Advanced Drilling and Well Technology

Advanced Drilling and Well Technology

Book Editors:

Bernt S. Aadnoy
University of Stavanger

Iain Cooper
Schlumberger

Stefan Z. Miska
University of Tulsa

Robert F. Mitchell
Halliburton

Michael L. Payne
BP

Society of Petroleum Engineers

Printed in the United States of America.

The paper used in this book meets the minimum requirements of ANSI/NSIO Z39.48-1992 (R1997). ⊗

ISBN 978-1-55563-145-1

09 10 11 12 13 14 15 16 / 9 8 7 6 5 4 3 2 1

Society of Petroleum Engineers
222 Palisades Creek Drive
Richardson, TX 75080-2040 USA

http://store.spe.org/
books@spe.org
1.972.952.9393

Foreword

For years, the prerequisite textbook for understanding fundamental drilling engineering principles has been *Applied Drilling Engineering* (Bourgoyne et al. 1986), better known to many readers as "The Big Red Drilling Book." This new drilling volume, *Advanced Drilling and Well Technology,* is meant to complement that classic textbook, and its genesis is twofold.

First, it was felt that there were many topics more advanced than the basic concepts covered in the original textbook, which very much focuses on the fundamentals of drilling engineering. To have included all aspects of drilling, particularly those that have appeared in the last 15 years, would have led to a single volume that would be both unfocused and unwieldy. Second, the drilling industry is at a turning point. With the average age of the current workforce in their late 40s, the next 10 years will see much of today's drilling expertise retired (although, we hope, still contributing to the growing body of knowledge!). Therefore, it was felt that we must develop a mechanism to capture some of that knowledge in a form that can be passed down to the younger community who will be drilling the wells of tomorrow. To this end, we have expanded upon *Applied Drilling Engineering* to provide more detail in certain areas (i.e., casing design, wellbore hydraulics, foam dynamics), and we have elaborated on the latest thinking that has taken advantage of developments in computer processing. For example, the sections on well-control modeling, geomechanics, and heat transfer all describe the fundamental science and governing equations behind many of the sophisticated simulators that are used to plan today's wells.

Major technology breakthroughs take a significant time to incubate before they are readily accepted as appropriate techniques. The first horizontal well was drilled in the 1920s (although some would say that it was actually *dug* in the previous century) but, as a technique, directional drilling only achieved economic viability in the 1980s. Indeed, directional drilling has been one of the most significant developments in the industry, and many of the sections within this book describe developments that have been enabled by directional drilling (such as geosteering) or that have led to refinements and improvements to the technique; the discussions on drillstring design and vibration highlight the issues associated with deviated and extended-reach wells, as well as the loads and shocks such trajectories place on the drilling equipment.

The most recent evolution (some would say revolution) of drilling has been the development of rotary steerable systems. These have led to higher rates of penetration, improved borehole quality, and access to previously unobtainable reserves. This is very much a technology that is still in its infancy, and there is no doubt that eventually we will see it overtaking motors, in a range of sizes and at a reduced cost, so that it becomes the ubiquitous system for both vertical and deviated wells. We do not include an explicit description of the variety of rotary steerable mechanisms and tools, but many of the advanced techniques that we describe are facilitated by this breakthrough technology and have implications for trajectory and the corresponding drillstring design.

We also discuss some of the newer drilling procedures that have appeared over the last few years and that have steadily gained a degree of acceptance that will only increase in the future. Indeed, this text is the first that has gathered together formal descriptions of the tools and techniques associated with coiled-tubing drilling, underbalanced and managed-pressure drilling, and casing drilling. There are other techniques that we have chosen not to include for the sake of space or because of the immaturity of the technology. To this end, we do not describe the parallel well design and drilling techniques of steam-assisted gravity drainage (SAGD) or other dedicated heavy-oil-recovery techniques. Nevertheless, as unconventional reservoirs become a more significant component of total hydrocarbon production, these techniques will be addressed in future texts.

It is often said that the "easy" oil has been drilled. This is only partially true, and many of the techniques we describe in this book address methodologies for increasing the recovery factor from existing reservoirs. Nevertheless, we have seen a shift toward more complex reservoirs, and the chapters on high-pressure/high-temperature applications and deepwater operations highlight some of the difficulties we face in drilling in such environments. We will continue to face ever more harsh conditions and will have to continually adapt both the processes and the equipment used to drill efficiently and to produce safely from these wells. We feel that the material in this book provides the necessary background for understanding the pertinent issues when designing a well for ever-increasing water depths or when encountering significantly overpressured zones.

There have been enough significant drilling technology developments in the last 15 years to fill two or three new textbooks. Detailed discussions of some of these developments will no doubt appear in a future volume. The reliability and accuracy of well surveying has undergone significant improvements over the last 10 years. The increasing precision of sensors and the plurality of these sensors in the drillstring has led to a whole suite of new uncertainty models and analyses that mean more-complex trajectories in multiwell programs can be drilled to significantly increase reservoir contact. Space dictates that the discussion of those uncertainty models and their role and implementation in well planning and real-time well redesign will have to wait for another day.

Similarly, many new logging-while-drilling technologies have appeared in the last few years and continue to increase both the speed and the precision of drilling. Seismic while-drilling technology has enabled the drillbit to be placed on the seismic map with an accuracy that can enable significant course corrections to be made and hazards to be avoided. Similarly, refined electromagnetic, nuclear, and acoustic measurements and the complementary improvements in telemetry rates have illuminated the borehole around and behind the bit to the degree that simple geometric steering has been ubiquitously replaced by geology-based steering for the complex wellbore trajectories necessitated by the more challenging environments faced by today's driller.

Now is a time more than any other where we in the drilling community have benefitted from the rapid progress in other industries. The developments in signal processing in the telecommunications world have been applied to the logging- and measurement-while-drilling environment to extend the application of the triple and quad combo (and more) from mere formation evaluation to enable not only the reservoir engineer but also the drilling engineer to make decisions in real time that will have a material benefit to both the quality of the borehole and the ultimate production from the reservoir. We have included detailed descriptions of some, but not all, of these measurements in the chapter on Wellbore Measurements: Tools, Techniques, and Interpretation, focusing on those that we feel have added the most significant value to the driller and that have had the most exposure in the literature in terms of helping maximize reservoir contact and ultimate production while also assisting in minimizing drilling trouble time.

Pore pressure while drilling was often seen to be the "holy grail" of drilling measurements. The recent development of real-time formation pressure while-drilling tools has led to a whole new area of interpretation that allows a much deeper understanding of the nature of drilling fluid behavior and well response during drilling, in addition to giving us that most fundamental of values. A variety of tools and techniques exist, and we attempt to cover all of the currently employed tools as well as a detailed analysis of the supercharging phenomenon, which has the most significant impact on a reliable interpretation of the measurements.

This would not be an "advanced" textbook unless we were allowed to try and speculate on some of the drilling systems (or should we say rock destruction mechanisms?) that may be used in the future. Therefore, in Chapter 9 we take the liberty of describing some of the techniques that appear in the literature and in conferences periodically that have yet to obtain wide acceptance in the field (such as laser and electropulse drilling), but that could provide the next significant breakthrough in terms of rate of penetration when we are able to fully deploy their effects downhole.

This book would not have been possible without many people sacrificing significant amounts of time during one of the busiest periods the drilling industry has ever seen. The editors would like to personally thank all of the authors, illustrators, and reviewers (and their families) for their patience, dedication, quality of work, and passion that they continually brought through the 6 years it has taken to develop the volume you now have in your hands. Finally, we must acknowledge John Thorogood, who had the initial vision of a textbook that covered the more recent breakthroughs in drilling technology and processes, and who assembled the editorial team and indeed many of the original authors in a conference room in Amsterdam some 6 years ago. We believe that we have compiled a book that lives up to that vision and can now take its place on your shelf alongside "The Big Red Drilling Book" (currently in revision).

Bernt S. Aadnoy
Iain Cooper
Stefan Z. Miska
Robert F. Mitchell
Michael L. Payne

Acknowledgments

The editors would like to thank all of the hard-working authors who contributed to this fine volume, as well as their employers who allowed them to participate in this venture. We also would like to thank the SPE staff for all of their efforts to organize this material into a book. Finally, we would like to thank Schlumberger for providing the image used on the cover.

Contents

Chapter 1

The Well-Construction Process

Claire Davy, EITICAT, and **Bernt S. Aadnoy,** University of Stavanger

This book has been compiled principally to lead well engineers through advanced drilling and well-design technology. The well-construction process provides the fundamental backbone to support such well design and completion design. We also will introduce the new technology that is applied in the design of wells and in the drilling process.

Advanced Drilling and Well Technology is aimed at the application of fundamental well-design techniques, but it also addresses far more complex challenges; the SPE textbook *Fundamentals of Drilling Engineering* (Mitchell 2009) holds a wealth of information on the fundamentals. The objective of this book is to capture the vast development that has taken place in drilling during the past decade to address meeting the requirements of designing more-challenging wells and completions. Significant advances have been seen in deepwater drilling, underbalanced drilling, logging while drilling, and pressure and temperature measurements during drilling, as a few examples. These two SPE books are closely tied together and will be considered complementary.

In short, this book will link theory to practical applications and includes detailed case studies where appropriate. Although some derivations of fundamentals are required, the idea is to refer to *Fundamentals of Drilling Engineering* to a large extent.

This chapter sets out the framework of the well-construction process into which, in practice, the design techniques are drawn one at a time at the appropriate points. Often, several alternative ways to design a well will be reviewed at the outset. Once the selection has narrowed the choice to a limited number of designs, thorough checks will be made on each design to assess viability and determine the value each design offers in terms of both direct outcomes and opportunities (particularly to address long-term life-cycle issues); the risks will also be assessed.

1.1 This Chapter in the Context of This Book

We will begin by describing the steps that are normally carried out in well design—working on the principle of designing geared to the well objectives, which usually requires bottom-up design. The reader is then systematically walked through each step. As the steps are recounted, the text points to example topics that the designer should investigate more deeply by referring to particular parts of the book. Toward the end of the chapter are examples of how to address a variety of different challenges facing well designers. Some examples point to how the topics set out in this book can be used.

This first chapter links the reader into the various topics covered in this book, giving instructions for locating pertinent information. It also points to the interplay between topics, guiding well designers to areas in the book on related issues, possible allied effects, and technologies that can be used during various processes.

These introductions are performed using the basis of a typical high-level holistic well-engineering management system, which is intended to provide support to the well-design and completion-design processes.

1.2 Overview of a Well Project

The elements of well construction are considered in the larger perspective, which includes project management, opportunity and risk assessment, and economic evaluation.

The overview of a well project includes all the high-level processes of well design, implementation, reporting, and feedback. These processes form part of the well-delivery process and the decisions made therein, [more fully described in Okstad (2006)], which are the outcome of sound project management.

The well-delivery process encompasses all the activities required to get from start to finish and is the responsibility of the drilling team. The process spans from the first concept of need for a well through developing that concept into a detailed design, constructing the well, maintaining it during its life, and, finally, abandoning it. The process noted here provides an extremely simplified version for the purpose of guiding engineers new to advanced well design. Experienced engineers will know that many of the components of the process as noted here are in fact carried out in an iterative manner and that the results of one calculation or inference often affect others. A highly skilled well-design engineer is able to determine when the outcome of one finding will impact other aspects of advanced well design and can take appropriate follow-up action. Responsibilities within the drilling team for carrying out design and planning, approving this work, and overseeing the implementation on-site are normally defined in the well-engineering management system, as are the responsibilities at the interfaces of external information input (e.g., by geologist or reservoir engineer), and at the interfaces for reporting to external parties such as regulatory bodies.

The well-delivery process has been viewed and defined from many perspectives—especially in the late 1990s—in order to fine tune and highlight the effectiveness of the parts of the process that were significantly impacting the key performance indicators (Kelly 1994; Robins and Roberts 1996; Dudouet and DeGuillaume 1995; Dupuis 1997; Morgan et al. 1999). However, in practice, the detail needed in almost every case of well design and completion design is different from one well to another. The main uses of the descriptions of business processes, at whatever level of detail is needed, are to ensure that matters requiring follow-up are properly flagged, that input (at the required accuracy) is provided by all necessary parties, and that an audit trail exists to support the process and enable tracking of decisions made.

The framework that guides the reader through this chapter refers to a management system that supports this well-delivery process and the subprocesses of well and completion design, well construction, and completion placement, right through to a responsible handover of the well for production, and all processes underpinned throughout by well integrity-assurance processes.

1.2.1 Advanced Well Design. Excellent advanced well design is the thread that sustains a successful well-delivery process. The process of advanced well design is similar to that used at a basic level; however, there are two principal differences as the design proceeds:

- Alternative ways of fulfilling the needs are constantly considered.
- The designer has to push out of the conventional *box* and sometimes beyond the comfort zone of both the designer and his or her peers. Design *rules* that may have been set previously must be challenged consistently. We must always revert to the objectives we are trying to achieve and not remain bound by traditional design methodology.

To be skilled in advanced well design, the designer must first become conversant with the technologies that are available and not be put off by perceived risk. Every risk that is presented can be minimized. As designers investigate further into experiences with certain technologies of interest, they will become more able to state for their management what real risks still exist with that technology and what opportunities others have realized.

1.3 Objectives, Needs, Assurance, and Resources

The first step in advanced well design is establishing the boundary conditions: What are we aiming to do? What is the arena in which we will work? How can we be sure to achieve the objectives? And what tools, equipment, and skilled people do we have at our disposal?

The well will be requested to fulfill specific corporate objectives. It will also have to meet design constraints, or *needs,* such as conveying fluids of a particular composition or accessing a reservoir at a point displaced from the well's surface location. Assurance that the well design meets these objectives and needs must be established to enable release of finances for the project. To estimate the authorization for expenditure (AFE), the specific resources required to carry out the work program must also be established.

To commence the journey along the well-delivery process, we begin with what is normal, good practice at the start: to define the objectives of the well. The objectives will depend on the purpose of the well that is to be constructed, and the environment within which it is set.

A well may be needed to fulfill a number of objectives depending on its purpose—it may need to meet the requirements of wildcat exploration, field appraisal, development, or even redevelopment:

- For data only or for injection, production, or observation
- To fulfill license commitment requirements

The well activity could be envisaged in several varying environments:

- Onshore or offshore
- New basin or known geology
- Sweet crude, sour crude, gas, water, or steam
- Situated in hot desert conditions, in temperate climate, or in arctic conditions
- Located in an area of strict environmental compliance or under various regulatory regimes including those that are not yet fully matured
- The well may be for a major oil company or for a much smaller independent

All of these varying requirements place different demands on the well. However, the processes of well design can still follow a similar path.

1.3.1 Objectives. *The Starting Point.* The best practice is to focus on the target objective. This includes repeat sessions with the client to clarify well objectives. Often, the client's link for a well's purpose to corporate strategies is not fully annotated initially, and as the well engineer seeks answers to questions, there is a firming up of strategies and objectives by the client's team, which often consists of members of several different functions (e.g., reservoir engineers and geologists). Specific objectives for the drilling program and for the completion program will become evident as this questioning process proceeds.

The well engineer must establish from these corporate objectives any specific requirements or limitations to well design:

- Final wellbore size
- Completion intervals
- Formation evaluation requirements
- Perforation strategies and sand control

Approved corporate objectives should be documented with sign-off by all relevant parties. This provides well designers with their proper remit and a good starting point.

Examples that will be more fully explained in the book will tend to hone in on wells using advanced well design and new technologies to meet the objectives of the well within the constraints imposed.

Final Wellbore Size. Today, many more options can be considered for final wellbore size. Traditionally, final casing or liner sizes were set between 4.5 and 7 in.; these sizes permitted tubing strings to be contained within the casing or liner and to have a packer set therein, isolating the annulus between the tubing and the casing or liner. Production technologists could specify the ideal tubing size to suit all the anticipated production fluids and capacities throughout the well's life. The casing or liner size was established from the required tubing size, but the tubing string size was often restricted by a non-optional final well-bore size. Exploration wells aiming to log only the borehole could even run out of workable hole size when casings had to be set to close off troublesome zones.

Advancements to provide more options recently have been demonstrated. These have been driven by many significant needs, including the requirement for bigger-bore completions [e.g., for tubing with an outside diameter of $9\frac{5}{8}$ in. (244 mm) for high-rate wells].

The requirement for too many casing seats traditionally has forced the designer into a smaller-than-practicable final wellbore size, or into having very large surface casing sizes. Several technologies have recently come to the forefront to address these challenges: expandable casings, managed-pressure drilling, and casing drilling.

Expandable casings have the potential to offer an entirely different way of setting out casing configurations within the well, enabling the reservoir wellbore size that is ideal for production to be realized. Expandable casings can even enable focus on optimizing the hole size without compromising reservoir wellbore size (e.g., where as extended-reach wells require an ideal hole size in the high-angle portion, or require having a trouble zone cased off early).

Being able to push casing seats down farther is one of the many positive attributes of the managed-pressure drilling technique, which is described in Chapter 9.3. In some cases, this technique is combined with casing drilling to ensure the attainment of the casing-seat depth. The technique of casing drilling is also described in Chapter 9.2, and it has been used in innovative ways to overcome loss of final wellbore size caused by hole problems.

Several well examples have revolutionized completion design, with the production packer now being seated in production casing above a sealed liner top, together with production-well kill philosophies that do not require kill fluid in the annulus close to the reservoir. These examples in turn reduce the casing or liner size needed across the reservoir and are well illustrated in the over-pressured gas wells of the southern North Sea.

Monobore completions are now accepted practice in some areas where corrosion of the casings is unlikely to occur during the production life of the well. The principle of desiring the ideal tubing-size specification thus becomes even more important because it is now possible to obtain this.

Completion Intervals. Are the zones to be produced all to be open from the outset? Is there a likelihood of differing pressure regimes from one zone to another? Is the reservoir pressure predictable before the zones are penetrated by drilling? Will the pore pressure be detectable at the time of drilling?

The answers to these questions and the certainty of each will lead well designers to ensure that they have contingent plans in place to avoid wasting time (and money) in responding to a possible outcome for which there is no established plan.

Formation Evaluation Requirements. The biggest advance in this respect is the use of underbalanced-drilling techniques, which can offer not only the benefit of reducing impairment but also the allowance of well testing while drilling (see Chapter 9.1).

There is considerable reward for designers in minimizing he critical path time for rig operations. As these are heightened further, innovation is creeping in to remove activities from the critical path, including formation evaluation activities. Openhole logging tools have become crammed into shorter units to enable combination runs to replace several runs of logging tools and to become a norm. Likewise, longer coring runs are achievable in one attempt. More focus is placed on enlightening the customer with the cost of information gained in these ways to enable compromises to be settled upon. Logging results derived from measurement-while-drilling techniques are now widely accepted as definitive surveys for formation evaluation purposes by oil companies and regulatory bodies, although there is controversy over the outcome readings. Unfortunately, it is still necessary to spend time on a gyro survey to define directional paths of wells that may later have other wells nearby or are to be sidetracked in or near a reservoir.

Perforation Strategies and Sand Control. The first design choice would be to perforate or to have an openhole completion of sorts. Perforation technology has moved ahead, and the choice for selection now spans a wide range.

Once the well designer has information about pore pressures and reservoir characterization, including the extent and thickness of interbeds and the nature of the reservoir rock and its permeability mechanisms, agreement can be reached on how perforation intervals will be chosen—what log they may need to be based upon, the intensity of perforation (shots per foot), and the phasing. From mud types and prediction of formation-invasion characteristics and perforation-channel constraints on production, the desired depth of penetration and type of perforation charges can be selected. These must be compatible with the technique by which they will be introduced and discharged.

The whole completion sequence will have to be considered in detail at this point and *what if* scenarios worked through to ensure that the well can be used in the way required.

Openhole completions can have screens or slotted pipe set across them, or simply can have raw sand-face exposed. The use of expandable sand screens and sandpacks have changed the options for completion extensively.

1.3.2 Define the Needs of the Well. Differentiating the needs of a well at an early stage can simplify much of the planning. If the final product to be conducted through the well is to be gas, then this immediately calls for a requirement for gas integrity of the well. This also can lead to constraints on well configurations, tubular connections, and kick tolerances permitted by some oil companies' in-house well-design policies.

If the fluids that are to be conveyed will have significant carbon dioxide or sour gas content, then this normally implies further metallurgy constraints to the materials that will be used.

The production magnitude, lifetime production profile, and changes of fluids must be established. Often, this will be linked closely to the economics of the project that this well supports.

For example, if the well is to be an exploration well to collect geological samples and log data only and no fluid is to be flowed, then the size of the final conduit could well be quite small (6 in., or 152 mm). However, if the well must produce at a high flow rate, a maximum-size reservoir conduit will be most favored. When considering if this is the requirement, it also needs to be established whether a high off-take will be sustainable by the reservoir over a considerable period. Related facts to be established are the viability (both practical and economic) of re-entering the well to change tubing size or completion configuration as the well life proceeds and the supply from the reservoir changes in pressure and, perhaps, rate.

Similarly, the shape of the well must be understood by all in light of the forecasts of hydrocarbon deliverability (e.g., with a lengthy horizontal well) and the susceptibility of wells to water ingress over time.

The key to good well design at this stage is having already thought through remedial plans to cope with the changes of fluids flowing into the well during the well's predicted life cycle. The key to success for a company is having all the stakeholders agree on such plans, and the likelihood of their being needed at this stage. Several iterations made on such points at this stage will provide huge gains for the advanced well-design engineer (and the company), both in the subsequent design process and in the longer term.

A similar set of discussions is needed to establish the completion design and to challenge it. It must be established whether any form of smart technology is to operate or be resident in this well at any time in the well's life. Will downhole monitoring systems be required? Are downhole flow-control devices envisaged? Would the field development benefit from special adaptations or special configurations of wells? Could the well be expected to have a junction in it?

A basis of design for both the completion and the well is normally drafted at this stage to capture the assumptions thus far and to provide a repository for more detail as further data on the materials or formation properties are collected.

The basis of design can be used as the first formal statement showing that well integrity is to be established for the life of the well. The document will name the conditions likely to be encountered and point out how each is to be met in a way that shows that basic philosophies supporting well integrity are maintained. For example, most oil companies adopt a two-barrier policy, and demonstration of the intent of meeting this policy can guide the development of designs and will certainly influence material and equipment selections and operational sequencing.

Data Sourcing, Collation, and Analysis. Once the basic requirements for the well have been established, as many data as possible are assembled to define the environment within which this well must be constructed. Data sought are typically:

- Likely and available surface location, including note of restrictions
- Weather, water depth (if any), and any constraints implied thereby
- Subsurface data including
  - A geological prognosis
  - Geological information showing regional stresses
  - Pore pressure/depth profiles, reservoir fluid type
  - Fracture pressure/depth profiles
  - Temperature/depth profiles
  - Nearby well paths

○ The nature of each of the geological formations and their drill ability, well-bore stability performance at various hole angles and azimuths

○ Mud programs from past wells and their success

For each of the components of work in data collation and analysis, the advanced methodologies in this book can be considered. Guidance on when these advanced methodologies may be used appropriately is indicated in **Table 1.1.**

First Pass: Planning the Well. The interpretation of these data with a view to meeting well objectives (established earlier) is used to define (ideally), in the following order:

1. Completion details
2. Well paths
3. Formation characteristics that will impact well design
4. Casing-seat placements in certain formations or at or below specific depths
5. Casing configurations: sizes, weights, and connections

The overall well design is thereby generated, incorporating

- Completion design
- Well profile and direction
- Casing design: production casing or liner and intermediate and surface strings

For each of the components of work in planning the well, the advanced technologies can be considered. Guidance on when these advanced technologies may be used appropriately is given in **Table 1.2.** The pore pressure and fracture pressure plot with depth are the keys to determining which enabling technology could add value by reference to **Table 1.3.**

TABLE 1.1— DATA COLLATION AND ANALYSIS TOPICS IN THIS BOOK		
Data Collation and Analysis Topic		Chapter for Reference
Wellbore stability	Chap. 5.1	Geomechanics and Wellbore Stability
	Chap. 6.1	Images While Drilling
	Chap. 6.2	Geosteering
Pore pressure	Chap. 5.2	Normalization and Inversion Methods

TABLE 1.2—ESSENTIAL WELL-PLANNING TOPICS IN THIS BOOK		
Well-Planning Topic		Chapter for Reference
Casing seat placements Casing configurations Tubing selection	Chap. 2	Advanced Casing and Tubing Design
Completion details	Chap. 3.4	Loads, Friction and Buckling
Well paths—trajectory design	Chap. 3.1	Advanced Drillstring Design
Well-control modeling	Chap. 4.2	Well-Control Modeling

TABLE 1.3—ENABLING-TECHNOLOGIES TOPICS IN THIS BOOK		
Enabling-Technologies Topic		Chapter for Reference
Using underbalanced-drilling fluid hydrostatics	Chap. 9.1	Underbalanced-Drilling Operations
Using at-balanced-drilling fluid hydrostatics	Chap. 9.3	Managed-Pressure Drilling
Using foam as drilling fluid	Chap. 4.3	Foam Drilling
Casing while drilling	Chap. 9.2	Casing While Drilling

1.3.3 Assurance to Achieve the Design. Once the well design is outlined, it will be important to ensure that the complete drilling program can support the requirements to fulfill the objectives of the well. Assurance that the well can have continued integrity will become progressively more of a concern to satisfy the regulatory authorities.

Management will need to be presented with several ways to design the well, with the cost for each set out and a list of the strengths, weaknesses, opportunities, and threats presented by each. Assurance that the designed well can be put in the ground will need to be justified. The more non-standard and the more deviations from traditional technology that are proposed, the greater the depth of study to prove the project viable. Identification of the opportunities that the design offers is highly valuable at this stage. The reservoir engineers should be encouraged to share their vision of how the well will continue to drain the field during its life, including production rates, fluid types, and whether water influx or sand production will likely occur after a specific number of years.

Certain parts of this book provide guidance on how this can be addressed within an advanced design approach. **Table 1.4** points to the following components:

- Directional program
- Drilling, workover, and completion fluids
- Drilling practices

Considerations for Special Environmental Conditions. When data are initially collated, it will become evident whether special environmental conditions exist. Examples that will require a complete additional set of considerations are deepwater or high-pressure/high-temperature subsurface conditions.

For each of these special environments, there are complete chapters in this book to guide the engineer as to extra considerations and different approaches that must be made for such advanced well design as per **Table 1.5**.

TABLE 1.4—FURTHER WELL-PLANNING TOPICS IN THIS BOOK		
Well-Planning Topic		Chapter for Reference
Drillpipe	Chap. 3.1	Advanced Drillstring Design
	Chap. 3.2	Drillstring Dynamics
Mud and completion-fluids program	Chap. 4.1	Wellbore Hydraulics
Directional program—geosteering wellbore measurements: tools, techniques and interpretation	Chap. 6.2	Geosteering
	Chap. 6.5	Drilling Vibration
Formation evaluation program	Chap. 6.4	Formation Pressure While Drilling

TABLE 1.5—SPECIALIST ENVIRONMENTAL CONDITION TOPICS IN THIS BOOK		
Specialist Environmental Condition Topic		Chapter for Reference
Deepwater wells	Chap. 7	Deepwater Drilling
High-pressure/high-temperature wells	Chap. 8	High-Pressure/High-Temperature Well Design and Drilling

1.3.4 Resources to Meet These Requirements. Resources to enable the designed well to be put in place are next determined by the advanced well-design engineer. In particular, it is determined what is available and what is appropriate with respect to

- Drilling rigs or workover units
- Service equipment
- Access routes to supply the drilling rig and their limitations
- Local assistance, support, and any constraints
- Expert personnel to deliver the required objectives

New technologies have emerged within the provision of many services required for constructing and servicing wells. Examples are set out in Chapter 9.4, "Coiled Tubing Drilling" and Chapter 9.5, "Novel Drilling Techniques."

The complete well design and the resources to be used are normally documented in a drilling program. Appendices of this program provide the outline of how special components of the program such as the mud or the directional program will be executed. For operations on the United Kingdom Continental Shelf (UKCS), the drilling program is referred to as the Health and Safety Executive (HSE) Submission. As in many countries, it must be submitted to the regulatory authority for approval or acceptance prior to permission being granted for the oil company to construct the well.

The detailed procedures for all the planned activities then are compiled in a detailed drilling program. The well engineer performs significant planning with all of the required parties and also organizes formal risk assessments that address the additional risks created by the interactivity of the various contributors (which are over and above the risk assessments that each service provider performs in-house on the activities for which each is wholly responsible). It is important not to miss the risks presented by bringing together different teams of service providers: No two well designs are identical; thus, a myriad of risks are presented for which minimization, mitigation, and responses need to be thought through formally before activity on-site.

A typical management system for companies operating on the UKCS formalizes what in fact tends to work well anywhere. Excerpts of flow charts from this management system are provided in **Figs. 1.1 through 1.4** to show how the design process is rigorously integrated; the charts include definite steps and checks by the well examiner for well integrity, and risk assessment activities are injected as the design develops right from the basis of design through the well-drilling program and on through the development of detailed procedures. Well design and completion design are addressed in the same rigorous manner.

The management system goes on to point out in particular the steps in which well-integrity concerns would be identified as the well is constructed (or as the completion is put in the ground), as reporting and handover of the product well to the customer is achieved, and as review occurs.

1.4 Implementation

1.4.1 Delivery of the Advanced Well Design. Enabling on-site delivery of the design requires several key systems to be in place. The size of the project to which this well contributes can vary greatly. Where the well

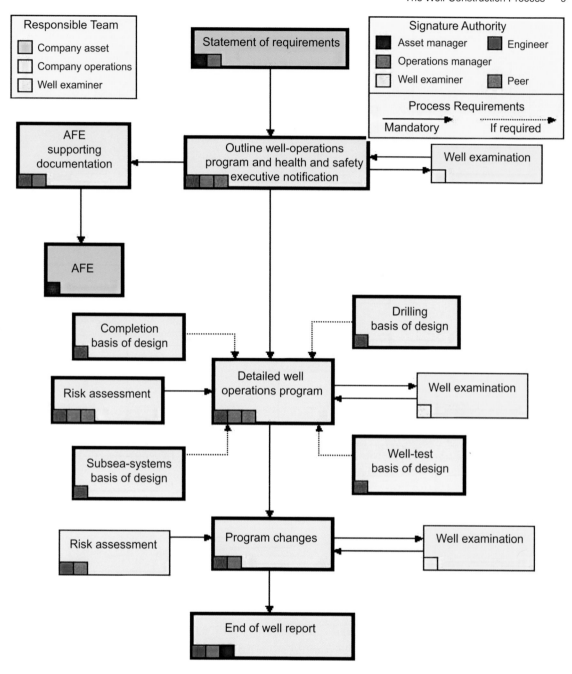

Fig. 1.1—Overall well-planning process—document and signature requirements (courtesy of Chris Dykes International).

is a lone exploration well, systems will be streamlined and likely customized. Where the well is one of many, it often requires support by a corporate set of systems that also support, for example, the production operations of a platform.

There are, however, systems of importance that are noted here as ideally required to some scale, whatever the project. The systems needed to support the successful construction of the designed well comprise, as a minimum, systems that perform the following tasks:

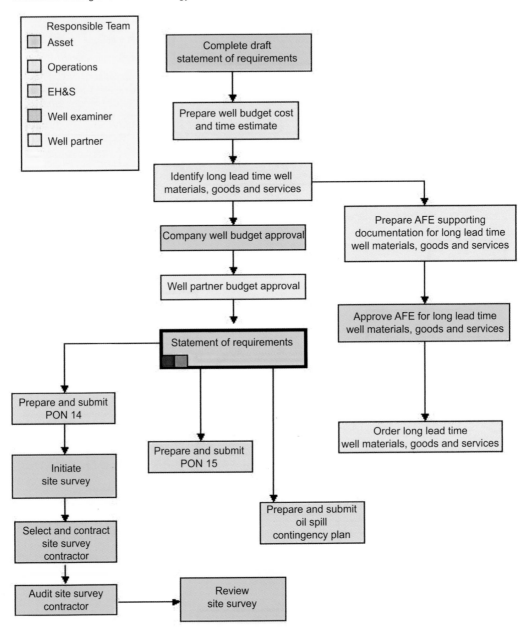

Fig. 1.2—Well-planning process—preplanning phase (courtesy of Chris Dykes International).

- Clearly define and document responsibilities and accountabilities
- Publish policies and procedures, particularly to ensure safety at site
- Monitor for actual operations: *as-built* vs. planned to detect and manage change
- Alert for variations in local data that may compromise the design
- Enable effective communication
- Ensure follow-through on authoritative directions

Clearly Define and Document Responsibilities and Accountabilities. A meeting to *drill the well on paper* can be effective in achieving these definitions. Usually, responsibilities and accountabilities will be documented in the management system.

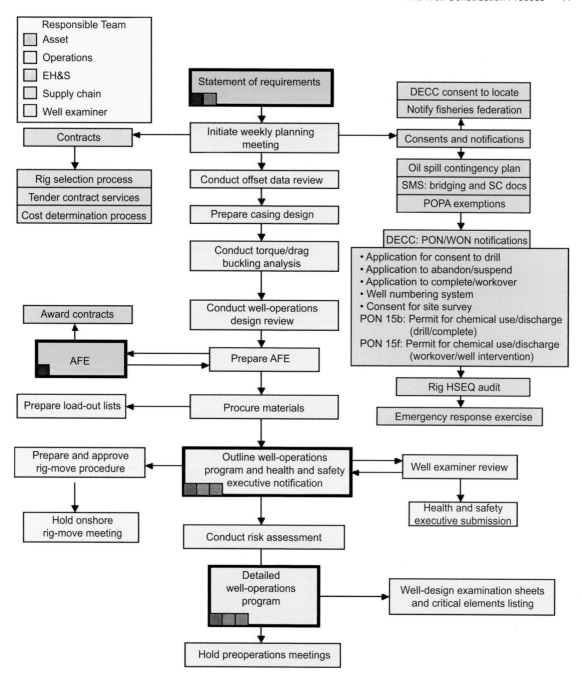

Fig. 1.3—Well-planning process—detailed planning phase (courtesy of Chris Dykes International).

Publish Policies and Procedures, Particularly to Ensure Safety. Well standards will exist to define policies affecting the construction process and the outcome well. These are normally broad enough to encompass completions and completion installation operations, well cleanup and well testing, and handover of wells to the customer.

Monitor Actual Operations: As-Built vs. Planned to Detect Change and Manage It. Formal documents explaining needed changes to the program are drawn up and channeled through the approval process before gaining the approval to enact the change. Following are examples of the need for such change:

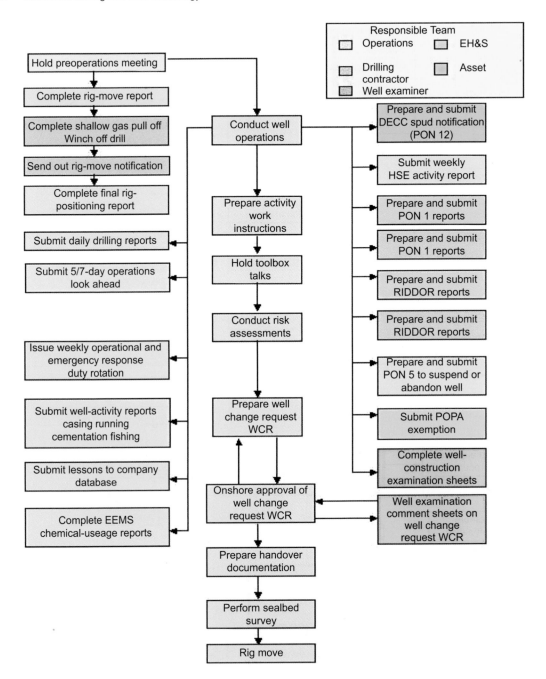

Fig. 1.4—Well-planning process—program (courtesy of Chris Dykes International).

- Cement job fails
- Stuck casing must be set high
- Actual geology or pore pressure is not as predicted

Alert for Variations in Local Data That May Compromise the Design. Variations that occur on site must not be permitted to compromise design. For example, geology picks coming in too high or not at all will affect the casing setting depth.

Enable Communication. Plans, including execution responsibilities, need to be relayed from town to key personnel at site. From the well-site the results of monitoring are fed back to town by the contractors daily, weekly, or monthly.

Ensure Follow-Through on Authoritative Directions. The oil company representative at the wellsite is the focus and co-coordinator for overseeing communications and for ensuring that the oil company directives are implemented.

1.5 Reporting and Feedback

1.5.1 What Is Reported? Minimum reporting requirements are very strict. They are used for the following purposes:

- To meet obligatory responsibilities and satisfy regulatory authorities
- To inform others in the site team and oil company, which is ultimately translated into being for safety or commercial purposes

It is mandatory that all details demonstrating well integrity and compliance to the approved program are reported.

How Does Reporting Occur? Reporting can occur in a number of formats and frequencies from daily drilling reports to weekly summaries, and, after the well is finished, the facts will be recorded in an end of well report. Reports are by the oil company drilling supervisor or representative to the oil company office responsible for the license. They also may originate from the service company and drilling contractor.

1.6 How Best To Use Advanced Technologies in the Overall Advanced Well Design

Using advanced technologies in the overall advanced well design requires:

1. Concept options to be generated
2. A robust business case for the preferred choice
3. A sound deployment framework
4. Excellence in project management

This book provides the well-design engineer with information on the new technologies that are available for well construction and gives examples of how those technologies can be used in advanced well design. Engineers will find these examples useful in taking their capabilities beyond the basics.

As the engineer follows specific examples, it is also useful to consider what other innovative applications there might be for that particular technology. In the evolving energy industry, creativity and responsible deployment of new technologies will provide the competitive edge.

This book serves only to introduce the technologies and some examples of their application. Further research to gain full familiarity with the particular technology of interest is highly recommended.

A caution is needed here about incorporating new technology into well design. The extent to which new technology deployment can benefit from good project management cannot be overstated. The upside is that new technology implementation can improve systems performance because so much attention has to be paid to detail. The downside is that if such attention and a rigorous approach to risk management are not made, the application of the new technology may be doomed. The preparation of proposals needs, therefore, to incorporate an analysis of the strengths, weaknesses, opportunities and threats (SWOTs) for each option so that these are thoroughly considered prior to a management challenge.

Case Study. When addressing case studies, the development of design options broadly follows the steps set out in this chapter. Reference should be made by the designer to the relevant technology in this book and in published papers such as those found in the SPE literature. For each well-design option indicated, the activities of well construction, including the completion, should be set out by the designer; each engineer is likely to end this exercise with slightly different outcomes.

Objective. To illustrate the value that a horizontal well would have compared to a traditional vertical well, for an oil company that is unfamiliar with (and apprehensive about) horizontal wells.

Identifying Boundary Conditions. Reservoir Plan. The sand trend is in the north/south direction within the reservoir. What is the optimum direction for a horizontal well in this reservoir? **Fig. 1.5** is a sketch of a plan at reservoir level, with North oriented up the page, showing braided sands.

Assessment.

1. A horizontal well with east/west orientation is needed to transect all three north/south channels of sand.
2. The well can commence at any point on the surface.

Concepts Considered. The design engineer is to consider only two concepts (vertical and horizontal) in order to compare these wells. However, during the design process it may become evident that with an open view to new technology, other options could be evaluated (e.g., expandables, multilaterals, or underbalanced drilling). Verification of the well objectives can be carried to ensure that they are properly formulated in the context of the boundary conditions. This provides a sound basis from which several concepts of well-design solutions begin to emerge.

Here, the analysis is directed in particular at the traditional vertical well vs. a horizontal well; however, local factors may cause some of these comparisons to differ when put into practice. This analysis is needed to generate a business case, particularly in the instance that this may be the first venture for some companies away from tradition. **Table 1.6** provides a comparison of a traditional vertical well and a horizontal well.

The conceptual designs showing simplistic well architecture and costs for each of the two approaches are generated as a function of the time/depth curve. Because the horizontal well is a longer drilled depth, it is no surprise that the time estimated for drilling is greater. Let us say that the duration to drill the horizontal well came out as the length known as P50, corresponding to the most likely time required. Further computations using a risked set of outcomes could enable generation of other points corresponding to P90 and P10.

According to one set of figures, the most likely outcome would be approximately 140 days for the long horizontal well, but considering both the possible accelerating of events and the delaying of progress, this well could take as little as 125 days or as much as 170 days. In contrast, the vertical well is compared at approximately 120 days but has a much tighter range for P10 to P90, from approximately 115 to 130 days.

There are several matters that such an analysis permits us to confirm. Apart from the horizontal well taking more time than the vertical well, it is observed that there is a greater risk attached to the horizontal well duration; this is denoted by the wider spread at P10 and P90. It is important that similar analysis is undertaken on the possible production and incremental liberated reserves to check that the benefits of these convincingly outweigh the cost concerns and added risk.

This process has the capability to prove the case one way or another and can be quite powerful when inserted into a business case. In this way, a large number of drilling/reservoir options can be screened to identify the best economic and technically feasible solutions. The design engineer is expected to produce

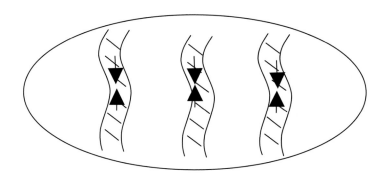

Fig. 1.5—Drainage points for three braided sands.

TABLE 1.6—COMPARISON OF VERTICAL AND HORIZONTAL WELLS	
Traditional Vertical Well	Horizontal Well
• Is simple and relatively inexpensive	• Longer zone in the pay sands
• Well design is easily accepted as a proven method with a low cost, but is it a good value?	• Greater confidence in penetrating the reservoir
• May not intersect any of the pay sand channel	• Concern over perceived risks of achieving the well
	• Higher initial well cost and interventions
	• Accesses more reserves
	• Reservoir depletion is less understood (e.g., well may have a shorter life if cones water and less of a choke effect around the wellbore)
	• Increased production rate is likely to accelerate production

such analyses to screen the available design options and prove the case for the application of new vs. traditional technology.

Project Management. Excellence in project-management techniques is required to facilitate and benefit from the use of the new technologies.

Proposals for the application of new technology are ideally supported by information to present the state of the art, minimize risk, and maximize value. Typical studies that can be undertaken to provide this information include:

- Identification of the new technology benefits where they apply to exploration wells
- Technology gap analysis at the conceptual stages of a project to demonstrate which technologies have the potential to add value
- The position of each new technology to be used in a project—set out in the technology deployment plan to support the field development plan or the field redevelopment plan
- Technical assessment of the integration of new/new and new/traditional technologies to achieve the best-value well over a well's entire life cycle
- The cost and potential value generated by the new technology: for example, how the correct application of the new technology can
  - ○ Improve project economics
  - ○ Liberate reserves of either bypassed oil or previously unattainable reserves
- Status of the maturity of each technology, naming examples in which deployment has occurred in similar circumstances
- The degree of reliability of the equipment components
- The level of confidence with which the technology can be deployed in the particular setting for which it is intended—onshore/offshore, local country, or basin
- The amount of planning that is needed to successfully deploy the technology
- The root causes that have underpinned existing success stories
- The background and the underlying cause of relevant failures of this technology
- Contributing factors likely to aid success or to undermine the deployment

- A catalog of the man-hours required to effectively achieve the volume of planning required to bring this technology deployment to reality without cost overruns

The well engineer will find that it is useful to get a feel for the above matters so that they are well equipped to respond to customer requests for use of newer technologies in a realistic manner. This book is intended as a good starting point for both those considering such implementation and also to support those who have deployed certain newer technologies and can benefit from prompts, knowledge and methodologies that apply to other technologies as yet untried. It is hoped that the information provided can help with preparing the businesses cases to bring each to fruition. Keeping updated with results and findings from other industry colleagues and other SPE colleagues will complement what you find in this book.

Acknowledgments

Figs. 1.1 through 1.4, which show the diagrams of the well engineering management system, are courtesy of Chris Dykes International, and a special acknowledgment goes to Chris Dykes for his kind permission to reprint these.

References

Dudouet, M. and DeGuillaume, J. 1995. Main Contractor's Viewpoint of the Lekhwair Turnkey Project: Case History. Paper SPE 29417 presented at the SPE/IADC Drilling Conference, Amsterdam, 28 February–2 March.

Dupuis, D. 1997. True Line of Business: Drilling Project Management. Paper SPE 37633 presented at the SPE/IADC Drilling Conference, Amsterdam, 4–6 March.

Kelly, J.H. 1994. Teamwork: The Methodology for Successfully Drilling the Judy/Joanne Development. Paper SPE 28862 presented at the European Petroleum Conference, London, 25–27 October.

Mitchell, R.M. (ed). 2009 (in progress). *Fundamentals of Drilling Engineering*. Richardson, Texas: SPE.

Morgan, D.R., Willis, J.C., and Lindley, B.C. 1999. Development and Implementation of a Process That Fosters Organizational Learning. Paper SPE 52773 presented at the SPE/IADC Drilling Conference, Amsterdam, 9–11 March.

Okstad, E.H. 2006. Decision Framework for Well Delivery Processes Application of Analytical Methods to Decision Making. Doctoral thesis, Department of Petroleum Engineering and Applied Geophysics, Norwegian University of Science and Technology, Trondheim, Norway.

Rayfield, M.A., Johnson, G.A., McCarthy, S., and Clayton, C. 2008. Parallel Appraisal and Development Planning of the Pluto Field. Paper SPE 117004 presented at the SPE Asia Pacific Oil and Gas Conference and Exhibition, Perth, Australia, 20–22 October.

Robins, K.B. and Roberts, J.D. 1996. Operator/Contractor Teamwork is the Key to Performance Improvement. *SPEDC* **11** (2): 98–103. SPE-29333-PA. DOI: 10.2118/29333-PA.

Chapter 2

Casing Design

David Lewis, Blade Energy Partners and **Richard A. Miller,** Viking Engineering

2.1 Introduction to Casing Design

This casing design chapter is intended for readers that have a background in engineering studies and have a working knowledge of casing and drilling operations.

The chapter will cover the following subjects:

- Tubular failure criteria and theories of strength
- Casing load identification and the estimation of design loads
- Design approaches
- Connections, environmental considerations, and materials
- Special problems in casing design

The following subjects will not be discussed in detail because it is assumed that the reader has a basic understanding of these subjects:

- Casing and hole size selection
- Casing seat selection
- Tubing design, which is beyond the scope of this chapter (however, many of the principles used for casing can be applied for tubing)

Material covered in Bourgoyne et al. (1986) will not be repeated in detail, for the sake of brevity.

2.1.1 Design Process. The process of mechanical design of tubulars is one of matching strength or resistance to load, subject to several constraints. Regardless of the design methodologies, limit states, or failure criteria used, the goal of all design processes is the same—to ensure with sufficient confidence that strength exceeds all loading during the entire service life of the tubular.

One typical constraint imposed in the selection of tubulars is selection from a set of standard outside diameter (OD) or American Petroleum Institute (API) sizes. Once the required casing shoe depths are determined from fracture and pore pressure distributions and hole stability analysis, the loads on each string are estimated, and a standard tubular that matches the load is selected. Starting with the desired hole size at the bottomhole location, the size sequence of the casing strings is established, and the weight and grade of the each string are selected to satisfy the design load requirements. Thus, it is almost a habit to begin with, say, a 20-in. (508-mm) string, follow it with a 13⅜-in. (340-mm) string, and follow it with a 9⅝-in. (244.5-mm) string, and finally a 7-in. liner. Not surprisingly, these have become the *standard* sizes, in turn encouraging more use of these sizes, and thus feeding this cycle.

2.1.2 The Goal of Design. Suppose that the shoe depths for a well are specified. Casing design then involves selecting the size, weight, and grade of the strings and connections to be run to the shoe depths. In essence, this is a constrained optimization problem. The goal of the design process is to specify strings such that their capacity exceeds all loads during service life. The constraints to the design process include

- Minimum acceptable design factors for a given load
- Restrictions on the diameter of the first string that passes through the wellhead
- Desired bottomhole size and minimum casing internal dimensions to accommodate completion equipment and enable economic production rates
- Running and cementing clearances for subsequent strings

The casings in a well are load-bearing structures. The goal is to design the composite structure such that it can safely withstand all of the loads that may be placed on it during its intended service life. The load placed on the tubular should not exceed the strength of the tubular resisting that particular load type. This can be stated in formula format as follows:

$$\text{Load} < \text{Resistance.} \qquad\qquad\qquad\qquad\qquad\qquad\qquad\qquad\qquad\qquad\qquad (2.1)$$

The outcome of a design process should be a safe, reliable, and cost-effective design. Safety and reliability are addressed differently by different design approaches.

Casing Design Considerations. A proper design must consider more than just the casing pipe body OD, weight, and grade. Other important considerations are the connection and the environment exposed to the casing. Environmental issues include corrosive fluids, both formation and injected fluids, and brittle failure due to sulfide stress cracking (SSC).

Design End Result. The final design should include the following:

- Size, weight, and grade
- Connection type
- Material requirements
- Inspection requirements for quality assurance and quality check (QA/QC)
- Operational procedures and precautions

Oilfield Tubulars. A casing or liner is usually run for the following reasons:

- To provide major structural support for the wellhead and other wellbore tubulars
- To maintain wellbore stability
- To isolate formations
- To control well pressures during drilling, production, and intervention

Principal Load Types. Casing load types and causes of loads are summarized in **Table 2.1.** The loads will be defined in more detail later.

Loads can be treated independently or as some combination of providing major structural support, maintaining well bore stability, isolating formations, and controlling well pressure in all phases of the well life.

TABLE 2.1—LOAD TYPES AND CAUSES	
Load Type	Cause
Axial tension/compression	Running, slackoff or pickup after cementing, changes in temperature and pressure
Pressure	Burst or collapse due to kick, pressure test, shut-in, cementing, injection, circulation, evacuation, or other operations
Bending	Bending load due to buckling or doglegs
Torsion and shear	Normally not considered in casing design unless the liner will be rotated while cementing, or if the casing is used for drilling

Some loads are intentionally applied, while others are accidental. Intentional loads are loads that will happen with a high degree of certainty, such as pressure tests and running loads. A kick load or tubing leak may not happen and therefore is an accidental load. Intentional and accidental loads are both possible and therefore must be considered. The relevance of intentional and accidental loads becomes more important when considering reliability-based design.

Design Process. Once a design approach is selected, the process of design usually progresses in the following steps:

- Identify all possible load scenarios and estimate the load parameters.
- Calculate the principal loads at every point in the string: axial force, internal and external pressure, bending stresses, and possibly torsion.
- Calculate the strength of the pipe to resist the loads.
- Check the design and make refinements.

Uncertainty is involved in the design process because loads describe future events. Some of the categories of uncertainty are

- Load parameters and magnitudes
- Casing strength
- Failure mode and consequence

2.2 Theories of Strength

In engineering structures, strength is the only attribute of the structure that can be controlled. The design process is one of comparing the loads likely to be placed on the structure to the resistance of the structure to that particular load. The strength or resistance of the structure is a response to an applied load and differs for different load types or modes. For example, resistance to bending is different from resistance to internal pressure.

This section presents different theories of strength used to estimate the resistance of casing to the principal load types. These include the standard API strength theories and some alternative methods currently under consideration for API *Bull. 5C3* (*Bull. 5C3* 1999) (*ISO/TR 10400* 2007). This chapter focuses only on the pipe body. Connection strength is discussed in Section 2.5, and material selection is discussed in Section 2.6.

2.2.1 Common Yield-Based Strength Equations. Bourgoyne et al. (1986) describes the common strength equations defined by API *Bull. 5C3* (*ISO/TR 10400* 2007). These equations are summarized here for completeness.

Tension—API Pipe Body Yield.

$$F_{ten} = \sigma_Y \times A_p. \quad\quad\quad\quad\quad\quad\quad\quad\quad\quad\quad\quad\quad\quad\quad\quad\quad\quad\quad\quad\quad\quad (2.2)$$

The tension rating is compared to the calculated axial force applied to the casing.

Internal Pressure—API Minimum Internal Yield Pressure (MIYP).

$$p_{API} = 0.875\left(\frac{2\sigma_Y t}{d_o}\right). \quad\quad\quad\quad\quad\quad\quad\quad\quad\quad\quad\quad\quad\quad\quad\quad\quad\quad\quad (2.3)$$

The MIYP is compared to the calculated differential pressure across the pipe body, which is Δp or $p_i - p_o$. The equation has its derivation from thin-walled pressure vessel theory with a 0.875 reduction factor for manufacturing variations in the wall thickness.

External Pressure—API Collapse. The API collapse equations are complex and consider four distinct regimes. The governing collapse regime is determined by the casing diameter-to-thickness ratio (*D/t*). The API collapse rating is compared to the equivalent collapse pressure p_c.

$$p_c = p_o - \left(1 - \frac{2}{(d_o/t)}\right)p_i. \qquad\qquad\qquad (2.4)$$

Yield Collapse.

$$P_{YP} = 2\sigma_{YPa}\left[\frac{(D/t)-1}{(D/t)^2}\right] \quad \text{for } \frac{D}{t} \leq \left(\frac{D}{t}\right)_{YP} \qquad\qquad (2.5)$$

$$\left(\frac{D}{t}\right)_{YP} = \frac{\sqrt{(A-2)^2 + 8\left(B + \dfrac{C}{\sigma_{YPa}}\right)} + (A-2)}{2\left(B + \dfrac{C}{\sigma_{YPa}}\right)}$$

$A = 2.8762 + 0.10679 \times 10^{-5}\sigma_{YPa} + 0.2131 \times 10^{-10}\sigma_{YPa}^2 - 0.53132 \times 10^{-16}\sigma_{YPa}^3$

$B = 0.026233 + 0.50609 \times 10^{-6}\sigma_{YPa}$

$C = -465.93 + 0.030867\sigma_{YPa} - 0.10483 \times 10^{-7}\sigma_{YPa}^2 + 0.36989 \times 10^{-13}\sigma_{YPa}^3$

Plastic Collapse.

$$P_p = \sigma_{YPa}\left[\frac{A}{(D/t)} - B\right] - C \quad \text{for } \left(\frac{D}{t}\right)_{YP} \leq \frac{D}{t} \leq \left(\frac{D}{t}\right)_{PT}. \qquad (2.6)$$

$$\left(\frac{D}{t}\right)_{PT} = \frac{\sigma_{YPa}(A-F)}{C + \sigma_{YPa}(B-G)}$$

$$F = \frac{46.95 \times 10^6\left(\dfrac{3(B/A)}{2+(B/A)}\right)^3}{\sigma_{YPa}\left(\dfrac{3(B/A)}{2+(B/A)} - (B/A)\right)\left(1 - \dfrac{3(B/A)}{2+(B/A)}\right)^2}$$

$$G = F\frac{B}{A}$$

Transition Collapse.

$$P_T = \sigma_{YPa}\left[\frac{F}{(D/t)} - G\right] \quad \text{for } \left(\frac{D}{t}\right)_{PT} \leq \frac{D}{t} \leq \left(\frac{D}{t}\right)_{TE}. \qquad (2.7)$$

$$\left(\frac{D}{t}\right)_{TE} = \frac{2 + B/A}{3B/A}$$

Elastic Collapse.

$$P_E = \frac{46.95 \times 10^6}{(D/t)[(D/t)-1]^2} \quad \text{for } \frac{D}{t} > \left(\frac{D}{t}\right)_{TE}. \qquad (2.8)$$

Yield Strength Adjusted for Tension. The API equations use an adjusted yield strength to account for the reduction in collapse resistance under axial tension.

$$\sigma_{YPa} = \left(\sqrt{1 - 0.75(\sigma_z/\sigma_Y)^2} - 0.5\sigma_z/\sigma_Y\right)\sigma_Y. \qquad\qquad (2.9)$$

Triaxial Stress. Triaxial stresses due to the combination of internal pressure, external pressure, axial force, and bending moment are compared to the material minimum yield strength. The von Mises equivalent (VME) stress equation is used to compute triaxial stresses in thick-walled cylinders.

$$\sigma_{VME} = \sqrt{\sigma_z^2 + \sigma_h^2 + \sigma_r^2 - \sigma_z\sigma_h - \sigma_z\sigma_r - \sigma_r\sigma_h + 3\left(\tau_1^2 + \tau_2^2 + \tau_3^2\right)}, \dots\dots\dots\dots\dots\dots \text{(2.10)}$$

where

$$\sigma_z = \frac{F_a}{\pi\left(r_o^2 - r_i^2\right)} \pm \sigma_b,$$

$$\sigma_h = -\left[\frac{r_i^2 r_o^2\left(p_o - p_i\right)}{\left(r_o^2 - r_i^2\right)}\right]\frac{1}{r^2} + \left[\frac{\left(p_i r_i^2 - p_o r_o^2\right)}{\left(r_o^2 - r_i^2\right)}\right] \quad \text{from Lamé's equations, and}$$

$$\sigma_r = \left[\frac{r_i^2 r_o^2\left(p_o - p_i\right)}{\left(r_o^2 - r_i^2\right)}\right]\frac{1}{r^2} + \left[\frac{\left(p_i r_i^2 - p_o r_o^2\right)}{\left(r_o^2 - r_i^2\right)}\right] \quad \text{from Lamé's equations,}$$

and τ_1, τ_2, and τ_3 are shear stresses that are usually assumed to be zero in tubular design because there are no torsional shear stresses or shear forces to give rise to horizontal shear stresses.

2.2.2 Alternative Internal Pressure Equations. The API MIYP is conservative. Numerous tests have shown that the true ductile rupture limit is much greater than suggested by the API equation, by 20 to 40% depending on the D/t ratio. Several other burst strength theories are therefore presented in this section.

Capped-End Yield. The capped-end yield rating uses thick-walled pressure vessel theory and eliminates several of the conservative assumptions that underlie the API MIYP. Capped-end yield is derived by placing Lamé's thick-wall derivations for hoop and radial stresses into the VME equation. The derivation is lengthy and is presented here in an abbreviated form.

Formulas for axial stress, hoop stress, and radial stress are placed into the VME equation. The three shear stresses are set equal to zero, and the triaxial stress is found on the inside surface, the point of maximum stress.

$$\sigma_{VMEr_i}^2 = \left[\frac{3\left(p_i - p_o\right)^2\left(A_o\right)^2}{\left(A_o - A_i\right)^2}\right] + \left[\frac{\left(p_o A_o - p_i A_i\right)}{A_s} + \sigma_z\right]^2, \quad \dots\dots\dots\dots\dots\dots \text{(2.11)}$$

where $A_o = \pi r_o^2$ and $A_i = \pi r_i^2$.

The equation is written in this form to show the internal components within the VME equation after the insertion of Lamé's equations. The last term of the expression is effective tension, which can be written as

$$\frac{\left(p_o A_o - p_i A_i\right)}{A_o - A_i} + \sigma_z = \frac{\left(p_o A_o - p_i A_i\right)}{A_p} + \frac{F_a}{A_p} \pm \sigma_b.$$

Defining an *effective* tension T_{eff},

$$T_{eff} = F_a - p_i A_i + p_o A_o,$$

VME can be written as

$$\sigma_{VMEr_i}^2 = 3\Delta p^2\left(\frac{A_o}{A_p}\right)^2 + \left(\frac{T_{eff}}{A_p} \pm \sigma_b\right)^2.$$

When a pipe is cap-ended or when a pipe is continuously restricted in place that will not allow any axial movement (i.e., is cemented in place), $T_{eff} = 0$. Equating σ_Y to σ_{VME} results in the classical capped-end yield equation for burst:

$$P_{\text{capped end}} = \frac{4}{\sqrt{3}} \sigma_Y \frac{t(d_o - t)}{d_o^2}. \quad \dots \dots \dots \dots \dots \dots \dots \dots \dots \dots \dots \dots \quad (2.12)$$

Because the capped-end burst derivation does not assume a *thin wall* like the API equation for burst, it is a better estimate of the burst capacity than the API MIYP equation. However, even this equation does not correctly estimate the true or *ultimate* burst capacity of the pipe because it uses the VME and Lamé's equations, which are applicable only in the elastic regime. Therefore, it only estimates the pressure at which the cross section first begins to yield at the inner radius.

Ultimate Burst or Rupture Limit. When the internal pressure reaches the capped-end pressure limit, the innermost fiber has just yielded. All the other fibers of the pipe are still within the elastic limit and have not lost their ability to withstand load. As the internal pressure continues to increase, outer fibers begin to yield from the inside diameter (ID) toward the OD, until at some maximum internal pressure, the entire cross section has yielded and becomes fully plastic. Burst can occur beyond this point. Therefore, this pressure is more indicative of the true rupture limit of the pipe and thus represents the limit load. Tests have indicated that the true ductile burst pressure is very close to this limiting pressure. **Fig. 2.1** illustrates this process.

Klever-Stewart Rupture Limit. The Klever-Stewart rupture limit is based on theoretical and experimental work by Klever and Stewart (1998), and it is currently adopted by the joint API/ISO committee on Oil Country Tubular Goods (OCTG) performance properties. The model proposed in its most general form is

$$p_{B,K\text{-}S} = K_n K_T \sigma_{\text{ult}} \frac{2(t_{\min} - m_f t_n)}{d_o - (t_{\min} - m_f t_n)}, \quad \dots \dots \dots \dots \dots \dots \dots \dots \dots \dots \quad (2.13)$$

where K_n, the index correction factor, is given by

$$K_n = \left(\frac{1}{2}\right)^{1+n} + \left(\frac{1}{\sqrt{3}}\right)^{1+n}$$

and K_T, the tension correction factor, is given by

$$K_T = \sqrt{1 - \left(\frac{T_{\text{eff}}}{T_{\text{UTS}}}\right)^2}$$

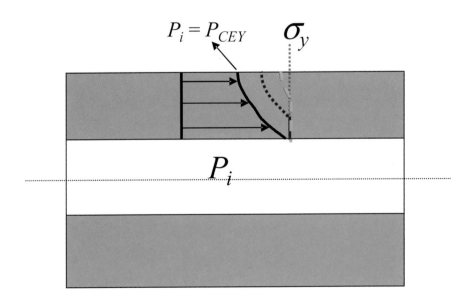

Fig. 2.1—Progression of yield as pressure increases beyond capped-end yield.

Note the similarity of this equation (barring the many constants and corrections) to the thin-walled pressure vessel equation used by API, in that the limit is proportional to wall thickness and material strength, and inversely proportional to the OD.

The term K_n is an experimentally determined correction term to account for the nonelastic behavior of materials. It is essentially a power-law-type fit to the stress-strain curve, and it incorporates the true strain of the material. The value of n should be experimentally determined for each material grade, but Klever and Stewart provide a simple curve fit for n, based on the yield strength:

$$n = 0.169 - 0.000882 \ \sigma_Y/1000$$

K_T is a tension correction factor that accounts for the impact of effective tension T_{eff} on the ductile rupture limit. In general, both T_{eff} and the ultimate tensile strength T_{UTS} are functions of diameter and wall thickness. The tension effect is ignored here: therefore, $K_T = 1$. Test data have indicated that this is a conservative assumption.

σ_{ult} is the ultimate strength of the material, obtained from uniaxial tensile strength tests on specified samples. In the absence of actual measured data, the API-specified minimum ultimate strength can be used (as this is the minimum tensile strength). t_{min} is the minimum measured wall thickness for the tubular. t_n and m_f are the flaw depth and flaw factor, respectively, to account for cracks and flaws in the tubular. They derive from the fracture mechanics-based crack propagation limits of different grades. They should be experimentally based. Klever and Stewart suggest that $m_f = 1$ can be used for high-fracture-toughness materials, but $m_f = 2$ is more appropriate for low-fracture-toughness materials. From the experimental data presented by Klever and Stewart (1998), it appears that for a flaw size of 5 to 6% of wall thickness, the ductile rupture limit is between 95 and 100% of the limit without flaws. This suggests that a factor of safety of 1.1 can account for strength degradation due to undetected flaws.

It should be noted that the equations presented above are the limit state functions and are used when material and geometric property data are available. When nominal and minimum properties are used, as is usual in design, *ISO/TR 10400* (2007) has provided a design equivalent of the above equations with appropriate derating factors, and these should be used instead of the limit state. The purpose of showing the limit states above is to illustrate the ultimate load-carrying capacity of the tubular.

Hill's Fully Plastic Burst Limit. The Hill rupture limit (Hill 1950) is obtained from classical mechanical analysis of a thick-walled capped-end cylinder subjected to internal pressure, under the assumption of elastic to perfectly plastic material behavior that obeys the von Mises yield criterion. The internal pressure at which the entire wall has plasticized based on the VME stress is the Hill rupture limit and is given by

$$p_{B,\text{Hill}} = \frac{2}{\sqrt{3}} \sigma_Y \ln\left(\frac{d_o}{d_o - 2t}\right). \ \dots\dots\dots\dots\dots\dots\dots\dots\dots\dots\dots\dots\dots\dots\dots \ (2.14)$$

Other versions of the Hill equation may use minimum wall and ultimate strength.

2.2.3 Alternative Collapse Strength Equations.

Collapse is one of the more complex strength functions in tubular design. The mode of collapse failure is strongly related to the d_o/t ratio of the pipe. For thin shells with a high d_o/t ratio, collapse is similar to buckling instability. For thicker-walled pipes, collapse is controlled by yield strength.

Collapse strength can be estimated based on several methods. API collapse is based on tests on 2,488 specimens of K55, N80, and P110 over a wide range of d_o/t ratios and ovality, manufacture and process tolerances, material imperfections, etc. Timoshenko derived a relationship for collapse limit state that includes the standard geometric and material parameters plus ovality and axial stress. A third collapse relationship is the Tamano equation for ultimate collapse strength, being selected by the API/ISO as an alternative collapse equation.

API collapse strength theory has served the industry well over the years. The collapse strength has an empirical basis and a reliability-based minimum rating; thus, it is likely to be fairly accurate. Despite the robustness of the API equations, it should be noted that the equations are based on tests conducted more than 40 years ago. Manufacturing processes, tolerance control, and material performance have all improved over this time.

Timoshenko Collapse. The theoretically derived Timoshenko collapse equation includes ovality and axial stress. It is not often used in design but is included here for completeness:

$$p'_{external} - p'_{internal} = \frac{P_y + P_e\left(1 + \frac{3\phi d_o}{2t}\right) - \sqrt{\left[P_y + P_e\left(1 + \frac{3\phi d_o}{2t}\right)\right]^2 - 4P_y P_e}}{2}, \qquad (2.15)$$

where

$$P_y = \Gamma \frac{2\sigma_y t}{d_o}$$

$$P_e = \frac{2E}{1 - v^2 \left(\frac{t}{d_o}\right)^3}$$

$$\phi = \frac{2\left(d_{o,\,max} - d_{o,\,min}\right)}{d_{o,\,max} + d_{o,\,min}} \quad \text{(This is the definition of ovality.)}$$

$$\Gamma = \sqrt{1 - 3\left(\frac{\sigma_z}{2\sigma_Y}\right)^2} - \left(\frac{\sigma_z}{2\sigma_Y}\right) \qquad (2.16)$$

$$\sigma_z = \frac{F_a}{A_p} + \sigma_b \qquad (2.17)$$

Tamano Collapse. The Tamano equation is an ultimate collapse strength or limit state for collapse. It attempts to derive a suitable interaction between the elastic collapse limit that applies for high d_o/t ratios and the yield collapse limit that applies for low d_o/t ratios. Corrections for eccentricity, ovality, and residual stress are also included.

The original Tamano equation was found to diverge from test data in some cases, particularly in combined loading situations. The API/ISO committee generalized the Tamano equation to improve collapse predictions. They introduced separate functions to account for imperfections in the yield range, the transition range, and the elastic range of pipe collapse. The resulting generalized collapse limit state equation is known as the Klever-generalized Tamano.

$$p_{c\,ult} = \frac{\left(p_e + p_y\right) - \sqrt{\left(p_e - p_y\right)^2 + 4p_e p_y H_{ult}}}{2\left(1 - H_{ult}\right)} \qquad (2.18)$$

The term H_{ult} is a correction factor proposed by Tamano to account for ovality, ϕ, eccentricity, ε, and the normalized residual stress, $\bar{\sigma}_r$. These are defined as

$$H_{ult} = 0.127\phi + 0.0039\varepsilon - 0.440\bar{\sigma}_r + h_n, \quad h_n = 0.055 \text{ for N80, 0 for all other grades,}$$

$$\phi = 100\frac{\left(d_{o,\,max} - d_{o,\,min}\right)}{d_{o,\,average}}$$

$$\varepsilon = 100\frac{\left(t_{max} - t_{min}\right)}{t_{average}}$$

$$\bar{\sigma}_r = \frac{\sigma_{residual}}{\sigma_Y}$$

$$p_e = k_{els} \frac{2E}{\left(1-v^2\right)} \frac{1}{m\left(m-1\right)^2},$$

where k_{els} is an elastic limit correction factor, taken to be 1.089 in the limit state;

$$m = \frac{d_{o,\,average}}{t_{average}},$$

rather than the ratio of the nominal values as in Tamano:

$$p_y = k_{yls} 2\sigma_{ye} \frac{m-1}{m^2}\left(1+\frac{1.5}{m-1}\right),$$

where k_{yls} is a yield limit correction factor, taken to be 0.9911 in the limit state.

In the p_y equation, it is customary to use the yield strength derated for the presence of real axial tension (or compression).

$$\sigma_a = \frac{F_a}{\pi} \frac{m^2}{d_o^2\left(m-1\right)}$$

$$\sigma_{ye} = \frac{1}{2}\left(\sqrt{4\sigma_Y^2 - 3\sigma_z^2} - \sigma_z\right)$$

The equations presented above are the limit state functions and are used when material and geometric property data are available. When nominal and minimum properties are used for design, *ISO/TR 10400* (2007) has provided a design equivalent of the above equations with appropriate derating factors, which are used instead of the limit state.

2.2.4 Probabilistic Strength Theories. *Introduction.* The strength estimation methods discussed previously all result in a deterministic value for strength—that is, a single value. This single value is based on numerous assumptions, most of them on the conservative side. Examples are minimum yield strength and a 12.5% allowable reduction in wall thickness. However, in a deterministic estimate of strength, it is implicitly assumed that if the load exceeds the calculated (deterministic) strength, the pipe will fail. In traditional design approaches, this possibility is minimized by separating the applied load from the calculated strength by some predetermined margin or safety factor.

Since the parameters that determine the strength of a pipe are variable, its true limits are uncertain. Yield strength, wall thickness, diameter, and other governing parameters are all random variables. Deterministic design considers the *minimum* strength of a pipe in the absence of quantified information on the variables.

However, it is possible to quantify the strength variability through measurement and testing and use the distributions of strength in design. This is the method of probabilistic strength determination.

Probabilistic Determination of Strength. In probabilistic methods, every variable is considered a random variable with a specified distribution. For a given strength equation, an uncertainty model is developed for each variable, preferably based on actual measurements and statistical modeling of the data. Measurements of the typical variables that appear in most strength equations are readily available, such as yield strength, wall thickness, and OD. The various individual distributions are then combined together using a Monte Carlo method or other analytical statistical methods (first-order or second-order uncertainty propagation methods) according to the rating equation, resulting in a probabilistic distribution of strengths.

This method is illustrated by finding the thin-walled pressure vessel internal yield pressure for a 9⅝-in., 47-lbm/ft, L80 casing string. The thin-walled pressure vessel strength is the same as the API strength (Eq. 2.2), but without the specified 12.5% wall reduction. Table 2.1 gives the nominal values and the mean μ and standard deviation σ' for the distribution of yield strength, OD, and wall thickness, obtained from measurements from a particular manufacturer. Then, from first-order propagation of uncertainty, the mean and standard deviation of the function are given by

$$\mu\left[f\left(x_i\right)\right] = f\left[\mu(x_i)\right], \quad\dotfill \quad (2.19)$$

$$\sigma'\left[f\left(x_i\right)\right] = \sqrt{\sum_{i=1}^{n}\left[\frac{\partial f\left(x_i\right)}{\partial x_i}\right]^2\left[\sigma'\left(x_i\right)\right]^2}, \quad \ldots\ldots\ldots\ldots\ldots\ldots\ldots\ldots\ldots\ldots\ldots \quad (2.20)$$

where x_i = the variables in the strength function.

Using these equations, the mean and standard deviation of the thin-walled pressure vessel strength function can be calculated. The nominal value, mean, and standard deviation of the strength also are shown in **Table 2.2**. **Fig. 2.2** shows the distribution schematically. The API value (calculated as per Eq. 2.2) is also shown on the distribution. For the provided statistics of d_o, t, and σ_Y, the API rating underestimates the calculated distribution of the strength.

The above example illustrates the impact of probabilistic consideration of strength. However, a low probability of failure is sought in design, so the simple first-order method shown above may not accurately reflect the probability of failure when the tail of the strength distribution is compared to a load. The method may still be used to

TABLE 2.2—PROBABILISTIC BARLOW STRENGTH DETERMINATION			
Barlow equation:	$P_{\text{Barlow}} = \sigma_Y \dfrac{2t_{\text{nom}}}{d_{o,\text{nom}}}$		
Casing:	9⅝ in., 47 lbm/ft L80		
Parameter	Nominal Value	Mean	Standard Deviation
d_o (in.)	9.625	9.635	0.003
Wall thickness (in.)	0.472	0.479	0.003
Yield strength (psi)	80,000	87,000	2,751
Thin-walled pressure vessel (psi)	7,846	8,650	279
API value (psi)	6,865	—	—

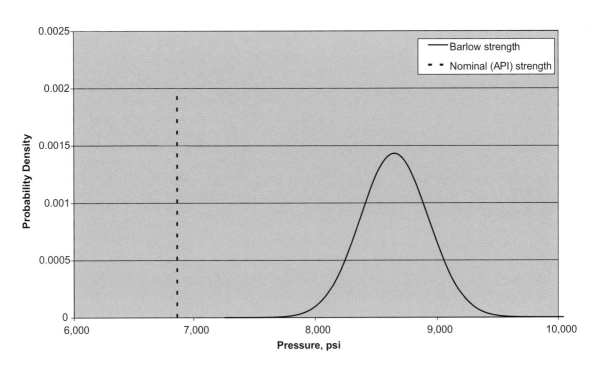

Fig. 2.2—Distribution of the thin-walled pressure vessel burst strength of a 9⅝-in., 47-lbm/ft, L80 casing.

get a reasonable estimate of the failure probability or for sensitivity analyses. Care must be taken to properly estimate the lower-tail probability of strength and, if load is also considered probabilistically, the upper-tail probability of the load. These types of analyses may require more sophisticated statistical tools and techniques.

2.3 Casing Standard Loads

This section discusses load assumptions used in casing design. Load assumptions for dry-tree wells, including land and platform wells, are discussed in detail. A section on subsea wells or wet-tree wells shows the differences between the two types of wells.

A load case is the description of internal pressure, external pressure, and temperature over the length of a casing string at a point in time. The load case describes an event that occurs in the life of the well, such as a kick or a frac job down the production casing. A framework for creating loads is provided here; however, there is no industry consensus for defining a unified set of load cases.

2.3.1 Burst Loads—Drilling and Production. The burst load at any depth is defined as follows:

$$p_{burst} = p_i - p_o. \dotfill (2.21)$$

The burst internal load (p_i) for casing is the surface pressure plus the hydrostatic pressure of the fluid in the casing. The load can be a planned load like a pressure test or an unplanned but possible load such as a kick or tubing leak.

The burst backup or outside load (p_o) depends on the fluid on the outside of the casing. The external pressure assumption can vary depending on the particular load assumption. The pressure profile can be mud hydrostatic, mud base fluid, pore pressure, base fluid density of cement, or a combination of several components. It is reasonable to use pore pressure backup in the open hole, below the previous casing shoe. A conservative design approach would use the base fluid gradient in the cased-hole annulus, above the previous shoe, with zero pressure at the wellhead. Nonzero mudline pressures for subsea wells deserve careful consideration.

Drilling Kick. Drilling kick is normally the critical drilling burst load. The worst-case kick internal load is the maximum anticipated surface pressure (MASP) during a kick plus a gas gradient to the shoe of the casing. Section 2.3.5 describes the calculation process for MASP. Gas is normally assumed as the worst case, resulting in the highest possible surface pressure. Gas is often assumed for kicks in oil wells since there is usually associated gas with the oil. In a worst-case situation, where a kick is shut in, it is possible for gas to migrate to surface and displace liquids from the wellbore, resulting in a gas gradient to surface from bottom-hole pressure.

$$p_i = MASP + \rho_i \times Z \times c. \dotfill (2.22)$$

Here, the conversion factor c is 0.052 psi/(ft × lbm/gal) for US oilfield units and 9.8 (kPa × L)/Kg × m for metric units.

External pressure depends on the assumed fluids outside the casing of interest. As an example, the load assumptions could be defined as follows. The mud gradient is used to the top of cement (TOC), assuming a short time frame between setting the casing and applying the load. The cement mix water gradient for cemented pipe in pipe assumes that the cement is set and the density has reverted to mix water. Pore pressure is a reasonable assumption for the pressure below the previous shoe.

$$p_o = \rho_{MW} \times Z \times c \text{ (mud weight gradient above the TOC)}, \dotfill (2.23)$$

$$p_o = p_{TOC} + (Z - Z_{TOC}) \times \rho_{water} \times c \text{ (cemented pipe in pipe)}, \dotfill (2.24)$$

$$p_o = \text{pore pressure below the previous shoe.} \dotfill (2.25)$$

A circulating temperature profile could be used for this load case to consider the effect of elevated temperature at the top of the string. Alternatively, the geothermal temperature gradient may result in higher axial tension, corresponding to the condition of shutting the kick in for a long time.

Pressure Test. The pressure test internal load is the test pressure plus the mud gradient to the shoe. The normal assumption is that the internal mud weight is the mud weight in the hole when the casing was run and set. If the plug was displaced with a different-density fluid, the fluid density should be revised.

$$p_i = p_{surf} + \rho_i \times Z \times c. \quad\dots\dots\dots\dots\dots\dots\dots\dots\dots\dots\dots\dots\dots\dots\dots\dots\dots \quad (2.26)$$

External pressure depends on the assumed fluids outside the casing of interest. Eqs. 2.23 through 2.25 could be used, for example. Alternatively, using the mud base fluid density in Eq. 2.23 would yield a more conservative pressure profile. The geothermal temperature profile is typically applied to pressure test loads.

Production Tubing Leak. The tubing leak internal load is the shut-in tubing pressure (SITP) plus a packer fluid gradient to the packer depth (assumes leak occurs at the top of the well). The highest SITP normally occurs early in the life of the well. For simplicity, the load may be extended to the casing shoe even though the load physically ends at the packer. Production tubing leak is normally the critical production burst load for wells that are not stimulated.

$$p_i = SITP + \rho_i \times Z \times c. \quad\dots\dots\dots\dots\dots\dots\dots\dots\dots\dots\dots\dots\dots\dots\dots\dots\dots \quad (2.27)$$

External pressure depends on the assumed fluids outside the casing of interest. Eqs. 2.23 through 2.25 could be used, for example. The mud base fluid gradient is usually assumed because the time frame for production loads can extend for many years after the mud is left behind in the casing. Original mud weight is not assumed because solids settling could occur.

Both a producing temperature profile and the geothermal gradient may be considered for tubing leak loads (hot tubing leak and cold tubing leak). The elevated production temperatures result in higher compression forces, which exacerbate bending stresses due to buckling. The geothermal temperature profile results in the highest tension forces at the top of the string.

This load condition, although named *tubing leak,* is actually any pressure that is placed on the production casing (most likely from a tubing leak). However, this pressure can result from packer or hanger leaks.

Production Stimulation. For stimulation down the casing, the internal load is the maximum stimulation pressure plus the stimulation fluid gradient to the shoe. Stimulation down the tubing is considered by a tubing leak load case, where the internal pressure profile is the maximum stimulation pressure acting on the packer fluid column.

External pressure is the same as described in the previous Production Tubing Leak section. An injection temperature profile accompanies these pressure profiles, resulting in the highest tension forces.

2.3.2 Collapse Loads—Drilling and Production.
The collapse external or outside load (p_o) for casing depends on the load assumption. Typical collapse loads are fluid hydrostatic pressure like cement, mud, and pore pressure. The collapse backup or internal load (p_i) depends on the fluid or lack of fluid on the inside of the casing. The collapse load at any depth is defined by Eq. 2.4.

Geothermal temperatures are typically used for collapse loads. This temperature assumption results in the highest string tension, which lowers collapse resistance.

Cement Collapse—Drilling and Production. Cement collapse applies to both drilling and production casing. The internal load is the hydrostatic pressure of the displacement fluid. This load can become an issue for large-OD casing that is set deep with a light displacement fluid.

$$p_i = Z \times \rho_{DispFluid} \times c \quad\dots\dots\dots\dots\dots\dots\dots\dots\dots\dots\dots\dots\dots\dots\dots\dots\dots \quad (2.28)$$

The external load is the unset cement and mud hydrostatic pressure on the outside of the casing.

$$p_o = Z \times \rho_{MW} \times c + (Z - Z_{TOC}) \times \rho_{Cmt} \times c \quad\dots\dots\dots\dots\dots\dots\dots\dots\dots\dots\dots \quad (2.29)$$

Drilling Collapse. The drilling collapse load assumes lost returns resulting in a fluid level drop to some depth. The depth may be determined by an arbitrary rule, such as 50% of the current casing shoe depth or one-third of the deepest subsequent open hole, or it could be calculated by balancing a zone of depleted pressure with the current mud weight (lost returns). Internal pressure above the fluid level is zero; the internal pressure below the fluid level is the mud weight gradient.

$$p_i = \text{zero (above the fluid level)} \quad \dotfill \quad (2.30)$$

$$p_i = (Z - Z_{FL}) \times \rho_{MWi} \times c \text{ (below the fluid level)} \quad \dotfill \quad (2.31)$$

External pressure depends on the fluids outside the casing of interest and is typically the original mud weight and cement density outside the casing, though a full column of mud may be used for the sake of simplicity.

Production Casing Evacuation Collapse. Production casing can be exposed to severe collapse loads up to zero internal pressure as a worst case. If it can be determined that full evacuation will never occur, an abandonment pressure other than zero may be assumed at the perforations. The internal pressure can be based on a packer fluid column height that will result in the design abandonment pressure at the perforations.

$$p_i = \text{zero (for full evacuation or above the fluid level)} \quad \dotfill \quad (2.32)$$

$$p_i = p_{\text{abandonment}} - (Z_{\text{perfs}} - Z) \times \rho_{MWi} \times c \text{ (below fluid level)} \quad \dotfill \quad (2.33)$$

External pressure is the same as for the Drilling Collapse section.

Salt Loading Collapse. Salt loading collapse is a special case. It can be a very severe collapse load and is included as a design load only if salt loading is expected. The internal pressure is the same as defined in the Drilling Collapse section for drilling casing or the Production Casing Evacuation Collapse section for production casing.

As a first pass, a uniform external load may be approximated by using an overburden gradient of 1 psi/ft (23 kPa/m) or higher for the salt section. However, nonuniform pressure considerations also should be included. Many engineers take an interaction diagram approach between uniform and nonuniform pressures. Pore pressure is normally assumed above and below the salt section.

2.3.3 Tension Loads. Typical tension loads include overpull while running casing and bump plug while cementing. Other tension loads that may apply include tension for setting the slips after the cement has set and tension effects due to stimulation temperature cool-down. Additionally, the highest tension loads may occur in conjunction with one of the previously described load cases, possibly during a stimulation event.

Prior to the cement setting up, the axial force can be calculated using the free-body diagram method described in Bourgoyne et al. (1986). For a string with no size or weight crossovers, the axial force at any depth is equal to

$$F_a = (Z_{\text{shoe}} - Z) \times w_{\text{air}} + p_{i,\text{shoe}} A_i - p_{o,\text{shoe}} A_o, \quad \dotfill \quad (2.34)$$

where w_{air} is the weight in air of the casing in lbm/ft (kg/m). The pressure terms in Eq. 2.34 are evaluated at the bottom of the string regardless of the depth of interest for axial force. Once both ends are fixed, by the hanger at the top and by cement at the bottom, then any change in pressure or temperature causes a change in tension.

$$\Delta F_a = \alpha E A_p \Delta T + 2v (\Delta p_i A_i - \Delta p_o A_o) \quad \dotfill \quad (2.35)$$

For uncemented sections of casing, the change in temperature ΔT and the change in pressure Δp is the average change over the uncemented interval. For casing depths axially constrained by cement, changes in force are due to changes in temperature and pressure at that particular depth. There is also a nonlinear force associated with buckling that can be added to Eq. 2.35.

A size or weight crossover in the uncemented portion of a string complicates the use of Eq. 2.35. This statically indeterminate problem can be solved by allowing each section to elongate or contract without constraint due to pressure and temperature changes. Then, the overall change in force for the uncemented section is found by solving for the restoring force required to return the string to its initial constrained length. The change in force due to the exposed piston area must also be included.

Bump Plug Tension—Drilling and Production.   This is a tension load caused by the pressure × area force while the cement is not set. The axial force in Eq. 2.34 is increased by the incremental pressure used to bump the plug.

Running Overpull—Drilling and Production.   This is the tension that can be pulled above the hanging weight of the casing in mud if the casing gets stuck while running in the hole. The assumption is that the casing is almost to bottom when it gets stuck.

Typical overpull designs include a fixed pull above string weight such as 100,000 lbf (445 kN) for large-OD casing. Smaller casing may be designed to string weight plus a percentage of joint strength, 15% for example, to avoid a design driven by overpull requirements. Variations on overpull calculations include using air weight instead of buoyed weight, which is more conservative.

If the casing is run in a directional well with high drag and overpull is a problem, the pickup tension can be modeled using a torque-and-drag model to better represent expected loads.

Set Slips Tension—Drilling and Production.   Additional tension may be applied to the uncemented casing above the TOC after the cement is set. This operation can limit buckling due to temperature increase while drilling below the casing or during thermal injection. The initial axial force in Eq. 2.34 would include the incremental overpull for all depths above the TOC.

2.3.4 Wet-Tree Wells—Subsea Wells.
Wet trees are located at the mudline of the seabed. Casing is run and landed in the subsea wellhead, and annuli are normally not accessible at surface. This difference changes the assumptions for backup for casing.

The same loads as described for dry-tree wells apply for subsea wells with the few changes in backup pressure. Several external pressure options exist for burst loads:

- Original mud weight
- A column of mud balancing pore pressure at the previous shoe (if the TOC is below the previous shoe)
- The original hydrostatic pressure at the wellhead on a column of base fluid
- Seawater gradient down to the wellhead and base fluid gradient from the wellhead to the TOC
- Zero pressure at the wellhead and base fluid gradient from the wellhead to the TOC

It is difficult to know with certainty what the backup pressure is in a subsea well. This uncertainty typically leads to using a lower-pressure, more conservative profile.

2.3.5 MASP Determination.
MASP for kick can be determined in several ways. These include frac at shoe and gas to surface, limited kick size and intensity, or frac at shoe and water to surface.

Frac at Shoe, Gas to Surface.   This method is used extensively in the industry for several reasons. It is conservative in that the casing is fully evacuated of mud and is filled with gas, the loads and assumptions are easy to understand, and calculations are straightforward and easily duplicated.

MASP from frac at shoe and gas to surface involves several steps.

- The first is to estimate the gas gradient. The gas gradient should be calculated based on methane at the highest bottomhole pressure (BHP) in the deepest openhole section below the current shoe. Methane will normally be the least dense gas, resulting in a higher MASP, hence more conservative. If a calculated gas gradient is not readily available, then a common rule of thumb is 0.1 psi/ft (2.3 kPa/m) for hole depths above 10,000 ft (3048 m) and 0.15 psi/ft (3.4 kPa/m) for greater hole depths.
- Use the fracture pressure at the shoe and subtract off the gas gradient to surface to determine the frac-at-shoe MASP.
- Use the highest BHP in the openhole section and subtract off the gas gradient to surface to determine the gas-to-surface MASP.

- The *minimum* of frac-at-shoe MASP or gas-to-surface MASP becomes the design MASP. The reason for this is that either method can limit the MASP depending on the frac-at-shoe pressure and the BHP. There is no need to design for a MASP that is not physically possible.
- If drilling liners are set below the casing, the highest MASP must be used, based on the BHP of the various openhole sections.

Limited Kick Size and Intensity. The limited kick size and intensity method can reduce the MASP when compared to the more conservative gas to surface method. However, modeling kick pressures is much more involved.

- An appropriate and representative kick volume and intensity is required.
- Realistic circulating pressures are difficult to calculate. Kick models that assume the gas remains in a single bubble are not realistic for synthetic muds. Kick models that accurately handle a distributed gas bubble are sophisticated and may not be readily available for a typical well design.
- The risk remains that an actual kick may occur with loads higher than the design load.

Frac at Shoe, Water to Surface. The MASP is determined in a manner similar to frac at shoe, gas to surface. This method assumes that it is possible to pump water down the annulus and frac the shoe and therefore limit the surface pressure. Practical issues include having enough water available to pump for extended periods of time. This may be possible for an offshore location, but may be a problem for some land operations.

Example—MASP Determination—Metric Units. **Fig. 2.3** shows the steps needed to determine MASP in metric units. The example shows a case in which the gas to surface is MASP and a case where frac at shoe determines MASP.

Example—MASP Determination—Field Units. **Fig. 2.4** shows the steps needed to determine MASP in field units. The example shows a case in which the gas to surface is MASP and a case in which frac at shoe determines MASP.

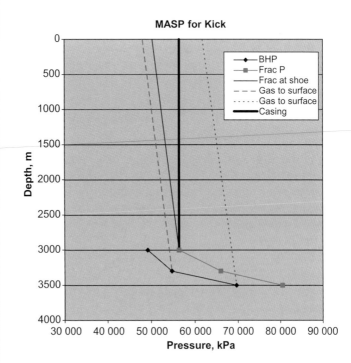

Given that casing is set at 3000 m, using this data set, determine the MASP for the next hole section to 3300 m.

Depth (m)	BHP (kPa/m)	Frac P (kPa/m)	BHT (°C)	Gas Grad. (kPa/m)
0				
3,000	16.4	18.8	116	2.00
3,300	16.6	20.0	121	2.05
3,500	19.9	23.0	138	2.27

Frac at shoe = 3000 * 18.8 = 56 400 kPa
Psurf = 56 400 – 3000 * 2.05 = 50 250 kPa

Gas to surf. BHP = 3300 * 16.6 = 54 780 kPa
Psurf = 54 780 – 3300 * 2.05 = 48 015 kPa

MASP = lesser of 50 400 or 48 015 kPa, therefore **MASP = 48 015 kPa**

Given the same data set, determine the MASP for the next hole section to 3500 m.

Frac at shoe = 3000 * 18.8 = 56 400 kPa
Psurf = 56 400 – 3000 * 2.05 = 50 250 kPa

Gas to surf. BHP = 3500 * 19.9 = 69 650 kPa
Psurf = 69 650 – 3500 * 2.27 = 61 705 kPa

MASP = lesser of 50 250 or 61 705 kPa, therefore **MASP = 50 250 kPa**

Fig. 2.3—MASP determination: metric units.

Given that casing is set at 10,000 ft, using this data set, determine the MASP for the next hole section to 11,000 ft

Depth (ft)	BHP (psi)	Frac P (psi)	BHT (°F)	Gas Grad. (psi/ft)
0				
10,000	7,100	8,200	240	0.089
11,000	7,950	9,570	250	0.0923
11,500	10,100	11,600	280	0.0999

Frac at shoe = 8,200 psi Psurf = 8,200 − 10,000 * 0.0888 = 7,312 psi Gas to surf. BHP = 7,950 psi Psurf = 7,950 − 11,000 * 0.0923 = 6,935 psi MASP = lesser of 7,312 or 6,935 psi, therefore **MASP = 6,935 psi**

Given the same data set, determine the MASP for the next hole section to 11,500 ft.

Frac at shoe = 8,200 psi Psurf = 8,200 − 10,000 * 0.0888 = 7,312 psi Gas to surf. BHP = 10,100 psi Psurf = 10,100 − 11,500 * 0.0999 = 8,951 psi MASP = lesser of 7277 or 8,951 psi, therefore **MASP = 7,312 psi**

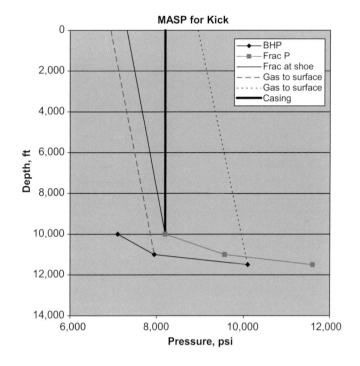

Fig. 2.4—MASP determination: field units.

2.3.6 SITP. SITP should be based on BHP at the perforations less a methane gradient to surface. If the reservoir fluid can be accurately characterized, then a more representative gradient can be used. This may significantly lower the SITP, especially for an oil reservoir.

2.3.7 Drilling Loads (for Drilling Casings and Liners). Drilling loads describe events that occur prior to well completion. Some dual-purpose strings are exposed to both drilling and production loads. **Table 2.3** is a summary of possible drilling load cases.

2.3.8 Production Loads (for Production Casings and Liners). Production loads are applied to casing strings and liners that either contact produced and injected fluids or are directly outside the tubing string. **Table 2.4** lists possible production loads.

2.4 Design Approaches for Casing

The process of design should ensure that the load-bearing structure is capable of withstanding all possible service life loads. *All design approaches, regardless of their details, strive to ensure this.*

   Three standard design approaches are discussed here:

- Working stress design
- Limit states design
- Reliability-based design

2.4.1 Data Required for Design. Before commencing on a design exercise, relevant data for design need to be gathered. The following summary lists data that are needed for detailed casing design.

- Directional survey
- Pore pressure and fracture gradient

TABLE 2.3—DRILLING LOADS FOR DRILLING CASINGS AND LINERS

Load Type	Load Condition	Internal Pressure	External Pressure	Temperature Profile
Burst	Kick	MASP on a gas (or fluid) gradient	Mud or base fluid gradient above previous shoe, pore pressure in open hole	Circulating or geothermal
Burst	Pressure test	Test pressure on the internal fluid gradient	Mud or base fluid gradient above previous shoe, pore pressure in open hole	Geothermal
Collapse	Drilling collapse	Zero to top of fluid, internal fluid gradient to shoe	Mud gradient or mud and cement gradient	Geothermal
Collapse	Cement collapse	Displacement fluid gradient	Mud and cement gradient to casing shoe	Geothermal
Tension	Bump plug	Displacement pressure plus bump margin above fluid gradient	Mud and cement gradient to casing shoe	Geothermal
Tension	Running overpull	Mud gradient	Mud gradient to casing shoe	Geothermal

TABLE 2.4—PRODUCTION LOADS FOR PRODUCTION CASINGS AND LINERS

Load Type	Load Condition	Internal Pressure	External Pressure	Temperature Profile
Burst	Tubing leak	SITP on packer fluid gradient	Base fluid gradient above previous shoe, pore pressure below	Producing and geothermal
Burst	Pressure test or stimulation down casing	Surface pressure on internal fluid gradient	Base fluid gradient above previous shoe, pore pressure below	Geothermal for pressure test. Stimulation for stimulation load
Burst	Stimulation through tubing	Surface pressure above packer fluid gradient	Base fluid gradient above previous shoe, pore pressure below	Stimulation
Collapse	Cement collapse	Displacement fluid gradient	Mud and cement gradient to casing shoe	Geothermal
Collapse	Production collapse	Zero pressure **or** packer fluid balancing abandonment pressure	Mud gradient or mud and cement gradient	Geothermal
Tension	Bump plug	Displacement pressure + bump margin above displacement fluid gradient	Mud and cement gradient to casing shoe	Geothermal
Tension	Running overpull	Mud gradient	Mud gradient to casing shoe	Geothermal

- Temperature profiles
- Casing size, type, and setting depth
- Mud weights
- Cement tops and densities
- Reservoir pressure and depth
- Produced fluid and injection fluid densities
- Packer fluid density
- H_2S and CO_2 concentration
- Maximum pressure loads

In some cases the data are readily available, and in other cases they must be estimated. Since design loads are based on these data, they should be as accurate as possible. If there is uncertainty in the data, casing may be underdesigned or overdesigned. Underdesigned casing can result in a failure with serious consequences.

While overdesigned casing is unlikely to fail, there is an economic impact that the designer is obliged to consider. In some instances, overdesigned casing can unnecessarily increase risk, like running a very heavy string near the design limits of elevators and slips.

2.4.2 Working Stress Design. Working stress design (WSD) is the most commonly applied approach in casing design, and it has a long history in the oil industry and many other industries.

Basic Approach. In working stress design, tubulars are designed to a *working stress*. The working stress is the ratio of design strength to a design safety factor that typically exceeds unity. Thus, the design check in WSD may be stated as

$$\text{Load} \leq \frac{\text{Design strength}}{\text{Safety factor}}. \quad\dots\dots\dots\dots\dots\dots\dots\dots\dots\dots\dots\dots\dots\dots\dots\dots \text{(2.36)}$$

It is customary to multiply the safety factor by the load, and the resulting design load is sometimes referred to as the *factored* load. This results in the familiar design check:

$$\text{Load} \times \text{Safety factor} \leq \text{Design strength}. \quad\dots\dots\dots\dots\dots\dots\dots\dots\dots\dots\dots\dots\dots\dots \text{(2.37)}$$

In WSD, the design strength is always the minimum strength of the material. Further, elastic criteria are used for design strength. Therefore, in WSD, exceeding the design strength may imply that the onset of yielding is possible rather than predict that a catastrophic failure is imminent.

Recommended Design Factors in WSD. The purpose of the safety factor in WSD is to account for uncertainties in strength and load estimation, and any other uncertainties, by separating the applied load from the resistance of the material to that load.

The typical safety factors used in applying the above design checks are listed in **Table 2.5.** These values are empirical and are based on experience and tradition. Several companies have different factors or ranges of factors for different load combinations or time periods in the life of the well. Documented evidence of the basis of these safety factors is rarely available.

Limitations of WSD. Despite the overwhelming acceptance of WSD in the industry and the advantage of its simplicity, there are several limitations to the approach.

In WSD, safety factors are recommended across the mode of loading and are usually independent of the applied load case. The burst safety factor is the same for a kick load or a pressure test load. Since the uncertainties and variability for different load cases are different, WSD results in safety-factor-consistent designs, but not risk-consistent designs. Thus, if a safety factor of 1.1 in burst is satisfied for both pressure test and kick, the risk of failure for the kick load is usually far lower than it is for a pressure test because the design kick load has a very low probability of occurrence. For simple wells, this may lead to a gross overdesign of the casing.

A more serious implication of the risk inconsistency of WSD is that for complex wells such as high-pressure/high-temperature (HP/HT) or deepwater wells, it is often difficult to satisfy the standard safety factors within geometric constraints. Engineers then tend to compromise the recommended safety factors by being more specific about the loads. However, it is unclear what additional risk is being taken in compromising the safety factors. Intuitively, reducing the safety factor is equivalent to increasing the risk of failure, but it is difficult to quantify the increased risk.

TABLE 2.5—TYPICAL SAFETY FACTORS	
Minimum Safety Factors—Casing	Pipe Body
VME	1.25
Axial	1.3–1.6
Burst (MIYP)	1.0–1.25
Collapse	1.0–1.1

WSD loads are estimated as if they are certain to occur. It is common practice to estimate a high end of the expected load to account for the possible uncertainty in the load. This tends to invite overdesign for infrequent loads.

Finally, due to the *masking* nature of the safety factors, it is impossible to assess the relative risk inherent in using different safety factors. As a result, the ability to consider the risk-consequence relationship is lost. For instance, the consequences of failure in a complex, expensive well near a populated area are much greater than those in an inexpensive well in the middle of a desert. It can be argued that higher safety factors should be used for the complex well, and lower safety factors are acceptable for the cheaper and simpler well. However, in practice, it is easier to satisfy the safety factors in the cheaper and simpler well but difficult to do so in the complex well. This practice has the illogical result of accepting a higher risk for projects with higher consequences of failure.

2.4.3 Limit States Design. In limit states design, the load-bearing structure is designed to a limit load rather than the working stress. Two types of limits are often addressed:

- Ultimate limit states, which address the ultimate and catastrophic failure of structures. Examples of this are rupture and parting.
- Serviceability limit states, which address the compromise of the serviceability of the structure, even though ultimate failure has not been reached. Examples of this are buckling and deflection beyond permissible limits.

In the oil industry, limit states are already used for the design of offshore structures and pipelines. For casing and tubing design, *ISO/TR 10400* (2007) attempts to bring limit states into the design process.

In limit states design for casing, the strength side in Eq. 2.37 is replaced with a limit state. For burst, the Klever-Stewart rupture limit in Eq. 2.13 or the Hill limit in Eq. 2.14 can be used. Limit states design continues to use design safety factors because the uncertainties in load estimation, load occurrence probability, and failure model must still be considered. A factor of 1.0 is appropriate for ductile rupture, provided that the minimum wall and imperfection threshold (based on inspection technique) are used in the calculation. It is important to note that usage of limit state equations such as Klever-Stewart and Klever-generalized Tamano usually requires measurement and quantification of material properties and geometric parameters to be useful in predicting a limit state. This is unlike WSD, where minimum properties and geometric parameters are used directly. When measurements are not available, minimum properties will have to be used in the limit state equations. *ISO/TR 10400* (2007) provides design equivalents of all the limit state equations contained in the standard.

In practice, a dual-design methodology of working stress design for normal loads and limit states design for infrequent or survival-type loads may be appropriate.

2.4.4 Reliability-Based Design Approaches. *Basic Approach in Reliability-Based Design.* In reliability-based design, the uncertainty, frequency of occurrence, and variability in all of the characteristic variables that define load and strength are considered explicitly. Every variable is considered a random variable with a specified distribution.

Variables on the strength side include yield strength, wall thickness, and OD. Load variables include BHP, fracture pressure, fluid density, kick frequency, etc. In general, the true variability and uncertainty in these parameters may not conform to a specific probability distribution, and the true distributions must be respected. The strength distribution is easier to quantify, since most of the strength parameters are measurable and available. Section 2.2.4 includes an example of the determination of strength distribution analytically using a first-order reliability method.

The load side is harder to quantify. Both variability in the load magnitude and probability of occurrence of the load have to be considered. For instance, it is customary to use a gas gradient to surface from the fracture strength at the shoe for a well control load. However, the probability of occurrence of this load is not 100% as implied in a deterministic design. The probability of occurrence of this load is dependent upon operational practice as well as *Mother Nature* and is an unlikely event. Moreover, the extreme load conditions assumed in design may never have been observed in practice. Therefore, statistical modeling of load distribution is often difficult and may require application of extreme-value theories.

Suppose $R(\tilde{x})$ is the distribution of the strength from a limit state, and $Q(\tilde{x})$ is the distribution of the load effect, where $(\tilde{x})$ are the random variables that determine the strength and load effect. Then a limit state function for the pipe may be written as

$$g(\tilde{x}) = R(\tilde{x}) - Q(\tilde{x}). \quad\dots\dots\dots\dots\dots\dots\dots\dots\dots\dots\dots\dots\dots\dots\dots\dots\dots \quad (2.38)$$

Eq. 2.38 is known as the g function. With the load and strength distributions known, the probability that the load will exceed strength can be determined easily. This is the probability of failure, and its complement is reliability. In terms of the limit state function, the probability of failure P_f is

$$P_f = P\left[g(\tilde{x}) < 0\right]. \quad\dots\dots\dots\dots\dots\dots\dots\dots\dots\dots\dots\dots\dots\dots\dots\dots\dots \quad (2.39)$$

Depending upon the load and strength, the two distributions may have an interference area, as shown in **Fig. 2.5.** The interference area is the probability of failure P_f for that particular load and strength. The complement of this $(1 - P_f)$ is the reliability of that particular design. This approach is sometimes referred to as *quantitative risk analysis* (QRA).

Since interference of tails is of interest, the statistical methods used to estimate the probability of failure should accurately model the tail probabilities. Using Monte Carlo simulations to find P_f may not be adequate (or practical) if probability of failure less than 10^{-3} is sought. In general, 10^{-x} reliability requires at least 10^{x+2} trials for a Monte Carlo simulation to be representative.

Resistance-Based Probabilistic Design. In the absence of a properly modeled load distribution, it is prudent to use the resistance distribution and a deterministic load representing a worst-case load. Then, in Eq. 2.38, load is deterministic. The probability of failure thus determined is indicative of a worst case or upper limit of the probability of failure.

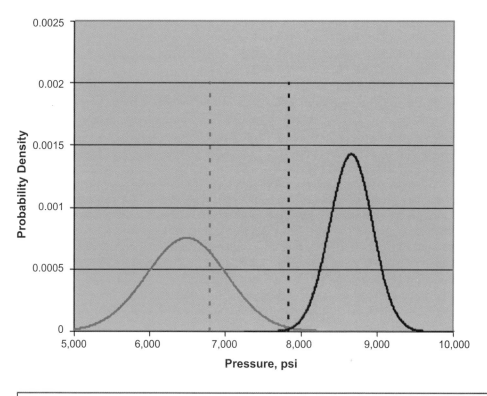

Fig. 2.5—Example of interference between the burst load and casing strength distributions.

Load and Resistance Factor Design (LRFD). Using QRA or a classical reliability-based design approach is computation- and statistics-intensive. It may not be practical for routine casing design. The LRFD approach retains the probabilistic nature of QRA while utilizing familiar equations and design checks to perform the design. The design factors are precalibrated to the failure probabilities. Hence, LRFD is a deterministic design process but has all the probabilistic capabilities built into it.

LRFD first recognizes that load-side uncertainties, dictated by operational practice, instrumentation error, and natural uncertainties, are intrinsically different from strength-side uncertainties, dictated by manufacturing process and quality control. Rather than use a single factor of safety that has to account for these vastly different types of uncertainties, LRFD uses individual load factors (L_f) and resistance factors (R_f). The resistance factors are calibrated to the uncertainty in the strength as well as the uncertainty in the limit state being used to determine strength. The load factors are calibrated to the uncertainty of each load, the deterministic design check being used, and some preselected target failure probabilities. The process of calibration encompasses the entire tubular population, usage distribution, and design envelope.

Once calibrated load and resistance factors are determined, they can be used in a deterministic design check. A desired target probability of failure is selected, and the appropriate load and resistance factors are obtained, depending upon the load type and the class of material. If a particular design satisfies this design check, the design has a probability of failure lower than the target used in selecting the factors. Thus, in terms of the load effect and resistance, the design check may be stated as

$$L_f \times Q(\overline{x}) \leq R_f \times R(\overline{x}), \dots\dots\dots\dots\dots\dots\dots\dots\dots\dots\dots\dots\dots\dots\dots\dots\dots\dots \text{(2.40)}$$

where ($\overline{x}$) = deterministic parameters used in a design check equation, Q = load effect, and R = strength in the chosen deterministic equation.

2.5 Design of Connections

Tubulars used in most wells in the oil industry have threaded connections. Two major types of connections are API connections and proprietary connections. API connections are manufactured according to specifications and tolerances provided by the API. Proprietary or premium connections are designed and manufactured by commercial manufacturers. Specifications for proprietary connections are determined by each manufacturer.

Studies have shown that most tubular failures are related to connections; estimates of 85 to 95% have been reported (Schwind 1998). Typically, connections receive little attention when designing tubulars. Some reasons for this include the large number of connections available and lack of understanding of connection performance. Connections must withstand the same loads applied to the pipe body, and often the connection is weaker in tension, compression, pressure, or bending.

The VME stress limit curve of the pipe is the ideal goal for a connection. Connections that match or exceed this limit under different combinations of pressure and axial load are said to be *transparent.* Unfortunately, this ideal is harder to realize than may be expected. For any connection, therefore, it is important to quantify departure from the ideal, through testing, analysis, or a combination of both. This is far more complex for a connection than it is for the pipe body due to the load effect on a connection, the need to consider both mechanical integrity and leak resistance, and the dependence of connection performance on the load and the load path or sequence.

2.5.1 API Connections. The original API tubular specifications were established in the early 20th century with the intent of standardizing pipe sizes and connections so that material from one mill/user could be assembled with material from other sources. The original API specifications thus focused on establishing interchangeability.

As deeper wells were drilled and loads became higher, leak resistance and tensile capacity became more important. Improvements in API connections came about after research by steel companies and operators. Makeup procedures, development of API thread compound, connection coating, upset pin ends, and other improvements helped to improve API connection performance.

API connections include 8 round, buttress thread and coupled (BTC), and extreme line (X-Line). API 8 round refers to 8 threads per inch and round shape of the thread crest and root. These connections are available in long and short couplings, referring to the length of the threads. LTC (long thread and coupled) and STC (short thread and coupled) have different axial and leak resistance performance. BTC couplings include regular and special clearance. X-Line connections are a pin-by-box connection with a metal-to-metal seal. X-Line connections are not very common and have been replaced for the most part by proprietary connections.

Operational limits for API connections vary by well type and loads and by operator's guidelines. General guidelines from published literature indicate that API connections can provide reliable service for the majority of wells drilled and completed (Klementich 1995).

The internal pressure resistance for a given pipe and connection is the lowest of the internal yield pressure of the pipe, the internal yield pressure of the coupling, or the internal pressure leak resistance at the connection critical cross section. The internal leak resistance pressure is based on the interface pressure between the pipe and coupling threads resulting from makeup and the internal pressure itself, with stresses in the elastic range. Thread tolerances, surface treatment, pipe *dope* application (and type), and tension all can affect leak resistance.

A leak path exists in both 8 round and BTC connections due to the thread-cutting process. This helical path is the void between the 8 round thread root and crest and the void between the BTC stab flanks. The leak path must be plugged with solid particles in the thread compound. API thread compound consists of base organic grease with lead, graphite, and other solids to provide lubrication between the threads to prevent galling during makeup and to plug the helical path. One of the problems with thread compound is deterioration with time and temperature, resulting in loss of sealability through the thread leak path. High temperature (>250°F or 121°C) can cause the compound to evaporate, dry out, and shrink. Gas can penetrate the organic grease, and the base grease can react with well fluids, resulting in loss of the seal.

API 8 Round. Connection leak resistance and joint strength equations for API 8 round connections are given in API *Bull. 5C3* (1999). Variables for coupling internal yield pressure include yield strength, coupling OD, pitch diameter, thread length, taper, and thread height. Joint strength variables include cross-sectional area, OD and ID of pipe, yield strength, and ultimate strength. The round thread is designed to provide a seal at both the stab flank and load flank. The voids at the root and crest are filled with thread compound to prevent a spiral leak path.

A tension load applied to an 8 round connection tends to open the stab flank, creating a leak path. The effect of tension on leak resistance was demonstrated by Thomas and Bartok (1941). Test data showed that if a safety factor of 1.6 was honored for tension, API connections did not leak at the applied pressure load.

As the tension load approaches 50 to 62.5% of API tension rating, the leak resistance for 7-in.-OD and larger connections should be derated. The effect of tension on leak resistance varies by grade, OD, and wall thickness. In general, the leak resistance decreases dramatically as the OD increases from 7 to 20 in. (Schwind 1990).

API BTC. Buttress threads are designed to resist high-axial-tension and -compression loading. Connection leak resistance and joint strength equations for API BTC connections are also given in API *Bull. 5C3* (1999). Factors that affect leak resistance for BTC connections are stab flank clearance, contact pressure, and coupling yield. The combination of minimum stab flank clearance and Teflon-impregnated thread compound prevents leaks through the spiral leak path.

More details on API connections, including diagrams and rating equations, can be found in Bourgoyne et al. (1986).

2.5.2 Proprietary Connections. The connection loads generated by wells with higher pressure, increased depth, and higher temperature may require connections with higher capacity than API connection. Alternatively, tight running clearances may dictate a connection geometry that is flush or slightly larger than the pipe OD. For these applications, the industry has developed proprietary or premium connections. These connections are designed to increase either the performance or the geometric transparency with respect to the pipe body. The cost of premium connections ranges from two to five times the cost of API connections.

Typical proprietary connections have a metal-to-metal seal that is energized during makeup. Some connections have more than one seal or backup elastomer seals. It is difficult or impossible to test multiple seals under field conditions.

Some of the problems with connections in the past include lack of sealability from external pressure and low efficiency in compression loading. Designing a connection for internal pressure and tension ratings of the pipe body is fairly easy. Designing a connection that is efficient in external pressure and compression is more of a challenge.

Because proprietary connections are developed by companies in competition with one another, there are no industry standards or specifications. Therefore, the users of proprietary connections need to be aware of the limitations.

Operating Envelopes. One of the main issues with premium connections is that of determining what the performance limits are for a given connection, size, weight, and grade. Although finite-element analysis is a valuable tool used in connection design, physical testing is the only way to establish the operating envelope with reasonable certainty. Testing involves a number of connection samples and a test program to establish an operating envelope. **Fig. 2.6** is a typical performance envelope showing test points of combined pressure and axial force. It is common to use the pipe body VME ellipse plot with connection test loads superimposed to show how connection performance compares to the pipe body. Once loads that meet the test criteria and failure loads are known, an operating envelope can be developed. It is also possible to have a gas envelope and a liquid envelope superimposed on the same diagram.

Typical requirements for a test program may include some combination of the following:

- Make-and-break tests to identify galling tendencies
- Gas and liquid sealability

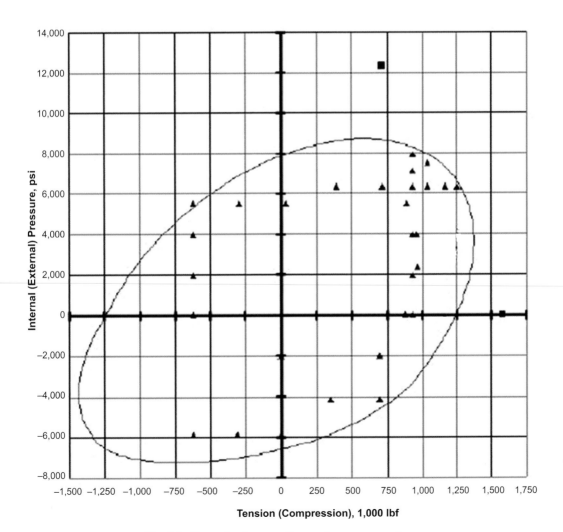

Fig. 2.6—Example of connection performance envelope.

- Maximum pressure and tension load
- Maximum pressure and compression load
- Thermal cycling
- External pressure for leak resistance from the annulus
- Minimum and maximum makeup torque
- Minimum and maximum thread interference
- Minimum and maximum seal interference
- Loading the connection to failure

An operating envelope needs to be established in a planned manner that simulates the downhole conditions that a connection may be exposed to. The connection supplier may have test data to adequately define an operating envelope to compare against design loads. Service conditions outside the experience envelope may require a test program. A connection will be qualified on the basis of these loads and the expected operating conditions, keeping in mind the potential for performance to be dependent on the sequence of applied loads. The result is a fit-for-purpose connection that meets the requirements of a particular well or field.

An option that is usually more costly and time-consuming is to qualify a connection to have the same capacity as the pipe body. This requires defining a test program that covers the extremes of pressure and tension. Some limitations of this method are that the results for a given connection, size, weight, and grade are not necessarily scalable to other sizes, weights, and grades for the same connection design. Each connection must be treated independently until analysis and testing defines the operating envelope.

Fig. 2.6 shows a performance envelope for an example proprietary connection, with test loads indicated as triangles. This envelope includes test points in all four quadrants of pressure and force combinations. The square data points indicate loads to failure.

Connection Qualification Standards. Several attempts have been made to standardize connection testing. These include API *RP 37* (*RP 37* 1980) in 1958. This standard was not widely used due to testing requirements that were extreme, took considerable time, and were costly.

A joint industry project (DEA-27) was organized in 1986 through the Drilling Engineering Association (DEA) to address problems with high clearance and flush joint connections (Payne and Schwind 1999). The test program was straightforward and dealt with tension and internal pressure loads. Testing led to improved connection designs. The original test program did not include external pressure testing. An addendum to the DEA-27 test program to include external pressure tests was created. Test results showed that flush connections tended to leak when external pressure was applied. Since the connections were designed to hold internal pressure, it is not surprising to find that leaks occurred at pressures as low as 40–50% of pipe body collapse (Payne and Schwind 1999). Testing showed that external pressure can deform the pins, and subsequent internal pressure tests failed. This highlighted that the load sequence can affect connection performance.

An API task force addressed a connection standard testing procedure in 1985. The result of the work was API *RP 5C5* (1990). This standard was published in 1990. A fairly large number of test specimens are required for some of the service classes designated by *RP 5C5* (1990). For this reason, testing to *RP 5C5* (1990) requirements is costly. Material requirements and thermal cycling and seals testing were included in this standard.

An international standard, *ISO 13679—Procedures for Testing Casing and Tubing Connections* (*ISO 13679* 2002), has been developed to address connection testing issues. Experience from previous connection testing standards has been incorporated in this standard.

This standard divides test severity into four test classes, with Connection Application Level (CAL) IV being the most severe for gas service at elevated temperature. CAL I class is tested at ambient temperature with liquid for least severe applications. Eight test samples for CAL IV and three for CAL I are required for *ISO 13679* testing, compared to as many as 27 for an API *5C5* test. CAL II and CAL III are service classes between classes IV and I and are intended for gas and liquid service at less severe conditions than CAL IV. Depending upon the severity of the application, the connection is tested for internal as well as external pressure, with and without axial load and bending, and in different directions along the VME curve, in order to assess the impact of load sequence. Some tests such as bending and external pressure are optional or not required for CAL II and III.

One of the strengths of *ISO 13679* is the documentation of connection geometry and performance data. This document fully defines the connection and requires the manufacturer to define the loads the connection can sustain. The service load envelope (SLE) defined by the manufacturer can be verified using *ISO 13679* procedures. In addition to confirming the SLE, an estimate of the limit load envelope (LLE) can be determined by testing to failure. A cross section drawing and a test load in graphical form (VME plot) is to be provided by the manufacturer. Quality control procedures for the manufacture of test specimens shall be documented and consistent with procedures for connections manufactured for well service.

Detailed load steps and the sequence of loading are to be documented. The order of loads applied may affect seal integrity. Also, tolerance extremes along with makeup extremes are tested.

Test Steps Summary. An example test loading program may consist of pressure and tension load points that progress around the VME ellipse several times in counterclockwise (CCW) and clockwise (CW) directions. Referring to **Fig. 2.7,** the outer ellipse is pipe body VME and the inner ellipse is 95% VME. The sequence in the testing is summarized here:

- Start at Load Point 1 with 0 pressure and tension for 95% VME. Progress CCW through Points 2–9 by increasing pressure up to the maximum and adjusting the tension for 95% VME. Hold times are from 5 to 15 minutes except the Point 2 hold time of 60 minutes. At Point 9, switch from internal to external pressure and progress to Point 14.
- Travel CW from Point 14 to Point 10. Change from external pressure to internal pressure and progress CW back to Point 1.
- Reverse the loading again, tracing CCW from Point 1 to Point 9. Change from internal to external pressure and travel from Point 10 to Point 14.
- Switch from external pressure to internal pressure. Finish moving CCW to Points 1 and 2.

At the end of testing, a connection test report and a test load envelope are to be prepared.

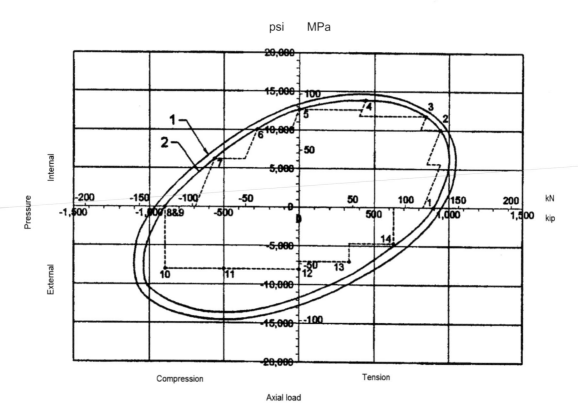

Fig. 2.7—Example of test load envelope (*ISO 13679:2002*).

2.6 Material Selection Considerations in Casing Design

The selection of material is one of the most important and most often overlooked parts of a complete casing design. In its simplest form, the selection of material is the specification of the material grade. Additionally, consideration of metallurgy, heat treatment, the environment in which the material is exposed, and inspection criteria are all part of material selection.

Material selection invariably involves a fairly detailed consideration of corrosion and metallurgy. Moreover, several metallic and alloy components are involved in a complete well design, including casing, tubing, drillpipe, tool joints, surface equipment, line pipe, liner hangers, and other auxiliary and special-purpose equipments. Such a broad treatment of the subject is beyond the scope of this text. An attempt is made to address the key concepts and fundamentals of material selection, with specific focus on casing design. A detailed and useful review of oil industry metallurgy and corrosion is provided by Craig (1993), and is recommended for further reading.

Material considerations are usually a far greater concern for other oil industry components such as surface equipment and tubing than they are for casing. Environmental exposure for drilling casing strings is usually accidental and of short duration, such as in the instance of a kick or a tubing leak. Sections of production casings and liners below the packer and the production tubing are continuously exposed to the produced fluid, which can contain corrosive elements. The production tubing by casing annulus can contain brines or other possible corrosive fluids for long durations. Stimulation with acid or CO_2 injection can expose the tubing and production casing/liner to corrosive environments for a moderately extended period. In general, packer fluids and stimulants are treated with inhibitors to minimize corrosion. Although the external environment outside the pipe is a consideration, it is not a significant concern for casings.

2.6.1 Oil Industry Materials. Alloy steel is the most common oil industry material. Several grades of steel are used in design and are classified by API from H40 to Q125. API *Spec. 5CT* (1998)/*ISO 11960:2004* lists the various grades, with their corresponding chemistries and mechanical properties. All the standard API grades are low-carbon (< 0.5%) steels with chromium, molybdenum, and manganese as key alloying elements. In addition to yield strength variations permissible, tensile strength and hardness are key properties of interest. The difference in properties is obtained by a combination of chemistry and heat treatment. The heat treatment of oil industry tubulars varies widely, but in general consists of the following two steps:

- Cooling to room temperature, either normalized in still air or oil or quenched. Quenching, or rapid cooling with water, leads to a finer microstructure and hence higher strength, but also reduces ductility.
- Tempering, or raising the temperature back to a critical temperature (usually the lower critical temperature of the material), to regain ductility. This is especially important for quenched alloys.

Both seamless and electric-resistance-welded (ERW) manufactured tubulars are available. Each has its own advantages and disadvantages. Many designers prefer seamless because ERW pipe has a weld seam, but a carefully quality-controlled and inspected ERW pipe with adequate heat relief of the weld will perform in a manner similar to seamless pipe. The benefits of lower eccentricity, more-uniform wall thickness, and better surface finish offered by ERW can be of great advantage in some design situations.

Beyond the standard API grades, several non-API grades are often recommended by designers, depending upon the loads and expected corrosion problems. These include high-collapse-strength casing, low-alloy steels, and nickel- or cobalt-based corrosion-resistant alloys (CRAs). Tubing and production casing are among the tubulars often considered for CRA applications because they are most likely to be exposed to corrosive environments over a long duration.

One of the most important classifications of API grades with implications on material selection is the classification into groups. **Table 2.6** shows the grouping of the different grades. Of all the groups, Group 2 has the greatest control on chemistry, heat treatment, and mechanical properties, in an attempt to improve its resistance to SSC.

Hardness is another important property indicative of the ability of a material to perform when exposed to corrosive environments. It is a nondestructive measure of the ultimate strength of the material and is

TABLE 2.6—API GROUPING OF CASING AND TUBING GRADES		
Group	Grades Applicable to	Comments
1	H40, J55, K55, N80	Limited control on chemical composition. Not normalized unless required by purchaser.
2	M65, L80 (including 9Cr and 13Cr), C90, C95, T95	Most stringent control of chemistry and mechanical properties. Restricted chemistry. Specified hardness maximum and allowable hardness variation. More detailed heat treatment.
3	P110	Limited control on chemical composition.
4	Q125	Controlled chemistry and properties. No hardness maxima but control on hardness variation.

indicative of the brittleness or, conversely, toughness of the material. Toughness of a material may be understood to be its ability to resist propagation of a crack or a fracture under conditions of stress. Glass is as strong as many steels in tensile strength, but is not tough at all in comparison to steel. This is because even a very small crack in glass tends to propagate very quickly under an imposed stress. Thus, in addition to tensile strength and stiffness, toughness is an important determinant of the strength of a material.

Hardness is measured using several different scales. The main difference between the scales relates to the method of application of the hardness indentation and the load at which it is applied. The common scale used in the oil industry is the Rockwell C scale, and a number known as the HRC measures the hardness on the Rockwell C scale. Typically, HRC ≤ 22 is desirable when exposure to corrosive environments is likely. (The other common scales are Rockwell B, which usually is applied to softer materials with lower tensile strengths (57–110 ksi), and the Brinnell scale, applied for field testing or rapid testing of redundant items.

Another common test to measure the toughness of a material is the Charpy impact test, wherein a specimen with a notch is subjected to an impact from a pendulum hammer. Upon impact, a part of the energy is absorbed by the specimen as it fails, and the hammer rises to a lower height than its release height. The difference in heights is the energy absorbed by the specimen and is measured as the *impact energy* in ft-lbf or J. For some grades, the minimum impact energy is specified.

2.6.2 Corrosion. Oil industry tubulars are invariably subjected to corrosive environments. The goal is to select the metallurgy and corrosion inhibition approach such that the tubulars continue to be able to withstand the loads, both cyclical and static. Corrosion can be categorized in several different ways. For casing, the following categorization is convenient:

- Uniform corrosion
- Pitting corrosion
- Environmentally assisted cracking

Uniform corrosion, in which corrosion results in a generalized and uniformly distributed loss of material at a constant rate, is an idealized view of corrosion. This is typically the design basis too, especially for production tubing where corrosion is predicted in terms of *mils* or mm per year. Pitting is a common corrosion mechanism, the attack and severity being dependent upon the environment. The pits form and corrode rapidly, causing failure in very short times. Once pitting is initiated, it becomes the anode, and the unpitted surroundings act as a cathode. The anodic reaction reduces the pH of the environment locally, causing further corrosion, until failure. Environmentally assisted cracking is the most important corrosion mechanism in the selection of materials. Stress corrosion cracking (SCC; cracking in the presence of tensile stress and corrosive environment), SSC (cracking in the presence of H_2S), and chloride stress cracking (cracking in the presence of chlorides) are all classes of environmentally assisted cracking. Environmentally assisted cracking is usually a consequence of the atomic hydrogen resulting from the electrochemical corrosion reaction infusing into the metal matrix, leading to embrittlement damage. As a result, the material fails at a stress far below the ductile limits of the pipe. A crack or imperfection in the presence of a deleterious environment and a tensile stress is the most potent combination for this failure mechanism.

The Corrosion Cell. Corrosion is an electrochemical process and involves the flow of electrons. It therefore requires an electrochemical cell consisting of one surface that discharges electrons and another that receives it. The anode is responsible for the discharge of electrons. The reaction at the anode is an oxidation reaction, which results in the metal going into positive metal ions and releasing electrons in the process:

$$M \rightarrow M^+ + e$$

In the case of steels, the metal involved is iron, and the reaction is typically

$$Fe \rightarrow Fe^{+2} + 2e, Fe^{+2} \rightarrow Fe^{+3} + e$$

The released electrons are received by the cathode, resulting in a reducing reaction

$$2H^+ + 2e \rightarrow H_2$$

Anode and cathode are quite often part of the same string. Anodes are located at surface cracks, stress concentrations, scratches, etc. A cathode is readily provided by exposed steel or surface impurities. The environment hastens the reaction. Also, depending upon the environment and the pressure loading, the free atomic hydrogen can infuse into the metal matrix before it becomes molecular, causing hydrogen embrittlement and damage.

Corrosion can be mitigated by breaking the chemical reaction. There are four typical approaches:

- Using coatings to change the metal-solutions reaction interface and provide a temporary barrier to the reaction
- Using chemical inhibitors that slow down the chemical reaction
- Using cathodic protection by changing the corrosion cell such that the metal that needs protection becomes a cathode at the expense of a sacrificial anode
- Changing the metallurgy to provide resistance to corrosion

Of these, chemical inhibitors remain the most popular approach with oil industry tubulars. When this does not work, designers tend to raise the alloying content, typically chrome or nickel, to increase the resistance to corrosion.

2.6.3 Environmental Considerations in Corrosion of Casing.
The most common environmental considerations in the selection of materials for casing are chloride, CO_2, and H_2S. Temperature, pressure, and pH are also part of the definition of the environment and influence its impact.

Chlorides and CO_2. Produced fluids often contain CO_2 and chlorides, usually in combination. The typical corrosion progress begins with pitting, followed by SCC at the tip of the pits. Both of these mechanisms are usually combated by using high-chromium alloys.

The rate of corrosion in a CO_2 environment is a function of the partial pressure of CO_2, temperature, and duration of exposure. A classical method for determining the rate of corrosion to be expected in the presence of CO_2 was given by de Waard and Milliams (1975):

$$\log R = 8.78 - \frac{2320}{(T+273)} - 5.55 \times 10^{-3}T + 0.67 \log p_{CO_2}, \quad \dots\dots\dots\dots\dots\dots\dots\dots\dots \quad (2.41)$$

where
  R = rate of corrosion, mil/yr
  T = temperature, °C
  p_{CO2} = partial pressure of CO_2, psi

This equation does not take the grade into consideration, nor does it consider the effect of the formation of a protective carbonate layer on the metal during CO_2 corrosion. It therefore gives conservative values for the corrosion rate, but it is a useful check in determining the need for resistant grades.

Increasing the Cr content usually helps in combating CO_2 corrosion. The threshold Cr level is usually 12% to achieve resistance to CO_2 corrosion, which explains the 13Cr alloys in the API tubular grades.

Chlorides enhance SCC of the steels. At high temperatures, the SCC impact of chlorides worsens. At temperatures above 350°F, normal steels can no longer be used, and nickel-based CRAs are required. Also, the presence of O_2 and free sulfur, and high partial pressures of CO_2 and H_2S, may prompt the application of CRAs. Duplex steels containing both ferrite and austenite microstructures offer good resistance to chloride stress cracking. However, as the H_2S partial pressure increases, the usefulness of the material diminishes since presence of H_2S with CO_2 can cause severe pitting corrosion in duplex. In such a case, nickel-based alloys may be required.

Hydrogen Sulfide and SSC. Oil industry tubular selection is most critically affected by the presence of H_2S. Although H_2S also causes pitting corrosion, the bigger concern is that it causes catastrophic SSC. The tendency for SSC is dependent upon the partial pressure of H_2S, the solution pH, temperature, presence and size of initial imperfections, and the material composition or alloying. Low-temperature, high-stress conditions lead to the most severe susceptibility to SSC. Unfortunately, this means that SSC is a major concern higher up the string and often the worst near the surface.

Fig. 2.8 shows the effect of temperature on SSC of tubulars. Low-strength tubulars are less susceptible to SSC. As the strength increases, so does susceptibility to SSC. As the temperature increases, SSC tendency decreases, and higher-strength material can be used.

Fig. 2.9 shows the effect of the concentration of H_2S on the SSC tendency of different oilfield tubular grades. As the figure shows, the behavior of normalized and tempered steels is worse than that of quenched and tempered steels. Once again, higher strength leads to lower tolerable levels of H_2S before SSC becomes a concern.

A general rule of thumb for carbon and low-alloy steels is to maintain hardness below HRC 22. The acceptable hardness limits for sour service for oil industry tubulars are specified by the National Association of Corrosion Engineers (NACE) in NACE-MR-01-75 (NACE *Standard MR-01-75* 2003). Detailed chemistry requirements are specified in the standard. The impact of temperature on SSC susceptibility is also discussed in this standard. One important observation from the standard is that higher-strength tubulars can be used deeper in a well where the temperature is higher. For instance, P110 can be used at temperatures in excess of 175°F without risking SSC, provided that the material is not subsequently cooled after exposure to H_2S.

2.6.4 Design for Environment. Design and selection of materials is subject to several complex considerations beyond the scope of this chapter. When environmental considerations dictate material selection, it is

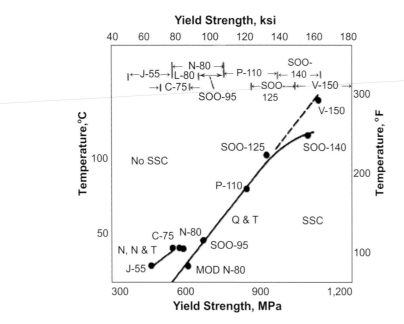

Fig. 2.8—Effect of temperature on sulfide stress cracking of casing and tubing material (Kane and Greer 1977).

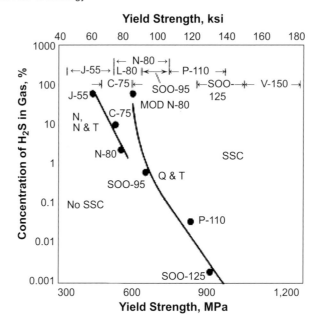

Fig. 2.9—Effect of H_2S on sulfide stress cracking of casing and tubing material (Kane and Greer 1977).

important to involve the expertise of metallurgists and corrosion engineers. This section reviews the approaches used to assist in the selection of materials.

General Rule of Thumb Approaches. Most material selection exercises are generalized rules of thumb. We have already discussed some rules of thumb. A summary that includes them follows:

- For CO_2 corrosion, if the partial pressure of CO_2 is less than 7 psia, corrosion is not likely; if it is between 7 and 30 psia, moderate corrosion is likely; and if it is greater than 30 psia, severe corrosion is likely (Eq. 2.41 may also be used).
- Maintaining hardness below HRC 22 improves SSC resistance and sour-service capability of carbon steel alloys.
- As temperature, pressure, CO_2, chloride, and H_2S concentrations increase, the selection progresses as
  - Plain carbon steels (H40–C75)
  - Sour-service carbon steels (L80, C90, C95, and depending upon temperature, P110)
  - 13% Cr with single-phase martensitic structure (this is still an API grade)
  - Duplex steels, with two-phase structure (alloying with Cr > 13%, with Ni), especially to combat CO_2 effects
  - Nickel alloys, to combat both CO_2 and H_2S environmentally assisted cracking
- Using a higher factor of safety in normal *ductile* design reduces the possibility of SCC (by limiting the stress level on the tubular).

Fig. 2.10 provides an approximate guideline for selection of materials as a function of the partial pressure of CO_2 and H_2S, provided that no chlorides are present. The NACE *Standard MR-01-75* (2003) is a very useful reference for material selection and should be consulted in detail when sour service is being considered.

Fracture Mechanics Approach. The strength of engineering materials, as measured by yield strength, is far lower than the strength suggested by the chemical bonding between individual atoms in the material. For instance, typical steels have chemical bond strength in the range of E/5 to E/10, where E is the modulus of elasticity. Yet, when used as load-bearing structures, their strength is startlingly lower than this limit, around 0.002E to 0.01E. This deterioration in strength is thought to be caused by imperfections and cracks.

Surface imperfections are always present in manufactured materials. In the presence of certain environments and stress, the cracks and imperfections have a potential to grow until they reach catastrophic proportions and

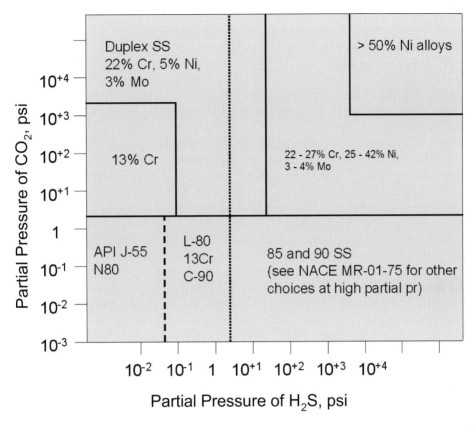

Fig. 2.10—Guideline for material selection in the presence of H_2S and CO_2 (after Sumitomo 2008).

lead to failure. Although standard design takes care of pressure, axial force, bending, temperature, and buckling stresses, it does not take into account the impact of imperfections on the integrity of the tubular.

Fracture mechanics is the study of the propagation of cracks and methods to mitigate against such propagation. Indeed, in engineering design, a primary obsession has been the search for methods to escape from the consequences of cracks and crack propagation. For centuries, this has been achieved by ensuring that structures remain in compression to keep cracks closed.

The diagram in **Fig. 2.11** shows the type of crack under consideration for tubular design. Some of the assumptions are as follows:

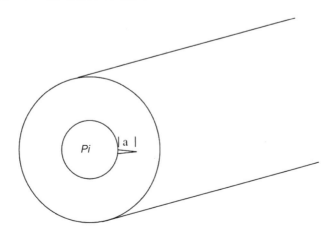

Fig. 2.11—Diagram of idealized crack in Oil Country Tubular Goods (OCTG).

- Crack propagation is perpendicular to the applied load.
- The pipe is infinitely long with an infinitely long longitudinal crack.
- Internal pressure acts on the faces of the cracks.
- The crack depth dimension is *a*.

A detailed review of fracture mechanics is beyond the scope of this document. Enough background is provided to understand the implications for casing design. Miannay (1997) provides an enriching review of the details of fracture mechanics.

Basics of Fracture Mechanics Applied to Design. In fracture mechanics, a crack is assumed to exist, and its behavior under load and exposure to environment is investigated. If a crack exists, then any load that tends to *open* it up results in excessive stress at the crack tip. The crack opens up in response to this load in order to relieve the strain energy that is building at the tip. However, for the crack to open up, it has to overcome the surface energy required to break the chemical bond strength or cohesive stress that keeps the material in one piece. Thus, propagation of a crack depends upon the competition between the energy relief provided by opening up and the energy required to open up. For lengths of crack smaller than a *critical* length, it is energetically more efficient for the crack to not propagate, which is the desirable condition. When the crack length approaches or exceeds the critical length, it is energetically more efficient for the crack to grow catastrophically, leading to failure.

The property that indicates the critical length of cracks is the fracture toughness of the material. Glass, which has nearly the same tensile strength as steel, has very low fracture toughness, and the critical length of a crack even in the presence of a small tensile load is very small, on the order of microns. Glaziers use this property when they *cut* class by creating a small surface crack and applying tension, usually in the form of a small bending moment. The crack propagates through the glass plate at the crack location, resulting in a clean break. For some steels, the critical crack length can be meters long. Achieving adequate fracture toughness without sacrificing strength is one of the key challenges for materials scientists and metallurgists.

In oil industry tubulars, surface imperfections always exist. Unless inspected, surface imperfections or defects cannot be detected. Even when inspected, there is a limit to the smallest defect depth that can be detected by current inspection methods. A typical detection limit is 5% of the wall thickness.

The effect of a load on the crack depends upon the load and crack orientation. A crack can be stressed in one of three modes, as shown in **Fig. 2.12.** For casing strings, the most common mode is *Mode 1,* or the opening mode.

When a material with a crack is loaded according to Mode 1, the stress just ahead of the crack tip increases. The stress intensity factor, K_1, is a measure of the stress state ahead of the crack tip. The principal stresses

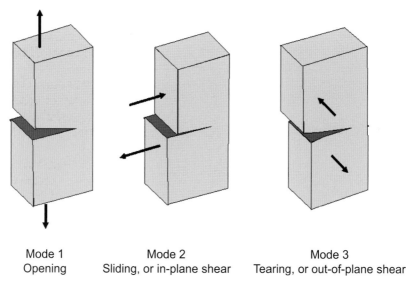

Mode 1	Mode 2	Mode 3
Opening	Sliding, or in-plane shear	Tearing, or out-of-plane shear

Fig. 2.12—Modes in which a crack is stressed.

ahead of the crack tip can be expressed as functions of the stress intensity factor, K_1. The stress intensity depends upon the load, far-field stress, and crack geometry. One of the most basic expressions for stress intensity factor for an infinite plate with a crack is

$$K_1 = \sigma \sqrt{\pi a}, \quad \dots\dots\dots\dots\dots\dots\dots\dots\dots\dots\dots\dots \quad (2.42)$$

where

σ = far-field stress

a = defect depth

Note that the unit of stress intensity is stress times root length ($\text{psi}\sqrt{\text{in.}}$ or $\text{kPa}\sqrt{\text{m}}$). The K_1 values for several typical geometries and loading types are listed in standard references [i.e., Miannay (1997)].

For Mode 1, the principal stresses in the vicinity of the crack (at any distance r from the crack tip and polar angle θ) can be found in terms of the stress intensity:

$$\sigma_1 = \frac{K_1}{\sqrt{2\pi r}} \cos\frac{\theta}{2}\left(1 + \sin\frac{\theta}{2}\right) \quad \dots\dots\dots\dots\dots\dots\dots\dots\dots\dots \quad (2.43)$$

$$\sigma_2 = \frac{K_1}{\sqrt{2\pi r}} \cos\frac{\theta}{2}\left(1 - \sin\frac{\theta}{2}\right) \quad \dots\dots\dots\dots\dots\dots\dots\dots\dots\dots \quad (2.44)$$

$$\sigma_3 = v\left(\sigma_1 + \sigma_2\right) \quad \dots\dots\dots\dots\dots\dots\dots\dots\dots\dots\dots\dots\dots \quad (2.45)$$

The singularity at the origin is resolved by assuming a plastic zone around the crack tip, and limiting the elastic fracture mechanics analysis to the region beyond this zone. The size of the zone can also be determined by the above equations. For instance, for plane stress, the radius of the plastic zone is given by

$$r_{\text{plastic}} = \frac{1}{6\pi} \frac{K_1^2}{\sigma_Y^2}. \quad \dots\dots\dots\dots\dots\dots\dots\dots\dots\dots\dots \quad (2.46)$$

Equations for K and principal stresses for different crack geometries and loading types are given in textbooks on fracture mechanics (Miannay 1997).

The material resistance to crack propagation depends upon its metallurgy and the impact of the environment in which it is used. When the environment is sour (i.e., in the presence of H_2S), the material resistance is known as K_{1SSC}, a critical K_1. When the stress reaches K_{1SSC}, the crack reaches its *critical* length in the material for the given load. Once this limit is exceeded, the crack grows catastrophically until failure. Thus, K_{1SSC} is a direct measure of the fracture toughness of the material. It can be thought of as a material property analogous to yield strength. However, its value is dependent on both metallurgy and the environment. Just like yield strength, K_{1SSC} has a statistical distribution for a given material and environment.

The fracture mechanics approach to design is to ensure that the application of the material is such that

$$K_1 < K_{1SSC}. \quad \dots\dots\dots\dots\dots\dots\dots\dots\dots\dots\dots\dots\dots\dots \quad (2.47)$$

A failure assessment diagram (FAD) is used to determine the safety of a structure for a given defect depth. **Fig. 2.13** illustrates a FAD for SSC. It plots the ratio of the applied K_1 to K_{1SSC} on the y axis, against the ratio of the applied load to the limit load of the structure on the x axis. The safe region is the one where the combined stress is less than the limiting condition. The FAD takes into account both the fracture mechanics and ductile loading. The presence of the defect reduces the allowable load that can be placed on the structure. When the defect is of critical length, no load can be placed on the structure without risking failure.

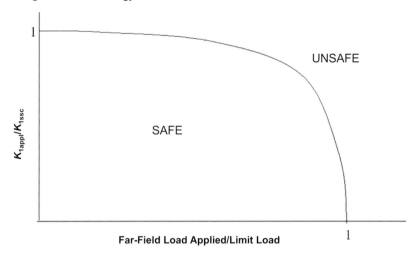

Fig. 2.13—Failure assessment diagram.

As mentioned, K_{ISSC} is dependent upon the material as well as the environment. For SSC, the environment is characterized by the partial pressure of H_2S and the temperature. K_{ISSC} is estimated through laboratory experiments that include double-cantilever beam tests and slow strain rate tests under the expected conditions of environment. Several measurements may be necessary to characterize the statistical spread of K_{ISSC}.

The stress intensity K_1 (the load side) is usually calculated from a suitable model using fracture mechanics theory. It assumes a given flaw geometry and loading condition due to internal pressure and tension. Several assumptions are made to simplify the mathematics of the estimation of stress intensity. Finite-element analyses also can be used to estimate the stress intensity. Solving for K_1 is especially complicated for real cracks in real materials.

Since K_1 depends on flaw geometry, controlling the defect size through tight inspection can increase the allowable internal pressure load for the same temperature and H_2S partial pressure. Thus, the material application can be tailored to the inspection level. By inspecting to a smaller defect size, the stress intensity at a given load can be reduced, and thus the pipe can be applied at a higher load level. The approach can be formalized in terms of *pipe application levels* based on defect size and environment.

ISO/TR 10400 Fracture of Pipe Body. Very few design approaches formally apply fracture mechanics to the selection and inspection of materials. Fracture of the pipe body in the presence of crack-like imperfection is addressed in *ISO/TR 10400* (2007). The same method as described above is used with some slight differences in nomenclature. The material fracture toughness is referred to as K_{mat} or K_{leac}. K_{leac} refers to the fracture toughness for environmental assisted cracking (subscript *eac*). This is a function of both the material and the environment in which the material is exposed. The value of K_{leac} will change depending on the environment, such as H_2S, temperature, pH, and water chemistry.

The fracture limit state function defines the FAD shown in Fig. 2.13. The equation used in *ISO/TR 10400* (2007) is shown in Eq. 2.48. The equation cannot be solved explicitly for the internal pressure; rather, an iterative solution is sought.

$$\left(1 - 0.14 L_r^2\right)\left(0.3 + 0.7 e^{-0.65 L_r^6}\right) = \frac{p_{iF}\left(\dfrac{d_o}{2}\right)^2 \sqrt{\pi a}}{\left[\left(\dfrac{d_o}{2}\right)^2 - \left(\dfrac{d_o}{2} - t\right)^2\right] K_{leac}} \left[2G_0 - 2G_1\left(\dfrac{a}{\dfrac{d_o}{2} - t}\right)\right.$$

$$\left. + 3G_2\left(\dfrac{a}{\dfrac{d_o}{2} - t}\right)^2 - 4G_3\left(\dfrac{a}{\dfrac{d_o}{2} - t}\right)^3 + 5G_4\left(\dfrac{a}{\dfrac{d_o}{2} - t}\right)^4\right], \quad \ldots\ldots\ldots (2.48)$$

where

$$L_r = \frac{\sqrt{3}}{2} \left(\frac{p_{iF}}{\sigma_Y} \right) \left(\frac{\dfrac{d_i}{2} + a}{t - a} \right)$$

a = defect depth
K_{Ieac} = fracture toughness of the material
L_r = load ratio, applied load/limit load
p_{iF} = internal pressure at the fracture

The parameters G_0–G_4 in **Table 2.7** are obtained exactly following the methodology in API *RP 579* (2000).

ISO/TR 10400 (2007) provides equations for the lower bounds for K_{Ieac} values. One equation is for sweet systems with no detectable presence of H_2S. The lower detectable limit of H_2S using currently available technology is around 1 ppm. The other equation is for sour systems. These equations are based on data in published literature.

The results for these two equations are tabulated in the *ISO/TR 10400* (2007) Annex E tables. The values for sweet-fracture capped end are similar to values for ductile-rupture capped end, except for some of the higher grades. The tabulated values for sour-fracture capped end are conservative, on the basis of published data. This may require measuring K_{Ieac} to determine a more representative value for fracture design.

The preferred method to determine K_{Ieac} for a given material is to expose the material to field environmental conditions and measure K_{Ieac}. This is done in a laboratory under controlled conditions using existing standard test methods.

ISO/TR 10400 Ductile Rupture of Pipe Body. A fracture mechanics approach for including material imperfections in the ductile rupture equation has been suggested in *ISO/TR 10400* (2007). *ISO/TR 10400* uses the Klever-Stewart ductile rupture formula, which includes a penalty for eccentricity in the pipe as well as penalty for the presence of a flaw or imperfection. Both penalties reduce the critical wall thickness. It is recommended that the minimum wall penalty be based on the minimum possible wall thickness, which is 12.5% of the wall thickness for standard API monogrammed casing. If a probabilistic design is desired, then the mean and standard deviation of measured wall thickness is used. For imperfections, if design is deterministic, then the maximum-size imperfection that can pass undetected through an inspection unit is used for the imperfection penalty. If a probabilistic design is desired, the depth of imperfection is still the same as for deterministic design, but the method uses a statistical representation of the frequency of imperfection. The frequency of occurrence of an imperfection has a significant impact on the probability of rupture at a given pressure.

TABLE 2.7—VALUES OF G_0–G_4 FOR FAILURE ASSESSMENT DIAGRAM

d/t or d_{wall}/t	a/t	G_0	G_1	G_2	G_3	G_4
4	0.0	1.120 000	0.682 000	0.524 500	0.440 400	0.379 075
4	0.2	1.242 640	0.729 765	0.551 698	0.458 464	0.392 759
4	0.4	1.564 166	0.853 231	0.620 581	0.503 412	0.427 226
10	0.0	1.120 000	0.682 000	0.524 500	0.440 400	0.379 075
10	0.2	1.307 452	0.753 466	0.564 298	0.466 913	0.398 757
10	0.4	1.833 200	0.954 938	0.676 408	0.539 874	0.454 785
20	0.0	1.120 000	0.682 000	0.524 500	0.440 400	0.379 075
20	0.2	1.332 691	0.763 153	0.569 758	0.470 495	0.401 459
20	0.4	1.957 764	1.002 123	0.702 473	0.556 857	0.467 621
40	0.0	1.120 000	0.682 000	0.524 500	0.440 400	0.379 075
40	0.2	1.345 621	0.768 292	0.572 560	0.472 331	0.402 984
40	0.4	2.028 188	1.028 989	0.717 256	0.566 433	0.475 028
80	0.0	1.120 000	0.682 000	0.524 500	0.440 400	0.379 075
80	0.2	1.351 845	0.770 679	0.573 795	0.473 108	0.403 649
80	0.4	2.064 088	1.042 414	0.724 534	0.571 046	0.478 588

The imperfection penalty implicit in the Klever-Stewart formula is based on fracture mechanics. The penalty is a generalization of the Klever-Stewart formula, which incorporates the effect of flaws. The imperfection penalty is based on the energy at the crack tip rather than the stress intensity associated with brittle burst. The approach used is as follows:

- Measure the crack tip energy J_1 for a representative sample of pipes.
- Calculate the applied J load for different pipes using finite-element analysis.
- Evaluate the pressure at which $J = J_1$, which is the limit load for the given defect.

This is identical to the use of K_1 and K_{1SSC}. The two approaches are equivalent since the crack tip energy is related to the stress intensity.

The imperfection penalty is applied in the rupture equation as a reduction in the wall thickness. If a is the maximum depth of flaw that can pass inspection undetected, and K_a is the correction factor based on the fracture mechanics approach used in *ISO/TR 10400* (2007), then t_{dr}, the wall thickness in the ductile rupture limit equation, is

$$t_{dr} = t_{min} - K_a a \dotfill (2.49)$$

Concluding Remarks on Materials Selection. A basic understanding of API grades, corrosion mechanisms, and fracture mechanics is useful in determining material suitability for the design environment. The API grade designation provides information on material strength, chemical composition, and mechanical properties. The presence of H_2S may restrict using some API grades. Corrosion is a function of both the downhole environment and material chemistry. Laboratory test data and field experience can provide insight into matching the material with the environment. Fracture mechanics also can be used to match materials with a specific environment and stress threshold.

Though the advanced well engineer should be competent in material selection, a metallurgical expert should be consulted for critical design issues.

2.7 Casing Special Problems

This section discusses other issues not normally addressed in routine casing design. Topics include thermal effects, buckling, annular pressure buildup (APB), wellhead growth, casing wear, and thermal well design.

2.7.1 Thermal Loading and Its Implications. The temperature of the wellbore is a function of its operating conditions. Wellbore temperatures change during various operations such as drilling, circulation, cementing, shut-in, production, or injection. Casing strings and the contents of the wellbore annuli respond to these temperature changes, defined as the difference between the temperatures during an initial or quiescent condition and some final operating condition. For example, a tubing string is installed during well completion. The packer may be set and the tubing hung off when the wellbore fluids are close to geothermal temperature. When the well starts producing, the tubing is heated. The change in temperature causes thermal expansion of the tubing. The tubing may buckle, depending on the tubing weight, the internal and external fluids, and the packer constraints on tubing movement. Simultaneously, the fluid in the tubing annulus expands and causes a pressure increase known as APB. These phenomena, which are the result of temperature changes in the wellbore, have a serious impact on the loads imposed on the casing strings in particular and the integrity of the wellbore in general. This section addresses some of the more serious problems that arise from thermally induced loads on the casing strings in the wellbore.

Temperature Prediction Approaches. Wellbore heat transfer is a mature discipline, and there exist standardized techniques to compute the temperatures in a wellbore. From the point of view of casing design and analysis, it is necessary to determine the temperatures in the wellbore during production, injection, drilling, and shut-in. The temperatures during cementing usually can be obtained as a special case of circulation followed by shut-in.

The nature of heat transfer in a well during production or injection through a tubing is discussed by Ramey (1962) in his classic paper. The heat transfer during circulation and drilling is addressed by Raymond (1969). Despite the passage of time, and notwithstanding the development of sophisticated computer programs that determine wellbore temperatures, these papers are classic works, since they describe the salient physics underlying wellbore heat transfer. A detailed discussion of wellbore heat transfer is beyond the scope of this section. The work by Hasan and Kabir (2002) includes a discussion of all issues concerning wellbore heat transfer and includes an extensive bibliography.

Temperatures are typically determined by using a dedicated wellbore thermal simulator. Given a well configuration, the simulator calculates the temperature distribution in the wellbore by using analytical solutions or numerical methods for different operational circumstances. The simulator treats the wellbore as an axisymmetric finite-difference grid (Wooley 1980). Heat transfer from the flow stream to the wellbore is calculated by using correlations for convective heat transfer. Natural convection in the annuli may be modeled by using correlations given by Dropkin and Somerscales (1965). Heat transfer in the tubing and cement sections is modeled by conduction. Heat transfer in the formation is calculated by solving the 2D heat conduction equations. Suitable boundary conditions are imposed far away from the wellbore and at the wellbore-formation interface. The output of a wellbore thermal simulator consists of the temperature in each string and annulus of the wellbore as a function of depth. In addition, the pressures in the flow stream are also calculated and provided as an output.

Fig. 2.14 shows example results of wellbore heat-up while producing at three different rates. These results were obtained using a commercially available industry standard wellbore thermal simulator. Only the temperature of the produced fluid is shown, but the temperature of all tubulars, annuli, and produced fluids are available as results.

A linear geothermal temperature gradient is assumed above. The production temperature curves are based on production of 629 BOPD (100 m³/d), 1,258 BOPD (200 m³/d), and 3,145 BOPD (500 m³/d).

Fig. 2.15 is an example of cool-down in the lower part of the well and heat-up in the upper section due to circulation while drilling. Also, example temperature difference calculations are shown for the uncemented portion of casing assuming the TOC at 5,906 ft (1800 m).

Geothermal Temperature Estimation. In thermal analysis, geothermal temperature often is the reference for calculating temperature differences. The geothermal temperature should be based on temperature logs and temperature data from offset wells or the general region. Geothermal temperature is often characterized by a geothermal temperature gradient, in °F per 100 ft (°C per 100 m), considered to be a property of the

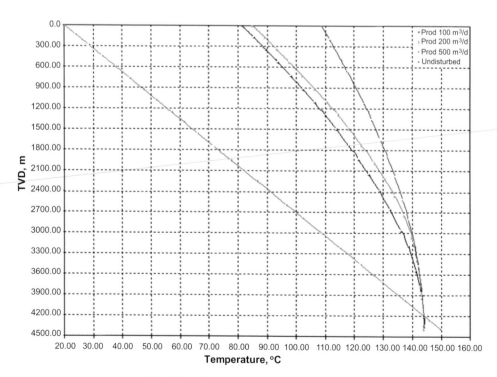

Fig. 2.14—Example of production heat-up.

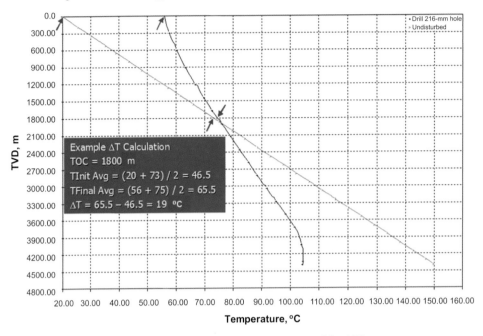

Fig. 2.15—Example of temperature change while drilling.

formations. Geothermal temperature gradients vary widely across the world and typically lie between 1.2°F per 100 ft (2.2°C per 100 m) and 1.8°F per 100 ft (3.3°C per 100 m).

For subsea wells, geothermal temperature estimation must include the temperature reduction from ambient as a function of water depth as well as the rapid increase in temperature to *normal* geothermal temperature below the mudline. A typical approach is given here:

- From surface to mudline, the temperature gradient characteristic of the waters in the region is used.
- The temperature at mudline is the temperature of water at the water depth to mudline. For most deepwater wells, the temperature at mudline is essentially constant at approximately 40°F (4°C).
- Below the mudline, a higher-than-normal temperature gradient is assumed such that the temperature at 1,000 ft (305 m) below the mudline is the same as it would be if it were a land well.
- At a depth greater than 1,000 ft (305 m), the normal geothermal temperature gradient is used.

Geothermal temperature estimation is critical in casing design, especially for deepwater and HP/HT wells. Therefore, care must be exercised in obtaining accurate temperature data. If the gradients have to be estimated, it is recommended that sensitivities be conducted over a range of expected geothermal temperature gradients. Additionally, the geothermal gradient may vary over the total well depth depending on lithology (e.g., through a thick salt body).

2.7.2 Steel Yield Strength Deration. For most steels, the yield strength reduces with increasing temperature. This thermal yield derating should be reflected in design. For casing, the rate of decrease in yield strength with temperature is a function of the grade under consideration and can be obtained experimentally. For design purposes, a linear thermal yield deration of 0.03% per °F above ambient (0.054% per °C) is used, though it may vary from 0.02 to 0.05% per °F (0.04 to 0.09% per °C).

Thermal yield deration of tubulars made of corrosion-resistant alloys is not as straightforward. Both chemistry and manufacturing process can significantly impact the shape and magnitude of deration. Data on the specific alloy may be required to accurately describe the derating curve.

In design, the derated yield strength is used in place of minimum yield strength in the strength calculations to account for the temperature effect. The temperature at the depth of interest is used for each load condition.

In limit state functions, ultimate strength is often used instead of yield strength. Ultimate strength of steels is not a strong function of temperature within the practical range of wellbore temperatures (< 500°F or 260°C). Therefore, it is not appropriate to derate ultimate strength for temperature.

2.7.3 Buckling and Post-Buckling Behavior. Fig. 2.16 shows a wellbore with uncemented sections of casing. Depending on operational conditions, the uncemented sections of casing strings can buckle. For example, the circulating drilling mud cools the lower sections of the wellbore and heats the upper section of the wellbore, resulting in the circulating temperature profile shown in Fig. 2.15. Depending on the depth of the drillstring and casing strings, geothermal temperature gradient, and flow rate, the 9⅝-in. (244.5-mm) liner may experience cooling while the 16-in. (406.4-mm) casing may experience heating. The uncemented portion of the casing may buckle if temperature increases during circulating mud are sufficient. Similarly, during production, the uncemented strings may experience greater temperature increases. If buckling occurs while drilling, it can promote casing wear due to contact between the rotating drillpipe tool joints and the exposed helix on the inside of the casing. Further, buckling creates additional bending stresses in the casing. If there is a potential for buckling, it must be considered during the design stage. Neglecting to do so can result in problems of well integrity.

Buckling of casing strings can also occur due to nonthermal loads such as formation subsidence or pressure changes in the annuli. It is important to understand the causes of buckling and determine whether the buckled state of the tubular can compromise the integrity of the string.

An unsupported casing string buckles when it experiences an *effective* compressive load greater than its critical buckling load. If the tubular is in an inclined hole, the tubular initially buckles into a sinusoidal shape. The load at which the string buckles is known as the critical sinusoidal buckling load. If a compressive load greater than the sinusoidal buckling load is applied, the pitch of the buckled tubular decreases. If the load is

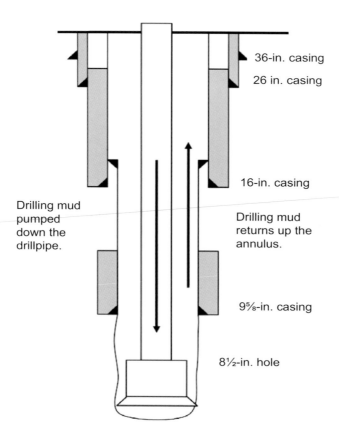

36-in. casing

26 in. casing

16-in. casing

Drilling mud pumped down the drillpipe.

Drilling mud returns up the annulus.

9⅝-in. casing

8½-in. hole

Fig. 2.16—Wellbore with uncemented casing.

gradually increased, there occurs a point when the tubular changes from a sinusoidal to a helical shape. The load at which this occurs is known as the critical helical buckling load.

Buckling of drillpipe and casing/tubing in constrained holes has been extensively studied by the drilling community. Many technical papers have contributed to the overall understanding of buckling and elastic stability (Mitchell 1982 and 1988; Sparks 1984; Hammerlindl 1980). These works address buckling of constrained tubulars in straight holes, curved holes, and inclined holes, and they discuss the effects of tubular weight on the buckling loads.

Effective Buckling Force. The axial force at any depth is given by Eq. 2.34. This is a true axial force that accounts for the weight of the casing, buoyancy, and any changes in force from the initial condition due to changes in temperature or pressure. Using the convention that tensile loads are positive and compressive loads are negative, the effective force is found by adjusting the true force in the string by internal and external pressure terms.

$$F_{eff} = F_a - p_i A_i + p_o A_o \quad \dots\dots\dots\dots\dots\dots\dots\dots\dots\dots\dots\dots\dots\dots\dots \quad (2.50)$$

The effective force is not a true force in that it cannot be measured with a strain gauge or weight indicator. Rather, it is useful for predicting buckling tendencies. The casing string has a tendency to buckle if the effective force is less than a critical buckling force.

$$F_{eff} < F_{cr} \rightarrow \text{buckling} \quad \dots\dots\dots\dots\dots\dots\dots\dots\dots\dots\dots\dots\dots\dots\dots \quad (2.51)$$

It is convenient to set the critical buckling force equal to zero, such that a negative or compressive effective force implies buckling.

Critical Buckling Loads. The critical buckling load is a function of the casing geometry, the hole angle and curvature, the clearance between the casing and the hole section, and the fluids inside and outside the casing.

Vertical Hole (Hole Inclination Less Than 15°). The critical sinusoidal buckling load in a vertical hole is given by the classical Euler buckling limit for a slender column:

$$F_{cr, \sin} = \frac{4\pi^2 EI}{L_{uc}^2}, \quad \dots\dots\dots\dots\dots\dots\dots\dots\dots\dots\dots\dots\dots\dots\dots \quad (2.52)$$

where

$F_{cr, \sin}$ = critical sinusoidal buckling load

I = moment of inertia of the tubular; $I = \frac{\pi}{64}\left(d_o^4 - d_i^4\right)$

The critical helical buckling load in a vertical hole is given by (Lubinski et al. 1962)

$$F_{cr, hel} = -1.94\sqrt[3]{EIw_b^2}, \quad \dots\dots\dots\dots\dots\dots\dots\dots\dots\dots\dots\dots\dots\dots \quad (2.53)$$

where

$F_{cr, hel}$ = critical helical buckling load, lbf
w_b = buoyed weight per unit length of the tubular, lbm/ft

$w_b = \left[w_{air} + 0.052\left(\rho_i A_i - \rho_o A_o\right)\right]$

w_{air} = air weight per unit length of the tubular, lbm/ft
Inclined Hole (Hole Inclination Greater Than 15°). The critical sinusoidal buckling load in a straight inclined hole is given by (Dawson and Paslay 1984)

$$F_{cr, \sin} = 2\sqrt{\frac{EIw_b \sin\varphi}{12r_c}}, \quad \dots\dots\dots\dots\dots\dots\dots\dots\dots\dots\dots\dots\dots\dots \quad (2.54)$$

where

φ = hole inclination, radians, and

$r_c = \dfrac{(\text{hole ID}) - (\text{tubing OD})}{2}$, the radial clearance.

The critical helical buckling load in an inclined hole is given by (Chen et al. 1990)

$$F_{cr,\text{hel}} = \sqrt{\frac{8EIw_b \sin\varphi}{12r_c}} = \sqrt{2}F_{cr,\sin}. \dotfill (2.55)$$

Curved Hole. The critical sinusoidal and helical loads are obtained by solving the following quadratic equation (He and Kyllingstad 1993):

$$F_{cr}{}^4 = \left(\frac{\beta EI}{r_c}\right)^4 (F_n)^2, \dotfill (2.56)$$

where

$F_n = \left[\left(F_{cr}\varphi' + w_b \sin\bar{\varphi}\right)^2 + \left(F_{cr}\vartheta' \sin\bar{\varphi}\right)\right]^{1/2}$,

β = 4, for sinusoidal buckling load,

β = 8, for helical buckling load,

φ' = build rate of the curved section,

ϑ' = walk rate of the curved section, and

$\bar{\varphi}$ = average hole inclination of the curved section.

Assumption in Casing Design. In casing design, it is convenient to assume that the critical buckling load is zero, which implies that the string buckles for any effective compressive force. This is a conservative assumption, especially for casing in high doglegs and at high inclination. Where needed, it is recommended that the above critical buckling limits be used.

Ways To Alleviate Buckling. As described in Section 2.7.3, the tubular is said to be buckled whenever $F_{\text{eff}} < F_{cr}$. This may happen only over the lower portion of the unsupported pipe. A buckling neutral point is often defined, above which the pipe is in effective tension and therefore cannot buckle.

Note that in determination of buckling tendency, the effective force is used rather than the real force. The real force acting on the tubular is the force that would be measured by an imaginary load cell placed across the axial cross section of a tubular immersed in a fluid. The effective force is a fictitious quantity defined by Eq. 2.50. Buckling can be alleviated by increasing the effective force. The real axial force can be increased by pulling additional tension prior to setting wellhead slips or by holding pressure while waiting on cement, keeping in mind that this latter operation may run the risk of forming a microannulus between the casing and cement. Alternatively, increasing external pressure may alleviate buckling, a practice that allows running tools through buckled tubing strings.

Buckling also can be addressed by raising the cement top above the depth where the effective force is zero. The effective force may remain below the critical buckling force below the TOC; however, the cement provides lateral support and prevents the pipe from otherwise buckling.

Post-Buckling Behavior. When the tubular is buckled, additional bending stresses are created in it. This bending stress is a function of the compressive load that causes buckling. The maximum bending stress occurs at the OD of the tubular (Lubinski et al. 1962).

$$\sigma_b = \pm\frac{d_o r_c}{4I} F_{\text{eff}} \dotfill (2.57)$$

The bending stress has the ± sign to indicate that the bending stress can be compressive (−) or tensile (+). The bending stress is ultimately used to determine the VME stress in the tubular. The VME stress must be computed using both maximum compressive and tensile bending stress to determine the worst case.

The bending stress is a function of the effective force acting on the tubular. The force across the length of a tubular can vary; therefore, bending stress can vary across the length also. In considering the effects of post-buckling stresses, the problem must be examined to determine the location of the maximum bending stress, usually the point of maximum compressive stress. This usually can be determined by inspection of the wellbore geometry.

When a tubular is buckled helically, it is in contact with the constraining hole. The compressive load on the string creates a normal contact force F_n between the buckled string and the constraining hole. This normal contact force is given as (Mitchell 1982)

$$F_n = \frac{r_c F_{eff}^{\,2}}{4EI}. \dots\dots\dots\dots\dots\dots\dots\dots\dots\dots\dots\dots\dots (2.58)$$

The normal contact force is actually a force per unit length of contact. This force must be multiplied by the contact length in order to determine the total contact force between the string and the hole.

It is common to calculate an effective dogleg severity (DLS) for helically buckled pipe as a function of the geometry, stiffness, and effective compressive force. The DLS is a useful characterization of buckling, especially in estimating wear when drilling through the buckled section. The DLS is given by

$$\text{DLS} = C_n \frac{r_c \left| F_{eff} \right|}{2EI}. \dots\dots\dots\dots\dots\dots\dots\dots\dots\dots\dots (2.59)$$

The conversion coefficient C_n is equal to 68,755 using the US oilfield unit system where r_c is in inches, yielding DLS in degrees per 100 ft. In SI units with r_c in mm, C_n is equal to 1,718,900, yielding DLS in degrees per 30 m.

2.7.4 APB. APB is a consequence of the difference in the unconstrained volume change of a fluid and the volume change allowed by its container. **Fig. 2.17** illustrates the mechanism of APB. The fluid volume change may be caused by thermal expansion or the addition or removal of fluid. The annulus changes its volume in response to volume changes in the fluid and thermal expansion of the tubulars according to Lamé's equations,

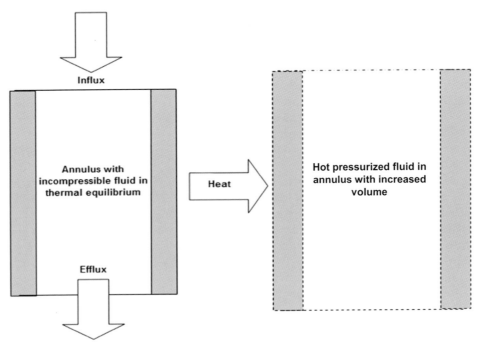

**Volume change equilibrates pressure increase
and stresses in container**

Fig. 2.17—Mechanics of APB.

all while maintaining mechanical equilibrium. By knowing the magnitude of the fluid volume change and the mechanical response of the annulus to pressure and temperature changes, the magnitude of APB in one or more of the annuli can be computed. The details of this mechanism are described in several publications (MacEachran and Adams 1994; Oudeman and Bacarreza 1995). The magnitude of APB influences the design of casing strings and also determines strategies to relieve pressure in the annuli.

Annuli differ from other well components in that they are not the result of purposeful design. Rather, they are a consequence of tubular design and the well construction process. Therefore, the ability of an annulus to withstand loads is evaluated at the end of the design process. **Fig. 2.18** shows the different kinds of annuli in a wellbore. The primary annulus, Type I, is formed by the production tubing and casing. It is bounded on the top and bottom by the wellbore seal assembly and completion hardware, including packers and seals. In addition, there may be an annular safety valve, gas lift valves, and related equipment depending on the nature of the well. The secondary annuli can be of two kinds—Types II and III. The Type II annulus is formed by adjacent casings or liners. It is bounded at the top by the wellhead seal assembly and at the bottom by cement. The cement top in this instance is above the shoe of the outer string of the annulus. The Type III annulus is similar, except that its bottom is open to the formation. The cement top lies below the shoe of the outer casing string, either by design or accident. Regardless of annulus type, the well design process quantifies the magnitude of APB that may occur during the life of the well. If the estimated annular pressures exceed a certain limit, then suitable mitigation strategies are employed. In land and platform completed wells, the primary and secondary annuli are usually accessible, allowing excess fluid pressure to bleed off. However, in

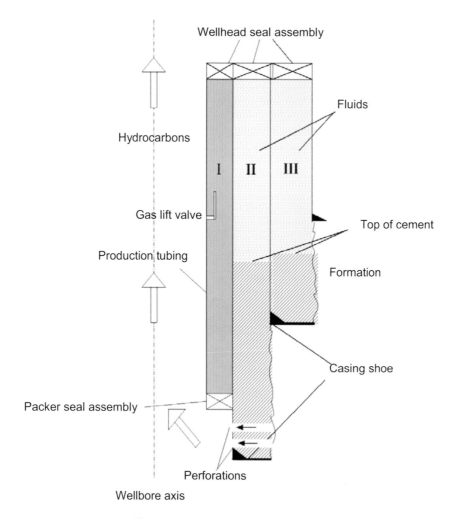

Fig. 2.18—Types of annuli in a wellbore.

subsea wells, secondary annuli cannot be accessed, and strategies to relieve excess annular pressure must be incorporated into well design.

Determination of APB. The calculation of APB in a closed elastic container is based on comparing the unconstrained volume expansion of the fluid due to temperature change and the annular volume that is available for such expansion. If the temperature change is small, the unconstrained volume expansion is merely the product of the isobaric volumetric coefficient of thermal expansion α_f, the temperature change ΔT, and the original volume of fluid V_f. However, the actual volume of expansion available for the expanding fluid is less than $\alpha_f(\Delta T)$ V_f. The resulting pressure change in the fluid is typically related to these two quantities by the isothermal bulk modulus B_f, which is commonly referred to as the fluid compressibility. Mathematically, this is given by

$$\Delta p_{APB} = B_f \frac{\Delta V_f - \Delta V_a}{V_f}. \quad\dots\dots\dots\dots\dots\dots\dots\dots\dots\dots\dots\dots\dots\dots\dots\dots\dots\dots\dots (2.60)$$

Here, ΔV_a is the change in volume of the elastic container. This approach usually works well when the initial temperature and pressure distributions in the fluid do not vary significantly over the volume of the container and the temperature change of the fluid is small. In a typical deepwater wellbore annulus, neither of these conditions is satisfied. Neglecting the nonlinear pressure/volume/temperature (PVT) behavior of annular fluids can significantly underestimate the magnitude of APB (Ellis et al. 2002).

Halal and Mitchell (1994) propose a method that accounts for the effect of PVT behavior on annular pressures. The method is explained by considering water in a rigid container. Assume that water is initially at 14.7 psi (101 kPa) and that its temperature increases from 70 to 80°F (21 to 27°C). Since the container is rigid, its volume change is zero. As a result, the density of water does not change. The resulting pressure change in the container is the pressure required to maintain the density of water constant over the temperature change, shown as line AB in **Fig. 2.19.**

This simple example illustrates several points:

1. The net change in the volume of the fluid is equal to the change in the volume of the container.
2. Though the net volume change of the fluid depends on the volume change of the container, it is calculated independently. The effect of volume change due to elasticity of the container is incorporated by accounting for the density change in the fluid that accompanies the pressure change.

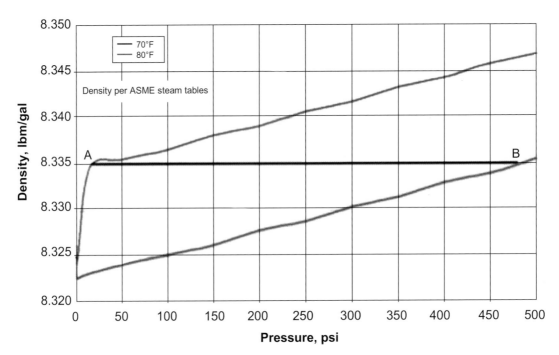

Fig. 2.19—Density of water as a function of pressure.

3. The elastic volume change of the container can be calculated independently. The volume change of the container is determined as a function of the unknown final pressure of the fluid and the temperature change in the walls of the container.
4. Equating the net volume change of the fluid and the container automatically satisfies mechanical equilibrium. As a result, this approach can be used for connected multiple volumes in a typical wellbore.
5. The calculation of fluid volume change depends only on the state equation that describes its density as a function pressure and temperature. Since volume changes are additive, the net volume change of individual fluids can be computed and added. This is especially useful for annuli with a gas cap.

The example described in Fig. 2.19 assumes a rigid container. However, wellbore annuli are bounded by elastic casing strings. Regardless of whether strings are bounded by other annular fluids, by formation, or by cement, it is reasonable to assume that the volume changes in the annuli due to pressure and temperature changes are linear functions of Δp and the temperature changes of the casing strings that bound each annulus. The Lamé equations can be used to describe the volume change of an annulus due to these pressure and temperature changes. In shortened form, the net volume change of an annulus ΔV_{ann} can be expressed by

$$\Delta V_{ann} \approx K'\Delta p, \quad \dotfill \quad (2.61)$$

where K' is the reciprocal of the annular stiffness. Unlike the example described in Fig. 2.19, the fluid density does not remain constant in an elastic container. It is therefore convenient to work with the net fluid volume change in the container as the fluid moves from its initial (P_i, T_i) state to its final condition. This is depicted in **Fig. 2.20.**

Fig. 2.20 is a slight modification of Fig. 2.19, albeit a schematic representation. The origin of the pressure (x axis) has been shifted from zero to p_i. Thus, the x axis now represents the change in pressure with respect to the initial value. On the y axis, density has been replaced by change in volume. Since mass is conserved in the container as the fluid is heated, it can be shown that the net volume change of the fluid is given by

$$\Delta V_f(p,T) = \int_{annulus} \frac{\Delta\rho(p,T)}{\rho_i(p_i,T_i)+\Delta\rho(p,T)}\, dV. \quad \dotfill \quad (2.62)$$

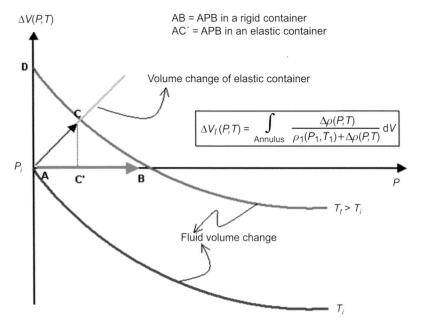

Fig. 2.20—Calculation of APB in an elastic container.

The integration in Eq. 2.62 is performed over the length of the annulus. Since the initial temperature of the fluid is known, the lower T_i curve can be obtained as function of pressure. Note that the volume change represented by the lower curve is negative since the fluid is compressed as the pressure increases isothermally. The final pressure of the fluid at temperature T_f is unknown. However, the volume change of the fluid at this temperature can be computed as a function of pressure. This is indicated by the upper T_f curve. Eq. 2.61 predicts that the volume of the casing changes linearly with change in pressure of the fluid, depicted by the solid gray line AC in Fig. 2.20. Since the volume change of the fluid and the container must be equal when the fluid temperature is T_f, the APB is determined by the intersection of the red curve and line AC.

For a perfectly rigid container, the volume change of the container is represented by the horizontal line AB. Elasticity of the tubulars tends to reduce APB. If the casing strings bounding the annulus are unconstrained by external pressure, then APB follows the line AC. If APB also exists outside these casings, then the line illustrating APB will fall somewhere between AB and AC.

These two observations make the calculation of an APB calculation straightforward, even when considering that fluid density varies over the length of the annulus. The fluid volume change calculation in Eq. 2.62 requires the integration of a function that varies with depth. In practice, fluid PVT behavior can be described by curve fitting the data to a polynomial that is linear in temperature and parabolic in pressure (Zamora et al. 2000). In terms of specific gravity, the equation has the form

$$SG(p,T) = (a_0 T + b_0) + (a_1 T + b_1)p + (a_2 T + b_2)p^2. \quad\dots\dots\dots\dots\dots\dots\dots\dots \text{(2.63)}$$

Eq. 2.63 is useful for synthetic-based drilling muds. Water-based muds can use the PVT behavior described by the American Society of Mechanical Engineers (ASME) steam tables.

If T_f is less than T_i, the fluid in the annulus cools and the pressure change is negative. The approach shown for calculating APB can be used for this reverse APB without modification.

APB Mitigation. If the magnitude of APB is sufficient to threaten well integrity, then the casing must be designed to either accommodate or relieve fluid expansion. The simplest mitigation strategy for APB is surface relief of the pressure in any given annulus through wellhead valves. Surface access is not always possible, particularly in deepwater wells. As a result, mitigation strategies that reduce or control the APB within acceptable limits must be considered. Depending on the criticality of the well, more than one mitigation strategy may be advisable. Some of the available mitigation strategies are vacuum-insulated tubing, syntactic foams, rupture disks, nitrogen gas cushions, and open shoes (Payne et al. 2003).

In most cases, mitigation focuses on preventing casing collapse. Outward APB failures could cause a connection leak or possibly a pipe body rupture. Either of these failures would not be detected in a producing well, for the annular fluid release is small and wellbore integrity is maintained. However, inward collapse failures cause concern regardless of the particular string. Collapse of a large-diameter casing can result in a nonuniform mechanical load on inner strings, causing a cascading failure that may compromise production casing pressure integrity or impinge on the flow conduit. The consequences of either type of APB failure should be considered, but typically the threat of collapse failures dictates mitigation strategy.

Vacuum-Insulated Tubing. Vacuum-insulated tubing (VIT) is used to isolate hot production fluids from the rest of the well. VIT completions were used by BP in the Marlin deepwater wells after a high-profile APB failure (Ellis et al. 2002; Gosch et al. 2004). A joint of VIT consists of concentric tubing strings welded at each end. The annular space between the tubulars is evacuated so that the heat transfer from the tubing to the annulus is minimized. Any heat that escapes into the annulus is due to radiation across the evacuated space between the inner and outer walls of the VIT or through conduction at the tubing connections.

Compared to the pipe body, the connections of the VIT have a much lower thermal resistance. As a result, the heat transfer across the VIT connections can be higher than across the VIT pipe body, even if coupling insulators are installed. Natural convection in the annulus can circulate heat that conducts through connections, increasing the overall heat transfer. To diminish convective heat transfer, highly viscous fluid gels may be placed in the annulus, preventing circulation. This combination of VIT and a high-viscosity packer fluid can effectively limit the APB in HP/HT wells.

While VIT can be an extremely effective APB mitigation strategy, its use is offset by its high cost. The VIT technology is still largely proprietary, and its design and use are currently on a case-by-case basis. Additionally, the presence of a vacuum complicates tubing design in that internal and external pressure loads must be considered without a backup pressure.

Syntactic Foams. Syntactic foams belong to a class of materials known as cellular solids, and they are characterized by an internal porous structure. The pore spaces usually are reinforced with glass or carbon fiber glass beads. The bulk properties of the cellular solid are a function of the porous structure and the reinforcing material in the pore spaces (Gibson and Ashby 1997). The behavior of syntactic foam is determined principally by its crush pressure and its compression ratio as shown in **Fig. 2.21.**

This figure shows the response of syntactic foam when it is subjected to hydrostatic pressure. The volumetric strain (shown on the *x* axis) increases linearly until the crush pressure is reached. At this point, the foam modules begin to crush catastrophically until all the pore spaces either have collapsed or are filled with the invading fluid. When this happens, crushing ceases. The volumetric strain at this point is known as the compression ratio, which dictates the total fluid volume relief provided by the foam. It is important to note that the crush pressure of the foam is a function of temperature. Because these foams usually are made of materials known as thermoplastics, the crush pressure falls rapidly below the glass transition temperature.

By adjusting the chemistry of the matrix resin and the glass beads that constitute the foam, its properties can be tailored such that different sections of foam modules crush at predetermined temperatures and pressures. When the foam modules crush in the sealed annulus and create space for the thermally expanding fluid, the pressures in the annulus are relieved. The technology to design foam modules for APB mitigation is still proprietary and must be done on a case-by-case basis.

Nitrogen Gas Cushion. Sometimes a column of nitrogen is placed at the top of the annulus. The column acts like a shock absorber in that the compressibility of gas is significantly greater than for liquid. Therefore, depending on the initial pressure at the wellhead and the volume of gas, the magnitude of APB in the presence of a column of gas is reduced.

Mitigation with nitrogen has its limitations, primarily in placement risk. Nitrogen is typically placed as a foamed spacer ahead of the primary cement job. If partial or lost returns occur while cementing, then there is no guarantee that the nitrogen will be placed in the annular space. Also, the initial nitrogen pressure at placement may be very high in deep wells with heavy mud weights. The compressibility of nitrogen is much lower at high pressures; thus, the dense gas cannot counteract as much mud fluid expansion. Finally, there is risk that the nitrogen may coalesce and rise in the annulus. If the annulus is trapped, then the nitrogen bubble has

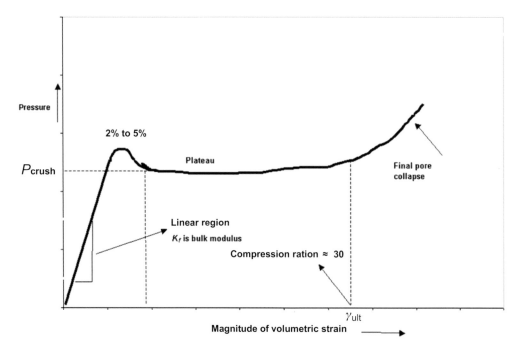

Fig. 2.21—Typical behavior of syntactic foam.

little room to expand, resulting in the migration of high BHPs back to the mudline. With deep strings and high mud weights, the possibility of gas migration raising mudline pressures may outweigh any benefit of APB mitigation. Nitrified spacers have been bullheaded down the external sideof casing to overcome placement issues; however, bullheading cannot be used for all well situations. Design calculations for nitrogen cushions should consider the impact of placement risk.

Open Shoes. While an open casing shoe may not be considered as an APB mitigation measure, common well construction practice is to leave the shoe of the previous casing string uncemented when possible. If the openhole section below the annulus not subsequently plug due to barite settling, cement channeling, or hole instability, then the pressure in the annulus is limited by the leakoff pressure at the shoe. This is illustrated in **Fig. 2.22.**

The fluid leakoff volume is calculated as shown in the figure. This figure is similar to Fig. 2.20, except that the *x* axis refers to the pressure at the bottom of the annulus in the openhole section. Although open shoes do provide a *relief valve* for APB, there are some key limitations that should be considered. Leaving the previous shoe open may not always be possible due to zonal isolation requirements or by geological and stability considerations. Also, open shoes add the additional risk of formation fluid influx. During production, annular fluids can expand and leak off to the open shoe. When the well is shut in, temperatures drop to the initial conditions, and a fluid influx may enter at the open shoe. If the fluid influx is a hydrocarbon, the safety of the wellbore may be compromised. In some cases, depending upon the temperature and pressure in the annulus, hydrocarbon gas influx can also cause gas hydrates, which can subsequently seal off the open shoe.

For these reasons, an open shoe by itself may not be an adequate mitigation strategy. In design, it is prudent to consider open shoes as closed to assess the risk of APB-induced collapse or burst failure.

Rupture Disks. Rupture disks are another available mitigation option for APB. Rupture disks are diaphragm disks run in a special casing sub. They open or rupture at a predesigned pressure, creating communication between the two annuli. The disk can provide a leak path for annular fluids to the formation, or it can provide a means of equalizing extreme pressure differences across a string.

Both burst and collapse rupture disks can be used, depending upon the dictating failure mode for the string under consideration. The rupture pressure is selected so that the disk opens before differential pressures exceed a casing rating. They may be used even if one of the annuli is cemented, provided that there is adequate standoff for the disk membrane to open unimpeded. Manufacturing tolerances can provide pressure relief within 5% of nominal ratings, which eliminates the uncertainty associated with relying on an outer casing to mechanically rupture before APB exceeds an inner-casing collapse.

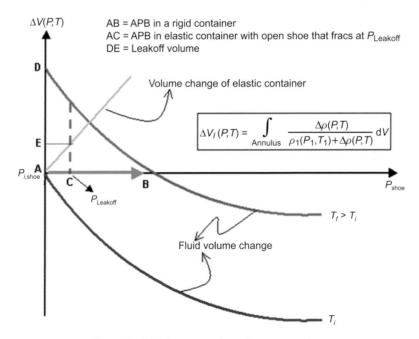

Fig. 2.22—APB in an annulus with an open shoe.

Designers of rupture disks must comprehend the other purposes of the string. For example, it is unacceptable for the rupture disk to compromise the burst strength of the pipe that has been considered in standard casing design, especially if margins of safety in standard design are low. Rupture disks operate on differential pressure across the disk; thus, multiple APB scenarios should be considered in which some annuli are sealed and others bleed pressure. Redundant disks are typically used, whether through multiple disks in a sub or multiple subs in a string. The latter redundancy may be important when considering the variety of possible APB scenarios of leakoff or barite settling.

APB Summary. There are many reasons why APB may not occur in a well. Shoes may remain open, providing a leakoff point for fluids. Connections may not hold high pressures. Gas entrained in wellbore fluids may reduce APB. Casing performance may exceed conservative ratings. These factors may explain why APB failures historically have been rare.

However, the high cost of a failure in a deepwater well lends credibility to pursuing mitigation even where the likelihood of a failure is low. The failure at Marlin caused the field to be shut in, resulting in significant financial losses due to workover operations and deferred production. Other observed failures have led to premature well abandonment. Though the particulars of APB loading scenarios may be debated, economics dictate that a relatively inexpensive and reliable mitigation strategy should be pursued for deepwater wells.

Not all mitigation methods discussed here are appropriate for all situations. VIT may not be compatible with all completion strategies. Syntactic foam requires significant up-front engineering and may be difficult to incorporate in a single well or small development. Nitrogen can be ineffective for deep annuli with heavy muds. The narrow window between adjacent casing ratings may diminish the utility of rupture disks. In pursuing an appropriate APB mitigation strategy, it is important to consider multiple options during the well planning phase. It can be especially challenging to design APB mitigation for an existing well nearing completion.

2.7.5 Wellhead Motion (WHM). WHM is caused by the resultant of forces exerted by uncemented casing strings that terminate at the wellhead. Though WHM can occur during different stages of well construction, the term generally refers to the motion caused during production.

WHM during production in platform wells is a consequence of temperature changes in the various casing strings that are hung off at the wellhead. WHM is primarily governed by the temperature distribution along different strings, their initial temperatures, axial stiffness of the strings, frictional forces due to differential motion between the conductor and formation, and frictional forces at steel-cement interfaces of the inner tubulars. To a lesser extent, WHM is affected by APB, buckling of the relatively limber inner strings, and axial drag between them. Apart from impacting the design of platform facilities, WHM causes a redistribution of loads exerted at the wellhead by the various strings that are hung off.

In addition to the motion caused by thermal expansion of the strings during production, the wellhead also moves during the well construction process. This motion is a result of the axial loads imposed on the wellhead by the hanging weight of the inner strings. Typically, the wellhead is installed on the conductor or the surface casing. As the inner strings are hung off at the wellhead and cemented, the wellhead moves downward incrementally. Simultaneously, each successive operation causes a redistribution of the loads exerted on the wellhead by the various strings. The extent of downward motion and redistribution of loads can be estimated by knowing the sequence of drilling operations, the weight of the strings, and the mechanisms by which the strings are hung off at the wellhead. Typically, these loads are compressive, and they result in a downward movement of the wellhead. Thermal forces exerted on the wellhead by the expanding casing strings during production result in an upward motion of the wellhead. While in principle the net motion of the wellhead is the algebraic sum of the downward motion of the wellhead during the well construction process and the upward motion caused during production, it is customary to refer to the motion caused during production as the final WHM. This is a consequence of the operational difficulties involved in tracking the motion of the wellhead during the well construction process. It is convenient to treat the location of the wellhead at the time of well completion as the initial location and to measure any growth with reference to this initial point. Therefore, while the exact magnitude of the downward motion of the wellhead during the well construction process is not an important quantity by itself, its value is necessary to determine the loads exerted by each string at the wellhead before the well is set on production.

This section describes the basis of a simple model known as the elastic spring model to calculate WHM and thermal stresses created due to thermal expansion of the strings. This model assumes the following:

1. All casing and tubing strings are tied to the wellhead and cannot move independently.
2. The inner strings undergoing thermal expansion behave like a set of elastic springs connected in parallel to a rigid wellhead that can move up or down, and to a rigid foundation, as shown in **Fig. 2.23.** Since the stiffness and length change in each string is different, the wellhead moves such that resulting forces on it are in equilibrium.
3. The upward force on the wellhead is transmitted to the conductor through the wellhead. The force on the conductor is resisted by the frictional force of the formation. The final WHM is a function of the relative magnitudes of the force due to the inner strings and the friction between the conductor and the formation.
4. If a string is well cemented in both the inner and outer annuli, the assumption is that its uncemented length is solely responsible for thermal elongation.
5. Temperature increase is determined with respect to the undisturbed temperature profile. Alternatively, a cementing profile can be used for each string.
6. The movement of the wellhead while landing each of the casing/tubing strings is determined. However, the initial position for determining wellhead growth due to thermal expansion is its position after all strings are landed.

Thermal Growth. A tubular of cross-sectional area A_p and length L_{uc} fixed between rigid supports is shown in **Fig. 2.24.**

If the tubular is subjected to a temperature rise ΔT during production, it tends to elongate by a length ΔL_T according to the coefficient of thermal expansion α. Now assume that the rigid support AB in Fig. 2.24 moves a distance ΔL, to a new position A′B′. Since only a portion of the thermal strain $\Delta L_T / L_{uc}$ is relieved, a thermal force is created in the tubular.

$$\Delta F = K(\Delta L_T - \Delta L), \dots\dots\dots\dots\dots\dots\dots\dots\dots\dots\dots\dots\dots\dots\dots\dots\dots\dots\dots (2.64)$$

where

K = axial stiffness of the string, given by $K = \dfrac{EA_p}{L_{uc}}$.

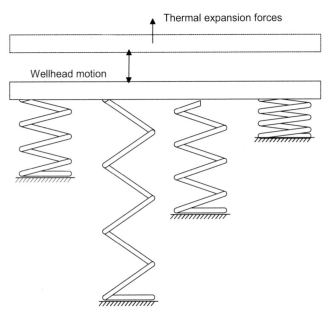

Fig. 2.23—Elastic spring model for wellhead motion.

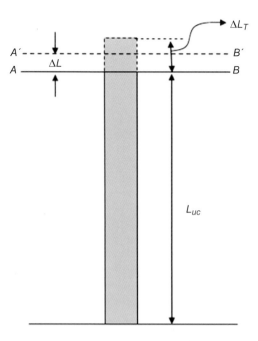

Fig. 2.24—Creation of thermal stress.

In this discussion, the term ΔL_T is referred to as the unrestrained growth of the string, while ΔL is the actual growth. The difference between the unrestrained and actual growths is responsible for the generation of thermal stress and WHM.

Assume that a total of N strings are hung off at the wellhead. Let L_i and $A_{p,i}$ denote the uncemented length and cross-sectional area of the ith string that is hung off at the wellhead. Also, let ΔT_i denote the average temperature change across the uncemented length of the ith string. The WHM due to thermal expansion of the N strings hung off at the wellhead is then given by

$$\Delta L_{\text{thermal}} = \frac{\sum\limits_{i=1}^{N} K_i \Delta L_i}{\sum\limits_{i=1}^{N} K_i}. \quad\dots\dots\dots\dots\dots\dots\dots\dots\dots\dots\dots\dots\dots\dots\dots\dots (2.65)$$

Each stiffness K_i is determined for the particular string's geometry. In the $\Delta L_{\text{thermal}}$ equation, ΔL_i is the unconstrained thermal expansion of the ith string. The force in the ith string as a result of WHM is given by

$$F_i = K_i(\Delta L_i - \Delta L_{\text{thermal}}) + F_i^0. \quad\dots\dots\dots\dots\dots\dots\dots\dots\dots\dots\dots\dots\dots\dots\dots\dots (2.66)$$

F_i^0 is the initial load in the ith string. The overall $\Delta L_{\text{thermal}}$ falls between the minimum and maximum ΔL_i. Thus, the equation for F_i predicts that WHM will apply tension to some strings and compression to others, subject to the force balance constraint that the sum of all forces is zero.

$$\sum_{i=1}^{N} F_i = 0 \quad\dots\dots\dots\dots\dots\dots\dots\dots\dots\dots\dots\dots\dots\dots\dots\dots\dots\dots\dots\dots (2.67)$$

The wellhead moves incrementally downward due to loads imposed during the well construction process. Typically, by the time the well is completed, the hanging weight of the inner strings creates a compressive load on the outer strings. At the wellhead, the inner strings are in tension while the conductor and surface casing are in compression. The downward motion of the wellhead when the inner strings are hung off can be calculated by the physical model described above. The resulting movement of the wellhead due to the preloads is given by

$$\Delta L_{\text{preloads}} = \frac{\sum\limits_{i=1}^{N} F_i^0}{\sum\limits_{i=1}^{N} K_i}. \quad \dots\dots\dots\dots\dots\dots\dots\dots\dots\dots\dots\dots\dots\dots\dots\dots \quad (2.68)$$

The net movement of the wellhead is then given by

$$\Delta L_{\text{net}} = \Delta L_{\text{thermal}} - \Delta L_{\text{preloads}}. \quad \dots\dots\dots\dots\dots\dots\dots\dots\dots\dots\dots\dots\dots\dots \quad (2.69)$$

Therefore, the net growth as given in the ΔL_{net} equation is smaller when the strings are pretensioned, and the initial position of the wellhead is taken at some point before one or more of the pretensioned strings is installed.

Inelastic Effects at the Formation Interface. The discussion so far assumes that only uncemented sections of strings that terminate at the wellhead participate in WHM. From a mechanical force equilibrium standpoint, the upward motion of the wellhead during thermal expansion can be viewed as the result of a balance between the tendencies of the inner strings to push the wellhead up countered by the axial tension generated in the stiff and relatively cooler outer strings. This assumes that sections of the inner strings below the top of cement and the section of the conductor below the formation interface are rigid and motionless, implying that the net upward force due to the inner strings must be resisted by the frictional force between the motionless conductor and the formation.

In general, the goal is to ensure that the conductor continues to be under compressive stress when the well is thermally loaded and that the forces at the wellhead are redistributed. The conductor, sometimes in concert with the surface casing, is the structural support member and is intended to be in compression. The structural requirements are not met when the conductor goes into tension.

Fig. 2.25 illustrates the forces acting on the conductor. The thermal force F_{inner} on the wellhead is generated by $N-1$ strings. If these strings behave elastically, the force on the wellhead due to the inner strings is given by

$$F_{\text{inner}} = \sum_{i=2}^{N} K_i \Delta L_{T,i} - \Delta L \sum_{i=2}^{N} K_i. \quad \dots\dots\dots\dots\dots\dots\dots\dots\dots\dots\dots\dots\dots \quad (2.70)$$

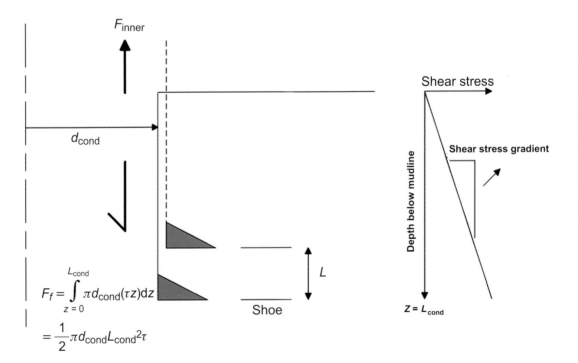

Fig. 2.25—Forces on the conductor.

$\Delta L_{T,i}$ is the unconstrained length change in each string due to temperature change. For convenience, this can be rewritten as

$$F_{inner} = a - b\Delta L, \quad \dots\dots\dots\dots\dots\dots\dots\dots\dots\dots\dots\dots\dots \quad (2.71)$$

where

$$a = \sum_{i=2}^{N} K_i \Delta L_{T,i}$$

$$b = \sum_{i=2}^{N} K_i$$

The frictional force on the conductor is denoted by F_f and is calculated by assuming that the shear stress on the conductor increases linearly with a gradient of τ over the length L_{cond}, which is the length of the conductor in contact with the formation that is capable of resisting the force due to the inner strings. Ideally, this length should be the same as the length of the conductor between the mudline and the shoe. The frictional force on the conductor is given by

$$F_f = \int_{L=0}^{L_{cond}} (\pi d_o dL) \tau L$$
$$= \frac{1}{2} \pi d_o \tau L_{cond}^2 \quad \dots\dots\dots\dots\dots\dots\dots\dots\dots\dots\dots\dots\dots \quad (2.72)$$

From Fig. 2.25, the equilibrium of the conductor is described by

$$F_{inner} = F_f$$

$$a - b\Delta L = (\pi/2)d_{cond}\tau(L_{cond} - \Delta L)^2. \quad \dots\dots\dots\dots\dots\dots\dots\dots\dots \quad (2.73)$$

Setting $\mu = (\pi/2)d_{cond}\tau$ in Eq. 2.73 and recognizing that L_{cond} is much greater than ΔL, the equation can be simplified as

$$a - b\Delta L \approx \mu L_{cond}^2. \quad \dots\dots\dots\dots\dots\dots\dots\dots\dots\dots\dots\dots\dots \quad (2.74)$$

Solving for L_{cond} yields

$$L_{cond} \approx \sqrt{\frac{a - b\Delta L}{\mu}}. \quad \dots\dots\dots\dots\dots\dots\dots\dots\dots\dots\dots\dots\dots \quad (2.75)$$

This equation has two unknowns, ΔL and L_{cond}. The constants a and b are known for a given well configuration. Although the true L_{cond} is not known, assuming that the entire length of the buried conductor resists WHM is sufficient to indicate the order of magnitude of the WHM and the forces in the conductors.

Once the temperature changes in the casing strings are determined, the WHM can be calculated easily. However, WHM is notorious in that predictions are seldom close to what is observed in the field. Surprisingly, theoretical calculations are not necessarily conservative. There have been instances when observed WHM has far exceeded the calculated values. Equally numerous are instances when the predictions are far in excess of the observed values of WHM. This is largely due to a poor and inadequate model of the inelastic effects that manifest at the conductor-formation interface. The analytical method presented here can give a rough estimate of WHM, but a more robust finite-element model with sophisticated formation interaction may be required to improve accuracy. This can be especially important for high-temperature wells.

Preload and Liftoff. The method of latching the strings to the wellhead can have an effect on WHM. In some wells, the wellhead base plate is merely landed on the conductor or drive pipe rather than welded. The inner strings are hung off at the wellhead as usual. In such configurations, it is important to consider the effect of preexisting loads in the strings (i.e., the loads prior to the creation of the thermal forces).

At the conclusion of well completion, all strings except the conductor and surface casing typically exert tensile forces at the wellhead. The conductor and surface casing are in compression. When the well starts producing, the thermal forces begin to offset the preexisting loads in the various strings. As tensile forces in the inner strings reduce, compressive force in the conductor is simultaneously relieved. The compressive force applied to the conductor may reach zero, resulting in the wellhead lifting off of the base plate. Beyond this point, the conductor no longer affects WHM. The net axial stiffness in Eq. 2.65 is reduced due to the absence of the conductor. As a result, the WHM can be greater than expected. Therefore, it is important to track the WHM and the forces in the strings as a function of the temperature increase in each string.

In summary, WHM is a complicated phenomenon, despite seeming simplicity. The nonlinear effects at the conductor-formation interface, the details of wellhead construction, and the procedures used in landing the strings during well construction have significant effects. These effects can be only partially modeled, partly due to numerical difficulties in modeling the formation-casing interface, and partly due to the inability to track the effects of the various well construction processes on the wellhead and string forces. As a result, order-of-magnitude estimates often must be deemed sufficient. If the orders of magnitude are not appropriate, as may be the case in critical wells, more-detailed solution approaches must be undertaken on a case-by-case basis.

2.7.6 Casing Wear. Casing wear is a localized phenomenon caused by the interaction of a rotating drillstring, the inner surface of a casing, and abrasive hardfaced materials. The magnitude of the contact pressure between the rotating string and the casing surface is one of the factors influencing the severity of wear. The severity of wear can be quantified by the volume of steel removed from the contacting surfaces or the localized reduction in the wall thickness at different points of the casing surface. The reduction in wall thickness affects the internal and external pressure ratings of the tubular. Further, the wear site may become a preferential point for corrosion. It is important to understand the causes of casing wear and identify methods to prevent or mitigate it. If wear is inevitable, a method to assess the remaining internal and external pressure capacities of the worn tubular becomes important.

Fig. 2.26 shows the interaction between the casing and the drillstring. The causes of casing wear and the physical mechanisms are well documented in the oilfield literature (Bradley 1975; Fontenot and McEver 1975; Bradley and Fontenot 1975; True and Weiner 1975; Lewis 1968). The principal conclusions of these studies indicate that wear due to casing-drillpipe interaction shows a qualitative change in the wear mechanism, from abrasive to adhesive wear, at a threshold contact pressure of approximately 250 psi (Williamson 1981). Abrasive wear occurs when a rough hard surface slides on a softer surface, such as tool joint hardbanding against a casing inside diameter, and plows a groove in it. Adhesive wear occurs when two surfaces are in rubbing contact, with or without lubrication. Adhesion or bonding occurs at the *sharp* contacts of the interface, and fragments are pulled off one surface to adhere to another. Adhesive wear is the more severe mechanism of the two. In general, the volume worn away depends on the contact load between the tool joint and casing surface, the hardness of the surface being worn away, the contact length between the two surfaces, and a wear coefficient.

Estimation of Wear. The following explains the basis of a model to calculate casing wear caused on the inner surface of a casing string by the tool joints on the drillstring (White and Dawson 1985; Archard 1953). The adhesive wear model compares the energy required to remove a certain amount of material to the total work done. The wear efficiency is defined as

$$\kappa = \frac{\text{Energy absorbed in wear}}{\text{Total mechanical work done}},$$

$$= \frac{VH}{\mu F_n x} \quad\quad\quad\quad\quad\quad\quad\quad\quad\quad\quad\quad\quad\quad\quad\quad\quad (2.76)$$

where
  κ = wear efficiency,
  V = volume of metal removed from the worn surface,

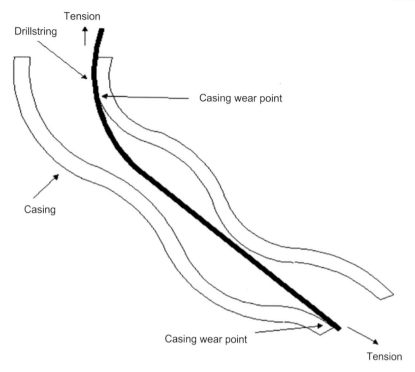

Fig. 2.26—Preferential casing wear due to string interaction.

TABLE 2.8—WEAR PROPERTIES OF CASING GRADES				
Mud Type	Casing Grade	Wear Efficiency, κ	K/H (in.²/lbf)	Hardness, H (psi)
	K55	0.0001	3.60E–10	277,778
Water	N80	0.00023	8.10E–10	283,951
	P110	0.00063	1.40E–10	450,000
	K55	0.0006	2.20E–10	272,727
Oil	N80	0.0012	3.90E–10	307,692
	P110	0.0017	4.20E–10	404,762

H = Brinnell hardness,
μ = coefficient of friction between the wearing surfaces,
x = distance of sliding contact, and
F_n = normal contact force between the surfaces.

The distance of sliding is the length of the worn area and represents the distance traveled by the tool joint. The equation for κ can be rearranged to find an expression for the volume worn, provided that the magnitude of the normal force and the wear efficiency are known. The values of κ and H for different casing grades in the presence of water and oil-based mud are shown in **Table 2.8** (White and Dawson 1985).

Let v_z represent the rate of penetration in the hole section being drilled and L_h define the length of this hole section. The sliding distance for a single tool joint can be calculated on the basis of drillstring rotation speed ω. If r_{TJ} is the radius of the tool joint, the sliding distance for a single tool joint is

$$x_o = \frac{L_h}{v_z} \omega r_{TJ}. \qquad\qquad\qquad\qquad\qquad\qquad\qquad\qquad\qquad\qquad\qquad (2.77)$$

If p is the tool joint pitch, or the length of a joint of drillpipe, the total sliding distance for all tool joints is given by

$$x_T = \frac{L_h^2}{v_z p} \omega r_{TJ}. \quad \dots\dots\dots\dots\dots\dots\dots\dots\dots\dots\dots\dots\dots\dots\dots\dots \quad (2.78)$$

The sliding distance per unit length of hole drilled is given by

$$x = \frac{L_h}{v_z p} \omega r_{TJ}. \quad \dots\dots\dots\dots\dots\dots\dots\dots\dots\dots\dots\dots\dots\dots\dots\dots \quad (2.79)$$

Substituting this value of sliding distance into Eq. 2.76 yields the volume of wear per unit length.

$$V = \frac{\kappa \mu \omega r_{TJ} F_n L_h}{H v_z p}. \quad \dots\dots\dots\dots\dots\dots\dots\dots\dots\dots\dots\dots\dots\dots \quad (2.80)$$

This is the classical expression for volume of material removed in adhesive wear and forms the basis of most wear estimation algorithms. The primary difference between models is the empirical basis of the factors used. The normal force is related to the string axial force F_a.

$$F_n = \frac{F_a}{R} L_{TJ}, \quad \dots\dots\dots\dots\dots\dots\dots\dots\dots\dots\dots\dots\dots\dots\dots\dots \quad (2.81)$$

where
L_{TJ} = length of the tool joint, and R = radius of curvature, given by $R = \dfrac{100 \times 180}{\pi \, \mathrm{DLS}}$ for US oilfield units or

$R = \dfrac{30 \times 180}{\pi \, \mathrm{DLS}}$ for SI units (DLS in degrees per 100 ft or degrees per 30 m).

Depth of Wear Groove. **Fig. 2.27** shows the geometry of the wear groove. The depth δ can be calculated for an assumed area of material removal depicted by the shaded region. From the geometry of the figure, the area of the shaded region A_w is given by

$$A_w = r_{TJ}^2 (\theta - \sin\theta \cos\theta). \quad \dots\dots\dots\dots\dots\dots\dots\dots\dots\dots\dots\dots \quad (2.82)$$

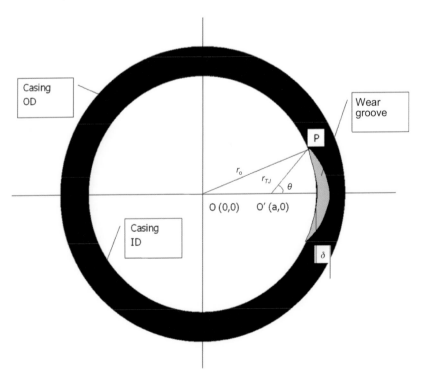

Fig. 2.27—Wear groove geometry.

The angle θ is obtained from the triangle OO'P, where the dimension a is defined such that

$$\theta = \pi - \cos^{-1}\left(\frac{a^2 + r_{TJ}^2 - r_o^2}{2ar_{TJ}}\right); \ a = r_o - (r_{TJ} - \delta). \ \dots\dots\dots\dots\dots\dots\dots (2.83)$$

The depth of the wear groove is obtained by setting A_w equal to V from Eq. 2.80 and solving for δ. The depth of the wear groove is the effective *local* reduction in wall thickness and is responsible for reduction in the performance ratings of the string.

Internal Pressure Rating. If casing wear occurs despite the use of appropriate hardfacing, the performance properties of the worn tubulars must be reassessed. The internal pressure ratings of a tubular vary inversely with the diameter-to-thickness ratio D/t, thus implying that the burst rating is approximately a linear function of the wall thickness. The conventional oilfield procedure of accounting for the burst rating of a worn tubular consists of derating the API MIYP of the tubular in proportion to the wall loss. For example, if the depth of the worst crescent or groove in the casing indicates wall loss of 10%, the internal pressure rating is 90% of the original value. Since the API MIYP is a yield-based rating rather than a failure limit, considerable casing wear may be accommodated before a true burst scenario is reached.

Two approximate solutions have been proposed to incorporate wear into burst ratings (Bradley 1975). The first solution uses the Lamé equations to calculate the stress in the casing. The second solution is based on the solution for the stress distribution in a cylinder with an eccentric bore (Timoshenko and Goodier 1970). Based on these two solutions, the ratio of the worn casing burst pressure to the unworn burst pressure, p/p_o, is a linear function of the ratio of the remaining wall thickness to the original wall thickness. This linear derating is identical to the convention used for MIYP.

Wear due to drillpipe rotation differs from wear due to wireline tripping (Bradley 1975). Since wireline wear tends to create regions of stress concentration with sharp and localized changes in geometry, the internal pressure rating must be derated for the remaining wall thickness and then divided by 1.4, a stress concentration factor (Roark 1954).

External Pressure Rating. Nippon Steel investigated the effect of internal crescent wear grooves on the collapse of casing (Kuriyama et al. 1992). The investigation was based on theoretical analysis, full-scale experiments, and finite-element analyses of three tubular products: 5½ 17.00 N80, 7 29.00 N80, and 7 29.00 P110. Wear grooves to simulate wall losses from 20 to 45% of nominal wall were machined on the inner surfaces of the specimens. The worn specimens were tested in a collapse test chamber, where the length-to-diameter ratio exceeded 6. Results of finite-element and theoretical analyses were compared with the experimental data. Examination of the results indicates that the experimental results and theoretical predictions exhibit identical trends, with the experimental data falling slightly less than theoretical predictions.

The results indicate that the external pressure rating decreases linearly with remaining wall thickness. **Fig. 2.28** illustrates this point. The ratio of the collapse pressure of the worn casing to that of the unworn casing is proportional to the ratio of the remaining wall thickness to the original wall thickness. This simply means that if 20% of the wall has been worn away, then the collapse rating is 80% of that of the original casing. It would be overly conservative to recalculate the collapse from API equations based on D/t with the reduced wall thickness.

The Nippon algorithm for derating the collapse performance is based on modeling the worn casing as a cylinder with an eccentric bore as shown in **Fig. 2.29.** The derating procedure consists of the following steps:

1. Calculate the eccentricity (distance between the centers) and the internal radius, r_i in Fig. 2.29, by using the minimum wall thickness in the worn section and nominal or average properties for the unworn section of the casing circumference.
2. Calculate the external pressure required to initiate yielding on the inner surface of the unworn casing by using the Lamé hoop stress formula.

$$p_{o,\text{ID yield}} = 2\sigma_Y \frac{\left(\dfrac{d_o}{t} - 1\right)}{\left(\dfrac{d_o}{t}\right)^2} \ \dots\dots\dots\dots\dots\dots\dots\dots\dots\dots\dots\dots\dots\dots\dots\dots\dots (2.84)$$

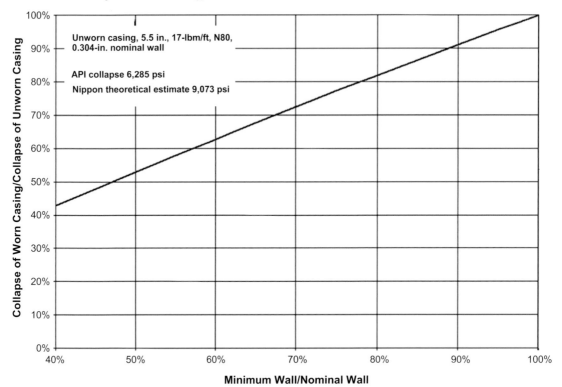

Fig. 2.28—Effect of localized wall reduction on collapse for 5½-in., 17-lbm/ft, N80 casing.

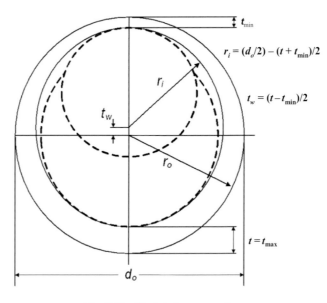

Fig. 2.29—Model of worn casing.

3. Calculate the external pressure required to yield the ID of the worn casing, by using the dimensions described in Fig. 2.29.

$$P_{o,\text{ ID yield, worn}} = \sigma_Y \left(\frac{r_i^2 + r_o^2}{2r_o^2} \right) \left[\frac{\left(r_o^2 + r_i^2 - t_w^2 \right)^2 - 4r_i^2 r_o^2}{\left(r_o^2 - r_i^2 \right)^2 - \left(r_i - 2t_w \right)^2} \right] \quad \dots\dots\dots\dots\dots\dots\dots\dots (2.85)$$

4. Calculate K_{wear}, the ratio of the pressure in Step 2 to the pressure in Step 3.

$$K_{wear} = \frac{p_{o,\text{ID yield}}}{p_{o,\text{ ID yield, worn}}} \quad \dots\dots\dots\dots\dots\dots\dots\dots\dots\dots\dots\dots\dots\dots\dots\dots\dots\dots \quad (2.86)$$

5. Calculate the collapse pressure of the worn casing from the nominal $p_{collapse}$ rating.

$$p_{worn} = \frac{p_{collapse}}{K_{wear}} \quad \dots\dots\dots\dots\dots\dots\dots\dots\dots\dots\dots\dots\dots\dots\dots\dots\dots\dots\dots \quad (2.87)$$

In general, the Nippon study implies that the external pressure rating calculated by direct proportioning based on the remaining wall thickness is a reasonable and conservative estimate of the collapse strength. The Tamano collapse (Eq. 2.18) also can be used for the deration calculations because it is similar in form to the equation used in the Nippon study.

2.7.7 Thermal Casing Design. Steam-cycling wells require casing designed to handle the extreme temperature changes due to the injection of superheated steam and subsequent cool-down during production cycles. The casing needs to be able to survive multiple cycles of heat-up and cool-down during its life. It is desirable to fully cement casing strings to surface to prevent unsupported casing from buckling and to insulate the casing to reduce heat losses into formations above the reservoir. In some cases, it may not be practical or possible to cement casing to surface. Uncemented casing will be discussed later.

Fully Cemented Casing. Assuming the casing is fully cemented and is constrained at the wellhead, the casing is not allowed to elongate due to thermal expansion. The thermal stresses result in axial forces within the casing.

$$\Delta F_{thermal} = A_p E \alpha \Delta T \quad \dots\dots\dots\dots\dots\dots\dots\dots\dots\dots\dots\dots\dots\dots\dots\dots\dots\dots \quad (2.88)$$

This relationship assumes that the stresses are below the yield strength of the casing.

The design limit or required casing strength is based on the change in temperature between the steam at injection quality and geothermal temperature at surface. Due to the high thermal stresses and forces, thermal casing design differs from conventional casing design in that it utilizes both the tensile and the compression capacity of the casing. This increases the elastic range to $\sim 2\sigma_Y$. Depending on the temperature increase, allowing the casing to yield may still be required. The general rule of thumb for yield strength required is

$2\sigma_Y = 207\Delta T$ in US oilfield units, with σ_Y in psi.

$2\sigma_Y = 2.57\Delta T$ in SI units, with σ_Y in MPa. $\dots\dots\dots\dots\dots\dots\dots\dots\dots\dots\dots\dots\dots\dots\dots$ (2.89)

This is illustrated in **Fig. 2.30** as the Holliday (Shell) method of thermal casing design. The concept is as follows:

- The *as cemented* tension at (1) is the starting condition.
- As steam is injected the temperature increases causing compression from (1) to (2). At (2), the yield strength is exceeded and the casing continues to yield. Strain or work hardening can take place, depending on the material.
- Steady-state steam injection conditions exist from (2) to (3), where the casing reaches constant temperature.
- At (3), steam injection stops and the production cycle starts. The casing cools and begins to behave as an elastic material. Reduced temperature results in the tendency to shorten the casing. Since the casing is fixed, it cannot shorten and the tensile stress increases.
- At (4), the yield strength is exceeded and the casing starts to yield in tension.
- The casing continues to yield in tension until steam injection starts again at (5) and the tensile stress is reduced.
- The thermal cycle is repeated. The stress-strain path may change slightly, depending on how the material properties change with temperature and strain.

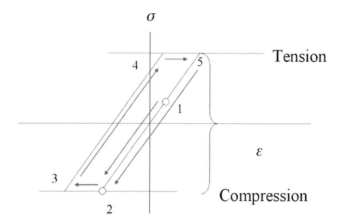

Fig. 2.30—Holliday concept of thermal casing design.

Some other issues associated with this concept of thermal casing design include the following:

- Stress effects from casing internal and external pressure must also be considered.
- Some casing material work hardens due to plastic deformation, which affects the σ and ε behavior of future cycles.
- The material properties are affected by increased temperature.
- Some materials exhibit a Bauschinger effect, where yielding in compression tends to reduce the tensile yield stress at the other end of the cycle.
- The combined effect of these factors can reduce the previous rule of thumb to a usable stress window of around 1.4 to 1.6 times yield strength.

An interesting aspect in the steam injection cycle is that K55 casing achieves a yield strength comparable to L80 material after a few thermal cycles (Maruyama et al. 1990). This is thought to be caused by work hardening and strain aging.

Work Hardening Materials. **Fig. 2.31** shows the cyclic history for a work hardening material such as K55-grade casing. The first cycle path from (1) to (2) shows the elastic response to compression increasing due to increasing temperature. At (2), the casing starts to yield. Plastic deformation takes place from (2) to (3). Path (3) to (4) shows stress relaxation while the temperature is held constant. Cooling starts at (4), and path (4) to (5) shows the elastic response to increasing tension due to cooling and is parallel to the compression path. At (5), the casing starts to yield in tension at a stress less than the expected pipe body yield tension. This is thought to be due to the Bauschinger effect of reduced tensile yield after yielding in compression. Finally, (6) is the large residual tensile load remaining at cool-down.

The second cycle starts at (6) and is offset by the residual tensile load remaining. Path (6) to (3) is the compression load due to heat-up progressing elastically. The rest of the second cycle follows the first cycle, except that the residual stress (6') is slightly higher than that in the previous cycle. This may have been caused by additional work hardening of the material. Subsequent cycles will essentially repeat the path for the second cycle.

Non-Work-Hardening Materials. Casing grades L80 and C95 behave in a similar manner with respect to the heat-up and compression as shown in Fig. 2.31. However, they behaved elastically through the cool-down phase without evidence of yielding.

Uncemented Casing. When casing is unsupported in thermal wells, the casing is able to elongate with increased temperature above the top of cement. If the casing is fixed at the surface, this results in helically buckled casing. The buckled casing may be tolerated, provided that the helix does not prevent tool passage and the bending stresses do not cause connection failures. Long sections of unsupported casing in thermal wells tend to behave like cemented casing after they have become fully buckled. The fully buckled casing has wall contact from top to bottom, and any additional thermal compressional stress follows the stress-vs.-strain

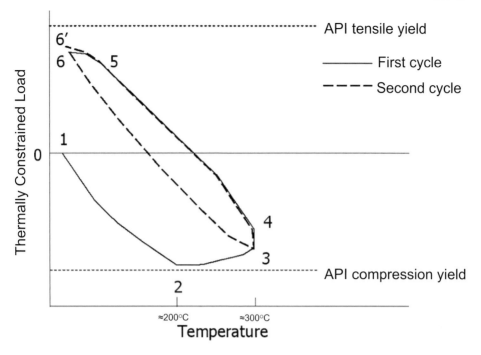

Fig. 2.31—Cyclic thermal loading for a work-hardening material.

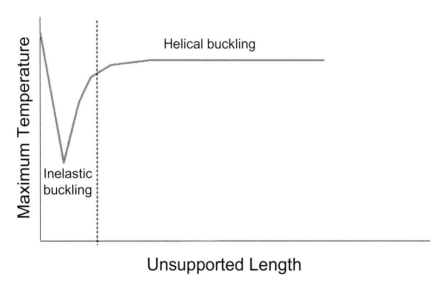

Fig. 2.32—Unsupported casing effect.

curve shown in Fig. 2.30. When the yield stress is exceeded, the casing yields in a manner similar to that of cemented casing. One difference is the length that was absorbed in the process of buckling the casing, resulting in less yielding required to handle the total thermal stress.

Short unsupported sections due to a poor cement job result in high buckling loads that are likely to buckle the casing. A short section may be around 100 ft (30 m), for example, depending on the casing and hole size. The problem with short sections is that inelastic buckling takes place, resulting in high bending loads well in excess of the yield strength. **Fig. 2.32** shows this concept, where the allowable maximum temperature is reduced significantly for short unsupported sections.

Other considerations for uncemented casing include pretensioning the casing after the cement is set. This moves the starting point up the stress-strain curve, and there is more room for compression loads before the casing yields due to temperature increase.

Nomenclature

a = defect depth, in.

a_i = coefficient; $i = 0,1,2$ are obtained from density measurements in tests

A_i = inside surface area, in.^{2}

A_o = outside surface area, in.^{2}

A_p = cross-sectional area of the casing, in.^{2}

A_w = worn area, in.^{2}

B_f = isothermal bulk modulus

B_i = coefficient; $i = 0,1,2$ are obtained from density measurements in tests

C = conversion constant (0.052 for lbm/gal and ft)

C_n = conversion constant

d_{cond} = conductor diameter, ft

d_o = nominal outside diameter of tubular, ft

D = nominal diameter of tubular, ft

E = modulus of elasticity of steel, 3×10^7 psi

F_a = axial force, lbf

F_{cr} = effective force, lbf

$F_{cr, hel}$ = critical helical buckling load, lbf

$F_{cr, sin}$ = critical sinusoidal buckling load, lbf

F_{eff} = critical force, lbf

F_f = friction force, lbf

F_i = force in the ith string, lbf

F_{inner} = thermal force on the wellhead, lbf

F_i^o = pre-existing load in the string, lbf

F_n = normal contact force, lbf

F_{ten} = tension force, lbf

g = function

H = Brinnell hardness

H_{ult} = correction factor

I = moment of inertia, in.^{4}

k = wear efficiency

K' = reciprocal of annular stiffness

K = axial stiffness of string

K_1 = stress intensity factor

K_{1SSC} = threshold stress intensity factor (material resistance)

K_a = correction factor

K_{els} = elastic limit correction factor

K_{leac} = fracture toughness of the material

K_n = the index correction factor

K_T = the tension correction factor

K_{wear} = ratio of external pressure to yield the ID to the external pressure to yield the ID of the worn casing, psi

K_{yls} = yield limit correction factor

L = length, ft

L_{cond} = conductor length, ft

L_f = load factor

L_h = length of hole section, ft

L_r = load ratio, applied load/limit load

L_{TJ} = length of tool joint, ft

L_{uc} = length of uncemented section of the casing, ft

m_f = flaw factor

p_{API} = API minimum internal yield pressure, psi

$p_{B,\,Hill}$ = Hill rupture limit pressure, psi

$p_{B,\,K\text{-}S}$ = Klever-Stewart rupture limit pressure, psi

p_c = collapse pressure, psi

$p_{collapse}$ = collapse pressure, psi

$p_{c\,ult}$ = ultimate collapse limit pressure, psi

p_i = internal pressure, psi

p_o = external pressure, psi

$p_{o,ID\,yield}$ = external pressure to yield the ID, psi

$p_{o,ID\,yield,worn}$ = external pressure to yield the ID of the worn casing, psi

p_{surf} = surface pressure, psi

p_{worn} = collapse pressure of worn casing, psi

p_{YP} = yield collapse pressure, psi

P = pressure, psi

P_{burst} = burst pressure, psi

P_{CO2} = partial pressure of CO_2, psia

P_e = elastic collapse pressure, psi

P_E = elastic collapse pressure, psi

P_f = probability of failure

P_{iF} = internal pressure at the fracture, psi

$P_{i,\,shoe}$ = initial pressure at the casing shoe

$P_{Leakoff}$ = leakoff pressure at the casing shoe

P_p = plastic collapse pressure, psi

P_{shoe} = pressure at the casing shoe

P_T = transition collapse pressure, psi

P_y = yield pressure, psi

Q = load effect

r = radius, in. or ft

r_c = radius clearance, ft

r_i = inside radius, in.

r_o = outside radius, in.

$r_{plastic}$ = radius of the plastic zone, in.

r_{TJ} = radius of tool joint, in.

R = strength in the chosen deterministic equation

R_f = resistance factor

SG = specific gravity

t = nominal wall thickness, in.

t_{dr} = wall thickness in a ductile rupture limit, in.

t_{min} = minimum measured wall thickness, in.

t_n = flaw depth, in.

T = temperature, °F or °C

T_{eff} = effective tension, lbf

T_f = final fluid temperature, °F or °C

T_i = initial fluid temperature, °F or °C

T_{UTS} = ultimate tensile strength, lbf

v = poisson's ratio

V = volume, in.^{3} or ft^3

V_f = original fluid volume, in.^{3}

w_{air} = air weight of casing per unit length, lbm/ft or kg/m

w_b = buoyed weight per unit length of tubular, lbm/ft

$(\overline{x})$ = deterministic parameters used in a design check equation

Z = depth, ft

α = coefficient of thermal expansion (6.5 – 6.9E^{-6}/F or 11.7 – 12.4E^{-6}/C)

β = has the value of 4 for sinusoidal buckling load and has the value of 8 for helical buckling (unit-less)

γ_{ult} = ultimate strain

δ = depth of wear groove, in.

ΔF = thermal force created in tubulars, lbf

ΔF_a = change in axial force, lbf

$\Delta F_{thermal}$ = axial force due thermal stress in tubulars, lbf

ΔL = change in length, ft

$\Delta L_{thermal}$ = change in length due to temperature, ft

ΔL_T = elongation in length, ft

ΔP = change in pressure, psi

ΔP_{APB} = change in annular pressure build, psi

ΔT = change in temperature, °F

ΔV_{ann} = the net change in volume of an annulus, ft^3

ΔV_a = change in volume of the elastic container, ft^3

ε = eccentricity

θ = polar angle with respect to the x axis (in the crack plane), radians

θ' = build rate of the curved section, deg/100 ft

μ = mean

ρ_i = gas gradient based on BHP and BHT, psi/ft

ρ_{MW} = mud weight, lbm/gal

ρ_{MWi} = current mud weight, lbm/gal

ρ_W = density of water at 70°F and 14 psi (i.e. 8.33 ppg), lbm/gal

σ' = standard deviation

σ = far-field stress, psi

σ_b = bending stress, psi

σ_h = hoop yield stress, psi

σ_r = radial yield stress, psi

$\overline{\sigma}_r$ = normalized residual stress, psi

σ_{ult} = ultimate yield stress, psi

σ_{VME} = von Mises equivalent yield stress, psi

σ_{VMEri} = inside von Mises equivalent yield stress, psi

σ_Y = yield stress, psi

σ_{Ya} = adjusted yield stress, psi

σ_z = axial yield stress, psi

τ_1 = shear stress, psi

τ_2 = shear stress, psi

τ_3 = shear stress, psi

ϕ = ovality

φ = hole inclination, radians

φ' = walk rate of the curved section, deg/100 ft

$\overline{\varphi}$ = average hole inclination of the curved section, radians

ω = rate of rotation of the drill string, rev/min

Acknowledgements

The authors gratefully acknowledge Viking Engineering and Blade Energy Partners for their support in the writing of this chapter.

References

Archard, J.F. 1953. Contact and Rubbing of Flat Surfaces. *Journal of Applied Physics* **24**: 981–988. DOI: 10.1063/1.1721448.

Bourgoyne, A.T., Millheim, K.K., Chenevert, M.E., and Young, F.S. Jr. 1986. *Applied Drilling Engineering*, Textbook Series, SPE, Richardson, Texas **2**.

Bradley, W.B. 1975. Experimental Determination of Casing Wear by Drill String Rotation. *Journal of Engineering for Industry: Transactions of the ASME, Series B* **97**: 464–471.

Bradley, W.B. and Fontenot, J.E. 1975. The Prediction and Control of Casing Wear. *JPT* **27** (2): 233–245. SPE-5122-PA. DOI: 10.2118/5122-PA.

Bull. 5C3, Formulas and Calculation for Casing, Tubing, Drill Pipe, and Line Pipe Properties. 1999. API, Washington, DC.

Chen, Y.-C., Lin, Y.-H., and Cheatham, J.B. 1990. Tubular and Casing Buckling in Horizontal Wells. *JPT* **42** (2): 140–141, 191. SPE-19176-PA. DOI: 10.2118/19176-PA.

Craig, B.D. 1993. *Practical Oilfield Metallurgy and Corrosion*, second edition. Tulsa: PennWell Corporation.

Dawson, R. and Paslay, P.R. 1984. Drillpipe Buckling in Inclined Holes. *JPT* **36** (10): 1734–1738. SPE-11167-PA. DOI: 10.2118/11167-PA.

de Waard, C. and Milliams, D.E. 1975. Carbonic Acid Corrosion of Steel. *Corrosion* **31** (5): 177.

Dropkin, D. and Somerscales, E. 1965. Heat Transfer by Natural Convection in Liquids Confined by Two Parallel Plates Which Are Inclined at Various Angles With Respect to The Horizon. *Journal of Heat Transfer* **87**: 77–84.

Ellis, R.C., Fritchie, D.G. Jr., Gibson, D.H., Gosch, S.W., and Pattillo, P.D. 2002. Marlin Failure Analysis and Redesign; Part 2, Redesign. Paper SPE 74529 presented at the IADC/SPE Drilling Conference, Dallas, 26–28 February. DOI: 10.2118/74529-MS.

Fontenot, J.E. and McEver, J.E. 1975. A Laboratory Investigation of the Wear of Casing Due to Drillpipe Tripping and Wireline Running. *Journal of Engineering for Industry: Transactions of the ASME, Series B* **97**: 445–456.

Gibson, L. and Ashby, M.F. 1997. *Cellular Solids: Structure and Properties*, second edition. Cambridge, UK: Cambridge University Press.

Gosch, S.W., Horne, D.J., Pattillo, P.D., Sharp, J.W., and Shah, P.C. 2004. Marlin Failure Analysis and Redesign; Part 3, VIT Completion With Real-Time Monitoring. *SPEDC* **19** (2): 120–128. SPE-88839-PA. DOI: 10.2118/88839-PA.

Halal, A.S. and Mitchell, R.F. 1994. Casing Design for Trapped Annulus Pressure Buildup. *SPEDC* **9** (2): 107–114. SPE-25694-PA. DOI: 10.2118/25694-PA.

Hammerlindl, D.J. 1980. Basic Fluid and Pressure Forces on Oilwell Tubulars. *JPT* **32** (1): 153–159. SPE-7594-PA. DOI: 10.2118/7594-PA.

Hasan, A.R. and Kabir, C.S. 2002. *Fluid Flow and Heat Transfer in Wellbores*. Richardson, Texas, USA: Society of Petroleum Engineers.

He, X. and Kyllingstad, A. 1993. Helical Buckling and Lockup Conditions for Coiled Tubing in Curved Wells. *SPEDC* **10** (1): 10–15. SPE-25370-PA. DOI: 10.2118/25370-PA.

Hill, R. 1950. *The Mathematical Theory of Plasticity*. Oxford, UK: Oxford University Press.

ISO 13679:2002, Petroleum and Natural Gas Industries—Procedures for Testing Casing and Tubing Connections. 2002. Geneva, Switzerland: ISO.

ISO/TR 10400:2007, Petroleum and Natural Gas Industries—Equations and Calculations for the Properties of Casing, Tubing, Drill Pipe and Line Pipe Used as Casing or Tubing, edition 1. 2007. Geneva, Switzerland: ISO.

Kane, R.D. and Greer, J.B. 1977. Sulfide Stress Cracking of High-Strength Steels in Laboratory and Oilfield Environments. *JPT* **29** (11): 1483–1488; *Trans.*, AIME, **263**. SPE-6144-PA. DOI: 10.2118/6144-PA.

Klementich, E.F. 1995. Unraveling the Mysteries of Proprietary Connections. *JPT* **47** (12): 1055–1059; *Trans.*, AIME, **299**. SPE-35247-PA. DOI: 10.2118/35247-PA.

Klever, F.J. and Stewart, G. 1998. Analytical Burst Strength Prediction of OCTG With and Without Defects. Paper SPE 48329 presented at the SPE Applied Technology Workshop on Risk Based Design of Well Casing and Tubing, The Woodlands, Texas, 7–8 May. DOI: 10.2118/48329-MS.

Kuriyama, Y., Tsukano, Y., Mimaki, T., and Yonezawa, T. 1992. Effect of Wear and Bending on Casing Collapse Strength. Paper SPE 24597 presented at the SPE Annual Technical Conference and Exhibition, Washington DC, 4–7 October. DOI: 10.2118/24597-MS.

Lewis, R.W. 1968. Casing Wear. *Drilling*: 48–56.

Lubinski, A., Althouse, W.S., and Logan, J.L. 1962. Helical Buckling of Tubing Sealed in Packers. *JPT* **14** (6): 655–670; *Trans.*, AIME, **225**. SPE-178-PA. DOI: 10.2118/178-PA.

MacEachran, A. and Adams, A.J. 1994. Impact on Casing Design of Thermal Expansion of Fluids. *SPEDC* **9** (3): 210–216. SPE-21911-PA. DOI: 10.2118/21911-PA.

Maruyama, K., Tsuru, E., Ogasawara, M., Inoue, Y., and Peters, E.J. 1990. An Experimental Study of Casing Performance Under Thermal Cycling Conditions. *SPEDE* **5** (2): 156–164. SPE-18776-PA. DOI: 10.2118/18776-PA.

Miannay, D.P. 1997. *Fracture Mechanics*. Mechanical Engineering Series. New York: Springer.

Mitchell, R.F. 1982. Buckling Behavior of Well Tubing: The Packer Effect. *SPEJ* **22** (5): 616–624. SPE-9264-PA. DOI: 10.2118/9264-PA.

Mitchell, R.F. 1988. New Concepts for Helical Buckling. *SPEDE* **3** (3): 303–310; *Trans.*, AIME, **285**. SPE-15470-PA. DOI: 10.2118/15470-PA.

NACE *Standard MR-01-75, Material Requirements Standard. Metals for Sulfide Corrosion Cracking and Stress Corrosion Cracking Resistance in Sour Oilfield Environments*. 2003. Houston: NACE.

Oudeman, P. and Bacarreza, L.J. 1995. Field Trial Results of Annular Pressure Behavior in High Pressure/High Temperature Well. *SPEDC* **10** (2): 84–88. SPE-26738-PA. DOI: 10.2118/26738-PA.

Payne, M.L. and Schwind, B.E. 1999. A New International Standard for Casing/Tubing Connection Testing. Paper SPE 52846 presented at the SPE/IADC Drilling Conference, Amsterdam, 9–11 March. DOI: 10.2118/52846-MS.

Payne, M.L., Pattillo, P.D., Sathuvalli, U.B., Miller, R.A., and Livesay, R. 2003. Advanced Topics for Critical Service Deepwater Well Design. Presented at the Deep Offshore Technology Conference (DOT03), Marseille, France, 19–21 November.

Ramey, H.J. Jr. 1962. Wellbore Heat Transmission. *JPT* **14** (4): 427–435; *Trans.*, AIME, **225**. SPE-96-PA. DOI: 10.2118/96-PA.

Raymond, L.R. 1969. Temperature Distribution in a Circulating Drilling Fluid. *JPT* **21** (3): 333–341; *Trans.*, AIME, **246**. SPE-2320-PA. DOI: 10.2118/2320-PA.

Roark, R.J. 1954. *Formulas for Stress and Strain*, third edition, 370. New York: McGraw-Hill. *RP 5C5, Recommended Practice for Evaluation Procedures for Casing and Tubing Connections*, second edition. 1980. Washington, DC: API.

RP 37, Recommended Practice Proof-Test Procedure for Evaluation of High-Pressure Casing and Tubing Connection Designs, 2nd edition. 1980. Washington, DC: API.

RP 579, Recommended Practice Fitness for Service, 1st Edition, 2000. Washington, DC: API.

Schwind, B.E. 1990. Equations for Leak Resistance of API 8-Round Connections in Tension. *SPEDE* **5** (1): 63–70. SPE-16618-PA. DOI: 10.2118/16618-PA.

Schwind, B.E. 1998. Mobil Qualifies Three Tubing/Casing Connection Product Lines. *Hart's Petroleum Engineer International* (November): 59–62.

Sparks, C.P. 1984. The Influence of Tension, Pressure and Weight on Pipe and Riser Deformations and Stresses. *Journal of Energy Resources and Technology* **106** (1): 46–54.

Spec. 5CT, Specification for Casing and Tubing, sixth edition. 1998. Washington, DC: API.

Sumitomo Products for the Oil and Gas Industries. Sumitomo Metals, http://www.sumitomometals.co.jp/e/business/sm-series.pdf. Downloaded 2 March 2009.

Thomas, P.D. and Bartok, A.W. 1941. Leak Resistance of Casing Joints in Tension. Paper presented at the 22nd API Annual Meeting on Standardization of Oilfield Equipment, San Francisco, November.

Timoshenko, S. and Goodier, J.N. 1970. *Theory of Elasticity*, third edition, 70–71, 198–202. Singapore: McGraw-Hill Education.

True, M.E. and Weiner, P.D. 1975. Optimum Means of Protecting Casing and Drillpipe Tool Joints Against Wear. *JPT* **27** (2): 246–252. SPE-5162-PA. DOI: 10.2118/5162-PA.

White, J.P. and Dawson, R. 1985. Casing Wear: Laboratory Measurements and Field Predictions. *SPEDE* **2** (1): 56–62. SPE-14325-PA. DOI: 10.2118/14325-PA.

Williamson, J.S. 1981. Casing Wear: The Effect of Contact Pressure. *JPT* **33** (12): 2382–2388. SPE-10236-PA. DOI: 10.2118/10236-PA.

Wooley, G.R. 1980. Computing Downhole Temperatures in Circulation, Injection, and Production Wells. *JPT* **32** (9): 1509–1522. SPE-8441-PA. DOI: 10.2118/8441-PA.

Zamora, M., Broussard, P.N., and Stephens, M.P. 2000. The Top Ten Mud-Related Concerns in Deepwater Drilling. Paper SPE 59019 presented at the SPE International Petroleum Conference and Exhibition in Mexico, Villahermosa, Mexico, 1–3 February. DOI: 10.2118/59019-MS.

SI Metric Conversion Factors

bbl	× 1.589 873	E – 01 = m^3
ft	× 3.048*	E – 01 = m
°F	(°F – 32)/1.8	= °C
gal	× 3.785 412	E – 03 = m^3
in.	× 2.54*	E + 00 = cm
in.^{2}	× 6.451 6*	E + 00 = cm^2
lbf	× 4.448 222	E + 00 = N
lbm	× 4.535 924	E – 01 = kg
psi	× 6.894 757	E + 00 = kPa

*Conversion factor is exact.

Chapter 3

Advanced Drillstring Design

3.1 Advanced Drillstring Design—Jackie E. Smith, Stress Engineering Services

3.1.1 Introduction. Drillpipe is by far the largest part of a drillstring. Its purpose is to support the bit and bottomhole assembly and to provide a means to pull the bit out of the hole. It provides a means to rotate the bit or act as supporting means for a downhole motor to rotate the bit. The drillstring also acts as a conduit for drilling fluid.

The selection process for the drillpipe consists of strength considerations, size considerations, and cost. Strength is normally the first item considered, then size, and, finally, cost.

Strength refers to several properties, including the pipe's ability to pull the drillstring out of the hole (tension capacity) and to transmit torque to the bit (torque capacity). There are other strength considerations such as internal pressure from the drilling fluid, bending in directional holes, fatigue, external pressure, compressive loads, and buckling.

Pipe size and tool joint size are driven by hydraulics, fishability, and elevator hoisting capacity. Other considerations relating to size include buckling strength, bending stresses, fatigue resistance, and external pressure.

Items affecting cost are pipe availability, drillpipe features that enhance the pipe's performance, and features that enhance the pipe's usable life. It is easily seen that none of these items—strength, size, or cost—stands alone; each affects, and is affected by, the other two.

3.1.2 Drillpipe Specifications. Drillpipe specifications have historically been defined by the American Petroleum Institute (API). However, with proprietary tool joints and proprietary pipe sizes, weights, and grades, and with the introduction of standards from the International Organization of Standards (ISO), there are many widely used drillpipe assemblies that are not officially covered by API specifications.

Drillpipe, or a drillpipe assembly, as shown in **Fig. 3.1.1,** is made from three different components: the pipe body, a tool joint pin, and a tool joint box. The components are joined by friction welding.

The pipe body is made in standard sizes, weights, grades, and upset configurations. **Table A-1,** located in the Appendix to this chapter, shows a list of many sizes and weights available. The size denotes the pipe's outside diameter (OD) in inches. There are nine commonly used sizes ranging from 2⅜ to 6⅝ in. The weight is the *nominal* weight given in lb/ft and is an indicator of wall thickness. There are a few exceptions, but in most cases, the calculated weight per foot of the pipe does not equal the nominal weight.

The nominal weight is not directly calculated from the OD and wall thickness. The *actual* weight of drillpipe assemblies includes the weight of the tool joints and is shown in many drillpipe performance tables. **Table A-2** in the Appendix shows the actual weight, along with other properties, of a few commonly used assemblies. **Table A-3** in the Appendix shows the properties of selected pipe sizes.

The strength of the drillpipe body is calculated based on the OD and internal diameter (ID) of the pipe and the material yield strength. Because of wear, the OD of the pipe body decreases as the pipe is used. Wear and erosion on the pipe's inside surfaces are not considered in strength calculations. Pipe is classified, based on

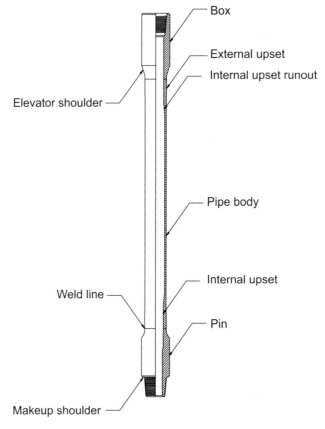

Fig. 3.1.1—Drillpipe assembly.

the amount of wear, as New, Premium, Class 2, or Class 3. New pipe has no wear. For service load calculations, the dimensions of New pipe are rarely used, except for New drillstrings and in some critical service applications such as in deep water, where wall uniformity is verified through frequent inspections. Premium pipe has up to 20% of the specified wall thickness worn away. Service load calculations for Premium pipe are made assuming uniform wear of 20% of the specified wall thickness on the outside surface of the pipe. Similarly, Class 2 pipe has up to 30% of the specified wall thickness worn away, and the service loads are calculated accordingly. Pipe with less than 70% of the specified wall thickness is classified as Class 3 pipe. Values for hook load capacity for the different wear classes can be found in *ISO/NP 10407-1* (under development 10/5/07) and API *RP 7G* (1995). The color coding applied to identify drillpipe class after it has been inspected is shown in **Fig. 3.1.2.**

Drillpipe Body. Drillpipe bodies are seamless chrome-molybdenum steel tubes with chemistries approximately equal to American Society for Testing and Materials (ASTM) 4127–4133 or an ISO 25CrMo4. Carbon content is normally at approximately 0.27% to 0.33%. The chemistry of the pipe and its subsequent heat treating is designed to obtain a specific yield strength, toughness, and ductility. There are four commonly used strength ranges called grades. **Tables 3.1.1 and 3.1.2** show the specified properties for the four grades of pipe. Other specialty grades are available with higher yield strengths or enhanced H_2S resistance.

The pipe is heat treated by austenitizing, then rapidly cooling with a water quench, normally from the OD only. The quench produces a Martensitic microstructure that is very hard and brittle. The pipe is tempered by reheating to a precise temperature and held at that temperature for a specific time to obtain the strength, toughness, and ductility properties shown in Tables 3.1.1 and 3.1.2. The tempering temperatures are generally around 1,000°F for Grade S and up to around 1,250°F for Grade E. The higher the pipe strength, the lower the tempering temperature. API and ISO specifications include tensile and yield strength requirements and impact strength requirements. Fracture toughness is very important for drillpipe. The most common failure

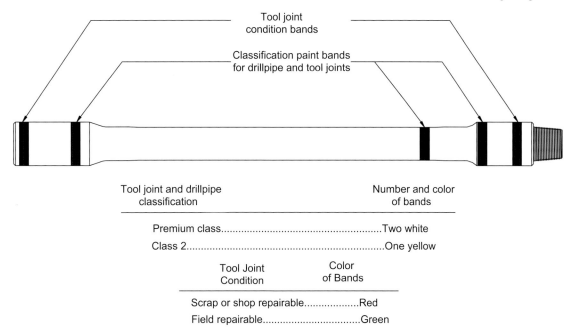

Fig. 3.1.2—Drillpipe and tool joint color code identification. (From *RP 7G, Recommended Practice for Drill Stem Design and Operating Limits*, 1998. Reproduced courtesy of the American Petroleum Institute.)

TABLE 3.1.1—DRILLPIPE BODY PROPERTIES

Material	Yield Strength (psi) Minimum	Maximum	Tensile Strength (psi) Minimum	Elongation (%) Minimum
Grade E	75,000	105,000	100,000	0.5
Grade X	95,000	125,000	105,000	0.5
Grade G	105,000	135,000	115,000	0.6
Grade S	135,000	165,000	145,000	0.7

TABLE 3.1.2—DRILLPIPE BODY CHARPY IMPACT VALUES

	SR[a]	Test Temperature	Minimum Average Absorbed Energy (J) Specimen Size (mm × mm) 10 × 10	10 × 7.5	10 × 5	Minimum Specimen Absorbed Energy (J) Specimen Size (mm × mm) 10 × 10	10 × 7.5	10 × 5
Grades E, X, G, S		21 ± 3 °C	54	43[b]	30[c]	47	38[b]	26[c]
Grades E, X, G, S	SR 20	−10 ± 3 °C	41	33[b]	27[d]	30	24[b]	20[d]

(a) Supplementary requirement. Requirement that may be added to the specification by the purchaser.

(b) Based on 80% of full size (10 × 10 mm)

(c) Based on 55% of full size (10 × 10 mm)

(d) Based on 67% of full size (10 × 10 mm)

mode of drillpipe is fatigue. A fatigue crack forms in the pipe and propagates through the pipe wall, allowing the drilling fluid, under pressure, to escape. This is called a *washout* and is detected on the rig floor by a drop in standpipe pressure. Pipe with good fracture toughness can be pulled out of the hole before the crack propagates to the point of pipe separation.

Upset. The heat associated with welding the tool joint to the drillpipe body changes the material properties of the pipe body and tool joint at the weld. The post-weld heat-treatment process can bring the yield strength of the heat-affected zone of the pipe and tool joint up to only approximately 80,000 to 100,000 psi.

To compensate for the loss of strength in the tool joint and higher-strength drillpipe, the drillpipe body is upset on each end to increase the wall thickness. Three upset configurations are used depending on the size of the tool joint that will be attached: internal upset (IU), internal-external upset (IEU), and external upset (EU). These upset configurations are shown in **Fig. 3.1.3.**

As shown in the figure, the wall thickness of IU pipe is increased by decreasing the pipe's ID. IEU pipe has a decreased ID and increased OD. EU pipe has an increased OD. There is some increase in the OD on IU upsets and some decrease in the ID on EU upsets to allow for machining after attaching tool joints.

The external surface of the upset is controlled by a die that is fit around the pipe during upsetting. The internal surface is controlled only by a punch inside the pipe the diameter of the finished upset ID. The punch has a very slight taper so that it can be removed after the pipe is upset. The transition region between the upset ID and pipe body ID on IU and IEU pipe is an uncontrolled surface and in the past was often the location of fatigue cracks because of stress raisers. Drillpipe manufacturers have developed processes that produce smooth transition zones over lengths of several inches that have very nearly eliminated fatigue cracks in this region.

Tool Joints. Tool joints are the threaded connections that are welded to the upset drillpipe (see Fig. 3.1.1). Tool joints are rotary shouldered connections, so named because when made up, they tighten against a shoulder. In the made-up condition, the dominant load in the threads is in the axial direction—tension in the pin, compression in the box.

Tool joints are defined by their OD, ID, and thread form. Other features of tool joints that must be specified are pin tong length, box tong length, box hardbanding type and configuration, and pin hardbanding type and configuration. Hardbanding is not normally used on all tool joints but is usually specified for the box tool joints.

Tool Joint OD and ID. The selection of tool joints is based on balancing the requirements for torsional strength and for minimizing drilling-fluid pressure loss through the bore. The torsional strength of the tool joint is a function of the pin ID and box OD. Torsional strength is limited by the tensile load the pin can withstand at the last engaged thread (LET) and by the compressive load the box can withstand in the counterbore. The LET is the last pin thread closest to the makeup shoulder that is engaged with the box. Eq. 3.1.14 contains an area term, A_C, which is the smaller of the cross-sectional area of the pin ¾ in. from the makeup shoulder or the box ⅜ in. from the makeup shoulder.

OD. The OD of a tool joint is a compromise between torsional strength, fishability, and elevator hoisting capacity. And there could be cases in which the tool joint OD is picked to limit the pressure loss in the annulus, thus minimizing the equivalent circulation density (ECD).

Fishability. The annular clearance between the tool joint OD and wall of the hole must be large enough for an overshot fishing tool to *swallow* the box. Fishing tool manufacturers publish overshot sizes and performance properties. It is important to retain the ability to use a full-strength overshot if there is a reasonable chance for having drillstring problems or stuck pipe.

Torsional Strength. The torsional strength of the box is a function of the OD because the critical area of the box is the box counterbore section between the threads and makeup shoulder. See **Fig. 3.1.4.**

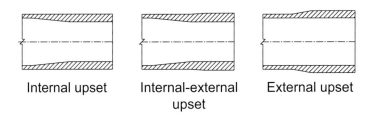

Internal upset   Internal-external   External upset
                             upset

Fig. 3.1.3—Drillpipe upsets.

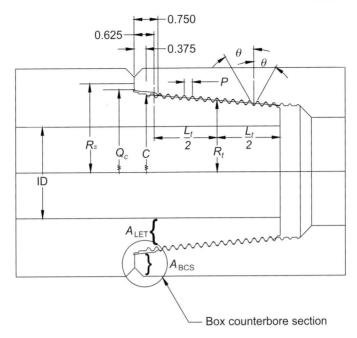

Fig. 3.1.4—Thread form nomenclature.

Elevator Hoisting Capacity. The elevator hoisting capacity is dependent on the projected area of the elevator shoulder as seen later in Eq. 3.1.49. Fig. 3.1.1 shows the elevator shoulder of the box.

ID. As the bore diameter decreases, the torsional strength increases until A_{LET} becomes greater than A_{BCS}. Generally speaking, however, the torsional strength is controlled by the pin ID; the smaller the ID, the greater the torsional strength. The makeup torque value in Table A-2 includes a P or B. This denotes whether the connection is pin-weak or box-weak. P is pin-weak, so the torsional strength of the connection is a function of the pin ID. B is box-weak, so the torsional strength of the connection is a function of the box OD.

Pressure Loss. **Fig. 3.1.5** shows the internal configuration of a conventional tool joint pin and box. There are a number of places in the tool joint that create a pressure loss as the drilling fluid is circulated down the string. The pressure loss varies inversely with the pin ID—the smaller the ID, the greater the pressure loss, and there are additional pressure losses with each change in ID. The power required to pump the drilling fluid down the pipe is a major concern because of the cost of power and because hydraulic horsepower consumed in moving the fluid is not available at the bit. The compromise is to select a bore diameter that will give adequate torsional strength at a pressure loss that can be tolerated.

Threads. Not counting proprietary connections, there are seven rotary shouldered connection types and nine thread forms (IADC 1992). **Table A-4,** located in the Appendix of this chapter, lists the properties of the thread forms used on rotary shouldered connections. **Fig. 3.1.6** contains drawings of each thread form.

The thread form nomenclature is based on the crest width. Thread form V-0.050 has a crest width of 0.050 in.; thread form V-0.040 has a crest width of 0.040 in.; and so on. In the numbered connection (NC) nomenclature, V-0.038R is indicative of the root radius, which is 0.038 in.

For pipe sizes up to 5½ in., the most widely used connections on drillpipe are the NCs. They were adopted in 1958 by API as an improvement over internal flush (IF) threads because the NC threads have a radiused root while the IF threads have a flat root. The NC threads are more fatigue-resistant, and while rarely done on tool joints, it is easier to cold roll the thread roots. NC threads are interchangeable with IF threads. Many people still identify tool joints by the IF nomenclature: 3½ IF instead of NC38 or 4½ IF instead of NC50. There are many IF connections specified and used on drillpipe today, but none have a flat thread root.

For 5½ and 6⅝ in. pipe, full hole (FH) tool joints are used more than any other. API Reg connections are used on rock bits and many bottom hole assembly components such as drill collars, stabilizers and other special tools. The FH and API Reg thread forms have a deeper thread and are used on heavier members to decrease bearing stress on the thread flank during stabbing and makeup.

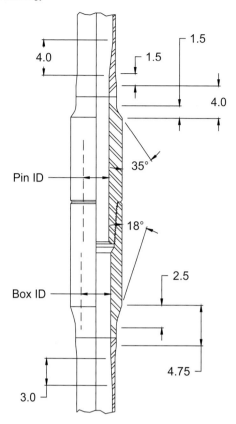

Fig. 3.1.5—Internal configuration of tool joint.

The H90 was designed for use on heavy-wall bottom hole assembly components. The 90° thread form is easier to stab and is less likely to incur damage when stabbing. The 90° thread form also has a larger radial force component on the thread flank as shown in the cosine term later in Eq. 3.1.14. Slimline H90 threads were designed for small diameter tool joints. The 90° thread form give the threads a shallow, low profile thread.

API Reg and H90 are rarely used on drillpipe. Reg is widely used on bottomhole assembly components. The thread taper on larger Reg connections is greater to aid in stabbing heavy bottomhole assembly member pins into the boxes hanging in the slips.

Tool Joint Metallurgy and Material Properties. Tool joints made to API or ISO standards all have the same yield strength regardless of the pipe grade to which they are attached. Other specifications such as Canada's *IRP Volume 1* (2008) specifies a lower tensile strength for increased H_2S resistance.

Tool joints are made from a chrome-molybdenum steel similar to that used for the drillpipe body, but the relatively complex shape of tool joints, compared to the cylindrical shape of the pipe body, increases their susceptibility to quench cracking. They are machined to their finished shape except for threading before heat treating and are quenched at a slower rate. The slower quench rate along with the heavy tong section requires a higher carbon content to get the needed hardness. Other elements may be added to give the tool joints the desired strength, toughness, and ductility. Most manufacturers have developed proprietary chemistries and heat-treating procedures that meet their design requirements.

API tool joints are heat treated to a strength of 120,000 psi minimum yield, regardless of the strength of the pipe to which they are welded. Most drillpipe is manufactured to API specifications. There are some exceptions in which a lower strength is specified for H_2S service, and there have been strings of pipe manufactured with higher-strength tool joints. Also, some deep drilling operations have required custom drillpipe products in which the tool joints have been manufactured from higher-strength materials including up to 140,000-psi-minimum-yield-strength material.

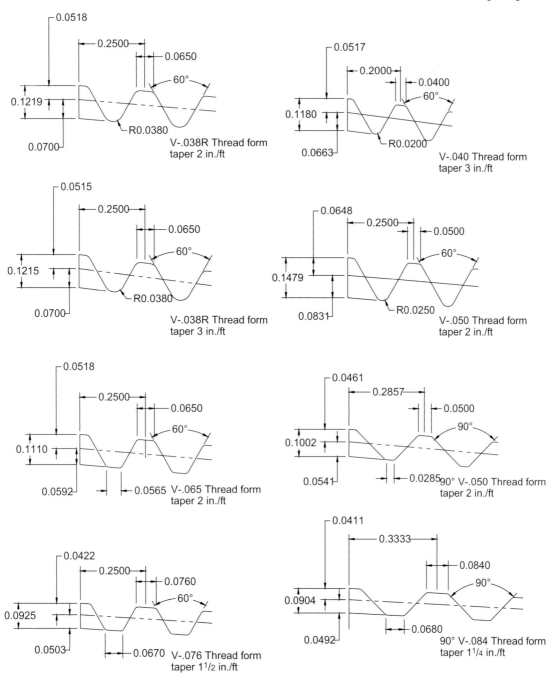

Fig. 3.1.6—Rotary shouldered connection thread forms.

Tool Joint Torsional Strength. Tool joint torsional strength is the result of frictional forces on the makeup shoulder and thread flanks. The equation used to calculate the torsional strength of a tool joint and other rotary shouldered connections, often referred to as the Jack Screw formula, was derived by Farr (1957). The derivation of the equation is as follows:

The forces acting on the screw thread are shown in **Figs. 3.1.7 and 3.1.8.** The total axial force exerted must equal the algebraic sum of the axial components of the friction forces and the forces normal to the thread surface. Hence,

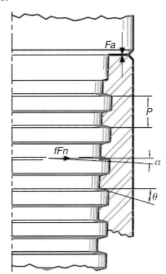

Fig. 3.1.7—Tool joint thread (Farr 1957). Reprinted courtesy of the American Society of Mechanical Engineers.

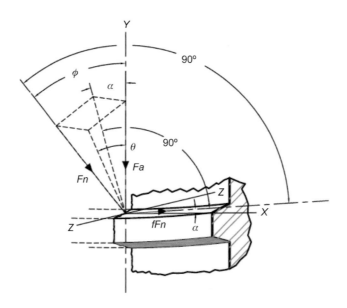

Fig. 3.1.8—Forces on tool joint thread flank (Farr 1957). Reprinted courtesy of the American Society of Mechanical Engineers.

$$F_a = F_n \left(\cos \phi - f \cdot \sin \alpha \right), \quad \dotfill \quad (3.1.1)$$

from which

$$F_n = \frac{F_a}{\cos \phi - f \cdot \sin \alpha}. \quad \dotfill \quad (3.1.2)$$

When a torque T is applied, the joint will make up p inches in one revolution. If a small amount of makeup, Δp, is considered, so that the axial force may be considered constant, the work done is given by

$$W_a = F_a \left(\Delta p \right) = \text{useful work} \quad \dotfill \quad (3.1.3)$$

$$W_f = f \cdot F_n \left(\frac{\Delta p}{\sin \alpha} \right) = \frac{f \cdot F_a (\Delta p)}{\sin \alpha (\cos \phi - f \sin \alpha)} = \text{work in overcoming thread friction} \quad \ldots \ldots \ldots \quad (3.1.4)$$

$$W_s = f \cdot F_a 2\pi R_s \left(\frac{\Delta p}{p} \right) = \text{work in overcoming shoulder friction} \quad \ldots \ldots \ldots \ldots \ldots \quad (3.1.5)$$

$$W = 2\pi T \left(\frac{\Delta p}{p} \right) = \text{work applied to joint.} \quad \ldots \ldots \ldots \ldots \ldots \ldots \ldots \ldots \ldots \ldots \quad (3.1.6)$$

Now,

$$W = W_a + W_f + W_s, \quad \ldots \ldots \ldots \ldots \ldots \ldots \ldots \ldots \ldots \ldots \ldots \ldots \ldots \ldots \ldots \quad (3.1.7)$$

and, by substitution,

$$T = \frac{F_a \cdot p}{2\pi} + \frac{F_a \cdot f \cdot p}{2\pi \sin \alpha (\cos \phi - f \sin \alpha)} + R_s f F_a. \quad \ldots \ldots \ldots \ldots \ldots \quad (3.1.8)$$

But

$$\tan \alpha = \frac{p}{2\pi R_t}, \text{ or } \frac{p}{2\pi} = R_t \tan \alpha. \quad \ldots \ldots \ldots \ldots \ldots \ldots \ldots \ldots \ldots \quad (3.1.9)$$

Substituting in the second term on the right gives

$$T = \frac{F_a p}{2\pi} + \frac{F_a R_t f}{2\pi \sin \alpha (\cos \phi - f \sin \alpha)} + R_s f F_a. \quad \ldots \ldots \ldots \ldots \ldots \ldots \ldots \quad (3.1.10)$$

The angle ϕ is dependent on the lead angle and the included angle of the threads. From Fig. 3.1.7 and Fig. 3.1.8, it is found that

$$\cos \phi = \frac{1}{\sqrt{1 + \tan^2 \theta + \tan^2 \alpha}}. \quad \ldots \ldots \ldots \ldots \ldots \ldots \ldots \ldots \ldots \ldots \quad (3.1.11)$$

Using this expression and noting that α is usually quite small in tool joints, it may be shown that

$$\cos \alpha (\cos \phi - f \sin \alpha) = \cos \theta, \text{ approximately.} \quad \ldots \ldots \ldots \ldots \ldots \ldots \quad (3.1.12)$$

The expression for the torque is written finally:

$$T = F_a \left(\frac{p}{2\pi} + \frac{R_T f}{\cos \theta} + R_S f \right) \quad \ldots \ldots \ldots \ldots \ldots \ldots \ldots \ldots \ldots \ldots \ldots \quad (3.1.13)$$

or

$$T_{TJ} = \frac{Y_{TJ} \times A_C}{12} \left[\frac{p}{2\pi} + f \left(\frac{R_T}{\cos \theta} + R_S \right) \right]. \quad \ldots \ldots \ldots \ldots \ldots \ldots \ldots \quad (3.1.14)$$

A_C is the smaller of the cross-sectional area at the last engaged thread of the pin or the cross-sectional area of the box ⅜ in. from the makeup shoulder. Eq. 3.1.20 is used to calculate the makeup torque. For tool joints, makeup torque is usually 60% of the torsional strength of the connection.

The torsional strength of a rotary shouldered connection is achieved by the moment from the frictional forces on the makeup shoulder and thread flanks and by the axial deformation of the pin base and box counterbore section. With double-shoulder connections, there is an additional shoulder at the nose of the pin that creates a second axial force that combines with the frictional forces on its contacting surface and adds to the frictional force on the thread flanks (*RP 7G* 1995). There is also the work done in compressing the pin nose.

$$T_{DSC} = \frac{Y_{TJ} A_C}{12}\left[\frac{p}{2\pi} + f\left(\frac{R_T}{\cos\theta} + R_S\right)\right] + \frac{Y_{TJ} A_N}{12}\left[\frac{p}{2\pi} + f\left(\frac{R_T}{\cos\theta} + R_N\right)\right] \quad \dots\dots\dots\dots\dots \quad (3.1.15)$$

The assumption is made for double-shoulder connections that yielding occurs simultaneously in the pin nose and critical area. Manufacturing tolerances of the pin length and box depth could cause yielding of the pin nose to occur slightly before or slightly later than yielding in the critical area. Based on the success of double-shoulder connections, this assumption apparently has not been a problem. Like single-shoulder connections, makeup torque for a double-shoulder connection is a percentage of its torsional strength–again, usually 60%.

Torsional stress in the tool joint tong area normally is not calculated because of the relative size of the tool joint compared to the pipe, and because the torsional strength of the threaded area is much lower than that of the tong area. The torsional stress of the tool joint threads is often lower than that of the pipe body. This is not necessarily a detriment, however, because as long as the torque applied to a tool joint (the drilling torque) is less than the makeup torque, there are no increases in stresses in the pin or box. When drilling torque exceeds makeup torque, the connection makes up downhole and often results in stretched pins, belled boxes, or twistoffs.

Example. Calculate the torsional strength of a 6⅝ × 3¼-in. NC50 tool joint. Refer to Fig. 3.1.5. The values needed to calculate the torsional strength can be found in the following publications:

- API *RP 7G* (1995)
- API *Spec. 7* (1998)
- *ISO 10424-2:2007* (2007)
- *IADC Drilling Manual* (1992)
- Manufacturer's product information

$$A_{LET} = \frac{\pi}{4}\left[\left(c - 2\cdot Ded - \frac{T_{pr}}{96}\right)^2 - ID_{TJ}^2\right], \quad \dots\dots\dots\dots\dots\dots\dots\dots\dots\dots\dots\dots \quad (3.1.16)$$

where $c = 5.042$ in., $D_{ed} = 0.070$ in., $T_{pr} = 2.000$ in./ft, and $ID_{TJ} = 3.250$ in.

$$A_{LET} = \frac{\pi}{4}\left\{\left[5.042 - (2)(0.070) - \frac{2}{96}\right]^2 - 3.250^2\right\} = 10.415 \text{ in.}^2$$

$$A_{BCS} = \frac{\pi}{4}\left[OD_{TJ}^2 - \left(Q_c - \frac{3\cdot T_{pr}}{96}\right)^2\right], \quad \dots\dots\dots\dots\dots\dots\dots\dots\dots\dots\dots\dots \quad (3.1.17)$$

where $OD_{TJ} = 6.625$ in. and $Q_c = 5.313$ in.

$$A_{BCS} = \frac{\pi}{4}\left\{6.625^2 - \left[5.313 - \frac{(3)(2)}{96}\right]^2\right\} = 12.824 \text{ in.}^2$$

$$A_{LET} < A_{BCS} \therefore A_C = 10.415 \text{ in.}^2$$

$$\frac{p}{2\pi} = \frac{0.250}{2\pi} = 0.040 \text{ in.}$$

$$R_T = \frac{2c - (L_{PC} - 0.625)\dfrac{T_{pr}}{12}}{4}, \quad \dots\dots\dots\dots\dots\dots\dots\dots\dots\dots\dots\dots \quad (3.1.18)$$

where $L_{PC} = 4.500$ in.

$$R_T = \frac{(2)(5.042) - (4.500 - 0.625)\dfrac{2.000}{12}}{4} = 2.360$$

$$R_S = \frac{OD_{TJ} + Q_C}{4} = \frac{6.625 + 5.313}{4} = 2.925 \text{ in.} \quad \dots\dots\dots\dots\dots\dots\dots\dots\dots\dots\dots \quad (3.1.19)$$

$$T_{TJ} = \frac{(120,000)(10.415)}{12}\left\{0.040 + 0.080\left[\frac{2.359}{\cos(30)} + 2.925\right]\right\} = 51,217 \text{ ft-lbf}$$

Makeup torque is usually 60% of the torsional strength.

$$T_{MU} = (0.60)(51,217) = 30,730 \text{ ft-lbf} \quad \dots\dots\dots\dots\dots\dots\dots\dots\dots\dots\dots\dots\dots\dots \quad (3.1.20)$$

Tool Joint Tensile Capacity. The maximum tensile capacity of a tool joint is equal to the product of its cross-sectional area at the last engaged thread of the pin and the yield strength of the material—most often 120,000 psi. Tool joint material properties are shown in **Table 3.1.3.**

$$P_{TJ} = Y_{TJ} \cdot A_{LET} \quad \dots\dots\dots\dots\dots\dots\dots\dots\dots\dots\dots\dots\dots\dots\dots\dots\dots\dots\dots\dots \quad (3.1.21)$$

The tensile capacity of the tool joint will be decreased if the makeup torque exceeds a value called T4. T4 is the makeup torque at which pin yield and shoulder separation occur simultaneously with an externally applied tensile load and is given by the equation

$$T4 = \left(\frac{Y_{TJ}}{12}\right)\left(\frac{A_{LET}A_{BCS}}{A_{LET} + A_{BCS}}\right)\left[\frac{p}{2\pi} + f\left(\frac{R_T}{\cos\theta} + R_S\right)\right]. \quad \dots\dots\dots\dots\dots\dots\dots\dots\dots \quad (3.1.22)$$

API *RP 7G* (1995) and *ISO/NP 10407-1* (under development 10/5/07) use the *pyramid curve* shown in **Fig. 3.1.9** to visually examine the relative loads in the tool joint.

The curve contains a triangular shape, the *pyramid*, that contains information about the tool joint and a parabolic curve that contains information about the pipe body. The left leg of the triangle, from the origin to T4, represents the tensile load that results in shoulder separation of the tool joint at a given makeup torque without yielding the pin. If, for example, the makeup torque was 24,000 ft-lbf, the shoulders would separate at a tensile load of 1,061,250 lbf, but pin yielding would not occur until the tensile load reached 1,249,800 lbf. The right leg of the triangle, from T4 to 51,217 ft-lbf on the horizontal axis, represents separation of the shoulder accompanied by yielding of the pin. In the case of 5 in. 19.50-lbm/ft S135 drillpipe with 6⅝ × 3¼-in. NC50 tool joints with a makeup torque of 30,730 ft-lbf and with the drilling torque not exceeding the makeup torque,

$$P_o = \frac{12(A_{LET} + A_{BCS})T_{MU}}{A_{BCS}\left[\dfrac{p}{2\pi} + f \cdot \left(\dfrac{R_T}{\cos\theta} + R_S\right)\right]}. \quad \dots\dots\dots\dots\dots\dots\dots\dots\dots\dots\dots\dots\dots \quad (3.1.23)$$

TABLE 3.1.3—TOOL JOINT MATERIAL PROPERTIES			
Yield Strength (psi)		Tensile Strength (psi)	Elongation
Minimum	Maximum	Minimum	Minimum
120,000	165,000	140,000	13%

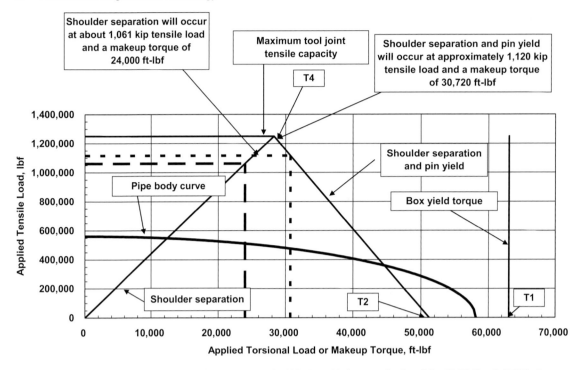

Fig. 3.1.9—Tool joint pyramid curve for 6 ⅝ × 3 ¼-in. NC50 tool joint attached to 5-in. 19.50-lbm/ft S135 pipe.

P_o is the tension required to separate the tool joint shoulders after T_{MU} is applied and is represented by the line from the origin to the apex at T4. T_{MU} is the torque that is applied to the tool joint before tension is applied, the makeup torque. Do not use Eq. 3.1.23 if T_{MU} is greater than T4 because P_o will be greater than P_{TJ}.

$$P_{T4T2} = \left(A_{LET} + A_{BCS}\right)\left\{Y_{TJ} - \frac{12 \cdot T_{MU}}{A_{LET}\left[\frac{p}{2\pi} + f\left(\frac{R_T}{\cos\theta} + R_S\right)\right]}\right\} \quad \dots\dots\dots\dots\dots\dots\dots \quad (3.1.24)$$

P_{T4T2} is the tension required to yield the pin after T_{MU} is applied and is represented by the line from T4 to T2.

$$T1 = \left(\frac{Y_{TJ}}{12}\right)\left\{A_{BCS}\left[\frac{p}{2\pi} + f\left(\frac{R_T}{\cos\theta} + R_S\right)\right]\right\} \quad \dots\dots\dots\dots\dots\dots\dots\dots\dots\dots\dots\dots \quad (3.1.25)$$

T1 is the torsional strength of the tool joint box and is represented by a vertical line at that value on the *x* axis.

$$T2 = \left(\frac{Y_{TJ} \cdot A_{LET}}{12}\right)\left[\frac{p}{2\pi} + f\left(\frac{R_T}{\cos\theta} + R_S\right)\right] \quad \dots\dots\dots\dots\dots\dots\dots\dots\dots\dots\dots\dots \quad (3.1.26)$$

T2 = torsional strength of the tool joint pin

$$T4 = \left(\frac{Y_{TJ}}{12}\right)\left(\frac{A_{LET}A_{BCS}}{A_{LET} + A_{BCS}}\right)\left[\frac{p}{2\pi} + f\left(\frac{R_T}{\cos\theta} + R_S\right)\right] \quad \dots\dots\dots\dots\dots\dots\dots\dots \quad (3.1.27)$$

T4 is the makeup torque at which pin yield and shoulder separation occur simultaneously with an externally applied tensile load.

$$T_{\text{PB}} = \frac{2 \cdot J_{\text{PB}}}{12 \cdot \sqrt{3} \cdot \text{OD}_{\text{PB}}} \cdot \sqrt{Y_{\text{PB}} - \frac{P_Q^2}{A_{\text{PB}}^2}} \quad \dots \dots \dots \dots \dots \dots \dots \dots \dots \dots \dots \dots \dots \quad (3.1.28)$$

T_{PB} is the allowable torque to the pipe body when a hanging weight or tensile load P_Q is present.

Tool Joint-Drillpipe Weld. Tool joints are attached to drillpipe by friction welding. Two friction-weld processes are commonly used: continuous drive and inertia drive. Both welds are made by converting mechanical energy from a rotating tool joint into heat energy by pushing the weld face of the tool joint against the end of the pipe held stationary and then stopping it to produce the weld. **Fig. 3.1.10** is a schematic of the process. The continuous-drive method rotates the tool joint at a constant speed of approximately 500 rev/min while holding the tool joint against the pipe at bearing pressures approaching 20,000 psi. Friction heats the tool joint and the pipe up to temperatures below melting while the force applied extrudes the softer hot material radially outward and inward, forming *ram's horns*. At the correct time, rotation is stopped, the tool joint is held against the pipe for a brief time, and the weld is complete.

The inertia-drive process grips the tool joint in a free-spinning flywheel turning 1,500 to 2,000 rev/min and pushes it against the stationary drillpipe, at the same bearing pressure as in the continuous drive weld process. The kinetic energy of the spinning flywheel is converted to frictional heat. When the energy of the flywheel is very nearly consumed, the flywheel is stopped, the tool joint is held against the pipe for brief time, and the weld is complete. After welding, the *ram's horns* are removed by machining, and then discontinuities between the pipe and tool joint are smoothed by grinding.

In both processes, the temperature is below that of melting—both are solid-state welds with good metallurgical properties and small grain size.

Tool joints are welded to the drillpipe after the tool joints and the drillpipe have been heat treated. The welding process reaustenitizes both members near the weld faces, and the subsequent slow cooling normalizes them. To regain the strength lost during welding, the weld is heat treated by induction heating and then quenching with either air or a polymer. Air quenching is normally carried out by directing air jets into the weld zone from a ring around the OD of the weld and a wand in the ID of the weld. In most cases, the polymer quench is on the OD only. The welding and heat-treating process produce a heat-affected zone approximately 1 in. wide on both sides of the weld. The width and profile of the heat-affected zone depends on the equipment used in the post-weld heat treatment. API specifications say that the tensile strength of the weld zone— the material yield strength times the total cross-sectional area—must be 1.10 times the tensile strength of the pipe body. Because of the chemistry of the pipe body and tool joint, the material strength at the weld is between 80,000 and 100,000 psi. With the added cross-sectional area from upsetting, the weld strength is greater than the strength of the pipe body even with S135 and higher-strength pipe.

Hardbanding. Drillpipe is often hardbanded (sometimes referred to as hardfacing) to inhibit tool joint wear, to inhibit casing wear, and to decrease the coefficient of friction between the tool joint and casing. See **Fig. 3.1.11.** Hardbanding is usually applied on the box but is often also applied on the pin. It is not applied on the pin only. A problem with applying hardfacing to the pin is the loss of tong space. Box hardbanding covers approximately 3 in. adjacent to the elevator shoulder and extends about ¾ in. down the elevator shoulder. There are sometimes three or four fingers equally spaced around the elevator shoulder to prevent *mud cutting* or erosion of the elevator shoulder. Sometimes the hardfacing is raised above the surface of the tool joint tong area. When casing wear is an issue, the material used for hardbanding is formulated to minimize casing wear. When casing wear is not a problem, tungsten carbide is often used to minimize tool joint wear. Currently, most prudent operators and drilling contractors specify casing-friendly hardbanding for

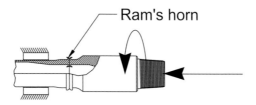

Fig. 3.1.10—Friction welding tool joint to drillpipe.

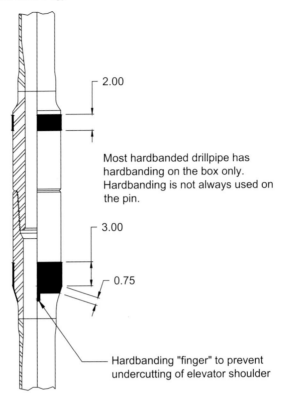

2.00

Most hardbanded drillpipe has
hardbanding on the box only.
Hardbanding is not always used on
the pin.

3.00

0.75

Hardbanding "finger" to prevent
undercutting of elevator shoulder

Fig. 3.1.11—Tool joint hardbanding.

their drillpipe. Tungsten carbide hardbanding is still used on heavy weight drillpipe (HWDP) and bottom hole assembly components that are primarily in the open hole and pose a much reduced risk of generating high casing wear.

The application of hardbanding produces a heat-affected zone in the tool joint, but this does not cause a problem.

Thread Compound. Thread compound is used on rotary-shouldered connections to prevent galling, to help provide a seal for drilling fluid, and to provide consistent frictional properties. The threads and makeup shoulder of either the pin or box are coated before making the connection up. The compound usually is applied to the box because the box threads are easy to get to when the pipe is in the slips.

Thread compound is a grease-based material that contains a percentage of finely powdered metallic solids or environmentally friendly solid. The metals primarily used in thread compounds are copper and zinc. At one time lead was widely used, but the dangers associated with lead and the environmental restrictions have minimized its use in compounds. The solids form a layer between the contacting surfaces of the tool joint that prevents galling, serves as a gasket to contain the internal pressure in the drillpipe, and provides a consistent coefficient of friction.

Thread compounds are formulated to provide a coefficient of friction between the pin and the box of 0.08 to approximately 0.12. The makeup torque tables in API *RP 7G* (1995), in *ISO/NP 10407-1* (under development 2007), and in Table A-2 of this chapter's Appendix contain makeup torque values calculated with a coefficient of friction of 0.08. Thread compounds with the higher frictional properties allow the tool joints to be made up to higher torques and provide additional torsional strength.

API *RP 7A1* (1992) defines a standardized test for thread compound friction factors. In this context, friction factor is not the same as coefficient of friction. The friction factor is a multiplier for the calculated makeup torque of a tool joint. When using a thread compound with a friction factor of 1.1, the published makeup torque will be multiplied by 1.1. For example, the makeup torque for a 6¼ × 3 NC46 shown in Table A-2 is 23,795 ft-lbf. Using a thread compound with a friction factor of 1.1 would allow the connection to be made up to 26,174 ft-lbf. Thread compound manufacturers usually put the friction factor value on the product label.

3.1.3 Drillpipe Service Loads.

Drillpipe service loads include tensile loading, torsional loading, internal pressure, compressive loading, external pressure, and bending. As stated at the beginning of the chapter, drillpipe is selected on the basis of its strength, size, and cost. The primary strength consideration is the pipe's tensile capacity or the ability to support the weight of the string. There are three size considerations: optimizing drilling fluid flow rate and hole cleaning, fishability, and elevator hoist capacity.

3.1.4 Pipe Body Loads and Stresses. *Tensile Loads.*

The required tensile strength of drillpipe is calculated by determining the total hanging weight or force required to pull the pipe out of the hole and adding a safety factor or *margin of overpull* to the value. For straight, vertical holes, the hanging weight is the sum of the weights of the bottomhole assembly and the weight of all the drillstring members above the bottomhole assembly times the buoyancy factor of the drilling fluid. For directional holes, the calculations are a little more complex because of the angle of the hole and frictional drag. The required tensile capacity of the pipe is equal to what the pipe has to pull out of the hole plus the margin of overpull.

The force required to pull drill collars out of the hole is

$$P_{DC} = \left[W_{DC1} \cdot \left(\cos \alpha_1 + \mu \cdot \sin \alpha_1 \right) + W_{DC2} \cdot \left(\cos \alpha_2 + \mu \cdot \sin \alpha_2 \right) \right] \cdot k_B \quad \dots\dots\dots\dots\dots \quad (3.1.29)$$

$$k_B = \left(1 - \frac{\rho_M}{\rho_S} \right) \quad \dots\dots\dots\dots\dots\dots\dots\dots\dots\dots\dots\dots\dots\dots\dots\dots\dots\dots\dots\dots\dots\dots \quad (3.1.30)$$

The weight of the drillstring is the length of all the drillpipe times its buoyed unit weight. (Note: In this context, *drillstring* is defined as drillpipe only. *Drillstring* does not include drill collars or other components such as motors, jars, or stabilizers, for example. *Drillstem* in this context is all components from the swivel to the bit.) The length of the drillstring is the hole depth minus the total length of drill collars and other drillstem members.

$$W_{DPA} = k_B \cdot L_{DPA} \cdot w_{DP}. \quad \dots\dots\dots\dots\dots\dots\dots\dots\dots\dots\dots\dots\dots\dots\dots\dots\dots\dots\dots\dots\dots \quad (3.1.31)$$

The force required to pull all drillstring members out of the hole is calculated in Eq. 3.1.32, where MOP is the margin of overpull. The MOP is the additional allowed tension to be applied to free stuck pipe. It is the difference between the maximum allowable tension and the calculated hook load (Azar and Samuel 2007).

$$P_{string} = W_{DPA1} \left[\left(\cos \alpha_1 + \mu \sin \alpha_1 \right) + W_{DPA2} \cdot \left(\cos \alpha_2 + \mu \sin \alpha_2 \right) \right] \cdot k_B + P_{BHA} + MOP \quad \dots\dots\dots \quad (3.1.32)$$

The tensile stress in the pipe from the hanging weight is

$$\sigma_{TPB} = \frac{F_{string}}{A_{PB}}. \quad \dots\dots\dots\dots\dots\dots\dots\dots\dots\dots\dots\dots\dots\dots\dots\dots\dots\dots\dots\dots\dots\dots\dots\dots\dots \quad (3.1.33)$$

The tensile stress is in the axial direction.

Also, there is a tensile stress induced in the tool joint pin from makeup. The tensile stress in the pin will be increased by the hanging weight of the pipe below the pin, as indicated by the first term in this equation:

$$\sigma_{LET} = \frac{P_{DS}}{\frac{\pi}{4} \left(OD_{TJ}^2 + ID_{TJ}^2 \right)} + \sigma_{MU} \quad \dots\dots\dots\dots\dots\dots\dots\dots\dots\dots\dots\dots\dots\dots\dots\dots\dots\dots \quad (3.1.34)$$

3.1.5 Torsional Loads.

The action of rotating the bit induces a torsional load on the drillpipe. When the bit is rotated by a topdrive or the rotary table, the torsional load is the sum of the bit torque and the frictional drag from rotating the pipe in the hole. When the bit is rotated with a motor, the pipe must counteract the torque of the motor. Torsional stress in the pipe body is given by the following equation:

$$\tau_{PB} = \frac{12 \cdot T \cdot OD_{PB}}{2 \cdot J_{PB}}. \quad \dots\dots\dots\dots\dots\dots\dots\dots\dots\dots\dots\dots\dots\dots\dots\dots\dots \quad (3.1.35)$$

Torsional stress is a shear stress at 45° from the axis.

3.1.6 Internal Pressure.

$$\sigma_P = \frac{P_I \cdot OD_{PB}}{0.875 \cdot 2 \cdot \tau_{PB}} \quad \dots\dots\dots\dots\dots\dots\dots\dots\dots\dots\dots\dots\dots\dots\dots\dots\dots \quad (3.1.36)$$

Internal pressure produces a tangential stress maximum on the outside surface.

Pipe Bending and Dogleg Severity. Drillpipe will always have an axial load, either tension or compression, and will often be subjected to a bending load because of hole curvature. Because the tool joints have a larger diameter than the pipe, the pipe body will not exactly follow the curvature of the hole and, depending on the axial load, may not even touch the wall of the hole. The procedure for calculating the bending stresses in the pipe as a function of axial load and hole curvature is not covered in this chapter. Instead, an approximation is used based on the assumption that the drillpipe body does exactly follow the curvature of the hole. A calculation procedure for more accurately calculating bending stresses around doglegs or in build zones is given by Lubinski (1988).

Bending stresses using this assumption are calculated as follows:
Radius of curvature:

$$\rho = \frac{100 \cdot 180}{HC \cdot \pi}. \quad \dots\dots\dots\dots\dots\dots\dots\dots\dots\dots\dots\dots\dots\dots\dots\dots\dots\dots\dots \quad (3.1.37)$$

Bending stress:

$$\sigma_{BPB} = \frac{E \cdot OD_{PB}}{2 \cdot \rho}. \quad \dots\dots\dots\dots\dots\dots\dots\dots\dots\dots\dots\dots\dots\dots\dots\dots\dots \quad (3.1.38)$$

The bending stress is in the axial direction maximum on the outside surface. The more rigorous calculations can be found in API *RP 7G* (1995) and *ISO/NP 10407-1* (under development 10/5/07).

When drilling in angled holes and when the coefficient of friction between the drillstring and the hole is greater than the cotangent of the friction angle, the drillstring must be pushed to apply weight to the bit. The drillstring is thus subjected to compressive loads and buckling. Buckling is addressed in Chapter 3.4 of this book.

3.1.7 Drillpipe Selection. *Pipe Strength.*

Pipe selection usually begins with picking a pipe that can handle the weight of the string. The hole depth and diameter will be known along with planned bottomhole assemblies and mud weights. This is sometimes an iterative process wherein heavier pipe or larger pipe is required at the top of the string, but the amount must be calculated.

As an example, select the drillpipe for a 16,000-ft total depth (TD) hole. The hole is vertical for 3,000 ft, then builds to 40° at a build rate of 5° per 100 ft. The hole profile is shown in **Fig. 3.1.12.** The required bit weight is 20,000 lbf, and the drill collar size is 7¼ × 2¼. The hole size at TD is 8½ in. The coefficient of friction between the pipe and hole wall is 0.3. Mud weight is 12 lbm/gal. Pipe size will be 5 in. A transition string of 10 joints of heavyweight pipe will be run above the drill collars.

3.1.8 Drill Collar Length.

Drill collar weight is calculated using Eq. 3.1.39.

$$W_{DC} = \frac{\pi}{4}\left(OD_{DC}^2 - ID_{DC}^2\right) \cdot \rho_S \cdot L_{DC} \quad \dots\dots\dots\dots\dots\dots\dots\dots\dots\dots\dots\dots\dots \quad (3.1.39)$$

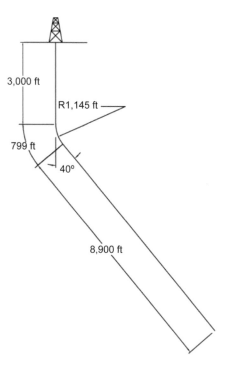

3,000 ft

R1,145 ft

799 ft

40°

8,900 ft

Fig. 3.1.12—Hole profile for drillpipe selection example.

$$\rho_S = 490 \ \frac{\text{lbm}}{\text{ft}^3}$$

$$L_{DC} = 31 \ \text{ft}$$

$$W_{DC} = \frac{\pi}{4}\left(7.25^2 - 2.25^2\right) \cdot \frac{(490) \cdot (31)}{144} = 3935.3 \ \text{lbm}$$

$$W_{DC} = 3935.3 \ \text{lbm}$$

or

$$W_{DC} = 126.9 \ \frac{\text{lbm}}{\text{ft}}$$

Determine the total length of the drill collars [where BW = bit weight, lbf, and NP = neutral point (equal to 70%)]:

$$L_{DC} = \frac{\text{BW}}{\text{NP} \cdot k_B \cdot W_{DC} \cdot (\cos\alpha - \mu \cdot \sin\alpha)} \quad \dots\dots\dots\dots\dots\dots\dots\dots \quad (3.1.40)$$

Set the neutral point at 0.70 times the length of the collars from the bottom and solve for total drill collar length. The neutral point is the axial location on the drill collars at which the axial load is zero. The top of the drill collar string is in tension, the bottom is in compression, and there is a point at which the load goes from tension to compression.

$$k_B = \frac{\rho_S - \dfrac{1728 \cdot W_M}{231}}{\rho_S}, \quad \dots\dots\dots\dots\dots\dots\dots\dots\dots\dots\dots\dots\dots\dots \quad (3.1.41)$$

where k_B = buoyancy factor, α = hole angle, and μ = coefficient of friction, and 1728/231 converts lbm/gal to lbm/ft³.

$$k_b = \frac{0.283 - \dfrac{(1728) \cdot (12)}{231}}{0.283}$$

$$k_B = 0.82$$

$$L_{DC} = \frac{20,000}{(0.8)(0.82)(126.9)(\cos 40 - 0.3 \cdot \sin 40)} \quad \dots\dots\dots\dots\dots\dots\dots\dots\dots\dots\dots\dots\dots \quad (3.1.40)$$

$$L_{DC} = 481 \text{ ft}$$

With a collar length of 31 ft, 16 collars are needed. Sixteen collars are 5⅓ stands, or 496 ft.

$$L_C = 496 \text{ ft}$$

The force to pull the drill collars out of the lateral section of the hole is

$$L_{DC} = (496)(126.9)(0.82)\left[\cos(40) + 0.3 \cdot \sin(40)\right] = 49,490 \text{ lbf}.$$

A transition string of heavyweight drillpipe is often placed directly above the drill collars in vertical holes because it has been found to reduce fatigue failures in drillpipe just above the collars. Further to this, it has been found that fewer fatigue failures occur when the bending strength ratio (BSR) of adjoining drillstring members is less than 5.5. BSR in this case is defined as the ratio of the bending section moduli of the two members (e.g., between the topmost drill collar and the pipe just above it). For severe drilling conditions such as hole enlargement, corrosive environment, or hard formations, a BSR of less than 3.5 helps to reduce the frequency of fatigue failures.

Section modulus of drill collar:

$$Z_{DC} = \frac{\pi}{32 \cdot OD_{DC}}\left(OD_{DC}^4 - ID_{DC}^4\right) \quad \dots\dots\dots\dots\dots\dots\dots\dots\dots\dots\dots\dots\dots \quad (3.1.42)$$

$$Z_{DC} = 37.065 \text{ in.}^3$$

Transition string BSR:

$$Z \geq \frac{37.065}{5.5} = 6.739.$$

The recommended section modulus of the transition string is > 6.74 in.³.

Calculate the section modulus of 5-in. heavyweight pipe.

From **Table 3.1.4**,

$$OD_{HW} = 5 \text{ in.}$$

$$ID_{HW} = 3 \text{ in.}$$

$$Z_{HW} = \frac{\pi}{32 \cdot 5}\left(5^4 - 3^4\right) = 10.68 \text{ in.}^3$$

Ratio of drill collar section modulus to pipe section modulus:

TABLE 3.1.4—HEAVYWEIGHT DRILLPIPE*			
Size (in.)	Pipe ID (in.)	Tool Joint	Weight (lbm/ft)
$3^1/_2$	$2^1/_4$	NC38	23.4
4	$2^9/_{16}$	NC40	29.9
$4^1/_2$	$2^3/_4$	NC46	41.1
5	3	NC50	50.1
$5^1/_2$	$3^3/_8$	$5^1/_2$ FH	57.6
$6^5/_8$	$4^1/_2$	$6^5/_8$ FH	71.3

* Personal communication with D. Brinegar, Smith International, Houston (2004).

$$\text{Ratio} = \frac{Z_{DC}}{Z_{HW}} = \frac{37.07}{10.68} = 3.47.$$

This falls below the recommended BSR of 5.5.

A heavyweight string length of 10 joints is a good rule of thumb. The weight of 10 joints of 5-in. heavyweight pipe from Table 3.1.4 is

$$W_{HW} = 10 \cdot L_{HW} \cdot w_{HW}$$

$$L_{HW} = 31 \text{ ft}$$

$$w_{HW} = 50.1 \frac{\text{lbm}}{\text{ft}}$$

$$W_{HW} = (10)(31)(50.1) = 15,531 \text{ lbm}$$

Force to pull heavy weight out of hole:

$$P_{HW} = (15,531)(0.82)\left[\cos(40) + 0.3 \cdot \sin(40)\right] = 12,212 \text{ lbf}.$$

Total length of drill collars and heavy weight:

$$L_{DC} + L_{HW} = 496 + 310 = 806 \text{ ft}.$$

Drillpipe length in deviated section of hole:

$$8,900 - 806 = 8,094$$

Calculate the force required to pull the drillpipe out of the deviated section of the hole. Find approximate drillpipe adjusted weight. (Tool joint OD and ID have not yet been determined; therefore, the weight of the drillpipe at this time must be estimated.) From Table A-2 for 5-in. drillpipe, the adjusted weight is 23.08 lbm/ft.

$$P_{DP} = (23.08)(8094)(0.82)\left[\cos(40) + 0.3 \cdot \sin(40)\right]$$
$$= 146,885 \text{ lbf}$$

Force to pull pipe out of build zone—hanging weight and frictional load below build zone:

$$W_H = 49,490 + 12,212 + 146,885 = 208,587 \text{ lbf}.$$

Calculate the lateral force on the tool joints. The lateral force on the tool joints can be calculated from Eq. 44 in Lubinski (1988b) (see also Van Vlack 1989; Jones 1992).

Radius of curvature of hole from Eq. 3.1.37,

$$\rho = \frac{(180)}{HC \cdot \pi} = \frac{(100)(180)}{(5)(\pi)} = 1,146 \text{ ft} \quad \dots\dots\dots\dots\dots\dots\dots\dots\dots\dots\dots\dots \quad (3.1.43)$$

$$F_{TJL} = \frac{W_H \cdot L}{\rho} = \frac{(208,587)(31)}{1,146} = 5,642 \text{ lbf} \quad \dots\dots\dots\dots\dots\dots\dots\dots\dots\dots\dots \quad (3.1.44)$$

There is 799 ft of pipe in the build zone, which is

$$\frac{799}{31} = 26 \text{ joints of pipe.}$$

One set of tool joints per pipe comes to a total lateral force of

$$(5,345)(26) = 138,970 \text{ lbf}.$$

This lateral force times the coefficient of friction is

$$(138,970)(0.3) = 41,691 \text{ lbf}.$$

It takes a 41,691-lbf tensile load to pull the pipe through the build zone at the maximum hole depth.

Force to pull pipe out of vertical portion of hole:

$$(23.08)(3000)(0.82) = 56,777 \text{ lbf}.$$

Total force to pull string out of hole:

$$F_{string} = 49,490 + 12,212 + 146,885 + 41,691 + 56,777 = 307,055 \text{ lbf}. \quad \dots\dots\dots\dots\dots\dots \quad (3.1.45)$$

Calculate the hook load capacity of Premium class 5-in. 19.50-lbm/ft S135 drillpipe. The wall thickness for New 5-in. drillpipe from Table A-1 is 0.362 in. The wall thickness for Premium 5-in. drillpipe is

$$(0.8)(0.362) = 0.290 \text{ in.}$$

The ID of 5-in. pipe from Table A-1 is 4.276 in. The OD of Premium 5-in. pipe used for calculations is

$$OD_P = 4.276 + (2)(0.290) = 4.855 \text{ in. OD}$$

Tensile capacity of pipe:

$$T_P = \frac{\pi}{4}(4.855^2 - 4.276^2)(135,000) = 560,559 \text{ lbf}.$$

Multiply this number by a design factor of 0.9 and subtract a margin of overpull value of 100,000 lbf. The margin of overpull is the additional tensile load, above the tensile load required to pull the pipe out of the hole, that could be used if the pipe becomes stuck.

Available tensile capacity of the pipe:

$$T_P = (0.9)(560,559) - 100,000 = 404,503 \text{ lbf}.$$

The calculated force required to pull the pipe out of the hole is 306,055 lbf. In terms of tensile requirements, 5-in. 19.50-lbm/ft S135 pipe is sufficient for this drilling program, but is it the right size pipe for the hole size? The hole size at TD is 8½ in.

The pipe must also be selected as a good match for the hole size and for drilling hydraulics. Selection of the pipe size based on hole diameter depends on the ability to fish the pipe and the desired annular velocity of the drilling fluid.

3.1.9 Fishing the Pipe. When drillpipe fails or *parts* (*twists off* is a common term), the portion of the string left in the hole must be *fished* out of the hole. Sometimes when pipe gets stuck, the pipe above the *stuck* point is backed off and the pipe left in the hole must be fished.

If, in the case of the parted pipe, the tool joint fails, the top of the fish will have a tool joint box looking up. If the pipe body fails, the fish will have the fractured end of the pipe body looking up. If the failed pipe body cannot be grabbed by the fishing tools, the pipe body will be milled away, leaving a tool joint at the top of the fish.

In these cases, fishing tools must grab the tool joint, and the drillstring selection process must allow for this. There must be enough annular clearance between the tool joint and the hole wall for the fishing tool to enclose the tool joint. Fishing-tool manufacturers publish information on the required annular clearance of their tools. For 8½-in.-diameter holes, tools to fish 7-in.-diameter tool joints are readily available. Some tools are available for tool joint diameters of 7¼ in., but they may be limited in strength.

The pipe selected in this exercise is a 5-in. 19.50-lbm/ft S135 pipe. The tool joint for this pipe will have an OD, as shown in Table A-2, of 6⅝ in., which can be fished in an 8½-in. hole.

3.1.10 Annular Velocity of Drilling Fluid. Bringing the cuttings generated by drilling to the surface requires a certain drilling-fluid velocity. The velocity is usually between 100 and 400 ft/min, depending on the properties of the drilling fluid and the cuttings, as well as the hole size and orientation. For the drilling program described above, assume a minimum annular velocity of 200 ft/min to bring the cuttings to the surface and a maximum of 350 ft/min to prevent hole erosion. Pump flow rate is 350 gal/min.

$$\text{Annular area} = \frac{\pi}{4}\left(D_H^2 - D_{DP}^2\right) = \frac{\pi}{4}\left(8.5^2 - 5^2\right) = 37.11 \text{ in.}^2, \qquad (3.1.46)$$

$$\text{Annular velocity} = \left(\frac{350 \text{ gal}}{37.11 \text{ in.}^2}\right)\left(\frac{231 \dfrac{\text{in.}^3}{\text{gal}}}{12 \dfrac{\text{in.}}{\text{ft}}}\right) = 182 \frac{\text{ft}}{\text{min}}. \qquad (3.1.47)$$

Thus, 5-in. drillpipe in an 8½-in.-diameter hole will yield the desired annular velocity.

An additional parameter often considered in drilling hydraulics is the ECD. This becomes a factor when pressure at the bottom of the hole required to move the fluid at the desired velocity becomes great enough to damage the formation.

3.1.11 Pressure Loss in the Drillpipe. The hydraulic horsepower at the bit is an important drilling parameter, and efforts are made to maximize the pump flow rate times the drilling fluid pressure at the nozzles of the drill bit (Lubinski 1988b). The maximum pressure available at the bit is

$$P_{bit} = P_{standpipe} - P_{drillpipe} - P_{annulus} \qquad \cdots\cdots\cdots\cdots\cdots\cdots\cdots\cdots\cdots\cdots\cdots\cdots\cdots\cdots (3.1.48)$$

In this relationship, P_{bit} is the pressure loss of the drilling fluid from being pushed through the bit nozzles. $P_{standpipe}$ is the drilling fluid pressure as it enters the drillstring. $P_{drillpipe}$ is the pressure loss of the drilling fluid as it is being pumped down the drillpipe. $P_{annulus}$ is the pressure loss of the drilling fluid as it is being pumped to the surface. **Fig. 3.1.13** shows the estimated pressure loss through the pipe and up the annulus of 12 lbm/gal drilling fluid in an 8½-in.-diameter hole for different drillpipe and tool joint sizes.

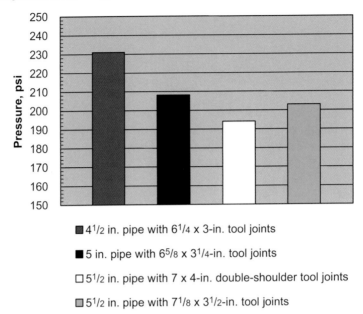

Fig. 3.1.13—Estimated pressure loss required to move 350 gal/min of 12-lbm/gal drilling fluid through 1,000 ft of pipe and annulus of 8½-in. hole.

Fig. 3.1.13 illustrates the hydraulic advantage of using the largest possible drillpipe and the largest possible tool joint ID. Other factors such as availability and cost of tools required to handle pipe do not always allow picking pipe size on hydraulic merits alone.

3.1.12 Elevator Hoist Capacity. Eq. 3.1.49 is used to calculate the projected area of contact of the tool joint elevator shoulder and the elevator. Dimensions for the ID of the elevator can be found in manufacturer's catalogs. The elevator ID for 5 in. pipe is 5¼ in. The projected area of contact times the bearing stress is

$$P_E = \frac{\pi}{4} \left[OD^2_{TJ} - \left(C_E + f_W \right)^2 \right] \cdot \sigma_B,$$

$$P_E = \frac{\pi}{4} \left[6.625^2 - \left(5.25 + \frac{3}{32} \right)^2 \right] \cdot 65,000 = 536,383 \text{ lbf.} \quad \dots\dots\dots\dots\dots\dots\dots\dots \quad (3.1.49)$$

In this example, a bearing stress of 65,000 psi is used. This is the recommended bearing stress from elevator manufacturers. Higher bearing stresses are sometimes used, but the wear rate of the elevator latch and hinge increases with the higher stresses.

This tool joint/elevator combination is capable of hoisting 536,383 lbf. The string weight from Eq. 3.1.45 is 307,055 lbf.

3.1.13 Slip Capacity. Slips grip the uppermost joint of pipe and keep the drillstring from falling in the hole when it is disconnected from the kelly or top drive. The slips grip the pipe with a radial force that is produced by the action of the pipe weight and a tapered slip bowl. See **Fig. 3.1.14.** Only in rare cases is drillpipe crushed by excessive slip loading; however, slip crushing is a major concern with landing strings used in setting casing and lowering BOPs and wellhead equipment in offshore deepwater applications.

The slips are fitted with dies that grip the pipe. These dies have small teeth that leave marks in the pipe that act as stress raisers. Fatigue failures in drillpipe often originate in slip marks.

Referring to Fig. 3.1.14, the safe load to prevent slip crushing can be calculated using the following derivation (Reinhold and Spiri 1959).

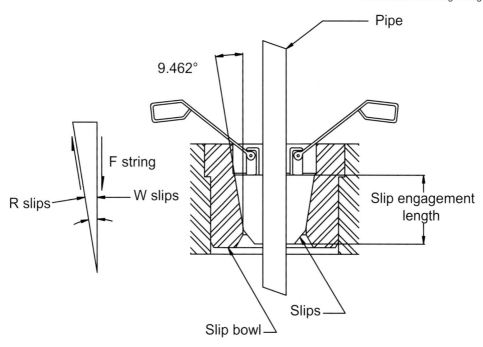

Fig. 3.1.14—Slips and slip bowl.

From Fig. 3.1.14, the following two equations must be satisfied for equilibrium of the slip:

$$F_{string} = R_{slips} \left(\sin \alpha + \mu \cos \alpha \right), \quad \dotso \dotso \quad (3.1.50)$$

$$W_{slips} = R_{slips} \left(\cos \alpha - \mu \sin \alpha \right). \quad \dotso \dotso \quad (3.1.51)$$

Eliminating R_{slips} from these equations gives

$$W_{slips} = F_{string} \frac{\left(1 - \mu \cdot \tan \alpha\right)}{\tan \alpha + \mu} = F_{string} \frac{1}{\tan \left(\alpha + \psi\right)}, \quad \dotso \quad (3.1.52)$$

where $\tan \alpha = \dfrac{1}{6}$ (API standard taper angle = 9.462°), and $\mu = 0.2$ is the coefficient of friction of two metal surfaces coated with thread compound (Sathuvalli et al. 2002). Note: This is not the value used in the Spiri-Reinhold paper (Reinhold and Spiri 1959)—they used 0.06.

ψ = friction angle = $\tan^{-1} \mu.$

Calculate the string weight that causes yielding in the pipe:

$$\left(\frac{F_{string}}{A_{PB}} \right)_{ID} = Y_{PB} \left[\frac{2}{1 + \left(1 + \dfrac{2 \cdot OD_{PB}^2}{OD_{PB}^2 - ID_{PB}^2} \dfrac{K \cdot A_{PB}}{A_L}\right)^2 + \left(\dfrac{2 \cdot OD_{PB}^2}{OD_{PB}^2 - ID_{PB}^2} \dfrac{K \cdot A_{PB}}{A_L}\right)^2} \right]^{\frac{1}{2}}. \quad \dotso \quad (3.1.53)$$

This equation is derived from the von Mises-Hencky theory, which defines the beginning of yield for a triaxial stress state in the pipe body:

$$Y_{DP} = \frac{1}{\sqrt{2}} \left[\left(\sigma_2 - \sigma_1\right)^2 + \left(\sigma_3 - \sigma_1\right)^2 + \left(\sigma_3 - \sigma_2\right)^2 \right]^{0.5}, \quad \dots\dots\dots\dots\dots\dots \quad (3.1.54)$$

$$\sigma_1 = \text{axial stress} = \frac{F_{string}}{A_{PB}}.$$

Both the radial and hoop stresses can be calculated from basic equations for an externally pressured thick-wall cylinder. Radial stress is zero at the inside radius. Hoop stress is maximum at the inside radius:

$$\sigma_3 = -\frac{2 \cdot P_{slips} \cdot OD_{DP}^2}{OD_{DP}^2 - ID_{DP}^2}, \quad \dots\dots\dots\dots\dots\dots\dots\dots\dots \quad (3.1.55)$$

where

$$P = \text{external pressure on pipe} = \frac{W_{slips}}{A_{slips}} \quad \dots\dots\dots\dots\dots\dots\dots\dots \quad (3.1.56)$$

And from Eq. 3.1.52,

$$W_{slips} = F_{string} \cdot K, \quad \dots\dots\dots\dots\dots\dots\dots\dots\dots\dots\dots\dots\dots\dots \quad (3.1.57)$$

$$\psi = \tan^{-1} 0.2 = 11.31,$$

$$K = \text{Lateral Load Factor} = \frac{1}{\tan\left(a + \psi\right)}$$

$$K = \frac{1}{\tan\left(9.462 + 11.31\right)} = 2.636.$$

A_{slips} = surface area of pipe in contact with slips.
Length of slips = 16 in.

$$A_{slips} = 16 \cdot \pi \cdot 5 = 251.3 \text{ in.}^2$$

Calculate the allowable string weight for the example pipe. From Eq. 3.1.53,

$$F_{string} = (5.275)(135,000)\left[\frac{2}{1 + \left(1 + \frac{(2)(5)^2}{5^2 - 4.276^2}\frac{(2.636)(5.275)}{251.3}\right)^2 + \left(\frac{(2)(5)^2}{5^2 - 4.276^2}\frac{(2.636)(5.275)}{251.3}\right)^2} \right]^{\frac{1}{2}}$$

$$= 566,203 \text{ lbf.} \quad \dots\dots\dots\dots\dots\dots\dots\dots\dots\dots\dots\dots \quad (3.1.58)$$

Eq. 3.1.45 says that the maximum tensile load the pipe will see is 307,055 lbf. This means that the pipe will not yield under the action of the slips.

3.1.14 Drillpipe and Tool Joint Torsional Strength. Torque required to rotate drill collars in lateral section of hole:

$$T_{DC} = \frac{w_{DC} \cdot L_{DC} \cdot k_B \cdot \mu \cdot \sin(\alpha) \cdot \frac{OD_{DC}}{2}}{12} = \frac{(126.9)(496)(0.82)(0.3)\sin(40)(7.25)}{(2)(12)}$$

$$= 3007 \text{ ft-lbf.} \quad \dots\dots\dots\dots\dots\dots\dots\dots\dots\dots\dots\dots \quad (3.1.59)$$

Torque required to rotate heavyweight pipe in lateral section of hole:

$$T_{HW} = \frac{w_{HW} \cdot L_{HW} \cdot k_B \cdot \mu \cdot \sin(\alpha) \cdot \dfrac{OD_{HW}}{2}}{12} = \frac{(50.1)(310)(0.82)(0.3)\sin(40)(6.625)}{(2)(12)}$$

$$= 678 \text{ ft-lbf.} \quad \dotfill \quad (3.1.60)$$

Torque required to rotate drillpipe in lateral section of hole:

$$T_{DP} = \frac{w_{DP} \cdot L_{DP} \cdot k_B \cdot \mu \cdot \sin(\alpha) \cdot \dfrac{OD_{DP}}{2}}{12} = \frac{(23.87)(8094)(0.82)(0.3)\sin(40)(6.625)}{(2)(12)}$$

$$= 8433 \text{ ft-lbf.} \quad \dotfill \quad (3.1.61)$$

The torque to rotate the pipe is 8,433 ft-lbf. This does not include the bit, nor does it include bottomhole assembly members that could increase rotating torque.

From Eq. 3.1.20, we see that the makeup torque for the tool joints is 30,730 ft-lbf. The torsional strength of the pipe body is

$$TS_{PB} = \frac{0.577 \cdot Y_{PB} \cdot J_{PB}}{\dfrac{OD_{PB}}{2}}, \quad \dotfill \quad (3.1.62)$$

$$J_{PB} = \frac{\pi}{32}\left(OD_{PB}^4 - ID_{PB}^4\right) = \frac{\pi}{32}\left(4.855^4 - 4.276^4\right) = 21.733 \text{ in.}^4 \quad \dotfill \quad (3.1.63)$$

Note that in the calculation of J_{PB}, the pipe OD is the OD for Premium class pipe.

$$TS_{PB} = \frac{(0.577)(135,000)(21.733)}{\dfrac{4.855}{2} \cdot 12} = 58,114 \text{ ft-lbf.}$$

The maximum allowable drilling torque with this drillstring should be kept under 30,000 ft-lbf. If the drilling torque is expected to go higher than the makeup torque of the tool joints, the makeup torque can be increased, but only after careful consideration. Increasing the makeup torque may decrease the allowable tensile load, as seen on the pyramid curve in Fig. 3.1.9. Increasing the makeup torque also can increase the risk of fatigue failures in the tool joint and the risk of shoulder damage on some tool joint types.

Earlier in this example the drillpipe weight was based on the selection of tool joint OD and ID, which were not known. The calculations confirm that 5 in. 19.50-lbm/ft S135 drillpipe with a 6⅝ × 3¼-in. tool joint is adequate for this drilling program.

3.1.15 Drillpipe Failures. Drillpipe failures are generally caused by one of the following listed phenomena. As is often the case, these items can be related. Fatigue failures frequently begin in corrosion pits. Corrosion damage is accelerated by high stresses resulting from service loads.

Fatigue. Fatigue is the most common failure mode of drillpipe and usually occurs in the pipe body. As the drillpipe is rotated, any bending that takes place in the pipe because of hole curvature, buckling, whirl, or other causes results in a cyclic stress reversal that can lead to a fatigue failure.

In addition to corrosion pits, other stress raisers such as slip marks or dings and bruises from handling the pipe can also be initiation points. The failures occur where the bending stresses are highest, which is usually near the tool joints. Slip marks are on the box end, and, as expected, more fatigue failures occur on the box end than the pin end. Fatigue cracks sometimes occur at mid-length when the pipe has been run in compression and has buckled.

Tool joint fatigue failures on drillpipe are rare but do occasionally occur. Heavier-wall bottomhole assembly members will experience more fatigue-related connection failures because of the bending moment induced in the connection by the heavy wall.

Corrosion. Corrosion causes pitting on the surface of the pipe and becomes a stress raiser. Fatigue cracks often begin at corrosion pits (Jones 1992).

Excessive Service Loads. Very few drillpipe failures occur solely from excessive service loads—either tension, torque, or pressure. The final failure mode of a fatigue crack could be considered excessive loading because of the reduction in cross-sectional area of the member as the crack propagates.

Environment in Which the Pipe Is Used. The presence of free atomic hydrogen, high material strength, and high tensile stress can result in brittle fractures. This topic is also discussed in Chap. 2.

Wear. Rotating the pipe in the hole continually subjects the tool joints and pipe body to wear. Wear can reduce the cross-sectional area of the members to the point that they can no longer support the loads applied.

Frictional Heat. Heat checking occurs when, rotating the pipe through doglegs, the hanging weight of the string and the severity of the dogleg create very high lateral loads. (In the example worked through in this chapter, Eq. 3.1.44, the lateral load on the tool joints when pulling the string out of the hole was 5,642 lbf.) High rotary speeds coupled with high radial loads can sometimes create enough heat for the surface of the tool joint to reach the austenitizing temperature. Rapid cooling by the drilling fluid then forms untempered Martensite, and small cracks can form. There have been cases in which frictional heat caused the pipe body to go into the plastic state. API *RP 7G* (1995) and *ISO/NP 10407-1* (under development 5 October 2007) are both good references for drillpipe failures.

Nomenclature

A_{BCS} = cross-sectional area of tool joint box counterbore

A_C = smaller of the cross-sectional areas of tool joint box counterbore or pin at last engaged thread

A_{LET} = cross-sectional area of tool joint pin at last engaged thread

A_N = cross-sectional area of pin nose on double-shoulder tool joints

A_{PB} = cross-sectional area of drillpipe body

A_{slips} = area of contact of slips on drillpipe

c = pitch diameter of tool joint

C_E = elevator bore diameter

D_{DP} = drillpipe body outside diameter

Ded = dedendum of tool joint thread

D_H = measured depth of drilled hole

E = modulus of elasticity

f = coefficient of friction between contacting thread flanks and contacting makeup shoulders

f_W = elevator wear factor

F = force

F_a = axial force on tool joint thread

F_n = normal force on tool joint thread

F_{string} = force to pull drillstring out of hole

HC = hole curvature, degrees/100 ft

ID_{DC} = inside diameter of drill collar

ID_{HW} = inside diameter of heavyweight drillpipe

ID_{TJ} = inside diameter of tool joint pin

J_{PB} = polar moment of inertia of drillpipe body

k_B = buoyancy factor

K = lateral load factor in slip capacity equations

L_{DC} = length of drill collar

L_{DPA} = length of drillpipe assembly

L_{HW} = length of heavyweight drillpipe

L_{PC} = length of tool joint pin

NP = neutral point

OD_{DC} = outside diameter of drill collar

OD_{HW} = outside diameter of heavy weight

OD_{PB} = outside diameter of drillpipe body

OD_{TJ} = outside diameter of tool joint

p = distance in work equation *(force)•(distance)*

P = axial load

$P_{annulus}$ = drilling fluid pressure loss in annulus

P_{BHA} = axial load created by bottomhole assembly weight and drag

P_{bit} = drilling fluid pressure loss through rock bit nozzles

P_{DC} = axial load created by drill collar weight and drag

P_{DP} = tensile capacity of drillpipe body

$P_{drillpipe}$ = drilling fluid pressure loss through drillpipe or drillstem

P_E = axial load supported by elevators

P_I = internal pressure

P_o = tension required to separate the tool joint shoulders after T_{MU} is applied; represented by the line from the origin to the apex at T4 (Fig. 3.1.9)

P_Q = tensile load on drillpipe body

P_{slips} = axial load supported by slips

$P_{standpipe}$ = drilling fluid pressure loss in standpipe

P_{string} = axial load created by drillstring weight and drag

P_{T4T2} = tension required to yield the pin after T_{MU} is applied

P_{TJ} = tool joint tensile capacity

Q_c = diameter of box counterbore

R_S = mean shoulder radius

R_{slips} = radial force on slips normal to surface in contact with slip bowl

R_N = mean radius of pin nose on double-shoulder tool joints

R_T = mean thread radius of tool joint

T = torque

T1 = torsional strength of tool joint box

T2 = torsional strength of tool joint pin

T4 = makeup torque at which pin yield and shoulder separation occur simultaneously with an externally applied tensile load

T_{DC} = torque required to rotate drill collars

T_{DP} = torque required to rotate drillpipe

T_{HW} = torque to rotate heavy weight

T_{MU} = makeup torque

T_P = allowable drillpipe body tensile capacity

T_{PB} = allowable drillpipe body torsional capacity

T_{pr} = tool joint thread taper

TS_{PB} = torsional strength of pipe body

T_{TJ} = torsional strength of tool joint

w_{DC} = weight per foot of drill collar

w_{DP} = weight per foot of drillpipe

w_{HW} = weight per foot of heavyweight pipe

W = work applied to tool joint

W_a = work in axial direction applied to tool joint

W_{DC} = weight of drill collar

W_{DC1} = weight of drill collar string 1

W_{DC2} = weight of drill collar string 2

W_{DPA} = buoyed weight of drillpipe in drillstem

W_{DPA1}, W_{DPA2} = buoyed weight of additional drillpipe sizes in drill stem

W_{DS} = weight of drillstring (drillpipe only)

W_{DS1} = weight of drillstring 1

W_{DS2} = weight of drillstring 2

W_f = work in overcoming thread friction of tool joint

W_{HW} = buoyed weight of heavyweight drillpipe

W_s = work in overcoming tool joint shoulder friction

W_{slips} = force from slips normal to drillpipe surface

Y_{PB} = material yield strength of drillpipe body

Y_{TJ} = material yield strength of tool joint

Z = section modulus of drillstem element

Z_{DC} = section modulus of drill collar

Z_{HW} = section modulus of heavyweight pipe

α = helix angle of tool joint thread; angle of drilled well; angle of slips in slip bowl

Δp = small axial displacement to tool joint pin into tool joint box during makeup tightening

θ = half the included angle of tool joint thread

μ = coefficient of friction

ρ = radius of curvature of hole

ρ_M = density of drilling fluid

ρ_S = density of steel

σ_1, σ_2, σ_3 = principal stresses acting on respective planes

σ_B = bending stress

σ_{BPB} = bending stress in drillpipe body

σ_{LET} = axial stress at last engaged thread of tool joint pin

σ_{MU} = axial stress at last engaged thread of tool joint pin from makeup torque

σ_P = tangential stress in drillpipe body from internal pressure

σ_{TPB} = tensile stress in drillpipe body

τ_{PB} = shear stress in pipe body

ϕ = angle between force normal to thread flank and tool joint axis

ψ = friction angle from coefficient of friction between slips and slip bowl

References

Azar, J.J. and Samuel, G.R. 2007. *Drilling Engineering.* Tulsa: PennWell Books.

Farr, A.P. 1957. Torque Requirements for Rotary Shouldered Connections and Selection of Connections for Drill Collars. Paper 57-Pet-19 presented at the ASME Petroleum-Mechanical Engineering Conference, Tulsa, 22–25 September.

GrantPrideco. 2009. 5-7/8″ Drill Pipe Buckling. http://www.grantprideco.com/drilling/products/DrillPipe/5_78.asp.

IADC. 1992. *IADC Drilling Manual.* Houston: IADC Publications.

Industry Recommended Practice for the Oil and Gas Industry; IRP Volume 1 2008; Enform, 1538 – 25th Ave. NE Calgary AB T2E 8Y3.

IRP Volume 1-2008. Industry Recommended Practice for the Oil and Gas Industry–Critical Sour Drilling, fifth edition. 2008. Calgary, Alberta: Enform.

ISO/NP 10407-1. Petroleum and Natural Gas Industries—Rotary Drilling Equipment—Part 1: Drill Stem Design And Operating Limits, first edition. Under development. 10/5/07. Calgary, Alberta: ISO, TC 67/SC 4.

ISO 10424-2:2007. Petroleum and Natural Gas Industries—Rotary Drilling Equipment—Part 2: Threading and Gauging of Rotary Shouldered Thread Connections, first edition. Calgary: ISO, TC 67/SC 4.

ISO 11961:1996. Petroleum and Natural Gas Industries—Steel Pipes for Use as Drill Pipe—Specification, first edition. Calgary: ISO, TC 67/SC 5.

Jones, D.A. 1992. *Principles and Prevention of Corrosion*. New York City: Macmillan.

Lubinski, A. 1988a. Fatigue of Range 3 Drill Pipe. In *Developments in Petroleum Engineering, Vol. 2*, ed. S. Miska. Houston: CRC, Gulf Publishing Company.

Lubinski, A. 1988b. Maximum Permissible Dog-Legs in Rotary Bore Holes. In *Developments in Petroleum Engineering, Vol. 1*, ed. S. Miska. Houston: CRC, Gulf Publishing Company.

Reinhold, W.B. and Spiri, W.H. 1959. Why Does Drill Pipe Fail in the Slip Area? *World Oil* Oct.: 100–115.

RP 7A1, Recommended Practice for Testing of Thread Compound for Rotary Shouldered Connections, first edition. 1992. Washington, DC: API.

RP 7G, Recommended Practice for Drill Stem Design and Operating Limits, 16th edition. 1998. Washington, DC: API.

Sathuvalli, U.B., Payne, M.L., Suryanarayana, P.V., and Shepard, J. 2002. Advanced Slip Crushing Considerations for Deepwater Drilling. Paper SPE 74488 presented at the IADC/SPE Drilling Conference, Dallas, 26–28 February. DOI: 10.2118/74488-MS.

Spec. 5D, Specification for Drill Pipe, fifth edition. 2001. Washington, DC: API.

Spec. 7, Specification for Rotary Drill Stem Elements, 39th edition. 1998. Washington, DC: API.

Van Vlack, L.H. 1989. *Elements of Material Science and Engineering,* 6th edition. Addison-Wesley Series in Metallurgy & Materials Engineering, Upper Saddle River, New Jersey: Prentice Hall.

Appendix—Drillpipe and Tool Joint Properties

	TABLE A–1—DRILLPIPE SIZES AND WEIGHTS (API *RP 7G*; Grant Prideco 2009)						
	New Drillpipe Body Nominal Dimensions			Premium Class OD and Wall Thickness for Calculating Pipe Properties		Class 2 OD and Wall Thickness for Calculating Pipe Properties	
Pipe Size and Weight	OD (in.)	Wall Thickness (in.)	ID (in.)	OD (in.)	Wall Thickness (in.)	OD (in.)	Wall Thickness (in.)
$2^3/_8$ in., 4.85 lb/ft	2.375	0.190	1.995	2.299	0.152	2.261	0.133
$2^3/_8$ in., 6.65 lb/ft	2.375	0.280	1.815	2.263	0.224	2.207	0.196
$2^7/_8$ in., 6.85 lb/ft	2.875	0.217	2.441	2.788	0.174	2.745	0.152
$2^7/_8$ in., 10.40 lb/ft	2.875	0.362	2.151	2.730	0.290	2.658	0.253
$3^1/_2$ in., 9.50 lb/ft	3.500	0.254	2.992	3.398	0.203	3.348	0.178
$3^1/_2$ in., 13.30 lb/ft	3.500	0.368	2.764	3.353	0.294	3.279	0.258
$3^1/_2$ in., 15.50 lb/ft	3.500	0.449	2.602	3.320	0.359	3.231	0.314
4 in., 11.85 lb/ft	4.000	0.262	3.476	3.895	0.210	3.843	0.183
4 in., 14.00 lb/ft	4.000	0.330	3.340	3.868	0.264	3.802	0.231
4 in., 15.70 lb/ft	4.000	0.380	3.240	3.848	0.304	3.772	0.266
$4^1/_2$ in., 13.75 lb/ft	4.500	0.271	3.958	4.392	0.217	4.337	0.190
$4^1/_2$ in., 16.60 lb/ft	4.500	0.337	3.826	4.365	0.270	4.298	0.236
$4^1/_2$ in., 20.00 lb/ft	4.500	0.430	3.640	4.328	0.344	4.242	0.301
5 in., 16.25 lb/ft	5.000	0.296	4.408	4.882	0.237	4.822	0.207
5 in., 19.50 lb/ft	5.000	0.362	4.276	4.855	0.290	4.783	0.253
5 in., 25.60 lb/ft	5.000	0.500	4.000	4.800	0.400	4.700	0.350
$5^1/_2$ in., 19.20 lb/ft	5.500	0.304	4.892	5.378	0.243	5.318	0.213
$5^1/_2$ in., 21.90 lb/ft	5.500	0.361	4.778	5.356	0.289	5.283	0.253
$5^1/_2$ in., 24.70 lb/ft	5.500	0.415	4.670	5.334	0.332	5.251	0.291
$5^7/_8$ in., 23.40 lb/ft	5.875	0.361	5.153	5.731	0.289	5.658	0.253
$5^7/_8$ in., 26.30 lb/ft	5.875	0.415	5.045	5.709	0.332	5.626	0.291
$6^5/_8$ in., 25.20 lb/ft	6.625	0.330	5.965	6.493	0.264	6.427	0.231
$6^5/_8$ in., 27.70 lb/ft	6.625	0.362	5.901	6.480	0.290	6.408	0.253

*From *RP 7G, Recommended Practice for Drill Stem Design and Operating Limits,* 1998. Reproduced courtesy of the American Petroleum Institute.

TABLE A–2—EXAMPLES OF TOOL JOINT AND DRILLPIPE ASSEMBLY PROPERTIES FOR SELECTED PIPE SIZES

Pipe Size, Weight, and Grade	Tool Joint	OD (in.)	ID (in.)	Tool Joint Torsional Strength (ft-lbf)	TJ Tensile Strength (ft-lbf)	Makeup Torque (ft-lbf)	Tool Joint/ Pipe Body Torsional Ratio	Pin Tong Space (in.)	Box Tong Space (in.)	Adjusted Weight (lbm/ft)	Minimum Tool Joint OD for Premium Class (in.)	Capacity (US gal/ft)	Displace- ment (US gal/ft)
$2^3/_8$ 4.85 G105	NC26	$3^3/_8$	$1^3/_4$	6,900	313,700	4,125B	1.03	9	10	5.55	$3^7/_{32}$	0.160	0.085
$2^7/_8$ 10.40 S135	NC31	$4^3/_8$	$1^5/_8$	16,900	623,800	10,167P	0.81	9	11	12.00	$4^1/_{16}$	0.184	0.184
$3^1/_2$ 13.30 E75	NC38	$4^3/_4$	$2^{11}/_{16}$	18,100	587,300	10,864P	0.97	10	$12^1/_2$	14.30	$4^1/_2$	0.311	0.219
4 14.00 G105	NC40	$5^1/_4$	$2^{13}/_{16}$	23,500	711,600	14,092P	0.72	9	12	15.55	5	0.447	0.238
$4^1/_2$ 16.60 G105	NC46	$6^1/_4$	3	39,700	1,048,400	23,795P	0.92	9	12	19.59	$5^{19}/_{32}$	0.582	0.300
5 19.50 S135	NC50	$6^5/_8$	$3^1/_4$	51,217	1,551,700	30,730P	0.41	9	12	23.08	$6^5/_{16}$	0.726	0.353
$5^1/_2$ 21.90 S135	$5^1/_2$ FH	$7^1/_2$	3	87,200	1,925,500	52,302P	0.96	10	12	28.21	$6^{15}/_{16}$	0.893	0.364
$5^7/_8$ 23.40 S135	XT57	7	$4^1/_4$	94,300	1,208,700	56,500B	0.89	10	15	26.46	$6^{19}/_{32}$	1.059	0.341
$6^5/_8$ 27.70 S135	$6^5/_8$ FH	8	5	73,700	1,448,900	44,196P	0.54	10	13	30.64	8	1.394	0.392

TABLE A–3—PROPERTIES OF SELECTED PIPE SIZES

Pipe Size, Weight, and Grade	Upset	Pipe Torsional Strength (ft-lbf)	Pipe Tensile Strength (lbf)	Wall Thickness (in.)	Nominal ID (in.)	Pipe Body Area (in.^{2})	Pipe Body Moment of Inertia (in.^{4})	Pipe Body Polar Moment of Inertia (in.^{4})	Internal Pressure (psi)	Collapse Pressure (psi)
$2^3/_8$ 4.85 G105	EU	6,700	136,900	0.190	1.995	1.304	0.784	1.568	14,700	16,800
$2^7/_8$ 10.40 S135	EU	20,800	385,800	0.362	2.151	2.858	2.303	4.606	29,747	33,997
$3^1/_2$ 13.30 E75	EU	18,600	212,200	0.368	2.764	2.829	0.445	9.002	13,800	15,771
4 14.00 G105	IU	32,600	313,900	0.330	3.340	2.989	0.639	12.915	15,159	17,325
$4^1/_2$ 16.60 G105	IEU	43,100	462,800	0.337	3.826	4.407	9.610	19.221	13,761	15,727
5 19.50 S135	IEU	74,100	712,100	0.362	4.276	5.275	14.269	28.538	17,105	19,548
$5^1/_2$ 21.90 S135	IEU	91,300	786,800	0.361	4.778	4.597	19.335	38.670	12,679	17,722
$5^7/_8$ 23.40 S135	IEU	105,500	844,200	0.361	5.153	4.937	23.868	47.737	10,825	16,591
$6^5/_8$ 27.70 S135	IEU	137,300	961,600	0.362	5.901	5.632	35.040	70.080	7,813	14,753

TABLE A–4—PROPERTIES OF TOOL JOINT THREAD FORMS

Thread Form	Threads per Inch	Taper (in./ft)
V-.038R	4	2
NC23 through NC50	4	3
NC56 through NC77		
V-.040	5	3
$2^3/_8$, $2^7/_8$, $3^1/_2$, $4^1/_2$ Reg		
$3^1/_2$, $4^1/_2$ FH[a]		
V-.050	4	2
$6^5/_8$ Reg, $5^1/_2$, $6^5/_8$ FH	4	3
$5^1/_2$, $7^5/_8$, $8^5/_8$ Reg		
V-.065	4	2
$2^3/_8$, $2^7/_8$, $3^1/_2$, 4, $4^1/_2$ SH[b]and WO[c]		
$2^7/_8$, $3^1/_2$ XH[d]		
$5^1/_2$, $6^5/_8$ IF[e]		
90°-V-.050	$3^1/_2$	2
H90		
90°-V-.084	3	$1^1/_4$
SLH 90[f]		
V-.076	4	$1^1/_2$
PAC[g], OH[h]		

(a) FH was originally a designation meaning 'full hole' to enhance the product description implying that the tool joint had a large bore.

(b) SH was originally a designation meaning 'slim hole' to enhance the product description implying that the tool joint had a small OD.

(c) WO was originally a designation meaning 'wide open' to enhance the product description implying that the tool joint had a large bore.

(d) XH was originally a designation meaning 'extra hole' to enhance the product description implying that the tool joint had a large bore.

(e) IF was originally a designation meaning 'internal flush' to enhance the product description implying that the tool joint had a large bore.

(f) SL was originally a designation meaning 'slim line' to enhance the product description implying that the tool joint had a small OD.

(g) Connections developed by Philip A. Cornell using a V-0.076 thread form.

(h) OH was originally a designation meaning 'open hole' to enhance the product description implying that the tool joint had a large bore.

SI Metric Conversion Factors

ft	$\times$ 3.048*	E − 01 = m
ft³	$\times$ 2.831 685	E − 02 = m³
ft-lbf	$\times$ 1.355 818	E + 00 = J
ft/min	$\times$ 5.08*	E − 03 = m/s
°F	$\times$ (°F − 32)/1.8	= °C
°F	$\times$ (°F + 459.67)/1.8	= K
gal	$\times$ 3.785 412	E − 03 = m³
gal/min	$\times$ 2.271 247	E − 01 = m³/h
in.	$\times$ 2.54*	E + 00 = cm
in.²	$\times$ 6.451 6*	E + 00 = cm²
in.³	$\times$ 1.638 706	E + 01 = cm³
lbf	$\times$ 4.448 222	E + 00 = N
lbm	$\times$ 4.535 924	E − 01 = kg
lbm/gal	$\times$ 1.198 264	E + 02 = kg/m³
lbm/ft³	$\times$ 1.601 846	E + 01 = kg/m³
mL	$\times$ 1.0*	E + 00 = cm³
psi	$\times$ 6.894 757	E + 00 = kPa

*Conversion factor is exact.

3.2 Drillstring Vibrations—**P.D. Spanos,** Rice University; **N. Politis,** BP America; **M. Esteva,** Rice University; and **M. Payne,** BP America

3.2.1 Introduction. Drillstring vibrations are extremely complex because of the random nature of a multitude of factors such as bit/formation interaction, drillstring/wellbore interaction, and hydraulics. They involve several phenomena that render the analysis quite challenging. Three primary modes of vibration are present while drilling: axial, torsional, and lateral. Related to these are phenomena including bit bounce, stick/slip, and whirling, respectively.

Drillstring vibrations can be induced by external excitations such as bit/formation interaction (Dunayevsky et al. 1993). In these cases, the tuning of the excitation source to a natural frequency of the drilling assembly or its components may yield destructive motions. Self-excited vibrations are also present downhole (Finnie and Bailey 1960). Vibrations may also be caused by the flow in the drillstring annulus (Paidoussis et al. 2007). The dynamic behavior of the drillstring can be either transient (nonstationary) or steady-state (stationary).

Drillstring vibrations directly affect the drilling performance because the various assembly components may experience premature wear and damage (Dykstra et al. 1994; Macdonald and Bjune 2007; Mason and Sprawls 1998), and the rate of penetration (ROP) decreases because part of the drilling energy needed to cut the rock is *wasted* in vibrations (Dareing et al. 1990; Macpherson et al. 1993; Wise et al. 2005). Further, vibrations can cause interference with measurement-while-drilling (MWD) tools (Lear and Dareing 1990). Finally, vibrations often induce wellbore instabilities that can worsen the condition of the well and reduce the directional control and the overall quality of the wellbore (Dunayevsky et al. 1993).

Drill collars and adjacent drillpipes have been recognized for many years as the components subjected to the most-harmful vibrations. Thus, the bottomhole assembly (BHA) not only influences the overall dynamic response of the assembly, but it is also the location of most failures (Dareing 1984b; Gatlin 1957). Therefore, vibration mitigation requires understanding the dynamic behavior of the BHA. Note, however, that downhole vibrations can be a valuable source of information that provides insight into bit wear, formation properties, and drillstring/wellbore interactions. It has also been suggested that they can be used as a potential seismic source (Booer and Meehan 1993; Poletto and Bellezza 2006). Further, drillstring vibrations have been considered as a means of enhancing the drilling effectiveness by increasing the available power at the bit (Dareing 1985).

3.2.2 Axial Vibrations. Axial vibrations of a drillstring involve motions of its components along its longitudinal axis. Drillstrings are subject to both static and dynamic axial loadings. The classical buckling theory provides the maximum static weight-on-bit (WOB) that the assembly can sustain without buckling (Lubinski et al. 1962; Mitchell 2008). It yields a certain operational static axial constraint corresponding to the applied WOB that fulfills the appropriate criteria (i.e., safety factor, ROP requirements). On the other hand, dynamic axial loads on the drilling assembly originate primarily from bit/formation interactions. They induce time-dependent fluctuations of the weight applied to the bit, and are rather erratic.

Historically, drillstring axial modes of vibration have been observed in the field, together with torsional modes, before their lateral counterparts (Finnie and Bailey 1960) because they can travel from the bottom of the well to the surface, whereas lateral vibrations are usually trapped below the neutral point (Inglis 1988). Drillstring axial dynamic behavior was the object of pioneering theoretical investigations of downhole vibrations in the early 1960s (Bailey and Finnie 1960).

Severe axial vibrations develop quite often when drilling with roller-cone (RC) bits, owing to their type of interaction with the formation. Specifically, the multilobe pattern generated by tricone bits at the bottom of the well is a major source of axial excitations for vertical or near-vertical wells in which the drillstring/borehole interactions are limited and the effective damping is reduced (Skaugen 1987). In the most severe cases, the axial vibrations can be observed at the surface because they may induce bouncing of the kelly and whipping of drawworks cables (Dareing 1983, 1984a).

Axial vibrations can be detrimental (Paslay and Bogy 1963) or beneficial (Dareing 1985) to drilling because they affect the WOB and consequently the ROP. Axial vibrations can be especially harmful to drilling if they have large-amplitude oscillations as in the case of the tuning of the drillstring natural frequencies with excitation frequencies of three cycles per bit revolution (Dareing 1984a). Further, the wellbore does not directly restrain the axial displacement of drill collars, resulting in large-amplitude oscillations. This may make the bit start bouncing off the formation, rendering the rock-breakage process erratic, and thereby reducing the overall ROP. Axial vibrations also have indirect consequences because of downhole coupling mechanisms that induce, for instance, significant lateral displacements (Dunayevsky et al. 1993; Shyu 1989). Existing literature reviews on this subject include those by Chevallier (2000), Dykstra (1996), Payne (1992), Sengupta (1993), Spanos et al. (1995), and Spanos et al. (2003).

An important feature of axial vibrations is the temporary liftoff of the drill bit from the formation, known as the bit-bounce phenomenon. Analytical investigations of bit bounce have been reported by Paslay and Bogy (1963) and Spanos et al. (1995). Downhole MWD tools have enabled the detection of wide and frequent WOB fluctuations that are sometimes not discernible at the surface. In extreme cases, the axial load of the drill bit vanishes rapidly and more or less periodically (Cunningham 1968; Deily et al. 1968; Vandiver et al. 1990; Wolf et al. 1985), together with the torque on bit (TOB) (Besaisow and Payne 1988). These instances correspond to a liftoff of the drill bit, and the process is accordingly called bit bounce.

Two primary causes are considered for the drill-bit liftoff. The first is the irregularities in the formation surface, sometimes resulting from drilling with a tricone bit (Sengupta 1993), which can generate a three-lobe pattern downhole. A second postulated source is the frequency tuning of mud pressure with the axial natural frequencies of the drilling assembly (Cunningham 1968). Bit bounce has many consequences including low ROP, excessive fatigue of the downhole drillstring components, and eventually well damage (Nicholson 1994).

The effects of the axial vibrations on drilling performance have led to the analytical investigation of drillstring axial vibrations because the late 1950s (Paslay and Bogy 1963; Bogdanoff and Goldberg 1958, 1961; Bradbury and Wilhoit 1963) when the first MWD data-recording devices became available. The similarities between the propagation of axial and torsional waves in drilling assemblies have encouraged their joint examination to determine, for instance, the natural frequencies of drilling assemblies (Bailey and Finnie 1960).

Although the wave equation can be used for some analytical solutions, the increasing availability of computers has fostered the use of numerical discretization methods, such as finite elements, to study the axial-vibration modes (Skaugen 1987; Dareing and Livesay 1968). The influence of downhole equipment such as shock subs, which are designed to absorb some of the axial energy, has also been investigated (Kreisle and Vance 1970).

A Continuous Axial-Vibration Model. This section discusses the mathematical representation of the axial behavior of the drillstring. The equation governing the undamped axial motion $\xi(x, t)$ of a linear elastic bar is a second-order partial-differential equation (Dareing 1984b; Bailey and Finnie 1960; Bradbury and Wilhoit 1963; Craig 1981) called the undamped classical wave equation,

$$\frac{\partial^2 \xi(x,t)}{\partial x^2} = \frac{1}{c^2} \frac{\partial^2 \xi(x,t)}{\partial t^2}, \dots\dots\dots\dots\dots\dots\dots\dots\dots\dots\dots\dots\dots\dots\dots (3.2.1)$$

whose general solution involves the superposition of terms of the form

$$\xi_n(x,t) = \left(A_n \sin \frac{\omega_n}{c} x + B_n \cos \frac{\omega_n}{c} x \right)$$
$$\times \left(C_n \sin \omega_n t + D_n \cos \omega_n t \right), \quad n = 1, 2, \dots, \dots\dots\dots\dots\dots (3.2.2)$$

where A_n, B_n, C_n, and D_n are constants and $\omega_n x/c$ are dimensionless parameters (Bailey and Finie 1960). The constants A_n, B_n, C_n, and D_n are determined by imposing the boundary and initial conditions. The axial wave velocity c can be expressed in terms of Young's modulus, E, and the density, ρ, of the material as

$$c^2 = \frac{E}{\rho}. \dots\dots\dots\dots\dots\dots\dots\dots\dots\dots\dots\dots\dots\dots\dots\dots\dots\dots\dots\dots\dots (3.2.3)$$

The equation of motion for the axial vibrations of a drillstring of cross-sectional area A_s, accounting for damping and subjected to an external forcing function, can be described by this second-order hyperbolic equation (Sengupta 1993; Bronshtein and Semendyayev 1997; Chin 1994):

$$\rho\frac{\partial^2 \xi}{\partial t^2}+c_a\frac{\partial \xi}{\partial t}-E\frac{\partial^2 \xi}{\partial x^2}+\rho g_z = g_a\left(x,t,\xi,\frac{\partial \xi}{\partial x},\frac{\partial \xi}{\partial t}\right), \quad \dots\dots\dots\dots\dots\dots\dots\dots\dots\dots \quad (3.2.4)$$

where c_a is a damping factor, g_z is the acceleration by gravity, and g_a is the external axial force per unit mass applied on the drillstring.

In many cases, such as in the presence of nonlinearities or arbitrary forcing functions in time and space, a closed-form solution is difficult, or even impossible, to obtain. In these cases, alternative procedures based on numerical techniques can prove useful, including finite differences (Bathe 1982; Thomson and Dahleh 1997), boundary elements (Brebbia et al. 1984; Burnett 1987; Chen and Zhou 1992), and finite elements (Bathe and Wilson 1976; Khulief and Al-Naser 2005; Melakhessou et al. 2003; Przemieniecki 1968; Reddy 1993).

A Bit-Bounce Model. The bouncing of the bit corresponds to the intermittent lift of the drilling assembly off the formation. This phenomenon relates primarily to tricone bits because they tend to create a pattern on the surface of the rock that may result in large-amplitude longitudinal vibrations of the BHA.

Spanos et al. (1995) presented a model that considers the coupling of axial and torsional vibrations of the BHA submitted to an excitation originating from the rock surface. This representation relied on a sinusoidal angular variation of the elevation of the surface without radial variation. Further, a quarter-sine radial variation established the continuity of the surface in its center; that is,

$$S(r,\phi)=\begin{cases} S_0 \sin\left(\frac{r}{\Delta r_b}\frac{\pi}{2}\right)\sin(3\phi), & 0 \le r \le \Delta r_b \ \text{ and } \ 0 \le \phi \le 2\pi \\[2mm] S_0 \sin(3\phi), & \Delta r_b \le r \le r_b, \end{cases} \quad \dots\dots\dots\dots\dots\dots\dots \quad (3.2.5)$$

where r_b is the radius of the borehole, Δr_b is a smaller radius than r_b, and r and ϕ are the radial and angular coordinates, respectively. This equation produces the surface shown in **Fig. 3.2.1.**

Then, the axial model was combined with a torsional one, in an effort to capture the coupling of these two vibration modes during a liftoff period of the bit. Finally, formulating conditions for liftoff of the bit and for resuming contact permitted us to conduct numerical analysis of the phenomenon. The special aspects of the solution are shown in Fig. 3.2.1.

Condition for Liftoff. When the drill bit moves in contact with the formation at a certain time, its axial displacement after the timestep, because of free vibration, can be calculated from the governing equation of motion by setting the excitation equal to zero. If this displacement is greater than the corresponding value of the profile elevation, then the drill bit is no longer in contact with the formation.

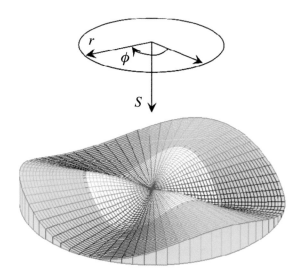

Fig. 3.2.1—The surface model of Spanos et al. (1995) relies on a sinusoidal, radially constant variation of the altitude. A quarter-sine radial variation of the surface, in the center, allows continuity in the profile. Reprinted courtesy of the American Society of Mechanical Engineers.

Condition for Resuming Contact. When the drill bit is not in contact with the formation at a certain time, and the displacement calculated from the free-vibration equation for the next time is less than the corresponding value of the profile elevation, the drill bit will be intercepted by the profile. To calculate the intermediate value of the time at which the drill bit comes in contact with the formation, an interpolation scheme such as in the Newton-Raphson technique can be applied.

Criterion of Bit Motion. When the drill bit is in contact with the formation, its rotational motion is governed by the motive torque from the stiffness of the drillstring above and the resisting TOB from the formation. The drill bit accelerates when the former is greater than the latter, and vice versa. But the TOB is a function of the WOB, which again depends on the position of the drill bit. Thus, the axial and torsional vibrations become coupled at the bit when the latter is in contact with the formation; an iterative scheme is necessary to solve the simultaneous equations of axial and torsional vibrations.

Frequency Dependence of Damping Matrices. While deriving the expression of the damping matrix, it is assumed that the matrix depends only on the dominant frequency of vibration. In evaluating the expression at a certain step, the dominant mode must be assumed before solving the system of equations. After solving it, the dominant mode can be checked from the modal expansion of the displacement vector. In case of disagreement, the steps are to be repeated.

The analytical model thus affords the option of a meaningful investigation of the complex phenomenon of bit liftoff. It is evident that the detection of the critical rotary speeds depends on the discretization scheme. Priority should be given to the scheme that can capture, at least approximately, all the resonant frequencies falling within the operating range.

Note that in this approach the formation surface has only one spatial frequency that corresponds to a three-lobe formation. This represents the most widely accepted formation-surface profile in case of drilling with tricone bits. The feature of having a polyharmonic profile is not needed in modeling the interaction among the axial vibrations of the drillstring, the bit liftoff, and the lobe amplitude modulation. Thus, the frequency of excitation is only three times the frequency of the rotary speed, and higher multiples are neglected. Nevertheless a polyharmonic profile can be incorporated, in which case, of course, the optimum discretization scheme has to be changed accordingly.

The model can be enhanced to reduce the extent of physical idealization. An obvious extension is the incorporation of the mud pressure fluctuation and of the bending vibrations. The model can also be improved by introducing the formation impedance. Additionally, the friction from the sidewall can be included. Because the solution is based on numerical integration, it can also accommodate nonlinear friction models. The lobe-amplitude-modulation model can be developed by considering the cone geometry and the details of the bit/formation interaction. For adequate calibration, the model must be tested for drillstrings with more-complicated geometric properties and for drilling under several operating conditions and formations. It can then be used for real-time feedback to the driller.

The interested reader can find a detailed description of this modeling approach in Sengupta (1993) and Spanos et al. (1995).

3.2.3 Torsional Vibrations. Downhole measurements show that the application of a constant rotary speed at the surface does not necessarily translate into a steady rotational motion of the drill bit. In fact, the downhole torsional speed typically shows large-amplitude fluctuations in time. This discrepancy in the rotational speed of the drillstring is a result of the large torsional flexibility of the drilling assembly (Skaugen 1987) and the presence of torsional-vibration modes.

Drillstring torsional vibrations remained undetected for a long time, perhaps because of the large inertia of the rotary table. The table acts almost as a clamped connection at the top that strongly attenuates torsional vibration modes traveling upward from the drill bit. Nevertheless, significant dynamic torque fluctuation can be experienced at the rig floor (Dareing 1984a).

The torsional-vibration modes of drilling assemblies can be classified in two categories: transient and steady-state. Transient vibrations correspond to localized variations of drilling conditions, as encountered, for example, when the lithology changes. Steady-state vibrations may also develop at least for an extended time segment.

Drillstring torsional vibrations, like their axial counterparts, may hinder drilling. Torsional vibrations can lead to excessive loadings resulting in equipment wear, joint failure, or damage to the drill bit (Brett 1992;

Elsayed et al. 1997). Previous literature reviews on this subject include those by Dykstra (1996), Payne (1992), Spanos et al. (2003), Kotsonis (1994), Leine (1997), and van den Steen (1997).

An important class of torsional vibrations is associated with the stick/slip behavior of the drill bit. Stick/slip is a self-excited torsional vibration induced by the nonlinear relationship between the friction-induced torque and the angular velocity at the bit (Jansen and van den Steen 1995). It can produce rotational speeds as great as 10 times the nominal rotary speed, as well as total standstill or even reverse motion of the bit. The stick/slip phenomenon has been observed to occur during as much as 50% of the drilling time (Dufeyte and Henneuse 1991). An early investigation of the friction-induced torsional vibrations and stick/slip of the drill bit has been reported in Belokobyl'skii and Prokopov (1982). Since then, the stick/slip behavior of drill bits has been examined extensively, both analytically and experimentally.

The drill bit might come to a standstill because of a sudden WOB increase or combined effects of significant drag, a tight hole, severe doglegs, and keyseatings. The static friction that must be overcome to start the drill-bit rotation again can be significantly higher than the normal Coulomb friction acting on the drilling assembly. Because the rotary table or the top-drive system is constantly turning, the drillstring stores torsional energy and twists up. When the available torque can finally overcome the static friction, the stored energy is suddenly released, and the drill bit comes loose. As the elastic torsional energy of the assembly decreases, so does its rotational speed. Then, the drill bit ultimately comes to a standstill again, and the whole cycle repeats itself. Therefore, the stick/slip is a phenomenon in which the excitation forces are produced by the motion of the system itself and the drillstring-downhole friction (Cull and Tucker 1999; Leine 2000; Lin and Wang 1991; Narasimhan 1987). However, the phenomenon requires certain drilling conditions before it can develop. Stick/slip of the bit will not occur ordinarily if the drillstring is shorter than a critical length (Lin and Wang 1991; Dawson et al. 1987). The critical length of the assembly is a function of the rotary speed of the string, the dry friction, and the system's viscous damping (Lin and Wang 1990). During stick/slip, assuming a constant rotary speed, the longer the drilling assembly is, the more severe the torsional vibrations are. As the rotary speed approaches the critical speed, the stick/slip frequency approaches the torsional natural frequency of the drillstring. Novel models to investigate the self-excited vibrations of the drillstring have been presented in Khulief et al. (2007), Navarro-López and Cortés (2007), and Richard et al. (2004, 2007).

Stick/slip may result in extensive bit wear, backward rotation, severe shock loadings of the drillstring, fatigue, and eventually failure of drilling equipment (Dufeyte and Henneuse 1991; Smit 1995). It also decreases the ROP by 25%, typically, perhaps because of the nonlinear relationship between the drilling rate and the rotational speed of the drill bit (van den Steen 1997). Also, the whipping and high-speed rotation of the drill bit in the slipping phase can generate severe axial and lateral vibrations of the BHA that may result in drillstring-connection failures. At the surface, the stick/slip phenomenon is sometimes characterized by a groaning noise and sawtooth-like variations, of large amplitude, of the applied torque (van den Steen 1997; Dufeyte and Henneuse 1991; Kyllingstad and Halsey 1988).

Possible remedies include greater drillstring stiffness, higher BHA inertia, increased rotational speed, and a reduced difference between static and dynamic frictions (van den Steen 1997; Dawson et al. 1987). MWD tools make it possible to detect the stick/slip phenomenon and identify its severity while drilling, thereby allowing real-time adoption of remedial measures. Control of the rotational behavior of a drilling assembly can be achieved by varying the rotary speed or the WOB, modifying mud properties (to alter downhole friction), and changing the type of drill bit or the configuration of the BHA (Smit 1995). In a parallel direction, an increasingly popular solution to stick/slip introduced in the 1980s is to increase the drillstring damping by means of an active damping system. This system reduces the torque fluctuations and torsional drillstring vibrations affecting the stick/slip conditions. The underlying concept is to reduce the amplitude of the downhole rotational vibrations using a closed circuit that provides torque feedback. The feedback is used by the rotational drive, which slows down the rotary rate when the torque increases and speeds it up when the torque decreases. This system is often called an impedance-control system or soft-torque system (van den Steen 1997; Jansen and van den Steen 1995; Smit 1995; de Vries 1994; Dekkers 1992; Javanmardi and Gaspard 1992). A related procedure is given by Tucker and Wang (1999). Note that the effectiveness of H_∞ control (methods used in control theory to synthesize controllers achieving robust performance) in suppressing stick/slip oscillations has been studied by Serrarens et al. (1998). Further, an optimal-state feedback control that can be effective in suppressing stick/slip oscillations once they are initiated has been suggested by Yigit and Christoforou (2000). Finally, a control approach based on modeling-error compensation has been proposed by Puebla and Alvarez-Ramirez (2008).

The wave equation has been used to describe the torsional behavior of drilling assemblies (Bailey and Finney 1960; Bradbury and Wilhoit 1963) since the early 1960s. Another approach assuming a friction-induced torsional drillstring-vibration mechanism that leads to self-excited downhole vibration-related phenomena and, particularly, the stick/slip of the bit (Belokobyl'skii and Prokopov 1982) was introduced in the early 1980s. Fourier transforms have also been used to compute torsional resonant frequencies (Halsey et al. 1986). Further, the problem has been modeled as a single-degree-of-freedom system (van den Steen 1997; Lin and Wang 1991; Dawson et al. 1987; Kyllingstad and Halsey 1988), a multiple-degree-of-freedom system (Zamanian et al. 2007), and a continuous system (Brett 1992; Belayev et al. 1995).

A Continuous-Torsional-Vibration Model. Torsional drilling vibrations are primarily associated with polycrystalline-diamond-compact (PDC) bits (Chin 1994) because these bits are associated with high downhole friction coefficients that can result in stick/slip oscillations of the bit.

The governing differential equation for the torsional vibration of the assembly is similar to that governing its axial vibrations. This explains why early investigations often considered the two vibration modes simultaneously (e.g., see Bailey and Finnie 1960; Bogdanoff and Goldberg 1958, 1961; Dareing and Livesay 1968). The torsional behavior of the assembly is described by

$$\rho J \frac{\partial^2 \phi}{\partial t^2} - JG \frac{\partial^2 \phi}{\partial x^2} = g_T(x, \phi, t), \quad \dots\dots\dots\dots\dots\dots\dots\dots\dots\dots\dots\dots\dots \quad (3.2.6)$$

where J is the polar moment of inertia of the considered cross section of drillstring, ϕ is the angular displacement of that section, G is the shear modulus of the drilling-assembly material, and $g_T(x, \theta, t)$ is the torsional applied load. The product JG quantifies the torsional stiffness of the system. A rigorous derivation of Eq. 3.2.6 can be found in Craig (1981). As in the case of continuous axial vibrations, Eq. 3.2.6 may be solved by using numerical techniques. A simple finite-element approach to solve this equation can be found in Raftoyiannis and Spyrakos (1997).

A Stick/Slip Model. This section presents an approach to analytical modeling of the stick/slip phenomenon; it relies primarily on the model presented by Dawson et al. (1987). This approach uses a single-degree-of-freedom representation of the drillstring in which a massless torsional spring of stiffness k models the entire length of the drilling assembly. The rotary table drives the system at the surface at a constant speed Ω. Therefore, the equation of motion is

$$I\ddot{\phi} + c_r \dot{\phi} + F(\dot{\phi}) + k\phi = k\Omega t, \quad \dots\dots\dots\dots\dots\dots\dots\dots\dots\dots\dots\dots\dots \quad (3.2.7)$$

where ϕ is the angular displacement of the BHA, c_r is the coefficient of viscous damping, k is the torsional stiffness of the drillstring, I is the mass moment of inertia with respect to the rotation axis, and $F(\dot{\phi})$ denotes the friction-induced forces. Eq. 3.2.7 may be normalized by the moment of inertia, yielding

$$\ddot{\phi} + 2\zeta\omega_0\dot{\phi} + f(\dot{\phi}) + \omega_0^2\phi = \omega_0^2\Omega t, \quad \dots\dots\dots\dots\dots\dots\dots\dots\dots\dots\dots\dots \quad (3.2.8)$$

where

$$\omega_0 = \sqrt{k/I} \text{ and } \zeta = \frac{c_r}{2\sqrt{kI}}. \quad \dots\dots\dots\dots\dots\dots\dots\dots\dots\dots\dots\dots\dots\dots \quad (3.2.9)$$

Assume $f(\dot{\phi})$ of the form

$$f(\dot{\phi}) = \begin{cases} f_1 - \dfrac{f_1 - f_2}{V_0}\dot{\phi} & 0 \le \dot{\phi} < V_0 \\ f_2 & V_0 \le \dot{\phi} \end{cases}; \quad \dots\dots\dots\dots\dots\dots\dots\dots\dots\dots \quad (3.2.10)$$

that is, a piecewise linear model with a change at $\dot{\phi} = V_0$. The parameters f_1, f_2, and V_0 in Eq. 3.2.10 depend on the physical characteristics of the drilling assembly. Note that this simple model considers the reduction of the friction when the system switches from a static to a kinetic state.

Next, to investigate the stick/slip behavior of the assembly, one considers two distinct time segments. The first phase extends from $t = 0$ to $t = t_1$, which is the time at which the angular velocity of the drillstring equals V_0. During this time, the bit is in the slipping phase; its corresponding equation of motion is

$$\ddot{\phi}_1 + 2\zeta\omega_0\dot{\phi}_1 + f_1 - \frac{f_1 - f_2}{V_0}\dot{\phi}_1 + \omega_0^2(\phi_1 - \Omega t) = 0, \quad \dots\dots\dots\dots\dots\dots \quad (3.2.11)$$

or, equivalently,

$$\ddot{\phi}_1 + 2\zeta_1\omega_0\dot{\phi}_1 + \omega_0^2\phi_1 = \omega_0^2\Omega t - f_1, \quad \dots\dots\dots\dots\dots\dots\dots \quad (3.2.12)$$

where the introduction of the damping ratio ζ_1 simplifies the notation. Note that

$$\zeta_1 = \zeta - \frac{f_1 - f_2}{2V_0\omega_0}. \quad \dots\dots\dots\dots\dots\dots\dots\dots\dots \quad (3.2.13)$$

For the slipping phase, the equation of motion as described in Eq. 3.2.12 is subject to the initial conditions

$$\begin{cases} \phi_1(t)\big|_{t=0} = -f_1/\omega_0^2 \\ \dot{\phi}_1(t)\big|_{t=0} = 0 \end{cases} \quad \dots\dots\dots\dots\dots\dots\dots\dots \quad (3.2.14)$$

The solution of Eq. 3.2.12 depends on the value of ζ_1. Specifically, for $\zeta_1 < 1$, the solution of the equation of motion is given by (Dawson et al. 1987; Craig 1981)

$$\phi_1(t) = \left(c_1 \sin\omega_d t + c_2 \cos\omega_d t\right)e^{-\zeta_1\omega_0 t} - \frac{f_1 + 2\zeta_1\omega_0\Omega}{\omega_0^2} + \Omega t, \quad \dots\dots\dots\dots \quad (3.2.15)$$

where

$$\omega_d = \sqrt{1 - \zeta_1^2}\,\omega_0, \quad c_1 = \frac{\Omega(2\zeta_1^2 - 1)}{\omega_0}, \text{ and } c_2 = \frac{2\zeta_1\Omega}{\omega_0}.$$

Similarly, for $\zeta_1 < 1$, the solution of Eq. 3.2.12 is

$$\phi_1(t) = \left(c_1'e^{\sqrt{\zeta_1^2-1}\omega_0 t} + c_2'e^{-\sqrt{\zeta_1^2-1}\omega_0 t}\right)e^{-\zeta_1\omega_0 t} - \frac{f_1 + 2\zeta_1\omega_0\Omega}{\omega_0^2} + \Omega t, \quad \dots\dots\dots \quad (3.2.16)$$

where

$$c_1' = \frac{\Omega(2\zeta_1^2 + 2\zeta_1\sqrt{\zeta_1^2-1} - 1)}{2\omega_0\sqrt{\zeta_1^2-1}}, \text{ and } c_2' = \frac{\Omega(-2\zeta_1^2 + 2\zeta_1\sqrt{\zeta_1^2-1} + 1)}{2\omega_0\sqrt{\zeta_1^2-1}}.$$

Taking the first time derivative of $\phi_1(t)$ in Eq. 3.2.15 or 3.2.16 yields the rotational speed of the BHA, $\dot{\phi}_1(t)$. By definition, the time t_1 is the instant at which $\dot{\phi}_1(t) = V_0$. Therefore, taking the first time derivative of Eq. 3.2.15 or 3.2.16 and solving for the smallest roots of either equation yields t_1. Then, substituting t_1 back in this equation yields the value of the displacement of the BHA.

Next, one is interested in the sticking phase of the pattern, for which the equation of motion of the BHA may be written

$$\ddot{\phi}_2 + 2\zeta\omega_0\dot{\phi}_2 + \omega_0^2\phi_2 = \omega_0^2\Omega t - f_2, \quad \dots\dots\dots\dots\dots\dots \quad (3.2.17)$$

with the initial conditions

$$\phi_2(t)\big|_{t=0} = \phi_1(t)\big|_{t=t_1} \text{ and } \dot{\phi}_2(t)\big|_{t=0} = V_0. \quad \dots\dots\dots\dots\dots\dots \quad (3.2.18)$$

The solution of Eq. 3.2.17 is (Dawson et al. 1987)

$$\phi_2(t) = e^{-\zeta\omega_0 t}\left(c_3 \cos\omega_b t + c_4 \sin\omega_b t\right) - \frac{f_2 + 2\zeta\omega_0\Omega}{\omega_0^2} + \Omega t \quad\ldots\ldots\ldots\ldots\ldots\ldots \quad (3.2.19)$$

where

$$\omega_b = \sqrt{1-\zeta^2}\,\omega_0,$$

$$c_3 = \dot{\phi}_1(t_1) + \frac{f_2 + 2\zeta\omega_0\Omega}{\omega_0^2}, \quad \text{and} \quad c_4 = \frac{1}{\omega_b}\left[V_0 - \Omega + \zeta\omega_0\left(\dot{\phi}_1(t_1) + \frac{f_2 + 2\zeta\omega_0\Omega}{\omega_0^2}\right)\right].$$

As for the slipping phase, taking the first time derivative of Eq. 3.2.19 and solving for the smallest root of $\dot{\phi}_2(t) = V_0$ yields the time t_2 at which the rotational velocity of the BHA equals V_0 again. Beyond t_2, Eq. 3.2.11 governs the BHA motion again.

After the BHA is stuck, its velocity remains null and the magnitude of the displacement ϕ increases linearly until $\phi = -f_1/\omega_0^2$. Therefore, the torque originating from the twisting of the drillstring is $T = k\phi$, where k is the torsional stiffness that Eq. 3.2.7 introduced.

3.2.4 Lateral Vibrations. Lateral vibrations, also known as transverse, bending, or flexural vibrations, are widely recognized as the leading cause of drillstring and BHA failures (Vandiver et al. 1990; Chin 1988; Mitchell and Allen 1985). Paradoxically, the impact of drillstring lateral modes of vibration remained unrecognized for a considerable period of time because most lateral vibrations do not travel to the surface, even in vertical wells (Chin 1994). Furthermore, lateral vibrations are dispersive and of frequencies higher than those of their torsional counterparts. Accordingly, they attenuate rapidly while propagating toward the surface (Payne et al. 1995). Therefore, they are difficult to detect on the basis of surface measurements alone. Developments of downhole measurement techniques, especially MWD tools, have helped capture the significance of these vibrations and their impact on equipment failures (Vandiver et al. 1990).

Various downhole mechanisms can induce lateral oscillatory modes, including primarily bit/formation and drillstring/borehole interactions. WOB fluctuations may also give rise to lateral instabilities because of linear axial/lateral coupling (Vandiver et al. 1990). Also, the initial curvature of the BHA can result in lateral vibrations (Vandiver et al. 1990; Payne et al. 1995).

Many studies have addressed the harmful effects of lateral vibrations in drilling systems (Vandiver et al. 1990; Chin 1988; Mitchell and Allen 1985, 1987; Allen 1987; Burgess et al. 1987; Close et al. 1988; Dubinsky et al. 1992; Rogers 1990). Lateral vibrations can cause severe damage to the borehole wall (Mason and Sprawls 1998; Jansen 1992), affect the drilling direction (Millheim and Apostal 1981), and result in precessional instabilities (Chin 1994). Moreover, lateral vibrations may initiate borehole/formation patterns resulting in axial and torsional drill-bit vibrations (Dareing 1984b). Despite their inherent damaging nature, flexural vibrations can be used in a positive manner by providing directional control at the bit and by increasing the ROP (Chin 1994; Kane 1984). Previous literature surveys on this subject include Chevallier (2000), Dykstra (1996), and Payne (1992).

An important subset of lateral vibrations is the *whirling* of the BHA. Whirling is a condition where the instantaneous center of rotation moves about the bit face as the bit rotates (Warren et al. 1990; Vandiver et al. 1990; Brett et al. 1990), and it can be forward, backward, or chaotic. Several factors can induce whirling. Mass imbalance, such as that created by MWD tools, or an initially bent BHA, together with high compressive loads generated by WOB, can produce some eccentricity of the drilling assembly. As drill collars rotate, this eccentricity induces a dynamic imbalance. The center of mass is then sent off the centerline of the drillstring, resulting in drill-collar whirl. The magnitude of the centrifugal force acting, in a D'Alembert sense, at the center of mass of the collar is proportional to the initial eccentricity, the square of the rotation rate, and the mass of the collar (Vandiver et al. 1990; Kotsonis 1994). Numerical solutions of the *eigen problem* for whirling modes for a range of whirl speeds and an attempt to correlate the numerical predictions with observation by using a small experimental rig have been reported in Coomer et al. (2001). Furthermore, the tuning of the rotation rate of the drill collars with their natural frequencies can make a part of the BHA bow out of its natural shape and perform forward synchronous whirl. Maximum bending deflections occur at rotary rates close to the lateral natural frequencies of the drill collars, referred to as the critical speed. This critical rotary speed is modified by fluid-added mass, stabilizer clearance, and stabilizer friction, of which the respective influences are complex (Jansen 1992). The amplitude of vibrations resulting from bit whirl increases with the formation strength for both PDC and RC bits. Additional factors affecting the whirl behavior include viscous-fluid damping, gravity, coupling of axial with torsional and lateral vibration modes, and wall contact (Jansen 1991).

Bit whirl is quite harmful for drag bits with PDC inserts (Warren et al. 1990; Brett et al. 1990). Tricone bits, on the other hand, do not suffer as extensively from whirling because they penetrate the formation, thereby reducing their sideways motion (Kotsonis 1994). Drillstring components experiencing whirling cause a series of problems. Whirling is a major cause of reduced ROP and early failure of downhole equipment (Mason and Sprawls 1998), it contributes significantly to drill-collar wear and connection fatigue, and whirl-induced lateral displacements of the assembly can result in severe and repeated contacts with the borehole wall, which can lead to surface abrasion of the drilling equipment and deterioration of the wellbore condition.

Most of the BHA operates in compression, making it a region where buckling and whirling are likely to occur. Severe whirling occurrences can be observed on the rig floor by the lateral motion of the traveling block and the whipping of the drawworks. BHA whirling, nonetheless, remains difficult to detect, while fatigue accumulates and eventually results in failure of the equipment.

The forward synchronous whirl or forward whirl of a drill collar occurs when that section rotates around the borehole with the same direction as the drillstring rotation generated by the rotary table (Jansen 1991). During forward whirl, the same side of the collar is in continuous contact with the borehole wall. The mechanism is therefore a cause of flat spots on collar joints (Vandiver et al. 1990). Forward whirl may develop during normal drilling operations. It is usually induced by an out-of-balance mass, although it is unlikely to develop if the mass eccentricity is less than the stabilizer clearance (Jansen 1991). Also, the friction resulting from stabilizer/borehole contact introduces instability at certain whirling frequencies, resulting in nonsynchronous, self-excited, large-amplitude vibrations (Van Der Heijden 1993). Thus, depending on the downhole conditions and drilling parameters, stable forward whirl may develop or evolve into other whirling patterns. For instance, the repeated impacts of the collars on the borehole wall can gradually transform forward whirl into backward whirl (Jansen 1992).

The best-known kind of nonsynchronous collar whirl is backward whirl. It may occur during normal operations, and it develops when the instantaneous center of rotation of a drillstring section lies between the center of the mass of the collar and the borehole wall. More specifically, the whirl qualifies as backward if the instantaneous center of rotation travels around the borehole in a direction opposite to the driving rotation. Backward whirl can originate from the friction between the stabilizers and the wellbore if this exceeds structural and hydrodynamic damping forces (Shyu 1989). This leads to backward rolling or slipping motions of the stabilizers that, in turn, can produce a self-excited backward whirling motion of the drill collars (Jansen 1991). Further, collars can drive the whirl if the slip is positive, or resist it if the slip is negative. The case of extreme backward whirl is called *pure backward whirl*. This is the rolling without slipping of the drill collars on the inside of the wellbore in the direction opposite to that imposed by the rotary table (Jansen 1992). The friction-induced transition from forward to backward whirl leads to perfect backward whirl at the limit of zero mass eccentricity (Van Der Heijden 1993). Backward whirl is a significant threat to drilling assemblies because it superimposes on the forward rotary speed, thereby inducing fluctuating bending moments with periodic changes of sign (Jansen 1991). These strong moment fluctuations give rise to high-amplitude bending-stress cycles. Thus, the fatigue life of the drill-collar connections can be shortened significantly when the bending-stress cycles accumulate at a rate much greater than the rotary speed (Vandiver et al. 1990). The backward whirl of drill collars at a frequency close to one of the natural frequencies of the assembly may lead to wall contact and can produce drill-collar precession—that is, a backward rolling motion of the drill collars along the borehole wall.

In practice, synchronous whirl cannot develop if the clearance at the stabilizers exceeds the eccentricity of the center of mass of the collars. When forward whirl is impossible, collars may either whirl backward or in an irregular fashion (Kotsonis 1994). Extreme cases of nonperiodic behaviors are called chaotic whirl, because the motion then depends strongly on initial conditions. The irregular motion is induced by nonlinear fluid forces, stabilizer clearance, and interactions with the borehole wall. Chaotic motions also can develop for low values of stabilizer friction (Jansen 1992). Finally, chaotic whirling of drilling components may comprise minor components of randomness (Kotsonis and Spanos 1997).

Lateral vibrations of drillstrings, and modeling techniques for this problem, have been the focus of several publications since the mid-1960s. The two common ones involve closed-form solutions and finite-element discretization. The closed-form solutions have been the basis of early analyses (Lichuan and Sen 1993). However, the great complexity of the problem has limited their applicability. Fortunately, the versatility of finite-element analysis and the advent of computers have facilitated the consideration of several parameters involved in the problem. To date, several aspects of drillstring behavior related to lateral vibrations have been studied. Typical studies address determination of natural frequencies (Chen and Géradin 1995; Christoforou and Yigit 1997; Frohrib and Plunkett 1967), critical-bending-stress calculations (Mitchell and Allen 1985; Plunkett 1967; Spanos et al. 1997), stability analysis (Vaz and Patel 1995), and prediction of lateral displacements of drilling assemblies (Dykstra 1996; Yigit and Christoforou 1998). Further, some analyses have identified critical failure parameters and the conditions that trigger a transfer of energy between lateral and rotational modes of vibration (Yigit and Christoforou 1998).

A Continuous-Lateral-Vibration Model. For a continuous model, the Euler-Bernoulli beam theory is considered, and the small-slopes assumption is adopted. In this regard, the following hypotheses are made. The plane sections that are normal to the beam axis before deformation remain plane and normal to the beam axis after deformation; this implies that the deformations are a result of bending only. Also, the beam is considered elastic, and Hooke's law relates the stresses and the strains. The Euler-Bernoulli equation is

$$\rho\frac{\partial^2 u}{\partial t^2} + \frac{\partial^2}{\partial x^2}\left(EI_z\frac{\partial^2 u}{\partial x^2}\right) = g(x,t), \quad \dots\dots\dots\dots\dots\dots\dots\dots\dots\dots\dots\dots\dots\dots \quad (3.2.20)$$

where $u(x, t)$, ρ, E, and I_z are the lateral displacement, the mass density, the modulus of elasticity, and the relevant moment of inertia of the cross section of the beam, respectively, and $g(x, t)$ denotes the external loading, x refers to the axis of the beam, and t indicates time.

Then, the effect of the axial forces is taken into account. The beam is subjected to an axial force P that is considered positive when tensile. The axial force induces an additional moment that modifies the shear-moment relationship and leads to the following differential equation:

$$\rho\frac{\partial^2 u}{\partial t^2} + \frac{\partial^2}{\partial x^2}\left(EI_z\frac{\partial^2 u}{\partial x^2}\right) - P\frac{\partial^2 u}{\partial x^2} = g(x,t). \quad \dots\dots\dots\dots\dots\dots\dots\dots\dots\dots\dots\dots \quad (3.2.21)$$

The solution of these equations by means of numerical techniques such as finite elements has also been widely used in different engineering applications. Exhaustive information about this technique can be found in Przemieniecki (1968), Reddy (1993), and Raftoyiannnis and Spyrakos (1997).

A Whirling Model. Many studies have approached the whirling phenomenon by adopting a 2D, single lumped mass representation of the assembly (Vandiver et al. 1990; Kotsonis 1994; Jansen 1992). The whirling motion of drillstrings relates to torsional vibrations in the sense that whirling requires rotation of the assembly to develop. The whirling of BHA components involves motions of these components about the borehole central axis. Previous analyses (Shyu 1989; Kotsonis 1994; Jansen 1992, 1991; Kotsonis and Spanos 1997) have characterized the whirling of a certain location in the BHA. That is, these approaches pursue a single-degree-of-freedom representation of the BHA. Because a primary consequence of whirling is the fatigue failure of drilling equipment, the location that is investigated along the drillstring is ordinarily midway between two stabilizers, where the lateral deflection is the largest.

Assuming a constant rotary speed, the equations of motion of a point equidistant between two stabilizers may be written as (Lee 1993)

$$m\ddot{y} + c_w\dot{y} + k_w y = me_0\Omega^2\cos(\Omega t), \quad \dots\dots\dots\dots\dots\dots\dots\dots\dots\dots\dots\dots\dots\dots \quad (3.2.22)$$

and

$$m\ddot{z} + c_w\dot{z} + k_w z = m e_0 \Omega^2 \sin(\Omega t), \quad \dots\dots\dots\dots\dots\dots\dots\dots \text{(3.2.23)}$$

where y and z are the lateral coordinates introduced in **Fig. 3.2.2**, m is the equivalent mass of the collar, c_w is the damping coefficient, k_w is the equivalent lateral stiffness of the collar, e_0 is the eccentricity of the center of mass, and Ω is the rotational speed of the drilling assembly. Eqs. 3.2.22 and 3.2.23 describe the planar motion of the considered point accounting for inertial and damping forces. An elastic restoring term is also considered. Note that the excitation term on the right-hand side of Eqs. 3.2.22 and 3.2.23 incorporates the force arising from the rotation of the unbalanced element.

A derivation of the equations of motion of a particular point along the drillstring accounting for fluid forces, stabilizer clearance, borehole contact, gravity, and linear and parametric coupling has been provided by Kotsonis (1994). Specifically, introducing the dimensionless polar coordinates (r, θ) of the center of the collar section at the considered location along the drilling assembly, the equations of motion may be written as

$$\beta_w \ddot{r} - \beta_w r \dot{\theta}^2 + \lambda \frac{\Omega}{\omega} r (\beta_w - 1)\left(2\dot{\theta} - \lambda \frac{\Omega}{\omega}\right) + \beta_w \Gamma \dot{r} + \Re[F_k]$$

$$= \varepsilon \frac{\Omega^2}{\omega} \cos(\theta - Z) - \chi \sin(\theta) + \Re[C_{PL}], \quad \dots\dots\dots\dots\dots \text{(3.2.24)}$$

and

$$\beta_w r \ddot{\theta} + 2\beta_w \dot{r}\dot{\theta} - 2\lambda \frac{\Omega}{\omega}\dot{r}(\beta_w - 1) + \beta_w \Gamma r\left(\dot{\theta} - \lambda \frac{\Omega}{\omega}\right) + \Im[F_k]$$

$$= \varepsilon \frac{\Omega^2}{\omega} \sin(\theta - Z) - \chi \cos\theta + \Im[C_{PL}], \quad \dots\dots\dots\dots\dots \text{(3.2.25)}$$

where $\Re[.]$ and $\Im[.]$ denote the real and imaginary parts of the quantity $[.]$, respectively, and the exact definition of the other terms introduced is referred to in Kotsonis (1994) and Kotsonis and Spanos (1997). From these studies, it is seen that the degree of complexity increases significantly when the analysis incorporates more factors. However, one may be interested in the investigation of the whirling behavior of a collar section—or even of an entire drillstring—as opposed to a single location along the assembly. To pursue this objective, the physical assembly must be modeled by a multiple-degree-of-freedom system (Lee and Kim 1986).

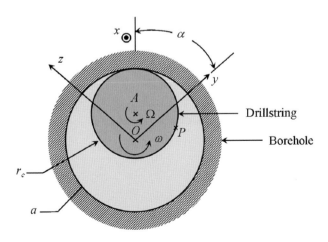

Fig. 3.2.2—Coordinate system for whirling investigation, after Shyu (1989). Reproduced courtesy of the Massachusetts Institute of Technology.

3.2.5 Coupled Vibrations. Multiple vibration modes can occur simultaneously while drilling. These include bit bounce, stick/slip, and forward and backward whirl, together with linear and parametric coupling between axial, torsional, and lateral vibrations. Coupling between the axial forces at the bit and the lateral oscillations of the drillstring, for instance, can originate from the initial curvature of the BHA. Vandiver et al. (1990) proposed a practical analogy to understand linear-coupling mechanisms: "Linear coupling is easy to visualize by taking a thin ruler or a piece of paper, giving it a slight curve, and then pressing axially on the ends. The object responds by additional bending in the plane of the initial curvature." Further, to explain parametric coupling mechanisms, Dunayevsky et al. (1993) proposed to visualize the dynamic coupling of axial vibrations with lateral deflections of a drillstring by inducing "a snaked motion in a vertically hanging rope by moving its end up and down at a particular frequency."

Although coupling mechanisms, as described by the preceding analogies, may appear easy to examine, they drastically increase the complexity of drillstring vibrations. For instance, the coupling between axial forces and lateral displacements induces time-dependent bending moments. Further, whirling produces curvature that, in turn, is needed for linear coupling to occur. Accordingly, BHA whirl and linear coupling usually develop simultaneously (Vandiver et al. 1990).

The drill bit plays an important role in coupling mechanisms because it converts the axial vibrations to torsional vibrations, or equivalently, relates the WOB to the TOB. Torsional vibrations help suppress axial and lateral vibrations, because the time-dependent angular velocity does not allow the other kinds of oscillations enough time to build up (van den Steen 1997). Similarly, the bouncing of the bit suppresses low-frequency torsional oscillations. Further, a comparison between linear and nonlinear models for coupling of axial and torsional vibrations has been made in Sampaio et al. (2007).

Coupling mechanisms are also crucial in generating certain vibration modes. For instance, the coupling between axial and lateral vibrations has been studied in Shyu (1989), Al-Hiddabi et al. (2003), and Trindade et al. (2005). **Table 3.2.1** provides a summary of downhole excitation mechanisms and their respective influence on axial-, torsional-, and lateral-vibration modes. This table reflects the intricate nature of downhole vibrations by showing that most sources result in significant vibrations in more than one mode. Further, **Table 3.2.2** summarizes the most important characteristics of the three primary drillstring modes.

As Chin (1994) has emphasized, analyzing axial, torsional, and lateral vibrations of drillstrings separately may appear to be of limited interest because it is the combined effect that is important in practice. However, he has also asserted that the separate analysis of each of these mechanisms is crucial to "ensure physical integrity and to minimize the possibility of formulation errors." In this context, the examination of individual vibration modes is an intermediate stage that expedites the elucidation of the intricate nature of drillstring vibrations.

3.2.6 Paradigm of Nonlinear Drillstring Modeling: Lateral Vibrations. The preceding sections have provided a mostly qualitative conspectus of vibration encountered in the drilling process. In this section and the ensuing Sections 3.2.7 and 3.2.8, approaches are outlined for determining transient and steady-state vibrations of drilling systems. For this purpose, a generic problem dealing with lateral vibrations of a BHA is used as a vehicle for providing paradigms of accounting for nonlinearities resulting from the borehole, and capturing deterministic and erratic (stochastic) features in the vibration pattern.

TABLE 3.2.1—DRILLSTRING-EXCITATION SOURCES, AFTER BESAISOW AND PAYNE (1988), SHOWING THAT THEY USUALLY INDUCE SIMULTANEOUSLY SEVERAL VIBRATION MODES		
Source	Primary Motion	Secondary Motion
Mass imbalance or bent pipe	Lateral	Axial/torsional/lateral
Misalignment	Lateral	Axial
Tricone bit	Axial	Torsional/lateral
Loose drillstring	Axial/torsional/lateral	–
Rotational walk	Lateral	Axial/torsional
Asynchronous walk or whirl	Lateral	Axial/torsional
Drillstring whip	Lateral	Axial/torsional

TABLE 3.2.2—SUMMARY OF MAJOR DRILLING VIBRATION MODES			
	Axial	Torsional	Lateral
Alternative name	Longitudinal	Rotational	Bending, flexural, transverse
Related phenomena	Bit bounce	Stick/slip, torque reversals	Whirling
Coupling mechanisms	Bit boundary condition	Bent drillstring components	–

Equation of Motion. The dynamic behavior of a BHA requires a multiple-degree-of-freedom approach. To this end, a finite-element model accounting for three displacements, and three rotations assigned at each node, can be used to model the BHA. In order to limit the computational requirements, and to facilitate the interpretation of the numerical results, only in-plane vibrations are examined.

An Euler-Bernoulli beam finite-element model that consists of two-node linear elements with two degrees of freedom assigned at each node, a lateral displacement u, and a rotation θ is derived for the description of the BHA dynamics. Then the resulting system of equations of motion is given by

$$\left[M\right]\ddot{u}(t)+\left[C\right]\dot{u}(t)+\left[K\right]u(t)=\mathbf{g}(t), \dotfill (3.2.26)$$

where $\left[M\right]$, $\left[C\right]$, and $\left[K\right]$ are the system mass, damping, and stiffness matrices, respectively; $\mathbf{g}(t)$ is the vector of the excitation applied to the system; and $\mathbf{u}$, $\dot{\mathbf{u}}$, and $\ddot{\mathbf{u}}$ are the vectors of displacement, velocity, and acceleration, respectively.

The element stiffness can be defined to account for the effect of a constant axial force, following the technique explained by Przemieniecki (1968). The applied load P is considered positive if tensile, resulting in the element-stiffness matrix

$$\left[\mathbf{k}_e\right]=\begin{bmatrix} \dfrac{12EI_z}{l_e^3}+\dfrac{6P}{5l_e} & & Symmetric & \\[2ex] \dfrac{6EI_z}{l_e^2}+\dfrac{P}{10} & \dfrac{4EI_z}{l_e}+\dfrac{2Pl_e}{15} & & \\[2ex] \dfrac{-12EI_z}{l_e^3}-\dfrac{6P}{5l_e} & \dfrac{-6EI_z}{l_e^2}-\dfrac{P}{10} & \dfrac{12EI_z}{l_e^3}+\dfrac{6P}{5l_e} & \\[2ex] \dfrac{6EI_z}{l_e^2}+\dfrac{P}{10} & \dfrac{2EI_z}{l_e}-\dfrac{Pl_e}{30} & \dfrac{-6EI_z}{l_e^2}-\dfrac{P}{10} & \dfrac{4EI_z}{l_e}+\dfrac{2Pl_e}{15} \end{bmatrix} \dotfill (3.2.27)$$

where E is the modulus of elasticity, I_z is the relevant moment of inertia of the cross section, and l_e is the length of the element.

A consistent-mass matrix is used to represent the BHA mass. The formulation allows the consideration of added-mass effects because of the presence of drilling mud inside and outside of the drillstring. Specifically,

$$\left[\mathbf{m}_e\right]=\begin{bmatrix} \dfrac{13M_t}{35}+\dfrac{6\rho I_z}{5l_e} & & Symmetric & \\[2ex] \dfrac{11M_t l_e}{210}+\dfrac{\rho I_z}{10} & \dfrac{M_t l_e^2}{105}+\dfrac{2\rho I_z l_e}{15} & & \\[2ex] \dfrac{9M_t}{70}-\dfrac{6\rho I_z}{5l_e} & \dfrac{13M_t l_e}{420}-\dfrac{\rho I_z}{10} & \dfrac{13M_t}{35}+\dfrac{6\rho I_z}{5l_e} & \\[2ex] \dfrac{-13M_t l_e}{420}+\dfrac{\rho I_z}{10} & \dfrac{-M_t l_e^2}{140}-\dfrac{\rho I_z l_e}{30} & \dfrac{-11M_t l_e}{210}-\dfrac{\rho I_z}{10} & \dfrac{M_t l_e^2}{105}+\dfrac{2\rho I_z l_e}{15} \end{bmatrix}, \dotfill (3.2.28)$$

where ρ and M_t are the density and the total mass of the element, respectively. The total mass of the element is the sum of the drillstring mass M, the added mass of the drilling fluid inside the tubular M_i, and the added mass of the annulus M_a. This means

$$M_t = M + M_i + C_M M_a, \dots\dots\dots\dots\dots\dots\dots\dots\dots\dots\dots\dots\dots\dots\dots\dots \quad (3.2.29)$$

where the added-mass coefficient C_M is a function of the vibration frequency ω, the mud properties, and the dimensions of the BHA and wellbore. For more details on the specific expression of this coefficient, see Payne (1992).

Finally, Rayleigh damping can be assumed for the BHA model. That is,

$$[\mathbf{C}] = \alpha_d [\mathbf{M}] + \beta_d [\mathbf{K}], \dots\dots\dots\dots\dots\dots\dots\dots\dots\dots\dots\dots\dots\dots\dots\dots \quad (3.2.30)$$

where the parameters α_d and β_d can be selected such that the damping ratios of particular modes are in the span of reported damping ratios, which range from 0.01% to as high as 0.65%, depending on well conditions and whether added lubricants are used (Brakel 1986).

Wellbore Consideration. The wellbore is modeled as shown in **Fig. 3.2.3.** The BHA lateral displacement is unconstrained for lateral displacement within the clearance α. For node lateral displacements exceeding α, an additional spring is activated to account for the contact with the wall. This system is referred to in the literature as a setup spring (Crandall 1961). Thus, a Hertzian contact law described by the equation

$$F_i[u_i(t)] = \begin{cases} k_2[u_i(t) + \alpha] & for \quad u_i \leq -\alpha \\ 0 & for \quad -\alpha \geq u_i \leq \alpha \quad i = 1,3,5,\dots,N-1 \\ k_2[u_i(t) - \alpha] & for \quad u_i \geq -\alpha \end{cases} \dots\dots\dots\dots\dots\dots \quad (3.2.31)$$

is assumed (see **Fig. 3.2.4**). The symbol k_2 denotes the rock Hertzian stiffness coefficient that directly affects the nonlinearity of the system, and N is the total number of degrees of freedom of the finite-element model. Incorporating Eq. 3.2.31 into Eq. 3.2.26 yields the nonlinear system of equations

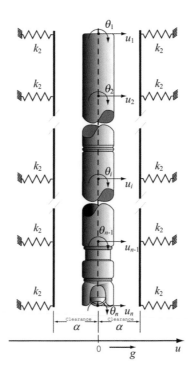

Fig. 3.2.3—BHA finite-element model for lateral vibrations considering wellbore contact (Spanos et al. 2002). Reprinted courtesy of the American Society of Mechanical Engineers.

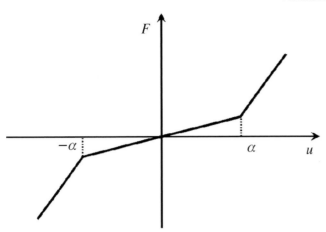

Fig. 3.2.4—Typical force-displacement diagram resulting from the adoption of a piecewise-linear model that accounts for the drillstring/formation interaction.

$$[\mathbf{M}]\ddot{\mathbf{u}}(t)+[\mathbf{C}]\dot{\mathbf{u}}(t)+[\mathbf{K}]\mathbf{u}(t)+\mathbf{F}[\mathbf{u}(t)]=\mathbf{g}(t). \quad \dots\dots\dots\dots\dots\dots\dots\dots\dots\dots\dots \quad (3.2.32)$$

3.2.7 Paradigm of Deterministic Response Determination: Lateral Vibrations. Obviously, the complexity of the dynamic system described by Eq. 3.2.32 renders its closed-form solution impossible to derive. In this section, techniques for the computation of transient as well as steady-state response are presented.

Transient-Response Algorithm. Note that a variety of numerical integration schemes are available for treating nonlinear structural dynamics problems such the one described by Eq. 3.2.32. Explicit methods, such as the Euler forward algorithm, use the equilibrium conditions at the previous timestep t_n to compute the response at time $t_{n+1} = t_n + \Delta t$. Accordingly, explicit algorithms are noniterative and easy to implement. Implicit approaches, on the other hand, which include the Euler backward, Newmark-β, Wison-θ, and Houbolt algorithms, use equilibrium conditions at the current timestep t_{n+1} to interpolate the solution. Implementing an implicit method is, as expected, more complex than implementing an explicit scheme. The choice of a procedure, however, also requires considering the stability characteristics of the various algorithms.

The Newmark-β method is a one-step method applicable to systems of linear or nonlinear differential equations. The initial-value problem involving Eq. 3.2.32, and the initial conditions, are first expressed at discrete time locations:

$$\begin{cases} [\mathbf{M}]\ddot{\mathbf{u}}_t + [\mathbf{C}]\dot{\mathbf{u}}_t + [\mathbf{K}]\mathbf{u}_t + \mathbf{F}(\mathbf{u}_t) = \mathbf{g}_t \\ \mathbf{u}_{t=0} = \mathbf{u}_0 \\ \dot{\mathbf{u}}_{t=0} = \dot{\mathbf{u}}_0 \end{cases} , \quad \dots\dots\dots\dots\dots\dots\dots\dots\dots\dots \quad (3.2.33)$$

which are located Δt seconds apart. This set of equations is first solved in its discrete form. Specifically, the solution is obtained by enforcing static equilibrium, including the effects of inertia and damping forces. Note that the nonlinearity in Eq. 3.2.33 relates to the stiffness term only.

The Newmark-β method is based on a Taylor-series expansion up to the second derivatives of the system displacements and velocities. It results in expressions of the displacements and velocities in the form (Kardestuncer et al. 1987)

$$\begin{aligned} \mathbf{u}_{t+\Delta t} &= \mathbf{u}_t + \Delta t \dot{\mathbf{u}}_t + \left(\frac{1}{2}-\beta\right)\Delta t^2 \ddot{\mathbf{u}}_t + \beta \Delta t^2 \ddot{\mathbf{u}}_{t+\Delta t} \\ \dot{\mathbf{u}}_{t+\Delta t} &= \dot{\mathbf{u}}_t + \Delta t \dot{\mathbf{u}}_t + \left(1-\gamma\right)\Delta t \ddot{\mathbf{u}}_t + \gamma \Delta t \ddot{\mathbf{u}}_{t+\Delta t} \end{aligned} , \quad \dots\dots\dots\dots\dots\dots\dots \quad (3.2.34)$$

where γ and β are parameters of the integration algorithm. The values adopted for these parameters will determine the nature of the approach. **Table 3.2.3** summarizes some of the most popular parameter settings in the family of Newmark integration schemes. **Fig. 3.2.5** shows a flowchart for the complete procedure applied to a nonlinear multiple-degree-of-freedom system with equations of motion represented as in Eq. 3.2.33. Obviously, given all the system parameters and the excitation representation in Eq. 3.2.33, this algorithmic procedure can be used to determine the vibration response in any specific drilling problem.

Steady-State Response. Clearly, the preceding transient response algorithm can be used to integrate the equation of motion *long enough* to determine the steady-state vibration solution. Nevertheless, in this section, a systematic approach is presented for circumventing the computationally costly time-domain integration to determine the steady-state response of a drillstring to monochromatic harmonic excitation. It is based on the work conducted by Payne (1992). The model is based on finite-element discretization that is capable of accommodating even frequency-dependent inertia and damping characteristics.

Similar to Eq. 3.2.26, a finite-element approach of the BHA dynamic response under monochromatic excitation leads to

$$[\mathbf{M}]\ddot{\mathbf{x}} + [\mathbf{C}]\dot{\mathbf{x}} + [\mathbf{K}]\mathbf{x} = \mathbf{F}\cos(\omega t) \qquad (3.2.35)$$

In this equation, $[\mathbf{M}]$ and $[\mathbf{K}]$ are the mass and stiffness matrices, respectively, defined in Section 3.2.6. Further, $\mathbf{F}$ is the excitation vector and $\ddot{\mathbf{x}}$, $\dot{\mathbf{x}}$, and $\mathbf{x}$ are the acceleration, velocity, and displacement vectors of the BHA, respectively. The parameters of the cosine function are the forcing frequency ω and time t.

BHA models that do not include damping have been proposed previously. By their very nature, they are incapable of predicting the response of a BHA at its natural frequencies. A discussion of generalized BHA damping presented by Apostal et al. (1990) involved Rayleigh damping, structural damping, and viscous damping. In this model, empirical formulas based on drillstring damping experiments (Payne 1992) are used to obtain the damping ratio by use of

$$\zeta = af_v^b, \qquad (3.2.36)$$

where f_v is the vibration frequency in hertz, and the coefficients a and b are functions of the mud density (specific gravity), ρ_{mud}. The expressions for these empirical coefficients are

$$a = (0.601)\rho_{mud}^{8.75} \qquad (3.2.37)$$

and

$$b = (0.15) - (1.026)\rho_{mud} \qquad (3.2.38)$$

The variety of excitations that act on an operating BHA include bit forces, drill collar mass imbalance, stabilizer loads, and drillpipe kinematics (Besaisow and Payne 1988). When using this model, a monochromatic harmonic excitation can be considered. This excitation is used because the capability to determine the

TABLE 3.2.3—NEWMARK INTEGRATION APPROACHES			
Procedure	Type	β	γ
Average acceleration (trapezoidal rule)	Implicit	1/4	1/2
Linear acceleration	Implicit	1/6	1/2
Fox and Goodwing (royal road)	Implicit	1/12	1/2
Purely explicit	Explicit	0	0
Central difference	Explicit	0	1/4

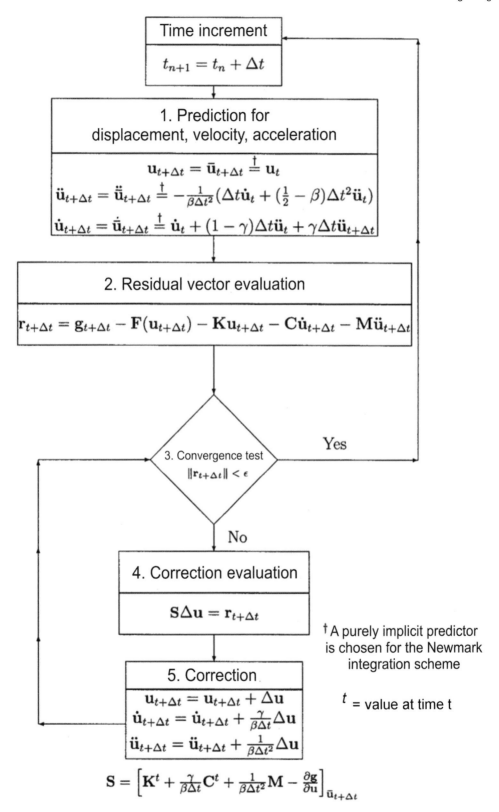

Fig. 3.2.5—Purely implicit Newmark-β direct integration scheme, after Kardestuncer et al. (1987). Reproduced from *Finite Element Handbook* with permission of the publisher, McGraw-Hill.

response of a dynamic model of the drillstring to this kind of excitation may expedite the study of the effects of any of the more specific sources of excitation described in the foregoing.

Boundary conditions of concern for lateral BHA vibrations include those at the stabilizers and the bit. The stabilizers are represented by pins that restrict lateral displacement but not rotation. The bit is represented by an excitation node at which a lateral force is applied to the model.

The solution approach for determining the BHA dynamic response must be carefully considered, because the frequency-dependent, added-mass coefficient and empirical damping function lead to the following revision of Eq. 3.2.35:

$$[\mathbf{M}(\omega)]\ddot{\mathbf{x}} + [\mathbf{C}(\omega)]\dot{\mathbf{x}} + [\mathbf{K}]\mathbf{x} = \mathbf{F}\cos(\omega t) \qquad \dots\dots\dots\dots\dots\dots\dots\dots\dots \quad (3.2.39)$$

It is noted that this equation symbolically represents time-domain convolutions in the inertia and damping terms. In this context, the steady-state solution, $\mathbf{x}(t)$, satisfies the equation

$$\left\{[\mathbf{K}] - \omega^2[\mathbf{M}(\omega)] + i\omega[\mathbf{C}(\omega)]\right\}\mathbf{X} = \mathbf{F}, \qquad \dots\dots\dots\dots\dots\dots\dots\dots \quad (3.2.40)$$

where i is $\sqrt{-1}$ and $\mathbf{X}$ is the complex amplitude of $\mathbf{x}(t)$. This equation can be solved with matrix inversion. However, matrix inversion does not provide any explicit information regarding the natural frequencies and mode shapes for the BHA, because the associated eigenvalue is not solved.

The representation of the transfer function of a dynamic system by modal superposition is well established for a variety of multiple-degree-of-freedom dynamic systems and is used in this model. The primary assumption of this technique is that the physical damping results in uncoupled modal equations of motion. Adopting this mathematical form for the damping terms, the steady-state displacement response for the system can be determined by

$$\mathbf{x}(t) = \sum_{r=1}^{l}\left[\left(\frac{\phi_r\phi_r^T\mathbf{F}}{K_r}\right)\left(\frac{1}{\sqrt{\left(1-r_r^2\right)^2+\left(2\zeta_r r_r\right)^2}}\right)\cos\left(\omega t - \alpha_r\right)\right], \qquad \dots\dots\dots\dots\dots \quad (3.2.41)$$

where the response is summed over the l modes of vibration ϕ_r, K_r is the modal stiffness, r_r is the ratio of the forcing frequency ω to the natural frequency of the mode ω_r, and ζ_r is the corresponding modal damping ratio. Note that fewer than the total number of system modes may be used in the summation in Eq. 3.2.41; the relative magnitude of the excitation frequency with respect to the natural frequencies usually suggests which modes are to be kept.

To implement this method, an eigenvalue problem must first be solved to obtain natural frequencies and normal modes for the BHA. In this case, the frequency dependence of the mass matrix $[\mathbf{M}(\omega)]$ indicates that the solution of the eigenvalue problem must be repeated at each frequency of interest to determine the frequencies and mode shapes accurately. Subsequently, this frequency-dependent modal information is used in Eq. 3.2.41. To avoid recalculation of the modes at each frequency, an alternative approach is adopted in which the transfer function, Eq. 3.2.41, is rewritten in the form

$$\mathbf{x}(t) = \sum_{r=1}^{l}\left[\left(\frac{\phi_r\phi_r^T\hat{\mathbf{F}}}{K_r}\right)\left(\frac{1}{\sqrt{\left(1-r_r^2\right)^2+\left(2\zeta_r r_r\right)^2}}\right)\cos\left(\omega t - \alpha_r\right)\right] \qquad \dots\dots\dots\dots\dots \quad (3.2.42)$$

Here, the forcing vector $\hat{\mathbf{F}}$ is equal to the original excitation force vector $\mathbf{F}$ plus an out-of-balance force vector to account for rotor eccentricities,

$$\mathbf{F}' = \omega^2\left([\mathbf{M}(\omega)] - [\mathbf{M}(\omega_m)]\right)\mathbf{X}, \qquad \dots\dots\dots\dots\dots\dots\dots\dots \quad (3.2.43)$$

where ω_m is the median frequency of interest, from which modal information is developed. For example, if the dynamic behavior in the frequency band of 0 to 50 Hz is of interest, the eigenvalue problem is formulated and solved at the median frequency of 25 Hz. Then, using iterative techniques, Eq. 3.2.42 is solved at all other frequencies of interest, with the out-of-balance force calculated by Eq. 3.2.43.

The requirement that the BHA remains within the wellbore is a nonlinear lateral effect that is easily understood but has not been accounted for by the previous equations. The iterative solution discussed in the preceding paragraph can address the nonlinear constraint imposed by the wellbore on the BHA if we introduce a minor modification to Eq. 3.2.42

$$\mathbf{x}(t) = \sum_{r=1}^{l} \left[\left(\frac{\phi_r \phi_r^T \left(\hat{\mathbf{F}} - F^* \right)}{K_r} \right) \left(\frac{1}{\sqrt{\left(1 - r_r^2\right)^2 + \left(2\zeta_r r_r\right)^2}} \right) \cos\left(\omega t - \alpha_r\right) \right], \quad \dots\dots\dots\dots (3.2.44)$$

where the contact restoring force to correct the response for the excessive displacement at the jth node is given by the equation

$$F^* = \left(\mathbf{X}_0^j - R_C\right) / \sum_{r=1}^{l} \left(\frac{\phi_r^j}{K_r \sqrt{\left(1 - r_r^2\right)^2 + \left(2\zeta_r r_r\right)^2}} \right). \quad \dots\dots\dots\dots (3.2.45)$$

In this equation, R_C denotes the local radial clearance and $\mathbf{X}_0^j$ is the initial displacement of the jth node, at which the maximum violation of the wellbore constraint occurs. The displacement of the jth node in the rth mode of vibration is denoted by ϕ_r^j.

To demonstrate the preceding solution approach, consider in **Fig. 3.2.6** a steel BHA consisting of 43 ft (13.1 m) of 8-in. (203.2-mm)-outside-diameter (OD) drill collars and 100 ft (30.48 m) of 6.5-in. (171.45-mm)-OD drill collars, with five stabilizers, each 5 ft (12.52 m) in length. It is assumed that the BHA operates in a 12.25-in. (311.15-mm)-diameter hole with mud with a specific gravity of 1.20. Using a spacing of

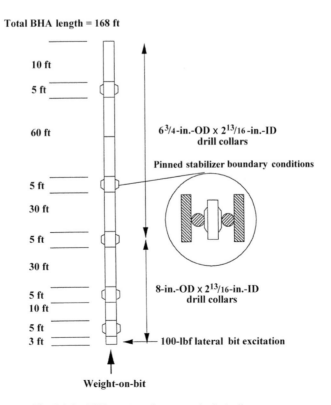

Fig. 3.2.6—BHA system for numerical studies.

approximately 2.5 ft (0.76 m), the model for the BHA involves 69 elements and 69 nodes with 131 degrees of freedom for lateral displacements and rotations. Stabilizers are treated as pins that restrict lateral displacement but allow rotation. A monochromatic harmonic excitation is applied laterally at the bit; the nominal magnitude of the force is 100 lbf (444.8 N).

Because of the large amount of information generated by a BHA frequency response calculation, results are presented herein in an abbreviated form. **Fig. 3.2.7** shows the first three eigenvectors as representative modal information.

Fig. 3.2.8 shows the BHA maximum von Mises stress response predicted through the use of different model formulations and solution techniques. The solid curve in Fig. 3.2.8 represents a predicted response for which the added-mass magnitude $C_M(\omega)$ term is not included in the system model. The importance of an accurate added-mass description in BHA dynamic analysis is seen in the shift between this response curve and the dashed or dotted response curves; the latter show responses for which the $C_M(\omega)$ term is included. The dashed curve depicts the response calculated through solution of the eigenvalue problem at each frequency of interest and application of Eq. 3.2.29. The dotted curve shows the response predicted by the proposed

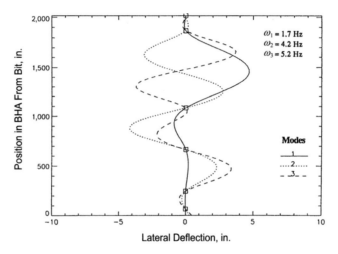

Fig. 3.2.7—Natural vibration Modes 1–3 for BHA case study.

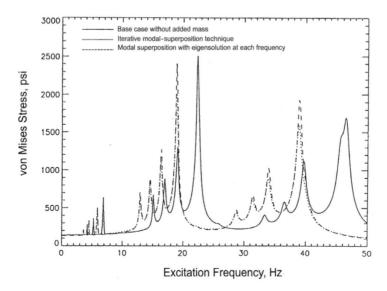

Fig. 3.2.8—Maximum BHA von Mises stress vs. excitation frequency for three solution techniques.

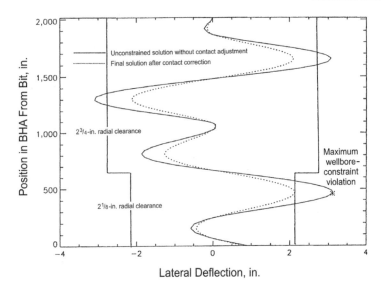

Fig. 3.2.9—Wellbore-contact correction of BHA displacement response.

iterative technique for which the eigenvalue problem is solved once at the median frequency of interest and Eqs. 3.2.42 and 3.2.43 are subsequently applied. In each of these cases, only 30 modal summations are used.

To illustrate the effect of introducing the nonlinear constraint imposed by the wellbore on the BHA, **Fig. 3.2.9** shows two calculated lateral-displacement responses along the BHA. The solid curve represents the unbounded response for which no consideration is given to the wellbore constraint. The response corrected for wellbore contact is represented by the dotted curve. It is worth mentioning that the computational time is minimally increased when the wellbore contact is introduced.

3.2.8 Paradigm of Stochastic Response Determination: Lateral Vibrations. It is clear from certain of the preceding comments of this chapter that drillstring vibrations are not completely understood, and to a great extent, they exhibit uncertainty. Interestingly, despite the inherent stochastic nature of many drilling variables, most vibration response models to date have pursued solely deterministic approaches. Nevertheless, examples of probabilistic analyses in a drilling context include computing the probability of encountering stuck pipes, and freeing them, using multivariate statistical analyses (Hempkins et al. 1987; Howard and Glover 1994; Shivers and Domangue 1993); generating authorizations for expenditures (Newendorp 1984; Peterson et al. 1995); and examining the fatigue life of downhole equipment (Dale 1989; Ligrone et al. 1995). Also, Murtha (1997) has presented some aspects of Monte Carlo simulations as employed in an oil-industry application. Literature reviews on this subject include Chevallier (2000) and Kotsonis (1994).

From a historical perspective, Bogdanoff and Goldberg (1958) presented in the late 1950s the first probabilistic approach to drillstring vibrations. They investigated the unknown stress distribution in a drilling assembly and argued that a formulation in terms of random variables was useful. Their mathematical model applied to axial and torsional vibrations. It modeled WOB and TOB using zero-mean normally distributed processes that were weakly stationary, mean-square continuous processes of second order. In other words, drilling was assumed sufficiently steady to consider the means and variances of the WOB and TOB as constant. The model yielded statistical information on shear stresses along the drillstring that could then be compared to postulated critical values. However, appropriate field data, such as power spectral densities, were unavailable, and only a qualitative analysis was performed. In a subsequent publication (Bogdanoff and Goldberg 1961), the same authors reaffirmed the appropriateness of the probabilistic treatment of drilling vibrations. They presented a model for axial force and torque characterization that statistically accounted for the influence of the lack of straightness of the well. Also, the model examined the buckling state in the compressive region of the assembly. No numerical example was presented because there were insufficient pertinent data available. Bogdanoff and Goldberg (1961) concluded their study by stating, "The very essence of a statistical analysis of the dynamical behavior of a drillpipe is the accommodation of uncertainty. Not only is it appropriate when the loads are known in a statistical sense, but it also is less sensitive than a deterministic analysis to uncertainty in boundary conditions, and the precise values of physical constants.... Therefore, its range of application in the present case and in highly complex problems is substantial and the advantages it offers in such problems should not be overlooked."

Skaugen (1987) studied the quasirandom nature of axial and torsional drilling vibrations. The uneven formation strength, the random rock breakage, and the amplification of these by modal coupling were assumed to result in quasirandom vibrations. On the basis of downhole measurements, the approach modeled axial displacements with sinusoids and a superimposed random component. The statistical-analysis results yielded significantly smaller displacement amplitudes than previous deterministic approaches.

Random-vibrations theory also has been applied to lateral drilling vibrations. In the 1990s, Kotsonis (1994) and Kotsonis and Spanos (1997) investigated the coupling of axial and lateral vibrations resulting from random excitations. A power spectral density with peaks located at three and six times the rotational speed represented those forcing functions, and the analysis described the whirling behavior of the assembly.

A stochastic approach in the analysis of lateral vibrations has been taken (Chevallier 2000; Politis 2002; Spanos et al. 2002), resulting in the development of techniques for the probabilistic treatment of drillstring vibrations. In the context of the pioneering work by Bogdanoff and Goldberg, a major thesis of this chapter is that a stochastic approach is quite appropriate for drillstring-vibration analysis. Obviously, in adopting a stochastic approach to drillstring vibrations, the excitation $\mathbf{g}_t$ in Eq. 3.2.33 must be described in a stochastic context. This ordinarily will involve probability distributions, such as the normal distribution for individual deviates of the excitation. Further, it will involve power spectra for the frequency content of the process as an ensemble (Roberts and Spanos 1990). In terms of stochastic vibration determination, two versatile tools of analysis are available: Monte Carlo simulation and statistical linearization. Both will be discussed in the ensuing sections.

Monte Carlo Approach for Stochastic Response Determination. Simulation in a broad sense involves performing sampling experiments on the model and the excitation of a system. The Monte Carlo method is a numerical method used to solve mathematical problems by random sampling (Spanos and Zeldin 1998).

Applied to the problem of drillstring vibrations, a Monte Carlo-based solution technique can be understood as involving several steps. First, the problem is expressed in terms of random variables. Considering a deterministic system, this stage requires the probabilistic characterization of excitation mechanisms. Next, these excitation mechanisms are quantified in terms of probability-density functions (or power spectra). Then, synthesis of random variables compatible with these probability distributions (or power spectra) is required, with each resulting in the generation of a particular, deterministic time history for the excitation mechanism of interest. Then, the response of the linear or nonlinear system to each deterministic excitation is determined. Finally, statistical information is retrieved from these data.

The approximation of the power spectral densities of excitation mechanisms is an important step in the Monte Carlo analysis of drilling-vibration problems. From a design point of view, several parameters, such as the maximum axial or lateral forces that a bit is required to sustain, can be used to define the characteristic magnitude and frequency content of the corresponding spectral densities.

The lateral excitation of the BHA depends strongly on the kind of drill bit used. The drill bit is chosen on the basis of the formation type of the well. Brakel (1986) has presented spectra of lateral excitations induced by the RC bit and the PDC bit. They most noticeably differ in their frequency content, because the excitation induced by the RC bits tends to possess significant levels of energy over the entire range of 0–50 rad/s, whereas PDC bits produce excitations with frequency content centered at the rotational driving frequency. This is because of the fact that the RC bit crushes the rock by impact of the protruding elements, which are uniformly distributed on rotating cones, as compared to the PDC bit in which the rock crushing involves protruding elements that are stationary with respect to the drillstring and thus induce excitation associated primarily with the rotary speed of the bit. In this context, excitations triggered by RC bits are modeled by band-limited white noise, as shown in **Fig. 3.2.10,** part (b). Further, the lateral excitations $\mathbf{g}(t)$ induced by PDC bits are modeled as the output resulting by passing white noise $\mathbf{w}(t)$ through a second-order filter:

$$\ddot{\mathbf{g}}(t) + \psi \dot{\mathbf{g}}(t) + \eta \mathbf{g}(t) = \mathbf{w}(t), \quad \dotfill (3.2.46)$$

(a) (b)

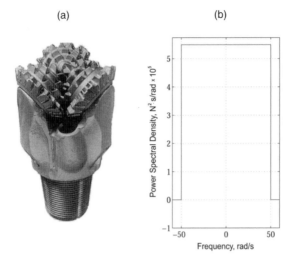

Fig. 3.2.10—(a) Typical RC bit; (b) RC-bit excitation spectrum (Spanos et al. 2002). Reprinted courtesy of the American Society of Mechanical Engineers.

with parameters ψ and η that are selected on the basis of field data. It is worth mentioning that the numerical implementation can also be performed by means of a digital filter where the values of the simulated load g_n, at each timestep Δt, can be obtained with

$$g_n = \sum_{i=1}^{p} a_i g_{n-i} + \sum_{j=1}^{q} b_j w_{n-j}, \quad \dots\dots\dots\dots\dots\dots\dots\dots\dots\dots\dots\dots \quad (3.2.47)$$

where w is white noise and the coefficients a_i and b_j can be obtained using various techniques; an extensive study on this approach can be found in Spanos and Zeldin (1998).

A spectrum corresponding to the signal generated by Eq. 3.2.46 is shown in **Fig. 3.2.11b.** For both kinds of bits, a zero-mean stationary Gaussian process is assumed.

Clearly, upon synthesizing by means of Eq. 3.2.47 realizations of the bit excitation, the time-domain integration algorithm in the Transient Response Algorithm section can be used repeatedly to determine an ensemble of the system response, which can be used to deduce distributional and spectral information of the system response.

Statistical Linearization for Stochastic Response Determination. In addition to Monte Carlo simulation, the method of statistical linearization is another versatile tool for nonlinear stochastic dynamic analysis. In the context of drilling-vibration problems, the equations of motion are linearized by replacing the original set of the governing nonlinear equations by an equivalent set of linear equations (Lin and Cai 1995; Roberts and Spanos 1990). Within this technique, two approaches permit solving the resulting set of equations: (1) the spectral matrix procedure in the frequency domain and (2) the covariance matrix procedure in the time domain.

Specifically, an equivalent set of linear equations is sought in the form

$$[\mathbf{M}]\ddot{\mathbf{u}}(t) + [\mathbf{C}]\dot{\mathbf{u}}(t) + ([\mathbf{K}] + [\mathbf{K}^{eq}])\mathbf{u}(t) = \mathbf{g}(t), \quad \dots\dots\dots\dots\dots\dots\dots\dots\dots\dots \quad (3.2.48)$$

where the nonlinear term $\mathbf{F}[\mathbf{u}(t)]$ in Eq. 3.2.32 is replaced by the equivalent linear system-stiffness matrix $[\mathbf{K}^{eq}]$ (Roberts and Spanos 1990), which is determined by the minimization of the mathematical expectation of the Euclidean norm of the error

$$\boldsymbol{\varepsilon} = \mathbf{F}(\mathbf{u}) - [\mathbf{K}^{eq}]\mathbf{u}. \quad \dots\dots\dots\dots\dots\dots\dots\dots\dots\dots\dots\dots\dots\dots \quad (3.2.49)$$

$$E\left\{\|\boldsymbol{\varepsilon}\|_2^2\right\} = E\left\{\boldsymbol{\varepsilon}^T \boldsymbol{\varepsilon}\right\} = \text{minimum}, \quad \dots\dots\dots\dots\dots\dots\dots\dots\dots\dots \quad (3.2.50)$$

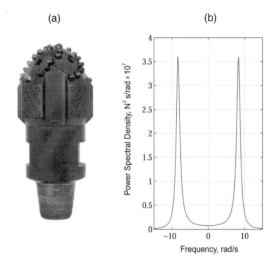

Fig. 3.2.11—(a) Typical PDC bit; (b) PDC-bit excitation spectrum (Spanos et al. 2002). Reprinted courtesy of the American Society of Mechanical Engineers.

where E and T denote the operators of mathematical expectation and transposition, respectively. This involves minimization in a least-squares sense:

$$\frac{\partial}{\partial k_{ij}^{eq}} = E\left\{\boldsymbol{\varepsilon}^T \boldsymbol{\varepsilon}\right\} = 0 \quad i, j = 1, 2, \ldots, n. \quad \dots\dots\dots\dots\dots\dots\dots\dots\dots\dots\dots\dots\dots \quad (3.2.51)$$

Because of the linearity of the mathematical expectation, Eq. 3.2.51 can be written as

$$\frac{\partial}{\partial k_{ij}^{eq}}\left[\sum_{i=1}^{N} E\{\varepsilon_i^2\}\right] = 0 \quad \overset{\text{expanding}}{\Rightarrow} \quad E\{u_j F_i\} = \sum_{r=1}^{N} k_{ir}^{eq} E\{u_r u_j\}, \quad j = 1, 2, \ldots, N. \quad \dots\dots\dots\dots \quad (3.2.52)$$

Then, the approximation of Gaussian displacement $\mathbf{u}$ is made to derive the elements of $[\mathbf{K}^{eq}]$:

$$k_{ij}^{eq} = E\left\{\frac{\partial F_i}{\partial u_j}\right\} = \frac{1}{\sqrt{2\pi}\sigma_i} \int_{-\infty}^{+\infty} \frac{\partial F_i}{\partial u_i} e^{-\frac{u_i^2}{2\sigma_i^2}} \, \mathrm{d}u_i. \quad \dots\dots\dots\dots\dots\dots\dots\dots\dots \quad (3.2.53)$$

For the adopted piecewise-linear contact law of Eq. 3.2.31, Eq. 3.2.53 yields

$$k_{ii}^{eq} = k_2 \operatorname{erfc}\left(\frac{\alpha}{\sqrt{2}\sigma_i}\right), \quad \dots\dots\dots\dots\dots\dots\dots\dots\dots\dots\dots\dots\dots\dots\dots \quad (3.2.54)$$

where $\operatorname{erfc}(\cdot)$ is the complementary error function. Next, with Eqs. 3.2.48 and 3.2.54, the statistics of the response are sought.

The Spectral-Matrix Solution Procedure. The transfer function of the equivalent linear system is

$$\mathbf{H}(\omega) = \left(-\omega^2 [\mathbf{M}] + i\omega[\mathbf{C}] + [\mathbf{K}] + [\mathbf{K}^{eq}]\right)^{-1}. \quad \dots\dots\dots\dots\dots\dots\dots\dots\dots \quad (3.2.55)$$

Then, assuming a stationary excitation $\mathbf{g}(t)$,

$$\mathbf{S}_{\mathbf{uu}}(\omega) = \mathbf{H}(\omega)\mathbf{S}_{\mathbf{gg}}(\omega)\mathbf{H}(\omega)^*, \quad \dots\dots\dots\dots\dots\dots\dots\dots\dots\dots\dots\dots\dots \quad (3.2.56)$$

where $\mathbf{S}_{\mathbf{gg}}$ is the power spectral density matrix of the excitation, and $(\cdot)^*$ denotes the complex conjugate. Next, the standard deviation of the ith degree of freedom can be determined by the equation

$$\sigma_i = \left(\int_{-\infty}^{+\infty} [\mathbf{S}_{\mathbf{uu}}(\omega)]_{ii} \, \mathrm{d}\omega\right)^{\frac{1}{2}} \quad \dots\dots\dots\dots\dots\dots\dots\dots\dots\dots\dots\dots\dots \quad (3.2.57)$$

Clearly, this involves a cyclic procedure for the determination of σ and $[\mathbf{K}^{eq}]$. The flowchart of **Fig. 3.2.12** shows the solution procedure as applied to the drilling problem.

The Covariance-Matrix Solution Procedure. A state space formulation is followed. The state vector $\mathbf{z}(t)$ is defined as

$$\mathbf{z}(t) = [\mathbf{u}(t)\ \dot{\mathbf{u}}(t)\ \mathbf{g}(t)\ \dot{\mathbf{g}}(t)]^T. \quad\dots\dots\dots\dots\dots\dots\dots\dots\dots\dots\dots\dots\dots\dots \quad (3.2.58)$$

The linearized system can be rewritten in the first-order matrix equation form

$$\dot{\mathbf{z}}(t) = \mathcal{F}\,\mathbf{z}(t) + \mathbf{h}(t), \quad\dots\dots\dots\dots\dots\dots\dots\dots\dots\dots\dots\dots\dots\dots\dots \quad (3.2.59)$$

where

$$\mathcal{F} = \begin{bmatrix} \mathbf{O} & \mathbf{I} & \mathbf{O} & \mathbf{O} \\ -[\mathbf{M}]^{-1}([\mathbf{K}]+[\mathbf{K}]^{eq}) & -[\mathbf{M}]^{-1}[\mathbf{C}] & -[\mathbf{M}]^{-1} & \mathbf{O} \\ \mathbf{O} & \mathbf{O} & \mathbf{O} & \mathbf{I} \\ \mathbf{O} & \mathbf{O} & -\eta\mathbf{I} & -\psi\mathbf{I} \end{bmatrix} \quad\dots\dots\dots\dots\dots\dots \quad (3.2.60)$$

and

$$\mathbf{h}(t) = [\mathbf{O}\ \mathbf{O}\ \mathbf{O}\ \mathbf{w}]^T. \quad\dots\dots\dots\dots\dots\dots\dots\dots\dots\dots\dots\dots\dots\dots \quad (3.2.61)$$

Considering the covariance matrices $\mathbf{V}$ and $\boldsymbol{\omega}_f$ of the vectors $\mathbf{z}(t)$ and $\mathbf{h}(t)$, respectively, are given by

$$\mathbf{V} = E[\mathbf{z}(t)\ \mathbf{z}^T(t)] \quad\dots\dots\dots\dots\dots\dots\dots\dots\dots\dots\dots\dots\dots\dots\dots\dots\dots \quad (3.2.62)$$

and

$$\boldsymbol{\omega}_f(t,\tau) = E[\mathbf{h}(t)\ \mathbf{h}^T(\tau)] = \mathbf{R}\ \delta(t-\tau), \quad\dots\dots\dots\dots\dots\dots\dots\dots\dots\dots \quad (3.2.63)$$

and recalling that the stationary $\mathbf{V}$ satisfies the Lyapunov matrix equation (Roberts and Spanos 1990)

$$\mathcal{F}\mathbf{V}^T + \mathbf{V}\mathcal{F}^T + \mathbf{R} = \mathbf{O}, \quad\dots\dots\dots\dots\dots\dots\dots\dots\dots\dots\dots\dots\dots\dots \quad (3.2.64)$$

one can solve for $\mathbf{V}$ numerically. This procedure is shown in the flowchart of **Fig. 3.2.13.**

To illustrate the preceding solution procedures, again a BHA model is considered. The BHA consists of drillpipes, a drill collar, and two stabilizers as shown in **Fig. 3.2.14**. The lateral excitation force is applied on the free drilling bit. The drillpipes are represented by 16 linear elements of equal length. No compressive force is applied on the drillpipes. The drill collar is represented by 64 linear elements of equal length, and a compressive force of 10 kN is applied on it. All the parts of the BHA are made of the same material, with a modulus of elasticity of 207 GPa and a density of 7833 kg/m³. A clearance of 10 cm between the drillstring and the wellbore is assumed. Finally, Rayleigh damping is assumed, and modal damping is selected between 2% and 10% for the first five modes.

The solution of the nonlinear equations of motion is obtained by use of the method of statistical linearization as described in the Statistical Linearization section. Further, relevant Monte Carlo simulations of the problem are conducted. Time histories compatible with the spectra corresponding to both types of drill bits are first generated by means of an auto-regressive moving average (ARMA) algorithm (Spanos and Zeldin 1998). A set of 300 time histories, each of which possesses more than 1,200 data points, is generated. The generated time histories are then used as excitations for the system, and the governing equations of motion are solved using a Newmark-β direct-integration scheme (Bathe and Wilson 1976). The statistical properties of the response are computed as ensemble averages over the simulated responses.

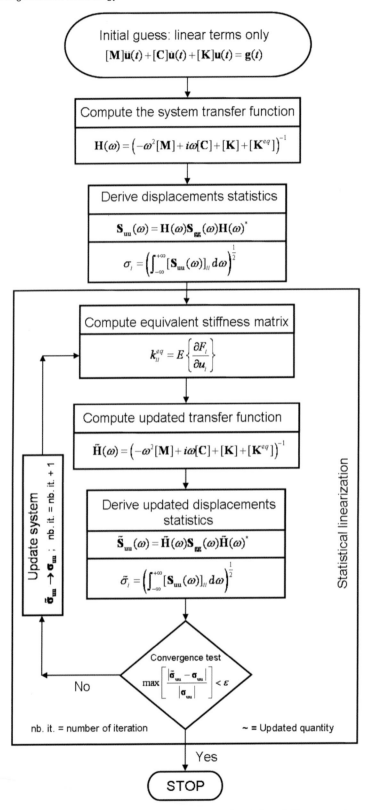

Fig. 3.2.12—Flowchart of equivalent linearization using the spectral-matrix solution procedure (Spanos et al. 2002). Reprinted courtesy of the American Society of Mechanical Engineers.

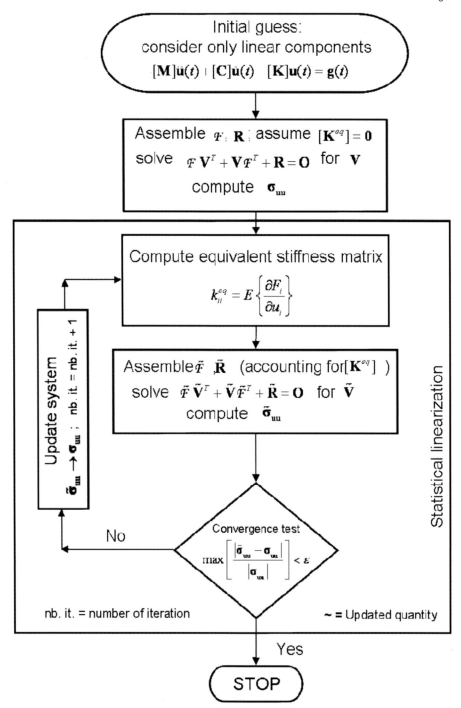

Fig. 3.2.13—Flowchart of statistical linearization using the covariance-matrix solution procedure (Spanos et al. 2002). Reprinted courtesy of the American Society of Mechanical Engineers.

Results for the numerical simulations regarding RC- and PDC-bit BHAs are shown in **Figs. 3.2.15 and 3.2.16,** respectively. The results are normalized on the abscissa with respect to the BHA stiffness, and on the ordinate by the clearance α. The figures show that for both kinds of bits, the solutions obtained by statistical linearization are in good agreement with the numerical results of the relevant Monte Carlo simulations. Further, the requisite computational time for the statistical linearization is significantly lower than that for the Monte Carlo simulations.

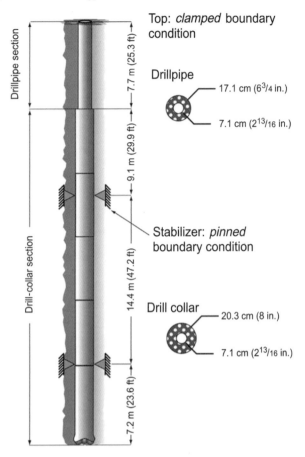

Fig. 3.2.14—Drillstring model considered for the numerical example (Spanos et al. 2002). Reprinted courtesy of the American Society of Mechanical Engineers.

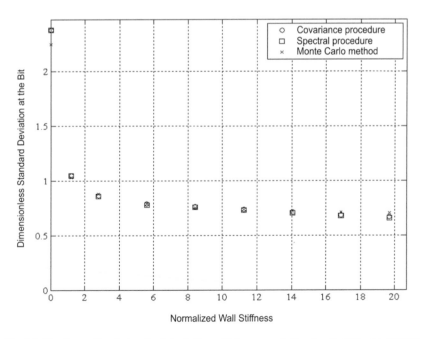

Fig. 3.2.15—Normalized standard deviation of bit vs. normalized wall stiffness; RC bit.

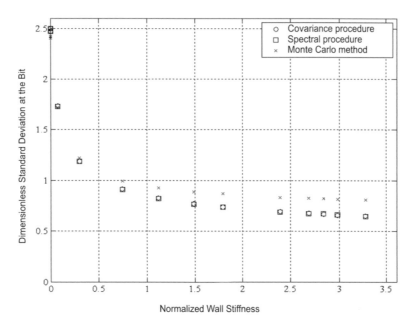

Fig. 3.2.16—Normalized standard deviation of bit vs. normalized wall stiffness; PDC bit.

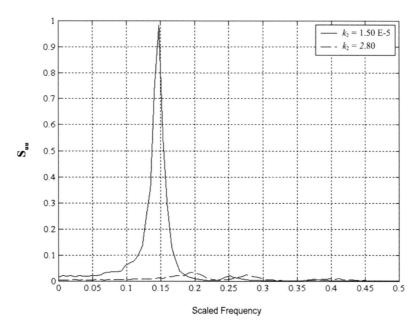

Fig. 3.2.17—Power spectrum of the response at the bit; RC bit.

The power spectra of the response of the bit for the RC-bit type of excitation and for the PDC-bit type of excitation are shown in **Figs. 3.2.17 and 3.2.18,** respectively. They both possess a peak close to the natural frequency of the system. However, in the case of the PDC bit, another peak that corresponds to the rotary frequency of the drill bit is observed. Further, they both capture the effects of the nonlinearity. Additional information and data regarding the stochastic approach delineated in the preceding can be found in references (Politis 2002; Spanos et al. 2002).

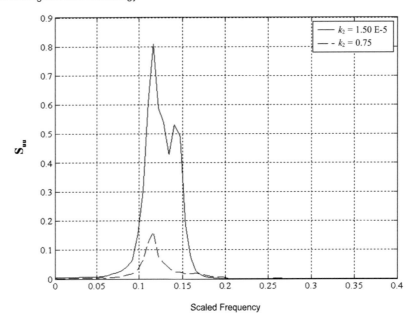

Fig. 3.2.18—Power spectrum of the response at the bit; PDC bit.

It is seen that the method of statistical linearization provides reliable and quite efficient results regarding this particular drilling-vibration problem.

3.2.9 Concluding Remarks. Clearly, downhole drilling vibrations are quite important for the drilling process, and the related issues should be addressed early in the drillstring-design process. We hope that this chapter provides a panoramic view of the pertinent physical phenomena and of the available tools of analysis.

It is expected that stochastic dynamics (Lin 1976; Roberts and Spanos 1990), system reliability (Ditlevsen and Madsen 1996; Thoft-Christensen and Murotsu 1986), optimization-based analysis (McCann and Suryanarayana 2001), and other novel approaches for vibration detection, isolation, and control (Christoforou and Yigit 2003; Viguié et al. 2008; Yigit and Christoforou 2006) will continue to receive increasing attention because they constitute a potent tool for achieving increased reliability and efficiency in drilling operations in the presence of drillstring vibrations. The ongoing developments in data acquisition (i.e., instrumentation, high-speed telemetry, distributed measurements) will continue refining the predictive capacity of simulation-based analyses of drillstring vibrations. Thus, as they evolve and are calibrated with more-sophisticated physical data sets, they will become more widely accepted as design tools in the drilling community.

Nomenclature

a = coefficient of damping ratio empirical equation in the steady-state lateral vibrations model
a_i = digital filter coefficient for load simulation
A_n = coefficient of axial wave equation general solution, L, m, ft
b = coefficient of damping ratio empirical equation in the steady-state lateral vibrations model
b_j = digital filter coefficient for load simulation
B_n = coefficient of axial wave equation general solution, L, m, ft
c = axial wave velocity, L/t, m/s, ft/s
c_1 = displacement solution coefficient for time interval 1 in stick/slip model
c_1' = displacement solution coefficient for time interval 1 in stick/slip model
c_2' = displacement solution coefficient for time interval 1 in stick/slip model

c_2 = displacement solution coefficient for time interval 1 in stick/slip model

c_3 = displacement solution coefficient for time interval 2 in stick/slip model

c_4 = displacement solution coefficient for time interval 2 in stick/slip model

c_a = damping coefficient in the equation of motion of the axial vibration model, m/L^{3}t, Kg/m^{3}s

c_w = damping coefficient for the whirling model, m/t, Kg/s

c_r = damping coefficient of stick/slip model, m L^2/t, Kg m^2/s rad

C_n = coefficient of axial wave equation general solution, L, m, ft

C_M = added mass coefficient in FE lateral vibration model

C_{PL} = coupling term in the whirling model, 1/t^2, 1/s^2

$[\mathbf{C}]$ = damping matrix for the FE lateral vibration model (units vary depending on degree of freedom)

D_n = coefficient of axial wave equation general solution, L, m, ft

e = Euler number

e_0 = eccentricity of center of mass in the whirling model, L, in

E = Young's modulus of elasticity, m/Lt^2, psi, Pa

f = normalized friction induced force in the stick/slip model, 1/t^2, 1/s^2

f_1 = friction parameter in the stick/slip model, 1/t^2, 1/s^2

f_v = vibration frequency in the steady-state lateral vibrations model, 1/t, Hz

f_2 = friction parameter in the stick/slip model, 1/t^2, 1/s^2

F = friction induced force in the stick/slip model, mL^2/t^2, N m

F_i = wellbore interaction force at the i-th degree of freedom, mL/t^2, N, lbf

F_k = equation of motion term in the whirling model, 1/t^2, 1/s^2

$\mathbf{F}$ = forcing vector in the steady-state lateral vibration model, mL/t^2, N, lbf

$\hat{\mathbf{F}}$ = forcing vector in the steady-state lateral vibration model, mL/t^2, N [lbf]

$\mathbf{F}^*$ = contact restoring force vector in the steady-state lateral vibration model, mL/t^2, N [lbf]

g = forcing term in the equation of motion of the lateral vibration model, mL/t^2, N [lbf]

g_a = forcing term in the equation of motion of the axial vibration model, mL/t^2, N [lbf]

g_n = digital filer output for load simulation, mL/t^2, N [lbf]

g_T = forcing term in the equation of motion of the torsional vibration model, mL/t^2, N [lbf]

$\mathbf{g}$ = forcing vector in the FE lateral vibration model

$\mathbf{g}_t$ = forcing vector at time t in the Newmark-β method

$\dot{\mathbf{g}}$ = first time derivative of the forcing vector in the FE lateral vibration model

$\ddot{\mathbf{g}}$ = second time derivative of the forcing vector in the FE lateral vibration model

G = shear modulus, m/Lt^2, psi [Pa]

$\mathbf{h}$ = state excitation vector in the covariance matrix procedure

$\mathbf{H}$ = transfer function matrix of the equivalent linear system

$\mathbf{H}^*$ = complex conjugate of the transfer function matrix of the equivalent linear system

I_z = cross section area moment of inertia in the FE lateral vibration model, L^4, m^4 [ft^4]

$\mathbf{I}$ = identity matrix

I = mass moment of inertia in the stick/slip model, m L^2, Kg m^2

J = cross section area polar moment of inertia, L^4, m^4 [ft^4]

k = torsional stiffness in the stick/slip model, mL^2/t^2, N m/rad

k_w = stiffness coefficient for the whirling model, m/t^2, N/m

k_{ij}^{eq} = diagonal element of the equivalent linear system stiffness matrix

k_{ij}^{eq} = (ij)-th element of the equivalent linear system stiffness matrix

k_{ir}^{eq} = (ir)-th element of the equivalent linear system stiffness matrix

k_2 = stiffness of the wellbore contact law, m/t^2, N/m [lbf/ft]

K = modal stiffness of the steady-state lateral vibration model

$[k_e]$ = elemental stiffness matrix for the FE lateral vibration model (units vary depending on degree of freedom)

$[\mathbf{K}]$ = global stiffness matrix for the FE lateral vibration model (units vary depending on degree of freedom)

$[\mathbf{K}^{eq}]$ = equivalent stiffness matrix in the statistical linearization technique (units vary depending on degree of freedom)

l = number of modes of vibration in the steady-state lateral vibration model

l_e = finite element length, L, m [ft]

m = equivalent mass of collar in the whirling model, m, Kg [slug]

$[\mathbf{m}_e]$ = elemental mass matrix for the FE lateral vibration model (units vary depending on degree of freedom)

M = drillstring finite element mass in the FE lateral vibration model, m, Kg [slug]

M_a = elemental added mass of annulus in the FE lateral vibration model, m, Kg [slug]

M_i = elemental mass of interior fluid in the FE lateral vibration model, m, Kg [slug]

M_t = elemental total mass in the FE lateral vibration model, m, Kg [slug]

$[\mathbf{M}]$ = global mass matrix in the FE lateral vibration model (units vary depending on degree of freedom)

n = element number in a time sequence

N = number of degrees of freedom in the FE lateral vibration model

$\mathbf{O}$ = zero matrix

p = digital filter order parameter for load simulation

P = drillstring axial force, mL/t², N [lbf]

q = digital filter order parameter for load simulation

r = dimensionless polar coordinate in the whirling model

r_b = radius of the borehole, L, in.

r_c = radius of the drillcollar, L, in.

r_r = ratio of forcing frequency and the r-th natural frequency

$\mathbf{r}_t$ = residual vector at time t in the Newmark-β method

$\dot{r}$ = first time derivative of the dimensionless polar coordinate in the whirling model, 1/t, 1/s

$\ddot{r}$ = second time derivative of the dimensionless polar coordinate in the whirling model, 1/t², 1/s²

R_C = local radial clearance in the steady-state lateral vibration model, L, in.

$\mathbf{R}$ = matrix term in Lyapunov equation

S = axial elevation in the bit-bounce model, L, m [ft]

S_0 = maximum axial elevation in the bit-bounce model, L, m [ft]

$\mathbf{S}_{gg}$ = power spectral density matrix of the excitation

$\mathbf{S}_{uu}$ = cross-spectrum matrix of the response

t = time, t, s

t_2 = time at which the sticking phase ends, t, s

u = lateral displacement in the continuous lateral vibration model, L, ft [m]

$\mathbf{u}_0$ = initial displacement vector in the Newmark-β method

u_i = i-th displacement in the FE lateral vibration model

u_j = j-th displacement in the FE lateral vibration model.

u_r = r-th displacement in the FE lateral vibration model

$\dot{\mathbf{u}}_0$ = initial velocity vector in the Newmark-β method

$\mathbf{u}_t$ = displacement vector at time t in the Newmark-β method

$\dot{\mathbf{u}}_t$ = velocity vector at time t in the Newmark-β method

$\ddot{\mathbf{u}}_t$ = acceleration vector at time t in the Newmark-β method

$\bar{\mathbf{u}}_t$ = displacement vector predictor at time t in the Newmark-β method

$\dot{\bar{\mathbf{u}}}_t$ = velocity vector predictor at time t in the Newmark-β method

$\ddot{\bar{\mathbf{u}}}_t$ = acceleration vector predictor at time t in the Newmark-β method

$\mathbf{u}$ = lateral displacement vector in the FE lateral vibration model

$\dot{\mathbf{u}}$ = lateral velocity vector in the FE lateral vibration model

$\ddot{\mathbf{u}}$ = lateral acceleration vector in the FE lateral vibration model

V_0 = velocity parameter in the stick/slip model, 1/t, rad/s

$\mathbf{V}$ = covariance matrix of the state vector

w_n = excitation simulated by digital filter white noise

$\mathbf{w}$ = excitation simulated vector by digital filter white noise

x = axial position along drillstring, L, ft [m]

$\mathbf{x}$ = displacement vector in the steady-state lateral vibration model

$\dot{\mathbf{x}}$ = velocity vector in the steady-state lateral vibration model

$\ddot{\mathbf{x}}$ = acceleration vector in the steady-state lateral vibration model

$\mathbf{X}_0^j$ = initial displacement of the j-th node in the steady-state lateral vibration model

$\mathbf{X}$ = complex amplitude of the displacement vector in the steady-state lateral vibration model

y = displacement in the whirling model, L, in.

$\dot{y}$ = velocity in the whirling model, L/t, in/s

$\ddot{y}$ = acceleration in the whirling model, L/t^2, in./s^2

z = displacement in the whirling model, L, in.

$\dot{z}$ = velocity in the whirling model, L/t, in./s

$\ddot{z}$ = acceleration in the whirling model, L/t^2, in./s^2

$\mathbf{z}$ = state space vector used in the covariance matrix solution procedure

$\dot{\mathbf{z}}$ = time derivative of state space vector

Z = equation of motion term in the whirling model

α = clearance parameter for wellbore contact law, L, in.

α_d = Rayleigh damping parameter

α_r = phase shift in the steady-state lateral vibration model

β = Newmark-β method parameter

β_w = added mass coefficient in the whirling model

β_d = Rayleigh damping parameter

γ = Newmark-β method parameter

Γ = equation of motion term in the whirling model, 1/t, 1/s

δ = Dirac delta function

Δ = increment

ε = equation of motion term in the whirling model, 1/t, 1/s

$\boldsymbol{\varepsilon}$ = error vector used in the statistical linearization technique

ε_i = i-th element of the error vector

ζ = damping ratio used in the stick/slip model

ζ_1 = damping ratio for notation simplification used in the stick/slip model

ζ_r = damping ratio of the r-th mode in the steady-state lateral vibration model

η = digital filter parameter for load simulation

ξ = axial displacement in the axial vibration model, L, m [ft]

ξ_n = n-th axial displacement term in the axial vibration model solution, L, m [ft]

θ = dimensionless polar coordinate in the whirling model.

$\dot{\theta}$ = first time derivative of the dimensionless polar coordinate in the whirling model, 1/t, 1/s

$\ddot{\theta}$ = second time derivative of the dimensionless polar coordinate in the whirling model, 1/t^2, 1/s^2

λ = equation of motion term in the whirling model, 1/t^2, 1/s^2

ρ = mass density, m/L^3, Kg/m^3 [slug/ft^3]

ρ_{mud} = mud density, m/L^3, Kg/m^3 [slug/ft^3]

σ_i = standard deviation of the i-th degree of freedom

σ_i^2 = variance of the i-th degree of freedom

τ = time shift, t, s

ϕ = angular displacement for continuous and stick/slip model

ϕ_1 = angular displacement in time interval 1 of the stick/slip model

ϕ_r = r-th mode of vibration of the steady-state lateral vibration model

ϕ_r^j = displacement of the j-th node in the r-th mode of the steady-state lateral vibration model

$\dot{\phi}$ = angular velocity in the stick/slip model, 1/t, rad/s

$\dot{\phi}_1$ = angular velocity in time interval 1 of the stick/slip model, 1/t, rad/s

$\ddot{\phi}$ = angular acceleration in the stick/slip model, 1/t^2, rad/s^2

$\ddot{\phi}_1$ = angular acceleration in time interval 1 of the stick/slip model, 1/t^2, rad/s^2

χ = equation of motion term in the whirling model, 1/t^2, 1/s^2

ψ = digital filter parameter for load simulation

ω = natural frequency of the whirling model, 1/t, rad/s

ω_0 = natural frequency of the stick/slip model, 1/t, rad/s

ω_b = damped frequency of sticking phase, 1/t, rad/s

ω_d = damped frequency of slipping phase, 1/t, rad/s

ω_n = n-th axial natural frequency, 1/t, rad/s

ω_m = median frequency of the steady-state lateral vibration model, 1/t, rad/s

ω_r = natural frequency of the r-th mode in the steady-state lateral vibration model, 1/t, rad/s

$\boldsymbol{\omega}_f$ = covariance matrix of the state excitation vector

Ω = rotary table speed of the stick/slip model, 1/t, rad/s

$\mathcal{F}$ = matrix term in the first-order matrix equation of the covariance matrix procedure

$\mathfrak{R}$ = real part of a quantity

$\mathfrak{I}$ = imaginary part of a quantity

Subscripts

i = i-th vector element

ii = matrix diagonal element

Superscripts

j = j-th node

T = transposition

$*$ = complex conjugate

References

Al-Hiddabi, S.A., Samanta, B., and Seibi, A. 2003. Non-Linear Control of Torsional and Bending Vibrations of Oilwell Drillstrings. *Journal of Sound and Vibration* **265** (2): 401–415. DOI: 10.1016/S0022-460X(02)01456-6.

Allen, M.B. 1987. BHA Lateral Vibrations: Case Studies and Evaluation of Important Parameters. Paper SPE 16110 presented at the SPE/IADC Drilling Conference, New Orleans, 15–18 March. DOI: 10.2118/16110-MS.

Apostal, M.C., Haduch, G.A., and Williams, J.B. 1990. A Study To Determine the Effect of Damping on Finite-Element Based, Forced-Frequency-Response Models for Bottomhole Assembly Vibration Analysis. Paper SPE 20458 presented at the SPE Annual Technical Conference and Exhibition, New Orleans, 23–26 September. 10.2118/20458-MS.

Bailey, J.J. and Finnie, I. 1960. An Analytical Study of Drill-String Vibration. *Trans.,* ASME, *Journal of Engineering for Industry* **82B** (May): 122–128.

Bathe, K.-J. 1982. *Finite Element Procedures in Engineering Analysis.* Englewood Cliffs, New Jersey: Prentice-Hall.

Bathe, K.-J. and Wilson, E. 1976. *Numerical Methods in Finite Element Analysis.* Englewood Cliffs, New Jersey: Civil Engineering and Engineering Mechanics Series, Prentice-Hall.

Belokobyl'skii, S. and Prokopov, V. 1982. Friction-Induced Self-Excited Vibrations of Drill Rig With Exponential Drag Law. *International Applied Mechanics* **18** (12): 1134–1138. DOI: 10.1007/BF00882226.

Belayev, A., Brommundt, E., and Palmov, V.A. 1995. Stability Analysis of Drillstring Rotation. *Dynamics and Stability of Systems: an International Journal* **10** (2): 99–110.

Besaisow, A.A. and Payne, M.L. 1988. A Study of Excitation Mechanisms and Resonance Inducing Bottomhole-Assembly Vibrations. *SPEDE* **3** (1): 93–101. SPE-15560-PA. DOI: 10.2118/15560-PA.

Bogdanoff, J.L. and Goldberg, J.E. 1958. A New Analytical Approach to Drill Pipe Breakage. *Proc.,* ASME Petroleum Mechanical Engineering Conference, Denver, 21–24 September.

Bogdanoff, J.L. and Goldberg, J.E. 1961. A New Analytical Approach to Drill Pipe Breakage II. *Trans.,* ASME, *Journal of Engineering for Industry* **83** (May): 101–106.

Booer, A.K. and Meehan, R.J. 1993. Drillstring Imaging—An Interpretation of Surface Drilling Vibrations. *SPEDC* **8** (2): 93–98. SPE-23889-PA. DOI: 10.2118/23889-PA.

Bradbury, R. and Wilhoit, J. 1963. Effect of Tool Joints on Passages of Plane Longitudinal and Torsional Waves Along a Drill Pipe. *Trans.*, ASME, *Journal of Engineering for Industry* **85** (May): 156–162.

Brakel, J.D. 1986. Prediction of Wellbore Trajectory Considering Bottom Hole Assembly and Drillbit Dynamics. PhD dissertation, University of Tulsa, Tulsa, Oklahoma.

Brebbia, C.A., Telles, J.C.F., and Wrobel, L.C. 1984. *Boundary Element Techniques*. Berlin: Springer-Verlag.

Brett, J.F. 1992. The Genesis of Torsional Drillstring Vibrations. *SPEDE* **7** (3): 168–174. SPE-21943-PA. DOI: 10.2118/21943-PA.

Brett, J.F., Warren, T.M., and Behr, S.M. 1990. Bit Whirl—A New Theory of PDC Bit Failure. *SPEDE* **5** (4): 275–281; *Trans.*, AIME, **289**. SPE-19571-PA. DOI: 10.2118/19571-PA.

Bronshtein, I.N. and Semendyayev, K.A. 1997. *Handbook of Mathematics*. Berlin: Springer.

Burgess, T.M., McDaniel, G.L., and Das, P.K. 1987. Improving BHA Tool Reliability With Drillstring Vibration Models: Field Experience and Limitations. Paper SPE 16109 presented at the SPE/IADC Drilling Conference, New Orleans, 15–18 March. DOI: 10.2118/16109-MS.

Burnett, D.S. 1987. *Finite Element Analysis From Concepts to Applications*. Reading, Massachusetts: Addison-Wesley.

Chen, G. and Zhou, J. 1992. *Boundary Element Methods*. London: Academic Press.

Chen, S.L. and Géradin, M. 1995. An Improved Transfer Matrix Technique as Applied to BHA Lateral Vibration Analysis. *Journal of Sound and Vibration* **185** (1): 93–106. DOI: 10.1006/jsvi.1994.0365.

Chevallier, A.M. 2000. Nonlinear Stochastic Drilling Vibrations. PhD dissertation, Rice University, Houston, Texas.

Chin, W.C. 1988. Why Drill Strings Fail at the Neutral Point. *Petroleum Engineer International* **60** (May): 62–67.

Chin, W.C. 1994. *Wave Propagation in Petroleum Engineering*. Houston: Gulf Publishing Company.

Christoforou, A.P. and Yigit, A.S. 1997. Dynamic Modelling of Rotating Drillstrings With Borehole Interactions. *Journal of Sound and Vibration* **206** (2): 243–260. DOI: 10.1006/jsvi.1997.1091.

Christoforou, A.P. and Yigit, A.S. 2003. Fully Coupled Vibrations of Actively Controlled Drillstrings. *Journal of Sound and Vibration* **267** (5): 1029–1045. DOI: 10.1016/S0022-460X(03)00359-6.

Close, D.A., Owens, S.C., and Macpherson, J.D. 1988. Measurement of BHA Vibration Using MWD. Paper SPE 17273 presented at the IADC/SPE Drilling Conference, Dallas, 28 February–2 March. DOI: 10.2118/17273-MS.

Coomer, J., Lazarus, M., Tucker, R.W., Kershaw, D., and Tegman, A. 2001. A Non-Linear Eigenvalue Problem Associated With Inextensible Whirling Strings. *Journal of Sound and Vibration* **239** (5): 969–982. DOI: 10.1006/jsvi.2000.3190.

Craig, R.R. Jr. 1981. *Structural Dynamics: An Introduction to Computer Methods*. New York: John Wiley and Sons.

Crandall, S.H. 1961. Random Vibration of a Nonlinear System With a Set-Up Spring. Technical Report No. AFOSR709, Accession No. AD0259320, Massachusetts Institute of Technology, Cambridge, Massachusetts (June 1961).

Cull, S. and Tucker, R. 1999. On the Modelling of Coulomb Friction. *Journal of Physics A: Math. Gen.* **32** (11): 2103–2113. DOI: 10.1088/0305-4470/32/11/006.

Cunningham, R.A. 1968. Analysis of Downhole Measurements of Drill String Forces and Motions. *Trans.*, ASME, *Journal of Engineering for Industry* **90** (May): 208–216.

Dale, B.A. 1989. Inspection Interval Guidelines To Reduce Drillstring Failures. *SPEDE* **4** (3): 215–222. SPE-17207-PA. DOI: 10.2118/17207-PA.

Dareing, D.W. 1983. Rotary Speed, Drill Collars Control Drillstring Bounce. *Oil & Gas Journal* **81** (23): 63–68.

Dareing, D.W. 1984a. Drill Collar Length Is a Major Factor in Vibration Control. *JPT* **36** (4): 637–644. SPE-11228-PA. DOI: 10.2118/11228-PA.

Dareing, D.W. 1984b. Guidelines for Controlling Drill String Vibrations. *Journal of Energy Resources Technology* **106** (June): 272–277.

Dareing, D.W. 1985. Vibrations Increase Available Power at the Bit. *Trans.*, ASME, *Journal of Energy Resources Technology* **107** (March): 138–141.

Dareing, D.W. and Livesay, B.J. 1968. Longitudinal and Angular Drillstring Vibrations With Damping. *Trans.*, ASME, *Journal of Engineering for Industry* **90B** (4): 671–679.

Dareing, D.W., Tlusty, J., and Zamudio, C. 1990. Self-Excited Vibrations Induced by Drag Bits. *Trans.*, ASME, *Journal of Energy Resources Technology* **112** (March): 54–61.

Dawson, R., Lin, Y., and Spanos, P. 1987. Drill String Stick-Slip Oscillations. *Proc.*, 1987 SEM Spring Conference on Experimental Mechanics, Houston, 14–19 June, 590–595.

Deily, F.H., Dareing, D.W., Paff, G.H., Ortolff, J.E., and Lynn, R.D. 1968. Downhole Measurements of Drill String Forces and Motions. *Trans.*, ASME, *Journal of Engineering for Industry* **90** (May): 217–225.

Dekkers, E. 1992. The "Soft Torque Rotary" System. PhD dissertation, University of Twente, Enschede, The Netherlands.

de Vries, H.M. 1994. The Effect of Higher Order Resonance Modes on the Damping of Torsional Drillstring Vibrations. PhD dissertation, University of Twente, Enschede, The Netherlands.

Ditlevsen, O. and Madsen, H.O. 1996. *Structural Reliability Methods*. New York: John Wiley and Sons.

Dubinsky, V.S.H., Henneuse, H.P., and Kirkman, M.A. 1992. Surface Monitoring of Downhole Vibrations: Russian, European and American Approaches. Paper SPE 24969 presented at the European Petroleum Conference, Cannes, France, 16–18 November. DOI: 10.2118/24969-MS.

Dufeyte, M-P. and Henneuse, H. 1991. Detection and Monitoring of the Slip-Stick Motion: Field Experiments. Paper SPE 21945 presented at the SPE/IADC Drilling Conference, Amsterdam, 11–14 March. DOI: 10.2118/21945-MS.

Dunayevsky, V.A., Abbassian, F., and Judzis, A. 1993. Dynamic Stability of Drillstrings Under Fluctuating Weight on Bit. *SPEDC* **8** (2): 84–92. SPE-14329-PA. DOI: 10.2118/14329-PA.

Dykstra, M. 1996. Nonlinear Drill String Dynamics. PhD dissertation, University of Tulsa, Tulsa, Oklahoma.

Dykstra, M.W., Chen, D.C.-K., Warren, T.M., and Zannoni, S.A. 1994. Experimental Evaluations of Drill Bit and Drill String Dynamics. Paper SPE 28323 presented at the SPE Annual Technical Conference and Exhibition, New Orleans, 25–28 September. DOI: 10.2118/28323-MS.

Elsayed, M.A., Dareing, D.W., and Vonderheide, M. 1997. Effect of Torsion on Stability, Dynamic Forces, and Vibration Characteristics in Drillstrings. *Trans.*, ASME, *Journal of Energy Resources Technology* **119** (March): 11–19.

Finnie, I. and Bailey, J.J. 1960. An Experimental Study of Drill-String Vibration. *Trans.*, ASME, *Journal of Engineering for Industry* **82B** (May): 129–135.

Frohrib, D. and Plunkett, R. 1967. The Free Vibrations of Stiffened Drill Strings With Static Curvature. *Trans.*, ASME, *Journal of Engineering for Industry* **89** (February): 23–30.

Gatlin, C. 1957. How Rotary Speed and Bit Weight Affect Rotary Drilling Rate. *Oil & Gas Journal* **55** (May): 193–198.

Halsey, G.W., Kyllingstad, A., Aarrestad, T.V., and Lysne, D. 1986. Drillstring Torsional Vibrations: Comparison Between Theory and Experiment on a Full-Scale Research Drilling Rig. Paper SPE 15564 presented at the SPE Annual Technical Conference and Exhibition, New Orleans, 5–8 October. DOI: 10.2118/15564-MS.

Hempkins, W.B., Kingsborough, R.H., Lohec, W.E., and Nini, C.J. 1987. Multivariate Statistical Analysis of Stuck Drillpipe Situations. *SPEDE* **2** (3): 237–244; *Trans.*, AIME, **283**. SPE-14181-PA. DOI: 10.2118/14181-PA.

Howard, J.A. and Glover, S.B. 1994. Tracking Stuck Pipe Probability While Drilling. Paper SPE 27528 presented at the SPE/IADC Drilling Conference, Dallas, 15–18 February. DOI: 10.2118/27528-MS.

Inglis, T.A. 1988. *Directional Drilling,* Petroleum Engineering and Development Studies Vol. 2. London: Graham and Trotman.

Jansen, J.D. 1991. Non-Linear Rotor Dynamics as Applied to Oilwell Drillstring Vibrations. *Journal of Sound and Vibrations* **147** (1): 115–135. DOI: 10.1016/0022-460X(91)90687-F.

Jansen, J.D. 1992. Whirl and Chaotic Motion of Stabilized Drill Collars. *SPEDE* **7** (2): 107–114. SPE-20930-PA. DOI: 10.2118/20930-PA.

Jansen, J. and van den Steen, L. 1995. Active Damping of Self-Excited Torsional Vibrations in Oil Well Drillstrings. *Journal of Sound and Vibration* **179** (26): 647–668. DOI: 10.1006/jsvi.1995.0042.

Javanmardi, K. and Gaspard, D. 1992. Soft Torque Rotary System Reduces Drillstring Failures. *Oil & Gas Journal* **90** (41): 68–71.

Kane, J. 1984. Dynamic Lateral Bottom Hole Forces Aid Drilling. *Drilling* **11** (August): 43–47.

Kardestuncer, H., Norrie, D.H., and Brezzi, F. eds. 1987. *Finite Element Handbook*. New York: McGraw-Hill.

Khulief, Y. and Al-Naser, H. 2005. Finite Element Dynamic Analysis of Drillstrings. *Finite Elements in Analysis and Design* **41** (13): 1270–1288. DOI: 10.1016/j.finel.2005.02.003.

Khulief, Y.A., Al-Sulaiman, F.A., and Bashmal, S. 2007. Vibration Analysis of Drillstrings With Self-Excited Stick-Slip Oscillations. *Journal of Sound and Vibration* **299** (3): 540–558. DOI: 10.1016/j.jsv.2006.06.065.

Kotsonis, S.J. 1994. Effects of Axial Forces on Drillstring Lateral Vibrations. MS thesis, Rice University, Houston, Texas.

Kotsonis, S.J. and Spanos, P.D. 1997. Chaotic and Random Whirling Motion of Drillstrings. *Trans., ASME, Journal of Energy Resources Technology* **119** (4): 217–222.

Kreisle, L.F. and Vance, J.M. 1970. Mathematical Analysis of the Effect of a Shock Sub on the Longitudinal Vibrations of an Oilwell Drill String. *SPEJ* **10** (4): 349–356. SPE-2778-PA. DOI: 10.2118/2778-PA.

Kyllingstad, A. and Halsey, G.W. 1988. A Study of Slip/Stick Motion of the Bit. *SPEDE* **3** (4): 369–373. SPE-16659-PA. DOI: 10.2118/16659-PA.

Lear, W.E. and Dareing, D.W. 1990. Effect of Drillstring Vibrations on MWD Pressure Pulse Signals. *Journal of Energy Resources Technology* **112** (2): 84–89. DOI: 10.1115/1.2905727.

Lee, C.-W. 1993. Vibration Analysis of Rotors, in *Solid Mechanics and Its Applications*. Dordrecht, The Netherlands: Kluwer Academic Publishers.

Lee, C.W. and Kim, Y.D. 1986. Finite Element Analysis of Rotor Bearing Systems Using a Modal Transformation Matrix. *Journal of Sound and Vibration* **111** (3): 441–456.

Leine, R.I. 1997. Literature Survey on Torsional Drillstring Vibrations. Internal report No. WFW 97.069, Division of Computational and Experimental Mathematics, Eindhoven University of Technology, Eindhoven, The Netherlands.

Leine, R.I. 2000. Bifurcations in Discontinuous Mechanical Systems of Filippov-Type. PhD dissertation, Eindhoven University of Technology, Eindhoven, The Netherlands.

Lichuan, L. and Sen, G. 1993. Studies on Some Problems of Vibrations and Stability of Drill Collar. In *Proc. of the 11th International Modal Analysis Conference: 1–4 February 1993, Kissimmee, Florida*, 567–571.

Ligrone, A., Botto, G., and Calderoni, A. 1995. Reliability Methods Applied to Drilling Operations. Paper SPE 29355 presented at the SPE/IADC Drilling Conference, Amsterdam, 28 February–2 March. DOI: 10.2118/29355-MS.

Lin, Y.K. 1976. *Probabilistic Theory of Structural Dynamics*. Melbourne, Florida: Krieger Publishing.

Lin, Y.K. and Cai, G.Q. 1995. *Probabilistic Structural Dynamics: Advanced Theory and Applications*. McGraw-Hill Education (ISE Editions), New York.

Lin, Y-Q. and Wang, Y-H. 1990. New Mechanism in Drillstring Vibration. Paper OTC 6225 presented at the Offshore Technology Conference, Houston, 7–10 May.

Lin, Y-Q. and Wang, Y-H. 1991. Stick-Slip Vibration of Drill Strings. *Trans., ASME, Journal of Engineering for Industry* **113** (February): 38–43.

Lubinski, A., Althouse, W.S., and Logan, J.L. 1962. Helical Buckling of Tubing Sealed in Packers. *JPT* **14** (6): 655–670; *Trans.*, AIME, **225**. SPE-178-PA. DOI: 10.2118/178-PA.

Macdonald, K.A. and Bjune, J.V. 2007. Failure Analysis of Drillstrings. *Engineering Failure Analysis* **14** (8): 1641–1666. DOI: 10.1016/j.engfailanal.2006.11.073.

Macpherson, J.D., Mason, J.S., and Kingman, J.E.E. 1993. Surface Measurement and Analysis of Drillstring Vibrations While Drilling. Paper SPE 25777 presented at the SPE/IADC Drilling Conference, Amsterdam, 23–25 February. DOI: 10.2118/25777-MS.

Mason, J. and Sprawls, B. 1998. Addressing BHA Whirl: The Culprit in Mobile Bay 1998. *SPEDC* **13** (4): 231–236. SPE-52887-PA. DOI: 10.2118/52887-PA.

McCann, R.C. and Suryanarayana, P.V.R. 2001. Horizontal Well Path Planning and Correction Using Optimization Techniques. *Journal of Energy Resources Technology* **123** (3): 187–193. DOI: 10.1115/1.1386390.

Melakhessou, H., Berlioz, A., and Ferraris, G. 2003. A Nonlinear Well-Drillstring Interaction Model. *Journal of Vibration and Acoustics* (January): 46–52.

Millheim, K.K. and Apostal, M.C. 1981. The Effect of Bottomhole Assembly Dynamics on the Trajectory of a Bit. *JPT* **33** (12): 2323–2338. SPE-9222-PA. DOI: 10.2118/9222-PA.

Mitchell, R.F. 2008. Tubing Buckling—The Rest of the Story. *SPEDC* **23** (2): 112–122. SPE-96131-PA. DOI: 10.2118/96131-PA.

Mitchell, R. and Allen, M. 1985. Lateral Vibration: The Key to BHA Failure Analysis. *World Oil* **200** (March): 101–104.

Mitchell, R.F. and Allen, M.B. 1987. Case Studies of BHA Vibration Failure. Paper SPE 16675 presented at the SPE Annual Technical Conference and Exhibition, Dallas, 27–30 September. DOI: 10.2118/16675-MS.

Murtha, J. 1997. Monte Carlo Simulation: Its Status and Future. *JPT* **49** (4): 361–370. SPE-37932-PA. DOI: 10.2118/37932-PA.

Narasimhan, S. 1987. A Phenomenological Study of Friction Induced Torsional Vibrations of Drill Strings. MS thesis, Rice University, Houston.

Navarro-López, E.M. and Cortés, D. 2007. Avoiding Harmful Oscillations in a Drillstring Through Dynamical Analysis. *Journal of Sound and Vibration* **307** (1–2): 152–171. DOI: 10.1016/j.jsv.2007.06.037.

Newendorp, P.D. 1984. A Strategy for Implementing Risk Analysis. *JPT* **36** (10): 1791–1796. SPE-11299-PA. DOI: 10.2118/11299-PA.

Nicholson, J.W. 1994. An Integrated Approach to Drilling Dynamics Planning, Identification, and Control. Paper SPE 27537 presented at the IADC/SPE Drilling Conference, Dallas, 15–18 February. DOI: 10.2118/27537-MS.

Paidoussis, M.P., Luu, T.P., and Prabhakar, S. 2007. Dynamics of a Long Tubular Cantilever Conveying Fluid Downwards, Which Then Flows Upwards Around the Cantilever as a Confined Annular Flow. *Journal of Fluids and Structures* **24** (1): 111–128. DOI: 10.1016/j.jfluidstructs.2007.07.004.

Paslay, P.R. and Bogy, D.B. 1963. Drill String Vibrations Due to Intermittent Contact of Bit Teeth. *Trans., ASME, Journal of Engineering for Industry* **85B** (May): 187–194.

Payne, M.L. 1992. Drilling Bottom-Hole Assembly Dynamics. PhD dissertation, Rice University, Houston.

Payne, M.L., Abbassian, F., and Hatch, A.J. 1995. Drilling Dynamic Problems and Solutions for Extended-Reach Operations. In *Drilling Technology 1995*, PD-Volume 65, ed. J.P. Vozniak, 191–203. New York: ASME.

Peterson, S.K., Murtha, J.A., and Roberts, R.W. 1995. Drilling Performance Predictions: Case Studies Illustrating the Use of Risk Analysis. Paper SPE 29364 presented at the SPE/IADC Drilling Conference, Amsterdam, 28 February–2 March. DOI: 10.2118/29364-MS.

Plunkett, R. 1967. Static Bending Stresses in Catenaries and Drill Strings. *Trans., ASME, Journal of Engineering for Industry* **89B** (1): 31–36.

Poletto, F. and Bellezza, C. 2006. Drill-Bit Displacement-Source Model: Source Performance and Drilling Parameters. *Geophysics* **71** (5): 121–129. DOI: 10.1190/1.2227615.

Politis, N.P. 2002. An Approach for Efficient Analysis of Drill-String Random Vibrations. MS thesis, Rice University, Houston.

Przemieniecki, J.S. 1968. *Theory of Matrix Structural Analysis*. New York: McGraw-Hill Book Company.

Puebla, H. and Alvarez-Ramirez, J. 2008. Suppression of Stick-Slip in Drillstrings: A Control Approach Based on Modeling Error Compensation. *Journal of Sound and Vibration* **310** (4–5): 881–901. DOI: 10.1016/j.jsv.2007.08.020.

Raftoyiannis, J. and Spyrakos, C. 1997. *Linear and Nonlinear Finite Element Analysis in Engineering Practice*. Pittsburgh, Pennsylvania: ALGOR.

Reddy, J.N. 1993. *An Introduction to the Finite Element Method*, second edition. New York: McGraw-Hill Science/Engineering/Math.

Richard, T., Germay, C., and Detournay, E. 2004. Self-Excited Stick-Slip Oscillations of Drill Bits. *Comptes Rendus Mecanique* **332** (8): 619–626. DOI: 10.1016/j.crme.2004.01.016.

Richard, T., Germay, C., and Detournay, E. 2007. A Simplified Model To Explore the Root Cause of Stick-Slip Vibrations in Drilling Systems With Drag Bits. *Journal of Sound and Vibration* **305** (3): 432–456. DOI: 10.1016/j.jsv.2007.04.015.

Roberts, J.B. and Spanos, P.D. 1990. *Random Vibration and Statistical Linearization*. New York: John Wiley and Sons.

Rogers, W. 1990. Drill Pipe Failures. In *Drilling Technology Symposium 1990: Presented at the Thirteenth Annual Energy-Sources Technology Conference and Exhibition, New Orleans, January 14–18, 1990*, Volume 27, ed. R.L. Kastor and P.D. Weiner, 89–93. New York: ASME.

Sampaio, R., Piovan, M., and Lozano, G.V. 2007. Coupled Axial/Torsional Vibrations of Drill-Strings by Means of Non-Linear Model. *Mechanics Research Communications* **34** (September): 497–502. DOI: 10.1016/j.mechrescom.2007.03.005.

Sengupta, A. 1993. Dynamic Modeling of Roller Cone Bit Lift-Off in Rotary Drilling. PhD dissertation, Rice University, Houston.

Serrarens, A.F.A., van de Molengraft, M.J.G., Kok, J.J., and van den Steen, L. 1998. H$_\infty$ Control for Suppressing Stick-Slip in Oil Well Drillstrings. *IEEE Control Systems* **18** (2): 19–30. DOI: 10.1109/37.664652.

Shivers, R.M. III and Domangue, R.J. 1993. Operational Decision Making for Stuck-Pipe Incidents in the Gulf of Mexico: A Risk Economics Approach. *SPEDC* **8** (2): 125–130. SPE-21998-PA. DOI: 10.2118/21998-PA.

Shyu, R-J. 1989. Bending Vibration of Rotating Drill Strings. PhD dissertation, MIT Department of Ocean Engineering, Cambridge, Massachusetts.

Skaugen, E. 1987. The Effects of Quasi-Random Drill Bit Vibrations Upon Drillstring Dynamic Behavior. Paper SPE 16660 presented at the SPE Annual Technical Conference and Exhibition, Dallas, 27–30 September. DOI: 10.2118/16660-MS.

Smit, A. 1995. Using Optimal Control Techniques To Dampen Torsional Drillstring Vibrations. PhD dissertation, University of Twente, Enschede, The Netherlands.

Spanos, P.D., Sengupta, A.K., Cunningham, R.A., and Paslay, P.R. 1995. Modeling of Roller Cone Bit Lift-Off Dynamics in Rotary Drilling. *Journal of Energy Resources Technology* **117** (3): 197–207. DOI: 10.1115/1.2835341.

Spanos, P.D., Payne, M.L., and Secora, C.K. 1997. Bottom-Hole Assembly Modeling and Dynamic Response Determination. *Journal of Energy Resources Technology* **119** (3): 153–158. DOI: 10.1115/1.2794983.

Spanos, P.D. and Zeldin, B. 1998. Monte Carlo Treatment of Random Fields: A Broad Perspective. *Applied Mechanics Reviews* **51** (March): 219–237.

Spanos, P.D., Chevallier, A.M., and Politis, N.P. 2002. Nonlinear Stochastic Drill-String Vibrations. *Journal of Vibration and Acoustics* **124** (4): 512–518. DOI: 10.1115/1.1502669.

Spanos, P.D., Chevallier, A.M., Politis, N.P., and Payne, M.L. 2003. Oil and Gas Well Drilling: A Vibrations Perspective. *Shock & Vibration Digest* **35** (March 2003): 85–103. DOI: 10.1177/0583102403035002564.

Thoft-Christensen, P. and Murotsu, Y. 1986. *Application of Structural Systems: Reliability Theory*. New York: Springer-Verlag.

Thomson, W.T. and Dahleh, M.D. 1997. *Theory of Vibration With Applications*, fifth edition. Upper Saddle River, New Jersey: Prentice-Hall.

Trindade, M.A., Wolter, C., and Sampaio, R. 2005. Karhunen-Loève Decomposition of Coupled Axial/Bending Vibrations of Beams Subject to Impacts. *Journal of Sound and Vibration* **279** (3–5): 1015–1036. DOI: 10.1016/j.jsv.2003.11.057.

Tucker, W.R. and Wang, C. 1999. On the Effective Control of Torsional Vibrations in Drilling Systems. *Journal of Sound and Vibration* **224** (1): 101–122. DOI: 10.1006/jsvi.1999.2172.

van den Steen, L. 1997. Suppressing Stick-Slip-Induced Drillstring Oscillations: A Hyperstability Approach. PhD dissertation, University of Twente, Enschede, The Netherlands.

Van Der Heijden, G.H. 1993. Bifurcation and Chaos in Drillstring Dynamics. *Chaos, Solitons & Fractals* **3** (2): 219–247. DOI: 10.1016/0960-0779(93)90068-C.

Vandiver, K.J., Nicholson, J.W., and Shyu, R.-J. 1990. Case Studies of the Bending Vibration and Whirling Motion of Drill Collars. *SPEDE* **5** (4): 282–290. SPE-18652-PA. DOI: 10.2118/18652-PA.

Vaz, M.A. and Patel, M.H. 1995. Analysis of Drill Strings in Vertical and Deviated Holes Using the Galerkin Technique. *Engineering Structures* **17** (6): 437–442. DOI: 10.1016/0141-0296(95)00098-R.

Viguié, R., Kerschen, G., Golinval, J.-C., McFarland, D.M., Bergman, L.A., Vakakis, A.F., and van de Wouw, N. 2009. Using Passive Nonlinear Targeted Energy Transfer To Stabilize Drill-String Systems. *Mechanical Systems and Signal Processing: Special Issue on Nonlinear Structural Dynamics* **23** (January): 148-169. DOI: 10.1016/j.ymssp.2007.07.001.

Warren, T.M., Brett, J.F., and Sinor, L.A. 1990. Development of a Whirl-Resistant Bit. *SPEDE* **5** (4): 267–274; *Trans.*, AIME, **289**. SPE-19572-PA. DOI: 10.2118/19572-PA.

Wise, J.L., Mansure, A.J., and Blankenship, D.A. 2005. Hard-Rock Field Performance of Drag Bits and a Downhole Diagnostics-While-Drilling (DWD) Tool. *GRC Transactions* **29** (September).

Wolf, S.F., Zacksenhouse, M., and Arian, A. 1985. Field Measurements of Downhole Drillstring Vibrations. Paper SPE 14330 presented at the SPE Annual Technical Conference and Exhibition, Las Vegas, Nevada, USA, 22–26 September. DOI: 10.2118/14330-MS.

Yigit, A.S. and Christoforou, A.P. 1998. Coupled Torsional and Bending Vibrations of Drillstrings Subject to Impact With Friction. *Journal of Sound and Vibration* **215** (1): 167–181. DOI: 10.1006/jsvi.1998.1617.

Yigit, A.S. and Christoforou, A.P. 2000. Coupled Torsional and Bending Vibrations of Actively Controlled Drillstrings. *Journal of Sound and Vibration* **234** (1): 67–83. DOI: 10.1006/jsvi.1999.2854.

Yigit, A.S. and Christoforou, A.P. 2006. Stick-Slip and Bit-Bounce Interaction in Oil-Well Drillstrings. *Trans.,* ASME, *Journal of Energy Resources Technology* **128** (4): 268–274. DOI: 10.1115/1.2358141.

Zamanian, M., Khadem, S.E., and Ghazavi, M.R. 2007. Stick-Slip Oscillations of Drag Bits by Considering Damping of Drilling Mud and Active Damping System. *Journal of Petroleum Science and Engineering* **59** (3–4): 289–299. DOI: 10.1016/j.petrol.2007.04.008.

SI Metric Conversion Factors

$1 \text{ GPa} = 145.038 \text{ kip/in.}^2$
$1 \text{ kg/m}^3 = 1.94 \times 10^{-3} \text{ slug/ft}^3$
$1 \text{ kN} = 224.809 \text{ lbf}$

3.3 Jars and Jarring Dynamics—Åge Kyllingstad, National Oilwell Varco

Jarring is the process of freeing stuck strings by means of a jar. A jar is a telescopic hammer tool that is placed in the string above the stuck point. Optionally, a jar accelerator can be used together with a jar to enhance the jarring energy and impact force.

3.3.1 Description of Jarring Tools. There are several types of jars briefly described below. A common feature of all types is the free stroke in which the upper part (hammer) is accelerated until it hits the lower part (anvil).

Mechanical jars have a mechanical latch mechanism with a preset release force. Except for one particular type allowing the release force to be adjusted by torque, the release force cannot be adjusted downhole. The jar fires (moves from its latched position into the free stroke) as soon as the overpull exceeds the jar release force.

In contrast, the release force of *hydraulic jars* has no preset value but is determined by the actual overpull at the jar. This is accomplished by a hydraulic mechanism and a so-called metering stroke in which oil is forced to flow through a small orifice. At the end of the metering stroke, the oil bypasses the orifice, causing the hydraulic resistance to drop to a very low value (free stroke). The metering stroke represents a delay time, allowing the jarring operator to set the desired overpull. This flexibility and the wide operating range of jar release force are major advantages of hydraulic jars. Disadvantages are risks of unintended jar firing during normal drilling operations, long metering times for low overpull, overheating the oil during repeated jarring, and risk of overloading the jar during the metering stroke.

Hydromechanical jars represent hybrid jars that combine initial mechanical release (to avoid uncontrolled firing) with hydraulic action to provide flexibility and adjustable release force **(Fig. 3.3.1).**

Jars can be either *single-acting* or *double-acting*. The latter type can be used both upward for creating tension impact loads and downward for creating compression impact loads. A schematic view of a double-acting hydraulic jar is shown in Fig. 3.3.1.

The drilling industry also distinguishes between *drilling jars* and *fishing jars*. While the former type is designed for being a standard member of a drillstring, the latter is designed for shorter time intervals during so-called fishing operations (freeing and retrieving stuck string sections left in the well).

A *jar accelerator*, also called a jar intensifier, is an optional tool commonly used in jarring. It acts as an elastic spring and increases the jarring energy and impact force. The spring medium can be steel (*mechanical

Hydraulic Jar

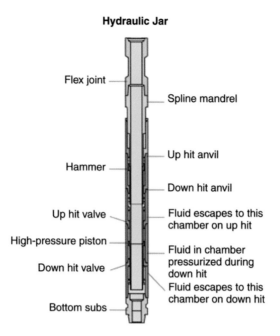

Fig. 3.3.1—Schematic view of a double-acting hydraulic jar (http://www.glossary.oilfield.slb.com/DisplayImage. cfm?ID=331). Image copyright Schlumberger. Used with permission.

accelerator), compressible oil (*hydraulic accelerator*), or nitrogen (*gas accelerator*). Jar accelerators are placed above the jar, normally with a few drill collars between the jar and the accelerator. Jar accelerators also can be single-acting or double-acting. Their ability to enhance the jarring impact force is discussed in more detail below.

Typical static spring characteristics for a single-acting mechanical accelerator are shown in **Fig. 3.3.2**.

3.3.2 Jarring Dynamics. The jarring cycle can be divided into the following phases:

1. Loading phase
2. Acceleration phase
3. Impact phase
4. Resetting phase

The second and third phases are discussed in some details here, while the two remaining phases will be discussed rather briefly.

Acceleration Phase. For simplicity we shall here make the following assumptions:

- The string above the collars (on top of the jar) is long and uniform, implying that we can neglect reflections of the wave resulting from the sudden change of tension across the jar.
- The collar section above the jar is short, so that the acceleration time is much longer than the acoustic round-trip time in the collar section.
- The distance from the jar to the stuck point is also short, so that transient motion of the jar anvil (lower part of the jar) can be neglected during the acceleration phase.

With these assumptions, we can use a lumped-mass approximation for the acceleration phase. The equation of motion for the hammer mass M_j (upper part of jar plus the drill collars above it) is

$$M_j \frac{dv}{dt} = F_0 - m_p c v_j - F_j. \quad \dots\dots\dots\dots\dots\dots\dots\dots\dots\dots\dots\dots\dots\dots\dots\dots\dots\dots\dots \quad (3.3.1)$$

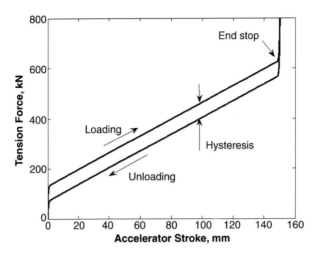

Fig. 3.3.2—Typical spring characteristics of a single-acting mechanical accelerator.

Here, $M_j = m_j l_j$ is the total mass of the jarring section, v_j is the jar hammer speed (positive upwards), F_0 is the prejarring force (actual overpull) just before start of the acceleration phase, m_p is the mass per unit length of drillpipe, c is the sonic speed, and F_j is the jar friction force. Because drillpipe is not truly uniform but has regularly spaced tool joints, c is slightly lower than the sonic speed in completely uniform steel pipes. The product $m_p c$ is often called the characteristic impedance for longitudinal waves. The two first terms on the right side of Eq. 3.3.1 represent the pull force from the pipe. It is easily verified that the solution with zero start speed is

$$v_j = \frac{F_0 - F_j}{m_p c}\left[1 - \exp\left(-\frac{m_p c}{M_j}t\right)\right] = v_0\left[1 - \exp(-t/\tau)\right]. \quad \dots\dots\dots\dots\dots\dots (3.3.2)$$

This speed approaches the contraction speed, $v_0 = (F_0 - F_j)/(m_p c)$, when the time is much greater than the time constant τ defined by $\tau = M_j/(m_p c)$. The acceleration phase ends when the free stroke of the jar is reached—that is, when

$$\int_0^{t_1} v\,dt = v_0 t_1 + v_0 \tau \exp(-t_1/\tau) = s_j. \quad \dots\dots\dots\dots\dots\dots\dots\dots (3.3.3)$$

This equation implicitly determines the acceleration time t_1 and the impact speed $v_1 = v(t_1)$.

Numerical Example.

F_0 = 600 kN   overpull
F_j = 100 kN   jar friction force
l_j = 30 m   length of jarring section
m_j = 161 kg   specific weight of jarring section (6.75 × 2.25 in. drill collars)
m_p = 32 kg/m   specific weight of 5-in. drillpipe
c = 5200 m/s   sonic speed in drillpipe
s_j = 0.15 m   free stroke of jar

These input values give

v_0 = 3.00 m/s   contraction speed (speed limit)
τ = 0.029 s   time constant
t_1 = 0.054 s   acceleration time
v_1 = 2.55 m/s   impact speed (at end of acceleration phase)

Note that the acceleration time is much longer than the acoustic round-trip time in the jar hammer (collars), which is $t_j = 2l_j/c = 0.011$ seconds. This is one reason why the lumped-mass approximation applies for the acceleration phase. Another reason is that the finite transition time from prejarring to the acceleration phase represents a high damping that effectively dampens internal vibrations in the jarring hammer.

If an accelerator is placed between the jarring mass section and the pipe section, the equations of motion become

$$M_j \frac{dv_a}{dt} = F_a(t) - F_j \quad \dots\dots\dots\dots\dots\dots\dots\dots\dots\dots\dots\dots\dots\dots\dots \quad (3.3.4)$$

$$M_a \frac{dv_j}{dt} = F_0 - m_p c v_j - F_a(t). \quad \dots\dots\dots\dots\dots\dots\dots\dots\dots\dots\dots \quad (3.3.5)$$

Here, F_a represents the variable tension force across the accelerator. It is often modeled as a linear spring without damping, but real accelerators have both hysteresis (internal friction) and end stops that reduce their efficiency.

Analytic solutions of the above equations exist if the spring force is linear, but because of their complexity they will not be given here. Instead, numerical solutions are plotted for three different cases (see **Fig. 3.3.3**). The free jar stroke (area below the curves) is the same for all cases.

The ideal accelerator has an infinitely soft spring with no end stop or internal friction. It keeps the acceleration constant because the force does not drop as the jar hammer moves and the accelerator contracts. In contrast, the real accelerator has a significant drop in tension force. Actually, the real accelerator does not start to contract before the hammer speed has reached a level at which the pipe force equals the unloading curve of the accelerator. Finally, the case with a defective accelerator is essentially the same as using no accelerator, except that the accelerator mass is added to the effective jarring hammer mass. Notice that the lower speed curve approaches the contraction speed of 3.0 m/s exponentially, as predicted by the simple theory above.

Impact Phase. The impact phase starts when the jar hammer hits the jar anvil. The dynamics for this phase are quite different from the acceleration phase because the jarring hammer can no longer be treated as a lumped mass. To simplify matters, we shall here make the following assumptions:

- The jarring hammer (jar and collar section above the jar) is uniform, with no cross section changes causing reflections.
- The fish (collar section between the jar and the stuck point) is also uniform.
- The fish is at rest at the start of the impact phase.
- The stuck point is absolutely fixed.

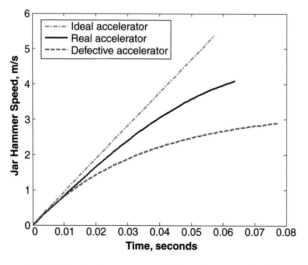

Fig. 3.3.3—Jar hammer speed for various accelerators.

When the jar hammer hits the anvil, two tension waves are generated; one propagates downward toward the stuck point, and the other upward. It can be seen that the magnitude of these tension waves is the same and equal to

$$F_1 = \frac{m_j m_f}{m_j + m_f} c \cdot v_j. \qquad (3.3.6)$$

Here, m_j and m_f are masses per unit length of jarring hammer and fish, respectively, and v_j is the speed of the jarring mass just before impact, hereafter called the impact speed.

When the tension wave propagating downward hits the stuck point, it is reflected with the same amplitude and sign as the primary incident wave. Neglecting the relatively small preimpact overpull from the jar friction F_j, the net tension force acting on the stuck point becomes

$$2F_1 = \frac{2m_j m_f}{m_j + m_f} c \cdot v_j. \qquad (3.3.7)$$

When the wave propagating upward reaches the upper end of the jarring hammer section, it is reflected. An accelerator acts as a nearly free end of the jar hammer section, so that the tension wave is reflected back with a sign shift. With no accelerator, the reflection is not total because the reflection coefficient depends on the weight ratio of the two sections. In both cases, the net tension force (sum of primary and reflected waves) is very much reduced, and a compression wave is transmitted into the fish section toward the stuck point. The magnitude of the reflected and transmitted wave is

$$F_2 \approx -\frac{2m_j}{m_j + m_f} F_1 \qquad (3.3.8a)$$

for the case with an accelerator, and

$$F_2 = -\frac{m_j - m_p}{m_j + m_p} \cdot \frac{2m_j}{m_j + m_f} F_1 \qquad (3.3.8b)$$

with no accelerator.

This compression wave is now propagating downward and, to a large extent, it neutralizes the primary tension wave. The time difference between the primary tension wave and this reflected compression wave equals the acoustic round-trip time in the jarring hammer. The duration time of the primary impact force, therefore, is

$$\tau = \frac{2l_j}{c}. \qquad (3.3.9)$$

With the above simplifications, the primary impact force is a square pulse with an amplitude of $2F_1$ and a duration of τ. In real life, due to nonperfect string geometry, wellbore friction, and a nonideal stuck point, the impact force at the stuck point tends to be broader and lower in magnitude.

The remaining impact phase is characterized by a complex mixture of waves traveling up and down in the bottomhole assembly, with partial reflection and energy leakage in both ends. These reflections cause the tension waves to dampen out, and the secondary impacts will never reach the level of the primary impact. After approximately 1–2 seconds, the entire string has a static overpull, which is reduced slightly by the jar extension and a possible motion of the fish. If the block is moving and stretches the string also when the jar fires, the post-jarring overpull can be higher than the prejarring overpull.

3.3.3 Resetting and Loading Phases. To reset a jar, the overpull must be turned into compression by lowering the string and reduce the hook load below the slackoff weight of the free string. If the hook load is monitored as a function of block height, then jar closing and resetting of the jar is indicated as a flat part of the curve at which the hook load is nearly constant. When the jar is reset, the string and accelerator can be stretched again to store new energy for a new jarring cycle.

3.3.4 Jarring Geometry Optimization. The above analysis of jarring dynamics has shown that the jarring hammer speed depends on many factors, such as applied overpull (prejarring force), jar friction, jarring hammer mass, and accelerator stiffness and performance, and also that the theoretical impact force is proportional to the speed and mass per unit length of the jarring hammer. These facts suggest that a short but heavy collar section is better than a longer section. This may not always be the case, mainly because real sticking forces are not concentrated in one point but are distributed over some length. In differential sticking, the sticking force can be distributed over tens of meters. In this case, mass momentum (hammer mass times hammer speed) and impact duration can be more important than the impact force peak amplitude. A long and heavy hammer section producing a moderate speed but a high momentum $(M_j \cdot v_j)$ and a long action time $(2l_j/c)$ could therefore be more capable of freeing a differentially stuck string than a shorter and harder-hitting hammer section. For this reason, the jarring hammer should not be shorter than two to three collar lengths.

Nomenclature

c = sonic speed, m/s
F_a = accelerator force, N
F_j = jar friction force, N
F_0 = prejarring force, N
F_1 = primary impact force, N
F_2 = magnitude of reflected impact force, N
l_j = length of jar hammer section, m
m_j = specific mass of jar hammer section, kg/m
l_j = length of jar hammer section, m
m_f = specific mass of fish section (below jar), kg/m
m_p = specific mass of drill pipes, kg/m
M_a = accelerator mass, kg
M_j = jar hammer mass, kg
s_j = free stroke of jar, m
t = time variable, s
t_1 = time of impact, s
v = jarring mass speed, m/s
v_a = accelerator speed, m/s
v_0 = contraction speed, m/s
v_1 = impact speed, m/s
v_j = jar hammer speed, m/s
τ = jar speed time constant, s

General Reference

Diagram of Jar. Schlumberger, www.glossary.oilfield.slb.com/DisplayImage.cfm?ID= 331. Downloaded 29 December 2008.

3.4 Loads, Friction, and Buckling—Robert F. Mitchell, Halliburton

3.4.1 Rigorous Torque and Drag Analysis: Introduction. While there have been many papers written on drillstring analysis, papers based on the fundamental analysis of torque and drag models have been relatively sparse. The original paper on this concept was presented by Johancsik et al. (1973). The formulation published by Sheppard et al. (1987) represents the standard way that the industry does torque and drag modeling today.

The basis for torque and drag analysis consists of three principal features:

1. The drillstring trajectory is specified, usually as constant-curvature segments.
2. Bending moments are neglected.
3. The drillstring force is assumed to be tangent to the trajectory.

Because the bending moments and shear forces are not present in this formulation, these models are often called *soft-string* models. These three features represent a significant simplification of the drillstring-equilibrium problem. Only a single force and a single torque need to be determined to solve the equilibrium equations. While simplified, this model has been found to be very useful; for example, see Lesso et al. (1989).

In the full drillstring-equilibrium problem, the drillstring forces, moments, and displacements are all unknown. The drillstring is constrained to lie within the wellbore, but the locations of contact between the drillstring and the wellbore are also unknown. Further, these calculations must be formulated as *large displacement*. By this we mean that 3D geometry plays a large role in the formulation of the drillstring-equilibrium problem. However, it does not mean that large strains are part of the formulation. Drillstring components are typically steel, which will not endure large strains without failure.

Modern formulation of large-displacement elastic-drillstring problems usually start with the chapter on the *elastica* in the classic textbook by Love (1944). One of the cleanest presentations of this material is the paper by Nordgren (1974). One of the more well-known papers on drillstring-model formulation is by Walker and Friedman (1977). A comprehensive paper on formulation, unfortunately often overlooked, is the work of Ho (1986). Ho followed this paper with a simplification for soft-string models (Ho 1988), using a curvilinear coordinate system based on the Serret-Frenet equations for a curve in space (Zwillinger 1996).

In this study, we will apply the large-displacement equilibrium equations, as presented by Nordgren (1974), to the curvilinear coordinate system used by Ho (1988), specialized to constant-plane curvature. We will first derive the equilibrium equations for a soft-string model, demonstrate its failure to solve all equilibrium equations, and show analytic solutions for curvature in a vertical plane. Finally, we show that including shear forces allows the solution of all equilibrium equations, again with analytic solutions in a vertical plane.

The Curvilinear Coordinate System. If the position of the drillstring is given as $\bar{u}(s)$, where s is the arc length of the curve (i.e., measured depth), then the unit tangent $\bar{t}(s)$ to the curve $\bar{u}(s)$ is given by

$$\bar{t}(s) = \frac{d\bar{u}(s)}{ds} = \bar{u}'(s), \quad \dots\dots\dots\dots\dots\dots\dots\dots\dots\dots\dots\dots\dots \quad (3.4.1)$$

where we have used ' to indicate derivative with respect to s. The derivative of the tangent vector is

$$\bar{t}'(s) = \kappa(s)\bar{n}(s), \quad \dots\dots\dots\dots\dots\dots\dots\dots\dots\dots\dots\dots\dots \quad (3.4.2)$$

where $\kappa(s)$ is the curvature, and $\bar{n}(s)$ is the unit normal to the curve. The third coordinate is called the binormal vector $\bar{b}(s)$, defined by

$$\bar{b}(s) = \bar{t}(s) \times \bar{n}(s), \quad \dots\dots\dots\dots\dots\dots\dots\dots\dots\dots\dots\dots \quad (3.4.3)$$

where $\times$ is the vector cross product. The triad $[\bar{t}(s), \bar{n}(s), \bar{b}(s)]$ forms a moving coordinate system along the drillstring trajectory. The last two derivatives we need to complete the definition of the coordinate system are

$$\bar{n}'(s) = -\kappa(s)\bar{t}(s) + \tau(s)\bar{b}(s)$$
$$\bar{b}'(s) = -\tau(s)\bar{n}(s), \quad \dots\dots\dots\dots\dots\dots\dots\dots\dots\dots\dots\dots\dots\dots \quad (3.4.4)$$

where $\tau(s)$ is the torsion of the curve. The torsion of the curve is not to be mistaken for the mechanical torsion of the drillstring. Instead, torsion is a measure of the helical nature of the curve. For instance, a constant-pitch helix has constant torsion, while a plane curve has zero torsion. These equations are called the Serret-Frenet equations (Zwillinger 1996).

Constant-curvature planar arcs are commonly used to define wellbore trajectories by means of the *minimum-curvature* method (Sawaryn and Thorogood 2005). This method simplifies the coordinate equations because $\bar{b}(s)$ and $\kappa(s)$ are constant and $\tau(s)$ is zero. For each segment i,

$$\bar{u}(s) = R\bar{t}_i \sin[\kappa_i(s-s_i)] + R\bar{n}_i\{1 - \cos[\kappa_i(s-s_i)]\} + \bar{u}_i$$
$$\bar{t}(s) = \bar{t}_i \cos[\kappa_i(s-s_i)] + \bar{n}_i \sin[\kappa_i(s-s_i)]$$
$$\bar{n}(s) = -\bar{t}_i \sin[\kappa_i(s-s_i)] + \bar{n}_i \cos[\kappa_i(s-s_i)] \quad \dots\dots\dots\dots\dots\dots\dots\dots\dots \quad (3.4.5)$$
$$\bar{b}(s) = \bar{t}_i \times \bar{n}_i$$

Here, $\bar{t}_i$ is the tangent vector at the beginning of the segment, $\bar{n}_i$ is the normal vector at the beginning of the segment, R is the radius of curvature of the segment, $\kappa_i = 1/R$ is the curvature of the segment, and s_i is the measured depth at the beginning of the segment. A potential defect in this model is that $\bar{n}(s)$, $\bar{b}$ and κ may be discontinuous where two segments are joined.

The Force and Moment Balance in a Drillstring. The change in pipe force $\bar{F}$, resulting from applied-load vector $\bar{w}$, is given by the following equation:

$$\frac{d\bar{F}}{ds} + \bar{w} = \bar{0}, \quad \dots\dots\dots\dots\dots\dots\dots\dots\dots\dots\dots\dots\dots\dots\dots\dots\dots \quad (3.4.6)$$

where $\bar{w}$ is force per unit length of the drillstring. The change in moment $\bar{M}$ resulting from applied-moment vector $\bar{m}$ and pipe force $\bar{F}$ is given by the following equation:

$$\frac{d\bar{M}}{ds} + \bar{t} \times \bar{F} + \bar{m} = \bar{0}. \quad \dots\dots\dots\dots\dots\dots\dots\dots\dots\dots\dots\dots\dots \quad (3.4.7)$$

The Loads on a Drillstring. What composes the load vector $\bar{w}$? The load vector $\bar{w}$ is

$$\bar{w} = \bar{w}_{bp} + \bar{w}_{st} + \bar{w}_c + \bar{w}_d + \Delta\bar{w}_{ef}, \quad \dots\dots\dots\dots\dots\dots\dots\dots\dots\dots \quad (3.4.8)$$

where the various terms will be described in the following text. We define the buoyant weight $\bar{w}_{bp}$ of the pipe as

$$\bar{w}_{bp} = [w_p + (\rho_i A_i - \rho_o A_o)g]\bar{i}_z, \quad \dots\dots\dots\dots\dots\dots\dots\dots\dots\dots\dots \quad (3.4.9)$$

and the pressure-area terms, known as the stream-thrust terms, F_{st}, group together:

$$\bar{F}_{st} = [(p_o + \rho_o v_o^2)A_o - (p_i + \rho_i v_i^2)A_i]\bar{t}$$
$$\bar{w}_{st} = \frac{d\bar{F}_{st}}{ds} \quad \dots\dots\dots\dots\dots\dots\dots\dots\dots\dots\dots\dots\dots\dots\dots\dots \quad (3.4.10)$$

The term $\Delta\bar{w}_{ef}$ is caused by complex flow patterns in the annulus. For many cases of interest, this term is zero, particularly for static fluid and for narrow annuli without pipe rotation. Because of the advanced nature of the computation of this term, this term will be neglected for the remaining discussion. The remaining terms are the mechanical-force terms. If the drillstring contacts the wellbore, there is a contact force w_c perpendicular to the wellbore, as shown in **Fig. 3.4.1.**

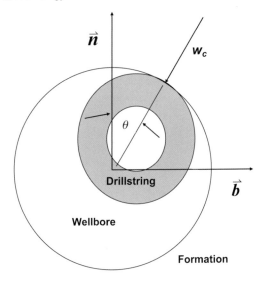

Fig. 3.4.1—Contact-force angle.

Note that w_c lies in the $\bar{n}$-$\bar{b}$ plane at angle θ with respect to the $\bar{n}$ vector. There is no contact force in the tangent-vector direction, because the contact force is perpendicular to the wellbore:

$$\bar{w}_c = w_c(\cos\theta\,\bar{n} + \sin\theta\,\bar{b}). \quad\quad\quad (3.4.11)$$

A second force, friction, is associated with the contact force. The Coulomb friction model is particularly simple in concept. If two surfaces in contact with a normal force F_N are sliding relative to each other, the friction force points in the direction opposite to the motion and has the magnitude of the product of the contact force and the dynamic coefficient for friction μ_f. The Coulomb friction relationship is shown in **Fig. 3.4.2.**

If the drillstring is sliding, there will be a friction-drag force $\bar{w}_d$ tangent to the wellbore and pointing in the direction opposite the sliding. The friction force may act in either direction, depending on whether the pipe is being run into or out of the hole.

The equation for sliding friction is

$$\bar{w}_d = \pm\mu_f w_c \bar{t}, \quad\quad\quad (3.4.12)$$

where the choice of + or − depends on the direction of sliding. If the string is sliding into the hole, then the negative sign holds. If the string is pulling out of the hole, the positive sign is used. The applied couple per unit length associated with this drag force is given by

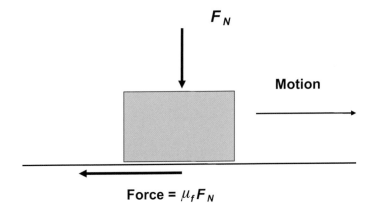

Fig. 3.4.2—Coulomb friction force.

$$\vec{m} = \pm \mu_f w_c \vec{t} \times r_o(-\cos\theta\,\vec{n} - \sin\theta\,\vec{b})$$
$$= \pm \mu_f r_o w_c (\sin\theta\,\vec{n} - \cos\theta\,\vec{b}). \quad\dots\dots\dots\dots\dots\dots\dots\dots\dots\dots\dots \quad (3.4.13)$$

When the drillstring is rotated, the friction force is no longer oriented axially, but is now applied opposite of the direction of rotation, in the $\vec{n}$-$\vec{b}$ plane, as shown in **Fig. 3.4.3.**

The equation for clockwise pipe rotation is

$$\vec{w}_d = \mu_f w_c (\sin\theta\,\vec{n} - \cos\theta\,\vec{b}). \quad\dots\dots\dots\dots\dots\dots\dots\dots\dots\dots\dots \quad (3.4.14)$$

If m_t is the applied torque per unit length in the tangential direction, then the applied torque can be inferred from Fig. 3.4.3 (in the orientation of this drawing, the tangent vector $\vec{t}$ points into the paper):

$$m_t = -\mu_f w_c r_o \quad\dots\dots\dots\dots\dots\dots\dots\dots\dots\dots\dots\dots\dots\dots\dots \quad (3.4.15)$$

The Mechanical Response of a Drillstring. Next, we will model the drillstring as an elastic solid material. Because a solid material can develop shear stresses, we formulate $\vec{F}$ in the following way:

$$\vec{F} = F_a\vec{t} + F_n\vec{n} + F_b\vec{b}, \quad\dots\dots\dots\dots\dots\dots\dots\dots\dots\dots\dots\dots \quad (3.4.16)$$

where F_a is the axial force, F_n is the shear force in the normal direction, and F_b is the shear force in the binormal direction. If we consider Eq. 3.4.10 with the equilibrium Eq. 3.4.6, we can group the stream-thrust terms with the axial force to define the effective force F_e:

$$F_e = F_a + F_{st}$$
$$= F_a + (p_o + \rho_o v_o^2)A_o - (p_i + \rho_i v_i^2)A_i. \quad\dots\dots\dots\dots\dots\dots\dots\dots \quad (3.4.17)$$

The casing moments for a circular pipe are given by

$$\vec{M} = EI\kappa\,\vec{b} + M_t\vec{t}, \quad\dots\dots\dots\dots\dots\dots\dots\dots\dots\dots\dots\dots\dots\dots \quad (3.4.18)$$

where EI is the bending stiffness and M_t is the axial torque. The first term in Eq. 3.4.18 may seem peculiar. The situation is illustrated in **Fig. 3.4.4.**

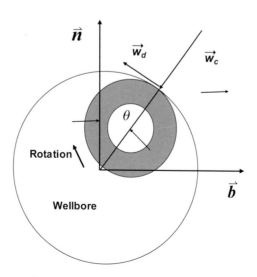

Fig. 3.4.3—Friction force due to rotation.

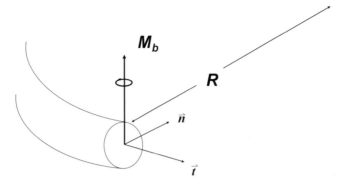

Fig. 3.4.4—Bending moment perpendicular to *t-n* plane.

For a round pipe, the bending moment points in the direction perpendicular to the plane and is proportional to the curvature of the pipe. In the figure we can see that the pipe curves in the $\vec{t}$-$\vec{n}$ plane, so that the steel on the inside of the curve is compressed, while the steel on the outside of the curve is extended. The forces associated with these displacements generate a moment that is proportional to the curvature ($1/R$) of the pipe and points perpendicular to the plane.

The moment equilibrium equation. Eq. 3.4.7 gives, using Eqs. 3.4.16 through 3.4.18,

$$EI\left(\frac{d\kappa}{ds}\vec{b} - \kappa\tau\,\vec{n}\right) + \frac{dM_t}{ds}\vec{t} + (M_t\kappa - F_b)\vec{n} + F_n\vec{b} + \vec{m} = 0, \quad \dots \dots (3.4.19)$$

where we have used the Serret-Frenet equations (Zwillinger 1996), Eqs. 3.4.2 and 3.4.4.

For a constant-curvature wellbore, the first term in Eq. 3.4.19 is constant, which means that the bending-moment term vanishes:

$$\frac{dM_t}{ds}\vec{t} + (M_t\kappa - F_b)\vec{n} + F_n\vec{b} + \vec{m} = 0. \quad \dots \dots (3.4.20)$$

For a constant-curvature trajectory, the bending moment defined in Eq. 3.4.18 may be discontinuous at survey locations. Further, the bending moment does not even appear in the equilibrium Eq. 3.4.20. These considerations suggest that the minimum-curvature trajectory is not an exact representation of a real drillstring configuration.

General Equilibrium Equations for Arbitrary Trajectory. If we substitute load equations into the equilibrium equations, assuming a general-curvature trajectory, we get

$$\frac{d}{ds}F_e - \kappa F_n + w_{bp}\left(\vec{t}\cdot\vec{i}_z\right) + \vec{w}_d\cdot\vec{t} = 0$$

$$\frac{d}{ds}F_n + F_e\kappa - F_b\tau + w_{bp}\left(\vec{n}\cdot\vec{i}_z\right) - w_c\cos\theta + \vec{w}_d\cdot\vec{n} = 0$$

$$\frac{d}{ds}F_b + F_n\tau + w_{bp}\left(\vec{b}\cdot\vec{i}_z\right) - w_c\sin\theta + \vec{w}_d\cdot\vec{b} = 0 \quad \dots \dots (3.4.21)$$

$$\frac{d}{ds}M_t + \vec{m}\cdot\vec{t} = 0$$

$$-ET\,\kappa\tau + M_t\kappa - F_b + \vec{m}\cdot\vec{n} = 0$$

$$EI\frac{d\kappa}{ds} + F_n + \vec{m}\cdot\vec{b} = 0$$

Note that τ and $d\kappa/ds$ vanish for a minimum-curvature trajectory.

The Torque-and-Drag Drillstring Model. A form of drillstring analysis called *torque-and-drag* analysis makes special assumptions that simplify the analysis and have been found useful for modeling real drillstrings. These assumptions are that the drillstring lies along the wellbore trajectory, and that the bending moment and shear forces may be neglected. These assumptions treat the drillstring like a cable or a chain, with no bending stiffness, and this model is sometimes called a *soft-string* model. The full analysis then is called a *stiff-string* analysis.

If we apply these assumptions to Eq. 3.4.21, we get

$$\frac{d}{ds}F_e + w_{bp}\left(\vec{t} \cdot \vec{i}_z\right) + \vec{w}_d \cdot \vec{t} = 0$$

$$F_e\kappa + w_{bp}\left(\vec{n} \cdot \vec{i}_z\right) - w_c\cos\theta + \vec{w}_d \cdot \vec{n} = 0$$

$$w_{bp}\left(\vec{b} \cdot \vec{i}_z\right) - w_c\sin\theta + \vec{w}_d \cdot \vec{b} = 0 \quad \dots\dots\dots\dots\dots\dots\dots\dots \quad (3.4.22)$$

$$\frac{d}{ds}M_t + \vec{m} \cdot \vec{t} = 0$$

$$M_t\kappa + \vec{m} \cdot \vec{n} = 0$$

$$\vec{m} \cdot \vec{b} = 0.$$

We immediately can see that this model is approximate, because only four of the six equilibrium conditions can be satisfied in general. First, Eq. 3.4.22 (line 4) and Eq. 3.4.22 (line 5) may not be mutually consistent with the specified applied-moment vector $\vec{m}$, and Eq. 3.4.22 (line 6) may not be satisfied by $\vec{m}$ either. In practice, Eqs. 3.4.22 (line 1) through 3.4.22 (line 4) are solved, and the remaining equilibrium equations are ignored. Note that there is the possibility of developing other *flavors* of torque-and-drag analysis by using other subsets of the equilibrium equations.

Drag Calculations. For the drag calculations, we will assume that the pipe is being run into the hole. If we assume linear Coulomb friction to model the drag force on the pipe, then w_d is given by Eq. 3.4.12. For this assumption, the drag force will be pointing in the negative s direction. To modify these equations for pulling out of the hole, we need only change the sign of the friction coefficient μ_f. The equations modeling drag calculations are the following:

$$\frac{dF_e}{ds} + w_{bp}t_z + w_d = 0$$

$$F_e\kappa + w_{bp}n_z - w_c\cos\theta = 0$$

$$w_{bp}b_z - w_c\sin\theta = 0 \quad \dots\dots\dots\dots\dots\dots\dots\dots\dots\dots\dots \quad (3.4.23)$$

$$M_t = \text{constant}$$

$$M_t\kappa + \mu_f w_c r_0\cos\theta = 0$$

$$-\mu_f w_c r_0\sin\theta = 0.$$

Given Eq. 3.4.5, the parameters t_z, n_z, and b_z can now be evaluated:

$$t_z(s) = \vec{t}(s) \cdot \vec{i}_z$$

$$n_z(s) = \vec{n}(s) \cdot \vec{i}_z \quad \dots\dots\dots\dots\dots\dots\dots\dots\dots\dots\dots\dots\dots\dots \quad (3.4.24)$$

$$b_z = (\vec{t}_i \times \vec{n}_i) \cdot \vec{i}_z.$$

We can now solve Eq. 3.4.23 (line 2) and Eq. 3.4.23 (line 3) to get the contact force w_c and angle θ:

$$w_c = \sqrt{(F_e\kappa_i + w_{bp}n_z)^2 + w_{bp}^2 b_z^2}$$

$$\theta = \tan^{-1}\left(\frac{w_{bp}b_z}{F_e\kappa_i + w_{bp}n_z}\right) \quad \dots\dots\dots\dots\dots\dots\dots\dots\dots\dots \quad (3.4.25)$$

The presence of the effective force F_e in the contact-force equation is due to the curvature of the wellbore. The product of the axial force and the curvature is known as the *capstan* force because when rope was wrapped around a ship's capstan, the tension in the rope created contact force, and the resulting friction between the rope and the capstan was used to lift loads.

The resulting differential equation for the force F_e becomes

$$\frac{dF_e}{ds} + w_{bp}t_z - \mu_f \sqrt{\left(F_e\kappa_i + w_{bp}n_z\right)^2 + w_{bp}^2 b_z^2} = 0 \qquad \text{(3.4.26)}$$

Eq. 3.4.26 is a first-order differential equation that can be solved only in general by numerical methods. There is, however, one specific assumption that does allow analytic solutions: that the circular arc lies in a vertical plane. This assumption makes b_z equal to zero, resulting in the following:

If

$$F_e\kappa_i + w_{bp}n_z > 0$$

$$\frac{dF_e}{ds} + w_{bp}(t_z - \mu_f n_z) - \mu_f\kappa_i F_e = 0, \qquad \text{(3.4.27)}$$

else

$$\frac{dF_e}{ds} + w_{bp}(t_z + \mu_f n_z) + \mu_f\kappa_i F_e = 0.$$

The solution to Eq. 3.4.27 (line 1) is

$$F_e(s) = \left[F_e^i - W(s_i)\right]\exp\left[\mu_f\kappa_i(s - s_i)\right] + W(s)$$

$$W(s) = \frac{w_{bp}}{\kappa_i(1 + \mu_f^2)}\left[(1 - \mu_f^2)n_z + 2\mu_f t_z\right], \qquad \text{(3.4.28)}$$

where F_e^i is the value of F_e at $s = s_i$. The solution to Eq. 3.4.27 (line 2) is obtained with the appropriate sign changes to Eq. 3.4.28.

Torque Calculations. If we adopt the friction equation for pipe rotation, the equilibrium equations now have the form

$$\frac{dF_e}{ds} + w_{bp}t_z = 0$$

$$F_e\kappa_i + w_{bp}n_z - w_c\cos\theta - \mu_f w_c\sin\theta = 0$$

$$w_{bp}b_z - w_c\sin\theta + \mu_f w_c\cos\theta = 0 \qquad \text{(3.4.29)}$$

$$\frac{d}{ds}M_t - \mu_f w_c r_0 = 0$$

$$0 = 0$$

$$0 = \kappa M_t.$$

The first equilibrium equation is easily solved by integration:

$$F_e = F_e^i - w_{bp}\left[\vec{u}(s) - \vec{u}(s_i)\right]\cdot\vec{i}_z. \qquad \text{(3.4.30)}$$

Using Eqs. 3.4.29 (line 2) and 3.4.29 (line 3), we can solve for the contact angle θ and then evaluate the contact force:

$$\theta = \tan^{-1}\left(\frac{w_{bp}b_z}{F_e\kappa_i + w_{bp}n_z}\right) + \tan^{-1}\mu_f$$

$$\qquad \text{(3.4.31)}$$

$$w_c = \sqrt{\frac{\left(F_e\kappa_i + w_{bp}n_z\right)^2 + w_{bp}^2 b_z^2}{1 + \mu_f^2}}$$

Eq. 3.4.29 (line 4) must be solved using numerical methods, because of the complex form of Eq. 3.4.31 (line 2). As in the case of the drag calculation, the assumption that the circular arc lies in a vertical plane allows analytical solutions. This assumption makes b_z equal to zeiro, resulting in

$$\theta = \tan^{-1}\mu_f$$

$$w_c = \frac{|F_e\kappa_i + w_{pb}n_z|}{\sqrt{1+\mu_f^2}} \quad \dots\dots\dots\dots\dots\dots\dots\dots\dots\dots\dots\dots\dots\dots\dots\dots\dots \quad (3.4.32)$$

$$\frac{dM_t}{ds} = \frac{\mu_f r_o |F_e\kappa_i + w_{pb}n_z|}{\sqrt{1+\mu_f^2}}.$$

The solution to Eq. 3.4.32 (line 3), taking the case $F_e\kappa_i + w_{bp}n_z > 0$, is

$$\left(M_t - M_t^i\right)\frac{\sqrt{1+\mu_f^2}}{\mu_f r_o} = \int_{s_i}^{s} F_e\kappa_i\,ds + w_{bp}\int_{s_i}^{s}\bar{n}(s)ds\cdot\bar{i}_z, \quad \dots\dots\dots\dots\dots\dots\dots\dots\dots \quad (3.4.33)$$

where

$$\int_{s_i}^{s} F_e\kappa_i\,ds = F_e^i\kappa_i\left(s - s_i\right) - w_{bp}\left\{\bar{t}_i\sin\kappa_i\left(s - s_i\right) + \bar{n}_i\left[1 - \cos\kappa_i(s - s_i)\right]\right\}\cdot\bar{i}_z$$

$$\int_{s_i}^{s}\bar{n}(s)ds = \frac{1}{\kappa_i}\left[\bar{t}(s) - \bar{t}_i\right]. \quad \dots\dots\dots\dots\dots\dots\dots\dots\dots\dots\dots\dots\dots\dots \quad (3.4.34)$$

Torque and Drag With Shear Forces. As we have seen, the single-force soft-string drillstring model is approximate because all of the moment equations cannot be satisfied. These problems can be solved by retaining the shear forces F_n and F_b.

If we substitute load Eqs. 3.4.8 and 3.4.14 into the equilibrium Eqs. 3.4.2 and 3.4.3, assuming a minimum-curvature trajectory, we get

$$\frac{d}{ds}F_e - \kappa F_n + w_{bp}\left(\bar{t}\cdot\bar{i}_z\right) + \bar{w}_d\cdot\bar{t} = 0$$

$$\frac{d}{ds}F_n + F_e\kappa + w_{bp}\left(\bar{n}\cdot\bar{i}_z\right) - w_c\cos\theta + \bar{w}_d\cdot\bar{n} = 0$$

$$\frac{d}{ds}F_b + w_{bp}\left(\bar{b}\cdot\bar{i}_z\right) - w_c\sin\theta + \bar{w}_d\cdot\bar{b} = 0 \quad \dots\dots\dots\dots\dots\dots\dots\dots\dots \quad (3.4.35)$$

$$\frac{d}{ds}M_t + m_t = 0$$

$$M_t\kappa - F_b + m_n = 0$$

$$F_n + m_b = 0.$$

Drag Calculations With Shear. For the drag calculations, we will assume that the pipe is being run into the hole. If we assume linear Coulomb friction to model the drag force on the pipe, then $\bar{w}_d$ is given by Eq. 3.4.12. For this assumption, the drag force will be pointing in the negative s direction. To modify these equations for pulling out of the hole, we need only change the sign of the friction coefficient μ_f. The equations modeling drag calculations are the following:

$$\frac{dF_e}{ds} - \kappa F_n + w_{bp}t_z - \mu_f w_c = 0$$

$$\frac{d}{ds}F_n + F_e\kappa + w_{bp}n_z - w_c\cos\theta = 0$$

$$\frac{d}{ds}F_b + w_{bp}b_z - w_c\sin\theta = 0 \quad\quad \dots\dots\dots\dots\dots\dots\dots\dots (3.4.36)$$

$$M_t = 0$$

$$-F_b + \mu_f r_o w_c \sin\theta = 0$$

$$F_n - \mu_f r_o w_c \cos\theta = 0.$$

Given Eq. 3.4.5, the parameters t_z, n_z, and b_z now can be evaluated:

$$t_z = \vec{t}(s)\cdot\vec{i}_z$$

$$n_z = \vec{n}(s)\cdot\vec{i}_z \quad\quad \dots\dots\dots\dots\dots\dots\dots\dots\dots\dots\dots\dots (3.4.37)$$

$$b_z = \vec{b}(s)\cdot\vec{i}_z$$

and we can solve Eq. 3.4.36 (line 5) and Eq. 3.4.36 (line 6) to get the shear forces F_n and F_b, the contact force w_c, and angle θ:

$$F_n = \mu_f r_o w_c \cos\theta$$

$$F_b = \mu_f r_o w_c \sin\theta$$

$$\mu_f r_o w_c = \sqrt{F_n^2 + F_b^2} \quad\quad \dots\dots\dots\dots\dots\dots\dots\dots\dots\dots\dots (3.4.38)$$

$$\theta = \tan^{-1}\left(\frac{F_b}{F_n}\right).$$

The equilibrium equations now become

$$\frac{d}{ds}F_e - \kappa F_n + w_{bp}t_z - \frac{1}{r_o}\sqrt{F_n^2 + F_b^2} = 0$$

$$\frac{d}{ds}F_n + F_e\kappa + w_{bp}n_z - \frac{F_n}{r_o\mu_f} = 0 \quad\quad \dots\dots\dots\dots\dots\dots\dots\dots\dots (3.4.39)$$

$$\frac{d}{ds}F_b + w_{bp}b_z - \frac{F_b}{r_o\mu_f} = 0$$

$$M_t = 0.$$

Because b_z is a constant for the minimum-curvature trajectory, Eq. 3.4.39 (line 3) has the analytic solution

$$F_b = F_b^0 \exp(\frac{s}{r_o\mu_f}) + r_o\mu_f w_{bp}b_z \quad\quad \dots\dots\dots\dots\dots\dots\dots\dots (3.4.40)$$

Because $r_o\mu_f$ is typically small relative to measured depth s, the exponential term in Eq. 3.4.40 becomes either very large (running in) or very small (pulling out changes the sign of μ_f). Therefore, the most reasonable choice for F_b^0 is zero. This will result in a discontinuity in F_b at survey points. We have already observed that the bending moment also may be discontinuous at survey points, so this is another deficiency of the minimum-curvature trajectory.

Eq. 3.4.39 is a system of first-order differential equations that only can be solved in general by numerical methods. One specific assumption (that the circular arc lies in a vertical plane) does allow analytic solutions.

This assumption makes b_z terms equal to zero. If we take the suggested special case $F_b^0 = 0$, then the remaining differential equations become

If

$F_n > 0,$

then

$$\frac{dF_e}{ds} - \left(\kappa + \frac{1}{r_o}\right)F_n + w_{bp}t_z = 0,$$

else .... (3.4.41)

$$\frac{dF_e}{ds} - \left(\kappa - \frac{1}{r_o}\right)F_n + w_{bp}t_z = 0$$

and

$$\frac{d}{ds}F_n + F_e\kappa + w_{bp}n_z - \frac{F_n}{r_o\mu_f} = 0.$$

The solution to Eq. 3.4.41 (lines 1 and 3) is

$$F_e(s) = \frac{\alpha_1 F_m(s,\alpha_1) - \alpha_2 F_m(s,\alpha_2)}{\alpha_1 - \alpha_2}$$

$$F_n(s) = \frac{\alpha_1\alpha_2}{\alpha_1 - \alpha_2}\left[F_m(s,\alpha_1) - F_m(s,\alpha_2)\right]$$

$$F_m(s,\alpha_1) = \left[F_m^i(\alpha_1) - W(s_i,\alpha_1)\right]\exp\left[-\kappa(s-s_i)/\alpha_1\right] + W(s,\alpha_1)$$

$$F_m(s,\alpha_2) = \left[F_m^i(\alpha_2) - W(s_i,\alpha_2)\right]\exp\left[-\kappa(s-s_i)/\alpha_2\right] + W(s,\alpha_2) \quad\quad\quad\text{...................} \quad (3.4.42)$$

$$W(s,\alpha_1) = \frac{-w_{bp}}{\kappa_i\left(1+\alpha_1^2\right)}\left[\left(1-\alpha_1^2\right)n_z + 2\alpha_1 t_z\right]$$

$$W(s,\alpha_2) = \frac{-w_{bp}}{\kappa_i\left(1+\alpha_2^2\right)}\left[\left(1-\alpha_2^2\right)n_z + 2\alpha_2 t_z\right]$$

$$\alpha_1 = \frac{-1}{2\mu_f\kappa_o r_o}\left(1+\sqrt{1-4\kappa\kappa_o r_o^2\mu^2}\right)$$

$$\alpha_2 = \frac{-1}{2\mu_f\kappa_o r_o}\left(1-\sqrt{1-4\kappa\kappa_o r_o^2\mu^2}\right)$$

$$\kappa_o = \kappa + 1/r_o.$$

In real wellbores, $1/r_0 \gg \kappa$, so the exponents in Eq. 3.4.42 can be written, to good approximation, as

$$\alpha_1 \approx \frac{1}{\mu_f}, \quad \frac{\kappa}{\alpha_1} \approx \kappa\mu_f$$

.... (3.4.43)

$$\alpha_2 \approx \frac{r_o\mu_f\kappa}{4}, \quad \frac{\kappa}{\alpha_2} \approx \frac{4}{r_o\mu_f}$$

The exponential term using α_2 must be removed from this solution, for the same reason that we chose the initial condition on F_b to be zero. The remaining constant term is evaluated from initial conditions:

$$F_m^i(\alpha_1) = \left[F_e^i(\alpha_1 - \alpha_2) + \alpha_2 W(s_i,\alpha_2)\right]/\alpha_1$$

.... (3.4.44)

$$F_m^i(\alpha_2) = W(s_i,\alpha_2),$$

where F_e^i is the value of F_e at $s = s_i$, and F_n^i is the value of F_n at $s = s_i$.

The solution to Eq. 3.4.41 (lines 2 and 3) is the same as Eq. 3.4.41 (lines 1 and 3) except that $\kappa_o = \kappa - 1/r_o$ instead of $\kappa + 1/r_o$.

Torque Calculations With Shear. If we adopt the friction equation for pipe rotation, the equilibrium equations now have the form

$$\frac{dF_e}{ds} - \kappa F_n + w_{bp} t_z = 0$$

$$\frac{dF_n}{ds} + F_e \kappa + w_{bp} n_z - w_c \cos\theta - \mu_f w_c \sin\theta = 0$$

$$\frac{dF_b}{ds} + w_{bp} b_z - w_c \sin\theta + \mu_f w_c \cos\theta = 0 \qquad\qquad\qquad \cdots \cdots \cdots \cdots \cdots \cdots \cdots \quad (3.4.45)$$

$$F_n = 0$$

$$F_b = \kappa M_t$$

$$\frac{d}{ds} M_t - \mu w_c r_0 = 0.$$

The first equilibrium equation is easily solved by integration:

$$F_e = F_e^i + w_{bp}[u_z(s) - u_z(s_i)]. \qquad \cdots \cdots \cdots \cdots \cdots \cdots \cdots \cdots \cdots \cdots \quad (3.4.46)$$

Using Eq. 3.4.45 (lines 2, 3, 5, and 6), we can solve for the contact angle θ and then evaluate the contact force:

$$\theta = \tan^{-1}\left(\frac{w_{bp} b_z}{F_e \kappa_i + w_{bp} n_z}\right) + \tan^{-1} \mu_f$$

$$+ \sin^{-1}\left[\frac{\mu_f \varepsilon}{\sqrt{1+\mu_f^2}} \frac{F_e \kappa + w_{bp} n_z}{\sqrt{\left(F_e \kappa + w_{bp} n_z\right)^2 + \left(w_{bp} b_z\right)^2}}\right] \qquad \cdots \cdots \cdots \cdots \cdots \cdots \quad (3.4.47)$$

$$w_c = \frac{\mu_f \varepsilon}{1+\bar{\mu}_f^2} + \frac{\sqrt{\left(1+\bar{\mu}_f^2\right)\left(F_e \kappa + w_{bp} n_z\right)^2 + \left(1+\mu_f^2\right)\left(w_{bp} b_z\right)^2}}{1+\bar{\mu}_f^2}$$

$$\varepsilon = \kappa r_o, \quad \bar{\mu}_f = \mu_f \sqrt{(1-\varepsilon^2)} \approx \mu_f$$

Eq. 3.4.45 (line 6) must be solved using numerical methods because of the complex form of Eq. 3.4.47 (line 2). As in the case of the drag calculation, the assumption that the circular arc lies in a vertical plane allows analytical solutions. This assumption makes b_z terms equal to zero, resulting in

$$\theta = \tan^{-1}\mu_f - \sin^{-1}\left(\frac{\kappa \mu_f r_o}{\sqrt{1+\mu_f^2}}\right)$$

$$w_c = \left|\frac{F_e \kappa + w_{bp} n_z}{\cos\theta + \mu_f \sin\theta}\right| = \frac{\left|F_e \kappa + w_{bp} n_z\right|}{\sqrt{1+\mu_f^2(1-\kappa^2 r_o^2)}} \qquad\qquad \cdots \cdots \cdots \cdots \cdots \cdots \cdots \cdots \cdots \cdots \quad (3.4.48)$$

$$\cong \frac{\left|F_e \kappa + w_{bp} n_z\right|}{\sqrt{1+\mu_f^2}} \quad \text{since } \kappa r_o \ll 1$$

$$\frac{dM_t}{ds} = \frac{\mu_f r_o \left|F_e \kappa + w_{bp} n_z\right|}{\sqrt{1+\mu_f^2}}.$$

The solution to Eq. 3.4.48 (line 3), taking the case $F_e \kappa + w_{bp} n_z > 0$, is

$$\left(M_t - M_t^i\right)\frac{\sqrt{1+\mu_f^2}}{\mu_f r_o} = \int_{s_i}^{s} F_e \kappa\, ds + w_{bp}\int_{s_i}^{s} \bar{n}(s)\, ds \cdot \bar{i}_z, \qquad\dots\dots\dots\dots\dots\dots\dots\dots\dots\dots\dots (3.4.49)$$

where

$$\int_{s_i}^{s} F_e \kappa\, ds = F_e^i \kappa\left(s - s_i\right) - w_{bp}\left\{\bar{t}_i \sin\kappa\left(s - s_i\right) + \bar{n}_i\left[1 - \cos\kappa\left(s - s_i\right)\right]\right\}\cdot\bar{i}_z$$

$$\dots\dots\dots\dots\dots\dots\dots (3.4.50)$$

$$\int_{s_i}^{s} \bar{n}(s)\, ds = \frac{1}{\kappa}\left[\bar{t}(s) - \bar{t}_i\right].$$

Conclusions. The exact vector formulation of torque-and-drag analysis has revealed some defects in the conventional formulation and has provided solutions to these defects. In the soft-string analysis where shear forces are neglected, two of the moment-balance equations are not satisfied in general. These equations can provide a measure of the validity of these assumptions for particular cases. A second defect has been identified in the equations for rotating friction. The term $\sqrt{1+\mu_f^2}$ needs to be included in the contact-force calculation. Because this term is missing in the conventional formulation, contact forces will be overestimated, with subsequent error in the torque calculations. Finally, a new full-contact solution has been developed that does not neglect the shear forces in the drillstring. This formulation, which is not substantially more complex than the conventional model, satisfies all of the equilibrium equations.

3.4.2 Analysis of Tubing Buckling: Introduction.

A problem almost unique to the petroleum industry is the problem of tubing buckling. The meaning of *buckling* in the mechanical engineering community almost exclusively concerns the determination of elastic stability loads [i.e., the conditions governing the stability of structural elements, such as bars, plates, shells, and columns (Timoshenko and Gere 1961)].

In the petroleum industry, the buckling of tubing refers to the equilibrium configuration of tubing loaded above its *critical load*. In the mechanical engineering literature, this is often referred to as *post-buckling* equilibrium. A simple example illustrates the ideas of critical load and equilibrium.

Fig. 3.4.5 shows a simply supported beam, with length ℓ, axial load P, a midspan vertical load Q, and bending stiffness EI. The equilibrium displacement at the midpoint of the beam is given by

$$\delta = \frac{Q\ell^3}{16EI}\frac{\left(\tan u - u\right)}{u^3}$$

where $\qquad\dots\dots\dots\dots\dots\dots\dots\dots\dots\dots\dots\dots\dots\dots\dots\dots\dots\dots\dots\dots\dots\dots\dots\dots (3.4.51)$

$$u = \frac{\ell}{2}\sqrt{\frac{P}{EI}}$$

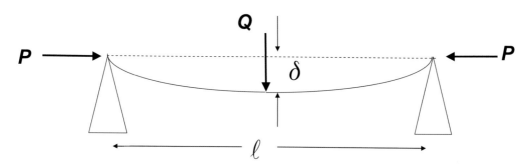

Fig. 3.4.5—Buckling analysis of a centrally loaded beam.

See Timoshenko and Gere (1961) for more details about the derivation of this solution. Note that Eq. 3.4.51 predicts infinite displacement for u equal to $\pi/2$. One interpretation of this fact is that the beam has no resistance to deflection for the axial load P_{crit}, where

$$P_{crit} = \frac{\pi^2 EI}{\ell^2}. \dots\dots\dots\dots\dots\dots\dots\dots\dots\dots\dots\dots\dots\dots\dots\dots\dots\dots\dots\dots\dots\dots\dots\dots\dots (3.4.52)$$

This value of P is called the critical load. The critical load is dependent on the boundary conditions of the problem. For instance, the critical load increases by a factor of 4 if the end conditions are cantilevered instead of simply supported. If we rearrange Eq. 3.4.51 to find the load required to give a specified displacement δ, then

$$Q = \delta \frac{16EI}{\ell^3} \frac{u^3}{\tan u - u} \dots\dots\dots\dots\dots\dots\dots\dots\dots\dots\dots\dots\dots\dots\dots\dots\dots\dots\dots\dots\dots\dots\dots\dots\dots (3.4.53)$$

and, at the critical load, the value of Q required to give displacement δ is zero.

Fig. 3.4.6 shows the deflection as a function of u. The dotted vertical line indicates $\pi/2$, and we see that the displacement becomes very large near this line.

Fig. 3.4.7 shows the value of Q required to produce a unit displacement. At u equal to $\pi/2$, we see that Q is zero.

The meaning of Figs. 3.4.6 and 3.4.7 is clear for P less than the critical load. Do these figures have any meaning for P greater than the critical load? Fig. 3.4.6 seems to indicate that we have negative displacement for a positive applied load, a nonsense result. But is this the right interpretation?

Fig. 3.4.8 shows Fig. 3.4.5 redrawn for P greater than critical load. In this figure, we can see that the force Q is now a constraining force, balancing the displacements caused by the axial force P. If we impose a fixed displacement, as in Fig. 3.4.7, then the force Q must change sign to become a constraining force, as we see in Fig. 3.4.7 for P greater than critical load.

Plane buckling, as illustrated in this example, almost never occurs in practice because the tendency to buckle out of the plane is too strong. In a wellbore, however, the pipe is constrained on all sides and can find an equilibrium configuration for supercritical loads. This problem was originally posed and solved by Lubinski et al. (1962). They assumed that the equilibrium configuration of a weightless rod constrained by a cylindrical wellbore was a constant-pitch helix. The differential equations describing this problem are (Mitchell 1988)

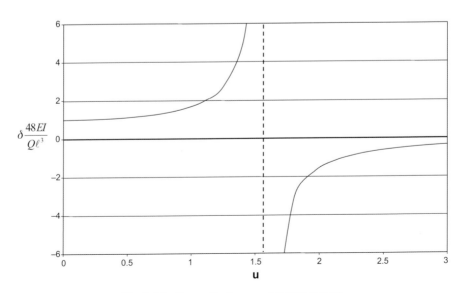

Fig. 3.4.6—Beam displacement for fixed load.

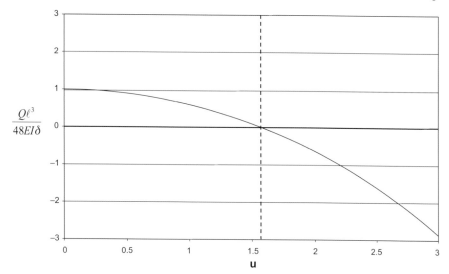

Fig. 3.4.7—Beam load for fixed displacement.

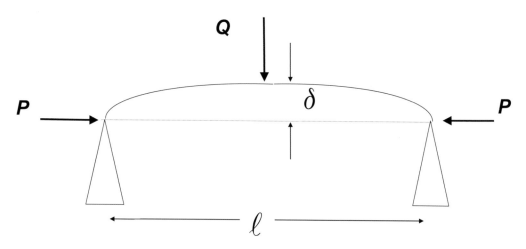

Fig. 3.4.8—Displacements for supercritical P.

$$EI \frac{d^4}{ds^4} u_1 + \frac{d}{ds}\left(P \frac{d}{ds} u_1 \right) + \frac{w_n}{r_c} \sin \theta = 0$$

$$EI \frac{d^4}{ds^4} u_2 + \frac{d}{ds}\left(P \frac{d}{ds} u_2 \right) + \frac{w_n}{r_c} \cos \theta = 0, \quad \dotfill \quad (3.4.54)$$

where u_1 is the displacement in the 1 coordinate direction, u_2 is the displacement in the 2 coordinate direction, w_n is the contact force, and r_c is the radial clearance between the tubing and the wellbore. The coordinate system is illustrated in **Fig. 3.4.9.**

Displacements for a constant-pitch helix are given by

$$u_1 = r_c \cos(\theta)$$

$$u_2 = r_c \sin(\theta)$$

$$\theta = \beta s, \quad \dotfill \quad (3.4.55)$$

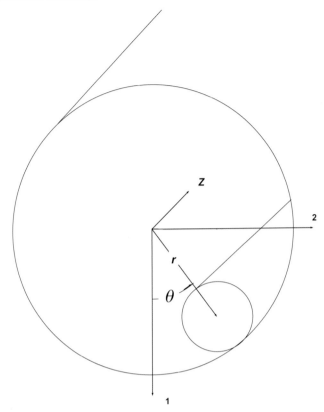

Fig. 3.4.9—Coordinate system for buckling analysis.

where β is a constant related to the pitch $\wp$ of the helix by

$$\wp = \frac{2\pi\sqrt{1-r_c^2\beta^2}}{\beta} \approx \frac{2\pi}{\beta}, \quad \dots\dots\dots\dots\dots\dots\dots\dots\dots\dots\dots\dots\dots\dots\dots\dots\dots\dots\dots\dots\dots \text{(3.4.56)}$$

where the approximate equation assumes that $\wp^2 \gg 4\pi^2 r_c^2$. Lubinski et al. (1962) formulated their solution in terms of pitch $\wp$, but we have chosen an alternative formulation for convenience. If we substitute Eq. 3.4.55 into Eq. 3.4.54, we find only the following result:

$$r_c EI\beta^2(\alpha^2 - \beta^2) = w_n \quad \dots\dots\dots\dots\dots\dots\dots\dots\dots\dots\dots\dots\dots\dots\dots\dots\dots\dots\dots\dots\dots\dots\dots \text{(3.4.57)}$$

where α is defined as (Timoshenko and Gere 1961)

$$\alpha = \sqrt{\frac{P}{EI}}. \quad \dots\dots\dots\dots\dots\dots\dots\dots\dots\dots\dots\dots\dots\dots\dots\dots\dots\dots\dots\dots\dots\dots\dots\dots\dots \text{(3.4.58)}$$

For a positive contact force, we require that $\beta^2 < \alpha^2$. Note that β can be either positive or negative in Eq. 3.4.57. This means that the helix can be either right-hand or left-hand; neither is preferred. What is interesting about Eq. 3.4.57 is that β is not determined by equilibrium, other than being constrained to be smaller than α in absolute value.

Lubinski et al. (1962) determined $\wp$ by using the method of virtual work. In terms of β, their result is

$$\beta_{LW} = \pm\sqrt{\frac{P}{2EI}}. \quad \dots\dots\dots\dots\dots\dots\dots\dots\dots\dots\dots\dots\dots\dots\dots\dots\dots\dots\dots\dots\dots\dots\dots \text{(3.4.59)}$$

The two solutions correspond to a right-hand helix and a left-hand helix, with pitch

$$\wp = \sqrt{\frac{8\pi^2 EI}{P}}, \quad \dots\dots\dots\dots\dots\dots\dots\dots\dots\dots\dots\dots\dots\dots\dots\dots\dots\dots \quad (3.4.60)$$

where we have again assumed that $\wp^2 \gg 4\pi^2 r_c^2$. We can now solve for the contact force, using Eqs. 3.4.57 and 3.4.59 (Mitchell 1986):

$$w_n = r_c EI(\alpha^2 - \beta_{LW}^2)\beta_{LW}^2 = \frac{r_c P^2}{4EI}. \quad \dots\dots\dots\dots\dots\dots\dots\dots\dots\dots\dots\dots \quad (3.4.61)$$

We can see that the contact force is positive, as we required.

Cheatham and Pattillo (1984) give a convincing development of the method of virtual work. Nevertheless, there is something mysterious about this method, which is supposed to be equivalent to the equilibrium equations, yet produces a specific value for β when the equilibrium equations do not.

Boundary Conditions. In the first example problem, we commented that the boundary conditions had a strong effect on the displacement of the beam. It would seem reasonable to expect that boundary conditions would have an equally strong impact on the pitch of a helically-buckled pipe.

A comprehensive analysis of tubing buckling consistent with boundary conditions at a packer or a centralizer has never been done in a completely satisfactory way. Many attempts have been made to connect this constant-pitch solution to a beam-column solution that brings the pipe from the wellbore wall to the packer. The following conditions must be met where the two solutions join:

1. Wellbore contact
2. Wellbore tangency
3. Continuity of curvature
4. Continuity of shear tangent to the wellbore
5. Positive contact force between pipe and wellbore
6. All pipe displacements within the borehole

Case 1: The Cantilever Packer. What are the appropriate boundary conditions for a helically buckled pipe? One common application is tubing sealed in packers. For this boundary condition, we propose that a beam-column solution to the buckling equations brings the pipe from a centralized position, tangent to the wellbore centerline, to a point tangent to the wellbore wall. This problem was originally posed in Mitchell (1982). The following equations satisfy these conditions (Mitchell 1982):

$$u_{1b} = r_c\{S[\alpha s - \sin(\alpha s)] + Y[\cos(\alpha s) - 1]\}/\delta$$
$$u_{2b} = r_c\varepsilon_o\{Y[\sin(\alpha s) - \alpha s] - X[\cos(\alpha s) - 1]\}/\delta$$
$$S = \sin(\alpha\Delta s_o)$$
$$Y = 1 - \cos(\alpha\Delta s_o) \quad \dots\dots\dots\dots\dots\dots\dots\dots\dots\dots\dots\dots\dots\dots \quad (3.4.62)$$
$$X = \alpha\Delta s_o - S$$
$$\delta = \alpha\Delta s_o S - 2Y$$

where ε and s_o are undetermined.

Fig. 3.4.10 shows the shape of the beam-column solution. The following boundary conditions must be satisfied where the beam-column solution (Eq. 3.4.62) connects to the constant-pitch solution:

$$u'_{2b}(s_o) = r_c\beta = r_c\alpha\varepsilon$$
$$u''_{1b}(s_o) = -r_c\beta^2$$
$$u''_{2b}(s_o) = 0 \quad \dots\dots\dots\dots\dots\dots\dots\dots\dots\dots\dots\dots\dots\dots\dots\dots \quad (3.4.63)$$
$$u'''_{2b}(s_o) = -r_c\beta^3.$$

These conditions represent continuity of slope, continuity of curvature in the 1 and 2 direction, and continuity of shear in the 2 direction. In Eq. 3.4.63, lines 1 and 2 restrict ε:

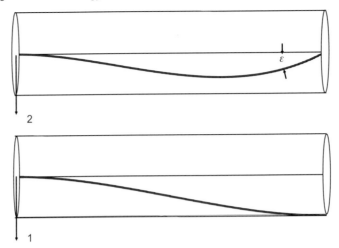

Fig. 3.4.10—Cantilever packer boundary conditions.

$$\varepsilon = \sqrt{\frac{\cos(\alpha s_o) - 1}{\delta}} = \frac{\beta}{\alpha}. \quad\quad\quad\quad\quad\quad\quad\quad\quad\quad\quad\quad\quad\quad\quad (3.4.64)$$

Eq. 3.4.64 has real values for αs_o in the range, $(0, 2\pi]$. Lines 3 and 4 require that

$$\alpha s_o \cos(\alpha s_o) - \sin(\alpha s_o) = 0$$
$$s_0 \sin(\alpha s_0) = 0. \quad\quad\quad\quad\quad\quad\quad\quad\quad\quad\quad\quad\quad\quad\quad\quad\quad\quad (3.4.65)$$

In Eq. 3.4.65, lines 1 and 2 state that the packer boundary condition can be satisfied only for $s_0 = 0$. In general, the packer boundary condition is consistent with a constant-pitch helix only for $\beta = 0$. More-complex versions of this boundary condition also fail to connect with a constant-pitch helix (Sorenson and Cheatham 1986).

Perhaps the solution is a helix with variable pitch? If we take Eq. 3.4.55 with θ an unknown function of s, then the contact force can be eliminated from the equilibrium Eq. 3.4.54, producing the following equation in θ only:

$$[\theta''' - 2(\theta')^3 + \alpha^2 \theta']' = 0. \quad\quad\quad\quad\quad\quad\quad\quad\quad\quad\quad\quad\quad (3.4.66)$$

Performing the integration

$$\omega'' - 2\omega^3 + \alpha^2 \omega - \frac{1}{2}C = 0,$$

where

$$\omega = \theta'$$
$$C = 2\omega_0'' - 4\omega_0^3 + 2\alpha^2 \omega_0 \quad\quad\quad\quad\quad\quad\quad\quad\quad\quad\quad\quad\quad (3.4.67)$$

Eq. 3.4.67 can be multiplied by ω' and integrated:

$$(\omega')^2 - \omega^4 + \alpha^2 \omega^2 - C\omega - D = 0,$$

where

$$D = (\omega_0')^2 - \omega_0^4 + \alpha^2 \omega_0^2 - C\omega_0. \quad\quad\quad\quad\quad\quad\quad\quad\quad\quad (3.4.68)$$

Eq. 3.4.68 is solved by separation of variables:

$$\int ds = \int \frac{d\omega}{\sqrt{\omega^4 - \alpha^2 \omega^2 + C\omega + D}} \quad \dots\dots\dots\dots\dots\dots\dots\dots \quad (3.4.69)$$

This equation can be solved in general (Gradshteyn and Ryzhik 2000), but, for our purposes, consider this special solution so that Eq. 3.4.69 can be factored (Mitchell 2002):

$$C = 0$$
$$D = \frac{\alpha^4}{4}. \quad \dots\dots\dots\dots\dots\dots\dots\dots\dots\dots\dots\dots\dots\dots\dots\dots\dots\dots \quad (3.4.70)$$

Using these values of C and D, one possible integral of Eq. 3.4.69 gives

$$\omega(s) = \tfrac{\sqrt{2}}{2}\alpha \tanh\left[\tfrac{\sqrt{2}}{2}\alpha(s - s_0) + \varphi_p\right] = \theta'(s)$$
$$\theta(s) = ln\left[\frac{\cosh\left[\tfrac{\sqrt{2}}{2}\alpha(s - s_0) + \varphi_p\right]}{\cosh(\varphi_p)}\right], \quad \dots\dots\dots\dots\dots\dots\dots\dots \quad (3.4.71)$$

where φ_p is a constant to be determined.

We can now reexamine the boundary conditions given in Eq. 3.4.63, being generalized for a variable-pitch helix:

$$u'_{2b}(s_o) = r_c\theta'_o = r_c\alpha\varepsilon$$
$$u''_{1b}(s_o) = -r_c\theta'^2_o$$
$$u''_{2b}(s_o) = r_c\theta''_o \quad \dots\dots\dots\dots\dots\dots\dots\dots\dots\dots\dots\dots\dots\dots\dots \quad (3.4.72)$$
$$u'''_{2b}(s_o) = r_c\theta'''_o - r_c\theta'^3_o.$$

Eq. 3.4.72 (line 1), together with Eq. 3.4.72 (line 2), restricts ε:

$$\varepsilon = \sqrt{\frac{\cos(\alpha s_o) - 1}{\delta}} = \tfrac{\sqrt{2}}{2}\tanh\phi_p. \quad \dots\dots\dots\dots\dots\dots\dots\dots \quad (3.4.73)$$

Eq. 3.4.72 (line 4) is now satisfied identically, while the numerical solution of Eq. 3.4.72 (line 3) requires that $\alpha s_o = 5.94992\dots$, which then requires that $\phi_p = 0.235586\dots$ because of Eq. 3.4.73.

If we recheck Eq. 3.4.62 (lines 1 and 2), we find that it violates a boundary condition not previously discussed: all valid solutions to the beam-column equations must lie within the wellbore.

As shown in **Fig. 3.4.11,** the solution to Eq. 3.4.62 lies outside the wellbore over part of the range of s. The solution to this problem is to formulate a two-segment beam-column solution, as shown in **Fig. 3.4.12.**

This problem is considerably more difficult to solve than the single-segment problem, and it must be solved numerically. A discussion of this solution is given in the Appendix. There are multiple solutions to this problem, but only one lies within the wellbore:

$$\alpha\Delta s_0 = 3.817290\dots$$
$$\alpha\Delta s_1 = 1.425441\dots$$
$$\varepsilon_0 = \pm 0.536227\dots$$
$$\varepsilon_1 = \pm 0.681518\dots$$
$$\theta_1 = \pm 0.897795\dots$$
$$\phi_p = 1.996967\dots$$

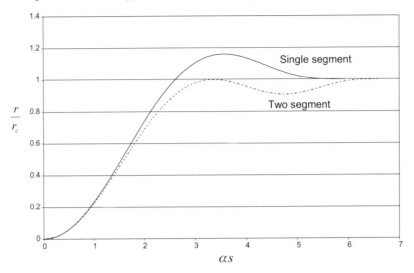

Fig. 3.4.11—Single-segment beam-column solution violates wall constraint.

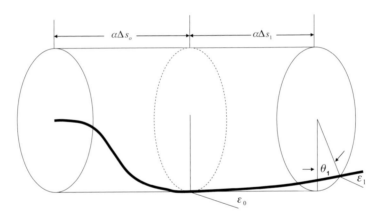

Fig. 3.4.12—Two-segment boundary condition.

Because of symmetry, we can change the sign of θ_1, ε_0, and ε_1 to obtain the helix with the opposite rotation. In Fig. 3.4.11 we can see that this solution stays within the wellbore.

Case 2: The Centralizer. We have noted that boundary conditions have a strong effect on the solution. Having satisfied one type of boundary condition, which new boundary condition should we consider now? The case of pinned-end conditions should give the solution most distinct from the cantilever case. Further, the pinned-end case most closely models the way a casing centralizer works, so there is utility in solving this problem. For this boundary condition, we propose that a beam-column solution to the buckling equations brings the pipe from a centralized, moment-free position to a point tangent to the wellbore wall. The following equations satisfy these conditions:

$$u_{1b} = r_c\{[\alpha s - \sin(\alpha s)] + [\cos(\alpha s_o) - 1]\alpha s\}/\mu$$
$$u_{2b} = r_c\varepsilon\{\alpha s_o[\sin(\alpha s) - \alpha s] + [\alpha s_o - \sin(\alpha s_o)]\alpha s\}/\mu$$
$$\mu = \alpha s_o \cos(\alpha s_o) - \sin(\alpha s_o), \quad \dots\dots\dots\dots\dots\dots\dots\dots\dots \quad (3.4.74)$$

where ε and s_o are undetermined.

Fig. 3.4.13 shows the shape of the beam-column solution. The boundary conditions given in Eq. 3.4.72 must be satisfied where the beam-column solution (Eq. 3.4.74) connects to the full-contact solution. Eq. 3.4.72 (line 1), together with Eq. 3.4.72 (line 2), restricts ε:

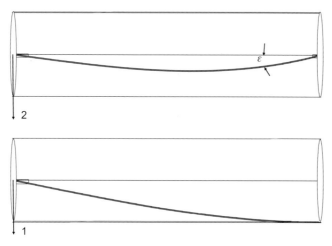

Fig. 3.4.13—Centralizer boundary conditions.

$$\varepsilon = \sqrt{\frac{-\sin\alpha s_o}{\mu}} = \frac{\sqrt{2}}{2}\tanh(\phi_c) \quad \dots\dots\dots\dots\dots\dots\dots\dots\dots\dots\dots\dots\dots\dots\dots\dots\dots\dots \quad (3.4.75)$$

Eq. 3.4.75 has real values for αs_o in the range $(0, \pi)$. If we again assume the solution given in Eq. 3.4.71, we find that it also satisfies the boundary conditions, with the numerical solution of Eq. 3.4.72 (line 3) requiring that $\alpha s_o = 2.505309\ldots$, which then requires that $\varphi_c = 0.819652\ldots$ because of Eq. 3.4.75.

The single-segment beam-column solution satisfies all the boundary conditions but fails to satisfy positive contact force, as shown in **Fig. 3.4.14.**

A solution using two beam-column segments, as illustrated in Fig. 3.4.12, was found to satisfy all the boundary conditions by using the following parameters:

$$\alpha\Delta s_0 = 2.468666\ldots$$
$$\alpha\Delta s_1 = 1.912386\ldots$$
$$\varepsilon_0 \quad = \pm 0.466547\ldots$$
$$\varepsilon_1 \quad = \pm 0.687209\ldots$$
$$\theta_1 \quad = \pm 1.181881\ldots$$
$$\phi_c \quad = 2.124768\ldots$$

These parameters also connect to a variable-pitch solution:

$$\theta(s) = \pm\ln\left[\frac{\cosh\left[\frac{\sqrt{2}}{2}\alpha(s-\Delta s_o-\Delta s_1)+\phi_c\right]}{\cosh(\phi_p)}\right]+\theta_1$$

$$\theta'(s) = \pm\frac{\sqrt{2}}{2}\alpha\tanh\left[\frac{\sqrt{2}}{2}\alpha(s-\Delta s_o-\Delta s_1)+\phi_c\right] \quad \dots\dots\dots\dots\dots\dots\dots\dots\dots\dots\dots\dots\dots \quad (3.4.76)$$

As stated previously, the $\pm$ sign indicates that either a right-hand or left-hand helix is allowed. As shown in Fig. 3.4.14, the two-segment solution provides a positive contact force.

Buckling Results. Fig. 3.4.15 shows the behavior of the pitch of both the cantilever and the centralizer solutions. In both cases, the pitch converges to the Lubinski et al. pitch (1962). Part of the conventional wisdom about tubing buckling is that the Lubinski et al. solution (1962) applies only *far away* from the boundary conditions. This analysis demonstrates for the first time that, for plausible-boundary conditions, the conventional wisdom is correct. Furthermore, the analysis provides a method for calculating the magnitude of *far*

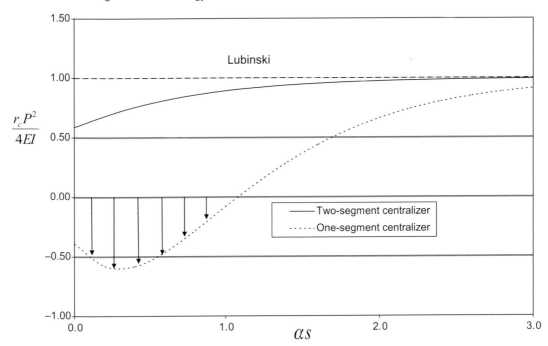

Fig. 3.4.14—Contact force is negative for one-segment centralizer.

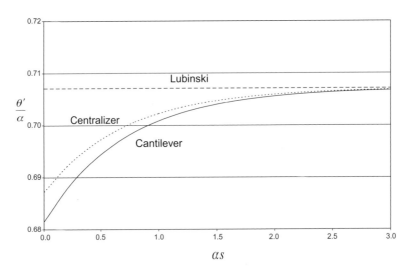

Fig. 3.4.15—Transisition to constant-pitch helix for both boundary conditions.

away. For instance, the transition section from the boundary to 99% of the Lubinski et al. solution (1962) for the two-segment cantilever has length Δs:

$$\alpha\Delta s = \alpha\Delta s_0 + \alpha\Delta s_1 + 2.082 - \sqrt{2}\,\phi_p$$
$$= 4.501, \quad\dots\dots\dots\dots\dots\dots\dots\dots\dots\dots\dots\dots\dots\dots \quad (3.4.77)$$

where the appropriate values of $\alpha\Delta s_o$, $\alpha\Delta s_1$, and φ_{p2} are chosen for the cantilever. A similar calculation for the centralizer solution gives

$$\alpha\Delta s = \alpha\Delta s_0 + \alpha\Delta s_1 + 2.082 - \sqrt{2}\,\phi_c$$
$$= 3.458. \quad\dotfill (3.4.78)$$

Using the data from **Table 3.4.1**, we find that $\alpha \approx 0.0371$ in.$^{-1} \approx 0.445$ ft^{-1}. For these values, the transition length is approximately 10.1 ft for a cantilever and approximately 7.8 ft for the centralizer.

There are three principal results of interest from helical-buckling analysis: tubing-length change, bending stresses, and contact forces.

Tubing-Length Change. Tubing-length change is determined by integrating the buckling *strain* over the buckled length. The buckling strain (i.e., the tubing-length change per unit length) is given by the following formula:

$$\varepsilon_b = -\tfrac{1}{2}\left(r_c\theta'\right)^2 \quad\dotfill (3.4.79)$$

As already discussed, the differences between Eq. 3.4.71, Eq. 3.4.76, and Lubinski et al. solution (1962) are negligible over most of the buckled length, so the Lubinski et al. length-change result is still essentially correct:

$$\Delta L_b = -\frac{r_c^2 \Delta P^2}{8EIw}. \quad\dotfill (3.4.80)$$

We can calculate a correction factor for Eq. 3.4.80 to evaluate the exact effect of the variable-pitch solution:

$$\Delta L_{bcorr} = \int \varepsilon_b \, ds - \Delta L_b = \tfrac{1}{2}r_c^2 \int \theta'' ds. \quad\dotfill (3.4.81)$$

Eq. 3.4.81 can be integrated to give

$$\Delta L_{bcorr} = \tfrac{1}{2}r_c^2 \beta_{LW}\left[1 - \tanh(\phi)\right]. \quad\dotfill (3.4.82)$$

The effect of Eq. 3.4.82 is small. Using the example problem data from Table 3.4.1, we see that the buckled-length change is 46.1 in., while the correction factor is 0.0012 in.

Bending Stresses. Bending-stress results are more interesting. For the beam-column solutions, the bending stresses are given by

$$M_i = EIr_c u_i'' \quad i = 1, 2. \quad\dotfill (3.4.83)$$

The total bending moment is therefore

$$M = EIr_c \sqrt{(u_1'')^2 + (u_2'')^2}. \quad\dotfill (3.4.84)$$

The total bending moment for the full-contact solution is given by

$$M = EIr_c \sqrt{(\theta')^4 + (\theta'')^2}. \quad\dotfill (3.4.85)$$

TABLE 3.4.1—DATA FOR LUBINSKI'S SAMPLE PROBLEM (Lubinski et al. 1962)	
P	66,320 lbf
E	30×10^6 psi
I	1.61 in.^{4}
r_c	1.61 in.

The maximum bending stress is related to the maximum bending moment:

$$\sigma_b = \frac{M d_o}{2I}, \quad \dots\dots\dots\dots\dots\dots\dots\dots\dots\dots\dots\dots\dots\dots\dots\dots\dots\dots\dots\dots\dots\dots\dots\dots \quad (3.4.86)$$

where d_o is the tubing outside diameter. **Figs. 3.4.16 and 3.4.17** show the maximum bending moment in the beam-column section and the full-contact section for the cantilever packer and the centralized pipe.

Note that in each case, the beam-column bending moment exceeds the full-contact bending moment, slightly for the cantilever packer and by 20% for the centralized pipe. While a real packer or centralizer will not behave exactly as these idealized models, these results suggest that bending stresses near the packer may exceed the predictions of the Lubinski et al. model (1962) by as much as 20%.

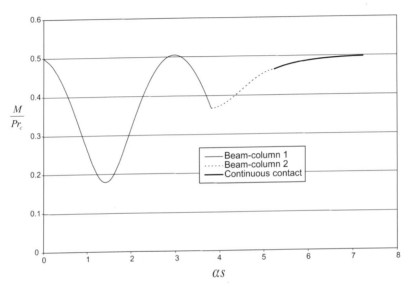

Fig. 3.4.16—Bending moment for two-segment cantilever packer.

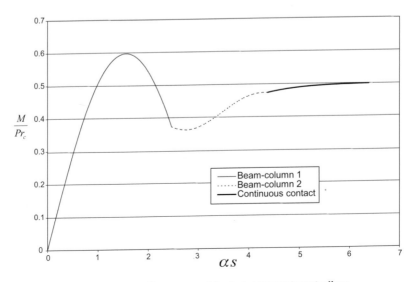

Fig. 3.4.17—Bending moment for two-segment centralizer.

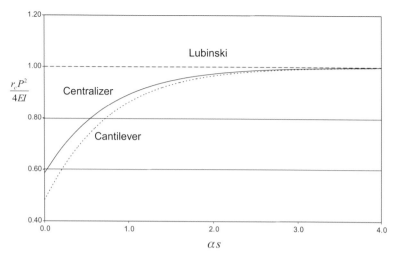

Fig. 3.4.18—Contact force for cantilever and centralizer.

Contact Forces. Once we have analytic solutions for θ', it is relatively easy to determine the contact force for the buckled pipe:

$$w_n = \frac{r_c P^2}{4EI}(10T^4 - 12T^2 + 3)$$

$$T = \tanh\left[\frac{\sqrt{2}}{2}\alpha(s - s_o) + \phi_p\right]. \quad \dotfill \quad (3.4.87)$$

The contact force for the cantilever packer and the contact force for the centralized pipe are shown in **Fig. 3.4.18.** The result rapidly converges to the contact force for the Lubinski et al. result (1962).

Conclusions and Comments. The missing piece of the helical-buckling problem has always been the transition from the packer to the fully-developed helix. Beam-column solutions were tried, but they could not be connected to a constant-pitch helix while satisfying necessary mechanical constraints. The discovery of variable-pitch solutions finally allows a completely valid answer to this problem.

This chapter provides solutions to two different boundary conditions: a cantilevered boundary and a pinned-end boundary. These solutions correspond to tubing sealed in a packer and a centralized casing, respectively. Both solutions converge to the Lubinski et al. solution (1962) *far from the boundary.* In addition, this distance can be quantified, so that an objective test can be performed regarding real buckling problems.

Buckled-length change is an important calculation in tubing design. The impact of the boundary on this calculation has also been quantified, though this effect can be shown to be small.

The most significant result is the bending moments generated in the beam-column section leading up to the fully-developed helix. While the cantilever solution gives only a 1% increase over the Lubinski et al. solution (1962), the centralizer solution gives a 20% increase. Both of these boundary conditions represent *ideal* rather than actual conditions. A real packer has some compliance, and a real centralizer has some resistance to tilt. If an engineer has specific information on the actual boundary condition, comparison to these special results should be useful in evaluating actual bending stresses.

Finally, this calculation points the way to solving another lingering question about helical buckling: What happens in tapered strings? The only commonly used buckling model for tapered strings uses *stacked* Lubinski solutions (Hammerlindl 1977). Again, the conventional wisdom is that these solutions are valid *far from* where the different-sized pipes are joined. But, the actual behavior of these buckled pipes near the point of junction is not yet known, so we cannot verify the conventional wisdom.

Nomenclature

A_i = cross-sectional internal flow area of the pipe, L^2, in.^{2}

A_o = total cross-sectional area of the pipe, including internal flow area, L^2, in.^{2}

b_z = component of the binormal vector in the $\bar{i}_z$ coordinate direction

$\bar{b}$ = binormal vector of the tubing displacement

$\bar{b}'$ = derivative of the binormal vector with respect to arc length s

C = constant of integration for buckling equation

d_o = tubing outside diameter, L, in.

D = constant of integration for buckling equation

E = Young's modulus, psi

F = force, mL/t^2, lbf

F_a = axial force, mL/t^2, lbf

F_b = shear force in the binormal direction, mL/t^2, lbf

F_e = effective force, mL/t^2, lbf

F_n = shear force in the normal direction, mL/t^2, lbf

F_m = special force definition to simplify notation, mL/t^2, lbf

F_{st} = magnitude of the stream-thrust force, mL/t^2, lbf

F_N = normal force, mL/t^2, lbf

$\bar{F}$ = pipe force (i.e., tubing internal force), lbf

$\bar{F}_{st}$ = stream-thrust force, mL/t^2, lbf

g = acceleration of gravity, L/t^2, lbf/lbm

$\bar{i}$ = coordinate unit vector

$\bar{i}_z$ = coordinate unit vector in the downward direction

I = moment of inertia of tubing, L^4, in.^{4}

ℓ = length of simply supported beam, L, ft

m_t = applied torque per unit length in the tangential direction, mL/t^2, in.-lbf/in.

$\bar{m}$ = applied moment-vector per unit length, mL/t^2, in.-lbf/in.

M = total bending moment, in.-lbf

M_i = bending moment in i direction, in.-lbf

M_t = axial torque, mL^2/t^2, in.-lbf

$\bar{M}$ = change in moment, mL/t^2, in.-lbf/in.

n_z = component of the normal vector in the $\bar{i}_z$ coordinate direction

$\bar{n}$ = unit normal vector of the tubing displacement

$\bar{n}_i$ = normal vector at the beginning of segment i

$\bar{n}'$ = derivative of the normal vector with respect to the arc length

p = fluid density, m/L^3, lbm/in.^{3}

p_i = fluid density inside the pipe, m/L^3, lbm/in.^{3}

p_o = fluid density outside the pipe, m/L^3, lbm/in.^{3}

P = buckling force, lbf

P_{crit} = critical buckling force, mL/t^2, lbf

Q = midspan vertical load in example problem, mL/t^2, lbf

r_c = tubing-casing radial clearance, L, in.

r_o = pipe outside radius, L, in.

R = radius of the curvature of the segment, L, in.

s = arc length of the curve (i.e., measured depth), L, in.

s_i = measured depth, segment i

S = parameter in beam-column equations

t_z = component of the tangent vector in the $\bar{i}_z$ coordinate direction

$\bar{t}_i$ = unit tangent vector of the tubing displacement, segment i

$\bar{t}$ = unit tangent vector of the tubing displacement

$\vec{t}'$ = derivative of the tangent vector with respect to arc length = $\dfrac{\mathrm{d}\vec{t}}{\mathrm{d}s}$

T = term in contact force equation

$\bar{u}$ = displacement vector of the tubing, L, in.

$\bar{u}_i$ = displacement vector of the tubing, segment i, L, in.

$\vec{u}'$ = derivative of the displacement vector with respect to arc length = $\dfrac{\mathrm{d}\bar{u}}{\mathrm{d}s}$

u_i = displacement component in the i coordinate direction, L, in.

u_z = displacement component in the $\bar{i}_z$ coordinate direction, L, in.

u_1 = tubing displacement, coordinate direction 1, L, in.

u_2 = tubing displacement, coordinate direction 2, L, in.

v = average flow stream velocity, L/t, in./s

v_i = average flow stream velocity inside pipe, L/t, in./s

v_o = average flow stream velocity outside pipe, L/t, in./s

w = load per unit length, m/t², lbf/in.

w_c = contact load, m/t², lbf/in.

w_n = normal contact load between the tubing and casing, lbf/ft

w_p = weight of the pipe in air, M/L, lbm/in.

$\bar{w}$ = applied load vector, lbf/in.

$\overline{W}_{bp}$ = buoyant weight of the pipe, m/t², lbf/in.

$\overline{W}_c$ = contact force vector, m/t², lbf/in.

$\overline{W}_d$ = the drag load on the pipe, lbf/in.

$\overline{W}_{st}$ = stream-thrust load, m/t², lbf/in.

W = special force definition to simplify notation, mL/t², lbf

X, Y, Z = parameters in beam-column equations

α = coefficient in parameter in beam-column equations solutions, L⁻¹, ft⁻¹

α_1 = coefficient in torque-drag analytic solution, dimensionless

α_2 = coefficient in torque-drag analytic solution, dimensionless

β = coefficient in solutions, L⁻¹, ft⁻¹

β_{LW} = Lubinski's coefficient in solutions, L⁻¹, ft⁻¹

δ = parameter in beam-column equations

ΔL_b = buckled length change, L, in.

ΔL_{bcorr} = buckled length change correction, L, in.

Δs = length of the transition section from the boundary to 99% of the Lubinski buckling solution for the two-segment cantilever, L, in.

Δs_0 = beam-column solution length for segment 1, L, ft

Δs_1 = beam-column solution length for segment 2, L, ft

ε = slope in beam-column solution, dimensionless

ε_0 = slope in beam-column solution for segment 1, dimensionless

ε_1 = slope in beam-column solution for segment 2, dimensionless

ε_b = buckling strain

θ = rotation angle of the pipe center about the center of the wellbore, radians

θ_1 = angle in beam-column solution, radians

κ = curvature, L⁻¹, 1/in.

κ_i = curvature of segment i, L⁻¹, 1/in.

μ = parameter in beam-column equations

μ_f = friction coefficient

$\bar{\mu}_f$ = modified friction coefficient

$\wp$ = pitch of a helix, L, ft

ϕ_c = unknown constant in centralizer solution, dimensionless

ϕ_p = unknown constant in packer solution, dimensionless

ξ = dimensionless length

ρ = fluid density, m/L³, lbm/in.³

ρ_i = density of fluid inside pipe, m/L^3, lbm/in.^{3}
ρ_o = density of fluid outside pipe, m/L^3, lbm/in.^{3}
σ_b = maximum bending stress, m/t^2-L, psi
τ = geometric torsion of a curve, L-1, 1/in
ψ = parameter in beam column equation, dimensionless
$\omega = \dfrac{d\theta}{ds}$, 1/L, 1/in.

Superscript
$'$ = derivative with respect to the arc length $\left(\dfrac{d}{ds}\right)$

References

Cheatham, J.B. Jr. and Pattillo, P.D. 1984. Helical Postbuckling Configuration of a Weightless Column Under the Action of an Axial Load. *SPEJ* **24** (4): 467–472. SPE-10854-PA. DOI: 10.2118/10854-PA.

Gradshteyn, I.S. and Ryzhik, I.M. 2000. *Table of Integrals, Series, and Products*. London: Elsevier Science & Technology.

Hammerlindl, D.J. 1977. Movement, Forces, and Stresses Associated With Combination Tubing Strings Sealed in Packers. *JPT* **29** (2): 195–208; *Trans.*, AIME, **263**. SPE-5143-PA. DOI: 10.2118/5143-PA.

Ho, H.-S. 1986. General Formulation of Drillstring Under Large Deformation and Its Use in BHA Analysis. Paper SPE 15562 presented at the SPE Annual Technical Conference and Exhibition, New Orleans, 5–8 October. DOI: 10.2118/15562-MS.

Ho, H.-S. 1988. An Improved Modeling Program for Computing the Torque and Drag in Directional and Deep Wells. Paper SPE 18047 presented at the SPE Annual Technical Conference and Exhibition, Houston, 2–5 October. DOI: 10.2118/18047-MS.

Johancsik, C.A., Dawson, R., and Friesen, D.B. 1973. Torque and Drag in Directional Wells—Prediction and Measurement. *JPT* **36** (6): 987–992. SPE-11380-PA. DOI: 10.2118/11380-PA.

Lesso, W.G., Mullens, E., and Daudey, J. 1989. Developing a Platform Strategy and Predicting Torque Losses for Modeled Directional Wells in the Amauligak Field of the Beaufort Sea, Canada. Paper SPE 19550 presented at the SPE Annual Technical Conference and Exhibition, San Antonio, Texas, 8–11 October. DOI: 10.2118/19550-MS.

Love, A.E.H. 1944. *A Treatise on the Mathematical Theory of Elasticity*, fourth edition. New York: Dover Books.

Lubinski, A., Althouse, W.S., and Logan, J.L. 1962. Helical Buckling of Tubing Sealed in Packers. *JPT* **14** (6): 655–670; *Trans.*, AIME, **225**. SPE-178-PA. DOI: 10.2118/178-PA.

Mitchell, R.F. 1982. Buckling Behavior of Well Tubing: The Packer Effect. *SPEJ* **22** (5): 616–624. SPE-9264-PA. DOI: 10.2118/9264-PA.

Mitchell, R.F. 1986. Simple Frictional Analysis of Helical Buckling of Tubing. *SPEDE* **1** (6): 457–465; *Trans.*, AIME, **281**. SPE-13064-PA. DOI: 10.2118/13064-PA.

Mitchell, R.F. 1988. New Concepts for Helical Buckling. *SPEDE* **3** (3): 303–310; *Trans.*, AIME, **285**. SPE-15470-PA. DOI: 10.2118/15470-PA.

Mitchell, R.F. 2002. Exact Analytical Solutions for Pipe Buckling in Vertical and Horizontal Wells. *SPEJ* **7** (4): 373–390. SPE-72079-PA. DOI: 10.2118/72079-PA.

Nordgren, R.P. 1974. On Computation of the Motion of Elastic Rods. *Journal of Applied Mechanics* **41**: 777–780.

Press, W.H., Flannery, B.P., Teukolsky, S.A., and Vetterling, W.T. 1992. *Numerical Recipes in Fortran 77: The Art of Scientific Computing*, second edition, 382–386. Cambridge, UK: Cambridge University Press.

Sawaryn, S.J. and Thorogood, J.L. 2005. A Compendium of Directional Calculations Based on the Minimum Curvature Method. *SPEDC* **20** (1): 24–36. SPE-84246-PA. DOI: 10.2118/84246-PA.

Sheppard, M.C., Wick, C., and Burgess, T.M. 1987. Designing Well Paths To Reduce Drag and Torque. *SPEDE* **2** (4): 344–350. SPE-15463-PA. DOI: 10.2118/15463-PA.

Sorenson, K.G. and Cheatham, J.B. Jr. 1986. Post-Buckling Behavior of a Circular Rod Constrained Within a Circular Cylinder. *Journal of Applied Mechanics* **53**: 929–934.

Timoshenko, S.P. and Gere, J.M. 1961. *Theory of Elastic Stability*. New York City: McGraw-Hill Companies.

Walker, B.R. and Friedman, M.B. 1977. Three-Dimensional Force and Deflection Analysis of a Variable Cross-Section Drillstring. *Journal of Pressure Vessel Tech.* **99**: 367–373.

Zwillinger, D. ed. 1996. *CRC Standard Mathematical Tables and Formulae*, 30th edition, 321–322. Boca Raton, Florida: CRC Press.

Appendix—Two-Segment Beam-Column Solutions

The section over interval Δs_o may be described by Eq. 3.4.62 for the cantilever problem, or by Eq. 3.4.74 for the centralizer problem. The second section is given by the following equation:

$$u_{1b2} = \psi_1(s) + \cos\theta_1 \psi_3(s) - \varepsilon_1 \sin\theta_1 \psi_4(s)$$

$$u_{2b2} = \varepsilon_o \psi_2(s) + \sin\theta_1 \psi_3(s) + \varepsilon_1 \cos\theta_1 \psi_4(s), \quad\cdots\cdots\cdots\cdots\cdots\cdots\cdots\cdots \text{(A-1)}$$

where the $\psi_i(s)$, $i = 1 \ldots 4$, are given by

$$\psi_1(s) = 1 - \frac{S(\xi - \sin\xi) - Y(1 - \cos\xi)}{\delta}$$

$$\psi_2(s) = \xi - \frac{(D+Y)(\xi - \sin\xi) - Z(1 - \cos\xi)}{\delta}$$

$$\psi_3(s) = \frac{S(\xi - \sin\xi) - Y(1 - \cos\xi)}{\delta}$$

$$\psi_4(s) = \frac{X(1 - \cos\xi) - Y(\xi - \sin\xi)}{\delta} \quad\cdots\cdots\cdots\cdots\cdots\cdots\cdots\cdots \text{(A-2)}$$

$$\xi = \alpha(s - \Delta s_o)$$

$$S = \sin(\xi_1)$$

$$X = \xi_1 - S$$

$$Y = 1 - \cos(\xi_1)$$

$$Z = \xi_1 \cos(\xi_1) - S$$

$$\delta = \xi_1 S - 2Y$$

$$\xi_1 = \alpha\Delta s_1.$$

The $\psi_i(s)$ were constructed so that

$$
\begin{aligned}
& u_{1b}(\Delta s_o) = u_{1b2}(\Delta s_o) \\
1. \quad & u'_{1b}(\Delta s_o) = u'_{1b2}(\Delta s_o) \\
& u_{2b}(\Delta s_o) = u_{2b2}(\Delta s_o) \\
& u'_{2b}(\Delta s_o) = u'_{2b2}(\Delta s_o).
\end{aligned}
\quad\cdots\cdots\cdots\cdots\cdots\cdots\cdots\cdots \text{(A-3)}
$$

Eq. A-2 must satisfy these boundary conditions:

$$
\begin{aligned}
& u''_{1b}(\Delta s_o) = u''_{1b2}(\Delta s_o) \\
2. \quad & u''_{2b}(\Delta s_o) = u''_{2b2}(\Delta s_o) \\
& u'''_{2b}(\Delta s_o) = u'''_{2b2}(\Delta s_o),
\end{aligned}
\quad\cdots\cdots\cdots\cdots\cdots\cdots\cdots\cdots \text{(A-4)}
$$

in addition to matching the solution for Eq. 3.4.71 or Eq. 3.4.76 at $s = \Delta s_o + \Delta s_1$. The boundary conditions are solved by finding the appropriate values of ε_o, ε_1, θ_1, $\alpha\Delta s_o$, and $\alpha\Delta s_1$. The boundary conditions (Eq. A-4) can be solved analytically to give ε_0, ε_1, and θ_1, so that the problem is reduced to finding $\alpha\Delta s_0$ and $\alpha\Delta s_1$. This problem was solved by scanning over the likely range of these two variables, seeking candidate solutions. The candidate solutions are then converged numerically using Broyden's method (Press et al. 1992).

SI Metric Conversion Factors

ft $\times$ 3.048* $E - 01 = m$
in. $\times$ 2.54* $E + 00 = cm$
in.^{2} $\times$ 6.451 6* $E + 00 = cm^2$
in.^{3} $\times$ 1.638 706 $E + 01 = cm^3$
lbf $\times$ 4.448 222 $E + 00 = N$
lbm $\times$ 4.535 924 $E - 01 = kg$

*Conversion factor is exact.

Chapter 4

Advanced Wellbore Hydraulics

4.1 Wellbore Hydraulics—Ramadan Ahmed, University of Oklahoma, and Stefan Miska, University of Tulsa

4.1.1 Introduction. During drilling of oil and gas wells, drilling fluid is circulated from surface to the bottomhole through the drillstring, the bit nozzles, and returns to surface in the annular region between the borehole and the drillstring. **Fig. 4.1.1** shows a schematic of the drilling fluid path in the wellbore. A drilling fluid has a number of important functions. It removes cuttings from the bottom of the hole, holds cuttings and weight material in suspension when circulation is interrupted, controls subsurface pressure, and transmits hydraulic horsepower.

For conventional drilling applications, we use water-based mud (WBM), oil-based mud (OBM), and synthetic-based mud (SBM). In addition, aerated fluids and foam are frequently used for underbalanced drilling (UBD) applications. For the purpose of hydraulics analysis, drilling fluids are generally classified as Newtonian and non-Newtonian fluids. Newtonian fluids such as water and mineral oil exhibit a direct proportionality between shear stress and shear rate under laminar flow conditions. Thus, the stress is expressed as

$$\tau = \mu \dot{\gamma}, \dotfill (4.1.1)$$

where μ is the viscosity of the fluid, which is a function of temperature and pressure, and $\dot{\gamma}$ is the shear rate.

The relationship between the shear stress and shear rate is used as the main rheological classification method for drilling fluids. Accordingly, different rheological models (constitutive equations) such as Bingham plastic, power-law, and Herschel-Bulkley [yield power-law (YPL)] models have been developed to represent correctly this relationship and perform wellbore hydraulics analysis. The YPL model is expressed as

$$\tau = \tau_y + K\dot{\gamma}^m, \dotfill (4.1.2)$$

where τ and τ_y are shear and yield stresses, respectively, K is the consistency index, and m is the fluid behavior index. At present, this model is preferable because it combines the effects of the Bingham ($\tau = \tau_y + K\dot{\gamma}$) and power-law ($\tau = K\dot{\gamma}^m$) fluid models. Applying the YPL model, the apparent viscosity of drilling fluids can be expressed as

$$\eta = \frac{\tau_y}{\dot{\gamma}} + K\dot{\gamma}^{m-1} \dotfill (4.1.3)$$

Often, drilling fluids show shear-thinning behavior (i.e., viscosity reduction resulting from increased shear rate) and fall into the non-Newtonian fluid category. For these fluids, the apparent viscosity, η, is used instead of the viscosity of the fluid in Eq. 4.1.1. The apparent viscosity is defined as the ratio of shear stress to the corresponding shear rate. It is a function of shear rate. For shear-thinning fluid, the apparent viscosity decreases as the shear rate increases.

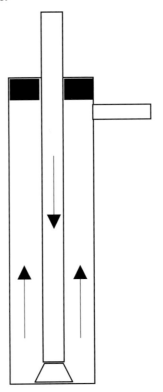

Fig. 4.1.1—Schematic of the drilling fluid path in conventional drilling.

Many drilling fluids, particularly bentonite suspensions, are not well approximated by either the Bingham plastic or power-law models. Majorities of drilling fluids, including bentonite suspensions, are well represented by the YPL model. It is important to note that rheologies of some non-Newtonian polymeric fluids that show Newtonian behavior at low shear rates are very difficult to describe using the YPL model. According to Eq. 4.1.3, a plot of viscosity vs. shear rate on a log-log scale gives a straight line for power-law fluids. A similar plot for fluids with yield stress forms a curve on a log-log scale. The yield stress usually is explained in terms of an internal structure that is capable of preventing deformation of a fluid element for values of shear stress less than the yield value. All rheological models presented above predict infinite viscosity at vanishing shear rate. However, polymer-based drilling fluids usually exhibit a constant viscosity (Newtonian behavior) at low shear rates.

Some drilling fluids, including bentonite mud, show time-dependent rheological behavior. Time-dependent fluids are those for which the shear stress is a function of both the magnitude of shear rate and shear rate history. These fluids usually are classified into two groups: thixotropic and rheopectic fluids. Under isothermal conditions, thixotropic fluid exhibits a reversible decrease in shear stress with time at a constant shear rate, whereas rheopectic fluid shows a reversible increase in shear stress with time at a constant rate of shear **(Fig. 4.1.2)**.

Complex drilling fluids such as invert-emulsion and polymeric fluids exhibit both viscous and elastic properties (i.e., viscoelastic properties). Viscoelastic fluids are those that show partial elastic recovery upon the removal of a deforming shear stress. Such materials exhibit properties of both fluids and elastic solids.

4.1.2 Rheological Characterization. Wellbore hydraulic calculations require rheological parameters of the fluid. These parameters normally are obtained using viscometric measurements that present shear stress and shear rate at the same known points in the viscometer. Different types of viscometers have been developed to determine rheological properties of fluids.

Rotational Viscometer. Couette viscometers (concentric-cylinder rotational viscometers) became the most popular due to their operational simplicity and mechanical reliability. Standard Couette viscometers in the field use a rotating cup (rotor) and stationary inner cylinder (bob) suspended with a torsion wire (coil) or a rotating inner cylinder suspended with a coil and stationary cup **(Fig. 4.1.3a)**. During the measurement, the

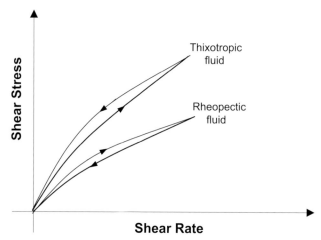

Fig. 4.1.2—Flow curves of time-dependent fluids.

sample fluid stays in the annular gap. The cup rotates at the desired speed and continuously deforms the test fluid. The coil experiences a torque due to the viscous resistance of the fluid, which is measured by the defection of the coil from the neutral position. Assuming isothermal laminar flow conditions and neglecting the contributions of the end sections, the shear stress acting at the inner cylinder wall is given by

$$\tau = \frac{\Gamma}{2\pi R_i^2 L_b}, \quad\dots\dots\dots\dots\dots\dots\dots\dots\dots\dots\dots\dots\dots\dots\dots\dots\dots\dots\dots\dots\dots\dots \text{(4.1.4)}$$

where Γ is measured torque.

To simplify the viscometric flow analysis, Couette viscometers usually are designed to have a narrow annular gap. For that reason, viscometers with radius ratios ($\kappa = R_i/R_o$) of greater than 0.99 are considered as a narrow slot (i.e., linear velocity distribution) in the determination of the shear rate, and the following equation can be used for estimating the shear rate at the inner or outer cylinder:

$$\dot{\gamma} = \frac{\omega R_i}{R_o - R_i} \quad\dots\dots\dots\dots\dots\dots\dots\dots\dots\dots\dots\dots\dots\dots\dots\dots\dots\dots\dots\dots\dots\dots\dots \text{(4.1.5)}$$

For wide-gap viscometers ($0.50 \le \kappa \le 0.99$), the narrow-slot approximation method is invalid. In this case, for standard Couette-type oilfield viscometers (i.e., rotating cup and stationary inner cylinder), the shear rate at the inner cylinder is estimated as

$$\dot{\gamma} = \frac{2\omega}{n*(1 - \kappa^{2/n*})}, \quad\dots\dots\dots\dots\dots\dots\dots\dots\dots\dots\dots\dots\dots\dots\dots\dots\dots\dots\dots \text{(4.1.6)}$$

where $n*$ is the slope of a logarithmic plot of the measured torque (Γ) vs. angular velocity (ω). Thus,

$$n* = \frac{dLn\Gamma}{dLn\omega} = \frac{dLn\tau}{dLnN}, \quad\dots\dots\dots\dots\dots\dots\dots\dots\dots\dots\dots\dots\dots\dots\dots\dots\dots \text{(4.1.7)}$$

where N is rotational speed in rev/min. The above analysis is not valid if the flow in the viscometric is under turbulent conditions, involves secondary flow such as Taylor vortices, or exhibits wall slip. Therefore, viscometric measurements should be carefully examined. Data points obtained under turbulent or secondary flow conditions must be removed from the data set before determining the rheological parameters. For Newtonian fluids and narrow-gap viscometers with the inner cylinder rotating, the following equation is presented to calculate the Taylor number:

$$\text{Ta} = \frac{\rho^2 \omega^2 (R_o - R_i)^3 R_i}{\mu^2} \quad\dots\dots\dots\dots\dots\dots\dots\dots\dots\dots\dots\dots\dots\dots\dots\dots \text{(4.1.8)}$$

When the Taylor number (Ta) is greater than the critical Taylor number (3,400), the flow becomes unstable and results in secondary flows (i.e., Taylor vortices). The formation of Taylor vortices in the viscometer dissipates additional energy and causes an increase in the measured torque. For non-Newtonian fluid in narrow-gap viscometers, the above equation can be applied using the apparent viscosity. However, for non-Newtonian polymeric solutions, the critical Taylor number is expected be higher than 3,400 (Macosko 1994).

Flow in Couette viscometers with a rotating outer cylinder is stable until the onset of turbulence at a Reynolds number of 50,000, where the Reynolds number is defined as (Macosko 1994)

$$\mathrm{Re}_T = \frac{\rho\omega(R_o - R_i)R_o}{n*} \qquad \dots\dots\dots\dots\dots\dots\dots\dots\dots\dots\dots \quad (4.1.9a)$$

Example 1. The following viscometric data (**Table 4.1.1**) were obtained from 6% bentonite suspension using an oilfield viscometer with a bob radius of 1.7245 cm and a cup radius of 1.8415 cm. Assuming YPL fluid, determine the rheological parameters of the fluid.

Solution. The diameter ratio of the viscometer is 1.7245/1.8415 = 0.9364. Hence, the shear rate should be calculated using Eq. 4.1.6. To apply this equation, we first need to estimate the slope of a logarithmic plot of the measured stress vs. rotational speed (Fig. 4.1.3b). After curve fitting the data points with a second-degree polynomial

$$Ln(\tau) = A(Ln(N))^2 + BLn(N) + C, \qquad \dots\dots\dots\dots\dots\dots\dots\dots\dots \quad (4.1.9b)$$

the slope of the curve can be calculated as $n* = 2ALn(N) + B$. For this particular case $A = 0.0462$; $B = 0.0632$; $C = 1.1721$. **Table 4.1.2** presents the shear rate obtained using Eq. 4.1.6. To determine the YPL model parameters (m, K, and τ_y), the data are presented in Fig. 4.1.3c as $Ln(\tau - \tau_y)$ vs. $Ln(\dot{\gamma})$. The correct value of the yield stress (τ_y) is obtained by varying its value until all the data points lie on a straight line. The results presented in Fig. 4.1.3b are obtained using yield stress value of 2.34 Pa (4.9 lbf/100 ft²). The slope and y-intercept of the straight line are used to determine the values of m and $Ln(k)$, respectively. Hence, $m = 0.61$ and $K = 0.39$ Pas$^{0.61}$ (0.82 lbfs$^{0.61}$/100 ft²).

Pipe Viscometer. Often pipe viscometers show better reliability and accuracy than rotational viscometers. However, pipe viscometers are relatively expensive and not convenient for field applications. As a result, they are commonly used for research purpose and in-line viscosity measurement. A standard pipe viscometer system (**Fig. 4.1.4**) has flow rate and pressure loss measuring instrumentations. To obtain reliable and

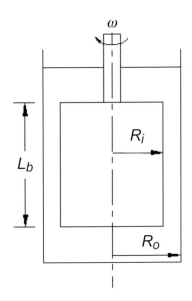

Fig. 4.1.3a—A stationary-cup Couette-type viscometer.

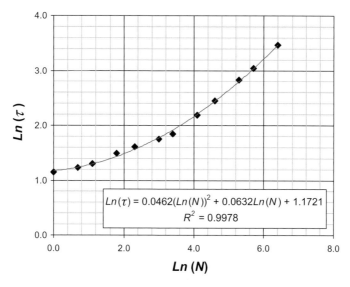

Fig. 4.1.3b—Logarithmic plot of shear stress vs. rotational speed.

Inside the plot:
$$Ln(\tau) = 0.0462(Ln(N))^2 + 0.0632Ln(N) + 1.1721$$
$$R^2 = 0.9978$$

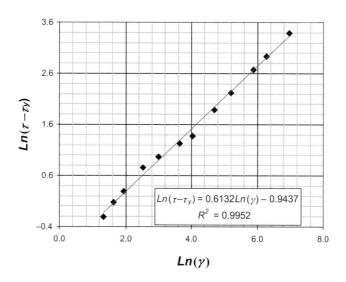

Fig. 4.1.3c—$Ln(\tau-\tau_y)$ vs. $Ln(\dot{\gamma})$.

Inside the plot:
$$Ln(\tau-\tau_y) = 0.6132Ln(\gamma) - 0.9437$$
$$R^2 = 0.9952$$

TABLE 4.1.1—VISCOMETRIC DATA OBTAINED FROM AN OILFIELD VISCOMETER											
Speed, N (rev/min)	1	2	3	6	10	20	30	60	100	200	300	600
Stress, τ (lbf/100 ft^2)	6.59	7.14	7.68	9.33	10.43	12.08	13.18	18.67	24.17	35.17	43.97	66.57

accurate measurements, these types of viscometers must have sufficiently long entrance and exit sections. This is mainly to establish a fully developed laminar flow condition in the test section. For power-law fluid, Collins and Schowalter (1963) carried out flow experiments to determine the entrance length in the pipe. Using their result, we propose the following correlation to estimate the entrance length (X_D):

$$X_D = (-0.126n + 0.1752)D \cdot \text{Re}, \quad \dots\dots\dots\dots\dots\dots\dots\dots\dots\dots\dots\dots\dots\dots\dots \quad (4.1.9c)$$

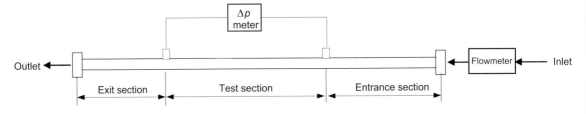

Fig. 4.1.4—Pipe viscometer system.

N (rev/min)	Ln(N)	τ (lbf/100 ft²)	τ (Pa)	Ln(τ)	n*	$\dot{\gamma}$ (1/s)	Ln($\dot{\gamma}$)	Ln($\tau-\tau_y$)
1	0.000	6.59	3.15	1.15	0.06	3.79	1.33	−0.21
2	0.693	7.14	3.42	1.23	0.13	5.11	1.63	0.07
3	1.099	7.68	3.68	1.30	0.16	6.94	1.94	0.29
6	1.792	9.33	4.47	1.50	0.23	12.58	2.53	0.75
10	2.303	10.43	4.99	1.61	0.28	20.05	3.00	0.97
20	2.996	12.08	5.78	1.75	0.34	38.46	3.65	1.24
30	3.401	13.18	6.31	1.84	0.38	56.66	4.04	1.38
60	4.094	18.67	8.94	2.19	0.44	110.65	4.71	1.89
100	4.605	24.17	11.57	2.45	0.49	181.91	5.20	2.22
200	5.298	35.17	16.84	2.82	0.55	358.45	5.88	2.67
300	5.704	43.97	21.05	3.05	0.59	533.78	6.28	2.93
600	6.397	66.57	31.87	3.46	0.65	1056.42	6.96	3.39

TABLE 4.1.2—CALCULATED VALUES OF EXPONENT n* AND SHEAR RATE

where n, D, and Re are the power-law index, pipe diameter, and Reynolds number, respectively. This correlation is valid only for n values greater than 0.2. The exit section length generally is shorter than the inlet section.

In order to analyze the viscometric flow, let us consider a short segment in the test section of the viscometer (Fig. 4.1.4) with diameter D and length ΔL. In this case, the flow rate through the segment is calculated from the velocity profile applying the continuity equation as

$$Q = 2\pi \int_0^R v(r)r\,dr, \quad\dots\dots\dots\dots\dots\dots\dots\dots\dots\dots (4.1.10)$$

where $v(r)$ is the axial velocity profile in the pipe viscometer **(Fig. 4.1.5)**. Integrating by part Eq. 4.1.10 and assuming that $v(R) = 0$ (i.e., a no-slip condition at the pipe wall), we get

$$Q = -\pi \int_0^R r^2 \frac{dv}{dr}\,dr \quad\dots\dots\dots\dots\dots\dots\dots\dots\dots\dots (4.1.11)$$

We know that the velocity gradient (shear rate), $\dfrac{dv}{dr}$, is a function of shear stress. It can be shown that for steady-state flow of fluid with constant density, the momentum balance yields the following expression:

$$\frac{\tau(r)}{r} = \frac{\tau_w}{R}, \quad\dots\dots\dots\dots\dots\dots\dots\dots\dots\dots\dots\dots\dots\dots (4.1.12)$$

where τ_w is the shear stress at the pipe wall, given by

$$\tau_w = \frac{R}{2}\frac{\Delta p}{\Delta L} \quad\dots\dots\dots\dots\dots\dots\dots\dots\dots\dots\dots\dots\dots (4.1.13)$$

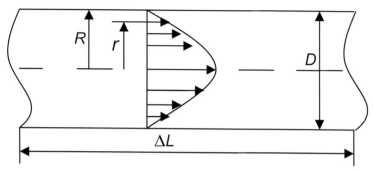

Fig. 4.1.5—Velocity profile in pipe flow.

Changing variables, Eq. 4.1.11 becomes

$$Q = -\pi \int_0^{\tau_w} \left(\frac{R}{\tau_w}\right)^3 \frac{dv}{dr} \tau^2 d\tau \quad \dots \dots \dots \dots \dots \dots \dots \dots \dots \dots \dots \quad (4.1.14)$$

Eq. 4.1.14 represents a general relationship between flow rate and shear stress. Denoting $\frac{dv}{dr} = f(\tau)$, differentiating with respect to τ_w, and upon rearrangement, we get

$$\frac{d\left(Q\tau_w^3\right)}{d\tau_w} = -\pi R^3 f(\tau_w)\tau_w^2 \quad \dots \dots \dots \dots \dots \dots \dots \dots \dots \dots \quad (4.1.15)$$

Hence, the shear rate at the pipe wall is

$$f(\tau_w) = \left(-\frac{dv}{dr}\right)_R = \dot{\gamma}_w = \frac{1}{\pi R^3 \tau_w^2} \frac{d\left(Q\tau_w^3\right)}{d\tau_w} \quad \dots \dots \dots \dots \dots \dots \dots \quad (4.1.16)$$

or

$$\dot{\gamma}_w = \frac{1}{\pi R^3} \tau_w \frac{dQ}{d\tau_w} + \frac{3Q}{\pi R^3} \quad \dots \dots \dots \dots \dots \dots \dots \dots \dots \dots \quad (4.1.17)$$

Because $\frac{Q}{\pi R^3} = \frac{2U}{D}$, Eq. 4.1.17 can be written in terms of mean velocity (U) and pipe diameter (D) as

$$\dot{\gamma}_w = \frac{\tau_w}{4} \frac{d\left(\frac{8U}{D}\right)}{d\tau_w} + \frac{3}{4}\left(\frac{8U}{D}\right) \quad \dots \dots \dots \dots \dots \dots \dots \dots \dots \quad (4.1.18)$$

Note that the following is true:

$$\tau_w \frac{d\left(Ln\tau_w\right)}{d\tau_w} = \left(\frac{8U}{D}\right)\frac{d\left(Ln\frac{8U}{D}\right)}{d\left(\frac{8U}{D}\right)} \quad \dots \dots \dots \dots \dots \dots \dots \quad (4.1.19)$$

From Eq. 4.1.19, we get

$$\frac{d\left(\frac{8U}{D}\right)}{d\tau_w} = \frac{\frac{8U}{D}}{\tau_w} \frac{d\left(Ln\frac{8U}{D}\right)}{d\left(Ln\tau_w\right)} \quad \dots \dots \dots \dots \dots \dots \dots \dots \quad (4.1.20)$$

Substituting Eqs. 4.1.20 to 4.1.18, we obtain

$$\dot\gamma_w = \frac{1}{4}\left[3+\frac{d\left(Ln\frac{8U}{D}\right)}{d\left(Ln\tau_w\right)}\right]\left(\frac{8U}{D}\right) \quad\dotfill\quad (4.1.21)$$

Introducing the flow behavior index (N), the above equation can be written as follows:

$$\dot\gamma_w = \left(\frac{3N+1}{4N}\right)\frac{8U}{D}, \quad\dotfill\quad (4.1.22)$$

where the flow behavior index, N, is expressed as:

$$N = \frac{d\left(Ln\tau_w\right)}{d\left(Ln\frac{8U}{D}\right)} \quad\dotfill\quad (4.1.23)$$

Pipe viscometer data (flow curve) are normally presented in terms of wall shear stress vs. nominal Newtonian shear rate ($8U/D$) on a logarithmic plot as schematically depicted in **Fig. 4.1.6a.** The flow behavior index N is the slope of the curve at a given shear stress. Once the value of the flow behavior index is known, the corresponding wall shear rate can be determined using Eq. 4.1.22 to plot the flow curve (wall shear stress vs. shear rate).

In general, we can rewrite the wall shear stress in terms of nominal Newtonian shear rate as follows:

$$\tau_w = K'\left(\frac{8U}{D}\right)^N, \quad\dotfill\quad (4.1.24)$$

where K' is the so-called generalized consistency index, which is a function of nominal Newtonian shear rate. If the log-log plot of wall shear stress vs. nominal Newtonian shear rate forms straight line, then we have a power-law fluid. In other words, the flow behavior index is constant and equal to the fluid behavior index, n. In other words, for power law fluids, $n = N$. Thus,

$$\tau = K\left(-\frac{dv}{dr}\right)^n \quad\dotfill\quad (4.1.25)$$

While K' and N are closely related to K and n, they are not the same.

Example 2. The data presented in **Table 4.1.3** come from a pipe viscometer with an inner diameter of 0.5 in. using 6% bentonite suspension, which has a specific gravity of approximately 1. Determine the YPL model parameters of the fluid.

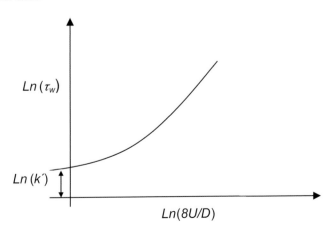

Fig. 4.1.6a—Wall shear stress vs. nominal Newtonian shear rate.

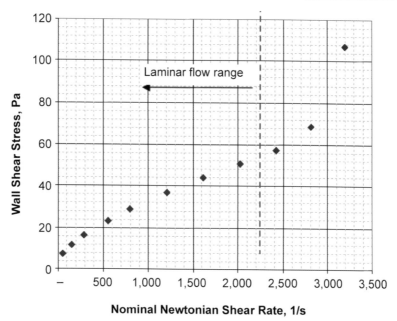

Fig. 4.1.6b—Wall shear stress vs. nominal Newtonian shear rate.

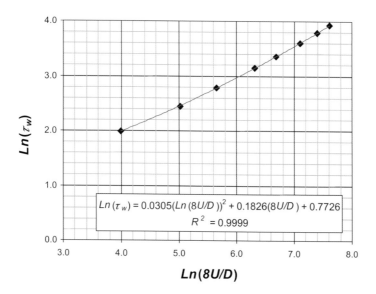

Fig. 4.1.6c—Flow curve: $Ln(\tau_w)$ vs. $Ln(8\,U/D)$.

Solution. To determine the rheological parameters, first the nominal Newtonian shear rate and wall shear stress should be calculated from the flow rate and pressure gradient (Eq. 4.1.13), respectively. Analysis of pipe viscometer data is presented in **Table 4.1.4.** The flow curve (Fig. 4.1.6b) can be plotted to screen out measurements that fall outside the laminar flow range. This can be done by careful examination of the trend of the flow curve for a sharp increase in wall shear stress as the nominal Newtonian shear rate increases. In this case, the last three data points in Fig. 4.1.6b appear to be out of the laminar flow range. Therefore, it is recommended to exclude these data points in the calculations of the rheological parameters. This is our first assumption; it should be verified using the Reynolds number when the correct rheological parameters are obtained. The remaining data points are presented in a logarithmic plot (Fig. 4.1.6c) to

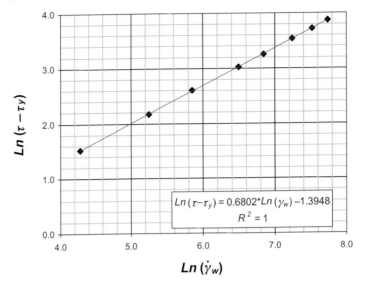

$$Ln(\tau-\tau_y) = 0.6802*Ln(\dot{\gamma}_w) - 1.3948$$
$$R^2 = 1$$

Fig. 4.1.6d—$Ln(\tau_w-\tau_y)$ vs. $Ln(\dot{\gamma}_w)$.

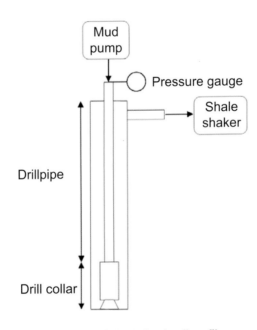

Fig. 4.1.6e—Schematic of well profile.

establish a relationship between N and $(8U/D)$ applying the polynomial curve fitting techniques used in Example 1. Accordingly, $N = 2*0.0305Ln(8U/D) + 0.1826$. After obtaining this relationship, the wall shear rate can be calculated using Eq. 4.1.22. The rheological parameters can be determined by plotting Lh $(\tau_w - \tau_y)$ vs. $Ln(\dot{\gamma}_w)$, as shown in Fig. 4.1.6d. Thus, $m = 0.68$, $Ln(K) = -1.3948$, and $K = 0.25$ Pas$^{0.68}$ (0.52 lbfs$^{0.68}$/100 ft^2). The results (Fig. 4.1.6d) are obtained using a yield stress value of 2.77 Pa (5.79 lbf/100 ft^2). In Table 4.1.4, the Reynolds number for each flow rate is calculated from these parameters. The Reynolds numbers for high flow rates (i.e., 10.17 gal/min and 8.96 gal/min, respectively) are more than 2,100, indicating non-laminar flow conditions. Therefore, other data points can be considered as laminar flow; this means that our previous assumption is valid.

TABLE 4.1.3—PIPE VISCOMETER DATA FOR 6% BENTONITE SUSPENSION										
Flow Rate (gal/min)	10.17	8.96	7.71	6.43	5.15	3.85	2.54	1.77	0.91	0.48	0.17
dp/dL [H₂O/in.]	3.43	2.21	1.84	1.63	1.41	1.18	0.92	0.75	0.52	0.37	0.24

TABLE 4.1.4—ANALYSIS OF PIPE VISCOMETER DATA FOR 6% BENTONITE SUSPENSION											
Q (gal/min)	dp/dL (Pa/m)	U (m/s)	$8\,U/D$ (/s)	T_w (Pa)	$Ln(8\,U/D)$	$Ln(\tau_w)$	N	$\dot\gamma_w$ (/s)	$Ln(\dot\gamma_w)$	$Ln(\tau_w - \tau_y)$	Re
10.17	33683.78	5.07	3191.90	106.95	8.07	4.67	0.67	3576.51	8.18	4.65	3041.83
8.96	21633.03	4.46	2810.32	68.68	7.94	4.23	0.67	3161.07	8.06	4.19	2555.45
7.71	18025.62	3.84	2418.42	57.23	7.79	4.05	0.66	2732.89	7.91	4.00	2079.77
6.43	15971.82	3.20	2017.38	50.71	7.61	3.93	0.65	2292.81	7.74	3.87	1620.74
5.15	13864.68	2.57	1616.28	44.02	7.39	3.78	0.63	1850.29	7.52	3.72	1193.46
3.85	11606.21	1.92	1207.06	36.85	7.10	3.61	0.62	1395.61	7.24	3.53	795.83
2.54	9059.11	1.27	797.59	28.76	6.68	3.36	0.59	936.05	6.84	3.26	445.52
1.77	7347.25	0.88	554.05	23.33	6.32	3.15	0.57	659.41	6.49	3.02	265.95
0.91	5095.58	0.45	284.81	16.18	5.65	2.78	0.53	348.63	5.85	2.60	101.84
0.48	3624.32	0.24	151.50	11.51	5.02	2.44	0.49	191.11	5.25	2.17	39.90
0.17	2305.05	0.09	53.77	7.32	3.98	1.99	0.43	71.91	4.28	1.52	7.98

4.1.3 Flow in Pipes and Circular Tubes. For steady isothermal flow of incompressible fluid in pipe, the axial momentum balance yields the relationship between shear stress distribution in the pipe and pressure gradient as

$$\tau = \frac{r}{2}\frac{dp}{dz} \quad\quad\quad\quad\quad\quad (4.1.26)$$

Hence, the wall shear stress at the wall can be calculated as

$$\tau_w = \frac{R}{2}\frac{dp}{dz}, \quad\quad\quad\quad\quad\quad (4.1.27)$$

where R is the radius of the pipe. The Fanning friction factor (friction factor) for pipe flow is defined as

$$f = \frac{\tau_w}{\frac{1}{2}\rho U^2} \quad\quad\quad\quad\quad\quad (4.1.28)$$

Newtonian Fluid. Laminar Flow. For laminar flow of Newtonian fluid in pipe with diameter D, the relationship between the pressure gradient and mean flow velocity can be obtained by combining the momentum (Eq. 4.1.26), constitutive (Eq. 4.1.1), and continuity (Eq. 4.1.10) equations and applying the boundary conditions. Thus,

$$\frac{dp}{dz} = \frac{32\mu U}{D^2} \quad\quad\quad\quad\quad\quad (4.1.29)$$

The pressure loss can be expressed in terms of friction factor as

$$\frac{dp}{dz} = \frac{2f\rho U^2}{D}, \quad\quad\quad\quad\quad\quad (4.1.30)$$

where

$$f = \frac{16}{Re} \quad \dots\dots\dots\dots\dots\dots\dots\dots\dots\dots\dots\dots\dots\dots\dots\dots\dots\dots\dots \quad (4.1.31)$$

and

$$Re = \frac{\rho U D}{\mu} \quad \dots\dots\dots\dots\dots\dots\dots\dots\dots\dots\dots\dots\dots\dots\dots\dots\dots \quad (4.1.32)$$

If the Reynolds number in a straight circular pipe is less than 2,100, then the flow is laminar. Consequently, the wall shear stress can be determined directly from the wall rate ($8\,U/D$) as

$$\tau_w = \mu \frac{8U}{D} \quad \dots\dots\dots\dots\dots\dots\dots\dots\dots\dots\dots\dots\dots\dots\dots\dots \quad (4.1.33)$$

Turbulent Flow. When the flow is turbulent, the pressure loss is calculated using Eq. 4.1.30. However, the turbulent friction factor correlations must be used to estimate the friction factor, f. It is important to note that the friction pressure loss under turbulent flow conditions is strongly affected by the pipe wall roughness. For Newtonian flow in smooth and rough pipes, the Colebrook equation presented here can be applied to determine the friction factor:

$$\frac{1}{\sqrt{f}} = -4\log\left[\frac{\varepsilon/D}{3.7} + \frac{1.255}{Re\sqrt{f}}\right], \quad \dots\dots\dots\dots\dots\dots\dots\dots\dots \quad (4.1.34)$$

where ε is the absolute roughness of the pipe wall.

Power-Law Fluid. *Laminar Flow.* The pressure loss for laminar pipe flow of power-law fluids is determined by applying the momentum equation presented previously (Eq. 4.1.27) as

$$\frac{dp}{dz} = \frac{4\tau_w}{D} \quad \dots\dots\dots\dots\dots\dots\dots\dots\dots\dots\dots\dots\dots\dots\dots\dots\dots \quad (4.1.35)$$

Integrating Eq. 4.1.14 for power law ($\tau = K\dot{\gamma}^n$), after some rearrangements the wall shear stress is expressed as

$$\tau_w = K\left(\frac{3n+1}{4n}\frac{8U}{D}\right)^n \quad \dots\dots\dots\dots\dots\dots\dots\dots\dots\dots\dots\dots \quad (4.1.36)$$

Eq. 4.1.36 is valid when the flow is laminar. That means that the Reynolds number defined below must be less than the critical Reynolds number (i.e., 2,100).

$$Re = \frac{D^n U^{n-2} \rho}{8^{n-1} K} \quad \dots\dots\dots\dots\dots\dots\dots\dots\dots\dots\dots\dots\dots\dots\dots \quad (4.1.37)$$

Turbulent Flow. The rheological complexity of non-Newtonian fluids coupled with random motion of fluid particles (turbulent eddies) make the mathematical treatment of turbulent non-Newtonian flow very difficult. As a result, our knowledge of non-Newtonian turbulent flow is largely restricted to semiempirical correlations. Based on experimental investigation and semitheoretical analysis, Dodge and Metzner (1959) developed a friction factor correlation for turbulent flow of non-Newtonian fluids through smooth pipes. For power-law fluids, the correlation is presented in this form:

$$\frac{1}{f^{0.5}} = \frac{4}{n^{0.75}}\log\left[Re f^{(1-n/2)}\right] - \frac{0.4}{n^{1.2}} \quad \dots\dots\dots\dots\dots\dots \quad (4.1.38)$$

For turbulent flow in rough pipe, the friction factor can be estimated using a similar equation proposed by Szilas et al. (1981):

$$\frac{1}{\sqrt{f}} = -4.0\log\left[\frac{\varepsilon/D}{3.7} + \frac{10^{-\beta/2}}{Re(4f)^{(2-n)/2n}}\right] \quad \dots\dots\dots\dots\dots\dots \quad (4.1.39)$$

and

$$\beta = 1.51^{1/n} \left(\frac{0.707}{n} + 2.12 \right) - \frac{4.015}{n} - 1.057, \quad \dots \dots \dots \quad (4.1.40)$$

where n is the power-law index.

Example 3. A drillstring is composed of 3,000 ft of 3½-in. drillpipe [inside diamtmeter (ID)= 2.6 in.] and a short drill collar. Determine pressure losses in the drillpipe when the mud flow rates are 100 and 300 gal/min. Consider the drillpipe as smooth. The drilling fluid used in the particular case best fits the power law model with a consistency index of 1.04 lbfs$^{0.6}$/100 ft^2 (0.50 Pas$^{0.6}$) and a fluid behavior index of 0.6. The density of the mud is 8.33 lbm/gal (1000 kg/m^3).

Solution 1. Mud flow rate = 100 gal/min: The mean flow velocity in the pipe is 1.84 m/s. If we assume laminar flow conditions in the pipe, the wall shear stress can be determined by applying Eq. 4.1.36:

$$\tau_w = K \left(\frac{3n+1}{4n} \frac{8U}{D} \right)^n = 14.1 \text{ Pa}$$

The Reynolds number can be estimated using Eq. 4.1.37:

$$\text{Re} = \frac{8\rho U^2}{\tau_w} = 1926$$

Hence, our laminar flow assumption is verified. The pressure loss can be calculated using Eq. 4.1.35 as

$$\Delta p = \frac{4\tau_w}{D} \Delta L = 777529.1 \text{ Pa } (112.8 \text{ psi}).$$

Solution 2. For 300 gal/min mud flow, the mean velocity is 5.52 m/s. Assuming laminar flow conditions; the wall shear stress becomes 27.16 Pa. The corresponding Reynolds number is 8965.6, which indicates turbulent flow conditions. Thus, the pressure loss must be determined using the Fanning friction factor (Eq. 4.1.38). When Re = 8965.6, the value of f becomes 0.0057. Applying Eq. 4.1.28, the correct wall shear stress is obtained as

$$\tau_w = \frac{1}{2} f \rho U^2 = 86.85 \text{ Pa}$$

Using Eq. 4.1.35, the pressure loss is determined as:

$$-\Delta p = \frac{4\tau_w}{D} \Delta L = 4806237 \text{ Pa } (697 \text{ psi})$$

YPL Fluid. *Laminar Flow.* By integrating Eq. 4.1.14, the exact analytical solution for steady, isothermal pipe flow of YPL fluid under laminar conditions can be obtained. After rearrangements, the equation can be expressed as:

$$\frac{8U}{D} = \frac{(\tau_w - \tau_y)^{(1+1/m)}}{K^{1/m} \tau_w^3} \left(\frac{4m}{3m+1} \right) \left[\tau_w^2 + \frac{2m}{1+2m} \tau_y \tau_w + \frac{2m^2}{(1+m)(1+2m)} \tau_y^2 \right] \quad \dots \dots \dots \quad (4.1.41)$$

Eq. 4.1.41 requires numerical solution to determine the shear stress (τ_w) for a given mean velocity. However, an explicit equation for wall shear stress can be obtained when the ratio of yield stress to the wall shear stress, τ_y/τ_w, is small.

The generalized Reynolds number for flow of YPL fluid in a circular tube is given by

$$\text{Re} = \frac{\rho U^{2-N} D^N}{K' 8^{N-1}} \quad \dots \dots \dots \dots \dots \dots \quad (4.1.42)$$

and, differentiating Eq. 4.1.41:

$$\frac{1}{N} = \frac{(1-2m)\tau_w + 3m\tau_y}{m(\tau_w - \tau_y)} + \frac{2m(1+m)\left[(1+2m)\tau_w^2 + m\tau_y\tau_w\right]}{m(1+m)(1+2m)\tau_w^2 + 2m^2(1+m)\tau_w\tau_y + 2m^3\tau_y^2} \quad \dots\dots\dots\dots \quad (4.1.43)$$

The generalized consistency index, K', is expressed as

$$K' = \frac{\tau_w}{\left(\frac{8U}{D}\right)^N} \quad \dots\dots\dots\dots\dots\dots\dots\dots\dots\dots\dots\dots\dots\dots\dots\dots\dots\dots\dots\dots\dots \quad (4.1.44)$$

The wall shear stress in Eqs. 4.1.43 and 4.1.44 must be obtained from Eq. 4.1.41 at the actual nominal Newtonian shear rate ($8U/D$). The critical Reynolds number for the transition from laminar to turbulent flow condition for YPL fluids is approximately 2,100. The critical Reynolds number **(Fig. 4.1.7)** slightly increases with the flow behavior index, N. Some drilling fluids with strong viscoelastic behaviors, such as polymer-based fluids, have a tendency of delaying this transition.

Turbulent Flow. Dodge and Metzner (1959) performed extensive experimental investigation and semitheoretical analysis for turbulent flow of non-Newtonian fluids through smooth pipes. After systematic manipulation of the mean velocity profile and application of conservation equations, they developed a general form of friction factor correlation for all time-independent and nonelastic fluids. The correlation is given by

$$\frac{1}{f^{0.5}} = \frac{4}{N^{0.75}} \log\left[\mathrm{Re} f^{(1-N/2)}\right] - \frac{0.4}{N^{1.2}} \quad \dots\dots\dots\dots\dots\dots\dots\dots\dots \quad (4.1.45)$$

The validity of the correlation has been verified with some non-power-law fluids (clay suspensions and polymeric gels). The graphic form of this equation is presented in Fig. 4.1.7. The solid curves represent the original data, while the dotted curves show the anticipated pattern of friction factor curves based on Eq. 4.1.45. Special care should be taken when this equation is used beyond the solid lines.

By analyzing Eqs. 4.1.34 and 4.1.45, Reed and Pilehvari (1993) proposed a modified form of the Colebrook equation (Eq. 4.1.34) to estimate the friction factor in rough pipe for YPL fluid by introducing the equivalent diameter concept. However, to the best of our knowledge, no experimental verification has yet been published. The modified form for the friction factor in rough pipes is

$$\frac{1}{\sqrt{f}} = -4.0\log\left[\frac{\varepsilon/D_{\mathrm{eff}}}{3.7} + \frac{1.26^{N-1.2}}{\left(\mathrm{Re} f^{(1-N/2)}\right)^{N-0.75}}\right], \quad \dots\dots\dots\dots\dots\dots\dots \quad (4.1.46)$$

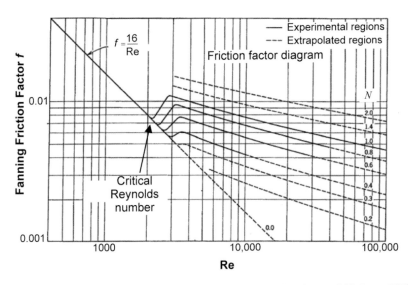

Fig. 4.1.7—Fanning friction factor vs. generalized Reynolds number (Dodge and Metzner 1959). Reprinted with permission from the American Institute of Chemical Engineers.

where the effective diameter, D_{eff}, is defined as

$$D_{eff} = \frac{4N}{3N+1} D \quad\quad\quad\quad\quad\quad\quad\quad\quad\quad\quad\quad\quad (4.1.47)$$

Example 4. Consider again Example 3 with another drilling fluid that best fits the YPL model with a consistency index of 1.044 lbfs$^{0.5}$/100 ft^2 (0.5 Pas$^{0.5}$), a fluid behavior index of 0.5, and yield stress of 10.44 lbf/100 ft^2 (5.0 Pa). Calculate drillpipe pressure losses at flow rates of 100 gal/min and 300 gal/min.

Solution 1. The mean flow velocity in the pipe is 1.84 m/s for a mud flow rate of 100 gal/min. Assuming laminar flow conditions, the wall shear stress of $\tau_w = 20.33$ Pa is numerically obtained from Eq. 4.1.41. The Reynolds number can be estimated using Eq. 4.1.42 as

$$\mathrm{Re} = \frac{8\rho U^2}{\tau_w} = 1330.$$

The flow is laminar, and the pressure loss can be calculated using Eq. 4.1.35 as

$$\Delta p = \frac{4\tau_w}{D} \Delta L = 1.13 \text{ MPa (163 psi)}.$$

Solution 2. For 300 gal/min mud flow, assuming laminar flow conditions, the wall shear stress becomes 33.48 Pa. The corresponding Reynolds number is 7273, suggesting turbulent flow conditions. Hence, the pressure loss should be calculated based on the Fanning friction factor using Eq. 4.1.45. This equation required the evaluation of the flow behavior index (N) using the laminar flow based wall shear stress (i.e., 33.48 Pa). Applying Eq. 4.1.43, we obtain an N value of 0.49. Subsequently, the value of f becomes 0.0052. For the turbulent flow, the wall shear stress can be calculated as:

$$\tau_w = \frac{1}{2} f \rho U^2 = 78.90 \text{ Pa},$$

and the friction pressure loss is:

$$\Delta p = \frac{4\tau_w}{D} \Delta L = 4.37 \text{ MPa (633 psi)}.$$

4.1.4 Flow in Annulus. In drilling applications, annular pressure loss analysis is very critical in determining the bottomhole pressure. The drillpipe can be concentric or eccentric, depending on the wellbore configurations. Eccentricity and diameter ratio (i.e., the ratio of diameter of the drillpipe to diameter of the wellbore) are very important geometric parameters of the annular flows. Typical drillpipe diameters, borehole diameters, and diameter ratios are presented in **Table 4.1.5.** For conventional drilling applications, the diameter ratio ranges from 0.4 to 0.6. Other special drilling applications such as casing drilling and coiled tubing may have diameter ratios that are out of the regular range.

Concentric Annulus. The equation of motion for steady incompressible flow in a concentric annulus of uniform cross section can be written using the cylindrical coordinate system (r, θ, z) as

$$\frac{\partial p}{\partial z} + \frac{1}{r} \frac{\partial}{\partial r} \left(r \tau_{rz} \right) = 0 \quad\quad\quad\quad\quad\quad\quad\quad\quad\quad (4.1.48)$$

TABLE 4.1.5—TYPICAL DRILLPIPE AND HOLE SIZES		
Pipe Diameter (in.)	Hole Diameter (in.)	Diameter Ratio
5	12^{1}/4	0.4082
4^{1}/2	8^{3}/4	0.5143
3^{1}/2	6^{3}/4	0.5185
2^{7}/8	6^{1}/4	0.4600
2^{3}/8	5^{1}/2	0.4318

For a constant pressure gradient (i.e., constant dp/dL), the above equation can be integrated to give the stress distribution in concentric annuli as

$$\tau_{rz} = -\frac{R_o}{2}\frac{dp}{dz}\left(\frac{r}{R_o} - \beta^2 \frac{R_o}{r}\right), \quad \dots\dots\dots\dots\dots\dots\dots\dots\dots\dots\dots\dots\dots \quad (4.1.49)$$

where $\beta = \lambda/R_o$ and λ represents the location in the annulus where the value of stress, τ_{rz}, vanishes. Eq. 4.1.49 clearly shows the presence of nonlinear shear stress distribution in the annulus. Furthermore, for a concentric annulus, the shear stresses at the inner and outer pipes are not identical. For the purpose of hydraulic calculations, we introduce the average wall shear stress ($\overline{\tau}_w$) that can be obtained by applying the axial momentum balance as

$$(p_i + p_o)\overline{\tau}_w = p_i\tau_{wi} + p_o\tau_{wo} = \frac{\pi}{4}(D_o^2 - D_i^2)\frac{dp}{dz}, \quad \dots\dots\dots\dots\dots\dots \quad (4.1.50)$$

where p_i and p_o are wetted perimeters calculated as $p_i = \pi D_i$ and $p_o = \pi D_o$. After simplification, Eq. 4.1.50 yields expressions for the average wall shear stress in terms of wall shear stresses or friction pressure gradient:

$$\overline{\tau}_w = \frac{\tau_{wo}D_o + \tau_{wi}D_i}{D_o + D_i} \quad \dots\dots\dots\dots\dots\dots\dots\dots\dots\dots\dots\dots\dots \quad (4.1.51)$$

or

$$\overline{\tau}_w = \frac{dp}{dz}\frac{D_{\text{hyd}}}{4}, \quad \dots\dots\dots\dots\dots\dots\dots\dots\dots\dots\dots\dots\dots\dots \quad (4.1.52)$$

where D_{hyd} is the hydraulic diameter of the annulus. By analogy to pipe flow, the friction factor for annular flow can be defined using the average wall shear stress as

$$f = \frac{\overline{\tau}_w}{\frac{1}{2}\rho U^2} \quad \dots\dots\dots\dots\dots\dots\dots\dots\dots\dots\dots\dots\dots\dots\dots \quad (4.1.53)$$

Newtonian Fluid. *Laminar Flow.* For Newtonian laminar flow, Lamb (1945) developed an analytical solution for Eq. 4.1.48. Applying this solution, it is possible to predict the pressure gradient directly from the mean velocity as

$$\frac{dp}{dz} = \frac{32\mu U}{D_L^2}, \quad \dots\dots\dots\dots\dots\dots\dots\dots\dots\dots\dots\dots\dots\dots \quad (4.1.54)$$

where D_L is the Lamb diameter given by

$$D_L^2 = D_o^2 + D_i^2 - \frac{D_o^2 - D_i^2}{Ln\frac{D_o}{D_i}} \quad \dots\dots\dots\dots\dots\dots\dots\dots\dots\dots \quad (4.1.55)$$

The above equation is valid when the flow condition is laminar. For the sake of consistency with pipe flow, we define the Reynolds as

$$\text{Re}_{\text{ann}} = \frac{D_{\text{eq}}U\rho}{\mu}, \quad \dots\dots\dots\dots\dots\dots\dots\dots\dots\dots\dots\dots\dots\dots\dots\dots \quad (4.1.56)$$

where the equivalent diameter, D_{eq}, is given by

$$D_{\text{eq}} = \frac{D_L^2}{D_{\text{hyd}}} \quad \dots\dots\dots\dots\dots\dots\dots\dots\dots\dots\dots\dots\dots\dots\dots \quad (4.1.57)$$

Then, the relationship between friction factor and Reynolds number can be established as

$$f = \frac{16}{\text{Re}_{\text{ann}}} \quad \dots\dots\dots\dots\dots\dots\dots\dots\dots\dots\dots\dots\dots\dots\dots\dots \quad (4.1.58)$$

Turbulent Flow. Experimental results suggest that laminar flow becomes unstable when the Reynolds number, defined by Eq. 4.1.56, exceeds the 2,100 limit. For turbulent flow, a pressure loss expression can be obtained by inserting Eq. 4.1.52 into Eq. 4.1.53. Thus, we get

$$\frac{dp}{dz} = \frac{2f\,\rho U^2}{D_{\text{hyd}}} \quad \dots\dots\dots\dots\dots\dots\dots\dots\dots\dots\dots\dots\dots\dots\dots\dots\dots\dots\dots \quad (4.1.59)$$

The Colebrook equation (Eq. 4.1.34) can be adopted to determine the friction factor in the annulus as

$$\frac{1}{\sqrt{f}} = -4.0\log\left[\frac{\varepsilon/D_{\text{eq}}}{3.7} + \frac{1.255}{\text{Re}_{\text{ann}}\sqrt{f}}\right], \quad \dots\dots\dots\dots\dots\dots\dots\dots\dots\dots\dots\dots\dots \quad (4.1.60)$$

where ε is the absolute roughness of the annulus. The roughness is assumed to be zero for relatively smooth pipes and annuli. The roughness of commercial steel pipe is approximately 50 μm. For a smooth annulus, the above equation reduces to

$$\frac{1}{\sqrt{f}} = -4.0\log\left[\frac{1.255}{\text{Re}_{\text{ann}}\sqrt{f}}\right] \quad \dots\dots\dots\dots\dots\dots\dots\dots\dots\dots\dots\dots\dots\dots\dots \quad (4.1.61)$$

Power-Law Fluid. *Laminar Flow.* Laminar flow of non-Newtonian fluid in concentric annuli has been studied by several investigators, including Fredrickson and Bird (1958). Different hydraulic calculation procedures have been developed to predict the annular pressure loss. Commonly used procedures are the exact method developed by Fredrickson and Bird (1958) and the approximate solution proposed by Whittaker (1985).

Exact Method. An analytical solution for laminar flow of power-law fluid in a concentric annulus has been presented by Fredrickson and Bird (1958) as

$$Q = \frac{n}{3n+1}\pi R_o^{\frac{3n+1}{n}}\left(\frac{1}{2K}\frac{dp}{dz}\right)^n\left[\left(1-\beta^2\right)^{1+\frac{1}{n}} - \kappa^{1-\frac{1}{n}}\left(\beta^2-\kappa^2\right)^{1+\frac{1}{n}}\right], \quad \dots\dots\dots\dots\dots\dots \quad (4.1.62)$$

where κ is the diameter ratio (i.e., $\kappa = R_i/R_o$) shown in **Fig. 4.1.8.** The parameter β in Eq. 4.1.62 is obtained numerically by solving the following integral equation:

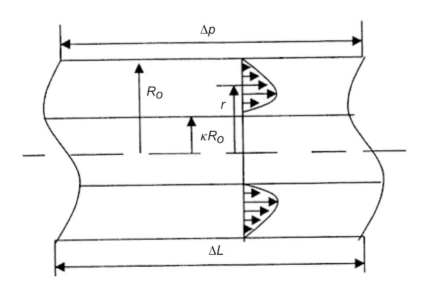

Fig. 4.1.8—Axial velocity profile for concentric annular flow.

$$\int_{\kappa}^{\beta}\left(\frac{\beta^2}{\xi}-\xi\right)^{\frac{1}{n}}d\xi = \int_{\beta}^{1}\left(\xi-\frac{\beta^2}{\xi}\right)^{\frac{1}{n}}d\xi \quad \dots\dots\dots\dots\dots\dots\dots\dots\dots\dots \quad (4.1.63)$$

The value of the dimensionless radius, ξ, ranges from κ to 1, where $\xi = r/R_o$. The parameter β indicates the location of the maximum velocity in the annulus in terms of dimensionless radius. Hanks and Larsen (1979) presented tabulated values for this parameter. For practical applications, the values of β can be estimated by (Chen et al. 2007)

$$\beta = 0.9904 \cdot \kappa^{0.4141} \cdot n^{0.01238} \quad \dots\dots\dots\dots\dots\dots\dots\dots\dots\dots \quad (4.1.64)$$

Fredrickson and Bird (1958) proposed the following expression to calculate the Reynolds number in the annulus.

$$\mathrm{Re}_{ann} = \frac{D_o^n U^{2-n}\rho(1+\kappa)}{2^{n-3}K(1-\kappa^2)^{n+1}}\Omega^n, \quad \dots\dots\dots\dots\dots\dots\dots\dots\dots\dots \quad (4.1.65)$$

where the dimensionless flow rate, Ω, is given by

$$\Omega = \frac{n\Gamma(1-\kappa)^{\frac{2n+1}{n}}}{2n+1} \quad \dots\dots\dots\dots\dots\dots\dots\dots\dots\dots \quad (4.1.66)$$

The value of the parameter Γ is obtained from **Fig. 4.1.9**. The annular flow is considered laminar when the Reynolds number is less than 2,100.

Approximate Method. Another way of calculating the laminar pressure loss in the annulus for power-law fluid is to use the Exlog solution (Whittaker 1985). According to this method, the pressure loss is calculated as

$$\frac{dp}{dz} = \left(\frac{4K}{D_{hyd}}\right)\left(\frac{8UG}{D_{hyd}}\right)^n, \quad \dots\dots\dots\dots\dots\dots\dots\dots\dots\dots \quad (4.1.67)$$

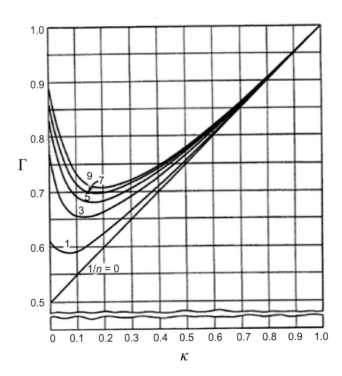

Fig. 4.1.9—Parameter Γ as a function of K and $1/n$ (Fredrickson and Bird 1958). Reprinted with permission from *Industrial and Chemical Engineering*. Copyright 1958 American Chemical Society.

where the parameters G, Z, and Y are given by

$$G = \frac{\left(1 + \dfrac{Z}{2}\right)[(3-Z)n+1]}{n(4-Z)} \dots\dots\dots\dots\dots\dots\dots\dots\dots\dots\dots\dots\dots \quad (4.1.68)$$

$$Z = 1 - \left[1 - \left(\frac{D_i}{D_o}\right)^Y\right]^{\frac{1}{Y}} \dots\dots\dots\dots\dots\dots\dots\dots\dots\dots\dots\dots\dots \quad (4.1.69)$$

$$Y = 0.37 n^{-0.14} \dots\dots\dots\dots\dots\dots\dots\dots\dots\dots\dots\dots\dots\dots\dots \quad (4.1.70)$$

The relevant Reynolds number for the flow is expressed as

$$\mathrm{Re}_{ann} = \frac{D_{eff}^n \rho U^{2-n}}{8^{n-1} K}, \dots\dots\dots\dots\dots\dots\dots\dots\dots\dots\dots\dots\dots \quad (4.1.71)$$

where D_{eff} is the effective diameter of the annulus given by

$$D_{eff} = \frac{D_{hyd}}{G} \dots\dots\dots\dots\dots\dots\dots\dots\dots\dots\dots\dots\dots\dots\dots \quad (4.1.72)$$

The Reynolds number presented in Eq. 4.1.71 must be less than 2,100 to have laminar flow conditions in the annulus.

Turbulent Flow. Laminar flow calculation methods presented in the previous section have also established procedures for determining the pressure loss under turbulent flow conditions.

Fredrickson and Bird Method. This approach recommends the following equation for determining the pressure loss in the annulus:

$$\frac{dp}{dz} = f \frac{\rho U^2 (1+\kappa)}{R_o (1-\kappa^2)}, \dots\dots\dots\dots\dots\dots\dots\dots\dots\dots\dots\dots\dots \quad (4.1.73)$$

where f is the Newtonian friction factor developed for smooth pipe (Eq. 4.1.61).

Exlog Solutions. According to this method, the friction factor is estimated applying the Dodge and Metzner (1959) equation for power-law fluids as

$$\frac{1}{f^{0.5}} = \frac{4}{n^{0.75}} \log\left[\mathrm{Re} f^{(1-n/2)}\right] - \frac{0.4}{n^{1.2}} \dots\dots\dots\dots\dots\dots\dots\dots\dots \quad (4.1.74)$$

After determining the friction factor, the pressure loss is calculated using Eq. 4.1.59.

YPL Fluid. *Laminar Flow.* A number of studies (Laird 1957; Hanks 1979) have been conducted to develop hydraulic models for annular flow of YPL fluids. Laird (1957) presented an exact solution for laminar flow of Bingham fluid in a concentric annulus. The solution requires an iterative procedure to determine plug boundaries. Hanks (1979) studied laminar flow of YPL fluids in a concentric annulus and developed design charts for the computation of the flow rate in terms of pressure drop, or vice versa. To date, however, no analytical solution for flow of YPL fluid in a concentric annulus has been published. Available methods are direct numerical solution and narrow-slot approximation.

Narrow-Slot Approximation. This approach represents the annulus with an equivalent narrow rectangular slot that has width w and height h, defined as

$$h = \frac{D_o - D_i}{2} \dots\dots\dots\dots\dots\dots\dots\dots\dots\dots\dots\dots\dots\dots\dots \quad (4.1.75)$$

and

$$w = \frac{\pi(D_o + D_i)}{2} \dots\dots\dots\dots\dots\dots\dots\dots\dots\dots\dots\dots\dots\dots \quad (4.1.76)$$

Hence, the equivalent narrow slot has the same cross-sectional area as the annulus. This method is reasonably accurate when the diameter ratio $D_i/D_o > 0.3$. The exact analytical solution for steady isothermal flow of YPL fluid in a narrow slot is expressed as

$$Q = \frac{wh^2(\tau_w - \tau_y)^{\frac{m+1}{m}}}{2K^{\frac{1}{m}}\tau_w^2}\left(\frac{m}{1+2m}\right)\left(\tau_w + \frac{m}{m+1}\tau_y\right) \qquad \dots \qquad (4.1.77)$$

The above equation can be adopted for the annulus using expressions presented in Eqs. 4.1.75 and 4.1.76. Thus,

$$\frac{12U}{D_o - D_i} = \frac{(\overline{\tau}_w - \tau_y)^{\frac{m+1}{m}}}{K^{\frac{1}{m}}\overline{\tau}_w^2}\left(\frac{3m}{1+2m}\right)\left(\overline{\tau}_w + \frac{m}{m+1}\tau_y\right) \qquad \dots \qquad (4.1.78)$$

The pressure loss can be determined from the average shear stress by applying Eq. 4.1.52. The Reynolds number of the flow is calculated as

$$\text{Re}_{ann} = \frac{12\rho U^2}{\tau_y + K\dot{\gamma}_w^m}, \qquad \dots \qquad (4.1.79)$$

where the wall shear rate is given by

$$\dot{\gamma}_w = \frac{1+2N}{3N}\frac{12U}{D_o - D_i} \qquad \dots \qquad (4.1.80)$$

and the value of the flow behavior index, N, is calculated from the following equation:

$$\frac{3N}{1+2N} = \left(\frac{3m}{1+2m}\right)\left[1 - \left(\frac{1}{1+m}\right)\frac{\tau_y}{\overline{\tau}_w} - \left(\frac{m}{1+m}\right)\left(\frac{\tau_y}{\overline{\tau}_w}\right)^2\right] \qquad \dots \qquad (4.1.81)$$

The average shear stress in Eq. 4.1.81 must be obtained from the laminar flow equation (Eq. 4.1.78).

Example 5. A drillstring (Fig. 4.1.6e) composed of 14,000 ft of 4½ in. drillpipe (ID = 3.826 in.) and 1,000 ft of 6¼ in. drill collar (ID = 2.5 in.) is used to drill an 8¾ in. well. A 12.5-lbm/gal YPL fluid having a consistency index of 0.21 lbfs$^{0.6}$/100 ft^2 (0.10 Pas$^{0.6}$), a fluid behavior index of 0.6, and yield stress of 14.6 lbf/100 ft^2 (7.00 Pa) has been used. Neglect the effect of temperature on fluid rheology. The drillpipe, drill collar, and wellbore walls are considered smooth. Determine annular pressure loss at a flow rate of 300 gal/min.

Solution. Flow in the annulus formed by the drillpipe and the wellbore: At 300 gal/min, the mean annular velocity is 0.66 m/s. Assuming laminar flow conditions, Eq. 4.1.78 can be solved numerically to obtain wall shear stress. For this part of the annulus, the wall shear stress becomes 10.73 Pa. Then, the Reynolds number can be calculated as:

$$\text{Re}_{ann} = \frac{12\rho U^2}{\overline{\tau}_w} = 738.$$

The Reynolds number indicates laminar flow conditions, and the wall shear stress obtained from Eq. 4.1.78 can be used to calculate the pressure loss as:

$$\Delta p_{pipe\text{-}zone} = \frac{4\overline{\tau}_w}{D_{hyd}}\Delta L = 1.70 \text{ MPa (246 psi)}$$

In the collar zone, the annular velocity is 1.0 m/s. Again assuming laminar flow conditions and applying Eq. 4.1.78, the wall shear stress becomes 11.16 Pa. The Reynolds number is:

$$\text{Re}_{ann} = \frac{12\rho U^2}{\overline{\tau}_w} = 1600$$

Since $\text{Re}_{ann} < 2,100$, the flow is laminar, and the pressure loss is calculated from the wall shear stress as

$$\Delta p_{\text{Collar}-\text{zone}} = \frac{4\bar{\tau}_w}{D_{\text{hyd}}} \Delta L = 0.21 \, \text{MPa} \, (31 \, \text{psi})$$

The annular pressure is the sum of the pressure loss in the drillpipe and drill collar zones. Hence,

$$\Delta p_{\text{ann}} = \Delta p_{\text{Collar}-\text{zone}} + \Delta p_{\text{Collar}-\text{zone}} = 1.91 \, \text{MPa} \, (277 \, \text{psi})$$

Direct Numerical Solution. **Fig. 4.1.10** shows a typical velocity profile for the flow of YPL fluid in a concentric annulus. Due to the yield stress, there will be an unshared portion of the fluid, which moves as a solid plug that has a ring shape with inner and outer radii of a and b, respectively. The plug velocity, U_p, is constant, and the shear stresses at the plug boundaries are equal to the yield stress, τ_y.

In sheared Region I, the momentum balance for any ring **(Fig. 4.1.11)** with the outer radius (r) and inner radius (R_i) is given by

$$\Delta p \pi \left(r^2 - R_i^2\right) = 2\pi \Delta L \left(\tau_{w,i} R_i - \tau r\right), \quad \dotfill \quad (4.1.82)$$

where $\tau_{w,i}$ is the shear stress at the pipe wall and $\Delta p = p_1 - p_2$. After simplification, Eq. 4.1.82 can be written in this form:

$$\tau r = \frac{-\Delta p \left(r^2 - R_i^2\right)}{2\Delta L} + \tau_{w,i} R_i \quad \dotfill \quad (4.1.83)$$

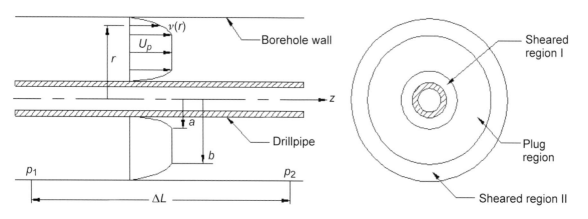

Fig. 4.1.10—Laminar flow of YPL fluid in concentric annulus.

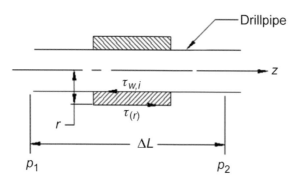

Fig. 4.1.11— Shear stresses in Region I.

The shear stress at the inner plug boundary (i.e., $r = a$) is equal to the yield stress. Hence, Eq. 4.1.84 is written for the inner plug boundary as

$$\tau_y a = \frac{-\Delta p \left(a^2 - R_i^2 \right)}{2 \Delta L} + \tau_{w,i} R_i \qquad (4.1.84)$$

Since the velocity gradient, dv/dr, is positive in this region, the constitutive equation can be expressed as

$$\tau = \tau_y + K \left(\frac{dv}{dr} \right)^m \qquad (4.1.85)$$

An expression for the velocity gradient can be obtained by combining Eqs. 4.1.83 and 4.1.85. Thus,

$$\frac{dv}{dr} = \left(c_{1,i} r + \frac{c_{2,i}}{r} - c_{3,i} \right)^{1/m}, \qquad (4.1.86)$$

where

$$c_{1,i} = \frac{-\Delta p}{2 \Delta L K} \qquad c_{2,i} = \frac{R_i}{K} \left(\tau_{w,i} + \frac{\Delta p R_i}{2 \Delta L} \right) \qquad c_{3,i} = \frac{\tau_y}{K}$$

If we assume a no-slip condition at the wall, an expression for the velocity profile can be obtained by integrating Eq. 4.1.86. Hence, the velocity profile in Region I (i.e., for $R_i \le r \le a$) is given by

$$v_1(r) = \int_{R_i}^{r} (c_{1,i} r + c_{2,i} / r - c_{3,i})^{1/m} dr \qquad (4.1.87)$$

The velocity at the plug boundary (i.e., $r = a$) can be determined as

$$U_p = v_1(a) = \int_{R_i}^{a} (c_{1,i} r + c_{2,i} / r - c_{3,i})^{1/m} dr \qquad (4.1.88)$$

Similarly, the momentum balance in sheared Region II (i.e., $b \le r \le R_o$) can be written for any ring (**Fig. 4.1.12**) with the outer radius R_o and inner radius r as

$$\Delta p \pi (R_o^2 - r^2) = 2 \pi \Delta L (\tau_{w,o} R_i - \tau r) \qquad (4.1.89)$$

The velocity gradient, dv/dr, is negative in this region; therefore, the constitutive equation is expressed as

$$\tau = \tau_y + K \left(-\frac{dv}{dr} \right)^m \qquad (4.1.90)$$

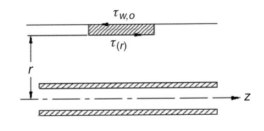

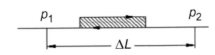

Fig. 4.1.12— Shear stresses in Region II.

An expression for the velocity gradient, which is obtained by combining Eqs. 4.1.89 and 4.1.90, is given as

$$-\frac{dv}{dr} = \left(c_{1,o} r + \frac{c_{2,o}}{r} - c_{3,o} \right)^{1/m}, \dotfill \quad (4.1.91)$$

where

$$c_{1,o} = \frac{\Delta p}{2\Delta L K} \qquad c_{2,i} = \frac{R_o}{K}\left(\tau_{w,o} - \frac{\Delta p R_o}{2\Delta L} \right) \qquad c_{3,o} = \frac{\tau_y}{K}$$

By integrating Eq. 4.1.91 and applying a no-slip condition at the wall, the velocity profile in sheared Region II is expressed as

$$v_{\mathrm{II}}(r) = \int_r^{R_o} (c_{1,o} r + c_{2,o}/r - c_{3,o})^{1/m}\, dr \dotfill \quad (4.1.92)$$

The velocity at the plug boundary (i.e., $r = b$) can be determined as

$$U_p = v_{\mathrm{II}}(b) = \int_b^{R_o} (c_{1,o} r + c_{2,o}/r - c_{3,o})^{1/m}\, dr \dotfill \quad (4.1.93)$$

The plug velocity, U_p, is constant. Hence, Eqs. 4.1.88 and 4.1.93 can be combined to match the velocity profiles as

$$\int_b^{R_o} \left(c_{1,o} r + c_{2,o}/r - c_{3,o}\right)^{1/m} dr = \int_{R_i}^a \left(c_{1,i} r + c_{2,i}/r - c_{3,i}\right)^{1/m} dr \dotfill \quad (4.1.94)$$

The overall momentum balance for the flow can be written as

$$\Delta p\, \pi\left(R_o^2 - R_i^2\right) = 2\pi\Delta L\, (\tau_{w,o} R_o + \tau_{w,i} R_i) \dotfill \quad (4.1.95)$$

After simplification, we obtain

$$\tau_{w,o} = \beta - \tau_{w,i}\frac{R_i}{R_o} \dotfill \quad (4.1.96)$$

and

$$\beta = \frac{\Delta p\left(R_o^2 - R_i^2\right)}{2\Delta L R_o} \dotfill \quad (4.1.97)$$

Similarly, the momentum balance for the plug region can be written as

$$\Delta p\, \pi(b^2 - a^2) = 2\pi\Delta L\, (\tau_y a + \tau_y b) \dotfill \quad (4.1.98)$$

Eq. 4.1.98 can be further simplified to obtain an expression for the plug thickness $(b - a)$:

$$b - a = \frac{2\tau_y}{\Delta p/\Delta L} \dotfill \quad (4.1.99)$$

It is worthwhile to note that the plug thickness has a physical constraint (i.e., $0 \le b - a < R_o - R_i$), which limits its value. This physical constraint must be used to avoid imaginary solutions. To obtain a relationship between pressure drop and volumetric flow rate, the plug velocity and the velocity profiles of the sheared regions need to be determined using Eqs. 4.1.87 and 4.1.92. However, these two equations require values of wall shear stresses and inner and outer radii of the plug, which are also unknown. In order to obtain the unknowns (a, b, $\tau_{w,i}$, and $\tau_{w,o}$), a system of four equations (Eqs. 4.1.84, 4.1.94, 4.1.96 and 4.1.99) needs to be established. This system is nonlinear and requires an iterative procedure with numerical integration to get the unknowns. After determining the values of wall shear stresses and inner and outer radii of the plug, the continuity equation can be applied to compute the flow rate numerically as

$$Q = 2\pi \int_{R_i}^{R_o} v(r)r\mathrm{d}r = 2\pi \left[\int_{R_i}^{a} v_{\mathrm{I}}(r)r\mathrm{d}r + \int_{a}^{b} U_p r\mathrm{d}r + \int_{b}^{R_o} v_{\mathrm{II}}(r)r\mathrm{d}r \right] \quad \dots\dots\dots\dots\dots\dots \quad (4.1.100)$$

Turbulent Flow. The narrow-slot approximation can be extended to determine friction pressure under turbulent flow conditions. However, the method has not been experimentally verified. Using this approach, the pressure loss gradient is calculated from the friction factor as

$$\frac{\mathrm{d}p}{\mathrm{d}z} = \frac{2f\rho U^2}{D_o - D_i}, \quad \dots\dots\dots\dots\dots\dots\dots\dots\dots\dots\dots\dots\dots \quad (4.1.101)$$

where f is estimated by adopting the pipe friction factor correlation as

$$\frac{1}{f^{0.5}} = \frac{4}{N^{0.75}} \log\left[\mathrm{Re}_{\mathrm{ann}} f^{(1-n/2)} \right] - \frac{0.4}{N^{1.2}} \quad \dots\dots\dots\dots\dots\dots\dots \quad (4.1.102)$$

The Reynolds number and flow behavior index in Eq. 4.1.102 must be calculated according to the procedure presented in Eqs. 4.1.78 through 4.1.82.

Eccentric Annulus. Experimental investigations (Hansen et al. 1999; Wang et al. 2000; Silva and Shah 2000) indicated that reduction in frictional pressure loss due to pipe eccentricity can be substantial. Determining the annular flow behavior of drilling fluids in eccentric annuli is quite important in well planning and hydraulic program development. A number of analytical and numerical studies (Piercy et al. 1933; Snyder and Goldstein 1965; Jonsson and Sparrow 1965; Guckes 1975; Mitshuishi and Aoyagi 1973; Haciislamoglu and Langlinais 1990; Fang et al. 1999; Hussain and Sharif 2000; Escudier et al. 2002) were conducted to develop a relationship between flow rate and pressure loss for eccentric annular flows. One of the earliest studies was that of Piercy et al. (1933), who presented an analytical solution that relates the pressure loss the flow rate. This and other studies (Wang et al. 2000; Silva and Shah 2000; Piercy et al. 1933) indicated significant variation of circumferential wall shear stresses.

The equation of motion in the Cartesian coordinate system for isothermal, incompressible, and fully developed steady laminar flow of viscous fluid in an eccentric annulus can be expressed as

$$-\frac{\partial p}{\partial z} + \frac{\partial}{\partial x}\left[\mu(\dot{\gamma})\frac{\partial v}{\partial x} \right] + \frac{\partial}{\partial y}\left[\mu(\dot{\gamma})\frac{\partial v}{\partial y} \right] = 0, \quad \dots\dots\dots\dots\dots\dots \quad (4.1.103)$$

where v represents the local axial velocity. For YPL fluids, the apparent viscosity is defined as the ratio of shear stress to the shear rate. Thus,

$$\mu(\dot{\gamma}) = \frac{\tau}{\dot{\gamma}} = \frac{\tau_y}{\dot{\gamma}} + K\dot{\gamma}^{m-1}, \quad \dots\dots\dots\dots\dots\dots\dots\dots\dots\dots \quad (4.1.104)$$

where m is the fluid behavior index. The shear rate for purely axial flow is given by

$$\dot{\gamma} = \left[\left(\frac{\partial v}{\partial x}\right)^2 + \left(\frac{\partial v}{\partial y}\right)^2 \right]^{\frac{1}{2}} \quad \dots\dots\dots\dots\dots\dots\dots\dots\dots\dots \quad (4.1.105)$$

Since the governing equations are nonlinear partial differential equations, it is very difficult to obtain an analytical solution. The numerical procedure is rather complex and computationally intensive. As a result, most investigators (Guckes 1975; Mitshuishi and Aoyagi 1973; Haciislamoglu and Langlinais 1990; Fang et al. 1999; Hussain and Sharif 2000; Escudier et al. 2002) applied complex numerical procedures. Guckes (1975) developed a series of dimensionless plots that are applicable for power-law and Bingham plastic fluids. The plots were obtained using finite-difference solutions and covered a broad range of fluid properties, diameter ratios, eccentricities, and flow rates. Haciislamoglu and Langlinais (1990) indicated that the effect of eccentricity on frictional pressure loss can be significant. More recently, an extensive numerical study on the effects of eccentricity on the velocity field and wall shear stress distributions was carried out by Fang et al. (1999).

Newtonian Fluid. Laminar Flow. For Newtonian fluid, an analytic solution (Piercy et al. 1933) is available to determine the pressure loss in an eccentric annulus

$$\frac{dp}{dz} = -\frac{8\mu Q}{\pi}\left[R_o^4 - R_i^4 - \frac{4E^2M^2}{B-A} - 8E^2M^2\sum_{i=1}^{\infty}\frac{ie^{-i(\beta+\alpha)}}{\sinh(iB-iA)}\right]^{-1}, \quad\dots\dots\dots\dots\dots \text{(4.1.106)}$$

where

$$M = (F^2 - R_o^2)^{\frac{1}{2}}, \quad F = \frac{R_o^2 - R_i^2 + E^2}{2E}, \quad A = \frac{1}{2}Ln\frac{F+M}{F-M}, \text{ and } \quad B = \frac{1}{2}Ln\frac{F-E+M}{F-E-M},$$

where E is the offset distance between the centers of the pipe and the borehole. Eq. 4.1.106 revealed that eccentricity has a very strong influence on pressure loss. At a constant pressure loss gradient, a small increase in eccentricity of the inner pipe can significantly increase the flow rate. This is due to a considerable increase in local velocities in the wider gap region.

Turbulent Flow. A convenient way of predicting the turbulent friction factor in an eccentric annulus is to adopt the pipe equations based on the effective diameter as proposed by Jones (Jones and Leung 1981). Subsequently, the pressure loss can be predicted as

$$\frac{dp}{dz} = \frac{2f\rho U^2}{D_o - D_i} \quad\dots\dots\dots\dots\dots\dots\dots\dots\dots\dots\dots\dots\dots\dots \text{(4.1.107)}$$

and

$$\frac{1}{\sqrt{f}} = -4.0\log\left[\frac{1.255}{\text{Re}_{ann}^e\sqrt{f}}\right], \quad\dots\dots\dots\dots\dots\dots\dots\dots\dots\dots\dots \text{(4.1.108)}$$

where the Reynolds number in an eccentric annulus, Re_{ann}^e, is expressed as

$$\text{Re}_{ann}^e = \frac{\rho U D_{eff}^e}{\mu} \quad\dots\dots\dots\dots\dots\dots\dots\dots\dots\dots\dots\dots\dots\dots \text{(4.1.109)}$$

and

$$D_{eff}^e = D_{hyd}\frac{16}{\phi}, \quad\dots\dots\dots\dots\dots\dots\dots\dots\dots\dots\dots\dots\dots\dots \text{(4.1.110)}$$

where ϕ is a hydraulic parameter calculated as

$$\phi = 16(a+b) \quad\dots\dots\dots\dots\dots\dots\dots\dots\dots\dots\dots\dots\dots\dots \text{(4.1.111)}$$

and geometric parameters a and b are estimated as

$$a = a_0e^3 + a_1e^2 + a_2e + a_3, \quad b = \alpha_0e^3 + \alpha_1e^2 + \alpha_2e + \alpha_3, \quad\dots\dots\dots\dots \text{(4.1.112)}$$

where a_0, a_1, a_2, a_3, α_0, α_1, α_2, and α_3 are regression coefficients, the values of which are dependent upon the diameter ratio as presented in **Table 4.1.6.** The diameter ratio (radius ratio) and dimensionless eccentricity are expressed as $\kappa = D_i/D_o$ and $e = E/(R_o - R_i)$, respectively.

Power-Law Fluid. To our knowledge, an analytical solution for power-law fluid in an eccentric annulus does not exist. However, a number of studies (Haciislamoglu and Langlinais 1990; Fang et al. 1999; Hussain and Sharif 2000; Escudier et al. 2002) have published numerical results. Numerical procedures have been very complex and computationally intensive. Approximate models have been developed for drilling applications. These models require the velocity distribution in an eccentric annulus to be systematically approximated by an equivalent velocity field. The most common approximation procedures are the narrow-slot model

TABLE 4.1.6—EQUATIONS FOR REGRESSION COEFFICIENTS	
$a_0 = -2.8711\kappa^2 - 0.1029\kappa + 2.6581$	$a_0 = 3.0422\kappa^2 + 2.4094\kappa - 3.1931$
$a_1 = 2.8156\kappa^2 + 3.6114\kappa - 4.9072$	$a_1 = -2.7817\kappa^2 - 7.9865\kappa + 5.8970$
$a_2 = 0.7444\kappa^2 - 4.8048\kappa + 2.2764$	$a_2 = -0.3406\kappa^2 + 6.0164\kappa - 3.3614$
$a_3 = -0.3939\kappa^2 + 0.7211\kappa + 0.1503$	$a_3 = 0.2500\kappa^2 - 0.5780\kappa + 1.3591$

(Tao and Donovan 1955; Vaughn and Grace 1965; Iyoho and Azar 1981; Uner et al. 1989), equivalent pipe method (Kozicki et al. 1966), concentric annuli approach (Luo and Peden 1990), and correlations (Haciis-lamoglu and Langlinais 1990) based on numerical results. The narrow-slot model neglects the effect of curvature and treats the eccentric annulus as if it were a slot of variable width. For eccentric annuli with low radius ratios and high eccentricities, the narrow-slot model performs poorly (Haciislamoglu and Langlinais 1990).

Laminar Flow. *Equivalent Pipe Model.* Kozicki et al. (1966) developed a generalized hydraulic equation for laminar flow of generalized fluid in ducts of arbitrary shape. Šesták et al. (2001) (see also Ahmed et al. 2006) verified the performance of this model by comparing model predictions with experimental measurements that are obtained from an eccentric annulus with a diameter ratio of 0.538. The equivalent pipe model adopts the shear rate equation for pipe flow and a similar expression for flow in slots to have the same form of the generalized shear rate equation, which is applicable for other duct geometries. Accordingly, the average shear rate for an eccentric annulus can be expressed as

$$\bar{\gamma}_w = \left[\frac{a}{n} + b\right]\left(\frac{8U}{D_{hyd}}\right), \dots\dots\dots\dots\dots\dots\dots\dots\dots (4.1.113)$$

where a and b are geometric parameters presented in Eq. 4.1.112, which is valid for $0 \le e \le 95\%$, $0.2 \le n \le 1.0$, and $0.2 \le \kappa \le 0.8$. The pressure loss can be determined from the friction factor as

$$\frac{dp}{dz} = 2f\frac{\rho U^2}{D_{hyd}} \dots\dots\dots\dots\dots\dots\dots\dots\dots\dots\dots (4.1.114)$$

The friction factor is calculated as

$$f = \frac{2^{3n+1}}{Re_{ann}^*}\left(\frac{a}{n} + b\right)^n \dots\dots\dots\dots\dots\dots\dots\dots\dots\dots (4.1.115)$$

and

$$Re_{ann}^* = \frac{\rho U^{2-n} D_{hyd}^n}{K} \dots\dots\dots\dots\dots\dots\dots\dots\dots\dots (4.1.116)$$

The flow regime in an eccentric annulus can be characterized using the Reynolds number (Luo and Peden 1990)

$$Re_{ann}^e = \frac{8\rho U^2}{K\bar{\gamma}_w^n} \dots\dots\dots\dots\dots\dots\dots\dots\dots\dots\dots (4.1.117)$$

The laminar flow is expected to be unstable when $Re_{ann}^e > 2,100$.

Correlation-Based Model. Haciislamoglu and Langlinais (1990) presented an accurate correlation for flow of power-law fluid in an eccentric annulus based on numerical simulation results. The correlation is valid for fluid with behavior index ranging from 0.4 to 1.0. It relates the pressure loss in an eccentric annulus to a concentric one as

$$\left(\frac{dp}{dL}\right)_e = \left(1 - 0.072\kappa^{0.8454}\frac{e}{n} - 1.5e^2\kappa^{0.1852}\sqrt{n} + 0.96e^3\kappa^{0.2527}\sqrt{n}\right)\left(\frac{dp}{dL}\right)_c, \dots\dots\dots\dots (4.1.118)$$

where $\left(\dfrac{\mathrm{d}p}{\mathrm{d}L}\right)_e$ and $\left(\dfrac{\mathrm{d}p}{\mathrm{d}L}\right)_c$ are pressure loss gradients in eccentric and concentric annuli, respectively. Eq. 4.1.118 is valid for eccentricities and diameter ratios ranging from 0 to 0.95 and 0.3 to 0.9, respectively.

Turbulent Flow. Hydraulic models for turbulent flow in an eccentric annulus are very limited. A correlation developed by Haciislamoglu and Langlinais (1990) has been adopted for turbulent flow as (Zamora et al. 2005)

$$\left(\frac{\mathrm{d}p}{\mathrm{d}L}\right)_e = \left(1 - 0.048\kappa^{0.8454}\frac{e}{n} - 0.67e^2\kappa^{0.1852}\sqrt{n} + 0.28e^3\kappa^{0.2527}\sqrt{n}\right)\left(\frac{\mathrm{d}p}{\mathrm{d}L}\right)_c \quad\cdots\cdots\cdots\cdots (4.1.119)$$

YPL Fluid. *Laminar Flow.* For YPL fluids, the most common approximation procedures, such as the equivalent pipe and correlations models, require estimation of the flow behavior index, which is used instead of the power-law index in the model. However, in some cases, the values of flow behavior index can be lower than 0.2. This may result in poor predictions of pressure losses and calculation errors. Hence, in order to obtain stable solutions, the concentric annuli model is preferred.

Concentric Annuli Model. The laminar flow of power-law and Bingham fluids in an eccentric annulus was modeled by Luo and Peden (1990) as an infinite number of concentric annuli with variable outer radius. This procedure has been extended for YPL fluids. Accordingly, a series of sectors of concentric annuli with variable outer radius are considered to represent an eccentric annulus as demonstrated in **Fig. 4.1.13.**

The radius of each concentric annulus (Fig. 4.1.13b) is determined by equating area A (Fig. 4.1.13a) to A^* (Fig. 4.1.13b) so that the corresponding shaded sectors have the same local mean velocities for the same flow rates. The advantage of this model over the narrow-slot model is that it is able to include the effect of curvature (i.e., $\tau_{w,i} \neq \tau_{w,o}$). However, the model still neglects circumferential wall shear stress variations in a given sector. The narrow-slot approximation (Eq. 4.1.78) or exact numerical method (Eqs. 4.1.84, 4.1.94, 4.1.96 and 4.1.99) presented previously can be applied in this model to calculate the mean flow velocity in each sector of a concentric annulus (i.e., shaded region of Fig. 4.1.13b). Extensive comparisons of model predictions (obtained using the exact numerical method) with published numerical results (Guckes 1975; Haciislamoglu and Langlinais 1990) for power-law fluids indicated a maximum discrepancy of 30%. Generally, the model underpredicts the pressure loss. In the limiting case of a concentric annulus, the model predictions are found to be in agreement with numerically obtained results. To reduce these discrepancies, a matching correlation, which is only a function of annular geometry, has been introduced to determine a corrected pressure loss:

$$\left(\frac{\mathrm{d}P}{\mathrm{d}L}\right)_{corrected} = \frac{1}{\Phi(\kappa,e)}\left(\frac{\mathrm{d}p}{\mathrm{d}L}\right)_{model}, \quad\cdots\cdots\cdots\cdots\cdots\cdots\cdots\cdots\cdots\cdots (4.1.120)$$

where $\Phi(\kappa,e) = \kappa^{0.27e}$. After introducing this correlation, the maximum discrepancy between model prediction and published results (Guckes 1975; Haciislamoglu and Langlinais 1990) has been reduced to approximately ±8%.

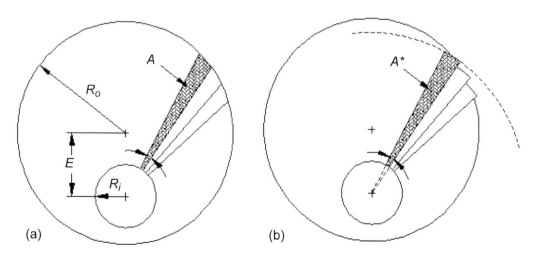

Fig. 4.1.13—Geometries of eccentric annulus (a) and equivalent annulus with series of concentric annuli (b).

Nomenclature

a = geometric parameter
A = curve fitting coefficient
b = geometric parameter
B = curve fitting coefficient
c = velocity gradient paramter
C = curve fitting coefficient
D = diameter
e = relative eccentricity
E = absolute eccentricity
f = friction factor
G = parameter
h = height
K = consistency index
K' = generalized consistency index
L = length
m = fluid behavior index
n = power-law index
n^* = slope
N = flow behavior index
p = pressure
Q = flow rate
r = radial distance from the center
R = radius
Re = Reynolds number
Re^e = Reynolds number in an eccentric annulus
Ta = Taylor number
U = mean flow velocity
v = local fluid velocity
w = width
X = length
z = z-axis
Z = hydraulic parameter
β = modified pressure loss
Δ = difference
ε = absolute roughness
$\dot{\gamma}$ = shear rate
Γ = torque
$\bar{\dot{\gamma}}$ = average shear rate
η = apparent viscosity
θ = θ-axis
κ = diameter or radius ratio
λ = location where shear stress vanishes
μ = viscosity
ξ = dimensionless radius
ρ = density
τ = shear stress
τ_y = yield stress
$\bar{\tau}_w$ = average wall shear stress
ω = rotational speed
Ω = dimensionless flow rate

Subscripts

ann = annulus
b = viscometer bob
D = entrance
eff = effective
eq = equivalent
hyd = hydraulic
i = inner
I = Region I
II = Region II
o = outer
p = plug
rz = stress notation
T = turbulence
w = wall
y = yield
Y = hydraulic parameter

References

Ahmed, R., Miska, S.Z., and Miska, W.Z. 2006. Friction Pressure Loss Determination of Yield Power Law Fluid in Eccentric Annular Laminar Flow. *Wiertnictwo Nafta Gaz* **23** (1): 47–53.

Chen, Z., Ahmed, R.M., Miska, S.Z., Takach, N.E., Yu, M. and Pickell, M.B. 2007. Rheology and Hydraulics of Polymer (HEC)-Based Drilling Foams at Ambient Temperature Conditions. *SPEJ* **12** (1): 100–107. SPE-94273-PA. DOI: 10.2118/94273-PA.

Collins, M. and Schowalter, W.R. 1963. Behavior of Non-Newtonian Fluids in the Entry Region of a Pipe. *AIChE Journal* **9** (6): 804–809. DOI: 10.1002/aic.690090619.

Dodge, D.W. and Metzner, A.B. 1959. Turbulent Flow of Non-Newtonian Systems. *AIChE Journal* **5** (2): 189–204. DOI: 10.1002/aic.690050214.

Escudier, M.P., Oliveira, P.J., and Pinho, F.T. 2002. Fully Developed Laminar Flow of Purely Viscous Non-Newtonian Liquids Through Annuli Including the Effects of Eccentricity and Inner-Cylinder Rotation. *International Journal of Heat and Fluid Flow* **23** (1): 52–73. DOI: 10.1016/S0142-727X(01)00135-7.

Fang, P., Manglik, R.M., and Jog, M.A. 1999. Characteristics of Laminar Viscous Shear-Thinning Fluid Flows in Eccentric Annular Channels. *Journal of Non-Newtonian Fluid Mechanics* **84** (1): 1–17. DOI: 10.1016/S0377-0257(98)00145-1.

Fredrickson, A.G. and Bird, R.B. 1958. Non-Newtonian Flow in Annuli. *Industrial & Engineering Chemistry.* **50** (3): 347–352. DOI: 10.1021/ie50579a035.

Guckes, T.L. 1975. Laminar Flow of Non-Newtonian Fluids in an Eccentric Annulus. *Trans., ASME, Journal of Engineering Industry* **97**: 498–506.

Haciislamoglu, M. and Langlinais, J. 1990. Non-Newtonian Flow in Eccentric Annuli. *Journal of Energy Resources Technology* **112** (3): 163–169.

Hanks, R.W. 1979. The Axial Laminar Flow of Yield-Pseudoplastic Fluids in a Concentric Annulus. *Industrial & Engineering Chemistry Process Design and Development* **18** (3): 488–493. DOI: 10.1021/i260071a024.

Hanks, R.W. and Larsen, K.M. 1979. The Flow of Power-Law Non-Newtonian Fluids in Concentric Annuli. *Industrial and Engineering Chemistry Fundamentals* **18** (1): 33–35. DOI: 10.1021/i160069a008.

Hansen, S.A., Rommetveit, R., Sterri, N., Aas, B., and Merlo, A. 1999. A New Hydraulics Model for Slim Hole Drilling Applications. Paper SPE 57579 presented at the SPE/IADC Middle East Drilling Technology Conference, Abu Dhabi, 8–10 November. DOI: 10.2118/57579-MS.

Hussain, Q.E. and Sharif, M.A.R. 2000. Numerical Modeling of Helical Flow of Viscoplastic Fluids in Eccentric Annuli. *AIChE Journal* **46** (10): 1937–1946. DOI: 10.1002/aic.690461006.

Iyoho, A.W. and Azar, J. 1981. An Accurate Slot-Flow Model for Non-Newtonian Fluid Flow Through Eccentric Annuli. *SPEJ* **21** (5): 565–572. SPE-9447-PA. DOI: 10.2118/9447-PA.

Jones, O.C. and Leung, J.C.M. 1981. An Improvement in the Calculation of Turbulent Friction in Smooth Concentric Annuli. *Journal of Fluids Engineering* **103** (4): 615–623.

Jonsson, V.K. and Sparrow, E.M. 1965. Results of Laminar Flow Analysis and Turbulent Flow Experiments for Eccentric Annular Ducts. *AIChE Journal* **11** (6): 1143–1145. DOI: 10.1002/aic.690110635.

Kozicki, W., Chou, C.H., and Tiu, C. 1966. Non-Newtonian Flow in Ducts of Arbitrary Cross-Sectional Shape. *Chemical Engineering Science* **21** (8): 665–679. DOI: 10.1016/0009-2509(66)80016-7.

Laird, W.M. 1957. Slurry and Suspension Transport—Basic Flow Studies on Bingham Plastic Fluids. *Industrial & Engineering Chemistry* **49** (1): 138–141. DOI: 10.1021/ie50565a041.

Lamb, H. 1945. *Hydrodynamics*, sixth edition. New York: Dover Publications.

Luo, Y. and Peden, J.M. 1990. Flow of Non-Newtonian Fluids Through Eccentric Annuli. *SPEPE* **5** (1): 91–96. SPE-16692-PA. DOI: 10.2118/16692-PA.

Macosko, C.W. 1994. *Rheology: Principles, Measurements, and Applications*, first edition. New York: Wiley-VCH.

Mitshuishi, N. and Aoyagi, Y. 1973. Non-Newtonian Fluid Flow in an Eccentric Annulus. *Journal of Chemical Engineering of Japan* **6** (5): 402–408.

Piercy, N.A.V., Hooper, M.S., and Winny, H.F. 1933. Viscous Flow Through Pipes With Core. *London Edinburgh Dublin Philosophical Magazine, and Journal of Science* **15**: 647–676.

Reed, T.D. and Pilehvari, A.A. 1993. A New Model for Laminar, Transitional and Turbulent Flow of Drilling Muds. Paper SPE 25456 presented an the SPE Production Operations Symposium, Oklahoma City, Oklahoma, 21–23 March. DOI: 10.2118/25456-MS.

Šesták, J.Í., Žitný, R., Ondrušová, J., and Filip, V. 2001. Axial Flow of Purely Viscous Fluids in Eccentric Annuli: Geometric Parameters for Most Frequently Used Approximate Procedures. In *3rd Pacific Rim Conference on Rheology*. Montreal: Canadian Group of Rheology.

Silva, M.A. and Shah, S.N. 2000. Friction Pressure Correlations of Newtonian and Non-Newtonian Fluids Through Concentric and Eccentric Annuli. Paper SPE 60720 presented at the SPE/ICoTA Coiled Tubing Roundtable, Houston, 5–6 April. DOI: 10.2118/60720-MS.

Snyder, W.T. and Goldstein, G.A. 1965. An Analysis of Fully Developed Laminar Flow in an Eccentric Annulus. *AIChE Journal* **11** (3): 462–467. DOI: 10.1002/aic.690110319.

Szilas, A.P., Bobok, E., and Navratil, L. 1981. Determination of Turbulent Pressure Loss of Non-Newtonian Oil Flow in Rough Pipes. *Rheologica Acta* **20** (5): 487–496. DOI: 10.1007/BF01503271.

Tao, L.N. and Donovan, W.F. 1955. Through Flow in Concentric and Eccentric Annuli of Fine Clearance With and Without Relative Motion of the Boundaries. *Trans. ASME* **77** (November): 1291–1301.

Uner, D., Ozgen, C., and Tosum, I. 1989. Flow of a Power-Law Fluid in an Eccentric Annulus. *SPEDE* **4** (3): 269–272. SPE-17002-PA. DOI: 10.2118/17002-PA.

Vaughn, R.D. and Grace, W.R. 1965. Axial Laminar Flow of Non-Newtonian Fluids in Narrow Eccentric Annuli. *SPEJ* **5** (4): 277–280; *Trans.*, AIME, **234**. SPE-1138-PA. DOI: 10.2118/1138-PA.

Wang, H., Su, Y., Bai, Y., Gao, Z., and Zhang, F. 2000. Experimental Study of Slimhole Annular Pressure Loss and Its Field Applications. Paper SPE 59265 presented at the IADC/SPE Drilling Conference, New Orleans, 23–25 February. DOI: 10.2118/59265-MS.

Whittaker, A. 1985. *Theory and Application of Drilling Fluid Hydraulics*. The EXLOG Series of Petroleum Geology and Engineering Handbooks. Boston, Massachusetts: International Human Resources Development Corporation.

Zamora, M., Roy, S., and Slater, K. 2005. Comparing a Basic Set of Drilling Fluid Pressure-Loss Relationships to Flow-Loop and Field Data. Paper AADE-05-NTCE-27 presented at the AADE National Technical Conference and Exhibition, Houston, 5–7 April.

SI Metric Conversion Factors

gal/min × 2.271 247   E – 01 = m³/h
in.   × 2.54*   E + 00 = cm

*Conversion factor is exact.

4.2 Well-Control Modeling—Rolv Rommetveit, eDrilling Solutions

4.2.1 Introduction. The development of a gas kick from the start of inflow until the gas is circulated out through the choke manifold is a complex interaction of many different subprocesses. This interaction and external factors such as mud and gas properties, reservoir conditions, and drilling and control procedures determine the character of the kick.

A gas kick in oil-based mud (OBM) will be more complex than in water-based mud (WBM) due to the high solubility of the gas in the base oil. As long as gas is dissolved, no gas expansion takes place, and the influx is partly *hidden* in the mud. When gas flashes out from the mud, the rest of the kick will develop more rapidly since both gas flashing and free-gas expansion take place simultaneously. It is almost impossible to consider all relevant factors related to the kick process by conventional methods.

Computer models have been developed in order to describe and analyze kick phenomena. Some of these codes are based on advanced, dynamic mathematical models describing the kick process. However, in order to verify the models, they need to be compared to real data from experiments. These can be used for verifying the different submodels involved, such as the free-gas transport model, dissolved-gas transport model, and frictional-pressure-loss model, as well as the hydrodynamic-well-flow model itself.

Of great importance are data from full-scale kick experiments performed under controlled conditions. In order to use a kick model as an engineering tool with confidence, it should be tested against full-scale data. If the simulator is to be used as a tool for decision-making support during well-control situations, this is a must.

This chapter will discuss well-control modeling. It will focus on a correct physical representation of the well-control process. The following aspects are discussed: assumptions and limitations, kick multiphase flow models and modeling, kick tolerance, gas migration, kicks in horizontal wells, well control in OBM, high-pressure/high-temperature (HP/HT) well control, deepwater well-control modeling, kick with lost circulation, hydrate formation during kicks, and modeling of complex well-control situations.

4.2.2 Assumptions and Limitations. Well control is a very complex phenomenon. In order to describe the complete process and produce meaningful results from the modeling, it is necessary to simplify. The main simplifying assumptions are as follows: The fluid flow in the well is considered a combined single- and two-phase flow problem. We restrict ourselves to one spatial dimension (i.e., position along flowline, s). The temperature, T, is for most applications assumed to be known along the flowline. The drift-flux approximation is used; the two-fluid formulation with separate momentum conservation equations for each phase is simplified to a combined momentum conservation equation as well as a slip relation. The complete limitations and assumptions used in advanced well-control modeling and simulations today are given by Rommetveit (1989).

4.2.3 Multiphase Flow Modeling of Kicks. The governing equations are those expressing conservation of mass and momentum. The energy conservation equation is for most purposes excluded from the modeling.

Conservation of mud mass:

$$\frac{\partial}{\partial t}\left[A(1-\alpha)\rho_1\right] = -\frac{\partial}{\partial s}\left[A(1-\alpha)\rho_1 v_1\right] + A\dot{m}_g \quad\dots\dots\dots\dots\dots\dots (4.2.1)$$

Conservation of mass of free gas:

$$\frac{\partial}{\partial t}\left(A\alpha\rho_g\right) = -\frac{\partial}{\partial s}\left(A\alpha\rho_g v_g\right) - A\dot{m}_g + q_g \quad\dots\dots\dots\dots\dots\dots (4.2.2)$$

Conservation of mass of dissolved gas:

$$\frac{\partial}{\partial t}\left[A(1-\alpha)x_{dg}\rho_1\right] = -\frac{\partial}{\partial s}\left[A(1-\alpha)x_{dg}\rho_1 v_1\right] + A\dot{m}_g \quad\dots\dots\dots\dots (4.2.3)$$

Conservation of mass of formation oil:

$$\frac{\partial}{\partial t}\left[A(1-\alpha)x_{fo}\rho_1\right] = -\frac{\partial}{\partial s}\left[A(1-\alpha)x_{fo}\rho_1 v_1\right] + q_{fo} \quad\dots\dots\dots\dots (4.2.4)$$

Conservation of total momentum:

$$\frac{\partial}{\partial t}\left[A(1-\alpha)\rho_1 v_1 + A\alpha\rho_g v_g\right] = -\frac{\partial}{\partial s}(Ap) - Af_1 - Af_2$$

$$+ A\left[(1-\alpha)\rho_1 + \alpha\rho_g\right]g\cos\theta$$

$$-\frac{\partial}{\partial s}\left[A(1-\alpha)\rho_1 v_1^2 + A\alpha\rho_g v_g^2\right] \quad \dots\dots\dots\dots\dots\dots\dots\dots \text{(4.2.5)}$$

Here, t is time, A is flowline cross-sectional area, α is void fraction, ρ_1 is mud density, v_1 is flow velocity of mud, $\dot{m}_g$ is rate of gas dissolution, ρ_g is gas density, v_g is flow velocity of gas, q_g is inflow of free gas, q_{fo} is influx of formation oil, x_{dg} is mass fraction of dissolved gas, x_{fo} is mass fraction of formation oil, p is pressure, f_1 is the frictional pressure loss term, and f_2 is the localized pressure loss term.

Eqs. 4.2.1 through 4.2.5 constitute a system of five equations for 13 unknowns: α, ρ_1, v_1, $\dot{m}_g$, ρ_g, v_g, q_g, q_{fo}, x_{dg}, x_{fo}, p, f_1, and f_2. To close the system, we need eight more equations, which we refer to as submodels. The general functional relationships are given here:

Mud density:

$$\rho_1 = \rho_1\left(p, T, x_{dg}, x_{fo}\right) \quad \dots\dots\dots\dots\dots\dots\dots\dots\dots\dots \text{(4.2.6)}$$

Gas density:

$$\rho_g = \rho_g\left(p, T\right) \quad \dots\dots\dots\dots\dots\dots\dots\dots\dots\dots\dots\dots \text{(4.2.7)}$$

Free-gas velocity:

$$v_g = v_g\left(p, T, x_{dg}, x_{fo}, \alpha, s, v_{mix}\right) \quad \dots\dots\dots\dots\dots\dots \text{(4.2.8)}$$

Gas influx:

$$q_g = q_g\left(p, T, s\right) \quad \dots\dots\dots\dots\dots\dots\dots\dots\dots\dots\dots \text{(4.2.9)}$$

Formation oil influx:

$$q = q_{fo}\left(p, T, s\right) \quad \dots\dots\dots\dots\dots\dots\dots\dots\dots\dots\dots \text{(4.2.10)}$$

Rate of gas dissolution:

$$\dot{m}_g = \dot{m}_g\left(p, T, x_{dg}, x_{fo}, \alpha, v_1, v_g, s\right) \quad \dots\dots\dots\dots\dots \text{(4.2.11)}$$

Frictional pressure loss:

$$f_1 = f_1\left(p, T, x_{dg}, x_{fo}, \alpha, v_1, v_g, s\right) \quad \dots\dots\dots\dots\dots \text{(4.2.12)}$$

Localized frictional pressure loss:

$$f_2 = f_2\left(p, T, x_{dg}, x_{fo}, \alpha, v_1, v_g, s\right) \quad \dots\dots\dots\dots\dots \text{(4.2.13)}$$

Here, $v_{mix} = \alpha v_g + (1-\alpha)v_l$ is the mixture velocity.

4.2.4 Simulators. The first gas kick computer simulator was developed by LeBlanc and Lewis (1968). From this pioneering simplistic model, gas-kick and well-control phenomena have been modeled with gradually more realistic models (Hoberock and Stanbery 1981; Nickens 1987; Ekrann and Rommetveit 1985; Podio and Yang 1986; White and Walton 1990; Choe and Juvkam-Wold 1996; Santos 1989; Rommetveit 1994). Some operators have their own in-house models. Other companies use commercial advanced well-control simulators for their evaluations of well control.

Nickens (1987) developed the first dynamic kick simulator for kicks in WBM. In parallel, a dynamic model that also accounted for gas solubility and kicks in OBM was developed by Ekrann and Rommetveit (1985). This model was upgraded gradually with results from extensive experimental and theoretical research on aspects like kicks in OBM, kicks in horizontal wells, slimhole kicks, HP/HT kick development, kicks with lost circulation, kicks in deep water, and kicks with hydrates. It was verified with results from full-scale kick experiments in an inclined test well (Rommetveit and Olsen 1989; Rommetveit and Vefring 1991). White and Walton (1990) implemented results from gas-migration tests in vertical and inclined geometries into their model. This model was also further developed for several years.

Advanced kick simulators gradually are used more and more in well-control planning and sensitivity evaluations for challenging wells. Real-time kick tolerance evaluations for critical wells are now under development. Advanced kick models also are integrated with advanced hydraulic models into real-time modeling systems for managed-pressure, dual-density, and standard drilling systems (Rommetveit et al. 2006a, 2006b, 2007; Bjørkevoll et al. 2008).

4.2.5 Gas Migration and Slip. The rate at which free gas rises up the wellbore is a key parameter in the development of a gas kick in a well. Studies on gas/liquid two-phase flow in annuli have focused on both Newtonian and non-Newtonian fluids. For Newtonian fluids, mechanistic models have been developed for vertical flows by Caetano (1986) and improved by Lage and Time (2000). A mechanistic model for the flow behavior of two-phase mixtures in horizontal or slightly inclined fully eccentric annuli has been developed by Lage et al. (2000), and this model is verified by extensive experiments. The model is composed of a procedure for flow pattern prediction and a set of independent models for calculating gas fraction and pressure drop in stratified, intermittent, dispersed-bubble, and annular flow. These models developed for Newtonian fluids cannot be applied directly to non-Newtonian fluids and drilling muds.

In general, the flow of mud-hydrocarbon mixtures upward in a wellbore is a very complex and transient multiphase flow process. The flow of free gas will, in general, take place in the bubble or slug flow regime. Annular flow will exist only if the influx is allowed to move uncontrolled and expand near the surface.

Experimental studies have been conducted specifically to improve the understanding of gas rise velocities in drilling and kick control situations [i.e., annulus flow geometry and non-Newtonian fluid (Johnson and White 1991; Hovland and Rommetveit 1992; Johnson and Cooper 1993; Rader et al. 1975; Santos and Azar 1997)]. Most of these studies are performed for vertical and deviated wells. Some studies have been done under accurately controlled conditions in flow loops.

Transition between bubble and slug flow will depend on the non-Newtonian properties of the mud-influx mixture. While this transition is at approximately 25% gas volume fraction for Newtonian fluids, it may take place at less than 5% in the non-Newtonian fluids (Johnson and White 1991).

The slug flow regime will give a much higher gas-slip velocity than the dispersed-bubble regime.

It was found that the gas velocity in the slug flow regime can be characterized by the Zuber-Findlay model (Zuber and Findlay 1965) as follows:

$$v_g = C_0 v_{mix} + v_0 \dots\dots\dots\dots\dots\dots\dots\dots\dots\dots\dots\dots\dots\dots\dots\dots\dots\dots\dots\dots\dots \quad (4.2.14)$$

Here, C_0 is a distribution factor and v_0 the gas slip velocity in stagnant mud. The two parameters C_0 and v_0 have been found to be a function of fluid viscosity, flow geometry [outside diameter (OD) of inner pipe], rotation of inner pipe, and well inclination.

Full-scale tests and experiments in vertical and inclined wells have also been performed (Rader et al. 1975; Rommetveit and Olsen 1989; Steine et al. 1996). The full-scale experiments performed by Rommetveit and Olsen (1989) considered both OBM and WBM. The well was a 2000-m-long and 60°-inclined research well.

A total of 24 different gas kick experiments were performed. In the experiments, parameters such as mud density, gas concentration, gas type, mud flow rate, injection depth, and mud type were varied, and large quantities of data were acquired and analyzed. Gas slip and migration were measured by means of five accurate pressure and temperature sensors spaced out in the well.

Slimhole Drilling. One experimental study has been performed in a full-scale test well (Steine et al. 1996). Comparison between simulated gas migration velocities and data from the experiments show that tool joints tend to slow down gas slip in slimhole geometries. However, gas rise velocities are predicted well by a state-of-the-art simulator when the drillpipe is slick with no external upset (Rommetveit et al. 1996).

4.2.6 Kicks in OBM. ***Phase Properties of Mud and Hydrocarbon Influxes.*** After an influx in a well, mixing will take place between the mud and the influx. Hydrocarbon influxes generally will range from dry gas (methane) through volatile oil and condensate to heavy oil.

Hydrocarbon gas solubility in the oil phase of the drilling mud is several orders of magnitude larger than in the water phase. Hence, OBM will behave significantly differently from WBM in the case of a kick.

After an influx, the mixture conditions of mud and influx will move toward thermodynamic equilibrium. This equilibrium condition will depend on pressure and temperature (depth) in the well. A fraction of the hydrocarbon influx will then dissolve in the mud. The major part of this will dissolve in the base oil. The dissolution process is not instantaneous, but is governed by molecular diffusion.

The volume of gas influx changes when it goes into solution. In general, the volume of the gas influx will shrink when the gas molecules move from the free state into the dissolved state. This effect is reduced with increasing pressure.

An influx of volatile oil in WBM will release free gas because of the pressure reduction when it is pumped upward in the well. This free gas will expand according to the real gas law.

Influx of volatile oil in OBM will mix completely with the base oil, and a new *pseudobase oil* with pressure/volume/temperature (PVT) properties different from those of the volatile oil will be created. The bubblepoint and volume of free gas released will be different from those of the volatile oil alone.

Influx of a dry gas in OBM will be soluble in the base oil. The bubblepoint of the mixture will govern where flashing of free gas will take place in the well. Gas or volatile oil in OBM will have the following special characteristics:

- As long as the influx is dissolved, it will stay rather concentrated. The influx will not be distributed over a larger portion of the mud by slippage, and a high gas peak can be expected when the gas reaches the surface.
- The expansion from such a system can be much more violent than from a gas-WBM system due to the combined effect of large expansion and violent flashing near the surface.

Gas Diffusion in Overbalance. If a well that is drilled with OBM overbalanced through a gas formation is left without circulation for an extended period of time, gas from the formation will diffuse through the mud-invaded zone and filter cake and accumulate in the drilling fluid. In the case of OBM at HP/HT conditions, methane is infinitely soluble in the oil, and substantial amounts of gas can be dissolved in the mud. The rather shallow mud-invaded zone in the near-wellbore region created when using OBM will enhance the process. The mechanisms for gas diffusion to take place are thoroughly discussed in Bradley et al. (2002).

Consequences. The rate of methane diffusing into a 1000-m HP/HT horizontal well with OBM has been estimated.[*] The results indicate that if the well is left without circulation for a number of days, a significant amount of gas will diffuse into the wellbore; see **Fig. 4.2.1.** The influx of methane will displace mud out of the well. This flow could easily be masked by other phenomena such as thermal expansion of the drilling fluid.

Consequences With Respect to Well Control. The key consequence is that it is entirely possible to diffuse a sufficient quantity of gas into the wellbore to cause a kick despite the drilling fluid being appropriately overbalanced and good operational practice being followed. There could be sufficient volumes of gas diffused into the well to underbalance the well or induce a kick while running in hole or when circulation is started. As a minimum consequence, substantial volumes of trip gas, together with significant gas cutting of the drilling fluid, could be expected.

Consequences From Changing Mud Properties. The quantity of gas eventually dissolved in the mud will impair the carrying capacity of the mud. This may cause precipitation of cuttings, weighting material, and

---

[*]Personal communication with B. Aas. 2006. Stavanger: Rogaland Research.

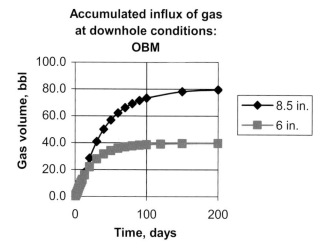

Fig. 4.2.1—Accumulated gas volume diffused into the well as a function of time (Rommetveit et al. 2003).

even viscosifying agents such as clays. A layer of low-density/low-viscosity fluid on the high side of the wellbore and a corresponding layer of high-density/high-viscosity fluid on the low side may develop. If the well inclination through the gas-bearing formation is close to horizontal, migration of the upper layer due to buoyancy is unlikely. However, on resumption of mud circulation, the lower-density layer may flow preferentially compared to the higher-density layer on the low side. As the lower-density fluid moves higher up in the well, well-control problems may result.

There are examples in the North Sea of HP/HT and near-HP/HT wells exhibiting severe operational problems that may be explained by this phenomenon. Examples include severe barite sag leading to well-control problems and loss of hole section, as well as complex well-control problems during openhole completion running. Factors that suggest gas diffusion as a potential mechanism include significant exposed gas-bearing formations, substantial periods of time with little or no circulation, high gas levels, or influxes occurring on recommencement of circulation, despite following company-recognized procedures and having drilling fluid within specification.

4.2.7 Kicks in Horizontal Wells. Studies on kick control in horizontal wells so far have been relatively few (Santos 1991; Wang et al. 1994; Currans et al. 1993; Vefring et al. 1995b). Santos (1991) performed computer simulation studies on the circulation out of a kick in a perfectly horizontal well. He assumed that the horizontal section is truly 90° and that the entire kick volume remains in the horizontal section during the influx period. He further assumed that the kick was distributed as a uniform gas/mud mixture extending from the drill bit back to a point in the annulus defined by the pit gain and gas void fraction (user-specified). Some of his arguments are direct consequences of the simplified assumptions used in his simulations.

The actual well-control operations are more complicated to implement for horizontal wells when the wait-and-weight method is used to circulate out of the kick. When the bottomhole pressure is held constant, which is a basic aim of all well-control operations, the pump pressure schedule becomes more complex due to unique well trajectories of horizontal wells. This requires an extension to the existing standard kick sheet as discussed by Currans et al. (1993). They also have discussed some other practical aspects of good well-control procedures in high-angle and horizontal wells, such as kick avoidance and minimizing swabbing.

A perfectly horizontal well (90°) is an exception. In many field situations, near-horizontal wells are drilled, and the actual well is an inclined hole either upward or downward, undulating in some cases and possibly with washout sections. In these cases, the buoyancy of the gas may cause it to become trapped in certain parts of the horizontal well.

On the basis of extensive experimental work (see below), a well-control simulator was further developed to handle kicks in high-angle and horizontal wells (Vefring et al. 1995a). The important new developments included

- Coupling the fluid flow in the horizontal section exposed to the reservoir with the influx from the formation
- New models for the gas-slip velocities in near-horizontal wells
- Models for three different gas-removal mechanisms.

Experiments. Laboratory experiments have been performed in order to investigate phenomena related to kicks in horizontal wells (Aas et al. 1993). The experiments were designed to provide both qualitative and quantitative knowledge of the removal of gas kicks in horizontal and near-horizontal wells. Specifically, the mechanisms associated with the removal of gas bubbles in the following three typical situations were studied:

- Upward-inclined end of the well
- Upward-inclined sections and local tops in the trajectory
- Out-of-gauge sections.

The mechanisms are as follows:
Removing Gas From End of Well. The removal of gas influx from the bottom end of an upward-inclined well was found to be dependent on the distance of the bit from the bottom end of the well. If the bit was placed sufficiently near the end of the well, the gas was mixed with the jets from the nozzles and displaced behind the bit. The gas then was transported downstream or accumulated behind the bit as a standing bubble, depending on flow parameters.
Upward-Inclined Sections and Local Tops. The gas could be removed by one or more of the following three different mechanisms, depending on the flow rates.

1. At sufficiently high flow rates, the gas is transported away essentially in the form of one big bubble.
2. At lower flow rates, the turbulent action from the liquid will tear off small bubbles from the end of the stationary bubble. The gas is then transported away in the form of entrained bubbles.
3. At even lower flow rates, without sufficient degree of turbulence in the mud, no free gas is transported away. The removal of gas will be by solution since the kill mud will be undersaturated at downhole conditions. But the solution process is found to be quite slow and dynamic. It can take on the order of 10 hours.

Out-of-Gauge Sections. It was found that gas trapped in out-of-gauge sections could be removed by solution and entrainment. Both of these two mechanisms depend upon the in-situ flow conditions and are interrelated. The solution of gas is a dynamic process. For the solution to be significant, the flow has to be turbulent. For the entrainment to be significant, the flow velocity must exceed a critical value. However, the entrained bubbles may or may not be transported out of the out-of-gauge section, depending upon the washout section length and degree of turbulence. Entrainment also has been found to increase the rate of gas removal by solution significantly.
Gas Removal. Three mechanisms by which the gas is removed are

- Gas transported as big bubbles
- Gas dissolved in the mud and then transported away
- Gas entrained in the fluid as small bubbles and transported away.

The critical parameters that govern the complete removal of gas are (1) gas rise velocity, (2) mass transfer rate from free gas to dissolved gas, and (3) rate of entrainment.
Full-Scale, High-Pressure Experiments. Full-scale experiments of gas kicks and gas removal from a horizontal well have been performed (Rommetveit et al. 1995). The experiments were performed in a comprehensive test setup composed of a 200-m horizontal well (flow loop), separate gas injection system, mud circulation system, instrumentation system, and data acquisition system. The horizontal well was constructed on surface in order to study kicks in horizontal wells. The 200-m-long well had an inside diameter of 9¼ in. The well was equipped with 5-in. drillstring and a 50-m-long drill collar section. The drillstring assembly

could be rotated. The last 50 m of the well was inclined upward by 4° from horizontal for part of the experiments. The wellbore was designed for pressures up to 170 bar, as well as full, realistic mud and gas flow rates.

Based on the results from these experiments, new pressure-loss models as well as gas-transport equations for horizontal well control have been developed (Rommetveit et al. 1998).

4.2.8 HP/HT Well Control. *Introduction.* The drilling of HP/HT wells poses special challenges compared to standard wells:

- High pressures and temperatures impact mud properties in a dynamic way and can have effects on well control.
- Small margins between pore and fracture pressures will prevail in sections of the well.
- The conditions are above the critical point for the gas/oil/condensate influx, which means that the hydrocarbon influx is infinitely soluble in the base oil of the mud.
- Hydrocarbon influx will totally mix with the base oil in OBM, and infinite amounts of gas can dissolve in the mud.
- Drilling of inclined and horizontal wells will make the consequences of barite sag serious.
- Significant quantities of gas can diffuse into a horizontal section of a well if OBM is used, even if the well is overbalanced.

The frequency of well-control incidents is higher than one per HP/HT well, and an increasing number of these take place during completion.

The Physics of an HP/HT Well. Maintaining the control of the well at all times is a question of understanding the physics of the well during changing conditions and using this knowledge to optimize the design of the well, to develop sound drilling procedures, and to handle unexpected situations during drilling in an optimal way.

Traditional and well-proven drilling and well-control practices and rules of thumb have been developed by *trial and error*. In some cases, these represent optimal solutions. However, when the drilling situation differs significantly from the traditional, old rules may not apply, and one will need to analyze the problems scientifically to revise the practices. Transient computer models with the correct physics built into them should be used to develop new procedures and practices for these wells.

The main physical parameters and interactions are discussed below with special focus on HP/HT impacts.

Drilling Mud Composition. Drilling mud is a mixture of many components with different properties. The various components react differently to pressure (p) and temperature (T). The gas solubility and mixture properties of more-complex hydrocarbon influxes with the mud will vary significantly. The most common components are water, base oil, and weight materials (solids). Other components are chemically active and mix or dissolve into the primary components of the mud for certain time periods during specific operations on the well. The duration of the time period may depend on the operations (for example, drill cuttings or kick fluid).

Drilling Mud Density. The density of the drilling mud is dependent on both p and T. The density distribution of mud in a well will vary with the temperature distribution in the wellbore. The temperature distribution depends on formation virgin temperature and drilling history.

The active mud volume in an HP/HT well may change significantly when circulation starts or stops, even when there is no lost circulation or influx. The reason for the changes may be one or more of the following effects:

- Mud expands or contracts due to temperature variations.
- Mud expands or is compressed due to pressure variations.
- The diameter of casing and openhole sections increases or decreases (ballooning).

Drilling Mud Rheology. The rheological properties of drilling fluids are often approximated to be independent of pressure and temperature. In many cases, this is a good approximation. For shallow wells, the temperature changes are not so large; hence, the rheological variations with temperature are small. Many wells have a large gap between pore pressure and fracture pressure, so errors in the estimation of the dynamic circulation pressure are of little consequence for well integrity or kick probability.

However, for wells with small margins between pore and fracture pressure, careful evaluations and analysis of the effects of temperature and pressure on rheology, wellbore hydraulics, and kick probability are needed.

There is also a difference between OBM and WBM in the pressure and temperature dependence.

Thermophysical Properties. Calculations of dynamic temperature profiles require information about specific heat and thermal conductivity of the materials represented in the well: well fluid, steel, formation, cement, water, and mud. Except for the well fluids, these properties may be obtained from several literature sources (Corre et al. 1984; Green 1984). Few or no data are available for HP/HT drilling fluids that are mixtures of several components with different properties.

Temperature Effects (Thermal Energy Transport). Mud temperature at a given position in the well changes rapidly, depending on the ongoing drilling operation. The temperature approaches the geothermal temperature during long stationary periods. When circulation starts, the lower part of the annulus will be cooled by cold mud from the drillstring, and the upper part of the annulus will be heated since hot mud is flowing up the annulus. Mud density as well as mud rheology at a given position will change rapidly in this phase. This normally results in change in the total mud volume in the hole.

Note that mud properties and well temperature influence each other mutually. Heat transfer and frictional heating depend on mud properties, which depend on mud temperature.

Pressure Effects. Pressure variations can be larger in HP/HT wells than in standard wells for several reasons:

- The hydrostatic pressure varies more, due to mud density changes caused by thermal effects. The mud density distribution will change due to drilling with different circulation rates, stationary periods, etc.
- The frictional pressure changes caused by temperature-driven rheology variations along the wellbore.
- Rheology changes also can induce flow regime transitions between laminar flow and turbulence, mainly in the drill collar section of the annulus. Turbulence will create higher frictional pressure losses.
- Surge and swab pressures can be more critical due to two factors. First, viscosities in the deepest and hottest parts of the wellbore may be higher (temperature-driven effects); second, the time-dependent gel strength increases with temperature for some muds.
- Mud rheology is not only temperature- and pressure-dependent, but also dependent on shear history. A rapid peak in the bottomhole pressure has been observed as gels are being broken during startup of pumps. This effect is more critical in HP/HT wells, due to small margins and higher viscosities.

Phase Properties of Mud and Hydrocarbon Influxes. After an influx in a well, mixing will take place between the mud and the influx. Hydrocarbon influxes generally will range from dry gas (methane) through volatile oil and condensate to heavy oil.

Influx of a dry gas in OBM at HP/HT conditions will be infinitely soluble in the base oil. The bubblepoint of the mixture will govern where in the well one will have free gas flashing from the mud.

Dynamic Sagging of Barite. It has been demonstrated in tests that all muds sag during shear flow. This is usually called dynamic sagging, to distinguish it from the sagging that may take place in fluids at rest.

Weight material typically will sag out of the drilling fluid in long, highly inclined sections of a well, when the circulation rate is low enough for the flow to be laminar and the drillstring is not rotated or is rotated only slowly. High turbulence and vigorous whipping of the drillstring will resuspend precipitated material.

The effect of dynamic sagging may become pronounced in wells with long horizontal sections. A significant loss of weight material from the mud may cause serious problems for the pressure control when the lighter mud reaches sections with small inclination.

Effects of Hydrocarbons on Mud Properties. Mud properties will be affected by hydrocarbon influx. The effects will be significantly larger in OBM compared to WBM due to the high solubility and mixing properties of hydrocarbons in OBM.

Hydrate Formation in the Well. Hydrate formation may take place in HP/HT wells, in particular with increasing water depths. This can create a difficult well-control situation. Evaluating the potential for hydrate formation, as well as investigating various well-control strategies vs. the hydrate-formation probability, now can be done using well-control models that account for the dynamic temperature effects on the hydrate probability (Petersen et al. 2001).

Laboratory Characterization of Mud and Mud-Influx Properties. In order to perform a thorough evaluation of the well-control aspects of HP/HT drilling, knowledge of the physical properties of the fluids involved is of significance.

Also, if advanced transient well control and thermohydraulic simulators are used, realistic input of fluid properties as part of the modeling will greatly enhance the relevance of the work.

Rheology (Rommetveit and Bjørkevoll 1997; Bjørkevoll et al. 2003) and density of drilling fluids used as well as the PVT properties of influx and influx/drilling fluid mixtures (Gard 1986; O'Bryan et al. 1988) should be tested vs. p and T. Muds used in HP/HT drilling also should be evaluated for static and dynamic barite sag.

Examples from HP/HT Wells. In spring 2001, BP Aberdeen was in the planning process of the Devenick HP/HT well in the North Sea (Rommetveit et al. 2003). The well was classified as an HP/HT well with a reservoir temperature of 150°C and a planned total depth (TD) at 4613 m true vertical depth (TVD). A 1000-m-long horizontal section through the reservoir was planned. One of the concerns was an expected high-pressure zone just above the reservoir formation, which could cause well-control incidents and had a direct impact on the casing design. The well was initially planned with OBM, but a switch to formate brine was found necessary at a later stage.

BP's motivation for performing the modeling work was as follows:

- There was a need for evaluating the combination of HP/HT and horizontal well control procedures.
- Confirm best-practice procedures.
- Identify any specific well-control risks.
- Evaluate casing design and kick tolerances.
- Address the change of fluid system from OBM to brine (initially motivated from the planned switch to brine during completion and the increasing number of kick incidents during completion).
- Improve crew training and make it more relevant.
- Address the concern for gas diffusion in OBM in the long horizontal section.

Thermohydraulic Modeling. An important issue for avoiding well-control incidents is to have good knowledge of the effective bottomhole pressure. Both hydrostatic and frictional pressures will depend on the present well conditions, which will change according to the operational state. **Fig. 4.2.2** shows the simulated equivalent circulating density (ECD) for the cesium/potassium formate mud that was planned for use in the Devenick HP/HT well. From this figure, the frictional pressures for various flow rates can be determined. In addition, it shows how the total static mud weight tends to decrease during a drill string connection due to a net temperature increase in the well. This can be related to a fingerprint that is expected

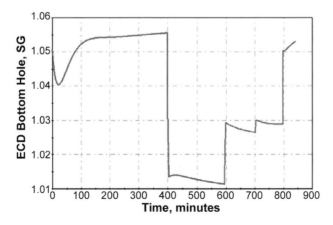

Fig. 4.2.2—ECD for a 1.62-SG formate mud in the 8½-in. section. First, the rate is 425 gal/min, circulating until steady state. Then, the pump is shut off before it increases to 80, 120, and 425 gal/min (Rommetveit et al. 2003).

to be observed in the pit tanks where a slight expansion should be seen. It is of importance to be able to distinguish such temperature-driven fingerprints from real influx situations, and in that respect, dynamic modeling can be a valuable tool.

When comparing field results with the study results it became clear that specifying accurate inlet and outlet mud temperatures was very important. This can be achieved by using information from offset wells or by using the model during the operation with updated information. In addition, very accurate modeling results can be achieved if measured PVT data of the mud are available and used in the modeling process. Using this, accurate predictions of the expected ECDs and fingerprints can be achieved.

In the Devenick study, both OBM and brine were modeled and compared with respect to ECDs, temperature distribution, and swab pressures. In **Fig. 4.2.3,** the bottomhole pressure for various pump rates while pulling out of the hole is shown. There is a clear need for pumping out of the hole to maintain an overbalance.

Kick Modeling. Focuses in the well control studies were

- Kick tolerances
- Undetected kicks
- Kill methods
- Comparing kick behavior in OBM vs. brine
- Surface flow behavior

In OBM, the incident of an undetected kick can cause problems. Since the gas influx dissolves in the mud, there will be no signals from the pit tank that an influx is being transported toward the surface. However, at some stage, free gas will occur, and a sharp increase in the pit levels can be seen. The kick is first detected when it is almost at surface, and the drilling crew have a very short time to activate well-control procedures. **Figs. 4.2.4 and 4.2.5** show the situation when a 4-bbl undetected kick is circulated up before cementing the 9⅞-in. casing.

Another issue in the study was to keep focus on the possible change of fluid system (from OBM to brine) and the impact this had on well control behavior. A kick taken in OBM will dissolve and remain at bottom if circulation is not resumed. However, the situation is quite different if a free-gas kick is taken in a brine mud. Even during closed-in conditions, gas migration will take place and lead to increased well pressures. **Fig. 4.2.6** shows the situation. In this case, the rig crew must initiate well-control procedures very quickly in order to avoid fracturing the casing shoe.

Influxes taken in brine typically lead to larger pressures and gas volumes at surface during the well kill compared to that seen for an OBM. An example is shown in **Figs. 4.2.7 through 4.2.10,** which compare two well-control scenarios and assumed a perfect kill situation.

From these figures, we can also observe that the surface pressures can vary due to geometrical changes. As the influx leaves the horizontal section and migrates upward, a sharp increase in surface pressures can be seen. After a while, it enters a region with larger volumes, and the kick length is shortened, which temporarily reduces the surface pressures. These are purely geometrical effects.

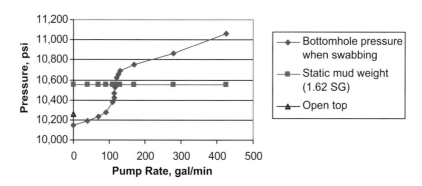

Fig. 4.2.3—The importance of maintaining circulation during swabbing operations to omit underbalanced conditions (Rommetveit et al. 2003).

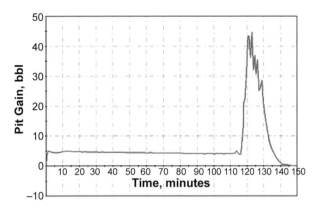

Fig. 4.2.4—A 4-bbl kick is taken when running the 9⅞-in. casing. Circulation rate 300 gal/min. No expansion in pit before the influx boils out of solution (Rommetveit et al. 2003).

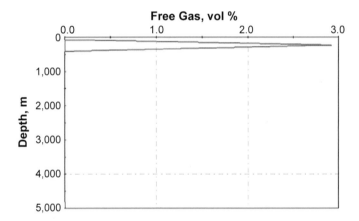

Fig. 4.2.5—Position of influx when the sharp increase in pit gain is observed, caused by presence of free gas (Rommetveit et al. 2003).

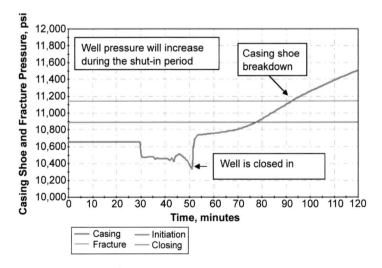

Fig. 4.2.6—In this case, a 20-bbl kick has been taken in brine. Note the pressure buildup even at closed-in conditions (Rommetveit et al. 2003).

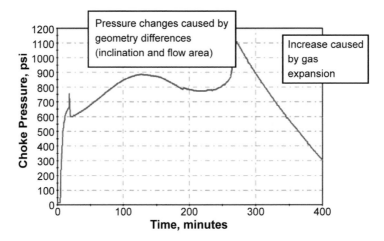

Fig. 4.2.7—Choke pressure development for a 100-bbl kick in OBM. The kick is taken in the horizontal section (Rommetveit et al. 2003).

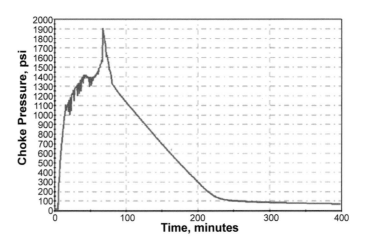

Fig. 4.2.8—Choke pressure development for a 100-bbl kick in brine. Note significantly larger pressures (Rommetveit et al. 2003).

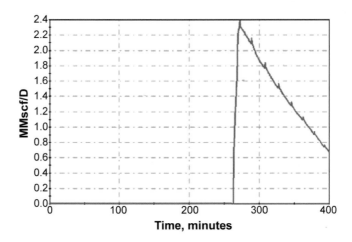

Fig. 4.2.9—Gas flow rate out for the 100-bbl kick in OBM (Rommetveit et al. 2003).

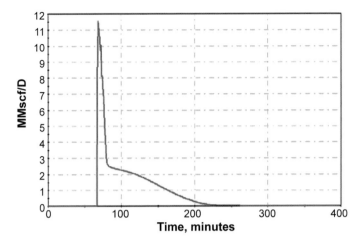

Fig. 4.2.10—Gas flow rate out for the 100-bbl kick in brine. Note that the surface flow rates are significantly larger and bottom-up time is significantly lower (Rommetveit et al. 2003).

Design Modifications. The initial modeling work focused on providing kick tolerances for validating the planned casing design. Due to the expected high-pressure zone just above the formation, some uncertainty existed as to whether it was possible to drill through this high-pressure formation without setting the 9⅞-in. casing first. There were also doubts about the strength of the formation around the 13⅜-in. casing shoe. Through close collaboration, the modeling tools were used during the drilling operation, using actual field data to assess the feasibility of setting the casing deeper. The results indicated clearly that this would have involved unacceptable risks, and as a result the 9⅞-in. casing was not deepened.

Initial kick tolerance results showed that the single-bubble model approach used by BP was too conservative. Hence, it was decided to use the more advanced dynamic kick simulator for updating the tolerances when necessary and to base the casing design on this. As the drilling operation proceeded, better data became available and kick tolerances were updated. In **Fig. 4.2.11,** an example of a kick tolerance curve is given.

Modifications to Procedures. The kick model was used to validate existing procedures. For instance, well-control simulations showed that there were no benefits in using the wait-and-weight kill method when drilling the 12¼-in. hole section. Swab calculations confirmed the need for maintaining circulation to minimize the risk of swabbing.

Simulation results showed that the rig crew had to be very careful and avoid undetected influxes in OBM. If a kick in brine occurred, well-control procedures had to be initiated quickly because well pressure would build up rapidly during shut-in due to fast migration rates.

It also became clear that the existing volumetric well-control procedures were not adequate because they did not account for inclination effects and hole geometry changes. Existing procedures could, in fact, make the well go underbalanced. Hence, new procedures were developed and verified with the kick model.

Modeling Support During Drilling Operation. During drilling operations, kick tolerances were recalculated. The model was calibrated with the most recent data for survey, depths, mud weights and rheology, temperature, and formation integrity testing (FIT). During critical phases of the operation, kick tolerances were updated to provide input for the future decisions.

Operational Success With No Well-Control Problems. Devenick was a major operational success for BP, being the first HP/HT horizontal well drilled. The effort put into well-control planning, the development of new procedures, increasing the understanding of HP/HT well-control phenomena, and crew training resulted in a successful well. The rig crews felt well prepared and were knowledgeable on what to expect from fluid behaviors in the well. They confidently and successfully applied complex kick prevention procedures, avoiding any major well-control events.

Modular dynamics formation tester (MDT) log results showed that the horizontal reservoir section had been drilled successfully with an overbalance of only 200 psi—a significant achievement considering the swab risk in horizontal hole sections with heavyweight mud.

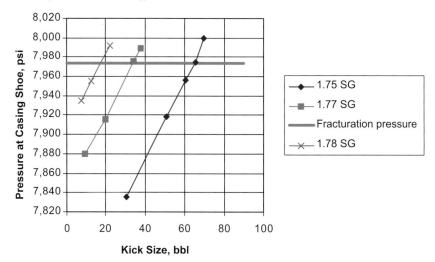

Fig. 4.2.11—Casing shoe pressure for different kick sizes using reservoir pressures of 1.75, 1.77, and 1.78 SG (Rommetveit et al. 2003).

4.2.9 Kicks in Slimhole Wells. Well control and hydraulics in slimhole wells have been studied in full-scale experiments (Steine et al. 1996). An advanced kick simulator was used to model the kicks (Rommetveit et al. 1996).

The simulator predicted the pressure losses very well in the part of the annulus that was concentric and without tool joints. In the section with tool joints, the predictions were reasonable. In the eccentric part of the annulus, the pressures were overpredicted by the simulator.

The predictions of gas-rise velocity were very good in the cases with laminar flow combined with rotation and with slightly turbulent flow. In the upper part of the well, the simulator overpredicted the gas-rise velocity. The most obvious explanation for this is the tool joints.

In a slimhole geometry without eccentricity and tool joints, both pressure losses and gas rise velocities are predicted very well by the simulator. Effects from tool joints and eccentricity influence the results in a slimhole geometry to a greater extent than in a traditional well geometry.

4.2.10 Kick With Lost Circulation. To simulate drilling problems when operating within narrow margins between formation and fracture pressure can cause great savings through improved planning. This is especially the case for HP/HT and deepwater wells. Lost circulation (i.e., underground flow) has in many cases proved to be a significant contributor to complex well-control problems and drilling delays. Well control of underground flow situations has been discussed by Wessel and Tarr (1991).

One advanced kick simulator has been expanded to include simulation of massive lost circulation (Petersen et al. 1998). This model allows interactive investigations of how to kill an influx while losses to formation take place. Kill procedures can thus be optimized through simulations.

The fracturing process is modeled in a simple way. Only three parameters define the fracturing: the fracture initiation pressure, the fracture propagation pressure, and the fracture closing pressure. The flow into (and out of) the fracture is determined by the flow needed to maintain the fracture pressure at the fracture position. A user-defined fraction of fluid lost to the formation determines the amount of fluid reentering the well.

Results are presented below from a simulation of a representative deepwater well with a complex well-control situation developing with underground blowout.

Fig. 4.2.12 shows a schematic well drawing with a fracture *arrow* on the left side of the casing shoe. The lines, one in the drillstring and two in the open hole, are mud fronts.

Consider a 5106-m-deep offshore well at a water depth of 904 m. The casing shoe is set at 2650 m with a diameter of 8.83 in. The 2456-m openhole section has a diameter of 8.54 in. The bit is as at the bottom, which is 6 m into the reservoir of rich natural gas. The drillstring has an OD of 4.23 in. and an inside diameter of

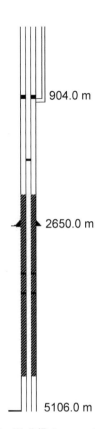

904.0 m

2650.0 m

5106.0 m

Fig. 4.2.12—Well (Petersen et al. 1998).

4 in. The drill collar is 150 m long with an inside diameter of 2.81 in. and an OD of 6.5 in. The well has two chokelines of 904 m with inside diameters of 2.5 in. each. See Fig. 4.2.12 (only one chokeline is shown).

The initial mud density was 1.95 specific gravity (SG). The reservoir pressure gradient is 2.079 SG, and at the casing shoe the fracture initiation, propagation, and closing pressures are 2.59 SG, 2.404 SG and 2.338 SG, respectively. This simulation has a recovery ratio of 0.5 (i.e., half of what goes into the fracture will reenter the annulus as the fracture closes).

The initial pump rate was 1500 L/min. The reservoir overpressure caused the pit gain to rise to 7 m³ within 3.5 minutes of simulated time. Shutting in the well took another minute, and in that time the pit gain had risen to 10.8 m³. The shut-in casing pressure (SICP) indicated a kill mud density of 2.234 SG. At this point, the pressure at the casing shoe has already surpassed the fracture propagation pressure.

We pump the kill mud at 500 L/min with the choke valve open enough (14%) to assure that the bottomhole pressure remains higher than the reservoir pressure. Apparently this opening is a little too small, because the pressure at the casing shoe passes the fracture initiation pressure at 11 minutes 25 seconds. The pump pressure drops to nothing in approximately 5 minutes, and the pit volume drops rapidly **(Fig. 4.2.13)**. The indications are clear that the well is having lost-circulation problems.

The pump rate is increased to 2500 L/min to get the well under control. (**Fig. 4.2.14** shows the free-gas distribution just before increasing the pump rate.) The fracture pressure combined with the pressure due to hydrostatic column in the openhole section must be higher than the reservoir pressure.

This is done to move as much mud into the openhole section as possible to increase the weight of the hydraulic column between the reservoir and the casing shoe in order to stop the flowing reservoir.

To avoid the gas influx in the casing section, the choke is kept at the relatively small aperture. This makes the majority of the gas escape into the fracture.

The gas distribution in the well a few minutes after the pump rate increase is illustrated in **Fig. 4.2.15**. The influx is already in the open hole, and the fracture pressure reduces the bottomhole pressure, thus causing the rate of gas influx to increase. It is not until the kill mud reaches the bottom hole (after 32 minutes 20 seconds)

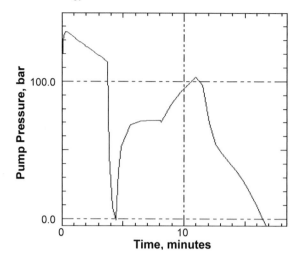

Fig. 4.2.13—Pump pressure (Petersen et al. 1998).

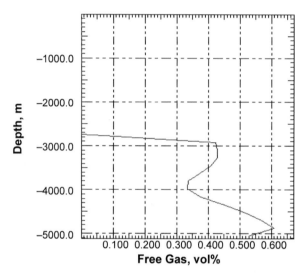

Fig. 4.2.14—Free gas at 16 minutes 43 seconds (Petersen et al. 1998).

that the flow from the reservoir starts declining. The pump pressure increases as the well fills with kill mud, indicating control over the reservoir has been reached (**Fig. 4.2.16**). It is now time to handle the lost-circulation situation. Apparently the new hydrostatic column combined with the fracture propagation pressure is enough to produce a bottomhole pressure higher than the reservoir pressure. The pump rate is thus reduced to 2000 L/min. Keeping an eye on the behavior of the pump pressure, the pump rate is reduced gradually to 500 L/min over a period of about a half hour (Fig. 4.2.16). The mud flow rate out has been this order of magnitude, ~500 L/min, since fracturing.

As the pump rate decreases to 500 L/min, the gas influx that has passed by the fracture starts entering the choke. The initial gas flux out reaches 78 000 L/min (78 minutes 30 seconds) but diminishes to half 50 minutes later. At a gas flow rate out of 22 000 L/min, the choke is opened to reduce the pressure at the casing shoe. The choke changes stepwise from 14 to 31% (205 minutes 35 seconds). This causes the situation at the casing shoe to change from a loss to a recovery condition after approximately 3 hours simulated time. Mud and gas begin leaving the fracture at the rate needed to keep the pressure at the casing shoe at the proper fracture pressure. As the gas flow rate out reaches 50 000 L/min and the mud rate out goes 80% higher than the mud rate in, the choke is reduced to 25% (249 minutes); see **Figs. 4.2.17 and 4.2.18.** The gas flow rate out peaks at

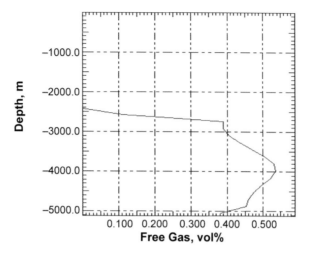

Fig. 4.2.15—Free gas at 19 minutes 30 seconds (Petersen et al. 1998).

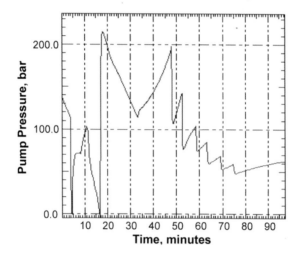

Fig. 4.2.16—Pump pressure (Petersen et al. 1998).

70 000 L/min and quickly drops while the mud flow rate out drops. We know that we are still in a fracture situation due both to the mud flow rates and due to the situation with the pump pressure (Fig. 4.2.17). The pump pressure sees the fracture pressure as the only boundary condition with no gas in the openhole section, so it stays calmly at a constant pressure **(Fig. 4.2.19).**

The choke is opened to 27% after 7 hours simulated time, causing the mud and gas flow rates to jump. After 535 minutes, the choke is reduced to 26% to decrease the discrepancies in the mud flow rates. Since we in general do not know when the fracture closes, it is wise to keep the flow rates as close as possible. The larger the flow rate out is with respect to the flow rate in, the larger the pressure drop in the annulus when the fracture closes. If the pressure drop is large enough, the bottomhole pressure could drop below the reservoir pressure, leading to another kick/fracture situation.

As the time passes on, the choke opening is changed with the aim in mind to keep the mud flow rates nearly the same, but with the flow rate out slightly higher. After 13 hours, the fracture empties and closes, causing a sudden pressure drop in the well. The pump pressure drops approximately 10 bar (Fig. 4.2.19). The remaining gas moves out while the choke opening gradually increases to 100%. **Fig. 4.2.20** shows the pit gain throughout this case, and **Fig. 4.2.21** shows the choke pressure. The *recoverable* mass in the fracture is shown in **Fig. 4.2.22.**

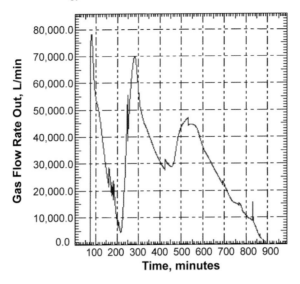

Fig. 4.2.17—Gas-flow rate out (Petersen et al. 1998).

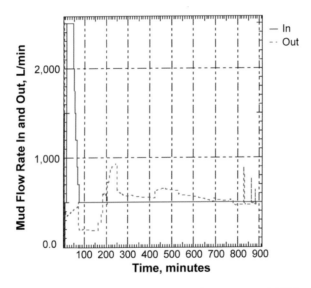

Fig. 4.2.18—Mud-flow rates in and out (Petersen et al. 1998).

4.2.11 Deepwater Well-Control Modeling. During a kick event, gas might migrate past the blowout preventer (BOP) and into the riser prior to kick detection. Some gas could also get trapped in the BOP stack during the shut-in procedure. If this gas is not handled in the appropriate way, gas will be released on Rotary Kelly Bushing (RKB), and an accident could be the result. The Zapata Lexington incident (Shaughnessy 1986; Gonzalez et al. 2000), with a gas explosion and five fatalities as a consequence, is believed to have occurred because

- Trapped gas in the BOP was circulated out at a very high rate.
- The diverter element in the riser was closed

Tests were performed by Amoco to evaluate the behavior of gas in a deepwater riser (Shaughnessy 1986; Gonzalez et al. 2000). Lloyd et al. (2000) evaluated how to handle gas in a deepwater riser.

Gas in a deepwater riser as a result of a kick has been studied by the use of advanced kick simulations (Nes et al. 1998). The following scenarios were simulated:

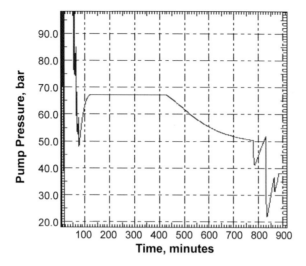

Fig. 4.2.19—Pump pressure (Petersen et al. 1998).

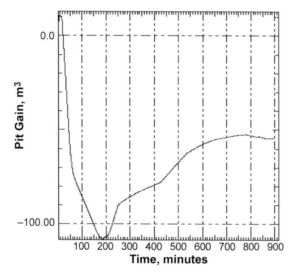

Fig. 4.2.20—Pit gain (Petersen et al. 1998).

- Gas migrating up the riser
- Pumping with gas in the riser
- Swabbed (concentrated) vs. drilled (distributed) kicks
- To what extent gas would be trapped in the mud

If handled properly, the gas in the riser, and trapped gas in the BOP, can be safely transported to the surface and out of the diverter. These results confirmed earlier tests performed by Amoco (Shaughnessy 1986) and published in 2000 (Gonzalez et al. 2000).

Hydraulics and well-control tests were performed in an ultra deepwater well (2741 m water depth) (Rommetveit et al. 2005). The simulated gas migration vs. that observed is shown in **Fig. 4.2.23. Fig. 4.2.24** shows that the decrease in pressure when gas reaches the surface is accurately reproduced by the well-control model. This shows that the gas had been distributed approximately as predicted by the model. For comparison, if all gas had moved as a single bubble, pressure would have dropped by more than 30 bar when the bubble reached the surface.

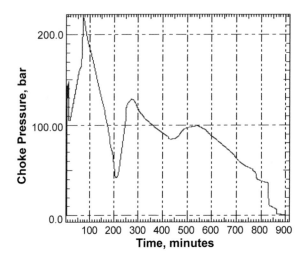

Fig. 4.2.21—Choke pressure (Petersen et al. 1998).

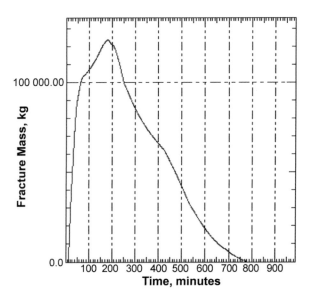

Fig. 4.2.22—Recoverable fracture mass (Rommetveit 2005; Petersen et al. 1998).

The pressure at one memory sensor vs. simulated pressure during the migration and subsequent circulation of gas in the riser is discussed below.

4.2.12 Kick With Hydrates. An advanced dynamic kick simulator has been expanded to include the determination of the potential for hydrate formation (Petersen et al. 2001). Using dynamic temperature simulation, detailed PVT computations of the hydrocarbon influx on the component level, and an advanced hydrate formation program, it is possible to obtain the *distance* in temperature from hydrate formation throughout the well at any time during a simulated operation.

The simulator includes code that takes into consideration the effect of hydrate-inhibitor chemicals such as salts and alcohols. Thus, it is possible to make several trial runs with differently inhibited muds to compare the dangers of running into hydrates.

The main purpose of the hydrate formation module is to combine many effects of well control in order to determine the chance of hydrate formation during well operations. Most hydrate reports are investigations into hydrates formation under carefully controlled conditions. This information is added into dynamic oil/gas

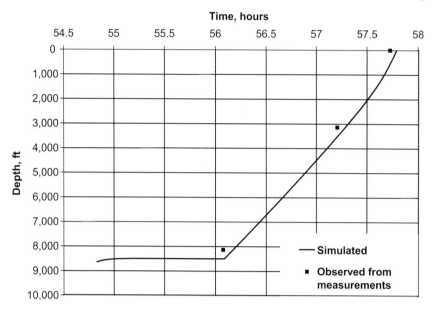

Fig. 4.2.23—Position of gas front during migration and subsequent circulation (Rommetveit 2005).

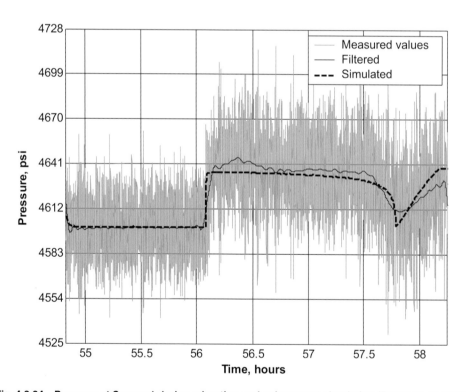

Fig. 4.2.24—Pressure at Sensor 1 during migration and subsequent circulation (Rommetveit 2005).

well operations where pressures, temperatures, and hydrocarbon compositions can vary dramatically within hours of operation.

4.2.13 Well-Control Modeling of Dual-Density Drilling Systems. The evolution of increasingly complex drilling systems for deep and ultradeep drilling conditions, such as dual-density drilling systems with subsea pumps, has necessitated development of safe well-control procedures for all possible scenarios. One model was further developed for the *Subsea Mudlift System* and used in the development of well-control procedures (Choe and Juvkam-Wold 1996, 1997).

Another such model was developed for the *Controlled Mud Pressure* (CMP) *System* (Rommetveit et al. 2006a). There has been special focus on studies of well control and early kick detection. Results show that an influx can be detected very early. A detailed procedure for handling an influx in a safe manner has been developed and verified with simulations. Kick detection when drilling and circulating at normal rates is expected to be superior to that of conventional drilling because return flow is indirectly measured very accurately by the CMP pump. The observed power consumption will effectively be a sensitive flow rate sensor and can serve as part of an accurate kick-detection system, provided that noise and other effects that influence power consumption are under sufficiently accurate control.

4.2.14 Integrated Well-Control and Thermohydraulic Modeling. Early well-control simulation tools did not consider dynamic temperature, but only did computations using the assumed *background* temperature profile (Rommetveit 1994). As offshore wells moved to deeper water, it became clear that some important transient effects could be simulated only by using a dynamic temperature model.

A general dynamic model for any flow-related operation (including complex well-control situations) during well construction and interventions has been developed (Rommetveit et al. 2006b). The model is a basis for a new generation of support tools and technologies needed in today's environment with advanced well designs, challenging drilling conditions, and need for fast and reliable real-time decision support. The solution method uses a *divide and conquer* scheme, which computes the flow in each well segment separately and then solves for the appropriate flow in the junctions. This simplifies simulating complex flow networks, such as multilateral wells or jet subs. The flexibility allows incorporation of additional pumps in the flow loop, as in dual-gradient systems (Rommetveit et al. 2006a). The model includes dynamic 2D temperature calculations, covering the radial area affecting the well and assuming radial symmetry in the vicinity of the well. Other features include flexible boundary conditions (which include drilling and tripping), non-Newtonian frictional pressure loss, transient well-reservoir interaction, slip between phases, and advanced PVT relationship. This approach gives higher flexibility, improved accuracy, reduced numerical diffusion, and increased computational speed.

The way the evolution of the state of the well is computed, the pressure computations and the heat/thermal computations are offset from each other. That is, they are not computed simultaneously. This simplifies the computation greatly.

4.2.15 Real-Time Drilling Operations Modeling Including Managed-Pressure Drilling. *Real-Time Decision Support.* An advanced dynamic-flow and well-control model is integrated into a decision support system with real-time simulations, diagnosis, and what-if evaluating possibilities. Real-time well-control and kick-tolerance evaluations are facilitated by linking the flow model to the drilling data recorded.

The flow model is also linked with other drilling subprocess models [torque and drag, rate of penetration (ROP), well stability, and pore pressure], constituting a real-time integrated drilling simulator (see **Fig. 4.2.25**) (Rommetveit et al. 2007).

Managed-Pressure Drilling. The advanced dynamic flow and temperature model was used during managed-pressure drilling (MPD) on an HP/HT well both for real-time control of choke setting and for offline simulations prior to the actual operations (Bjørkevoll et al. 2008). The two model variants used are described in some detail in the following sections and by **Fig. 4.2.26,** where color codes are used to distinguish the two.

Real-Time Model. The model ran with input from the rig, to update continuously the choke pressure setpoint, which was input to the automatic choke system. The main features of the real-time model are as follows:

- Dynamic mass transport, which means that changes in boundary conditions and temperature profile will propagate through the system rather than jumping straight to a new state.
- Pressure- and temperature-dependent density. For the actual mixture of Cs/K formate, calculations are based on laboratory measurements.
- Pressure- and temperature-dependent rheology. Laboratory data are adjusted automatically to match rig measurements done during operations, and interpolated to calculated pressure and temperature in each grid cell along the flow trajectory.

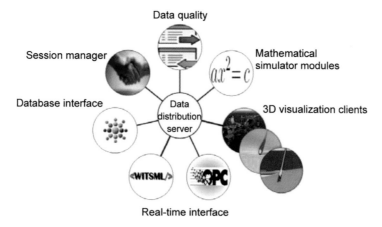

Fig. 4.2.25—Infrastructure for the Real-Time Integrated Drilling System.

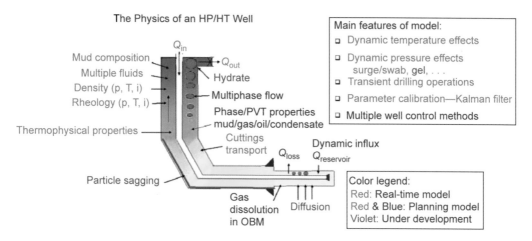

Fig. 4.2.26—Schematic overview of the model, both the offline variant used for planning and the online variant used for automated choke control.

- Frictional pressure loss is calculated on the basis of fitting a three-parameter rheology model to rheology, and using published methods for handling laminar, transitional, and turbulent flow.
- Cuttings load is taken into account.
- Rotation is taken into account when calculating frictional pressure loss and heat transfer.
- Multiple fluids are tracked through the system, and calculated pressures will change gradually, depending on details of rheology, density, inside diameters and ODs, angle of inclination, and operational parameters.
- A 2D detailed dynamic temperature model is seamlessly integrated with the 1D dynamic mass transport model. The temperature model takes into account formation properties of different lateral layers, casing/liner layers with cement and other materials behind, and fluid properties and operational parameters. Virgin formation temperature and mud temperature in are input parameters.

The real-time model can handle various operational sequences such as

- Circulation at different pump rates, including ramping the pump up and down, with and without rotation
- Static periods
- Drilling
- Tripping in and out
- Displacements with different fluids or one fluid with different densities

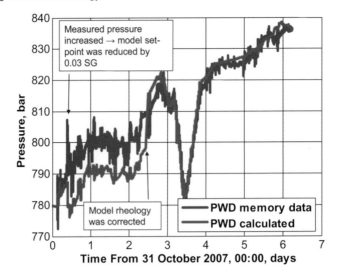

Fig. 4.2.27—Drilling Well 2, model/PMWD comparison.

Example Well. There was a deviation between pressure measurement while drilling (PMWD) and model pressure during the first part of drilling, and this turned out to be due to erroneous model configuration, most importantly wrong rheology input. The model was restarted with correct configuration, and after having stabilized, calculation kept within 2 to 3 bar deviation from PMWD for the rest of the run; see **Fig. 4.2.27.** This included pump stops (e.g., shortly after 3 days) in which choke pressure was increased automatically by more than 20 bar to compensate for loss of friction, but still with measured PMWD within approximately 3 bar from calculation.

Offline Model. The real-time model is derived from a more comprehensive offline model that was used for planning of operations. The offline model adds the following features, among others:

- Inflow of hydrocarbons or water from a reservoir
- Dynamic two-phase transport of fluids up the annulus
- Batch simulation of long predefined sequences, which has been used for
  - Testing and validation of the model
  - Preparing for upcoming operations
  - Giving input to procedures
  - Post-analysis

The offline model was used for detailed simulations of different operations before the start of the first MPD well, for testing and validation of the model, and for detailed adjustments and validation of operational working including well control procedures.

Automated Drilling System. The transient flow and temperature model is currently linked to a drilling-rig-control system. This will facilitate real-time kick-tolerance evaluations and automatic kill operations assisted by an advanced transient-flow and well-control model.

Nomenclature

A = cross-sectional area
C_0 = distribution factor
f_1 = frictional pressure loss term
f_2 = localized pressure loss term
$\dot{m}$ = mass transfer rate
$\dot{m}_g$ = gas dissolution rate

p = pressure
q = gas influx rate
Re = Reynolds number
s = distance
$\tilde{S}_{int}$ = interface area per length
t = time
T = temperature
v = velocity
v_0 = gas slip velocity
v_{mix} = mixture velocity
x_{dg} = mass fraction of dissolved gas
x_{fo} = mass fraction of formation oil
α = gas void fraction
ΔM = gas mass transfer during a timestep
Δt = timestep size
θ = hole inclination
v = kinetic viscosity
ρ = density
τ = contact time

Subscripts

g = gas
l = liquid
fo = formation oil

References

Aas, B., Bach, G.F., Hauge, H.C., and Sterri, N. 1993. Experimental Modeling of Gas Kicks in Horizontal Wells. Paper SPE 25709 presented at the SPE/IADC Drilling Conference, Amsterdam, 23–25 February.

Bjørkevoll, K.S., Rommetveit, R., Aas, B., Gjeraldstveit, H., and Merlo, A. 2003. Transient Gel Breaking Model for Critical Wells Applications With Field Data Verification. Paper SPE 79843 presented at the SPE/IADC Drilling Conference, Amsterdam, 19–21 February. DOI: 10.2118/79843-MS.

Bjørkevoll, K.S., Molde, D.O., Rommetveit, R., and Syltøy, S. 2008. MPD Operation Solved Drilling Challenges in a Severely Depleted HP/HT Reservoir. Paper SPE 112739 presented at the IADC/SPE Drilling Conference, Orlando, Florida, 4–6 March. DOI: 10.2118/112739-MS.

Bradley, N.D., Low, E., Aas, B., Rommetveit, R., and Larsen, H.F. 2002. Gas Diffusion—Its Impact on a Horizontal HP/HT Well. Paper SPE 77474 presented at the SPE Annual Technical Conference and Exhibition, San Antonio, Texas, 29 September–2 October. DOI: 10.2118/77474-MS.

Caetano, E.F. 1986. Upward Vertical Two-Phase Flow Through an Annulus. PhD dissertation, University of Tulsa, Tulsa.

Choe, J. and Juvkam-Wold, H.C. 1996. Well Control Model Analyzes Unsteady State, Two-Phase Flow. *Oil and Gas Journal* **94** (49): 68–77.

Choe, J. and Juvkam-Wold, H.C. 1997. A Modified Two-Phase Well-Control Model and Its Computer Applications as a Training and Educational Tool. *SPECA* **9** (1): 14–20. SPE-37688-PA. DOI: 10.2118/37688-PA.

Corre, B., Eymard, R., and Guenot, A. 1984. Numerical Computation of Temperature Distribution in a Wellbore While Drilling. Paper SPE 13208 presented at the SPE Annual Technical Conference and Exhibition, Houston, 16–19 September. DOI: 10.2118/13208-MS.

Currans, D., Brandt, W., Lindsay, G., and Tarvin, J. 1993. The Implications of High Angle and Horizontal Wells for Successful Well Control. Paper presented at the IADC European Well Control Conference, Paris, 2–4 June.

eControl: Functional Design Specification. Document 100501-19591-IZ-SA06-0102, Aker MH, Kristiansand, Norway (December 2007).

Ekrann, S. and Rommetveit, R. 1985. A Simulator for Gas Kicks in Oil-Based Drilling Muds. Paper SPE 14182 presented at the SPE Annual Technical Conference and Exhibition, Las Vegas, Nevada, 22–26 September. DOI: 10.2118/14182-MS.

Gard, J. 1986. PVT Measurements of Base Oils. Research report, PRC K-36/86, Rogaland Research, Stavanger.

Gonzalez, R., Shaughnessy, J.M., and Grindle, W.D. 2000. Industry Leaders Shed Light on Drilling Riser Gas Effects. *Oil and Gas Journal* **98** (29).

Green, D.W. and Perry, R.H. 1984. *Perry's Chemical Engineers' Handbook*, sixth edition. New York: McGraw-Hill.

Hoberock, L.L. and Stanbery, S.R. 1981. Pressure Dynamics in Wells During Gas Kicks: Part 2—Component Models and Results. *JPT* **33** (8): 1367–1378. SPE-9822-PA. DOI: 10.2118/9822-PA.

Hovland, F. and Rommetveit, R. 1992. Analysis of Gas-Rise Velocities From Full-Scale Kick Experiments. Paper SPE 24580 presented at the SPE Annual Technical Conference and Exhibition, Washington, DC, 4–7 October. DOI: 10.2118/24580-MS.

Johnson, A.B. and Cooper, S. 1993. Gas Migration Velocities During Gas Kicks in Deviated Wells. Paper SPE 26331 presented at the SPE Annual Technical Conference and Exhibition, Houston, 3–6 October. DOI: 10.2118/26331-MS.

Johnson, A.B. and White, D.B. 1991. Gas-Rise Velocities During Kicks. *SPEDE* **6** (4): 257–263. SPE-20431-PA. DOI: 10.2118/20431-PA.

Lage, A.C.V.M. and Time, R.W. 2000. Mechanistic Model for Upward Two-Phase Flow in Annuli. Paper SPE 63127 presented at the SPE Annual Technical Conference and Exhibition, Dallas, 1–4 October. DOI: 10.2118/63127-MS.

Lage, A.C.V.M., Rommetveit, R., and Time, R.W. 2000. An Experimental and Theoretical Study of Two-Phase Flow in Horizontal or Slightly Deviated Fully Eccentric Annuli. Paper SPE/IADC 62793 presented at the IADC/SPE Asia Pacific Drilling Technology, Kuala Lumpur, 11–13 September. DOI: 10.2118/62793-MS.

LeBlanc, J.L. and Lewis, R.L. 1968. A Mathematical Model of a Gas Kick. *JPT* **20** (8): 888–898; *Trans.*, AIME, **243**. SPE-1860-PA. DOI: 10.2118/1860-PA.

Lloyd, W.L., Andrea, M.D., and Kozicz, J.R. 2000. New Considerations for Handling Gas in a Deepwater Riser. Paper SPE 59183 presented at the IADC/SPE Drilling Conference, New Orleans, 23–25 February. DOI: 10.2118/59183-MS.

Nes, A., Rommetveit, R., Hansen, S. et al. 1998. Gas in a Deep Water Riser and Associated Surface Effects Studied With an Advanced Kick Simulator. Presented at IADC Deep Water Well Control Conference, Houston, 26–27 August.

Nickens, H.V. 1987. A Dynamic Computer Model of a Kicking Well. *SPEDE* **2** (2): 159–173; *Trans.*, AIME, **283**. SPE-14183-PA. DOI: 10.2118/14183-PA.

O'Bryan, P.L., Bourgoyne, A.T., Monger, T.G., and Kopeck, D.P. 1988. An Experimental Study of Gas Solubility in Oil-Based Drilling Fluids. *SPEDE* **3** (1): 33–42; *Trans.*, AIME, **285**. SPE-15414-PA. DOI: 10.2118/15414-PA.

Petersen, J., Rommetveit, R., and Tarr, B.A. 1998. Kick With Lost Circulation Simulator, a Tool for Design of Complex Well Control Situations. Paper SPE 49956 presented at the SPE Asia Pacific Oil and Gas Conference and Exhibition, Perth, Australia, 12–14 October. DOI: 10.2118/49956-MS.

Petersen, J., Bjørkevoll, K.S., and Lekvam, K. 2001. Computing the Danger of Hydrate Formation Using a Modified Dynamic Kick Simulator. Paper SPE 67749 presented at the SPE/IADC Drilling Conference, Amsterdam, 27 February–1 March. DOI: 10.2118/67749-MS.

Podio, A.L. and Yang A.-P. 1986. Well Control Simulator for IBM Personal Computer. Paper SPE 14737 presented at the SPE/IADC Drilling Conference, Dallas, 10–12 February. DOI: 10.2118/14737-MS.

Rader, D.W., Bourgoyne, A.T., and Ward, R.H. 1975. Factors Affecting Bubble-Rise Velocity of Gas Kicks. *JPT* **27** (5): 571–584. SPE-4647-PA. DOI: 10.2118/4647-PA.

Rommetveit, R. 1989. A Numerical Simulation Model for Gas-Kicks in Oil Based Drilling Fluids. PhD dissertation, University of Bergen, Bergen, Norway.

Rommetveit, R. 1994. Kick Simulator Improves Well Control Engineering and Planning. *Oil and Gas Journal* **92** (34): 64–71.

Rommetveit, R. and Bjørkevoll, K.S. 1997. Temperature and Pressure Effects on Drilling Fluid Rheology and ECD in Very Deep Wells. Paper SPE 39282 presented at the SPE/IADC Middle East Drilling Technology Conference, Bahrain, 23–25 November. DOI: 10.2118/39282-MS.

Rommetveit, R. and Olsen, T.L. 1989. Gas Kick Experiments in Oil-Based Drilling Muds in a Full-Scale Inclined Research Well. Paper SPE 19561 presented at the SPE Annual Technical Conference and Exhibition, San Antonio, 8–11 October. DOI: 10.2118/19561-MS.

Rommetveit, R. and Vefring, E.H. 1991. Comparison of Results From an Advanced Gas Kick Simulator With Surface and Downhole Data From Full Scale Gas Kick Experiments in an Inclined Well. Paper SPE 22558 presented at the SPE Annual Technical Conference and Exhibition, Dallas, 6–9 October. DOI: 10.2118/22558-MS.

Rommetveit, R., Bjørkevoll, K.S., Bach, G.F. et al. 1995. Full Scale Kick Experiments in Horizontal Wells. Paper SPE 30525 presented at the SPE Annual Technical Conference and Exhibition, Dallas, 22–25 October. DOI: 10.2118/30525-MS.

Rommetveit, R., Nes, A., Steine, O.G., Harries, T.W.R., Maglione, R., and Sagot, A. 1996. The Applicability of Advanced Kick Simulators to Slim Hole Drilling. Presented at the IADC Well Control Conference for Europe, Aberdeen, 22–24 May.

Rommetveit, R., Time, R.W., and Bjørkevoll, K.S. 1998. Large Scale Experiments of Non-Newtonian Two Phase Flow in Horizontal Annuli With Relevance to Well Control in Horizontal Wells. Presented at the 8th International Conference on Multiphase Flow, Cannes, France, 18–20 June.

Rommetveit, R., Fjelde, K.K., Aas, B. et al. 2003. HP/HT Well Control: An Integrated Approach. Paper OTC 15322 presented at the Offshore Technology Conference, Houston, 5–8 May.

Rommetveit, R., Bjørkevoll, K.S., Gravdal, J.E. et al. 2005. Ultra-Deepwater Hydraulics and Well Control Tests With Extensive Instrumentation: Field Tests and Data Analysis. *SPEDC* **20** (4): 251–257. SPE-84316-PA. DOI: 10.2118/84316-PA.

Rommetveit, R., Bjørkevoll, K., Petersen, J. et al. 2006a. A Novel, Unique Dual Gradient Drilling System for Deep Water Drilling, CMP, Has Been Proven by Means of a Transient Flow Simulator. IBP1400_06, presented at the Rio Oil and Gas Expo and Conference, Rio de Janeiro, 11–14 September.

Rommetveit, R., Bjørkevoll, K.S., Petersen, J., and Frøyen, J. 2006b. A General Dynamic Model for Flow Related Operations During Drilling, Completion, Well Control and Intervention. IBP1373_06, presented at the Rio Oil and Gas Expo and Conference, Rio de Janeiro, 11–14 September.

Rommetveit, R., Bjørkevoll, K.S., Halsey, G.W. et al. 2007. eDrilling: A System for Real-Time Drilling Simulation, 3D Visualization and Control. Paper SPE 106903 presented at the Digital Energy Conference and Exhibition, Houston, 11–12 April. DOI: 10.2118/106903-MS.

Santos, O.L.A. 1989. A Dynamic Model of Diverter Operations for Handling Shallow Gas Hazards in Oil and Gas Exploratory Drilling. PhD dissertation, Louisiana State University, Baton Rouge, Louisiana.

Santos, O.L.A. 1991. Well-Control Operations in Horizontal Wells. *SPEDE* **6** (2): 111–117. SPE-21105-PA. DOI: 10.2118/21105-PA.

Santos, O.L.A. and Azar, J.J. 1997. A Study on Gas Migration in Stagnant Non-Newtonian Fluids. Paper SPE 39019 presented at the Latin American and Caribbean Petroleum Engineering Conference, Rio de Janeiro, 30 August–3 September. DOI: 10.2118/39019-MS.

Shaughnessy, J.M. 1986. Test of Effect of Gas in a Deepwater Riser. Internal memorandum, Amoco Production Company.

Steine, O.G., Rommetveit, R., Maglione, R., and Sagot, A. 1996. Well Control Experiments Related to Slim Hole Drilling. Paper SPE 35121 presented at the SPE/IADC Drilling Conference, New Orleans, 12–15 March. DOI: 10.2118/35121-MS.

Vefring, E.H., Wang, Z., Gaard, S., and Bach, G.F. 1995a. An Advanced Kick Simulator for High Angle and Horizontal Wells—Part I. Paper SPE 29345 presented at the SPE/IADC Drilling Conference, Amsterdam, 28 February–2 March. DOI: 10.2118/29345-MS.

Vefring, E.H., Wang, Z., Rommetveit, R., and Bach, G.F. 1995b. An Advanced Kick Simulator for High Angle and Horizontal Wells—Part II. Paper SPE 29860 presented at the SPE Middle East Oil Show, Bahrain, 11–14 March. DOI: 10.2118/29860-MS.

Wang, Z., Peden, J.M., and Lemanczyk, R.Z. 1994. Gas Kick Simulation Study for Horizontal Wells. Paper SPE 27498 presented at the SPE/IADC Drilling Conference, Dallas, 15–18 February. DOI: 10.2118/27498-MS.

Wessel, M. and Tarr, B.A. 1991. Underground Flow Well Control: The Key to Drilling Low-Kick-Tolerance Wells Safely and Economically. *SPEDE* **6** (4): 250–256; *Trans.*, AIME, **291**. SPE-22217-PA. DOI: 10.2118/22217-PA.

White, D.B. and Walton, I.C. 1990. A Computer Model for Kicks in Water- and Oil-Based Muds. Paper SPE 19975 presented at the SPE/IADC Drilling Conference, Houston, 27 February–2 March. DOI: 10.2118/19975-MS.

Zuber, N. and Findlay, J.A. 1965. Average Volumetric Concentration in Two-Phase Flow System. *ASME Journal of Heat Transfer* **87**: 453–468.

SI Metric Conversion Factors

bbl	× 1.589 873	$E - 01 = m^3$
gal/min	× 2.271 247	$E - 01 = m^3/h$
in.	× 2.54*	$E + 00 = cm$

*Conversion factor is exact.

4.3 Foam Drilling—Ramadan Ahmed, University of Oklahoma, and Stefan Miska, University of Tulsa

4.3.1 Introduction. The need for technologies to reduce cost and improve recovery from existing hydrocarbon reserves is well known (Kuru et al. 1999). One of the most effective methods of cost reduction relies on improvements in drilling technologies. Particularly, the development of underbalanced drilling technology is beneficial for drilling partially depleted reservoirs and reentry wells. During conventional (overbalanced) drilling, mud filtrate penetrates the near-wellbore formation because drilling fluid pressure becomes higher than the pore pressure. Mud filtrate penetration alters near-wellbore pore fluid properties (*Underbalanced Drilling Manual* 1997). As a result, well productivity decreases significantly. Underbalanced drilling technology is often preferred to minimize problems associated with formation damage, lost circulation, and differential sticking (Culen et al. 2003; Devaul and Coy 2003; Santos et al. 2003). In field applications, many different techniques are available for achieving underbalanced conditions. These mostly involve circulating low-density fluids such as aerated mud and foams. However, flow behavior of foam is complex. Using foam or other low-density fluid alone does not always guarantee underbalanced conditions. Excessive frictional pressure losses may result in overbalanced conditions even with foams (*Underbalanced Drilling Manual* 1997). Therefore, hydraulics of aerated fluids and drilling foams should be well understood to accurately predict downhole pressure and minimize drilling problems.

In addition to drilling, foam is widely used in many industrial applications. In the petroleum industry, foams are used as drilling and completion fluids in low-pressure and water-sensitive formations, and as a displacing agent in conjunction with waterfloods. Some of these applications involve flowing of foam through pipes, annuli, and porous media. Other applications of foam include cementing and fracturing. Foamed cements are introduced and show some advantages over the conventional cements. Due to their compressibility and bubble structure, foamed cements enable internal deformation without cracking when they are exposed to thermal and mechanical stresses. Nonetheless, performing foam cement jobs requires additional equipment such as a foam generator, among others (Green et al. 2003).

In some applications, such as established gas-producing areas, conventional drilling poses downhole pressure risk. Foam drilling techniques allow drilling to proceed at or below formation pore pressure. This results in higher rate of penetration (ROP) and limited or no formation damage. Besides this, foam drilling is applied to counter problems with formation water and loss of circulation in boreholes that have formation-damage-sensitive zones. In short, this technology has strong potential to provide significant benefits including increased productivity, high ROP, better hole cleaning, reduced operational difficulties associated with drilling in low-pressure reservoirs, and improved formation evaluation while drilling. However, in-depth understanding of foam structure, texture, rheology, stability, and hydraulics is necessary to reduce risks and costly operations.

Compared with traditional drilling muds, foam is an unstable fluid system. Drilling foam contains a high volume fraction of gas phase (air or other gases such as nitrogen). The gas phase exists as dispersed bubbles separated by thin liquid films. It is a structured gas-liquid dispersion **(Fig. 4.3.1)** with typical bubble dimensions between 10 μm and 1 cm. The liquid component of drilling foam contains water, surfactant, and in some cases additive polymers [hydroxylethylcellulose, polyanionic cellulose (PAC), xanthan gum, and carboxymethylcellulose]. The function of the additive polymer is to improve the stability and control the rheology of foam. The surfactant usually makes up a small amount of the liquid phase (0.5 to 2% by volume). The generation of foam requires the transfer of mechanical energy to surface free energy. Surfactants play a great role in the generation of foam and maintaining the stability of the interface. Therefore, the type of surfactant and its concentration is very important in designing of a foam program. Addition of polymers allows the formation of foams with considerable stability. In spite of this, foams are metastable systems that will always decay over time to achieve a lower energy configuration.

Traditionally, gas volume fraction or foam quality is used as a mean of classification for foams (Ahmed et al. 2003a). Dry foams are characterized by higher amounts of gas phase and coarse bubbles. **Fig. 4.3.2** presents the pattern of relative viscosity of foam (i.e., relative to the liquid phase, which is $[\mu_f(\dot{\gamma})/\mu_L(\dot{\gamma})]$ at different foam qualities. When the quality reaches 98%, foam turns into a mist. There is an essential difference between foam and mist. In a mist, the liquid phase is dispersed in the gas phase as small liquid droplets. As a result, mist is treated as a regular gas-liquid dispersion. The liquid droplets are free and do not possess a structure.

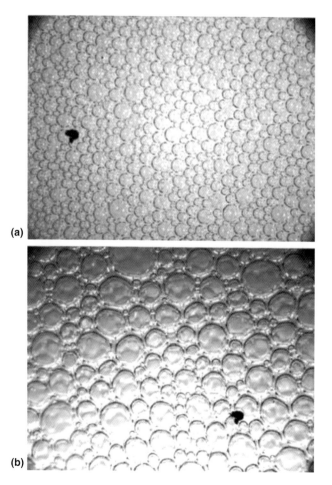

(a)

(b)

Fig. 4.3.1—Structures of low- and high-quality foams: (a) Γ = 70% and (b) Γ = 90%.

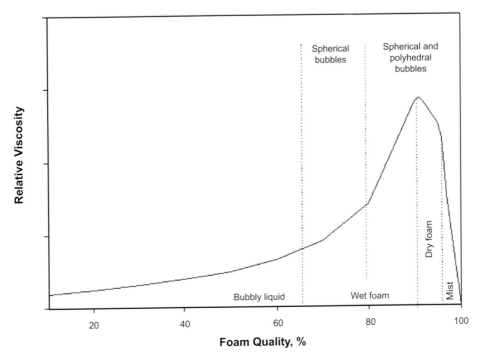

Fig. 4.3.2—Relative viscosity at different foam qualities.

In drilling operations, foam may be generated at surface and then injected to the drillpipe, or its components can be injected separately because foam generation occurs as the mixture flows down the drillpipe, bottomhole assembly (BHA), and drill bit (Lyons et al. 2000). Perhaps the major advantage of foam, compared with other drilling fluids, is that it offers more flexibility over equivalent circulating density (ECD) control. It is well known that proper wellbore pressure management is important not only for pore pressure control but also for wellbore stability. Controlling borehole pressures is of particular importance in deepwater drilling operations where the difference between pore pressure and formation fracture pressure is very small. Accurate wellbore pressure control is achieved by varying the proportions of liquid and gas phases (i.e., injection rates), base fluid rheology, and backpressure at the surface. In addition, foam is an environmentally friendly drilling fluid because it contains only a small fraction of liquid phase to be treated at the surface. Moreover, the polymers and surfactants, which are used to generate foam, are biodegradable.

Foam is a compressible non-Newtonian fluid with a complex structure that is strongly influenced by its quality, liquid-phase composition, and pressure. Foam flow analysis is difficult because many variables such as quality, liquid-phase viscosity, wall slip, pressure, and temperature are affecting the flow behavior. Moreover, methods of foam generation, degree of foam equilibration, shear history, foam texture, and type of surfactant and its concentration play great roles in determining the flow behavior. Selection of appropriate foam quality, injection rates, and type and concentration of polymers and surfactants is critical for achieving the desired wellbore hydraulics in terms of wellbore pressure and cuttings transport. Foam density (gravitational component) and viscosity (friction pressure losses) effects mostly control the wellbore pressure distribution. The kinetic energy changes are usually small compared to the gravitational and friction pressure components. Cuttings transport depends upon many factors; however, again, foam density and viscosity effects are the controlling parameters for a given foam flow rate and wellbore geometric configuration.

4.3.2 Foam Characteristics. In order to make an engineering analysis, relevant properties of foam should be described properly. Hence, foam needs to be characterized like other materials to determine its features (rheology, stability, and compressibility), composition (quality, expansion ratio, density), and structure (bubble size and texture). A systematic characterization of foam properties is necessary to provide effective methods for solving foam drilling problems. Therefore, the next topics will focus on foam characterization.

Physical Properties of Foam. *Foam Quality.* Foam quality is the volumetric fraction of the gas phase at a given temperature and pressure, and mathematically expressed as

$$\Gamma = \frac{V_g}{V_f} = \frac{V_g}{V_g + V_L} \quad \dots\dots\dots\dots\dots\dots\dots\dots\dots\dots\dots\dots\dots\dots\dots\dots\dots \quad (4.3.1)$$

Based on their quality, foams usually are classified as bubbly liquid, wet foam, and dry foam (Fig. 4.3.2). Bubbly liquids are formed at low qualities (when Γ is approximately less than 60% for aqueous foams). In the intermediate quality range (i.e., 60 to 94%), wet foam is generated. Usually, dry foams are observed when quality is greater than 94%. However, these ranges may vary slightly depending on the composition of the foam, pressure, and temperature. Generally, dry foams can flow only if the gas bubbles are deformed (*Underbalanced Drilling Manual* 1997).

Compressibility. Hydraulic and rheological analysis of foam flow requires determination of foam compressibility. Isothermal compressibility of foam is given mathematically by

$$C_f = -\frac{1}{V_f}\frac{dV_f}{dp}, \quad \dots\dots\dots\dots\dots\dots\dots\dots\dots\dots\dots\dots\dots\dots\dots\dots \quad (4.3.2)$$

where V_f is volume of foam, which is given by $V_f = V_g + V_L$. Neglecting liquid-phase compressibility and gas-phase solubility, the foam compressibility can be written in terms of quality and gas-phase isothermal compressibility as

$$C_f = \Gamma C_g \quad \dots\dots\dots\dots\dots\dots\dots\dots\dots\dots\dots\dots\dots\dots\dots\dots\dots\dots \quad (4.3.3)$$

Equations of State—Pressure/Volume/Temperature (PVT). A simple relationship between the PVT of foam can be developed by neglecting the gas phase solubility and assuming the liquid phase as incompressible. At the present, there are limited numbers of foam PVT equations (Morrison and Ross 1983; Lord 1981) that are used for designing purposes. For ambient and intermediate pressure ranges, the PVT behavior of foams can be predicted from PVT properties of the two phases. At high pressures, there is a possibility of departure from the traditional equation of state. This phenomenon can be attributed to, first, the mass transfer between the gas and liquid phases (solubility and evaporation), and, second, structural and fluid property variations resulting from changes in temperature and/or pressure. As a result, there is no single equation of state that predicts the exact PVT behavior of foams at any pressure and temperature. Morrison and Ross (1983) derived an equation of state for foam from the Laplace equation. This equation requires parameters such as pressure inside the gas bubble, interfacial surface area, and surface tension, which are difficult to determine in the field. As a result, practical application of this equation is very limited. Lord (1981) presented an equation of state for foam based on the real gas equation and an assumption that the liquid phase is incompressible. This equation expresses specific volume of foam, v_F, as a function of pressure, temperature, liquid phase specific volume (v_L), and gas mass fraction (W_g). Thus,

$$v_F = \frac{b^* p + a^*}{p}, \quad \dots\dots\dots\dots\dots\dots\dots\dots\dots\dots\dots\dots\dots\dots\dots\dots \quad (4.3.4)$$

where

$$a^* = \frac{W_g z R_g T}{M} \quad \dots\dots\dots\dots\dots\dots\dots\dots\dots\dots\dots\dots\dots\dots\dots\dots \quad (4.3.5)$$

and

$$b^* = (1 - W_g)v_L, \quad \dots\dots\dots\dots\dots\dots\dots\dots\dots\dots\dots\dots\dots\dots\dots\dots \quad (4.3.6)$$

where v_L is the specific volume of the liquid phase and z is gas compressibility factor. The relationship between gas mass fraction and foam quality is given by

$$W_g = \frac{\rho_g \Gamma}{\rho_g \Gamma + \rho_L (1 - \Gamma)} \quad \dots\dots\dots\dots\dots\dots\dots\dots\dots\dots\dots\dots\dots\dots \quad (4.3.7)$$

Eq. 4.3.4 is valid if the change in gas solubility is negligible. Lourenço (2002) experimentally investigated PVT properties of aqueous foams at different temperatures. Results suggested that model predictions (Lord 1981) deviate slightly from the measurement as the pressure increases.

Example 1. Water and air form 95%-quality foam at surface conditions. It is required to calculate foam quality at downhole conditions of $p = 500$ psia and $T = 200°F$.

Solution. At surface conditions, $p = 14.7$ psia, $T = 60°F$, and $\rho_L = 62.4$ lbm/ft³. The real gas law states

$$p = z\rho_g R_g T / M,$$

where the gas compressibility factor, z, is approximately 1 for air in this particular case. The universal gas constant, R_g, is 10.7 psia·ft³/lbmol/°R, and $M = 28.97$ lbm/lbmol. Using the above equation, the density of gas at surface conditions is estimated as

$$\rho_g = \frac{pM}{zR_g T} = \frac{14.7 \text{ psia} \times 28.97 \text{ lbm / lbmol}}{10.7 \text{ psia} \bullet \text{ ft}^3 / \text{lbmol} / °R \bullet (60 + 459.7) \text{ }°R}$$

$$= 0.077 \text{ lbm/ft}^3$$

Similarly, gas density at downhole conditions is

$$\rho_g = 2.052 \text{ lbm/ft}^3$$

If we neglect changes in gas solubility due to pressure change, the gas mass fraction at surface and downhole conditions is the same. Applying Eq. 4.3.7 at the surface conditions,

$$W_g = \frac{0.077 \text{ lbm / ft}^3 \bullet 0.95}{0.077 \text{ lbm / ft}^3 \bullet 0.95 + 62.4 \text{ lbm / ft}^3 \bullet (1 - 0.95)}$$

$$= 0.0228$$

Rearranging Eq. 4.3.7, an expression for quality can be obtained as a function of gas mass fraction and density of the two phases:

$$\Gamma = \left[\frac{\rho_g}{\rho_L} \left(\frac{1}{w_g} - 1 \right) + 1 \right]^{-1}$$

Applying the above equation, the quality at downhole conditions is

$$\Gamma = \left[\frac{2.052 \text{ lbm/ft}^3}{62.4 \text{ lbm/ft}^3} \left(\frac{1}{0.0228} - 1 \right) + 1 \right]^{-1}$$

$$= 0.4149$$

Stability and Drainage. Stability of foam is critical during underbalanced drilling operations. The stability depends upon a complex relationship between the surfactant concentration, foam quality and texture, and liquid-phase rheology. Liquid-phase rheology and surface tension predominantly control the stability and drainage behavior of foams under static condition. Generally, foam becomes unstable when the liquid film that separated the gas bubbles drains due to gravity and coalescence. As the film gets thinner due to the drainage, the probability of bubble rupturing becomes very high. The liquid-phase viscosity adversely affects the process of drainage by hindering the flow of liquid phase in the film. This may be accomplished through addition of polymers to the liquid phase. The coalescence process mainly occurs due to the pressure difference created by variations in bubble size (curvature). When two foam bubbles with different sizes share a common border, there will be a transfer of mass from the small bubble to the larger one due to the pressure difference.

This phenomenon can be theoretically examined using the Laplace equation that gives the pressure differential between the inside and outside of a gas bubble:

$$\Delta P = P_i - P_o = \frac{4\sigma}{r_b} \quad \dots\dots\dots\dots\dots\dots\dots\dots\dots\dots\dots\dots\dots\dots\dots\dots\dots \quad (4.3.8)$$

It is apparent from the Laplace equation that the smaller the radius of a bubble, the greater the internal pressure. This means that small bubbles have a relatively higher internal pressure than the large ones. Consequently, small bubbles merge into large ones. The merging increases the diameter of the large bubbles.

Morrison and Ross (1983) studied the interactions between different bubble geometries and explained the limitations of gas and liquid volumes from a surface-tension point of view. Foam stability was investigated by considering the amount of potential energy it requires to maintain different textures/structures. They found that the diffusion of gas phase through the liquid films and rupturing of the films mainly control the drainage process. In addition to the effect of gravity and coalescence, the method of foam generation considerably affects the drainage phenomenon. Rand and Kraynik (1983) investigated the foam generation method effect on drainage time. They observed that drainage time increases with the increase in foam generation pressure and decrease in foam bubble size. Similar results have been reported by other investigators (Harris 1985).

Fundamental methods for measuring foam stability for both static and dynamic foams are summarized by Nishioka et al. (1996). They compared the advantages and disadvantages of conical- and cylindrical-type stability-measuring apparatuses. One of the simplest methods for assessing foam stability at ambient conditions is to measure the amount of foam drainage with time. This method consists of

1. Preparing a given volume ($V_L = 100$ mL) of foaming solution
2. Mechanically mixing using an appropriate blender that generates homogeneous foam at approximately constant temperature
3. Pouring the foam quickly into a graduated cylinder and recording initial foam volume (V_f)
4. Recording the volume of liquid phase drained V_D with time
5. Measuring the mass of foam (m_F) left in the mixer.

After completion of the test, fractional drainage (the ratio of volume of liquid drained to the total liquid volume in the cylinder) at a given time is calculated as

$$F_d(t) = \frac{V_D(t)}{V_L - m_F / \rho_L}. \quad \dots\dots\dots\dots\dots\dots\dots\dots\dots\dots\dots\dots\dots\dots\dots\dots \quad (4.3.9)$$

Half-life is the time that corresponds to 50% drainage. The foam quality can be approximately determined by

$$\Gamma = 1 - \frac{V_L - m_F / \rho_L}{V_F}. \quad \dots\dots\dots\dots\dots\dots\dots\dots\dots\dots\dots\dots\dots\dots\dots\dots \quad (4.3.10)$$

For successful drilling operations, it is necessary that foam exhibits high stability in the presence of large amounts of contaminants such as formation water and crude oil. Rojas et al. (2001) investigated the effect of oil and salt on the stability of aqueous foam. This study indicated that the stability of foam is highly dependent on physicochemical properties of the surfactant, contaminates, and their interaction. To evaluate the effect of salt on the stability of foam, different concentrations of monovalent salt were investigated. The result (**Table 4.3.1**) suggested that the addition of salts considerably reduces the stability of aqueous foam.

Foam stability under the presence of monovalent and divalent salts (sodium chloride and calcium chloride), different foaming agents, and different polymer (PAC) concentrations were studied by Argillier et al. (1998). The foam stability was quantified on the basis of liquid-phase drainage measurements. The foam was poured into a funnel, which extended into a graduated cylinder, and the drained liquid was recorded as a function of time. **Fig. 4.3.3** shows foam stability measurements for high-quality (~85%) foam. The measurements indicated the presence of three drainage regimes. In the beginning, the drainage rate tended to increase with time (Regime I). After some time, the rate reached a value and stayed approximately constant for a time (Regime

TABLE 4.3.1—EFFECT OF SALT ON THE STABILITY OF FOAMS (Rojas et al. 2001)		
Foam Type	Salt Concentration	Half-Life (min)
Aqueous foam	0.5%	64
	1.0%	4
Polymer-based foam	0.5%	349
	1.0%	322

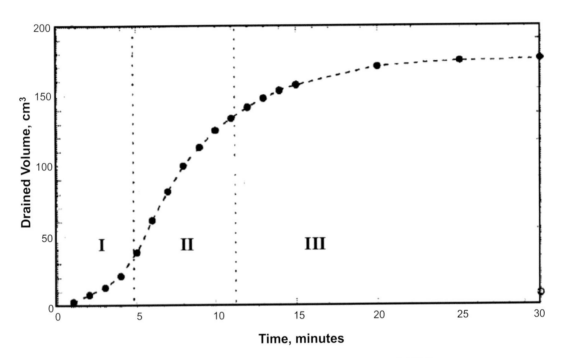

Fig. 4.3.3—Drained liquid volume vs. time (Argillier et al. 1998).

II). Eventually, the rate decreased asymptotically to zero (Regime III). Similar results were reported by Lourenço (2002), indicating the existence of the three drainage regimes.

According to the results of Rojas et al. (2001), the effect of crude oil on the stability of foam is insignificant when crude oil concentration is less than 10% **(Fig. 4.3.4).** The results clearly showed the influence of crude oil American Petroleum Institute (API) gravity on foam stability. Heavy crude oils showed higher foam stability than light oils. Heavy crude oils would be expected to emulsify more slowly than the others do, and this would be expected to retard the rate of foam lamella breakdown.

Foam Texture and Bubble Size. Several investigators (Harris 1985; Lourenço et al. 2004; David and Marsden 1969) studied foam texture and bubble size distribution. David and Marsden (1969) measured bubble size and bubble size distribution under a microscope. Experimental results **(Fig. 4.3.5)** indicated that both mean bubble diameter and bubble size distribution are influenced by foam quality. As the quality increases, the average bubble size and range of bubble sizes increase. Bubble size distributions of low-quality foams showed a sharp peak near the mean diameter.

In some cases, bubble size may considerably affect rheological properties of foam. Foam flows are characterized by bubble deformation that has significant effect on rheology. Bubble size is considered one of the parameters that affect the degree of bubble deformation. In steady homogeneous shear flow, small changes in bubble shape from sphericity depend upon a capillary number. The capillary number for aqueous foam is given by

$$Ca = \frac{r_b \mu_L \dot{\gamma}}{\sigma}, \quad \dots\dots\dots\dots\dots\dots\dots\dots\dots\dots\dots\dots\dots\dots\dots\dots\dots\dots\dots \quad (4.3.11)$$

where r_b, σ, and $\dot{\gamma}$ denote average bubble radius, interfacial tension, and shear rate, respectively. The capillary number compares viscous and interfacial forces. When foam is continuously deformed, the

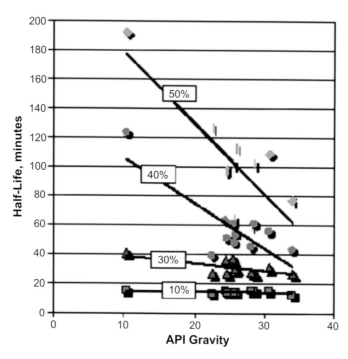

Fig. 4.3.4—Effect of the crude oil API gravity on foam stability at different oil concentrations (Rojas et al. 2001).

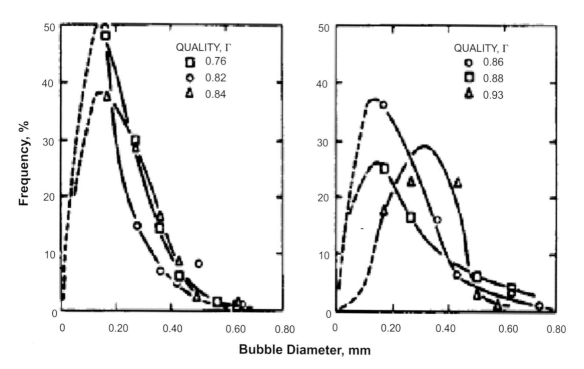

Fig. 4.3.5—Bubble size distributions of foams (David and Marsden 1969).

viscous forces tend to distort the bubbles, while the interfacial forces oppose the distortion and tend to favor sphericity.

Harris (1985) investigated the effects of texture, liquid-phase viscosity, and quality on rheology of foams using a recirculating flow loop. The rheology of foams was found to be affected by texture to a lesser extent. Results suggested that rheology of foam is strongly influenced by the quality and liquid phase viscosity. Test foams generated at higher shear rates, surfactant concentrations, and pressures had smaller mean bubble size (fine texture). Fine-textured foams showed better stability than the coarse ones **(Fig. 4.3.6)**.

An extensive experimental study (Lourenço 2002) was conducted using a nonrecirculating (single-pass) flow loop. Flow behaviors of foams in pipe viscometers (2-, 3-, and 4-in. nominal diameters) were investigated. Foam was generated by flowing an air/water/surfactant mixture through a partially closed ball valve that was placed upstream of the pipe viscometers. The valve opening position (differential pressure across the valve) was used to control foam generation. Foams with different flow properties were generated by regulating the valve. Differential pressure across the valve and the 4-in. pipe viscometer were carefully monitored while keeping all other test parameters constant. Significant change in pressure loss was observed in the 4-in. pipe as the valve opening position changed **(Fig. 4.3.7)**. It is apparent from the experimental results that flow properties of foam can be affected considerably by the differential pressure across the foam-generating valve.

Foam Structure and Bubble Shape. Foam quality is the fundamental descriptive parameter for a macroscopic structure of foam. As the foam quality increases from zero to unity, first a bubbly liquid will be formed at low foam qualities. However, a critical transition occurs as foam quality increases to the rigidity transition, where it develops its structure and becomes rigid. This critical value differs slightly, depending on the liquid-phase compositions. For water-based foams, the rigidity transition occurs at approximately 63% quality (Ahmed et al. 2003a). The bubble structures of low-quality foams appear spherical.

Further increase in foam quality exhibits peculiar rheological properties when the quality exceeds 80%. The structure of bubbles begins to change from a spherical to polyhedral configuration. The bubbles deform against their neighbors, but they remain separated by thin films of liquid phase that keep the bubbles from rupturing. When such high-quality foams are subjected to small shear deformation, they exhibit strong elastic response and yield stress (Princen and Kiss 1989; Pal 1999).

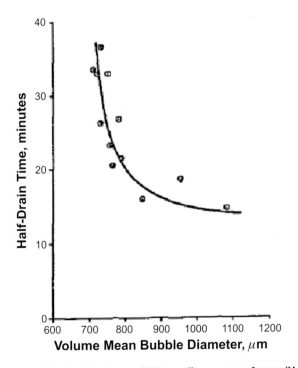

Fig. 4.3.6—Static drain times of 70%-quality aqueous foams (Harris 1985).

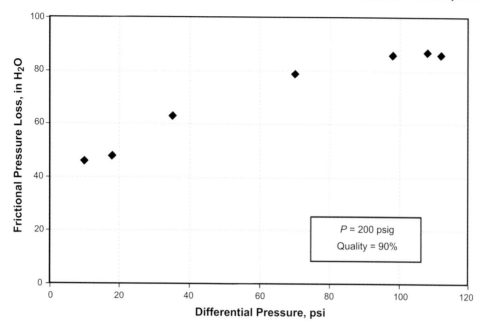

Fig. 4.3.7—Pressure loss in the 4-in. pipe vs. differential pressure across foam-generating valve (data from Lourenço 2002).

As the foam quality approaches 95% (dry foam limit), the bubbles acquire an increasingly polyhedral shape. Due to the structural change, the foam system attains the highest viscosity when the quality reaches the dry foam limit. Previous studies (Debrégeas et al. 2001; Gopal and Durian 1998) on foam rheology have found that polyhedral bubbles predominantly appear when the foam quality is between 88 and 95%. Additional increase in the quality beyond the dry foam limit moderately decreases the viscosity until the foam stability limit is reached. Further increase in quality beyond the stability limit convert the dry foam into mist, resulting in a drastic viscosity reduction (*Underbalanced Drilling Manual* 1997; Okpobiri and Ikoku 1986). The stability limit of water-based foam under ambient conditions is approximately 97%. Addition of viscosifiers into the liquid phase increases this limit (*Underbalanced Drilling Manual* 1997).

Foam Breaking. The simplest technique of foam breaking is to allow the foam to degrade in an open pit. However, because of costs, space, and environmental limitations, offshore operations have used various advanced techniques to facilitate breaking of the foam. Commonly used foam-breaking mechanisms in drilling applications are classified as mechanical, chemical, or combined. Chemical foam breakers (defoamers) accelerate the foam-decaying process significantly. Chemical foam breaking is accomplished by spraying a suitable defoamer on top of foam in the return line. For effective breaking, the foam should be mixed thoroughly with the defoamer. Dilution of the defoamer with a solvent may assist the process of mixing and, hence, breaking. However, care should be taken during dilution because adding a solvent may reduce the concentration of the defoamer and its breaking strength. Therefore, it is important to conduct an experiment to determine the effectiveness of the defoamer. The experiment can be performed by making small batches of foam using a laboratory mixer and spraying a defoamer onto the foam. Several types of mechanical foam-breaking systems exist in the field. The simplest method is that of allowing the foam to drain by gravity and decay naturally in an open pit. This natural foam-breaking process is very slow; hence, it requires large amounts of pit volume and space. In some instances, this process is accelerated by mechanical agitation and spraying of a liquid phase. Other mechanical foam-breaking systems such as the hydrocyclone apply strong centrifugal forces to separate the liquid phase from the gas. Injection of defoamer upstream of the hydrocyclone significantly improves the efficiency of separation (*Underbalanced Drilling Manual* 1997).

Filtration Properties of Foam. One of the major advantages of underbalanced drilling (UBD) is that it minimizes or prevents formation damage by reducing the filtration process because the pressure in the drilling fluid is less than the formation pressure. However, one of the main drawbacks of UBD is the difficulty in maintaining underbalanced conditions all the time. Short-term excursions into overbalanced conditions may result in considerable formation damage. On the basis of laboratory experiments, Herzhaft et al. (2000) suggested that in some tight reservoirs, permeability impairments due to the penetration of UBD fluids can be worse than that of correctly designed conventional fluid. They experimentally investigated filtration properties of various aqueous foams with a modified API filtration cell. This study noted the importance of characterization of filtration properties of foams to select the less damaging fluid. Experiments were conducted according to the classical API filtration test procedures (7 bar pressure, 30 minutes of filtration) with a standard API filtration cell (Cell A) and special filtration cell having a volume of 1 L (Cell B). Several liquid-phase formulations with solids were investigated including various clays and weighting agents. The result **(Fig. 4.3.8)** showed variations in filtration with different liquid phase formulations and solids concentrations:

   i. BM1 + 80 g/L solid A (Cell A)
  ii. BM1 + 40 g/L solid A (Cell A)
 iii. BM1 + 40 g/L solid A (Cell B)
 iv. BM1 + 60 g/L solid A (Cell A)
  v. BM1 + 40 g/L solid B (Cell A)
 vi. BM1 + 40 g/L solid A + 1 g/L PAC (Cell B).

The initial foaming solution, BM1, was composed of 1% surfactant and tap water with pH value of 10. Solids A and B were bentonite and nonswelling clay particles simulating drilled solids, respectively. In addition to fluid pressure penetration, formation damage may occur due to chemical effects that should be considered in designing drilling foams.

   Foam Rheology and Wall Slip. Rheological properties of foam are important for pressure loss calculation and cuttings transport prediction. Equations of motion are based on well-known principles of conservation of mass, momentum, and energy. However, they cannot be solved without the constitutive equation that relates the shear stress with the resulting shear rate. If the shear stress vs. shear rate is expressed in a graphical form, we call this relationship the rheological curve (flow curve).

   Einstein (1906) and Hatschek (1911) pointed out that foam should be treated as a single-phase fluid with viscosities significantly greater than either phase. Drilling foams are complex mixtures of gas and liquid, the rheological and hydraulic properties of which are influenced by several flow and fluid parameters such as foam quality, liquid-phase viscosity, foam texture, pressure, and temperature. Some of these parameters are uncontrollable. As a result, foam flow measurements usually exhibit a higher degree of scattering and

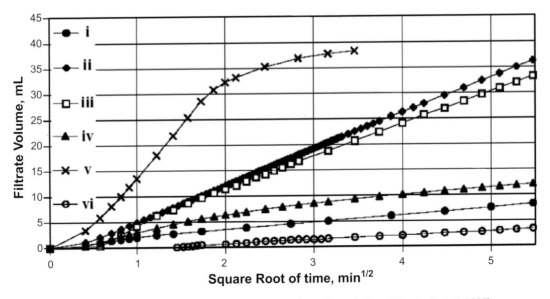

Fig. 4.3.8—Static filtration properties of various foam formulations (Herzhaft et al. 2000).

randomness. In our experimental investigation, we found that it is practically impossible to make two foams that have exactly the same texture (bubble size and bubble size distribution).

Drilling foam rheological properties are largely influenced by foam quality and liquid-phase viscosity. Temperature and pressure affect foam rheology by regulating foam quality and liquid-phase viscosity. For instance, increasing the pressure at a given temperature significantly reduces the volume occupied by the gaseous phase, indirectly reducing the foam quality. At constant pressure, increasing the temperature obviously decreases the viscosity of the liquid phase. Consequently, the foam viscosity decreases as the temperature increases.

Wall Slip. Foam-rheology measurements present considerable experimental difficulties due to the phenomenon of wall slip. Different flow curves are obtained when foam is tested with viscometers with different pipe sizes. Rheology experiments conducted using rotational viscometers that have different dimensions also indicate the presence of wall slip (Yoshimura and Prud'homme 1988). Therefore, the effect of wall slip on rheological measurements should be eliminated to determine a true relationship between shear rate and shear stress.

Wall slip is normally considered one of the most important characteristics of foam flows. Slip occurs in foam flows because of the displacement of the gas phase away from the wall (Heller and Kuntamukkula 1987). Therefore, a convenient description of the wall-slip mechanism is based on the existence of a thin liquid film that does not itself slip, but wets the wall and lubricates the foam flow. Because of the wall slip, a higher foam flow rate is normally observed compared to that which would occur if the slip were not present. **Fig. 4.3.9** presents typical velocity and shear stress distributions of foam flow in pipe viscometers. The thickness of the film, δ_s, is of course greatly exaggerated in this figure. For aqueous foam flow in smooth pipe, the effective film thickness, which is calculated from typical experimental data, is in the range of 10 μm (Kraynik 1988).

An experimental investigation (Thondavadi and Lemlich 1985) performed using rough pipe viscometers indicated that the wall slip can be reduced or eliminated by the use of rough surfaces. A similar study was conducted by Saintpere et al. (1999) to develop foam-rheology characterization procedures. Rheology tests were conducted with a parallel-plate viscometer. Wall slip was minimized by applying *grooved* surfaces on the viscometer walls. Roughness protrusions have a tendency to reduce the effective film thickness. Slip velocity can be expressed in terms of effective film thickness, wall shear stress, and liquid-phase viscosity (Kraynik 1988):

$$u_s = \frac{\delta_s \tau_w}{\mu_L} \quad \dots \dots \dots \dots \dots \dots \dots \dots \dots \dots \dots \dots \dots \dots \dots \dots \dots \quad (4.3.12)$$

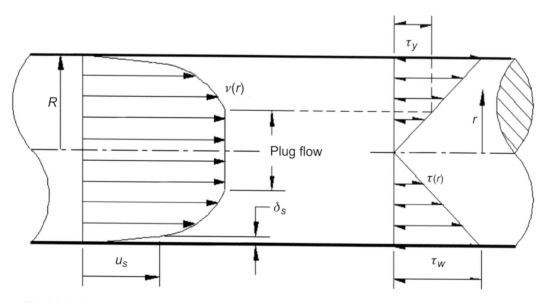

Fig. 4.3.9—Velocity and shear stress distributions in pipe flow for high-quality foam with yield stress.

The above equation can be extended for non-Newtonian liquid film if apparent viscosity is used instead of the viscosity.

Rheology Measurement and Wall Slip. The application of pipe viscometers for the measurement of rheological properties of foam is well established. Pipe viscometers are used to determine wall slip and rheological parameters such as consistency index (K_f) and flow behavior index (n_f). A standard procedure for estimating wall slip involves determination of foam rheology using pipe viscometers that have different internal diameters. It is accepted that without wall slip, different viscometer diameters should give a single flow curve for a given fluid if the temperature and pressure in the viscometers are maintained constant. Because of wall slip, several foam flow curves are often obtained from viscometers with different pipe sizes. **Fig. 4.3.10** shows typical viscometric data (uncorrected flow curves) of foam that indicate the presence of wall slip. The dotted curve in this plot shows the pattern of flow curve after the wall slip correction. The gap between the corrected and uncorrected curves displays the contribution of the wall slip to the nominal Newtonian shear rate.

To obtain the true rheology of foam, the contribution of wall slip to the nominal wall shear rate should be removed systematically. This is done using different methods: the Mooney postulate, which assumes that the slip velocity is directly proportional to wall shear stress (Mooney 1931), or the Oldroyd-Jastrzebski method, which assumes that slip velocity (u_s) depends on both wall shear stress (τ_w) and pipe diameter (Jastrzebski 1967). Both procedures introduced proportionality constants or slip coefficients that relate the slip velocity with flow parameters. Gardiner et al. (1998) and Jastrzebski (1967) discussed these two procedures in detail. The Oldroyd-Jastrzebski method gives reasonable slip velocity, while the Mooney postulate sometimes gives inaccurate predictions. Beyer et al. (1972) used the Mooney postulate and determined slip velocities of foams for qualities ranging from 0.75 to 0.98. The measurements suggested that at a given wall shear stress, the slip velocity decreases as the quality increases because the effective film thickness diminishes as the liquid volume fraction decreases. Princen et al. (1980) drew a similar conclusion after presenting an analytical solution to the liquid film thickness of extremely-high-quality foams and emulsions. However, Eq. 4.3.12 apparently shows that the slip velocity depends not only on the effective film thickness but also on the resulting wall shear stress, which is relatively less for low-quality foams. The product of the effective film thickness and wall shear stress may increase as the quality increases, resulting in a higher slip velocity.

According to the Oldroyd-Jastrzebski method, the slip velocity is expressed as a function of wall shear stress, pipe diameter, and slip coefficient:

$$u_s = \beta \frac{\tau_w}{D} \qquad\qquad\qquad\qquad\qquad\qquad\qquad\qquad\qquad\qquad\qquad (4.3.13)$$

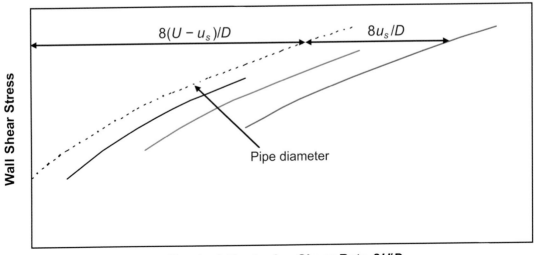

Fig. 4.3.10—Flow curves of foam with wall slip.

This method suggests a three-step procedure to determine the value of the slip coefficient, β:

1. Obtain a plot of Newtonian wall shear rate, $8U/D$, vs. $1/D^2$ at constant wall shear stress, τ_w.
2. Determine the slip coefficients from slopes of the straight lines. The value of β is determined by dividing the slope of a least-square line by $8\tau_w$.
3. Develop a functional relationship between the slip coefficient and wall shear stress by curve fitting the plot of β vs. τ_w.

In addition to the wall shear stress, the slip coefficient is influenced by foam quality and liquid-phase viscosity (Gardiner et al. 1998; Özbayoğlu et al. 2002). Once the slip coefficient is determined as a function of wall shear stress, the measured volumetric flow rate, Q_m, from pipe flow experiments needs to be corrected to the value it would have in the absence of the *slip* by the following equation:

$$Q_c = Q_m - \frac{\pi \beta \tau_w D}{4} \quad\text{..} \quad (4.3.14)$$

The corrected flow rate represents the true shear flow that corresponds to the resulting wall shear stress or frictional pressure gradient. The corrected flow rate is used to calculate the true nominal Newtonian shear rate, $8(U - u_s)/D$, which is useful to determine the true rheological properties of foam.

Foam Rheology Models. A number of investigators have studied foam rheology and developed models. Both empirical and mathematical modeling approaches were considered to predict foam rheology as a function of quality and liquid-phase viscosity. There have been several experimental studies (Beyer et al. 1972; Lourenço et al. 2004; Özbayoğlu et al. 2002; Khade and Shah 2004; Chen et al. 2005a) conducted on drilling-foam rheology, covering a wide range of foam qualities, liquid-phase compositions, temperatures, and pressures. Although there are differences in the results of these investigations, it can be deduced that the rheology of foam primarily depends on quality, liquid-phase viscosity, and texture (method of foam generation). Experimental results (Beyer et al. 1972) show approximately a 10% difference in frictional pressure loss due to secondary effects of pressure (pressure effects excluding the change in quality) on the rheology.

Aqueous Foams. These types of drilling foams are made from compressed gas, water, and surfactants. Low-quality aqueous foams (i.e., $\Gamma < 0.54$) do not show a rigid or tight bubble structure. They drain very rapidly. Mitchell (1971) measured rheology of low-quality aqueous foams using pipe viscometers. This study indicated that low-quality aqueous foams behave as Newtonian fluids:

$$\mu_f = \mu_L (1 + 3.6\Gamma), \quad\text{..} \quad (4.3.15)$$

where μ_f and μ_L denote foam and liquid phase viscosities, respectively. High-quality aqueous foams do not flow as freely as the low-quality foams. Flow of high-quality aqueous foam exhibits bubble interference, deformation, and non-Newtonian behaviors such as shear thinning and yield stress (Mitchell 1971). As the foam quality increases from a low quality range to a high one, a structural change occurs as the quality approaches a rigidity transition, which is approximately 0.63 for aqueous foams (Holt and McDaniel 2000). This critical value differs slightly, depending on the liquid-phase composition. Above this critical value, the foam becomes rigid and structurally ordered. The structure of foam bubbles appears spherical or polyhedral. Considering foam as a Newtonian fluid, Mitchell (1971) also developed the following empirical equation to estimate the high viscosity quality ($0.54 < \Gamma < 0.96$) of aqueous foams. Thus,

$$\mu_f = \mu_L \frac{1}{1 - \Gamma^{0.49}} \quad\text{..} \quad (4.3.16)$$

Sanghani and Ikoku (1983) experimentally studied rheology of high-quality aqueous foams ($0.67 < \Gamma < 0.96$) using a concentric annular viscometer. They proposed a power-law-type rheology model and presented charts for effective viscosity of foam in annular flow as a function of average wall shear rate. Martins et al. (2001) determined relationships between power-law fluid parameters (consistency index, K, and flow behavior index, n) and foam quality:

$$n_f = 0.82 \left(\frac{1-\Gamma}{\Gamma} \right)^{0.52} \quad \dots\dots\dots\dots\dots\dots\dots\dots\dots\dots\dots\dots \quad (4.3.17)$$

and

$$K_f = 0.081 \left(\frac{1-\Gamma}{\Gamma} \right)^{-1.59}, \quad \dots\dots\dots\dots\dots\dots\dots\dots\dots\dots \quad (4.3.18)$$

where the unit of K_f is Pa·s^{n_f}. The above equations are proposed for high-quality aqueous foams ($0.6 < \Gamma < 0.95$). Li and Kuru (2003) developed empirical correlations using the experimental values of Sanghani and Ikoku (1983). For $0.67 \leq \Gamma \leq 0.915$, these correlations are expressed as

$$K_f = 0.3543 e^{3.516\Gamma} \quad \dots\dots\dots\dots\dots\dots\dots\dots\dots\dots\dots\dots \quad (4.3.19)$$

and

$$n_f = 1.2085 e^{-1.9897\Gamma} \quad \dots\dots\dots\dots\dots\dots\dots\dots\dots\dots\dots \quad (4.3.20)$$

For $0.915 \leq \Gamma \leq 0.98$, the values of K_f and n_f are determined by

$$K_f = -102.8175\Gamma + 103.2723 \quad \dots\dots\dots\dots\dots\dots\dots\dots\dots\dots \quad (4.3.21)$$

$$n_f = 2.5742\Gamma - 2.1649 \quad \dots\dots\dots\dots\dots\dots\dots\dots\dots\dots\dots \quad (4.3.22)$$

Recently, Özbayoğlu et al. (2002) performed a comparative study (**Fig. 4.3.11**) to investigate the predictive ability of the available aqueous-foam rheology models. They conducted extensive flow experiments to determine flow curves of high-quality (70 to 90%) aqueous foams. Analysis of the flow curves indicated that foam behaves like a pseudoplastic fluid with a negligible yield stress and shear thinning effect. The flow behavior and consistency indices vary nonlinearly with foam quality. The result illustrated that the power-law model can better characterize the rheology of foams at 70 and 80% qualities, whereas the rheology of foam at 90% quality best fits the Bingham plastic model. Experimental results were compared with predictions of available

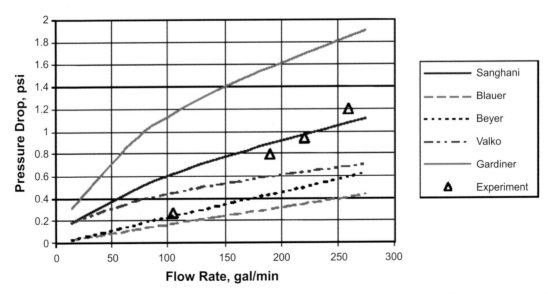

Fig. 4.3.11—Model comparison with experimental data for 4-in. pipe with 80%-quality foam (Özbayoğlu et al. 2002).

rheology models. Significant differences were observed between model predictions and pressure loss measurements.

Polymer-Based Foams. Very thick and stable polymer-based foams can be generated by adding a polymer viscosifier to the liquid phase. The rheology of polymer-based foams greatly depends on the liquid-phase rheology and foam quality. Several investigators (Mitchell 1971; Reidenbach et al. 1986; Cawiezel and Niles 1987) studied the flow properties of polymer-based foams at different qualities and viscosifier concentrations. Results indicated that the yield power-law model can best describe the rheological behavior of these foams. A recent study (Khade and Shah 2004) on guar-foam rheology suggested that the flow behavior index of low-quality ($\Gamma < 0.5$) polymer-based foams is the same as the flow behavior index of the liquid phase, while the fluid-consistency index increases approximately linearly with the foam quality. Thus,

$$n_f = n_L \qquad (4.3.23)$$

and

$$K_f = K_L(1+3.6\Gamma) \qquad (4.3.24)$$

In addition to quality and liquid phase rheology, rheologies of high-quality polymer-based foams may be affected considerably by other factors such as foam texture, composition, and wall slip. Consequently, rheology studies of polymer-based foams often are limited to a specific polymer type. Khade and Shah (2004) presented empirical correlations for rheologies of guar-based foams as

$$n_f = n_L(1+C_1\Gamma^\xi) \qquad (4.3.25)$$

and

$$K_f = K_L e^{C_2\Gamma+C_3\Gamma^2}, \qquad (4.3.26)$$

where C_1, C_2, C_3, and ξ are empirical constants (**Table 4.3.2**) that depend on the composition of the liquid phase.

Most recently, experimental investigations (Chen et al. 2007) on polymer-based drilling foams were carried out in a large-scale flow loop (**Fig. 4.3.12**) that permitted foam flow through different pipe sections (2 in., 3 in., and 4 in.) and an annular test section (6 in. × 3.5 in.). Rheology tests were performed with foams that had different polymer (hydroxylethylcellulose) concentrations and 1% commercial surfactant. During the experiments, frictional pressure losses across the pipe and annular sections were measured for different gas/liquid flow rates, polymer concentrations, and foam qualities. Significant rheological variations were observed (**Fig. 4.3.13**) between aqueous foams containing no polymers and polymer-thickened foams. Besides this, experimental data from pipe sections showed three distinct flow curves, indicating the presence of wall slip. The slip coefficient decreases as the foam quality or polymer concentration increases.

On the basis of experimental investigation of the behavior of foamed polymer solutions in a large-scale vertical tube, Valkó and Economides (1997) developed the principle of volume equalization to describe the rheology of polymer-based foams. This technique uses the specific-volume expansion ratio, ε, as the additional parameter to represent the gas volume fraction. This quantity is defined as the ratio of the liquid density to the foam density, which varies along the flow path because of the change in pressure. The specific-volume expansion ratio at a given temperature and pressure is given by

TABLE 4.3.2—EMPIRICAL CONSTANTS AT DIFFERENT GUAR CONCENTRATIONS				
Concentration	C_1	C_2	C_3	ξ
20 lbm per 1000 gal	−2.10	−1.99	8.97	7.30
30 lbm per 1000 gal	−0.15	−2.38	8.88	6.51
40 lbm per 1000 gal	−0.66	−0.49	5.62	5.17

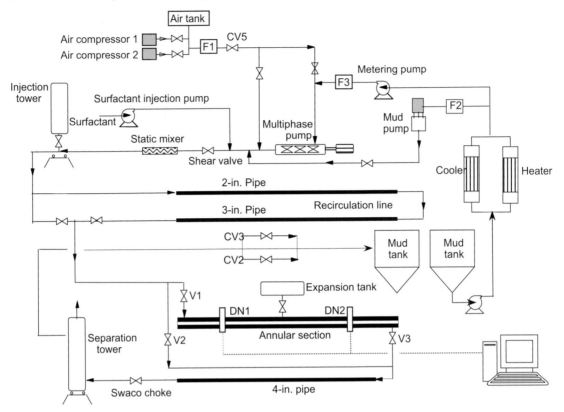

Fig. 4.3.12—Schematic of test facility (Chen et al. 2007).

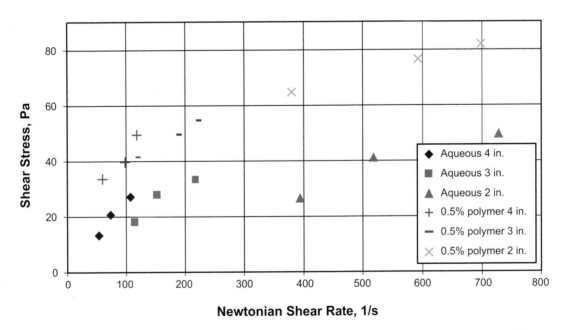

Fig. 4.3.13—Flow curve of 80%-quality foam with different polymer concentrations (Chen et al. 2007).

$$\varepsilon = \frac{\rho_L}{\rho_f} \quad \dots \quad (4.3.27)$$

The principle of volume equalization is derived from an invariance requirement. It assumes that for a straight duct flow of constant cross section, both compressible and incompressible flows possess the invariance property. This means that the loss of mechanical energy is proportional to the kinetic energy; in other words the friction factor is constant (Valkó and Economides 1997). A constitutive equation that provides the required invariance is called the volume-equalized equation. For instance, the volume-equalized power-law equation can be written as

$$\frac{\tau}{\varepsilon} = K_{VE} \left(\frac{\dot{\gamma}}{\varepsilon} \right)^{n_f} \quad \dots \quad (4.3.28)$$

One of the advantages of this approach is that when the volume-equalized wall shear stress is plotted against the volume-equalized nominal Newtonian shear rate on a log-log scale, the result is a straight line for a wide range of foam qualities and pressures.

Foam-Rheology Measurement. Rheology characterization of foam requires the selection of a convenient apparatus to determine the rheology accurately. Different types of foam viscometers are used for rheology characterization. Pipe viscometers are commonly used because measurements are relatively less affected by the drainage effect. Pipe viscometers allow dynamic measurement of rheology as the foam flows through the viscometer. However, careful test setup and measurement is required to maintain repeatability of the measurements. Assumptions made during the development of viscometric equations should not be violated. The concept of a pipe viscometer is developed on the assumption of incompressible isothermal laminar flow conditions. The physical properties of components of foam such as density and viscosity should not change during the measurement. Maintaining incompressible flow conditions seems unrealistic, but it is possible to minimize the effect of foam expansion during rheology measurement. This can be achieved by minimizing the ratio of pressure differential across the viscometer to the static pressure (i.e., $\Delta P/P \ll 1$). In addition to this, a flow visualization port is important to verify the homogeneity of the foam.

To measure wall slip, a series of pipe viscometers with different diameters is required. The pipes can be assembled in parallel or in series. Both approaches have their own advantage and drawback. When viscometers are arranged in series, maintaining exactly the same foam quality through each pipe section is difficult because of foam expansion. The test foam expands as it flows from one test section to another due to the reduction in static pressure resulting from significant frictional pressure loss across the test sections and pipe fittings. Especially if the pipe arrangement is from the narrowest pipe to the widest pipe, the expansion effect may considerably affect the measurement. With parallel pipe viscometers, it is possible to control static pressure in each test section, but it is difficult to keep exactly the same foam composition. In both cases, the rheology measurements are made by measuring the differential pressure across test sections while keeping other test parameters such as gas- and liquid-phase mass flow rates, pressure, and temperature constant. Measured differential pressure and other test parameters are used to calculate the nominal Newtonian shear rate ($8U/D$) and wall shear stress. The procedure presented previously (i.e. Oldroyd-Jastrzebski method) can be applied to determine the wall slip.

Once the slip velocity is determined for a given data set, the generalized flow behavior index (n') and generalized consistency index (K') can be obtained by plotting wall shear stress against slip-corrected nominal Newtonian shear rate on a log-log scale (**Fig. 4.3.14a**). If the data points fit a single straight line, the foam is considered a power-law fluid; otherwise, another constitutive relation such as the yield power-law model may be used.

To get in-depth understanding of n' and K', let us consider a fully developed viscometric pipe flow (Fig. 4.3.9) under steady-state conditions with a velocity profile, $v(r)$. The volumetric flow rate due to the shearing of the fluid is expressed as

$$Q_{sh} = 2\pi \int_0^{R-\delta_s} (v - u_s) r \, dr, \quad \dots \quad (4.3.29)$$

where δ_s is the slip layer thickness, which can be neglected for simplicity as being much less than R. Integrating by parts and assuming that $v(R) - u_s = 0$, we get

$$Q_{sh} = -\pi \int_{O}^{R} r^2 \frac{d(v - u_s)}{dr} dr \quad \dots\dots\dots\dots\dots\dots\dots\dots\dots\dots\dots \quad (4.3.30)$$

We know that $d(v - u_s)/dr$ is a function of shear stress, τ. From conservation of linear momentum, for steady-state pipe flow of fluid with constant density, we can write

$$\frac{\tau(r)}{r} = \frac{\tau_w}{R} \quad \dots\dots\dots\dots\dots\dots\dots\dots\dots\dots\dots\dots\dots\dots\dots\dots\dots\dots\dots \quad (4.3.31)$$

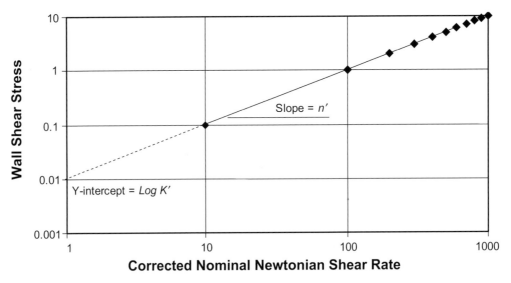

Fig. 4.3.14a—Typical flow curve of foam on a log-log plot.

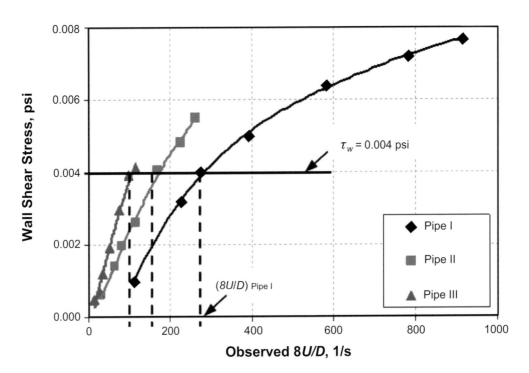

Fig. 4.3.14b—Wall shear stress vs. observed 8*U/D* for different pipe sizes.

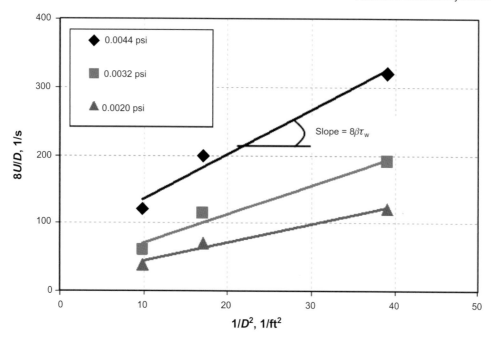

Fig. 4.3.14c—Nominal Newtonian shear rate, $8U/D$, vs. $1/D^2$.

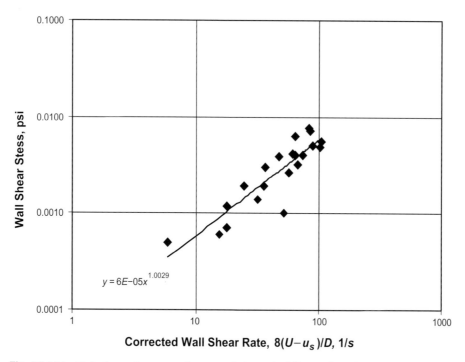

Fig. 4.3.14d—Wall shear stress vs. slip-corrected nominal Newtonian shear rate, $8(U - u_s)/D$.

Assuming negligible slip layer thickness (Yoshimura and Prud'homme 1988), δ_s, at the wall, shear stress variation in the slip layer can be neglected. Hence, the shear stress near the pipe wall $(R - \delta_s)$, τ_w^*, is approximately given by

$$\tau_w^* \approx \tau_w = \frac{R}{2} \frac{\Delta P}{\Delta L} \quad \ldots\ldots\ldots\ldots\ldots\ldots\ldots\ldots\ldots\ldots\ldots\ldots\ldots\ldots \quad (4.3.32)$$

After changing variables, Eq. 4.3.30 takes the form

$$Q_{sh} = -\pi \int_0^{\tau_w} \left(\frac{R}{\tau_w}\right)^3 \frac{d(v-u_s)}{dr} \tau^2 d\tau \quad \dots\dots\dots\dots\dots\dots\dots\dots\dots \quad (4.3.33)$$

Denoting $d(v-u_s)/dr = f(\tau)$, and differentiating with respect to τ_w and upon rearrangements, we obtain

$$\frac{d\left(Q_{sh}\tau_w^3\right)}{d\tau_w} = -\pi R^3 f\left(\tau_w\right)\tau_w^2 \quad \dots\dots\dots\dots\dots\dots\dots\dots\dots \quad (4.3.34)$$

Hence, the shear rate near the pipe wall $(R-\delta_s)$ is

$$f\left(\tau_w\right) = \left[-\frac{d(v-u_s)}{dr}\right]_R \quad \dots\dots\dots\dots\dots\dots\dots\dots\dots \quad (4.3.35)$$

or

$$\dot{\gamma}_w = \frac{1}{\pi R^3} \tau_w \frac{dQ_{sh}}{d\tau_w} + \frac{3Q_{sh}}{\pi R^3} \quad \dots\dots\dots\dots\dots\dots\dots\dots\dots \quad (4.3.36)$$

Note that

$$\frac{Q_{sh}}{\pi R^3} = \frac{U-u_s}{R} = \frac{2(U-u_s)}{D} \quad \dots\dots\dots\dots\dots\dots\dots\dots\dots \quad (4.3.37)$$

Therefore, Eq. 4.3.36 takes the form

$$\dot{\gamma}_w = \frac{\tau_w}{4} \frac{d\left[\dfrac{8(U-u_s)}{D}\right]}{d\tau_w} + \frac{3}{4}\left[\frac{8(U-u_s)}{D}\right] \quad \dots\dots\dots\dots\dots\dots\dots\dots\dots \quad (4.3.38)$$

Note that the following is true:

$$\tau_w \frac{d\left(\log \tau_w\right)}{d\tau_w} = \left[\frac{8(U-u_s)}{D}\right] \frac{d\left[\log \dfrac{8(U-u_s)}{D}\right]}{d\left[\dfrac{8(U-u_s)}{D}\right]} \quad \dots\dots\dots\dots\dots \quad (4.3.39)$$

From Eq. 4.3.39 we get

$$\frac{d\left[\dfrac{8(U-u_s)}{D}\right]}{d\tau_w} = \frac{\dfrac{8(U-u_s)}{D}}{\tau_w} \frac{d\left[\log \dfrac{8(U-u_s)}{D}\right]}{d\left(\log \tau_w\right)} \quad \dots\dots\dots\dots\dots \quad (4.3.40)$$

Substituting Eq. 4.3.38 to Eq. 4.3.36, we get the expression for wall shear rate:

$$\dot{\gamma}_w = \frac{1}{4}\left\{3 + \frac{d\left[\log \dfrac{8(U-u_s)}{D}\right]}{d\left(\log \tau_w\right)}\right\}\left[\frac{8(U-u_s)}{D}\right] \quad \dots\dots\dots\dots\dots \quad (4.3.41)$$

Therefore, if we plot the wall shear stress (τ_w) against the corrected nominal Newtonian shear rate $[8(U-u_s)/D]$ on a log-log scale (Fig. 4.3.14a), the generalized flow behavior index (n') can be obtained from the slope of the plot (flow curve) at a given corrected shear rate. Hence, the generalized flow behavior index is expressed as

$$n' = \frac{d(\log \tau_w)}{d\left[\log \frac{8(U - u_s)}{D}\right]} \quad \dots \dots \dots \dots \dots \dots \dots \dots \dots \dots \dots \dots \dots \quad (4.3.42)$$

Substituting Eq. 4.3.40 to Eq. 4.3.39 gives

$$\dot{\gamma}_w = \left(\frac{3n' + 1}{4n'}\right) \frac{8(U - u_s)}{D} \quad \dots \dots \dots \dots \dots \dots \dots \dots \dots \dots \dots \quad (4.3.43)$$

In general,

$$\tau_w = K' \left[\frac{8(U - u_s)}{D}\right]^{n'} \quad \dots \dots \dots \dots \dots \dots \dots \dots \dots \dots \dots \dots \quad (4.3.44)$$

Eq. 4.3.44 is an equation of tangents to the flow curve (log-log scale). If the curve fits a straight line on the log-log plot, we have a power-law fluid. In other words $n = n'$ and

$$\tau_w = K_f \left[-\frac{d(v - u_s)}{dr}\right]^n \quad \dots \dots \dots \dots \dots \dots \dots \dots \dots \dots \dots \quad (4.3.45)$$

It is worthwhile to note the difference between K_f and K'.

Example 2. The experimental data in **Table 4.3.3** have been obtained from three pipe viscometers, Pipe I, II, and III, with internal diameters of 1.92, 2.90, and 3.82 in., respectively. The viscometers are arranged in series to determine the foam rheology and wall slip. Foam quality was approximately 80% under the experimental conditions ($T = 85°F$ and $P = 100$ psia). Assume the flow as laminar and establish a relationship between the slip coefficient and wall shear stress, then determine the rheological parameters of the test foam.

Solution. The relationship between the slip coefficient and wall shear stress can be established by applying the Oldroyd-Jastrzebski method. In order to apply this procedure, the experimental data need to be presented in the form of observed Newtonian shear rate ($8U/D$) vs. wall shear stress as shown in **Table 4.3.4**. This information is used to determine the observed $8U/D$ in each pipe that corresponds to the same wall shear stress as presented in Fig. 4.3.14b.

Table 4.3.5 can be obtained by applying the technique presented in Fig. 4.3.14b for different wall shear stresses. The graphically obtained $8U/D$ values are presented in Fig. 4.3.14c as a function of $1/D^2$. This is to estimate the slip coefficient from the slope of the regression line at a given wall shear stress (as shown in Fig. 4.3.14c).

Graphically obtained slip coefficient values are correlated to the wall shear stress by

$$\beta = 8 \times 10^6 \tau_w^2 - 46062 \tau_w + 226.15,$$

TABLE 4.3.3—EXPERIMENTAL DATA FROM THREE PIPE VISCOMETERS			
Flow Rate (gal/min)	Measured Pressure Loss Gradient (psi/ft)		
	Pipe I	Pipe II	Pipe III
20.0	0.0250	0.0099	0.0063
41.0	0.0800	0.0232	0.0088
50.0	0.1000	0.0323	0.0151
71.2	0.1252	0.0433	0.0239
105.0	0.1600	0.0670	0.0373
140.9	0.1800	0.0795	0.0490
165.0	0.1925	0.0910	0.0515

TABLE 4.3.4—OBSERVED NOMINAL NEWTONIAN SHEAR RATE VS. WALL SHEAR STRESS

Observed 8U/D (1/s)			Wall Shear Stress (psi)		
Pipe I	Pipe II	Pipe III	Pipe I	Pipe II	Pipe III
110.9	32.2	14.1	0.001	0.0006	0.0005
227.3	66.0	28.9	0.0032	0.0014	0.0007
277.1	80.4	35.2	0.0040	0.0019	0.0012
394.7	114.6	50.1	0.0050	0.0026	0.0019
582.3	169.0	73.9	0.0064	0.0040	0.0030
781.0	226.7	99.2	0.0072	0.0048	0.0039
914.7	265.4	116.1	0.0077	0.0055	0.0041

TABLE 4.3.5— OBSERVED NOMINAL NEWTONIAN SHEAR RATE AT A GIVEN WALL SHEAR STRESS

D (ft)	$1/D^2$ (1/ft^2)	8U/D (1/s)		
		$\tau_w = 0.0044$ psi	$\tau_w = 0.0028$ psi	$\tau_w = 0.0012$ psi
0.160	39.06	320	190	120
0.242	17.12	200	115	70
0.318	9.87	120	60	38

where the wall shear stress and coefficient are given in psi and ft^2/psi·s, respectively. From this equation, the slip coefficient can be determined for a given wall shear stress. Using this procedure and Eq. 4.3.13, the slip velocity can be estimated to calculate the slip-corrected nominal shear rate, $8(U - u_s)/D$. Fig. 4.3.14d presents a logarithmic plot of wall shear stress vs. the corrected nominal shear rate. The data points lie approximately on a straight line, indicating power-law fluid rheology with an n_f value of approximately one (i.e., Newtonian fluid). The consistency index, K_f, is 6×10^{-5} lbf·s/in.^{2} (0.864 lbf-sec per 100 ft^2). The yield stress is negligible.

In the past, a limited number of studies (Minssieux 1974; Marsden and Khan 1966) were conducted to determine the rheology of foam using Couette-type viscometers. However, the issue of drainage and the phenomenon of wall slip restricted the usage of Couette viscometers for foam-rheology applications. Recently, a number of foam-rheology studies (Washington 2004; Pickell 2004; Lauridsen et al. 2002; Pratt and Dennin 2003; Chen et al. 2005b) have been conducted using Couette-type (rotational) viscometers. Chen et al. (2005b) used a flow-through Couette viscometer for measuring rheology of aqueous foams. Viscometer cups and rotors were roughened to minimize wall slip. Test foam was generated by mechanical mixing of surfactant solutions with the gas phase. During the experiment, the foam continuously flows through the viscometer (**Fig. 4.3.15**). However, an optimum foam flow rate needs to be determined in order to get reliable and repeatable measurements from a flow-through Couette viscometer. Determining the optimum flow rate is important to allow enough flow through the viscometer to minimize the effect of foam drainage on the measurements, but flow rate should not be too high, because it would allow the axial flow to interfere with the viscometer readings. Reasonable foam rheology measurements were obtained with this apparatus. Once the appropriate flow rate through the viscometer is maintained, torque measurements can be made at different rotational speeds.

A standard flow-through Couette viscometer (**Fig. 4.3.16**) uses a rotating inner cylinder suspended with a torsion wire and stationary cup. Consider the flow of foam confined between two roughened concentric cylinders with the inner cylinder rotating. For isothermal laminar flow condition, when the inner cylinder is rotating at a constant angular velocity (ω) and the cup is held stationary, the wall shear stress acting on the rotor is given by

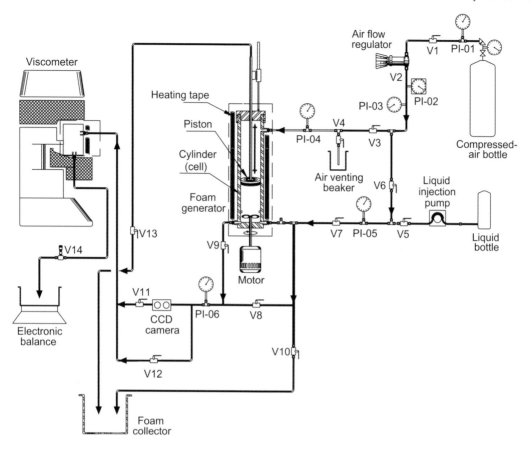

Fig. 4.3.15—Schematic of the foam generator/viscometer.

$$\tau = \frac{T_m - T_e}{2\pi R_i^2 L_b}, \quad\ldots\ldots\ldots\ldots\ldots\ldots\ldots\ldots\ldots\ldots\ldots\ldots\ldots\ldots\ldots\ldots\ldots\ldots\ldots\ldots\ldots\ldots \text{(4.3.46)}$$

where T_m is measured torque, while T_e is additional amount of torque due to the end effect and mechanical friction. At higher shear stress readings, the contributions of the end effect and mechanical friction to the measured torque are minimal. Hence, the wall shear stress can be expressed approximately as

$$\tau \approx \frac{T_m}{2\pi R_i^2 L_b} \quad\ldots\ldots\ldots\ldots\ldots\ldots\ldots\ldots\ldots\ldots\ldots\ldots\ldots\ldots\ldots\ldots\ldots\ldots\ldots\ldots\ldots\ldots\ldots\ldots \text{(4.3.47)}$$

For low shear measurement, the torque due to the end effect and mechanical friction may become comparable with the measurement. In this case, the viscometer needs to be calibrated using standard fluid to correct the measurements.

In viscometric flow analysis, usually the gap is considered as a narrow slot (i.e., linear velocity distribution). With this assumption, the following equation can be used for calculating the shear rate:

$$\dot{\gamma} \approx \frac{\omega R_i}{R_o - R_i} \quad\ldots\ldots\ldots\ldots\ldots\ldots\ldots\ldots\ldots\ldots\ldots\ldots\ldots\ldots\ldots\ldots\ldots\ldots\ldots\ldots\ldots\ldots\ldots\ldots \text{(4.3.48)}$$

The narrow-slot approximation is valid for $R_i/R_o > 0.99$. To get reasonable foam rheology measurements, the gap between the cup and rotor should not be in the range of bubble size (i.e., 50 to 300 µm). For wide-gap viscometers ($0.50 \leq R_i/R_o \leq 0.99$), the narrow-slot approximation is invalid. In this case, the correct shear rate is given by (Macosko 1994)

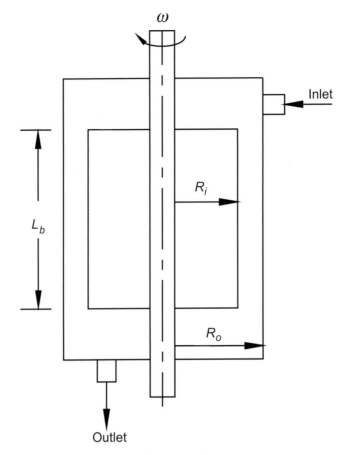

Fig. 4.3.16—Schematic of flow-through Couette-type viscometer.

$$\dot{\gamma} = \frac{2\omega}{n^*\left(1 - k^{2/n^*}\right)}, \quad \dots \dots \dots \dots \dots \dots \dots \dots \dots \dots \dots \dots \dots \dots \dots \dots \quad (4.3.49)$$

where n^* is the slope of a logarithmic plot of torque T_m vs. angular velocity, ω. Thus,

$$n^* = \frac{d \ln T_m}{d \ln \omega} \quad \dots \dots \dots \dots \dots \dots \dots \dots \dots \dots \dots \dots \dots \dots \dots \dots \dots \dots \dots \quad (4.3.50)$$

Eq. 4.3.49 can be simplified by taking n^* as unity (i.e., assuming Newtonian fluid). However, this simplification is only valid when n^* is constant and close to one. After measuring shear stress values for different shear rates from a Couette viscometer, the flow curve can be obtained by plotting the data on a log-log scale.

Time-Dependent Rheological Behavior. During foam drilling operations, in order to reduce drillstring pressure losses and injection pressure requirements, the components of foam are injected directly into the drillpipe without prefoaming. Foam generation occurs as the mixture flows down the drillpipe, BHA, and drill bit (Lyons et al. 2000). As the foam evolves from an aerated fluid state to a fully equilibrated state, it shows time-dependent rheological behavior, which is a function of properties of the fluid and flow parameters. A series of experiments were conducted (Pickell 2004) to verify the hypothesis that foam viscosity increases with added energy. A flow-through Couette viscometer was coupled with a foam generator. Foam was generated by mechanical agitation of a surfactant solution in the presence of air. This study showed that foam viscosity increases with added energy while average bubble size decreases with the mixing time. Eventually the average bubble size reaches a limiting value, which is a function of its composition, temperature, and pressure.

4.3.3 Hydraulics of Foam Drilling. Wellbore hydraulics has long been recognized as one of the most important considerations to implement efficient, safe, and economical procedures to meet the objectives of underbalanced drilling operations. An important task during well planning and designing is obtaining reliable predictions of frictional pressure losses, downhole pressure, and equivalent circulating density. Using low-density drilling foam alone does not guarantee underbalanced conditions. The frictional pressure loss for foam can be comparable to the gravitational pressure gradient. This can result in a downhole pressure that exceeds the pore pressure even when the hydrostatic head of the fluid does not.

Several investigators (Lourenço 2002; Rand and Kraynik 1983; Harris 1985; Nishioka et al. 1996; Rojas et al. 2001; Argillier et al. 1998; Lourenço et al. 2004; David and Marsden 1969; Ahmed et al. 2003b) studied the flow behaviors of foams. Hydraulic optimization of underbalanced drilling with foam is defined as the process of selecting the best combination of annular backpressure, gas/liquid flow rate, and bit nozzle sizes that would maximize drilling rate and ensure effective cuttings transport while keeping the circulating bottomhole pressure to a minimum. During foam drilling, a certain backpressure needs to be applied to maintain the stability of foam close to the surface. Generally, 96% quality is considered the upper limit for aqueous foam, above which foam becomes unstable and turns into a mist (Kuru et al. 2005). This limit depends on the composition of the liquid phase, particularly the polymer concentration. For polymer-based foams, this limit can be as high as 99% (*Underbalanced Drilling Manual* 1997).

Hydrostatic Pressure Distribution in the Annulus. The bottomhole pressure due to the static foam column can be analyzed in terms of backpressure and quality at the surface. The governing equation for pressure distribution in a static column of any fluid in a vertical channel is given by

$$dp = \rho_f g dh \quad \dots\dots\dots\dots\dots\dots\dots\dots\dots\dots\dots\dots\dots\dots\dots\dots\dots\dots\dots\dots (4.3.51)$$

From Eq. 4.3.4, the density of foam can be expressed as

$$\rho_f = \frac{p}{b*p + a*} \quad \dots\dots\dots\dots\dots\dots\dots\dots\dots\dots\dots\dots\dots\dots\dots\dots\dots\dots (4.3.52)$$

Combining Eqs. 4.3.51 and 4.3.52 and integrating, we get a relationship between bottomhole pressure and depth for isothermal condition:

$$a*\ln\frac{p_b}{p_s} + b*(p_b - p_s) = gH \quad \dots\dots\dots\dots\dots\dots\dots\dots\dots\dots\dots\dots (4.3.53)$$

The hydrostatic pressure gradient can be estimated using Eqs. 4.3.52 and 4.3.53. **Fig. 4.3.17** presents hydrostatic pressure gradient as a function of depth for different return qualities at the surface with standard backpressure (i.e., fully opened choke).

Example 3. Consider a foam drilling operation in which air and water are used as the gas and liquid phases, respectively. Calculate the anticipated static bottomhole pressure at a depth of 5,000 ft if the surface pressure is 120 psia and foam quality, Γ_s, is 0.95. Assume the temperature is constant at 80°F.

Solution. At surface conditions, $p_s = 120$ psia, $T = 539.7°R$, and $\rho_L = 62.4$ lbm/ft³. The real gas law states

$$p = z\rho_g R_g T / M$$

The universal gas constant, R_g, is 10.7 psia·ft³/lbmol·°R, and $M = 28.97$ lbm/lbmol. Assuming the gas compressibility factor to be 1, the density of gas at the surface conditions is estimated as

$$\rho_g = \frac{pM}{R_g T} = \frac{120 \text{ psia} \times 28.97 \text{ lbm/lbmol}}{10.7 \text{ psia} \bullet \text{ft}^3/\text{lbmol}/°R \bullet 539.7 °R}$$

$$= 0.602 \text{ lbm/ft}^3$$

If we neglect changes in gas solubility due to pressure changes, the gas mass fractions at surface and downhole conditions are the same. Applying Eq. 4.3.7 at the surface condition,

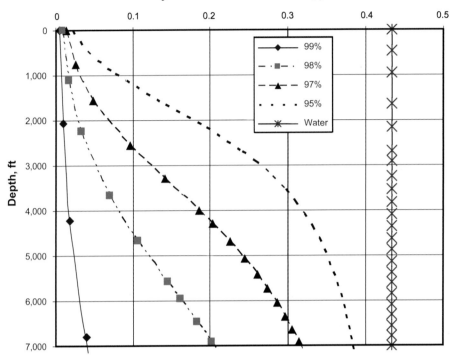

Fig. 4.3.17—Annular hydrostatic pressure gradient distributions for different return foam qualities.

$$W_g = \frac{0.602 \text{ lbm} / \text{ft}^3 \bullet 0.95}{0.602 \text{ lbm} / \text{ft}^3 \bullet 0.95 + 62.4 \text{ lbm} / \text{ft}^3 \bullet (1 - 0.95)}$$

$$= 0.1528$$

According to Eq. 4.3.53, the parameters $a*$ and $b*$ are required to determine the anticipated static bottomhole pressure. Applying Eq. 4.3.5,

$$a* = \frac{W_g R_g T}{M} = \frac{0.1528 \bullet 10.7 \text{ psia} \bullet \text{ft}^3 / \text{lbmol} / {}^{\circ}R \bullet 539.7 \, {}^{\circ}R}{28.97 \text{ lbm} / \text{lbmol}}$$

$$= 30.46 \text{ psia} \bullet \text{ft}^3 / \text{lbm} = 4386.24 \text{ psfa} \bullet \text{ft}^3 / \text{lbm}$$

Similarly, the parameter $b*$ is obtained from Eq. 4.3.6 as

$$b* = \frac{1 - W_g}{\rho_L} = \frac{1 - 0.1528}{62.4} \text{ ft}^3/\text{lbm}$$

$$= 0.0136 \text{ ft}^3/\text{lbm}$$

Eq. 4.3.53 is written for field units as

$$a* \ln \frac{p_b}{p_s} + b* (p_b - p_s) = \frac{gH}{g_c} = H,$$

where g_c is a dimensional constant. Introducing all known variables into the above equation, we obtain

$$4386.24 \bullet \ln \frac{p_b}{17280} + 0.0136 (p_b - 17280) = 5000$$

Solving the above expression iteratively, the static bottomhole pressure p_b is 344.86 psia.

Friction Pressure Loss Calculations in Pipe and Annulus. *Using Yield Power-Law Rheology Model.* The previous sections have placed us in a position to predict the rheology of the foam. The first step in the hydraulic calculation procedure is to select appropriate rheology models based on the composition of the foam and the ranges of flow parameters. In this section, a foam hydraulic model is developed by assuming foam as a yield power-law fluid and considering slip at the wall. For both pipe and annular flow, the frictional pressure gradient under isothermal flow conditions can be expressed as

$$\frac{dP}{dL} = \frac{4\bar{\tau}_w}{D_H}, \dotfill (4.3.54)$$

where $\bar{\tau}_w$ is the average wall shear stress.

Pipe Flow. For isothermal pipe flow, the relationship between mean velocity and wall shear stress is given by

$$\frac{8(U - u_s)}{D} = \frac{(\tau_w - \tau_y)^{(1+1/m)}}{K_f^{1/m} \tau_w^3} A * \left(\tau_w^2 + B * \tau_y \tau_w + C * \tau_y^2 \right), \dotfill (4.3.55)$$

where

$$A* = \frac{4m}{3m+1} \qquad B* = \frac{2m}{1+2m} \qquad C* = \frac{2m^2}{(1+m)(1+2m)}$$

Eq. 4.3.55 can be solved numerically for τ_w if the values of U, τ_y, K_f, m, and u_s are known. However, the slip velocity, u_s, is a function of τ_w and β; therefore, development of a reliable model for slip velocity prediction represents one of the most important tasks in foam hydraulics. This information is critical for accurate prediction of the frictional pressure loss because contribution of the wall slip to the mean flow is significant in most cases. When an accurate correlation for the slip coefficient is unavailable, hydraulic predictions can be made by assuming a no-slip condition at the wall. Then the rheological parameters (m, K_f, and τ_y) should be obtained from plots of wall shear stress vs. uncorrected shear rate.

To determine if the flow is laminar or turbulent, we need to calculate the Reynolds number. Blauer et al. (1974) investigated flow behavior of foam in pipe and capillary viscometers. They concluded that the friction pressure losses for foam flow can be determined using friction factor–Reynolds number correlations or charts that are developed for a single-phase fluid. Hence, the Dodge and Metzner (1959) chart presented in **Fig. 4.3.18** can be applied for estimating the friction factor, where the Reynolds number is expressed as

$$\mathrm{Re} = \frac{8\rho_f U^2}{(\tau_y + K_f \dot{\gamma}_w^m)} \dotfill (4.3.56)$$

The wall shear rate, $\dot{\gamma}_w$, is obtained by

$$\dot{\gamma}_w = \frac{3n'+1}{4n'} \left(\frac{8U}{D} \right) \dotfill (4.3.57a)$$

The term $(3n' + 1)/4n'$ in Eq. 4.3.57a is calculated as*

$$\frac{3n'+1}{4n'} = \frac{3m+1}{4mC_c} \dotfill (4.3.57b)$$

The parameter C_c is expressed as

$$C_c = (1 - x)(1 + B * x + C * x^2), \dotfill (4.3.57c)$$

where $x = \tau_y / \tau_w$. Using the friction factor, the wall shear stress is determined as

---

*Personal communication with S. Z. Miska. 2004. Tulsa: University of Tulsa.

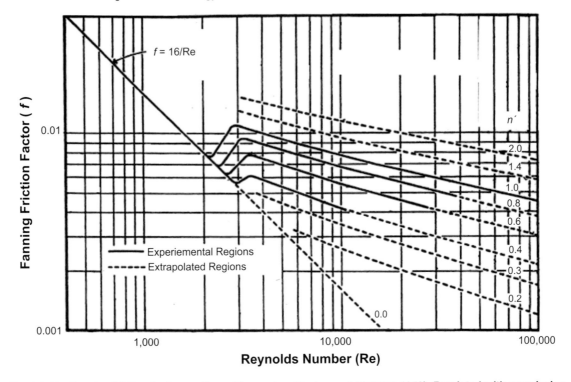

Fig. 4.3.18—Fanning friction factor vs. Reynolds number (Dodge and Metzner 1959). Reprinted with permission from the American Institute of Chemical Engineers.

$$\tau_w = \frac{1}{2} f \rho U^2 \quad \dots\dots\dots\dots\dots\dots\dots\dots\dots\dots\dots\dots\dots\dots\dots\dots\dots\dots\dots\dots\dots\dots \quad (4.3.58)$$

The parameter C_c is a function of the wall shear stress, which is unknown. Thus, an iterative procedure is required to determine the wall shear stress for a given flow rate or mean velocity.

Annular Flow. An exact analytical solution for laminar concentric annular flow of yield power-law fluids is presented by Hanks (1979). The analytical solution requires an iterative procedure that involves numerical integration. Hanks presented useful engineering design charts and calculation procedures for predicting the frictional pressure loss from the flow rate or vice versa. The design charts were computed from the exact numerical solution.

Another method for estimating frictional pressure loss is modeling a concentric annulus as a narrow slot. Reasonable predictions can be obtained using the narrow-slot approximation for boreholes that have higher pipe-to-borehole-diameter ratios (i.e., $D_i/D_o > 0.3$). An analytical solution using the narrow-slot approximation is expressed as

$$\frac{12U}{D_o - D_i} = \frac{(\tau_w - \tau_y)^{(1+1/m)}}{K_f^{1/m} \tau_w^2} \left(\frac{3m}{2m+1} \right) \left(\tau_w + \frac{m}{1+m} \tau_y \right) \quad \dots\dots\dots\dots\dots\dots\dots\dots\dots \quad (4.3.59)$$

The Reynolds number for annular flow is expressed as

$$\text{Re} = \frac{12\rho U^2}{(\tau_y + K_f \dot{\gamma}_w^m)}, \quad \dots\dots\dots\dots\dots\dots\dots\dots\dots\dots\dots\dots\dots\dots\dots\dots\dots\dots \quad (4.3.60)$$

where the shear rate at the wall is obtained by

$$\dot{\gamma}_w = \frac{1+2n'}{3n'} \left(\frac{12U}{D_o - D_i} \right) \quad \dots\dots\dots\dots\dots\dots\dots\dots\dots\dots\dots\dots\dots\dots\dots\dots \quad (4.3.61)$$

The relationship between wall shear stress and n' is given by

$$\frac{3n'}{1+2n'} = \left(\frac{3m}{1+2m}\right)\left(1 - \frac{1}{1+m}x - \frac{m}{1+m}x^2\right), \dots\dots\dots\dots\dots\dots\dots\dots \quad (4.3.62)$$

where $x = \tau_y/\tau_w$.

Using Volume-Equalized Rheology Model. As discussed previously, often difficulties arise when trying to characterize the rheological behavior of dynamic foams. As foam flows through a pipe or annulus, the pressure decreases and the foam expands, resulting in higher flow velocity and quality. Since foam rheology is greatly affected by the quality, the rheological parameters may vary significantly within a short distance. Different rheological models that account for foam quality change can be found in the literature. The volume-equalized model is one of the most convenient rheology models. According to this model, volume-equalized flow data of foam at various pressures can be reduced to a master flow curve **(Fig. 4.3.19)**, relating wall shear stress to the nominal shear rate with the help of the expansion ratio, ε. For low-quality foams, the method of volume equalization has been verified by various authors (Mooney 1931; Saintpere et al. 2000).

Modeling of Pressure Traverse Along the Wellbore. For incompressible fluids, it is possible to estimate both frictional pressure loss and hydrostatic pressure drop independently and then the overall pressure drop. This approach is not valid for a compressible fluid such as foam because the frictional pressure loss and hydrostatic pressure drop are coupled through pressure-dependent foam qualities (densities). Hence, to properly design a foam hydraulic program, one needs to know the anticipated pressure profile in the wellbore, foam linear flow velocities, quality, etc. along the wellbore and drillstring. In particular, it is very important to predict the anticipated bottomhole pressure, pressure changes across the drill bit, and injection pressure at surface for different foam flow rates. Determining an accurate hydraulic model is necessary for better downhole pressure control and improvement in the efficiency of foam drilling operations. The hydraulic model should incorporate key features of foam such as compressibility, wall slip, and change in rheology. Taking these into consideration, momentum balance for steady-state annular flow can be written as

$$A\,dp + d(\beta_f \dot{m} U) + \rho_f g\cos\alpha \Delta LA + \pi(D_o \tau_{w,o} + D_i \tau_{w,i})\Delta L = 0. \dots\dots\dots\dots\dots\dots\dots \quad (4.3.63)$$

The momentum correction factor is defined as

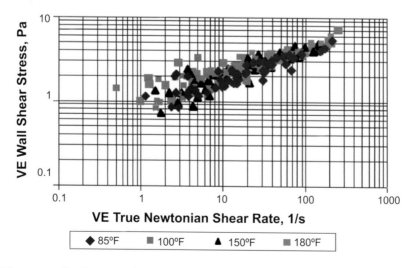

Fig. 4.3.19—Volume-equalized master flow curve for 60–90% foams at different temperature and pressure (Lourenço et al. 2004).

$$\beta_f = \frac{2\pi\rho_f \int\limits_{D_i/2}^{D_o/2} v^2 r\, dr}{\dot{m}U} \quad \cdots\cdots\cdots\cdots\cdots\cdots\cdots\cdots\cdots\cdots\cdots\cdots\cdots \quad (4.3.64)$$

After substituting the inner and outer wall shear stresses with the mean wall shear stress, Eq. 4.3.63 can be rearranged to have the following form:

$$\frac{dP}{dL} + \frac{d(\beta_f \dot{m}U)}{A\,dL} + \rho_f g\cos\alpha + \frac{4\bar{\tau}_w}{D_H} = 0, \quad \cdots\cdots\cdots\cdots\cdots\cdots\cdots\cdots\cdots \quad (4.3.65)$$

where D_H is hydraulic diameter, α is hole inclination angle (from vertical), and β_f is the momentum correction factor. The value of this factor depends upon the shape of the velocity profile. For Newtonian pipe flow under laminar condition, the momentum correction factor is 4/3. The value of β_f for laminar pipe flow of a power-law fluid is given by

$$\beta_f = \frac{(1/n+3)^2}{(1/n+1)^2}\left(1 - \frac{2}{1/n+3} + \frac{1}{1/n+2}\right) \quad \cdots\cdots\cdots\cdots\cdots\cdots\cdots \quad (4.3.66)$$

According Eq. 4.3.66, the momentum correction factor is a weak function of the flow behavior index, n. For shear-thinning fluids, the value of the momentum correction factor ranges from 1.00 to 1.33. It is difficult to obtain a similar expression for annular flows **(Fig. 4.3.20)**. As the first approximation, β_f values of pipe flow may be used for annular flows.

Neglecting the variation of β_f in a computational segment and integrating Eq. 4.3.65 between two neighboring points i and $i + 1$ (upstream and downstream, respectively) as presented in Fig. 4.3.20, we obtain

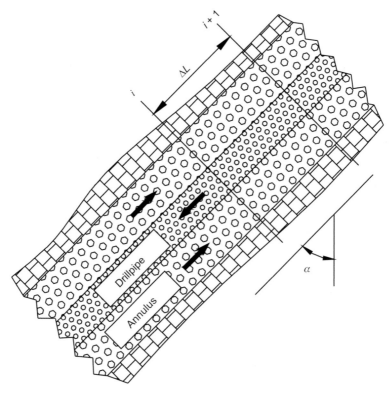

Fig. 4.3.20—Flow through drillpipe and annulus.

$$P_i - P_{i+1} + \frac{\dot{m}}{A}\beta_f(U_i - U_{i+1}) + \bar{\rho}_f g \cos\alpha \bullet \Delta L + \frac{4\bar{\tau}_w}{D_H}\Delta L = 0, \quad \dots\dots\dots\dots \quad (4.3.67)$$

where

$\dot{m}$ = mass flow rate of foam, which is constant along the wellbore for steady-state flow conditions,

A = flow cross-sectional area,

$\bar{\rho}_f$ = average foam density between point i and $i + 1$,

U_i = average velocity at point I,

and

$\bar{\tau}_w$ = average-mean wall shear stress between point i and $i + 1$.

Various iterative techniques can be used to determine the pressure traverse, and hence, the desired foam velocity, quality, apparent viscosity, etc. along the wellbore. We propose the following approach, assuming that the pressure at the top of the hole ($P_i = P_1$) in the annulus is known. This is simply an atmospheric pressure if the foam is not discharged to the pits through the surface choke. If we further assume that the upstream pressure, P_{i+1}, is known, Eq. 4.3.67 can be solved for ΔL. This is our first estimate of length of the wellbore segment (ΔL_i) that corresponds to the pressure difference ($\Delta P_i = P_i - P_{i+1}$) in the computational domain. Thus,

$$\Delta L_i = \frac{\left[(P_{i+1} - P_i) + \beta_f \frac{\dot{m}}{A}(U_{i+1} - U_i)\right]D_H}{4\bar{\tau}_w + \bar{\rho}_f g D_H \cos\alpha} \quad \dots\dots\dots\dots\dots\dots \quad (4.3.68a)$$

The differential pressure in the computational segment, ΔP_i, should be small enough to neglect the effect of expansion on properties of foam. Moreover, it is useful to limit the segment length so that $\Delta L_i < \Delta L_{max}$, where ΔL_{max} is the maximum length of the computational segment. The calculations should be repeated for other computational segments until the total segment length ($\Delta L_1 + \Delta L_2 + \cdots + \Delta L_n$) is greater than the wellbore length, L. Finally, for the last segment, upstream pressure needs to be varied systematically to obtain the following condition:

$$\sum_{i=1}^{n}\Delta L_i \approx L, \quad \dots\dots\dots\dots\dots\dots\dots\dots\dots\dots\dots\dots \quad (4.3.68b)$$

where n is the number of computational segments and P_{n+1} is the bottomhole pressure.

The foam mass flow rates, velocities, average foam density, and average-mean wall shear stress are calculated as shown below.

1. Mass flow rates of liquid and gas phases are

$$\dot{m}_L = \rho_L Q_L \quad \dots\dots\dots\dots\dots\dots\dots\dots\dots\dots\dots\dots\dots\dots \quad (4.3.69)$$

$$\dot{m}_g = \rho_{g,std} Q_{g,std}, \quad \dots\dots\dots\dots\dots\dots\dots\dots\dots\dots\dots\dots \quad (4.3.70)$$

where the quantities $Q_{g,std}$ and $\rho_{g,std}$ are flow rate and density of the gas phase at standard conditions. The liquid phase is assumed incompressible.

2. Foam density at a given point can be calculated as

$$\rho_{f,i} = \frac{P_i}{a + bP_i}, \quad \dots\dots\dots\dots\dots\dots\dots\dots\dots\dots\dots\dots \quad (4.3.71)$$

where

$$a = \frac{w_g RT}{M_g} \quad \dots\dots\dots\dots\dots\dots\dots\dots\dots\dots\dots\dots\dots \quad (4.3.72)$$

$$b = \frac{B^0 w_g RT}{M_g} + (1 - w_g)\frac{1}{\rho_L} \quad \dots\dots\dots\dots\dots \quad (4.3.73)$$

$$w_g = \frac{\dot{m}_g}{\dot{m}_g + \dot{m}_L}, \quad \dots\dots\dots\dots\dots\dots\dots\dots\dots\dots\dots\dots\dots\dots\dots\dots\dots \quad (4.3.74)$$

where

T = temperature (absolute)

and

B^0 = modified second viral coefficient of the gas.

3. Mean foam velocity at a given point is determined by

$$U_i = \frac{\dot{m}}{A\rho_{f,i}} \quad \dots\dots\dots\dots\dots\dots\dots\dots\dots\dots\dots\dots\dots\dots\dots\dots\dots\dots \quad (4.3.75)$$

4. Average foam density from point i to $i + 1$ is

$$\bar{\rho}_f = \frac{1}{b^2(P_{i+1} - P_i)}\left[b(P_i - P_{i+1}) + a \ln\left(\frac{a + bP_{i+1}}{a + bP_i}\right)\right] \quad \dots\dots\dots\dots\dots\dots \quad (4.3.76)$$

It should be noticed that the quantities a and b in Eqs. 4.3.71 and 4.3.76 are dependent not only on pressure but also on temperature. However, in a first iteration, we assume that the temperature is constant along the computational segment, ΔL_i. In the second iteration, the temperature may be updated using the geothermal temperature gradient. The calculations should be repeated until the desired convergence on the values of ΔL_i is obtained. After determining average foam density, we may now calculate the average-mean foam velocity as

$$\bar{U} = \frac{\dot{m}}{A\bar{\rho}_f} \quad \dots\dots\dots\dots\dots\dots\dots\dots\dots\dots\dots\dots\dots\dots\dots\dots\dots\dots\dots \quad (4.3.77)$$

The average-mean wall shear stress can be calculated using Eq. 4.3.59 if the average-mean foam velocity and the values of rheological parameters are known in the computational segment.

4.3.4 Cuttings Transport With Foam. Foam has good cuttings-transport ability even in the laminar flow regime. The gaseous phase in foam contributes to foam quality and helps the liquid phase to form a relatively stable lamellae structure. This structure is able to hold drilled cuttings and prevents them from falling. The use of drilling foams is increasing because foams exhibit properties that are desirable for good cuttings transport. Drilling foam has a good potential for taking the place of conventional drilling fluids. Cuttings transport with conventional drilling fluid systems has been studied for horizontal and inclined wellbore configurations. Experiments have been conducted by numerous investigators to determine the optimum flow rates needed to avoid the problems that are created by insufficient cleaning or excessive flow rates. However, there is still a significant lack of information when foam is used for drilling purposes.

The increasing use of foam for drilling has created the need for a better understanding of cuttings transport with foam. Previous investigators (Krug and Mitchell 1972; Okpobiri and Ikoku 1986; Guo et al. 1995) used different methods for determining the minimum gas and liquid injection rates that are required to transport cuttings in vertical wells using foam. Krug and Mitchell (1972) recommended 1.5 ft/sec annular velocity as the minimum velocity required at the bottom of the hole for effective cuttings transport in vertical wells.

Okpobiri and Ikoku (1986) developed a semiempirical model for predicting the minimum gas and liquid injection rates for foam and mist drilling operations, taking into account the frictional pressure losses caused by cuttings, pressure drop across bit nozzles, and particle settling velocities. They observed an increase in friction pressure losses with an increase in solid mass flow rate when foam drilling operations were performed in a laminar flow region in which the foam qualities varied between 55 and 96%. This approach, however, is limited to only vertical wells. Guo et al. (1995) recommended that a critical cuttings concentration should be specified at the bottom of the hole to determine the minimum foam velocity for efficient cuttings transport.

One of the most basic functions of drilling fluid is to transport cuttings out of the borehole. This requires designing a fluid system that has better cuttings transport capabilities.

Carrying Capacity of Foams. During foam drilling, rock cuttings mix with the fluid and form a suspension at the bit. As the fluid flows through the annulus, different forces such as gravity, buoyancy, and hydrodynamic forces act on cuttings particles. These forces have a tendency of affecting the trajectory and motion of the particles in the fluid. The particles slip in the fluid due to the actions of these forces. As a result, they do not have the same velocity as the fluid. The slip velocity depends on the properties of the fluid and suspended particles, and it has a negative impact on the cuttings transport.

Generally, a solid particle falling in a fluid under the action of gravity accelerates until the buoyancy and drag forces just balance the gravitational force; then it continues to fall at constant velocity (settling velocity), which is given by

$$v_s = \sqrt{\frac{4gd_p(\rho_p - \rho_f)}{3\rho_f C_D}}, \quad \dots \dots \dots \dots \dots \dots \dots \dots \dots \dots \dots \dots \dots \quad (4.3.78)$$

where the particle drag coefficient, C_D, is given by

$$C_D = \frac{24}{\text{Re}_p} + \frac{6}{1 + \text{Re}_p^{0.5}} + 0.4, \quad \dots \dots \dots \dots \dots \dots \dots \dots \dots \dots \dots \quad (4.3.79)$$

where Re_p is the particle Reynolds number. Eq. 4.3.79 can be valid for Newtonian and non-Newtonian fluids if the definition of the particle Reynolds number is the same in both cases (Dedegil 1987). Hence, it is necessary to define the particle Reynolds number in a more general form as

$$\text{Re}_p = \frac{\rho_f v_s^2}{\tau}, \quad \dots \dots \dots \dots \dots \dots \dots \dots \dots \dots \dots \dots \dots \quad (4.3.80)$$

where τ is the shear stress, which is determined by the rheological model of the fluid at a representative shear rate, v_s/d_p (Dedegil 1987). The effect of cuttings concentration on settling velocity needs to be considered when calculating the slip velocity of a single particle, because the cuttings' effect is significant at high concentrations. Therefore, the slip velocity obtained from Eq. 4.3.78 needs to be modified by the hindered settling factor to account for hydrodynamic interference and particle collision. For solids volume fractions between 0.001 and 0.4, this factor is given by (Govier and Aziz 1972)

$$f_s = e^{-5.9c}, \quad \dots \dots \dots \dots \dots \dots \dots \dots \dots \dots \dots \dots \dots \quad (4.3.81)$$

where c is in-situ cuttings concentration. For fluid with yield stress, some of the fine particles may not settle through a static fluid due to the yield stress effect. For a spherical particle, the force required to overcome the yield strength of a static fluid is $\pi \tau_y d_p^2$. This means that there is a critical particle size, d_c, below which the particles do not settle. The critical diameter can be estimated by (Bourgoyne et al. 1986)

$$d_c = \frac{6\tau_y}{g(\rho_s - \rho_f)} \quad \dots \dots \dots \dots \dots \dots \dots \dots \dots \dots \dots \dots \dots \quad (4.3.82)$$

Herzhaft et al. (2000) experimentally investigated solids-carrying capacity of different-quality foams. Sedimentation tests were carried out using a vertical transparent polyvinyl chloride (PVC) cylinder of 35 mm diameter and 240 mm length. Spherical glass beads were used as cuttings particles. Settling velocities of glass beads with different diameters (2 to 10 mm) were tested. The beads were colored in order to visualize their trajectory during the experiment. Three different-quality polymer-based foams, the rheological properties of which are given in **Table 4.3.6**, were used in the experiment. The base fluid was 0.3% PAC solution. The result indicated that the settling velocity of the particles reduces as the quality increases. **Figs. 4.3.21a and 4.3.21b** compare the measured settling velocity with predictions of Eq. 4.3.78. Theoretical predictions are slightly less than the measured data; however, the predictions are still encouraging. The settling behavior of a solid particle in structured fluids such as foam is different from that in ordinary fluids. Eq. 4.3.78 can be used to estimate settling velocities of cuttings for optimization of hole cleaning. For a vertical well, if the in-situ cuttings concentration is known, then the cuttings transport ratio can be estimated by

$$F_T = \frac{U - f_s v_s}{U} \quad \dots \dots \dots \dots \dots \dots \dots \dots \dots \dots \dots \dots \dots \quad (4.3.83)$$

TABLE 4.3.6—RHEOLOGICAL PROPERTIES OF TEST FOAMS

Fluid	τ_y (Pa)	K (Pa·s^m)	m
PAC (3 g/L)	0.00	0.50	0.51
Foam ($\Gamma = 0.84$)	4.70	3.50	0.47
Foam ($\Gamma = 0.90$)	6.10	4.33	0.47
Foam ($\Gamma = 0.96$)	10.10	5.50	0.45

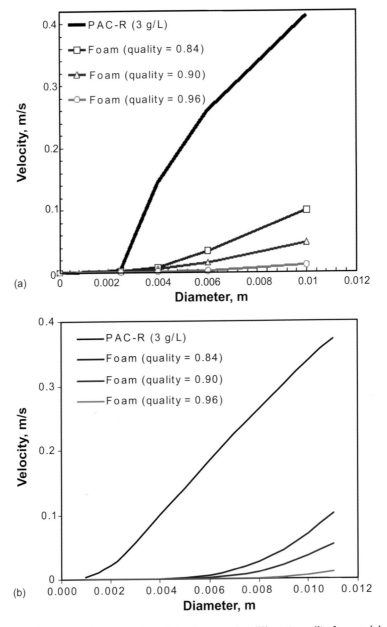

Fig. 4.3.21—Settling velocity as a function of particle diameter for different-quality foams: (a) measured (Herzhaft et al. 2000) and (b) predicted from Eq. 4.3.78.

Saintpere et al. (2000) evaluated carrying capacities of foams and conventional fluids using a small-scale experimental setup with an inclined pipe section **(Fig. 4.3.22)**. Aqueous solutions of PAC and xanthan gum were used as conventional fluids. Different-quality foams were generated using these solutions as base fluids. In order to simulate drilled cuttings, glass balls with different diameters were used. The angle of inclination was varied from horizontal to vertical during the experiments. **Fig. 4.3.23** presents the percentage of cuttings removed as a function of angle of inclination at different dimensionless circulating times. Results show the existence of inclination angles between 30 and 45° that are difficult to clean. Inclinations around 0 and 80° have relatively higher cuttings-transport rates than inclinations between 30 and 45°.

Empirical Cuttings Transport Studies. Martins et al. (2001) conducted extensive experiments to determine the cuttings-bed erosion capacity of drilling foams in horizontal wells. The research involved foaming-agent selection, rheological characterization, and development of a flow loop to test the erosion capacity at high angles of inclination. Results presented in **Fig. 4.3.24** show that the increase in gas-injection rate considerably improves bed erosion. Higher liquid-injection rates have better bed-erosion capacity than lower injection rates at a given gas flow rate. Sensitivity of equilibrium bed height for changes in angle of inclination was investigated. Experimental results **(Fig. 4.3.25)** suggest that bed erosion is significantly less for 45 and 75° inclinations. An empirical correlation that expresses dimensionless equilibrium cuttings bed height (h/D_o) as a function of the Reynolds number and power-law exponent, n, for horizontal configuration was developed. Hence,

$$\frac{h}{D_o} = a_o - b_o \, \mathrm{Re}^{c_o} \, n^{d_o}, \quad \dots\dots\dots\dots\dots\dots\dots\dots\dots\dots\dots\dots\dots\dots\dots\dots \quad (4.3.84)$$

where the coefficients a_o, b_o, c_o, and d_o are empirical constants.

Recently, cuttings transport and hydraulic investigations with aqueous foam were conducted by Özbayoğlu el al. (2003) using a large-scale flow loop (4.5 × 8 in. test section) under ambient temperature and pressure conditions. Foam quality was varied from 70 to 90%. Inclination angles ranged from 70 to 90° from vertical. A mathematical model was developed for predicting frictional pressure losses and cuttings transport in foam drilling. Model predictions were compared with experimentally measured data. Experimental results in terms of cuttings-bed cross-sectional area are presented **(Fig. 4.3.26)** as a function of average annular velocity.

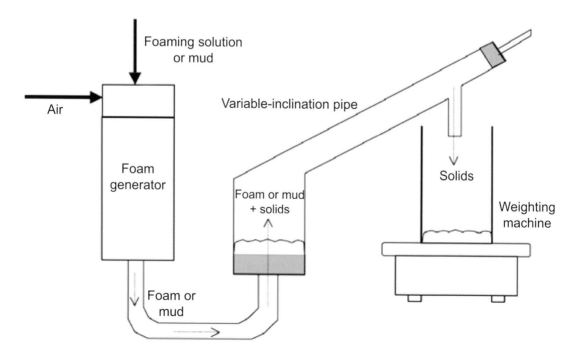

Fig. 4.3.22—Small-scale experimental setup (Saintpere et al. 2000).

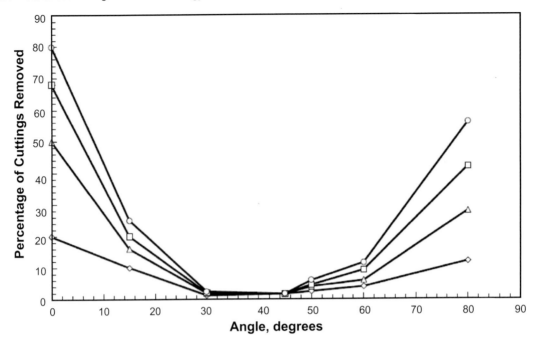

Fig. 4.3.23—Percentage of cuttings removed as a function angle of inclination from vertical (Saintpere et al. 2000).

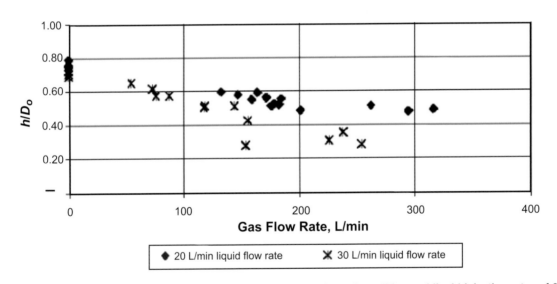

Fig. 4.3.24—Equilibrium bed height vs. gas-injection rate for horizontal condition and liquid injection rates of 20 L/min and 30 L/min (Martins et al. 2001).

Results indicate that the cuttings-bed area decreases as the average annular flow velocity increases at lower flow velocities (i.e., $U < 10$ ft/sec). At higher flow velocities, the bed area remains approximately the same as the velocity increases. The presence of a cuttings bed (shown in **Fig. 4.3.27**) within the wellbore was reported even at higher annular velocities. This observation was attributed to the high viscosity of the foam, which dampens the turbulent effects close to the cuttings-bed surface. With reduced turbulent effects, cuttings on the bed surface cannot be picked up efficiently.

Further analysis of the experimental data presented in Fig. 4.3.26 and **Figs. 4.3.28 through 4.3.30** also reveals that the effect of inclination on the equilibrium bed height is minimal when the ranges of foam quality and inclination angle are 70 to 90% and 80 to 90°, respectively.

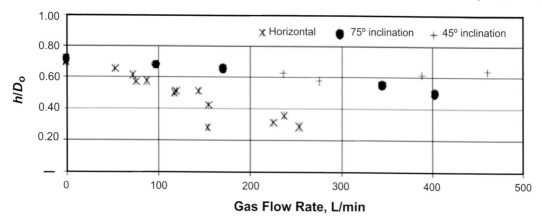

Fig. 4.3.25—Equilibrium bed height vs. gas-injection rate at three different inclination angles for liquid injection rate of 30 L/min (Martins et al. 2001).

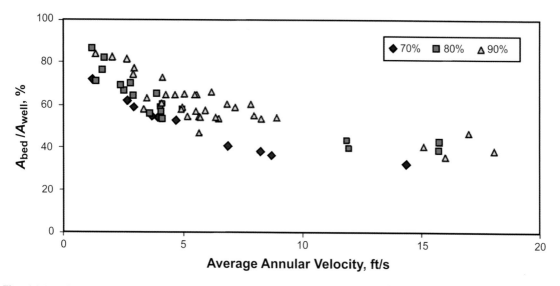

Fig. 4.3.26—Cuttings bed development as a function of foam flow rate (Özbayoğlu et al. 2005). Reprinted with permission from Elsevier.

Based on results of flow loop experiments, Özbayoğlu et al. (2003) developed an empirical correlation that describes the test variables in dimensionless form. Equilibrium cuttings-bed height is considered as a way of quantifying hole-cleaning performance for a successful drilling operation. The following variables are considered major independent drilling variables that control the formation of cuttings beds in the wellbore: inclination angle, feed cuttings concentration, fluid density, a term representing the apparent fluid viscosity, average velocity, and dimensions of the pipe and wellbore. Five dimensionless groups are identified as important variables from the results of the dimensional analysis. These variables are

1. Cuttings volumetric concentration, C_c
2. Inclination angle from vertical, α
3. Dimensionless bed area, $\dfrac{A_{bed}}{A_w}$
4. Reynolds number, $\mathrm{Re} = \dfrac{\rho U D}{\mu}$

Fig. 4.3.27—Cuttings bed formed in the test section (80%-quality foam at 500 gal/min) (Özbayoğlu 2002).

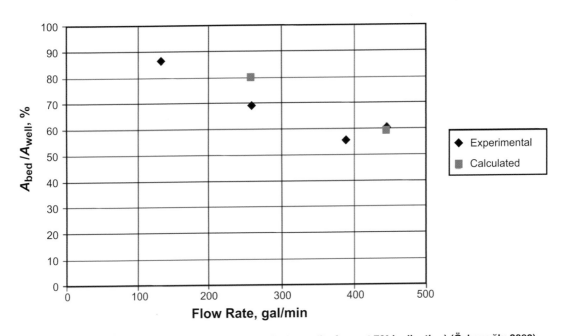

Fig. 4.3.28—Cuttings-bed area vs. foam flow rate (80%-quality foam at 70° inclination) (Özbayoğlu 2002).

5. Froude number, $Fr = \dfrac{U^2}{gD}$

The relation between the dimensionless bed area and the rest of the dimensionless groups is expressed as follows:

For $n' \geq 0.9$,

$$\frac{A_{bed}}{A_w} = 4.1232(C_c)^{0.0035}(\mathrm{Re})^{-0.2198}(\mathrm{Fr})^{-0.2164} \quad \dots \dots \dots \dots \dots \dots \dots \dots \dots \quad (4.3.85)$$

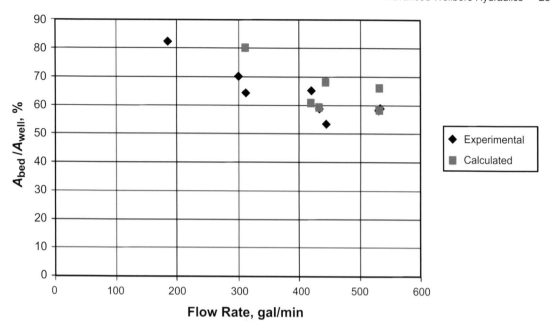

Fig. 4.3.29—Cuttings-bed area vs. foam flow rate (80%-quality foam at 80° inclination) (Özbayoğlu 2002).

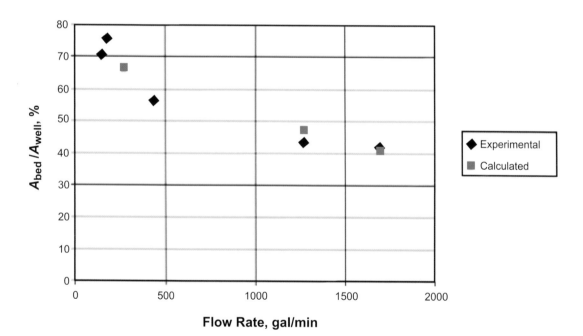

Fig. 4.3.30—Cuttings-bed area vs. foam flow rate (80%-quality foam at 90° inclination) (Özbayoğlu 2002).

For $0.6 < n' < 0.9$,

$$\frac{A_{bed}}{A_w} = 0.7115(C_c)^{0.0697}(Re)^{-0.0374}(Fr)^{-0.0681} \quad \dots \dots \dots \dots \dots \dots \dots \dots \dots \dots \dots \dots \dots \quad (4.3.86)$$

For $n' \leq 0.6$,

$$\frac{A_{bed}}{A_w} = 1.0484(C_c)^{0.0024}(Re)^{-0.1502}(Fr)^{-0.0646} \quad \dots \dots \dots \dots \dots \dots \dots \dots \dots \dots \dots \dots \dots \quad (4.3.87)$$

These empirical correlations are valid for highly deviated wells (i.e., $\alpha > 60°$). The inclination angle is removed from the correlations because flow loop experiments conducted at different inclinations (from horizontal to 65°) indicated that an inclination greater than 70° has little influence on the bed height. In other words, Özbayoğlu's correlations are valid for highly inclined wells.

Recently, an experimental study of cuttings transport with foam at intermediate angles has been conducted by Capo et al. (2006) in a full-scale flow loop **(Fig. 4.3.31)** under ambient temperature and pressure conditions. An anionic surfactant (concentration of 1 vol%) was used to generate aqueous foams. Air and tap water were used as gas and liquid phases, respectively. Tests were conducted to determine the effects of inclination angle, foam quality, foam velocity, and rate of penetration on cuttings transport. In-situ cuttings concentration and frictional pressure loss were measured. **Figs. 4.3.32 and 4.3.33** show in-situ cuttings concentration as a function of mean velocity for different rates of penetration.

In the effort to develop a convenient correlation for the cuttings bed area, the data were analyzed using dimensional analysis techniques. Dimensionless groups considered in the analysis were volume-equalized Reynolds number (Re_ε), Archimedes number (Ar), Froude number (Fr), ratio of solid density to foam density (s), and inclination angle (α). A correlation for the ratio of cuttings bed area to wellbore area as a function of the dimensionless groups is given by

$$\ln\left(\frac{A_b}{A_w}\right) = \alpha_1 Ar^{\alpha_2} Fr^{\alpha_3} Re_\varepsilon^{\alpha_4} s^{\alpha_5} \alpha^{\alpha_6} \quad \dots\dots\dots\dots\dots\dots\dots\dots\dots\dots\dots \text{(4.3.88)}$$

Equations for dimensionless groups and values for the empirical constants are presented in Capo et al. (2006).

Mechanistic Cuttings-Transport Models. Very limited numbers of foam cuttings-transport models are available for predicting in-situ cuttings concentration and optimization of hole cleaning. These models adopt cuttings-transport models developed for conventional drilling.

A transient cuttings-transport model for vertical wells has been presented by Li and Kuru (2004). The effects of key drilling parameters such as drilling rate, annular geometry, and influx rate on the efficiency of cuttings transport have been investigated. A number of hole-cleaning charts have been proposed to optimize cuttings transport. The model considered foam as a single phase (a homogeneous compressible mixture of liquid and gas). Assuming laminar flow condition and incompressible solid phase, material balance equations for foam and solid phases in a differential segment of the annulus can be given by

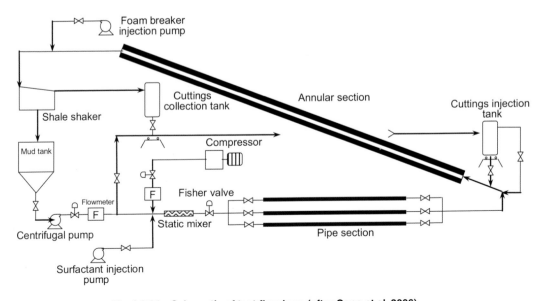

Fig. 4.3.31—Schematic of test flow loop (after Capo et al. 2006).

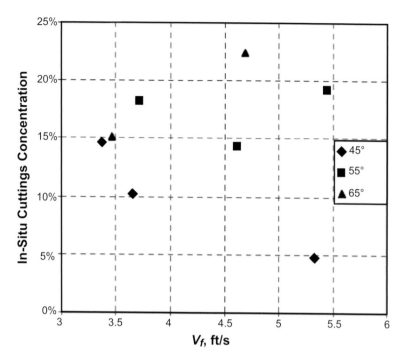

Fig. 4.3.32—In-situ cuttings concentration vs. foam velocity at different inclination angles (quality = 70% and ROP = 39 to 55 ft/hr) (after Capo et al. 2006).

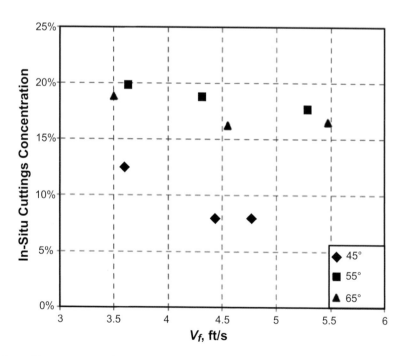

Fig. 4.3.33—In-situ cuttings concentration vs. foam velocity at different inclination angles (quality = 70% and ROP = 22 to 44 ft/hr) (after Capo et al. 2006).

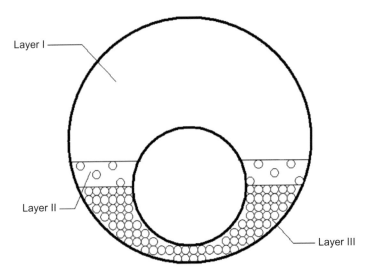

Fig. 4.3.34—Cross section of the wellbore (Özbayoğlu et al. 2003).

$$\frac{\partial(c_f \rho_f)}{\partial t} + \frac{\partial(c_f \rho_f u_f)}{\partial x} = s_f \qquad \dots \dots \dots \dots \dots \dots \dots \dots \dots \dots \dots \dots \dots \dots \dots \quad (4.3.89)$$

and

$$\frac{\partial(c)}{\partial t} + \frac{\partial(c\,u_s)}{\partial x} = 0, \qquad \dots \dots \dots \dots \dots \dots \dots \dots \dots \dots \dots \dots \dots \dots \dots \dots \dots \quad (4.3.90)$$

respectively, where s_f is a source term that represents reservoir fluid influx. In order to get the solution, the material balance equations were coupled with the momentum equation.

Özbayoğlu et al. (2003) developed a three-layer mechanistic model for foam cuttings transport. The three-layer modeling approach is preferred to describe the cuttings-transport phenomenon in horizontal and highly inclined wellbores. The model describes the flow of cuttings and drilling-fluid mixture in inclined annuli by a system of nonlinear equations that are obtained from the conservation equations and empirical correlations. The unknowns, such as moving bed layer thickness, velocities in each layer, in-situ cuttings concentration, and frictional pressure loss, will be determined by solving the system of equations numerically. Three distinct layers **(Fig. 4.3.34)** are considered in the development of this model. Additionally, the following assumptions are made to formulate the model:

1. Layer I is a fluid layer without cuttings and with uniform physical and chemical properties.
2. Layer II is a moving bed layer, which is a mixture of cuttings and fluid.
3. Layer III is a stationary cuttings-bed layer, which is uniformly compacted with constant porosity and negligible pore fluid flow.
4. Cuttings are assumed uniform, incompressible, and spherical.

To obtain a numerical solution, the wellbore is divided into small longitudinal grids. The length of each grid should be small enough to minimize the change in fluid velocity, pressure gradient, and fluid properties within the grid element. However, these flow parameters are considered varying from one grid to another grid because of the compressibility of foam. Therefore, the flow rate in a given grid ($i + 1$) is calculated using the flow parameters determined at the neighboring grid (i) as

$$Q_{i+1} = Q_i \left[(1 - \Gamma_i) + \frac{P_i T_{i+1}}{P_{i+1} T_i} \Gamma_i \right], \qquad \dots \dots \dots \dots \dots \dots \dots \dots \dots \dots \dots \dots \dots \quad (4.3.91)$$

where the indices i and $i + 1$ refer to the consecutive grid order. For a steady-state flow condition, the mass balance for fluid phase in a single grid can be expressed as (Özbayoğlu et al. 2003)

$$v_I A_I \rho_f + v_{II} A_{II} \rho_f \left(1 - C_{C,II}\right) = \bar{v} A_w \left(1 - C_C\right) \rho_f \quad\dots\dots\dots\dots\dots\dots\dots\dots\dots\dots\dots (4.3.92)$$

A similar equation can be written for solid-phase material balance. Thus,

$$v_{II} A_{II} C_{C,II} \rho_c = \bar{v} A_w C_C \rho_c, \quad\dots\dots\dots\dots\dots\dots\dots\dots\dots\dots\dots\dots\dots\dots\dots (4.3.93)$$

where v_{II} is the slurry velocity at the second layer, which is defined as

$$v_{II} = v_{II,f} - \frac{v_{slip} \rho_c C_{C,II}}{\rho_s} \quad\dots\dots\dots\dots\dots\dots\dots\dots\dots\dots\dots\dots\dots (4.3.94)$$

The in-situ slurry density, ρ_s, is given by

$$\rho_s = \rho_f \left(1 - C_{C,II}\right) + \rho_c C_{C,II} \quad\dots\dots\dots\dots\dots\dots\dots\dots\dots\dots\dots\dots (4.3.95)$$

The average fluid velocity in the wellbore, $\bar{v}$, and the feed cuttings concentration, C_c, on the right side of Eqs. 4.3.92 and 4.3.93, are then defined as

$$\bar{v} = \frac{Q_f + \lambda A_{bit} \text{ROP}}{A_w} \quad\dots\dots\dots\dots\dots\dots\dots\dots\dots\dots\dots\dots\dots\dots (4.3.96)$$

and

$$C_C = \frac{Q_c}{Q_f + Q_c}, \quad\dots\dots\dots\dots\dots\dots\dots\dots\dots\dots\dots\dots\dots\dots\dots\dots (4.3.97)$$

respectively, where λ is the correction factor for cuttings accumulation in the wellbore. Cuttings concentration, C_c, is not the in-situ cuttings concentration; it is rather cuttings volumetric rate relative to the total volume flow rate.

A free-body diagram of a wellbore section is presented in **Fig. 4.3.35** to develop momentum equations for each layer. If the fluid density and velocity are assumed constant for a given wellbore grid, then the momentum balance for the upper layer (Layer I) can be expressed as

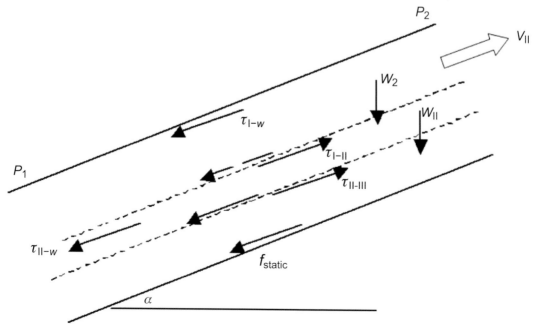

Fig. 4.3.35—Free-body diagram of a wellbore grid (Özbayoğlu et al. 2003).

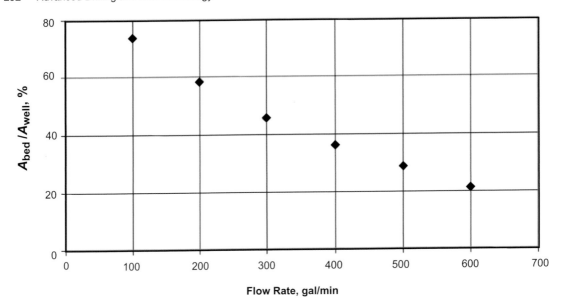

Fig. 4.3.36—Simulation result obtained with the model (Özbayoğlu 2002).

$$\frac{\Delta P}{\Delta L}A_{\mathrm{I}} - \tau_{\mathrm{I,II}}S_{\mathrm{I,II}} - \tau_{\mathrm{I,w}}S_{\mathrm{I,w}} - \rho_f gA_{\mathrm{I}}\sin\alpha = 0 \quad\text{(4.3.98)}$$

Similarly, the momentum equations for Layer II and Layer III are given by

$$\frac{\Delta P}{\Delta L}A_{\mathrm{II}} + \tau_{\mathrm{I,II}}S_{\mathrm{I,II}} - \tau_{\mathrm{II,III}}S_{\mathrm{II,III}} - \tau_{\mathrm{II,w}}S_{\mathrm{II,w}} - \rho_s gA_{\mathrm{II}}\sin\alpha = 0 \quad\text{(4.3.99)}$$

and

$$\frac{\Delta P}{\Delta L}A_{\mathrm{III}} + \tau_{\mathrm{II,III}}S_{\mathrm{II,III}} - F_{\mathrm{III\text{-}w}} - \rho_b gA_{\mathrm{III}}\sin\alpha = 0, \quad\text{(4.3.100)}$$

respectively, where $F_{\mathrm{III\text{-}w}}$ is the friction force (i.e., force per length) between the bed and the wellbore. Interfacial shear stresses that are presented in Eqs. 4.3.94, 4.3.95, and 4.3.96 are generally expressed as

$$\tau_{i-j} = f_{i-j}\frac{\rho_i\left(v_i - v_j\right)^2}{2}, \quad\text{(4.3.101)}$$

where indices i and j indicate the surfaces where the shear stress is acting.

A friction factor correlation developed by Televantos et al. (1979) is used to calculate the interfacial shear stress between the layer boundaries. Determination of the in-situ cuttings concentration in the second layer is very important for defining the mixture density and viscosity. At steady-state conditions, a convection-diffusion equation can be applied to obtain the in-situ cuttings concentration. Thus,

$$v_y\frac{\partial C}{\partial y} = N_C\frac{\partial^2 C}{\partial y^2}, \quad\text{(4.3.102)}$$

where

$$v_y = v_s\cos\alpha \quad\text{(4.3.103)}$$

A system of equations can be established when all the constitutive relations are substituted into the conservation equations. This system is nonlinear and requires an iterative procedure to get a numerical solution.

Details of the model and expressions for in-situ cuttings concentration and slip velocity are available in previous works (Özbayoğlu 2002 and Özbayoğlu et al. 2003). **Fig. 4.3.36** shows predictions of the model for a 5×2½-in. horizontal wellbore with eccentricity of 50% and ROP of 50 ft/hr. The following fluid and cuttings properties were assumed: mud weight of 8.4 lbm/gal, $K_f = 20$ cp equivalent, average cuttings size of 0.118 in., and specific gravity of 2.65.

Model predictions indicate strong influence of flow rate (i.e., the annular fluid velocity) on cuttings transport. Increase in the flow rate drastically prevents the bed buildup by improving the cuttings-carrying ability of the fluid. Especially when the flow becomes turbulent, cuttings are transported more effectively and the bed thickness reduces significantly.

Nomenclature

a = parameter used in viral equation of state
a^* = parameter used in the equation of state
a_o = empirical constant
A = flow cross-sectional area
A_{bed} = bed cross-sectional area
A_{bit} = drill bit cross-sectional area
Ar = Archimedes number
A_w = annular flow area
A^*, B^*, C^* = dimensionless fluid parameters
b = parameter used in viral equation of state
b^* = parameter used in the equation of state
b_o = empirical constant
B^0 = modified second viral coefficient
c = in-situ cuttings concentration
c_o = empirical constant
C_1, C_2, C_3 = empirical constant
Ca = capillary number
C_c = cuttings volumetric concentration
C_D = drag coefficient
C_f = compressibility of foam
c_f = concentration of foam
C_g = compressibility of gas phase
d_c = critical diameter
d_o = empirical constant
d_p = particle diameter
D = pipe diameter
D_H = hydraulic diameter
D_i = inner diameter of annulus
D_o = outer diameter of annulus
e = exponential function
f = friction factor
f_s = hindered settling factor
F_D = fractional drainage
F_T = cuttings transport ratio
F_{III-w} = friction force between the wellbore and stationary bed
Fr = Froude number
g = gravitational acceleration
g_c = dimensional constant
h = cuttings bed height
H = bottomhole depth
j = subscript indicating the surface where shear stress is acting

κ = radius ratio

K_f = foam consistency index

K_L = liquid phase consistency index

K_{VE} = volume-equalized consistency index

K' = generalized consistency index

L = length

L_b = viscometer rotor length

L_i = length of wellbore computational segment

L_{max} = maximum length of wellbore computational segment

m = flow behavior index for yield power-law fluids

m_F = mass of foam

$\dot{m}$ = mass flow rate

$\dot{m}_g$ = mass flow rate of gas phase

$\dot{m}_L$ = mass flow rate of liquid phase

M = molecular weight

M_g = gas molecular weight

n_f = foam flow behavior index

n_L = liquid phase flow behavior index

n' = generalized flow behavior index

n^* = slop of torque vs. angular velocity (log-log plot)

N_C = diffusion coefficient

p = static pressure

p_b = bottomhole pressure

p_s = annular pressure at the surface

P_i = pressure inside foam bubble

P_o = pressure outside foam bubble

Q_m = measured flow rate

Q_c = slip-corrected flow rate

$Q_{g,std}$ = gas flow rate at standard conditions

Q_L = liquid flow rate

Q_{sh} = flow rate due the shearing of the fluid

r_b = average bubble radius

R = pipe radius

R_i = viscometer rotor radius

R_o = viscometer cup radius

R_g = universal gas constant

Re = Reynolds number

Re_p = particle Reynolds number

Re_ε = volume-equalized Reynolds number

s = dimensionless group

S = wetted perimeter

$S_{ll,w}$ = wetted perimeter of moving layer in contact with borehole wall

t = time

T = absolute temperature

T_e = additional torque due to end effect and friction

T_m = measured torque

u_s = slip velocity

U = mean velocity

$\overline{U}$ = average velocity between point i and $i + 1$

v = local velocity

v_F = specific volume of foam

v_L = specific volume of the liquid phase

v_s = settling velocity

v_{slip} = slip velocity
$\bar{v}$ = average fluid velocity in the wellbore
V_D = volume of drainage
V_g = gas phase volume
V_f = foam volume
V_L = liquid phase volume
w = mass fraction
w_g = gas mass fraction
$\boldsymbol{x}$ = dimensionless parameter
z = gas compressibility factor
α = inclination angle from vertical
β = slip coefficient
β_f = momentum correction factor
δ_s = effective film thickness of slip layer
ε = specific-volume expansion ratio
$\dot{\gamma}$ = shear rate
$\dot{\gamma}_w$ = shear rate at the wall
Γ = foam quality
μ_L = liquid phase viscosity
μ_f = foam viscosity
ξ = empirical constant
ρ_f = density of foam
ρ_g = density of gas phase
$\rho_{g,std}$ = density of gas at standard condition
ρ_L = density of liquid phase
$\bar{\rho}_f$ = average foam density between point i and $i + 1$
σ = interfacial tension
τ_w = wall shear stress
$\tau_{w,o}$ = shear stress acting on the outer wall
$\bar{\tau}_w$ = average wall shear stress at point i
$\bar{\tau}_{\bar{w}}$ = average-mean wall shear stress between i and $i + 1$
τ^* = average wall shear stress in the slip layer
τ_y = yield stress
ω = angular velocity

References

Ahmed, A., Kuru, K., and Saasen, A. 2003a. Critical Review of Drilling Foam Rheology. *Transactions of the Nordic Rheology Society* **11** (Session 2).

Ahmed, R., Saasen, A., and Ergun, K. 2003b. Mathematical Modeling of Drilling Foam Flows. Paper 03-033 presented at the CADE/CAODC Drilling Conference, Calgary, 20–22 October.

Argillier, J.-F., Saintpere, S., Herzhaft, B., and Toure, A. 1998. Stability and Flowing Properties of Aqueous Foams for Underbalanced Drilling. Paper SPE 48982 presented at the SPE Annual Technical Conference and Exhibition, New Orleans, 27–30 September. DOI: 10.2118/48982-MS.

Beyer, A.H., Millhone, R.S., and Foote, R.W. 1972. Flow Behavior of Foam as a Well Circulating Fluid. Paper SPE 3986 presented at the SPE Annual Meeting, San Antonio, Texas, 8–11 October. DOI: 10.2118/3986-MS.

Blauer, R.E., Mitchell, B.J., and Kohlhaas, C.A. 1974. Determination of Laminar, Turbulent, and Transitional Foam Flow Losses in Pipes. Paper SPE 4885 presented at the SPE California Regional Meeting, San Francisco, 4–5 April. DOI: 10.2118/4885-MS.

Bourgoyne, A.T., Millheim, K.K., Chenevert, M.E., and Young, S.F. 1986. *Applied Drilling Engineering*. Textbook Series, SPE, Richardson, Texas **2**.

Capo, J., Yu, M., Miska, S.Z., Takach, N., and Ahmed, R. 2006. Cuttings Transport With Aqueous Foam at Intermediate Inclined Wells. *SPEDC* **21** (2): 99–107. SPE-89534-PA. DOI: 10.2118/89534-PA.

Cawiezel, K.E. and Niles, T.D. 1987. Rheological Properties of Foam Fracturing Fluids Under Downhole Conditions. Paper SPE 16191 presented at the SPE Production Operations Symposium, Oklahoma City, Oklahoma, 8–10 March. DOI: 10.2118/16191-MS.

Chen, Z., Ahmed, R.M., Miska, S. et al. 2005a. Rheology Characterization of Polymer Drilling Foams Using a Novel Apparatus. *Annual Transactions of the Nordic Rheology Society* **13**.

Chen, Z., Ahmed, R.M., Miska, S.Z. et al. 2005b. Rheology of Aqueous Drilling Foam Using a Flow-Through Rotational Viscometer. Paper SPE 93431 presented at the SPE International Symposium on Oilfield Chemistry, Houston, 2–4 February. DOI: 10.2118/93431-MS.

Chen, Z., Ahmed, R.M., Miska, S.Z., Takach, N.E., Yu, M., and Pickell, M.B. 2007. Rheology and Hydraulics of Polymer (HEC)-Based Drilling Foams at Ambient Temperature Conditions. *SPEJ* **12** (1): 100–107. SPE-94273-PA. DOI: 10.2118/94273-PA.

Culen, M.S., Al-Harthi, S., and Hashimi, H. 2003. Omani Field Tests Compare UBD Favorably to Conventional Drilling. *World Oil* **224** (4): 27–30.

David, A. and Marsden, S.S. Jr. 1969. The Rheology of Foam. Paper SPE 2544 presented at the Fall Meeting of the Society of Petroleum Engineers of AIME, Denver, 28 September–1 October. DOI: 10.2118/2544-MS.

Debrégeas, G., Tabuteau, H., and di Meglio, J.-M. 2001. Deformation and Flow of a Two-Dimensional Foam Under Continuous Shear. *Physical Review Letter* **87** (17): 178,305–178,309. DOI: 10.1103/PhysRevLett.87.178305.

Dedegil, M.Y. 1987. Particle Drag Coefficient and Settling Velocity of Particles in Non-Newtonian Suspensions. *ASME Journal of Fluids Engineering* **109** (3): 319–323.

Devaul, T. and Coy, A. 2003. Underbalanced Horizontal Drilling Improves Productivity in Hugoton Field. *World Oil* **224** (4): 33–36.

Dodge, D.W. and Metzner, A.B. 1959. Turbulent Flow of Non-Newtonian Systems. *AIChE Journal* **5** (2): 189–204. DOI: 10.1002/aic.690050214.

Einstein, A. 1906. A new determination of molecular dimensions. *Ann. Phys* **19**: 289–306.

Gardiner, B.S., Dlugogorski, B.Z., and Jameson, G.J. 1998. Rheology of Fire-Fighting Foams. *Fire Safety Journal* **31** (1): 61–75. DOI: 10.1016/S0379-7112(97)00049-0.

Gopal, A.D. and Durian, D.J. 1998. Shear-Induced Melting of an Aqueous Foam. Paper O29.02 presented at the March Meeting of the American Physical Society, Los Angeles, 16–20 March.

Govier, G.W. and Aziz, K. 1972. *The Flow of Complex Mixtures in Pipes*. New York: Van Nostrand Reinhold.

Green, K., Johnson, P.G., and Hobberstad, R. 2003. Foam Cementing on the Eldfisk Field: A Case Study. Paper SPE 79912 presented at the SPE/IADC Drilling Conference, Amsterdam, 19–21 February. DOI: 10.2118/79912-MS.

Guo, B., Miska, S., and Hareland, G. 1995. A Simple Approach to Determination of Bottom Hole Pressure in Directional Foam Drilling. *Proc.*, 1995 ASME-ETCE Conference, Houston, Drilling Technology PD-vol. 65, 329–338.

Hanks, R.W. 1979. The Axial Laminar Flow of Yield-Pseudoplastic Fluids in a Concentric Annulus. *Industrial and Engineering Chemistry Process Design and Development* **18** (3): 488–493. DOI: 10.1021/i260071a024.

Harris, P.C. 1985. Effects of Texture on Rheology of Foam Fracturing Fluids. *SPEPE* **4** (3): 249–257. SPE-14257-PA. DOI: 10.2118/14257-PA.

Hatschek, E. 1911. Die viskosität der dispersoide. *Kolloid-Z* **8**: 34–39.

Heller, J.P. and Kuntamukkula, M.S. 1987. Critical Review of the Foam Rheology Literature. *Industrial and Engineering Chemistry Research* **26** (2): 318–325. DOI: 10.1021/ie00062a023.

Herzhaft, B., Toure, A., Bruni, F., and Saintpere, S. 2000. Aqueous Foams for Underbalanced Drilling: The Question of Solids. Paper SPE 62898 presented at the 2000 SPE Annual Technical Conference and Exhibition, Dallas, 1–4 October. DOI: 10.2118/62898-MS.

Holt, R.G. and McDaniel, J.G. 2000. Rheology of Foam Near the Order-Disorder Phase Transition. *Proc.*, Fifth Microgravity Fluid Physics and Transport Phenomena Conference, NASA Glenn Research Center, Cleveland, Ohio, USA, CP-2000-210470, 1006–1027.

Jastrzebski, Z.D. 1967. Entrance Effects and Wall Effects in an Extrusion Rheometer During the Flow of Concentrated Suspensions. *Industrial and Engineering Chemistry Fundamentals* **6** (3): 445–453. DOI: 10.1021/i160023a019.

Khade, S.D. and Shah, S.N. 2004. New Rheological Correlations for Guar Foam Fluids. *SPEPF* **19** (2): 77–85. SPE-88032-PA. DOI: 10.2118/88032-PA.

Kraynik, A.M. 1988. Foam flows. *Annual Review of Fluid Mechanics* **20**: 325–357. DOI: 10.1146/annurev.fl.20.010188.001545.

Krug, J.A. and Mitchell, B.J. 1972. Charts Help Find Volume, Pressure Needed for Foam Drilling. *Oil and Gas Journal* **70** (February 1972): 61–64.

Kuru, E., Miska, S., Pickell, M., Takach, N., and Volk, M. 1999. New Directions in Foam and Aerated Mud Research and Development. Paper SPE 53963 presented at the Latin American and Caribbean Petroleum Engineering Conference, Caracas, 21–23 April. DOI: 10.2118/53963-MS.

Kuru, E., Okunsebor, O.M., and Li, Y. 2005. Hydraulic Optimization of Foam Drilling for Maximum Drilling Rate in Vertical Wells. *SPEDC* **20** (4): 258–267. SPE-91610-PA. DOI: 10.2118/91610-PA.

Lauridsen, J., Twardos, M., and Dennin, M. 2002. Shear-Induced Stress Relaxation in a Two-Dimensional Wet Foam. *Physical Review Letter* **89** (9). DOI: 10.1103/PhysRevLett.89.098303.

Li, Y. and Kuru, E. 2003. Numerical Modeling of Cuttings Transport With Foam in Vertical Wells. Paper presented at the Petroleum Society's Canadian International Petroleum Conference 2003, Calgary, June 10–12.

Li, Y. and Kuru, E. 2004. Optimization of Hole Cleaning in Vertical Wells Using Foam. Paper SPE 86927 presented at the SPE International Thermal Operations and Heavy Oil Symposium and Western Regional Meeting, Bakersfield, California, 16–18 March. DOI: 10.2118/86927-MS.

Lord, D.L. 1981. Analysis of Dynamic and Static Foam Behavior. *JPT* **33** (1): 39–45. SPE-7927-PA. DOI: 10.2118/7927-PA.

Lourenço, A.M.F. 2002. Study of Foam Flow Under Simulated Downhole Conditions. MS thesis, University of Tulsa, Tulsa, Oklahoma.

Lourenço, A.M.F., Miska, S.Z., Reed, T.D., Pickell, M.B., and Takach, N.E. 2004. Study of the Effects of Pressure and Temperature on the Viscosity of Drilling Foams and Frictional Pressure Losses. *SPEDC* **19** (3): 139–146. SPE-84175-PA. DOI: 10.2118/84175-PA.

Lyons, W.C., Guo, B., and Seidel, F.A. 2000. *Air and Gas Drilling Manual*, second edition, Chap. 10, 1–84. New York: Professional Engineering Series, McGraw-Hill.

Macosko, C.W. 1994. *Rheology: Principles, Measurements, and Applications*. New York: Advances in Interfacial Engineering Series, Wiley-VCH.

Marsden, S.S. and Khan, S.A. 1966. The Flow of Foam Through Short Porous Media and Apparent Viscosity Measurements. *SPEJ* **6** (1): 17–25; *Trans.*, AIME, **237**. SPE-1319-PA. DOI: 10.2118/1319-PA.

Martins, A.L., Lourenço, A.M.F., and de Sa, C.H.M. 2001. Foam Property Requirements for Proper Hole Cleaning While Drilling Horizontal Wells in Underbalanced Conditions. *SPEDC* **16** (4): 195–200. SPE-74333-PA. DOI: 10.2118/74333-PA.

Minssieux, L. 1974. Oil Displacement by Foams in Relation to Their Physical Properties in Porous Media. *JPT* **26** (1): 100–108; *Trans.*, AIME, **257**. SPE-3991-PA. DOI: 10.2118/3991-PA.

Mitchell, B.J. 1971. Test Data Fill Theory Gap on Using Foam as a Drilling Fluid. *Oil and Gas Journal* **69** (September 1971): 96–100.

Mooney, M. 1931. Explicit Formulas for Slip and Fluidity. *Journal of Rheology* **2** (2): 210–222. DOI: 10.1122/1.2116364.

Morrison, I.D. and Ross, S. 1983. The Equation of State of a Foam. *Journal of Colloid and Interface Science* **95** (1): 97–101. DOI: 10.1016/0021-9797(83)90076-0.

Nishioka, G.M., Ross, S., and Kornbrekke, R.E. 1996. Fundamental Methods for Measuring Foam Stability. In *Foams*, Vol. 57, ed. R.K. Prud'homme and S.A. Khan, Chap. 6, 275–285. New York: Surfactant Science Series, Marcel Dekker.

Okpobiri, G.A. and Ikoku, C.U. 1986. Volumetric Requirements for Foam and Mist Drilling Operations. *SPEDE* **1** (1): 71–88; *Trans.*, AIME, **281**. SPE-11723-PA. DOI: 10.2118/11723-PA.

Özbayoğlu, E. 2002. Cuttings Transport With Foam in Horizontal and Highly-Inclined Wellbores. PhD dissertation, University of Tulsa, Tulsa, Oklahoma.

Özbayoğlu, E.M., Kuru, E., Miska, S., and Takach, N. 2002. A Comparative Study of Hydraulic Models for Foam Drilling. *J. Cdn. Pet. Tech.* **41** (6): 52–61.

Özbayoğlu, E.M., Miska, S.Z., Reed, T., and Takach, N. 2003. Cuttings Transport With Foam in Horizontal and Highly Inclined Wellbores. Paper SPE 79856 presented at the SPE/IADC Drilling Conference, Amsterdam, 19–21 February. DOI: 10.2118/79856-MS.

Özbayoğlu, E.M., Miska, S.Z., Takach, N., and Reed T. 2005. Using Foam in Horizontal Well Drilling: A Cuttings Transport Modeling Approach. *Journal of Petroleum Science and Engineering* **46** (4): 267–282.

Pal, R. 1999. Yield Stress and Viscoelastic Properties of High Internal Phase Ratio Emulsions. *Colloid & Polymer Science* **277** (6): 583–588. DOI: 10.1007/s003960050429.

Pickell, M.B. 2004. Preliminary Studies of Aqueous Foam for Transient Rheological and Texture Properties Using a Flow-Through Couette Viscometer. MS thesis, University of Tulsa, Tulsa, Oklahoma.

Pratt, E. and Dennin, M. 2003. Nonlinear Stress and Fluctuation Dynamics of Sheared Disordered Wet Foam. *Physical Review. E* **67** (5). DOI: 10.1103/PhysRevE.67.051402.

Princen, H.M., Aronson, M.P., and Moser, J.C. 1980. Highly Concentrated Emulsions. II. Real Systems. The Effect of Film Thickness and Contact Angle on the Volume Fraction in Creamed Emulsions. *Journal of Colloid and Interface Science* **75** (1): 246–270. DOI: 10.1016/0021-9797(80)90367-7.

Princen, H.M. and Kiss, A.D. 1989. Rheology of Foams and Highly Concentrated Emulsions IV. An Experimental Study of the Shear Viscosity and Yield Stress of Concentrated Emulsions. *Journal of Colloid and Interface Science* **128** (1): 177–187. DOI: 10.1016/0021-9797(89)90396-2.

Rand, P.B. and Kraynik, A.M. 1983. Drainage of Aqueous Foams: Generation-Pressure and Cell-Size Effects. *SPEJ* **23** (1): 152–154. SPE-10533-PA. DOI: 10.2118/10533-PA.

Reidenbach, V.G., Harris, P.C., Lee, Y.N., and Lord, D.L. 1986. Rheological Study of Foam Fracturing Fluids Using Nitrogen and Carbon Dioxide. *SPEPE* **1** (1): 31–41; *Trans.*, AIME, **281**. SPE-12026-PA. DOI: 10.2118/12026-PA.

Rojas, Y., Kakadjian, S., Aponte, A., Márquez, R., and Sánchez, G. 2001. Stability and Rheological Behavior of Aqueous Foams for Underbalanced Drilling. Paper SPE 64999 presented at the SPE International Symposium on Oilfield Chemistry, Houston, 13–16 February. DOI: 10.2118/64999-MS.

Saintpere, S., Herzhaft, B., Toure, A., and Jollet, S. 1999. Rheological Properties of Aqueous Foams for Underbalanced Drilling. Paper SPE 56633 presented at the SPE Annual Technical Conference and Exhibition, Houston, 3–6 October. DOI: 10.2118/56633-MS.

Saintpere, S., Marcillat, Y., Bruni, F., and Toure, A. 2000. Hole Cleaning Capabilities of Drilling Foams Compared to Conventional Fluids. Paper SPE 63049 presented at the SPE Annual Technical Conference and Exhibition, Dallas, 1–4 October. DOI: 10.2118/63049-MS.

Sanghani, V. and Ikoku, C.U. 1983. Rheology of Foam and Its Implications in Drilling and Cleanout Operations. Paper ASME AO-203 presented at the Energy-Sources Technology Conference and Exhibition, Houston, 30 January–3 February.

Santos, H., Rosa, F.S.N., and Cunha, J.C. 2003. Field Case History Shows Merit of UBD in Northeastern Brazil. *World Oil* **224** (5): 38–42.

Televantos, Y., Shook, C.A., Carleton, A., and Street, M. 1979. Flow of Slurries of Coarse Particles at High Solids Concentrations. *Canadian Journal of Chemical Engineering* **57**: 255–262.

Thondavadi, N.N. and Lemlich, R. 1985. Flow Properties of Foam With and Without Solid Particles. *Industrial and Engineering Chemistry Process Design and Development* **24** (3): 748–753. DOI: 10.1021/i200030a038.

Underbalanced Drilling Manual. 1997. Chicago: Gas Research Institute.

Valkó, P. and Economides, M.J. 1997. Foam Proppant Transport. *SPEPF* **12** (4): 244–249. SPE-27897-PA. DOI: 10.2118/27897-PA.

Washington, A. 2004. Preliminary Studies of the Rheology of Foam Using Rotational Viscometer. MS thesis, University of Tulsa, Tulsa, Oklahoma.

Yoshimura, A. and Prud'homme, R.K. 1988. Wall Slip Corrections for Couette and Parallel Disk Viscometers. *Journal of Rheology* **32** (1): 53–67. DOI: 10.1122/1.549963.

SI Metric Conversion Factors

cp	$\times$	1.0*	E $-$ 03 = Pa·s
ft	$\times$	3.048*	E $-$ 01 = m
ft^3	$\times$	2.831 685	E $-$ 02 = m^3
ft/hr	$\times$	8.466 667	E $-$ 05 = m/s
°F	$\times$	(°F $-$ 32)/1.8	= °C
°F	$\times$	(°F + 459.67)/1.8	= K
in.	$\times$	2.54*	E + 00 = cm
in.^{2}	$\times$	6.451 6*	E + 00 = cm^2
lbf	$\times$	4.448 222	E + 00 = N
lbm	$\times$	4.535 924	E $-$ 01 = kg
lbm/ft^3	$\times$	1.601 846	E + 01 = kg/m^3
lbm mol	$\times$	4.535 924	E $-$ 01 = kmol
mL	$\times$	1.0*	E + 00 = cm^3
psi	$\times$	6.894 757	E + 00 = kPa

*Conversion factor is exact.

Chapter 5

Geomechanics

5.1 Geomechanics and Wellbore Stability—John Cook, Schlumberger, and Stephen Edwards, BP

5.1.1 Introduction. From the very beginnings of the oil industry, the mechanical behavior of the formation has played an important role. After all, to drill a well, a large volume of rock has to be broken up and removed. A great deal of effort has been spent in designing drill bits to do this efficiently and rapidly, without wearing out or breaking too quickly.

More-widespread applications of rock-mechanics (or geomechanics) ideas arrived relatively recently, from the late 1970s onwards. This was probably brought about by the increase in the numbers of deviated and horizontal wells being drilled, the move of the drilling industry to more and more extreme environments (high temperatures, high pressures, high tectonic stresses), some high-profile compaction and subsidence problems, and the development of new generations of drilling rigs (especially offshore) whose costs were so high that time spent working pipe or sidetracking after hole-instability problems could no longer be tolerated.

Now, after roughly 30 years of development, geomechanics plays an accepted role in the oil and gas industry. Many companies now require screening or auditing of all proposed projects to examine the potential for costly drilling or production problems arising from the response of the rock.

This chapter aims to provide a concise description of the science behind petroleum geomechanics, the ways of collecting the data needed to carry out studies and bring benefits to oilfield operations, and the approaches that have been found effective both in planning and in implementation. Some of the material is a recapitulation of that found in Mitchell (2007). Other books that provide fundamentals, background, and more-advanced treatments of rock mechanics are Jaeger and Cook (1979), Fjær et al. (1992), and Charlez (1991, 1997).

Because this is a drilling textbook, the focus will be on wellbore instability. This is a major cause of lost time and equipment during drilling; estimates of its total cost to the industry vary but figures of USD 2 to 5 billion per year are widely quoted. These costs are obvious for a deepwater offshore rig, where every hour spent working the pipe to get the bottomhole assembly (BHA) past a tight spot might cost USD 10,000 or more (or roughly USD 3 per second). The costs are still present for low-cost onshore rigs, but more diluted; 20% lost time on a single well may not add up to many dollars on such a rig, but each rig drills many wells, and 20% lost time on a multiwell field-development program can make or break the project.

For example, in the Cusiana field in Colombia, very severe wellbore-instability problems were largely responsible for millions of dollars of additional costs per well. Many of the approaches described later in this chapter were applied to solve this problem and reduce the typical time to drill a well by a factor of approximately three, with accompanying cost savings (Last et al. 1995). In the Tullich field in the North Sea, careful planning and data acquisition prevented instability problems that had troubled nearby developments, and allowed optimization of the locations of wellheads and manifolds to minimize drilling problems (Russell et al. 2003).

There are two significant factors that have helped in successes like these. The first is a move away from prevention of rock failure toward management of its consequences. In the early days of wellbore-stability prediction, modeling often showed that a well might have no stable mud-weight window (i.e., no mud-weight value that could be chosen to avoid both instability and lost circulation). Nevertheless, the well would be drilled successfully. It is now accepted that rock failure does not necessarily mean wellbore failure, that traditional rock-mechanics approaches can be unduly conservative, and that even if large volumes of cavings are produced, careful hole-cleaning procedures can mitigate the problem and allow casing to be run. The second factor is the rapid development of logging while drilling (LWD), and in particular the ability to monitor near-bit annular pressure and hole geometry in real time as the hole is being drilled. This allows us to treat the wellbore as a mechanical experiment on the rock, under the most relevant conditions possible, and to use the results of the experiment to improve drilling conditions in the current well and in planning for the next one.

5.1.2 What Is Wellbore Instability? Wellbore instability includes the following phenomena:

- Breakage of intact rock around the wellbore because of high stresses generated there by the in-situ stress conditions or by sudden temperature changes
- Loosening of already-fractured rock around the wellbore
- Growth of fractures from the wellbore into the formation, sometimes with significant loss of drilling fluid
- Softening and breakage of the rock because of interactions with the drilling fluid
- Squeezing of soft rocks, such as salt, into the wellbore
- Activation of pre-existing faults that intersect the wellbore

The first five of these are very common. The effects range from negligible, through expensive, to terminal. Minor instability, such as a small amount of rock breaking off the wall and falling into the well, is rarely a problem during drilling and can be allowed for easily in logging or cementing. If the instability becomes worse, and significant volumes of cavings are generated by mechanical failure, they can pack around the drilling assembly and lead to excessive torque, drag, and bottomhole pressure, followed by twistoffs, stuck pipe, lost drilling assemblies, and sidetracks. The cavings can, of course, also be removed safely from the hole by good drilling practices and hole-cleaning procedures. This still may leave the enlarged hole as a major problem for the petrophysicist or cementing engineer, however. Failure of a zone of already-fractured rock may generate so many cavings, so rapidly, that even good hole-cleaning procedures are overwhelmed. Fracture growth, or flow into existing fractures, can result in the loss of large quantities of expensive drilling fluids, followed by loss of pressure control of the well. Many rocks—shales in particular—interact with the water in drilling fluids on short or long time scales, generating large volumes of softened and swollen material, or of harder, more-coherent fragments. Both of these situations can delay or prevent the running and cementing of casing to the required depth.

There are, of course, other ways that some of these operational problems can arise. Tortuosity from the directional-drilling process, keyseating, poor hole cleaning (of cuttings rather than cavings), differential sticking, equipment failure, and junk in hole can all cause stuck pipe. It is important to identify the mechanism of stuck pipe before trying to solve the problem. This chapter will, however, focus on wellbore instability as the mechanism of stuck pipe. Understanding and predicting wellbore instability is one of the key areas of petroleum geomechanics. Both the drilling process and the subsurface conditions play significant roles in wellbore instability, and so in problematic wells or areas, it is vital that the drilling engineering team and the geology or subsurface team understand something of each other's disciplines and communicate effectively.

5.1.3 Basics of Mechanics. Geomechanics requires some understanding of basic mechanics—in other words, stress, strain, and material response. Because we aim to deal with subsurface phenomena, the treatment must be 3D from the start, and this can cause confusion. If the following discussion confuses rather than clarifies, other approaches can be found in Fjær et al. (1992), Davis and Selvadurai (1996, 2002), Priest (1993), and many other engineering texts.

Stress. Stress is given by force divided by the area over which the force is applied. Force is a vector and has an orientation as well as a magnitude. Similarly, the area over which the force is applied also has an orientation and magnitude. This means that stress is not such a simple quantity. **Fig. 5.1.1** illustrates this. In Fig. 5.1.1a, a column with cross-sectional area A is loaded in compression by a weight F. The cross section of the column is subjected to a normal stress of F/A. Because the force acts normal to the surface we have chosen, there is no shear stress on that surface. In Fig. 5.1.1b, the same load F hangs from a hook glued to a wall over

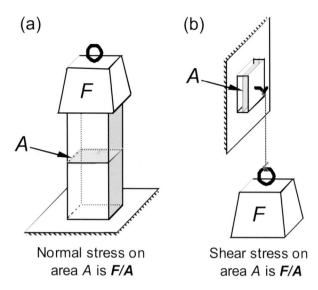

(a)   **(b)**

Normal stress on   Shear stress on
area *A* is ***F/A***   area *A* is ***F/A***

Fig. 5.1.1—Illustration of normal and shear stresses.

an area *A*. The indicated plane is subjected to a shear stress of *F/A*. The force acts parallel to the surface, so there is no normal stress acting on it. If in Fig. 5.1.1a we choose to examine a plane in the column that is not horizontal, there are components of force both normal and parallel to the surface, and so there are both normal and shear stresses acting on this plane.

This example illustrates two points that are true in general of stress:

- The relative magnitudes of shear and normal stress acting on a plane vary with the orientation of the plane.
- There are orientations of planes that have only normal stress acting on them.

As mentioned above, we must deal with 3D stresses. In general, a stress state has six independent components; the examples of Fig. 5.1.1 appear to have only one but this is because the others are all zero. Imagine a small cube drawn within the material of an engineering component—the wall of a drillpipe, say—or of a sandstone formation **(Fig. 5.1.2a)**. The loads imposed on the drillpipe by torque, mud pressure, or other sources, or on the sandstone by gravity and tectonics, generate stresses on the faces of the cube. Each opposite pair of faces will have a normal stress and two shear stresses acting on it, and the magnitudes of these are the six components mentioned above. (One normal stress and two shear stresses on three pairs of faces imply nine components rather than six, but rotational equilibrium means that only three of the shear stresses are independent.)

Now imagine drawing another cube in the same place but with a different orientation; the values of the six components will change, although the stress state itself does not. This is equivalent to changing the orientation of the plane in Fig. 5.1.1a; the loading on the system does not change, but the normal and shear stresses on the plane do. It can be shown that we can draw a cube in the material that has only normal stresses acting on its faces, just as we can find orientations of the plane in Fig. 5.1.1a with only normal stresses. The six components of the stress state are then the normal stresses acting on the three pairs of faces, and the orientation of the cube (Fig. 5.1.2b). The normal stresses in this geometry are called the principal stresses, and in petroleum geomechanics, as in most other branches of mechanics, specifying the values of the principal stresses and their orientations is the most common way of describing a stress state.

A quantity such as stress, that has rotational properties like this, is a variety of tensor, and the stress state is often referred to as the stress tensor. The cubes drawn in the material correspond to different sets of axes on which the components of the tensor are evaluated, and the process of finding the principal stresses is often called rotating the axes, or diagonalizing the stress. The directions of the principal stresses are called principal directions or principal axes; they are always mutually perpendicular.

The starting point in most geomechanics studies is the pre-existing or in-situ state of stress in the formations of interest. It should be clear from the preceeding discussion that in order to specify in-situ stress we

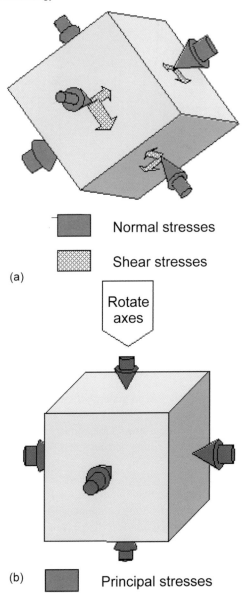

Normal stresses

Shear stresses

(a)

Rotate
axes

(b)   **Principal stresses**

Fig. 5.1.2—Components of the stress tensor, referred to (a) arbitrary axes and (b) principal axes.

need three magnitudes and some orientation information. Because in most areas of interest the overburden is a principal stress (i.e., one of the principal stresses is vertical), the in-situ stress state usually can be specified by three magnitudes and the direction of one of the horizontal stresses.

Around the wellbore, as we will see, the stress state may not be so simple, and complete specification of all six components is often needed. The units of stress are force/area; the most common units in the oil field are pounds per square inch (psi), and megapascals (MPa), where 1 MPa = 145.038 psi. Some areas use the unit *bar;* 1 bar is 0.1 MPa (approximately 1 standard atmosphere). In geomechanics, compressive stresses are denoted by positive numbers (in most other fields of mechanics, tensile stresses are given positive values). It is conventional in rock mechanics that the maximum compressive stress is called σ_1, the intermediate is σ_2, and the minimum compressive stress is σ_3.

Mohr's Circle. The Mohr's-circle diagram is a very common and useful way of illustrating the properties of stress. It is usually seen as a geometric construction involving two of the three principal stresses, and this is how we will start. The two principal-stress values are plotted along a horizontal line, and the point midway between them is found. A semicircle is then drawn centered on this point, with a diameter equal to the difference between the stresses. **Fig. 5.1.3a** shows this construction for principal-stress values of 10 and 40 MPa. Suppose in Fig. 5.1.3a that the 40-MPa principal stress lies along the *x*-axis in space, as shown, and the 10-MPa stress lies along the *y*-axis (the other principal stress must of course lie along the *z*-axis). All the planes that contain the *z*-axis map onto the circumference of the semicircle; the *y–z* plane maps to the 40-MPa point, and the *x–z* plane maps to the 10-MPa point. A plane at an angle *β* degrees to the *y–z* plane maps to a point 2*β* degrees around the circumference from the 40-MPa point, as shown in Fig. 5.1.3b. The Mohr's-circle construction tells us that the position of this point along the normal stress line gives the normal stress on that plane (in this case, approximately 28 MPa), and the vertical position gives the shear stress acting on that plane (in this case, approximately 14.5 MPa). The calculations that justify this are shown in Appendix A. Note that the two planes perpendicular to the principal-stress directions lie on the horizontal axis; they have no shear stress acting on them, as we expect.

Mohr's circle gives a graphical way of examining how normal and shear stresses on a plane vary with the orientation of that plane. This is particularly useful for rock mechanics, because, as we will see, the yield and failure of intact or fractured rock depends on these quantities. Mohr's circle allows us to predict when yield or failure will occur in an intact rock, and the orientation of the planes on which failure will occur, and also allows us to predict the stress levels under which further deformation will occur on a pre-existing fracture or fault.

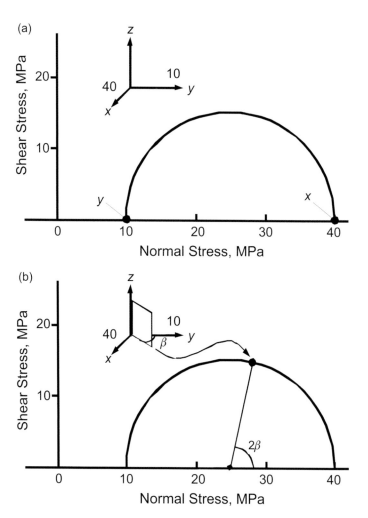

Fig. 5.1.3—Mohr's circle construction (a) for principal stresses of 10 and 40 MPa, and (b) with the point representing a plane at *β* degrees to the *y-z* plane.

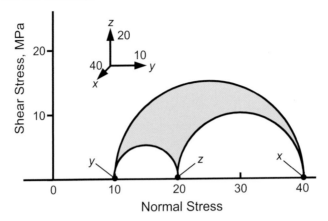

Fig. 5.1.4—Mohr's circle with all three principal stresses.

It is possible to include all three principal stresses in the Mohr's-circle diagram, and **Fig. 5.1.4** shows an example of this. The third principal stress is 20 MPa. A semicircle is drawn, following the same principles as in Fig. 5.1.3a, for each pair of principal stresses. Each semicircle represents the family of planes containing one principal-stress direction. The shaded area enclosed by the semicircles represents all other planes in the system; as in the 2D diagram, the position of the point along the horizontal axis gives the normal stress on the plane, and its height above the line gives the shear stress. The 3D diagram is particularly useful for examining the response of pre-existing faults and fractures to a change in the in-situ stress, caused by such factors as depletion, injection, and tectonism. The mapping of stress states into the 3D diagram is shown in Davis and Selvadurai (2002).

Effective Stress. If a material is porous, as are most rocks, its mechanical response is influenced not just by the stresses applied to it but also by the pressure of the fluid within its pores. If the applied stress and pressure change with time, the combined effects can be understood, for elastic materials at least, using poroelastic models, usually based on the work of Biot (1962). A great deal of effort has been devoted to obtaining mathematical solutions to Biot's equations that include the effects of temperature, anisotropy, chemical interactions, plasticity, and other factors, and these are useful in understanding the general principles of rock behavior. When stress and pressure are relatively steady, or when the permeability of the rock is so high that fluid-pressure gradients die away very rapidly, a simpler approach can be used, with the concept of effective stress.

The effective stress for a particular process, such as elastic deformation or failure, is the combination of stress and pore pressure that controls the process. Some of these processes will be discussed below, and the effective stress for each of them will be introduced.

Strain. Strain is a measure of the change in shape of a material in response to stress, or more formally the displacement gradient. Normal strains involve lengthening or shortening; the magnitude of the normal strain along a line in a material is defined as the change in length of the line divided by its original length. Shear strains result in changes in the angles between pairs of lines in the material; the magnitude of the (engineering) shear strain is the change in the angle between two mutually perpendicular lines in the material **(Fig. 5.1.5).** Most geomechanics only considers small (often called infinitesimal) strains—that is, magnitudes less than *approximately* 0.1. Large (often called finite) strains are not common in geomechanics but are important in some areas, for example in the description of salt movement. Their definition, notation, and relation to the stress tensor are complex and beyond the scope of this chapter; the interested (and committed) reader should consult advanced mechanics textbooks such as Billington and Tate (1980) or Spencer (1985). Strain has no units; as the definition shows, it is dimensionless.

Strain is a tensor quantity just as stress is, and as such it can be referred to different axis orientations. It is possible to find a set of axes where all strain components are normal rather than shear, and these are called the principal axes or directions; the normal strains, by analogy with stress, are called the principal strains. Wellbore-stability analysis, the focus of this chapter, rarely needs to calculate strains explicitly (because

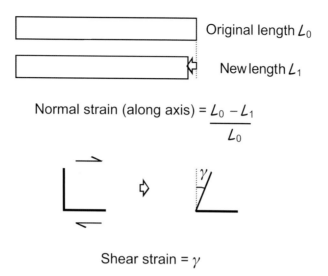

Normal strain (along axis) $= \dfrac{L_0 - L_1}{L_0}$

Shear strain $= \gamma$

Fig. 5.1.5—Simplified definitions of normal and shear strain.

rock-mechanics failure criteria are expressed in terms of stresses). The next section will mention strains in connection with elastic moduli, but after that we will not need them.

Material Response. Elasticity. The simplest link between stress and strain for a material is elastic behavior, where any strain or deformation is recovered on removal of the applied stress. The simplest form of elasticity is linear isotropic elasticity; linearity means that applying twice the stress generates twice the strain, and isotropy means that a given stress will generate the same strain levels whatever the orientation of the stress state relative to the material (i.e., the material looks the same in all directions).

If rock behavior is linear isotropic elastic, it can be described by two well-known elastic constants: Young's modulus and Poisson's ratio. Young's modulus relates stress and strain in a rod. If a compressive force F is applied along the axis of a rod with cross-sectional area A, the stress σ parallel to the axis is given by F/A (it is a normal stress and a principal stress in this case). If the initial length of the rod was L_0, and its new length is L_1, the normal strain ε parallel to the axis is $(L_0 - L_1)/L_0$ (using the compression-positive sign convention). The Young's modulus E is given by

$$E = \frac{\sigma}{\varepsilon} = \frac{F}{A} \frac{L_0}{\left(L_0 - L_1\right)} \quad \dots\dots\dots\dots\dots\dots\dots\dots\dots\dots\dots\dots\dots\dots\dots\dots\dots \quad (5.1.1)$$

and has the dimensions of stress.

The length of the cylinder decreases when the stress is applied, but its diameter increases, from d_0 to d_1, giving a lateral strain ε_l of $(d_0 - d_1)/d_0$ (note that this is negative for a diameter increase; it is a tensile strain). The Poisson's ratio v is the negative of the ratio of the lateral strain to the axial strain.

$$v = -\frac{(d_0 - d_1)}{d_0} \frac{L_0}{\left(L_0 - L_1\right)}. \quad \dots\dots\dots\dots\dots\dots\dots\dots\dots\dots\dots\dots\dots\dots\dots\dots \quad (5.1.2)$$

Thermodynamics dictates that Poisson's ratio lies between -1 and $+0.5$; in practice, it lies between 0 and 0.5, and between 0.2 and 0.4 for many rocks and other materials. It is dimensionless.

There are other ways of describing linear isotropic elasticity, each using two elastic constants. Apart from E and v, the other constants that are used are the bulk and shear moduli, the compressibility, and Lamé's parameters. Any two are enough to describe the behavior; the relationships between them are set out in Appendix B, together with the elastic stress/strain equations.

As mentioned before, elastic deformation responds both to applied stress and to pore-fluid pressure. This can be understood using an effective stress for elasticity. Effective stress will be denoted by σ' (rather than σ for total stress). The effective stress for elasticity, for a homogeneous linear material, is given by

$$\sigma' = \sigma - \left(1 - \frac{K_{\text{frame}}}{K_{\text{grain}}}\right) p_p, \quad \dots \dots \dots \dots \dots \dots \dots \dots \dots \dots \dots \dots \dots \dots \quad (5.1.3)$$

where K_{frame} and K_{grain} are the bulk moduli of the rock skeleton (in the absence of fluid) and the grains making up the rock, respectively, and p_p is the pore-fluid pressure. The term in parentheses is widely known as Biot's parameter, usually denoted by α. Biot's parameter approaches zero for stiff (usually strong) rocks and approaches unity for low-stiffness (usually weak) rocks. Because the bulk moduli can be obtained from sonic and mineralogical log interpretation, it is possible to use this approach to get more and more-precise values of α. This is not usually worthwhile, except for hard rocks that are to be hydraulically fractured. There are many uncertainties in geomechanical analysis, in the in-situ stresses and rock strength in particular, and efforts in a project should be concentrated on improving estimates of these rather than of α. For weak rocks, which is where most of our problems arise, it is reasonable practice to make a first assumption that $\alpha = 1$. (Elsewhere in this book, the symbol α is used to denote the coefficient of thermal expansion. It is so widely used in rock mechanics for Biot's coefficient, however, that we will do that in this chapter and will use α_T for the expansion coefficient.)

Unfortunately, rocks tend not to be isotropic linear elastic materials. Bedding and depositional fabric make them anisotropic; the almost universal presence of cracks and microcracks makes them nonlinear; common processes such as rock plasticity and wellbore instability are by definition nonelastic; and many formations are discontinuous (i.e., they are more or less fractured). Predicting the behavior [for example, the mud weight (MW) window] of rocks with discontinuities, anisotropy, or nonlinearity requires far more-complex models, and requires much more input data; most of the time this is not available and simple linear isotropic models are used instead. This is satisfactory in most situations, but in some cases it does not give a good prediction of behavior. In such cases, more-advanced modeling can help to illuminate the processes underlying the behavior, but it is not the entire solution. Progress (e.g., reduction of drilling problems) is more likely to be made when this theoretical illumination is combined with better diagnosis of the problems by data collection while drilling, and with modification of the drilling plan. This approach is discussed in more detail in Section 5.1.7.

Yield and Plasticity. When a rock or other material is stressed beyond its elastic limit, it may yield (i.e., undergo permanent, or plastic, deformation without breaking). When the material is unloaded, it does not return to its original shape. This happens very widely on a geological time scale, leading to folding, but can also happen in the shorter term. For example, many shales are brittle (i.e., they break abruptly with little or no plasticity) in conventional laboratory tests, but can yield and sustain high plastic strains without breaking when loaded very rapidly, as happens under the tooth of a roller-cone drill bit (Cook et al. 1991). Yield and plasticity in general are caused by shear stresses (i.e., they are favored by large differences between principal stresses). It follows that the action of yield or plasticity is to reduce shear stresses, and so to reduce the differences between principal stresses.

Yield and plasticity are important in the oilfield environment in

- Influencing the in-situ stress field
- Determining the stress field in and around salt
- Modifying the behavior of the tips of hydraulic fractures
- Modifying the stress field around wellbores and perforations

The last of these has received much attention, with the development of many analytic or numerical models for MW limits in the presence of plasticity. This is because the elastic models that we will deal with in this chapter have been found to be very conservative; they predict hole instability at a certain MW, but the hole can be drilled successfully below this MW. This is because plasticity strengthens the rock (in the same way that plasticity in a stainless steel strengthens), and also because it reduces the shear stresses in the borehole geometry.

Plasticity in a metal occurs when the shear stress in the metal is high enough to move imperfections in the crystal structure; this is expressed mathematically by yield criteria such as the Tresca criterion:

$$\sigma_1 - \sigma_3 > k. \quad \dots \dots \dots \dots \dots \dots \dots \dots \dots \dots \dots \dots \dots \dots \dots \dots \dots \quad (5.1.4)$$

This and other metal yield criteria such as the von Mises criterion are not sensitive to pressure; the metal responds only to the difference between the principal stresses, not to their absolute magnitude. Rocks behave differently; they usually become stronger as the mean stress level increases, and require a pressure-sensitive yield criterion. The most widely used is the Coulomb or Mohr-Coulomb (M-C) criterion, which is based on the frictional behavior of sliding surfaces. There are a number of ways of stating the M-C criterion. The most obvious physically is

$$\tau > S + \sigma'_n \tan\Phi, \qquad\qquad (5.1.5)$$

where τ is the shear stress needed to cause shear on a specified plane, S is the cohesion of the rock, σ'_n is the effective normal stress on the plane, and Φ is the angle of internal friction. The cohesion is the resistance to shear of the intact rock, and the second term is the enhancement to this resistance because of the friction caused by the effective normal stress on the plane. For intact rock there is no discontinuity on the plane, and so the conventional idea of friction is replaced by that of internal friction. Φ is typically 40 to 50° for sandstones, lower for shales.

The effective normal stress here is that appropriate to plasticity; it is not the same as the one we have already seen for elastic deformation. For plasticity, the effective stress is given by

$$\sigma' = \sigma - p_p. \qquad\qquad (5.1.6)$$

Note that there is no coefficient in front of the pore pressure.

Given a set of principal stresses, it is possible to find the shear and normal stresses on a particular plane (using Mohr's circle or the equations in Appendix A) and evaluate the M-C criterion to determine whether shear will occur on that plane. There is no guarantee, however, that this is the most highly stressed plane and that shear is not occurring in another orientation. There are various ways to overcome this difficulty. The first, which is less physical but very suitable for computation, is to rearrange the M-C criterion directly in terms of the effective principal stresses. This has a very simple form:

$$\sigma'_1 - N_\Phi \sigma'_3 > \text{UCS}, \qquad\qquad (5.1.7)$$

where UCS is the unconfined compressive strength (i.e., the compressive strength when the minimum effective stress is zero). The coefficient N_Φ does not have a commonly used name; it is given by

$$N_\Phi = \frac{1 + \sin\Phi}{1 - \sin\Phi}. \qquad\qquad (5.1.8)$$

For an internal-friction angle of 30°, $N_\Phi = 3$.

This arrangement of the M-C criterion allows one to say when yield will occur as the principal stresses are changed, but not the orientation of the plane on which it occurs. This can be found, however, using the Mohr's-circle construction. **Fig. 5.1.6a** shows the usual construction, but now an additional line is included that represents the yield criterion in Eq. 5.1.5 (with cohesion of 5 MPa and internal-friction angle of 30°). In Fig. 5.1.6b, the minimum principal stress has been decreased, so that the Mohr's circle now touches the yield-criterion line. On the plane that maps onto this point of contact, the shear and normal stresses satisfy the yield criterion. The angle labeled as 2β shows that the plane where this condition is satisfied lies at an angle β to the maximum principal stress.

Examination of Fig. 5.1.6 will show that, provided the internal-friction angle is greater than zero (i.e., the slope of the M-C-criterion line is positive), the angle β must be greater than 45°; in other words, the angle between the plane on which shear occurs and the maximum-stress direction must be less than 45°. This is a general principle and will be revisited in the discussions of failure and of understanding the stresses in the Earth's crust.

It is important to understand that the M-C criterion for yield is a model only; it is a workable approximation to the real behavior of rock, which is more complicated (as for elasticity). For example, the yield line in the Mohr's-circle diagram is frequently concave downward; the rate at which the shear strength increases with

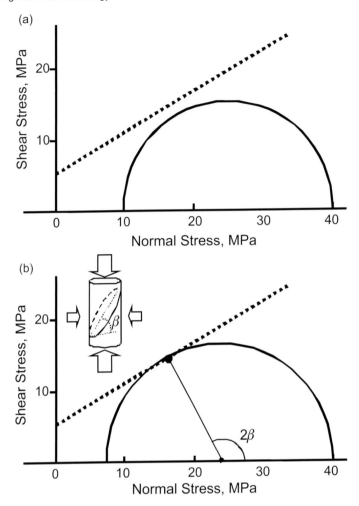

Fig. 5.1.6—The M-C criterion (dotted line) on the Mohr's circle diagram (a) before yield and (b) at the point of yield.

normal stress is much higher at low normal stresses than at high values. This will be discussed after the section on failure.

Failure. Elasticity and plasticity are straightforward to describe and define (although they can be difficult to model, of course). Failure, surprisingly, is more difficult. A piece of metal tested in tension may undergo small or large plastic strains before breaking into two pieces. A piece of copper pipe or wire can be bent into a right angle to fit its function; a large plastic strain is actually necessary for it to work properly. The same plastic strain in the structure of an aircraft would spell disaster. In the geomechanics world, a core of rock tested under confining pressure may break into two or more pieces and yet still support enough load to be a viable structure. So failure needs to be defined in terms of function; elastic and plastic deformations are characteristics of a material, but failure is a characteristic of an engineering structure or function. This difference is vital in the discussion of the prediction and consequences of wellbore instability.

Rock failures in compression and tension are very different. We will discuss compressive failure first. In spite of the complexities of defining failure discussed in the preceeding paragraph, most predictions of compressive failure in rock structures are made on the basis of the failure by shear of cylinders of rock under compression in laboratory tests. The failure point of such tests is usually taken as the maximum load supported by the sample, although it is quite common, as mentioned above, for the sample to continue to support a lower but still substantial load after failure. The most common type of test is carried out on a cylinder of rock under atmospheric pressure. The peak stress in such a test is the UCS. Because there is no lateral restraint in such a test, failure is often through axial splitting of the sample and is often violent, because of the sudden release of stored elastic energy. The so-called triaxial test is another common, but more complex, measurement. The core is surrounded by a flexible jacket, with two steel end plates, then placed in a pressure vessel. Confining pressure is applied by means of hydraulic oil, and then an additional axial stress is applied to deform and fail the sample. The axial and radial strains of the sample are usually measured. **Fig. 5.1.7** shows a typical output from a triaxial compression test on weak sandstone.

The axial-stress data in Fig. 5.1.7, starting from the left-hand corner, show an initial steep linear increase (elastic behavior). At approximately 75 MPa, the gradient decreases; this is the beginning of yield. At approximately 95 MPa, the stress drops abruptly; the peak stress here is usually regarded as the failure stress, or the strength. Note some further points about this figure:

- The test was continued beyond the peak stress, and the core continued to support a substantial stress (roughly half the peak stress in this case). This is called the residual strength, and it arises from the frictional strength of the fault that now crosses the sample, pressed together by a normal stress arising from the confining pressure. Some wellbore-stability models take this residual strength into account (Somerville and Smart 1991).
- The radial strain goes negative, slowly at first then more rapidly. The negative values mean that the sample diameter is increasing, at a rate given by Poisson's ratio. The rate increases when yield begins, because of the formation of microcracks in the sample.
- The volume strain is initially positive, then decreases again and becomes negative (i.e., the sample volume first decreases then increases, eventually exceeding the starting volume). This is known as dilatancy, and is again the consequence of microcracking, which introduces new open space into the sample.

In general, a higher confining pressure in such a test

- Increases the peak and residual stresses, at least until the sample begins to compact
- Inhibits microcracking, so that the onset of dilatancy is delayed until higher axial strains, or is suppressed altogether

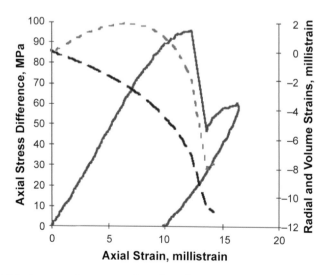

Fig. 5.1.7—Typical stress/strain curve for a core of weak rock tested in compression. In this case, the confining pressure (the pressure acting on the curved surface of the core) was 20 MPa. The horizontal axis is the axial strain in units of millistrain (10 millistrain = 1% shortening). The solid line is the additional stress acting on the flat faces of the core (the axial stress difference), plotted against the left vertical axis. The long- and short-dashed lines are the radial- and volume-strain curves respectively, plotted against the right vertical axis.

It is common practice to test a rock at a series of confining pressures, and then plot the peak stresses against confining pressure. This generates a failure envelope for the rock, such as the one shown in **Fig. 5.1.8.** This example illustrates both the higher strength of the rock as the confining pressure is increased and the reduced rate of increase at higher pressures. The M-C criterion shown for plasticity in Eq. 5.1.7 is also commonly applied to rock failure. The slope of the failure envelope can be used to estimate a value for N_ϕ or the angle of internal friction, which can then be used in predictions. Because the slope varies with confining pressure, it is wise to estimate its value at a pressure appropriate to the environment for which the prediction is being made.

As for elasticity and plasticity, compressive rock failure in reality can be much more complex than the behavior represented by the M-C criterion, and much more complex mathematical models can also be used to describe it. Complexities in the real behavior include the following:

- The effects of sample size. Larger samples tend to be weaker than smaller samples of the same rock.
- The effects of environment. Increasing water content in particular tends to reduce the strength of rocks, by attacking the cementation between the grains, or by lowering the resistance to crack propagation in the grains themselves (a form of stress corrosion).
- The intermediate stress. There is some evidence that the intermediate stress σ_2 also enters into the failure behavior. This is much more difficult to test experimentally, but a number of alternative failure criteria have been developed to deal with it, such as the Drucker-Prager and the Lade (Ewy 1999) criteria.
- Anisotropy of the rock can lead to failure by splitting along bedding planes. This can be very significant for drilling.
- The M-C criterion is derived from the geometry of the triaxial test. It does not necessarily apply well to other geometries, such as that of the wellbore.
- Finally, and perhaps most importantly, other failure modes. Tensile splitting under low confining stresses and compaction under high pressures are both important; compaction plays a major role in the production behavior of some oil reservoirs but is less important for drilling. These are the main reasons

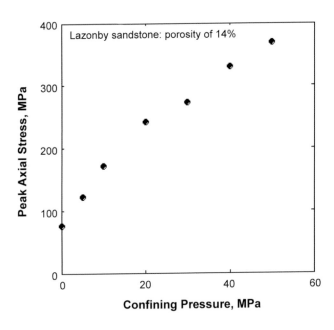

Fig. 5.1.8—Peak axial stress vs. confining pressure for seven triaxial compression tests on a sandstone. There are several ways to plot this type of data; the vertical and horizontal axes here represent the maximum and minimum compressive effective stresses, respectively (i.e., σ'_1 and σ'_3), so the slope of the line is N_ϕ from Eq. 5.1.7.

causing the yield or failure line in plots such as that in Fig. 5.1.8 to be concave downward. Tensile split-ting occurs at low confining pressures and is inhibited by small changes in confining pressure, leading to very rapid strengthening at low pressures (Ashby and Hallam 1986). Compaction occurs at high pressures—in general, the higher the porosity, the lower the pressure at which compaction begins (Wong et al. 1992)—and leads to complex behavior that shows itself as a downward curvature of the yield envelope. In some cases, the yield envelope curves all the way down to meet the pressure axis, forming a cap.

It is possible to construct predictive models incorporating most of these complexities, and each of them becomes significant in some environments. Nevertheless, simple models using the M-C criterion are the most commonly used. Reasons for this include

- Speed of computation
- Reduced input-data requirements
- The increasing ability to monitor the real state of the hole using LWD measurements
- The incompleteness of rock failure criteria as indicators of wellbore failure

Some of these reasons will be discussed in more detail later in this section, or elsewhere in the chapter.

Rock failure in tension is more straightforward than in compression. Rocks tend to be very weak in tension; most natural rock masses contain many pre-existing fractures, and so have zero tensile strength, and even in intact rocks, cracks propagate very easily. The accepted criterion for tensile failure is that the minimum effec-tive principal stress (i.e., the least compressive effective stress) becomes more negative than the tensile strength σ_t:

$$\sigma'_3 < \sigma_t. \qquad\qquad\qquad\qquad\qquad\qquad\qquad\qquad\qquad\qquad\qquad\qquad\qquad\qquad (5.1.9)$$

The tensile strength σ_t is numerically negative here, and may be zero.

Drained and Undrained Deformation. Both elastic and plastic deformation can change the volume of a sample of rock. Elastic deformation (by compressive stresses) always leads to a reduction in volume, but plastic deformation usually leads to an increase in volume (again, dilatancy), at least at low to moderate con-fining pressures. As mentioned before, plasticity under high confining pressures can lead to volume reduction (i.e., compaction).

Most of the volume decrease or increase is through changes in the pore volume of the rock (the volume of the grains does change, but this can be neglected except in very stiff rocks). If the fluid in the pores cannot move rapidly enough, reduction in pore volume leads to increasing fluid pressure, and dilatancy leads to decreasing fluid pressure. In a highly permeable rock, with a low-viscosity fluid, the fluid can move very rapidly, and any changes in fluid pressure are very short-lived. In a rock with low permeability, or with a high-viscosity fluid, the pressure changes may last a significant time. If the fluid movement and pressure decay are rapid compared to the rate of loading of the rock, the deformation is *drained*. If the fluid cannot move rapidly enough to eliminate the pressure changes, the deformation is *undrained*. Undrained deforma-tion can also be induced by barriers to fluid movement, such as a jacket around the sample. The variations in pore pressure caused by undrained loading can affect the mechanical response; for example, dilatancy in an undrained test leads to an increase in the effective confining pressure and so to an increase in the strength of the sample.

The major type of loading that we are concerned with in this chapter is, as we will see, the establishment of a stress concentration around the wellbore. The time scale for this is related to the rate of penetration (ROP) during drilling and to the bit size, and is not very rapid. In general, permeable rocks such as sand-stones, siltstones, and limestones undergo drained deformation during the drilling of the wellbore, and the pore-pressure changes can be neglected. (Note: It is important to distinguish between this effect of the defor-mation on the fluid pressure and the effects on the fluid pressure of fluid movement into or out of the well-bore.) Shales and mudstones, on the other hand, have such low permeabilities that their deformation around the wellbore is undrained. The pore-pressure variations induced by the changing stress around the well can persist for days or weeks and may influence the stability of the rock around the hole. This will be discussed further in Section 5.1.4.

Chemical Effects. Many rocks, shales in particular, interact with the water in drilling fluids on short or long time scales, generating large volumes of softened and swollen material, or of harder, more-coherent fragments. This is a very common problem and (until the use of oil-based drilling fluids became more common) was often regarded as the primary cause of wellbore instability. A vast amount of work has been connected to understand and prevent this type of instability, usually based on the contrasts in chemical activity between the ionic species present in the fluid and in the shale. It has proved to be very difficult to make quantitative predictions of chemical instability (for example, to predict the needed changes in mud density to prevent instability) or to predict the time before instability becomes intolerable. Part of this problem comes from the difficulty of characterizing the shales under investigation—they are exceptionally variable and intractable materials to work with—and part comes from the complexity of the coupling between the physical, mechanical, and chemical processes occurring during rock/fluid interaction. Because of the difficulty of quantitative prediction, problems of chemically induced instability are normally addressed by testing shale samples (sometimes under downhole conditions) against different drilling fluids developed by mud companies through chemical insights into the interaction mechanisms. A recent review is given by van Oort (2003); chemical effects will not be discussed in any more detail in this chapter.

This section has set out the basics of mechanics as applied to rocks. The next section will examine the response of a stressed rock to the drilling of a borehole.

5.1.4 Application to the Wellbore. Before the drill bit penetrates a region of rock, this region is subject to three in-situ principal stresses, which will generate shear stresses on planes within the rock. The rock is certainly strong enough to withstand these stresses; if this were not true, the rock would yield or fail until the differences between the in-situ stresses were too small to generate shear stresses exceeding the rock strength. When the drill bit has penetrated the rock, however, and passed some distance beyond our region of interest, a cylinder of rock has been replaced by a cylinder of fluid. Fluids cannot sustain shear stresses—effectively, they have zero shear strength. This difference between the original rock and the fluid leads to a redistribution of the stresses in the rock around the hole. This is known as a stress concentration and is common in mechanics (for example, around holes or cracks in stressed components).

Although the original rock could withstand the stresses within it, it may not be able to do this under the new distribution of stresses caused by the stress concentration, and it may yield, crack in tension, or fail. This is the major origin of wellbore instability. Understanding and predicting instability relies on calculation of the stress field around the hole, evaluation of the failure criteria, and in some cases evaluating the evolution of yield, damage, or failure around the hole.

Calculation of the stress field around a wellbore is not very difficult provided some specific conditions are met, but it rapidly becomes complicated if they are not. The conditions are as follows:

- The axis of the well lies along a principal-stress direction.
- The rock is permeable but there is no flow within the rock, nor in or out of the wellbore [e.g., fluid pressure is uniform within the formation, and there is an impermeable filter cake (in other words, the rock around the wellbore undergoes drained deformation)].
- The rock is continuous (i.e., not fractured), homogeneous, isotropic, and linearly elastic.
- The depth of interest is a long distance from the beginning or the end of the hole (i.e., the hole can be thought of as infinitely long).

We will discuss in detail the situation in which all these conditions are met, then relax the first condition, which is clearly important for deviated wells, and the second condition, which is important for shales and mudstones. Relaxing the third condition can introduce a great deal of complexity, which will be examined as needed. The fourth condition rarely needs to be relaxed, but when it must be (for example, to calculate failure around the bit), numerical modeling is needed. This situation is not covered.

Wellbore Stresses for Simple Geometry and Material. This assumes that all the simplifying conditions are met. Imagine a wellbore drilled in permeable, linear elastic rock, with a wellbore pressure p_w higher than pore pressure p_p, and a filter-cake-forming fluid in the well. It lies along one of the principal directions of the in-situ stress state in the formation—say, along the direction of σ_C, so that σ_A and σ_B are perpendicular to the well axis. We are describing the in-situ stresses like this to avoid using the subscripts 1, 2, and 3, which imply relative magnitudes, and which we will need later when we discuss failure. We will assume, though, that $\sigma_C > \sigma_A > \sigma_B$, so that this is the situation of a vertical well in a typical in-situ stress field. A long distance away from the wellbore, the principal stresses are undisturbed and lie along their in-situ directions. Because the well fluid cannot exert a shear stress on the wellbore wall, the only stress acting on the surface of the wellbore must be a principal stress, and the radial direction must therefore be a principal direction. This means the other principal stresses must lie parallel to the wellbore surface; for this simple geometry, symmetry demands that the other principal stresses lie axially and tangentially to the wellbore. In moving from the far field to the wellbore wall, the principal-stress directions rotate from their far-field values to values dictated by the wellbore geometry. Away from the well we can speak about σ_A, σ_B, and σ_C, but close to the well we generally speak about σ_r, σ_θ, and σ_z, the radial, tangential, and axial stresses, respectively. Tangential stress is often called hoop stress. We also have to specify the position around the wellbore, because the principal stresses vary around the perimeter; θ is the clockwise angle between the radial direction we are considering and the direction of σ_A. **Fig. 5.1.9** sums up this geometry.

The effective stresses around the hole are given by (Fjær et al. 1992)

$$\sigma_r' = \left(\frac{\sigma_A' + \sigma_B'}{2}\right)\left(1 - \frac{r_w^2}{r^2}\right) + \left(\frac{\sigma_A' - \sigma_B'}{2}\right)\left(1 - \frac{4r_w^2}{r^2} + \frac{3r_w^4}{r^4}\right)\cos 2\theta + \left(p_w - p_p\right)\frac{r_w^2}{r^2} \quad \cdots\cdots\cdots (5.1.10)$$

$$\sigma_\theta' = \left(\frac{\sigma_A' + \sigma_B'}{2}\right)\left(1 + \frac{r_w^2}{r^2}\right) - \left(\frac{\sigma_A' - \sigma_B'}{2}\right)\left(1 + \frac{3r_w^4}{r^4}\right)\cos 2\theta - \left(p_w - p_p\right)\frac{r_w^2}{r^2} \quad \cdots\cdots\cdots\cdots (5.1.11)$$

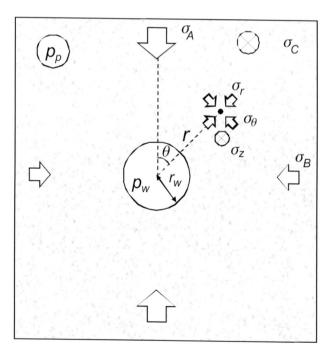

Fig. 5.1.9—Geometry for the wellbore stress concentration. The borehole, with radius r_w, lies along a principal-stress direction (in this case, the direction of σ_C). The order of stresses is $\sigma_C > \sigma_A > \sigma_B$. The stresses normal to the wellbore are σ_A and σ_B, the rock pore pressure is p_p, and the wellbore pressure is p_w. The angle around the wellbore is θ, and the normal stresses in the radial, tangential, and axial directions relative to the wellbore are σ_r, σ_θ, and σ_z, respectively. σ_r and σ_θ are principal stresses only at the wellbore wall and when $\theta = 0$, $\pi/2$, π, or $3\pi/2$.

$$\sigma_z' = \sigma_C' - 2v\left(\sigma_A' - \sigma_B'\right)\frac{r_w^2}{r^2}\cos 2\theta \quad \dotfill \quad (5.1.12)$$

$$\tau_{r\theta}' = \left(\frac{\sigma_A' - \sigma_B'}{2}\right)\left(1 + \frac{2r_w^2}{r^2} - \frac{3r_w^4}{r^4}\right)\sin 2\theta \quad \dotfill \quad (5.1.13)$$

The other two shear stresses, $\tau_{z\theta}$ and τ_{zr}, are zero. If the hole lies along one of the other principal-stress directions, the subscripts on the stresses on the right-hand sides of the Eqs. 5.1.10 through 5.1.13 should be changed to reflect this.

Remembering Eqs. 5.1.10 through 5.1.13 is rarely, if ever, necessary. Even for this most simple geometry, analysis is almost always carried out with a computer, and for deviated wells or more-complex materials, the equations are so lengthy that hand calculation is an invitation to make errors. Expressions for the maximum and minimum stresses around the hole are worth remembering, but, as will be seen, these are much simpler. The important points from the equations are as follows:

- The stress-concentration effect decreases rapidly away from the hole (because of the $1/r^2$ and $1/r^4$ terms).
- The stresses vary around the hole (because of the terms in θ).
- At the borehole wall, the radial effective stress is constant and equal to the overbalance.
- As summarized in **Fig. 5.1.10,** at the borehole wall, the tangential effective stress varies between a maximum value of $3\sigma_A' - \sigma_B' - (p_w - p_p)$ along the $\theta = \pi/2$, $3\pi/2$ directions (e.g., Position A in Fig. 5.1.10) and a minimum value of $3\sigma_B' - \sigma_A' - (p_w - p_p)$ along the $\theta = 0$, π directions (e.g., Position B in Fig. 5.1.10). As we will see, the maximum tangential stress often can be linked to breakouts, and the minimum tangential stress can be linked to drilling-induced fractures.

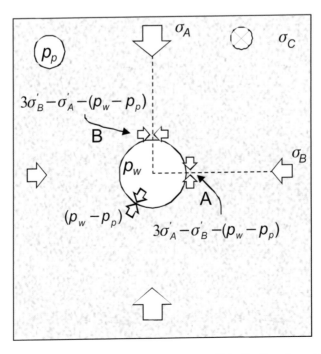

Fig. 5.1.10—Location and values of maximum and minimum tangential effective stresses, and the radial effective stress. For this discussion, we assume $\sigma_C > \sigma_A > \sigma_B$.

Wellbore Stresses for Deviated Wells. The equations for stresses around a deviated hole are too long to be written down conveniently or clearly. It is best to deal with the problem in the following stages, and to work in tensor notation, which we will not cover here (it is discussed briefly in Appendix A).

- The in-situ stress state is expressed in terms of three axes related to the wellbore geometry, usually the wellbore axis, a radius ending on the low side of the hole, and the vector normal to these two directions. The geographical orientation of the in-situ stress state is needed for this calculation, together with the deviation and azimuth of the wellbore.
- The equations of elasticity are solved to give the stress concentration around the wellbore. This is a more complex calculation than for the simple geometry because the far-field stresses referred to the wellbore-axis system now include shear stresses, and the shear stresses $\tau_z\theta$ and τ_{zr} are not necessarily zero.
- The principal stresses and directions are calculated by diagonalizing the stress tensor. This is made easier by the requirement that one principal-stress direction must still lie perpendicular to the wellbore surface, as for the simple geometry.

The key difference between the stress states for the deviated geometry and the simple geometry is that the two principal stresses that are parallel to the wellbore surface do not lie parallel to the well axis and perpendicular to it, respectively. This is illustrated in **Fig. 5.1.11** and leads to significant complications in the prediction and interpretation of failure patterns in deviated wellbores, obtained for example from image logs.

Low-Permeability Rocks and Flowing Formations. One of the assumptions in the above discussion is that the permeability of the rock is high. This means that fluid moves rapidly in response to pressure gradients within the rock. This is suitable for sandstones, siltstones, most carbonates, and other relatively permeable rocks (say, with permeability greater than 1 μDarcy) with water, gas, or light-to-medium oil in the pores. Unfortunately, most wellbore-instability problems occur in shales, which typically have extremely low permeability (on the order of 1 nDarcy, or 0.001 μDarcy), and here we cannot assume that deformation is drained, on a time scale of days or even weeks. The changes in stress around the wellbore can lead to significant changes in pore pressure, which persist for long periods, and should be taken into account in wellbore-stability calculations. The time-dependent nature of the stress state now means that the effective-stress approximation for elastic behavior can no longer be used, and the solution is much more complex. Solutions exist for a wellbore parallel to a principal-stress axis, when the other principal stresses are equal (Bratli et al. 1983) and when they are not (Detournay and Cheng 1988). The Bratli et al. (1983) solution is steady-state and, like solutions to the well-testing equations, assumes a boundary condition at a finite radius.

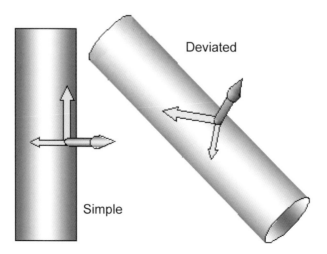

Fig. 5.1.11—Directions of the principal stresses at the wellbore wall for the simple geometry, and a possible set of directions for the deviated-well geometry.

The Detournay and Cheng (1988) solution is time-dependent, but the zero-time (completely undrained) solution can be used for the purposes of wellbore-stability prediction in shales.

More-Complex Material Properties. So far, we have considered rocks that are continuous, homogeneous, isotropic, and linearly elastic. Relaxing these conditions brings us to the boundaries of current knowledge in petroleum rock mechanics.

Modeling the stability of holes in fractured formations is well advanced in civil engineering for tunnels and mines, but here the formation is relatively easy to access, and there is usually plenty of time to collect detailed data; fracture distributions can be mapped and characterized if they are significant. In the oil field we do not have this luxury, and we need a generalized model for a fractured formation. Few models of this kind have been built (Santarelli et al. 1992b). Inhomogeneous formations (e.g., where one side of the borehole is in a different formation than the other side) present similar modeling problems. A few nonlinear elastic models have been generated and used (Santarelli 1987; Nawrocki et al. 1996), but they generally require more data than are easily available for operational use.

Anisotropic rocks have been studied in detail, and models are available to calculate the stresses around boreholes in such formations. We will restrict ourselves to transversely isotropic (TI) rocks, in which all directions lying in a single plane are equivalent, but directions lying out of that plane are not. This is by far the most common kind of anisotropy in oilfield rocks, because of the way that they are formed; the single plane generally corresponds to the bedding plane. The input data are more often available than, for example, those of nonlinear or fractured formations, because they can sometimes be generated from seismic or well-log data. Even for TI rocks, however, there are several levels of complexity, depending on the geometry (Aadnoy 1987):

- If the TI axis of the rock lies along a principal-stress direction, and the well axis is either parallel or perpendicular to the bedding plane, the stress-concentration solution is relatively straightforward and can be written down and calculated fairly easily.
- If the well is deviated with respect to the stress directions or the TI axis, the solution is much more complex.
- If the axis of the TI medium is not parallel to one of the axes of stress, and the well is also deviated, the solution is even more complex.

When the stresses around the hole have been obtained, a failure criterion appropriate for TI rocks must be used, which introduces one more level of difficulty. Several models for the stresses and failure modes around wellbores in TI rocks have been published, with different levels of this complexity (Aadnoy 1987, 1989; Amadei 1983; Willson et al. 1999; Ong and Roegiers 1993; Atkinson and Bradford 2001). Aadnoy (1987), Ong and Roegiers (1993), and Atkinson and Bradford (2001) deal with the complete problem, with general orientations of rock, stress field, and borehole; the others deal with special cases. Although the majority of rocks that we drill are anisotropic, and this property plainly plays an important role in much of the instability that is seen, it is still far from routine to use models such as those mentioned above to improve drilling performance. This is partly because of the intrinsic technical difficulty of dealing with anisotropic mechanics and partly because of our relatively poor understanding of borehole failure in anisotropic formations; but more importantly, it is because of the difficulty of acquiring good formation and stress data along the borehole trajectory to use in the models.

Plasticity around the hole has been studied very extensively. As one might expect, approaches range from simple to extremely complex. The most simple approach is for an isotropic rock, exhibiting perfect plasticity, with a borehole lying along one principal-stress direction, and with the other two principal stresses, normal to the borehole, being equal. Perfect or ideal plasticity is an idealized material response where the stress cannot exceed the yield stress—there is no hardening or strengthening—and the material can undergo infinite strain. The solution in this case is simple, relying on a combination of the stress-equilibrium equation in radial coordinates with the M-C criterion. The algebra is lengthy, however, and so the reader is referred to the literature (Somerville and Smart 1991; Bratli and Rinses 1981; Brady and Brown 1993).

There is an enormous array of more-complicated plasticity solutions in the literature, including the effects of hardening, anisotropic stresses, residual strength, deviated holes, other yield criteria, finite strains, and softening. The last of these is particularly interesting because it represents a step toward failure of the rock, by generating localized bands of intense deformation. Calculations like this are complex and time-consuming, and they require information on the nature of the material in the bands. It is possible to predict the onset and propagation of shear bands around the wellbore (Papanastasiou and Vardoulakis 1994), but this is not really feasible in the time- and data-constrained drilling environment. These approaches are used more widely in the completions area, for the prediction of sand production (van den Hoek et al. 2000).

This section has dealt with calculation of the stress field around the wellbore. The next section will examine how the stress field can cause failure of the rock around the hole, and potentially of the hole itself.

Wellbore Failure. In the section on material properties, the idea was introduced of failure as loss of function rather than as a material property. This is very important in discussing wellbore problems caused by rock failure. **Fig. 5.1.12** illustrates this for a few example situations. In each case, the diagram represents a cross section through a well; the dashed or solid circles represent the as-drilled hole size.

Hole 1 is a perfect in-gauge hole, such as might be drilled in a very strong rock under low in-situ stress conditions. Hole 2 has a pair of regions on either side where the stresses have exceeded the yield criterion for the rock and some plasticity has occurred (the factors controlling the occurrence and location of regions like this will be discussed later in this section). Although the rock is no longer behaving elastically, no faults or shear planes have appeared. Has the hole failed as a piece of engineering? Clearly not; this hole would not present a problem to the driller, the cementing engineer, or the petrophysicist. In fact, an appropriate logging tool might detect the plastic zones and so give more information about stress state or rock strength than for a hole resembling Hole 1.

In Hole 3, the deformation process seen in Hole 2 has become more severe; shear planes have formed, pieces of rock have dropped from the wellbore wall into the drilling fluid, and they have been swept up the annulus to the shale shakers. The rock has clearly failed, but has the wellbore? The BHA can still move up and down the hole, so from the driller's point of view, there is no serious problem. The cementing engineer and petrophysicist will have to take care, but once again an appropriate logging campaign will gather useful information. The only way this could be a serious problem is if a long section of failed rock suddenly falls out and drops down the well onto the top of the BHA.

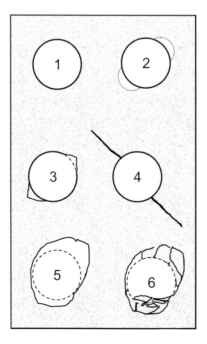

Fig. 5.1.12—Some examples of rock failure around the wellbore. See text for details.

In Hole 4, the rock around the hole has failed in tension, and a short fracture has formed on each side of the hole. It has not propagated far away from the hole, and so lost circulation has not occurred; reasons for this will be discussed subsequently in detail, but this behavior is favored by high mud density (high borehole pressure). Some fluid, however, has leaked into the fractures. Because the fracture volume grows as the hole advances, this would not be noticeable as a rapid mud loss. Over a few hundred meters of well, however, a few barrels of mud might have been lost into the fractures. Although the rock has failed, the borehole is still functional—there is no barrier to drilling or logging. Great care needs to be taken with this hole, however. If the mud pumps are switched off for some reason, or circulation rate is reduced, and the bottomhole pressure drops as a result, the fractures along the well may close, expelling fluid into the wellbore and up to the surface. Here it may be seen as the first sign of a kick, to which the response may be an increase in mud density. This, of course, will make the problem worse, and may lead to genuine lost circulation. Knowing that this kind of fracturing is happening, or even that it is possible in a given well geometry, is important in avoiding serious problems.

Finally, Holes 5 and 6 show situations where the rock has undergone gross failure around the hole; perhaps the in-situ stresses are very high, or the rock is very weak, but large volumes of rock are broken and many cavings are being generated. In Hole 5, the mud is performing well, good hole-cleaning practices are being followed, and the cavings have been swept out of the well. Although logging this hole would present some problems, and cementing will need some care, it would be possible to get the casing down to the section total depth (TD), and the well could be a reasonable success. In Hole 6, the hole has not been properly cleaned, cavings remain in the well, and the BHA or casing is unlikely to be able to be tripped past them, up or down. This well has failed. The key point of these examples is that the rock-mechanics aspects (strength, in-situ stresses, and wellbore-pressure regime) could have been identical for Holes 5 and 6. What has made the difference between a moderate success and a failure was drilling practice alone. It is important to keep sight of the fact that drilling practices have much more impact on success or failure of the hole than do better rock-mechanics models or even better input data. Nevertheless, rock-mechanics modeling can be a major factor in tilting the balance toward success, providing a strategic framework for drilling the well, tactics for problem avoidance and mitigation, and a rational basis for diagnosis if problems do emerge. We will now go on to examine how the stresses around the well lead to the kinds of rock failure seen in Fig. 5.1.12.

Wellbore Stresses and Simple Failure Modes. We will start with the simplest case, as usual. Eqs. 5.1.10 through 5.1.13 describe the effective stress state around a wellbore lying along one of the principal-stress directions, in an isotropic permeable rock, with a filter cake. Fig. 5.1.10 summarizes the values at the wall of the hole, which are what we need, and also mentions a restriction on the in-situ stress state, namely that $\sigma_C > \sigma_A > \sigma_B$. We will use the stress values near the wellbore wall in Eqs. 5.1.7 and 5.1.9 to predict shear and tensile failures of the rock, respectively.

The two most common types of rock failure around the borehole are

- Shear failure when the tangential stress is the maximum principal stress σ_1 and the radial stress is the minimum principal stress σ_3 (bear in mind that the shear failure criterion involves two principal stresses), which can lead to the well-known breakout geometry
- Tensile failure when the tangential stress is more tensile than the tensile strength of the rock

The M-C criterion is $\sigma_1' - N_\Phi \sigma_3' > \text{UCS}$. Incorporating the values of tangential and radial principal stresses from Position A in Fig. 5.1.10 in this criterion gives

$$3\sigma_A' - \sigma_B' - (N_\Phi + 1)(p_w - p_p) > \text{UCS} \quad \dots\dots\dots\dots\dots\dots\dots\dots\dots\dots\dots\dots \quad (5.1.14)$$

or

$$p_w < \frac{3\sigma_A' - \sigma_B' - \text{UCS}}{(N_\Phi + 1)} + p_p \quad \dots\dots\dots\dots\dots\dots\dots\dots\dots\dots\dots\dots\dots\dots\dots\dots\dots\dots \quad (5.1.15)$$

If the pressure in the wellbore falls below this value, the rock at Position A in Fig. 5.1.10 is predicted to undergo shear failure, and possibly form breakouts. Much of wellbore-stability planning is concerned with calculating this minimum mud pressure, in order to avoid shear failure. The minimum mud pressure needed is increased by a high difference between the principal stresses normal to the hole (σ_A and σ_B in this example), low rock strength (UCS), high pore pressure, and low rock internal-friction angle.

The failure mode described by Eq. 5.1.15 can lead to the breakout geometry shown in Hole 3 of Fig. 5.1.12 (the sheared material can also be lost from the lower breakout, of course). In this example, the breakouts lie on a diameter approximately 45° clockwise from vertical, and this implies that the maximum stress in the plane of the paper lies at approximately 45° counterclockwise from vertical (i.e., perpendicular to the line joining the two breakout lobes).

The criterion for tensile failure is given by Eq. 5.1.9:

$$\sigma_3' < \sigma_t$$

Incorporating this into the tensile failure criterion gives

$$3\sigma_B' - \sigma_A' - (p_w - p_p) < \sigma_t \quad \dots\dots\dots\dots\dots\dots\dots\dots\dots\dots\dots \quad (5.1.16)$$

or

$$p_w > 3\sigma_B' - \sigma_A' + p_p - \sigma_t \quad \dots\dots\dots\dots\dots\dots\dots\dots\dots\dots \quad (5.1.17)$$

If the wellbore pressure rises above this value, a crack is predicted to form at Point B in Fig. 5.1.10. We will discuss further propagation of the crack later. So tensile failure of the wellbore wall is promoted by high wellbore pressure, large difference between the principal stresses normal to the wellbore, low pore pressure, and low tensile strength. (Remember that in our compression-positive sign convention, σ_t is a negative number and becomes more negative if the rock is stronger.)

The fractures initiated at this pressure may or may not propagate away from the wellbore; this is discussed in the next section. Because the fracture initiation depends on the presence of the wellbore, these fractures are called drilling-induced fractures; they are not necessarily a problem and can be very useful in clarifying stress states and directions.

In these examples, the stress regime has been chosen so that these examples work correctly. In the more general case, because the three principal stresses vary with position around the wellbore, locating the positions of potential failure, and the pressures at which they occur, requires checking all the combinations of principal stresses all around the well. It is feasible to do this manually for a single depth in a simple well, but for a range of depths, or for deviated wells, it is time-consuming, and a computer should be used.

Other Failure Modes. The last section dealt with the two most common kinds of rock failure around the hole, shear failure from the combination of tangential and radial principal stresses and tensile failure from a reduced tangential principal stress. There are many other ways that the rock can fail. They are discussed below briefly, and some of them are treated in more detail in Section 5.1.8.

Other Shear-Failure Modes. Shear failure can also be induced by other combinations of the three principal stresses around the hole. For example, if the axial effective stress is the maximum and the radial effective stress is the minimum, and together they exceed the M-C criterion, the rock can fail with a geometry that does not generate breakouts immediately. The geometry of the predicted failure planes is shown schematically in **Fig. 5.1.13** (it is not so easy to define as are breakouts because similar patterns of shear planes should form at every point along the wellbore where the conditions are satisfied, and these patterns interfere with one another). Because the axial stress can vary with azimuth around the hole, this type of failure can be localized on opposite sides of a borehole, just as for conventional breakouts.

Similarly, other combinations of the three principal stresses can lead to different geometries of shear failure. The possibilities are cataloged and illustrated in Bratton et al. (1999) and can be recognized in some

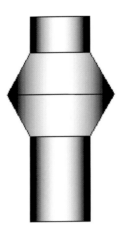

Fig. 5.1.13—Predicted geometry of shear planes around the well when failure is caused by axial stress as the maximum and radial stress as the minimum.

circumstances in borehole images. Calculating by hand whether each mode will occur is a lengthy process, but most software packages for wellbore-stability prediction will do this automatically.

Lost Circulation. The tensile failure criterion discussed above (Eq. 5.1.17) is for initiation of a drilling-induced fracture at the wellbore wall. There is nothing in this criterion about fracture propagation away from the wellbore. Fractures that do not propagate accept relatively small quantities of fluid, and do even this slowly, as the hole advances; they do not generally lead to lost-circulation incidents. Lost circulation occurs when a significant fracture propagates away from the wellbore, or when the wellbore intersects a conductive natural fracture, or when a drilling-induced fracture connects with a conductive natural fracture (there are other causes of lost circulation, of course, such as streaks of extremely high permeability, or vugular porosity, but these are not influenced by geomechanics). The theory and practice of hydraulic fracturing tells us that propagation occurs only (for a nonpenetrating fluid such as drilling fluid) when the pressure of the fluid in the fracture is above the minimum total stress in the formation. This is usually but not always the minimum horizontal stress. So drilling-induced fractures occur at a wellbore pressure given by Eq. 5.1.17 (or its equivalent for deviated wells), and fracture propagation occurs when wellbore pressure exceeds the minimum formation stress. Clearly, the propagation pressure can be above or below the initiation pressure.

If the initiation pressure is below the propagation pressure (which is favored by a large difference between the principal stresses perpendicular to the well), slowly increasing the wellbore pressure will lead to stable initiation of a fracture at the wellbore wall, followed by its growth, until the wellbore pressure reaches the fracture-propagation pressure, when large amounts of fluid will suddenly be lost. If the initiation pressure is above the propagation pressure, on the other hand, increasing the wellbore pressure will lead to initiation of a drilling-induced fracture followed by immediate propagation and loss of fluid. Under these conditions, it is unwise to use the higher initiation pressure as the lost-circulation pressure (i.e., to use it as the upper MW limit). Although a fracture cannot propagate unless it initiates first, formations are full of natural fractures that act as good starting points for hydraulic-fracture propagation and remove the need for initiation by a drilling-induced fracture. Even if a leakoff or casing-shoe test may indicate a high fracture-initiation pressure and an intact, uncracked interval of the formation, there is no guarantee that the next 100 ft of drilling, or indeed the next foot, will not intersect an unfavorably oriented fracture and lose its fluid because the pressure is higher than the minimum in-situ stress. The minimum in-situ stress is the safe and conservative value for the maximum wellbore pressure.

The differences between fracture initiation and propagation are discussed further in Section 5.1.6, with reference to interpretation of LOTs.

Having said all this, there are many examples of successful drilling with higher wellbore pressures, for example using lost-circulation additives, and especially in depleted reservoirs where the minimum in-situ stress has dropped along with the formation pressure. The ways in which this might happen, through use of drilling-fluid additives to plug fractures, are discussed in Sections 5.1.5 and 5.1.7.

Natural Fractures and Fractured Formations. Nearly all formations contain fractures, but mechanical behavior dominated by fractures is, fortunately, rare.

One case was mentioned in the preceding text in the context of lost circulation. If the wellbore meets a natural fracture that is propped open, perhaps because it has been sheared after opening, then whole mud can be lost down the fracture, and severe lost circulation is the result. Unless the well is drilled with total losses, the maximum wellbore pressure that can be applied in these circumstances is the formation pressure because this is the natural pressure in the fracture. If there is a risk of kicks or blowouts because of this, the well becomes very difficult to drill; either underbalanced drilling or mudcap drilling is needed.

If the well meets a natural fracture that is closed, mud loss will occur when the fracture opens. If the fracture is oriented perpendicular to the minimum in-situ stress, this will occur when the mud pressure equals this stress value. If the fracture is not oriented in this way, the mud pressure needed to open it will be higher.

Some formations are heavily fractured; as a rule of thumb, if the fracture spacing is comparable to the wellbore diameter, the behavior will begin to be dominated by the fractures. The chances then are that some of the fractures will be unfavorably oriented, and so the lost-circulation pressure will be the minimum in-situ stress. There are other consequences, however. Even if whole mud does not leak into the fracture network, filtrate may do so and, in an impermeable formation such as a shale, start to reduce the effective normal stress acting on the faces between blocks of rock around the wellbore. This reduces the frictional forces between them and the mechanical stability of the arrangement. Swabbing the well during a trip, or backreaming, or rough drilling, can then break up the arrangement of blocks, and allow them to fall into the well. This can often cause unmanageable instability problems. Under these circumstances, it is vital to minimize fluid invasion, by keeping the static mud density as low as possible consistent with stability in other areas, and keeping dynamic changes in wellbore pressure as small as possible.

Several authors have constructed numerical models for wellbore instability in fractured formations (Santarelli et al. 1992a, 1992b; Zhang et al. 1999; Chen et al. 2002). These are difficult models to generalize, because they rely on the input of specific fracture geometries. They all agree, however, on the importance of minimizing mud infiltration.

Bedding-Plane Effects. In fissile shales (i.e., shales that split along their bedding planes), a different failure mode emerges when the hole is nearly parallel to the bedding (Okland and Cook 1998). Increasing deviations through the shale may lead to gradual changes in the stability of the hole, but when the axis of the hole comes within a few degrees (10 to 15°) of being parallel to bedding, the hole is liable to catastrophic collapse. This is not predicted by the failure criteria discussed previously, but it seems to be similar to the roof-collapse failures observed in mines and tunnels. It appears to be very difficult to manage, and the best approach during planning is probably to ensure that the well never gets near this critical deviation through a susceptible shale bed (Edwards et al. 2004). If the geometry of the target reservoir requires this deviation, it is advisable to put a kink in the well to go through the shale at a higher deviation, accepting the additional torque and drag that this generates.

Thermal Effects. Rocks, just like other materials, expand when they are heated. In the wellbore wall, this expansion is largely prevented by the rest of the formation, and so the increase in temperature leads to an increase in compressive stress. It also leads to an increase in pore pressure, because the pore fluid also expands. If the rock is relatively permeable, this extra pore pressure drains away rapidly, but in a shale it can persist around the wellbore for hours to days.

The full solution of this coupled problem of fluid and heat flow is fairly complex, especially as the wellbore temperature (which is a boundary condition for the problem) is rarely constant in time. For the purposes of wellbore-stability prediction and management, however, we will as usual adopt a simple approach. Neglecting the pore-pressure effects, the additional tangential stress $\Delta\sigma$ generated by the temperature rise is

$$\Delta\sigma = \frac{E\,\alpha_T}{1-\nu}\Delta T, \qquad\qquad\qquad (5.1.18)$$

where E is Young's modulus, α_T is the coefficient of thermal expansion of the rock, v is Poisson's ratio, and ΔT is the temperature change.

Expansion coefficients of most solids are approximately $10^{-5}/°C$, and most Poisson's ratios are approximately 0.33, so this equation can be reduced to

$$\Delta\sigma \approx 1.5.10^{-5} E \Delta T. \quad\dots\dots\dots\dots\dots\dots\dots\dots\dots\dots\dots\dots\dots\dots (5.1.19)$$

In other words, 1°C generates the stress corresponding to 0.0015% strain in the rock. For a rock with a modulus of 10 GPa, a temperature change of 10°C generates 1.5 MPa of thermal stress. Eq. 5.1.19 tells us that thermal effects are unlikely to be important unless the temperature changes are large, and the rock is stiff (i.e., has a high Young's modulus). Large temperature changes can easily occur in a well, for example when circulation is resumed after a break of a few hours in a deep well. Cold mud from the top of the well is put suddenly in contact with hot rock at the bottom, reducing the compression there, and hot mud from the bottom of the well is put in contact with cold rock at the top, increasing the compression. Because the upper sections of the hole are usually cased, the dominant effect will be the cooling of the hot openhole sections at the bottom, and possible initiation of drilling-induced fractures there. These will not, however, lead to lost circulation unless the static MW is already greater than the fracture gradient.

If some of the formations are unusually stiff, or if the Geothermal gradient is unusually high, it is worthwhile to include thermal stresses in a wellbore stability analysis; most software packages have this capability, and very few additional data are needed. Although thermal effects are unlikely to initiate real problems that were not already present, they can help to explain features such as drilling-induced fractures or minor hole enlargement seen under unexpected conditions. There have been cases where the mud has been deliberately cooled to improve shear stability (Maury and Guenot 1995), but the biggest advantage that the authors found from doing this was to increase the operating life of the LWD tools in the hole, rather than improving wellbore stability.

5.1.5 State of Stress in the Subsurface. The state of stress in the Earth is a vital ingredient in wellbore-stability planning and other geomechanical projects. This section describes the origins and variation of the stress in the Earth. Methods of estimating the stress state in order to build a mechanical Earth model (MEM), will be covered in Section 5.1.6.

Origins of Stress. The state of stress in the subsurface has been the subject of much research and debate. Later in this chapter it will become clear that it is the key to the prediction of MWs required for wellbore stability. However, it is difficult to measure stress directly in the Earth (i.e., in situ). Some in-situ measurement techniques (e.g., hydraulic-fracturing stress measurement and the XLOT—Section 5.1.6) do exist. However, estimates of stress in the subsurface are more often based on indirect observations of the consequences of stress (e.g., movement of faults or deformation of boreholes) and on modeling.

Interest in the stresses in the Earth is widereaching and encompasses many areas outside of the oil and gas industry. Much work has come from the global plate-tectonics and earthquake-prediction community, who look at crustal-scale processes. Other areas where Earth stresses have been studied traditionally are the tunneling and mining industries where projects are usually in *hard* (i.e., crystalline, igneous, or metamorphic) rock environments. So, before delving into the wealth of ideas on the origins of subsurface stress it is important to step back and consider the geological settings in which we are primarily involved within the petroleum industry.

Hydrocarbons are generated from organic-rich material that is formed at the Earth's surface and deposited in sedimentary basins. Therefore, the vast majority of oil and gas wells are drilled in sedimentary basins. Sedimentary basins form where the Earth's crust has subsided below the elevation of surrounding land. The basin is filled by sediment eroded from the surrounding high ground, washed downhill, and deposited. When the Earth's outer rigid layer (lithosphere) is viewed as a whole, sedimentary basins typically form only a relatively thin veneer on top of the thicker, harder basement rocks.

Fig. 5.1.14 shows a cross section through the Earth's crust. The sedimentary basin is seen to be a relatively thin skin lying on top of the basement. Sedimentary-basin penetrations are typically only up to 4 or 5 km deep, whereas the Earth's lithosphere is typically on the order of several tens to 100 km thick.

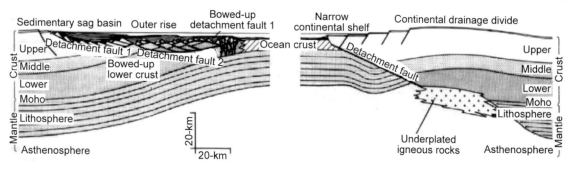

Fig. 5.1.14—Examples of sedimentary basins in a continental-margin setting. The thickness of the sedimentary basin is small in comparison to that of the lithosphere as a whole. From Busby and Ingersoll (1995).

It is the larger-scale deformation of the basement in response to lithospheric-scale processes such as plate tectonics that create the low spots in which sedimentary basins form. In a young basin, however, the sediments may sit quite passively within the low spot and be largely uncoupled from the larger-scale lithospheric processes going on around them. Characteristics of Earth stress seen in crustal-scale and hard-rock studies may therefore not be as relevant to these relatively young sedimentary environments, typical of many of the younger basins (mainly Tertiary—i.e., between 0 and 60 million years old) in which hydrocarbons are produced today. Examples include the Gulf of Mexico, much of the central and northern North Sea, the West Coast of Africa, the Nile delta, and much of the offshore Trinidad area. Where a basin sits passively on top of basement, the predominant force acting on the sediments is gravity. In the following section, we will consider how gravity and the weight of the sediments themselves generate the majority of the in-situ stresses that are experienced when drilling in these environments.

With the passing of time, sediments can be buried to great depth, will compact and harden, and can become more strongly coupled to the underlying basement. They can therefore become more intimately linked to the larger-scale tectonic forces that control the deformation of the lithosphere. When we find ourselves drilling for hydrocarbons in such basins, we are typically in older (pre-Tertiary) rocks such as in the southern North Sea or the onshore Permian Basin of the US, or in more-active tectonic settings such as the foreland fold and thrust basins of the Andes in Colombia or the strike/slip pull-apart basins associated with major shear zones such as the San Andreas fault in California. In such settings, although gravitational loading is likely to still contribute the majority of the in-situ stress, significant additional tectonic stress may also exist.

Gravitational Loading. The weight of the overburden is typically the largest source of stress in most sedimentary basins. As discussed in the preceding section, this is particularly true in basins that are relatively young and that are not in very active tectonic settings. Such basins are often referred to as *relaxed* basins, where the term *relaxed* is intended to mean that there is no compressional tectonic stress.

As sediments accumulate in a relaxed basin, they are loaded by the weight of the overlying material (other sediment and water). The weight of this overburden creates vertical stress. Horizontal stresses are generated as the sediment tries to deform sideways in response to the vertical stress but is constrained by the surrounding material (either the sides of the basin or other sediment).

As the sediment accumulates, it may be constrained from expanding sideways on both sides. An example of this is where sediment infills a rifted graben (**Fig. 5.1.15**). Once the graben has formed, the sides of the basin can be considered as fixed boundaries. An example of where this has occurred is in the Tertiary infilling of central and northern North Sea grabens.

Alternatively, as sediment is deposited and buried it may be free to expand basinward such as in a passive-margin setting (**Fig. 5.1.16**). In these settings, it is common to see basinward slumping accommodated on normal faults (also called growth faults in this setting) as the wedge of sediment collapses under its own weight. Examples of such basins occur along the US Gulf Coast and into the deepwater Gulf of Mexico.

These two different boundary conditions on sediment deformation may influence the way in which the stresses within the sediment develop in response to gravitational loading. In the case where the sediment is

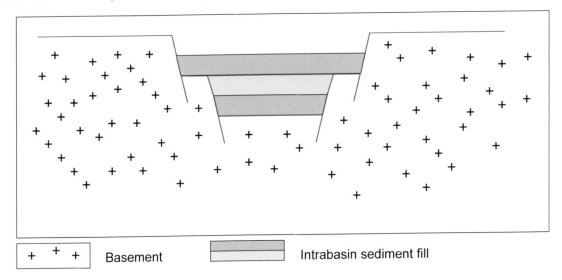

Fig. 5.1.15—Rift-bounded sedimentary basin. The sediment deposited and loaded after rifting will be confined later-ally, imposing approximately uniaxial strain conditions.

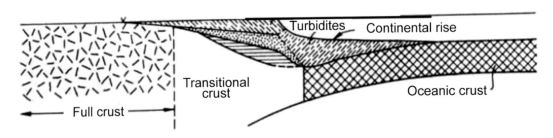

Fig. 5.1.16—Passive-margin sedimentary basin. The sediments are free to slump basinward under gravity to fill the space provided by the thinned continental crust and oceanic crust. Basinward slumping typically occurs on growth faults (normal faults) in response to vertical loading as more sediment is deposited. The magnitude of horizontal stress is therefore most likely controlled by the frictional strength of these growth faults (Busby and Ingersoll 1995).

constrained on both sides (and if we assume these features are much longer than they are wide such that movement in and out of the plane of Fig. 5.1.15 is negligible), a uniaxial-strain condition exists. Uniaxial strain literally means that strain (movement in response to applied stress) is able to occur only along one axis—in this case the vertical axis.

This condition forms the basis of the uniaxial-strain model (USM), traditionally one of the most commonly used models for estimating in-situ horizontal-stress magnitudes in petroleum geomechanics. In this model, the magnitude of horizontal stress is determined by the elastic properties of the rock, specifically Poisson's ratio (see Section 5.1.3). A rock with low Poisson's ratio, when loaded vertically and constrained on both sides, will transfer only a small amount of the load sideways to generate horizontal stress. A material with higher Poisson's ratio will transfer more of the load, generating a higher horizontal stress. A material with Poisson's ratio equal to 0.5 will transfer the entire load sideways such that the vertical and horizontal stresses are equal. The USM is presented in the following section.

A different model can be used to approximate the case where the sediment is free to slump downhill toward the basin (Fig. 5.1.16). The key control on the magnitude of horizontal stress in this case is the strength of the sediment. The vertical stress increases (as sediment is deposited) until it overcomes the strength of the sediment, which fails by faulting (or by moving on an existing fault), allowing movement of the material downward and basinward. The magnitude of the vertical stress (which promotes basinward movement) and the magnitude of the horizontal stress (which resists basinward movement) are in equilibrium, and the ratio of these magnitudes is controlled by the frictional strength of either the sediment itself or the fault gouge material. The model that describes this state of stress is therefore called the frictional-equilibrium model (FEQM), and it also is presented in the following section.

Both the USM and FEQM are simple gravitational-loading models. Under simple gravitational loading, the vertical stress is a principal stress and (as long as the rock has some finite strength or stiffness) it will be the maximum stress.

USM. The USM describes the distribution of stress in a setting such as that shown in Fig. 5.1.15. A more general form of this model is given by Thiercelin and Plumb (1994). However, if we assume that not only is there no horizontal strain but that the sediment behavior is also linear, isotropic, and elastic, the magnitude of both horizontal stresses can be expressed very simply as a function of the vertical stress, pore pressure, and the Poisson's ratio. In reality, these assumptions probably do not hold over geologic time. The basin boundaries may expand or contract, allowing horizontal strain to occur. Compacting and diagenetic processes that occur as rocks are buried and are subjected to pressure and temperature detract from the assumption of linear isotropic elasticity. The USM may therefore be more suited to describing relatively small changes in horizontal stress that occur over short periods of time such as during reservoir depletion. However, despite its shortcomings, the USM remains a commonly used framework (albeit often dependent on calibration from stress measurements or estimates from LOTs) for determining horizontal-stress magnitudes in the petroleum industry. Practical use of the USM is discussed in Section 5.1.6.

The most convenient way of considering the USM is in terms of effective stress. The effective stress has been defined in Eq. 5.1.3. Here we will assume that Biot's parameter $\alpha = 1$ such that effective vertical stress (σ'_v) is simply the total vertical stress (σ_v) minus the pore pressure:

$$\sigma'_v = \sigma_v - p_p \quad\text{...} \quad (5.1.20)$$

and the horizontal effective stress (σ'_h) is the total horizontal stress (σ_h) minus the pore pressure:

$$\sigma'_h = \sigma_h - p_p \quad\text{...} \quad (5.1.21)$$

The USM can then be expressed simply in terms of the ratio of the effective stresses (effective stress ratio) and Poisson's ratio (v):

$$\frac{\sigma'_h}{\sigma'_v} = \frac{v}{1-v} \quad\text{...} \quad (5.1.22)$$

Alternatively, the total horizontal stress can be given as

$$\sigma_h = \left(\frac{v}{1-v}\right)\sigma'_v + p_p \quad\text{...} \quad (5.1.23)$$

FEQM. The FEQM describes the state of stress in settings such as that shown in Fig. 5.1.16. It could be applied to determine the magnitude of the minimum stress in any active faulting environment provided the magnitude of the maximum stress is known. However, it is most easily applied in normal faulting environments (where gravitational loading is the only significant source of stress) as here the maximum stress is the vertical stress, which is usually well constrained (Section 5.1.6). It is also worth bearing in mind that in any active faulting environment, even when absolute stress magnitudes are not known, the FEQM can provide a useful method of constraining the ratio of the maximum and minimum stress if the frictional strength of the rock is known (Moos and Zoback 1990).

There are a number of criteria for describing rock failure that could be used in the FEQM. One of the most commonly used is the M-C criterion (already seen in Section 5.1.3), which is based on frictional sliding of two surfaces. We can use the M-C criterion here to give the magnitude of the minimum effective stress (in the simple gravitational-loading model, this is the horizontal stress in the plane perpendicular to the fault plane) as a function of the maximum (vertical) effective stress and the shear strength of the sediment. If new faults have to be created to allow the rocks to deform, then we must invoke a shear stress sufficient to overcome the cohesion of the intact rock plus the frictional strength of the rock. However, in most well-developed sedimentary basins in passive margin settings, growth faulting is already well established. It is therefore likely to be largely the frictional strength of the material in these fault planes (sometimes referred to as the residual strength or post-failure strength), with little or no contribution from cohesion, that controls the magnitude of the horizontal stress.

We can make use of expressions already given for the M-C criterion in Eq. 5.1.5. In Eq. 5.1.5, a difference between maximum and minimum effective stresses is expressed as a function of the internal friction and cohesion. Failure will occur on some plane when this condition is met. In the gravitational-loading model in Fig. 5.1.16, if we assume that the existing faults have formed at the appropriate angle (i.e., it is on these planes that the failure condition is met) and that there is no cohesion on these faults, then we can express the magnitude of the minimum total stress as

$$\sigma_h = \left[\frac{1 - \sin \Phi}{1 + \sin \Phi}\right] \sigma_v' + p_p \quad \dots\dots\dots\dots\dots\dots\dots\dots\dots\dots\dots\dots\dots\dots \quad (5.1.24)$$

The friction angle (Φ) in this case is that of the material in the existing fault zones. Because this parameter refers to post-failure properties, it is sometime called the residual friction angle. In a developed-growth faulting system, the fault-zone material will have already experienced significant shearing, which is likely to lead to weakening. The residual friction angle is therefore typically lower than the internal-friction angle of intact rock. The FEQM will be discussed further in Section 5.1.6.

Additional Sources of Stress. In the preceding sections, we considered scenarios where simple gravitational loading was the only source of stress in the subsurface. A sound understanding of these simple models is worthwhile because they are quite applicable in many (particularly young and relaxed) basins in which the petroleum industry operates. The gravitational models also provide a good benchmark against which to compare more-complex states of stress.

Understanding and predicting states of stress in simple gravitational-loading settings forms a large part of the workload of a wellbore-stability specialist. However, it is often the more-complicated environments where many of the industry's wellbore-instability problems occur. Hydrocarbon provinces with significant tectonic stress (e.g., Andes foreland fold and thrust basins of Colombia) and those where stress perturbations exist around salt (e.g., deepwater Gulf of Mexico) are two examples with very considerable wellbore-stability challenges.

Tectonic Stress. After gravitational loading, tectonic stress is probably the next-most-significant source of stress in the subsurface. The theory of plate tectonics describes the Earth's outer layer (or lithosphere) as individual plates that *float* on top of a weaker asthenosphere. Features such as compressional mountain belts, extensional rifts, subduction zones with associated vulcanism, and major shear zones (such as the zone in which the San Andreas fault sits) are clear evidence that these plates move. Ultimately, plate motions are linked to the cooling processes of the Earth, although exactly how the forces are transferred to the lithosphere is not well understood. Convection cells in the viscous material of the asthenosphere causing *basal drag* at the base of the lithosphere and upwelling of the asthenosphere at spreading centers leading to *ridge push,* are two possible sources of plate-driving forces.

Whatever the exact mechanism of plate-driving force, a stress field exists in the Earth's crust that is for the most part strongly aligned with the motion vectors of the plates. A classic example of this and how it can influence drilling activities is found in the foreland fold and thrust belt of the Colombian Andes.

The South American plate is moving approximately northwest relative to the Pacific plate (**Fig. 5.1.17**). Where these plates have collided in compression, the Andes Mountains have formed. The plate-driving forces have generated a strong NW/SE-oriented stress field that is clearly manifested in the boreholes drilled in the oil fields of the foothills. Borehole breakout and hole ovalization are observed in these wells. The orientation of this hole ovalization is very uniform and is clearly aligned NW/SE, consistent with the strong compressive tectonic stresses that have generated the adjacent mountains. **Fig. 5.1.18** shows borehole-breakout directions measured in the Cupiagua field. Wellbore deformation associated with this tectonic stress was a major cause of drilling problems such as packoff and stuck pipe in this and surrounding fields. As a result, these wells

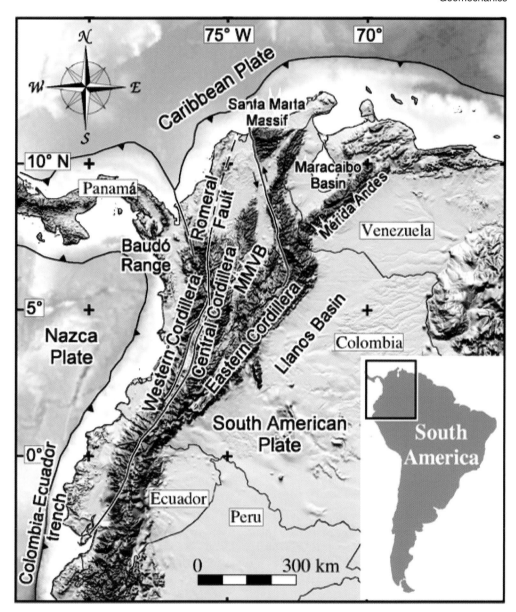

Fig. 5.1.17—Active NW/SE convergence of the Nazca plate and the South American plate has given rise to the creation of the Andes Mountains and generates high horizontal stresses in the surrounding crust. The foothills contain foreland basins that have been buried deeply, strongly lithified, and reactivated as fold and thrust belts. This process has resulted in a strong coupling of the sediments to the underlying tectonic basement so that they are also now subject to high horizontal tectonic stress (Gomez et al. 2005).

were extremely difficult, time-consuming, and expensive to drill (Last et al. 1995). It is not just areas in close proximity to mountain belts or other major tectonic features that are subject to tectonic stress. Tectonic plates are rigid bodies capable of transmitting stress over long distances. The World Stress Map (Reinecker et al. 2005) is a compilation of many thousands of stress measurements from around the world plotted within a plate-tectonic framework. The source of many of these stress measurements is earthquakes. These tend to occur mostly along plate boundaries, because this is where most of the deformation is accommodated. However, intraplate earthquakes are by no means rare. Other indications of stress such as borehole measurements also indicate a consistent pattern of stress in intraplate areas. In northwest Europe for example,

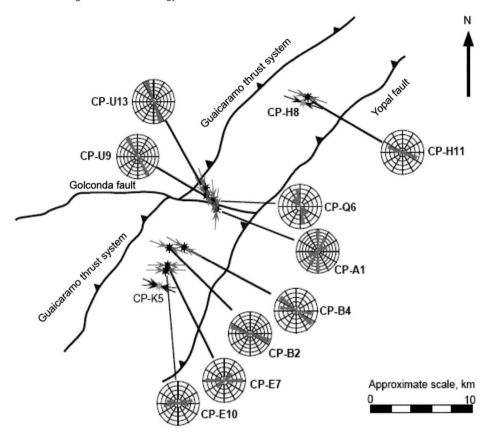

Fig. 5.1.18—Map of orientation of maximum horizontal stress in the Cupiagua field. Courtesy of GeoMechanics International.

the nearest major plate boundaries are the North Atlantic Ridge that pushes southeast and the northward colliding of the African plate into the Eurasian plate (giving rise to the Alps) to the south. The stress field in northwest Europe and the North Sea appears to be the resultant of these two plate-driving forces. **Fig. 5.1.19** shows an extract from the World Stress Map (Reinecker et al. 2005) showing northwest Europe and surrounding plate boundaries.

The influence of either plate-driving force appears to increase with proximity (i.e., the trend of the maximum horizontal stress swings gradually from north/south close to the Alps to a more northwest/southeast orientation close to the Mid-Atlantic Ridge).

Most estimates of maximum-horizontal-stress orientation in the World Stress Map come from either earthquakes or deep borehole measurements and are therefore from sampling at some significant depth in the crust. At shallow depths in more recent sediments (the Tertiary of the central and northern North Sea for example), the stress-orientation signature is less clear or even nonexistent. The tectonic stresses experienced at depth are not always transmitted strongly to the shallower rocks. It is therefore important when undertaking a geomechanical analysis for a petroleum engineering application to examine carefully the existing well data in the area because larger-scale regional tectonic stress trends may or may not be present at the depths of interest. Caliper- and image-log data are the best data sources for determining if there is a significant tectonic-stress component and if so what the orientation (and ultimately magnitude) of this stress is. This is discussed further in Section 5.1.6 on determining the state of stress.

Despite projects such as the World Stress Map and others in academia and the industry over the last 20 or so years, it is surprising how poorly constrained the state of stress remains in most sedimentary basins around the world. Even in areas considered to be major hydrocarbon-producing provinces, few or no data are publicly available.

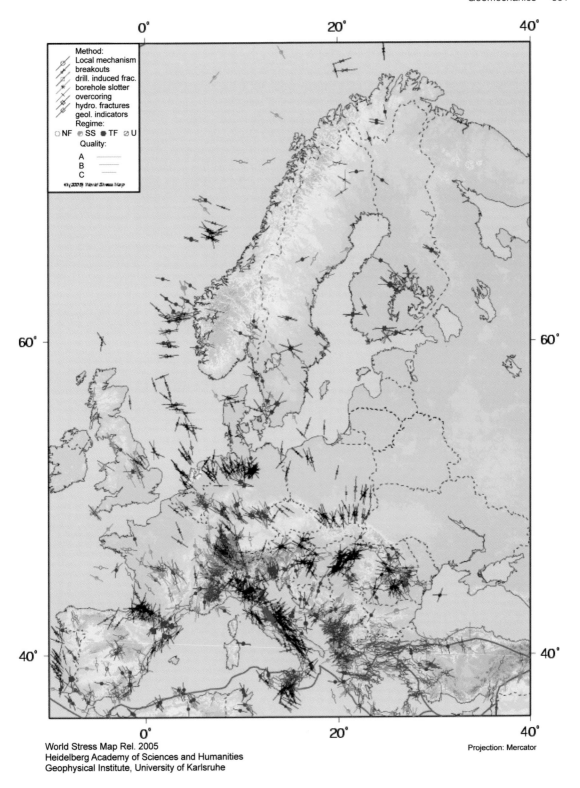

Fig. 5.1.19—Northwest Europe World Stress Map extract indicating the direction of the maximum horizontal stress. From the World Stress Map project (Reinecker et al. 2005).

Local Stress Perturbations. Large-scale patterns of stress in the Earth's crust are clearly visible in the World Stress Map and are somewhat predictable. At a smaller scale, local perturbations can occur. Some documented examples include stress reorientations around faults or other major structural features and around salt bodies.

Stress anomalies have been observed around major shear zones and faults in various parts of the world. These observations tend to be made in areas with a strong tectonic-stress component (probably because the effects of stress are more dramatic and noticeable in such areas) usually in strike/slip regimes (see Section 5.1.5), although there is no reason they should not occur in any stress regime.

Major strike/slip faults in the brittle part of the crust often move in a stick/slip fashion. Movement on such faults may be controlled by a few sticking points (rough spots or asperities on the fault surface), rather than by the stress being equally distributed over the entire fault surface. If this is the case, large stress concentrations will occur around the sticking points while other parts of the fault may be relatively free of stress or, in effect, be *shadowed* from the stress by the adjacent sticking point. The local distribution of stress around the sticking point will clearly concentrate as stress builds up in the system, whereas the sections of the fault in the stress shadow may be acting as free surfaces. A completely free surface transmits no shear stress, which demands that the orientations of the principal stresses in the adjacent areas reorient to become parallel and perpendicular to the free surface **(Fig. 5.1.20).** This combination of sticking points with their associated stress concentrations and free surfaces with their reoriented stress fields produces a complex pattern of stress that is difficult to predict and would require very careful and detailed analysis from existing well logs to understand and apply to practical wellbore-stability studies. Fortunately, these effects appear to be significant only in local areas adjacent to major tectonically active features and should not have a significant effect on wellbore stability in the vast majority of hydrocarbon provinces.

Many sedimentary basins at some time in the past have seen the generation of evaporite sequences. Thick accumulations of halite (sodium chloride—i.e., common salt) underlie sediments in many of the basins around the Atlantic rim, including the North Sea, offshore West Africa, offshore Brazil, and around the US Gulf Coast and Gulf of Mexico. Over long periods of time, salt has very low shear strength, and flows when

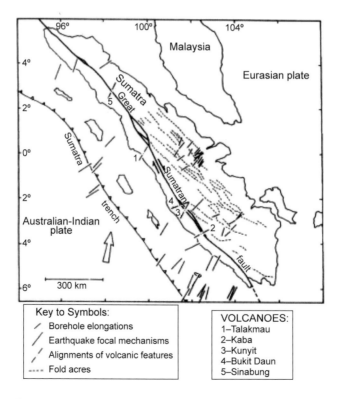

Fig. 5.1.20—Directions of maximum horizontal stress around the Great Sumatran fault. Stress concentrations around sticking points along the fault may be responsible for local variations in the stress direction (Minster and Jordan 1978).

subjected to a differential stress under some confining pressure. This occurs when salt is loaded by overlying sediment deposition. Unlike sedimentary rocks, salt is noncompressible and therefore does not compact as it is buried. Salt density is less than that of compacted sediment, and it therefore tends to flow upwards through the sediment if it is able to. Under the right conditions, salt diapirs will form. There are also many other geometries of mobile salt including salt sheets, salt dikes, and salt pillows. Subsidence of the overlying sediment occurs as salt is squeezed out, creating low spots that attract further deposition and loading. Thus, quite rapidly in geological terms, a dynamic system of sediment loading and salt movement develops.

Salt movement is sometimes referred to as halokinesis. The processes of halokinesis and associated sediment deformation are often collectively referred to as salt tectonics (Alsop et al. 1996). The term tectonics in this context has nothing to do with tectonic plates. Salt tectonics is a subject worthy of a dedicated discussion. Particularly in the deepwater Gulf of Mexico, it is increasingly recognized as playing a fundamental role not only in the geomechanics of drilling and completions but also in development of hydrocarbon provinces. Some 90% of BP's future prospects (at the time of writing) in the deepwater Gulf of Mexico are subsalt.

Despite its importance in some of the key hydrocarbon basins in the world, salt tectonics and associated geomechanics in the surrounding sediments are not well understood. What is clear is that the presence of salt, particularly in the diapiric mode, is likely to have an impact on the in-situ stress in the surrounding sediments. While the stress field within the salt is likely to be quite simple (because it behaves as a fluid, over time the stresses will tend toward isotropy), the distribution of stresses around the salt may be complex. The salt/sediment interface may act as a free surface (because the salt cannot sustain long-term shear stress) such that the orientations of the principal stresses are forced to be parallel and perpendicular to the surface. Where diapirism is active, an upward and outward stress will be applied to the surrounding rock. Similarly, if the salt body is passively sitting within the sediment, the buoyancy forces generated because of the lower density of the salt will also create additional stress. Around dome-shaped salt bodies, a radial pattern of stress is often generated. More-sophisticated numerical modeling of salt tectonic processes has shown that, in fact, a whole range of stress conditions can exist around the salt—including both increases and decreases in both horizontal- and vertical-stress magnitudes, and substantial rotation of the stress components away from vertical and horizontal. One of the first 3D MEMs was constructed in the late 1990s for the area around the Mungo diapir in the North Sea, based on finite-element methods and the free-surface boundary condition mentioned previously, and was used for drilling optimization in the field (Bratton et al. 2001). **Fig. 5.1.21** shows an example of numerical-modeling results around salt.

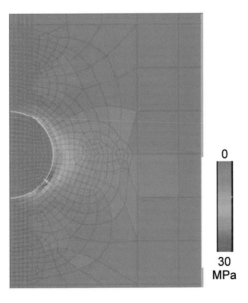

0

30
MPa

Fig. 5.1.21—Example of numerical modeling of stress around salt. The figure shows a contour plot of the difference between the minimum and maximum horizontal stresses around a spherical salt body (Fredrich et al. 2003).

Unfortunately, there are currently very few data sets around salt bodies with which to test these models. Good-quality caliper- or image-log information that could constrain stress orientation is scarce around salt bodies, partly because of the tendency for hole instability (and therefore the reluctance to run expensive logging tools, which may be lost) in near-salt and subsalt environments. Similarly, stress-magnitude data in the near-salt environment are scarce. Although there is much anecdotal evidence of perturbed stress magnitudes from low LOT values and of frequent occurrences of lost circulation around salt, there have been no rigorous compilations of such data within the industry to date with which to validate the modeling.

The phenomena discussed in the preceding paragraphs mean that stresses and material properties can vary rapidly and unpredictably around salt bodies. Holt et al. (2000) and Beacom et al. (2001) discuss some of these problems, and how team-based visualization can help in the diagnosis of problems.

Willson and Fredrich (2005) give a very comprehensive review of geomechanical problems associated with drilling in and near salt.

Thermal Stress. Temperature may play a role as a source of stress in the Earth in some circumstances. In a confined setting, an increase in temperature will result in an increase in compressive stress.

In most basins, temperature increases more or less linearly with depth, so the effect of temperature on the stress field, if it is important, is difficult to separate from the effect of increasing overburden. It might be expected that if all other things were equal, the stress magnitude in a hotter basin would be greater than that in a cooler basin (assuming uniaxial-strain boundary conditions in both). The increase in horizontal stress ($\Delta\sigma_h$) for a given increase in temperature under uniaxial strain conditions is dependent on the thermal-expansion coefficient (α_T) and the elastic properties (Young's modulus and Poisson's ratio) and is given by

$$\Delta\sigma = \frac{E\alpha_T}{1-v}\Delta T \quad\dotfill\quad (5.1.25)$$

Values of the thermal-expansion coefficient are not commonly measured on sedimentary rocks. The few published values that do exist are in the range of $0.1 \times 10^{-5}/°C$ to $5 \times 10^{-5}/°C$. As mentioned in Section 5.1.4, the effect of temperature could be significant for relatively stiff rocks in areas of high geothermal gradient. Currently, there is probably no sufficiently large database of adequate quality with which to test this model.

Thermally induced stress changes may be more important on the scale of the borehole or the reservoir, in the time frame of drilling and producing a field (Section 5.1.4). Circulating drilling fluid can have the effect of significantly heating or cooling in the area directly around the wellbore. Production of fluids from the reservoir may also have a significant heating effect on the overburden, and injected fluids may have a significant cooling effect on the reservoir.

Formation Pressure. All of the forces acting on the sedimentary basins in which we drill are borne by a combination of the sediment grains and the pore fluid that fills the gaps between these grains. We have already been introduced to this concept in the discussion on effective stress in Section 5.1.3. Here we will explore how formation pressure (often referred to as pore pressure) can evolve during deposition and burial and how it has a fundamental effect on the subsurface stress state and basin-scale geomechanics.

The Origins of Formation Pressure. At the surface of the basin where sediment is deposited (e.g., the seafloor if in a marine basin), there is no stress between the individual grains (i.e., there is zero effective stress), and the grains sit loosely together. The pressure in the formation fluid will be only that resulting from the weight of the overlying fluid (seawater if in a marine environment or atmospheric pressure if on land). Pressure has the same dimensions as stress (force divided by area), but it is a scalar. The force exerted by a pressure is the same in all directions, and there are no shear forces. The pressure exerted by the weight of a column of fluid is known as hydrostatic pressure (p) and is given by the product of the fluid density (ρ_f), gravity (g), and the height of the column (h):

$$p = \rho_f gh \quad\dotfill\quad (5.1.26)$$

During the process of sediment burial, as long as the pore fluid is in pressure communication with the surface, the formation pressure remains hydrostatic. Hydrostatic pressure is usually called *normal pressure* in the petroleum industry. Pressure communication between some depth in the basin and the surface requires that the fluid can flow between the grains to equalize the pressure. This is a function of the permeability of the sediment and also of the time allowed for pressure equalization to occur. If the sediment permeability is too low for a given burial rate, it will not allow the fluid to escape, and pressure communication will be lost. If pressure communication is lost as burial continues, the overburden load will increase but the formation fluid will no longer be able to escape from between the grains. As water is highly incompressible, the additional load of the increasing overburden is borne largely by the water and not by the more compressible grain-to-grain framework of the sediment. The formation fluid now bears the weight of part of the overlying sediment as well as the overlying fluid, and so the pressure is raised above hydrostatic values. Pressures above hydrostatic are referred to as overpressure or abnormal pressure.

The mechanism of overpressure generation just described is often called undercompaction or compaction disequilibrium. The fact that the sediment cannot compact when the pore fluid is not able to escape is quite readily observed in the log response when drilling through overpressured sections. An undercompacted sediment will have a lower effective stress and a higher porosity than its normally compacted equivalent, and this can be detected from several different logging measurements including sonic, density, and resistivity.

Undercompaction is by far the most significant mechanism of overpressure generation and the most commonly encountered in the major basins being developed today. The deepwater Gulf of Mexico is a good example of a basin where undercompaction is prevalent. Sedimentation in deepwater environments tends to be predominantly of fine-grained material, meaning the permeability of the sediment is low. Additionally in the Gulf of Mexico, deposition and burial rates are high because there is a large volume to fill and abundant sediment from the main river systems that feed the basin (e.g., the Mississippi). The combination of low-permeability sediment and high rates of deposition is ideal for the generation of overpressure. Indeed, this combination is so effective in parts of the deepwater Gulf of Mexico that overpressure is encountered almost at the seafloor throughout the entire thickness of sediment. On the other hand, there are many basins of similar age and depositional rate where normal pressures are encountered through most of the drilled overburden. These tend to be sand-rich basins, where the more-permeable and less-compactible sands allow the pressure to be bled off easily to the surface (or laterally) before any overpressure is generated. The way in which the basin is *plumbed* is an important factor in determining overpressure generation.

The other aspect of this plumbing problem is that if a formation is isolated from drainage but is connected (by a route that is permeable on a geological time scale) to a high-pressure zone, that formation will also accumulate a high pore pressure, independent of its state of compaction. This is often called repressurization, and it can be difficult to deal with.

Origins and prediction of overpressure are dealt with in more detail in Section 5.1.6.

The Influence of Formation Pressure on the Subsurface State of Stress. The overall state of stress in a basin is fundamentally linked to, and in a large part controlled by, the formation-fluid pressure. This is particularly true in a basin where in-situ stress is predominantly a result of gravitational loading. In such a basin, the maximum stress is the vertical stress. This is simply the stress resulting from the weight of the overlying material (sediment and fluid) and is controlled by formation pressure only to the extent that thick undercompacted sequences are lower in density, and this generates a lower vertical stress. The horizontal-stress magnitudes, however, are much more intimately linked to the pore pressure.

The pore pressure is a component of the total stress (Eq. 5.1.3) and exerts equal forces in all directions. Thus, in an overpressured setting, where the pore pressure bears much of the weight of the overburden, this weight is transmitted equally in all directions, such that the total horizontal-stress magnitudes are higher than those in the equivalent normally pressured setting. In the extreme case where the formation fluid bears the entire weight of the overburden, the pore pressure is equal to the total stress (there is zero effective stress), and all three total principal-stress (one vertical and two horizontal) magnitudes are equal. Settings that approach this extreme scenario are not uncommon in basins such as the deepwater Gulf of Mexico where very little dewatering has been possible during burial.

This relationship between the pore pressure and the total horizontal-stress magnitude is represented in both of the simple gravitational-loading models we examined earlier (the USM and the FEQM in Section 5.1.5). It is also very clearly observed in field data from overpressured basins from around the world. Data collected while drilling provide a reasonable first-order approximation both for pore pressure and for total horizontal-stress magnitude.

Field Data. **Fig. 5.1.22** shows a typical profile of pore pressure and fracture pressure from a post-well analysis. Because the well has already been drilled, there are various sources of field data to constrain the magnitude of both the pore pressure and fracture pressure.

Pore pressure is estimated by a variety of direct and indirect methods. Actual downhole pressure measurements are made routinely in permeable formations and provide the most accurate pressure values. However, other direct techniques such as background to connection gas ratios and flow checks on connections, and indirect techniques such as estimates from sonic logs, are typically all combined to give a reasonable estimate of pore pressure.

The fracture pressure is determined from LOTs and occurrences of lost circulation. The fracture pressure can often be taken as a reasonable first-order approximation to the minimum total horizontal-stress magnitude, for the following reasons. When the pressure within the wellbore is sufficiently high (if a high MW is used or if additional surface pressure is applied) it can open a hydraulic fracture in the wellbore wall (Section 5.1.4). In a normal-stress regime (see Section 5.1.5) where the vertical stress is the maximum stress, the fracture will open in the vertical plane perpendicular to the least-compressive stress. Where there is some horizontal-stress anisotropy, this is typically called the minimum horizontal stress (σ_h). Fracture initiation may have to overcome some tensile strength in addition to the stress concentration at the wellbore wall. However, the rock at the wellbore wall may already contain small cracks and fractures either naturally occurring or induced by the process of drilling the hole. Such cracks and fractures effectively remove the tensile strength

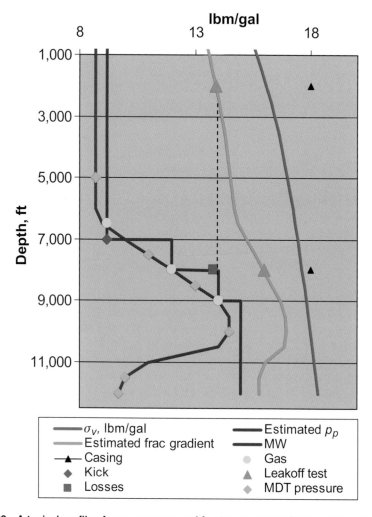

Fig. 5.1.22—A typical profile of pore pressure and fracture pressure from a post-well analysis.

and may provide a conduit for wellbore fluids to bypass the near-wellbore stress concentration. Once the fracture is beyond the near-wellbore stress concentration, the pressure required to propagate the fracture further is only slightly greater than σ_h. Thus, the fracture pressure seen at the surface may in fact often be only marginally higher than σ_h. As the wellbore pressure is reduced, the fracture closes. However, a mechanically closed fracture may remain hydraulically open such that during repressurization the fracture will reopen at a wellbore pressure approximately equal to σ_h. Thus, in an extended LOT or a repeated lost-circulation event, the fracture pressure may be very close to σ_h.

The exact relationship between various pressures during the LOT or lost-circulation event and σ_h is sufficiently important, yet inconsistently used, that it warrants a separate discussion (Section 5.1.6). For the time being, it is enough to say that on the basis of the above considerations and empirical field data, the fracture pressure trend typically derived from a series of LOTs and lost-circulation events is a reasonable first-order approximation to the trend of σ_h.

The data in Fig. 5.1.22 come from a well that was drilled through a normally pressured interval, a pressure ramp, and into a pressure regression. In order to normalize for depth, pressure values in Fig 5.1.22 are plotted as equivalent MW. Through the normally pressured section, the fracture pressure gradient (and by implication σ_h) can be seen to increase. This is mainly because of compaction and increased density generating a higher vertical-stress gradient. However, as the well enters the pressure ramp, a corresponding increase in fracture pressure (above that resulting from just an increasing vertical-stress gradient) is clearly seen.

This relationship between pore pressure, vertical stress, and fracture pressure has long been recognized, and many attempts to quantify it have been made. Hubbert and Willis (1957) are generally credited with the first attempt. They presented a form of the FEQM (in Eq. 5.1.24) that we have already discussed. The FEQM, the USM, and all subsequent attempts to describe the variation of fracture pressure (FP) and/or σ_h with pore pressure and vertical stress have followed the same basic format, using an equation of the form

$$FP = K\left(\sigma_v - p_p\right) + p_p \quad \dots\dots\dots\dots\dots\dots\dots\dots\dots\dots\dots\dots\dots \text{(5.1.27)}$$

K is typically referred to as either the matrix stress coefficient or the effective stress ratio. If FP is assumed interchangeable with σ_h, Eq. 5.1.27 can be rearranged for K to show that it is equal to the ratio of the horizontal to the vertical effective stress—hence the name effective stress ratio:

$$K = \frac{\left(\sigma_h - p_p\right)}{\left(\sigma_v - p_p\right)} \quad \dots\dots\dots\dots\dots\dots\dots\dots\dots\dots\dots\dots\dots\dots\dots\dots \text{(5.1.28)}$$

In the USM, K is a function of an elastic parameter, the Poisson's ratio v of the rock:

$$K = \frac{v}{\left(1-v\right)}, \quad \dots\dots\dots\dots\dots\dots\dots\dots\dots\dots\dots\dots\dots\dots\dots\dots\dots \text{(5.1.29)}$$

and in the FEQM, K is a function of a strength parameter, the internal friction angle Φ of the rock or of the fault material:

$$K = \left[\frac{1 - \sin\Phi}{1 + \sin\Phi}\right] \quad \dots\dots\dots\dots\dots\dots\dots\dots\dots\dots\dots\dots\dots\dots \text{(5.1.30)}$$

While both the USM and the FEQM provide a theoretical framework within which to predict horizontal-stress magnitudes and fracture pressures, there is typically a significant amount of uncertainty in the appropriate values of the parameters (v and Φ) that are required for these models. Additionally, it should be remembered that these models are very much simplifications of reality and the assumptions such as isotropic linear elasticity and uniaxial strain are rarely satisfied.

A more pragmatic approach to determining the appropriate value for K has been adopted in most cases in the drilling literature by simply backcalculating K from data sets where σ_h and/or FP, pore pressure, and vertical stress are known from field tests and measurements. Some of the more well-known relationships for K come from the Gulf Coast data sets and include a relationship between K and effective vertical stress from Matthews and Kelly (1967), a relationship between K and depth from Pennebaker (1968), two relationships between K (or actually a Poisson's-ratio value, which is used to compute K through Eq. 5.1.29) and depth below seafloor—one for the Gulf Coast and shelf and one for the deepwater Gulf of Mexico—from Eaton (1969, 1997), and a relationship between K and density from Christman (1973).

Breckels and van Eekelen (1982) developed an approach that does not backcalculate a K value, but simply establishes a relationship between depth and fracture gradient for various regions around the world. They were careful to emphasize that the data used to establish the relationships were representative of the fracture gradient, and not of drilling-induced fractures. Although this approach does not establish the kind of detailed profile usually employed in an MEM, it is a very useful first step.

Stress Regimes and Why They Are Important During Drilling. Most of the discussion so far in this section has concerned the subsurface state of stress in a predominantly gravitational-loading environment. In this environment, the vertical stress is the maximum stress. If faults develop in this situation, they are of a type known as normal faults—the fault plane is steeply inclined and the material above the fault plane moves downward (this nomenclature has nothing to do with the previous use of *normal* in the description of stresses and strains). Thus, the configuration where the vertical stress is the maximum stress σ_1, and the minimum stress σ_3 and the intermediate stress σ_2 are horizontal, is called a normal-faulting stress regime **(Fig. 5.1.23).** This nomenclature is after Anderson (1951), who was primarily describing field observations of faults. The term normal-faulting stress regime is used in petroleum geomechanics simply to define the relative order of the stress magnitudes; it does not necessarily mean that normal faults are actually present or active.

When horizontal stresses in addition to those generated by gravitational loading are applied, the relative ordering of the principal-stress magnitudes can change. Such additional horizontal stress is most often tectonic, but can also be associated with salt or local structural complexity. Tectonic stresses are predominantly horizontal such that in a tectonically active area one or both of the horizontal stresses may become greater than the vertical stress. The setting where one of the horizontal stresses is now σ_1, the vertical stress is now σ_2, and the other horizontal stress is σ_3 is referred to as a strike/slip faulting stress regime after the Andersonian convention. Faulting, if it occurs, will be strike/slip (Fig. 5.1.23). The setting where σ_1 and σ_2 are horizontal and σ_3 is vertical is referred to as a thrust-faulting, or reverse-faulting, stress regime (Fig. 5.1.23).

The stress regimes described in the preceding and shown in Fig. 5.1.23 are idealized end members in which the three principal stresses remain vertical and horizontal. This is probably a reasonable approximation for most basins. However, in complex tectonic settings, at shallow depths near steep surface topography, or around salt bodies, the principal stresses may be rotated out of the vertical and horizontal planes.

Understanding the relative magnitudes (and ideally the absolute magnitudes) of the three principal stresses is a prerequisite for any attempt at resolving the stresses onto the wellbore wall (Section 5.1.4), where the

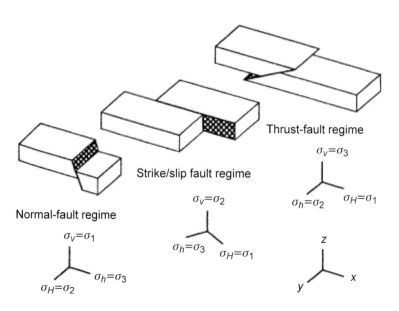

**Fig. 5.1.23—Andersonian stress regimes (Engelder 1993). From Engelder, Terry: *Stress Regimes in the Lithosphere.*
© 1993 Princeton University Press. Reprinted by permission of Princeton University Press.**

calculation of required MW is performed. In the last 20 years or so, as a greater number of directional wells have been drilled, the importance of the deviation angle and the azimuth of the well relative to the in-situ stress has been highlighted.

Some rules of thumb apply to picking the most stable hole trajectory for a given stress regime. These rules are based on trying to minimize the magnitude of stress difference between the two principal stresses (differential stress) acting at the wellbore wall. For example, in a normal-faulting stress regime, if we assume that the two horizontal stresses are more or less equal,

$$\sigma_3 = \sigma_2 < \sigma_1, \quad \dots\dots\dots\dots\dots\dots\dots\dots\dots\dots\dots\dots\dots\dots\dots\dots \quad (5.1.31)$$

then a vertical well will experience the least differential stress of any trajectory because it will be perpendicular to the plane containing σ_2 and σ_3, which are almost equal **(Fig. 5.1.24a).** However, in a case where one horizontal stress (σ_2) is slightly lower than the vertical stress (σ_1) but both σ_1 and σ_2 are significantly greater than σ_3, then

$$\sigma_1 = \sigma_2 > \sigma_3 \quad \dots\dots\dots\dots\dots\dots\dots\dots\dots\dots\dots\dots\dots\dots\dots\dots \quad (5.1.32)$$

also is a normal-faulting stress regime. Under this stress configuration, the most stable well is one drilled at a deviation of 60° in the NW/SE direction (Fig. 5.1.24b).

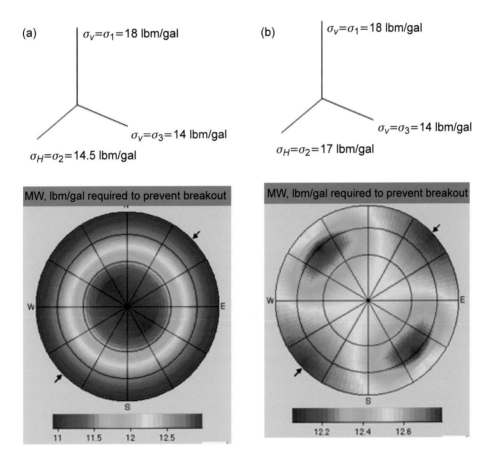

Fig. 5.1.24—The lower-hemisphere projections show the required mud weight as a function of wellbore azimuth and deviation under the stress fields shown and at a depth of 10,000 ft with a pore pressure of 10 lbm/gal and typical values for rock strength. For the stresses in Fig. 5.1.24a, the most stable well (the one that requires the lowest MW) is a vertical well. In Fig. 5.1.21b, the stress field is also that of a normal faulting stress regime, but the most stable well under these conditions is one drilled at a deviation of 60° in the NW/SE direction.

Picking the least stable hole (on the basis of differential stress rather than any considerations of such factors of drillability) is simpler—a hole drilled along the intermediate-stress direction will see the highest differential stress and potentially will suffer the most from wellbore instability. For example, a vertical well in a strike/slip regime has the most potential for trouble from high differential stresses (although in reality this would probably be the easiest well to drill and clean from an operational point of view). This argument also can be applied to oriented perforations in a completion.

In summary, to get an accurate wellbore-stability prediction, not only is the stress regime important but so too are the absolute magnitudes of the stresses. In fact, it is actually the full effective-stress tensor and the rock strength at any given point on the wellbore wall that determine the required MW, and even in simple scenarios, the result of combining all of these parameters is sometimes found to be not very intuitive. It is, therefore, strongly recommended that a full wellbore-stability calculation always be performed rather than relying on rules of thumb.

Determining the State of Stress. Determining the magnitudes and orientation of the initial in-situ stress state is a vital part of almost all geomechanics projects. It can be straightforward or can involve varying degrees of difficulty and uncertainty. The procedures for estimating the stress state, and the data sources required for this, are discussed in detail in Section 5.1.6.

The Influence of Depletion. As hydrocarbons are produced, the pore pressure in the reservoir drops (unless there is some pressure-support mechanism such as a large connected aquifer or strong compaction drive). This is called depletion. As the reservoir pore pressure drops, so too do the in-situ stress magnitudes. The total vertical stress is not affected, as the weight of the overburden remains unchanged. However, the effective vertical stress increases simply by the amount that the pore pressure has decreased (assuming Biot's parameter α is 1), and this will be important for stability calculations especially in the case of sand-production prediction. From a drilling perspective, the response of the horizontal-stress magnitudes to depletion is of more interest because of the effect on fracture pressure. This response, however, is somewhat more challenging to predict.

As with the problem of predicting virgin in-situ horizontal-stress magnitudes, there are broadly three choices available for predicting how σ_h will change with depletion. The choices are (1) to assume the response is elastic, (2) to assume that the system is in a state of frictional equilibrium and that the response will be governed by the frictional strength of bounding faults, and (3) to take an experimental approach by collecting in-situ measurements of horizontal stress at different stages of depletion and derive an empirical relationship between the change in pore pressure and the response of the horizontal-stress magnitude.

Under the elastic assumption, the simplest approach is to assume uniaxial strain conditions and linear isotropic elasticity. Let the ratio of the change in horizontal stress ($\Delta\sigma_h$) to a given change in pore pressure (Δp_p) be called the stress-depletion ratio A_{SDR}:

$$\frac{\Delta\sigma_h}{\Delta p_p} = A_{SDR} \quad\dots\dots\dots\dots\dots\dots\dots\dots\dots\dots\dots\dots\dots\dots\dots\dots\dots\dots \text{(5.1.33)}$$

Then, for linear isotropic elastic behavior

$$A_{SDR} = \alpha\frac{\left(1-2v\right)}{1-v} \quad\dots\dots\dots\dots\dots\dots\dots\dots\dots\dots\dots\dots\dots\dots\dots\dots \text{(5.1.34)}$$

Note that some authors use other ratios to characterize depletion behavior for example, the ratio of change in effective horizontal stress to change in effective overburden. The principles behind the numbers are all the same, but to avoid confusion, it is important to establish exactly what ratio is being used or plotted.

The implication of Eq. 5.1.34 is that if Poisson's ratio remains constant, then A will be constant. In other words, the decrease in horizontal stress with decrease in pore pressure will be linear.

During moderate depletion, strains are fairly small and the time scale over which the deformation takes place is short. Sand reservoirs can also be relatively isotropic (at least compared to layered shale formations). For these reasons, the assumptions of linear isotropic elasticity and uniaxial strain may be quite appropriate. On the other hand, there is also a growing body of microseismic reservoir-monitoring data that shows a clear correlation between increases in production and/or injection rates and microseismic activity. These data are from a limited number of fields and are probably not representative of every case. However, there is clear evidence that in at least some cases the reservoir is near the frictional-equilibrium limit. In such a case, the expression for the stress-depletion ratio A_{SDR} will depend on the stress regime. In the simplest case of gravitational loading (in a normal-faulting regime), it is given by

$$A_{SDR} = \alpha \frac{2 \sin \Phi}{1 + \sin \Phi} \quad \dots\dots\dots\dots\dots\dots\dots\dots\dots\dots\dots\dots\dots\dots\dots\dots \quad (5.1.35)$$

The implication of Eq. 5.1.35 is (as for the elastic assumption) that the rate of reduction in horizontal stress with pressure depletion will be linear.

There are relatively few fields around the world where a sufficiently good data set exists to determine the actual value of A_{SDR}. Ideally, an initial pore-pressure and σ_h measurement would be required, followed by subsequent pore-pressure and σ_h measurement after some depletion has occurred. Data sets including both measurements at both times are rare. Also, many reservoirs are not massive sands with uniform properties throughout, but may be layered and heterogeneous and not necessarily in pressure communication from layer to layer. In such a reservoir, pore pressures may vary considerably between layers, although it is rare to make detailed pressure measurements once the reservoir is on production (instead pressures are usually inferred from production pressures). Hydraulic fractures are typically tens to hundreds of feet in height and can therefore extend through many layers, giving rise to some uncertainty in the measurement of σ_h where different layers are depleted by different amounts.

For the above reasons, it is important to look carefully at the setting from which data are derived and the quality of the measurements. That being said, a handful of published data does exist where the authors have a reasonable degree of confidence in the quality of the measurements (Addis et al. 1994; Salz 1977; Teufel et al. 1993).

Table 5.1.1 presents available values of the stress-depletion ratio from published and other sources. From the values in Table 5.1.1 it can be seen that most of the values of A_{SDR} are in the range 0.5 to 0.7. Assuming Biot's parameter α is 1, this corresponds to Poisson's ratios between 0.2 and 0.3 if Eq. 5.1.34 is assumed to apply, or to internal-friction angles between approximately 20 and 30° if Eq. 5.1.35 is assumed to apply. These values are within a reasonable range.

The stress-depletion effect is clearly quite significant, with σ_h typically dropping by at least half of the magnitude of depletion. Given the close correlation between σ_h and fracture pressure, it is easy to see a potential problem if depleted reservoirs are to be drilled. In mature basins such as the Gulf Coast and shelf and the North Sea, drilling depleted reservoirs is becoming increasingly common as deeper reservoir intervals are targeted and existing intervals are worked over. **Fig. 5.1.25** shows a generic well profile containing depleted intervals to illustrate the problem. The MW must be kept above the pore pressure to prevent influx of

TABLE 5.1.1—STRESS-DEPLETION RATIOS		
Field/Location	Stress-depletion Ratio (A_{SDR})	Data Source
Vicksburg formation south Texas	0.53	(Salz 1977)
Waskom east Texas	0.46	(Whitehead et al. 1987)
Magnus field North Sea	0.68	(Addis et al. 1994)
Ekofisk Chalk North Sea	0.8	(Teufel et al. 1993)
Wytch Farm southern UK	0.65	(Addis 1997[*])
Gulf of Mexico shelf	0.63	BP[**]
Gulf of Mexico deep water	0.65	BP[†]

[*] Reprinted with permission from Elsevier.
[**], [†] Unpublished results from BP research.

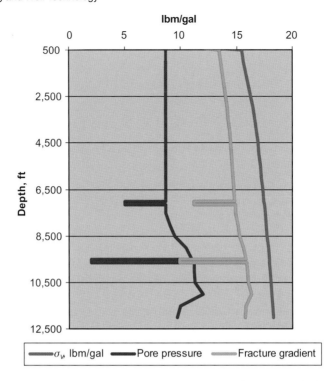

Fig. 5.1.25—A generic well profile containing depleted sand intervals. As the pressure in the sands is reduced, so too is the fracture gradient. This is a common problem when infill drilling operations take place on producing fields.

fluid from virgin-pressure intervals and to prevent hole collapse in the shale intervals. Ideally, the MW should also be kept below the fracture pressure to avoid lost circulation. The only way to do this is to set multiple strings of casing or liners to isolate the depleted intervals. Although the use of expandable liners has gone some way toward helping in these situations, wells like this often remain impractical because of the number of liners that would need to be set and the uncertainty in the exact location of depleted- and virgin-pressure sands.

Several large new fields where stacked pay layers exist have motivated extensive studies on the effect of depletion on future pressure and future drillability. Two different philosophies have developed to cope with the issue. The first philosophy is to avoid drilling depleted intervals where possible. In a new development plan, this means bottom-up development (i.e., predrilling to the deepest targets before starting to produce from the shallower targets). This is certainly the safe and conservative approach. However, in some cases the economic impact of such a predrill strategy is very large, because of the coupled effect of increased capital expenditure up front and the delay of first oil. This means that the shallower targets have to be developed first even though the operator knows that at some time in the future these intervals will have to be drilled when they are severely depleted. The reality is that many reservoir intervals today, and in the future, have to be drilled with MWs that exceed σ_h and the natural fracture pressure. Lost circulation is common in such cases, with thousands of barrels often being lost. From around 2000 onwards, however, solutions to the problem have been developed. Additives to the mud have been developed either to prevent the fractures from initiating (van Oort et al. 2003) or to rapidly plug and screen out the fractures, preventing further propagation (Morita et al. 1990; Alberty and McLean 2001; Aston et al. 2004). These approaches are best suited to sands, where the permeability and filtrate loss allows the additives to concentrate at the wellbore wall or within the fractures. These methods have been shown to be very effective if the mud additives are designed correctly for the particular setting and if they are applied in conjunction with appropriate drilling practices such as careful annular-pressure management to prevent pressure spikes and avoidance of backreaming, which can remove the additives from the wellbore wall and fracture mouths (see also Section 5.1.7).

5.1.6 Predrill Planning. *Introduction to Predrill Planning and the Concept of the MEM.* As the wells we drill become more challenging and more expensive, the importance of good planning has become increasingly critical. This has fostered a growing awareness among drillers of the importance of understanding the nature of the subsurface. A good modern drill plan has more than just formation tops and casing points. It should include all the predicted major subsurface features that could impact drilling operations. This includes faults, fractured zones, areas of high bedding dip or other structural complexity, weak or fissile zones, and any other known drilling hazard. This type of information comes mainly from integrating offset-well drilling and logging data with a seismic interpretation of the subsurface.

In addition to this description of the qualitative risks, the predrill plan also specifies the MW window. The MW window is bounded on the low side by either the minimum mud pressure required to prevent excessive shear failure of the wellbore wall (MW_{min}, also sometimes called collapse pressure) or the pore pressure of any permeable intervals, whichever of these is higher. On the high side, the MW should not exceed the fracture gradient (there are numerous definitions of fracture gradient, but in practice this is the MW at which excessive lost circulation occurs).

One of the core tasks of the geomechanics engineer is to define this MW window. The bulk of the work of defining the MW window is in calculating MW_{min}. This involves determining the rock strength, pore pressure, and stress magnitudes (closely related to fracture gradient) and how these vary with depth. Such a description of stress and strength in the Earth is often referred to as an MEM (Plumb et al. 2000). The MEM can then be used to calculate the MW window for a range of different well trajectories.

In the predrill planning stage, the MEM is based on seismic data and offset-well data. The accuracy of the MEM is therefore highly dependent on the quality and applicability of these data. In a wildcat exploration well, the MEM may contain significant uncertainties. For example, parameters such as pore pressure are only inferred from seismic velocity. Also, in a structurally complex area, even when offset-well data exist, parameters such as stress may vary significantly over short distances. So, although the predrill MEM should represent the best understanding of the properties of the region of interest at that time, it should be recognized that the MEM must be verified or updated as a new well is drilled and new information is acquired. In this section, we describe the building of the predrill MEM. In Section 5.1.7, we will discuss how the MEM is updated while drilling and how such updates can impact drilling operations.

An MEM can be a 1D, 2D, or 3D model. The 3D model is obviously the most desirable because it may cover a whole field (or more), and so the same model can be used for multiple well locations. 3D MEMs are built in the industry, but they are currently rare because they typically require significantly more effort and resources. The most common approach is to build a well-specific 1D MEM. These models are quite easily built, edited, and transferred as simple spreadsheet or text files, making them much easier to work with than their 3D equivalents.

Table 5.1.2 summarizes the main components of an MEM and the input data typically used to estimate these parameters. Each parameter is then discussed in more depth in the following sections.

Building the MEM. *Rock Strength.* A characteristic of rock (and similar porous/granular materials) that is not shared by all materials (e.g., metals) is that its strength increases with confining pressure. We therefore typically define rock strength by using some value of intrinsic or unconfined strength plus some description of how its strength increases with confining pressure. This has already been illustrated in Section 5.1.3 by using the parameters of cohesion and angle of internal friction. Cohesion is a measure of resistance to shear stress with zero confining stress (the intercept of the strength envelop on the shear stress axis of the Mohrs-circle diagram in Fig. 5.1.6). Internal-friction angle (IFA) is related to the slope of the strength envelope in Fig. 5.1.6–i.e., how much increase in strength occurs for a given increase in normal stress (confining pressure).

Describing rock strength in terms of cohesion is an intuitive approach within the framework of the Mohrs-circle diagram (Fig. 5.1.6). In practice, the parameter that is most commonly measured in the laboratory is the UCS, which has been described in Section 5.1.3. UCS and cohesion are different but related parameters. Both can be used to describe the intrinsic or unconfined strength of the rock. We will use UCS for the remainder of this discussion because it is the more widely used and quoted of the two in the oil industry.

UCS and IFA can be measured on core samples in laboratory rock mechanics tests. However, this presents a significant challenge in wellbore-stability work because core samples are almost always only available over a very limited interval of the reservoir. Most of the overburden is typically not cored. To address this issue, the relatively limited amount of data from mechanically tested core samples has been used to make

TABLE 5.1.2—THE MAIN COMPONENTS OF A MEM AND THE INPUT DATA TYPICALLY USED

MEM Component	Input Data Source
Rock- strength parameters	Typically from correlations with log-measured values such as compressional wave velocity, shear wave velocity, density (or parameters derived from them; e.g., shear modulus, Young's modulus), and clay content, which have been generated from laboratory measurement on core. Can also be obtained from measurements on core and in some cases from cuttings/cavings.
Vertical stress (σ_v)	Density logs, velocity-derived density (from sonic log, interval velocity, check shot).
Pore pressure (p_P)	Measurements from modular dynamics tester (MDT) and repeat formation test (RFT). Inferred from sonic, resistivity, density logs. Estimated from drilling indicators, kicks, flow checks, connection gas.
Minimum horizontal stress (σ_h)	Typically model-based, calibrated to LOTs, XLOTs, hydrofrac closure pressures, lost-circulation events.
Maximum-horizontal-stress direction	Borehole breakouts or drilling-induced tensile fractures observed from image (or caliper) logs. Earthquake focal mechanisms, World Stress Map.
Maximum-horizontal-stress magnitude	No direct measurement. Typically constrained from model-based breakout and fracture observations and/or frictional equilibrium.

correlations between laboratory-measured strength parameters and rock properties that can be measured downhole with logs. These correlations allow us to populate the MEM throughout the overburden.

Studies have identified correlations between rock strength and properties such as grain size, number of grain contacts and dominant grain type (quartz or clay mineral) (Plumb et al. 1992), porosity, and stiffness (Deere and Miller 1966). Most of the published work in this area is based on sands because reservoir cores are much more readily available than shale cores. For wellbore-stability work, however, we are primarily concerned with shales. If we make the assumption that shales are predominantly fine-grained materials in which the clay minerals are the load-bearing medium, we can significantly narrow the range of variables in the strength correlation. Although this assumption is clearly a simplification, which belies the range and complexity of fine-grained lithologies, it has allowed some quite-simple shale-strength correlations to be used successfully, as with Horsrud's correlation (Horsrud 2001), shown in **Fig. 5.1.26.**

In essence, both UCS and IFA tend to increase as the rock becomes more compact and better cemented. Therefore, log measurements that respond to such changes in the rock have potential for developing correlations with UCS. There are several logs that have this potential and have been used for strength correlations. They include density, porosity, and sonic. Both the density and the porosity logs are likely to pick up the general strength trends. However, there can be issues with this approach under circumstances in which the degree of cementation is variable. Two rocks with the same porosity (or density) may have very different strengths if one is strongly cemented and the other is uncemented or weakly cemented. The sonic log, because it is responding to the stiffness of the material, is more likely to detect the difference between an uncemented and a cemented rock, although where grains are closely interlocked at higher confining pressures this might not always be the case. In addition, the shear-wave velocity is probably also a better measure of rock strength than the compressional-wave velocity, becasue it is sampling only the grain-to-grain framework and is not affected by the fluid. In practice, as we will see in the following, it is actually the compressional-wave velocity V_p that is most commonly used to estimate shale strength. Although shear may be a better measure of rock strength, the fact that compressional-wave velocity is much more common (especially in while-drilling measurements) probably accounts for its widespread use.

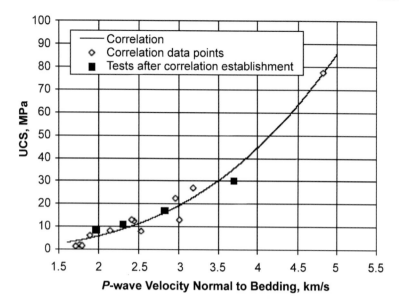

Fig. 5.1.26—UCS as a function of compressional velocity from laborator tests on North Sea shales (Horsrud 2001). The correlation shown is UCS = $0.77 V_p^{2.93}$.

There are few published correlations for shale strength (Horsrud 2001; Lal 1999; Lashkaripour and Dusseault 1993; Chang et al. 2006). Chang et al. (2006) provides a comprehensive compilation and review of these. Horsrud (2001) also gives a good summary of previous work in this area and goes on to present details of laboratory data from North Sea shales that demonstrate a good correlation between UCS and compressional-wave velocity. Fig. 5.1.26 shows Horsrud's correlation. This correlation, and modifications to it based on local data sets, are used routinely in wellbore-stability calculations around the world.

Correlations for IFA in shales are less well-developed. This may be partly because the range of possible values of IFA is smaller than those for UCS. IFA for shale is almost always in the range of 20°–45°, whereas UCS can vary from as small as 100 psi in a very weak rock to perhaps 20,000 psi in a very strong rock. There are also fewer data sets of IFA measurements on shales on which to base such correlations.

The physical background on how IFA varies as a function of log-derived properties is less intuitive than for UCS. Published correlations, however, are based on compressional-wave velocity that is at least convenient, because this also is the basis for the UCS correlation. One such published correlation from Lal (1999) is widely used

$$\sin \Phi = \frac{V_p - 1}{V_p + 1} \quad \dotfill \quad (5.1.36)$$

A more recent review of data from sedimentary rocks (Chang et al. 2006) offers some alternative correlations.

Pitfalls in Estimating Rock Strength From Log Measurements. Although the preceding correlations for UCS and IFA appear to be quite simple, some care must be taken to ensure that they are used properly. First, appropriate velocity measurements must be used as an input. Second, the limitations of the derived strength value must be understood.

Velocity is affected by both confining pressure and the frequency of the measured wave. The laboratory-measured velocities that form the basis of the correlations have been adjusted to correct for the effect of confining pressure, such that they should be equivalent to an in-situ sonic-log measurement. The frequency of the compressional wave is also in the same range as that of a sonic-logging tool. Velocities derived from other sources should be corrected to sonic-log equivalents in order to use them in these strength correlations. Seismic-wave velocities for examples (from either surface seismic or check-shot-type measurements) may be approximately 2 to 5% slower than the sonic-log equivalents.

When using sonic-log measurements, the hole angle also has to be considered if there is significant forma-tion anisotropy in the form of bedding or lamination. The correlations assume the measurement is made perpendicular to bedding. In a highly deviated well, however, the logging tool might be sampling a velocity closer to parallel to bedding. Where bedding anisotropy is strong, there might be a significant difference in bedding-parallel and bedding-perpendicular measurements. Anisotropy of up to 50% has been reported by Hornby et al. (2003) in strongly laminated shales **(Fig. 5.1.27)**.

One more significant potential error exists. The effect of pore fluid on compressional-wave velocity can be large. If the in-situ measurement is made in a hydrocarbon-bearing zone, it may be slower than if the pore fluid is water or brine. Gas has a very strong effect on the compressional-wave velocity. Because we are mainly considering shales in the wellbore-stability field, this is not often a problem other than in source rocks or unconventional settings.

Even when the appropriate input velocity is used, the limitations of the derived strength estimate should be understood. This is an estimate of strength for an intact rock in the direction perpendicular to bedding. If the actual in-situ rock is damaged (cracked, fractured, or fissile) it will likely behave as a much weaker material. Also, if there is a significant strength anisotropy, such as along fissile bedding planes, the behavior of the wellbore wall (particularly in a borehole drilled at a low angle to the plane of anisotropy) may be governed by the strength of the weak planes rather than by the *bulk* strength of the rock.

Vertical Stress (σ_v). Vertical stress is perhaps the most fundamental component of the MEM. It has a direct influence on pore-pressure and horizontal-stress magnitudes. In most (simple) settings, it is also (fortunately) quite straightforward and unambiguous to determine. In simple settings, vertical stress can usually be consid-ered a principal stress, equal to the weight per unit area of the overburden (where this weight can be estimated from density logs). Circumstances where this assumption breaks down include the following: areas of sig-nificant topography (mountainous areas), areas where strong tectonic forces or the presence of salt causes the principal stresses to rotate out of the vertical and horizontal planes, and areas where stress arching can occur (above a compacting reservoir for example). In most settings, however, the preceding simple assumption is

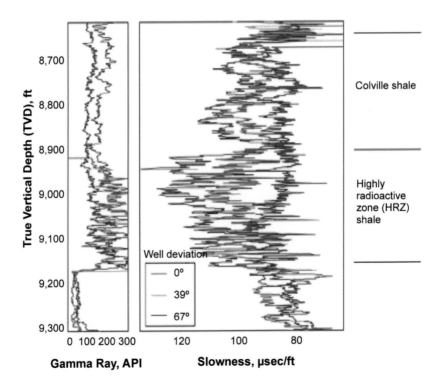

Fig. 5.1.27—From Hornby et al. (2003). Significant velocity anisotropy developed in the HRZ and overlying shales of the North Slope in Alaska. The HRZ is a highly organic, strongly laminated source rock.

reasonable and the term *overburden* is typically used interchangeably *with vertical stress.* We will treat this assumption as valid in the following discussion.

Perhaps because it appears to be rather a straightforward parameter, many rules of thumb have been created for estimating the overburden. For a long time, 1 psi/ft was thought to be a good enough estimate most of the time. In offshore environments with varying water depths, this is clearly inadequate, so numerous correlations have arisen that give overburden simply as a function of water depth and depth below seafloor. Although these correlations are an improvement on the 1 psi/ft model, they still fail to take account of lateral variations in compaction, presence of salt, or other lithologies. For this reason, other than in the most simple and laterally uniform settings, a site-specific vertical stress using the best local estimate of density available should yield the most accurate estimate of vertical stress.

Vertical stress (σ_v) is derived through integration of the density from surface to the depth of interest.

$$\sigma_v = \int_z^0 \rho g \times dz \quad \dots\dots\dots\dots\dots\dots\dots\dots\dots\dots\dots\dots\dots\dots\dots\dots\dots\dots\dots\dots\dots\dots\dots\dots \quad (5.1.37)$$

The density of the overburden can be estimated in a variety of ways. The following presents some of the more common methods.

Density Logs. Density measurements are available from both LWD and wireline logs. Using density measurements from an offset well is one of the most common approaches to estimating σ_v. This approach assumes that the density profile at the offset well is the same as that at the location of the MEM. This is often a good assumption if there are no large lateral variations in lithology or state of compaction.

Density logs are rarely run to surface. Typically, the density between surface and the top of the density log is estimated by curve extrapolation, which clearly provides some potential for error. The resulting error in the vertical stress will be most significant at shallow depths (**Fig. 5.1.28**). At greater depths, because the vertical stress is the integral of all densities above that depth, the error will be less.

Fig. 5.1.28a shows an example of a density log extrapolated to surface. Two alternative extrapolations are shown. One is a power-law extrapolation to an assumed mudline density. This type of density relationship with depth is typical of a compacting sediment in a young basin. The alternative is a simple straight-line extrapolation back to the seafloor. This might be more realistic in an older basin where the rocks have already

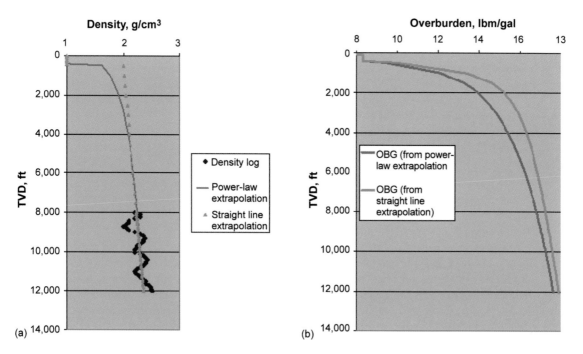

Fig. 5.1.28—Two possible extrapolations to surface of a density log and their effects on the overburden gradient (OBG).

been buried, compacted, and uplifted. In both examples, the density curve is extrapolated to the seafloor and a water density of approximately 1 g/cm^3 is used from seafloor to surface. As water is significantly less dense than rock or even unconsolidated sediment, water depth is a factor that has one of the largest effects on the vertical-stress profile.

The effect of the different extrapolation assumptions on the resulting vertical stress is shown in Fig. 5.1.28b. Each of the density profiles on the left has been integrated to compute a vertical stress (overburden). The choice of extrapolation model to use to surface can make a significant difference to the computed overburden, particularly at shallow depths. Clearly, it is useful either to have a good understanding of how to extrapolate the density data or to find some other way to estimate the density between the top of the density log and surface. In the absence of density measurements, sonic velocity data (from logs, seismic, or check shot) probably provides the best alternative.

One concern to keep in mind when working with density logs is that the measurement is very sensitive to hole size and must be quality controlled carefully before the values are used. Anomalously low density values can arise in enlarged-hole intervals.

Density From Sonic. Where direct density measurements do not exist, a *synthetic* or *pseudo* density can be computed from velocity. For most sand- and shale-type lithologies, a good correlation exists between density and velocity. Some well-known relationships between density and velocity exist, such as the Gardner correlation (Gardner et al. 1974):

$$\rho = aV_p^b, \quad \dots\dots\dots\dots\dots\dots\dots\dots\dots\dots\dots\dots\dots\dots\dots\dots\dots\dots\dots\dots\dots \quad (5.1.38)$$

where V_p is velocity in ft/sec and a and b are empirically derived constants [$a = 0.23$ and $b = 0.25$ in the original paper by Gardner et al. (1974), although these need to be calibrated for local rocks]. Often, a straight-line relationship between density and velocity also provides an adequate fit over limited intervals.

The exact correlation is actually a function of lithology and mineralogy. It should therefore be calibrated locally where both density and velocity have been measured in the same well. Care should be taken when extrapolating a correlation derived in one interval to an interval of significantly different depth or temperature, as clay diagenesis can lead to different density/velocity relationships (e.g., the smectite-to-illite conversion).

The advantage of using a velocity-derived density is that it can provide data all the way to surface, thus avoiding the need for curve extrapolation. Sonic logs, like density logs, are rarely run to surface. However, velocity data to surface can be obtained from check shots. Interval velocities from surface seismic can also be used where careful processing is applied. Care must be taken here that reasonable velocities can be extracted, particularly for very shallow and very deep intervals and for structurally complex settings. It may also be necessary to apply a frequency correction to seismic and check-shot velocities if they are to be used in a log-derived correlation to density.

Special Cases. In some settings, there are factors that can influence the overburden such that it cannot be derived simply by integrating the density of the overlying material. These settings include the following:

1. Areas of stress arching. Stress arches can develop where there are high horizontal stresses, or above depleting and compacting reservoirs or around salt. A stress arch can shield the underlying rock from some of the overburden weight such that the vertical stress below a stress arch is not given by integrating the density of the overlying material.
2. Areas of significant surface topography. In mountainous areas or in areas where significant bathymetric features exist, the weight of the topographic highs is not just supported by the material vertically below but is also partly supported by the adjacent material. This can cause significant variations in vertical stress, particularly at shallow depths. This should also be considered where large density discontinuities exist in the subsurface, such as around salt or igneous intrusions.
3. Drilling a deviated well under a changing water depth or density anomaly (e.g., salt). When estimating the overburden in a deviated well, it is important to integrate the density of the material lying vertically above any point in the well. If the water depth is changing as a function of horizontal well

step-out, or if the well passes under salt, this density profile will not be that measured in the actual well itself.

Formation Pressure (p_p). Formation pressure (the pressure of the fluid in the pore space of the rock) is a key component of the MEM, and is a vital concern for drilling safety and efficiency. The generation of pore-pressure values above hydrostatic has been touched on in Section 5.1.5, together with the influence of pore pressure on the state of stress.

There is a variety of different ways of generating a pore-pressure profile for a prospective well. Before discussing these, a distinction must first be made between two types of mechanism for generating excess pore pressures.

In Section 5.1.5, undercompaction of shale was discussed briefly. This is where a shale formation is loaded (by further burial) so rapidly that pore-fluid drainage cannot occur rapidly enough to transfer the load onto the solid framework of the rock. Load is transferred to the fluid instead, elevating its pressure, and the porosity of the shale remains (substantially) higher than normal as it is buried. Normal porosity in this context means the porosity that the shale would have at this level of overburden load (i.e., depth) with hydrostatic pore pressure. This type of overpressure is called undercompaction pore pressure, or compaction disequilibrium, or Type 1 overpressure, and is common in rapidly buried, young sediments. Its distinguishing feature is that pore pressure is strongly related to shale porosity.

A second type was also mentioned in Section 5.1.5, where the interval of interest is repressurized by a fluid from another formation (or, more rarely, by temperature or horizontal-stress changes). Because shale compaction is irreversible, this repressurization does not substantially increase the porosity of shales in the interval. This is called repressurization, or Type 2 overpressure; its distinguishing feature is that pore pressure is not strongly related to shale porosity.

The next sections deal with methods for building pore-pressure profiles for the MEM. It is important to discover, at an early stage, whether Type 1 or Type 2 overpressure is likely to be significant in the region of interest, through discussion with the geology and geophysics teams, and examination of offset-drilling experience.

Because of its importance in drilling safety, formation-pressure estimation has been investigated for many years, and there are many approaches and methods in the literature. This section will not attempt to cover them all, but to give an introduction to the principles. Mouchet and Mitchell (1989) provide a good handbook on the basics of practical-formation pressure estimation, and more recent developments can be found in Mitchell and Grauls (1998) and AADE (1998).

Undercompaction Overpressure. There are many well-established methods for predicting undercompaction, or Type 1, overpressure and generating a pore-pressure profile for the MEM. These are based on deviations from the normal compaction behavior or trend for the region of interest. Because shale porosity is difficult to measure unambiguously from log data, other properties such as density, resistivity, and especially sonic traveltime are used instead. The general approach is as follows (sonic velocity will be used as an example here, but resistivity or density could also be used).

- Accumulate log data from offset wells for the region of interest. These data should come from as much of the sedimentary column as possible, especially the shallower, younger regions.
- Locate the shale intervals (usually by taking gamma ray values above a locally determined value).
- Plot sonic traveltime (or its logarithm) for these intervals against TVD.
- Search for intervals where there is a clearly defined linear trend of traveltime decreasing with depth. The gradient of this line represents the normal compaction trend for the region.
- Extrapolate the normal-compaction trend over the depths of interest, and search for deviations from it. Deviations towards higher traveltimes than the normal compaction trend are likely to indicate higher porosities than those expected on the trend, and so indicate the presence of undercompaction overpressure.
- The pressure can be quantified by various methods, following one of the two methods below (Traugott 1997).

It is assumed in the undercompaction approach that the measured property depends only on effective stress (not, for example, on changes in lithology, or temperature, of fabric), that there is no repressurization, and that

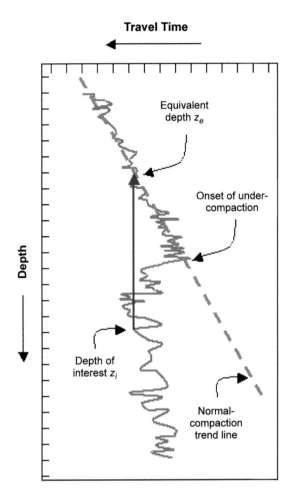

Fig. 5.1.29—Schematic of the vertical method for pore-pressure estimation in undercompacted environments.

the sediments have always been buried, never uplifted (so that the current effective stress in the shales is the maximum that they have ever encountered). Although these seem restrictive, undercompaction approaches are useful and very widely used. In addition, there are more-complex variations that relax one or more of the assumptions.

Vertical Methods. This is illustrated in **Fig. 5.1.29**, for shale sonic-traveltime data (red curve). First, draw a normal-compaction trend line (green dashed line) that fits the traveltime plot in the intervals where it is linear with depth. Then, find the measured value of the traveltime at the depth of interest z_i, where the pressure value is required. Next, use the normal trend line to find the equivalent depth (z_e) at which the measured value would normally be expected (vertical blue arrow). Then (because the traveltime depends on porosity, which depends on effective stress), the effective stress at the depth of interest is taken as equal to the effective stress at the equivalent depth (which is calculated as the difference between overburden and normal pore pressure). The pore pressure at z_i is then calculated as the difference between the overburden at z_i and the calculated effective stress there. The well-known *equivalent-depth method* is a vertical method (Foster 1966).

Horizontal Methods. This is illustrated in **Fig. 5.1.30**. First, as for the vertical methods, draw a normal-compaction trend line (green dashed line) that fits the traveltime plot in the intervals where it is linear with depth. Then find the measured value Δt of the traveltime at the depth of interest. At the same depth, find the expected value Δt_{normal} on the normal-compaction trend line. Use the measured and expected values in a

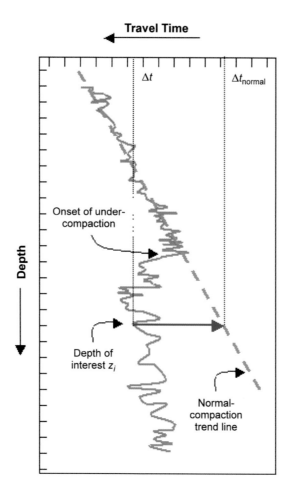

Fig. 5.1.30—Schematic illustration of the horizontal method for pore pressure estimation in undercompacted environments.

locally calibrated equation to find the effective stress at the depth of interest. The pore pressure at the depth of interest is then calculated as the difference between overburden and effective stress at the depth of interest, as for the vertical methods. Eaton's equation (Eaton 1975) is the best-known of the horizontal methods and perhaps the most widely used pore-pressure prediction approach overall:

$$\sigma' = \sigma'_{normal}\left(\frac{\Delta t_{normal}}{\Delta t}\right)^3 \quad \ldots\ldots\ldots\ldots\ldots\ldots\ldots\ldots\ldots\ldots\ldots\ldots\ldots\ldots\ldots\ldots \quad (5.1.39)$$

There are many variations on these two methods. These include different ways of plotting the depth data (e.g., logarithmic vs. linear) to obtain the normal-compaction trend line, different combinations of stress components to obtain the effective-stress value (e.g., the isotropic component of stress rather than the overburden), and different forms of the equation used in the horizontal methods (e.g., the value of the exponent in Eaton's equation is usually set by local experience). One of the attractions of the horizontal methods is that they can be expressed as a simple equation and can be easily coded into, for example, spreadsheets, as follows. Combining Eaton's equation with the effective-stress relationship gives, for sonic-log data,

$$p_p = \sigma_v - \left(\sigma_v - p_p^{normal}\right)\left(\frac{\Delta t_{normal}}{\Delta t}\right)^3, \quad \ldots\ldots\ldots\ldots\ldots\ldots\ldots\ldots\ldots\ldots\ldots\ldots \quad (5.1.40)$$

and for resistivity,

$$p_p = \sigma_v - \left(\sigma_v - p_p^{\text{normal}}\right)\left(\frac{R}{R_{\text{normal}}}\right)^{1.2} . \qquad \dots\dots\dots\dots\dots\dots\dots\dots\dots\dots\dots \quad (5.1.41)$$

The use of log data has been implicit in the discussion of undercompaction overpressure, but this is not the only source of data. Cuttings-density data can be used within the same framework, as can LWD sonic, resistivity, and density, and sonic traveltime derived from check shots and surface- and borehole-seismic data. This means that the undercompaction approach is well suited to being used on a fieldwide basis, to generate 3D pore-pressure maps (Sayers et al. 2002b), and also in real time while drilling. For example, collection of sonic traveltime or resistivity values from LWD measurements can lead to a revised normal compaction trend for the specific well being drilled, rather than for the offset wells, and can also be processed to give an up-to-date estimate of the pore pressure in the shales currently being drilled—which can be very useful if there is a known or unknown sand a few feet in front of the bit (Sayers et al. 2002a; Malinverno et al. 2004).

Problems arise in deepwater environments (Smith 2002) because the onset of overpressure may be so rapid (i.e., so close to the mud line, that there may be no detectable normal-compaction trend). This means that a direct relationship between effective stress and velocity must be sought. A model of this kind is presented by, for example, Dutta et al. (2001).

An interesting recent development has been the construction of a predrill pore-pressure cube for the entire northern Gulf of Mexico, using public and private check-shot data (Sayers et al. 2005). Tollefsen et al. (2006) report how the cube was used as the starting point for construction of a well in Vermillion Block 338. Combining the predrill model with real-time updating allowed the elimination of a liner string and other changes in the well, with estimated savings of USD 1.7 million.

Repressurization Overpressure. Repressurization, or Type 2, overpressure is much more difficult to predict. Shale compaction, because of increasing effective stress during burial and drainage, results in large changes in porosity and elastic stiffness, which are reflected in the large changes in traveltime or resistivity that are used in undercompaction methods. If the pore pressure is pumped up after compaction, however, or if the effective stress on the shale is reduced by other factors such as uplift, the compaction is not reversed. Shale porosity is a unique function of effective stress only when the effective stress is increasing. When it decreases, the porosity of the shale does increase but to a much smaller degree than it decreased, and the corresponding changes in traveltime or resistivity are also much smaller. So, a simple search for departures from a compaction trend cannot tell us about repressurization overpressure.

There are approaches that deal with this kind of overpressure. Bowers (1995) explicitly accounts for the loading and unloading processes and provides a log-based method for determining where the unloading begins. Doyle (1998) describes a comprehensive approach. This includes prediction both with the techniques described in the preceding and with a basin-modeling technique (tracking the evolution of pressure and porosity across the basin during its geological history), and also updating of the predictions during drilling using seismic-while-drilling and cuttings-based measurements. Standifird and Matthews (2005) give a more detailed description of the basin-modeling approach and its updating during drilling with revised formation tops and resistivity data.

Minimum-Horizontal-Stress Magnitude (σ_h). In Section 5.1.5, the state of stress in the subsurface was discussed. Some simple models of horizontal-stress magnitude were introduced. In this section, we discuss how they can be used in practice to estimate the minimum-horizontal-stress-magnitude (σ_h). When using such models for engineering applications, their limitations, and the simplifications they assume, should be kept in mind. The results of such models should be treated cautiously until the point where sufficiently good and plentiful field data can be collected to validate them. The use of such field data will be described.

Section 5.1.5 describes two simple models for σ_h (the USM and the FEQM), and the geological/basin boundary conditions under which these models might be applicable. These are both models where gravitational loading is the dominant (or only) source of stress. In more-complex settings (e.g., where tectonics plays a significant role), a more sophisticated model may be required, or more reliance on empirical calibration may be needed.

The USM. Despite its limitations (see Section 5.1.5), the USM remains in wide use in various forms. As shown in Eq. 5.1.22, it is theoretically the Poisson's ratio of the rock that controls the magnitude of σ_h for a given vertical stress (σ_v) and pore pressure (p_p). Poisson's ratio can be determined experimentally in the laboratory on core samples. However, it is rare to have core samples outside of the reservoir interval. Poisson's ratio can also be determined, where both compressional- and shear-wave-velocity measurements are available, from

$$
v = \left(\frac{\left(\dfrac{V_p}{V_s} \right)^2 - 2}{2\left(\dfrac{V_p}{V_s} \right)^2 - 2} \right) \quad \dotfill \quad (5.1.42)
$$

The Poisson's ratio in Eq. 5.1.42 is called the dynamic Poisson's ratio because it is derived from high-frequency (dynamic) deformations. The rate of these deformations is much higher than those that are applied to rock in situ, in the Earth, through geological loading. Under in-situ loading conditions, it is the static Poisson's ratio (that obtained in laboratory measurements over a period of minutes to hours) that is more applicable to geological conditions. Thus, some conversion from dynamic to static Poisson's ratio should be applied in order to use these log-based Poisson's-ratio values in the USM. In practice, it is very difficult to make a meaningful correction from dynamic to static. For this reason, the absolute magnitude of σ_h obtained through the USM is typically calibrated to measurements and observations from hydraulic fracturing and LOTs (more on this in the following). However, dynamic Poisson's ratios used with the USM may still give a sense of the contrast in σ_h between different lithologies. For example, adjacent lithologies with very similar dynamic Poisson's ratios might be expected to have a similar value of σ_h, whereas a stress contrast might be expected between adjacent lithologies with significantly different dynamic Poisson's ratios.

One important consideration when looking at dynamic Poisson's ratios is the pore-fluid effect. Gas, because it is very compressible, has a big impact on compressional-wave velocity and, thus, on dynamic Poisson's ratio. If gas is present, a fluid correction should be applied to extract the dry frame Poisson's ratio of all lithologies.

In practice then, σ_h derived from the USM needs to be corrected to the absolute magnitude of σ_h as obtained from some kind of measurement or test. The widely used Eaton fracture-gradient equation (Eaton 1969) is an example of this type of correction (in which fracture gradient and σ_h are assumed to be equivalent). Eaton's fracture-gradient equations (Eaton 1969) are purely empirical and are based on results of LOTs. There is one relationship for the Gulf Coast and Gulf of Mexico shelf (Eaton 1969) and another for the deepwater Gulf of Mexico (Eaton and Eaton 1997). Each relationship consists of back-calculated Poisson's ratio as a function of depth below the mudline. In addition to addressing the difference between static and dynamic Poisson's ratios, calibration to known values also addresses the nonelastic behavior (e.g., creep) of rock over geological time.

A standard LOT measures the resistance of a formation to fracturing. It is not in itself a direct measure of σ_h. The standard leakoff pressure (LOP), however, is often used as an approximation to σ_h. A more detailed discussion of the relationship between σ_h and the LOP is given in Section 5.1.4. For the purposes of this discussion, we will assume that a carefully selected LOP from a properly performed LOT is a reasonable first-order approximation to σ_h.

Because LOTs are almost always performed in shales, such an approach will provide a method of predicting σ_h for shales. Values of σ_h in sands can be determined from minifracs or datafracs performed before hydraulic-fracturing reservoir stimulation and are typically observed to be lower than in shales.

FEQM. As discussed in Section 5.1.5, in basins where active faulting is occurring, it may be that σ_h is not controlled by the elastic response to loading as in the USM assumption, but may instead be controlled by the frictional strength of appropriately oriented faults (Addis et al. 1994). In Eq. 5.1.24, it can be seen that in this case it is the residual-friction angle ($\Phi_{residual}$) that controls the magnitude of σ_h for a given σ_v and p_p. Values of $\Phi_{residual}$ can be measured in the laboratory. Values of between 11 and 20° have been reported (Wu et al. 1998) for weak shales. Higher values would be expected in more-sandy material and stronger (highly compacted or nonclastic) rocks. There have also been attempts to estimate friction-angle values from log measurements. However, in practice, as with the USM, quantitative predictions of σ_h using the FEQM rely on calibration to observations from stress measurements and LOTs.

Fig. 5.1.31 shows the theoretical variation in σ_h calculated by both the USM and the FEQM for assumed values of Poisson's ratio and $\Phi_{residual}$, respectively, under two different pore-pressure regimes (normally pressured and mildly overpressured). It should be noted that (for the purposes of simplification) the relationships

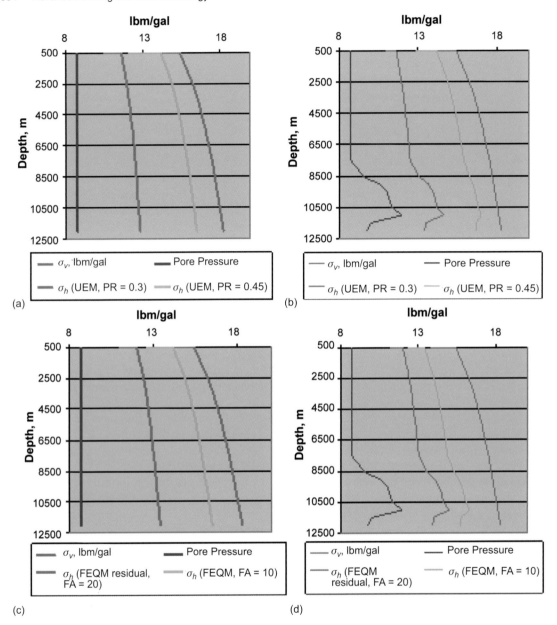

Fig. 5.1.31—Effects of variations in Poisson's ratio (PR) and friction angle (FA) on predicted stress in various models and environments. (a) The effect of the variation in PR from 0.3 to 0.45 on the calculated values of σ_h (using the USM) in a normally pressured setting. (b) The effect of the variation in PR from 0.3 to 0.45 on the calculated values of σ_h (using the USM) in a mildly overpressured setting. (c) The effect of the variation in $\Phi_{residual}$ on the calculated values of σ_h (using the FEQM) in a normally pressured setting. (d) The effect of the variation in $\Phi_{residual}$ on the calculated values of σ_h (using the FEQM) in a mildly overpressured setting.

shown in Fig. 5.1.31 assume that Poisson's ratio and residual-friction angle are constant with depth. In reality, as we see in the following section, these values usually appear to change with depth.

Calibrating Models of σ_h. From the above discussion, it is clear that commonly used models of σ_h rely on calibration to measurements and observations. The following is a discussion of how this can be achieved in practice.

As discussed in Section 5.1.1, the USM and the FEQM have the same form:

$$\sigma_h = K\sigma_v' + p_p, \quad\quad\quad\quad\quad\quad\quad\quad\quad\quad\quad\quad\quad (5.1.43)$$

where K is typically called the *matrix stress coefficient* or the *effective-stress ratio*. In the USM, $K = v/(1 - v)$, and in the FEQM, $K = (1 + \sin\Phi)/(1 - \sin\Phi)$.

In practice, because both models have the same form, we need only be concerned about seeking to calibrate K to the observed values of σ_h, regardless of which model we assume is most applicable. Although understanding the physical background to these models and how they may or may not apply to the setting in question is important, it can certainly avoid a lot of potential confusion if Poisson's ratio and friction angle are left out of the conversation, and a simple empirical K value is all we discuss.

Rearranging Eq. 5.1.43, we can see that K is simply the ratio of horizontal effective stress to vertical effective stress (hence the name effective-stress ratio):

$$K = \frac{\sigma'_h}{\sigma'_v}, \dots\dots\dots\dots\dots\dots\dots\dots\dots\dots\dots\dots\dots\dots\dots\dots\dots\dots\dots \quad (5.1.44)$$

where $\sigma'_h = \sigma_h - p_p$ and $\sigma'_v = \sigma_v - p_p$.

Vertical stress (σ_v) and pore pressure (p_p) are usually quite readily estimated over much of the interval depth. We can therefore determine K at any point where we have a measurement of σ_h. When we have determined several values of K, we can simply estimate the best empirical correlation to describe how K varies. K probably varies as a function of some mechanical property of the rock (for example, where the USM assumptions are correct, this would be the Poisson's ratio). Christman (1973) finds a good correlation between K and density. However, it is more common (at least in relatively simple gravitational-loading settings) for such correlations to describe how K varies with depth, as in the well-known Matthews and Kelly correlation (Matthews and Kelly 1967) and (indirectly) the Eaton equations (Eaton 1969).

Calibration of σ_h: An Example. The following example illustrates a typical situation, in which there is one existing well in a field, and fairly standard data are available. From these data, a relationship for K might be derived and be used to populate a simple MEM with σ_h. Here, we assume that K is a simple function of depth, which is reasonable for a simple gravitational-loading environment.

In **Fig. 5.1.32a,** p_p, σ_v, and some estimates of σ_h taken from LOTs in an existing exploration well are shown. Fig. 5.1.32b shows the values of K derived from these σ_h points plotted against depth. For reference, Eaton's Gulf of Mexico shelf K values are also shown in this plot. Two possible best-fit trend lines—a power-law fit and a linear fit—through these K values are shown. Both provide a reasonable fit across the depth of interest but diverge significantly at shallow depths (above 4,000 ft) and at greater depth (below 15,000 ft). Taking both of these trend-line fits and computing σ_h across the entire depth range yields the results presented in Fig. 5.1.32c. For reference, the Eaton Gulf of Mexico shelf trend is also plotted (Eaton 1969).

Note that the plots shown in Figs. 5.1.32a through 5.1.32c are based on LOT-derived values of σ_h and as such are likely to represent the relationship for shales. Because it is primarily shales with which we are concerned in wellbore-stability calculations, this is usually not an issue. However, where σ_h values for other lithologies are sought, different relationships would be needed.

Care must also be taken in extrapolating these types of relationships in complex geological and tectonic settings where effective-stress ratios may not be a simple function of depth.

Tectonic stress can potentially vary significantly across major geological features such as folds (a large anticlinal structure, for example, may be extensional at the crest and compressive in the core) and faults (tectonic stress that is transmitted through basement rocks may be absent in shallower rocks where they are decoupled by a detachment zone). Stress also varies in a notoriously unpredictable way around salt. In all of these settings, more-complex modeling (typically numerical, such as finite-element modeling) is likely to be required to predict values of σ_h.

Relationship Between LOPs and σ_h. The LOP is used routinely as an estimate of σ_h. The LOT, however, is not performed to measure σ_h and can therefore sometimes yield misleading results. The LOT typically is performed after drilling out the casing shoe in order to determine whether the cement job successfully isolated the casing annulus and to estimate the upper safe limit for MW or equivalent circulating density (ECD) to drill the next hole section.

To understand the relationship between the wellbore pressure measured during an LOT and the stresses in the rock surrounding the wellbore, it is instructive to consider the complete cycle of an XLOT. This is a procedure very similar to that of a hydraulic-fracturing stress measurement performed specifically to measure σ_h

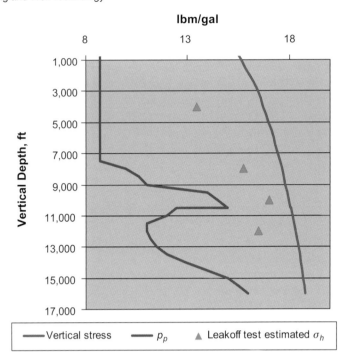

Fig. 5.1.32a—Input data to derive the stress model.

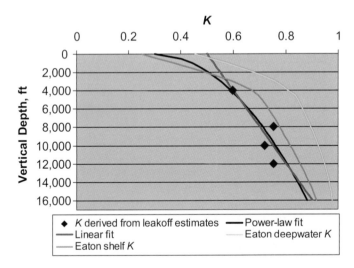

Fig. 5.1.32b—K values derived from the data in Fig. 5.1.32a, plotted as a function of depth. Two reasonable best-fit trends are presented, a linear trend and a power-law trend. Eaton's K values from both the Gulf of Mexico shelf and Gulf of Mexico deep water are plotted for reference.

in scientific boreholes (Hickman and Zoback 1983), or a minifrac (also called datafrac) performed to determine various parameters (including σ_h) required for the design of a large reservoir fracture-stimulation job (Economides and Nolte 1989). Although XLOTs are rarely performed in the industry, the first part of the test is the same as a standard LOT.

Fig. 5.1.33 represents a pressure-vs.-time (or vs.-volume, assuming a constant pumping rate during pressurization) record from an XLOT. Stage 1 is the initial pressurization once the well is shut in. The slope of the line is a function of the compressibility of the whole system (fluid, casing, pumping lines and equipment, and the rock exposed to the test). The LOP is usually defined as the point at which the pressure-buildup slope

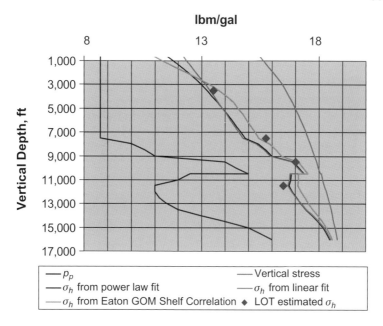

Fig. 5.1.32c—Values of σ_h derived from the various relationships of K with depth shown in Fig. 5.1.32b. The original leakoff-derived input data points are shown for reference.

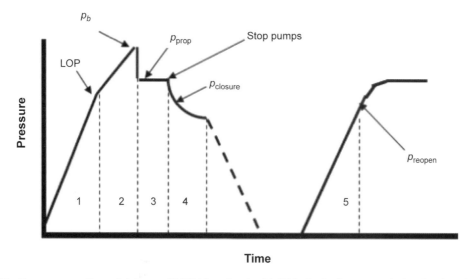

Fig. 5.1.33—Pressure vs. time plot for an XLOT. The standard LOT is typically stopped shortly after the LOP is seen.

deviates from linearity (Fig. 5.1.33). The LOP represents the point at which the system stiffness decreases, which under normal circumstances (assuming this is not mechanical failure of some part of the enclosing system such as cement or pump) is likely to be the initial opening of tensile fractures at the wellbore wall. How this relates to σ_h is discussed further later in this section.

In the standard LOT, this initial deviation from linearity is where pumping is stopped. In the XLOT, pumping continues and the pressure is typically seen to continue to increase (Stage 2) until the breakdown pressure (p_b) is reached. Breakdown is defined as the point at which the pressure actually drops, which indicates that the tensile fracture is growing at a faster rate (in terms of volume) than the rate at which the pumps are supplying fluid.

After some volume of fluid has been pumped into the fracture to ensure that it has propagated some distance from the wellbore wall (at fracture-propagation pressure p_{prop}, Stage 3 in Fig. 5.1.33), the pumps are stopped. During Stage 4, pressure bleeds off (either to the formation if the system is closed or back to surface in a flowback test) and the fracture closes. The pressure in the fracture just at the point where the fracture closes is called the fracture-closure pressure ($p_{closure}$) and is a good measure of the stress acting perpendicular to the fracture. This is the minimum compressive stress (usually σ_h).

In an XLOT, a second pressurization cycle is then performed, which reopens the fracture (Stage 5). The fracture created in Stages 1–4 is likely to remain hydraulically open (albeit mechanically closed) after Stage 4. In this case, it will be pressurized along its length in the second cycle (Stage 5) such that the reopening pressure (P_{reopen}) is again a good measure of σ_h.

Although $P_{closure}$ and P_{reopen} are the best-quality measures of σ_h, they are rarely obtained because the XLOT is rarely performed. The standard LOT, on the other hand, is performed at most casing shoes and therefore offers a much larger data set.

What then is the relationship between the standard LOP and σ_h? The answer depends very much on the nature of the borehole wall during the pressurization cycle. We can think about the nature of the borehole wall in terms of two end members:

1. An intact impermeable rock
2. A rock containing relatively long (one wellbore diameter or more), pre-existing, permeable cracks oriented perpendicular to σ_h

In the intact case (Case 1), the LOP will theoretically be equal to the breakdown pressure (where the breakdown pressure p_b is the pressure in the wellbore p_w in Eq.5.1.17 at breakdown), which is a function of both horizontal stresses and tensile strength and can be significantly higher than σ_h, depending on the relative magnitudes of all of these parameters. In this case, the LOP may resemble a breakdown pressure (Fig. 5.1.33) in that it is followed by a distinct pressure drop.

In the case of pre-existing cracks (Case 2), fluid can penetrate during pressurization to act on the sides of the crack, and the LOP is likely to be closer to σ_h. In this case, the LOT curve may approximate the fracture re-opening part of the curve in Fig. 5.1.33, and the LOP could be considered close to p_{reopen}, which as discussed is considered a good approximation to σ_h. This situation of long pre-existing cracks can arise naturally, of course, but in addition, large tensile fractures have been seen to form in the rathole below casing as a result of pressure surges during casing running and cementing.

In reality, many LOPs probably represent some process that falls in between these end members (Case 3). Although large pre-existing fractures are possible, completely intact rocks are probably rare. In addition to crack formation during natural rock deformation, the process of drilling a hole through rock may well generate small microfractures in the material surrounding the wellbore. In the case of short pre-existing cracks, fluid pressure can still act on the walls of the crack. Thus, if the LOP represents the point where these cracks start to open, it is likely to be a reasonable approximation of σ_h but may be more influenced by near-wellbore effects (e.g., getting pressure to the tip of the crack, solids content, fluid viscosity) and by the tensile strength or fracture toughness (Rummel and Winter 1983) of the rock, than in the end-member Case 2. In this case, some additional pressure may still be required for the fracture to propagate rapidly. This appears to be consistent with the typical shape of an LOT curve.

Fig. 5.1.34 is a schematic of LOT curves from the cases previously discussed (Case 1, Case 2, and the intermediate case, Case 3). The LOP from Case 1 may not be a good approximation to σ_h. The LOP from Case 2 should be a good approximation to σ_h, and the LOP from Case 3 may be a reasonable approximation to σ_h.

In summary, the relationship between LOP and σ_h is a complex function of in-situ stress and the nature of pre-existing cracks as well as the properties of the fluid during the LOT. Care should be taken when using LOT data to estimate σ_h. It is recommended that the original pressure/volume record and any other operational information be reviewed in the interpretation of LOTs.

Finally, in some circumstances useful information on the minimum stress can be obtained from drilling data, without any special operations. Edwards et al. (2002) describe calculation of σ_h values from annular-pressure-while-drilling measurements, detecting opening and closing of hydraulic fractures.

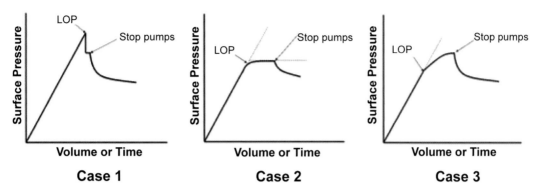

Fig. 5.1.34—Various types of LOTs, representing different downhole processes.

Horizontal-Stress Orientation. Where significant horizontal-stress anisotropy exists, the MW required for hole stability will be a function of the azimuth (as well as the deviation) at which the well is drilled. Horizontal-stress orientation can therefore be an important part of the MEM. In most relatively simple settings, we assume one principal stress is vertical (Section 5.1.5) such that the other two principal stresses are horizontal (and mutually perpendicular). By determining the direction of one of these horizontal principal stresses we have therefore defined the orientation of the complete stress tensor.

The direction of maximum horizontal stress (σ_{H_Az}) is usually the azimuth in which the resultant plate-tectonic driving forces act. Regional stress patterns mapped on the World Stress Map for the most part reflect this. Fig. 5.1.19 is an extract from the northwest European part of the World Stress Map (Reinecker et al. 2005). Here, σ_{H_Az} is seen to be largely determined by the present-day major plate boundaries, the Mid-Atlantic Ridge to the northwest and the African-Eurasian collision (forming the Alps) to the south. However, tectonic stress may not be present in shallower/younger sediments that are not strongly coupled to the underlying basement. Also, there may be local perturbations around geological structure or features such as salt. For these reasons, examination of actual well data within the field of study is recommended to determine the local stress field in the area and in the depth of interest. The World Stress Map, on the other hand, does still provide useful context and can provide some constraints in a rank wildcat situation where no close-offset-well data exist.

Well-log data have been used increasingly in the past 20 or so years to determine horizontal stress orientation. Borehole breakouts originally described by Leeman (1964) and subsequently investigated further by Babcock (1978) were shown to form on the walls of a vertical wellbore at an azimuth perpendicular to the maximum-horizontal-stress direction (**Fig. 5.1.35**). The use of oriented multiarm-caliper-log data to detect these breakouts was demonstrated by Bell and Gough (1982) and Plumb and Hickman (1985). More recently, wellbore images from both wireline (Zoback et al. 1985) and LWD tools (Bratton et al. 1999) have been used to identify breakouts.

When using four-arm-caliper data, care must be taken to distinguish breakouts from other features such as washouts and drilling-induced keyseats. A scheme for such differentiation (Plumb and Hickman 1985) is shown in **Fig. 5.1.36**. A common mistake is to confuse keyseating with breakout. Keyseating in this context is the erosion of one or more sides of the borehole wall by the BHA or drillpipe. In directional wells, the drillpipe tends to be in contact with either the top or bottom of the hole. As the drillpipe rotates it can erode the wellbore wall, causing elongation. Although this phenomenon is sometimes thought only to occur in wells deviated by more than 10°, it can in fact occur in a wellbore with any deviation at all. It is recommended that all high-side/low-side elongation be interpreted very carefully. A similar process of erosion can also occur on the sides of a hole where changes in azimuth occur. Therefore, any hole elongation that occurs in an area of high dogleg severity must also be treated cautiously when interpreting for breakout.

Image logs offer a potentially less ambiguous method of determining stress direction. These logs can actually create an image of a breakout on the borehole wall. The characteristic symmetrical nature of the breakout (i.e., that they form on both sides of the wellbore) helps to distinguish them from keyseat-type features. The

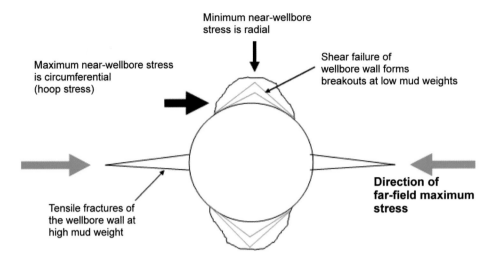

Fig. 5.1.35—Schematic cross section of a borehole showing a typical breakout aligned perpendicular to the direction of the maximum stress. For a vertical wellbore where horizontal-stress anisotropy exists, the breakout forms perpendicular to the azimuth of σ_H.

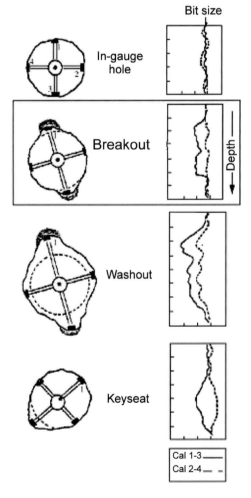

Fig. 5.1.36—Distinguishing breakout from other hole-geometry effects using four-arm-caliper data (Plumb and Hickman 1985).

fracture-bounded edges of the breakout are also distinct from features formed by erosion. **Fig. 5.1.37** shows examples of breakouts observed in different types of image log.

Drilling-induced tensile fractures (DITFs) are another commonly observed feature in image logs. As discussed in Section 5.1.4, while breakouts form because of compressive failure in the wellbore wall (at relatively low MWs), DITFs form because of tensile stress in the wellbore wall (at relatively high MWs). DITFs form perpendicular to the least compressive stress (i.e., at 90° to breakouts) and are therefore another useful indicator of horizontal stress orientation in a vertical wellbore. Because DITFs are quite narrow features, they are not typically detected by caliper logs (most caliper-log pads are wider than a DITF). However, they are often very clear in the higher-resolution (typically resistivity-based) image logs. Because they are typically axial fractures (i.e., parallel to the wellbore wall) they are quite distinctive and should not be easy to confuse with other features. **Fig. 5.1.38** shows examples of DITFs.

Breakouts and DITFs are both good indicators of horizontal-stress orientation in vertical wells (where the vertical stress is a principal stress). Where the wellbore is deviated, such that it is not parallel to one of the principal stresses, breakouts, DITFs, and other modes of failure can still occur, but they do not necessarily have a simple (parallel/perpendicular) relationship to the in-situ stress field. Interpretation of such features in deviated wellbores should be approached with care.

Observations of breakouts and DITFs are by far the most common methods used to determine stress orientations. Other core- and log-based methods do exist but typically require significantly more specialist analysis to yield robust results. These methods include the following.

1. Sonic log-based acoustic anisotropy (Franco et al. 2005). This method is based on azimuthal changes in shear- and compressional-wave velocity around a wellbore wall that are functions of anisotropic in-situ stress. Resolving velocity changes as a function both of azimuth around the well and of depth into the formation requires running a sonic tool in a specific mode. Specialist processing is also required to decouple stress-induced effects from intrinsic anisotropy resulting from rock fabric, such as bedding.
2. Numerous core-based methods. Clearly these all require the core to be oriented in some way. Core-based methods include anelastic-strain recovery (ASR) (Teufel 1983), differential strain analysis (DSA) (Strickland and Ren 1980; Ren and Roegiers 1983), and acoustic anisotropy (Ren and Hudson 1985).

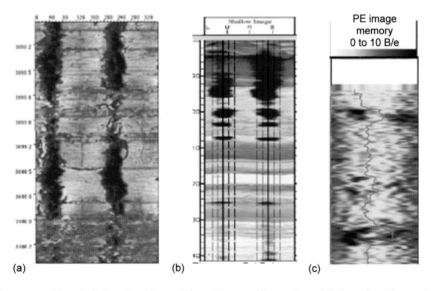

(a)   (b)   (c)

Fig. 5.1.37—Examples of borehole breakout from different types of image logs. (a) A breakout imaged using a wireline ultrasonic tool (UBI); courtesy of GeoMechanics International. (b) A breakout imaged using an LWD resistivity-at-bit (RAB) tool. (c) A breakout imaged from the photoelectric-factor data recorded on an LWD density tool (Greenwood et al. 2006). Example (b) is from Paper JJJ, 1999 by T. Bratton, T. Borneman, Q. Li, R. Plum, J. Rasmus, and H. Krabbe, presented at the 40th Annual SPWLA Conference, Oslo, Norway.

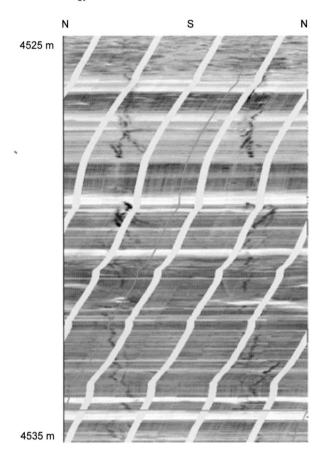

Fig. 5.1.38—Example of drilling-induced tensile fractures seen in a four-pad wireline resistivity (FMI) imaging log (light colors are higher resistivity). The fracture extends through mudstone and thin silty sands.

Maximum Horizontal-Stress Magnitude (σ_H). The maximum horizontal-stress magnitude σ_H is the most difficult parameter in the MEM to determine. Unlike σ_h, which can be directly measured by the hydraulic fracture and LOTs, there are no methods to measure σ_H directly. For this reason, σ_H typically has to be constrained through some model-based approach. Two common approaches are described in more detail below. These are based on frictional equilibrium and on the observation of breakouts and DITFs. A third approach, based on inversion of leakoff data from multiple wells, is described in "Normalization and Inversion Methods," Section 5.2 of this book.

Constraining σ_H From Frictional Equilibrium. The concept of frictional equilibrium in the crust has already been covered in Section 5.1.5. If we have determined all the other components of the MEM (as described above), we can use the frictional-equilibrium principle to constrain σ_H. In other words, for a given set of σ_v, σ_h, and p_p values, the upper limit of σ_H is determined by the frictional strength of the rock mass (usually the residual-friction angle on existing favorably oriented faults).

The range of possible stress magnitudes for a given frictional strength can be illustrated as a stress polygon (Zoback et al. 1985). The stress polygon displays the permissible magnitudes of horizontal stress (assuming frictional equilibrium) for a given value of σ_v, p_p, and $\Phi_{residual}$. There are three sectors to the polygon, each representing a different Andersonian faulting regime (Section 5.1.5).

Fig. 5.1.39 shows an example of two stress polygons (Zoback et al. 2003). Fig. 5.1.39a shows the permissible magnitudes of horizontal stress at a depth of 3 km assuming hydrostatic pore pressure, a value of 30° for $\Phi_{residual}$, and a known σ_v of 70 MPa. Fig. 5.1.39b shows the permissible magnitudes of horizontal stress for the same assumptions as in Fig. 5.1.39a, except that in this case, p_p is significantly higher (i.e., this is an overpressured environment). The increased pore pressure has the effect of greatly reducing the range of

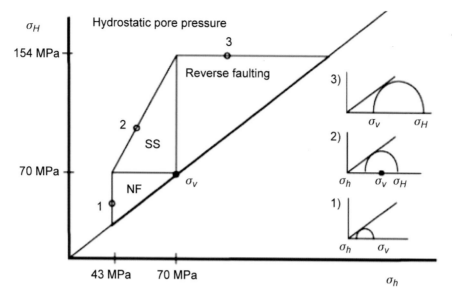

Fig. 5.1.39a—Stress polygon (Zoback et al. 2003). For a given vertical stress and frictional strength, the range of permissible horizontal stresses can be computed. The stress state can lie anywhere within the polygon. However, if it is known to be in an active state of faulting, the stress state will lie on one of the bounding lines of the polygon (depending on the relative magnitudes of the three stresses—i.e., the stress regime) such as the example stress states shown (1—normal faulting, 2—strike/slip faulting, and 3—reverse faulting). Reprinted with permission from Elsevier.

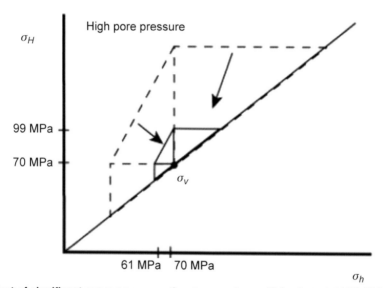

Fig. 5.1.39b—Effect of significant overpressure on the stress polygon (Zoback et al. 2003). This plot represent the same setting (vertical stress and rock strength) as Fig. 5.1.39a but with an increased pore pressure. Reprinted with permission from Elsevier.

possible stress values. This is consistent with the observation that in highly overpressured environments, we rarely see significant stress anisotropy. In both parts of Fig. 5.1.39, the range of values of σ_H can be seen that are allowed for a known value of σ_h.

Constraining σ_H From Observations of Wellbore Failure. When the stress concentration at the wellbore wall exceeds the rock strength, the rock in the wellbore wall will fail, either in compression or in tension. When we can observe such a failure in a wellbore and we know the wellbore pressure (MW) at the time of failure, we can estimate (assuming that we know all other components of the MEM), through some failure criterion and through the equations that describe stress at the wellbore wall (i.e., Equations 5.1.10–5.1.13), the magnitude of σ_H.

There are several different modes of stress-induced failure of the wellbore wall, depending on the relative magnitude of the in-situ stress, rock strength, and MW (Bratton et al. 1999). However, the most common are breakouts (shear failure of the wellbore wall at relatively low MWs) and DITFs (tensile failure of the wellbore wall at high MW). Assuming we identify these modes of failure correctly, we can use that observation to constrain σ_H.

Where a breakout is observed in a wellbore that was exposed to a known MW, we can conclude, on the basis of the wellbore-stability calculation (Section 5.1.4), that σ_H must have exceeded a certain value for the breakout to have formed. This places a lower limit on the magnitude of σ_H (i.e., σ_H must be at least this high for the breakout to have formed).

On the other hand, in the case where no breakout is observed, we can conclude that the magnitude of σ_H cannot exceed a certain value; i.e., we can derive an upper limit (σ_H cannot be higher than this, otherwise a breakout would have formed). Depending on the mechanism of breakout detection, the nonobservation of a breakout is perhaps more difficult to confirm than the positive observation of a breakout. For example, if only four-arm-caliper data are available, a wellbore that appears to be in gauge may actually have failed, but the failed rock may not have been removed from the wellbore wall. High-resolution image logs are a more reliable way of detecting failed or intact rock.

The observation of DITFs can also be used to constrain σ_H. As for breakouts, the assumption is that the wellbore pressure at which a DITF forms is a function of the stress concentration (and therefore of σ_H) at the wellbore wall (Eq. 5.1.10). Under this assumption, high values of σ_H relative to σ_h tend to promote the formation of DITFs. Therefore, as with the observation of breakout, where a DITF is observed for a given wellbore pressure, we can calculate the lower limit of σ_H. In other words, we can say σ_H must be at least this high for a DITF to have formed.

Although this method seems to be used quite widely, there are a couple of points that should be kept in mind. First, when we apply Eq. 5.1.10 to solve for σ_H, we are assuming that the boundary conditions behind Eq. 5.1.10 are applicable. One of these boundary conditions is that the wellbore wall is initially intact and impermeable. As discussed previously in this section, during the process of wellbore pressurization, fluid may penetrate and pressurize pre-existing cracks, which is not a process that is captured by Eq. 5.1.10. As we saw in the discussion of LOT interpretation, tensile fractures may actually propagate from the wellbore wall at a pressure closer to σ_h. This should be kept in mind when DITFs are observed in wells where the wellbore pressure used to drill (or pressure surges during trips) may have been close to or have exceeded σ_h. Second, if the assumptions behind Eq. 5.1.10 are valid, then the tensile strength of the rock should be included in the calculation. Estimates and measurements of tensile strength can vary quite widely, which adds some significant uncertainty to the σ_H estimate (although an assumption of zero tensile strength will still enable a lower bound to be estimated) for σ_H.

When using observation of breakouts and DITFs to estimate σ_H, it is important to know the pressure to which the wellbore has been exposed between being drilled and making the observation. This is usually approximated by the MW to which the hole has been exposed. However, transient pressures such as swab and surge that occur during normal drilling practices must also be considered (particularly for the case of DITFs). For example, a surge pressure that might occur if the wellbore packs off while pumping may be several hundred psi greater than the hydrostatic MW. It may be the surge pressure that is responsible for creation of a DITF, as opposed to the hydrostatic pressure. Likewise, if the wellbore experiences a significant reduction in pressure because of swabbing, resulting in a breakout, the calculation of σ_H must be based on the swabbing pressure in order to derive an accurate value. Ideally a time-based record of downhole pressures should be used to ensure that the appropriate wellbore pressure is used in the calculation.

Example of Constraining σ_H From Observation of Breakout and DITF. Well A is a vertical well in which a high-resolution image log was run, and breakouts were observed at 5,000 ft, 8,000 ft, and 10,000 ft. All components of the MEM except σ_H have been determined. These values are presented at the observed breakout depths in **Table 5.1.3**. The values in Table 5.1.3 can be inserted into Eq. 5.1.15 (where σ_A would be σ_H and σ_B would be σ_h) to determine the lower bound to σ_H (i.e., σ_H must be at least this great for breakouts to have formed). The calculated lower bounds to σ_H are included in Table 5.1.3, and the final estimated stress profiles are shown in **Fig. 5.1.40a.**

TABLE 5.1.3—MEM COMPONENTS AT DEPTHS OF OBSERVED BREAKOUTS IN VERTICAL WELL A,
AND CORRESPONDING CALCULATED LOWER BOUNDS TO σ_H

Depth of Observed Breakout (ft)	σ_v (lbm/gal)	p_p (lbm/gal)	σ_h (lbm/gal)	UCS (lbm/gal equivalent)	$\Phi°$	Minimum Pressure Experienced by Wellbore (lbm/gal)	Lower Bound to σ_H (lbm/gal)
5,000	16.99	8.70	12.77	15.41	25.00	9.00	15.54
8,000	17.63	9.10	13.28	14.45	27.00	9.60	15.92
10,000	17.98	11.20	14.52	13.49	30.00	11.80	17.60

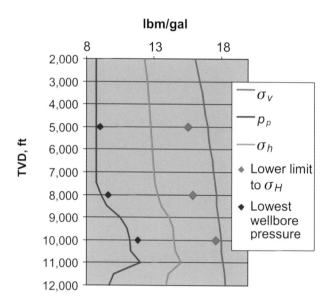

Fig. 5.1.40a—Stress profiles and calculated lower-bound values of σ_H from breakout observations in Well A.

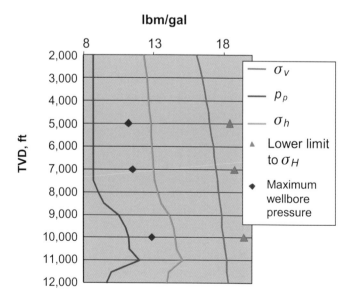

Fig. 5.1.40b—Stress profiles and calculated lower-bound values of σ_H from observed DITFs in Well B.

TABLE 5.1.4—INPUTS USED TO CALCULATE LOWER BOUND TO σ_H FROM OBSERVATIONS OF DITFs IN WELL B						
Depth of Observed DITF (ft)	σ_v (lbm/gal)	p_p (lbm/gal)	σ_h (lbm/gal)	Tensile Strength	Maximum Pressure Experienced by Wellbore (lbm/gal)	Lower Bound to σ_H (lbm/gal)
5,000	16.99	8.7	12.77	0	11.2	18.40
7,000	17.43	8.7	12.98	0	11.5	18.74
10,000	17.98	11.2	14.52	0	12.9	19.47

Well B is also a vertical well but is in a different part of the world than Well A. A high-resolution image log was also run in Well B, and DITFs were observed in three places. **Table 5.1.4** shows the MEM components at the depths of the observed DITFs. These values can be substituted into Eq. 5.1.16 (where σ_A would be σ_H and σ_B would be σ_h) to determine the minimum value of σ_H (i.e., σ_H must be at least this high for DITFs to have formed). The calculated lower bounds to σ_H are included in Table 5.1.4, and the final estimated stress profiles are shown in Fig. 5.1.40b.

The MW Window. Having built the MEM as described above, we are ready to calculate the MW window for a given well trajectory. In the most common case of a 1D MEM, the model could simply be a text file or spreadsheet containing all of the MEM parameters along the well trajectory in question, referenced to vertical depth. At this stage in the process, software is typically required to compute an MW window efficiently. The software is simply using Eqs. 5.1.10 through 5.1.17 to compute the required MWs. However, for a well oriented nonparallel to a principal stress, and where the calculation has to be done at multiple depths, it would be very time-consuming if performed by hand.

The number of depths at which to perform the calculation depends on the level of detail required. For a well profile with multiple changes of lithology, for example, more depths may be necessary (perhaps every 10 ft) than for the case where the interval modeled is predominantly shale (perhaps every 100 or more ft). Alternatively, we might want to look in detail over a shorter interval within a well where particular problems have occurred or where a high-resolution image log is available for calibration. In this case, we might want to go as far as performing the calculation every foot.

As described in the introduction to Section 5.1.6, the lower bound to the mud-weight window is defined by either the pore pressure of permeable intervals or the minimum MW (MW_{min}) required to prevent excessive shear failure, whichever of these is the greatest value. In a strong rock, MW_{min} can be lower than the pore pressure (**Fig. 5.1.41,** Zone 1). If this rock is also impermeable, then it can be drilled safely with a MW lower than pore pressure. Often, however, MW_{min} is higher than pore pressure, particularly in a tectonic setting or in a deviated well.

The upper bound to the MW window is the fracture gradient. Although fracture gradient can be defined in several ways, including the fracture-initiation pressure, the measured LOP, and the estimated minimum horizontal stress, in practice it is whatever the wellbore can sustain without taking significant losses. In some wells, the well-control and kick-tolerance environment is such that suffering minor losses that may be returned on connections (wellbore breathing/ballooning) is not a major concern. In other cases, any lost circulation may pose a threat to well control, and a more conservative fracture gradient might be used. Different fracture gradients also may be used for different lithologies. Sand fracture gradients are typically thought to be lower than shale fracture gradients in normal_faulting environments because the stiffer sand is thought to carry less horizontal stress. Often, however, the sand fracture gradient can in practice be increased through the appropriate use of lost-circulation materials that prevent hydraulic tensile fractures in sand from propagating.

Fig. 5.1.41a shows an example of a MW-window plot. Two different well trajectories are considered: a vertical well (Well A) and an extended-reach well with a long 60° tangent section (Well B). The geology is assumed to be laterally continuous (layer-cake-type model), so both trajectories penetrate the same rock properties at the same vertical depth. A simple 1D MEM can be used to calculate the MW window for both wells. The key inputs to the MEM are shown in Fig. 5.1.41b. The fracture gradient chosen in this case is the minimum horizontal stress in the shales. This is thought to be the most likely practical upper limit (above

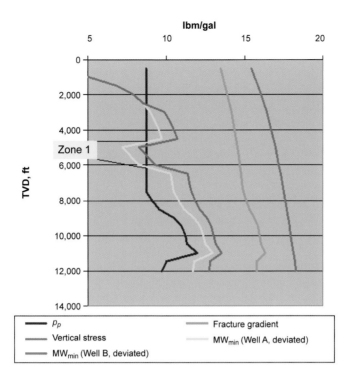

Fig. 5.1.41a—Mud-weight window for Wells A (vertical) and B (60° deviated) through the same MEM. The lower limit at any depth is given by either MW_{min} or the pore pressure (whichever is larger). The upper limit is given by the fracture gradient.

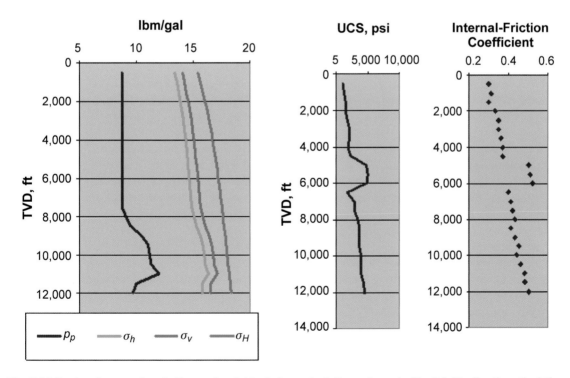

Fig. 5.1.41b—Input parameters to the mud-weight-window calculations shown in Fig. 5.1.41a. For the calculation shown, Poisson's ratio and Biot's coefficient are assumed to be constant and equal to 0.3 and 1, respectively.

which, fractures will propagate rapidly from the wellbore wall, inducing significant lost circulation) in this case.

In Well A, from surface to a depth of approximately 2,500 ft, the rock is strong enough to be drilled under-balanced (i.e., with a MW less than pore pressure). However, as there may be permeable zones within that interval, the pore pressure is the lower limit to the MW window. In most of the rest of the well, the lower limit is given by the calculated MW_{min}. Only in Zone 1, which is a particularly strong zone (for example, it may be a particularly well-cemented sandstone), does the pore pressure again become the lower limit to the MW window.

Well B is vertical down to 2,500 ft and is therefore identical to Well A. Below 2,500 ft Well B builds angle rapidly to 60°. The higher angle of the hole (in this normal-faulting regime) requires a higher MW to prevent shear failure.

The MW_{min} values shown in Fig. 5.1.41a are calculated to prevent any shear failure at all from occurring in the wellbore wall. In practice, this might be an overconservative way to determine the required MW. In reality, it depends on the requirements of the particular hole section. For example, some degree of shear failure in the wellbore wall can be tolerated without seriously interfering with drilling and casing operations. Thus, if the objective is just to drill and case this interval, we could use a lower MW without much risk. On the other hand, if the objective is to obtain high-quality formation-evaluation logs in this interval, we might well want to keep the hole as close to gauge as possible, as the quality of some logs is reduced by a rugose wellbore wall. Similarly, in a production interval, it may be beneficial to keep the hole as close to gauge as possible to optimize installation of the completion equipment. Unlike a simple string of casing, completion equipment can be more complex and troublesome to run into a rugose hole. A good cement job (uniform cement around all components) is also sometimes easier to achieve in a gauge hole. Similarly, some completion equipment, such as an expandable sand screen, relies on making uniform contact with the wellbore wall. In a rugose hole, problems may be more likely during production and compaction because of nonuniform loading of the completion equipment.

Calibrating and Sense Checking—Comparison to Offset-Well Caliper, Drilling Experience, and Other Factors. Before we use the MEM to make MW recommendations for future wells, it is a good idea to test the model results on existing wells to ensure that the calculated MWs are consistent with observations. Although this may seem like an obvious step, it is often overlooked. If the steps described in Section 5.1.6 to build an MEM have been followed, then we should have already included a large amount of well data in deriving the inputs to the MEM. However, either in these wells, or other wells not yet included, do the MWs predicted by the MEM make sense when compared to observations while the well was actually drilled? To answer this question, we need to read the drilling reports. Daily drilling or operations reports will usually describe when and where hole problems were encountered.

Reality-Check Examples. Suppose we have built an MEM based on data from Wells X and Y, which provided a good data set including good-quality LOT records, caliper logs, and image log with which to estimate stress orientations and magnitudes. Well Z is in the same field. Formation depths and pressures are slightly different at the Well-Z location, so we build a new 1D MEM for the Well-Z location. Well Z has a more limited set of logs, but it does include a density and sonic log from which to estimate rock strength and pore pressure. On the basis of our understanding of the geological setting of the field, it seems reasonable to assume that the stress regime at Location Z is comparable to that at Locations X and Y. So, we have an MEM with vertical stress, pore pressure, and rock strength specific to the Well Z location and a stress model derived in offset wells applied to the Well-Z location.

Case 1. We run the trajectory from Well Z through our MEM, and it calculates a minimum required MW to prevent excessive shear failure. However, when we read the drilling reports from Well Z, we discover that in the 12¼-in. hole section, they actually used a MW between 0.5 and 1 lbm/gal below that calculated by the MEM. We continue reading the drilling reports all the way to the end of the hole section and through the casing running and see no indications of hole problems. No image logs or caliper logs were run, so we have no direct indication of what happened to the hole. What does this mean? Some possible explanations are discussed below.

Explanation (a): An error in the MEM. It may be that there really is a fundamental difference between our MEM and reality. For example, our assumption that the stresses derived at Wells X and Y will apply to Well Z may be invalid. This is more likely to be the case in structurally complex and/or tectonic environments. Alternatively, our shale-pore-pressure estimate (based on the sonic and density logs) may contain uncertainties and error bars. The petrophysical models used to estimate pore pressure relate velocity to effective stress and are sensitive to processes such as diagenesis, cementation, uplift, and unloading. When these secondary processes are significant, the uncertainty on the shale-pressure estimates increases. Even when we have good pressure measurements in adjacent sands, shale pressures can still have considerable uncertainty because of lateral pressure-transfer processes, which affect sands and shales differently.

In short, each of the inputs to the MEM contains some uncertainty, and it requires some experience and engineering judgment to assess which, if any, of the inputs may be in error.

Explanation (b): Rock is *failing* but not *collapsing*. In the Introduction to this chapter, the difference between rock failure and well failure was discussed. The wellbore-stability model calculates the MW required to prevent the rock in the wellbore wall from failing. Typically, in fairly brittle rocks, once the rock has failed it is likely to become detached from the wellbore wall and fall (or be knocked by the BHA/drillstring) into the annulus, potentially creating drilling problems. In a more ductile rock (softer shales or high-porosity sands) or if the mechanical action of the BHA/drillstring on the wellbore wall is quite gentle, this post-failure removal of the rock may not occur and the well may appear to be intact.

Explanation (c): Hole is collapsing but does not impact drilling practices significantly. Again, as discussed in the chapter Introduction, we should bear in mind the difference between rock failure and hole failure. The wellbore wall may well be failing and producing cavings into the annulus. However, the degree to which this impacts the drilling process negatively depends on a number of factors. Wellbore instability may actually be occurring, and careful observation may detect cavings at the shakers or other evidence, but this may not be mentioned in a daily drilling report if there was no negative consequence to drilling operations. This may be the case in certain circumstances, including the following:

- In large hole sizes (e.g., 17½ in. or bigger) where a caving may still be small in relation to the size of the annulus (drillstring would be 6⅝-in. diameter at most).
- Where good hole cleaning is achieved. The power and surface-pressure ratings available on modern mud pumps enable very large flow rates, which combined with a well-kept mud rheology, can keep most hole sizes clean, even in the case where there is some degree of hole enlargement and cavings production.
- In near-vertical holes. In high-angle wells, hole rugosity (resulting from instability) can cause tight hole, packoff, and even stuck pipe during trips. In near-vertical holes, however, where most material is quickly removed because of good hole cleaning, quite significant rugosity can be tolerated on trips (depending on the BHA configuration or other factors) because the drillstring is quite free to move and is not pushed against the side of the hole.

In summary, there are many factors that affect whether hole instability impacts drilling. A certain amount of instability in one hole (with given hole-cleaning capacity, BHA type, and related factors) may cause no problem at all, but similar instability might be the cause of stuck pipe and sidetracks in another hole with a different combination of such factors as hole-cleaning capacity and BHA configuration.

Case 2. We run the trajectory from Well Z through our MEM and it calculates a minimum required MW to prevent excessive shear failure. However, when we read the drilling reports from Well Z, we discover that in the 12¼-in. hole section, they actually used a MW between 0.5 and 1 lbm/gal below that calculated by the MEM. We read through several days of operations that describe the process of drilling this interval, and we are surprised not to read of any hole problems. We read how the hole section (quite a long high-angle hole) reached TD in record time, a couple of days ahead of schedule, meaning the drilling crew would likely get a good bonus for their two weeks' work. If we were to read only to this point, we might conclude that our model was wrong. However, until the casing is at TD and cemented, there is potential for things to go wrong. In fact, many if not most cases of instability develop into significant operational problems only when tripping out of the hole. As we continue to read, we learn that after a short bottom-up circulation was performed at TD, they began to trip out of the hole as fast as they could. After pulling a few stands, the hole began to pull tight. They washed back down and tried again to pull out of the hole, but this time they packed off and nearly got stuck. It gets worse as we read ahead. The next week is spent fighting instability trying to get out of the hole and then back into it to clean up. They raise the MW in 0.5-lbm/gal increments and are soon in excess of the original recommended MW. With the high MW and the continual packoff, they accidentally cause a pressure surge high enough to break down the formation and begin losing mud. A further week is spent struggling between losses and instability until the hole is abandoned, sidetracked, and redrilled with the recommended MW.

In summary, hole instability does not always occur right at the bit while drilling. It can occur some hours or even days after the hole has been drilled. Although the rock may initially yield soon after it is drilled, it may be the multiple pressure fluctuations to which the well is exposed (e.g., pumps on and off, swab and surge because of pipe movement) that destabilize the yielded rock and create instability. Such an unstable hole section may have little impact on operations while the wellbore below it is drilled, because it only contains a drillstring. However, trying to trip out through such a section with several hundred feet of BHA equipment of much larger diameter than the drillstring may be a much more difficult operation.

Case 3. We run the trajectory from Well Z through our MEM and it calculates a minimum required MW to prevent excessive shear failure. When we read the drilling reports from Well Z, we discover that in the 12¼-in. hole section they used a MW equal to or higher than that calculated for the entire hole section. However, despite seeming to have used the appropriate MW, we are surprised to read that there were considerable hole problems. Although there is no mention of cavings or any explicit description of hole instability, there are numerous occasions where they experience tight hole, high torque and drag, and even stuck pipe.

On further investigation, we discover that an oriented four-arm-caliper tool was run in this hole, and we acquire the data. The caliper data show several sections of the well where the hole is enlarged. At first glance, it appears that our wellbore-stability model must be wrong. However, when we look at the caliper data relative to the trajectory of the hole, we see that most of the enlargement is on the high side and low side of the hole. In these places, the softer rock has been eroded by the action of the drillpipe (like a keyseat), and the stronger rock has remained in gauge. On tripping out of an enlarged hole and into an in-gauge hole, overpull was experienced that resulted in extra reaming, which probably just further enlarged the already enlarged hole section. Lower annular velocities in the enlarged sections meant that the hole was not being cleaned efficiently, and cuttings beds were building up. This was not instability but simply a hole-geometry and associated hole-cleaning problem.

In another section of the hole, we see that the enlargement is on the sides of the hole, and at first we think this really is shear failure and breakout. However, when we look closely, we see that this enlargement is at the exact point where there is a sharp change in the hole azimuth. The hole azimuth had been drifting out of range for the previous several hundred feet, and at this point some corrective action was taken to get back on track. This created a high dogleg and side forces on the side of the hole that were sufficient to erode the softer portions. This interval was also problematic on trips, but again this was not a hole-stability problem but a hole-geometry problem.

Case 4. We run the trajectory from Well Z through our MEM and it calculates a minimum required MW to prevent excessive shear failure. When we read the drilling reports from Well Z, we discover that in the 12¼-in. hole section they used a MW equal to or higher than that calculated for the entire hole section. However, despite seeming to have used the appropriate MW, we are surprised to read that there were considerable hole problems. The drilling report describes large volumes of cavings coming over the shakers very soon after the start of the interval. Several days are spent trying to drill farther, but neither mud-weight increases nor the pumping of sweeps to clean the hole seems to help. A day after increasing the MW, so that it is now 1 lbm/gal higher than the recommendations of the wellbore-stability model, the instability just seems to be getting worse and the hole is abandoned. The hole condition was too poor to run any wireline logs, so it at first appears that we will never know the cause of the instability. However, on reading the BHA description, we notice that an LWD density tool was run on this hole section, and we go back to look for those data. The operator was running the density tool only as a wireline replacement for a density measurement in possible pay sands they were anticipating toward the bottom of the hole. Although they were offered the wellbore images and oriented-ultrasonic-caliper data that the density tool records as it rotates, they declined because it was an additional cost. Well Z is sidetracked and redrilled with a higher MW—1.5 lbm/gal higher than the wellbore-stability model calculated. Other than this increase in MW, nothing else about the well is changed. The drilling reports read very similar to the report of the first attempt to drill and end with a stuck pipe and lost BHA. Fortunately, the LWD provider kept the raw data from the original hole in Well Z and is able to provide the density, photoelectric factor (PEF), and caliper data, which clearly show that all of the hole enlargement is occurring along bedding planes and around fault zones. On re-examination of the seismic, we see that because of the dip of the beds in this area, the well trajectory is almost parallel to the bedding. In addition, there are some indications of minor faulting, which had been overlooked originally. The well has clearly encountered a zone of fissile and fractured shale in which the fabric of the rock is playing a significant role in the instability. The pressure of the mud in cases like this does not act to support the wellbore wall because the mud tends to penetrate the network of weaknesses in the rock. The wellbore-stability model is a simple model of shear failure of intact rock and does not account for the processes at work here. Increasing MW does not solve this type of instability.

Cases 1–4 above describe some real-life examples of wellbore-stability-model predictions that were not a good match to the apparent operational experience. When building and testing these models, we clearly have to be aware of the whole picture. We must keep in mind the evolution of instability with time, the interaction of instability with the drilling equipment, other processes that can be significant in the mechanics of drilling, other mechanisms of instability not accounted for by the model used, and of course the possibility that the geological complexity of the subsurface is beyond what we can describe accurately from looking at an offset well drilled some miles away.

Identifying and Communicating Other Hazards. In the preceding sections, we have discussed how to build an MEM and calculate the safe MW window. However, as noted in the discussion on calibration and sense checking, even when we keep the MW within the safe window, we may still encounter other drilling hazards. These are typically disturbed zones or discontinuities, often associated with faults or highly deformed intervals. They might be highly fractured and fissile zones of rock that cannot be stabilized by applying MW. They might be zones containing natural open fractures to which lost circulation will occur at any MW higher than pore pressure. These are hazards for which a correct choice of MW may not be the deciding factor in success. They are more usefully described in a qualitative way (a description of what they are, the mechanism by which they are troublesome, the corresponding mitigation, and some estimate of the probability of encountering them), and it is important that the predrill plan still captures them.

A good drilling plan should try to identify and communicate the possible drilling hazards before drilling. If a hazard is anticipated, drilling practices can be adopted accordingly to minimize its impact. Given the inherent uncertainty in the nature of subsurface prediction, identifying drilling hazards is a difficult task. Ideally, input from a wide range of team members should be sought. In some cases, people familiar with drilling wells in the area can offer a good insight into the nature of certain hazards. Other times, in new areas, or where we do not have access to offset-well drilling data or experience, we have to rely more heavily on seismic interpretation and geological models. In all cases, we should seek the insight of a multidisciplinary group including geologist, geophysicist, petrophysicist, drilling engineer, and wellsite drilling personnel (wellsite geologists, mud loggers, drilling-performance/-optimization engineers).

When all the potential drilling hazards have been captured, they need to be documented and communicated. The important thing is to ensure that the knowledge distilled from the group is conveyed in a clear and concise way to the people who are at the forefront of operational decision making while the well is drilled. This includes both office- and rig-based staff. The drilling industry is fraught with stories of miscommunication or complete lack of communication either between subsurface and drilling teams or between the office-based team and the wellsite.

Visualization is an important part of communicating drilling hazards. There has been a trend in recent years to make use of 3D visualization. First, this can help to illuminate the relationship between the geology and the hazard and thus help the team to understand the root cause of the issue. Second, clear visualization is very helpful in communicating these hazards to a wider team. **Figs. 5.1.42a and 5.1.42b** show two examples of 3D drilling-hazard visualization. Models like this are typically built where some offset-well data exist. The well paths, geological model, and drilling problems encountered are displayed simultaneously. Planned well paths can be added to the display to help see where we might expect to encounter hazards.

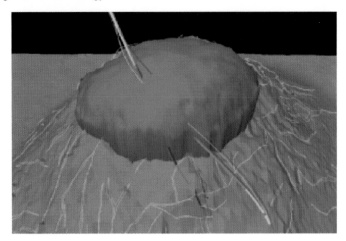

Fig. 5.1.42a—Example of 3D visualization, analysis, and communication of drilling hazards within a complex geological environment. The well paths shown intersect a salt diapir and are color-coded according to the drilling hazards they encountered, such as fractured fissile shale around the diapir, bedding-parallel drilling on the flanks of the diapir, and open natural fractures within the chalk at the base of the diapir (Bratton et al. 2001). Image copyright Schlumberger. Used with permission.

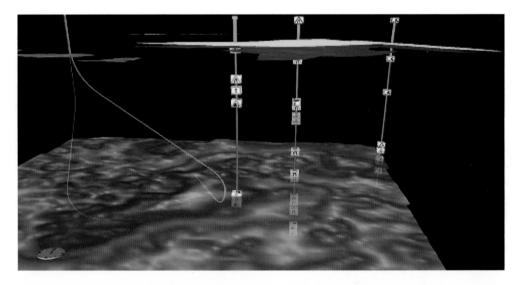

Fig. 5.1.42b—Example of 3D visualization of mapped well events in three vertical exploration and appraisal wells. The well paths are shown together with the major geological features such as formation tops and faults (for clarity, not all of these are shown here; they can be added and removed from the view by a simple mouse click). In this example, the drilling events in the existing wells are indicated by the different symbols. How these events relate in a geological context to the planned hook-shaped well paths shown in green can be seen clearly (Greenwood et al. 2006).

As a well is drilled, new data are acquired and the models may be updated. If this is done in real time, while the well is being drilled, it can help to clarify any new issues that are encountered. Because events can develop very rapidly while drilling, communication in this stage becomes critically important. Any tools that help with interpretation and communication while drilling can be extremely valuable. This type of real-time or while-drilling approach is discussed further in Section 5.1.7.

5.1.7 Wellbore Stability While Drilling: Modeling, Monitoring, Diagnosing, and Improving. The preceding sections in this chapter have described the mechanisms that control wellbore stability and have outlined the type of data and approach required for predrill wellbore-stability planning. Armed with this knowledge, in a perfect world, good planning should lead to prevention of wellbore-stability problems. In reality, however, the predrill understanding of the subsurface is never perfect. Despite considerable advances in seismic and other subsurface technology in recent years, it is very often still the case (particularly in exploration wells, but also to a significant extent in appraisal and development wells) that fundamental properties such as depth, lithology, pressure, nature of faults, and rock fabric carry a high uncertainty until the well is actually drilled.

Given this inherent predrill uncertainty, even in the best-planned wells, what can be done to mitigate potential wellbore-stability problems while drilling? It is worth pausing at this point in the discussion and taking a step back to think about this problem. It is certainly preferable to be able to predict a problem and to avoid its occurrence completely. In fact, some would argue that once a problem is encountered (when there has been failure to predict it), it is too late to be modeling and monitoring. With some problems, that may be true. However, with wellbore stability, problems tend to develop over a matter of hours or even days, so that early identification of a problem can enable remedial action in time to prevent serious consequences. So, although predrill planning is a very important part of the drilling process, the while-drilling analysis described in the following sections is equally, or sometimes more, important.

An approach that has proved to be effective is illustrated in **Fig. 5.1.43.** The first stage of the workflow consists of updating the wellbore-stability model with newly acquired data (e.g., LWD measurements) in conjunction with close monitoring for signs of instability. Modeling and monitoring provide the information required to recognize instability and diagnose the mechanism, which points to the appropriate remedial action to improve or manage the wellbore instability.

The phrase *monitoring wellbore stability while drilling* refers to the process of observation. Various pieces of information are available while drilling (cavings; drilling parameters such as torque, drag, and annular pressure; perhaps LWD calipers or even wellbore images) that give some indication of what the wellbore is actually doing. This is distinct from the phrase *modeling wellbore stability while drilling.* Modeling means running a piece of software that requires some input (e.g., rock strength and Earth stress/pore pressure) and provides an output, typically the MW required to prevent instability for a given depth and hole trajectory.

Modeling Wellbore Instability While Drilling. The predrill wellbore-stability model has been described in Section 5.1.6. When the well is drilled, we typically find that things are not exactly as we expected. For the wellbore-stability model to still be of use, the inputs must be updated.

In development scenarios where a lot of offset-well data are available to constrain the predrill input parameters, there may be little updating required. Perhaps minor changes to the depth of a particular formation top or small adjustments to pore pressure are required. On the other hand, in some exploration environments, there may be very little predrill information available. Drilling out of the base of salt in the deepwater Gulf of Mexico, for example, is often done almost *blind,* with a very large uncertainty in parameters such as pore pressure (and, hence, such parameters as effective stress and rock strength).

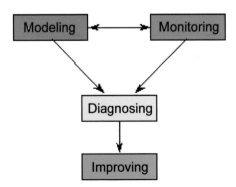

Fig. 5.1.43—A conceptual summary of the types of information used to address wellbore stability while drilling. Modeling, monitoring, diagnosing, and improving wellbore stability are each dealt with in turn in the following sections.

The wellbore-stability model is more sensitive to some inputs than to others. The magnitude of predrill uncertainty also varies from one parameter to another, as do the data available while drilling to update the model. So, although it is important to ensure all the inputs to the model are accurate, some inputs tend to be more important and easier to update than others. **Table 5.1.5** presents a summary of the input parameters that can be updated while drilling and the measurements available to make such updates. These are discussed in more detail in the following.

Pore Pressure. In abnormally pressured basins (this includes many basins being developed today), pore pressure is probably the most important parameter to update while drilling. Pore pressure is very often significantly higher or lower than predicted. Pressure ramps at depths that are different from those predicted are a common source of instability. Pore pressure has a large impact on the MW requirement for stability. If the pore pressure changes, it usually implies that some other inputs to the model (e.g., the magnitude of the horizontal stress) should also change. The change in required MW for a given change in pore pressure is a function of many parameters. However, in a typical environment, a 1-lbm/gal change in pore pressure can often mean a change in required MW of 0.5 lbm/gal or more.

In the predrill phase, pore pressure is predicted from offset-well data, seismic velocities, and/or basin modeling. While drilling, LWD measurements that can be used to estimate shale pressures are common. Sonic,

TABLE 5.1.5—DATA AVAILABLE FOR REAL-TIME UPDATING OF WELLBORE-STABILITY MODEL*

MEM Component	Data Available for Update	Typical Degree of Sensitivity of Stability Model	Ease of Update
Pore pressure (p_p)	Analysis of LWD sonic, resistivity, or density for shale. Real pressure measurements for sands. Gas, flow checks, kicks, etc.	Very sensitive. Required mud weight to prevent shear failure is strongly correlated with pore pressure.	Easy where appropriate logs are run. This is the most commonly updated while-drilling parameter for stability.
Rock strength	Lithology-based correlations from sonic, resistivity, and density. Wellsite strength tests on cuttings and cavings. Wellsite correlations between strength and clay content.	Moderately sensitive. Strength anisotropy— e.g., if an unexpectedly fissile shale is encountered—can have a large effect on the stability model.	Strength of isotropic rock is simple to update from logs, cuttings, or cavings measurements. Anisotropic strength is very difficult to update.
Vertical stress (σ_v)	LWD density measurement, or indirectly (e.g., from a sonic to density transform). Cuttings and cavings density may also be used if corrected for downhole conditions.	Largely insensitive. Because vertical stress is controlled by all the material above a depth, it changes very slowly. Long intervals of anomalous density are needed to make a significant change to σ_v.	Easy in principle, although software is typically not set up to accommodate changes because they are rarely thought to be significant.
Minimum horizontal stress (σ_h)	Same as for p_p because σ_h is a function of p_p. Additionally, leakoff tests and induced-fracture or lost-circulation events.	Most important effect of change in σ_h is on fracture gradient. Stability is moderately sensitive, depending on orientation of well path and stress field.	Easily updated through stress model where pore pressure is being updated. Leakoff tests are performed in most wells, but are often subject to interpretation difficulties.
Maximum horizontal stress (σ_H)	Requires caliper- or image-based observations of breakout for model-based update.	Moderately sensitive, depending on orientation of well path and stress field. Large changes from predrill estimates are unlikely except in tectonically complex area.	Difficult to update. Clear observation of breakout is rare while drilling. Also requires all other parameters to have been determined.

* Compare to Table 5.1.2—the predrill equivalent.

resistivity, and density logs can all be used to estimate shale pressure. Calculating shale pore pressure at the wellsite from LWD measurements (real-time pore-pressure analysis) is quite common today in overpressured basins. Most of the LWD providers offer this service, as do several independent companies who specialize in real-time pressure analysis.

The ability to take direct pressure measurements in permeable formations (typically sands) is a recent development in LWD technology. While in the past it was necessary to wait for wireline measurements, this new LWD technology offers a significant opportunity to sense-check pressure estimates while drilling. For wellbore stability, we are usually interested in shale pressure, which may not be the same as the sand pressure. However, the sand pressure measurements, when viewed within the geological context, can often put some constraints on the shale pressure estimates.

Rock Strength. Rock strength is another important parameter to update while drilling and can be estimated in a variety of ways from data available while drilling. Several relationships exist between strength parameters (UCS, cohesion, friction angle) and direct log-measured parameters (e.g., density and compressional- and shear-slowness) or parameters derived from log measurements (Young's modulus, shear modulus, porosity), some of which are available from LWD tools (Plumb et al. 1992; Horsrud 2001). Many of these are the same as the relationships used in the predrill stage.

While drilling, in addition to LWD data, several techniques exist to determine strength parameters from cuttings and cavings. These are rather specialized and not in widespread use. They include the following:

- Estimating rock strength as a function of clay mineralogy or cation-exchange capacity (CEC) (Leung and Steig 1992)
- Estimating rock strength as a function of plasticity-index-type tests (Kageson-Loe et al. 2004)
- *Direct* strength tests using indentation or scratching methods on cuttings or cavings (Schei et al. 2000)
- Ultrasonic measurements at the wellsite of the wave speed in small samples, from which strength parameters are estimated through correlations similar to those for log-measured wave speeds (Schei et al. 2000)

Vertical Stress. The vertical stress or overburden at any point is a function of the density of all the material above it. Therefore, local discrepancies in density between the predrill predicted value and the actual value tend to have a small effect on the total integrated result. The vertical stress is therefore fundamentally different from other properties such as strength and pore pressure and typically does not require significant updates while drilling.

The exceptions to this general rule are as follows:

- At shallow depth, where there are perhaps only a few hundred feet of material over which to integrate, a typical error in density prediction may lead to a significant overburden error. The problem is compounded by the typical lack of density or velocity measurements at shallow depth.
- If the density prediction is significantly wrong over a considerable depth, the cumulative effect may be significant. There are examples in the deepwater Gulf of Mexico where the presence of salt prevents accurate seismic velocities from being determined before drilling. In an exploration well, seismic velocities typically are used for the overburden prediction (by means of a velocity-to-density conversion). Thus, if the velocities are significantly wrong or are not known, there can be a large uncertainty in the overburden. **Fig. 5.1.44** shows an example based on a real data set where the overburden turned out to be significantly less compacted than was predicted. The highly undercompacted nature of the overburden meant that not only was the vertical stress significantly lower than predicted but also the pore pressure was significantly higher than predicted, severely narrowing the drilling window.

Minimum-Horizontal-Stress Magnitude (σ_h). σ_h is estimated predrill from a simple model correlation with effective vertical stress (Section 5.1.6) or empirically from estimates in offset wells based on LOTs, XLOTs, lost circulation, or hydraulic-fracturing tests. Similarly, while drilling the well, as pore pressure (and possibly vertical stress) is updated, a model-based σ_h should also be updated. Also, if good LOT or lost-circulation data are acquired while drilling, this should also be used to either confirm or update σ_h.

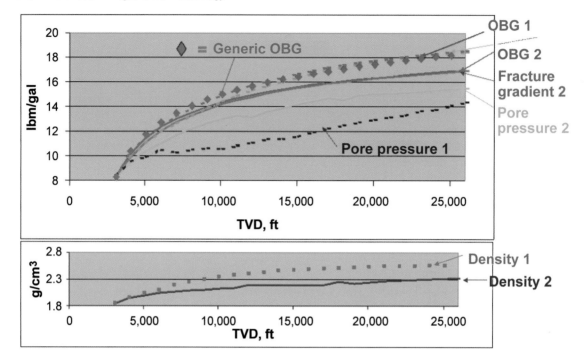

Fig. 5.1.44—Vertical stress profiles, based on a real deepwater Gulf of Mexico example. OBG is overburden gradient.

Examples of where σ_h changes significantly from the predrill estimate include the following:

- Where vertical stress and/or pore pressure change significantly from the predrill estimate, so too will σ_h. In the example shown in Fig. 5.1.44 (vertical-stress change), the magnitude of the horizontal stress, being a function of vertical stress and pore pressure, also changes. A more common example is where pore pressure alone changes. An unexpected pressure ramp or regression can have a significant impact on σ_h. In the simple models used to estimate σ_h (Section 5.1.6.), a 1-lbm/gal change in pore pressure would give rise to approximately a 0.5-lbm/gal change in σ_h.
- Changes in Lithology. In simple gravitational-loading models, σ_h is related to vertical effective stress by means of some property of the rock (Poisson's ratio or friction angle) that will vary from one lithology to another.
- Where significant components of subsurface stress are from sources other than gravitational loading (e.g., in tectonic settings, structurally complex settings, and around salt), changes in σ_h may be related to factors other than pore pressure and vertical stress. In these more-complex settings, updates to σ_h are likely to require some direct measurement of σ_h (from an extended LOT, for example).

Maximum-Horizontal-Stress-Magnitude (σ_H). Determining the orientation of the maximum horizontal stress has been discussed in Section 5.1.6. In most environments, it is unlikely to require updating while drilling (assuming it is known predrill). In complex tectonic environments or around salt, on the other hand, it may require updating while drilling. LWD image logs and oriented LWD calipers provide the possibility of identifying stress-direction indicators such as breakouts or drilling-induced fractures while drilling. LWD images and calipers are described further subsequently.

In both the predrill and while-drilling phases, the magnitude of the maximum horizontal stress (σ_H) is the most difficult parameter to determine. There is no direct way of *measuring* σ_H. Section 5.1.6 describes some of the ways in which σ_H is estimated through modeling, matched to observations of wellbore failure. This is also possible in the while-drilling phase, using LWD images. In complex tectonic environments, σ_H may be an important parameter that should be updated as the well is drilled. However, because of the difficulty of estimating this parameter, it appears to be rare that σ_H is updated while drilling.

Monitoring Wellbore Instability While Drilling. Monitoring methods for instability fall into two categories, direct and indirect. The former provide a direct observation of instability from an LWD image, LWD caliper, or from cavings. Indirect methods are those such as annular pressure or torque and drag that indicate a hole problem that might be the result of instability or might be caused by something else (i.e., poor hole cleaning).

Direct Methods: LWD Images. LWD imaging tools make an image of the wellbore wall. Several are commercially available today. These images are similar to some of the wireline images shown previously in Section 5.1.6. However, LWD images can provide a direct observation of the state of the borehole as the well is drilled. This offers the opportunity to use the information in the image to impact the way the well is drilled.

Images are generated by measuring a quantity that is sensitive to the differences between rock and drilling fluid. Density, resistivity, and gamma ray LWD imaging tools are all available. Typically, as the tool rotates, the sensor makes measurements around the circumference of the wellbore wall, enabling a full wellbore image to be made. These tools are described in more detail in Chapter 6.

Some LWD density and resistivity images are shown in the case study examples in the subsequent Diagnosing Wellbore-Instability Mechanisms section. Where the wellbore is enlarged beyond bit size, the tool reads the properties of mud rather than those of rock. The image can therefore show at a glance where the wellbore has failed or is enlarged for some reason, although care must be taken to avoid confusion between hole enlargement and tool e-centering. Because the image is also oriented, it is possible to determine where on the wellbore wall the enlargement has taken place, which can help determine the mechanism of the hole enlargement (see the subsequent Diagnosing Wellbore-Instability Mechanisms section). These tools also image geological features; this is typically the primary reason for running them. Geological features such as bedding, fractures, faults, and other rock fabrics can also play an important role in influencing the mechanism and extent of wellbore instability. (See Example 2 in the Diagnosis Case Studies.)

Transmitting LWD images from downhole to surface (i.e., real-time imaging) takes up telemetry bandwidth. Techniques such as image compression have made this easier. Typically, however, at the present time bandwidth is too narrow to transmit everything that is measured downhole. For that reason, the image is often available only after the tool is brought to surface and its memory is downloaded. Wells that are geosteered through the reservoir are currently the only wells where real-time images are used routinely. Clearly, in a well with stability issues, the sooner this information is available the better. For the wellbore-stability analyst, real-time transmission of data is therefore far preferable to memory data. Advances in measurement while drilling telemetry are bringing us closer and closer to the reality of real-time images as a standard logging deliverable.

Direct Method: LWD Calipers. Wellbore diameter is a key piece of information for stability analysis while drilling. The simple observation that the hole is enlarged in a particular area, coupled with other observations such as cavings, can be very powerful. Unlike wireline tools, LWD tools do not have *physical* calipers. In the LWD world, hole diameters are inferred from other measurements.

The most commonly used and perhaps most direct LWD caliper is the ultrasonic caliper (Maeso and Tribe 2001). A short ultrasonic pulse is aimed from the tool to the wellbore wall. A receiver positioned close by on the tool picks up the reflected signal as it bounces off the wellbore wall. The wave speed in the mud is known approximately (it has to be adjusted for density and temperature), so that the distance between the tool and the wellbore wall can be determined as simply the product of the wave speed and half of the traveltime. **Fig. 5.1.45** shows some examples of ultrasonic-caliper data presented in various ways.

Hole diameters can also be estimated from both LWD density and LWD resistivity measurements. These estimates rely on assuming either a density or resistivity for both the formation and the drilling fluid, and then modeling the diameter and geometry of the hole to fit the observed measurement. These computed calipers are a less direct estimate than the ultrasonic calipers and are subject to more uncertainty, which is discussed in more detail in Chapter 6.

These *direct* observations (calipers and images) of the state of the wellbore are probably the single most useful piece of information for wellbore-stability analysis while drilling, and their increasing real-time

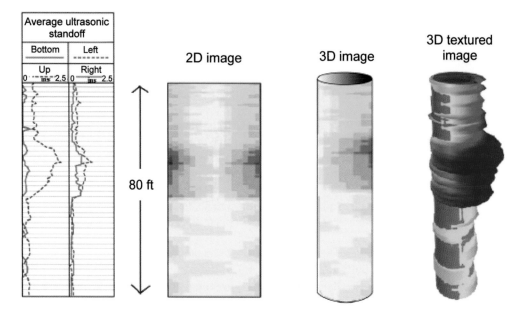

Fig. 5.1.45—Example of ultrasonic-caliper data displayed as raw data and as contoured 2D and 3D images (Maeso and Tribe 2001).

availability in the industry offers a huge opportunity for a step change in drilling performance in the future.

Resistivity. Along with gamma ray, resistivity is the most commonly run LWD tool. Most tools make at least two different resistivity measurements. Each measurement typically samples a different depth into the formation. In a shale, fluid invasion, particularly with oil-based mud (OBM), does not occur under normal conditions. Thus, both resistivities should read approximately the same value (if the hole is in gauge and the rock is approximately isotropic).

In a fractured/fissile shale, mud can penetrate the pre-existing weaknesses. Where this occurs using an OBM, an invasion profile (resistivity-curve separation) can be observed. Because of the large resistivity contrast between OBM and (water-bearing) formation, the shallow resistivity will read higher than the deep resistivity. **Fig. 5.1.46** shows some examples of this behavior.

An invasion profile in a shale is therefore very likely an early indication of instability. Resistivity curves can actually show anomalous features (similar to invasion) for a number of reasons. Hole enlargement is one such reason, which might also be an indication of instability. Intrinsic anisotropy (in a thinly bedded formation) can cause separation of the resistivity curves where the well penetrates the beds at an angle. Drilling-induced tensile hydraulic fractures are also a common cause of resistivity anomalies. These are usually accompanied by significant lost circulation.

Time-lapse resistivity (i.e., comparing resistivity from original logging runs to that on subsequent trips) is a very powerful tool for monitoring changes in hole condition. Where anomalous resistivity profiles are observed to change as a function of time, the cause is very likely hole stability or fracturing as opposed to some intrinsic anomaly such as anisotropy.

The simple description of resistivity behavior given here belies the fact that it is actually a very complicated function of such factors as anisotropy, hole enlargement, invasion, and fracturing. Given the current state of the science, for practical purposes it is perhaps best to use observations of resistivity anomalies in shale (curve separation) as an indication that something unusual is happening, which may very well be an early indication of instability. Such observations should be a red flag that triggers careful monitoring of other hole indicators that can be used in conjunction to diagnose the hole problem.

Time-Lapse LWD. The log that is typically delivered by the LWD vendor is a depth log. On a standard depth log, each time new footage is drilled the log measurements from the newly drilled piece of the well are appended to the previous log. Thus, under normal drilling conditions, the image log, caliper log, or resistivity

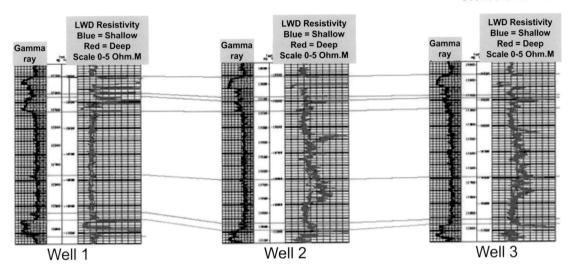

Fig. 5.1.46—Three Gulf of Mexico wells. Anomalous resistivity in Wells 2 and 3, associated with OBM invading pre-existing cracks/fractures in a fractured/fissile shale. Images courtesy of James Brenneke, BP.

log represents the state of the wellbore shortly after it is drilled (depending on distance behind bit and ROP). However, often during the course of drilling the hole, the BHA may be tripped in and out several times, and so it is possible to measure or image the wellbore at several times after the well has been drilled. This is called time-lapse LWD. It shows how the stability of the wellbore evolves through time, which can be a very powerful approach to monitoring the condition of the hole and understanding the causes of instability (Rezmer-Cooper et al. 2000). **Fig. 5.1.47** shows an example of time-lapse caliper data, and **Fig. 5.1.48** shows time-lapse image data.

The opportunity to image the wellbore wall soon after drilling and again at a later stage is something that is not available from wireline-logging measurements. Also, wireline tools are often not run in holes known to be unstable for fear of losing the tool. Here, again, LWD offers an opportunity to acquire data that would usually not be acquired.

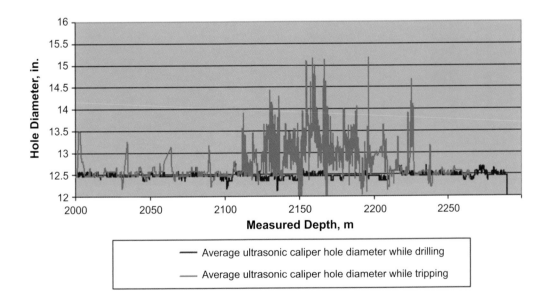

Fig. 5.1.47—Time-lapse caliper data, collected while drilling and while tripping.

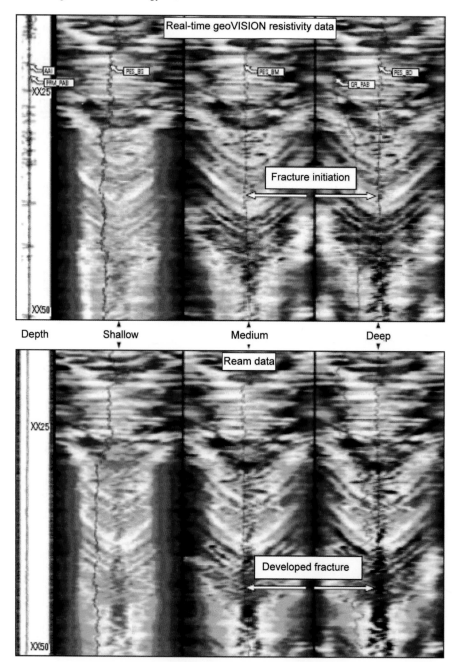

Fig. 5.1.48—Time-lapse image data for fracture identification. Drilling-induced fracture development can be seen between XX25 and XX50 in a calcareous-shale section. The upper image is a real-time resistivity image collected while drilling, showing the onset of a sequence of isolated fractures on the low side of the hole. The lower image, collected during reaming a few hours later, shows development of a single long fracture with increased width across the same depth interval. The images show shallow-, medium-, and deep-resistivity measurements from left to right (Inaba et al. 2003). Image copyright Schlumberger. Used with permission.

Although time-lapse information theoretically is available from most wells drilled with LWD imaging, logging companies have traditionally only provided the *while drilling* log and, therefore, typically do not have their surface software set up to provide the data in time-lapse format. It is, therefore, rare that such information is actually processed and analyzed.

Cavings. The terms cuttings and cavings are sometimes used rather loosely. Cuttings are strictly those pieces of rock that are cut by the bit or other hole-opening device (e.g., an underreamer). All other rock fragments are cavings. They are typically pieces of the wellbore wall that cave in at some point after the hole is made. **Fig. 5.1.49** illustrates typical cuttings. These are distinct from cavings shown in **Figs. 5.1.50 through 5.1.52.**

The observation of cavings during drilling is a simple but important piece of information in the effort to understand a wellbore-stability problem. The fact that cavings are seen at surface is a direct indication that some form of hole instability is occurring. In fact, this observation can often be the only piece of evidence that distinguishes a wellbore-stability problem from other hole problems (e.g., hole cleaning, keyseating).

Cavings can form at any time after the well is drilled. So, unlike cuttings, it is typically not possible to locate where they are coming from, just on the basis of when they appear at surface. It may take hours or days

Fig. 5.1.49—Typical shale cuttings from a polycrystalline-diamond-compact (PDC) bit in OBM at 17195 mud density (MD).

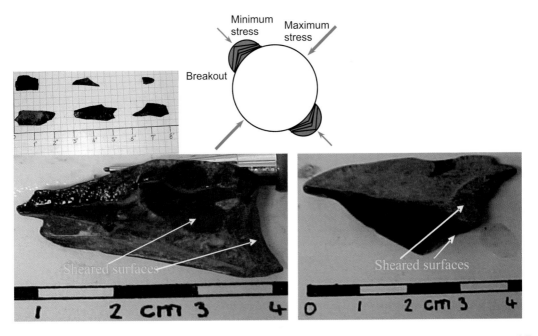

Fig. 5.1.50—Angular cavings arising from failure of intact rock at the wellbore wall (pictures courtesy of Tron Kristiansen, BP).

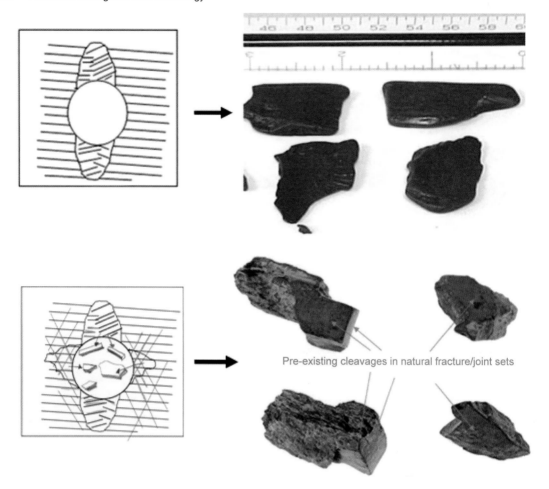

Fig. 5.1.51—Tabular-to-blocky cavings formed from instability dominated by pre-existing planes of weakness.

after drilling for them to develop. An interval that is producing cavings may then continue to produce cavings for several days or may stabilize after a shorter time. Where several distinct lithologies are open in a hole section, the lithology of the caving can reveal where in the well the instability is occurring. However, it is probably more often the case that shales encountered in a single hole section are more or less indistinguishable to the naked eye at surface. Micropaleontology is sometimes used to try to identify where in the open hole the cavings are coming from, but this is typically not used in near to real time. In summary, the observation of cavings typically tells us that there is a wellbore-stability problem, but it often does not tell us where the problem is.

Another important aspect of cavings is that they can also give an indication of the mechanism of instability. A range of mechanisms of instability was introduced in Section 5.1.4. Because some of these mechanisms are controlled by the intrinsic fabric of the rock, the morphology of the resulting cavings can often be characteristic of a particular mechanism. If the cavings can be used to determine the cause of the instability, the appropriate remedial action can be taken. Cavings morphology is thus potentially a very important piece of information. Three types of cavings—angular/splintery, tabular and blocky, and rubble—are described next.

Angular/Splintery. Failure of intact rock (where the stress at the wellbore wall exceeds the strength of the intact rock and there is no influence of any intrinsic fabric) typically yields angular and splintery cavings (Fig. 5.1.50). One face of the caving may be the wellbore wall and the others are newly formed fracture surfaces. Parallel faces and blocky or tabular forms are absent. When these are well preserved [typically when a good inhibitive water-based mud (WBM) or an OBM is used and when they are promptly removed from the

Fig. 5.1.52—Rubble/breccia cavings from fault and rubble zones.

wellbore], the sharp angular edges and fresh fracture surfaces are quite diagnostic. Increasing MW provides more support to the wellbore wall—effectively increasing the radial (minimum) effective stress and decreasing tangential (maximum) effective stress and, thus, inhibiting failure (Section 5.1.4).

Tabular and Blocky. Probably the most distinct and easy to recognize types of cavings are those that form where a strong planar fabric (such as in a fissile shale) provides the dominant influence on failure of the wellbore wall. These are typically seen in wells that drill close to parallel with the plane of weakness, for example parallel to bedding in a fissile shale. The tabular nature of these cavings (Fig. 5.1.51) is a result of the parallel planes of weakness (bedding planes, cleavage, or fracture sets). Blocky cavings may also form where two or more sets of pre-existing weaknesses intersect, such as in a fractured fissile shale (Fig. 5.1.51 lower left). Increasing the MW may or may not provide more support to the wellbore wall, depending on whether these planes of weakness are permeable. Field experience has shown that increasing MW is typically not effective at controlling this type of instability.

Rubble. The term *rubble zone* is often used by drillers (in the Gulf of Mexico in particular) when significant instability is encountered and the resulting cavings seen at the shakers have the appearance of rubble. The main characteristic of rubble is that the fragments are a variety of shapes, sizes, and forms and are somewhat random in nature. Volumes of this material tend to be large and overwhelm any cuttings being returned. Consequently, the overall appearance at the shakers is very similar to rubble that might be found at a building site. Most of the time, the source of the rubble is never observed directly. The precise origin of rubble zones is, therefore, often not known. However, in the Gulf of Mexico they typically appear to be adjacent to, or directly beneath, salt. There are several possible explanations for this. One is that the rubble zone is simply a large shear zone or fault zone that has accommodated the relative movement between the salt and surrounding rock. Drag zones and halos of intense deformation are observed around some salt bodies in the field. Another possible origin is the slumping of carapace material from the top of the salt body to the edges where it becomes buried by further sedimentation.

Rubble-like material is also seen in areas where there is no salt. These rubble zones are thought to be large faults. Faults that appear as a discrete line on seismic sections can in fact often be a zone of deformation, tens to hundreds of feet thick. Fault-zone material observed in outcrops is often described as breccia, which is a rubble-like material. Breccia can have a pervasive fabric where significant displacement accommodated by the fault has imparted an alignment to the fragments. Often, evidence of a weak fabric can be seen in rubble-zone material. It sometimes has a platy or scaly appearance. It is not as distinct as the fabric from a fissile shale but is not completely random.

It is difficult to find a textbook example of material from a rubble zone because each is different. Some rubble-zone material contains fragments that look very much like the angular and splintery cavings of Fig. 5.1.50. In other rubble zones, the material may tend more toward the tabular/blocky end of the spectrum if perhaps a fissile lithology is the source of the rubble or if the rubble is simply the result of several intersecting phases of faulting or fracturing. Fig. 5.1.52 shows some varied examples of cavings from rubble zones.

Field experience has shown that rubble zones are very difficult to deal with. Once the instability begins, it often just gets worse until the hole is lost. The reason for this is that MW increases are typically ineffective or are only temporarily effective at suppressing the instability because the mud does not provide the support to the wellbore wall (as explained in Section 5.1.4).

Indirect Indicators of Instability. While LWD calipers and images, as well as cavings, provide direct observations of instability, drilling parameters measured at the drill floor and downhole can serve as indirect indicators. These parameters are indicators of the general hole condition. Care must be taken in determining the cause of the hole condition, (i.e., wellbore instability or some other problem). These indirect observations alone are not sufficient to diagnose the problem completely as wellbore instability or otherwise. However, they are often the first sign of trouble and as such are an important piece of the puzzle.

Torque and Drag. The torque required to rotate the drillstring will tend to increase around an unstable section of hole. As cavings begin to pack off around the BHA and drillstring, extra torque is required to continue rotation. For the same reasons, the amount of drag increases through such an interval. These effects typically will be most noticeable in deviated wells, where cleaning of the unstable sections is more difficult. Instability can also generate *ledges* between stable and unstable intervals, and these are also often seen as tight spots.

Departures in torque and drag from the baseline trend can be an early indication of hole problems (Smirnov et al. 2003). **Fig. 5.1.53** is an example of a trip sheet that records the location of tight spots on each trip in and out of the hole. A persistent tight spot may well be the result of an area of hole instability, although there are several other possible causes (e.g., hole geometry, hole cleaning, excessive filter cake).

Annular Pressure While Drilling. The annular pressure sensor is run in many LWD BHAs today. It is more commonly referred to as pressure measurement while drilling (PMWD). The tool is described in more detail in Chapter 6.

PMWD provides one of the simplest yet most powerful pieces of downhole information available to the driller. PMWD data are used in a variety of ways but are particularly useful for monitoring wellbore stability for two reasons. First, PMWD measures the actual mud pressure to which the formation is subjected. Actual downhole pressure can differ considerably from the hydrostatic pressure exerted by the column of mud because of a variety of factors such as circulation and swab/surge. As the wellbore pressure often needs to be controlled carefully to manage wellbore instability, it is preferable to know the actual downhole pressure rather than simply the surface MW.

Second, when PMWD is monitored against time, it is an early and very effective indicator of hole deterioration. Any restrictions around the BHA (because of packoff in an unstable interval, for example) will be reflected immediately in the annular pressure below the restriction. **Fig. 5.1.54** shows an example of plotted PMWD vs. time. Here, when the pump is stopped during connections, cuttings and cavings fall toward the bottom of the hole and collect around the top of the BHA. This incipient packoff is noticed as a significant pressure spike every time pumps are turned back on. The use of PMWD data is discussed further along, in the Improving Wellbore Stability While Drilling section.

Diagnosing Wellbore-Instability Mechanisms While Drilling. The preceding section has discussed modeling and monitoring of wellbore stability while drilling. If it is determined that the wellbore is unstable, the next step should be to ask, "What is the mechanism of instability?" because this will determine the appropriate remedial action.

Diagnosing the mechanism of instability has been discussed in the Predrill Planning section. In the predrill stage, where offset-well data are available, wireline-log data (primarily image and caliper logs) are used together with wellbore-stability modeling for diagnosing hole instability. In the while-drilling phase, image and caliper data are again the keys to diagnosis of the problem. Although LWD image and caliper lack some

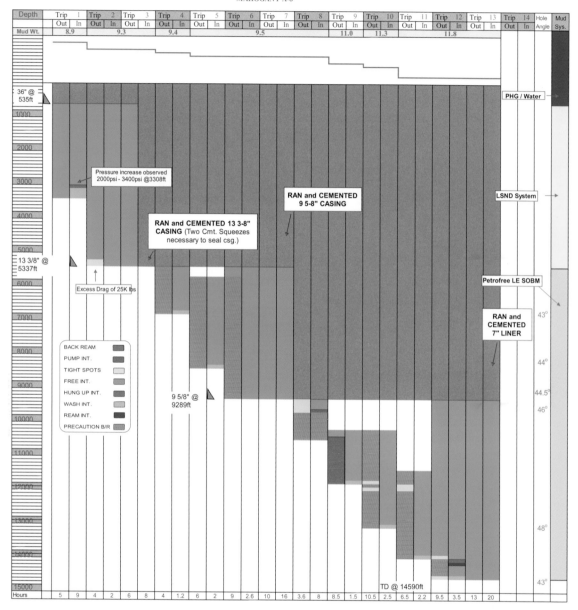

Fig. 5.1.53—Trip-sheet example (courtesy of Baroid Trinidad).

of the detail available from their wireline equivalents, information such as cavings and some of the indirect indicators of hole instability are more readily available in the while-drilling stage.

As in the predrill stage, identifying the relationship between the wellbore and the rock fabric is the key to understanding the mechanism of instability. The simple scheme for categorizing mechanisms of instability (Sections 5.1.6 and earlier in this section) is used again here. The differences in the way the measurements are made, and the timing of the measurements (soon after drilling in the LWD world and later in the wireline-logging world), must be kept in mind when analyzing LWD image and caliper data. However, the diagnostic wellbore features for each of these mechanisms that are observed in the data (e.g., symmetrical breakout, failure along bedding planes) are the same in real time as they are in predrill analysis.

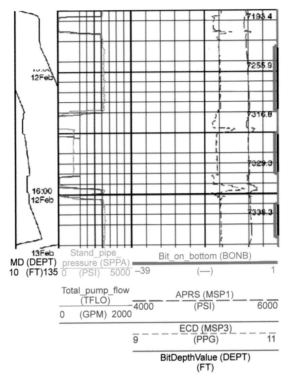

Fig. 5.1.54—An example of PMWD plotted against time. Here, when the pump is stopped during connections, cuttings and cavings fall toward the bottom of the hole and collect around the top of the BHA. This incipient packing-off is noticed as a significant pressure spike every time pumps are turned back on (Nonpuplished BP data; Schlumberger log display).

Table 5.1.6 summarizes the while-drilling data that can be used to diagnose mechanisms of instability. Diagnosing the cause of instability from the data available while drilling is challenging and is by no means routine. However, a combination of all the approaches listed in Table 5.1.6 and a certain amount of good luck can often be enough to understand what the wellbore is doing in time to have an impact on drilling operations.

Diagnosis Case Studies. Example 1 From Ecuador. **Fig. 5.1.55a** shows an example of a resistivity-at-bit (RAB) image from a wellbore that displays breakout on the side of the hole (Bratton et al. 1999). This is indicative of shear failure of intact rock. This example is from a near-vertical well. The breakouts are symmetrical and unambiguous. Breakouts like this are often seen in hard and moderately brittle rock. They are less commonly seen in softer, more-ductile rock.

Cavings from an equivalent section in an offset well in Ecuador are shown in Fig. 5.1.55b. The angular nature of these cavings and newly created fracture surfaces is indicative of this type of shear failure of intact rock.

Stability in this wellbore was modeled using a simple linear-elastic model that appears to have been quite appropriate for this rock. The results of the model are shown in Fig. 5.1.55c. The model predicts that breakouts will form for the MW being used.

In this example, a direct observation of instability, analysis of the cavings type, and a simple model of shear failure all point to the same conclusion that instability is a result of shear failure of intact rock, control of which requires an increase in MW.

Example 2 From Gulf of Mexico Shelf. **Fig. 5.1.56a** shows an azimuthal density image and associated ultrasonic caliper in a high-angle well in the Gulf of Mexico shelf (Edwards et al. 2004). This well, Wellbore 1 (WB1), was drilled close to parallel to bedding through fissile and fractured shale. Images of both the density and the PEF are shown. The PEF is very sensitive to the difference between formation pressure and mud density. It clearly indicates that failure of the wellbore wall is occurring along bedding planes (seen as low-amplitude sinusoids in this image) primarily on the high side and low side of the hole. The hole angle is

TABLE 5.1.6—SUMMARY OF WHILE-DRILLING DATA TYPES THAT CAN BE USED TO DIAGNOSE MECHANISMS OF INSTABILITY

Information Source	Shear Failure of Intact Rock	Shear Failure or Splitting of Impermeable Planes of Weakness	Failure Along Permeable Planes of Weakness	Disintegration of Rubble/Breccia Zones
LWD azimuthal density, resistivity, and gamma images	Symmetrical enlargement on opposite sides of the hole is diagnostic of breakout. Shear fractures at the breakout edge may be visible.	Image shows failure planes bounded by pre-existing planes of weakness (e.g., bedding planes). Tends to occur when wellbore is close to parallel to planes of weakness.	Resistivity image may detect penetration of drilling fluids into pre-existing planes of weakness. Otherwise, diagnostics same as for impermeable planes. Image probably cannot resolve.	Fracture-zone boundaries and individual contained fractures may be visible. Fractures crosscut bedding. High-resolution image may resolve rubble or brecciated fabric. Hole may be enlarged around entire circumference or preferentially on high and low sides.
Ultrasonic caliper	Oriented caliper may show hole ovalization on sides of hole.	Oriented calipers may show enlargements where wellbore intersects planes of weakness.	Oriented calipers may show enlargements where wellbore intersects planes of weakness.	Enlargement may be nondirectional or may occur on high side of hole. Enlargement is restricted to the rubble/breccia zones—e.g., may only occur in thin fault zone. Hole size is often very big—typically greater than breakout size in intact rock.
Cavings	Angular, newly created fracture surfaces	Predominantly tabular to blocky with parallel sides	Predominantly tabular to blocky with parallel sides	Rubble- or breccia-like cavings. May be a mixture of shapes and sizes characterized by being bounded by pre-existing planes of weakness.
Modeling	Intact-rock-shear -failure model (e.g., elastic/brittle). Real-time updated model shows mud-weight deficiency.	Difficult to model quantitatively. This mode can occur even when the intact rock model indicates that there is sufficient mud weight.	Stability models that determine required mud weight are inapplicable because mud does not provide sustained support for wellbore wall.	Stability models that determine required mud weight are not applicable because mud does not provide sustained support for the wellbore wall.
Resistivity, invasion	Oil-based mud does not invade intact shale.	Mud does not invade because planes of weakness are not preferentially permeable.	Resistivity shows invasion profile with OBM. Invasion can develop as soon as drilled, or may take some hours to develop. Higher mud weights can create more invasion.	Resistivity shows invasion profile with OBM. In highly rubbleized or brecciated zones, invasion is likely to develop as soon as the interval is drilled.

approximately 60°, and the bedding dip is approximately 12°, giving an attack angle of 18°. The LWD ultrasonic caliper is also shown in this figure. It can be seen that the borehole is overgauge in the interval that corresponds to the failed wellbore wall.

The image log shown in Fig. 5.1.56a is from the tool memory and was not studied in real time as the well was drilled. Had the image been studied in real time, the problem may have been diagnosed in time to prevent

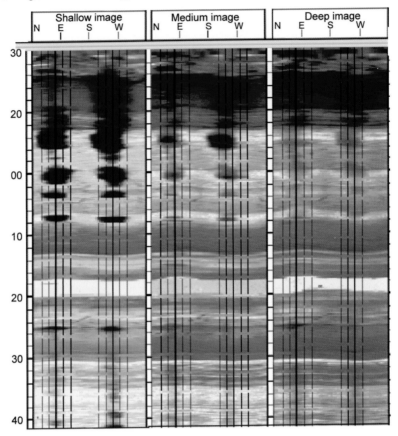

Fig. 5.1.55a—RAB image from Ecuador. From Paper JJJ, 1999 by T. Bratton, T. Bornemann, Q. Li, R. Plum, J. Rasmus, and H. Krabbe, presented at the 40th Annual SPWLA Conference, Oslo, Norway.

Fig. 5.1.55b—Cavings from breakouts in Ecuador well.

further drilling problems. The log shown in Fig. 5.1.56a was run some hours after the interval was drilled (a short trip occurred between drilling and logging this interval). However, as drilling proceeded, this interval became increasingly troublesome, ultimately leading to stuck pipe. The well was sidetracked and the interval was drilled again, Wellbore 2 (WB2) with a higher MW. The higher MW did not help. In fact, it may have

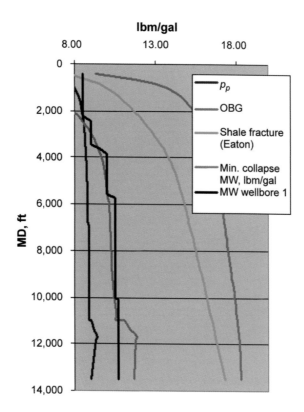

Fig. 5.1.55c—Results of a simple wellbore stability model show that insufficient mud weight was used in the lower portion of the well. This allowed shear failure of the wellbore wall, giving rise to the observed breakouts and angular cavings.

worsened the situation, with an even higher volume of cavings observed. The BHA was lost in this case, and no images were recovered. Fig. 5.1.56b shows the shale cavings recovered from this interval. They are distinctly tabular in nature, consistent with failure along bedding planes.

Although the BHA was lost together with memory data in WB2, real-time LWD data were recorded. Fig. 5.1.56c shows the LWD resistivity through this interval for both wellbores. Anomalously high resistivity for a shale (typically in the 1–1.3 $\Omega \cdot$ m range in this section) coupled with the separation of the resistivity curves is an indication that the OBM is penetrating the pre-existing fractures and fissile bedding planes. This is an indication of the type of instability occurring here and should have provided a warning that increasing the MW would not add more support to the wellbore wall.

Fig. 5.1.56d shows the results of wellbore-stability modeling performed for this interval. This simple model assumes an intact rock. The red curve labeled *collapse* is the calculated MW required to prevent excessive shear failure of the wellbore wall at the given hole angle. This calculation shows that inWB1, the MW should have been just sufficient to prevent instability. In WB2 there should have been more than enough MW to prevent instability. The fact that the model is inconsistent with the observations (given that there is a fairly high degree of confidence in the input parameters) is a strong indication that some mechanism other than simple shear failure of intact rock is dominant.

In this Gulf of Mexico shelf example, several pieces of information are used, within the geological context, to determine the location and mechanism of instability. Any one of these pieces of information alone would likely not have been sufficient to determine the nature of the instability unambiguously, but where several pieces of information point consistently to the same answer, a diagnosis can be made quite confidently. To summarize,

- LWD resistivity showed an anomalous profile through the unstable zone, indicating possible invasion of OBM into cracks or planes of weakness. This is an indication that MW may not be acting to support the wellbore wall and thus that increased MW may not make the well more stable.

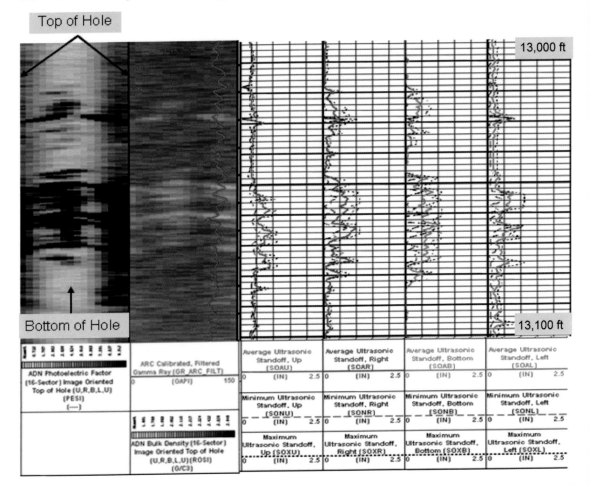

Fig. 5.1.56a—Data from an azimuthal LWD density tool with ultrasonic caliper. Tracks shown (from left to right): PEF image of the wellbore wall, density image of the wellbore wall, ultrasonic-measured standoff (i.e., distance between tool and wellbore wall) in four quadrants: up, right, bottom, left.

- LWD ultrasonic caliper confirmed this interval was overgauge and unstable.
- LWD PEF and density images confirmed the location of instability and indicated that failure was occurring along bedding planes. This interpretation was supported by the geological setting (i.e., knowledge that the well was being drilled at an angle of 18° to bedding).
- Shale cavings were characteristically tabular-to-blocky—an indication that pre-existing planes of weakness were a dominant control on failure.
- The disagreement between the simple wellbore-stability model and observations of instability was consistent with a more complex failure mechanism.

The solution to the stability problem in Example 2 is not as clear cut as the simple MW increase in Example 1. The possible approaches include the following:

- Drill the well at a higher angle to the weak bedding planes. Experimental and field evidence (Okland and Cook 1998) has shown that an angle of less than 20° between the wellbore axis and fissile bedding leads to instability. In Example 2, the well was redrilled (wellbore 3) at a higher angle (approximately 45° to bedding rather than the 18° used in Wellbores 1 and 2) and with MW reduced back to 10.8 ppg and it did not suffer any significant instability problems.
- Change the drilling fluid so that it does not enter the permeable planes of weakness but rather provides support to the wellbore wall. It may be possible to achieve this either by altering the mud chemistry

Fig. 5.1.56b—Tabular shale cavings from the unstable interval imaged in Fig. 5.1.56a. This type of caving is indicative of failure along pre-existing weak planes, in this case probably bedding planes.

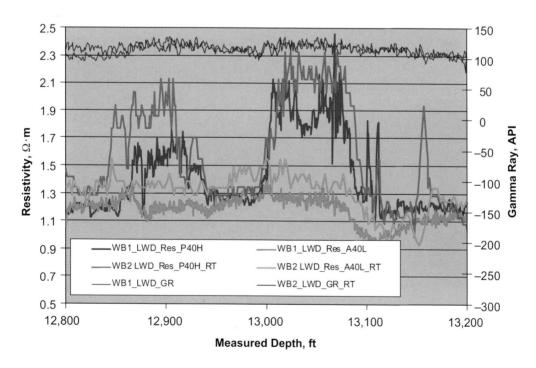

Fig. 5.1.56c—LWD resistivity curves from 2 wellbores through the interval shown in Fig. 5.1.56a. Wellbore 1 (WB1) was the first wellbore to penetrate this interval and used a mud weight of 10.7 ppg. After WB1 failed because of the instability in this interval, Wellbore 2 (WB2) was redrilled through the interval at the same azimuth and deviation but (probably mistakenly) using a higher mud weight of 11.7 ppg.

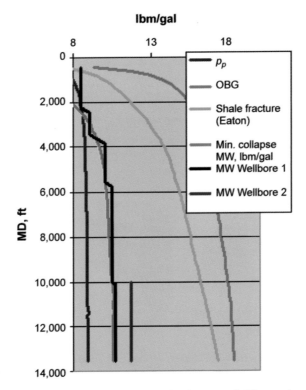

Fig. 5.1.56d—Result of simple shear-failure modeling assuming intact rock. The mud weights used in both well-bores (MW WB1 and MW WB2) were higher than the calculated minimum mud weight to prevent collapse (Min collapse MW) and should have been adequate to prevent this mechanism of failure.

such that the shale swells slightly, by reducing the permeability of the planes of weakness, or by adding crack-blocking additives. There is anecdotal evidence from other wells that indicates some success with this type of approach.

- A more experimental and so far untried approach would be to limit the instability by avoiding annular-pressure cycles. The mechanism of instability experienced in this example is to a large extent caused by the continuous pressure fluctuations in the annulus. With pumps on, the annular pressure is increased, forcing more fluid into the cracks/weaknesses. As pumps are turned off (or the well is swabbed by movement of the BHA), the annular pressure is lowered, allowing fluid to return to the annulus. As this process is repeated many times, the formation is destabilized. If on the other hand the pressure in the wellbore is kept constant or is gradually increased with time (i.e., so a positive pressure differential always exists between the wellbore and the formation), this destabilization might be avoided. One way this could be achieved is through continuous circulation. Experimental continuous-circulation tools are currently in existence, but it is not yet clear whether such a technology will become widely used.

Improving Wellbore Stability While Drilling. The preceding sections have discussed various ways of identifying and understanding hole-stability problems while drilling. This is of course the first step toward solving the problem. The next step is taking the appropriate action to cure or manage the problem. This section discusses how drilling practices affect wellbore instability and suggests ways in which stability can be improved or managed.

Mud Type, Formulation, and Density. As mentioned in Section 5.1.2, many wellbore-instability problems have their origin in the physicochemical interactions between the water-bearing components of the drilling fluid and the formation—usually shales or mudstones making up the overburden or appearing in streaks within the reservoir. OBMs generally produce far fewer wellbore-instability problems than do WBMs. On a new project, if wellbore instability is a possibility and environmental considerations and economics allow,

OBMs are a wise first choice. Knowledge (e.g., shale type) gained during drilling of the first few wells can then be used to assess whether WBMs can be used for subsequent wells, and to target their design.

Changing mud type from WBM to OBM is not something that is done in response to an unanticipated drilling problem on the current well; it requires a great deal of technical and financial planning. It is often done in a planned fashion, however, when there are known trouble zones that will be encountered in the well.

Changing the mud formulation (adding new components or keeping it within specification) is done both routinely and in response to problems. There are so many ways that muds can lose their specification, and so many types of additives, that most of this task is best left to a specialist mud engineer. For OBMs, it is especially important to keep the salinity and ionic species of the brine phase, and the fluid-loss characteristics, within specifications.

A relatively new area, where geomechanics has a strong influence on mud formulation, is that of additives for drilling through depleted reservoirs, where the fracture gradient has dropped. New wells are needed to access bypassed oil or deeper reservoirs, and they must pass through both the depleted reservoirs with low fracture gradient and the shales between the reservoirs, which may still require a high MW for stability. In principle, this could be accomplished by setting many casing strings, but in practice this is uneconomical, and so drilling-fluid additives and drilling techniques have been developed that allow the depleted sands to be drilled with high mud densities. The so-called *stress cage* effect is one approach to doing this (Alberty and McLean 2001). The principle behind the treatments or techniques is not well established at the time of writing but appears to be as follows:

- As fractures form around the wellbore they are plugged immediately by solids from the mud.
- The plugging prevents penetration of fluid from the wellbore down the fracture, so hydraulic propagation of the fracture by wellbore pressure is inhibited.
- The plugging material also wedges the fracture open, increasing the local hoop stress and so raising the initiation pressure for subsequent fractures.
- The plugging material can be added continuously, or as a spot treatment; it is sized to plug short fractures whose aperture can be calculated from the stress conditions and the rock elastic properties.

Mud density is a primary tool of the geomechanics engineer because it establishes the basis for our control of the stresses around the well. Monitoring it, keeping it within specification, and changing it appropriately in response to new circumstances in the drilling process are critical to the success of difficult wells. Traditionally, if excessive cavings were seen on the shakers, or if tight hole conditions were encountered, or one of the other classical symptoms of wellbore instability were seen, the response would be to increase the mud density. This is often the right thing to do; tradition is after all strongly based on experience. In the current era of fewer and more-expensive wells, however, the exceptions to the rule can lead to costly mistakes, and it is vital to diagnose the problem correctly before changing mud density. If the source of the problem is breakouts in weak but intact rock (the situation described in Section 5.1.5), then increasing mud density will help, by increasing the minimum principal effective stress at the borehole wall. If the rock is already fractured before the drill bit penetrates it, however, increased mud pressure may force fluid into the fractures, loosening and lubricating the assembly of rock fragments around the hole and promoting instability. The better response in this situation (identified by cavings monitoring) is to avoid borehole-pressure increases from any source (e.g., static MW, viscous sweeps, swab pressures, cuttings loading, drilled solids) and use good hole-cleaning practices to remove debris gently from the hole.

If the ROP rises, the cuttings loading of the annulus increases and may begin to generate high mud densities and wellbore pressures. In a well with a narrow MW window, this can cause losses, or pumping up and destabilization of fracture networks. Monitoring and restricting the ROP may be a good idea in such circumstances. High cuttings loadings may also cause problems if circulation is interrupted, especially in long deviated sections, where the cuttings may fall out of suspension and generate deep cuttings beds or avalanches.

Hole Cleaning. From the driller's point of view, the biggest problem that wellbore instability generates is probably the cavings in the hole; they increase torque and drag, packoff, the likelihood of stuck-pipe, and they raise the bottomhole pressure and overwhelm the surface facilities. If the rock around the hole is failing, then removing the debris is as important a task as preventing further failure. Hole cleaning, then, is a vital tool in the management of wellbore instability. It is, of course, a very mature subject, and there are many articles and papers written on detailed strategy and tactics for hole cleaning. It is important to remember the conditions of the rock around the hole, however, and make sure that the cleaning approach adopted does not make the source of the problem worse. Backreaming is a good example of this. In the right circumstances, it disturbs broken material lying on the bottom of the hole and stirs it up into the flowing mud. In the wrong circumstances, it breaks up the small cohesion of naturally fractured rock formations held together by the mud overbalance and generates large volumes of extra cavings. An example of this was seen in the Cusiana field (Last et al. 1995) where it was found that the volume of cavings seen at surface increased after backreaming; the source of the continuing instability was in fact the hole-cleaning technique that was being used to deal with the consequences of instability. In general, backreaming should not be used automatically, but only after consideration of what it might be doing to the remaining rock around the hole, especially if this is suspected to be heavily fractured.

Similar thought should be given to other hole-cleaning techniques, such as wiper trips (tripping a few stands of pipe out of the hole to dislodge and flush away cuttings and cavings from around the BHA), and low-/high-viscosity sweeps (pumping low-viscosity pills of mud to generate turbulence in the annulus and dislodge cuttings beds, followed by high-viscosity pills to suspend the solids and sweep them out of the well). Avoidance of the formation of cuttings and cavings beds through continuous good hole cleaning (e.g., through high pipe rotation speeds) is likely to be better for the condition of a fragile hole than is periodic intense cleaning.

Continuous monitoring of the shale shakers helps to determine the need for hole cleaning. For example, if there is a steady flow of cuttings-sized solids from the well, and this suddenly stops, even though ROP is maintained, it is safe to conclude that cuttings are accumulating somewhere. Prompt action will remove this accumulation and determine the origin of the change in hole-cleaning characteristics. If predrill planning indicates that instability should be occurring, but no cavings are seen on the shakers, a careful wiper trip or similar procedure may be worthwhile to confirm or refute expectations, and to clean up any downhole accumulations. Continuous monitoring is the key; it is important to record normal behavior in order to identify changes.

Borehole-Pressure Monitoring. The use of borehole-pressure-monitoring (or PMWD) data was discussed earlier in this section as an indicator of instability. It can be used in the same way to guide approaches to improving well stability. Continuous monitoring is again important, so that trends can be observed and sudden changes can be understood in the light of what is happening on the rig at the time. This is particularly important when the MW window is very narrow, or there is a borehole-pressure limit that must not be exceeded, as when drilling through fractured shales, for example (Bradford et al. 2000).

Trips. If instability is a problem on a well, tripping in and out of the hole should be performed with acute attention. Excessive swab and surge pressures below the bit can do a great deal of damage. If possible, a maximum trip velocity should be used, for which the calculated swab or surge pressures do not exceed relatively small values compared to the static mud pressure. This may mean long periods spent tripping, but in critical wells, time spent in this activity will save time spent in recovering from a tight or collapsed hole. As usual, this is particularly important when the formation is heavily fractured already; surge pressures are very effective in pumping up the fluid pressure in a fracture system, and swabbing is very effective in destabilizing the resulting assembly of loose blocks.

In difficult wells or sections, it may be worthwhile to take other actions to reduce the pressure excursions during tripping—for example, increasing the static MW during tripping out, shearing the mud to break its gel (both on surface in the tanks and downhole by pipe rotation), and even reducing static MW during tripping in (Bradford et al. 2000).

Limiting the trip speed in critical sections of the well is relatively common practice, but it should be remembered that the swab pressure beneath a drill bit is not limited to the depth interval immediately below the bit. The trip speed is often increased as soon as the bit is inside the casing (because the formation is thought to be protected), but the pressure excursions that this causes may well persist below the shoe, and cause damage there.

Because trips can be long and tedious, it is important that the whole rig crew know the importance of maintaining slow movements of the BHA, on night and weekend shifts and across crew changes, to avoid even longer and more tedious fishing work.

5.1.8 Difficult Situations. Vertical holes through intact rocks in a well-understood normal-stress environment should be relatively straightforward to drill, using the methods described above, even if the rock is weak and instability is present. Some environments, however, undoubtedly present much greater wellbore-instability challenges, and this section discusses them. Some of them have already been mentioned in previous sections, which will be referenced below.

Fractured Rocks. Instability problems in heavily fractured formations, and potential approaches to dealing with them, have been discussed in Section 5.1.4.

Bedding-Parallel Failure. Uncontrollable failure is often seen when drilling close to parallel to bedding in fissile, strongly bedded shales. Usually the only solution is to modify the well path. This environment is discussed in more detail in Sections 5.1.4 and 5.1.7.

Depleted Formations. Drilling through depleted formations is becoming increasingly common, together with the problems resulting from the reductions in fracture gradient that often accompany depletion. These problems are (at the time of writing) usually addressed through drilling-fluid additives, and this is discussed in more detail in sections 5.1.5 and 5.1.7.

Drilling Near Salt. Salt bodies such as diapirs are common in oilfield environments, and they generate drilling problems nearby because of their rapidly varying stress field, the presence of rubble zones arising from intense deformation during halokinesis, and the difficulties of characterizing them and their neighborhoods by seismic methods. Issues with drilling near salt are discussed in Section 5.1.5.

Drilling in Salt. Almost all the discussion in this chapter so far has related to pieces of rock breaking off and falling into the wellbore to cause problems. Salt usually behaves differently—it can undergo large, time-dependent, plastic strains without fracturing, under low differential stresses. In other words, it creeps, and in doing so it can close up the wellbore. Salt as referred to here is not just halite, or sodium chloride (although this is the most common evaporite mineral). It also includes sylvite (potassium chloride), carnallite (potassium magnesium chloride), and anhydrite (calcium sulfate).

Salts vary significantly in the rate at which they creep; composition, stress, temperature, and water content all influence this (Alsop et al. 1996). Some drilling in salt is essentially unaffected by hole closure; for example, the extended-reach wells on the Dieksand project passed through thousands of meters of the Haselgebirge salt without appreciable problems (Sudron et al. 1999). Some wells, however, close rapidly, on the timescale of the drilling process, and need constant backreaming to allow the BHA to move freely (Holt and Johnson 1986).

The bulk movement of salt is rapid on a geological timescale, but is still very slow on a human or drilling timescale. Poliakov et al. (1996) concluded that the Deborah number D_e (the ratio of viscoelastic relaxation time to viscous relaxation time) was very small ($\approx 3 \times 10^{-4}$) for salt diapirism, and so substantial shear stresses from large-scale movement of the salt are relaxed very rapidly. This means that to a first approximation, we can assume that the in-situ stress within a salt is close to isotropic; the shear stresses have been reduced by creep shear strain (it must be emphasized that this is only an approximation—salts can sustain small shear stresses, especially if they are still undergoing bulk motion).

In general, detailed laboratory characterization of a salt is needed in order to predict the closure rates of holes within it accurately, because the deformation parameters of the salt depend sensitively on its constitution, especially its water and clay contents. This characterization is conducted, for example, to assess the lifetimes of gas-storage caverns and cave and tunnel systems proposed for use as radioactive-waste storage.

Such detailed characterization is rare in the oil field, although it does happen (Maia et al. 2005). A less quantitative strategy is often adopted, based on local or global experience, for dealing with the problems that salt bodies present. For example, because the general principles of salt creep are well known (together with their approximate dependencies on stress and temperature), experience of the rate of closure from one well can be transferred to another under a different level of stress, or at a different temperature (provided the type of salt is unchanged). An important question to answer is "How long will this hole stay open under this MW?", where the meaning of *open* depends on the diameter of the drilling equipment that must pass through and the amount of reaming and redrilling that can be tolerated. Models for this exist, both simple (Barker et al. 1994) and complex (Maia et al. 2005; Carcione et al. 2006). Temperature is an important component of such models, because the creep of salt is thermally activated and so depends strongly on temperature. The effects of MW and in-situ stress levels influence the closure time through the role of stress in the creep rate; this is not an M-C relationship as for brittle rocks and other geomaterials, but is a power-law relationship.

Sometimes, the closure rate is not slow enough to allow trouble-free drilling. Options then include frequent reaming, or the use of unsaturated WBM to dissolve away the encroaching salt. The latter course may be difficult in long or mixed salt intervals, as solubilities change with salt type and with temperature, and it is easy to get substantial hole enlargement with its attendant hole-cleaning problems.

If the section is completed and cased, the trouble is, unfortunately, not necessarily over. The salt can continue to creep, eventually exerting the entire load of the overburden on the casing. Any asymmetry of loading, caused for example by residual anisotropy of the in-situ stress state, will accelerate casing collapse. Willson et al. (2003) discuss how to deal with this problem, and they include comparisons of collapse in nonuniform holes (in general, those drilled with WBMs) and uniform holes (drilled with OBMs or synthetic muds). The latter is significantly better; it may be possible to leave the annulus uncemented through the salt section in this case.

There are several other potential hazards when drilling through salt; Willson and Fredrich (2005) offer a very good review.

Underbalanced Drilling. In underbalanced drilling, the wellbore pressure is lower than the formation pressure, so formation fluids flow into the well from permeable zones. Specialist equipment is needed at surface to reduce the wellbore pressure to its required levels, and to handle the flow from the well safely and without environmental discharges. Underbalanced drilling can bring significant benefits in increasing ROP, minimizing differential sticking, and reducing casing costs and formation damage (Moore et al. 2004).

From a wellbore-instability perspective, underbalanced drilling brings some particular problems:

- The wellbore pressure is lower, and there is no support from the mud pressure, even in permeable formations. Because fluid is flowing into the well, the radial effective stress at the wellbore wall will be tensile, and this can encourage both shear and tensile failure of the rock.
- The wellbore-stability analyst must fight hard to get the adoption of a borehole-pressure recommendation for stability. The driving force behind a choice of borehole pressure in these wells is the fluid flow.
- Even if a recommended pressure is adopted, controlling bottomhole pressure within the required limits may be impossible. It is more difficult to predict pressure drops for the multiphase flow present in the annulus; influxes of produced fluid are occurring; and gas slugging may be possible, which imposes large pressure fluctuations on the wellbore wall.
- Fewer data are generally available from the well during underbalanced drilling. The high gas fractions in the drillpipe that are needed to control the annulus pressure can seriously reduce the bandwidth for LWD telemetry. It may also be more difficult to observe cavings morphology at surface because of the gas-handling equipment in place.
- The capacity of the fluid in the annulus to carry cuttings and cavings may be severely reduced, making hole cleaning much more difficult and so making it more likely that rock failure turns into a drilling problem.

For the rock-mechanical part of the problem, the same wellbore-stability models as used in conventional drilling planning can be used. Collection and use of stress and strength data proceed in just the same way. Some workers use more-complex mechanics models; for example, Parra et al. (2003) uses an undrained, poroelastic, coupled model, which predicts yielded zones in shale—but as always for more-complex models, a great deal more data are needed, and computation times can be long.

For the bottomhole-pressure part of the problem, detailed fluid-mechanical models of the well are needed. Saponja (1998) gives a clear summary of the procedures and problems.

Enlargement of the hole because of instability also impacts the hydraulics, and so a complex coupled problem can emerge. Hawkes et al. (2002) address this problem and illustrate it with an example of how hole enlargement in a shale stringer can effectively reduce or remove the available pressure/injection-rate window for underbalanced operations.

Finally, Guo (2001) discusses fluid-mechanical models, but also gives a simple formula for the potential temperature change resulting from gas expansion through the bit during underbalanced drilling. Although this is unlikely to lead to drilling problems, any local thermal contraction of the rock could lead to unexpected tensile cracking, which could be misinterpreted as drilling-induced fractures.

5.1.9 Other Aspects of Geomechanics. Geomechanics is used in other areas of oilfield operations, as well as in drilling. This section gives a brief overview of each.

Sand Production. When a well is put into production, the stresses around it may exceed the strength of the rock (just as for wellbore instability), and solid particles may be produced as well as hydrocarbons. This is called sand or solids production and is a very old problem, originally seen in water wells. It is found all over the world, although many of the traditional methods for dealing with it come from experiences in the Gulf of Mexico. It is a major economic problem, which often results in wells being choked back or taken out of production altogether. Consequences of uncontrolled sand production include bridging and blockage of wells, decline in productivity as the well fills up, collapse of completions because of cavity formation, elevated costs because of the need for environmentally safe disposal of oily sand, and especially erosion of everything in the production line, from downhole tubulars to chokes and surface flowlines. The last of these is particularly important for gas wells, where erosion is accelerated by the high production velocities, and the consequences of leakage from erosion of surface pipe elbows and other components are very serious.

There are a few significant differences from the drilling situation:

- Sand production is limited to the permeable reservoir intervals, usually sandstone or limestone, while most drilling problems are in the (essentially) impermeable shales.
- The direction of flow in the formation is, of course, out of the rock and into the well for production, while for drilling it is nearly always into the formation (or at least the pressure differential acts in that direction, even if the filter cake stops significant fluid flow).
- Wellbore instability is generally no longer a problem after casing has been installed. Sand production can be a problem immediately on completion of the well, or many years later when the produced water cut increases or the reservoir depletes.
- There are many ways in which wellbore stability can be improved while the well is being drilled. However, completions are often planned and components purchased, months or years in advance of the project, and there may be very limited space for maneuver if an unexpected problem arises. On the up side, more data are generally available for completions projects.

Prediction of sand production is conducted with techniques and equations similar to those seen earlier in the chapter. There is, however, an additional source of stress in the production environment. This is the drag force of the produced fluid on the grains of the reservoir rock, as it flows through the last few millimeters of formation and enters the perforation or wellbore. This is usually very small and not significant, for the production velocities typical of most wells. There are situations, however, where it is important:

- When the rock is extremely weak (i.e., essentially unconsolidated) or *beach sand* as it is sometimes called. In these cases, the drag force alone is enough to break the cementation of individual grains of sand and push them into the production flow stream. In many unconsolidated sands, the only forces holding the grains together are from capillary pressures in the thin films of water at grain contacts. These are easily broken by drag forces. A workable rule of thumb is that if the reservoir formation can be cored to make a small cylindrical plug (e.g., 1-in. diameter) that supports its own weight, the rock strength is enough to withstand drag forces in most situations.
- When the production velocity is not *typical*. This can arise if many of the perforations do not clean up properly, so that the entire flow of the well is concentrated on one or two small perforation tunnels, or it can arise at startup of a well after a shut-in period. If the well is opened up quickly, so the bottomhole pressure drops very rapidly, a very steep pressure gradient can be set up just inside the sandface, and this can lead to short-lived, very high fluid velocities and consequent sand production.
- When the fluid is very viscous, as in heavy oils. Even low flow velocities can then exert enough drag force on the rock grains to remove them.

Sand prediction involves calculating the maximum drawdown that can be applied to the sandface without producing sand, given the particular rock properties and geometry of the completion. It is usually based on log data (in particular, sonic velocities) but is often calibrated by reference to core testing (because the correlation between sonic properties and strength is poor when rock strengths are very low). It is sometimes possible to increase the maximum sand-free drawdown by changing the geometry of the completion. For example, phased (randomly oriented) perforation tunnels lie at all possible orientations perpendicular to the wellbore, and so sample the worst orientation for stability. Given an appropriate stress field, perforations can be oriented instead to sample the best orientation, and gain a good amount of extra allowed drawdown (Sulbaran et al. 1999). Another possibility is to selectively perforate only the stronger intervals in the reservoir section; provided the permeability is favorable, the higher drawdown this allows more than compensates for the smaller area open to flow.

Sand prediction of this kind must be made not just for the current state of the reservoir, but also for the future (because well operators would generally like their wells to be sand-free throughout the operating life). Two main factors influence the long-term performance:

- The change in stress state in the reservoir as it depletes. This may make things better or worse. It is necessary to know or calculate the stress-path coefficient of the reservoir to make predictions of this long-term behavior.
- The onset of water cut. This can either destroy cementation between grains, weakening the formation so that it fails under the in-situ stresses, or reduce capillary cohesion to zero, so previously stable banks and structures of loose sand are suddenly mobilized. There is also an indirect influence because when water cut increases, the drawdown on the well is often increased to keep oil levels up.

If these calculations show that sand is likely to be produced, and if this is likely to have a negative impact on well performance, then sand-control methods must be used. There is a wide range of these, and it is beyond the scope of this chapter to describe them in detail. They include gravel packing, stand-alone screens (i.e., tubular filters), frac and pack (a gravel pack combined with a hydraulic fracture), resin consolidation, expandable sand screens, and screenless completions. The installation of this kind of equipment is not easily reversible and so must be planned very carefully and executed diligently, and the components must continue to function properly for the lifetime of the well so they are very highly engineered and made of corrosion-resistant alloys. This means that sand-control completions can make up a substantial proportion of the total cost of a well.

Hydraulic Fracturing. Productivity of a reservoir interval depends on many factors, such as fluid viscosity, formation permeability, and the so-called skin around the well. Very often, the flow into a well is not enough to make it economical, and so the well is stimulated. Perhaps the most widespread form of stimulation is hydraulic fracturing. After the well is perforated, a viscous or filter-cake-forming fluid is pumped into the casing and perforation tunnels, using high-powered frac pumps. The hoop stress around the well or perforation is reduced to zero and below by the internal pressurization, as described by Eqs. 5.1.10 through 5.1.19 in Section 5.1.4. A fracture is initiated and then grows away from the well, often to distances of hundreds of meters on either side. The fracture is usually initiated with just a fluid, but shortly afterwards sand or proppant is also pumped in. This is carried by the viscous frac fluid far along the hydraulic fracture, and settles there, either by dehydration of the slurry through fluid flow into the formation, or by reduction (*breakage*) of the viscosity or gel strength of the carrier fluid. Breakage is usually carried out by chemical or thermal means. The remnants of the frac fluid are then flowed out of the well (*cleanup*) leaving the sand or proppant in the fracture, to act as a high-permeability pathway from the distant parts of the reservoir to the well.

Traditionally, hydraulic fracturing was used to stimulate gas wells, especially in low-permeability reservoirs, but its use has grown considerably in recent years and all kinds of wells are fractured now. High-permeability, weak sandstones are often fractured deliberately during installation of a gravel pack for sand control (frac and pack), giving high productivity and ensuring a good void-free gravel pack.

This sketch does not do justice to the great sophistication and complexity of hydraulic fracturing, which is a major worldwide business, mainly for gas wells but also for oil. The process of initiating and propagating the fracture has been the subject of many experimental and theoretical investigations, and many computer codes exist, in the business and academic worlds, for predicting the shape and extent of the fracture.

The initiation of the fracture is controlled by the Eqs. 5.1.10 through 5.1.19 as mentioned previously, and so fracture-initiation pressures, and the location and orientation of the initial frac, depend on the orientation of the hole relative to the in-situ stress field. The propagation pressure, however, depends primarily on the minimum in-situ stress, and the orientation of the frac when it has propagated a long way beyond the region influenced by the wellbore is generally perpendicular to the direction of this minimum in-situ stress.

Other key parameters of the fracture are its height, width, and length. Height is primarily controlled by the pumped fluid pressure and the in-situ stress level, and fractures are often designed to be contained within reservoir intervals by the higher stress levels expected within neighboring shale layers. Sonic logs are frequently run to measure Poisson's ratio through sand/shale sequences, for example, to assess the likely levels of stress variation (see Section 5.1.5) and its effect on fracture-height growth.

The width is controlled by fluid pressure and the Young's modulus of the formation. It increases as the fracture extends, but decreases again as pumping stops and the fracture begins to close. The proppant treatment (particle size and volume fraction in the fluid) must be designed so that an appropriate thickness of proppant pack is trapped in the fracture after closure, ideally without crushing, in order to achieve the right levels of fracture conductivity. The higher the formation permeability, the higher must be the fracture conductivity in order to influence the productivity of the well.

The length of the fracture is controlled by the geology, the volume of fluid pumped, and its efficiency (i.e., how much of the fluid leaks into the formation). As mentioned before, the length can reach hundreds of meters, especially in low-permeability rock where dehydration of the proppant slurry is slow. In more-permeable formations, a few tens of meters can be achieved before the slurry dehydrates; designers may aim for this dehydration so that, in a soft, low-modulus, high-permeability rock, the pumping pressure can be increased substantially to give extra fracture width and high conductivity.

There are many models for the growth of hydraulic fractures. Most are based on one of two assumed geometries. The Perkins-Kern-Nordgren model (Perkins and Kern 1961) is the most common and assumes that the cross section of the fracture, looking along its length, is an ellipse; the opening of the fracture decreases smoothly towards the upper and lower barriers. The Kristianovitch-Geertsma-DeKlerk model (Geertsma and de Klerk 1969) assumes that the cross section is a rectangle; the fracture opening is constant in height, and the upper and lower boundaries are marked by slip of the reservoir formation relative to its neighbors. Both models assume that the fracture is contained (i.e., its height growth is limited) and also that the fracture is planar. A number of much more complex models exist in the industry that relax some of these assumptions.

The picture given here of a two-winged fracture extending smoothly out from the well is model-based. Methods for mapping fracture growth have been developed recently, based on measurement and interpretation of the tilts (at the surface and in nearby wells) caused by the fracture, and also of the microseismic events that accompany it. Both these techniques have considerable difficulties in interpretation, but show that hydraulic-fracturing geometry is much more complex in reality than is assumed in the models (Wright et al. 1999).

This discussion has concerned fracturing for productivity increases, but the technique is also widely used for disposal of drilling wastes. Cuttings and cavings are ground up at surface, slurried, then pumped down the annulus of the current well or a nearby one, to grow a fracture and dispose of the solids. This is nearly always in an impermeable shale formation, and so there is very slow leakoff of the slurry fluid. This means that the fracture can grow very large and long, and care must be taken to understand where it is going (Moschovidis et al. 2000).

Reservoir Management. If hydrocarbons are produced from a reservoir without replacement by, for example, an aquifer drive or water injection, the formation pressure tends to drop. This changes the effective stresses acting on the formation, and this can have a number of effects.

Many formations compact, for example, when the effective stress on them is increased. The most notable example of this is Ekofisk in the North Sea, a thick chalk reservoir discovered in 1969. Production began in 1971. In 1984, it was realized that the seabed was subsiding, and measures had to be taken to raise the platform to a safe height above the waves. Rock-mechanics studies revealed that the chalk in the reservoir was compacting under the effects of increasing stress and water injection, and this precipitated a great deal of interest in compaction, and more generally in the mechanics of reservoirs (Chin et al. 1994; Sylte et al. 1999). Similarly, in the diatomite fields of California, compaction and the shear induced by it along bed boundaries caused damage to hundreds of wells (Fredrich et al. 1996).

The influence of production or injection is not limited to compaction and subsidence. In the Rangely field in Colorado, injection triggered minor earthquakes (Scholz 2002). In the Shearwater field in the North Sea, prediction of the effects of depletion was needed in order to make informed choices of completion equipment and sand-management options (Kenter et al. 1998). In a number of reservoirs, the dominant direction of fluid flow was found to be strongly affected by the influence of the in-situ stress field on fractures in the formation, allowing improved design of waterfloods and other secondary-recovery methods (Heffer et al. 1997).

Simple models exist for some aspects of this behavior. For example, surface subsidence as a consequence of compaction at the reservoir was modeled by Geertsma (1973); a more complex approach is used more recently by Segall (1992). Changes in horizontal stresses with depletion can be estimated using simple elastic equations or more-realistic failure equations (Addis et al. 1994), as described in Section 5.1.5. In general, however, these methods give approximate answers only. In recent years, several software packages have become available for modeling the mechanical response of the reservoir and overburden. These are usually based on finite-element approaches, and they often involve coupling between the mechanical response and the fluid-flow response of the formation. The deformation of the diatomite reservoir mentioned previously (Fredrich et al. 1996), for example, was studied using a large 3D fluid-flow model (a black-oil reservoir simulator with more than 90,000 gridblocks) coupled to a 3D finite-element mechanical model with nearly half a million nodes. The formations being studied were very weak, and so the simulations had to include their plastic response; this meant the simulations were complex and lengthy, but they gave good predictions of subsidence and well-failure locations (Fredrich et al. 1996). This type of coupled simulation is now available on small computers and is becoming widely used. The calculation time can still be rather long, and as with most simulations, the quality of the output depends on the quality of the input data; it is, however, a useful new tool in predicting the initial conditions and evolution of reservoirs (Samier et al. 2003; Onaisi et al. 2002; Stone et al. 2003).

The stress changes induced by production are also being used to indicate fluid movement and depletion. Changes in fluid type or pressure can alter the seismic properties of the reservoir significantly, and this can sometimes be seen on a repeated seismic survey. Because the reservoir is usually very thin, however, its influence on the overall seismic response can be small. If it induces stress changes in the overburden, however, these are present over much greater volume and affect the seismic-wave propagation to a correspondingly greater extent. This has been used for monitoring reservoir evolution (Kenter et al. 2004).

Nomenclature

a = constants in Gardner's correlation

A = area

A_{SDR} = stress depletion ratio ($\Delta\sigma_h/\Delta p_p$)

b = constants in Gardner's correlation

d_0, d_1 = initial and deformed diameters

D_e = Deborah number: the ratio of viscoelastic relaxation time to viscous relaxation time

E = Young's modulus

F = force

g = acceleration due to gravity

G = shear modulus

h = height of fluid column

$\mathbf{I}$ = identity matrix

k = constant in Tresca criterion

K_{frame} = bulk modulus of rock skeleton

K_{grain} = bulk modulus of rock mineral

K = matrix stress coefficient (effective-stress ratio)

l_{ij} = direction cosine

L_0, L_1 = initial and deformed lengths

MW_{min} = minimum MW

$\hat{\mathbf{n}}$ = unit vector normal to a surface, referred to the same axes as the stress components

N_Φ = coefficient in Mohr-Coulomb criterion

p = pressure

p_p = pore or formation pressure

p_p^{normal} = formation pressure on normal compaction trend

p_w = wellbore pressure

p_b = breakdown pressure in XLOT

P_{prop} = propagation pressure in XLOT

P_{closure} = closure pressure in XLOT

P_{reopen} = reopening pressure in XLOT

r = radius

r_w = wellbore radius

R = measured resistivity

R_{normal} = normal resistivity

S = rock cohesion

T = temperature

V_p = compressional wave velocity

V_s = shear wave velocity

x, y, z = Cartesian axes

z = depth; distance along the wellbore axis

$\mathbf{R}$ = rotation matrix

$\mathbf{T}$ = surface traction vector

α_T = thermal-expansion coefficient

α = Biot's parameter

β = compressibility; angle in Mohr's diagram

Δt = measured sonic traveltime

Δt_{normal} = normal sonic traveltime

ε = normal strain

ε_l = lateral strain

θ = azimuth angle around well, or angle to principal axis

λ = Lamé constant

ν = Poisson's ratio

ρ = rock density

σ = normal total stress

σ = total stress tensor

$\sigma_1, \sigma_2, \sigma_3$ = maximum, intermediate, and minimum compressive stress, respectively

σ_A, σ_B = principal stresses normal to wellbore axis

σ_C = principal stress parallel to wellbore axis

$\sigma_r, \sigma_\theta, \sigma_z$ = radial, hoop, and axial stresses, respectively

σ' = effective stress

σ'_h = effective horizontal stress

σ'_n = effective stress normal to failure plane

σ'_{normal} = effective stress on normal-compaction trend

σ'_v = effective vertical stress

σ_t = rock tensile strength

σ_v = vertical stress

σ_h = minimum horizontal stress

σ_H = maximum horizontal stress

σ_{H_Az} = azimuth of maximum horizontal stress

$\boldsymbol{\sigma}^T$ = transpose of the matrix of stress components

τ = shear stress

Φ = internal-friction angle

Φ_{residual} = residual-friction angle

Appendix A—Stress-Transformation Equations

Fig. 5.1.2 shows how the magnitudes of the components of stress change according to the orientation of the axes used to define them. This is because stress is a tensor quantity—specifically a second-rank tensor. (The rank refers to the number of indices on its components. A scalar can be thought of as a zero-rank tensor, and a vector as a first-rank tensor. The important property of a tensor quantity is that its components should transform under a rotation of coordinate axes in such a way as to keep its geometrical or physical meaning invariant.)

Consider a point within a medium, relative to x-, y-, and z-axes. The force acting on a surface normal to the x-axis produces a stress normal to the surface, σ_x, and a shear stress acting parallel to the surface. The shear stress can be decomposed into two components, acting along the y- and z-axes, τ_{xy} and τ_{xz}, respectively. The stresses acting on a surface normal to the y-axis are similarly σ_y, τ_{yx}, and τ_{yz}, and those on a surface normal to the z-axis are σ_z, τ_{zx}, and τ_{zy}. Putting these together gives the complete set of components of the stress tensor $\boldsymbol{\sigma}$ acting at the point:

$$\boldsymbol{\sigma} = \begin{pmatrix} \sigma_x & \tau_{xy} & \tau_{xz} \\ \tau_{yx} & \sigma_y & \tau_{yz} \\ \tau_{zx} & \tau_{zy} & \sigma_z \end{pmatrix} \quad \dots\dots\dots\dots\dots\dots\dots\dots\dots\dots\dots\dots \text{(A-1)}$$

Bold typeface represents multicomponent quantities such as vectors or matrices. τ_{xy} is the stress acting on the plane normal to the x-axis, in the direction of the y-axis. Because the material on which this stress is acting is at rest, there can be no net rotational forces acting, which implies

$$\begin{aligned} \tau_{xy} &= \tau_{yx} \\ \tau_{xz} &= \tau_{zx}, \quad \dots\dots\dots\dots\dots\dots\dots\dots\dots\dots\dots\dots\dots \text{(A-2)} \\ \tau_{zy} &= \tau_{yz} \end{aligned}$$

and so the stress tensor contains six independent numbers.

Other notations are also used for the components of the stress tensor; it is frequently written as

$$\boldsymbol{\sigma} = \begin{pmatrix} \sigma_{11} & \sigma_{12} & \sigma_{13} \\ \sigma_{21} & \sigma_{22} & \sigma_{23} \\ \sigma_{31} & \sigma_{32} & \sigma_{33} \end{pmatrix} \quad \dots\dots\dots\dots\dots\dots\dots\dots\dots\dots\dots\dots\dots\dots \text{(A-3)}$$

The rules governing how the values of the components of the stress tensor change as the axes rotate are simple to derive, at least in two dimensions. Consider the forces acting on a surface in a material referred to x-, y-, and z-axes, as in **Fig. A-1.** If the only forces acting in the material are in the x–y plane, we can treat this as a 2D case, with unit thickness. The material is at rest, so the forces acting on the surface AB must be balanced by the forces acting on the surfaces OA and OB. For example, the force in the y direction generated by σ_y acting on area OB and τ_{xy} acting on area OA must be balanced by the forces generated by σ_n and τ acting on area AB, and then resolved onto the y-direction. Note that the stresses must be converted to forces before their components in a particular direction can be calculated—stresses cannot be resolved directly. So, assuming OA = 1, and balancing forces along the x-direction,

$$\sigma_x + \tau_{yx} \tan\theta = \frac{\sigma_n}{\cos\theta} \cos\theta + \frac{\tau}{\cos\theta} \sin\theta, \quad \dots\dots\dots\dots\dots \text{(A-4)}$$

and in the y direction,

$$\tau_{xy} + \sigma_y \tan\theta = \frac{\sigma_n}{\cos\theta} \sin\theta - \frac{\tau}{\cos\theta} \cos\theta \quad \dots\dots\dots\dots\dots \text{(A-5)}$$

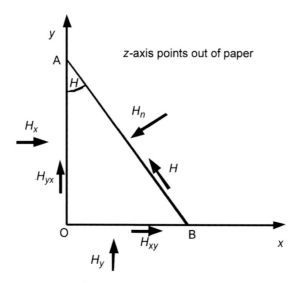

Fig. A-1—Geometry for the calculation of stress rotation in two dimensions.

Solving these two equations gives

$$\sigma_n = \sigma_x \cos^2 \theta + \sigma_y \sin^2 \theta + \tau_{xy} \sin 2\theta \quad \text{...................................... (A-6)}$$

and

$$\tau = \frac{1}{2}\left(\sigma_x - \sigma_y\right)\sin 2\theta - \tau_{xy} \cos 2\theta \quad \text{................................. (A-7)}$$

Note that it is possible to choose values of θ such that the shear stress τ on the plane AB is zero. The values are given by

$$\tan 2\theta = \frac{2\tau_{xy}}{\sigma_x - \sigma_y} \quad \text{.. (A-8)}$$

What this means is that the orientation of the plane AB can be chosen so that the only nonzero stress acting on it is a normal stress. There are two such planes given by this equation, whose normals are the principal directions of the stress state, which were referred to in Section 5.1.3, and the two normal stresses that emerge from this are the principal stresses. The principal-stress values are obtained by using Eq. A-8 in Eq. A-6:

$$\sigma_1 = \frac{1}{2}\left(\sigma_x + \sigma_y\right) + \sqrt{\tau_{xy}^2 + \frac{1}{4}\left(\sigma_x - \sigma_y\right)^2} \quad \text{.......................... (A-9)}$$

$$\sigma_2 = \frac{1}{2}\left(\sigma_x + \sigma_y\right) - \sqrt{\tau_{xy}^2 + \frac{1}{4}\left(\sigma_x - \sigma_y\right)^2} \quad \text{.......................... (A-10)}$$

We can use Eqs. A-6 and A-7 to see how Mohr's circle works. If we assume that σ_x and σ_y are now principal stresses, so that τ_{xy} is zero, we can calculate the normal and shear stresses σ_n and τ at an angle θ to the direction of σ_x:

$$\sigma_n = \sigma_x \cos^2 \theta + \sigma_y \sin^2 \theta \quad \text{.. (A-11)}$$

$$\tau = \frac{1}{2}\left(\sigma_x - \sigma_y\right)\sin 2\theta \quad \text{... (A-12)}$$

A few lines of algebra show that these are the coordinates of the point specified by the Mohr's-circle construction:

$$\sigma_n = \frac{1}{2}\left(\sigma_x + \sigma_y\right) + \frac{1}{2}\left(\sigma_x - \sigma_y\right)\cos 2\theta \quad\dots\dots\dots\dots\dots \text{(A-13)}$$

and

$$\tau = \frac{1}{2}\left(\sigma_x - \sigma_y\right)\sin 2\theta \quad\dots\dots\dots\dots\dots\dots \text{(A-14)}$$

The same principles apply to 3D stress states, but the algebra is lengthier. Let us start with a stress state represented by its components referred to x-, y-, and z-axes, as in Eq. A-1:

$$\boldsymbol{\sigma} = \begin{pmatrix} \sigma_x & \tau_{xy} & \tau_{xz} \\ \tau_{yx} & \sigma_y & \tau_{yz} \\ \tau_{zx} & \tau_{zy} & \sigma_z \end{pmatrix}$$

The process represented in Fig. A-1 is determination of the normal and shear stresses on an arbitrary plane. The 3D analog of this contains an additional complication, in that the orientation of the shear stress is no longer defined. All we can say is that it has to lie in the selected plane. This can be resolved by introducing the surface traction vector **T**, which is the limiting value of the ratio of force to the area on which it acts:

$$\mathbf{T} = \lim_{dA \to 0} \frac{d\mathbf{F}}{dA} \quad\dots\dots\dots\dots\dots\dots\dots\dots\dots \text{(A-15)}$$

Note that this is a vector, and its magnitude and direction depend on the orientation of the area A. Cauchy's equation (Davis and Selvadurai 1996) relates the traction on a given surface to the stress state and a unit vector normal to the surface:

$$\mathbf{T} = \boldsymbol{\sigma}^T \hat{\mathbf{n}}, \quad\dots\dots\dots\dots\dots\dots\dots\dots\dots\dots \text{(A-16)}$$

where $\boldsymbol{\sigma}^T$ is the transpose of the matrix of stress components and $\hat{\mathbf{n}}$ is the unit vector normal to the surface (referred to the same system of axes as the stress components). When **T** has been calculated from Eq. A-16, its components normal and parallel to the surface can be calculated, which are the normal stress and maximum shear stress acting on the surface:

$$\sigma_n = \mathbf{T}\hat{\mathbf{n}} \quad\dots\dots\dots\dots\dots\dots\dots\dots\dots\dots\dots \text{(A-17)}$$

$$\tau = \mathbf{T} - (\mathbf{T}\hat{\mathbf{n}})\hat{\mathbf{n}} \quad\dots\dots\dots\dots\dots\dots\dots \text{(A-18)}$$

The maximum shear stress τ is a vector. It can now be resolved onto a chosen pair of axes lying in the surface.

Finding the principal directions and stresses means finding the three surface orientations for which the traction vector is parallel to the surface normal vector; i.e., for which

$$\mathbf{T} = \boldsymbol{\sigma}^T \hat{\mathbf{n}} = \alpha\hat{\mathbf{n}} \quad\dots\dots\dots\dots\dots\dots\dots \text{(A-19)}$$

or

$$(\sigma = \alpha\mathbf{I})\,\hat{\mathbf{n}} = 0, \quad\dots\dots\dots\dots\dots\dots\dots\dots \text{(A-20)}$$

because the stress-components matrix is symmetric. $\mathbf{I}$ is the identity matrix. This has a unique solution if

$$\det(\boldsymbol{\sigma} - \alpha \mathbf{I}) = 0 \qquad \text{(A-21)}$$

This can be solved for three values of α, which are the principal stresses. Using these in Eq. A-20 gives three eigenvectors that are the principal directions.

The stress state represented by Eq. A-1 can also be represented on other sets of axes, say 1, 2, and 3. The relationship between the two sets of axes is expressed by direction cosines. For example, if the angles between the l-axis and the x-, y-, and z-axes are δ_{1x}, δ_{1y}, and δ_{1z}, the direction cosines for these axis pairs are

$$
\begin{aligned}
l_{1x} &= \cos \delta_{1x} \\
l_{1y} &= \cos \delta_{1y} \qquad\qquad\qquad\qquad\qquad\qquad\qquad\qquad\qquad \text{(A-22)}\\
l_{1z} &= \cos \delta_{1z}
\end{aligned}
$$

Putting together all the direction cosines for the two sets of axes gives a rotation matrix:

$$
\mathbf{R} = \begin{pmatrix} l_{1x} & l_{1y} & l_{1z} \\ l_{2x} & l_{2y} & l_{2z} \\ l_{3x} & l_{3y} & l_{3z} \end{pmatrix} \qquad\qquad\qquad\qquad\qquad \text{(A-23)}
$$

The components of the stress state referred to the new axes are then given by

$$\boldsymbol{\sigma}^* = \mathbf{R}\boldsymbol{\sigma}\mathbf{R}^T \qquad\qquad\qquad\qquad\qquad\qquad\qquad \text{(A-24)}$$

When, for example, the in-situ stress state is expressed with respect to the coordinate system of the wellbore, as in Section 5.1.4 for the stability analysis of deviated wells, Eq. A-24 is the equation that is used. The direction cosines l_{ij} come from the azimuth and deviation of the well relative to the in-situ stress orientations, and also from the position of the point of interest around the circumference of the well. Eq. A-24 represents six equations, each with nine terms on the right side, and it is typically evaluated at many points around and along the wellbore, which is a good reason to use software for this type of calculation.

Appendix B—Elastic Behavior and Relationships Between Elastic Constants

Eqs. 5.1.1 and 5.1.2 defined Young's modulus E and Poisson's ratio v, respectively, as the two independent elastic constants of a linear isotropic elastic material. The relationships between normal and shear stresses σ and τ and normal and shear strains ε and γ can be expressed in a number of ways, and this allows the elastic constants to be expressed in several different ways. There are always two independent numbers, however. One of the most useful ways in geomechanics is as follows:

$$
\begin{aligned}
E\varepsilon_x &= \sigma_x - v(\sigma_y + \sigma_z) \\
E\varepsilon_y &= \sigma_y - v(\sigma_x + \sigma_z) \\
E\varepsilon_z &= \sigma_z - v(\sigma_y + \sigma_x) \qquad\qquad\qquad\qquad\qquad\qquad \text{(B-1)}\\
G\gamma_{xy} &= \tau_{xy} \\
G\gamma_{xz} &= \tau_{xz} \\
G\gamma_{yz} &= \tau_{yz}
\end{aligned}
$$

Here G is the shear modulus; it is not independent of E and v, but given by

$$G = \frac{E}{2(1+v)} \qquad\qquad\qquad\qquad\qquad\qquad\qquad\qquad \text{(B-2)}$$

The first two lines of Eq. B-1 are used in the USM and similar models for horizontal stress (Section 5.1.5). For example, let σ_x and σ_y be equal horizontal stresses. In the USM, ε_x is assumed to be zero, so the first line of Eq. B-1 becomes

$$0 = \sigma_x - \nu(\sigma_x + \sigma_z)$$

$$\sigma_x = \frac{\nu}{1-\nu}\sigma_z \qquad\qquad \text{...} \quad \text{(B-3)}$$

For a porous material with a pore pressure, the stresses are replaced by effective stresses.

The other commonly used elastic constants are the bulk modulus K, the compressibility $\beta (= 1/K)$, and the Lamé constant λ (G is also called a Lamé constant). The Lamé constants are useful for expressing stress in terms of strain, as follows:

$$\sigma_x = (\lambda + 2G)\varepsilon_x + \lambda(\varepsilon_y + \varepsilon_z)$$

$$\sigma_y = (\lambda + 2G)\varepsilon_y + \lambda(\varepsilon_x + \varepsilon_z)$$

$$\sigma_z = (\lambda + 2G)\varepsilon_z + \lambda(\varepsilon_x + \varepsilon_y)$$

$$\tau_{xy} = G\gamma_{xy} \qquad\qquad \text{...} \quad \text{(B-4)}$$

$$\tau_{xz} = G\gamma_{xz}$$

$$\tau_{yz} = G\gamma_{yz}$$

The different elastic constants are interrelated as shown here:

$$E = 2G(1+\nu)$$

$$E = 3K(1-2\nu)$$

$$E = \frac{\lambda}{\nu}(1+\nu)(1-2\nu)$$

$$\nu = \frac{\lambda}{2(\lambda+G)} \qquad\qquad \text{...} \quad \text{(B-5)}$$

$$\nu = \frac{3K - 2G}{2(3K+G)}$$

$$K = \lambda + \frac{2}{3}G$$

References

AADE 1998. *Pressure Regimes in Sedimentary Basins and Their Prediction*. American Association of Drilling Engineers (AADE) Industry Forum, Lake Conroe, Texas, USA, 2–4 September.

Aadnoy, B.S. 1987. Continuum Mechanics Analysis of the Stability of Inclined Boreholes in Anisotropic Rock Formations. PhD dissertation, Norwegian Institute of Technology, Trondheim, Norway.

Aadnoy, B.S. 1989. Stresses Around Boreholes Drilled in Sedimentary Rocks. *J. Petrol. Sci. Eng.* **2** (4): 349–360. DOI: 10.1016/0920-4105(89)90009-0.

Addis, M.A. 1997. Reservoir Depletion and Its Effect on Wellbore Stability Evaluation. *International Journal of Rock Mechanics Mining Sciences* **34** (3–4): 423.

Addis, M.A., Last, N.C., and Yassir, N.A. 1994. Estimation of Horizontal Stresses at Depth in Faulted Regions and Their Relationship to Pore Pressure Variations. *SPEFE* **11** (1): 11–18. SPE-28140-PA. DOI: 10.2118/28140-PA.

Alberty, M.W. and McLean, M.R. 2001. Fracture Gradients in Depleted Reservoirs—Drilling Wells in Late Reservoir Life. Paper SPE 67740 presented at the SPE/IADC Drilling Conference, Amsterdam, 27 February–1 March 2001. DOI: 10.2118/67740-MS.

Alsop, G.I., Blundell, D.J., and Davison, I. ed. 1996. *Salt Tectonics*, 291–302. Bath, UK: Special Publication No. 100, The Geological Society.

Amadei, B. 1983. *Rock Anisotropy and the Theory of Stress Measurements*. Heidelberg, Germany: Springer-Verlag.

Anderson, E.M. 1951. *The Dynamics of Faulting and Dyke Formation With Applications to Britain*. Edinburgh, UK: Oliver and Boyd.

Ashby, M.F. and Hallam, S.D. 1986. The Failure of Brittle Solids Containing Small Cracks Under Compressive Stress States. *Acta Metallurgica* **34** (3): 497–510. DOI: 10.1016/0001-6160(86)90086-6.

Aston, M.S., Alberty, M.W., McLean, M.R., de Jong, H.J., and Armagost, K. 2004. Drilling Fluids for Wellbore Strengthening. Paper SPE 87130 presented at the IADC/SPE Drilling Conference, Dallas, 2–4 March. DOI: 10.2118/87130-MS.

Atkinson, C. and Bradford, I. 2001. Effect of Inhomogeneous Rock Properties on the Stability of Wellbores. *Proc.*, IUTAM Symposium on Analytical and Computational Fracture Mechanics of Non-Homogeneous Materials, Cardiff University, UK, 18–22 June.

Babcock, E.A. 1978. Measurement of Subsurface Fractures From Dipmeter Logs. *AAPG Bulletin* **62** (7): 1111–1126. Reprinted 1990 in *Formation evaluation II—log interpretation*, 457–472, ed. N.H. Foster and E.A. Beaumont. Tulsa: Treatise of Petroleum Geology Reprint Series No. 17, AAPG.

Barker, J.W., Feland, K.W., and Tsao, Y.-H. 1994. Drilling Long Salt Sections Along the U.S. Gulf Coast. *SPEDC* **9** (3): 185–188. SPE-24605-PA. DOI: 10.2118/24605-PA.

Beacom, L.E., Nicholson, H., and Corfield, R.I. 2001. Integration of Drilling and Geological Data To Understand Wellbore Instability. Paper SPE 67755 presented at the SPE/IADC Drilling Conference, Amsterdam, 27 February–1 March. DOI: 10.2118/67755-MS.

Bell, J.S. and Gough, D.I. 1982. The Use of Borehole Breakouts in the Study of Crustal Stress. In *Workshop on Hydraulic Fracturing Stress Measurements Proceedings*, 539–557, ed. M.D. Zoback and B.C. Haimson. US Geological Survey Open-File Report 82-1075, Reston, Virginia, USA.

Billington, E.W. and Tate, A. 1980. *The Physics of Deformation and Flow*. Dallas: McGraw-Hill.

Biot, M.A. 1962. Mechanics of Deformation and Acoustic Propagation in Porous Media. *J. Appl. Phys.* **33** (4): 1482–1498. DOI: 10.1063/1.1728759.

Bowers, G.L. 1995. Pore Pressure Estimation From Velocity Data: Accounting for Overpressure Mechanisms Besides Undercompaction. *SPEDC* **10** (2): 89–95. SPE-27488-PA. DOI: 10.2118/27488-PA.

Bradford, I.D.R., Aldred, W.A., Cook, J.M. et al. 2000. When Rock Mechanics Met Drilling: Effective Implementation of Real-Time Wellbore Stability Control. Paper SPE 59121 presented at the IADC/SPE Drilling Conference, New Orleans, 23–25 February. DOI: 10.2118/59121-MS.

Brady, B.H.G. and Brown, E.T. 1993. *Rock Mechanics for Underground Mining*, second edition. The Netherlands: Springer.

Bratli, R.K. and Rinses, R. 1981. Stability and Failure of Sand Arches. *SPEJ* **21** (2): 236–248. SPE-8427-PA. DOI: 10.2118/8427-PA.

Bratli, R.K., Horsrud, P., and Risnes, R. 1983. Rock Mechanics Applied to the Region Near a Wellbore. *Proc.*, 5th International Congress on Rock Mechanics, Melbourne, Australia, F1–F17.

Bratton, T., Bornemann, T., Li, Q., Plumb, R., Rasmus, J., and Krabbe, H. 1999. Logging-While-Drilling Images for Geomechanical, Geological and Petrophysical Interpretations. Paper JJJ presented at the 40th Annual Logging Symposium, SPWLA, Oslo, Norway.

Bratton, T., Edwards, S., Fuller, J. et al. 2001. Avoiding Drilling Problems. *Oilfield Review* **13** (2): 32–51.

Breckels, I.M. and van Eekelen, H.A.M. 1982. Relationship Between Horizontal Stress and Depth in Sedimentary Basins. *JPT* **34** (9): 2191–2199. SPE-10336-PA. DOI: 10.2118/10336-PA.

Busby, C.J. and Ingersoll, V.R. 1995. *Tectonics of Sedimentary Basins*. London: Blackwell Science.

Carcione, J.M., Helle, H.B., and Gangi, A.F. 2006. Theory of Borehole Stability When Drilling Through Salt Formations. *Geophysics* **71** (3): F31–F47. DOI: 10.1190/1.2195447.

Chang, C., Zoback, M.D., and Khaksar, A. 2006. Empirical Relations Between Rock Strength and Physical Properties in Sedimentary Rocks. *Journal of Petroleum Science and Engineering* **51** (3–4): 223–237. DOI: 10.1016/j.petrol.2006.01.003.

Charlez, P.A. 1991. *Rock Mechanics, Vol. 1 Theoretical Fundamentals*. Paris: Editions Technip.

Charlez, P.A. 1997. *Rock Mechanics, Vol. 2 Petroleum Applications*. Paris: Editions Technip.

Chen, X., Tan, C.P., and Detournay, C. 2002. The Impact of Mud Infiltration on Wellbore Stability in Fractured Rock Masses. Paper SPE 78241 presented at the SPE/ISRM Rock Mechanics Conference, Irving, Texas, USA, 20–23 October. DOI: 10.2118/78241-MS.

Chin, L.Y., Boade, R.R., Nagel, B., and Landa, G.H. 1994. Numerical Simulation of Ekofisk Reservoir Compaction and Subsidence: Treating the Mechanical Behavior of the Overburden and Reservoir. Paper SPE 28128 presented at Rock Mechanics in Petroleum Engineering, Delft, The Netherlands, 29–31 August. DOI: 10.2118/28128-MS.

Christman, S.A. 1973. Offshore Fracture Gradients. *JPT* **25** (8): 910–914. SPE-4133-PA. DOI: 10.2118/4133-PA.

Cook, J.M., Sheppard, M.C., and Houwen, O.H. 1991. Effects of Strain Rate and Confining Pressure on the Deformation and Failure of Shale. *SPEDE* **6** (2): 100–104; *Trans.*, AIME, **291**. SPE-19944-PA. DOI: 10.2118/19944-PA.

Davis, R.O. and Selvadurai, A.P.S. 1996. *Elasticity and Geomechanics*. Cambridge, UK: Cambridge University Press.

Davis, R.O. and Selvadurai, A.P.S. 2002. *Plasticity and Geomechanics*. Cambridge, UK: Cambridge University Press.

Deere, D.U. and Miller, R.P. 1966. *Engineering Classification and Index Properties for Intact Rock*. Technical Report, AFWL-TR-65-116, US Air Force Systems Command Air Force Weapons Lab, Kirtland Air Force Base, New Mexico, USA.

Detournay, E. and Cheng, A.H.-D. 1988. Poroelastic Response of a Borehole in a Non-Hydrostatic Stress Field. *International Journal of Rock Mechanics and Mining Science & Geomechanics Abstracts* **25** (3): 171–182. DOI: 10.1016/0148-9062(88)92299-1.

Doyle, E.F. 1998. Case Study—Comprehensive Approach to Formation Pressure Prediction and Evaluation on a Norwegian HPHT Well. In *Overpressures in Petroleum Exploration*, ed. A. Mitchell and D. Grauls, 149–155. Pau, France: Bull. des Centres de Recherches Exploration-Production Elf EP Memoire 22, Elf Aquitaine.

Dutta, N., Gelinsky, S., Reese, M., and Khan, M. 2001. A New Petrophysically Constrained Predrill Pore Pressure Prediction Method for the Deepwater Gulf of Mexico: A Real-Time Case Study. Paper SPE 71347 presented at the SPE Annual Technical Conference and Exhibition, New Orleans, 30 September–3 October. DOI: 10.2118/71347-MS.

Eaton, B.A. 1969. Fracture Gradient Prediction and Its Application in Oilfield Operations. *JPT* **21** (10): 1353–1360; *Trans.*, AIME, **246**. SPE-2163-PA. DOI: 10.2118/2163-PA.

Eaton, B.A. 1975. The Equation for Geopressure Prediction From Well Logs. Paper SPE 5544 presented at the SPE Annual Technical Conference and Exhibition, Dallas, 28 September–1 October. DOI: 10.2118/5544-MS.

Eaton, B.A. and Eaton, T.L. 1997. Fracture Gradient Prediction for the New Generation. *World Oil* **218** (10): 93–100.

Economides, M.J. and Nolte, K.G. ed. 1989. *Reservoir Stimulation*, second edition. Upper Saddle River, New Jersey: Prentice Hall.

Edwards, S.T., Bratton, T.R., and Standifird, W.B. 2002. Accidental Geomechanics—Capturing In-Situ Stress From Mud Losses Encountered While Drilling. Paper SPE 78205 presented at the SPE/ISRM Rock Mechanics Conference, Irving, Texas, USA, 20–23 October. DOI: 10.2118/78205-MS.

Edwards, S.T., Matsutsuyu, B., and Willson, S. 2004. Imaging Unstable Wellbores While Drilling. *SPEDC* **19** (4): 236–243. SPE-79846-PA. DOI: 10.2118/79846-PA.

Engelder, T. 1993. *Stress Regimes in the Lithosphere*. Princeton, New Jersey: Princeton University Press.

Ewy, R.T. 1999. Wellbore-Stability Predictions by Use of a Modified Lade Criterion. *SPEDC* **14** (2): 85–91. SPE-56862-PA. DOI: 10.2118/56862-PA.

Fjær, E., Holt, R.M., Horsrud, P., Raaen, A.M., and Risnes, R. 1992. *Petroleum Related Rock Mechanics*, second edition. Oxford, UK: Elsevier Science.

Foster, J.B. 1966. Estimation of Formation Pressures From Electrical Surveys—Offshore Louisiana. *JPT* **18** (2): 165–171. SPE-1200-PA. DOI: 10.2118/1200-PA.

Franco, J.L.A., de la Torre, H.G., Ortiz, M.A.M. et al. 2005. Using Shear-Wave Anisotropy To Optimize Reservoir Drainage and Improve Production in Low-Permeability Formations in the North of Mexico. Paper SPE 96808 presented at the SPE Annual Conference and Technical Exhibition, Dallas, 9–12 October. DOI: 10.2118/96808-MS.

Fredrich, J.T., Arguello, J.G., Thorne, B.J. et al. 1996. Three-Dimensional Geomechanical Simulation of Reservoir Compaction and Implications for Well Failures in the Belridge Diatomite. Paper SPE 36698 presented at the SPE Annual Technical Conference and Exhibition, Denver, 6–9 October. DOI: 10.2118/36698-MS.

Fredrich, J.T., Coblenz, D., Fossum, A.F., and Thorne, B.J. 2003. Stress Perturbations Adjacent to Salt Bodies in the Deepwater Gulf of Mexico. Paper SPE 84554 presented at the SPE Annual Technical Conference and Exhibition, Denver, 5–8 October. DOI: 10.2118/84554-MS.

Gardner, G.H.F., Gardner, L.W., and Gregory, A.R. 1974. Formation Velocity and Density—The Diagnostic Basis for Stratigraphic Traps. *Geophysics* **39** (6): 770–780. DOI: 10.1190/1.1440465.

Geertsma, J. 1973. Land Subsidence Above Compacting Oil and Gas Reservoirs. *JPT* **25** (6): 734–744. SPE-3730-PA. DOI: 10.2118/3730-PA.

Geertsma, J. and de Klerk, F. 1969. A Rapid Method Of Predicting Width And Extent Of Hydraulically Induced Fractures. *JPT* **21** (12):1571-1581. SPE 2458-PA. DOI: 10.2118/2458-PA.

Gomez, E., Jordan, T.E., Allmendinger, R.W., and Cardozo, N. 2005. Development of the Colombian Foreland-Basin System as a Consequence of Diachronous Exhumation of the Northern Andes. *GSA Bulletin* **117** (9): 1272–1292. DOI: 10.1130/B25456.1.

Greenwood, J., Bowler, P., Sarmiento, J.F., Willson, S., and Edwards, S. 2006. Evaluation and Application of Real-Time Image and Caliper Data as Part of a Wellbore Stability Monitoring Provision. Paper SPE 99111 presented at the IADC/SPE Drilling Conference, Miami, Florida, USA, 21–23 February. DOI: 10.2118/99111-MS.

Guo, B. 2001. Use of Spreadsheet and Analytical Models To Simulate Solid, Water, Oil and Gas Flow in Underbalanced Drilling. Paper SPE 72328 presented at the SPE/IADC Middle East Drilling Technology Conference, Bahrain, 22–24 October. DOI: 10.2118/72328-MS.

Hawkes, C.D., Smith, S.P., and McLellan, P.J. 2002. Coupled Modeling of Borehole Instability and Multiphase Flow for Underbalanced Drilling. Paper SPE 74447 presented at the IADC/SPE Drilling Conference, Dallas, 26–28 February. DOI: 10.2118/74447-MS.

Heffer, K.J., Fox, R.J., McGill, C.A., and Koutsabeloulis, N.C. 1997. Novel Techniques Show Links Between Reservoir Flow Directionality, Earth Stress, Fault Structure and Geomechanical Changes in Mature Waterfloods. *SPEJ* **2** (2): 91–98. SPE-30711-PA. DOI: 10.2118/30711-PA.

Hickman, S.H. and Zoback, M.D. 1983. The Interpretation of Hydraulic Fracturing Pressure Time Data for In-Situ Stress Determination. In *Hydraulic Fracturing Stress Measurements*, ed. M.D. Zoback and B.C. Haimson. Washington, DC: National Academic Press.

Holt, C.A. and Johnson, J.B. 1986. A Method for Drilling Moving Salt Formations—Drilling and Underreaming Concurrently. *SPEDE* **1** (4): 315–324. SPE-13488-PA. DOI: 10.2118/13488-PA.

Holt, J., Wright, W.J., Nicholson, H., Kuhn-de-Chizelle, A., and Ramshorn, C. 2000. Mungo Field: Improved Communication Through 3D Visualization of Drilling Problems. Paper SPE 62523 presented at the SPE/AAPG Western Regional Meeting, Long Beach, California, USA, 19–23 June. DOI: 10.2118/62523-MS.

Hornby, B.E., Howie, J.M., and Ince, D.E. 2003. Anisotropy Correction for Deviated-Well Sonic Logs: Application to Seismic Well Tie. *Geophysics* **68** (2): 464–471. DOI: 10.1190/1.1567214.

Horsrud, P. 2001. Estimating Mechanical Properties of Shale From Empirical Correlations. *SPEDC* **16** (2): 68–73. SPE-56017-PA. DOI: 10.2118/56017-PA.

Hubbert, M.K. and Willis, D.G. 1957. Mechanics of Hydraulic Fracturing. In *Petroleum Development and Technology. Transactions of the American Institute of Mining and Metallurgical Engineers*, Vol. 210, 153–168. Littleton, Colorado: AIME.

Inaba, M., McCormick, D., Mikalsen, T. et al. 2003. Wellbore Imaging Goes Live. *Oilfield Review* **15** (1): 24–37.

Jaeger, J.C. and Cook, N.G.W. 1979. *Fundamentals of Rock Mechanics*. London: Chapman and Hall.

Kageson-Loe, N.K., Stage, M.C., Christensen, H.F., and Havmoller, O. 2004. Application of Overburden Strength and Deformation Properties Derived From Drill Cutting I_p Data. *Proc.*, 6th NARMS Conference (Gulf Rocks 2004), Houston, Paper No. ARMA/NARMS 04-456.

Kenter, C.J., Schreppers, G.M.A., Blanton, T.L., Baaijens, M.N., and Ramos, G.G. 1998. Compaction Study for Shearwater Field. Paper SPE 47280 presented at SPE/ISRM Rock Mechanics in Petroleum Engineering (EUROCK 98), Trondheim, Norway, 8–10 July. DOI: 10.2118/47280-MS.

Kenter, C.J., van den Beukel, A.C., Hatchell, P.J. et al. 2004. Evaluation of Reservoir Characteristics From Timeshifts in the Overburden. Presented at Gulf Rocks 2004, 6th North American Rock Mechanics Symposium (NARMS): Rock Mechanics Across Borders and Disciplines, Houston, 5–9 June.

Lal, M. 1999. Shale Stability: Drilling Fluid Interaction and Shale Strength. Paper SPE 54356 presented at the SPE Asia Pacific Oil and Gas Conference and Exhibition, Jakarta, 20–22 April. DOI: 10.2118/54356-MS.

Lashkaripour, G.R. and Dusseault, M.B. 1993. A Statistical Study on Shale Properties: Relationships Among Principal Shale Properties. *Proc.,* Conference of Probabilistic Methods in Geotechnical Engineering, ed. K.S. Li and S.-C.R. Lo, Rotterdam, The Netherlands: Balkema.

Last, N., Plumb, R., Harkness, R., Charlez, P., Alsen, J., and McLean, M. 1995. An Integrated Approach to Evaluating and Managing Wellbore Instability in the Cusiana Field, Colombia, South America. Paper SPE 30464 presented at the SPE Annual Technical Conference and Exhibition, Dallas, 22–25 October. DOI: 10.2118/30464-MS.

Leeman, E.R. 1964. The Measurement of Stress in Rock; Part I, the Principles of Rock Stress Measurements; Part II, Borehole Rock Stress Measuring Instruments. *Journal of the South African Institute of Mining and Metallurgy* **65** (2): 45–114.

Leung, P.K. and Steig, R.P. 1992. Dielectric Constant Measurements: A New, Rapid Method To Characterize Shale at the Wellsite. Paper SPE 23887 presented at the SPE/IADC Drilling Conference, New Orleans, 18–21 February. DOI: 10.2118/23887-MS.

Maeso, C. and Tribe, I. 2001. Hole Shape From Ultrasonic Calipers and Density While Drilling—A Tool for Drillers. Paper SPE 71395 presented at the SPE Annual Technical Conference and Exhibition, New Orleans, 30 September–3 October. DOI: 10.2118/71395-MS.

Maia, C.A., Poiate, J.E., Falcão, J.L., and Coelho, L.F.M. 2005. Triaxial Creep Tests in Salt Applied in Drilling Through Thick Salt Layers in Campos Basin–Brazil. Paper SPE 92629 presented at the SPE/IADC Drilling Conference, Amsterdam, 23–25 February. DOI: 10.2118/92629-MS.

Malinverno, A., Sayers, C.M., Woodward, M.J., and Bartman, R.C. 2004. Integrating Diverse Measurements To Predict Pore Pressure With Uncertainties While Drilling. Paper SPE 90001 presented at the SPE Annual Technical Conference and Exhibition, Houston, 26–29 September. DOI: 10.2118/90001-MS.

Matthews, W.R. and Kelly, J. 1967. How To Predict Formation Pressure and Fracture Gradient From Electric and Sonic Logs. *Oil and Gas Journal* **65**: 92–106.

Maury, V. and Guenot, A. 1995. Practical Advantages of Mud Cooling Systems for Drilling. *SPEDC* **10** (1): 42–48. SPE-25732-PA. DOI: 10.2118/25732-PA.

Minster, J.B. and Jordan, T.H. 1978. Present-Day Plate Motions. *J. Geophysical Research* **83** (B11): 5331–5354. DOI: 10.1029/JB083iB11p05331.

Mitchell, A. and Grauls, D. eds. 1998. *Overpressures in Petroleum Exploration,* workshop proceedings. Pau, France: Bull. des Centres de Recherches Exploration-Production Elf EP Memoire 22, Elf Aquitaine.

Mitchell, R.F. ed. 2007. *Petroleum Engineering Handbook, Vol. 2—Drilling.* Richardson, Texas: SPE.

Moore, D.D., Bencheikh, A., and Chopty, J.R. 2004. Drilling Underbalanced in Hassi Messaoud. Paper SPE 91519 presented at the SPE/IADC Underbalanced Technology Conference and Exhibition, Houston, 11–12 October. DOI: 10.2118/91519-MS.

Moos, D. and Zoback, M.D. 1990. Utilization of Observations of Well Bore Failure To Constrain the Orientation and Magnitude of Crustal Stresses: Application to Continental, Deep Sea Drilling Project, and Ocean Drilling Program Boreholes. *J. Geophys. Res.* **95** (B6): 9305–9325. DOI: 10.1029/JB095iB06p09305.

Morita, N., Black, A.D., and Fuh, G.-F. 1990. Theory of Lost Circulation Pressure. Paper SPE 20409 presented at the SPE Annual Technical Conference and Exhibition, New Orleans, 23–26 September. DOI: 10.2118/20409-MS.

Moschovidis, Z., Steiger, R., Peterson, R. et al. 2000. The Mounds Drill-Cuttings Injection Field Experiment: Final Results and Conclusions. Paper SPE 59115 presented at the IADC/SPE Drilling Conference, New Orleans, 23–25 February. DOI: 10.2118/59115-MS.

Mouchet, J.P. and Mitchell, A. 1989. *Abnormal Pressures While Drilling: Origins, Prediction, Detection, Evaluation*. Paris: Elf EP-Editions, Editions Technip.

Nawrocki, P.A., Dusseault, M.B., and Bratli, R.K. 1996. Semi-Analytical Models for Predicting Stresses Around Openings in Non-Linear Geomaterials. In *EUROCK '96: Prediction and Performance in Rock Mechanics and Rock Engineering*, Volume 2, ed. G. Barla. Rotterdam, The Netherlands: Balkema.

Okland, D. and Cook, J.M. 1998. Bedding-Related Borehole Instability in High-Angle Wells. Paper SPE 47285 presented at SPE/ISRM Rock Mechanics in Petroleum Engineering, Trondheim, Norway, 8–10 July. DOI: 10.2118/47285-MS.

Onaisi, A., Samier, P., Koutsabeloulis, N., and Longuemare, P. 2002. Management of Stress Sensitive Reservoirs Using Two Coupled Stress-Reservoir Simulation Tools: ECL2VIS and ATH2VIS. Paper SPE 78512 presented at the Abu Dhabi International Petroleum Exhibition and Conference, Abu Dhabi, UAE, 13–16 October. DOI: 10.2118/78512-MS.

Ong, S.H. and Roegiers, J.-C. 1993. Influence of Anisotropies in Borehole Stability. *Intl. J. of Rock Mech. Min. Sci. and Geomech. Abstr.* **30** (7): 1069–1075. DOI: 10.1016/0148-9062(93)90073-M.

Papanastasiou, P.C. and Vardoulakis, I.G. 1994. Numerical Analysis of Borehole Stability Problem. In *Soil-Structure Interaction: Numerical Analysis and Modelling*, ed. J.W. Bull. London: Taylor and Francis.

Parra, J.G., Celis, E., and De Gennaro, S. 2003. Wellbore Stability Simulations for Underbalanced Drilling Operations in Highly Depleted Reservoirs. *SPEDC* **18** (2): 146–151. SPE-83637-PA. DOI: 10.2118/83637-PA.

Pennebaker, E.S. 1968. An Engineering Interpretation of Seismic Data. Paper SPE 2165 presented at the Fall Meeting of the Society of Petroleum Engineers of AIME, Houston, 29 September–2 October. DOI: 10.2118/2165-MS.

Perkins, T.K. and Kern, L.R. 1961. Widths of Hydraulic Fractures. *JPT* **13** (9): 937-949. SPE-89-PA. DOI: 10.2118/89-PA.

Plumb, R.A., Edwards, S., Pidcock, G., Lee, D., and Stacey, B. 2000. The Mechanical Earth Model Concept and Its Application to High-Risk Well Construction Projects. Paper SPE 59128 presented at IADC/SPE Drilling Conference, New Orleans, 23–25 February. DOI: 10.2118/59128-MS.

Plumb, R.A. and Hickman, S.H. 1985. Stress-Induced Borehole Elongation: A Comparison Between the Four-Arm Dipmeter and the Borehole Televiewer in the Auburn Geothermal Well. *Journal of Geophysical Research* **90** (B7): 5513–5521. DOI: 10.1029/JB090iB07p05513.

Plumb, R.A., Herron, S.L., and Olsen, M.P. 1992. Composition and Texture on Compressive Strength Variations in the Travis Peak Formation. Paper SPE 24758 presented at the SPE Annual Technical Conference and Exhibition, Washington, DC, 4–7 October. DOI: 10.2118/24758-MS.

Poliakov, A.N.B., Podladchikov, Y.Y., Dawson, E.C., and Talbot, C.J. 1996. Salt Diapirism With Simultaneous Brittle Faulting and Viscous Flow. In *Salt Tectonics*, ed. G.I. Alsop, D.J. Blundell, and I. Davison, 291–302. Bath, UK: Special Publication No. 100, Geological Society Publishing House.

Priest, S.D. 1993. *Discontinuity Analysis for Rock Engineering*. London: Chapman and Hall.

Reinecker, J., Heidbach, O., Tingay, M., Sperner, B., and Müller, B. 2005. *The World Stress Map Project*. http://www-wsm.physik.uni-karlsruhe.de/pub/home/index_noflash.html. Downloaded 3 July 2008.

Ren, N.K. and Hudson, P.J. 1985. Predicting In-Situ Stress State Using Differential Wave Velocity Analysis. *Proc., 26th Symposium on Rock Mechanics*, Rapids City, South Dakota, USA, 1235–1244.

Ren, N.K. and Roegiers, J.C. 1983. Differential Strain Curve Analysis: A New Method for Determining the Pre-Existing In-Situ Stress State From Rock Core Measurements. *Proc., 5th ISRM Congress on Rock Mechanics*, The Netherlands, Vol. 2, F117–127.

Rezmer-Cooper, I., Bratton, T., and Krabbe, H. 2000. The Use of Resistivity-at-the-Bit Measurements and Annular Pressure While Drilling in Preventing Drilling Problems. Paper SPE 59225 presented at the IADC/SPE Drilling Conference, New Orleans, 23–25 February. DOI: 10.2118/59225-MS.

Rummel, F. and Winter, R.B. 1983. Fracture Mechanics as Applied to Hydraulic Fracturing Stress Measurements. *Earthq. Predict. Res.* **2**: 33–45.

Russell, K.A., Ayan, C., Hart, N.J. et al. 2003. Predicting and Preventing Wellbore Instability Using the Latest Drilling and Logging Technologies: Tullich Field Development, North Sea. Paper SPE 84269 presented at the SPE Annual Technical Conference and Exhibition, Denver, 5–8 October. DOI: 10.2118/84269-MS.

Salz, L.B. 1977. Relationship Between Fracture Propagation Pressure and Pore Pressure. Paper SPE 6870 presented at SPE Annual Technical Conference and Exhibition, Denver, 9–12 October. DOI: 10.2118/6870-MS.

Samier, P., Onaisi, A., and Fontaine, G. 2003. Coupled Analysis of Geomechanics and Fluid Flow in Reservoir Simulation. Paper SPE 79698 presented at the SPE Reservoir Simulation Symposium, Houston, 3–5 February. DOI: 10.2118/79698-MS.

Santarelli, F.J. 1987. Theoretical and Experimental Investigation of the Stability of the Axisymmetric Wellbore. PhD dissertation, Imperial College, London.

Santarelli, F.J., Dahen, D., Baroudi, H., and Sliman, K.B. 1992a. Mechanisms of Borehole Instability in Heavily Fractured Rock Media. *Intl. J. Rock Mech. Min. Sci. & Geomech. Abstr.* **29** (5): 457–467. DOI: 10.1016/0148-9062(92)92630-U.

Santarelli, F.J., Dardeau, C., and Zurdo, C. 1992b. Drilling Through Highly Fractured Formations: A Problem, a Model, and a Cure. Paper SPE 24592 presented at the SPE Annual Technical Conference and Exhibition, Washington, DC, 4–7 October. DOI: 10.2118/24592-MS.

Saponja, J. 1998. Challenges With Jointed-Pipe Underbalanced Operations. *SPEDC* **13** (2): 121–128. SPE-37066-PA. DOI: 10.2118/37066-PA.

Sayers, C.M., den Boer, L., Nagy, Z., Hooyman, P., and Ward, V. 2005. Pore Pressure in the Gulf of Mexico: Seeing Ahead of the Bit. *World Oil* **226** (12).

Sayers, C.M., Hooyman, P.J., Smirnov, N. et al. 2002a. Pore Pressure Prediction for the Cocuite Field, Veracruz Basin. Paper SPE 77360 presented at the SPE Annual Technical Conference and Exhibition, San Antonio, Texas, USA, 29 September–2 October. DOI: 10.2118/77360-MS.

Sayers, C.M., Johnson, G.M., and Denyer, G. 2002b. Predrill Pore-Pressure Prediction Using Seismic Data. *Geophysics* **67** (4): 1286–1292. DOI: 10.1190/1.1500391.

Schei, G., Fjær, E., Detournay, E., Kenter, C.J., Fuh, G.-F., and Zausa, F. 2000. The Scratch Test: An Attractive Technique for Determining Strength and Elastic Properties of Sedimentary Rocks. Paper SPE 63255 presented at the SPE Annual Technical Conference and Exhibition, Dallas, 1–4 October. DOI: 10.2118/63255-MS.

Scholz, C.H. 2002. *The Mechanics of Earthquakes and Faulting*, second edition. Cambridge, UK: Cambridge University Press.

Segall, P. 1992. Induced Stresses Due to Fluid Extraction From Axisymmetric Reservoirs. *Pure and Applied Geophysics* **13** (3–4): 535–560. DOI: 10.1007/BF00879950.

Smirnov, N.Y., Lam, R., and Rau, W.E. 2003. Process of Integrating Geomechanics With Well Design and Drilling Operation. Paper AADE-030NCTE-28 presented at the AADE National Technology Conference, Houston, 1–3 April.

Smith, M.A. 2002. Geological Controls and Variability in Pore Pressure in the Deep Water Gulf of Mexico. In *Pressure Regimes in Sedimentary Basins and Their Prediction*, ed. A.R. Huffman and G.L. Bowers, 107–113. Tulsa: AAPG Memoir 76, American Association of Petroleum Geologists.

Somerville, J.M. and Smart, B.G.D. 1991. The Prediction of Well Stability Using the Yield Zone Concept. Paper SPE 23127 presented at Offshore Europe, Aberdeen, 3–6 September. DOI: 10.2118/23127-MS.

Spencer, A.J.M. 1985. *Continuum Mechanics.* London: Longman.

Standifird, W. and Matthews, M.D. 2005. Real Time Basin Modeling: Improving Geopressure and Earth Stress Predictions. Paper SPE 96464 presented at Offshore Europe, Aberdeen, 6–9 September. DOI: 10.2118/96464-MS.

Stone, T.W., Xian, C., Fang, Z. et al. 2003. Coupled Geomechanical Simulation of Stress Dependent Reservoirs. Paper SPE 79697 presented at the SPE Reservoir Simulation Symposium, Houston, 3–5 February. DOI: 10.2118/79697-MS.

Strickland, F.G. and Ren, N.K. 1980. Use of Differential Strain Curve Analysis in Predicting the In-Situ Stress State for Deep Wells. *Proc.,* 21st US Symposium on Rock Mechanics, Rolla, Missouri, USA, 523–532.

Sudron, K., Berners, H., Frank, U. et al. 1999. Dieksand 2: An Extended Well Through Salt, Increases Production From an Environmentally Protected Field. Paper SPE 52854 presented at the SPE/IADC Drilling Conference, Amsterdam, 9–11 March. DOI: 10.2118/52854-MS.

Sulbaran, A.L., Carbonell, R.S., and Lopez-de-Cardenas, J.E. 1999. Oriented Perforating for Sand Prevention. Paper SPE 57954 presented at the SPE European Formation Damage Conference, The Hague, 31
May−1 June. DOI: 10.2118/57954-MS.

Sylte, J.E., Thomas, L.K., Rhett, D.W., Bruning, D.D., and Nagel, N.B. 1999. Water Induced Compaction in
the Ekofisk Field. Paper SPE 56426 presented at the SPE Annual Technical Conference and Exhibition,
Houston, 3–6 October. DOI: 10.2118/56426-MS.

Teufel, L.W. 1983. Determination of In-Situ Stress From Anelastic Strain Recovery Measurements of Oriented Core. Paper SPE 11649 presented at the SPE/DOE Low Permeability Gas Reservoirs Symposium,
Denver, 14–16 March. DOI: 10.2118/11649-MS.

Teufel, L.W., Rhett, D.W., Farrell, H.E., and Lorenz, J.C. 1993. Control of Fractured Reservoir Permeability
by Spatial and Temporal Variations in Stress Magnitude and Orientation. Paper SPE 26437 presented
at the SPE Annual Technical Conference and Exhibition, Houston, 3–6 October. DOI: 10.2118/26437-
MS.

Thiercelin, M.J. and Plumb., R.A. 1994. A Core-Based Prediction of Lithologic Stress Contrasts in East
Texas Formations. *SPEFE* **9** (4): 251–258. SPE-21847-PA. DOI: 10.2118/21847-PA.

Tingay, M., Müller, B., Reinecker, J. et al. 2005. Understanding Tectonic Stress in the Oil Patch: The World
Stress Map Project. *The Leading Edge* **24** (12): 1276. DOI: 10.1190/1.2149653.

Tollefsen, E., Goobie, R.B., Noeth, S. et al. 2006. Optimize Drilling and Reduce Casing Strings Using Remote
Real-Time Well Hydraulic Monitoring. Paper SPE 103936 presented at the International Oil Conference
and Exhibition in Mexico, Cancun, Mexico, 31 August–2 September. DOI: 10.2118/103936-MS.

Traugott, M. 1997. Pore/Fracture Pressure Determinations in Deepwater. *World Oil*: *Deepwater Technology
Special Supplement* (August 1997): 68–70.

van den Hoek, P.J., Hertogh, G.M.M., Kooijman, A.P. et al. 2000. A New Concept of Sand Production Prediction: Theory and Laboratory Experiments. *SPEDC* **15** (4): 261–273. SPE-65756-PA. DOI: 10.2118/65756-
PA.

van Oort, E. 2003. On the Physical and Chemical Stability of Shales. *Journal of Petroleum Science and
Engineering* **38** (3–4): 213–235. DOI: 10.1016/S0920-4105(03)00034-2.

van Oort, E., Gradishar, J., Ugueto, G. et al. 2003. Accessing Deep Reservoirs by Drilling Severely Depleted
Formations. Paper SPE 79861 presented at the SPE/IADC Drilling Conference, Amsterdam, 19–21
February. DOI: 10.2118/79861-MS.

Whitehead, W.S., Hunt, E.R., and Holditch, S.A. 1987. The Effects of Lithology and Reservoir Pressure on
the In-Situ Stresses in the Waskom (Travis Peak) Field. Paper SPE 16403 presented at the Low Permeability Reservoirs Symposium, Denver, 18–19 May. DOI: 10.2118/16403-MS.

Willson, S.M. and Fredrich, J.T. 2005. Geomechanics Considerations for Through- and Near-Salt Well
Design. Paper SPE 95621 presented at SPE Annual Technical Conference and Exhibition, Dallas, 9–12
October. DOI: 10.2118/95621-MS.

Willson, S.M., Last, N.C., Zoback, M.D., and Moos, D. 1999. Drilling in South America: A Wellbore Stability Approach for Complex Geological Conditions. Paper SPE 53940 presented at the Latin American and
Caribbean Petroleum Engineering Conference, Caracas, 21–23 April. DOI: 10.2118/53940-MS.

Willson, S.M., Fossum, A.F., and Fredrich, J.T. 2003. Assessment of Salt Loading on Well Casings. *SPEDC*
18 (1): 13–21. SPE-81820-PA. DOI: 10.2118/81820-PA.

Wong, T.-F., Szeto, H., and Zhang, J. 1992. Effect of Loading Path and Porosity on the Failure Mode of
Porous Rocks. *Applied Mechanics Review* **45** (8): 281–293.

Wright, C.A., Weijers, L., Davis, E.J., and Mayerhofer, M. 1999. Understanding Hydraulic Fracture Growth:
Tricky but Not Hopeless. Paper SPE 56724 presented at the SPE Annual Technical Conference and Exhibition, Houston, 3–6 October. DOI: 10.2118/56724-MS.

Wu, B., Addis, M.A., and Last, N.C. 1998. Stress Estimation in Faulted Regions: The Effect of Residual Friction. Paper SPE 47190 presented at SPE/IADC Rock Mechanics in Petroleum Engineering, Trondheim,
Norway, 8–10 July. DOI: 10.2118/47210-MS.

Zhang, X., Last, N., Powrie, W., and Harkness, R. 1999. Numerical Modeling of Wellbore Behavior in Fractured Rock Masses. *J. Pet. Sci. Eng.* **23** (2): 95–115. DOI: 10.1016/S0920-4105(99)00010-8.

Zoback, M.D., Moos, D., Mastin, L., and Anderson, R.N. 1985. Well Bore Breakouts and In-Situ Stress.
Journal of Geophysical Research **90** (B7): 5523–5530. DOI: 10.1029/JB090iB07p05523.

Zoback, M.D., Barton, C.A., Brudy, M. et al. 2003. Determination of Stress Orientation and Magnitude in Deep Wells. *International Journal of Rock Mechanics and Mining Sciences* **40** (7–8): 1049–1076. DOI: 10.1016/j.ijrmms.2003.07.001.

5.2 Normalization and Inversion Methods—Bernt S. Aadnoy, University of Stavanger and Joannes Djurhuus, Statoil-Hydro

5.2.1 Introduction. Please observe that this chapter assumes knowledge of borehole stability as presented throughout Chapter 5.1. The details and derivations are therefore kept to a minimum. Also, the reader may refer to the references at the end of the chapter.

In this chapter we will present methods that have been particularly useful for borehole stability analysis. However, when performing modeling, there are certain principles to follow. Two of these are the following:

- Always use the same model when performing modeling and when the results are used to derive a prognosis for the next well. Consistency in modeling is required.
- Before modeling is performed, normalize all data to the same depth reference. Examples provided in this chapter are platform elevation, water depth, rock bulk density, pore pressure, and others.

In the following, various approaches will be presented, starting with a discussion about the use of pressures and pressure gradients.

5.2.2 Pressure and Pressure Gradients, Pore Pressure Example. In the drilling industry it is common to use gradients instead of actual pressures. Actually, relative density is used. If a drilling fluid has a density of 1.5 specific gravity (SG), this means that the density of the fluid is 1.5 times higher than that of water. The drilling fluid density is a key parameter during drilling because it relates to formation pressure, borehole stability, and so on. For that reason, other parameters such as pore pressure, overburden stress, fracture and collapse pressures, and others are converted to relative density for easy comparison with the mud weight. Please observe that this is allowable for steady-state processes only. If, for example, one has a transient process such as circulating out a kick, pressures should be used instead of gradients.

To understand similarities and differences between pressures and pressure gradients, we will study the formation pore pressure. In the following, a simple explanation of the origin of increased pore pressure will be given. There are, of course, many contributors to the generation of the pore pressure, such as tectonic effects and hydrocarbon conversion. However, in the opinion of the authors, the most dominating effect is that of buoyancy.

Water is the most abundant fluid in subsurface geology. The sediments usually are deposited in water (e.g., by rivers). Therefore, water exists in all rocks. **Fig. 5.2.1a** illustrates sediment deposited in water. Assuming full vertical communication upward, the pressure exerted at any depth is simply given by the weight of a water column. This scenario is actually the aquifer model used in geology.

The pressure exerted at the bottom of a water column is

$$p \text{ (bar)} = 0.098 d_{water} \text{ (SG)} z \text{ (m)}. \quad \dots \dots \dots \dots \dots \dots \dots \dots \dots \dots \dots \dots \dots \dots (5.2.1)$$

Here, the water density is given as relative to pure water, $d = \dfrac{\rho}{\rho_{water}}$. Seawater has a density of 1.03 SG.

Over geologic time, organic material has been deposited. Subjected to pressure and temperature over a long time, hydrocarbons have been formed. These are in general immiscible with water. Because these hydrocarbons often have a density less than that of water, they exhibit upward buoyancy and migrate toward the surface.

If vertical communication exists, the lighter fluid will migrate all the way to surface. However, imagine that a seal is placed somewhere in the rock as shown in Fig. 5.2.1b. This seal will capture the fluid. After some time, the volume underneath the seal is completely filled with oil. Let us assume the situation depicted in

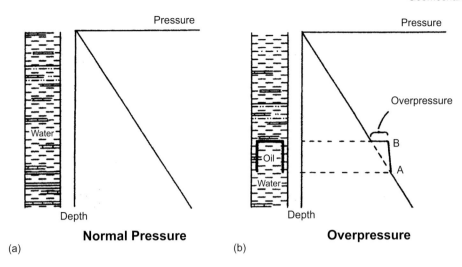

Fig. 5.2.1—Pressures vs. depth. (a) Normal pressure; (b) overpressure.

Fig. 5.2.1b. The seal or the cap is completely filled with oil. Both below and above the seal, normal pressure exists. The pressure at Point A is equal to

$$p_A = 0.098 d_{water} z_A \quad \text{..} \quad (5.2.2)$$

and is defined as a normal pore pressure. We will now compute the pressure at Point B. This is simply the pressure at Point A minus the pressure exerted by the oil column from A to B, or

$$p_B = 0.098(d_{water} z_A - d_{oil}(z_A - z_B)). \quad \text{...........................} \quad (5.2.3)$$

The overpressure at Point B is simply the pressure from Eq. 5.2.3 minus the normal pressure at that depth, or

$$\Delta p = 0.098\left(d_{water} - d_{oil}\right)\left(z_A - z_B\right). \quad \text{..........................} \quad (5.2.4)$$

Eq. 5.2.4 describes the basic pore pressure mechanism—namely, buoyancy. For an overpressure to develop, three conditions are required: a density contrast between the oil and the water, a height for a pressure to develop, and a caprock to confine the oil.

In the preceding we derived equations in terms of pressure. This is the physically correct measure to use. However, during drilling, one uses a drilling fluid of constant density. By using this density gradient, one takes out the depth dependence. Using this density as a reference simplifies the computation of pressures. The equivalent density of the pore pressure at any depth for the case of normal pressure is simply proportional to ρ_{water}. The equivalent density of the overpressure at Point B is simply

$$p_B = 0.098 d_{eq} z_B. \quad \text{..} \quad (5.2.5)$$

Equating this equation with Eq. 5.2.3, the equivalent density at depth B can be expressed as

$$\rho_{eq} = \rho_{oil} + \left(d_{water} - d_{oil}\right)\frac{z_A}{z_B}. \quad \text{...........................} \quad (5.2.6)$$

Fig. 5.2.2 shows the equivalent densities for the two cases of Fig. 5.2.1. Please observe that the overpressure shown in Fig. 5.2.2b results in a declining trend with depth, called a regression line. The slope of this regression line is directly governed by the density contrast between the water and the oil. Also, even if this

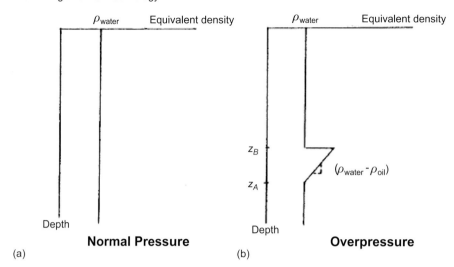

Fig. 5.2.2—Equivalent densities for (a) normal pressure and (b) overpressure.

equivalent gradient is decreasing, the pressure curve shown in Fig. 5.2.1b always increases with depth, provided that there is vertical communication.

Several observations should be pointed out:

- A regression on the equivalent gradient plot will usually result in an increase in pressure with depth when converted to pressure.
- The pore pressure might not be continuous with depth. A caprock may result in a discrete transition below and above.
- The average equivalent gradient is used for practical applications. Please observe that this is not identical to the local gradient or the derivative.

Fig. 5.2.3 shows a typical gradient plot as used in well construction. The two most important curves are the fracturing curve and the pore-pressure curve. In this case, the 9⅝-in. (24.45-cm) production casing is set above the top of reservoir. The 13⅜-in. (33.97-cm) casing must be set deep enough to avoid fracturing at the shoe.

The rock is clay from top to the 9⅝-in. (24.45-cm) shoe. We have only indirect ways of estimating pore pressure in impermeable rocks such as clays. The industry, however, uses this information as a true pore pressure, despite the questioning of its correctness. For critical wells, one may therefore adjust the pore pressure in clays, provided that there are no permeable zones that may cause well-control problems.

Below the 9⅝-in. (24.45-cm) casing is the sand reservoir. Here, we may obtain pore pressure readings from wellbore measurements. The slope of the regression line in the reservoir actually is caused by the density of the reservoir fluids, as discussed above. This example demonstrates the lack of pore pressure knowledge in impermeable rocks and the effects of buoyancy in permeable rocks.

5.2.3 Change of Depth Reference. The geologist often uses mean sea level (MSL) as a depth reference, whereas the driller uses the drill floor (RKB). Drilling and production platforms may have different elevations above sea level.

If we are using data from the same field or platform, we may use them without corrections. However, we often are compiling data from floating rigs, from fixed rigs with various drill floor elevations, and from wells with significant difference in water depths.

Obviously, using data of various origins may introduce significant errors. In this section we will show how to make the data consistent in a simple manner. The key is to select a reference system and to normalize all data to this common reference.

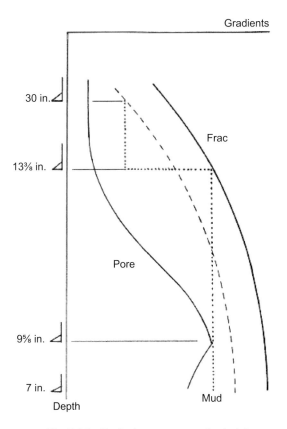

Fig. 5.2.3—Typical pressure gradient plot.

The data we are mainly considering at this point are the mud weight, the frac gradient, the overburden stress gradient, the horizontal stresses, and the pore pressure gradient. However, other pressure gradient data should also be corrected to a common reference.

Correcting to MSL. As already mentioned in the introduction, geologists often adjust the data to MSL. This has the advantage of removing the effects of various drill floor elevations (Aadnoy 1997).

Consider a scenario as shown in **Fig. 5.2.4.** A drilling rig has an elevation h_f from MSL to the drill floor. Assume that at a given depth z, the pressure is p. This can be either the static mud pressure, a frac pressure, or a pore pressure. Expressed as a gradient relative to the drill floor, this pressure can be expressed as

$$p = 0.098 d_{RKB} z.$$

We want to express this pressure relative to MSL. The pressure must be the same, but the gradient must be different due to a different reference level:

$$p = 0.098 d_{MSL} (z - h_f).$$

Equating the two equations results in an expression to correct the RKB data to an MSL reference:

$$d_{MSL} = d_{RKB} \frac{z}{z - h_f}. \quad \dots \dots \dots \dots \dots \dots \dots \dots \dots \dots \dots \dots \dots \dots \dots \dots \quad (5.2.7)$$

If we instead want to convert MSL data to RKB, we can write

$$d_{RKB} = d_{MSL} \frac{z - h_f}{z}. \quad \dots \dots \dots \dots \dots \dots \dots \dots \dots \dots \dots \dots \dots \dots \dots \dots \quad (5.2.8)$$

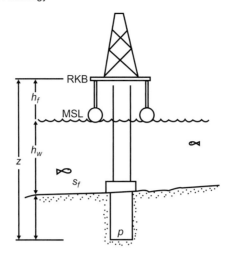

Fig. 5.2.4—Definition of references (Aadnoy 1997). Reprinted with permission from Elsevier.

In these examples, the reference depth z is chosen from the drill floor. Other reference levels can be used; the important factor is that the bottomhole pressure remains constant regardless of choice of reference level.

Correcting to Another Drill-Floor Level. One common problem is that parts of the field data are compiled during exploration drilling and parts during drilling from a production platform. Often the two have significantly different drill-floor elevations. **Fig. 5.2.5** illustrates the problem.

For future work, we are most likely to use the production platform drill floor as a reference (RKB2); we will define this as our reference level. A pressure p at depth z can be expressed as

$$p = 0.098 d_{RKB2}\, z = 0.098 d_{RKB1}(z - \delta h).$$

The correction equation is then

$$d_{RKB2} = d_{RKB1}\frac{z - \delta h}{z} \quad \dots\dots\dots\dots\dots\dots\dots\dots\dots\dots\dots\dots\dots\dots\dots\dots\dots\dots \quad (5.2.9)$$

or, if we want to use RKB1 as a reference,

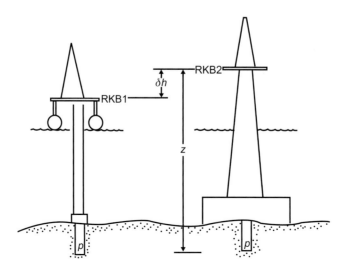

Fig. 5.2.5—Definition of various drill floor elevations (Aadnoy 1997). Reprinted with permission from Elsevier.

$$d_{RKB1} = d_{RKB2} \frac{z}{z - \delta h}. \quad \dots\dots\dots\dots\dots\dots\dots\dots\dots\dots\dots\dots\dots\dots\dots \quad (5.2.10)$$

Example 5.2.1. A leakoff pressure gradient of 1.5 SG is measured at a depth of 1000 m MSL (1025 m RKB). If the rig floor elevation is 25 m, determine the gradient from the drill floor.

$$d_{RKB} = 1.5 \frac{1025 - 25}{1025} = 1.46 \text{ SG}$$

5.2.4 Deepwater Geomechanic Model. Deepwater drilling has a high activity worldwide. In particular, in the Gulf of Mexico and offshore Brazil, the activity is high. In Europe, however, this is a fairly new activity. The first deepwater well was drilled offshore Norway in 1997. Since then, a number of wells have been drilled. On the UK side, deepwater drilling has taken place mainly along the Atlantic margin, close to the Shetland Islands and the Orkney Islands. A higher activity is foreseen towards the Faroe Islands, and also in Norway.

A general trend is that the deeper the water, the smaller the pressure margins during drilling. The average overburden weight is low, resulting in a low fracture gradient. This implies that in very deep waters exceeding 10,000 ft, water is the ultimate drilling fluid. To overcome this problem, several new techniques have been invented. In the Gulf of Mexico, a so-called dual-density approach was taken. This involves installing a pump at the wellhead to provide additional lift up the marine riser. Herrmann et al. (2001) show that by using this technique, a well can be drilled with fewer casing strings. Another approach has been taken in Norway by installing a buoyant wellhead up in shallower water. The buoyant wellhead is resting on a 20-in.-long (50.8-cm-long) marine riser and is placed on top of a buoyant element that is kept in position at a shallow depth by lines to the seafloor. The main advantage is that the well can be drilled with a standard drilling rig.

There are two major problems in the deepwater scenario. First, drilling problems such as circulation losses, hole collapse, and water inflow cause costly well operations. Sometimes stuck pipes, sidetracks, and tophole abandonment are the result. With narrow pressure margins, the setting depth of casing strings is an important issue.

The second class of problems relates to the design of the operation. The dual-density approach may be utilized to reduce the number of casing strings in a well, but it has a drawback of a costly pumping system installed at the wellhead, and it requires a drilling rig with deepwater capability. Wells using a shallow buoyant wellhead can be drilled with ordinary drilling rigs, provided they have dynamic positioning systems. Both concepts are feasible.

In this section we will focus on the geomechanics aspect of deepwater drilling, which is one of the most important issues, and more specifically on the modeling of and the accuracy of the fracture gradient. Webb et al. (2001) discusses another important issue–namely, to design the drilling fluid to heal mud losses. The latter topic will not be pursued in this book.

Fracturing Pressure in Deep Waters. Shallower sediments in deep water are often in a relaxed state. The overburden and the horizontal stresses are caused by compaction alone. Under such conditions, it is well known that the fracture pressure correlates with the overburden stress. Aadnoy et al. (1991) showed that the fracture pressure is approaching the overburden at shallow seabed penetration, on the basis of shallower North Sea data. Rocha and Bourgoyne (1996) performed a similar analysis and showed, using data from the US Gulf Coast and Brazil, that, on average, fracture pressure was equal to overburden stress. Their model is based on an exponential compaction model.

Aadnoy (1998) adapted the same philosophy but normalized all data to seabed, removing the effects of different water depths. Applying this method to deepwater wells, a good correlation was found. Also, a water-depth normalization method was developed based on the same model. As an estimate for the overburden stress, he used a model initially derived by Eaton (1969).

New Results. Aadnoy and Saetre (2003) presented a *worldwide* fracture model for deepwater applications. It was based on field data from Norway, the Gulf of Mexico, the Shetland Islands, Angola, and Nigeria. Also, it was based on Eaton's (1969) overburden stress model. Several new wells have been drilled offshore Norway recently, and the model has been revisited. Several new observations have been made. First, the lithology varies considerably from well to well, resulting in different overburden stresses. Second, Eaton's model overpredicts the overburden stress, at least offshore Norway. Based on this, the fracturing model will be modified.

Five deepwater wells offshore Norway were analyzed in detail. The lithology and the bulk densities were analyzed to provide an accurate overburden stress curve for each well.

Comparing the measured leakoff data to the overburden stress, it was found that the best correlation was obtained when the fracture prognosis was approximately 97% of the overburden stress. The best-fit fracture model for deep waters offshore Norway is

$$p_{wf} = 0.97\sigma_V, \quad \dotfill \quad (5.2.11)$$

where σ_V is the specific overburden stress for each well. This equation is valid for seabed penetrations between 403 and 2586 m. One assumption for application is normal pore pressure. **Fig. 5.2.6** shows one result of this correlation.

Normalization Methods. We propose to use Eq. 5.2.11 as a fracture prognosis for new wells offshore Norway. In other places, it is recommended to use the same equation as a fracturing estimate and to compare measured leakoff data to the model. If reference wells are drilled in the same area, data from these might be used for comparison as well.

In general, deepwater wells are drilled in different water depths. In order to compare these data they must be normalized to the same reference depth. Assuming a similar bulk density profile (overburden stress), Aadnoy (1998) derived the following normalization equations, which were further extended in Kaarstad and Aadnoy (2006).

The general form of the depth normalization equation is

$$D_{wf2} = D_{wf1} + \Delta h_w + \Delta h_f + \Delta D_{sb}$$

$$d_{wf2} = d_{sw} \frac{h_{w2}}{D_{wf2}} + \left(d_{wf1} \frac{D_{wf1}}{D_{wf2}} - d_{sw} \frac{h_{w1}}{D_{wf2}} \right) \frac{\int_{D_{sb2}} d_{b2}(D)dD}{\int_{D_{sb1}} d_{b1}(D)dD}. \quad \dotfill \quad (5.2.12)$$

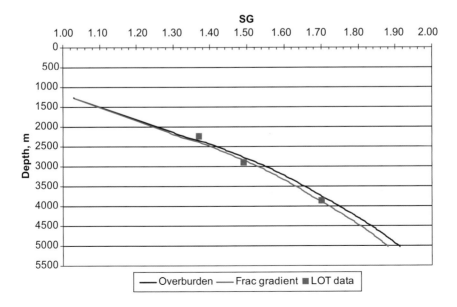

Fig. 5.2.6—Gradients of well A; water depth 1238 m (Kaarstad and Aadnoy 2006).

This equation requires detailed information about the bulk density profile and is applicable when significant reference data exist. However, simplifying assumptions often apply, and they can be categorized as follows:

Case 1: Different but Constant Bulk Densities. Applying constant bulk densities, the integrals are reduced, and the normalization equations can be expressed as follows:

$$D_{wf2} = D_{wf1} + \Delta h_w + \Delta h_f + \Delta D_{sb}$$

$$d_{wf2} = d_{sw}\frac{h_{w2}}{D_{wf2}} + \left(d_{wf1}\frac{D_{wf1}}{D_{wf2}} - d_{sw}\frac{h_{w1}}{D_{wf2}}\right)\frac{d_{b2}D_{sb2}}{d_{b1}D_{sb1}}. \quad\dots\dots\dots\dots\dots\dots\dots\dots\dots\dots\dots \quad (5.2.13)$$

Case 2: Constant Bulk Density. For wells in the same area, it may often be assumed that the bulk density is the same for the different wells. Eq. 5.2.12 is then further simplified, and the normalization equations become

$$D_{wf2} = D_{wf1} + \Delta h_w + \Delta h_f + \Delta D_{sb}$$

$$d_{wf2} = d_{sw}\frac{h_{w2}}{D_{wf2}} + \left(d_{wf1}\frac{D_{wf1}}{D_{wf2}} - d_{sw}\frac{h_{w1}}{D_{wf2}}\right)\frac{D_{sb2}}{D_{sb1}}. \quad\dots\dots\dots\dots\dots\dots\dots\dots\dots\dots\dots \quad (5.2.14)$$

These equations are used to normalize between varying water depths, platform elevations, and rock penetrations.

Case 3: Same Rock Penetration Below Seabed for Data and Prognosis Assuming Constant Bulk Density. When setting $\Delta D_{sb} = 0$, the following normalization equations result:

$$D_{wf2} = D_{wf1} + \Delta h_w + \Delta h_f$$

$$d_{wf2} = d_{wf1}\frac{D_{wf1}}{D_{wf2}} + \frac{d_{sw}\Delta h_w}{D_{wf2}}. \quad\dots\dots\dots\dots\dots\dots\dots\dots\dots\dots\dots\dots\dots\dots \quad (5.2.15)$$

Example of Use of Normalization Methods. Data normalization is an indispensable method to compare data sets with different references. Eq. 5.2.12 defines the general normalization equations used to compare pressures (e.g., overburden, leakoff pressure, in-situ stresses) with differences in bulk density, rig floor height, water depth, and depth of penetration. To demonstrate the application, we present two examples:

Example 5.2.2. The reference well is drilled in 400 m of water. These data will be used to derive a prognosis for a well in 1100 m water depth. Assume that the rig floor height, the bulk density, and the penetration depth remain unchanged. The following well data apply, using Case 3 above:

- Drill floor height: $h_f = 25$ m
- Total depth of well 1: $z_1 = 900$ m
- Water depth for well 1: $h_{w1} = 400$ m
- Leakoff pressure for well 1: $d_{wf1} = 1.5$ SG at 900 m depth
- Water depth for well 2: $h_{w2} = 1100$ m
- Seawater density: $d_{sw} = 1.03$ SG

The new depth reference is

$$D_2 = D_1 + \Delta h_w + \Delta h_f$$
$$= 900 \text{ m} + (1100 \text{ m} - 400 \text{ m}) + (25 \text{ m} - 25 \text{ m})$$
$$= 1600 \text{ m}$$

The prognosis for the leakoff pressure gradient of a similar well with 1100 m water depth is

$$d_{wf2} = 1.5 \text{ SG} \frac{900 \text{ m}}{1600 \text{ m}} + 1.03 \text{ SG} \frac{1100 \text{ m} - 400 \text{ m}}{1600 \text{ m}} = 1.29 \text{ SG}$$

at 1600 m depth. In this example, the increase of the water depth from 400 m to 1100 m resulted in a decrease in leakoff pressure gradient from 1.5 to 1.29 SG.

Example 5.2.3. Unless the wells are very close to each other, it is reasonable to assume that there are differences in the bulk densities. Changes in lithology have a significant effect on the overburden stress gradient. Therefore, the normalization should take into account differences in bulk density.

In this example we want to show the effect of differences in bulk density between the two wells. Case 1 applies here. We consider the same wells as in Example 5.2.2, with the following additional information:

- Bulk density gradient for reference well: $d_{b1} = 2.05$ SG
- Bulk density gradient for new well: $d_{b2} = 1.85$ SG

The new depth reference becomes

$$D_2 = D_1 + \Delta h_w + \Delta h_f + \Delta D_{sb}$$
$$= 900 + (1100 - 400) + 0 + 0.$$
$$= 1600 \text{ m}$$

The new leakoff pressure gradient is

$$d_{wf2} = 1.03 \text{ SG} \frac{1100 \text{ m}}{1600 \text{ m}} + \left(1.5 \text{ SG} \frac{900 \text{ m}}{1600 \text{ m}} - 1.03 \text{ SG} \frac{400 \text{ m}}{1600 \text{ m}} \right) \frac{1.85 \text{ SG}}{2.05 \text{ SG}} = 1.24 \text{ SG}.$$

We observe that the lower bulk density in Well 2 leads to a decrease in overburden stress, resulting in a lower leakoff prognosis. We also observe that water contributes significantly to the total overburden stress. The result is that with the same penetration depth, an increase in water depth gives a decrease in overburden stress and fracture pressure.

Fig. 5.2.7 shows an overburden stress curve for a land well. At surface, the gradient is 2.0 SG, and it increases with depth due to compaction. Assuming the same bulk density, a number of deepwater overburden stress curves are produced. For example, at 3000 m depth in 2000 m of water, the average overburden stress gradient is approximately 1.4 SG. The reason is that the overburden consists of 2000 m of water and 1000 m of rock. This example demonstrates why the overburden stress and the fracture strength decrease with increased water depth. At 3000 m of water and deeper, wells will probably be drilled with a mud close to seawater density.

5.2.5 Inversion Methods for In-Situ Stress Determination.
In this section we will present a powerful method to compute the magnitudes and directions of the principal in-situ stresses from multiple fracture (or leakoff) data. First a fracture model must be defined. Aadnoy and Chenevert (1987) presented the fracture equation in terms of different well directions.

Fracturing Theory. The principal stresses on the borehole wall are given by

$$\sigma_1 = p_w$$

$$\sigma_2 = \frac{1}{2}(\sigma_\theta + \sigma_z) + \frac{1}{2}\sqrt{(\sigma_\theta - \sigma_z)^2 + 4\tau_{\theta z}^2}$$

$$\sigma_3 = \frac{1}{2}(\sigma_\theta + \sigma_z) - \frac{1}{2}\sqrt{(\sigma_\theta - \sigma_z)^2 + 4\tau_{\theta z}^2} . \quad \dotsb \quad (5.2.16)$$

Overburden Stress Gradient, SG

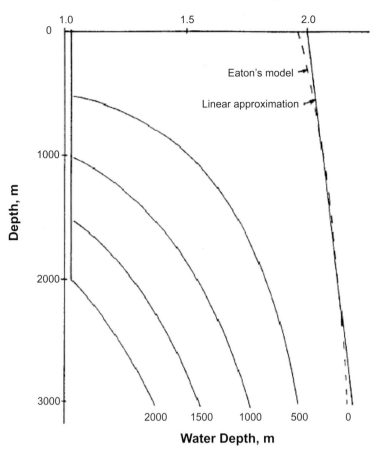

Fig. 5.2.7—Overburden stress for various water depths (Aadnoy 1998).

After calculating these principal stresses, the subscripts are rearranged such that index 1 always refers to the maximum compressive principal stress, 2 to the intermediate principal stress, and 3 to the least principal stress.

During drilling of petroleum wells, a so-called leakoff test is performed after each casing is set. This is to test out the strength of the hole below the casing to ensure that the strength is sufficient to handle the mud weights required for the next hole section and that the well can handle a pressure event such as a kick. These tests are the main input to the fracture prognosis. It might be stated that the pore pressure profile and the fracturing profile are the two most important parameters in the well.

Fracturing of the wellbore is initiated when the rock stress changes from compression to tension. By increasing the wellbore pressure, the hoop stress decreases accordingly. Therefore, fracturing occurs at high wellbore pressures. During well stimulation, while performing minifrac tests, or during extended leakoff operations, the wellbore is purposely fractured. During conventional leakoff testing, on the other hand, the borehole is usually only brought toward the fracturing stage.

The borehole will fracture when the minimum effective principal stress reaches the tensile rock strength σ_t. This is expressed as

$$\sigma_3' = \sigma_3 - p_p \leq \sigma_t. \quad\dotfill\quad (5.2.17)$$

Inserting Eq. 5.2.16 into Eq. 5.2.17, the critical tangential stress is given by

$$\sigma_\theta = \frac{\tau_{\theta z}^2}{\sigma_z - \sigma_t - P_p} + P_p + \sigma_t \quad \dots\dots\dots\dots\dots\dots\dots\dots\dots\dots\dots \quad (5.2.18)$$

Inserting the equation for the tangential stress, Eq. A-2, the critical borehole pressure is given by

$$P_w = \sigma_x + \sigma_y - 2(\sigma_x - \sigma_y)\cos(2\theta) + 4\tau_{xy}\sin(2\theta) - \frac{\tau_{\theta z}^2}{\sigma_z - \sigma_t - P_p} - P_p - \sigma_t. \quad \dots\dots\dots\dots \quad (5.2.19)$$

Differentiating Eq. 5.2.19, the position on the borehole wall where the fracture will arise is given by

$$\frac{dP_w}{d\theta} = 0 \rightarrow \tan(2\theta) = 2\frac{\tau_{xy}}{(\sigma_x - \sigma_y)}.$$

Thus the general fracturing equation is defined. It is valid for all cases, arbitrary directions, and anisotropic stresses.

Neglecting shear stresses of second order, and assuming zero tensile strength due to fissures or cracks, the following fracture equations result:

$$P_{wf} = 3\sigma_x - \sigma_y - P_p \text{ for } \sigma_x < \sigma_y, \text{ and } \theta = 90°$$

and

$$P_{wf} = 3\sigma_y - \sigma_x - P_p \text{ for } \sigma_y < \sigma_x, \text{ and } \theta = 0°. \quad \dots\dots\dots\dots\dots\dots\dots\dots\dots \quad (5.2.20)$$

These equations define the fracturing pressure for various conditions of the in-situ stresses. For a vertical hole, the following equation applies:

$$P_{wf} = 3\sigma_h - \sigma_H - P_p. \quad \dots\dots\dots\dots\dots\dots\dots\dots\dots\dots\dots\dots \quad (5.2.21)$$

Fig. 5.2.8 shows a typical pressure recording from a leakoff test. Usually, the deviation from the straight line is defined as the leakoff or the fracture initiation pressure.

Example 5.2.4. Assume that the following data exist for a well:

Overburden stress	σ_V	= 100 bar
Horizontal stress	σ_H	= σ_h = 90 bar
Pore pressure	P_p	= 50 bar
Borehole inclination	ϕ	= 40°
Borehole azimuth	ϑ	= 165°

Determine the fracture pressure for a vertical well and for the deviated well given above.

For the vertical well, the in-situ stresses relate directly to the borehole direction and become

$$\sigma_x = \sigma_y = 90 \text{ bar.}$$

The fracture pressure is determined directly by Eq. 5.2.21 and is

$$P_{wf} = 2\sigma_h - P_p = 2 \times 90 - 50 = 130 \text{ bar}$$

For the inclined well, the stresses must first be transformed to the orientation of the wellbore by Eqs. A-7 through A-12. The result is

$$\sigma_x = 94.13 \text{ bar}$$
$$\sigma_y = 90 \text{ bar}$$

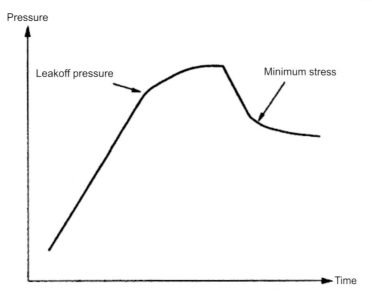

Fig. 5.2.8—Common interpretation of leakoff pressure.

$\sigma_z = 95.87$ bar

$\tau_{xz} = 4.92$ bar, $\tau_{yz} = \tau_{xy} = 0$.

The fracture pressure becomes

$$p_{wf} = 3 \times 90 - 94.13 - 50 = 125.9 \text{ bar}.$$

We observe that the fracturing pressure decreases with increased borehole inclination for this example.

The Inversion Technique. We will now use the fracture model to compute the magnitude of the in-situ stresses. The so-called inversion technique Aadnoy (1990) uses leakoff data to predict stresses in the formation, and it also predicts fracturing pressures for new wells. The following input parameters are needed: the fracture pressure, pore pressure, and overburden stress at each fracture location, as well as the directional data, which are the borehole azimuth and inclination. The following data result from the inversion technique.

Fig. 5.2.9 illustrates the method. Having two or more data sets, the inversion technique calculates the horizontal stress field that fits all data sets. In other words, a maximum and a minimum principal horizontal stress are calculated, along with their directions. This information is useful for a number of rock mechanical analyses. This section will focus on the application to wellbore fracturing operations.

The fracturing process is governed by Eq. 5.2.20. The two normal stresses are replaced by their transformation equations, and by rearranging, Eq. 5.2.20 now becomes

$$\frac{p_{wf} + P_p}{\sigma_v} + \sin^2 \phi = \left(3 \sin^2 \vartheta - \cos^2 \vartheta \cos^2 \phi\right)\frac{\sigma_k}{\sigma_V}$$

$$+ \left(3 \cos^2 \vartheta - \sin^2 \vartheta \cos^2 \phi\right)\frac{\sigma_l}{\sigma_V},$$

or, in short form,

$$p' = a\frac{\sigma_k}{\sigma_V} + b\frac{\sigma_l}{\sigma_V}. \quad \dotfill \quad (5.2.22)$$

The equation above has two unknowns, the horizontal in-situ stresses, called σ_k and σ_l. Having two data sets from two well sections with different orientation, one can determine these two unknown stresses. After calculating the stresses, the largest is redefined as σ_H and the smallest as σ_h.

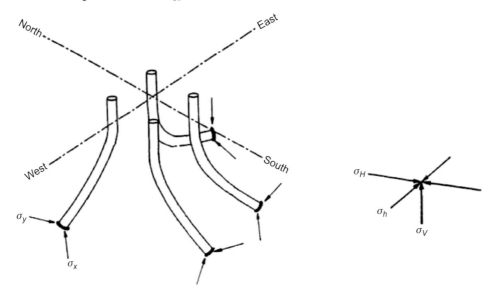

Fig. 5.2.9—Stresses acting on inclined boreholes are transformed from the in-situ stress field (Aadnoy 1997). Reprinted with permission from Elsevier.

The inversion technique takes advantage of the process described above. Often, we have many data from many wells. These will be used in the following to calculate the two horizontal in-situ stresses and their directions. Assume that we have many data sets. In matrix form, these data are

$$
\begin{bmatrix} p'_1 \\ p'_2 \\ p'_3 \\ \cdot \\ p'_n \end{bmatrix} = \begin{bmatrix} a_1 & b_1 \\ a_2 & b_2 \\ a_3 & b_3 \\ \cdot & \cdot \\ a_n & b_n \end{bmatrix} \begin{bmatrix} \sigma_k / \sigma_V \\ \sigma_l / \sigma_V \end{bmatrix}
$$

or, in short form,

$$
[p'] = [A][\sigma]. \quad\dotfill\quad (5.2.23)
$$

Here,

$$
a_i = 3\sin^2 \vartheta_i - \cos^2 \vartheta_i \cos^2 \phi_i
$$
$$
b_i = 3\cos^2 \vartheta_i - \sin^2 \vartheta_i \cos^2 \phi_i
$$
$$
p'_i = \frac{P_{wfi} + P_{pi}}{\sigma_{Vi}} \sin^2 \phi_i
$$

Eq. 5.2.23 is an overdetermined system of equations because there are many sets of data available to determine the two unknown stresses. For this general case, there will always be an error between the solution and some of the data sets. The unknown stresses must also be isolated by determining the inverse of the equation above. To solve these issues, the error between the model and the measurement is defined:

$$
[e] = [A][\sigma] - [p'].
$$

The squared error is

$$e^2 = [e]^T [e].$$

The error is minimized by requiring

$$\frac{\partial e^2}{\partial [\sigma]} = 0.$$

By performing the analysis shown above, the in-situ stresses are given by

$$[\sigma] = \left\{ [A]^T [A] \right\}^{-1} [A]^T [p'] . \dotfill (5.2.24)$$

At this stage, we observe that the equation for the stresses is too cumbersome for manual calculations. A computer program is required. Another issue not discussed so far is the determination of the direction of the in-situ stresses. Eq. 5.2.24 is computed in steps assuming a direction of the in-situ stresses from 0 to 90°. The direction at which the error is at a minimum value is the direction of one of the horizontal in-situ stresses. Aadnoy et al. (1994) presents a field case demonstrating the application of the inversion technique to determine the in-situ stresses. In the following section, we will use a numerical example to demonstrate the application of the inversion program.

Geological Aspects. Here, two scenarios will be presented to illustrate the fundamental application of the inversion technique.

In a *relaxed depositional environment,* we often neglect tectonic effects and assume that the horizontal in-situ stress field is due to rock compaction only. It is often called a hydrostatic or isotropic stress field in the horizontal plane. That implies the same horizontal stresses in all directions. If deviated boreholes are drilled, then there are no directional abnormalities for the same wellbore inclination, and the same leakoff value is expected in all geographical directions. Since the horizontal stresses in a relaxed depositional environment are lower than the overburden stress, the fracture gradient will decrease with hole angle as illustrated in **Fig. 5.2.10a.** This situation is relatively simple to analyze (i.e., by assuming a constant horizontal stress gradient for the field). However, even though relaxed environments exist, this ideal stress situation is rarely the case. Usually, a more complex stress situation exists.

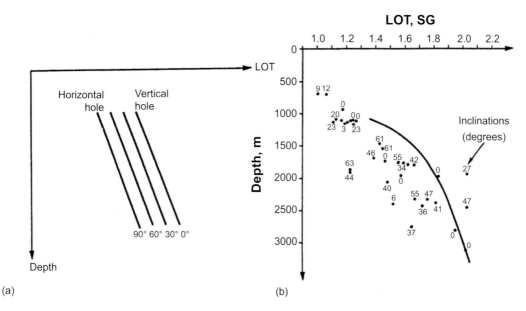

Fig. 5.2.10—Effects of stress anisotropy on Snorre leakoff data (Aadnoy 1997; reprinted with permission from Elsevier). (a) Expected leakoff behavior, relaxed depositional basin; (b) Snorre leakoff data vs. inclination.

The horizontal stress field usually varies with direction, and we have two different horizontal principal stresses. This stress state is called *anisotropic*. This may be caused by global geologic processes such as plate tectonics, or by local effects like salt domes, topography, or faults. The resulting stress state varies over the area. Fig. 5.2.10b shows an example from the Snorre field offshore Norway. Two immediate observations are that there is a considerable spread in the leakoff data, and there is no apparent trend with respect to wellbore inclination. It is obvious that the previously defined isotropic model is not useful for this case. The answer is that the stress state is different for many data points. By establishing a more complex stress model for Snorre, most of the data points shown were predictable with reasonable accuracy. The experience is that most oil fields exhibit anisotropic stress fields to some degree.

Example 5.2.5. In the following, an example is given to demonstrate some of the features and applications of the inversion technique. Imagine a field with three wells drilled, and with a fourth well under planning. The data for this case are listed in **Table 5.2.1**. The wells are shown in **Figs. 5.2.11a and 5.2.11b.**

First Example Calculation. To start the simulations, first let us use all data sets and estimate an average stress in the formation. The computer will generate the following results:

$$\sigma_H/\sigma_V = 0.864$$
$$\sigma_h/\sigma_V = 0.822$$
$$\beta_H = 44°.$$

The interpretation is as follows. The largest principal horizontal stress is 0.864 times the overburden stress, and its direction is 44° from north (northeast). The minimum principal horizontal stress is 0.822 times the overburden stress. However, please note that the data cover a considerable depth interval and a large geographical area. We therefore have to assess the quality of the simulation and evaluate if one stress model adequately describes this area. The program automatically provides a quality control. After the stresses have been computed, the program uses these as input data and provides a prediction of each input data set. If the measured and the predicted data are similar, the model is good. In contrast, a large discrepancy questions the validity of the stress model.

Comparison between the measured and the predicted leakoff data shows a rather poor correlation (see **Table 5.2.2**). Fowr practical applications, this difference should probably be within 0.05 to 0.10 SG. The conclusion at this stage is that a single stress model is not adequate for this field, and that the field has to be modeled with several submodels.

Second Example Calculation. Let us now study the stress state at approximately 1100 m depth, which is the 20-in. (50.8-cm) casing shoe location. By marking out data sets 1, 4, 7, and 10, and computing, the following stress state results:

TABLE 5.2.1—FIELD DATA (Aadnoy 1997[*])								
Data Set	Well	Casing (in.)	Depth (m)	P_{wf} (SG)	p_p (SG)	σ_v (SG)	ϕ (°)	ϑ (°)
1	A	20	1101	1.53	1.03	1.71	0	27
2		13⅜	1888	1.84	1.39	1.82	27	92
3		9⅝	2423	1.82	1.53	1.89	35	92
4	B	20	1148	1.47	1.03	1.71	23	183
5		13⅜	1812	1.78	1.25	1.82	42	183
6		9⅝	2362	1.87	1.57	1.88	41	183
7	C	20	1141	1.49	1.03	1.71	23	284
8		13⅜	1607	1.64	1.05	1.78	48	284
9		9⅝	2320	1.84	1.53	1.88	27	284
10	New	20	1100		1.03	1.71	15	135
11		13⅜	1700		1.19	1.80	30	135
12		9⅝	2400		1.55	1.89	45	135

[*] Reprinted with permission from Elsevier.

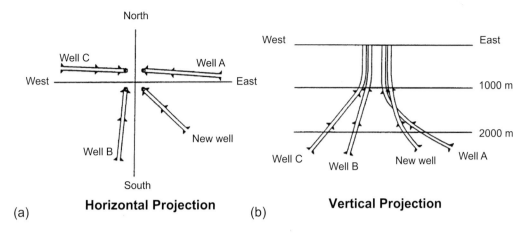

Fig. 5.2.11—Placement of the wells in the field (Aadnoy 1997). Reprinted with permission from Elsevier.

TABLE 5.2.2—SIMULATION USING ALL DATA									
Data set	1	2	3	4	5	6	7	8	9
Measured LOT (SG)	1.53	1.84	1.82	1.47	1.78	1.87	1.49	1.64	1.84
Predicted LOT (SG)	1.75	1.64	1.58	1.78	1.66	1.43	1.88	1.86	1.65

TABLE 5.2.3—SIMULATION USING DATA AT 1100-m DEPTH				
Data set	1	4	7	10
Measured LOT (SG)	1.53	1.47	1.49	
Predicted LOT (SG)	1.53	1.47	1.49	1.53

$$\sigma_H/\sigma_V = 0.754$$
$$\sigma_h/\sigma_V = 0.750$$
$$\beta_H = 27°.$$

Now we observe that the two horizontal stresses are nearly equal. This is expected because at this depth, no tectonics exist, and in a relaxed depositional environment, an equal (or hydrostatic) horizontal stress state is expected. The dominating mechanism here is compaction. Let us next evaluate the quality of the simulation by comparing input fracturing data with the modeled data.

A perfect match is seen in **Table 5.2.3,** and we will consider this simulation a correct assessment of the stress state at this depth. Also, a prediction for the new well is performed since we included data set 10 in the simulation.

Third Example Calculation. Now we want to investigate stresses at the 13⅜-in. (39.97-cm) casing shoe depths between 1607 and 1888 m. We use data sets 2, 5, 8, and 11 and obtain the following results:

$$\sigma_H/\sigma_V = 1.053$$
$$\sigma_h/\sigma_V = 0.708$$
$$\beta_H = 140°.$$

From **Table 5.2.4** we observe a rather poor comparison. The most probable reason is that the three input data sets are not consistent. One stress state will not adequately model all three locations. To further investigate this, we will therefore simulate several combinations of these data sets.

From **Table 5.2.5** we see that both simulation runs give a perfect match. This is always the case when using only two data sets, since there are two unknown in-situ stresses to be calculated. However, the prediction

TABLE 5.2.4—SIMULATION USING DATA BETWEEN 1600 AND 1900 m				
Data set	2	5	8	11
Measured LOT (SG)		1.84	1.78	1.64
Predicted LOT (SG)	1.73	1.37	1.31	0.77

TABLE 5.2.5—SIMULATION USING OTHER COMBINATIONS OF DATA			
Data set	2	5	11
Measured LOT (SG)	1.84	1.78	
Predicted LOT (SG)	1.84	1.78	1.95
Data set	5	8	11
Measured LOT (SG)	1.78	1.64	
Predicted LOT (SG)	1.78	1.64	1.71

of the first example calculation yields a too high leakoff value. The second example calculation seems more realistic.

Fourth Example Calculation. A final example is to model stresses at reservoir level. From Table 5.2.1 we use data sets 3, 6, 9, and 12. The results are

$\sigma_H/\sigma_V = 0.927$
$\sigma_h/\sigma_V = 0.906$
$\beta_H \quad = 77°.$

This is shown in **Table 5.2.6** and is considered a good representation of the stress fields at the reservoir level.

Discussion of Simulations. The previous examples demonstrate how the inversion technique can be used to estimate stresses and to perform predictions for new wells. We have shown a way to assess the quality of the simulation. A practical way to analyze fields is to first generate averages over larger depth intervals and areas, and then to investigate smaller intervals or parts using measurement and prediction

TABLE 5.2.6—SIMULATION AT RESERVOIR LEVEL				
Data set	3	6	9	12
Measured LOT (SG)	1.82	1.87	1.84	
Predicted LOT (SG)	1.82	1.87	1.84	1.86

TABLE 5.2.7—RESULTS OF SOME SIMULATIONS							
Run	Data Sets	Wells	Casing (in.)	σ_H/σ_V	σ_h/σ_V	β_H (°)	Comments
1	1–9	A, B, C	All	0.861	0.825	41	Local average
2	1, 4, 7	A, B, C	20	0.754	0.750	27	Good simulation
3	2, 5, 8	A, B, C	13⅜	1.053	0.708	50	Poor simulation
4	2, 5	A, B	"	0.891	0.867	13	Good simulation
5	2, 8	A, C	"	—	—	—	Poor simulation
6	5, 8	B, C	"	0.854	0.814	96	Good simulation
7	3, 6, 9	A, B, C	9⅝	0.927	0.906	77	Good simulation
8	2, 3	A	13⅜ to 9⅝	0.982	0.920	90	Poor simulation
9	5, 6	B	"	—	—	—	Poor simulation
10	8, 9	C	"	—	—	—	Poor simulation

	Casing				β_H	LOT New Well
Run	(in.)	Depth (m)	σ_H/σ_V	σ_h/σ_V	(°)	(SG)
2	20	1100–1148	0.754	0.75	44	1.53
6	13⅜	1607–1812	0.854	0.814	96	1.71
7	9⅝	2320–2423	0.927	0.906	90	1.86

TABLE 5.2.8—FINAL RESULTS OF FIELD SIMULATIONS (Aadnoy 1997*)

* Reprinted with permission from Elsevier.

comparison to assess the quality of each simulation. In this way stress models can be derived for the field.

The data given in Table 5.2.1 were simulated in a number of combinations. The results are listed in **Table 5.2.7.** To summarize, the resulting stress fields obtained in this example exercise are given in **Table 5.2.8.**

In Table 5.2.8 several observations can be made. The stress field is increasing with depth as expected. Furthermore, it is shown that the stress field is probably anisotropic. In particular, at reservoir level, the maximum horizontal stress apparently is approaching the overburden stress. The predicted stress fields are shown in **Fig. 5.2.12.**

5.2.6 Compaction Models. The compaction model was derived to account for pore pressure changes during depletion of a field. By introducing this model, we will have a tool to perform stress history calculations. If the pore pressure has changed over time, we can estimate what effect this will have on the fracturing pressure. One example is to normalize all leakoff data to a given reference pore pressure. If all data then form a trend, we may interpret this as that all data have the same origin. In this section, we will first define a simple compaction model, then use an example to illustrate the application.

Crockett et al. (1986) give a more general derivation of the so-called *backstress* concept, which is the influence of charged pore pressure on the fracturing pressure. Morita et al. (1988) also address the same problem. Aadnoy (1991) derives this concept in a simpler way. **Figs. 5.2.13a and 5.2.13b** show a rock before and after the pore pressure has been changed. Assuming that the overburden stress remains constant, and that no strain is allowed on the sides of the rock, we can calculate the changes in the horizontal rock stress. Because the overburden stress is constant and the pore pressure is lowered, for example, the rock matrix must take the load held by the initial pore pressure. This increased vertical matrix stress will, through the Poisson's ratio, also increase the horizontal stress. This horizontal stress increase is

$$\Delta\sigma = \Delta p_p \frac{1-2v}{1-v}. \quad\dots\dots\dots\dots\dots\dots\dots\dots\dots\dots\dots\dots\dots\dots\dots\dots \quad (5.2.25)$$

Inserting this matrix stress change into the general frac equations, the corresponding change in frac pressure can be calculated. The following equation results:

$$\Delta p_{wf} = \Delta p_p \frac{1-3v}{1-v}. \quad\dots\dots\dots\dots\dots\dots\dots\dots\dots\dots\dots\dots\dots\dots\dots \quad (5.2.26)$$

Example 5.2.6. As an example, **Table 5.2.9** shows some leakoff data and their associated pore pressure gradients. If we decide to normalize the data to the same pore pressure gradient, arbitrarily chosen to 1.80 SG, and with a Poisson's ratio of 0.25, the third entry looks like

$$p_{wf} - 1.98 = \frac{(1.80-1.44)(1-3\cdot0.25)}{1-0.25}$$

or

$$p_{wf} = 2.10 \text{ SG}.$$

The raw data and the pore-pressure-corrected data are shown in **Fig. 5.2.14.** It is observed that the raw data show hardly any correlation. The compaction-corrected data, on the other hand, fall nearly on a line, with a

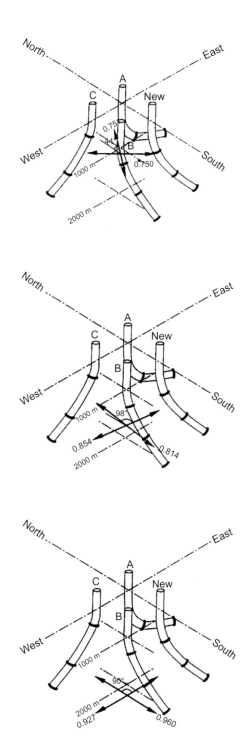

Fig. 5.2.12—Predicted in-situ stress fields from the simulations (Aadnoy 1997; reprinted with permission from Elsevier). (a) Predicted stress field at 1100–1148 m. (b) Predicted stress field at 1607–1812 m. (c) Predicted stress field at 2320–2428 m.

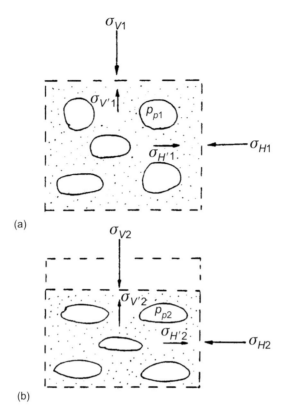

(a)

(b)

Fig. 5.2.13—Illustration of the compaction model: (a) The initial stress state. (b) After the pore pressure is reduced, the horizontal stress increases, while the overburden remains constant (Aadnoy 1991). Reprinted with permission from Elsevier.

TABLE 5.2.9—FIELD DATA (Aadnoy 1997*)		
Depth (m)	LOT (SG)	p_p (SG)
3885	2.10	1.79
3821	2.13	1.84
3818	1.98	1.44
3914	2.06	1.58
* Reprinted with permission from Elsevier.		

reasonable correlation. One interpretation is that the data are very similar, or have the same origin. The initial stress states for the four wells were possibly similar, but the variations observed in the frac pressures are mainly dependent on local variations in the pore pressures. In other words, the pressure depletion history is also reflected in the frac pressure.

Assuming that the vertical trend line of Fig. 5.2.14 (at 2.11 SG) reasonably well describes the frac pressure, the compaction equation can be expressed as

$$2.11 - p_{wf} = \frac{(1.80 - p_p)(1 - 3 \cdot 0.25)}{1 - 0.25}$$

or, rearranged the frac equation looks like

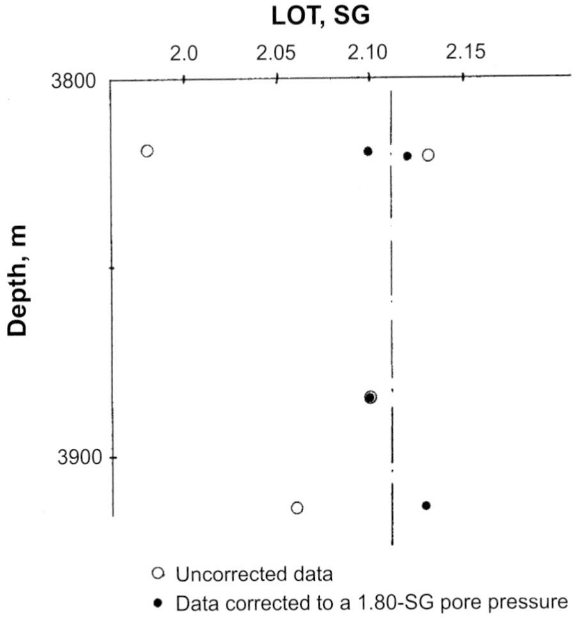

Fig. 5.2.14—Fracture data shown uncorrected and corrected with a compaction model. Reprinted with permission from Elsevier.

$$p_{wf} = 1.51 + \frac{1}{3}p_p$$

If Table 5.2.9 is used to reconstruct the data with this equation, the result is very close to the initial raw data.

Example 5.2.7. Another example is that in a field, the initial frac gradient during exploration drilling was 1.70 SG. After the field has been put on production, the overall reservoir pressure gradient declines from 1.5 to 1.3 SG. What is the expected change in the horizontal stresses and the frac pressure in the reservoir?

The change in the horizontal stress gradient would be

$$\Delta\sigma = \frac{(1.5 - 1.3)(1 - 2 \cdot 0.25)}{1 - 0.25} = 0.13 \text{ SG.}$$

If an infill well was drilled at a later stage, the estimated frac pressure gradient would be

$$p_{wf} = 1.7 - \frac{(1.5 - 1.3)(1 - 3 \cdot 0.25)}{1 - 0.25} = 1.63 \text{ SG.}$$

5.2.7 Bounds on In-Situ Stresses. Since wellbore stability analysis was introduced to the oil community in early 1980s, it has been observed that many analyses use poor in-situ stress analysis. The reported stress state may be outside the permissible range. A typical consequence of this is that the critical collapse curve may cross the fracture curve (see **Fig. 5.2.15a**). Adjusting the in-situ stresses, the frac and collapse curve are no longer crossing, as seen in Fig. 5.2.15b, which is physically correct. Earlier conclusions were that a well would be stable up to a certain inclination (e.g., 60°) but unstable above. We know today that this is not the case. Most wells can be drilled at any direction, provided that they are properly designed. Aadnoy and Hansen (2005) analyzed and defined the permissible range for the in-situ stresses.

Implementation of the bound has led to a significant improvement in borehole stability modeling. In the following section, a brief description is given, followed by two numerical examples.

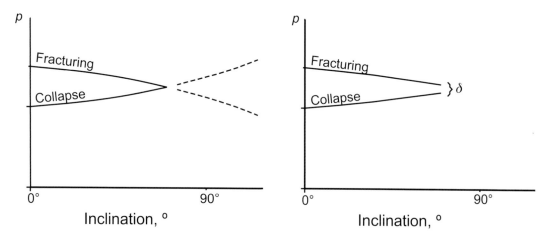

Fig. 5.2.15—Fracturing and collapse pressures vs. wellbore inclination. (a) Example of incorrect in-situ stress state that leads to physically incorrect results; (b) physically correct result (Aadnoy and Hansen 2005).

Knowledge of Instability Problems in a Well, or in Stable Borehole Sections. Some wells have known stability problems in given directions. Having such a case at hand, the critical frac and collapse pressure must be established. From these, the stability margin can be established along one of the principal in-situ stress directions. If the well is hardly drillable, a very small stability margin δ may result. Also, trouble-free borehole sections can be used to assess stress level limits. The stress magnitudes should be marked in Fig. 5.2.14 and used for prediction of similar wells. Assume that several stress data are established for a field or a well. Inserting these into Fig. 5.2.14 and making a straight line between the points will serve as the stress model. If only one data set exists, a straight line through the point and the (1, 1) point defines our stress model.

Having No Prior Knowledge of Stability Problems. The evolution of modern drilling technology has led to the observation that most boreholes can be drilled in any direction, given good planning and operational follow-up. However, the margins between success and failure are sometimes small. For this general scenario,

TABLE 5.2.10—GENERAL BOUNDS FOR IN-SITU STRESSES FOR VARIOUS STRESS STATES

Stress State	Upper Bound	Lower Bound
Normal fault	$\dfrac{\sigma_h}{\sigma_V}, \dfrac{\sigma_H}{\sigma_V} \leq 1$	$\dfrac{\sigma_h}{\sigma_V}, \dfrac{\sigma_H}{\sigma_V} \geq \dfrac{B+C}{A}$
Strike/slip fault	$\dfrac{\sigma_H}{\sigma_V} \leq \dfrac{A-C}{B}$	$\dfrac{\sigma_H}{\sigma_V} \geq 1$
	$\dfrac{\sigma_h}{\sigma_V} \leq 1$	$\dfrac{\sigma_h}{\sigma_V} \geq \dfrac{B+C}{A}$
Reverse fault	$\dfrac{\sigma_h}{\sigma_V}, \dfrac{\sigma_H}{\sigma_V} \leq \dfrac{A-C}{B}$	$\dfrac{\sigma_h}{\sigma_V}, \dfrac{\sigma_H}{\sigma_V} \geq 1$

In all entries $\sigma_H > \sigma_h$,
where $A = 7 - \sin\theta$, $B = 5 - 3\sin\theta$,
$C = [2p_p(1 + \sin\theta) + 2(\delta - \tau_c \cos\theta)]/\sigma_V$

a default model is proposed as follows. Letting the stability margin equal the cohesive strength in **Table 5.2.10,** the last argument of constant C vanishes, and it becomes $C = 2p_p(1 + \sin\theta)/\sigma_v$. For this case, the bound follows the solid lines of Fig. 5.2.15. The solid lines will therefore serve as our default stress model. In wildcat drilling, we may assume the lower bound as our model, but reevaluate it as soon as leakoff data become available, following the procedure outlined above.

Another advantage given by the model is the possibility to develop a continuous stress profile with depth, as opposed to current practices where one obtains discrete stress states at each casing shoe [leakoff test (LOT) measurement], and just extrapolates between these points. This will be demonstrated in the following examples.

This analysis is based on the relationship between the pressure margin and the in-situ stress tensor. From an operational perspective, this demonstrates that all wells have some directions that are more critical than others. Furthermore, to know these directions, an insight into the in-situ stresses and their directions is a requirement. All critical wells, such as long-reach wells, should therefore be analyzed to determine the pressure margin, which has a significant impact on the casing depth selection. In the following section, two examples from a field offshore Norway will be presented.

Example 5.2.8. In this example, we will investigate the in-situ stress bounds that give realistic wellbore stability analysis results. The following data are used from a field in the North Sea:

Depth: 1700 m
Overburden stress gradient: 1.8 SG
Pore pressure gradient: 1.03 SG
Cohesive rock strength: 0.2 SG
Friction angle of rock: 30°

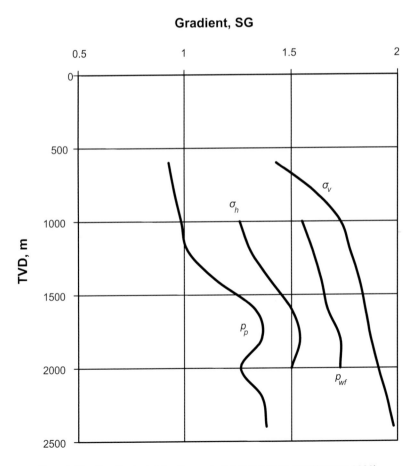

Fig. 5.2.16—Gradient plot for Example 5.2.9 (Aadnoy and Hansen 2009).

TABLE 5.2.11—IN-SITU STRESS BOUNDS AND FRACTURE PROGNOSIS FOR EXAMPLE 5.2.9*					
Depth (m)	p_p (SG)	σ_v (SG)	$\sigma_h/\sigma_v,$ $\sigma_H/\sigma_v,>$	σ_h (SG)	p_{wf} (SG)
1000	0.97	1.73	0.73	1.26	1.55
1200	1.02	1.78	0.74	1.31	1.60
1400	1.15	1.82	0.77	1.40	1.64
1600	1.33	1.85	0.81	1.50	1.67
1800	1.36	1.87	0.81	1.54	1.73
2000	1.26	1.91	0.78	1.50	1.73
* Parameters: $\delta = 0.1$ SG, $\theta = 35°$, $\tau_c = 0.5$ SG					

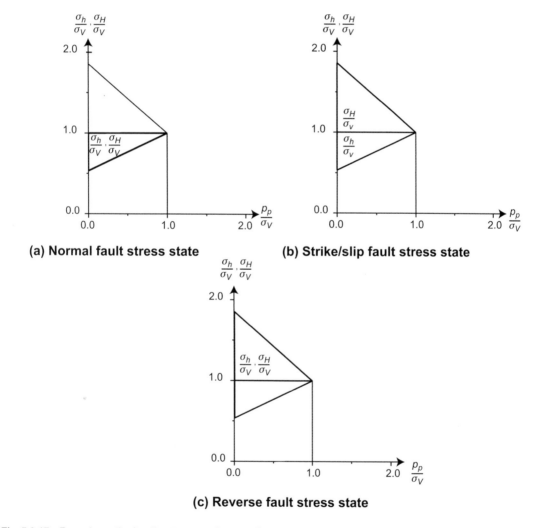

(a) Normal fault stress state   (b) Strike/slip fault stress state

(c) Reverse fault stress state

Fig. 5.2.17—Bounds on the in-situ stresses. Assumptions: $\theta = 30°$, $\delta = 0.866\tau_c$ (Aadnoy and Hansen 2005). (a) Normal fault stress state, (b) strike/slip fault stress state, (c) reverse fault stress state.

Wells are drilled at different inclinations at this depth. One horizontal section is drilled, confirming that it is possible to drill wells in any direction. From evaluation of well data, the minimum difference between the fracturing and collapse pressure is estimated as $\delta = 0.173$ SG. Inserting the data into Table 5.2.10, the following bounds are established, assuming a normal fault stress state:

$$1 \geq \frac{\sigma_h}{\sigma_V} \geq 0.80$$

$$1 \geq \frac{\sigma_H}{\sigma_V} \geq 0.80$$

$$\frac{\sigma_H}{\sigma_V} \geq \frac{\sigma_h}{\sigma_V}.$$

These conditions guarantee that the minimum difference between the critical fracturing pressure and the critical collapse pressure never exceeds δ. To determine the magnitude of the two horizontal in-situ stresses, one must use other methods. For this case, a number of LOT data were analyzed using the inversion technique.

Example 5.2.9. **Fig. 5.2.16** presents the overburden gradient and the pore pressure gradient for a production field in the North Sea. The intermediate casing is set in the depth interval 1100 to 1900 m. This interval is characterized by a pore pressure increase from 1.03 SG at the top to 1.4 SG near the bottom. Most of the interval consists of young clays. Well inclinations are low in most wells, resulting in a limited amount of data to develop a general model. It is difficult to establish stress magnitudes that provide a realistic frac prognosis over the entire interval.

Two problems were encountered. First, by changing the inclination toward horizontal in simulation studies, the aforementioned crossing of fracturing and collapse curves resulted, questioning the quality of earlier-used in-situ stress data. Second, LOT data exist only at the 1200-m depth and 1800- to 1900-m depths because these are the setting depths for the 18⅝-in. and 13⅜-in. (47.31-cm and 33.97-cm) casing strings. A well-optimization study suggested that the setting depth should be changed to somewhere in between these two depths. However, no data exist here.

The stress states at the top and bottom are different. Using either of these gives good results at one end but poor results at the other end. We will use the method of this chapter to produce a stress state that covers the entire depth interval. **Table 5.2.11** gives the bounds for the in-situ stresses. The minimum horizontal stress and the fracturing prognosis are shown in Fig. 5.2.16. **Fig. 5.2.17** shows the limits for the in-situ stresses that give realistic fracture and collapse curves for all wellbore orientations. The present example represents a normal fault stress state as shown in Fig. 5.2.17a.

The existing wells in the area have the 20-in. (50.8-cm) casing strings placed at approximately 1100 m, and the 13⅜-in. (33.97-cm) casings set at approximately 1800 m. Because LOTs are conducted at these depths, an estimate of the stress state can be established at these depths only.

Fig. 5.2.16 shows the resulting horizontal stress and the fracture gradient for a vertical well in the depth interval 1000 to 2000 m. Please observe that these are continuous throughout the entire depth interval. There is no longer a need to extrapolate between two casing points. Also observe that the fracture prognosis shows a continuous increase with depth. The results of Fig. 5.2.16 give a significant improvement to the stress analysis performed over the field for the past decade.

Nomenclature

d = pressure gradient defined as relative to water density
d_{b1} = bulk density gradient for reference well
d_{b2} = bulk density gradient for new well
d_{eq} = equivalent density related to the overpressure
d_{oil} = oil density gradient
d_{sw} = seawater density gradient
d_{water} = water density gradient
d_{wf} = leakoff pressure gradient
D_{sb} = depth of rock below seabed
D_{sb1} = depth of rock below seabed for reference well

D_{sb2} = depth below seabed for new well
D_{wf1} = leakoff pressure gradient for reference well
D_{wf2} = leakoff pressure gradient for new well
e = error
h = distance or height
h_f = distance from drill floor to sea level
h_w = water depth
I_1, I_2, I_3 = stress invariants
p = pressure
p' = normalized pressure to be used for the inversion technique
p_p = pore pressure
p_w = well pressure
p_{wf} = fracture pressure
s_f = seafloor reference level
z = true vertical depth
β_H = angle from north to the maximum horizontal stress
Δ = difference between two readings of a parameter
Δp = pressure difference
δ = stability margin
δh = difference in drill floor elevation between two platforms
θ = angle around borehole wall
v = Poisson's ratio
ρ = density
ρ_{water} = pure water density
σ = normal stress
σ_1 = maximum principal stress
σ_2 = medium principal stress
σ_3 = minimum principal stress
σ' = effective stress
σ_H = maximum horizontal stress
σ_h = minimum horizontal stress
σ_k = unknown horizontal in-situ stress
σ_l = unknown horizontal in-situ stress
σ_V = overburden stress
σ_t = tensile rock strength
$\sigma_r, \sigma_\theta, \sigma_z, \tau_{\theta z}$ = stresses on borehole wall
$\sigma_x, \sigma_y, \tau_{xy}$ = in-situ stresses transformed to xy directions
τ = shear stress
τ_c = cohesive formation strength
τ_{xy} = shear stress in y direction related to a surface normal to the x-axis
τ_{xz} = shear stress in z direction related to a surface normal to the x-axis
τ_{yz} = shear stress in z direction related to a surface normal to the y-axis
ϕ = borehole inclination
ϕ_i = borehole inclination for data set i
Φ = angle of internal friction for rock
ϑ = borehole azimuth
ϑ_i = borehole azimuth relative to the maximum horizontal stress for data set i

Superscripts

T = transpose
-1 = inverse

References

Aadnoy, B.S. 1990. Inversion Technique to Determine the In-Situ Stress Field From Fracturing Data. *Journal of Petroleum Science and Engineering* **4** (2): 127–141. DOI: 10.1016/0920-4105(90)90021-T.

Aadnoy, B.S. 1991. Effects of Reservoir Depletion on Borehole Stability. *Journal of Petroleum Science and Engineering* **6** (1): 57–61. DOI: 10.1016/0920-4105(91)90024-H.

Aadnoy, B.S. 1997. *Modern Well Design.* Houston: Gulf Publishing Company.

Aadnoy, B.S. 1998. Geomechanical Analysis for Deep-Water Drilling. Paper SPE 39339 presented at the IADC/SPE Drilling Conference, Dallas, 3–6 March. DOI: 10.2118/39339-MS.

Aadnoy, B.S. and Chenevert, M.E. 1987. Stability of Highly Inclined Boreholes. *SPEDE* **2** (4): 364–374. SPE-16052-PA. DOI: 10.2118/16052-PA.

Aadnoy, B.S. and Hansen, A.K. 2005. Bounds on In-Situ Stress Magnitudes Improve Wellbore Stability Analyses. *SPEJ* **10** (2): 115–120. SPE-87223-PA. DOI: 10.2118/87223-PA.

Aadnoy, B.S. and Saetre, R. 2003. New Model Improves Deepwater Fracture Gradient Values off Norway. *Oil and Gas Journal* **101** (5) 51–54.

Aadnoy, B.S., Soteland, T., and Ellingsen, B. 1991. Casing Point Selection at Shallow Depth. *Journal of Petroleum Science and Engineering* **6** (1): 45–55. DOI: 10.1016/0920-4105(91)90023-G.

Aadnoy, B.S., Bratli, R.K., and Lindholm, C. 1994. In-Situ Stress Modeling of the Snorre Field. *Proc.,* EUROCK 94: Rock Mechanics in Petroleum Engineering, Delft, The Netherlands, 871–878.

Crockett, A.R., Okusu, N.M., and Cleary, M.P. 1986. A Complete Integrated Model for Design and Real-Time Analysis of Hydraulic Fracturing Operations. Paper SPE 15069 presented at the SPE California Regional Meeting, Oakland, California, 2–4 April. DOI: 10.2118/15069-MS.

Eaton, B.A. 1969. Fracture Gradient Prediction and Its Application in Oilfield Operations. *JPT* **21** (10): 1353–1360; *Trans.,* AIME, **246**. SPE-2163-PA. DOI: 10.2118/2163-PA.

Herrmann, R.P., Smith, J.R., and Bourgoyne, A.T. 2001. Application of Dual-Density Gas Lift to Deepwater Drilling. *World Oil Deepwater Technology Supplement* (October 2001): 17–22.

Kaarstad, E. and Aadnoy, B.S. 2006. Fracture Model for General Offshore Applications. Paper SPE 101178 presented at the SPE Asia Pacific Oil and Gas Conference and Exhibition, Adelaide, Australia, 11–13 September. DOI: 10.2118/101178-MS.

Morita, N., Whitfill, D.L., Nygaard, O., and Bale, A. 1988. A Quick Method to Determine Subsidence, Reservoir Compaction, and In-Situ Stress Included by Reservoir Depletion. *JPT* **41** (1): 71–79. SPE-17150-PA. DOI: 10.2118/17150-PA.

Rocha, L.A. and Bourgoyne, A.T. 1996. A New Simple Method to Estimate Fracture Pressure Gradient. *SPEDC* **11** (3): 153–159. SPE-28710-PA. DOI: 10.2118/28710-PA.

Webb, S., Anderson, T., Sweatman, R., and Vargo, R. 2001. New Treatments Substantially Increase LOT/FIT Pressures to Solve Deep HTHP Drilling Challenges. Paper SPE 71390 presented at the SPE Annual Technical Conference and Exhibition, New Orleans, 30 September–3 October. DOI: 10.2118/71390-MS.

Appendix A—Borehole Stresses

Radial stress:

$$\sigma_r = p_w \quad \dots\dots\dots\dots\dots\dots\dots\dots\dots\dots\dots\dots\dots\dots\dots\dots \text{(A-1)}$$

Tangential stress:

$$\sigma_\theta = \sigma_x + \sigma_y - p_w - 2\left(\sigma_x - \sigma_y\right)\cos\left(2\theta\right) - 4\tau_{xy}\sin\left(2\theta\right) \quad \dots\dots\dots\dots\dots \text{(A-2)}$$

Axial stress, plane strain:

$$\sigma_z = \sigma_{zz} - 2v\left(\sigma_x - \sigma_y\right)\cos\left(2\theta\right) - 4v \cdot \tau_{xy}\sin\left(2\theta\right) \quad \dots\dots\dots\dots\dots\dots \text{(A-3)}$$

Axial stress, plane stress:

$$\sigma_z = \sigma_{zz} \quad \dotfill \quad \text{(A-4)}$$

Shear stress:

$$\tau_{\theta z} = 2\left(\tau_{yz} \cos\theta - \tau_{xz} \sin\theta\right) \quad \dotfill \quad \text{(A-5)}$$

$$\tau_{rz} = \tau_{r\theta} = 0 \quad \dotfill \quad \text{(A-6)}$$

Stress Transformation Laws.

$$\sigma_X = \left(\sigma_H \cos^2 \vartheta + \sigma_h \sin^2 \vartheta\right)\cos^2 \varphi + \sigma_v \sin^2 \varphi \quad \dotfill \quad \text{(A-7)}$$

$$\sigma_y = \left(\sigma_H \sin^2 \vartheta + \sigma_h \cos^2 \vartheta\right) \quad \dotfill \quad \text{(A-8)}$$

$$\sigma_{zz} = \left(\sigma_H \cos^2 \vartheta + \sigma_h \sin^2 \vartheta\right)\sin^2 \phi + \sigma_V \cos^2 \phi \quad \dotfill \quad \text{(A-9)}$$

$$\tau_{yz} = \frac{1}{2}\left(\sigma_h - \sigma_H\right)\sin\left(2\vartheta\right)\sin\phi \quad \dotfill \quad \text{(A-10)}$$

$$\tau_{xz} = \frac{1}{2}\left(\sigma_H \cos^2 \vartheta + \sigma_h \sin^2 \vartheta - \sigma_V\right)\sin\left(2\phi\right) \quad \dotfill \quad \text{(A-11)}$$

$$\tau_{xy} = \frac{1}{2}\left(\sigma_h - \sigma_H\right)\sin\left(2\vartheta\right)\cos\phi \quad \dotfill \quad \text{(A-12)}$$

Principal Stresses.

$$\sigma^3 - I_1\sigma^2 - I_2\sigma - I_3 = 0, \quad \dotfill \quad \text{(A-13)}$$

where

$$I_1 = \sigma_x + \sigma_y + \sigma_z \quad \dotfill \quad \text{(A-14)}$$

$$I_2 = \tau_{xy}^2 + \tau_{xz}^2 + \tau_{yz}^2 - \sigma_x\sigma_y - \sigma_x\sigma_z - \sigma_y\sigma_z \quad \dotfill \quad \text{(A-15)}$$

$$I_3 = \sigma_x\left(\sigma_y\sigma_z - \tau_{yz}^2\right) - \tau_{xy}\left(\tau_{xy}\sigma_z - \tau_{xz}\tau_{yz}\right) + \tau_{xz}\left(\tau_{xy}\tau_{yz} - \tau_{xz}\sigma_y\right) \quad \dotfill \quad \text{(A-16)}$$

$$\sigma_i = p_w \quad \dotfill \quad \text{(A-17)}$$

$$\sigma_{ik} = \frac{1}{2}\left(\sigma_\theta + \sigma_z\right) \pm \frac{1}{2}\sqrt{\left(\sigma_\theta - \sigma_z\right)^2 + 4\tau_{\theta z}^2} \quad \dotfill \quad \text{(A-18)}$$

For borehole stress, let

$$x = r$$
$$y = \theta$$
$$z = z.$$

Chapter 6

Wellbore Measurements: Tools, Techniques, and Interpretation

6.1 Images While Drilling—Tom Bratton and Iain Cooper, Schlumberger

6.1.1 Introduction. The ability to acquire logging-while-drilling (LWD) images (with both density and resistivity tools) enables the driller to truly *see* the downhole environment. In addition to clearly showing the interbedding of the formations and the dip of the bed, the images can be used to define fractures. In this chapter, we describe how such images can be used to determine both the fractures and failure mode. Using the image data, it is also possible to determine the maximum-stress direction. This information can then be used to correct the Earth model for more precise wellbore stability analysis (see Chapter 5.1), and in real time will enable the correct remedial action to be taken—for example, raising or lowering the mud weight so that a tolerable level of breakouts can be achieved.

However, it is the combinations of measurements that give rise to the most accurate drilling interpretations. Later in this chapter we highlight some examples of how combined measurements and time-lapse logging can enable drilling problems to be avoided. Indeed, combinations of caliper measurements (which provide a picture of the shape of the borehole) and image data can indicate the severity of formation breakout and the primary directions of failure.

While the quantitative images from single-depth resistivity tools and later density images were a significant technology breakthrough, the addition of multiple depths of investigation combined with surface dynamic measurements (e.g., torque, hookload, pump pressures), mud-logging measurements, and other downhole LWD measurements (internal and annular pressures while drilling, downhole weight and torque, etc.) enable a more complete picture of the borehole to be established while drilling.

6.1.2 The Importance of Downhole Measurements. The measurements of annular pressure and associated downhole and surface measurements are the *sight and touch* of modern-day drillers, enabling them to *see and feel* the dynamic motions of the drillstring, and the downhole behavior of the drilling fluid, so that optimal decisions can be made. Vibration and shock data, along with torque and weight-on-bit, can be used to modify drilling parameters for increased bit and bottomhole assembly (BHA) reliability and performance. The lifeblood of the drilling process is the drilling fluid, and downhole annular mud pressure is one of the most important pieces of information that the driller has to sense what is happening as the bit enters each new section of formation, or while running the bit into or out of the hole.

Downhole annular pressure measurements are being used in many drilling applications, including underbalanced, extended-reach, high-pressure/high-temperature, and, perhaps most significantly, deepwater wells (see Chapter 7). For example, because of shallow-water influx concerns in deepwater wells with narrow stability margins, effective mud weight differences of a few tenths of a pound per gallon can determine whether extra strings of casing are needed to protect shallow intervals. Accurate leakoff tests and/or formation integrity tests are essential to enable efficient management of the equivalent circulating density (ECD) within the safe pressure window (Rezmer-Cooper et al. 2000b).

In conjunction with other drilling parameters, real-time annular pressure measurements improve rig safety by helping avoid potentially dangerous well control problems by detecting gas and water influxes. These measurements are often used for the early detection of sticking, hanging or balling stabilizers, bit problems, and cuttings buildup, and for improved steering performance. While real-time pressure data are of significant value, the information from these measurements is also useful in planning the next well. However, the annular pressure measurement or ECD can only yield so much information on its own. The pressure data should be displayed in the context of the other drilling variables for useful diagnosis of drilling problems (Hutchinson and Rezmer-Cooper 1998).

Recent experiences with downhole pressure data have highlighted how trends in ECD can be used to anticipate drilling problems before they develop into serious events. In addition to real-time interpretation, post-event diagnosis is still necessary as not all drilling events or problems are straightforward or can be solved in real time.

Successful drilling requires that the pressure exerted by the drilling fluid stay within a tight mud-weight window defined by the pressure limits for wellbore stability. The lower limit is either the pore pressure in the formation or the limit for avoiding wellbore collapse. The upper pressure limit for the drilling fluid is the minimum that will fracture the formation. If the drilling fluid exceeds this pressure, there is a risk of creating or opening fractures, resulting in lost circulation and a damaged formation. Fracture gradients are determined from overburden weight and lateral stresses of the formation at depth, and from local rock properties. Density- and sonic-logging data can help provide estimates of rock strengths (Brie et al. 1998). Calculating offshore fracture gradients in deep water presents a special problem. The uppermost layers are replaced by a layer of water, which is less dense than rock. In these wells, the overburden stress is less than in a comparable onshore well of similar depth, and results in lower fracture gradients, as in general, fracture gradients decrease with increased water depth. Thus, increasing the water depth reduces the size of the margin between the mud weight required to balance formation pore pressures and that which will result in formation breakdown.

Once the wellbore stability pressure window has been determined, the driller has to do more than keep the drilling fluid between these limits. To correctly interpret the response of the downhole annular pressure measurement (and the corresponding ECD), it is important to appreciate the physical principles upon which it depends. The first is a static pressure due to the density gradients of the fluids in the borehole annulus—the weight of the fluid vertically above the sensor. The density of the mud column including the solids is typically called the equivalent static density, and the fluid densities are pressure- and temperature-dependent.

Second is a dynamic pressure related to pipe velocity (swab, surge, and drillpipe rotation), inertial pressures from string acceleration or deceleration when tripping, excess pressures to break mud gels, and the cumulative pressure losses in circulating fluids and solids to the surface. Flow past restrictions, such as cuttings beds or swelling formations, changes in hole geometry, and influxes and effluxes of fluids and solids to or from the annulus all contribute to the dynamic pressure. Efficient hole cleaning is vitally important in the drilling of directional and extended-reach wells, and optimized hole cleaning remains one of the major challenges. Although many factors affect hole cleaning ability, two important ones that the driller can control are pump flow rate and drillpipe rotation.

The pump flow rate is one of the most important parameters in effective hole cleaning. Inadequate flow results in increasing cuttings concentrations in the annulus. Cuttings accumulation can lead to a decrease in the annular cross-sectional area, and hence an increase in the ECD, ultimately leading to a packoff. The use of real-time annular pressure measurements allows early identification of an increasing ECD trend, caused by inefficient hole cleaning, and can help the driller avoid formation breakdown resulting from large pressure surges, or a costly stuck-pipe event. More details on the analysis of such measurements and their applications are given in Chapter 6.3. As we shall see, the additional information from the resistivity-at-the-bit images can give an early indication that drilling practices are less than optimal, and that remedial action is necessary.

LWD images can significantly enhance interpretation not only for petrophysicists and geologists, but also for drilling engineers. Invasion profiles are more easily understood when the rock geomechanics are considered. The key to distinguishing natural fractures from drilling-induced fractures is the addition of geomechanical modeling of the failure modes. The drilling engineer can determine the problematic failure modes and the drilling procedures that cause them by referring to the drilling mechanics data (e.g., the annular pressure while drilling). As images become available in real time, their value will further increase.

Resistivity curve separation is often a first indication at the wellsite that *something* is happening to the reservoir that does not fit a preconceived paradigm. Curve separation while drilling can occur because of formation fracturing caused by high ECDs.

A classification scheme based on 3D linear elasticity when used in conjunction with wellbore stability analysis can identify stress-induced artifacts in resistivity-at-the-bit images. This analysis can be used to help validate the strength and stress profiles used to plan future wellbores. In addition, petrophysical and geological interpretations are improved by correctly distinguishing natural formation properties from common drilling-induced artifacts. The use of image interpretation can be extended to conductive oil-based muds (OBM), which, although in their infancy, show much promise.

Real-time LWD data, annular pressure measurements, and resistivity-at-the-bit images can be used to

- Identify drilling hazards
- Improve geological and petrophysical interpretation
- Distinguish between naturally occurring and drilling-induced fractures
- Monitor dynamic formation failure and invasion
- Aid in the selection of remedial methods to optimize drilling operations, including reduction of tripping speeds to minimize hydraulic impact
- Give more accurate ECD management

The next step in combining traditional formation evaluation measurements, images, and drilling mechanics measurements is to use real-time images for even earlier information concerning potential formation breakdown. Drilling practices can then be modified in real time to minimize the impact of the pressure changes.

In this chapter we examine the physical processes associated with downhole hydraulics, but also add a new component to the measurement suite that can be used for interpretation: resistivity images, where the driller can now truly *see* the dynamic behavior of the borehole. However, to interpret the images we must first understand the nature of the wellbore stresses. The state of stress around the wellbore has a direct influence on wellbore stability, and ultimately drilling efficiency.

6.1.3 Wellbore Stresses. There are two sets of principal stresses important in the analysis of wellbore rock mechanics: far-field stresses and wellbore stresses. Far-field stresses exist in the formation far away from the wellbore and are not influenced by the borehole. However, wellbore stresses act on the formation at the mud-formation interface. These stresses are controlled by the mud density (and the corresponding ECD when pumping), in addition to the far-field stresses. **Fig. 6.1.1** illustrates these two sets of stresses.

Cartesian coordinates describe the far-field stresses: one stress is vertical (σ_v), and the two orthogonal stresses are horizontal. If the magnitudes of the two horizontal stresses are different, they are termed the minimum (σ_h) and maximum (σ_H) horizontal stresses. The direction of the horizontal stress completes the total description of the far-field stresses.

In a vertical well, a cylindrical coordinate system describes the wellbore stresses. Here, one stress is radial (σ_r), and the two orthogonal stresses are axial (σ_a) and tangential (σ_t). The axial stress is directed along the axis of the borehole, whereas the tangential stress is directed along the circumference of the borehole. The two plots in the lower left-hand corner of Fig. 6.1.1 illustrate how the magnitude of the far-field stresses changes as they approach the wellbore. The left plot illustrates a low mud density or ECD (with respect to the minimum horizontal stress), and the right plot illustrates a high mud density or ECD. In both plots, the minimum horizontal stress transforms into the radial stress, the maximum horizontal stress transforms into the tangential stress, and the vertical stress transforms into the axial stress.

Stresses in the Earth are generally compressive. The grains of the formation are forced together by, for example, the far-field vertical stress, which is created by the weight of the overburden. Tensile stresses act in the opposite direction, pulling the grains apart (see the right side of Fig. 6.1.1). These two stresses cause fundamentally different mechanisms of yield and failure. Shear failure is initiated by two orthogonal stresses with different magnitudes. Tensile failure is initiated by a single tensile stress. These two different mechanisms are commonly observed in wellbore images. Shear and tensile failure are independent. A formation can fail in shear without a tensile failure, fail in tension without a shear failure, or fail in both modes sequentially

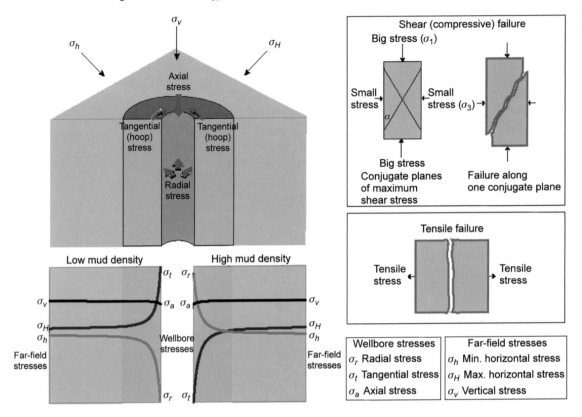

Fig. 6.1.1—Relationship of far-field stresses to wellbore stresses. Shear and tensile failure diagrams. Low mud density (or ECD) has low near-wellbore radial stresses, and high near-wellbore tangential stresses. High mud density (or ECD) has high near-wellbore radial stresses, and low near-wellbore tangential stresses (Rezmer-Cooper et al. 2000a).

or simultaneously. The geometry of the failure is dependent upon the two stresses causing the shear failure, and the single stress causing the tensile failure.

Failure Modes. There are six modes of shear failure and three modes of tensile failure for a vertical wellbore. **Fig. 6.1.2** shows a 3D representation. The modes are classified by the following (Bratton et al. 1999):

- Mode of origin (shear or tensile failure)
- Morphology (e.g., wide breakout, high-angle echelon)

The modes are as follows:

1. Shear failure—wide breakout. This is the mode more commonly called a *breakout*.
2. Shear failure—shallow knockout. The circumferential coverage is small and could be caused with a vertical fracture.
3. Shear failure—high-angle echelon. This mode makes high-angle fractures that cover up to a quarter of the borehole circumference.
4. Shear failure—narrow breakout. The annular coverage in this mode is typically less than 30°.
5. Shear failure—low-angle echelon.
6. Shear failure—deep knockout. This failure occurs in the vertical plane but is centered at the azimuth of the maximum horizontal stress.
7. Tensile failure—cylindrical. This mode is concentric with the borehole and is not visible on a wellbore image.
8. Tensile failure—horizontal. This mode creates horizontal fractures.
9. Tensile failure—vertical. This mode is exploited by hydraulic fracturing techniques.

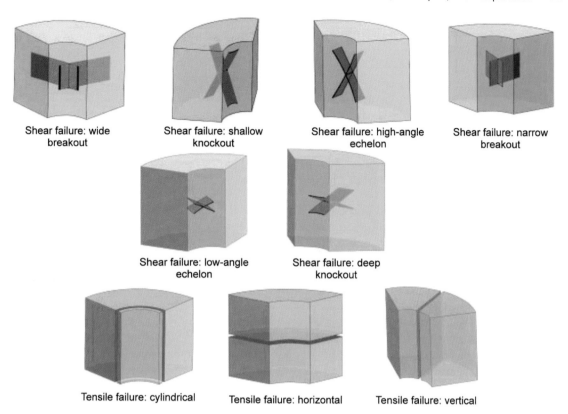

| Shear failure: wide breakout | Shear failure: shallow knockout | Shear failure: high-angle echelon | Shear failure: narrow breakout |

Shear failure: low-angle echelon   Shear failure: deep knockout

Tensile failure: cylindrical   Tensile failure: horizontal   Tensile failure: vertical

Fig. 6.1.2—Shear and tensile failure modes for vertical wellbores (Rezmer-Cooper et al. 2000a).

The stresses applied to the formation are controlled by two sources that are different in origin. The radial stress is caused by the pressure of the drilling mud, is largely controlled by the driller, and is generally dictated by the specific needs of the operation. This pressure can be measured in real time by the annular pressure-while-drilling sensor. The axial and tangential stresses are controlled by the far-field stresses.

The various modes of failure can therefore be associated with distinct mechanisms. This aids interpretation of the resistivity image.

6.1.4 Resistivity Images. The highest-resolution images available using LWD technology are recorded in the resistivity-at-the-bit tool. This tool makes electrode resistivity measurements in wells drilled with water-based (or conductive) mud. It may be run in any of several configurations, providing up to five resistivity measurements. Primarily, these measurements are intended for wellbore positioning and formation evaluation. An azimuthal orientation system uses the Earth's magnetic field as a reference to determine the angular position of the tool in the wellbore as the drillstring rotates. This allows the tool to make azimuthal resistivity measurements. Auxiliary measurements of radial and longitudinal shock and temperature are also made. The angular resolution as the tool rotates is approximately $6°$.

The tool provides three different types of resistivity measurement:

- Bit resistivity uses the bit and the lower few inches of the tool as a measure electrode. The resultant resistivity measurement has a measure point at the midpoint of this electrode, and with a vertical resolution equal to its length.
- Ring resistivity is an accurate, high-resolution, focused lateral measurement made by a 1.5-in.-tall, cylindrical electrode approximately 3 ft above the lower end of the tool. The investigation diameter of the ring is approximately 22 in. Alternatively, we could express this by saying that the depth of investigation is approximately 11 in.

- Three *button* electrodes provide focused, lateral resistivity measurements, and reside in a clamp-on sleeve. The buttons are longitudinally spaced to provide staggered depths of investigation. The tool records azimuthal data simultaneously at three depths of investigation (1, 3, and 5 in.).

Azimuthal button measurements are displayed as full-bore images of formation resistivity. These data are represented as *shallow*, *medium*, and *deep* images. Resolution is decreased by a factor of five in comparison to wireline imaging tools. All resistivity measurements can be acquired every five seconds. In addition to imaging fractures, stratigraphic features are sometimes visible on the at-bit resistivity images. For example, information has been obtained on slumped/folded sections, unconformities, and changes in lithology. The principal applications for resistivity images are structural geology, geosteering (see Chapter 6.2), and the diagnosis of borehole failure mechanisms, while the well is under construction.

6.1.5 Natural and Drilling-Induced Features. While drilling, it is important to distinguish natural features from those induced by the drilling process so that the drilling program can be modified to minimize the impact of the induced fractures. The left side of **Fig. 6.1.3** shows a sequence of shales and sands. The gamma ray is plotted in the depth track. The yellow shading extending from the gamma ray to the right edge of the track indicates a sand. The shallow resistivity measurement reads lower than the deep resistivity in the sands, while the measurements overlay in the shales. Resistivity curve separation allows estimation of petrophysical properties in what seems to be a typical invasion profile. The right side of Fig. 6.1.3 shows the corresponding resistivity image. The light colors indicate high resistivity, and the dark colors low resistivity. The orientation is typical for deviated wells, with the left edge of the track aligned with the top of the hole (the letters at the top of the image indicate the Upper, Right, Bottom, and Left quadrants of the borehole).

The addition of the image to the standard log curves confirms that the curve separation is not due to invasion by the drilling fluid, but due to borehole breakout. The shallow resistivity is strongly affected by the defects in the borehole wall, which are filled with conductive mud. However, the zone near the top of the interval is invaded; there is curve separation without an indication of breakout.

A geological analysis of borehole images includes the search for open natural fractures. Wrongly identifying drilling-induced fractures as natural fractures results in an optimistic forecast and could lead to incorrect remedial procedures being recommended for the drilling program. **Fig. 6.1.4** shows sinusoidal features that

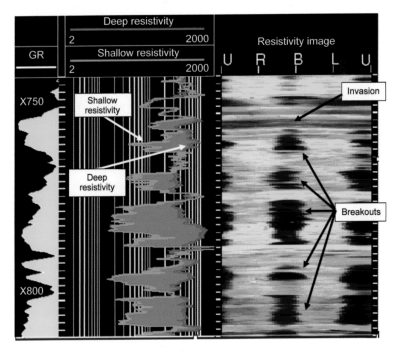

Fig. 6.1.3—Resistivity profiles and images for a sequence of shales and sands (Rezmer-Cooper et al. 2000a).

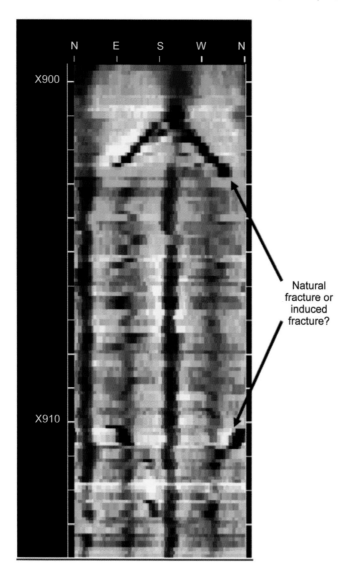

Fig. 6.1.4—Resistivity images of fractures (Rezmer-Cooper et al. 2000a).

could be drilling-induced or natural features. Geomechanical analysis will predict the depth, azimuth, and extent of drilling-induced fractures. The recognition of drilling-induced fractures and knowledge of their influence on logging measurements greatly improves the geological and petrophysical information, and also leads to a more efficient drilling process. Before we can deduce the nature of the features shown in Fig. 6.1.4, we must examine the borehole stresses.

Fig. 6.1.5 shows a plot of the wellbore stresses as a function of the ECD. In a vertical well, the radial stress increases with the ECD, the tangential stress decreases with the ECD, and the axial stress is independent of the ECD. Failure occurs when the wellbore stresses exceed the formation strength. Two classes of failure, shear and tensile, are commonly observed. Shear failure occurs when the shear stress exceeds the shear strength. Shear stress, proportional to the difference between the maximum and minimum wellbore stress, is highlighted in yellow. Tensile failure occurs when any stress becomes greater than the tensile strength of the formation. By convention, tensile stresses are negative, so any stress less than zero is tensile and is indicated by the hashing. The geometry of the failure changes with the ECD because the stresses that cause the failure change orientation. We examine in detail one of the failure modes, the wide breakout.

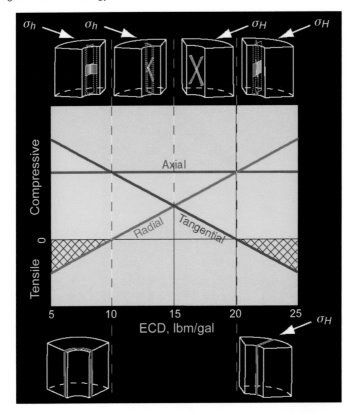

Fig. 6.1.5—Wellbore stresses as a function of the ECD (Rezmer-Cooper et al. 2000a).

Wide Breakouts. The wide breakout is a shear failure caused by a low ECD. The tangential stress is very large, while the radial stress is very small (Fig. 6.1.5). This large differential stress causes the failure. The failure occurs at a small angle to the maximum stress, the tangential stress, and is indicated by the green crosses in **Fig. 6.1.6**. Wide breakouts occur in the direction of the minimum horizontal stress. Shallow knockouts also occur in this direction. The failure modes due to a high ECD occur in the direction of the maximum horizontal stress. The resistivity image associated with a shear failure-wide breakout is given in **Fig. 6.1.7**. Drilling-induced features tend to be best imaged with the shallow button. As the depth of investigation increases, the feature goes away, which highlights that the mud-filled breakout is a near-wellbore event (Fig. 6.1.7).

Fig. 6.1.8 is an example of a vertical well drilled into a basin with unbalanced horizontal stresses. The maximum horizontal stress is approximately 20% greater than the minimum horizontal stress. Multiple failure modes are evident in this image. Wide breakouts are visible in the upper section. Offset from the wide breakout by 90° is a vertical fracture. Other fracture modes are also visible. To explain how one can have both breakouts and fractures in the same bit run we must examine the stress plots. The well was drilled with a static mud weight of 9.5 lbm/gal.

The stress plot in **Fig. 6.1.9a** shows the wellbore stresses at the central azimuth of the breakout. While the bit run was drilled with a static 9.5-lbm/gal mud, the ECD varied from approximately that value to nearly 12.5 lbm/gal. This range is shown in yellow. If failure occurs in the lower half of this range, the geometry of failure will be a wide breakout because the maximum stress is tangential, and the minimum stress is radial. If failure occurs in the upper half of this range, the geometry of the failure will be a shallow knockout, which looks very much like a vertical fracture, but occurs on the same azimuth as the breakout, the direction of the minimum horizontal stress.

The stress plot in Fig. 6.1.9b shows the wellbore stresses at the azimuth of the vertical fracture, the direction of the maximum horizontal stress. At this azimuth, the tangential stress is quite different, and much lower than the tangential stress in the minimum horizontal stress direction. Here, the tangential stress is negative, or

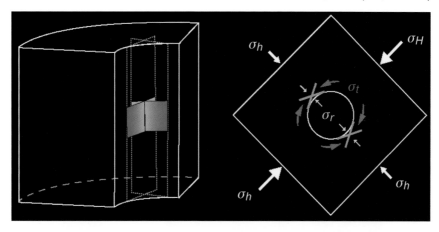

Fig. 6.1.6—Mechanism of a shear failure: wide breakout (Rezmer-Cooper et al. 2000a).

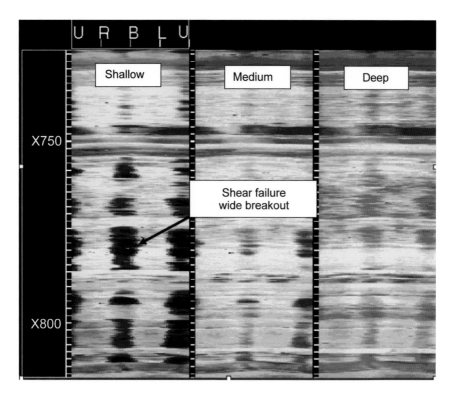

Fig. 6.1.7—Resistivity images of a shear failure: wide breakout. The resistivity images vary as a function of the depth of investigation (Rezmer-Cooper et al. 2000a).

tensile, in the upper half of the ECD range, causing a vertical fracture. Investigation of the image on an expanded vertical scale clearly highlights the different failure mode geometries.

Fig. 6.1.10 shows the interval where the formation is failing in the upper part of the ECD range. We would expect a shallow knockout in the direction of the minimum horizontal stress, and a vertical fracture in the direction of the maximum horizontal stress. The geometry of these two modes is identical, so they appear the same in the image. However, they are offset by 90°. In the upper part of the image, the shear failure mode changes to a wide breakout.

Fig. 6.1.11 shows the answer to the question posed by Fig. 6.1.4. The two sinusoids are drilling-induced fractures. They originate at the azimuth of the maximum horizontal stress, and extend away at a high angle,

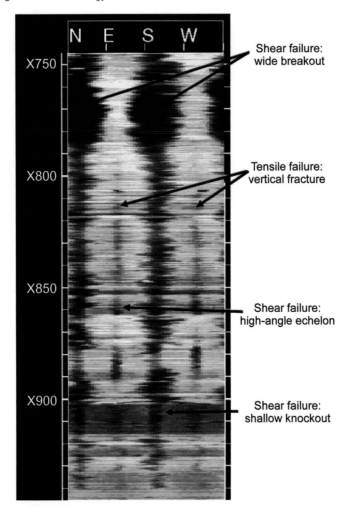

Fig. 6.1.8—Example resistivity images from a vertical well drilled into a basin with unbalanced horizontal stress (Rezmer-Cooper et al. 2000a).

at approximately 60°. The vertical fracture is much less obvious, with the high-angle echelon features more dominant.

Now that we can identify both the features and the mechanisms that appear on the resistivity image, we can start to combine the image with the drilling mechanics measurements to understand the causes behind the drilling-induced fractures. Time-lapse logging highlights the dynamic changes in the wellbore properties and can be vital in diagnosing the onset of wellbore problems.

6.1.6 Time-Lapse Logging. Curve separation in resistivity logs is often the first indication that something is happening in the reservoir that does not fit a preconceived paradigm. Curve separation can result from many causes:

- Anisotropy with high apparent formation dip
- Close proximity of tight streaks undetected in pilot wells
- Permeability variations in carbonate reservoirs
- Dynamic formation fracturing by heavy mud (high ECDs)

Tabanou et al. (1997) use real and simulated examples to illustrate the added value of 2-MHz multispacing resistivity logs during and after drilling, and subsequently evaluate the resistivity curve separation vs. time.

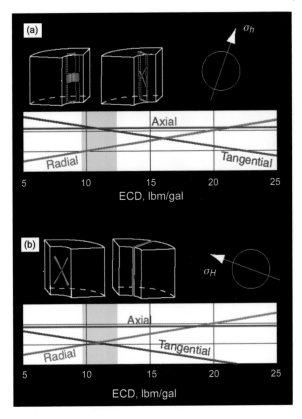

Fig. 6.1.9—Wellbore stresses: (a) at the central azimuth of the breakout shown in Fig. 6.1.8 and (b) at the azimuth of the vertical fracture shown in Fig. 6.1.8 (Rezmer-Cooper et al. 2000a).

They discuss one particular example from an offshore well drilled in the Gulf of Mexico with heavy OBM. The curve separation on the resistivity log was thought to be due to the heavy OBM fracturing the shales. Losses were observed while pumping by the driller. Fractures developed around the hole, most of them as *wings* propagating in the plane of the wellbore trajectory. As these fractures filled with OBM, they blocked the circulation of the 2-MHz currents, with the shallow phases affected more than the deep. With time, the fracture wings extended and affected all of the measurement depths.

Clearly, combining the annular pressure measurement with the above observations would have been beneficial in determining both the true leakoff pressure and the true effective mud weight at the depth of fracture. This would have enabled more precise modifications to be made to the mud or drilling program to minimize the losses, or perhaps even identified the potential for losses before they occurred. The next stage is to use images and annular pressure to identify the early onset of drilling-induced fractures, and to clearly distinguish them from naturally occurring fractures.

6.1.7 Horizontal Well Example. So far, we have identified the various failure mechanisms and highlighted how modeling can be used in conjunction with resistivity images to differentiate between drilling-induced features and naturally occurring features. However, there may still be confusion concerning the mechanism causing the fracture. Now we can look at combining the images and the annular pressure-while-drilling measurement. **Fig. 6.1.12** shows the trajectory of a horizontal well drilled in chalk, where the operator was searching for natural fractures (Rezmer-Cooper et al. 2000a). An axial fracture is faintly visible on the top and bottom of the horizontal wellbore **(Fig. 6.1.13)**. The image has been shifted 90° in this figure and is displayed *left, up, right, bottom,* so that the two wings of the axial fracture are more visible. Note that this fracture extends over a 1,000-ft interval. In addition, a few intervals show a much larger feature, but over a shorter 60-ft interval. To best interpret these data, all LWD measurements should be integrated.

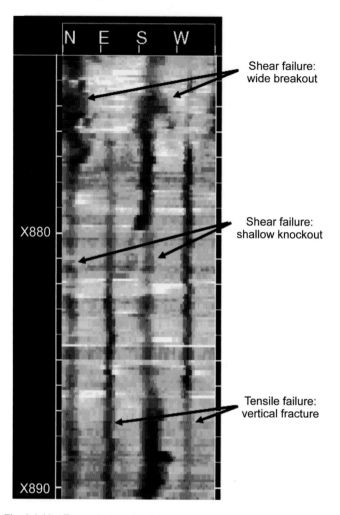

Fig. 6.1.10—Expanded scale of Fig. 6.1.8 (Rezmer-Cooper et al. 2000a).

The green line plotted on **Fig. 6.1.14** shows the depth of the bit as a function of time. The deep button sensor was 53 ft above the bit. Downhole annular pressure was recorded during this bit run, and Fig. 6.1.14 also displays (in white) the derived ECD during the run.

The LWD data normally presented are the data recorded by the sensor the first time it passes a new, deeper depth. During the first 1.75 hours of this time sequence, the well was drilled from X1,933 to X2,017 ft and imaged from X1,880 to X1,964 ft. The lower dashed line indicates the bottom of the wellbore during this time interval; the upper dashed line shows the sensor position when the bit is at the bottom of the well. The image shown in Fig. 6.1.14a was acquired within an hour of the bit penetrating the formation, and shows a very faint fracture. For the next 6 hours the BHA was raised and lowered a number of times. At approximately 8 hours, drilling continued, and the interval drilled 7 hours earlier was finally imaged (Fig. 6.1.14c). This section of the log shows a wide induced fracture in addition to the sought-after natural fractures, which appear as low-angle sinusoids. Fig. 6.1.14c was acquired while drilling between 7.75 and 8.75 hours. This time-lapse image clearly shows that the fracture was enlarged soon after drilling. However, the borehole was open 6 hours longer than the interval above X1,964 ft and the interval below X2,040 ft. The axial fracture seems to have extended during this time interval. Although the data typically presented are the data recorded by the sensor the first time it passes a new depth, the time-lapse data are also available, and we are often interested in what the image would show if we looked at the data during the working of the pipe (Fig. 6.1.14b). Note the dramatic change in the image between Fig. 6.1.14a and Fig. 6.1.14b while the pipe was worked up and down. It

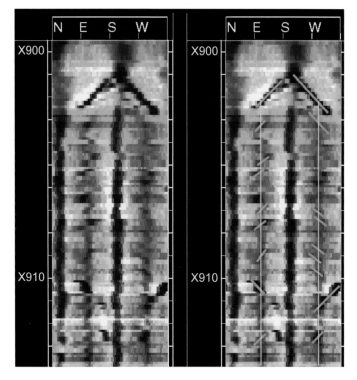

Fig. 6.1.11—Drilling-induced fractures (cf. Fig. 6.1.4) (Rezmer-Cooper et al. 2000a).

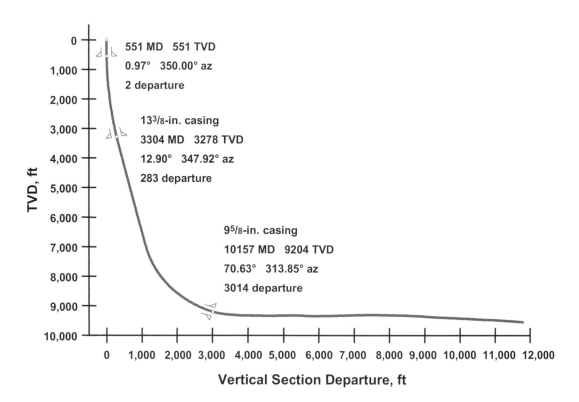

Fig. 6.1.12—Vertical section of well used for combined resistivity-at-the-bit and annular-pressure interpretation (Rezmer-Cooper et al. 2000a).

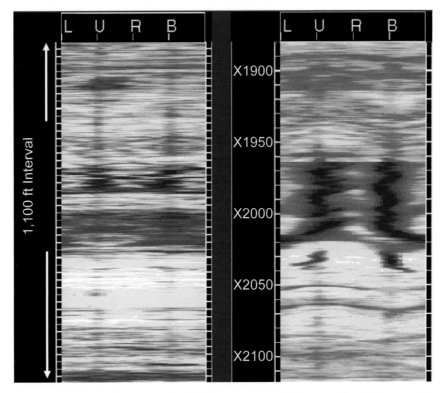

Fig. 6.1.13—Axial fracture in a horizontal well. Right images acquired 6 hours after left images (Rezmer-Cooper et al. 2000a).

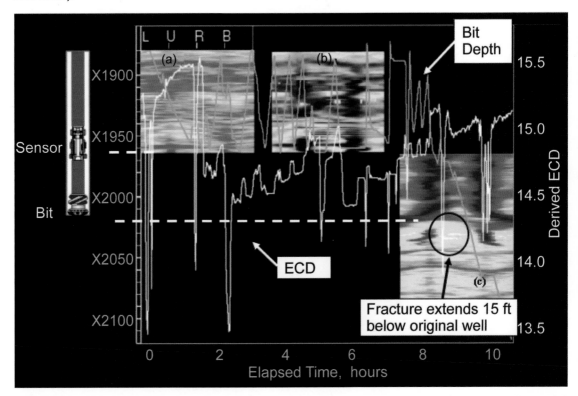

Fig. 6.1.14—Time-lapse images and annular pressure while drilling show growth of hydraulic fracture due to high ECDs. Depth of deep button resistivity sensor as a function of time. Central images acquired during working of the pipe (Rezmer-Cooper et al. 2000a).

appears that the fracture has been extended away from the wellbore from approximately X1,900 to X2,020 ft. It also appears that the fracture extended 15 ft beyond the bottom of the hole when it was created (from X2,020 to X2,035 ft) during the earlier stages of drilling.

The ECD varied between 13.5 and 15.5 lbm/gal during this bit run. The buildup in the ECD is clearly visible during the drilling of the top interval, and the ECD achieved at 1.5 hours was the maximum during the drilling of this section of the wellbore. Severe losses occurred in this well section every time the flow rate increased above a certain level. In this field, the tolerance between the pore pressure and the fracture gradient pressure is so small that the high flow rates and surge pressures led to high ECDs. The ECD can increase to the point where a hydraulic fracture is induced, as happened here. Eventually, the driller managed to reduce the ECD by working the pipe and circulating, and drilling resumed. The natural fractures that were the original focus of the imaging also appear in Fig. 6.1.14c; they appear as low-angle sinusoids.

Nomenclature

σ_a = axial stress
σ_h = minimum horizontal stress
σ_H = maximum horizontal stress
σ_r = radial stress
σ_t = tangential stress
σ_v = vertical stress

References

Bratton, T., Bornemann, T., Li, Q., Plumb, D., Rasmus, J., and Krabbe, H. 1999. Logging-While-Drilling Images for Geomechanical, Geological and Petrophysical Interpretations. *Trans.*, SPWLA 40th Annual Logging Symposium, Oslo, Norway, 30 May–3 June, Paper JJJ.

Brie, A., Endo, T., Hoyle, D. et al. 1998. New Directions in Sonic Logging. *Oilfield Review* **10** (1): 40–55.

Hutchinson, M. and Rezmer-Cooper, I.M. 1998. Using Downhole Annular Pressure Measurements to Anticipate Drilling Problems. Paper SPE 49114 presented at the SPE Annual Technical Conference and Exhibition, New Orleans, 27–30 September. DOI: 10.2118/49114-MS.

Rezmer-Cooper, I., Bratton, T., and Krabbe, H. 2000a. The Use of Resistivity-at-the-Bit Images and Annular Pressure While Drilling in Preventing Drilling Problems. Paper SPE 59225 presented at the IADC/SPE Drilling Conference, New Orleans, 23–25 February. DOI: 10.2118/59225-MS.

Rezmer-Cooper, I.M., Rambow, F.H.K., Arasteh, M., Hashem, M., Swanson, B., and Gzara, K. 2000b. Real-Time Formation Integrity Tests Using Downhole Data. Paper SPE 59123 presented at the IADC/SPE Drilling Conference, New Orleans, 23–25 February. DOI: 10.2118/59123-MS.

Tabanou, J.R., Bruce, S., Bonner, S., and Wu, P. 1997. Time Lapse Opens New Opportunities in Interpreting 2-Mhz Multispacing Resistivity Logs Under Difficult Drilling Conditions and in Complex Reservoirs. *Trans.*, SPWLA 38th Annual Logging Symposium, Houston, 15–18 June, Paper II.

SI Metric Conversion Factors

cycles/sec	× 1.0*	E + 00 = Hz	
ft	× 3.048*	E – 01 = m	
gal	× 3.785 412	E – 03 = m³	
in.	× 2.54*	E + 00 = cm	
lbm	× 4.535 924	E – 01 = kg	

*Conversion factor is exact.

6.2 Geosteering—W.G. Lesso Jr., Schlumberger

6.2.1 What Is Geosteering? It is a simple idea: take the trajectory design of a horizontal well and update the trajectory during drilling for optimum placement of the wellbore in the reservoir. Geosteering can be more precisely defined as "the drilling of a horizontal, or other deviated well, where decisions on well path adjustment are made based on real-time geological and reservoir data" (Peach and Kloss 1994).

In conventional directional and horizontal drilling, the well path is steered according to a predetermined geometric plan. The objective is to follow *the line* as closely as possible to a well defined target with equally well defined tolerances in three dimensional space. Geosteering is a departure from this convention. It is required when the reservoir target is ill defined, tolerances are tight, or the geology is so complicated as to make conventional directional drilling targeting impractical (see **Fig. 6.2.1**).

Geosteering implies continuous feedback of all available data into the model for the well path and reservoir. This directional drilling management is judged on the productivity of the well rather than adherence to curves on a plot or even amounts of dogleg severity. Potential gains in production must be balanced against additional drilling and formation evaluation costs. Proper characterization requires knowledge of where the well path is located, where the current trajectory plan will take the well path, and where the wellbore should go. Uncertainty in geological modeling and the need to maximize profitability require an interdisciplinary team approach.

Conventional directional drilling plots also do not adequately display geosteering decisions so that they can be understood. The emphasis in directional plotting has always been on maintaining a one-to-one aspect ratio between the vertical scale [true vertical depth (TVD)] and the horizontal scale (section or displacement). Horizontal steering changes are quite small in relation to drilling footage as reservoir thickness is usually much smaller than its length. A 10-ft-thick (3.05-m-thick) sand lens probably runs for several thousand feet. Ratios between 5:1 and 20:1 are common (see **Figs. 6.2.2 and 6.2.3**). It is important to distinguish these types of directional plots from those used in conventional directional drilling. They are *geosteering schematics*.

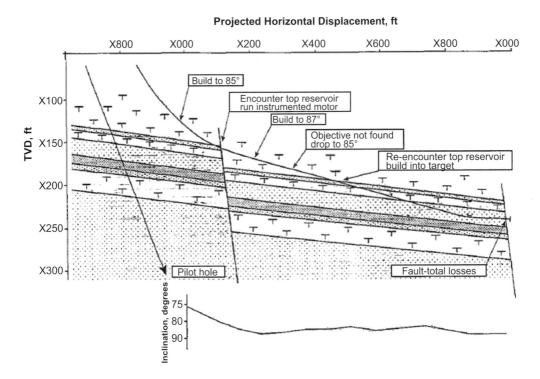

Fig. 6.2.1—An early example of geosteering. Correlation with marker beds when the wellbore is at a high inclination, but not horizontal, allows for changes and precise control of the final build into the target (Peach and Kloss 1994).

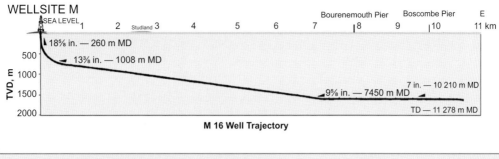

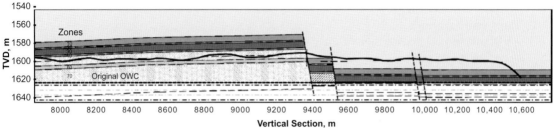

Fig. 6.2.2—An extended-reach geosteered well encountering substantial faulting and large changes in elevation or TVD near the reservoir. Note that on the trajectory plot at the top, the path looks like a simple horizontal well. It is only when the vertical scale (TVD) is greatly exaggerated in relation to the horizontal scale (displacement) on the bottom geosteering schematic, that the complicated nature of geosteering decision-making begins to appear (Meader et al. 2000).

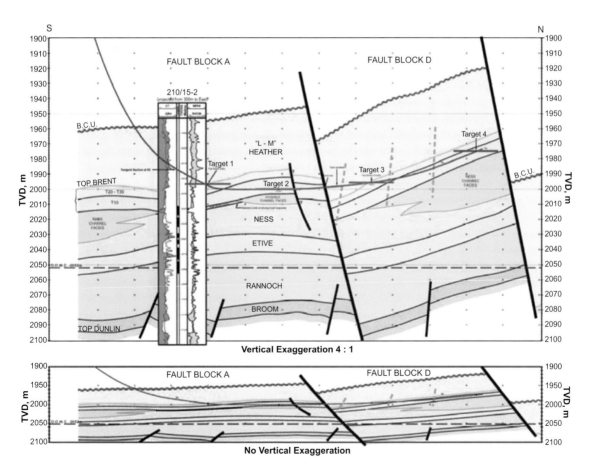

Fig. 6.2.3—A second geosteering schematic showing the detail that can be added (top). The 1:1 aspect ratio plot is below. Comments, log data, better lens definition, and trajectory options help in the placement decision process (Tribe et al. 2003b).

6.2.2 The Development of Geosteering. Geosteering was a natural development from the increased ease in drilling horizontal wells that occurred in the late 1980s with the introduction of steerable downhole motors. Operators found that they could land horizontal wells in large reservoir sections with huge increases in production as compared to a vertical or standard directional well. It was natural that horizontal projects were attempted in thinner reservoir sections. Several notorious cases occurred where the drainhole section ended up missing a 10-ft-thick (3.05-m-thick) sand altogether, resulting in very disappointing production. Groups from two disciplines, directional drilling and petrophysical formation evaluation, working independently in different parts of the world developed three techniques to resolve this issue. These approaches can be labeled as

1. Petrophysical modeling
2. Horizontal navigation
3. TVD, true vertical thickness (TVT), and true stratigraphic thickness (TST) analysis

Petrophysical modeling was developed by those modeling log responses in high wellbores angles. Resistivity logs are more sensitive to the angle at which they measure stratigraphic formation beds. They have additional character such as resistivity horns that, if understood, help in correlation or determination of placement within a bed.

Horizontal navigation originated in the directional drilling community. Correlations were made from offset wells and pilot holes that allowed for the plotting of bed boundaries on a vertical section directional plot. The proximity to a bed boundary or placement within the bed would be calculated by measuring distances on the plot. Directional projections allowing for various curve rates could then be used to determine footage before encountering or leaving a bed.

TVD, TVT, and TST analyses came from those wanting to strictly correlate log plots. Measured depth logs from high angle or horizontal wells recalculated on one of these indices allow placement decision makers to correlate offset well data with these logs much as they had done with vertical or standard directional wells.

A project team would use one of these techniques depending on the problems in the particular case, measurement data available, and the dominant group in the multidisciplinary team running the project.

It is not surprising that geosteering developed from both directional drilling and high-angle petrophysical log analysis. Both groups were sensitive to criticisms that the horizontal drainhole was not being placed in the intended formations. Geologists and log analysts are usually the economic buyers of a horizontal well, and the problems of high-angle resistivity data were a natural starting place for developing methods for improving the accuracy of horizontal well placement.

6.2.3 Petrophysical Modeling. Modeling petrophysical measurements grew out of the need to understand the anisotropic effects of intersecting formation stratigraphic boundaries in high-angle wells or at a low angle of incidence. These anisotropic effects are most pronounced in resistivity data. Basically, the tools will read one value when the tools are perpendicular to the stratigraphic beds (R_h) and a different value when they are parallel to the beds (R_v). The higher the ratio of R_v to R_h, the more anisotropy a resistivity sensor has in a formation.

All resistivity logging tools to date were developed to be run with the tool normal to bedding planes. While reservoir properties vary along the length of the tool, properties radially away from the tool are assumed to be constant. Boundary and thin-bed effects exist for many such logs. Additionally, environmental characteristics affect log data. Tools can be assumed to be centered in vertical wellbores, while they are most likely laying on the low side of the hole in horizontal wells. Cuttings buildups and conveyance of the logging tools are also different between vertical and high-angle wells. Most nuclear tools make measurements along preferential azimuths, increasing these effects.

But it is the anisotropic effects on resistivity devices that cause the most confusion in horizontal well evaluation. Deep investigation is a desired feature of resistivity logs in vertical wells to determine the true value of formation resistivity. Deep investigation for horizontal wells crosses multiple formation layers, which may result in anomalous features. While these features may complicate the evaluation, they may also improve the ability to geosteer the well by *seeing* the formations above and below the bit location. One unique aspect of the use of dual-propagation resistivity logging while drilling (LWD) logs in horizontal wells is the existence of a *polarization horn*, a spike of high resistivity as the tool crosses bed boundaries with sharp resistivity contrasts (see **Fig. 6.2.4**). In vertical wells, bed boundaries are horizontal and parallel to current

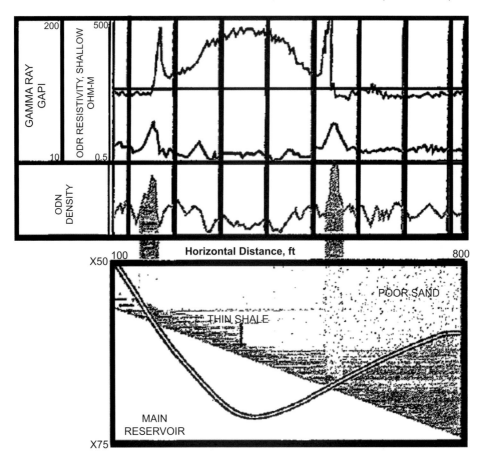

Fig. 6.2.4—Actual resistivity and gamma ray from a geosteered well where the resistivity contrast is high between the main reservoir and poor sand above it. Note the resistivity horns at the bed boundaries. Comparing these data with modeled calculations allows for easier determination of the bed boundaries (Meehan 1994).

loops. In horizontal wells, the bed boundaries cut the current loops and allow a charge buildup at the bed interface. This secondary field build is caused as the tool moves along the formation interface. This tool response in such formations can be modeled.

The response that individual LWD tools have in beds where the wellbore approaches horizontal can be modeled to show the expected resistivity values and any boundary effects. The process involves determining the anisotropic response for each type of resistivity measurement. Various service companies have done this, and these response models are usually quite complex. A resistivity model is often made from resistivity measurements in an offset well or a composite of several offset wells. Those resistivity data, usually multiple depths of investigation, are used to calculate a true resistivity, R_t. This is the actual resistivity of the formation. It is most closely approximated when a sensor package is nearly perpendicular to the formation beds, or R_h and the bed is infinitely thick. The response model for the tool used in the offset well(s) is needed to calculate R_v. Next, a stratigraphic column is defined for the offset well(s). The offset log interval where the horizontal well is expected to be placed is divided into discrete elements where R_t can be expected to be constant. These intervals can be less than 1 ft (0.30 m) in thickness up to 10 ft (3.05 m) thick. These constant R_t values result in a squared-off log showing step changes in resistivity. These R_t values are then used with the response models for the resistivity tools that are expected to be used in the planned horizontal well as shown in **Fig. 6.2.5**. Usually a table is calculated for a range of angles close to horizontal, say 85 to 95° or ±5° of incidence in one degree increments.

6.2.4 Navigation Computations. Geosteering navigation is based on adding formation top or structural data to directional plots. Log correlations are made between offset wells and/or pilot wells and the horizontal well.

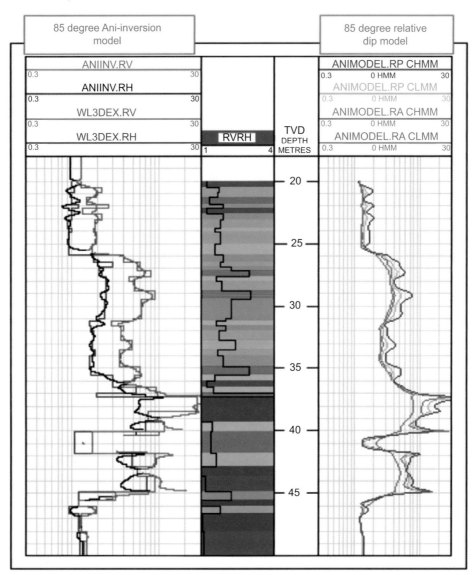

Fig. 6.2.5—An example of offset well resistivity modeling for use in geosteering. Original resistivity data are used to calculate values of vertical and horizontal true resistivity, shown as red and black curves respectively in the left track. These curves are then squared off to reflect a constant value for individual stratigraphic beds (blue and green curves). These R_v and R_h bed values and their ratio (center track) are then used to model a specific tool response at horizontal or near-horizontal well angles. In this case, the right-hand track shows the calculated tool responses for four resistivity depths of investigation curves at 85° (Page and Benefield 2002).

These measured depth pairs are then converted to 3D Cartesian coordinates using survey interpolation calculations so that the points can be plotted on the well traces on a vertical section plot. A line connecting the correlation among the wells effectively shows the formation boundary and apparent dip. Faults can be introduced and the bed boundaries shifted as an interpretation (see **Fig. 6.2.6**). Distances can be directly measured or calculated between the horizontal well and any bed boundary. Directional projections can be made during the drilling of the well based on current build/drop/turn rates to determine the distance in measured depth before a bed boundary is encountered. When a bed boundary is inadvertently encountered, directional projections for various curve rates can be used to calculate the footage necessary to drill before the bed is entered again. Each additional bed boundary determination adds to the correlation information and improves the structural model.

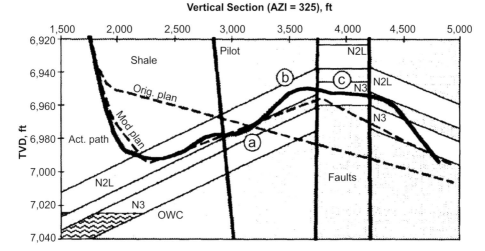

Fig. 6.2.6—Steering through a complex reservoir section using navigation techniques. Correlation and section calculations using data from a pilot well helped determine that the well exited the desired formation at point "a." Additional calculations showed that the well was approaching a faulted section, "b" and "c" (Lesso and Kashikar 1996).

6.2.5 Log Correlation Methods (TVD, TVT, and TST). Correlation of measured depth logs is only practical in vertical wells. As directional wells were drilled, logs were recalculated on a TVD index, effectively making the well appear vertical. Correlations can be made on directional wells up to approximately 70° of inclination. At higher angles and approaching horizontal or 90°, these TVD changes are small until ultimately all the log data are compressed into a single point or TVD. Additionally, the strike and dip of a bed has a larger effect on wells at higher angles and on different azimuths.

This led to the definition of a second log index calculation, TVT. TVT is defined as the vertical thickness of a bed along a constant azimuth or compass bearing. It is more difficult to use as the dip and strike of the bed must be known or interpreted. TST takes this process one step further. TST is the thickness of a bed as measured perpendicular to the bedding plane. It is the actual thickness of the bed independent of dip and strike. Its main disadvantage is that it is not workable over long intervals, but this diminishes when one is only looking at correlations within a single bed or series of closely linked beds. These differences between TVD, TVT, and TST are shown in **Fig. 6.2.7**.

The log correlation methods are usually used with petrophysical data that do not have significant anisotropic effects. They work well in simpler geosteering projects where only gamma ray data are used. They can be quite effective, and geosteering decision makers from a petrophysical background are experienced and comfortable with log correlation techniques.

6.2.6 Comprehensive Plots. The results of these three approaches to geosteering are usually summarized in a comprehensive plot of the data that were used in the geosteering process for the drilling of the horizontal section. These plots are constructed either in a log format (**Fig. 6.2.8**) or along a directional trajectory vertical section format (**Fig. 6.2.9**), and often a combination of the two, but one approach is dominant.

These plots will contain petrophysical data, dip calculations, images, lithology, drilling and mud-logging data, and trajectory calculations. The plots are a record of the well and useful in reservoir and production calculations as well as diagnosing production problems such as water coning later in the life of the well. They are not usually a good indication of the decisions in the actual geosteering process.

6.2.7 Real-Time Operations. Geosteering decisions in real time were initially recommendations that came from the analysis of log and trajectory data that were reformatted on various paper plots. Multiple revisions of these interpretations led to a lot of record keeping. It became apparent that these analyses needed to move from paper plots to video displays, and several service companies developed *geosteering screens*. They are usually divided into three parts: (1) The offset petrophysical data are usually shown in a vertical column with

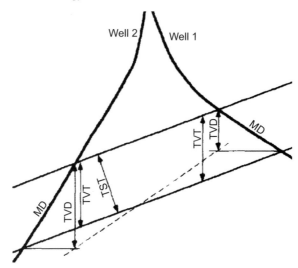

Fig. 6.2.7—The various depth index calculations that can be used in scaling log data for correlation between two wells. Measured depth, MD, only works well in vertical wells. TVD is used in standard directional wells. TVT and TST are used in horizontal geosteered wells where formation dip angle and strike make correlation difficult using TVD logs (Lesso and Kashikar 1996).

or without modeled data for anisotropic effects. This plot is usually displayed on the left-hand part of the screen. It is often scaled in TVD, although it could be in TVT or TST. (2) A vertical section display shows the current well trajectory and formation tops. (3) The current LWD log data, often with the offset modeled data *stretched* to match or correlate with the current measurements, is displayed on a vertical section index, which is the same scale used in the trajectory plot, usually below it. Thus, correlation lines could be drawn from markers in the offset data vertical column right to its point on the trajectory. A second line could then be drawn from this trajectory point up to the current log (with model) and match the same marker. This could only happen if the geological structure was mapped correctly in the trajectory plot and, thus, the LWD log data on the resulting vertical section index would correlate. Examples are shown in **Figs. 6.2.10 and 6.2.11.**

Geosteering screen software was set up to allow the worker to select marker correlation pairs on both the offset vertical log and the real-time LWD log. Using input layer-cake data on formation tops and thicknesses, these software applications would use various techniques to adjust the dips of the formations to obtain a unique correlation. The simpler models assumed uniform or parallel bed thickness in two dimensions without faulting. As these products matured, moer complex features such as 3D characteristics, pinchouts in formations, and fault models were added and could be used by the worker. This increased the number of possible solutions to the correlation picks and the need for more sophisticated interpretation by the geosteering coordinator.

As the technology matured, azimuthal measurements became available to the geosteering worker. Gamma ray or resistivity sensors were packaged in a preferential direction along a LWD collar. When a new (or unexpected) formation boundary was encountered, drilling could be stopped and the boundary investigated. The BHA would be oriented so that the sensors could measure the top and bottom of the hole. A test (see **Fig. 6.2.12**) may show that the gamma ray readings on the top of the hole are higher than on the bottom of the hole. This would indicate that the sand is below the wellbore and in order to re-enter the sand, the well path should be turned down and back into the sand.

6.2.8 Wellbore Imaging and Dip Calculations. The next tool to be applied to geosteering work was wellbore images. Resistivity and density variations obtained from azimuthal measurements are displayed as an image of the wellbore wall where color coding denotes the measured value (see **Figs. 6.2.13 and 6.2.14**). The smaller the angle between the azimuthal measurements, the finer the detail in the image and the larger the size of the data. Images were initially only available in recorded mode when the data was dumped from the tool on the surface after the bottomhole assembly run. This was necessary due to the large amount of data recorded. Advances in real-time measurement-while-drilling (MWD) telemetry later

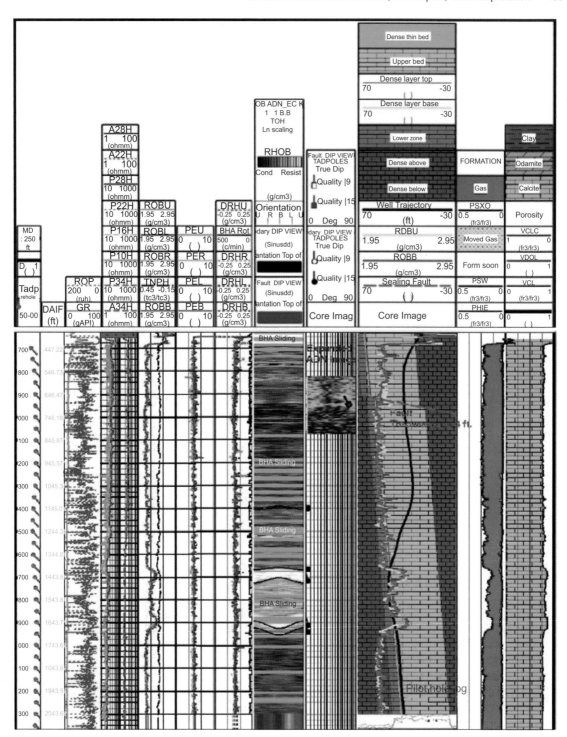

Fig. 6.2.8—A comprehensive log combining directional, drilling, petrophysical, dip, image, and lithology computations of a horizontal well section. Geosteering can require all of these data to make trajectory decisions. This complexity is difficult to manage without real-time software tools (Efnik et al. 1999).

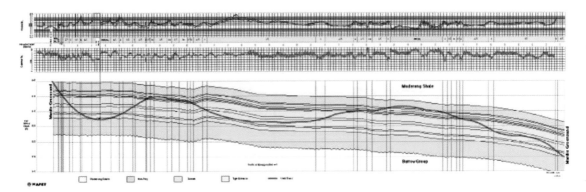

Fig. 6.2.9—Another comprehensive log, this time combining petrophysical data with directional trajectory and formation dip computations. Note the difference in the emphasis between this figure and Fig. 6.2.8 (Tippet and Beacher 2000).

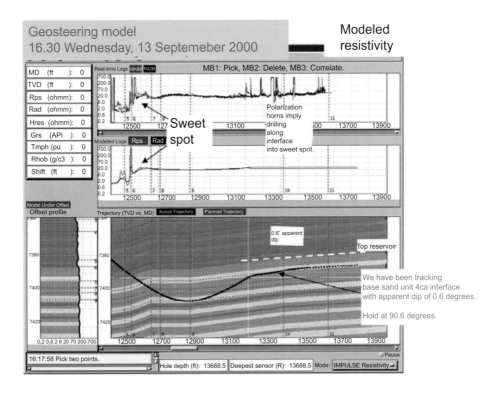

Fig. 6.2.10—An example of a real-time geosteering screen software product that can help with decision making on placing the well path. The formation vertical column is shown in the lower left. To the right of that is the trajectory section with the formation dip interpretation that was obtained by forcing a match between the modeled (middle plot) and actual LWD (upper plot) resistivity data (Tribe et al. 2003a).

permitted the transmission of images obtained from azimuthal measurements. This gave geosteering a better tool to determine the orientation and location of bed boundaries. Azimuthal sensors for resistivity and density data record data when the tool is rotating, without having to stop drilling to take a measurement. These images can be used to determine the dip of the changing formations encountered.

6.2.9 3D Visualization. Geosteering screens could not adequately demonstrate 3D structure such as drilling along the flank of an anticline. Apparent-dip calculations were made, but as 3D *designer well paths* were

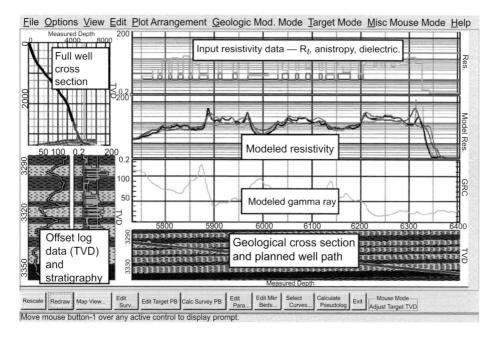

Fig. 6.2.11—A second example showing the modeled data with the vertical column and in the stretched log at the top. Both gamma ray and resistivity data were used in this analysis (Lott et al. 2000).

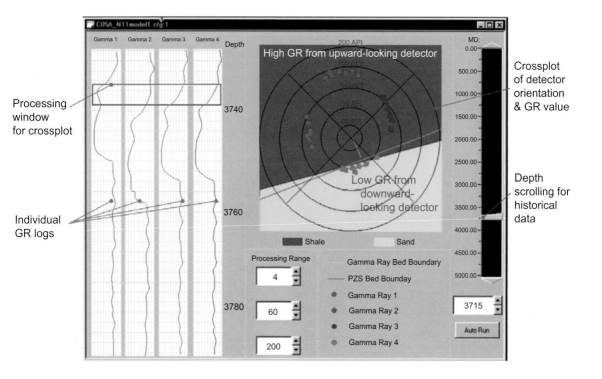

Fig. 6.2.12—This screen shows an azimuthal test being made with gamma ray data. The LWD tool in this case has four sensors at 90° spacings along the collar to measure up, down, left, and right. Those data are presented in a log format to the left, but it is the polar plot with gamma ray magnitude increasing from the center that reveals that lower counts are from the *down* sensor. This indicates that the sand is below and the well should be steered down (Phillips et al. 2000).

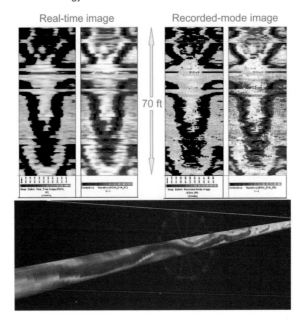

Fig. 6.2.13—The images can be rolled into a 3D view of the wellbore wall to help visualize the formation structure and dip in a geosteered well (Tribe et al. 2003b).

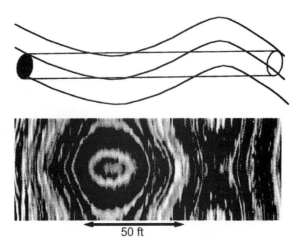

Fig. 6.2.14—A resistivity image detecting an inflection point in a nearly flat or zero-dip formation. The bull's-eye pattern shows the inflection point. The sinusoidal patterns on either side can be used to calculate relative dip. Trajectory data can relate this to actual formation dip (Rosthal et al. 1995).

developed and drilled with the advent of rotary steerable systems, it became increasingly difficult to visualize a well path with significant changes in azimuth and how it intersects formation boundaries. 3D cubes of the near-well volume were developed to address this issue. A *curtain plot* (**Fig. 6.2.15**), showing a curved surface that contained the wellbore or the drainhole section of the wellbore, would be defined in the cube. The 3D nature of the formation structure could then be shown on the curtain. These plots have gained limited use in real time. They are used mostly in the planning phases where they provide an overall understanding of this 3D problem and have thus had an indirect impact on geosteering decisions.

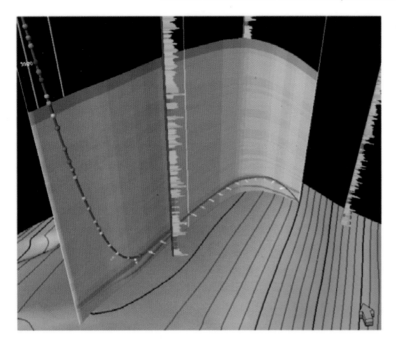

Fig. 6.2.15—A 3D curtain plot with a 3D designer horizontal trajectory. The curved surface shows the true 3D nature of the formation structure. These plots can usually be rotated and zoomed for different views. This allows for easier navigation calculations and geosteering decisions (Le Turdu et al. 2004).

Seismic data can also be used in geosteering work. In order to do this, the seismic data have to be translated from traveltime to a length (feet or meters). A reference plane or curtain is defined and the well path added. Usually a series of seismic horizons are added near the geosteered section of the well. An example is shown in **Fig. 6.2.16**. This is a valuable method when a high-resolution fine-scaled geological model has been defined. However, horizons picked directly on a 3D-seismic model may show a lot of uncertainty and may be subject to resolution problems when going from the large seismic scale to the finer log scale.

6.2.10 Look Ahead and Look Around. The information needed in geosteering decisions often leads workers to want to drill a little bit farther for that defining bit of data to clarify the decision. This is usually the wrong thing to do. It is best to stop drilling and determine the well position and make a trajectory decision before resuming drilling.

The ultimate tactic in geosteering is to understand the petrophysics, geology and structure ahead of the bit to deliver informed decisions on placement. Measurements to do this are along the BHA some distance back from the bit. This tends to keep geosteering in a passive/reactive mode making decisions only after the formation has been entered and seen by the sensors. It became highly desirable to develop tools that could look ahead of the bit and/or look around the wellbore. New LWD tools incorporating directional antennae and long measurement spacings have been developed that allow delineation of bed boundaries approximately 15 ft (4.57 m) away from the wellbore (see **Figs. 6.2.17 and 6.2.18**). These look-around tools are the first step in a new phase in geosteering. New interpretation software has been developed to translate these measurements into real-time structural maps. The processing is based on simple layer-cake models much as the first geosteering screens were. Based on these assumptions, distance and direction to bed boundaries can be displayed (**Fig. 6.2.19**).

6.2.11 The Geosteering Investigation. New tools will continue to be developed to help solve more difficult well placement problems. At its heart, geosteering remains a multidisciplinary investigation (see **Fig. 6.2.20**). The team must assimilate petrophysical, trajectory, and drilling data to determine the best fit in terms of the

Fig. 6.2.16—A large-scale visualization view of a geosteered well in seismic data. The well is contained in the vertical curtain display of seismic data, while a curved surface represents a seismic top near the horizontal section of the well. Usually these plots can be rotated and zoomed to show finer detail and aid in decision making (Mitra et al. 2004).

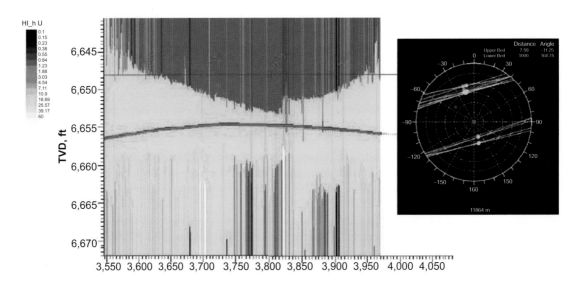

Fig. 6.2.17—Data from a look-around resistivity measurement showing a well path approaching a formation top in the vertical section view to the left. The azimuth view on the right shows the well placement within the formation with distance and direction data to both the top and the bottom (Omeragic et al. 2005).

wellbore location. A timely decision has to be made to adjust this placement as necessary without jeopardizing the stability of a horizontal well. This continues until a sufficient amount of the reservoir is drilled for the well to be viable for hydrocarbon production. A successful geosteered well has sufficient planning to allow for contingencies when formation surprises occur and has learning feedback loops to effectively expect the unexpected.

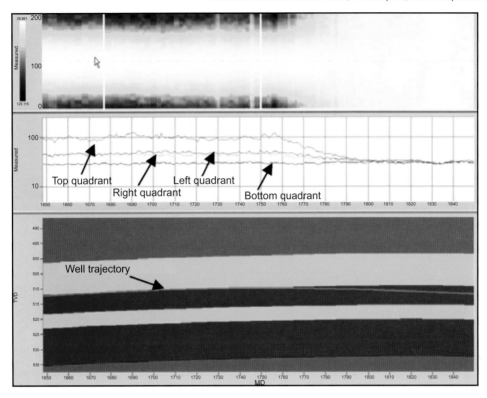

Fig. 6.2.18—A different display of look-around data with both image and log data combined with a trajectory section plot (Bittar et al. 2007).

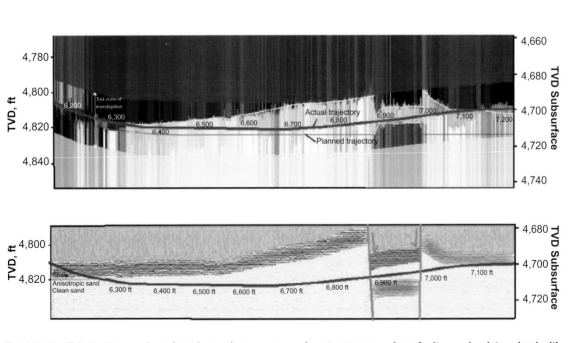

Fig. 6.2.19—This look-around section shows that more complex structures such as faults can be determined with simple layer-cake models delineating 15 ft (4.57 m) above and below the wellbore. The actual resistivity bed contrasts are seen in the upper plot, while the lower plot is an interpretation sketch. Image courtesy of Schlumberger.

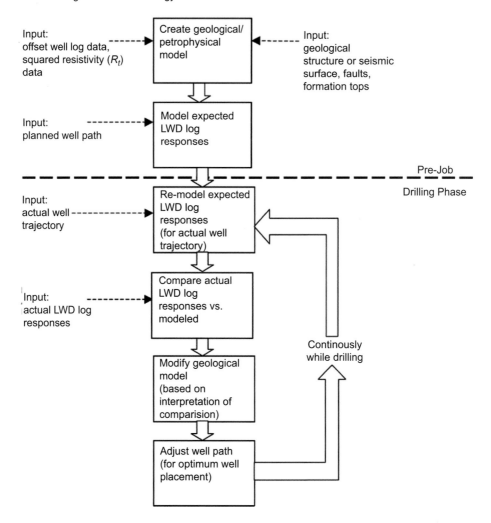

Fig. 6.2.20—A simple yet effective flow chart for the geosteering investigation (Lott et al. 2000).

Nomenclature

R_h = the resistivity value a tool measures when the plane that its sensors are reading is horizontal
R_t = the actual or true resistivity of a formation
R_v = the resistivity value a tool measures when the plane that its sensors are reading is vertical

References

Bittar, M., Klein, J., Beste, R. et al. 2007. A New Azimuthal Deep-Reading Resistivity Tool for Geosteering and Advanced Formation Evaluation. Paper SPE 109971 presented at the SPE Annual Technical Conference and Exhibition, Anaheim, California, 11–14 November. DOI: 10.2118/109971-MS.

Efnik, M.S., Hamawi, M., and Shamri, A. 1999. Using New Advances in LWD Technology for Geosteering and Geologic Modeling. Paper SPE 57537 presented at the SPE/IADC Middle East Drilling Technology Conference, Abu Dhabi, 8–10 November. DOI: 10.2118/57537-MS.

Lesso, W.G. Jr. and Kashikar, S.V. 1996. The Principles and Procedures of Geosteering. Paper SPE 35051 presented at the IADC/SPE Drilling Conference, New Orleans, 12–15 March. DOI: 10.2118/35051-MS.

LeTurdu, C., Bandyopadhyay, I., Ruelland, P., and Grivot, P. 2004. New Approach to Log Simulation in a Horizontal Drain—Tambora Geosteering Project—Balikpapan, Indonesia. Paper SPE 88448 presented at the SPE Asia Pacific Oil and Natural Gas Conference and Exhibition, Perth, Australia, 18–20 October. DOI: 10.2118/88448-MS.

Lott, S.J., Dalton, C.L., Bonnie, J.H.M., Roberts, M.J., and Cooke, G.P. 2000. Use of Networked Geosteering Software to Optimum High-Angle/Horizontal Wellbore Placement: Two UK North Sea Case Histories. Paper SPE 65542 presented at the SPE/CIM International Conference on Horizontal Well Technology, Calgary, 6–8 November. DOI: 10.2118/65542-MS.

Meader, T., Allen, F., and Riley, G. 2000. To the Limit and Beyond—The Secret of World-Class Extended-Reach Drilling Performance at Wytch Farm. Paper SPE 59204 presented at the IADC/SPE Drilling Conference, New Orleans, 23–25 February. DOI: 10.2118/59204-MS.

Meehan, D.N. 1994. Geological Steering of Horizontal Wells. *JPT* **46** (10): 848–852. SPE-29242-PA. DOI: 10.2118/29242-PA.

Mitra, P.P., Joshi, T.R., and Thevoux-Chabuel, H. 2004. Real Time Geosteering of High Tech Well in Virtual Reality and Prediction Ahead of Drill Bit for Cost Optimization and Risk Reduction in Mumbai High L-III Reservoir. Paper SPE 88531 presented at the SPE Asia Pacific Oil and Natural Gas Conference, Perth, Australia, 18–20 October. DOI: 10.2118/88531-MS.

Omeragic, D., Li, Q., Chou, L. et al. 2005. Deep Directional Electromagnetic Measurements for Optimal Well Placement. Paper SPE 97045 presented at the SPE Annual Technical Conference and Exhibition, Dallas, 9–12 October. DOI: 10.2118/97045-MS.

Page, G. and Benefield, M. 2002. Improved R_t and S_w Definition from the Integration of Wireline and LWD Resistivities, To Reduce Uncertainty, Increase Pay and Aid Reservoir Navigation. Paper SPE 78341 presented at the European Petroleum Conference, Aberdeen, 29–31 October. DOI: 10.2118/83968-MS.

Peach, S.R. and Kloss, P.J.C. 1994. A New Generation of Instrumented Steerable Motors Improves Geosteering in North Sea Horizontal Wells. Paper SPE 27482 presented at the SPE/IADC Drilling Conference, Dallas, 15–18 February. DOI: 10.2118/27482-MS.

Phillips, I.C., Paulk, M.D., and Constant, A. 2000. Real Real-Time Geosteering. Paper SPE 65141 presented at the SPE European Petroleum Conference, Paris, 24–25 October. DOI: 10.2118/65141-MS.

Rosthal, R.A., Young, R.A., Lovell, J.R., Buffington, L., and Arceneaux, C.L. 1995. Formation Evaluation and Geological Interpretation from the Resistivity-at-the-Bit Tool. Paper SPE 30550 presented at the SPE Annual Technical Conference and Exhibition, Dallas, 22–25 October. DOI: 10.2118/30550-MS.

Tippet, P.J. and Beacher, G.J. 2000. Integration of Subsurface Disciplines To Optimize the Development of the Mardie Greensand at Thevenard Island, Western Australia. Paper SPE 59410 presented at the SPE Asia Pacific Conference on Integrated Modeling for Asset Management, Yokohama, Japan, 25–26 April. DOI: 10.2118/59410-MS.

Tribe, I., Burns, L., Howell, P.D., and Dickson, R. 2003a. Precise Well Placement with Rotary Steerable Systems and LWD Measurements. *SPEDC* **18** (1): 42–49. SPE-82361-PA. DOI: 10.2118/82361-PA.

Tribe, I., Holm, G., Harker, S. et al. 2003b. Optimized Horizontal Well Placement in the Otter Field, North Sea Using New Formation Imaging While Drilling Technology. Paper SPE 83968 presented at Offshore Europe, Aberdeen, 2–5 September. DOI: 10.2118/83968-MS.

6.3 Pressure Measurement While Drilling—Iain Cooper, Schlumberger, Chris Ward, GeoMechanics International, and Peter Bern, BP

6.3.1 Introduction. This chapter covers the application of downhole pressure measurements to drilling operations at the rigsite. The majority of drilling downtime results from hole problems. These principally include those of lost circulation, formation fluid influx (or kick), hole collapse, differential sticking, and poor hole cleaning. Such events usually lead to time-consuming and expensive incidents such as lost mud and downhole tools, well-control incidents, stuck pipe, and stuck casing. It has been estimated that these problems account for approximately 10 to 15% of the drilling operations time in the North Sea. Pressure measurement while drilling (PMWD) has helped the industry gain invaluable understanding into the mechanisms of drilling problems and have aided in the implementation of the remedial solutions.

Most drilling hole problems occur when the downhole safe operating pressure limits are exceeded. These limits are defined by the pore, collapse, and fracture pressures. They are typically determined from offset well data, either by modeling or by certain measurements made while drilling the well, such as leakoff tests (LOTs) and formation integrity tests (FITs). If the stresses imposed during drilling are allowed to exceed safe pressure limits, hole problems are likely to occur. The pressure imposed is defined by the mud weight plus or minus any dynamic pressures resulting from pipe movement (e.g., swab/surge, rotation), from fluid flow (e.g., breaking the gel strength), or from restrictions (e.g., cuttings beds, packing-off, LOT, well control, managed-pressure drilling). The static mud weight traditionally is measured at surface, and the dynamic effects traditionally have been estimated using hydraulics models. Additional hole problems can occur if the conditions are insufficient to remove the drilled cuttings from the hole. Poor cuttings removal (hole cleaning) often results in excessive reaming times, packing off, and stuck pipe. PMWD enables the determination of real-time annular pressure losses. This can reveal important information about hole cleaning, barite sag, and swab and surge pressures.

PMWD is run on a variety of well configurations, but has most applications in the following:

- Extended-reach wells in which there is a narrow window between pore pressure and fracture gradient,
- Lost circulation, poor hole cleaning, or barite sag (Bern et al. 1996)
- Underbalanced drilling for monitoring circulating hydraulics, validating the hydraulics program, optimizing the gas injection rate, or optimizing the rate of penetration (ROP)
- Deepwater wells, in which the mud weight window is significantly narrower than in land or shallow-water operations, for early detection of well control issues, accurate control during well kills, and general management of the wellbore density
- High-pressure, high-temperature wells for understanding the influence of temperature on downhole drilling fluid density and rheology, managing lost circulation, barite sag, and well control (Leach and Quentin 1994)

6.3.2 Annular Pressure Measurements. The history of annular pressure measurements extends as far back as the 1950s, when pressures were recorded to verify simple hydraulics calculations. In the mid-1980s, Gearhart Industries provided recorded annular pressure sensors on their measurement-while-drilling (MWD) tools for measuring pressure drops. The first commercial real-time tools were developed in the mid-1990s. Since then, all the major MWD service companies have developed sensors for downhole pressure measurements while drilling. Quartz, strain, and resistor-based bellows gauges are three of the types of pressure gauges used in downhole tools today.

Measurement Specifications. For applications in medium and large boreholes, the sensors used in PMWD have been integrated into 6¾-, 8¼-, and 9½-in. collar sizes. For smaller hole sizes, the sensors are available as an addition to 4¾- and 3⅛-in. tool sizes. For coiled-tubing applications, 2⅛-in. sensors are also available, as are larger sizes.

Annular pressure sensors are typically available in a number of different ranges, depending upon the hole size or section of the well they are to be used in (typically, 5,000, 10,000, 20,000, and 25,000 psi). A typical resolution for an annular or internal pressure sensor is on the order of 1 psi.

Accuracy implies exactitude in measurement. The accuracy—strictly, the inaccuracy—of a measurement is a matter of degree, indicating how near our estimation, arrived at by means of an aid to measurement, comes to the true value, which is always unknown. The degree of inaccuracy is a matter of probability, assuming that the mode of measurement is theoretically correct, and the desired quantity is suitably isolated and defined.

The pressure at a certain true vertical depth (TVD) in the annulus is often expressed in terms of the equivalent mud weight (EMW) at that depth, and in oilfield units is given by

$$\text{EMW (ppg)} = \frac{P(\text{psi})}{0.052 \times \text{TVD (ft)}} \quad \dots\dots\dots\dots\dots\dots\dots\dots\dots\dots\dots\dots\dots\dots\dots\dots \quad (6.3.1)$$

The accuracy of the EMW calculation also depends on the sensor TVD accuracy, and this often can lead to the largest error in the EMW. Apart from the actual depth error, there can be a time synchronization error, particularly when the pipe movement is rapid (e.g., during tripping). Great care should be taken when evaluating EMW trip information. Although the PMWD sensor is normally some distance behind the bit, it is also useful to project this measurement to the bit because, unlike logging while drilling (LWD), PMWD responses usually are related to the current bit depth.

In the absence of hysteresis, the accuracy of each sensor is ±0.03% of the maximum pressure range of the sensor. The accuracy specification including hysteresis is ±0.1%. For example, a 5,000-psi pressure sensor will have an accuracy of ±1.5 psi (5 psi including hysteresis). A 20,000-psi sensor will have an accuracy of ±6 psi (20 psi including hysteresis).

Pressure sensors are very reliable, with tens of thousands deployed worldwide. Some of the failure modes that have been seen are as follows:

- Shock damage, usually seen as uncorrectable offsets. The shock levels that do this cause damage to the whole tool.
- Aging by cyclic pressure, from the modulator or an agitator. This shows as an increased offset on return to surface. This is the most troublesome failure, since it happens during the run and gives erroneous data. The error is typically less than 0.5% full scale, and arises only at combined high pressure and high temperature.
- Plugging of the sensor bore with mud. This is rare and gives an error in surface offset, without normally affecting the downhole measurement. The restriction tends to result in a smoothed annular pressure response. The annular pressure tends to remain higher than the bore pressure during the pumps-off period. Modifications to the design have made this failure mode relatively rare.

6.3.3 Equivalent Circulating Density. Successful drilling requires that the drilling fluid pressure stays within a tight mud-weight window defined by the pressure limits for well control and wellbore integrity or stability. The lower pressure limit is either the pore pressure in the formation (in which formation fluid can enter the wellbore in permeable zones) or the limit for avoiding wellbore collapse **(Fig. 6.3.1)**. If the annular pressure is less than the pore pressure, then formation fluid or gas could flow into the borehole, with the subsequent risk of a blowout at surface or underground. If the annular pressure is less than the collapse pressure, formation may spall into the well increasing the risk of pack-off and stuck pipe.

The upper pressure limit for the drilling fluid is the minimum that will fracture the formation. If the drilling fluid exceeds this pressure, there is a risk of creating or opening fractures, resulting in lost circulation and a damaged formation.

Pressures are often expressed as pressure gradients or equivalent fluid densities. The upper limit of the pressure window is usually called the formation fracture gradient, and the lower limit is called the pore pressure or collapse gradient.

The downhole annular pressure has three components:

1. The first is a static pressure due to the density gradients of the fluids in the borehole annulus: the weight of the fluid vertically above the pressure sensor. The density of the mud column including solids (such as drilled cuttings), and perhaps introduced formation fluids, is called the equivalent static density (ESD). The fluid densities are pressure- and temperature-dependent.
2. The second are dynamic pressures related to the drillstring velocity [swab/surge (Rudolf and Suryanarayana 1998) and rotation (McCann et al. 1995)], inertial pressures from the drillstring acceleration or deceleration while tripping, excess pressure to break mud gels, and the cumulative pressure losses required to move drilling fluids up the annulus. Flow past constrictions, such as cuttings beds or swelling formations; changes in hole geometry; and influxes and effluxes of liquids, gases, and solids all contribute to the dynamic pressure. The equivalent circulating density (ECD) is defined as the effective mud weight at a given depth created by the total hydrostatic (including the cuttings pressure) and dynamic pressures during circulation.
3. The third is related to restrictions in the hydraulic system. When the annulus becomes blocked due to excessive cuttings, cavings, and poor hole cleaning, the hole may pack-off, resulting in excessively

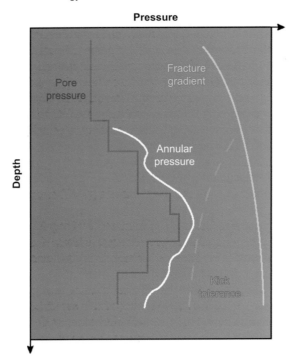

Fig. 6.3.1—The pressure window (Aldred et al. 1998). Image copyright Schlumberger. Used with permission.

high or low pressures below the restriction. In addition, there may be times when the annulus is shut in at surface (e.g., during an LOT, a well-control incident, or managed-pressure drilling) that can lead to higher or lower pressures than an open system.

The drilling fluid rheological properties (including viscosity, yield and gel strength) and dynamic flow behavior (laminar, transitional, or turbulent) can all affect the dynamic response of the annular pressure. The variation of the rheological properties with flow regime and temperature affects the total pressure measured downhole. Some of these parameters can be controlled by the driller, while others, such as downhole temperature, cannot, and furthermore, are not well modeled.

As the annular gap decreases, the effect of drillpipe pressure losses becomes more important. Annular pressure losses or the axial pressure drop depend on which part of the flow regime predominates when the rotation rate is changed **(Fig. 6.3.2)** (Aldred et al. 1998). At low flow rates, the pressure drop decreases with increasing rotation rate. At higher flow rates, the opposite effect is observed. It is also thought that the pipe eccentricity plays a role and that the effect of drillpipe rotation increases with hole angle. **Fig. 6.3.3** shows experimental data highlighting these effects (Hutchinson and Rezmer-Cooper 1998).

6.3.4 Data Display Formats. There are three modes in which PMWD data are recorded and displayed:

Memory Mode (Also Known as Recorded Mode). In this mode, the downhole tool is continually recording data into the tool memory. These data then can be retrieved when the tool is brought back to surface. Recorded-mode data tend to be sampled at a higher frequency than real-time data and as such are richer in detail, but more prone to noise. Recording gauges are programmable and can store data at required intervals from 1 second upward. Typical settings are from 5 to 20 seconds. It is important to realize that some short-timescale phenomena could be missed by an incorrect setting of the memory-mode recording interval. For example, fast transients due to swab/surge could be missed, or their peak pressure amplitudes could be distorted.

Real-Time Mode. In this format, the downhole pressure data are interfaced with a downhole MWD system (this can be either a mud pulse system, an electromagnetic system, or a system hardwired to the surface), and the real-time data are sent to surface every 30 seconds or so. With a mud pulse MWD system, the data can be

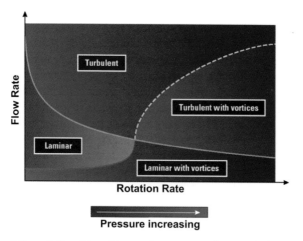

Fig. 6.3.2—Flow regimes and the relationship with annular pressure losses. At low flow rates, the pressure drop decreases with increasing rotation rate. At higher flow rates, the opposite effect is observed (Aldred et al. 1998). Image copyright Schlumberger. Used with permission.

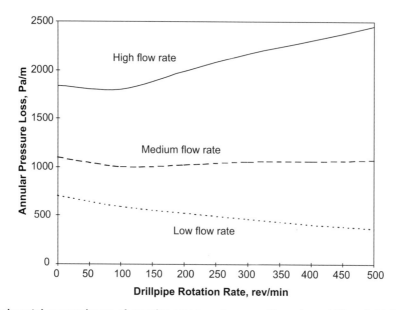

Fig. 6.3.3—Experimental comparisons of annular pressure losses with a clean drilling fluid (no solids) at low, medium, and high flow rates vs. drillpipe rotation rate (Hutchinson and Rezmer-Cooper 1998).

sent to surface only when the pump flow rate is above a certain threshold. Below this flow rate, no information can be sent to surface. There is no limitation with electromagnetic or hardwired systems.

Pumps-Off Mode. In this mode, the recorded-mode data are processed in the downhole tool, and a limited amount of information is sent to surface when pump circulation resumes for pulse-type MWD systems. This mode is of use during connections or LOTs. Either a continuous pumps-off recorded pressure or a limited set can be transmitted. Typical information from the limited set that is processed and transmitted to surface includes

- Maximum annular pressure (expressed in terms of ECD)
- Minimum annular pressure (expressed in terms of ECD)
- Average annular pressure during the connection
- Static mud weight, or equilibrated pressure (expressed as the ESD) if all transient fluid motion has ceased during the connection

As drilling progresses, additional joints of drillpipe must be connected to the drillstring at the surface to extend the reach of the drilling rig toward deeper objectives. During each pipe connection (or when the pumps are turned off in general), several factors contribute to the downhole pressure. Some of these factors are dynamic in nature, and the downhole pressure generally changes or fluctuates throughout the duration of each pipe connection (downhole pressure trace). Factors that may contribute to or affect the downhole pressure trace during a pipe connection (Ward and Andreassen 1997) include

1. Movement of the drillstring within the wellbore (rotation or reciprocation)
2. Temperatures and temperature gradients throughout the wellbore
3. Pressure gradients and propagation rates of pressure fronts throughout the wellbore
4. Mud viscosity, compressibility, and other static and dynamic fluid properties of the drilling mud, and their physical sensitivities to changes in temperature
5. Drilling mud weighting agents and loading of cuttings from drilling, and uniformity or nonuniformity of dispersal of both in the mud
6. Fluid flows into and out of the wellbore
7. Elastic and inelastic expansion of the wellbore and casing
8. Elastic expansion and elongation of the drillstring
9. Frictional pressure losses due to wellbore geometry and mud rheology (Haciislamoglu and Langlinais 1990; Haciislamoglu 1994)

A sample pressure trace during a connection is shown in **Fig. 6.3.4**.

6.3.5 Drilling Logs. The display of each of the three modes of PMWD data is critical to the usefulness of the information. Care must be taken with the setup of each of the presentation formats.

Typically, there are two distinct modes of display for drilling logs: time-based logs and depth-based logs. Depth-based logs are more associated with formation evaluation measurements, in which the characterization of the reservoir is of importance. For real-time drilling information and post-event interpretation of memory-mode data, a time-based format is more appropriate.

With both LWD and wireline formation evaluation data, a *triple combo* log presentation has the gamma ray data in track one, resistivity data in track two, and porosity data in track three. Presenting formation measurements out of context of one another would make lithological interpretation difficult. Similarly, it is important to interpret drilling information and measurements in the context of each other.

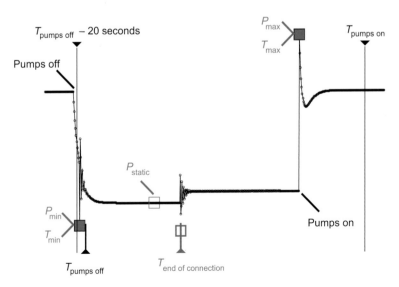

Fig. 6.3.4—Typical annular pressure trace during a connection. The static pressure (P_{static}), the maximum pressure (P_{max}) and the time it occurs (T_{max}), and the minimum pressure (P_{min}) and the time it occurs (T_{min}) are often sent to surface immediately after a connection.

One of the factors impeding annular pressure data interpretation in the industry today is the wide variety of formats used to present the data. More often than not, critical contextual information is missing, which prevents the ability to discriminate between the many different influences upon the annular pressure, and to determine from where in the borehole different drilling symptoms originate.

Pressure data should be presented in the context of many other drilling variables for an interpretation to be useful. For example, an annular pressure trace on its own tells you nothing more than the minimum and maximum pressures that occur during that time or depth interval. Other key measurements that are important to have are as follows:

- Pump flow rate
- Standpipe pressure
- Surface annular/casing pressure (if appropriate)
- Drillstring revolutions per minute
- Surface and downhole weight
- Surface and downhole torque
- Block velocity (i.e., ROP)
- Drilling fluid properties (mud weight in, Fann viscometer readings, gel strength, and temperature and pressure dependence if available)
- Total gas
- Cuttings volume rate
- Pit level
- Rig heave

It is also important that the measurements are made in the context of the previous drilling history in the region. Rig-based software tools are now available that permit the automatic evaluation of the rig state (i.e., tripping, circulating off bottom), which facilitates interpretation of the pressure traces. Drilling problems generally result in slower ROPs, and as a result data are sometimes compressed onto a depth scale. A time-based presentation is better suited for detailed analysis during problematic drilling intervals because many problems occur when the bit is off bottom.

It is also important to annotate the drilling logs with related information such as bit type, bottomhole assembly (BHA) design, mud rheology, reasons for tripping, and both BHA and bit condition after tripping.

Time-Based Log Format. **Fig. 6.3.5** shows a typical time-based log format, with the PMWD placed in the context of the other downhole and surface measurements. It is important to note that for real-time applications, the scales for the various parameters must be chosen with care to maximize the visibility of the expected signatures, particularly for that of the annular pressure, ECD, or EMW. For example, if the ESD is 12 lbm/gal, and the ECD is 14 lbm/gal, and the expected swab and surge pressures are approximately equivalent to ±1 lbm/gal, then the ECD probably should be plotted on a scale from 11 to 15 lbm/gal, so that the detailed pressure trace has a high sensitivity, and drilling anomalies can be easily visualized.

Depth-Based Log Format. A depth-based presentation is important for the assessment of drilling events in the context of the BHA position relative to the lithological boundaries. An example for the case of a washout is shown in **Fig. 6.3.6**.

6.3.6 Pressure Measurement Calibrations.
Master calibrations of the pressure sensors are performed over a range of temperatures using a dead-weight tester. At the location or wellsite, hydraulic tests using a hand pump are performed on the gauges before and after use in each well to verify calibrations.

It is important that the pressure gauges are calibrated in the shop before the tool is placed in the hole so that the correct pressures are recorded. Failure to do this will result in inaccurate pressure measurements, and potentially misleading interpretation.

6.3.7 Validation of Hydraulics Models and Cuttings Transport.
The value of PMWD lies in its ability to monitor pressures wherever the sensors are placed. It can also be used to validate predictive hydraulics models, and can refine the calculations from those models by enabling some of the unknown parameters to be refined or determined. Even though enhancements in computer power have led to more-complex hydraulics

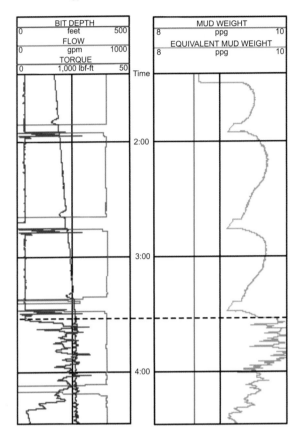

Fig. 6.3.5—Time-based log format.

models being developed (such as those described in Chapter 4), PMWD also can be used to understand the limitations of hydraulics models, particularly in allowing for the complex distributions of cuttings that can occur along the annulus, the effects of which typically are poorly handled by conventional hydraulics models (Luo et al. 1992, 1994).

An example of a comparison between the measured ECD and the predicted ECD using a model is shown in **Fig. 6.3.7.** It can be seen that in this case, the model consistently underpredicted the measured ECD. Examination of the PMWD data at zero flow rate revealed that the measured hydrostatic pressure was higher than the pressure based on surface mud weight. This difference was generally in the range of 0.1 to 0.25 lbm/gal. The difference can be explained on the basis of the cuttings accumulation in the annulus adding to the effective static mud weight. This explanation is consistent with the observation in Fig. 6.3.7 because the density difference is greater at higher ROPs, as can be seen in **Fig. 6.3.8.** The model does not take into account the contribution to the ECD made by the cuttings. In practice, it is better to use actual field data (where possible) to allow for the effect of the cuttings contribution to the mud weight. However, it is possible to make a rough estimate based on the cuttings feed rate, (which is driven by the ROP), flow rate, hole diameter, and transport ratio.

The differences between the PMWD and the mud weight data in Fig. 6.3.8 have been used to correct the model predictions on the basis of the actual downhole mud weight using the correlation shown in **Fig. 6.3.9.** The resulting corrections are now shown in **Fig. 6.3.10.** Measured and predicted ECDs now agree very well.

Many of the hydrodynamic responses of a pressure sensor to drilling parameters are not easily predicted, as illustrated above with the example of the cuttings. In the absence of prediction models for these effects, field procedures need to be developed that calibrate the response of the pressure sensor to each of the drilling control variables. In this way, it is then possible to discriminate the contribution that cuttings make to the ECD and determine whether the hole is being cleaned effectively.

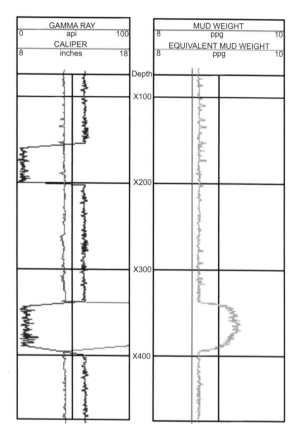

Fig. 6.3.6—Depth-based log format.

6.3.8 LOTs. With higher rig costs on many drilling projects, such as extended-reach and deepwater wells, time savings and precise measurements are critical. Accurate LOTs are essential to enable efficient management of the ECD within the pressure window, and the corresponding mud program.

An LOT is usually performed at the beginning of each well section, after the casing has been cemented, to test the integrity of the cement seal and to determine the fracture gradient below the casing shoe. In general, these tests are conducted by closing in the well at the surface or subsurface with the blowout preventer (BOP) after drilling out the casing shoe, and slowly pumping drilling fluid into the wellbore at a constant rate (typically 0.3 to 0.5 bbl/min), causing the pressure in the entire hydraulic system to increase. Downhole pressure buildup traditionally is estimated from standpipe pressure, but it can be monitored directly with pressure sensors in the annulus. If pressure measurements are made in the standpipe, then complex corrections must be made for the effects of temperature on mud density and for other effects on downhole fluid pressure. Pressures are recorded against the mud volumes pumped until a deviation from a linear trend is observed, indicating that the well is taking mud. This could be due to either failure of the cement seal or initiation of a fracture. The point at which the nonlinear response first occurs is the LOT pressure used to compute the formation fracture gradient. Sometimes the procedure is to stop increasing the pressure before the actual leakoff pressure is reached. In such cases, the planned hole section requires a lower maximum mud weight than the expected fracture pressure, and the LOT only pressures up to this lower value with no evidence of fracture initiation. This is called an FIT. If pumping continues beyond the fracture initiation point, the formation may rupture, pressure will fall, and the fracture could propagate. In some cases the test is taken past breakdown and a fracture is propagated, often called an extended LOT, for the purpose of accurately measuring stresses downhole.

Historically, pressures were measured at surface during a LOT, and a downhole pressure was calculated by adding the static mud weight, based on the surface mud weight, which is assumed to be homogeneous

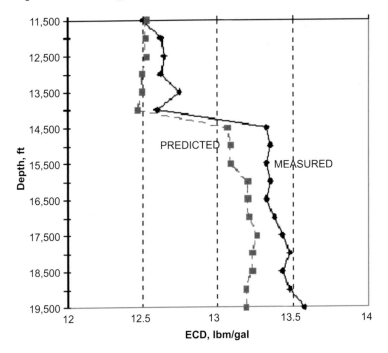

Fig. 6.3.7—Comparison between predicted and measured ECD.

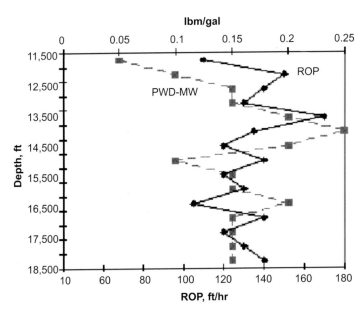

Fig. 6.3.8—Effect of ROP on the discrepancy between predicted and measured EMW.

throughout the fluid column. PMWD sees the same stress as the formation during an FIT and an LOT and is therefore a more accurate measure of the formation strength (Rezmer-Cooper et al. 2000).

Operators' standard practice is to circulate the well to an even mud weight (sometimes called *conditioning the mud*) before each FIT/LOT to decrease the uncertainty in interpretation. This time can be saved by recording the accurate downhole PMWD.

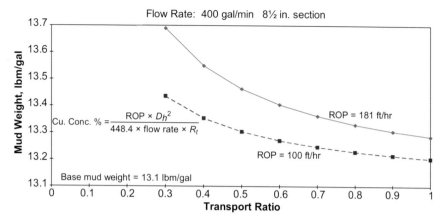

Fig. 6.3.9—Cuttings influence on mud weight. D_h = hydraulic diameter and R_t = transport ratio. The transport ratio is the measure of how fast the cuttings are carried in the mud compared to the annular fluid velocity. If no slip occurs the transport ratio is unity. The transport ratio can be regarded as the reciprocal of the number of bottoms-up that are required to bring the cuttings to surface.

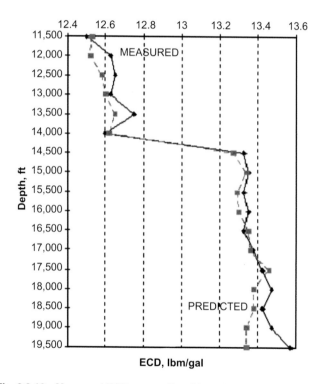

Fig. 6.3.10—Measured ECD vs. predicted based on true mud weight.

Alberty et al. (1999) highlight that it is sometimes difficult to pick the leakoff pressure from the buildup curve because there are other factors that can mask the pressure variation, such as

- Variations in the pump rate
- Changes in the mud gel strength
- Air trapped in lines
- Inaccurate pressure gauges
- Insufficient contrast between leakoff pressure and the mud weight used

It was recommended by Alberty et al. (1999) that two more easily identifiable points be picked. The first is the maximum pressure achieved (or the point of massive breakdown), which represents the maximum ECD to which the formation should be exposed. Minor losses may occur below this limit, with more severe losses at pressures exceeding this value. The second value they suggest picking is immediately after the pumps have shut down, the initial shut-in pressure, which represents the pressure at which any initiated fracture will collapse. The ESD should never exceed this value. Alberty et al. (1999) describe the difficulties in conducting quality FITs in the larger hole sections and in shallow casing shoes, and they give guidance on how best to perform the test.

In deep water, the mud density in the annulus can significantly vary, due to the pressure and temperature profile in the annulus. This is compounded by the use of synthetic muds, which are highly compressible. When drilling in deep water, the external temperature profile along the well path decreases from surface to seafloor. In deep, cold water, the temperature can get lower than the normal freezing point of water. Onshore, the external temperature typically increases with depth according to the geothermal gradient. The water temperature at the seafloor in a well drilled in 5,000 ft of water can easily be 100°F less than the rock temperature at a depth of 5,000 ft in an onshore well. The reduced external temperature profile has a significant effect on the circulating temperature for the whole well and will affect both the density and viscosity of the mud. The amount of cooling is strongly affected by the ratio of the water depth to the well depth; the deeper the well, the more chance the mud has to warm before it encounters the cool riser.

The cooler mud in a deepwater well will be denser and more viscous than the same mud in an onshore well. Mud density typically will increase with increasing pressure and decrease with increasing temperature. In water-based mud (WBM), the pressure effect is small, and the temperature dominates the behavior. The average mud weight is fairly close to the mud weight in. However, the effective density above the casing shoe can be considerably higher than the nominal mud weight in because of the deepwater cooling. This would lead to an incorrect estimate of the LOT pressure if a uniform mud density equivalent to that of the inlet density were assumed. On a land rig, the compressibility/expansivity effect is actually more noticeable, but the effective window between the pore pressure and the fracture gradient is likely to be considerably larger.

Indeed, the choice of mud type has been identified as being a crucial element in the accuracy of LOT interpretation. Leakoff testing with oil-based mud (OBM) or synthetic-based mud (SBM) is a more critical operation. If the formation is broken down during the test with OBM, the formation may not heal and regain the strength it had before the test. It is often preferable to perform an LOT with WBM, if possible, and then to displace it with the OBM for subsequent drilling (Alberty et al. 1999). LOT pressures obtained with WBM have been higher than those obtained with OBM (or SBM). The International Association of Drilling Contractors (IADC) deepwater guidelines report that this difference can be as high as 0.5 to 0.7 lbm/gal. **Fig. 6.3.11** shows how the downhole pressure, and the subsequent EMW, varies during an FIT. In this case, PMWD data indicates that the test downhole EMW reached an actual value of 1.952 SG.

A methodology using two downhole pressure points to calibrate the hydrostatic and compressibility offsets between surface and downhole is described by Rezmer-Cooper et al. (2000). A complete LOT profile then can be created at surface as soon as conventional pumping resumes and the pressure information is sent back to the surface, and it will yield a more accurate representation of the true formation breakdown potential because the effects of the pressure and temperature variations of the mud density and the borehole and casing compliance are now included. Using this method only requires two pumps-off measurements to be sent to surface on resumption of circulation after a LOT.

Analysis shows that the downhole annular pressure during the FIT/LOT is related to the standpipe pressure by a simple linear relationship:

$$P_{ann} = a + bP_{sp}, \qquad \dots\dots\dots\dots\dots\dots\dots\dots\dots\dots\dots\dots\dots\dots\dots \text{(6.3.2)}$$

where P_{ann} is the downhole annular pressure recorded during the LOT, P_{sp} is the standpipe pressure at the same time, and a and b are parameters (offset and gain) to be determined from the fit. The parameter a accounts for the hydrostatic offset between the two readings, and term b accounts for the mud compressibility/expansivity and for borehole and casing compliance. The borehole compliance term should be small during an LOT because only a small amount of formation typically is exposed. In general, the casing is an order of magnitude less compliant than the formation (Johnson and Tarvin 1993).

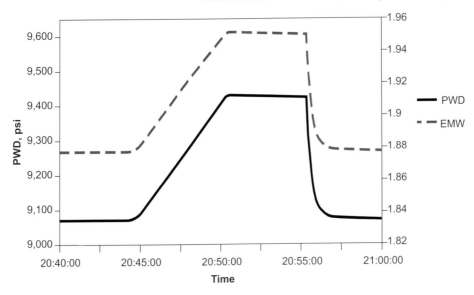

Fig. 6.3.11—Variation of EMW as a function of the downhole recorded-mode pressure during an FIT.

Fig. 6.3.12 shows the downhole annular pressure plotted against the standpipe pressure for an FIT undertaken in Marnock field in the central North Sea, 150 km due east of Aberdeen. The minimum and maximum downhole and surface pressures are determined and recorded, and then they are used to determine the parameters a and b. It is important to use the minimum and maximum values of pressure because a smaller spread could result in a significant error in determining the slope and intercept of the linear fit, thus inaccurately estimating the hydrostatic offset and compressibility/expansivity effects. Fig. 6.3.12 also shows the linear fit through the data from which the gain and offset parameters were calculated. Finally, the time of the peaks are matched, and the plot in **Fig. 6.3.13** shows the full recorded-mode data from downhole and the reconstituted downhole data based on the surface standpipe pressure. This highlights that the model agreement is very good, and an excellent reproduction of the downhole profile can be obtained from the surface data and limited downhole data sent to surface straight after the test.

Fig. 6.3.13 again highlights that the surface pressure measurement is noisier than the downhole pressure, as in the real-time LOT, and highlights the need for accurate surface instruments in addition to those downhole.

6.3.9 Riserless Operations and Shallow Water Flows. In many deepwater wells, the first casing or conductor pipe is usually 30 or 36 in. in diameter. The next hole section is typically 24 or 26 in. and is often drilled without a riser. In these wells, spent drilling fluid and cuttings are returned to the ocean floor around the wellhead. One of the biggest hazards drilling the riserless section is encountering a shallow water flow (SWF) (Hauser 1998). Standard operating practices in deepwater wells involve using a remotely operated vehicle (ROV) with a camera at the mud line to monitor flow coming out of the wellhead. At a connection, the driller holds the drillpipe stationary and turns off the pumps for a few minutes to allow fluid U-tubing oscillations to stabilize, and to observe whether there is flow at the wellhead. This information can be complemented by a pressure-measuring tool that is used to help identify, monitor, and mitigate SWFs. As the well flows, the weak unconsolidated sand flows into the wellbore, and the downhole pressure shows a characteristic increase due to the weight of the additional solids in the annulus.

According to a recent Minerals Management Services survey covering the last 15 years, SWF occurrences have been reported in approximately 60 Gulf of Mexico lease blocks involving 45 oil and gas fields or prospects. Problem water flow sands are typically found at depths from 950 to 2,000 ft, but some have been reported as deep as 3,500 ft below the seafloor. Frequently these problems are due to overpressurized and unconsolidated sands at shallow depths below the seafloor. They can lead to formation cave-in when uncontrolled water production occurs. If an influx is severe enough, wells can be lost due to continuous water flow. Extensive washouts can undermine the large casing that is the major support structure for the entire well.

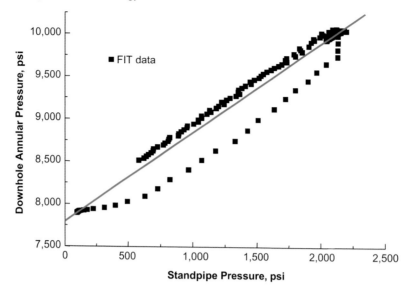

Fig. 6.3.12—Crossplot of downhole annular pressure and surface standpipe pressure during an FIT (Rezmer-Cooper et al. 2000).

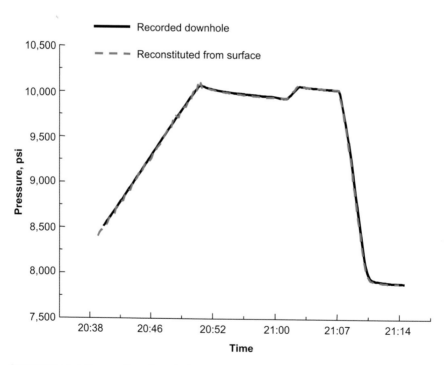

Fig. 6.3.13—Comparison between the downhole recorded annular pressure and the downhole pressure reconstituted from the surface data and a two-point calibration (Rezmer-Cooper et al. 2000).

In deepwater drilling, SWFs are commonly observed while drilling the riserless section. These can occur down to around 5,000 ft below the mud line in water depths more than approximately 1,500 ft. These are particularly common in areas of the Gulf of Mexico but have been observed in other deepwater areas such as the Norwegian North Sea. This section is normally drilled with unweighted seawater, and any interval penetrated with overpressure and permeability may begin to flow. It has also been proposed that SWFs may be initiated through supercharging of the formation, inducing a loss/gain situation. The sediments at this depth can be very unconsolidated, and the formation can quickly erode due to unrestricted flow. Because these sections typically are drilled without a riser, there are none of the usual surface indications of a flowing well (pit gain, gas, shut-in pressures). The flow can only be identified with the PMWD sensor and later confirmed with the ROV.

Such SWFs can pose no problem if they are of short duration and controlled. In an extreme case, continuing SWF problems led to the loss of the first Ursa template in the Gulf of Mexico through erosion around the structural casing and well collapse (Eaton 1999; Pelletier et al. 1999). After identifying the SWF situation, the PMWD sensor is used to help monitor and control the methods used to mitigate the flow. Several different methods have been attempted—setting casing above the flow zone and drilling ahead with weighted mud, allowing the well to flow while drilling and spotting heavy pills to balance the flow at total depth, drilling with a dual-density weighted mud system to balance the flow with no returns (often called "pump-and-dump"), and drilling at high ROPs with the weight of solids loading balancing the flow.

Monitoring the ECD helps the operator to assess both the depth and severity of the water flow, and to decide whether the flow is severe enough to stop drilling. Most conventional hydraulics models do not consider the effects of mud returns to the seafloor, and thus cannot accurately predict the expected ECD in these wells. A direct measurement of mud pressure solves this problem.

The following schematic examples show how to identify typical SWF situations with the PMWD sensor and some of the methods used to distinguish between a flowing and collapsed well (Ward 1998; Ward and Beique 1999).

Fig. 6.3.14 represents the typical situation while drilling through an SWF zone when drilling riserless. This graphic simplifies the cumulative effect of three distinct fluid columns that influence the measurement of the PMWD sensor in this situation. The first interval is the body of seawater between sea level and the seafloor. Except for the tiny effect of waves and tides, this pressure remains constant. Next is the interval between the seafloor and the SWF source. The annular pressure contributed by this interval is a combination of the sea-water drilling fluid plus the weight of drill cuttings returning up the annulus and the slurry produced from the SWF source. Besides the pressure profile itself, ROV observation while the well is static can confirm that SWFs are not pure formation water flows but rather a combination of formation water and unconsolidated

Fig. 6.3.14—Typical shallow water flow event when performing riserless drilling.

formation (sand and silt) that makes up a dense slurry. The lowermost interval is the new hole drilled below the source interval. In this example, it is where the PMWD sensor is situated, usually a short distance behind the bit. The annular pressure attributed to this interval is the column of the seawater drilling fluid plus the weight of suspended drill cuttings. In addition, pressure is added below the mud line while circulating due to dynamic pressure losses (i.e., ECD). This is usually a fairly small contribution in these riserless sections.

Changes in annular pressure, and therefore EMW, may be caused by many factors, including ECD pressures while circulating, SWF slurries, hole collapse and restrictions, and cuttings loads. It is important to understand that the point source measurement of the PMWD sensor reading is a cumulative reading of all the events occurring within the three zones previously described. Interpretation of changes in EMW must incorporate many of the variables occurring at or near the bit, as well as dynamic conditions that occur far above the bit in drilling situations.

Fig. 6.3.15 is a schematic example of a PMWD sensor depth log while drilling through a sequence of shaly and nonflowing *clean* zones. Many times the gamma ray tool measures a large washout, and the assumption that this interval is sand may be erroneous. For the purposes of clarity, we will refer to *clean* zones as any interval in which the gamma ray measurement approaches zero. This is the base case for interpretation of EMW in riserless drilling operations. A washout does not necessarily occur when drilling the clean zone, as evidenced by the caliper in the upper interval. The EMW is unaffected when no washout occurs. In the case of the lower clean zone, the hole washes out beyond the depth of investigation of the gamma ray sensor. A significant increase of 0.1 to 0.7 lbm/gal in EMW can occur when a clean zone washes out. However, this effect lasts only for the duration of the clean zone, and the EMW returns to its baseline soon after drilling the unconsolidated interval. In this case, the well is not flowing, and the section can be safely drilled.

An SWF can occur when the clean zone is charged above the normal seawater gradient. **Fig. 6.3.16** is a schematic of a PMWD sensor depth log through an SWF interval. In this depth-based example, the EMW

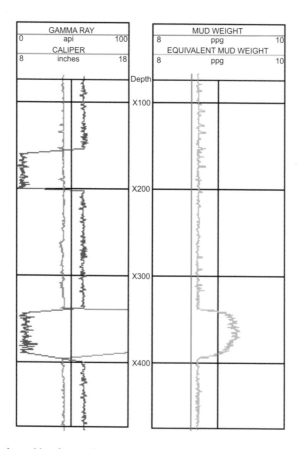

Fig. 6.3.15—Depth-based log format for a nonflowing washout when performing riserless drilling.

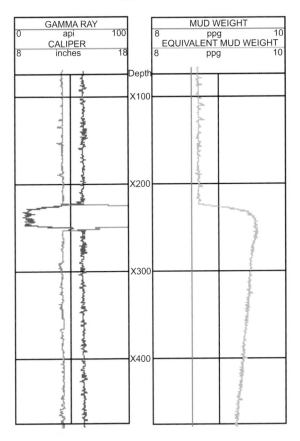

Fig. 6.3.16—Depth-based shallow water flow.

exhibits the sharp increase similar to the previous example but does not return to the baseline after the drilling through the interval. Instead, the well continues to flow while drilling ahead, and the EMW remains above the seawater mud weight gradient due to the weight of entrained solids in the SWF slurry. EMW typically reduces slowly with depth. Several theories partially explain this gradual reduction in EMW. First, as drilling continues, the interval between the flow zone and the bit accounts for more and more of the total annular pressure. The average density of the annular fluid in this interval is usually less than the combined density of the flowing slurry/drilling-fluid mixture. Second, the flow itself may be gradually diminishing with time, yet no known measurement of an SWF has been done to confirm this theory. Another explanation may be that the nature of the flow is changing and that decreasing amounts of solids are contained in the slurry. This particular signature of an SWF is consistent in the examples seen throughout the Gulf of Mexico and the North Sea.

The previous examples all related annular pressure and EMW to depth. Operators derive the most benefit from PMWD sensor information when the data are presented in a time format because much of the time is spent off bottom. The presentation of PMWD sensor data combined with synchronized surface data in a time format allows a drilling engineer to correlate downhole events with surface drilling data. Standard surface data measurements include standpipe pressure, flow in, ROP, and rotary revolutions per minute and torque. The quantity of data measured and displayed can overwhelm even those most experienced in PMWD interpretation, so great care must be taken to present only the most relevant data in real time.

The PMWD sensor schematic in **Fig. 6.3.17** represents a typical SWF event in time. Notice the abrupt increase in EMW when encountering the SWF. The sudden increase can occur in less than a minute and can exceed safety margins at the previous casing shoe. The gradual tapering of EMW is evident in time as well as depth. In the time presentation, however, EMW changes in different drilling operations help explain the hole conditions.

During a connection, both the pumps and rotary table stop. This results in the elimination of ECD. Typically the ECD is higher during the SWF due to the increased flow and rheology of the slurry. The frictional

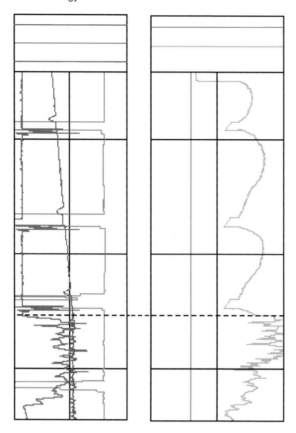

Fig. 6.3.17—Time-based shallow water flow. Dotted line shows onset of peak EMW.

pressure losses from the flowing zone will continue to exert on the sensor. (In this example, as well as others, the authors have expanded and simplified the PMWD sensor signature for clarity.) Well conditions are not static while a connection is made, as evidenced by the steady decrease of EMW during these intervals. This decrease may be attributed to solids fallout from the seawater drilling fluid, which has relatively little carrying capacity. When circulating and rotary drilling resume, the character of the EMW confirms the suspicion that cuttings were falling out of suspension during the connection. Cuttings are resuspended, and the EMW increases to a maximum value due to the increased cuttings load (Fig. 6.3.17). This load is quickly transported out of the hole and onto the seafloor, and the EMW begins to decrease as the annulus returns to a steady-state condition. The next connection starts the process all over again. The SWF initially aids in the removal of the cuttings by adding to the annular fluid velocity. Eventually, the zone may wash out, or the flow may diminish to such a degree that the trend may reverse and the EMW signature increases. Careful real-time observation with PMWD sensor data can minimize problems by alerting the operator to these conditions.

The PMWD sensor time log shown in Fig. 6.3.17 describes some other common signatures of SWFs. In this example, a flow is in progress, but the combined lifting capacity of the circulating drilling fluid and the flowing slurry are not enough to clean the hole properly. The resulting increasing maximum EMW trend is typical in a continuing SWF situation.

As the well is deepened below the flow zone, it becomes increasingly difficult to clean the well. The hole can often destabilize at this time, bridging and packing off the wellbore, typically when circulation is stopped. At this time, the flow of slurry and drilled cuttings exceeds the cleaning capacity of the well. At approximately 3:35 on this figure, immediately after a connection, the EMW increases and becomes erratic and spiky due to the restriction. This is correlated with a sharp increase in surface torque. At this stage, it is very difficult to drill ahead and common for the drillstring to become stuck.

The previous examples have described problems associated with SWFs occurring at a relatively shallow depth in reference to the subsea mud line. Not all problems in these intervals are associated with SWFs, as the PMWD depth log shown in **Fig. 6.3.18** illustrates. At no time is the well flowing in this example, but both the EMW and torque curves show dramatic increases in a section of the hole that is predominantly shale. This apparent hole failure results in a barely manageable hole-cleaning predicament. In this case, the collapse occurs during each connection. The sawtooth shape of the curves indicates that conditions improve while drilling the stand of pipe, but immediately deteriorate at a connection. However, the increased pump rate at X300 aids in hole cleaning and reduces the overall torque and EMW.

SWFs have proved to be a serious hazard when drilling the uppermost riserless sections of many deepwater wells. Historically, these flows have been difficult to deal with due to lack of surface and downhole information. Over the past few years, PMWD sensor information has increasingly been applied to this problem, leading to reduced risk in drilling these sections through more thoroughly informed decision making.

6.3.10 Wellbore Stability. In some wells, especially deviated and extended-reach, the window between the pore pressure and the fracture gradient may be small, and very accurate annular pressure information is essential to maintain operations within safe limits. Well-control requirements are such that circulation of an influx through the long choke and kill lines that run from the subsea BOP also imply a lower kick tolerance.

Fig. 6.3.19 is a modification of Fig. 6.3.1 to now include the effects of wellbore stability at two different wellbore inclinations: 30 and 70°. The influence of well deviation angle on the pressure window shows that managing the mud weight in extended-reach wells is made more difficult by annular pressure losses, which are inherently higher for wells with long horizontal sections.

6.3.11 Hole Cleaning. One of the primary uses of downhole pressure measurements has been to monitor hole cleaning. Efficient hole cleaning is vitally important in the drilling of directional and extended-reach wells, and optimized hole cleaning remains one of the major challenges. Although many factors affect hole

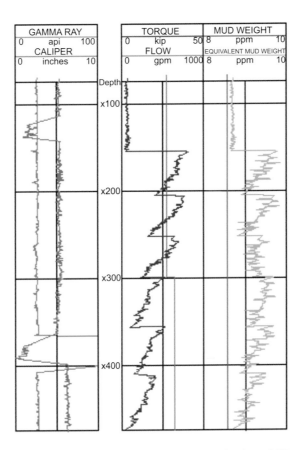

Fig. 6.3.18—Hole-cleaning issues in shallow riserless drilling.

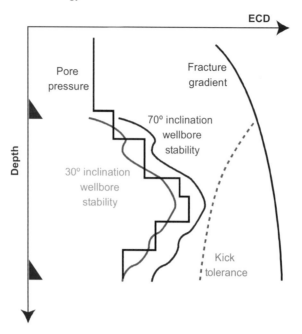

Fig. 6.3.19—Schematic representation of the decreasing pressure window as a function of well deviation (Hutchinson and Rezmer-Cooper 1998).

cleaning (as shown schematically in **Fig. 6.3.20**), two important ones that the driller can control are flow rate and drillpipe rotation.

6.3.12 The Effect of Flow Rate. The circulating mud flow rate is the most important parameter in determining effective hole cleaning. For fluids in laminar flow, fluid velocity alone cannot effectively remove cuttings from a deviated wellbore. Fluid velocity can disturb cuttings lying in the cuttings bed and push them up into the main flow stream. However, if the fluid has inadequate carrying capacity—yield point, viscosity, and density—then many of the cuttings will fall back into the cuttings bed. Mechanical agitation due to pipe rotation or backreaming can aid cleaning in such situations, but sometimes are inefficient and can even worsen the situation. Agitation that is too vigorous, such as rotating too fast with a bent housing in the motor, can have a detrimental effect on the life

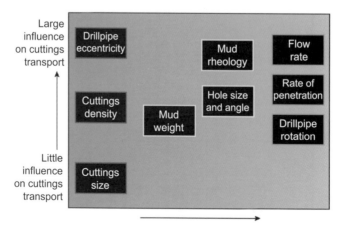

Fig. 6.3.20—Factors affecting hole cleaning. Some factors are under the control of the driller, such as surface mud rheology, rate of penetration, flow rate, and, to a lesser extent, hole angle. Others, including drillpipe eccentricity, and cuttings density and size, cannot be easily controlled (Aldred et al. 1998). Image copyright Schlumberger. Used with permission.

of downhole equipment. Inadequate flow results in increased cuttings concentrations in the annulus. A cuttings accumulation may lead to a decrease in annular cross-sectional area, and hence an increase in the ECD, ultimately leading to a plugged annulus, called a packoff. The use of real-time downhole annular pressure measurements allows early identification of an increasing ECD trend, caused by an increasing annular restriction, and helps the driller avoid formation breakdown resulting from large pressure surges or a costly stuck-pipe event.

An example shows how PMWD helps detect packoff (**Fig. 6.3.21**). The log shows that the annulus started to pack off at approximately 1:20. Other drilling parameters, such as increasing surface torque and variations in rotation rates, were becoming erratic. Standpipe pressure increased slightly. These warnings could have been interpreted as due to increased motor torque associated with an increase in surface torque. However, the large ECD increase confirmed that the mud flow was restricted around the BHA just above the annular pressure sensor. Based on the confirmation from PMWD, the driller reduced the mud flow rate and worked the pipe to prevent the ECD from exceeding the fracture gradient.

6.3.13 The Effect of Drillpipe Rotation. Another example demonstrates the effect of drillpipe rotation on hole cleaning in a horizontal section, 12¼-in hole. **Fig. 6.3.22** shows the annular pressure in white, rotary speed in red, and a rough approximation of produced cuttings seen at the surface in yellow. While sliding, cuttings fall out of suspension, causing a gradual drop in the effective mud density and resulting annular pressure. At approximately 20:20, pipe rotation resumes, resulting in an instantaneous increase in bottomhole pressure of roughly 50 psi, due to flow redistribution in the eccentric annulus and the pick up of cuttings.

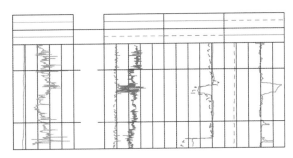

Fig. 6.3.21—Packing off. The driller responds in real time to an increase in ECD (red curve in Track 4) as the annulus packs off above the MWD tool (Hutchinson and Rezmer-Cooper 1998).

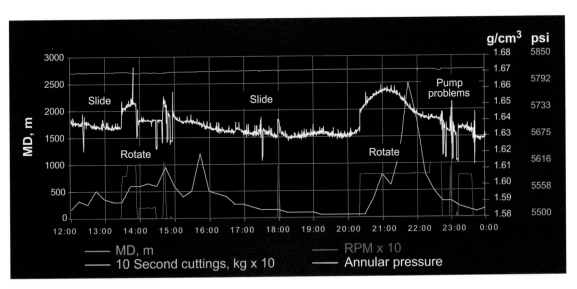

Fig. 6.3.22—Cuttings mobilization by a change in drillpipe rotation, and transport into less deviated portion of the well.

Then pressure continues to increase gradually as the cuttings are stirred up and become entrained in the flow, contributing to the effective density. Once the cuttings are fully mobilized, a pressure maximum is reached (with a corresponding maximum in the cuttings seen at surface after a delay (yellow peak) as the cuttings are circulated into a more vertical part of the well. The pressure then drops back to normal, with the corresponding decrease in cuttings seen at the surface.

In hole sections in which washouts are suspected or are known to occur, the immobile cuttings in the washout may not be seen by the PMWD tool until they are dislodged by the pipe movement. They then can pack off the BHA, causing large pressure surges. The PMWD tool may indicate a problem, but other indicators such as excess drag, overpull, and anomalous MWD readings should be observed as well. Backreaming may be necessary to improve hole conditions. Hole stability is necessary for avoiding trouble. An overgauge hole results in low annular accumulations, packoff, and a severe danger of stuck pipe. The correct mud weight for the hole angle, identification of early signs of hole instability, and minimization of out-of-gauge hole are essential.

Fig. 6.3.23 shows the effect of low mud rheology. There was an overloading of the shakers, packoffs, and excessive ECDs when the cuttings were lifted into the mud stream. As the yield point was increased from 15 to 25 lbf per 100 ft^2 during the day, this effect was mitigated, and then it disappeared.

It is often beneficial to circulate the hole prior to tripping to ensure that there will be no (or limited) sticking risk when pulling the BHA/bit out of the hole. **Fig. 6.3.24** shows the observed decrease in EMW measured by the PMWD tool as the hole was circulated clean. The annulus was loaded with cuttings, and on circulating the hole clean, the ECD reduced. Combined with the appropriate surface indication, in this case the material coming over the shakers, the PMWD tool indicated that the hole had indeed been cleaned.

It is important to recognize that the increase in ECD caused by drillpipe rotation will be related to the size of the cuttings bed. In the case of slide drilling, hole cleaning is less efficient, and cuttings tend to accumulate on the low side of the hole. Following long periods of slide drilling, it is recommended that the hole be circulated clean while rotating the drillpipe at normal rotary drilling revolutions per minute.

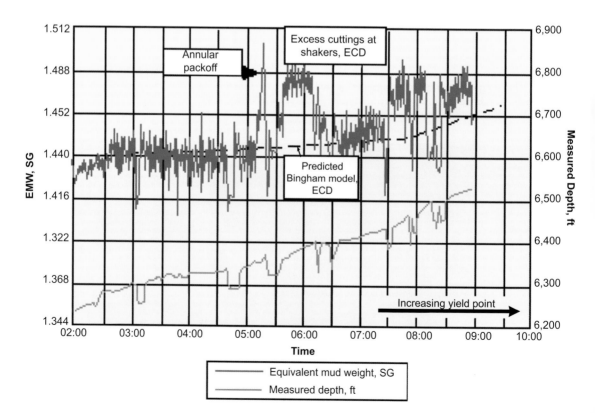

Fig. 6.3.23—PMWD equivalent mud weight vs. time (12¼-in. section, 55° inclination). Poor hole cleaning due to low mud rheology.

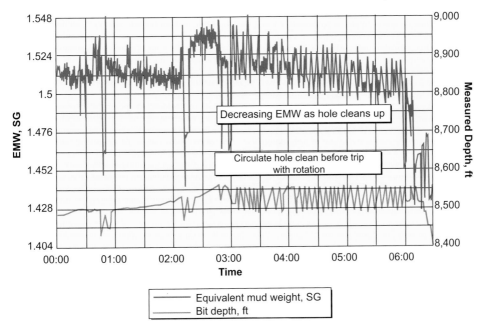

Fig. 6.3.24—EMW vs. time (2, 76°). EMW reduction as a result of hole cleaning prior to tripping.

Based on the assumption that no cuttings are removed from the well during the sliding phase, it is recommended that the well be circulated with rotation for 45 minutes for every 100 ft drilled in sliding mode. This figure is based on typical conditions for 12¼-in. and 8½-in. holes. Other operational issues are as follows:

- Hole cleaning is considerably worsened without pipe rotation, with the effect becoming more pronounced at higher angles.
- Circulation times before tripping out of the hole should be governed by cuttings returns and PMWD data. General guidelines of between 1.5 and 3 times bottom-up are given by the hole-cleaning charts, but often the amount of circulation can be reduced based on observation of PMWD trends.
- It is important to establish the downhole pressure trend when the hole is clean and to watch for an abnormal increase in the downhole pressure trend, which can indicate poor hole cleaning.
- Determine a maximum ROP to avoid overloading the annulus during sliding/rotary drilling.
- It is recommended that hydraulics and hole-cleaning simulations be run for all deviated wells. This enables mud properties and flow rates to be optimized to provide adequate hole cleaning in all sections of the well.
- Pulling through tight spots is acceptable, provided the pipe is free going down. Do not go immediately to the maximum overpull, but work up progressively, ensuring that the pipe is free to go down on every occasion.
- Backream only when absolutely necessary. Use models to determine the maximum allowable backreaming rate.

6.3.14 Plugging Below the Pressure Sensor. PMWD is an integrated measure of the effects in the borehole from the sensor to surface. Typically, the pressure sensor is located a short distance above the bit (from a few feet to a few tens of feet above the bit). Thus, if the annulus were to start packing off below the PMWD sensor, there would be no indication that the packoff was occurring from the PMWD sensor. However, if the standpipe pressure trace is not too noisy, there may be an indication on the pump pressure, manifested by an increase. An example is shown in **Fig. 6.3.25.** The fact that the PMWD tool sees nearly all packoff events shows that most are above the sensor, probably around the top of the BHA.

6.3.15 Gel Breaking. The thixotropic gel property of the drilling fluid is necessary to prevent cuttings and barite from dropping out of suspension during periods when the mud is not being circulated. During

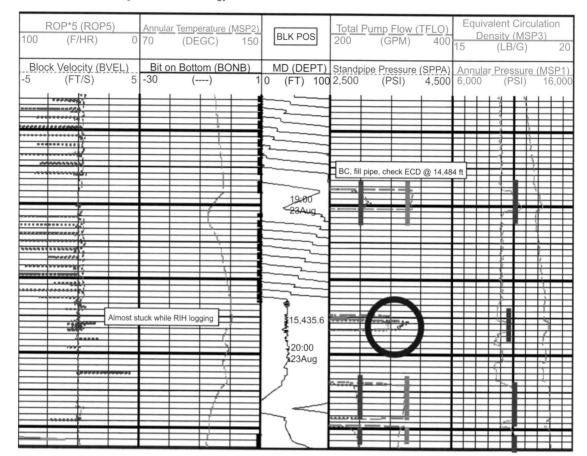

Fig. 6.3.25—Early detection of packoff below pressure sensor. Increase in standpipe pressure after running in hole (RIH) is not reflected on annular pressure sensor (ECD measurement).

times when the mud is not circulating, and even during connections, the mud can generate gel strength when in static conditions, and this requires significant pressure to break the gels after each connection. In some cases, this may be a sufficient pressure rise to initiate losses. Indeed, lost-circulation incidents are often blamed on the formation itself or mud, whereas PMWD data have shown that it can equally be drilling practices that are to blame (i.e., breaking gels by pumping instead of rotating first to shear the mud and break the gel). An example demonstrating the additional stress while breaking mud gels is shown in **Fig. 6.3.26** (Ward and Andreassen 1997). This example occurred while reaming down through the reservoir. In this case, a pressure surge over the mud weight up to twice the ECD was measured each time circulation was started after a connection.

6.3.16 Tripping Procedures. There are frictional pressure losses generated by fluid flow due to pipe movement—typically a positive surge pressure when running the pipe in, and negative swab pressures when pulling the pipe out. However, negative pressures due to fluid oscillations have also been seen when running in, and have been of sufficient magnitude to swab in an influx from the formation. Modeling these dynamic swab and surge pressures accurately is difficult, due to the uncertainties in geometry, downhole fluid properties, pipe acceleration effects, etc. Recorded-mode PMWD data can help determine the degree of inaccuracy in the models and can enable some degree of calibration.

An example of measured swab and surge pressures (Ward and Andreassen 1997) is shown in **Fig. 6.3.27**. It can be seen that the pipe movement can generate significant pressure transients. In some situations, this could be sufficient to cause hole or well-control problems.

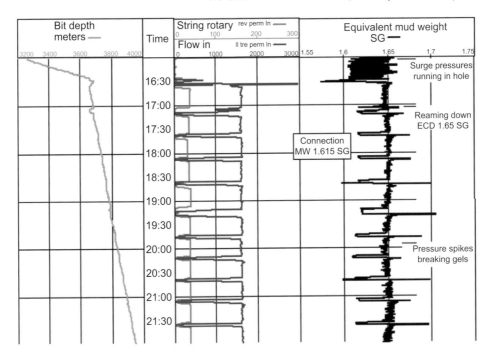

Fig. 6.3.26—Pressure spikes from breaking mud gels.

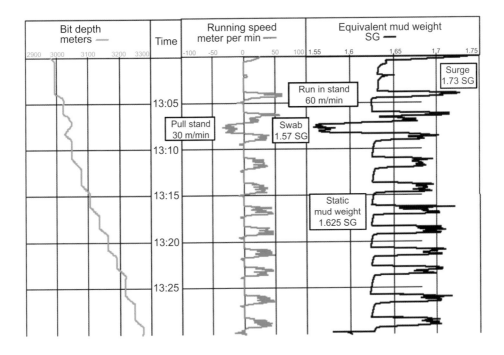

Fig. 6.3.27—Swab and surge pressures.

6.3.17 Well-Control Monitoring. The influx of formation fluid or gas into the wellbore due to or unexpected formation pressures is one of the most serious risks during drilling. The character of the influx will depend primarily upon the influx density, rate, and volume; the drilling fluid properties; and both borehole and drill-string geometry.

During gas kicks, the ECD response is dominated by the reduced density of the mud column, as heavier fluid is replaced by less dense gas or fluid, and by increased annular pressure losses due to friction and inertia when accelerating the mud above the gas influx. The reduced annular gap in slimmer annular geometries (e.g., casing drilling) can cause unique drilling problems.

Constant monitoring of all available drilling data is critical in detecting a downhole kick event. **Fig. 6.3.28** shows the PMWD response to a gas influx. When gas mixes with the drilling fluid, the density of the drilling fluid (and, therefore, the annular pressure) decreases. Fifty minutes after the ECD, shown in Track 3, started to decrease, a flow check confirmed that a gas influx had occurred. Note also that the increase in annular temperature, shown in Track 2, as the formation fluid, which is hotter than the circulating mud, warmed the borehole.

Pressure responses in both time and depth vary greatly from one well-control situation to another. However, similarities in the type of kick may result in a similar PMWD log. For example, in a saltwater kick **(Fig. 6.3.29)**, the influx occurs after the last connection is made, and both a reduction in EMW and an increase in the active pit totals are observed. This kick is recognized, drilling activities are stopped, and the well is shut in. The wellbore becomes a closed system, and the PMWD sensor records a pressure buildup curve that reaches equilibrium at the pore pressure of the invading formation. These kinds of recorded data have been used to dissect well-kill procedures and refine techniques used by rig crews. Normally, this information is not available from pulse MWD systems due to the slow circulation rates. However, the complete record of the kill from shut-in to circulating out the kick is available when the PMWD data are retrieved. Telemetry systems such as wired drillpipe and electromagnetic could be used in real time to obtain downhole pressure traces during well-control events.

In many cases, determining well-control problems from real-time pumps-on PMWD data is difficult in the absence of other data. However, when these data are combined with conventional mud-logging data, a clearer picture develops. **Fig. 6.3.30** demonstrates the use of gas hydrocarbon averages (percent) and pit volume totals (bbl) in the interpretation of PMWD data.

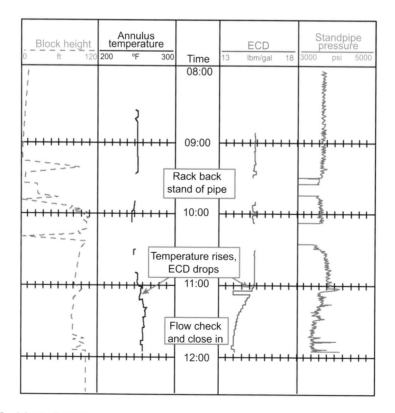

Fig. 6.3.28—PMWD response to a gas influx (Hutchinson and Rezmer-Cooper 1998).

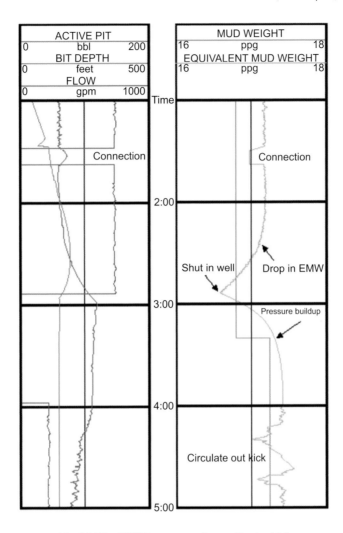

Fig. 6.3.29—PMWD response to a saltwater kick.

Pit volumes and gas-in-mud percentages indicate that the well is underbalanced over the three historical connections shown in this example. The length of time taken to complete each of the first two connections could account for the increase in flowback pit volumes and connection gas. Ambiguous surface data over several connections are common, especially in *loss/gain* situations. In fact, the rig crew did not raise the mud weight until after the third connection produced a sustained pit gain of approximately 10 bbl. An earlier response would have prevented this influx.

Well-control issues can appear during any rig operation, but especially dangerous is an influx while tripping. The log example shown in **Fig. 6.3.31** is an illustration of a kick occurring while tripping out of the hole. After the incident, the pore pressure was measured at 16.05 lbm/gal. This value was taken after shut-in from the stabilized PMWD buildup curve. Looking back, both the circulating EMW and the static EMW just before the short trip were greater than pore pressure. However, swab pressures from the short trip lowered the EMW below pore pressure, and the well took a gas influx. While tripping back in the hole, a decrease in pressure was noted as the PMWD sensor entered the swabbed-in light gas.

The mud-pulse PMWD service records only these kinds of events. There is no circulation throughout the short trip and none during the shut-in period of the well; also, the slow pump rates required to kill wells normally are not great enough to allow for real-time information to be transmitted. Future enhancements to this service will incorporate solutions for acquiring real-time data throughout all rig operations, whether circulating or not.

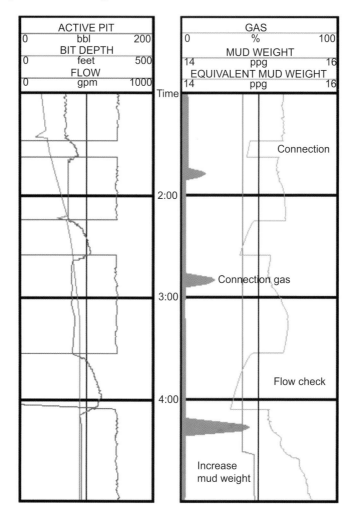

Fig. 6.3.30—Use of additional information to interpret a gas kick with PMWD.

6.3.18 Loss/Gain, Ballooning, and Breathing. The phenomenon, variously called borehole ballooning, breathing, loss/gain or wellbore storage, can result from drilling close to the fracture pressure. Slow mud losses are observed while drilling ahead, followed by mud returns after the pumps have been turned off, such as during a connection or flow check. Usually, any flows during these periods are a cause for concern because they may be mistaken as an influx of formation water, liquid hydrocarbons, or gas. The term *ballooning* originally implied that the wellbore diameter expands when circulation is started due to the additional ECD and contracts when circulation is stopped (Gill 1986, 1987, 1989). This explained the observed mud losses and gains and gave the analogy to blowing up and deflating a balloon. Some operators call this phenomenon breathing [i.e., the well takes (inhales) and returns (exhales) mud in response to turning the pumps on and off]. Others call it loss/gain due to the observed mud losses and gains. It now seems more likely that the phenomenon is due to fractures being opened and closed by the annular pressure fluctuations resulting from mud circulation and noncirculation (Bowman 1989; Holbrook 1989; Aadnoy 1996). Unfortunately the name *ballooning* has stuck.

As noted in the previous section, any influx from the formation can result in a well-control problem, the magnitude of which is dependent on its volume and composition. However, if the flow is due to mud returns, well control is not an issue. The question, then, is *how does one know unequivocally if it is an influx or if it is mud that was lost while drilling flowing back into the wellbore?* If the well is shut in, both situations typically show a pressure buildup.

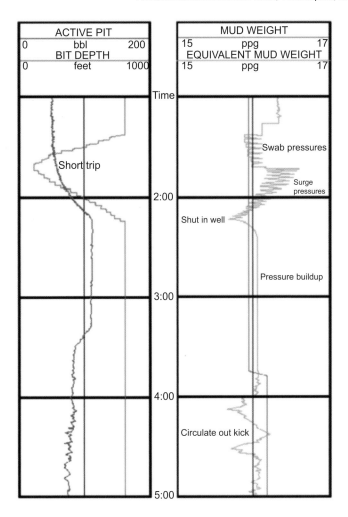

Fig. 6.3.31—PMWD response to kick while tripping.

This loss/gain situation has often been mistaken in the field for an influx of formation fluid. Misdiagnosis often leads to unwarranted well-control procedures that can be costly. One way to identify such a situation is with the PMWD signature during pumps-off periods. **Fig. 6.3.32** shows a series of three connections during a developing loss/gain situation. A normal connection is typically square-shaped when the pumps are stopped and started. When the pumps are off, the EMW is that of the whole annular column, in this case approximately 14.5 lbm/gal. During circulation, the EMW is quickly established at a constant value, in this example close to 15.5 lbm/gal. As the loss/gain develops, the PMWD connection signature changes. When the pumps are turned off, the EMW slowly decays to the static mud weight as mud bleeds back from the formation in a manner similar to a LOT, finally reaching the static mud weight at 4:00. When circulation is reestablished, the EMW slowly builds up to the ECD level as fractures are slowly refilled.

Loss/gain is a relatively common problem in deepwater wells due to the low overburden. If a loss/gain situation is misidentified as an influx, the normal response is to increase the mud weight which often leads to more ballooning and eventually total losses. The correct response is to decrease the mud weight, decrease the ECD (reduce the flow rate), or live with the losses and gains.

Mud losses while circulating are required for ballooning to occur, but these are partial and not total. Small losses that are continuous while drilling ahead may not be detected easily but can accumulate to a sizeable volume over a long drilling time. The losses must be into fractures that are contained within a limited fracture network and with little or no fluid leakoff into porosity. When the pumps are stopped, the annular pressure

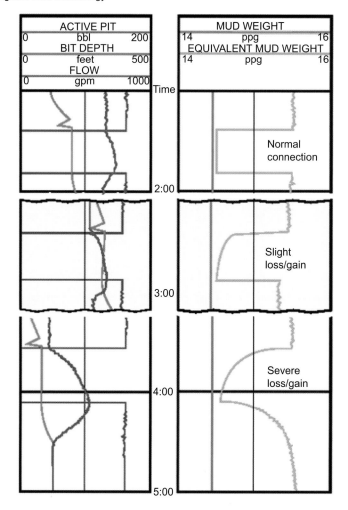

Fig. 6.3.32—PMWD signature during loss/gain events.

will fall, and mud lost to these fractures will flow back into the wellbore. Such returns are more noticeable than the losses because the return occurs rapidly and during a period when no flow is expected.

When ballooning occurs, the recorded PMWD response during the pumps-off period is diagnostic in that the pressure will fall sharply when the pumps are turned off, but at some point, the pressure will decay slowly to the static level due to mud flow from the loss zone. The static level will be reached when the mud return flow stops if the pumps are off long enough. The rate of the pressure decay should be related to the return flow rate, and it seems that it should be possible to predict the flowback volume from the pressure decay curve. In any event, the pressure will eventually reach, and then remain at, the static level until the pumps are restarted.

Once the pumps are restarted, the annular pressure should return to the level it was at before the pumps were turned off, assuming the flow rate is the same before and after the shutdown. There may be some delay in reaching the former pressure level (and ECD) as the fracture network is being recharged. In this case, not all the mud being pumped is coming up the wellbore at first. Some of the mud is refilling the system of fractures and being stored until the next time the pumps are turned off.

Fig. 6.3.33 shows the pressure changes during a normal connection, in the 8½-in. interval of a Gulf of Mexico well (Ward and Clark 1998). The mud in use was a 15.70-lbm/gal synthetic oil-based mud. In this case, the stand was drilled down to 15,856 ft, then reamed once before the pumps were turned off. During the pumps-off interval, the pit gained approximately 35 bbl, a normal flow from the lines when circulation is

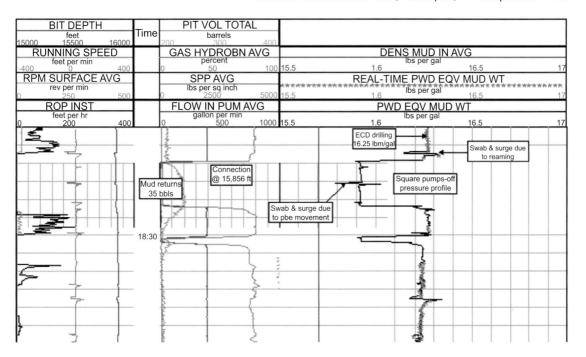

Fig. 6.3.33—Preballooning connection in an 8½-in. hole section. This has the normal *square well* signature.

stopped. The pumps were turned on again approximately 23 minutes later. When the pumps were turned off, there was a sharp pressure drop to near the static level. When the pumps were restarted, the pressure rose just as sharply to the level prior to reaming the pipe. This *square well* profile is characteristic of a normal connection with no signs of ballooning or flow. The EMW before and after the connection are the same at 16.26 lbm/gal. Reaming up and down caused a swab to 16.14 lbm/gal and surge to 16.32 lbm/gal before the pumps were stopped. The EMW dropped to 15.92 lbm/gal and stayed at that level, except for two times when the pipe was moved slightly. Once the pumps were turned on again, the EMW quickly increased back to 16.25 lbm/gal, the same as before the connection. Note that on Fig. 6.3.33, the effective downhole mud density is approximately 0.22 lbm/gal higher than that reported going in. The difference between the surface and downhole densities is due to the effect of temperature and pressure on the mud in the borehole.

Fig. 6.3.34 shows the connection at a depth of 16,679 ft. The mud weight in had been raised to 15.9 lbm/gal at this point, and the ECD was 16.48 lbm/gal. The swab and surge upon reaming up and down were similar to the previous example. When the pumps were turned off, the EMW fell sharply to 16.16 lbm/gal and then fell more gradually over the next 20 minutes to a near-static level of approximately 16.12 lbm/gal. When the pumps were restarted, the EMW rose quickly back to 16.47 lbm/gal, essentially the level before the connection. There were no reports at surface at this time, although the well was giving back 45 bbl, 10 bbl more than previously.

Fig. 6.3.35 shows a connection at 17,230 ft, at which ballooning was reported. The ECD was 16.42 lbm/gal as the stand was drilled down. When the pumps were stopped, the EMW fell rapidly to 16.37 lbm/gal and then gradually to 16.13 lbm/gal just before the pumps were restarted. After the pumps were restarted, the EMW eventually increased to 16.40 lbm/gal, or nearly the same as before the connection, but took more than 15 minutes even with a slightly higher flow rate. Ballooning was reported at this depth, and flow checks experienced returns of approximately 85 bbl. The gradual fall in pressure to the static level shown in Fig. 6.3.34 is indicative of returning mud flow with the pumps off. The return mud flow prevents the pressure from falling rapidly to the static level as shown in the preballooning well in Fig. 6.3.33.

Fig. 6.3.36 shows a connection at 17,696 ft, at which ballooning was severe just before a major lost-circulation event. Here, the slow buildup to the drilling ECD is obvious and can be observed with the real-time PMWD data. Also, the pressure does not reach static equilibrium during the 8-minute pumps-off period.

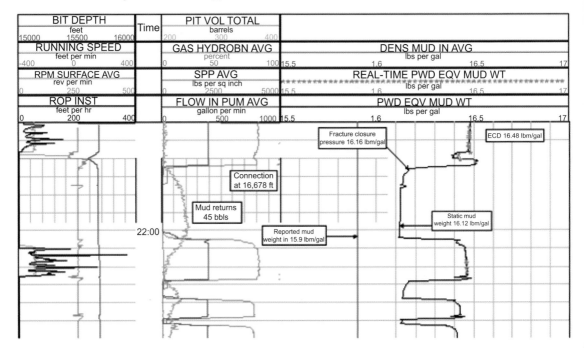

Fig. 6.3.34—Start of ballooning. Early indications of ballooning seen with PMWD, but not noticed at surface.

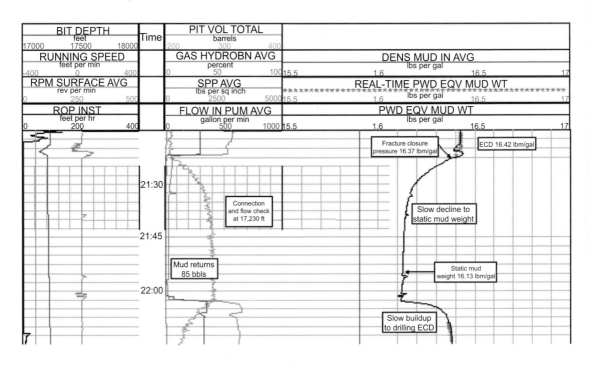

Fig. 6.3.35—Severe ballooning characterized by slow pressure decline and buildup curves.

How do the ballooning cases in Figs. 6.3.33 through 6.3.36 compare with a formation influx? Unpublished PMWD observations of influxes and kicks have shown that with a weighted mud (and ballooning always occurs with a weighted mud close to the fracture pressure), an influx will always be of a lower density than the drilling fluid, and in the case of gas, of a much lower density. Therefore, a drop in EMW characterizes the PMWD data

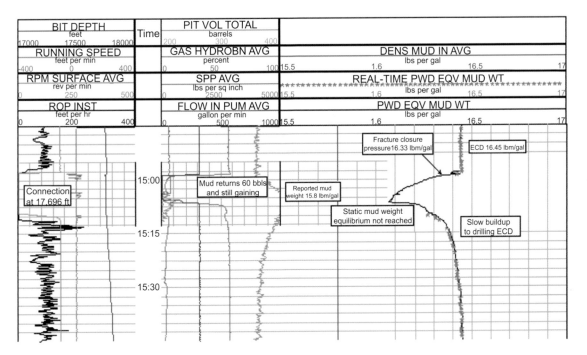

Fig. 6.3.36—Ballooning just before severe lost circulation. The quick connection does not allow static equilibrium to take place.

during an influx whether circulating, while static, or during tripping. A drop in ECD is also shown after the connection influx in Fig. 6.3.30. The magnitude of this drop depends on the relative densities of the influx and mud and the influx volume. In a ballooning well, the ECD always eventually goes back up to the same ECD as was observed before the static period. Had the pit gain been caused by an influx during the connection, then the ECD would not have increased back up to 16.45 lbm/gal but would have been somewhat less.

The connections made before ballooning was reported/observed on the rig raise questions about the shape of the initial section of the pumps-off portion of this pressure vs. time curve. There is a definite curvature in this section of the plot. This may be due to mud return flow that is overlooked or too small to be detected. If this is true, the onset of ballooning might be detectable via the PMWD tool before it becomes apparent on the rig floor. During ballooning, the PMWD signature is distinct, with a curved pressure decline when pumps are stopped and a slow curved pressure increase after pumps are restarted. In addition, on the decline curve there is typically a break in slope, which is interpreted to be the fracture closure pressure similar to that interpreted in many LOTs when pumping is halted.

A number of conclusions can be drawn from the preceding discussion of ballooning (Ward and Clark 1998):

- Ballooning is caused by slow mud losses during drilling and subsequent mud returns of equal volume when the pumps are turned off. The losses are confined within a limited fracture network in which little or no fluid leakoff into porosity is possible.
- Return mud flow will slow the pressure decline to the static level when the pumps are initially turned off.
- Mud losses will slow the return to the ECD level observed before the connection or the flow check after the pumps are turned on again.
- The ECD before and after the connection or the flow check will be the same, assuming the pump rate and rotary speed are the same. A lower ECD after the pumps are restarted is indicative of an influx.
- The downhole pressure response to turning the pumps off and on during a connection or a flow check is diagnostic for ballooning. A real-time PMWD tool measures this and can allow for correct decisions to be made.

6.3.19 Interpretation Guide. In this chapter, we have seen how the real-time monitoring of surface pressures alone can lead to potentially misinterpreted drilling events, primarily because of the nature of the fluid- and cuttings-filled column in the annulus, and the pressure window being navigated. Monitoring the ECD with downhole measurements certainly can assist in the interpretation of the complex events that can occur while drilling. However, a more complete picture of the events and potential causes can be achieved by combining all available information, both surface and downhole.

Real-time analysis of the causes, the signatures, and the parameters that are most appropriate to monitor is evolving, and not all possible events have yet been observed. However, certain common events do have repeatable signatures in both surface and downhole measurements. **Table 6.3.1** is a simple interpretation guide for some of the more common drilling events, highlighting how the ECD has changed most often in terms of its signature in a real-time depth-based (or, more usually, time-based) drilling log. Alongside the ECD signature for the event, the other indications or trends that may be seen in complementary measurements— such as surface pump pressure, surface and downhole torque, hookload, flow in, and flow out—are also

TABLE 6.3.1—BRIEF INTERPRETATION GUIDE FOR MOST COMMON EVENTS			
Event or Procedure	ECD Change	Other Indications	Comments
Mud gelation or pump startup	Sudden increase possible	Increase in pump pressure	Avoid surge by breaking rotation and slowing the pump
Cuttings pickup	Increase then leveling as steady state reached	Cuttings at surface	Increase may be more noticeable with rotation
Plugging annulus	Intermittent "spiky"surge increases	• Standpipe pressure • Surge increase? • Torque/rev/min fluctuations • High overpulls	Often an early warning for stuck pipe. Packoff may blow through before formation breakdown
Cuttings-load formation	Gradual increase	• Total cuttings expected not seen at surface • Increased torque • ROP decreases	If near plugging, may get pressure surge spikes
Plugging below sensor	Sudden increases as packoff passes sensor—none if packoff remains below sensor	• High overpulls • *Steady* increase in standpipe pressure	Monitor both standpipe pressure and ECD
Gas migration	Increase if well is shut in	Shut-in surface pressures increase linearly (approximately)	Take care if estimating gas migration rate
Running in hole	Increase—magnitude dependent on gap, rheology, speed, etc.	Monitor trip tank	Effect enhanced if nozzles plugged
Pulling out of hole	Decrease—magnitude dependent on gap, rheology, speed, etc.	Monitor trip tank	Effect enhanced if nozzles plugged
Making a connection	Decrease to static mud density	• Pumps on/off indicator • Pump flow rate lag	Watch for significant changes in static mud density
Barite sag	Decrease in static mud density or unexplained density fluctuations	High torque and overpulls	While sliding periodically or rotating wiper trip to stir up deposited beds, use a correct mud rheology
Gas influx	Decreases in typical-size hole	Increases in pit level and differential pressure	Initial increase in pit gain may be masked
Liquid influx	• Decreases if lighter than drilling fluid • Increases if influx accompanied by solids	Look for flow at mud line if relevant	Plan response if shallow water flow expected

given. In the last column are some comments on potential remedial action, or some caveats as to the complexity of the potential downhole event.

Nomenclature

a = hydrostatic pressure offset
b = bulk compressibility (includes mud compressibility and casing/openhole compliance effects)
P = pressure
P_{ann} = downhole annular pressure
P_{sp} = standpipe pressure

References

Aadnøy, B.S. 1996. Evaluation of Ballooning in Deep Wells. In *Modern Well Design*, second edition, Appendix B, 224–233. Rotterdam, The Netherlands: A.A. Balkema.

Alberty, M.W., Hafle, M.E., Mingle, J.C., and Byrd, T.M. 1999. Mechanisms of Shallow Waterflows and Drilling Practices for Intervention. *SPEDC* **14** (2): 123–129. SPE-56868-PA. DOI: 10.2118/56868-PA.

Aldred, W., Cook, J., Bern, P. et al. 1998. Using Downhole Annular Pressure Measurements To Improve Drilling Performance. *Oilfield Review* **10** (4): 40–55.

Bern, P.A., Zamora, M., Slater, K.S., and Hearn, P.J. 1996. The Influence of Drilling Variables on Barite Sag. Paper SPE 36670 presented at the SPE Annual Technical Conference and Exhibition, Denver, 6–9 October. DOI: 10.2118/36670-MS.

Bowman, G.R. 1989. Borehole Ballooning in Response to Gill. *Oil and Gas Journal* **87** (10 April 1989).

Eaton, L.F. 1999. Drilling Through Deepwater Shallow Water Flow Zones at Ursa. Paper SPE/IADC 52780 presented at the SPE/IADC Drilling Conference, Amsterdam, 9–11 March. DOI: 10.2118/52780-MS.

Gill, J.A. 1986. Charge Shales: Self-Induced Pore Pressures. Paper SPE 14788 presented at the SPE/IADC Drilling Technology Conference, Dallas, 9–12 February. DOI: 10.2118/14788-MS.

Gill, J.A. 1987. Well Logs Reveal True Pressures Where Drilling Responses Fail. *Oil and Gas Journal* **85**: 41–45.

Gill, J.A. 1989. How Borehole Ballooning Alters Drilling Responses. *Oil and Gas Journal* **87**: 43–51.

Haciislamoglu, M. 1994. Practical Pressure Loss Predictions in Realistic Annular Geometries. Paper SPE 28304 presented at the SPE Annual Technical Conference and Exhibition, New Orleans, 25–28 September. DOI: 10.2118/28304-MS.

Haciislamoglu, M. and Langlinais, J. 1990. Non-Newtonian Flow in Eccentric Annuli. *Journal of Energy Resources Technology* **112** (3): 163–169.

Hauser, B. 1998. Opening Remarks. Drilling Engineering Association (DEA) Shallow Water Flow Forum, The Woodlands, Texas, 24–25 June.

Holbrook, P. 1989. Discussion on Borehole Ballooning in Response to Gill. *Oil and Gas Journal* **87** (12 June 1989).

Hutchinson, M. and Rezmer-Cooper, I. 1998. Using Downhole Annular Pressure Measurements to Anticipate Drilling Problems. Paper SPE 49114 presented at the SPE Annual Technical Conference and Exhibition, New Orleans, 27–30 September. DOI: 10.2118/49114-MS.

Johnson, A.B. and Tarvin, J.A. 1993. Field Calculations Underestimate Gas Migration Velocities. *Oil and Gas Journal* **91** (46): 55–60.

Leach, C.P. and Quentin, K.M. 1994. Static and Circulating Kick Tolerance. Paper presented at the IADC Asia Pacific Well Control Conference, Singapore, 1–2 December.

Luo, Y., Bern, P.A., and Chambers, B.D. 1992. Flow-Rate Predictions for Cleaning Deviated Wells. Paper SPE 23884 presented at the SPE/IADC Drilling Conference, New Orleans, 18–21 February. DOI: 10.2118/23884-MS.

Luo, Y., Bern, P.A., and Chambers, B.D. 1994. Simple Charts To Determine Hole Cleaning Requirements in Deviated Wells. Paper SPE 27486 presented at the SPE/IADC Drilling Conference, Dallas, 15–18 February. DOI: 10.2118/27486-MS.

McCann, R.C., Quigley., M.S., Zamora, M., and Slater, K.S. 1995. Effects of High-Speed Pipe Rotation on Pressures in Narrow Annuli. *SPEDC* **10** (2): 96–103. SPE 26343-PA. DOI: 10.2118/26343-PA.

Pelletier, J.R., Ostermeir, R.M., Winker, C.D., Nicholson, J.W., and Rambow, F.H. 1999. Shallow Water Flow Sands in the Deepwater Gulf of Mexico: Some Recent Shell Experience. Paper presented at the International Forum on Shallow Water Flows, League City, Texas, 6–8 October.

Rezmer-Cooper, I.M., Rambow, F.H.K., Arasteh, M., Hashem, M.N., Swanson, B., and Gzara, K. 2000. Real-Time Formation Integrity Tests Using Downhole Data. Paper SPE 59123 presented at the IADC/SPE Drilling Conference, New Orleans, 23–25 February. DOI: 10.2118/59123-MS.

Rudolf, R.L. and Suryanarayana, P.V.R. 1998. Field Validation of Swab Effects While Tripping-In the Hole on Deep High Temperature Wells. Paper SPE 39395 presented at the IADC/SPE Drilling Conference, Dallas, 3–6 March. DOI: 10.2118/39395-MS.

Smith, M. 1998. Shallow Water Flow Physical Analysis. Paper presented at the IADC Shallow Water Flow Conference, Houston, 24–25 June.

Ward, C. 1998. Pressure-While-Drilling: Shallow Water Flow Identification. Paper presented at the Drilling Engineering Association (DEA) Shallow Water Flow Forum. The Woodlands, Texas, 24–25 June.

Ward, C. and Andreassen, E. 1997. Pressure-While-Drilling Data Improves Reservoir Drilling Performance. *SPEDC* **13** (1): 19–24. SPE 37588-PA. DOI: 10.2118/37588-PA.

Ward, C. and Beique, M. 1999. Pressure-While-Drilling Application for Drilling Shallow Water Flow Zones. Paper presented at the International Forum on Shallow Water Flows, League City, Texas, 6–8 October.

Ward, C. and Clark, R. 1998. Anatomy of a Ballooning Borehole Using Pressure While Drilling™ Tool. Paper presented at the Overpressures in Petroleum Exploration Workshop, Pau, France, 7–8 April.

SI Metric Conversion Factors

bbl	× 1.589 873	E – 01 =	m^3
ft	× 3.048*	E – 01 =	m
ft^2	× 9.290 304*	E – 02 =	m^2
°F	× (°F – 32)/1.8	=	°C
°F	× (°F + 459.67)/1.8	=	K
in.	× 2.54*	E + 00 =	cm
lbf	× 4.448 222	E + 00 =	N
lbm/gal	× 1.198 264	E + 02 =	kg/m^3
min	× 6.0*	E + 01 =	s
psi	× 6.894 757	E + 00 =	kPa

*Conversion factor is exact.

6.4 Formation Pressure While Drilling—Julian Pop, Paul Hammond, and Iain Cooper, Schlumberger

6.4.1 Introduction. The cost to drill complex wells has escalated significantly in the last few years. Numerous suites of logging-while-drilling (LWD) tools have focused on reducing the risks while drilling by assessing the formation pore pressure. An understanding of the pressure regime in the subsurface affects drilling safety and the overall nature of a casing-design program. Knowledge of the pore pressure enables an optimum mud program to be developed, so that wellbore-stability and well-control events can be avoided or minimized, and thus drilling can proceed at the optimum rate of penetration (Barriol et al. 2005). Typically, the mud weights are chosen so that the total pressure used in the wellbore is greater than the formation pressure, so that formation stresses can be controlled and well-control events avoided. However, as described in Chapter 9.1, we can see that wells are also drilled with the static mud weights below the formation pressure, or underbalanced. In deepwater drilling conditions the mud-weight window (the pressure regime between the formation pore pressure and/or stability and the formation fracture pressure) is quite narrow and can be difficult to define, particularly because other parameters also govern the selection of the mud properties, such as hole-cleaning ability and telemetry properties for the drillstring mud column. Therefore, the formation pressure is a key drilling parameter, and knowledge of it is key.

Typically, formation pore pressures have been estimated indirectly from predrill seismic measurements and from LWD measurements of resistivity and sonic properties. Formation resistivity depends directly on the porosity, the fluid in the pores, and its ionic strength. An increase in shale resistivity with depth, under normal compaction conditions, implies a reduction in porosity. Therefore, an anomalous formation pressure is associated with a shift in the normal compaction trend, indicated on a log by a reduction in resistivity (and an associated increase in porosity, **Fig. 6.4.1**) (Alford et al. 2005).

However, because of the indirect nature of these measurements, other factors can potentially mask changes in the normal compaction trend and can obfuscate the detection of abnormal pressures (Aldred et al. 1989). Factors that may complicate the inference are as follows:

- Organic deposits can increase resistivity.
- Borehole-geometry defects such as washouts or cavings can increase the error in the resistivity measurement.
- A nonuniform borehole temperature can affect the resistivity of the formation water.

Acoustic velocities, such as those used in seismic measurements, are a function of the porosity (i.e., the lower the acoustic velocity, the higher the porosity). Again, in normally compacted formations, the compaction increases with depth. Porosity, in turn, decreases with depth, and therefore sonic- and seismic-wave velocities generally increase with depth.

Overpressured zones are typically associated with sediments that have not compacted, so a deviation from this trend can be an indication of a potential drilling hazard. However, the depth resolution of seismic measurements is typically poor, and predrill hazards can be difficult to determine from acoustic measurements. Sonic measurements while drilling can refine the predrill information to give a more accurate indication of the risk of penetrating an overpressured zone **(Fig. 6.4.2)**.

Nevertheless, periodic calibration points are required for quantitative predictions. Until recently, leakoff tests, formation-integrity tests, and kick or influx events were used to ascertain the pressure limits, and these formed de facto calibration points.

However, measurements of formation pressure while drilling (FPWD) have revolutionized formation-pore-pressure estimation and now enable a wide range of drilling-efficiency and -quality improvements. For example, FPWD measurements can assist with the following:

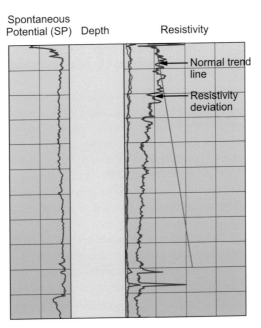

Fig. 6.4.1—Deviation in resistivity from the normal compaction trend may indicate anomalous formation pressures (Alford et al. 2005). Image copyright Schlumberger. Used with permission.

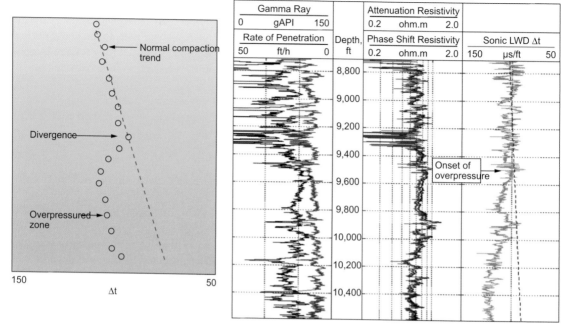

Fig. 6.4.2—The use of sonic measurements while drilling to identify potentially overpressured zones (Alford et al. 2005). Image copyright Schlumberger. Used with permission.

- Casing design. By enabling a better understanding of the near-wellbore pressure environment, planned casing points can be optimized. Significant cost savings can be achieved by avoiding premature casing runs and incorrect placement of the casing.
- Geosteering and geostopping. These decisions can be optimized to the real-time environment on the basis of FPWD measurements (Neumann et al. 2007). Quick decisions can eliminate the time wasted in drilling pressure-depleted formations and also can preserve virgin-pressure zones that have been scheduled for sidetrack development or completion. Detailed data on the local pressures and fluid mobility can help target the most productive zones and, when combined with models, aid in the determination of the optimal drainage length for horizontal wells.
- Highly faulted formations. Real-time pressure data can help with geosteering decisions made between the individual compartments.
- Carbonate reservoirs. The FPWD pressure and mobility data, when combined with surface measurements (such as pyrolysis or hydrogen-gas ratios), can aid in geosteering decisions that facilitate the placement of injectors above tar mats to avoid injecting into heavy-oil layers (Seifert et al. 2007).
- Determination of formation properties. Accurate determination of the formation pressure in the reservoir can make it possible to analyze both virgin and well-developed reservoirs. In virgin formations, pressure profiles can be combined with other LWD measurements to develop a full static model of the reservoir that can enable improved planning decisions with regard to future well development. In a developed reservoir, the pressure profiles can be used to understand the fluid movement in the reservoir at different levels within the formation.
- Dynamic reservoir modeling and simulation. Pressure profiles, defining gradients, and contact points can be combined with production histories (in addition to the static reservoir model) to develop a model for the dynamic pressures within the formation. These models are a key element in optimizing the total recovery from a particular reservoir and can determine the type and complexity of the completion system used for production.

The chapter will describe the basic theories of the tools, test sequences, and measurements of FPWD, will highlight how the significant phenomenon of supercharging (the increase in sandface pressure caused by the leakoff of drilling-fluid filtrate) can be allowed for in pressure interpretation, and ultimately will show how the measurements have been validated with post-drill measurements that are typically wireline-deployed.

6.4.2 Measurement Devices and Methodology. All major service companies provide an FPWD service, and there have been many recent publications that describe the details of the services. We will give the general principles of the measurement here. Details of particular tools and case studies are described in prior publications (Finneran et al. 2005; Fletcher et al. 2005; Pop et al. 2005b).

Any new measurement that is developed must provide robust, reliable, and repeatable readings. The various types of FPWD tools are somewhat similar to their wireline-deployed counterparts in that they are typically probe-style measurements based in collars or stabilizers.

The principle of operation is simple. Drilling is temporarily halted, and a probe is pushed into the formation, surrounded by a sealing device that can isolate the measurement from the borehole. The probe-type formation tester measures the pressure just inside the formation at the wellbore wall (the sandface) which is essentially the interface between the external mudcake and the formation. **Fig. 6.4.3** is a schematic representation of the pressure regimes across the formation.

A typical operational sequence will be described later in the section, but the designs are such that circulation can be maintained so that the mud and borehole can be conditioned, and the risk of differential sticking of the tool against the formation reduced. It is noted that if there is a drilling motor in the bottomhole assembly (BHA), the circulating mode can lead to vibrations, so the preference is to have the pumps off in a motor BHA configuration. Typical tool configurations are shown in **Fig. 6.4.4**.

The principal features of such tools are as follows (Pop et al. 2005b):

- A probe that is forced against the borehole wall with an elastomeric sealing element to aid with the quality of the seal
- Pistons that force the tool against the wall and that aid in the protection of mudcake formation in the vicinity of the probe from erosion due to mud circulation
- Collar/stabilizer designs that divert the flow away from the neighborhood of the probe and minimize the mud velocity in the neighborhood of the probe, resulting in less mudcake erosion and filtrate leakage into the formation

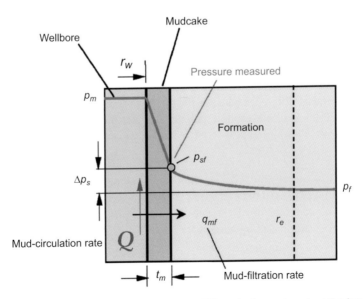

Fig. 6.4.3—Pressure regimes encountered by a probe-type FPWD tool when set against the formation wall. Depending on the mud filtration rate and the contrast in transmissibilities between the mudcake and the formation, the sandface pressure may not be close to the true (far-field) formation pressure. Mud circulation limits the thickness of the mudcake, thus promoting mud filtration, whereas under static wellbore conditions the mudcake will grow, eventually achieving maximum resistance (Pop et al. 2005b).

Fig. 6.4.4—Typical FPWD tools.

- Power source to allow the tool operate in pumps-on and pumps-off mode
- Electromechanical pretest system that allows precise control over the pretest rate and volume drawn into the tool
- Flowline and wellbore pressure gauges

The tools typically can be run in while-drilling mode in conjunction with rotary-steerable systems or after drilling. To minimize the time not drilling, and to avoid the potential risk of sticking, the total measurement time is minimized and is typically on the order of a few minutes to tens of minutes.

The depth at which the measurement is to be taken typically has been identified first by predrill and/or offset-well analysis and refined by other LWD measurements. Once the probe has been deployed and the measurement made, a signal is given and the probe is retracted. Real-time or post-event data are transmitted to surface, depending on the individual needs of the job.

6.4.3 Supercharging. The properties of both the formation and the mud will determine whether the sandface is an accurate representation of the true far-field pressure. If the filter cake is strong, it should form a good seal, and the measurement will give a more accurate representation of the formation pressure than a poorly sealing filter cake. If the seal is not ideal, then there is a pressure drop from the sandface to the far-field formation. When there is a significant difference between the sandface and the formation pressure, the formation is said to be *supercharged.* The magnitude of the supercharging is expressed by the following:

$$\Delta p_s = p_{sf} - p_f, \dotfill (6.4.1)$$

where p_{sf} is the sandface pressure, and p_f is the far-field formation pressure as per Fig. 6.4.3.

Other factors that may affect the sandface-pressure measurement are

- Mud type (oil-based, water-based, synthetic)
- Reservoir-fluid type
- Rock wettability
- Capillary pressure effects

The amount of supercharging depends upon the cumulative effects of the mud-filtration rate from the moment the formation was first drilled, and leakoff first occurred, and upon the ability of the formation to actually accommodate any filtrate. Phelps et al. (1984) have indicated that the amount of supercharging should be inversely proportional to the formation total mobility (i.e., inhibition of the mudcake seal against filtration increases the supercharging effect). For example, mud circulation that restricts the growth of the filter cake or drilling dynamics that erode or scrape off the filter cake could lead to an increase in supercharging.

Dynamic conditions when circulating the drilling fluid have a significant effect on the buildup of filter cake. This is well documented in the literature for both water-based and oil-based systems (Longeron et al. 1998). The properties in the wellbore for a given BHA configuration and circulation rate that ultimately will affect the degree of filtration are as follows:

- Mud type (solids particle-size distribution)
- Mud rheological parameters (yield stress, effective viscosity)
- Mud gel strength
- Formation mobility ratio between filtrate and the formation fluid

Pop et al. (2005b) highlight that supercharging can be associated with at least three phenomena:

1. Static effects. Rock wettability and capillary pressure when an immiscible mud filtrate invades the formation. Usually, this is of significance when small-volume pretests are used in the pressure-measurement process.
2. Pseudostatic effects. The overpressure existing after the mudcake has formed an effective seal cannot dissipate within the time frame of the formation test because of a very low formation mobility hindering leakoff.
3. Dynamic filtration. The degree of overpressure that exists in the near-wellbore region as a result of the mudcake not forming an effective seal. This can be caused by three factors:

- The mudcake has not fully formed after drilling. This is related to the formation permeability, the degree of overbalance while drilling, and the time between drilling and testing.
- There is mudcake erosion and growth inhibition directly attributable to the circulation rate.
- There are mechanical disturbances to the integrity of the mudcake (e.g., BHA vibrations leading to stabilizer gouging).

Experiments described by Pop et al. (2005b) have shown that the sandface pressure that is measured is influenced directly by the circulation rate, and that for measurement purposes there are benefits from performing the test with no circulation. However, the risk of sticking must also be considered when designing the test sequence.

The tools are designed so that the sealing elements and the probe achieve the best possible seal, and the effects of mud circulation can be minimized where possible, to avoid filtration leakage into the formation in the vicinity of the probe.

To allow for the effects of supercharging during tests, models have been developed that allow for the factors that influence supercharging (Chang et al. 2005). Models simulating the temporal development of the formation pressure and filtrate leakoff are valuable planning aids in understanding the measurements of formation pressure made while drilling, and they highlight that some of the experiences that are obtained with a wireline-deployed post-drill formation-pressure measurement may not necessarily be relevant in the while-drilling context, because of the effects of filtration rate.

We first outline the basic ideas behind the models used to simulate drilling-fluid-filtrate leakoff, formation pressure, and supercharging.

6.4.4 Modeling Fluid Loss. To predict supercharging, two elements must be combined: a model for the leakoff rate of drilling-fluid filtrate, from the instant of first penetration of the formation until the time of measurement, and a model for pressure and flow within the formation surrounding the wellbore, over the same period of time. The filtrate-leakoff model is itself a combination of a filter-cake-growth model and a wellbore-hydraulics calculation (Chang et al. 2005).

The main phenomena that must be captured in a model for while-drilling wellbore filtration are well known. After a very rapid pore-clogging phase, during the first minutes after a formation is drilled an external filter cake forms and grows in thickness, and the filtrate-leakoff rate decreases in response. During this quasistatic filtration period, for a given mud, the controlling factor on leakoff is the differential pressure between well-bore and formation. After some time, the leakoff rate stops decreasing and takes on an equilibrium dynamic-filtration value controlled by the shear stress exerted by the flowing mud at the surface of the cake, which in turn is controlled by the circulation rate of the mud, its rheology, and the geometry of the hole and drillstring. [This picture must be modified if conditions are such that leakoff rates are less than a critical value below which particles no longer accrete to the cake (Fordham et al. 1991).]

Further cake growth and decrease in leakoff can occur if the circulation rate decreases or the gap between the formation and BHA increases. Cake erosion and corresponding increase in leakoff rate can occur if the hydraulics conditions become more aggressive (higher circulation rates, narrower gap).

Standard drilling-hydraulics calculations give the wellbore pressure and the frictional-pressure-loss gradient (which is closely linked to the hydraulic shear stress), and so allow us to compute the main controls on cake growth and leakoff. As drilling proceeds, assuming for the moment that the drilling-fluid circulation rate stays constant in time, the hydraulic wall shear stress at a fixed depth first takes on large values because of the small gap for mud flow between drill bit and sandface. Shear stresses drop as collars and parts of the BHA with smaller outside diameters (ODs) arrive opposite the formation. They may rise again if there are any large-diameter elements in the BHA (stabilizers or, indeed, the FPWD tool), and finally they will drop to a low value once the drillpipe arrives.

This picture will be complicated by changes in circulation rate (e.g., at connections), working of the pipe, and mechanical scraping or plastering action of the drillstring. It is straightforward to include the first two in any simulation, but we do not at present know enough about filtration with mechanical action on the cake to model the latter process.

To model actual leakoff rates, it is also necessary to enter information about the rheology and filtration properties of the particular drilling fluid. Ideally, these data would come from on-site measurements on the actual fluid. While this may be practical for the rheology and static-filtration properties, the measurement of dynamic-filtration characteristics requires a more-complicated apparatus. For that reason, for the dynamic parameters, it may be necessary to rely on a library of typical values or to recognize that a significant uncertainty enters here and to simulate a range of cases.

The approach used to model filtrate-loss rates vs. time, and in changing hydraulic conditions, is that the filter cake grows or erodes at a finite rate until the leakoff rate becomes equal to that which would be measured in equilibrium dynamic filtration at the current hydraulic conditions. The cake growth rate, relevant if the current leakoff rate is greater than the appropriate equilibrium dynamic-filtration value, is taken to be that implied by static-filtration measurements at the current wellbore/formation differential pressure for a cake of the same material content. The erosion rate, relevant if the leakoff rate is less than the appropriate equilibrium dynamic value, is taken to be an independent parameter. Its value must be determined from experiments specially designed to probe cake behavior when mud-hydraulics conditions are stepped to more-aggressive values. In the present simulations, the erosion rate is zero unless the mud flow is turbulent.

In contrast, the model for processes within the formation is comparatively familiar. As in pressure-transient well testing, Darcy's law is combined with ideas of mass conservation and fluid compressibility. In situations that can be approximated as single-phase, such as invasion from an oil-based mud (OBM) into an oil-bearing zone with the properties of the two oils being similar, this leads to a simple diffusion equation for the pressure. In multiphase situations, such as invasion from water-based mud (WBM) into a hydrocarbon zone, a set of equations governing saturation(s) and pressure results. For simplicity, however, only single-phase situations will be discussed here.

The mathematics behind these models is as follows. The approach to filtration modeling taken here is strongly influenced by the model of Dewan and Chenevert (2001) and by the ideas of Fordham et al. (1991). To summarize, recall that Fordham et al. (1991) proposed that drilling-fluid filtration under constant circulating wellbore conditions can be understood in terms of a succession of phases:

1. An early-time quasistatic period, in which fluid loss and cake growth proceed in the same way as if the mud were not flowing, and in which the dominant control is the pressure difference between wellbore and formation
2. A transition period

3. A late-time equilibrium dynamic-filtration phase, in which cake growth has ceased, and the filtrate-loss rate takes on a constant value determined by the shear stress exerted by the flowing mud at the cake surface and is independent of the pressure difference between wellbore and formation and of the formation properties

During the equilibrium dynamic-filtration phase, it is believed that particles from the mud are no longer accreting to the cake [as opposed to the alternative view, which is not consistent with observation (Fordham et al. 1991), that an equilibrium is set up between accretion and removal]. It is further believed that accretion ceases because the filtrate-loss rate has fallen below a critical value that is dependent on the hydraulic shear stress exerted at the cake surface. To quote the abstract of Fordham et al. (1991), "The conjecture of a critical dynamic filtration rate ... means that invasion volumes should be independent of mud overbalance and of formation properties ... unless the formation permeability is so low that it limits the filtration flux to less than the critical flux." In that paper, the authors presented a simple mathematical description for filtrate-loss rates in each of the three phases listed in the preceding. The model in this paper simply reformulates those ideas into a format more suitable for inclusion in a simulator capable of predicting loss rates under time-dependent wellbore conditions.

The model of this paper differs from that of Dewan and Chenevert (2001) in a number of ways:

1. Here, concepts such as mudcake thickness and permeability, which are difficult to determine unambiguously from simple experiments, are avoided in favor of direct use of experimentally determined quantities, such as the pressure-dependent desorptivity.
2. In a similar vein, the requirement that the particle-size distribution in the mud be known is avoided by use of a simple critical flow-rate cutoff on the fluid-loss rate in dynamic filtration.
3. The model is formulated in terms of differential equations.

While these simplifications keep the model close to experimentally measurable quantities, there are disadvantages. For example, it cannot capture the slow decline in dynamic fluid-loss rates that is often observed and is conjectured to be the result of slow consolidation of the cake (Fordham et al. 1991) or a reduction in the mean size (and number) of particles accreting (Dewan and Chenevert 2001). The extent to which this is a serious issue depends on the time scale over which we need to model fluid-loss rates. A formation-pressure-measurement tool 50 ft behind the drill bit will first reach a target formation an hour or so after first exposure, and this may be a sufficiently short time that long-term cake-compaction effects are negligible. On the other hand, formation-pressure measurements could be taken days after first exposure, for example when a shallow formation is retested after drilling deeper formations. In this case, not only long-term cake compaction, but also the effects of tripping in and out, surge and swab, mechanical interactions between drillstring and cake, and long-term changes in mud properties, must be assessed and accounted for if they are likely to be significant.

In the following sections, the basic mathematical model is described first, then details are given of the *constitutive relations* (i.e., the functional forms used to capture the variation of mud-filtration properties in response to changing wellbore-hydraulics conditions). Where possible, the attempt has been made to keep the quantities used by the model to characterize the filtration properties of the fluid system close to those that are already routinely measured; for example, the key descriptor of static filtration is obtainable directly from a series of American Petroleum Institute (API) standard mud-filtration tests.

A formation flow model, assuming single-phase, radial, slightly compressible flow, is included as part of the model presented. A more elaborate formation model could be implemented, and this may be necessary for the correct prediction of supercharging pressures in situations where, for example, relative permeability and capillary pressure contributions are significant, or where filtrate and native-fluid viscosities are significantly different. In principle, there is no obvious difficulty in replacing the single-phase model with a two- or three-phase formation flow model, at least if we stay within the numerical framework described here.

Basic Equations. The total volume of filtrate lost into the formation per unit area of sandface, V, is related to the filtrate fluid volumetric loss rate per unit sandface area, q, through

$$\frac{dV}{dt} = q. \qquad \qquad (6.4.2)$$

We take time $t = 0$ to correspond to the instant at which the formation of interest is first drilled. Filtrate leakoff starts at this instant, and so

$$V(0) = 0. \qquad (6.4.3)$$

Contributions to invasion from fluid loss in the immediate vicinity of the working drill bit will be ignored. Some idea of the error involved can be obtained by comparing anticipated near-bit depths of invasion, using measured at-bit fluid-loss-rate values, against the radius of invasion computed from $V(t)$, $R_{\mathrm{inv}}(t) = r_w \left[1 + 2V(t) / \phi r_w \right]^{1/2}$. At-bit fluid losses could be included by suitable modification of Eq. 6.4.3.

Fluid loss from the drilling mud into the formation is usually associated with the growth of a filter cake of mud solids on the exposed surface of the formation. (We shall here neglect detailed treatment of any solids clogging the pore space of the formation.) The mass of solids in the filter cake, per unit sandface area, is denoted as M. This quantity can grow, through deposition of solids at a rate D, and can decrease as a result of erosion, at a rate E, so that

$$\frac{\mathrm{d}M}{\mathrm{d}t} = D - E. \qquad (6.4.4)$$

Here, the term *erosion* is used to refer to the process of removal of material, deposited at some more or less distant time in the past, from the mudcake by a newly imposed *strong* wellbore flow. It, and Eq. 6.4.4, should not be understood as implying that a state of dynamic equilibrium between solids deposition and removal exists once steady-state dynamic-fluid-loss conditions have been attained in a situation of constant-wellbore-flow conditions. In fact, the constant long-term fluid-loss rate in dynamic filtration is not the result of a dynamic balance but is believed to be a genuine cessation of deposition. Because there is no cake present at the instant the formation is first drilled, $M(0) = 0$.

It is convenient to introduce a new variable

$$V^*(t) = \frac{M(t)}{\kappa} \qquad (6.4.5)$$

to replace $M(t)$. Here V^* is the volume of filtrate per unit area that would have been lost had the current cake been built under constant conditions of differential pressure and hydraulic shear stress, and κ is the mass of mud solids per unit volume of filtrate. Writing $D^* = D/\kappa$ and $E^* = E/\kappa$, it follows that

$$\frac{\mathrm{d}V^*}{\mathrm{d}t} = D^* - E^*, \qquad (6.4.6)$$

subject to

$$V^*(0) = 0. \qquad (6.4.7)$$

Within the formation, assuming radial, single-phase, slightly compressible flow in a uniform medium, the pressure $p(r, t)$ satisfies

$$\frac{\partial p}{\partial t} = \frac{k}{\phi \mu c_t} \frac{1}{r} \frac{\partial}{\partial r} \left(r \frac{\partial p}{\partial r} \right), \qquad (6.4.8)$$

where k is the formation permeability, ϕ is the porosity, c_t is the total compressibility, and μ is the pore fluid viscosity. At the sandface, $r = r_w$, the formation flow must match the filtrate influx, and so

$$-\frac{k}{\mu} \frac{\partial p}{\partial r}(r_w, t) = q. \qquad (6.4.9)$$

At large distances the pressure tends to the undisturbed formation value, $p \to p_\infty$ as $r \to \infty$ for all t, and at the initial instant, $p(r, 0) = p_\infty$, $r \geq r_w$. For convenience, we denote the formation pressure at the sandface as

$p_{sf}(t) = p(r_w, t)$, and the pressure difference between the wellbore and the sandface (i.e., the pressure difference across the cake) as $p_{well}(t) - p_{sf}(t) = \Delta p(t)$. The pressure supercharging is simply $p_{sf}(t) - p_\infty$.

Eqs. 6.4.2, 6.4.6, and 6.4.8 are integrated forward in time to predict leakoff and supercharging. When doing so, it is necessary to specify as input the time variations of wellbore pressure, $p_{well}(t)$ and $dp_{well}(t)/dt$, and the drilling-fluid circulation. These, together with the wall shear rate and shear stress resulting from the circulating fluid and the local Reynolds number, are obtained, in the form of a set of tables of values vs. time, using a drilling-hydraulics package.

The next subsection outlines the constitutive relations.

Filtration Model. The mass rate of deposition of solids into the mudcake is related to q, the current filtrate loss rate, through

$$D^* = qa \quad\dotfill\quad (6.4.10)$$

and

$$a = \begin{cases} 1 & \text{if } q > q_{crit} \\ 0 & \text{otherwise.} \end{cases} \quad\dotfill\quad (6.4.11)$$

The role of the a term is to turn off accretion of solids into the cake should the filtrate-loss rate fall below the critical value q_{crit}. Provided the filtrate-loss rate exceeds the critical value, mud solids are fully incorporated into the filter cake at a rate proportional to the filtrate-loss rate. Cessation of deposition when the filtrate flux falls below a hydraulics-dependent value is the key idea of Fordham et al. (1991), while in the model of Dewan and Chenevert (2001), as fluid-loss rates decrease, only progressively smaller particles from the mud are allowed to adhere into the cake.

Little is known about mudcake erosion under conditions of intense wellbore-fluid circulation or mechanical interaction with drillstring hardware (indeed, it is also fair to say that we know too little about the consequences for fluid loss of cake compaction and plastering caused by mechanical interactions between the drillstring and the borehole wall), and we follow the lead of Dewan and Chenevert (2001), taking a purely heuristic approach.

The erosion process is here assumed to follow a linear kinetic law, with a rate constant (i.e., reciprocal time scale), λ. We assume that no erosion can occur unless the hydraulic shear stresses exerted by the flowing mud at the cake surface exceed a critical value, τ_{crit}. Furthermore, we assume that erosion drives the filter cake toward a state in which the equilibrium dynamic filtration fluid loss rate takes the value appropriate to the current wellbore hydraulics. Lastly, if there is no cake, no erosion can occur. Putting all these ideas together, the resulting expression for E is

$$E^* = \begin{cases} 0 & \text{if } V^* < 0 \\ \lambda b \max\left(0, V^* - V^*_{crit}\right) & \text{otherwise,} \end{cases} \quad\dotfill\quad (6.4.12)$$

where the factor b captures hydraulics wall-shear-stress threshold effects on erosion, and

$$b = \begin{cases} 1 & \text{if } \tau_w > \tau_{crit} \\ 0 & \text{otherwise.} \end{cases} \quad\dotfill\quad (6.4.13)$$

The value of V^*_{crit} will be discussed below, once we have linked filtrate-loss rate q to M and Δp. In essence, V^*_{crit} will take the value necessary for the fluid-loss rate to be equal to $q_{crit}(\tau_w)$ when $V = V^*_{crit}$. For simplicity, we propose that the threshold for cake erosion corresponds to the onset of turbulence in the mud flow past the cake. Thus, rather than setting a shear-stress threshold, we assume that erosion commences when the Reynolds number for the local mud flow exceeds the critical value for the transition to turbulence. Different erosion-onset criteria could be proposed, but they would not alter in essence the structure of the filtration model.

It is well known (Fordham et al. 1991) that the cumulative fluid-loss volume per unit area, in static filtration of drilling mud on a substrate of negligible resistance, varies with time as $V(t) = S(\Delta p)t^{1/2}$. The factor S is referred to as the desorptivity and is known from experiment to vary weakly with filtration pressure and to exhibit hysteresis if filtration pressure is decreased and subsequently increased. The associated fluid-loss rate in static filtration is $q(t) = \left[S(\Delta p) t^{-1/2} \right]/2$, and the deposited cake mass is $M(t) = \kappa S(\Delta p) t^{1/2}$. Combining these two expressions, to eliminate t in favor of the time-dependent quantity M, we obtain

$$q(t) = \frac{\kappa S^2(\Delta p)}{2M(t)}. \quad\dots\dots\dots\dots\dots\dots\dots\dots\dots\dots\dots\dots \quad (6.4.14)$$

This expression is a trivial identity for static-filtration data obtained under conditions of constant filtration pressure. The central modeling assumption of this section is as follows: Eq. 6.4.14 will be taken to hold true for the instantaneous values of the fluid-loss rate, cake-solids mass, and filtration pressure, at every instant in a wellbore filtration process. Thus we relate q, V^*, and Δp through

$$q = \frac{S^2(\Delta p)}{2V^*} \quad\dots\dots\dots\dots\dots\dots\dots\dots\dots\dots\dots\dots\dots\dots \quad (6.4.15)$$

at every instant, be it in quasistatic filtration with cake growth, in equilibrium dynamic filtration during which time no matter is being added to the cake, or during instants when the wellbore pressure or mud-circulation rate is changing.

The physical interpretation of this modeling assumption is as follows:

1. The cake is assumed to adjust instantaneously to any change in filtration-controlling parameters.
2. The state of the cake, and the fluid loss through it, is controlled by the mass of solids in the cake and the state of compaction of those solids (which, in turn, is controlled by the pressure drop across the cake).

Further insight can be obtained from consideration of the incompactible case. There, it is well known that $S \propto \sqrt{\Delta p}$, and the cake thickness T is simply proportional to V^*. As a result, Eq. 6.4.15 becomes $q \propto \Delta p / T$, which is the expected Darcy's-law relation.

Since the model assumes instantaneous adjustment of the filter cake to changing conditions, we must interpret it as being useful for predictions of the long-term features of the fluid-loss and cake-growth processes, and as being invalid should features on the time scales comparable with those for internal readjustments of the filter cake be of interest. Furthermore, it does not keep track of the small amounts of filtrate squeezed from the cake, or sucked into it, when the filtration pressure is changed; as a result, a small filtrate mass-balance error is incurred.

It follows at once, by simple rearrangement of Eq. 6.4.15, that the mass of solids in the cake at the end of a process of erosion must satisfy

$$V^*_{\text{crit}} = \frac{S^2(\Delta p)}{2q_{\text{crit}}(\tau_w)}. \quad\dots\dots\dots\dots\dots\dots\dots\dots\dots\dots\dots\dots \quad (6.4.16)$$

This specifies the last remaining parameter of the erosion model.

It now remains only to specify how the desorptivity and equilibrium dynamic-fluid-loss rate vary. First, since filter cakes are known to exhibit compaction hysteresis, it is necessary to introduce a bookkeeping variable, with the evolution equation, to permit this phenomenon to be tracked. Writing $\Delta p_{\text{max}}(t) = \max_{0 \le t' \le t} \left[p_{\text{well}}(t') - p_{sf}(t') \right]$, the maximum pressure difference that the filter cake has experienced up to the present time, then

$$\frac{d\Delta p_{\text{max}}}{dt} = \begin{cases} \max\left(0, \dfrac{dp_{\text{well}}}{dt} - \dfrac{dp_{sf}}{dt} \right) & \text{if } \Delta p \ge \Delta p_{\text{max}} \\ 0 & \text{otherwise.} \end{cases} \quad\dots\dots\dots\dots\dots \quad (6.4.17)$$

At the initial instant, $\Delta p_{\text{max}}(0) = p_{\text{well}}(0) - p_{sf}(0)$; the value of this quantity can be found from an early-time similarity solution.

For the pressure dependence of the desorptivity, we follow soil mechanics custom, and let

$$
S\left(\Delta p\right) =
\begin{cases}
S_{\text{ref}}\left(\Delta p_{\text{ref}}\right)\left(\dfrac{\Delta p}{\Delta p_{\text{ref}}}\right)^{n} & \text{if } \Delta p \geq \Delta p_{\text{max}} \\[3ex]
S_{\text{ref}}\left(\Delta p_{\text{ref}}\right)\left(\dfrac{\Delta p_{\text{max}}}{\Delta p_{\text{ref}}}\right)^{n}\left(\dfrac{\Delta p}{\Delta p_{\text{max}}}\right)^{n'} & \text{otherwise.} \quad\cdots\cdots\cdots\cdots\cdots\cdots\cdots\cdots\cdots\cdots\cdots
\end{cases}
\quad (6.4.18)
$$

For compactible filter cakes, $0 < n < 1/2$ and $n' \geq n$. If the cake does not decompact on reduction of Δp, then $n' = 1/2$.

We use a simple power-law increase of equilibrium dynamic-fluid-loss rate with circulating-mud wall shear stress (Hammond and Pop 2005). As a temporary expedient, to avoid the need for detailed consideration of drilling-fluid rheology at this stage, we let

$$
q_{\text{crit}}\left(\tau_w\right) = q_{\text{crit}}\left(\dot{\gamma}_{\text{ref}}\right)\left(\frac{\dot{\gamma}_w}{\dot{\gamma}_{\text{ref}}}\right)^{m}, \quad\cdots\cdots\cdots\cdots\cdots\cdots\cdots\cdots\cdots\cdots\cdots\cdots\cdots\cdots\cdots\cdots\cdots\cdots \quad (6.4.19)
$$

where $\dot{\gamma}$ is the shear rate in the flowing mud at the cake surface.

Example Supercharging Simulation. **Fig. 6.4.5** shows the results of a simulation of 29 hours of fluid loss into a 5-md formation and the associated supercharging, for a WBM with good dynamic-fluid-loss characteristics. Values of the drilling-fluid filtration and Bingham plastic-rheology parameters used in this and later simulations are given in **Table 6.4.1**. The drillstring geometry is generic and not intended to correspond in detail to any particular system; details are given in **Table 6.4.2**. **Fig. 6.4.6** shows the bit position, drilling fluid circulation rate, wellbore pressure, and other factors vs. time.

The formation is first penetrated by the drill bit at time $t = 0$. Pipe connections are made periodically, during which time circulation is stopped for 60 seconds; associated spikes are visible in Fig. 6.4.6 in the simulated wellbore pressure (caused by changes in the frictional contribution to wellbore pressure), and in Fig. 6.4.5 in the leakoff rate and sandface pressure (caused both by the changing wellbore pressure and by the growth and erosion of a small amount of static mudcake). Drilling proceeds, and at successive times the drill collars, the formation-pressure-measurement tool, a further section of collars, and eventually the drillpipe are opposite the formation in which the pressure is to be measured; this succession changes the gap between drillstring and sandface, and so affects the mean mud velocity in the annulus and then the filtrate-leakoff process. (Note, for example, the correlations between $\dot{\gamma}$ in Fig. 6.4.6 and the equilibrium dynamic-leakoff rate in Fig. 6.4.5; because the cake growth and erosion processes are comparatively slow for this fluid, the changes in $\dot{\gamma}$ are smoothed out in the actual leakoff rate.) Drilling pauses for 10 minutes for a first pressure measurement at $t = 1$ hour (the formation-pressure measurement is located 100 ft behind the bit, and the rate of penetration is 100 ft/hr). A further 100 ft is then drilled, and the string is then pulled back for a second formation-pressure measurement a little after 2 hours. **Fig. 6.4.7** shows a close-up of the leakoff rate and sandface pressure during this time. Drilling then continues ahead, at constant rate of penetration, for 22 hours. The string is then pulled back for a third extended formation-pressure measurement at approximately 26 hours.

The formation pressures measured 5 minutes after each arrival of the formation-pressure-measurement tool at the target formation are supercharged relative to the distant formation pressure by 0.842, 0.707, and 0.339 bar, respectively. This decrease in supercharging with time is a result of the generally decreasing filtrate-loss rate, which is itself the result of the combination of the downward steps in hydraulic shear stress as the drillstring-to-formation gap increases and the slow rates of cake growth and erosion for this mud. A closer look at the leakoff rate in Fig. 6.4.5 reveals two periods of quasistatic cake growth, roughly from $t = 0$ to $t = 2$ hours and from $t = 4$ hours to $t = 15$ hours, with periods of constant-leakoff-rate dynamic filtration between $t = 2$ and $t = 4$ hours, and from $t = 15$ to $t = 26$ hours. Toward the end of the simulation, from $t = 26$ hours onward, supercharging increases as a result of erosion of the cake caused by the turbulent mud flow around the collars (the flow is laminar when the drillpipe is opposite the formation).

6.4.5 When Will Supercharging Be Important? The simulations of the preceding subsection show that there can be quite a lot of temporal variation in the filtrate-leakoff rate because of changing

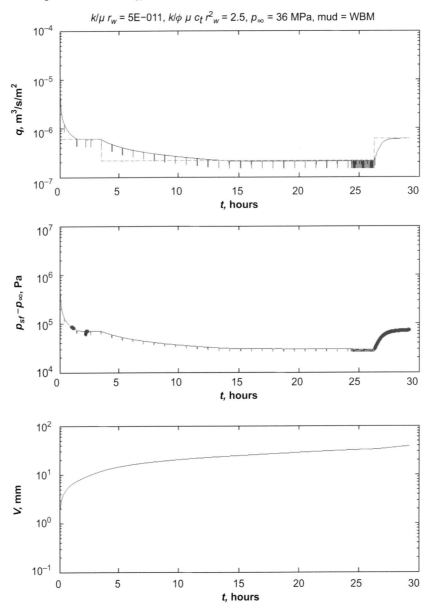

Fig. 6.4.5—Computed results from a coupled simulation of wellbore filtration and formation pressure. Top: filtrate-leakoff rate (solid, blue) and equilibrium dynamic-filtration-leakoff rate (broken, green). Middle: sandface pressure relative to the known formation pressure at great distances. The line is thickened at those times when the formation-pressure-measurement tool is opposite the target zone. Bottom: cumulative volume of filtrate. The formation permeability was 5 md, and the drilling-fluid parameters were as in Table 6.4.1.

wellbore-hydraulics conditions and cake growth and erosion. This frustrates any simple and accurate estimation of likely supercharging levels. But it is possible to make an approximate estimate, on the basis that the leakoff rate has taken its present value at all times since the start of drilling.

Using the well-known constant-rate solution for the formation-pressure diffusion equation, we obtain $p_{sf}(t) \approx p_{\infty} + (\mu r_w/2k)q(t)E_1(\phi\mu c_t r_w^2/4kt)$. The resulting approximation is acceptable in these cases, and the largest errors occur after rapid changes in leakoff rate. As further justification, we can estimate the time, t_{forget}, required after a step change in leakoff rate of size Δq for the sandface pressure to come within Δp of the value it would have had had the leakoff rate taken the post-change rate at all previous times. Using the

TABLE 6.4.1—DRILLING-FLUID FILTRATION AND RHEOLOGY PARAMETERS

	WBM	OBM
S_{ref} (Δp_{ref})	50×10^{-6} m/s$^{1/2}$ at 10^6 Pa	18×10^{-6} m/s$^{1/2}$ at 3.4×10^6 Pa
q_{crit} ($\dot{\gamma}_{ref}$)	3×10^{-7} m/s at 300 1/s	1.9×10^{-6} m/s at 270 1/s
n	0.3	0.3
n'	0.45	0.45
m	1	0.75
λ	8.33×10^{-4} 1/s	10^{-2} 1/s
μ_p	50×10^{-3} Pa·s	50×10^{-3} Pa·s
τ_y	3 Pa	3 Pa
ρ	1100 kg.m^{-3}	1100 kg.m^{-3}

TABLE 6.4.2—DRILLSTRING AND HOLE PARAMETERS

	Diameter (in.)	Length (ft)
Hole inside diameter	8.75	–
Bit outside diameter	8.5	1
Tool/collar outside diameter	6.75	215
Drillpipe outside diameter	5	–

previously quoted constant-leakoff-rate solution for the sandface pressure, it can be shown that t_{forget} is given approximately by $t_{forget} = t_0 / \left[\exp\left(2k\,\Delta p / \mu r_w \Delta q\right) - 1 \right]$. If the time after drilling t_0 is 1 hour, the formation permeability k is 10 md, the fluid viscosity is 1 mPa·s, and the wellbore radius is 0.1 m, then for $\Delta p = 10^4$ Pa and $\Delta q = 10^{-6}$ m/s, t_{forget} is approximately 10 minutes. For these parameter values at least, prechange conditions are largely forgotten after 10 minutes. Forgetting times are longer in lower permeabilities, or for larger changes in leakoff rate or more-stringent accuracy requirements (i.e., smaller Δp values). However, for measurements made several forgetting times after the last major change in leakoff conditions, it is reasonable to approximate the sandface pressure on the basis of the current leakoff rate.

Suppose we now determine that supercharging must be less than a particular level, p_{super}, for the measurements to be usable. Using the expression above to approximate the supercharging pressure, at any particular time after drilling we can divide the permeability/leakoff-rate plane into two regions. In one, the supercharge pressures are less than p_{super}; in the other, they exceed it. **Fig. 6.4.8** shows the result for $p_{super} = 1$ bar, and times after drilling of 1 hour, 1 day, and 10 days. Supercharging at a particular time after first drilling is important in the region of the k–q plane above the appropriate line—that is, low permeabilities and/or high loss rates.

Supercharging can be important in high permeabilities if the mud has high leakoff in dynamic filtration. Unfortunately, there is only weak correlation between easily measured static filtration data and the crossflow filtration behavior that is relevant here (Fordham et al. 1991). So without appropriate measurements having been made, we are unlikely to know exact values of q in practice; hence, it is important to simulate a range of possibilities when planning a job (underlining the value of a simple-to-use tool that enables a number of what-if scenarios to be investigated with ease), and/or to make some form of in-situ characterization of fluid-loss rates or appropriate mud properties.

The arguments of this subsection allow the likely supercharging pressure in a given situation to be estimated approximately. If a more precise prediction is required, or if the cake-building characteristics of the drilling fluid are such that fast adjustment of the cake to current hydraulic conditions may not be assumed, then detailed simulations can be performed. But whether a particular level of supercharging is important depends on the use to which the formation pressure measurements will be put. An example discussing the impact of supercharging on compartmentalization is given by Chang et al. (2005).

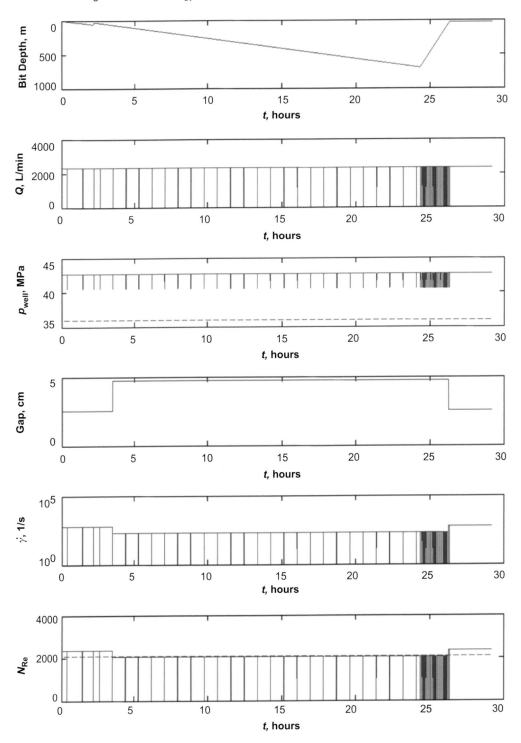

Fig. 6.4.6—Drilling and hydraulics inputs. From the top down: depths of bit below zone in which formation pressure is to be measured; drilling-fluid-circulation rate; wellbore pressure (solid, blue) and the (known) far-field formation pressure (broken, green); gap between drillstring and sandface at formation to be measured; shear rate for mud flow opposite formation; Reynolds number (solid, blue) for mud flow opposite formation and critical Reynolds number for onset of turbulence (broken, green). Drilling-fluid circulation is stopped for 60 seconds every time a connection is made, hence the downward spikes in wellbore pressure, wall shear rate, and other factors.

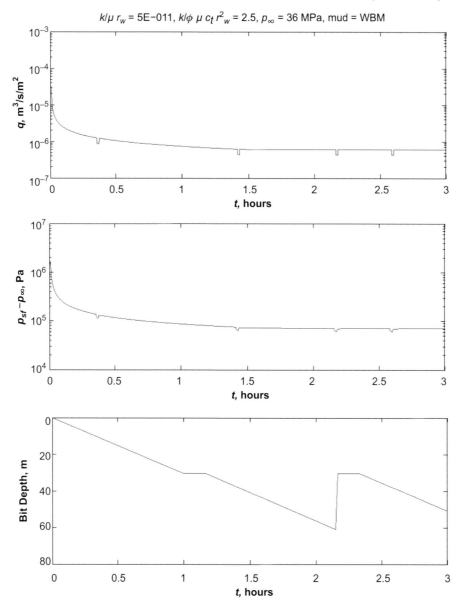

$k/\mu\ r_w$ = 5E−011, $k/\phi\ \mu\ c_t\ r^2_w$ = 2.5, p_∞ = 36 MPa, mud = WBM

Fig. 6.4.7—Close-up of the first 3 hours of the simulation shown in Fig. 6.4.5.

Planning for Supercharging. Coupled simulations of filtrate leakoff and formation pressure have shown that supercharging levels can vary by significant amounts as a well is drilled. Supercharging levels are shown to depend both on the formation permeability and on the filtrate-leakoff rate, and this in turn varies with hydraulic conditions in the wellbore (and the properties of the drilling fluid). As a result, prediction of supercharging while drilling, and especially its variation with time, requires a simulation framework within which the full history of fluid circulation and drilling operations can be captured. Fig. 6.4.8 may be used if a rough estimate of likely supercharging levels is required.

The simulation results show that high drilling-fluid-circulation rates and/or narrow BHA/formation clearances can lead to filter-cake erosion and increased filtrate-leakoff rates, and then to elevated supercharging pressures (or even supercharging pressures that increase with time). This possibility of significant (and nonmonotonic) variation of leakoff rate in time is the major difference between supercharging while drilling and supercharging when making wireline formation-pressure measurements. Use of detailed simulation is a

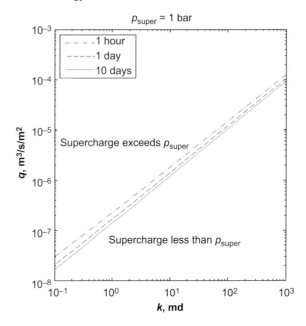

Fig. 6.4.8—Each diagonal line divides the formation permeability-equilibrium dynamic-fluid-loss rate plane into two regions, one when supercharging at the corresponding time after drilling is less than 1 bar, the other where it exceeds 1 bar. Large supercharge pressures occur for high loss rates and/or low formation permeabilities.

valuable means through which to develop insight into the characteristics and peculiarities of this new formation-pressure-measurement environment.

Subject to operational constraints, the sensitivity of leakoff rate and supercharging to wellbore hydraulic conditions can be exploited in job design to minimize the amount of supercharging at the time of measurement.

To simulate supercharging while drilling, it is necessary to know something about the dynamic filtration characteristics of the drilling fluid. Because measurements of these requires a special apparatus, accurate dynamic filtration-parameter values may not always be available. In that case, simulations must be performed using wisely chosen default values, so as to access worst-case scenarios.

Therefore, a capability to simulate filtrate leakoff and formation pressures in detail is a valuable aid in planning, understanding, and interpreting formation pressures measured while drilling.

6.4.6 Correcting Supercharging in FPWD Measurements. Now that we have an appreciation of how supercharging is affected by mud filtration, we describe methods for correcting for the supercharging effect so that accurate formation pressures can be estimated. The method described by Hammond and Pop (2005) consists of varying the wellbore pressure deliberately, measuring that and the sandface-pressure response, and then using a mathematical model, taking account of fluid flows within filter cake and formation to quantify the amount of sandface-pressure elevation. The true formation pressure then follows by subtracting this estimate of the supercharging from the actual measurement.

This method for estimating and removing the supercharging contribution to measured pressures allows accurate formation pressures to be obtained while drilling in circumstances where a good measurement was not previously possible. The time required to collect data to make the interpretation is sufficiently short that the method is compatible with operational constraints associated with formation-pressure measurements while drilling, such as avoidance of tool sticking. Errors are reduced because some of the necessary drilling-fluid-leakoff information is determined from the measurements made in situ rather than from separate laboratory measurements or databases.

The method differs from previous supercharging corrections as follows:

- The interpretation is based on detailed transient descriptions of the flow within the formation, of the filtrate-leakoff process, and of the filter-cake behavior.

- Changes are induced deliberately in the wellbore to perturb the leakoff process and to then extract information about leakoff rates in situ.

After the model derivation, data collected in a real well are then explored and used to demonstrate the new supercharging-estimation and -correction methodology.

6.4.7 Interpretation Methodology. The central idea of the new methodology is to use measurements of the sandface-pressure variations in response to deliberately induced changes in wellbore pressure to characterize the resistance to filtrate leakoff offered by the mudcake and, from this, to estimate the current filtrate-leakoff rate and, then, the amount by which the sandface pressure is elevated over values at great distances. This estimate of the supercharging pressure is then subtracted from the measured sandface pressure to give the true far-field formation pressure.

6.4.8 History Matching the Supercharging Estimation and Correction. The supercharging-estimation and -correction interpretation proposed here is a type of history matching. In the preceding subsection, we have described a forward model that allows the sandface pressure to be computed, given information about the drilling history, hydraulics, formation properties, and the characteristics of the drilling fluid. Symbolically, this model can be written as

$$p_{sf}(t) = \mathop{F}_{t'=-t_0}^{t'=t}\left[\text{mud properties}, p_\infty, k, \phi, \ldots, \text{hydraulics}(t')\right], \quad\quad\quad\quad\quad (6.4.20)$$

where F denotes a functional over all past times, and the formation was first exposed to leakoff at time $t = -t_0$. History matching is the process of selecting values for the various formation and drilling-fluid parameters appearing in the forward model, so as to minimize the discrepancy between measured and simulated sandface pressures at a number of times. That is, we seek parameters that minimize

$$\varepsilon = \sum_j \left| p_{sf}(t_j) - \mathop{F}_{t'=-t_0}^{t'=t_j}\left[\text{mud properties}, p_\infty, k, \phi, \ldots, \text{hydraulics}(t')\right]\right|^2 \quad\quad\quad\quad (6.4.21)$$

Among the parameters that can be determined in this way is p_∞, the true formation pressure. Wellbore-pressure variations can be imposed deliberately, in addition to any naturally occurring fluctuations, so as to ensure that the parameter determination is as well-posed as possible.

The preceding procedure is computationally costly, and so a faster method, based on an approximate calculation but within the same history-matching framework, has been developed. This approximate calculation also helps us to understand the limitations of the technique and the most appropriate types of wellbore-pressure variations to impose. The main components of the approximate calculation are a formation flow model linking leakoff rate to sandface pressure, some assumptions about the past history of filtrate leakoff and the duration of the measurement process relative to the time scale for changes in the filter cake, and a model relating filtrate-leakoff rate to the present state of the filter cake and the pressure difference across it. Each component is described in turn below.

It is well known that for single-phase, slightly compressible, radial flow in a homogeneous medium, the flow rate of fluid into the formation, $q(t)$, and the pressure within the formation at the wellbore wall, $p_{sf}(t)$, are linked by a convolution integral:

$$p_{sf}(t) = p_\infty + \int_{-t_0}^{t} R(t-t')q(t')dt' \quad\quad\quad\quad\quad\quad\quad\quad (6.4.22)$$

The impulse response R is given by

$$R(t) = \frac{\mu r_w}{k}\frac{4\kappa}{\pi}\int_0^\infty \frac{e^{-\kappa t u^2}}{u\left[J_1^2(u)+Y_1^2(u)\right]}du = \frac{\mu r_w}{k}\kappa R_D(\kappa t), \quad\quad\quad\quad (6.4.23)$$

where k is the formation permeability, ϕ its porosity, c_t the total compressibility of fluid and matrix, μ the pore-fluid viscosity, and for later convenience, we let $\kappa = k / (\phi \mu c_t r_w^2)$. J_1 and Y_1 are Bessel functions. The sandface-pressure response to a constant-unit-rate leakoff, $H(t)$, is obtained by integrating Eq. 6.4.23 to obtain

$$H(t) = \frac{\mu r_w}{k} \frac{4}{\pi} \int_0^\infty \frac{\left(1 - e^{-\kappa t u^2}\right)}{u^3 \left[J_1^2(u) + Y_1^2(u)\right]} du = \frac{\mu r_w}{k} H_D(\kappa t) \quad \dots \dots \dots \dots \dots \dots \quad (6.4.24)$$

In what follows, this expression is central to the numerical evaluation of the convolution integral (Eq. 6.4.22). Because only one parameter, κt, appears as an argument, it is possible to compute the convolution and expressions derived from it rapidly using precomputed lookup tables of R_D and H_D.

To make use of Eq. 6.4.22 to compute the sandface pressure, it is necessary to give the values of the leakoff rate at all times since the time when the formation was first drilled. It is not easy to do this with confidence, because even if the drilling-hydraulics history is measured and recorded, mechanical interactions between components of the BHA and the filter cake that may affect the leakoff rate are largely unknown.

Fortunately, some reasonable simplifying approximations can be made. Let us suppose that the measurement process starts at time $t = 0$, and that the wellbore pressure is deliberately varied from that time on. Also, introduce a quantity $q_{history}(t')$ that is equal to $q(t')$ for $-t_0 \le t' \le 0$. Its values for $0 \le t \le t'$ will be defined below. A trivial rearrangement of Eq. 6.4.22 then leads to

$$p_{sf}(t) = p_\infty + \int_{-t_0}^t R(t - t') q_{history}(t') dt''$$

$$+ \int_0^t R(t - t') \left[q(t') - q_{history}(t')\right] dt', \quad \dots \dots \dots \dots \dots \dots \quad (6.4.25)$$

which we can interpret as saying that the sandface pressure is the sum of the effects of past history (the first two terms) and recent events (the final term). It can be argued that a reasonable approximation to the first integral term in Eq. 6.4.25, which we denote as $p_{history}(t)$, can be had in wellbore-filtration contexts by assuming that the leakoff rate has taken a value equal to its current value at all times since the hole was opened to drilling. That is, we propose the approximation

$$q_{history}(t') = q(0) \text{ for } -t_0 \le t' \le t \quad \dots \dots \dots \dots \dots \dots \dots \dots \dots \dots \dots \dots \dots \dots \quad (6.4.26)$$

It then follows that for $t > 0$,

$$p_{history}(t) \approx p_\infty + \int_{-t_0}^t R(t - t') q(0) dt' = p_\infty + q(0) H(t + t_0) \quad \dots \dots \dots \dots \quad (6.4.27)$$

and

$$p_{sf}(0) \approx p_\infty + q(0) H(t_0) \quad \dots \dots \dots \dots \dots \dots \dots \dots \dots \dots \dots \dots \dots \dots \quad (6.4.28)$$

Now consider the modeling of fluid loss through the cake. The leakoff rate through a filter cake is approximately related to the pressure drop across it by

$$q = \frac{S^2(\Delta p)}{2V^*}, \quad \dots \dots \dots \dots \dots \dots \dots \dots \dots \dots \dots \dots \dots \dots \dots \dots \dots \dots \dots \dots \quad (6.4.29)$$

where V^* is the volume of filtrate per unit area that would have been lost had the current cake been built under constant conditions of differential pressure and hydraulic shear stress, $\Delta p(t) = p_{well}(t) - p_{sf}(t)$ is the difference between the current wellbore and sandface pressures, and $S(\Delta p)$ is termed the desorptivity. If we concern ourselves only with processes involving an already developed filter cake, and happening over time scales that are short compared to the time scale over which that cake grew, then it is permissible to treat V^* as a constant. In that case, the leakoff rate is a function only of the differential pressure across the cake. A good model for the pressure dependence of desorptivity is

$$S(\Delta p) = \begin{cases} S_{ref}(\Delta p_{ref})\left(\dfrac{\Delta p}{\Delta p_{ref}}\right)^n & \text{if } \Delta p \geq \Delta p_{max} \\[3ex] S_{ref}(\Delta p_{ref})\left(\dfrac{\Delta p_{max}}{\Delta p_{ref}}\right)^n\left(\dfrac{\Delta p}{\Delta p_{max}}\right)^{n'} & \text{otherwise} \end{cases}$$ . . . . (6.4.30)

If during the test, the cake is subjected to differential pressures that are less than the maximum differential pressure it has already experienced, then only the decompaction branch (the lower expression on the right-hand side of Eq. 6.4.30 is relevant, and we may write

$$S(\Delta p) = \alpha \Delta p^{n'}$$ . . . . (6.4.31)

Then from Eq. 6.4.29,

$$q(t) = \frac{\alpha^2 \left[\Delta p(t)\right]^{2n'}}{2V^*}$$ . . . . (6.4.32)

If the cake behaves in a nearly incompressible fashion during decompaction, then $n' \approx 1/2$. This completes the specification of the cake and fluid-loss model.

Finally, the cake and formation models are coupled, by combining Eqs. 6.4.25 and 6.4.32, and using Eq. 6.4.27, to obtain

$$p_{sf}(t) \approx p_\infty + \frac{\alpha^2 \left[p_{well}(0) - p_{sf}(0)\right]^{2n'}}{2V^*} H(t + t_0)$$
$$+ \int_0^t R(t - t')\frac{\alpha^2\left\{\left[p_{well}(t') - p_{sf}(t')\right]^{2n'} - \left[p_{well}(0) - p_{sf}(0)\right]^{2n'}\right\}}{2V^*}dt'.$$ . . . . (6.4.33)

While it is possible to use this expression as the basis of a parameter-fitting interpretation, numerical experiments suggest that it is better to work with the difference, $p_{sf}(t) - p_{sf}(0)$, rather than with $p_{sf}(t)$ itself. This reduces by one the number of unknown parameters that must be estimated (indeed, Eq. 6.4.36 below shows that within the present formulation p_∞ is not independent of the other parameters). Using Eq. 6.4.28, we obtain from Eq. 6.4.33,

$$p_{sf}(t) - p_{sf}(0) \approx \frac{\alpha^2\left[p_{well}(0) - p_{sf}(0)\right]^{2n'}}{2V^*}\left[H(t + t_0) - H(t_0)\right]$$
$$+ \int_0^t R(t - t')\frac{\alpha^2\left\{\left[p_{well}(t') - p_{sf}(t')\right]^{2n'} - \left[p_{well}(0) - p_{sf}(0)\right]^{2n'}\right\}}{2V^*}dt',$$ . . . . (6.4.34)

which on writing $A = \alpha^2 \mu r_w / (2kV^*)$ and $m = 2n'$, and using Eqs. 6.4.23 and 6.4.24, becomes

$$p_{sf}(t) - p_{sf}(0) = A\left[p_{well}(0) - p_{sf}(0)\right]^m \left\{H_D\left[\kappa(t + t_0)\right] - H_D(\kappa t_0)\right\}$$
$$+ A\kappa\int_0^t R_D\left[\kappa(t - t')\right]\left\{\left[p_{well}(t') - p_{sf}(t')\right]^m - \left[p_{well}(0) - p_{sf}(0)\right]^m\right\}dt'.$$ . . . . (6.4.35)

This equation is the basic time-domain convolution model; it links wellbore and sandface pressures, for while-drilling filtration with a compactible mudcake. Once the values of the three parameters (A, κ, m) in Eq. 6.4.35 have been determined, an estimate of the true formation far-field pressure follows from Eq. 6.4.28, as where

$$p_{\infty} = p_{sf}(0) - A \left[p_{well}(0) - p_{sf}(0) \right]^m H_D(\kappa t_0). \qquad \dots\dots\dots\dots\dots\dots\dots (6.4.36)$$

This completes the description of the basic supercharging-correction model.

In principle, given measurements of the wellbore and sandface pressures in suitable circumstances at a set of times (t_i), the values of the three unknown independent parameters $\{A, \kappa, m\}$ in Eq. 6.4.35 may now be determined using a standard optimization routine to minimize the sum of squared residuals $\sum_i \varepsilon^2(t_i)/\sigma_i^2$, where

$$\varepsilon(t) = p_{sf}(t) - p_{sf}(0) - A \left[\Delta p(0)^m \left\{ H_D \left[\kappa(t + t_0) \right] - H_D(\kappa t_0) \right\} \right.$$
$$\left. + \kappa \int_0^t R_D \left[\kappa(t - t') \right] \left[\Delta p(t')^m - \Delta p(0)^m \right] dt' \right] \qquad \dots\dots\dots\dots\dots (6.4.37)$$

and σ_i are estimates of the measurement errors.

In practice, it is not possible to estimate the value of m reliably in this way, because only modest variations in $\Delta p(t') - \Delta p(0)$ can be achieved safely. As a result, the value of m must be independently entered into the interpretation process on the basis of filtration measurements made on the drilling fluid. The comparative incompressibility of filter cakes on decompaction can be exploited when designing testing sequences so as to constrain the possible values of m.

Furthermore, as we shall argue in the following, the parameter κ influences Eq. 6.4.37 only very weakly. An adequate value can be obtained from the mobility measured when setting the tool measurement probe combined with a reasonable estimate of the porosity/compressibility product.

In consequence, only one unknown parameter, A, remains. The optimization procedure to determine its value is quick and straightforward.

In a practical application, the sandface and wellbore pressures will be measured at a discrete set of times. It is therefore necessary to specify a method for the numerical evaluation of the convolution integral in Eq. 6.4.37 given a table of values of $\Delta p(t)$ defined at a set of times (t_i), with $t_1 = 0$. Let us write $f(t') = \Delta p(t')^m - \Delta p(0)^m$, and let

$$I(t) = \kappa \int_0^t R_D \left[\kappa(t - t') \right] f(t') dt'. \dots\dots\dots\dots\dots\dots\dots\dots\dots (6.4.38)$$

Then, discretizing the range of integration, and replacing $f(t')$ on each panel of the mesh by the average of its values at the panel edges, we obtain

$$I(t_j) \approx \sum_{i=1}^{j-1} \kappa \int_{t_i}^{t_{i+1}} R_D \left[\kappa(t_j - t') \right] dt' \frac{f(t_{i+1}) + f(t_i)}{2}, \qquad \dots\dots\dots\dots\dots\dots (6.4.39)$$

but

$$\kappa \int_{t_i}^{t_{i+1}} R_D \left[\kappa(t_j - t') \right] dt' = \kappa \int_{t_i}^{t_j} R_D \left[\kappa(t_j - t') \right] dt' - \kappa \int_{t_{i+1}}^{t_j} R_D \left[\kappa(t_j - t') \right] dt'$$
$$= H_D \left[\kappa(t_j - t_i) \right] - H_D \left[\kappa(t_j - t_{i+1}) \right]; \qquad \dots\dots\dots\dots\dots\dots (6.4.40)$$

therefore,

$$I(t_j) \approx \frac{f(t_2)}{2} H_D \left[\kappa(t_j - t_1) \right] + \sum_{i=2}^{j-1} \frac{f(t_{i+1}) - f(t_{i-1})}{2} H_D \left[\kappa(t_j - t_i) \right], \qquad \dots\dots\dots (6.4.41)$$

where we have used $H_D(0) = 0$ and $f(0) = 0$ to drop a few terms. Using the method outlined in the preceding, we obtain

$$\varepsilon(t_j) = p_{sf}(t_j) - p_{sf}(0) - A \begin{bmatrix} \Delta p(0)^m \left\{ H_D\left[\kappa(t_j + t_0)\right] - H_D(\kappa t_0) \right\} \\ + \dfrac{\left[\Delta p(t_2)^m - \Delta p(t_1)^m\right]}{2} H_D\left[\kappa(t_j - t_1)\right] \\ + \displaystyle\sum_{i=2}^{j-1} \dfrac{\Delta p(t_{i+1})^m - \Delta p(t_{i-1})^m}{2} H_D\left[\kappa(t_j - t_i)\right] \end{bmatrix}. \quad \cdots\cdots\cdots \quad (6.4.42)$$

This completes the description of the numerical implementation of the interpretation. It remains to show that the dependence of Eq. 6.4.37 on κ is weak. This follows by rearranging Eq. 6.4.41 to obtain

$$I(t_j) \approx \frac{1}{2}\left\{ f(t_{j-1}) H_D\left[\kappa(t_j - t_{j-2})\right] - f(t_j) H_D\left[\kappa(t_j - t_{j-1})\right] \right\}$$
$$+ \sum_{i=1}^{j-1} \frac{f(t_{i+1})}{2}\left\{ H_D\left[\kappa(t_j - t_i)\right] - H_D\left[\kappa(t_j - t_{i+2})\right] \right\} \quad \cdots\cdots\cdots\cdots\cdots\cdots \quad (6.4.43)$$

and then using the large-argument asymptotic approximation

$$H_D(\tau) \approx (\pi/2)\log\tau. \quad \cdots\cdots\cdots\cdots\cdots\cdots\cdots\cdots\cdots\cdots\cdots\cdots\cdots\cdots \quad (6.4.44)$$

At this point, it will be seen that all but two of the appearances of the variable κ cancel and the remainder appear only in arguments of logarithms. Numerical experiments support the conclusion that the overall dependence on κ is weak.

6.4.9 Example Practical Application. To illustrate the method, we make use of the real example described by Hammond and Pop (2005). Data to test the interpretation, and explore wellbore filtration phenomena under realistic conditions, were collected during the drilling of a 12¼-in., 3,100-ft-deep vertical well. A water-based drilling fluid was used. The target formation, at a measured depth of 3,050 ft, was a water-bearing limestone, with permeability at approximately 1 md and with 16% porosity. Formation and wellbore pressures were measured using an FPWD tool, and drilling-fluid circulation rates and surface pressures were recorded. Formation pressures and pressure gradients had been determined previously in offset wells penetrating the target and adjacent formations using wireline formation-pressure tools. These measurements acted as ground truth for the present interpretation but were not revealed until after the interpretation presented here had been made.

A number of experiments relating to wellbore filtration and its effects on formation pressure were performed, somewhat complicating the sequence of operational events. In outline, the well was drilled until the formation-pressure-measurement tool was opposite the target formation, a period of approximately 14 hours of mud circulation followed, then the string was pulled to place the tool opposite a more shallow formation and to place the bit well above the target zone, giving a period of approximately 18.5 hours of static filtration (i.e., zero mud circulation) there. The string was then run in again to place the tool at the target, and a sequence of three periods of circulation was conducted; each period consisted of 1 hour of full circulation and 1 hour of zero circulation. During the last of these zero-circulation periods, at approximately 38 hours after the formation was first drilled, the well was shut in and the surface pressure was raised and then reduced in three steps, each lasting approximately 400 seconds. During this stepping sequence the tool was set on the formation, and the sandface- and wellbore-pressure data to be interpreted were collected.

A forward simulation was performed to give some insights into the possible behavior of the filtrate-leakoff rate and the sandface pressure. The formation was assumed to be normally pressured, and estimated formation properties and drilling-fluid-design parameter values were used (**Table 6.4.3**). Because

TABLE 6.4.3—DRILLING-FLUID PROPERTIES USED IN THE SIMULATION OF THE FIELD TEST	
	WBM
$S_{ref}\,(\Delta p_{ref})$	93×10^{-6} m/s$^{1/2}$ at 6.895×10^{5} Pa
$q_{crit}\,(\dot{\gamma}_{ref})$	3×10^{-7} m/s at 300 s^{-1}
n	0.3
n'	0.45
m	1
λ	8.33×10^{-4} s^{-1}
μ_p	8×10^{-3} Pa.s
τ_y	2 Pa
ρ	1114 kg/m^3

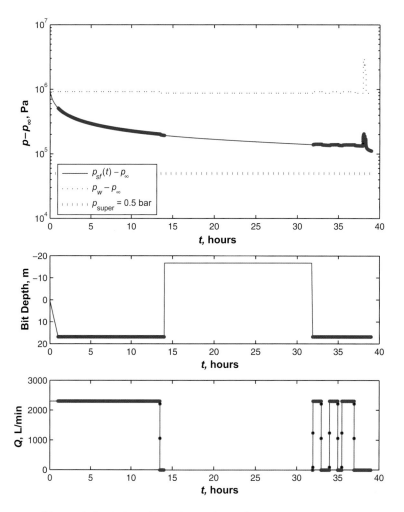

Fig. 6.4.9—Summary of forward simulation of filtration and sandface-pressure evolution for the field example.

drilling-fluid dynamic-filtration characteristics are not measured routinely, values for a similar laboratory mud were used.

Fig. 6.4.9 shows the simulated sandface pressure, **Fig. 6.4.10** summarizes the hydraulics opposite the target zone vs. time, and **Fig. 6.4.11** shows various filtration-related output quantities. As might be expected given

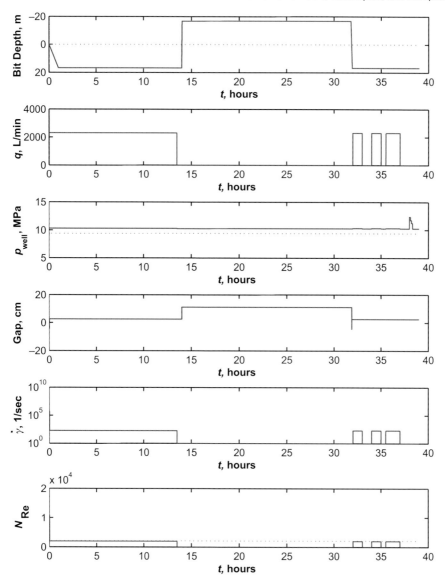

Fig. 6.4.10—Summary of hydraulics data for simulation of the field example.

the low formation permeability, some supercharging of the sandface pressure is predicted. Interestingly, as a result of the large hole size, the drilling fluid flow opposite the target formation is predicted never to be turbulent (on the basis of the Bingham plastic rheological model used, computed Reynolds numbers with a power-law model are greater), and as a result, no erosion of filter cake is predicted. This keeps predicted filtrate-leakoff rates low because cake deposited during the long static period is not removed when circulation is restarted. As a result, the predicted level of supercharging before pressure pulsing is near 1.3 bar, which is perhaps lower than might be expected. When the wellbore pressure is raised during the pulsing sequence, the supercharging increases to a maximum of approximately 2 bar. The other important point to note from Fig. 6.4.11 is that relatively little material is added to the filter cake during the pressure-pulsing sequence at $t = 38$ hours. This suggests that the basic assumption behind the interpretation—that the cake does not change much during the measurement process—may hold true.

Fig. 6.4.12 shows the actual measured sandface and wellbore pressures. Times are referenced to an arbitrary origin just before the probe-setting sequence began. The wellbore pressure is increased by approximately 20 bar at $t = 200$ seconds and is then stepped down at $t = 600$ seconds and $t = 1,000$ seconds. The probe

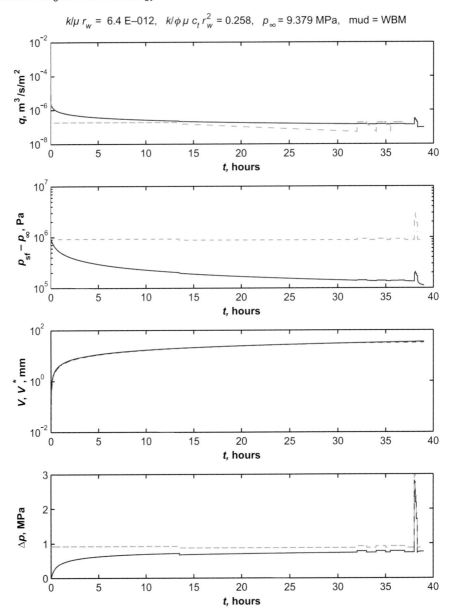

$k/\mu\, r_w$ = 6.4 E–012, $k/\phi\mu\, c_t\, r_w^2$ = 0.258, p_∞ = 9.379 MPa,   mud = WBM

Fig. 6.4.11—Summary of filtration-related outputs for simulation of the field example.

is in pressure communication with the formation from approximately $t = 100$ seconds until approximately $t = 1,300$ seconds. The large downward excursions in the probe pressure just after $t = 100$ seconds and $t = 700$ seconds are fluid extraction drawdowns. The final wellbore pressure step-down occurs after the probe had unset, and therefore the formation response to that event was not observed.

The most noticeable sandface-pressure responses to the wellbore-pressure variations that can be seen in the lower track of Fig. 6.4.12 are an upward spike at $t = 200$ seconds and a pair of smaller downward V-shaped features just after $t = 550$ seconds and $t = 1,000$ seconds. On the basis of calculations, not reported here, using a compactible-filter-cake model, the first of these is interpreted as the sandface-pressure response to fluid squeezed out of the cake as it compacts to a new, lower, void ratio when the wellbore pressure is increased. The two later events are interpreted similarly as the consequences of a small transient decrease in leakoff rate as the cake takes up fluid as it decompacts in response to the lower differential pressure. The decompaction events are smaller in magnitude than the compaction event because of cake-compaction hysteresis (the

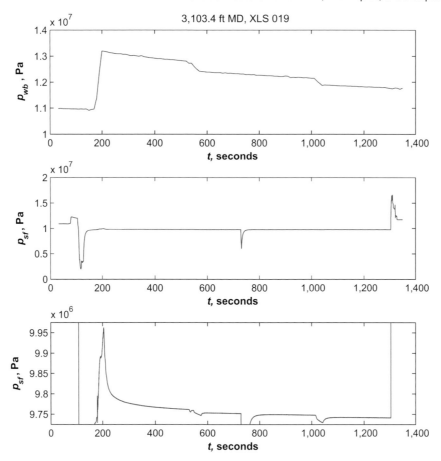

Fig. 6.4.12—Data from the field test. Measured wellbore pressure vs. time (upper track); measured sandface pressure (middle and bottom tracks, different pressure scales). The probe is set on the formation at approximately 75 seconds; the first drawdown is at 100 seconds and the second is at 700 seconds. The probe is unset at 1,300 seconds. MD = measured depth.

void-ratio change on decrease or removal of differential pressure is less than the void-ratio change when that differential pressure was applied).

These transient, finite-rate, cake-compaction-related events are not captured by the model of filter-cake behavior upon which the interpretation is based. We therefore edit them out from the data, in effect basing the interpretation on the long-term trends in the sandface pressure after the effects of these transient compaction flows have dissipated. **Fig. 6.4.13** shows a close-up of the measured sandface pressure from just before the first wellbore pressure step-down. Also shown on the plot are straight lines fitted to each section of data corresponding to each step of the wellbore-pressure staircase. These straight-line fits produce the edited data that are entered into the interpretation (**Fig. 6.4.14**). The same procedure also removes the probe drawdown transient.

The results of the parameter fitting are shown in **Fig. 6.4.15**. The estimated formation pressure is 9.5988 MPa. Taking the difference of the initial sandface pressure and the estimated formation pressure gives a value for the supercharging at the start of the interpreted sequence of 1.64 bar, which is of a magnitude similar to the values found in the forward simulation. The best estimate of the true far-field formation pressure, based on fitting a trend line through nonsupercharged wireline formation-pressure measurements in higher and lower zones, is 9.5437 MPa. A wireline pressure measurement in the same formation but in a different well gives 9.7216 MPa, which lies above the trend defined by the other formations and so may be supercharged. The interpreted formation pressure differs from the trend-line value by 0.55 bar. Given the liberties taken with the data to remove compaction-related features, this is a reasonable result. The comparison with the wireline

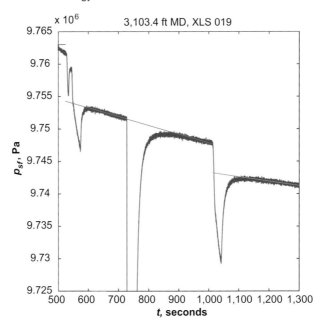

Fig. 6.4.13—Close-up of the measured sandface pressure as the wellbore pressure is stepped down (blue). The red and green lines are straight-line fits used to eliminate the probe drawdown and cake-compaction features from the data.

measurement in an adjacent well is not as good. Hence, further tests of the interpretation method should be made to assess its reliability.

The amount of supercharging affecting formation pressures measured while drilling can be estimated and quantified, both in theory and in practice, using the method described here. Values for the true far-field formation pressure can be obtained from sandface-pressure measurements when the method is applied. The method requires some changes to normal drilling operations, but these are limited to, for example, changes in the fluid circulation rate lasting for only minutes, at most, while the formation-pressure data are being collected.

It is very helpful to use a forward model to simulate the filtration process and sandface pressure, both when planning and when performing the interpretation. These simulations allow the engineer to explore likely system behavior and responses when planning the job, to assess the feasibility and suitability of the method, to check that the assumptions upon which it is based are likely to be satisfied, and to build some expectations about likely measured pressures and their behavior over time. The more that is known about the filtration properties of the drilling fluid, especially under dynamic conditions, the more accurately these simulations, and the better the testing sequence, can be designed.

6.4.10 Test and Pretest Sequences. Now that we have seen how we can allow for supercharging, let us look in more detail at the sequence of events that occurs during a typical formation-pressure measurement while drilling.

A typical fixed-mode operating sequence is given in **Fig. 6.4.16**, showing the probe extension, drawdown, buildup, and end of test (Pop et al. 2005a). Fixed-mode pretests are typically used (fixed drawdown volume) when the formation mobility is known to some degree of confidence. Given the limited information available in the *while-drilling* environment, and the wish to minimize the time stationary (or not actually drilling), it is not appropriate to perform the range of parameter settings that are often employed when using wireline-deployed formation-testing devices, and so much of the intelligence in determining appropriate pretest parameters is implemented directly into the tool.

Typical pretest phases that are desirable are as follows:

1. Initial phase. The point at which the sandface pressure goes below the formation pressure, and fluid from the formation flows into the tool, is identified.

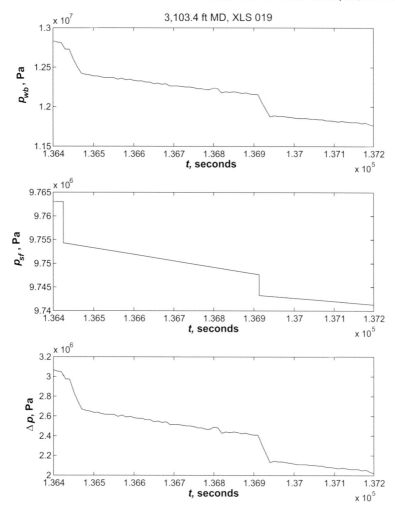

Fig. 6.4.14—Data input to the supercharging interpretation. The straight-line fits of Fig. 6.4.13 were used to create the sandface-pressure data in the middle track. The upper track is the wellbore pressure; the lower track is the differential pressure across the cake. The origin of the time axis has been shifted to reflect the time since the formation was first drilled.

2. Investigation phase. Formation is interrogated to obtain an initial estimate of its properties, essentially formation pressure and mobility.
3. Measurement phase. Best estimates of the formation pressure and mobility are obtained.

Before conducting the measurement phases, information acquired during the investigation, or during previous measurement phases, would be used to design optimal test sequences so that a stabilized sandface pressure is achieved at the end of each measurement phase, the total test time being subject to a time constraint.

Details of an optimized test sequence based on a time-limited pretest are described by Pop et al. (2005a), and they show that the problem can be solved as an optimization problem in the tool. Fixed-mode pretests typically perform adequately, but there are circumstances where they can be nonoptimal—first, when the overpressure is large or varies greatly, as can be the case in depleted reservoirs. (Chapter 5.1 addresses this and other difficult drilling situations like naturally fractured reservoirs. Such reservoirs can also be problematic and prone to difficult-to-interpret tests.) The second occurs where the formation is heterogeneous, and there is the potential for a wide range of formation permeabilities. The time-limited pretest can mitigate the problems in these types of reservoirs. The detailed sequence of steps is as follows (Pop et al. 2005a):

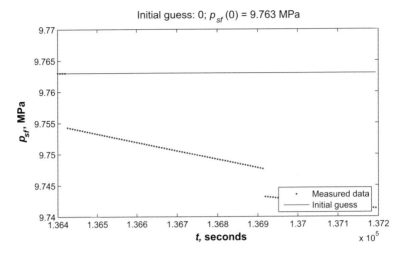

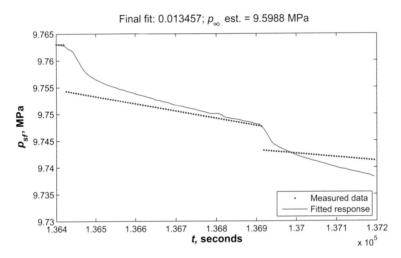

Fig. 6.4.15—Interpretation outputs including fitted sandface pressure.

1. Investigation phase
   a. Perform a controlled expansion of the flowline at a slow, constant rate—either continuously or in a series of measured steps.
   b. Recognize a point or range of points at which fluid is being extracted from the formation (i.e., the mudcake has been breached, the sandface pressure has gone below the formation pressure, and fluid flows from the formation into the tool).
   c. Terminate the drawdown after the extraction of a limited volume of fluid from the formation.
   d. Allow the probe pressure to reach a substantially stabilized sandface pressure in as short a time as possible.

2. Determine a first estimate of the formation pressure and a first estimate of the formation mobility from the first drawdown and buildup sequence.
3. Measurement Phase
   a. On the basis of the formation pressure and mobility just estimated, pretest volume and rate parameters are estimated. These are for a second pretest such that the probe pressure will be within some specified neighborhood of the stabilized sandface pressure at the end of the allotted time for the test.
   b. Perform a pretest using the parameters that have just been determined.

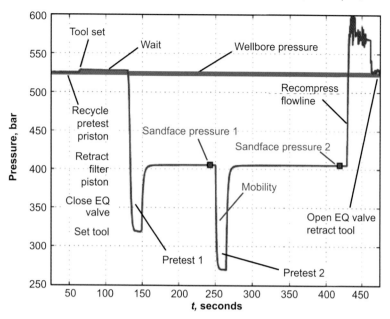

Fig. 6.4.16—Schematic of a typical fixed-mode operating sequence consisting of two pretests. The data are memory-mode data performed with an FPWD tool in a 2.5-md/cp formation, with pumps circulating.

If at the end of the investigation phase it is determined that the formation properties are such that a measurement phase is not warranted, the investigation phase proceeds to the end of the test period without interruption. However, if the pressure buildup has stabilized well before the end of the allotted test period, one or more additional measurement phases may be executed.

The investigation phase enables a quick estimate of the formation parameters, without prior knowledge of the particular conditions under which the test is to be performed, and is effectively a restriction on the amount of fluid that is withdrawn from the formation.

The constrained optimization to yield the measurement-phase pretest parameters effectively determines the measurement-phase pretest rate and duration such that at the end of the user-specified test period, the probe pressure is within a designated neighborhood of the stabilized sandface pressure and the maximum volume compatible with the previous requirement has been extracted from the formation. The mathematical details of the optimization, including the time to stabilization, the details of the tool-formation response model, and the selection of pretest parameters, are given in detail by Pop et al. (2005a).

6.4.11 Examples of Fixed-Mode and Time-Limited Pretests.
In this subsection, we will give a couple of examples of time-limited pretests to highlight the performance of the algorithm under a range of conditions (Pop et al. 2005a).

Low-Mobility Formation. The test sequence as determined by the algorithm for a 0.1-md/cp formation is given in **Fig. 6.4.17.** In this case, a stabilized pressure could not be reached in the time remaining after the investigation phase, and so no measurement was initiated. Under slightly different conditions (e.g., a smaller overbalance, a less-compressible flowline fluid, and a slightly more-mobile formation), a small-volume-measurement pretest could have been performed. This example highlights the difficulty of achieving stabilized sandface pressures in tight formations.

Intermediate-Mobility Formation. **Fig. 6.4.18** gives the expected response for a time-limited pretest for a formation of 1 md/cp. In this case, the initial-volume investigation phase is followed by a larger-volume measurement-phase pretest.

Moderate-Mobility Formation. In a moderately mobile formation of 20 md/cp, the pressure stabilization is very rapid and there is an opportunity for multiple measurement pretests. However, there is no advantage to having more than one measurement phase if the sandface pressures determined from the investigation and first measurement phase agree **(Fig. 6.4.19).**

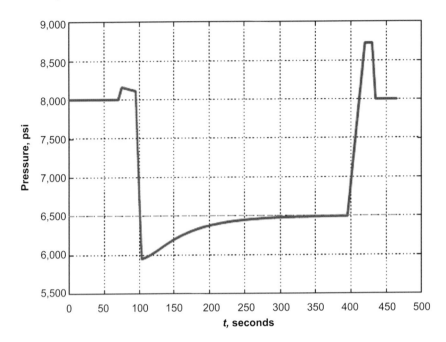

Fig. 6.4.17—The simulated time-limited pretest in a 0.1-md/cp formation, using a single pretest. The dashed magenta line represents the formation pressure.

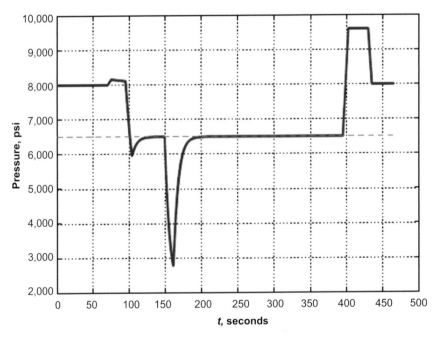

Fig. 6.4.18—The simulated time-limited pretest in a 1-md/cp formation, using a single pretest. The dashed magenta line represents the formation pressure.

High Overbalance. In this example, we contrast between a typical two-volume fixed-mode pretest and the time-limited pretest optimization under conditions where the overbalance is high (5,000 psi) and the formation permeability is relatively low (1 md/cp). **Fig. 6.4.20** shows that in the fixed-mode case, the combined volumes of the first and second pretest were insufficient to cause the probe pressure to go below the local formation pressure, and this is essentially a wasted test. **Fig. 6.4.21** shows that with the time-limited pretest,

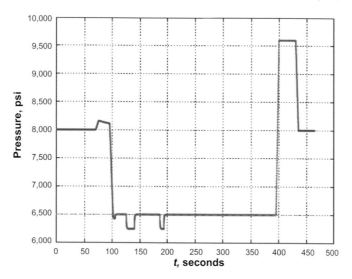

Fig. 6.4.19—The simulated time-limited pretest in a 20-md/cp formation. In this case, all phases of the test sequence reach stabilization before the criteria terminating the phases are achieved.

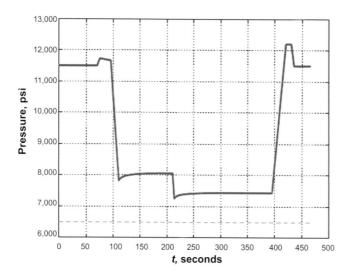

Fig. 6.4.20—A fixed-mode pretest when both volumes are insufficient to overcome the overbalance. This is typical of mature fields that have been depleted by production.

the investigation phase goes below the local sandface pressure without an excessive volume, with two successful measurement phases.

More-detailed examples of both synthetic and real case studies highlighting the importance of a time-limited field test are given in Pop et al. (2005b) and show that allowances for the unique environment that occurs in the while-drilling domain can be allowed for when making FPWD measurements.

6.4.11 Comparison With Post-Drill Data and Wireline Formation Testers.
Wireline-deployed tools have been the traditional method for formation-pressure testing. However, with more directional and horizontal wells being drilled, wireline technology becomes more difficult to deploy, typically necessitating the use of drillpipe or tractors. FPWD tools allow one to obtain estimates of formation pressure and fluid mobilities, without the risk of drillpipe-deployed tools, but these are sometimes perceived to be lower-quality measure-

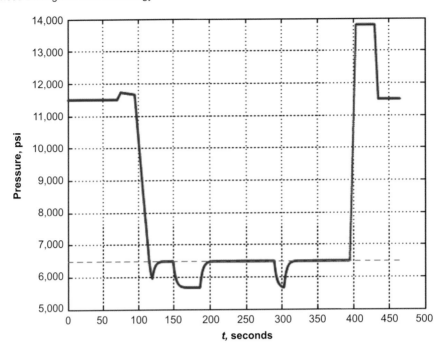

Fig. 6.4.21—This is the same situation as described in Fig. 6.4.20, but using the time-limited pretest.

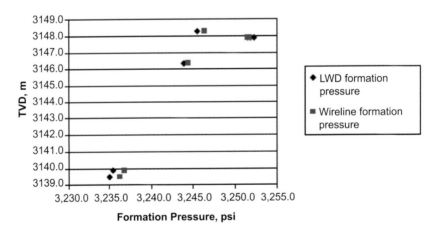

Fig. 6.4.22—Comparison of wireline and LWD data for a horizontal well (Seifert et al. 2007). TVD = true vertical depth.

ments because of the restrictions on the amount of data that can be transmitted during the drilling process. There have been numerous recent studies comparing the quality of wireline-deployed and while-drilling formation-pressure measurements, and they have all highlighted that under the right conditions, the quality of the FPWD measurements can be as good as the wireline versions (Seifert et al. 2007; Chang et al. 2005). This has led to the rapid acceptance of FPWD as a critical field-development measurement.

Fletcher et al. (2005) highlight that the similarity between the measurements can be on the order of the accuracy of the individual gauges. This is shown in **Fig. 6.4.22.** We note that when comparing wireline and LWD data, one must ensure that allowances are made for

- Differences between the measured depth and wireline depth
- Differing pad/probe orientations
- Individual sensor gauge accuracy

A more recent example also compares the wireline measurement with an FPWD measurement, and shows a match to within 1 psi and a fluid-gradient difference of less than 0.004 psi/ft (Mishra et al. 2007). The mobilities tested ranged from 4 to 1,100 md/cp, and tests were conducted with pumps both on and off. The S-shaped wells had tangents close to 70° deviation before dropping into the reservoir, and they necessitated the use of tough-logging-condition deployment of the wireline tool. The FPWD tool was run approximately 3 days after the equivalent wireline run. **Fig. 6.4.23** highlights that for one of the wells studied, the formation pressures estimated from each tool are nearly identical.

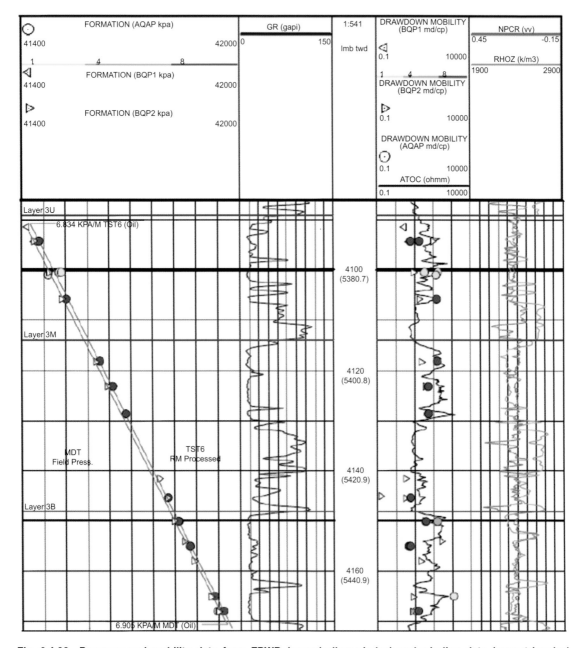

Fig. 6.4.23—Pressure and mobility data from FPWD (green/yellow circles) and wireline data (open triangles), highlighting nearly identical pressure gradients between each tool measurement (Mishra et al. 2007, © International Petroleum Technology Conference).

Nomenclature

b = nondimensional binary indicator for wall erosion

c_t = formation total compressibility, 1/Pa

D = rate of solids deposition, kg/m²/s

E = rate of erosion for solids in the filter cake, kg/m²/s

F = functional over all past times

$H(t)$ = sandface pressure response

H_D = dimensionless sandface pressure response

J_1 = Bessel function of the first kind

k = formation permeability, m²

m = shear-rate exponent in leakoff rate

M = mass of solids in the filter cake per unit sandface area, kg/m²

n = pressure exponent in desorptivity when differential pressure is greater than or equal to the maximum so far experienced (first loading, cake is normally consolidated)

n' = pressure exponent in desorptivity when differential pressure is less than the maximum experienced (unloading or reloading, cake is overconsolidated)

N_{Re} = Reynolds number

p_f = far-field formation pressure

$p(r,t)$ = formation pressure, Pa

p_m = mud pressure in wellbore, Pa

$p_{sf}(t)$ = sandface pressure, Pa

p_{well} = wellbore pressure, Pa

p_∞ = formation pressure at far distance, Pa

q = filtrate-leakoff rate, m³/s/m²

$q_{crit}\left(\dot{\gamma}_{ref}\right)$ = reference dynamic-fluid-loss rate, m³/s/m²

q_{mf} = mud filtration rate, m³/s/m²

Q = mud circulation rate, l/min

r = radial distance from borehole center

r_e = radius of influence, m

r_w = wellbore radius, m

R_D = nondimensional impulse response function

R_{inv} = radius of invasion

$S_{ref}\left(\Delta p_{ref}\right)$ = reference static desorptivity, m/s$^{1/2}$

t_m = filter cake thickness, m

T = cake thickness, m

V = filtrate-leakoff volume, m³/m²

Y_1 = Bessel function of the second kind

α = nondimensional binary indicator of accretion

$\dot{\gamma}$ = wall shear rate at sandface, 1/s

λ = rate constant for erosion of mudcake, 1/s

κ = mass of mud solids per unit volume of filtrate, kg/m³

λ = rate constant for filter-cake erosion, 1/s

μ = filtrate viscosity, Pa·s

μ_p = drilling-fluid plastic viscosity, Pa·s

ρ = drilling fluid density, kg/m³

σ = estimate of pressure measurement errors, Pa

$\tau_{crit.}$ = critical value for shear stress for erosion to occur, Pa

τ_w = wall shear stress at sandface, Pa

τ_y = drilling-fluid yield stress, Pa

ϕ = formation porosity

References

Aldred, W., Bergt, D., Rasmus, J., and Voisin, B. 1989. Real-Time Overpressure Detection. *Oilfield Review* **1** (3): 17–27.

Alford, J., Goobie, R.B., Sayers, C.V.M. et al. 2005. A Sound Approach to Drilling. *Oilfield Review* **17** (4): 68–78.

Barriol, Y., Glasser, K.S., Pop, J. et al. 2005. The Pressures of Drilling and Production. *Oilfield Review* **17** (3): 22–41.

Chang, Y., Hammond, P.S., and Pop, J.J. 2005. When Should We Worry About Supercharging in Formation Pressure While Drilling Measurements. *SPEREE* **11** (1): 165–174. SPE-92380-PA. DOI: 10.2118/92380-PA.

Dewan, J.T. and Chenevert, M.E. 2001. A Model for Filtration of Water-Base Mud During Drilling: Determination of Mudcake Parameters. *Petrophysics* **42** (3): 237–250.

Finneran, J.M., Green, C., Roed, H., Burinda, B.J., Mitchell, I.D.C., and Proett, M.A. 2005. Formation Tester While Drilling Experience in Caspian Development Projects. Paper SPE 96719 presented at the SPE Annual Technical Conference and Exhibition, Dallas, 9–12 October. DOI: 10.2118/96719-MS.

Fletcher, J., Seymour, G., Flynn, T., and Burchell, M. 2005. Formation Pressure Testing While Drilling for Deepwater Field Development. Paper SPE 96321 presented at the Offshore Europe Conference, Aberdeen, 6–9 September. DOI: 10.2118/96321-MS.

Fordham, E.J. and Ladva, H.K.J. 1989. Crossflow Filtration of Bentonite Suspensions. *Physico-Chemical Hydrodynamics* **11** (4): 411–439.

Fordham, E.J., Allen, D.F., and Ladva, H.K.J. 1991. The Principle of a Critical Invasion Rate and Its Implications for Log Interpretation. Paper SPE 22539 presented at the SPE Annual Technical Conference and Exhibition, Dallas, 6–9 October. DOI: 10.2118/22539-MS.

Hammond, P.S. and Pop, J.J. 2005. Correcting Supercharging in Formation-Pressure Measurements Made While Drilling. Paper SPE 95710 presented at the SPE Annual Technical Conference and Exhibition, Dallas, 9–12 October. DOI: 10.2118/95710-MS.

Longeron, D.G., Alfenore, J., and Poux-Guillaume, G. 1998. Drilling Fluids Filtration and Permeability Impairment: Performance Evaluation of Various Mud Formulations. Paper SPE 48988 prepared for presentation at the SPE Annual Technical Conference and Exhibition, New Orleans, 27–30 September. DOI: 10.2118/48988-MS.

Mishra, V.K., Pond, S., and Haynes, F. 2007. Formation Pressure While Drilling Data Verified With Wireline Formation Tester, Hibernia Field, Offshore Newfoundland. Paper IPTC 11249 presented at the International Petroleum Technology Conference, Dubai, 4–6 December. DOI: 10.2523/11249-MS.

Neumann, P.M., Salem, K.M., Tobert, G.P., Seifert, D.J., Dossary, S.M., Khaldi, N.A., and Shokeir, R.M. 2007. Formation Pressure While Drilling Utilized for Geosteering. Paper SPE 110940 presented at the SPE Saudi Arabia Technical Symposium, Dhahran, Saudi Arabia, 7–8 May.

Phelps, G.D., Stewart, G., and Peden, J.M. 1984. The Effect of Filtrate Invasion and Formation Wettability on Repeat Formation Tester Measurements. Paper SPE 12962 presented at the European Petroleum Conference, London, 22–25 October. DOI: 10.2118/12962-MS.

Pop, J., Follini, J.-M., and Chang, Y. 2005a. Optimized Test Sequences for Formation Tester Operations. Paper SPE 97283 presented at the Offshore Europe Conference, Aberdeen, 6–9 September. DOI: 10.2118/97283-MS.

Pop, J., Laastad, H., Eriksen, K.O., O'Keefe, M., Follini, J.-M., and Dahle, T. 2005b. Operational Aspects of Formation Pressure Measurements While Drilling. Paper SPE 92494 presented at the SPE/IADC Drilling Conference, Amsterdam, 23–25 February. DOI: 10.2118/92494-MS.

Seifert, D.J., Neumann, P.M., Dossary, S.M. et al. 2007. Characterization of Arab Formation Carbonates Utilizing Real-Time Formation Pressure and Mobility Data. Paper SPE 109902 presented at the SPE Annual Technical Conference and Exhibition, Anaheim, California, USA, 11–14 November. DOI: 10.2118/109902-MS.

SI Metric Conversion Factors

°API 141.5/(131.5+°API) = g/cm^3

bar	× 1.0*	E + 05 = Pa
cp	× 1.0*	E − 03 = Pa·s

ft	× 3.048*	E – 01 = m
ft²	× 9.290 304*	E – 02 = m²
ft³	× 2.831 685	E – 02 = m³
in.	× 2.54*	E + 00 = cm
lbm	× 4.535 924	E – 01 = kg
psi	× 6.894 757	E + 00 = kPa

*Conversion factor is exact.

6.5 Drilling Vibration—Chris Ward, GeoMechanics International

6.5.1 Introduction. Drillstrings are not stable, so downhole vibration is inevitable, but at a low level, it is harmless. However, severe downhole vibration can cause numerous problems such as bottomhole assembly (BHA) washouts, twistoffs, premature bit failure, accelerated failure of downhole equipment, excessive wear on tool joints, and damage to the topdrive and hoisting equipment. It also can lead to reduced rate of penetration (ROP) and hole enlargement. The financial losses incurred due to drilling dynamics are substantial and have been estimated to be on the order of 5 to 10% of drilling costs (Payne et al. 1995). Rising drilling costs, and increasingly complex and expensive downhole tools, make vibration mitigation and control a key issue in drilling optimization.

During drilling, various sources can excite the drillstring. The amplitude of the resultant drillstring vibrations will depend on the level (severity) of the excitation, the system damping, and the proximity of the excitation frequency to a natural frequency of the drillstring (Macpherson et al. 1993). When the frequency of any of the excitation sources is close to a natural frequency of the drillstring (axial, torsional, or lateral), then the string resonates and there is an amplification of vibration amplitude. If the amplitude levels are high, as they usually are at resonance in lightly damped conditions, then the drillstring will be subjected to fatigue loading that can result in localized or catastrophic failure (Reid and Rubia 1995). Vibration levels are generally highest at resonance, but high levels of vibration may exist in the drillstring, independent of drilling resonance, whenever a high level of excitation is present. Drilling with large-amplitude vibrations will result in accelerated drillstring failure. A study of drillstring failures (Hill et al. 1992) indicated that fatigue was the primary cause of the examined failures.

6.5.2 Types of Vibration. Vibration can induce three components of motion in the drillstring and bit **(Fig. 6.5.1)**: torsional, a motion causing twist/torque; lateral, a side-to-side motion; and axial, a motion along the drillstring axis. Combinations and interactions of these motions often lead to more complicated vibration motions. All dynamic motions and combinations can be detrimental to the bit and the drillstring.

Torsional Vibration (Stick/Slip). Torsional vibration is caused by periodic acceleration and deceleration of the bit and drillstring rotation, triggered by frictional torque on bit and BHA. It can cause large and damaging torsional vibrations in the string.

Due to the low torsional stiffness of drillstrings, instantaneous bit rotation is rarely the same as surface rotary speed. The string is continuously submitted to a light torsional pendulum effect due to friction at the bit and along the string. A non-uniform bit rotation develops in which the bit stops rotating momentarily at regular intervals, causing the string to periodically torque up and then spin free, accelerating the bit to high speeds (Pavone and Desplans 1994). As the torque is unwound, the bit speed then can slow to a rotation speed less than surface speed and in severe cases come to a stop or even reverse the rotation direction instantaneously. Because of this action, this mechanism is often called stick/slip vibration.

Stick/slip sets up primary torsional vibrations in the string with a frequency below 1 Hz, typically 0.05 to 0.5 Hz **(Fig. 6.5.2)**.

When severe, torsional vibration also can trigger lateral BHA vibration. It is usually associated with polycrystalline diamond compact (PDC) bits because of their higher friction, but can occur with rock bits and stabilizers. Stick/slip often can be observed at surface by large torque and/or revolutions-per-minute (rev/min) fluctuations.

Stick/slip-related problems include damage to PDC cutters due to increased impact loads, overtorqued connections or back-off, and topdrive stalling. Machine tests with rocks from hard to soft formations and different bits have shown that stick/slip leads to a 35% average ROP decrease (Dubinsky et al. 1992). Elf also

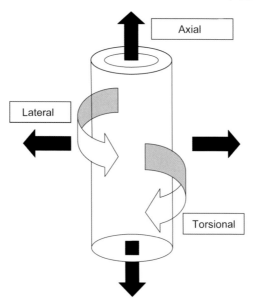

Fig. 6.5.1—The three main components of vibration motion: axial, lateral, and torsional (courtesy of Halliburton Sperry).

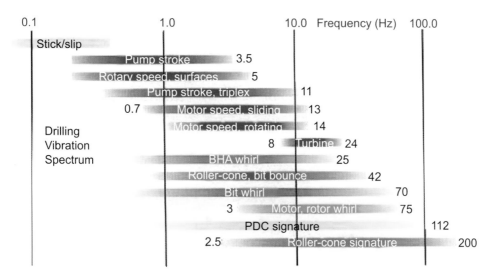

Fig. 6.5.2—Drilling-related spectral signatures (Macpherson et al. 2001).

concluded that stick/slip reduced ROP by 30 to 40% based on detailed analysis of field data (Payne et al. 1995).

Lateral Vibration (Whirl). Lateral vibration, or walk of the bit and/or BHA around the wellbore, can cause large and damaging stresses resulting from repeated impact of the bit and BHA with the wellbore. When severe, it also can trigger axial and torsional vibration. This mode of vibration is often called whirl and typically cannot be seen at the surface because the lateral bending vibration tends to dampen out along the string.

Bit whirl is the eccentric rotation of the bit about a point other than its geometric center and is caused by PDC bit-wellbore gearing, resulting from excessive side-cutting forces. As a result, the bit *walks* around the hole and produces unusual bottomhole patterns (**Fig. 6.5.3**). This vibration mechanism induces high-frequency

Fig. 6.5.3—A bottomhole pattern created by a PDC bit in soft rock. The five-lobed star pattern and a 1¼-in. (3.18-cm) overgauge hole suggest that the bit was whirling backward (Dykstra et al. 1994).

torsional and lateral vibrations of the bit in the range of 5 to 100 Hz, depending on the rotation speed and number of bit cutters. This can result in a large impact loading on the PDC cutters, resulting in their rapid failure.

BHA whirl is the walk of the BHA around the borehole caused by friction-driven gearing of stabilizers or tool joints with the wellbore. The mechanism can cause lateral and torsional vibrations of 5 to 20 Hz in the string, depending on the rotation speed and number of stabilizer blades. This results in repeated impacts of the BHA with the wellbore and often is responsible for stabilizer and tool joint damage **(Fig. 6.5.4)**.

Axial Vibration (Bit Bounce). Axial vibration is due to large weight-on-bit (WOB) fluctuations that cause the bit to repeatedly lift off bottom and then drop and impact the formation. This motion is often called bit bounce. When severe, it can trigger lateral BHA vibrations. This phenomenon is linked to the axial stiffness of the drillpipe and the mass of the BHA (Dubinsky et al. 1992) and is normally associated with rock bits. It can sometimes be identified at surface by severe axial shaking of the topdrive and hoisting equipment, especially in shallow vertical wells.

Bit bounce, which often occurs when drilling with tricone bits in hard formations, can cause axial bit vibration with a frequency range of 1 to 10 Hz. The resulting increase in impact loading can be damaging to the bit and BHA and lead to failure of downhole tools.

Other Types of Vibration. Other, more complicated and rarer types of vibration also exist. These are perhaps less well known and more difficult to recognize but result in equally damaging problems.

Bit chatter is caused by the individual bit teeth impacting the rock. It is usually a low-level vibration with high frequency, 50 to 350 Hz, depending on the rotation speed and number of bit teeth.

BHA-forced vibration is due to the resonance excitation of the BHA triggered by the other vibration mechanisms. This mechanism is always present at different magnitudes as *background* vibration even when all the other destructive mechanisms are absent. The failure caused by this non-severe vibration mechanism is often due to fatigue crack growth rather than rapid failure.

The dynamic component of axial load is caused primarily by bit-formation interactions, which result in fluctuations in the WOB (Dunayevsky et al. 1993). When these fluctuations are accounted for, the loss of mechanical stability becomes evident as rapidly growing lateral drillstring vibrations. This occurs in much the same way as inducing a snaked motion in a vertically hanging rope by moving its end up and down at a particular frequency. This phenomenon, which is associated with specific axial fluctuations, is called *parametric resonance*. It can occur for much smaller WOB values than the critical WOB obtained from a static-stability (buckling) analysis.

Furthermore, all of the above mechanisms can be coupled, in the sense that one can trigger the other. This is sometimes called *modal coupling* and typically has a frequency of 0 to 20 Hz. For example, bit whirl can be triggered by high bit speeds generated during stick/slip motion; stick/slip can cause lateral vibration of the BHA (parametric resonance) as the bit accelerates during the slip phase; and large lateral vibration of the

Fig. 6.5.4—Example of stabilizer damage caused by BHA whirl (courtesy of Halliburton Sperry).

BHA can in turn cause BHA-wellbore interaction and/or bit bounce. It often can be difficult to recognize the root cause of these vibrations.

6.5.3 Causes and Sources of Vibration. Vibration is caused by bit and string interactions with the rock under certain drilling conditions and in certain formation sequences (Dubinsky et al. 1992). Sources can excite vibrations directly, trigger other vibration mechanisms, or induce a resonance or natural harmonic into the drillstring. All of these can be damaging to downhole tools, bits, hole quality, and ROP.

During drillstring rotation, such mechanisms as imbalance, misalignment, bent pipe, drillstring walk about the inside of the wellbore diameter, or other geometric phenomena sometimes create excitations that are at the rotational frequency or multiples of the rotational frequency (Besaisow and Payne 1988). These mechanisms create forces and stresses that oscillate at frequencies of the excitation mechanisms that are multiples of applied rotation speed. When the excitation mechanisms' frequencies match one of the BHA natural frequencies, a resonance condition with growing stresses is generated. The speeds at which resonant conditions occur are called critical speeds.

The drillstring response to these excitation mechanisms is very complex, to a large extent due to the coupling that inherently exists between the motion of the bit and that of the drillstring. The frequency range measurements in the drillstring made at the surface can be divided into two distinct domains. A low-frequency domain (1 to 20 Hz) corresponding to excitation phenomena due to the rotary speed and its first harmonic and first axial, torsional, or bending natural modes of the drillpipe or BHA. This frequency range represents the drillstring failure risk domain; it exhibits the resonances of the drillstring as well as complex forms of behavior (e.g., BHA whirl, stick/slip). A higher-frequency domain (20 to 200 Hz) corresponds to the bit signature. Whether in pure rotary mode or drilling with a downhole motor, this *drilling efficiency* domain is valuable for ROP optimization.

For modeling, knowledge of the excitation mechanisms involved and their nature is required, along with good estimates of the BHA resonances from sophisticated models (Besaisow and Payne 1988). For field use, models can be used to define safe operating ranges of rev/min and weight-on-bit (WOB), or to change BHA configurations to those that minimize failure risk **(Fig. 6.5.5)**. A number of models have been developed to identify and avoid critical rotary speeds (Dykstra et al. 1995). Although some are quite advanced analytically, the accuracy of their predictions can be limited. This is particularly due to uncertainties in input data, especially with regard to boundary conditions such as hole size and shape. It is also due to assumptions that are made during the analysis procedure, the most important of which concern excitation magnitudes and locations. Despite their limitations, programs that analyze drillstring dynamics can be very worthwhile when used prudently. The programs can improve drilling performance significantly by comparing predictions with real-time monitoring information to adjust parameters on the fly. They can also help improve field drilling performance though post-analysis of previous bit runs.

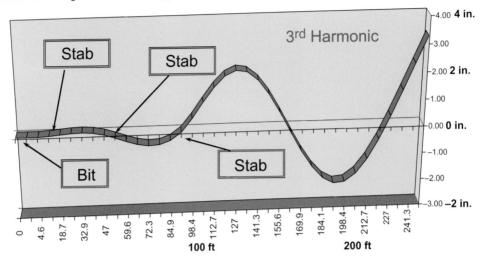

Fig. 6.5.5—An example of a model predicting BHA harmonics showing drillstring displacements and likely impact points. These models also can be used to predict critical rotary speeds and running parameters to avoid vibration hot spots. Third harmonic: Displacement along BHA in inches. Maximum displacement is BHA radio clearance in a 16-in. (40.64-cm) wellbore (courtesy of Halliburton Sperry).

Mass Imbalance. A rotating body is unbalanced when its center of gravity does not coincide with the axis of rotation (Dykstra et al. 1995). When the drillstring is rotated at one of its natural frequencies of lateral vibration, the deflection due to mass imbalance can be very large. This phenomenon is known as resonance, and rotary speeds at which resonance occurs are called critical speeds. Mass imbalance causes excitations primarily in the lateral direction that are on the order of the rotational frequency. This lateral excitation yields a smaller or secondary axial excitation that is on the order of twice the rotational frequency. Torsional excitations on the order of one or two times the rotational frequency are also induced. A bent-pipe mechanism is similar to the mass-imbalance mechanism (Besaisow and Payne 1988).

Misalignment. Misalignment of the drillstring, or buckling of the BHA, causes lateral excitations at the rotational frequency. Misalignment also causes axial and torsional excitations of twice the rotational frequency. Assuming that the drilled formation is not perfectly flat at the bottom, asymmetric rock strength contrasts can also induce lateral excitations.

Tricone Bit. The tricone rock bit causes excitations that are primarily three times the rotational frequency. Because the bouncing axial motion causes fluctuating levels of torque, a torsional excitation mechanism three times the rotational frequency is also induced. As the WOB is decreased and the bit-formation contact becomes less rigid, more harmonics from one to five times the rotational frequency can be seen.

Whirl. Another important excitation mechanism is the shaft whirl or drillstring walk mechanism. If slippage occurs, synchronous whirl vibrations at the rotational frequency are evident. The buckled sections of the BHA that are touching the hole, typically stabilizers, can also walk backward as a result of rubbing contact with the hole. Assuming no slippage, the walk rotational frequency is $[d_h/(d_h - d_d)]\omega$, where d_h is the hole diameter, d_d is the drill collar diameter, and ω is the rotation speed. Note that this mechanism occurs only with assemblies where the collar sections can buckle and touch the hole. Also, the backward walk mechanism will most likely occur in buckled sections below the neutral point.

BHA Resonance. Excitation of the drillstring in the vicinity of a natural frequency may lead to the development of an anomalously high-amplitude resonance and can induce severe vibration (Macpherson et al. 1993). This has long been recognized, and for example, the coincidence of the three-times bit rev/min signal and the first axial mode of many drillstrings has been noted. This is the underlying assumption for many drillstring dynamic models in which excitations at multiples of the bit rev/min are applied in harmonic analysis in order to predict rev/min operating windows (rev/min ranges for which the predicted vibration levels are low), sometimes called critical speeds.

Models using finite elements or wave and beam equations can be used to predict these resonances and to design BHAs less prone to resonance (Besaisow and Payne 1988). These simplified equations, however, do not predict all lateral modes that are important. Drillstring excitations may not occur at integer multiples of the rev/min; modeling boundary conditions may not be correct; the role of formations—or formation changes—appears critical in developing high-amplitude excitations; and lateral natural frequencies are often so closely spaced as to apparently close all operating windows. Therefore, avoidance of damaging resonance often depends on real-time monitoring.

6.5.4 Vibration Monitoring. The first attempts to record and process vibrations occurring at the surface and downhole took place in the early 1960s (Dubinsky et al. 1992). Vibration can be detected at surface through torque, rev/min, and standpipe pressure variations. It can be better detected downhole using measurements while drilling (MWD) from the monitoring of shocks, accelerations, torque, and WOB. Evidence for it can also be detected from post-bit-run inspection of damaged downhole equipment.

6.5.5 Tool Inspections. The nature of damage on downhole drilling components can be a direct indication of vibration source and mechanism. For example, localized wear on the tool joints is an indication of tool joints whirling around the wellbore, and damage on stabilizer blades is an indication of BHA whirl due to stabilizer-wellbore interaction.

6.5.6 Real-Time Vibration Modeling. Models that predict critical rotary speeds that excite lateral resonant vibrations have been introduced to aid BHA design and recommend operating parameters **(Fig. 6.5.6)** (Heisig and Neubert 2000). Some of these models have been adapted for real-time use. However, downhole data show that these models often have limited application in practice (Rewcastle and Burgess 1992). Contact between the drillpipe and wellbore dampens the transmission of vibrations in directional wells. In vertical wells, most models are highly sensitive to boundary conditions (e.g., BHA-wellbore contact points, hole size, bit-rock interaction), which results in limited predictive value. Running these models alongside real-time surface or downhole vibration monitoring can improve their value.

6.5.7 Surface Vibration Monitoring. Surface torque and rev/min oscillations can provide information on downhole vibrations (Dubinsky et al. 1992; Macpherson et al. 1993). In particular, this gives a good indication whether stick/slip vibrations are present, and whether drilling parameters can be adjusted in real time to avoid it. Surface torque, and to a lesser extent rev/min, tend to oscillate with a time period of several seconds. An alarm system, often in the form of *traffic lights*, can detect torsional stick/slip vibration and alert the driller.

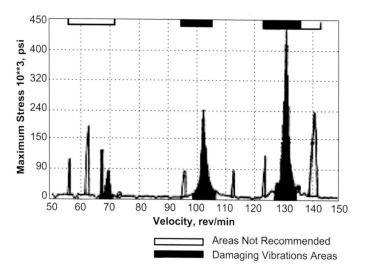

Fig. 6.5.6—Example of harmonic analysis for a 17½-in. (44.45-cm) hole. Note that all excitation mechanisms are included, with the primary mechanisms being highlighted as damaging vibration areas (Besaisow and Payne 1988).

Apart from torsional (stick/slip) and sometimes axial (bit bounce) vibrations, surface measurements of vibration are a poor indicator of the downhole vibration environment (Rewcastle and Burgess 1992). Lateral vibrations are dispersive, highly damped, and generally attenuated in the drillstring before reaching the surface. Lateral vibrations will, however, couple into the axial direction through either linear or parametric coupling, and are therefore sometimes detectable at the surface on axial vibration channels (Macpherson et al. 1993).

6.5.8 Downhole Vibration Monitoring. The first downhole MWD vibration devices had a simple shock sensor designed to monitor the cumulative transverse shock cycles on a particular tool component, principally to help determine tool fatigue levels (Alley and Sutherland 1991; Rewcastle and Burgess 1992; Cook et al. 1989). When a threshold was exceeded on the accelerometer (typically 25 G), a shock count was recorded and a cumulative shock history was used to determine the probability of failure through a statistical analysis. Even though it was possible to provide real-time shock information, this usually was not sufficient to diagnose the mode of vibration, and thus it was sometimes difficult to determine the correct course of action required to mitigate the vibration.

More recent tools continuously monitor BHA vibrations and diagnose the occurrence of vibration-related problems (Heisig et al. 1998). The type and severity of the diagnosed vibration are transmitted to the surface and displayed on the rig floor or a remote monitor. This allows the driller to take immediate remedial action by changing drilling parameters and, more importantly, to optimize the drilling process.

The low transmission bandwidth of MWD pulse telemetry does not permit the transmission of all the raw vibration data to the surface in real time. The downhole diagnosis of drilling dynamics data and transmission of diagnostic flags to the surface is the key to the problem. Recent advances in sensor technology and microelectronics have allowed the development of new downhole tools containing a full suite of dynamics sensors and high-speed data acquisition and processing systems (Heisig et al. 1998; Zannoni et al. 1993; Close et al. 1988). The sensors have been shown to have the capability to detect harmful BHA dynamic conditions such as whirling, lateral BHA shocks, stick/slip, and bit bounce by measuring changes in acceleration in the axial, tangential, and radial directions. They also can give advice and recommendations on the course of action for the driller to mitigate the vibration.

Fast-Fourier-transform (FFT) processing of high-frequency accelerations **(Fig. 6.5.7)** also can be used to determine the vibration frequencies. This can help to further diagnose the causes of vibration, and sometimes the excitation mechanisms and sources.

Bit manufacturers recently have developed easy-to-use, memory-mode, vibration-logging tools that have made obtaining relevant *at-bit* data feasible (Schen et al. 2005; Desmette et al. 2005; Roberts et al. 2005) for

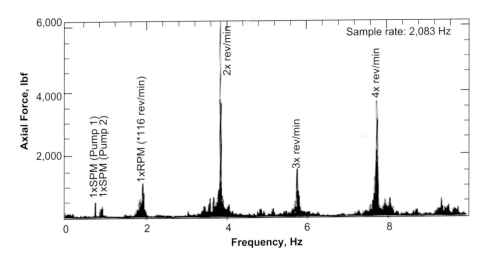

Fig. 6.5.7—Axial-force spectral signature prior to BHA fatigue failure (Macpherson et al. 1993).

assessing bit selections, design features, and running parameters. Typically, engineers have had only limited access to this type of data because MWD tools, when run, generally are placed well above the bit in the BHA, where dynamics can be significantly different from those at the bit.

6.5.9 Detecting Modes of Vibration. Each vibration mechanism has its own symptoms that assist in its detection. Sometime this can be seen at surface, always downhole with MWD and adequate measuring capabilities, and often from downhole equipment wear. Due to coupling between these mechanisms, severe downhole vibration often is accompanied by symptoms that belong to more than one mechanism. This makes the detection process more iterative.

Stick/Slip.
- Surface: characterized by surface torque, fluctuations in rev/min, and cyclicity; possible topdrive stalling; reduction in ROP
- Downhole: low-frequency torsional vibration
- Tool damage: PDC cutter impact damage; drillstring twistoff and/or washout; connections overtorqued or backed off

Bit and BHA Whirl.
- Surface: typically no direct observations; reduction in ROP
- Downhole: high-frequency downhole lateral and torsional vibration
- Tool damage: cutter and/or stabilizer damage; overgauge hole; increase in downhole torque

Bit Bounce.
- Surface: large surface vibration (shaking of hoisting equipment); large WOB fluctuations; reduction in ROP
- Downhole: large axial vibration
- Tool damage: bit damage (e.g., broken teeth, damaged bearing); BHA washout

Parametric Resonance and Modal Coupling.
- Surface: large WOB fluctuations; reduction in ROP
- Downhole: large lateral, torsional, and axial vibrations
- Tool damage: drillstring twistoff/washout

Rigsite Presentation. The cost of downhole vibration is now commonly recognized, and operators increasingly rely on surface and downhole information to reduce it.

The basic goal of real-time data analysis is to present the vibration data for a field supervisor or driller in a clear and simple form. Historically, MWD vibration severity levels are plotted against depth or time for the driller to monitor **(Fig. 6.5.8).** When high levels are observed, the driller can alter the operating parameters in an attempt to reduce them. Because of the complexity of downhole vibrations and the low bandwidth to transmit data, MWD systems that diagnose the vibration mechanism downhole and send up the result are now available. Some even have developed into some of the first *smart* advice systems used in the drilling industry, recommending to the driller which changes in rev/min and WOB to try to reduce the vibration and improve the ROP **(Fig. 6.5.9 and Table 6.5.1).**

6.5.10 Reducing Downhole Vibration. Vibration can be controlled while drilling by adjusting drilling parameters, particularly rev/min, WOB, and mud lubricity. Vibration risk also can be reduced by improving BHA/bit design and can be further mitigated using mechanical vibration dampeners such as shock subs, antiwhirl bits, and soft torque systems. Because vibration is difficult to eliminate completely, improvements to downhole tool reliability also have helped.

6.5.11 Reduced-Vibration Drilling Systems. A variety of different drilling systems and downhole drilling equipment has been developed over the years in an attempt to reduce downhole vibration.

Early on, shock subs placed directly behind the bit were developed to dampen vibration travel from the bit to the BHA. However, a shock sub, if incorrectly placed, rated, or worn to the point that it no longer meets specifications, can do more damage than it prevents (Macpherson et al. 1993). *Bendy bits* also perform the same function.

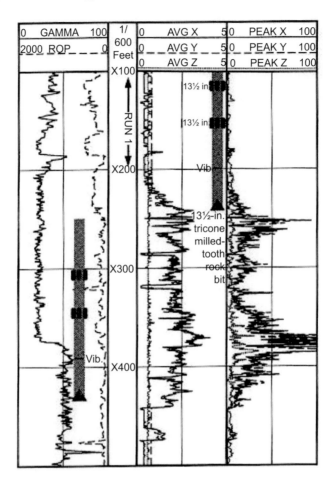

Fig. 6.5.8—Correlation of high lateral shocks with lithology (Zannoni et al. 1993).

A number of bits also have been designed to reduce specific types of vibration, such as antiwhirl, steering wheel, low-backrake, and long-gauge bits. They often have been very effective in reducing vibration generated from aggressive bits.

6.5.12 Mitigation of High Downhole Vibration. The avoidance of high vibration levels can be attempted in two ways: the BHA can be modeled and a harmonic analysis performed to predict the operating conditions, WOB, and rev/min that avoid resonant conditions, or the vibrations can be directly monitored while drilling to determine the optimum operating conditions (WOB, rev/min, and pump rate).

Stick/Slip. A soft torque system provides the most effective solution to this problem. The system consists of a topdrive (or rotary table) feedback mechanism that regulates surface torque fluctuation by altering rotary speed, thus enabling a more uniform bit rotation. It is essential that the system is regularly tuned to account for changing parameters such as depth and BHA configuration. Other possible remedial actions include increasing rev/min, lowering WOB (when stick/slip is due to bit-rock interaction), and increasing mud lubricity and use of roller reamers (when stick/slip is due to the BHA rubbing against the wellbore), all of which lead to reduced friction at the bit and BHA.

Bit Whirl. Antiwhirl bits have, in some cases, enabled PDC bits to drill into harder formations, but they have been less successful in highly interbedded formations. In situations where the geology is known, care must be taken when crossing formation interfaces where bit vibration and, hence, damage to the bit is likely. Also, bit whirl is likely when rotating off bottom or reaming because the bit is least constrained against

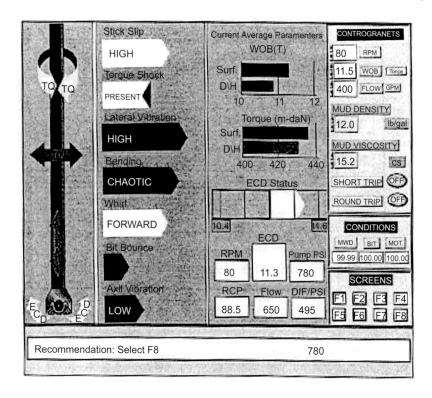

Fig. 6.5.9—Driller's advice screen (Dubinsky et al. 1992).

TABLE 6.5.1—RECOMMENDED DRILLING IMPROVEMENTS FOR DIFFERENT VIBRATION MECHANISMS		
	While Drilling	After Trip
Stick/slip	Increase rev/min and/or lower WOB	Add roller reamers
	Reduce the friction by increasing mud lubricity	Use a soft drive system (soft torque)
Bit bouncing	Increase or decrease the RPM slowly by steps	Add a shock sub
	Avoid cyclic variations of the WOB	Run a PDC bit
Whirling	Stop rotation and start with lower rev/min and greater WOB	Modify the BHA
	Ream the bottom of the hole	Add roller reamers
	Increase the mud viscosity	Add drillpipe protector
		Choose antiwhirl bit
		Change PDC bit profile

gearing with the wellbore. Therefore, with aggressive PDC bits, it is good practice to increase to full speed after tagging bottom.

BHA Whirl. Mud properties play an important role in BHA whirl. Roller reamers and nonrotating stabilizers provide a good solution when whirl is due to stabilizer-wellbore gearing. For the case of whirl due to tool joint-wellbore gearing, non-rotating drillpipe protectors can be effective.

Bit Bounce. Running a shock sub close to the bit can provide an effective solution. However, it is important to operate shock subs within their effective window of rev/min and WOB. At present, there is no guideline available on shock-sub selection. In the absence of a shock sub, changing of drilling parameters (e.g., WOB, rev/min) should be tried.

Parametric Resonance. Changing rev/min and/or reducing WOB are possible cures. Alternatively, dampening of WOB fluctuation by running a shock sub can provide an effective solution.

Nomenclature

d_d = drill-collar diameter
d_h = hole diameter
ω = rotation speed

References

Alley, S.D. and Sutherland, G.B. 1991. The Use of Real-Time Downhole Shock Measurements to Improve BHA Component Reliability. Paper SPE 22537 presented at the SPE Annual Technical Conference and Exhibition, Dallas, 6–9 October. DOI: 10.2118/22537-MS.

Besaisow, A.A. and Payne, M.L. 1988. A Study of Excitation Mechanisms and Resonance Inducing Bottom-hole-Assembly Vibrations. *SPEDE* 3 (1): 93–101. SPE-15560-PA. DOI: 10.2118/15560-PA.

Close, D.A., Owens, S.C., and Macpherson, J.D. 1988. Measurement of BHA Vibration Using MWD. Paper SPE 17273 presented at the IADC/SPE Drilling Conference, Dallas, 28 February–2 March. DOI: 10.2118/17273-MS.

Cook, R.L., Nicholson, J.W., Sheppard, M.C., and Westlake, W. 1989. First Real Time Measurements of Downhole Vibrations, Forces, and Pressures Used To Monitor Directional Drilling Operations. Paper SPE 18651 presented at the SPE/IADC Drilling Conference, New Orleans, 28 February–3 March. DOI: 10.2118/18651-MS.

Desmette, S., Will, J., Coudyzer, C., Richard, T., and Le, P. 2005. Isubs: A New Generation of Autonomous Instrumented Downhole Tool. Paper SPE 92424 presented at the SPE/IADC Drilling Conference, Amsterdam, 23–25 February. DOI: 10.2118/92424-MS.

Dubinsky, V.S.H., Henneuse, H.P., and Kirkman, M.A. 1992. Surface Monitoring of Downhole Vibrations: Russian, European, and American Approaches. Paper SPE 24969 presented at the European Petroleum Conference, Cannes, France, 16–18 November. DOI: 10.2118/24969-MS.

Dunayevsky, V.A., Abbassian, F., and Judzis, A. 1993. Dynamic Stability of Drillstrings Under Fluctuating Weight on Bit. *SPEDC* 8 (2): 84–92. SPE-14329-PA. DOI: 10.2118/14329-PA.

Dykstra, M.W., Chen, D.C.-K., Warren, T.M., and Azar, J.J. 1995. Drillstring Component Mass Imbalance: A Major Source of Downhole Vibrations. *SPEDC* 11 (4): 234–241. SPE-29350-PA. DOI: 10.2118/29350-PA.

Dykstra, M.W., Chen, D.C.-K., Warren, T.M., and Zannoni, S.A. 1994. Experimental Evaluations of Drill Bit and Drill String Dynamics. Paper SPE 28323 presented at the SPE Annual Technical Conference and Exhibition, New Orleans, 25-28 September.

Heisig, G. and Neubert, M. 2000. Lateral Drillstring Vibrations in Extended-Reach Wells. Paper SPE 59235 presented at the IADC/SPE Drilling Conference, New Orleans, 23–25 February. DOI: 10.2118/59235-MS.

Heisig, G., Sancho, J., and Macpherson, J.D. 1998. Downhole Diagnosis of Drilling Dynamics Data Provides New Level Drilling Process Control to Driller. Paper SPE 49206 presented at the SPE Annual Technical Conference and Exhibition, New Orleans, 27–30 September. DOI: 10.2118/49206-MS.

Hill, T.H., Seshadri, P.V., and Durham, K.S. 1992. A Unified Approach to Drillstem-Failure Prevention. *SPEDE* 7 (4): 254–260. SPE-22002-PA. DOI: 10.2118/22002-PA.

Macpherson, J.D., Jogi, P.N., and Vos, B.E. 2001. Measurement of Mud Motor Rotation Rates Using Drilling Dynamics. Paper SPE 67719 presented at the SPE/IADC Drilling Conference, Amsterdam, 27 February–1 March. DOI: 10.2118/67719-MS.

Macpherson, J.D., Mason, J.S., and Kingman, J.E.E. 1993. Surface Measurement and Analysis of Drillstring Vibrations While Drilling. Paper SPE 25777 presented at the SPE/IADC Drilling Conference, Amsterdam, 23–25 February. DOI: 10.2118/25777-MS.

Pavone, D.R. and Desplans, J.P. 1994. Applications of High Sampling Rate Downhole Measurements for Analysis and Cure of Stick-Slip in Drilling. Paper SPE 28324 presented at the SPE Annual Technical Conference and Exhibition, New Orleans, 25–28 September. DOI: 10.2118/28324-MS.

Payne, M.L., Abbassian, F., and Hatch, A.J. 1995. Drilling Dynamic Problems and Solutions for Extended-Reach Operations. In *Drilling Technology 1995, PD-Volume 65*, ed. J.P. Vozniak, 191–203. New York: ASME.

Reid, D. and Rubia, H. 1995. Analysis of Drillstring Failures. Paper SPE 29351 presented at the SPE/IADC Drilling Conference, Amsterdam, 28 February–2 March. DOI: 10.2118/29351-MS.

Rewcastle, S.C. and Burgess, T.M. 1992. Real-Time Shock Measurements Increase Drilling Efficiency and Improve MWD Reliability. Paper SPE 23890 presented at the SPE/IADC Drilling Conference, New Orleans, 18–21 February. DOI: 10.2118/23890-MS.

Roberts, T.S., Schen, A.E., and Wise, J.L. 2005. Optimization of PDC Drill Bit Performance Utilizing High-Speed, Real-Time Downhole Data Acquired Under a Cooperative Research and Development Agreement. Paper SPE 91782 presented at the SPE/IADC Drilling Conference, Amsterdam, 23–25 February. DOI: 10.2118/91782-MS.

Schen, A.E., Snell, A.D., and Stanes, B.H. 2005. Optimization of Bit Drilling Performance Using a New Small Vibration Logging Tool. Paper SPE 92336 presented at the SPE/IADC Drilling Conference, Amsterdam, 23–25 February. DOI: 10.2118/92336-MS.

Zannoni, S.A., Cheatham, C.A., Chen, C.-K.D., and Golla, C.A. 1993. Development and Field Testing of a New Downhole MWD Drillstring Dynamics Sensor. Paper SPE 26341 presented at the SPE Annual Technical Conference and Exhibition, Houston, 3–6 October. DOI: 10.2118/26341-MS.

Chapter 7

Deepwater Drilling

7.1 Deepwater Technology

7.1.1 The Deepwater Drilling Process—J.C. Cunha, University of Alberta

From ancient times men have used the word "earth" in two senses—to denote the soil and to name the planet which we inhabit. Although it was natural for man, the land dweller, to identify his planet with his domain of dirt and rocks, he did so out of ignorance of the world. Land is the lesser part of earth; it is water which is predominant on this planet (Brantly 1971).

Even though the amazing deepwater offshore developments are relatively recent, the search for offshore oil and gas, or at least the interest for such ventures, started almost simultaneously with the modern oil industry. Just 10 years after the Drake well, on 4 May 1869, Thomas F. Rowland from New York applied for a patent of a "submarine drilling apparatus" **(Fig. 7.1.1)**. Rowland's apparatus consisted of a platform and tender combination that almost 100 years later, in the 1940s and 1950s, would be the equipment predominantly used in offshore drilling activities.

Later in the same year (Brantly 1971) another inventor from the United States, Samuel Lewis, patented a "submarine drilling machine" that was composed of a drilling device placed on the top of a vessel that would be raised above the water and supported by six adjustable pillars touching the seafloor. This remarkable invention envisioned the basic principles of the modern jackup rigs that would be built one century later.

Although Rowland's and Lewis's designs have never been used, the oil industry started offshore exploration still in the nineteenth century, with many wells being drilled after 1897 off the coast of California at Summerland Beach, Santa Barbara, USA **(Fig. 7.1.2)**. Those wells were a mere continuation of the development of fields originally discovered onshore and later found to have part of their accumulation under the ocean. Nevertheless, the feasibility of those wells together with increasing technology developments indicated in those early years that part of the oil industry's future would certainly take place offshore.

In this chapter, we will concentrate on the deepwater drilling process. A brief evolution of offshore drilling will be presented, and then deepwater-well design aspects will be discussed, including descriptions of special equipment and nonconventional wells. Field examples will be provided.

Evolution of Offshore Drilling. Initially, offshore drilling was an extension of onshore activities; consequently, the first offshore wells were drilled near shore in shallow waters. To carry on those early drilling operations, causeways, piers, and small artificial islands were built near the shoreline. By 1910, the first successful overwater drilling in the United States took place at Ferry Lake in Caddo Parish, Louisiana (Chevron 2007). The rig used was set on pilings that could be floated out to the well by tugs and barges.

With the success obtained in the United States, where by 1920 drilling in lakes and offshore was a routine operation, other countries also started to develop their oil resources located offshore or under inland waters. In 1924, the exploration success already achieved on the shore of Lake Maracaibo, Venezuela, was repeated under the lake's water, with many discoveries being made in that year and in the following decades. A year later, wells were drilled over artificial islands in the Caspian Sea (AzerMSA 1999), Azerbaijan (then, USSR).

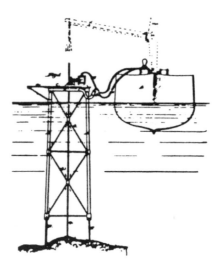

Fig. 7.1.1—Rowland's patent for a drilling tender: 4 May 1869.

Fig. 7.1.2—Offshore wells at Summerland Beach, Santa Barbara, USA, in 1899. Image courtesy of the University of Southern California Libraries, California Historical Society Collection, 1860–1960.

With the increasing demand for energy and the advances in technology, offshore equipment had an amazing development in the next decades, and by 1947, the first well "out of sight of land" was drilled 14.5 km offshore from a tender-supported fixed platform in the Gulf of Mexico. That well, located in Ship Shoal, coast of Louisiana, not only provided the first major oil discovery in the Gulf of Mexico but also represented the starting point for the modern offshore industry.

Two years later, various wells were drilled using the first moveable drilling rig, a pioneer submersible drilling barge designed by John Hayward. For the first time, the offshore oil industry was able to drill and test exploratory wells economically and, after conclusion of the job, move the rig to another location. A few years afterward, in 1953, a navy cargo craft named *Submarex* was adapted with a land rig cantilevered over her side, generating the first floating drilling vessel. The oil industry was taking its first steps toward deepwater development.

Currently, drilling operations are performed in water depths exceeding 3000 m. Oil production in water depths exceeding 2000 m already takes place in different parts of the world, with records being broken on a yearly basis. Next, we present a quick view on modern moveable drilling rigs and also on current deepwater production platforms. Although the emphasis of this chapter is on deepwater technology, rigs for shallow waters are also mentioned to better emphasize the evolution of the industry toward the more challenging deepwater environment.

Offshore Moveable Drilling Rigs. Moveable rigs originally were developed envisioning exploration drilling, because in addition to being cheaper than fixed platforms, they also could be used in multiple locations. Nowadays, moveable rigs may be used in either exploration or development drilling, depending on the strategy used to develop a field.

A *drilling barge* is a moveable rig used only in shallow waters, normally less than 50 m. Its design makes it appropriate to drill in lakes, rivers, and canals. It is not suitable for severe water movement normally found on open sea. It is towed from one location to another by tugboats.

Jackups **(Fig. 7.1.3)** are the most used offshore drilling rigs, comprising more than 50% of the worldwide fleet. They are used in shallow waters up to 110 m. The jackup is towed to the drilling location where the "legs" are lowered to the seafloor. Then the platform is raised above the water to the drilling position.

A *submersible rig* **(Fig. 7.1.4)** is used in shallow waters, normally less than 40 m. Currently it is the least used offshore rig, with less than 10 still in activity (Marine Drilling Rigs 2003). It has a submersible barge

Fig. 7.1.3—Jackup rig. Courtesy of Noble Drilling Services.

Fig. 7.1.4—A submersible drilling unit. Courtesy of Keppel Offshore & Marine.

and a piled platform. During transportation between locations, the barge is filled with air, making the vessel float. Once on location the barge is ballasted with water and sinks, resting on the bottom.

Semisubmersible rigs (**Fig. 7.1.5**) are the most widely used offshore rig to drill in water depths exceeding 100 m. Semisubmersibles have two or more pontoons that form the hull of the vessel. The drilling platform, storage area, and quarters are mounted on top of the hull. When moving from one location to another, the ballast tanks are empty, which allows the rig to float while being tugged. After being positioned on the new location, the tanks are filled with water, submerging the lower part of the structure. Since the submersible rig is designed for deeper waters, the lower hull will not rest on the seafloor. There are two different systems that can be used to keep the platform in place; one uses huge anchors that, together with the submerged portion of the vessel, keep it stable. Another, more modern, system is the dynamic-positioning system, which uses thrusters and a sophisticated navigation system to keep the vessel stable during drilling operations.

With the constant search for new oil and gas accumulation in deepwater areas, successive water-depth drilling records have been broken in recent years. Currently, there are semisubmersible rigs capable of drilling in water depths in excess of 2400 m.

Drillships (**Fig. 7.1.6**) are the ultimate deepwater drilling vessel. Even though the first drillships were not designed for an ultradeepwater environment, today the most challenging wells in deep water are drilled using modern drillships.

Drillships use a positioning system with multiple anchors or dynamic propulsion (thrusters), or a combination of both. Typically, drillships for deep and ultradeep water use dynamic-positioning systems and, although carrying larger payloads than semisubmersibles, have inferior motion characteristics.

Drilling operations in water depths greater than 3000 m using drillships are already feasible. With deepwater records being broken continually in recent years, it is virtually impossible to predict a depth limit for such operations.

Fig. 7.1.5—Semisubmersible rig for deepwater drilling. Courtesy of Noble Drilling Services.

Deepwater Development Systems. Much more challenging than drilling is the production of oil fields located in deepwater environments. Currently, oil fields located under water depths greater than 2000 m are already being produced; however, such developments are far from being considered routine in the industry. Extremely expensive, those developments are feasible only for large reserves with highly productive wells. The following is an overview of the most common production systems for deep water **(Fig. 7.1.7)**.

Fixed platforms normally can be installed in water depths up to 500 m. The jacket rests on the seafloor, and a deck is placed on the top, providing space for equipment, a drilling rig (which may be removed after the end of drilling operations), personnel quarters, and production facilities.

A *compliant tower* is a narrow, flexible tower. The deck on the top can have drilling and production equipment. Unlike conventional platforms that are designed to resist forces, compliant towers will flex with the action of wind, wave, and current, which makes them suitable for deeper waters. It can be used in water deeper than 800 m.

A *tension-leg platform* is a floating platform that will be kept in place by tensioned tendons with top and bottom segments used to attach it to the structure and seafloor, respectively. The tensioned tendons will limit vertical movement, allowing TLPs to be used in waters deeper than 1400 m.

Fig. 7.1.6—Drillship for ultradeep water. Courtesy of Transocean.

Deepwater Development Systems

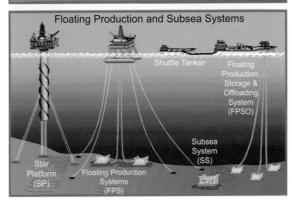

Fig. 7.1.7—Deepwater development systems [courtesy of the US Department of the Interior (Minerals Management Service 2008)].

A *mini-TLP* is equivalent to a small TLP. Its relatively low cost allows it to be used in the development of small deepwater reserves that would be uneconomical if produced with conventional platforms.

A *Spar platform* is a tall vertical cylinder held in place by mooring lines or tethers, as shown in Fig. 7.1.7. The cylindrical structure has spiral flanges to reduce vortex shedding in strong currents. Spar platforms currently are being used in water depths greater than 1600 m, and existing technology indicates that their use can be extended to water depths greater than 2000 m.

Subsea systems are used to produce single or multiple wells. The production, through a manifold and pipeline system, will be directed to a distant production facility. These systems are used currently in water depths greater than 1500 m (Minerals Management Service 2008).

A *floating production system* consists of a semisubmersible unit that may be equipped with drilling and production equipment and be kept in place by either a mooring or a dynamic-positioning system. It is used to produce subsea wells that will have their oil transported to the surface through production risers. It can be used in a large range of water depths, varying from less than 200 m to more than 2000 m.

A *floating production, storage, and offloading system* (FPSO) consists of a large tanker moored to the seafloor. FPSOs collect the production from nearby wells and periodically offload it to a carrier tanker. An FPSO may be used as a temporary production system while another type of platform is designed and built. It can also be used for marginally economic fields, avoiding the cost of pipeline infrastructure.

7.1.2 Deepwater Drilling Fluids—Rosana Lomba, Petrobras

Introduction. Complex deepwater and ultradeepwater drilling projects require renewed technological support aiming at the minimization of borehole problems and increased well productivity. In this context, chemical and physical properties of drilling fluids may determine the success of a drilling operation. Among other factors, low seabed temperatures, highly unconsolidated formations, low fracture pressures, shallow water flows and shallow gas flows, and a narrow operational margin between pore pressure and fracture gradient bring one's attention to the need for specially designed drilling fluids to ensure a successful operation. Also, the increasing number of depleted reservoirs demands the availability of aqueous and/or nonaqueous lightweight drilling fluids as an alternative to aerated fluids.

Drilling-Fluid Systems. Liquid drilling fluids may be classified into two main categories: aqueous and nonaqueous systems. The lithology, expected downhole pressures and temperatures, well geometry, formation-evaluation requirements (logging and fluid sampling), and environmental restrictions will determine the choice of one system over the other.

Aqueous Systems. **Tables 7.1.1 and 7.1.2** show examples of water-based drilling-fluid formulations. Typical additives to the systems are polymers, bridging solids, lubricants, weighting agents, salts, and defoamers. Each one performs a specific function regarding the required properties.

Nonaqueous Systems. Nonaqueous drilling fluids are water-in-oil emulsions. Different base oils may be used to formulate the system. Internal olefins, normal paraffins, isoparaffins, ester, and ether are commonly used base oils. The internal phase is a brine of low water activity to ensure chemical stabilization of shales. The emulsion is stabilized by a surfactant that keeps the water droplets from coalescing. Wetting agents are added to the system to make the solids oil-wet. Fluid-loss-control additives, viscosifiers, and weighting agents are other components of the system. **Table 7.1.3** presents a typical nonaqueous-drilling-fluid formulation with a 70:30 oil/water ratio.

TABLE 7.1.1—TYPICAL AQUEOUS-DRILLING-FLUID COMPOSITION FOR SHALE INHIBITION		
Additive	Function	Concentration (wt%)
Seawater	Continuous phase	As needed
Soda ash	Calcium remover	0–0.2
Xanthan gum	Viscosifier	0.15–0.45
Polyacrylamide (PAC) polymer	Viscosifier	0.15–0.45
PAC polymer	Fluid-loss reducer	0.3–0.45
NaCl	Shale inhibitor	5–31
Cationic polymer	Clay-swelling inhibitor	1.5–2.0
Triazine	Biocide	0–0.1
Caustic soda	pH controller	0–0.45
Detergent	Bit-balling preventer	0.02–0.1
Barite	Weighting agent	As needed

TABLE 7.1.2—TYPICAL AQUEOUS-DRILLING-FLUID COMPOSITION

Additive	Function	Concentration
Water	Continuous phase	As needed
NaCl	Shale inhibitor or weighting agent	2.8–31 wt%
Defoamer	Foam preventer	0.1 vol%
Xanthan gum	Viscosifier	0.2–0.55 wt%
Modified starch	Fluid-loss reducer	1.7–2.3 wt%
Magnesium oxide	pH buffer	0.25–0.45 wt%
Calcium carbonate	Bridging material or weighting agent	10–14 wt%
Lubricant	Lubricity agent	2–3 vol%
Triazine	Biocide	0–0.15 wt%

TABLE 7.1.3—TYPICAL NONAQUEOUS-DRILLING-FLUID COMPOSITION WITH A 70:30 OIL/WATER RATIO

Additive	Function	Concentration
Base oil	Continuous phase	0.67 vol%
Primary emulsifier	Emulsifier	2.5 wt%
Calcium oxide	Alkalinity	1.5 wt%
NaCl-saturated brine	Internal phase	0.34 vol%
Secondary emulsifier	Fluid-loss reducer	0.4–1.7 wt%
Organophilic clay	Viscosifier	1–1.2 wt%
Rheology modifier	Viscosifier	0.5–0.6 wt%
Wetting agent	Wettability	As needed
Barite	Weighting agent	As needed

Lightweight and Noninvasive Fluids in Deep Water. The occurrence of a narrow operational margin between pore pressure and fracture gradient has been commonly associated with some drilling problems, such as loss of circulation and well-control events. The use of lightweight fluids introduces the possibility of being able to drill ultradeepwater wells successfully, which has a significant impact on exploration activity. In addition, some deepwater reservoirs are already depleted, and there is a demand for lightweight fluids that are capable of avoiding circulation losses and minimizing formation damage.

From those operational perspectives, currently, development efforts are focused on two different topics: (1) the development of a dual-gradient drilling (DGD) system based on lightweight fluids and (2) the formulation of noninvasive drilling fluids.

DGD Systems Based on Lightweight Fluids. In the last few years, the industry has been addressing the use of lightweight fluids with great interest. Technology reviews and the development of equipment, materials, and computer simulators through several joint-industry projects and independent research groups are under way. One of the most beneficial ways of using lightweight fluids in a deepwater operation consists of managing the narrow pressure window.

For achieving the DGD condition, the industry has been focusing on the development of systems based on two distinct conceptual approaches: the use of lightweight fluids, which is the main focus of the present development efforts, and mechanical lifting, which consists of a pumping system to lift the mud from the seafloor up to the surface. The use of lightweight fluids consists of injecting hollow spheres or gas at the bottom of the marine riser to maintain the pressure in the subsea wellhead equal to the hydrostatic pressure of the seawater at the same water depth. Such a DGD system has one effective fluid gradient between the surface and the seafloor, and another one within the subsea well. Consequently, the effective mud weight at the previous casing is less than the effective mud weight at the current drilling depth.

Unlike the mechanical lifting systems, the use of lightweight fluids for achieving a dual-gradient condition consists of diluting the mud returns at the seafloor, through the injection of low-density materials, to decrease the density of the drilling fluids. From the point of injection upward to the surface, the density of the drilling fluid will be less than the effective mud density below the seafloor.

The DGD condition requires that the mud flowing through the riser exert hydrostatic pressure equivalent to that of the seawater at the seafloor. So, the diluting material must be added at a proper concentration to decrease the density of the drilling fluid to a value that fulfills this DGD requirement. Besides injecting the diluting material, it is also necessary to separate the material from the drilling fluid at surface. After this separation, the drilling fluid is processed in order to maintain the proper physical and chemical properties and is pumped back to the well through the drillstring. If hollow spheres are used as the diluting material, which generates an incompressible lightweight fluid, they will be reused after being extracted at surface. However, despite all recognized advantages of a DGD system based on this diluting material, this innovative drilling process requires new equipment and new operational procedures that are still being developed.

The gas lifted riser DGD system, which consists of injecting gas at the blowout preventer (BOP) level to reduce the marine-riser annular density down to the seawater density, requires much less development effort than the hollow-sphere system to be implemented in the field because of the improvements already introduced by underbalanced-drilling technology in recent years. The primary idea is to combine nitrogen injection with a high-pressure concentric-casing riser, which reduces the inside diameter of the riser and, consequently, the pumping volumetric requirements. The annular space between the outer riser and internal casing is filled with seawater to prevent collapse of the outer marine riser. A high-pressure rotating control head located at the top of the concentric riser seals the annular space, and the return stream is directed to an automated three-phase separator, which enables the continuous separation of the mixture components for recirculation, sampling, storage, flaring, or disposal.

The gas lifted riser DGD system involves the use of a closed and pressurized circulating system. On floating rigs, the key point consists of defining the best location for the rotating piece of equipment that holds pressure. This issue is critical, but, based on previous experiences, placing the rotating control head at the top of the inner riser is supposed to be the best option for this particular application. The low strength of the conventional drilling riser in terms of collapse and holding internal pressure is an important aspect to be considered in this analysis. However, besides the concentric riser concept, another very promising alternative is being developed—the surface BOP. This alternative not only will solve the difficulties associated with the strength of the conventional drilling riser but also will combine nicely with the slender well concept, presenting a significant potential for reducing overall drilling costs.

Noninvasive Fluids. Filtration control is a major criterion for fluids design. Avoiding fluid losses and minimizing formation damage depend on a minimum interaction between drilling fluids and the drilled rock. The adequacy of the drilling fluid to reach these goals is normally evaluated by the execution of conventional static and dynamic filtration experiments.

Fluid invasion into productive zones has been widely recognized as detrimental to well productivity. Filtrate and solids invasion can cause irreversible formation damage and permeability reduction. Drilling fluids are formulated to avoid excessive fluid penetration into productive zones. Nondamaging acid-soluble solids (for example, calcium carbonate) are usually added to drilling fluids in order to promote pore plugging and minimize fluid penetration. Also, specific polymers are used that reduce fluid invasion by means of surface chemistry and viscosity effects. The development of less-invasive nondamaging fluid formulations requires the knowledge of filtration mechanisms of solids-containing polymeric solutions in porous media.

The research has been carried out in various ways: the search for the understanding of the relative importance of factors contributing to filtration, the evaluation of solid additives to prevent loss of circulation, the evaluation of polymers to minimize fluid invasion into porous media, and the development and evaluation of noninvasive-fluid formulations. Theoretical and experimental studies on static and dynamic filtration of water-based drilling fluids have been carried out to evaluate the effects of the following: fluid type; solids shape, size, and concentration; polymer type and concentration; rock permeability; and applied differential pressure.

Regarding bridging materials, granular, laminated, and fiber-like solids act quite differently during filtration through a high-permeability unconsolidated porous medium, confirming the importance of shape effects on filtration mechanisms. Also, an increase in solids concentration does not necessarily lead to lesser invasion into the medium. Particle-size distribution and particle shape seem to be the major factors governing fluid invasion.

Commercial noninvasive-fluid formulations based on physicochemical mechanisms or surface interactions between additives and the permeable rock may be recommended for some specific applications, depending on rock type, downhole conditions, and drilling scenario.

Mud-Weight Program: Safety-Related Requirements. In case of riser disconnection, the "riser safety margin" establishes that the planned mud weight be such that the mud column together with the seawater column contains the exposed formation pressure. On the other hand, "kick tolerance" is defined as the maximum mud-weight gradient to control an influx without fracturing the more-fragile formation (often assumed to be the last exposed casing shoe). The "tripping safety margin" quantifies the changes in bottomhole pressures resulting from surge and swab effects, and it is normally expressed as an equivalent mud weight (lbm/gal). Although these requirements should be taken into consideration when defining the mud-weight program, the deepwater drilling scenario sometimes makes them difficult to be followed.

Gas Solubility in Nonaqueous Fluids. The most important operational aspects regarding an undesired formation-gas influx into the well, known as a kick, are the detection of the influx and its controlled circulation out of the well. The time taken to detect the kick is directly proportional to its size or initial volume inside the well, and this factor will dictate the pressures encountered during the well-control procedures. The maximum pressure at the surface, at the BOP (located at the mudline), and at the last casing shoe during the kick circulation are of particular interest to avoid gas leakage resulting from equipment failure or formation fracture, which can lead to a dangerous situation called a blowout.

The drilling scenario in deep and ultradeep water is quite complex, and the risk of blowouts should be mitigated. The interaction between formation gas and the drilling fluid under downhole pressure and temperature conditions is a very important issue concerning the safety of the drilling operation. It determines well-control procedures, and an unsuccessful operation could lead to severe environmental damages and losses of equipment and human lives.

In order to overcome the technical and environmental restrictions, synthetic liquids have been applied in the formulation of drilling fluids. Synthetic nonaqueous drilling fluids show excellent lubricity and shale-stabilization characteristics and may be the unique option when drilling in high-pressure/high-temperature (HP/HT) environments. Concerning the deepwater and ultradeepwater scenario, use of synthetic-based drilling fluids makes a well-control situation even more complex because of the inherent difficulties and higher costs involved in drilling offshore wells and because of the peculiar thermodynamic interaction between this kind of fluid and the formation gas. A better understanding of the interaction between formation gas and a synthetic-based drilling fluid during a kick detection and circulation situation would contribute to safe and economical drilling of offshore wells.

Several contributions have been published regarding gas dissolution in conventional oil-based drilling fluids (O'Brien 1981; Thomas et al. 1984; O'Bryan 1983; O'Bryan et al. 1988; O'Bryan and Bourgoyne 1990). Fewer works have focused on other organic fluids (Berthezene et al. 1999; Bureau et al. 2002).

Pressure/volume/temperature (PVT) measurements of mixtures of methane and organic liquids (linear paraffin and ester), currently applied in drilling-fluid systems for deepwater and ultradeepwater drilling, are presented. The measurement of thermodynamic properties of the methane/liquid mixtures, such as bubblepoint pressure, solubility, oil formation volume factor, gas formation volume factor, and liquid density, was performed for different temperatures. The results showed that the precise accounting of formation-gas solubility in the fluid under downhole conditions and during the kick circulation is a very important issue for drilling deepwater and ultradeepwater wells safely **(Figs. 7.1.8 through 7.1.12)** (Silva et al. 2004).

Hydrates. The potential for hydrate formation in drilling fluids increases with the increase in water depth. The low seabed temperatures and high pressures provide the proper conditions for hydrate formation and growth, especially in aqueous drilling fluids. Once the formation of hydrate starts, growth may be quite rapid and a solid mass may be formed in the wellbore, kill and choke lines, and/or inside the BOPs. The solid mass may block fluid circulation or even be strong enough to prevent drillstring movement.

To avoid the problem, drilling fluids are formulated either to inhibit or to delay hydrate formation. Another approach involves preventing the growth of hydrates once they are first formed.

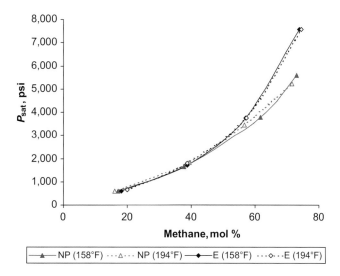

Fig. 7.1.8—Bubblepoint pressure as a function of methane molar fraction in n-paraffin (NP) and in ester (E).

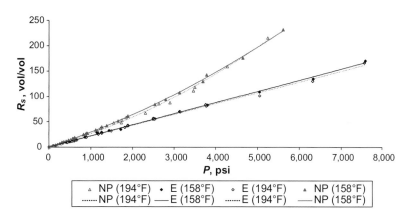

Fig. 7.1.9—Solubility of methane in n-paraffin (NP) and in ester (E).

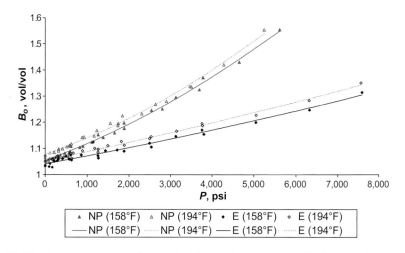

Fig. 7.1.10—Formation volume factor of oil for methane in n-paraffin (NP) and in ester (E).

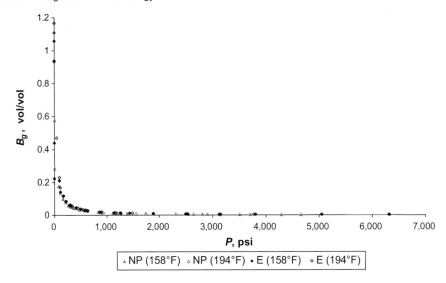

Fig. 7.1.11—Formation volume factor of gas for methane in n-paraffin (NP) and in ester (E).

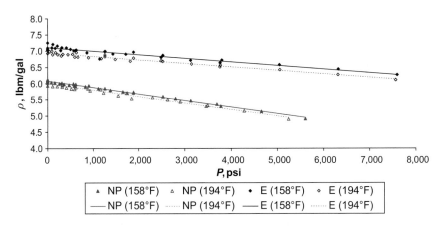

Fig. 7.1.12—Density of the saturated fluid for methane in n-paraffin (NP) and in ester (E).

Thermodynamic Inhibitors. Methanol, ethanol, glycols, and salts are commonly used additives for preventing hydrate formation in drilling mud. The combinations and product concentrations are established depending upon temperature and pressure conditions and the required degree of inhibition. Pressure/temperature diagrams are obtained experimentally to define the more-suitable mud composition for a given set of drilling conditions. Typical concentrations are 20% NaCl and 10% ethylene glycol (Kotkoskie et al. 1992; Ebeltoft et al. 1997).

Kinetic Inhibitors. Kinetic inhibitors are used to delay hydrate formation under certain conditions by delaying the appearance of the critical nuclei. Kinetic additives are copolymers or surfactants used in low concentrations so that they have only limited effect on fluid properties. The performance depends on the required subcooling.

Antiagglomerants. These kinds of additives are used to avoid the growth of hydrates once they begin to form. They are also called crystal modifiers. They may slow the rate of hydrate formation and/or prevent the agglomeration process.

Temperature-Dependent Fluid Rheology. The low temperatures encountered at the seabed (below 5°C) directly affect fluid rheology and lead to severe gel development of the drilling fluid. Moreover, the hydrostatic pressures at the bottom of the riser are usually very high, depending on fluid density and water depth.

The rheological properties of aqueous drilling fluids increase with the decrease in temperature and are affected only slightly by pressure. On the other hand, the rheological behavior of synthetic nonaqueous drilling fluids is affected by both temperature and pressure.

Significant differences may be found in computed equivalent circulating density (ECD) compared with an ECD calculated without considering rheology dependence on temperature and pressure (Davison et al. 1999). The variations may cause well-control problems, especially with a narrow operational margin between pore pressure and fracture gradient.

The reduced downhole rheology and reduced ECD often result in poor wellbore cleaning, barite sag, or other problems related to low rheology at bottomhole static and circulating temperatures. Adjusting the rheology upward can result in excessive ECD values and gel strengths.

Lost-circulation problems caused by the adverse increase of viscosity and ECD may result in well-control problems, high operational costs, and severe environmental damages. Therefore, new drilling-fluid systems should show an improved, flat rheological profile over a wide range of temperature and pressure. This behavior would allow for the maintenance of higher viscosities for improved cuttings-carrying capacity and barite-sag prevention without affecting ECD adversely (van Oort et al. 2004).

7.1.3 Deepwater Cementing—Cristiane Richard de Miranda, Petrobras

Introduction. There are several key differences in performing cementing jobs in deepwater wells compared with cementing jobs in onshore and shallow-water wells. The main differences are the following: very low temperatures; existence of different temperature gradients for the sea and for the formation; narrow operational window between pore pressure and fracture gradient that can result in major fluid losses when drilling, running casing, and cementing; possible occurrence of shallow water- or gas-flow problems; and formation and destabilization of gas hydrates.

The cement-slurry design and cementing operation must be engineered and executed in a manner that appropriately recognizes the specific problems of deepwater wells. Proper design and execution can reasonably assure good zonal isolation, structural support, and long-term durability of the cement sheath.

Temperature. (See Zhongming and Novotny 2003; Ward et al. 2001, 2003; Romero and Touboul 1998; Calvert and Griffin 1998.) The bottomhole circulating temperature (BHCT) of vertical onshore wells and of shallow-water wells can be determined using the data presented in American Petroleum Institute (API) Specification 10 (*RP 10B* 1997), and it is used to properly design cement slurries for cementing operations.

In contrast to onshore wells, the temperature found in deepwater wells is affected by a great number of factors that are not included in the API correlations. The main factors influencing the temperature in deepwater conditions are ocean-current velocities, sea temperature, presence or absence of a riser, and heat of hydration of the cement slurry. Because the temperature used to design and test cement slurries greatly influences their properties (e.g., thickening time, compressive-strength development, rheology, and fluid loss), it is necessary to determine the BHCT and the bottomhole static temperature in a manner that properly accounts for the unique conditions encountered in deepwater wells.

Small changes in temperature can cause significant alterations in the cement-slurry properties, such as thickening time and compressive-strength development. In practice, this can then lead to a cement-slurry design that does not match the required conditions. For example, if the BHCT was overestimated, then the slurry reaction may be slowed too much (overly retarded), resulting in an increased risk of gas or water flow from shallow formations and in an excessive waiting-on-cement (WOC) time. In a deepwater operation, where rig time is very expensive, this wasted WOC time results in a large, unnecessary expense. An accelerated cement slurry can bring serious problems if it sets before its complete displacement.

A numerical heat-transfer simulator is recommended to predict the temperature expected during circulation and the temperature achieved because of the heat recovery after the circulation is stopped. In general, an appropriate numerical heat transfer simulator should consider sea-current velocities, sea temperatures, well-deviation-survey data, rheological parameters and densities of the fluids (drilling fluid, washes, and cement slurry), heat of hydration of the cement slurry, thermal properties of the formations, more than one temperature gradient, circulation rate and time, temperature of the injected fluid, presence or absence of a riser, and insulation of the riser.

Because of the complexities found in deepwater wells, the document ISO *10426-3* (2003) recommends the use of a numerical simulator or the use of measurements from an offset well so the conditions of temperature and pressure placed on the cement slurry in the laboratory tests can be as close as possible to the actual conditions found in a deepwater well. Matching the temperature and pressure schedule of the laboratory tests to the actual deepwater-well conditions will improve the engineer's ability to design the cement slurry properly for a particular well.

Shallow Water Flow (SWF). In water depths greater than 152 m (500 ft), especially in depths greater than 305 m (1,000 ft), water can flow from shallow, overpressured formations, which can compromise the hydraulic integrity of the tophole section. This influx of water will result in a poor cement isolation, which in turn can result in problems such as buckling or shear of the casing, pressure/fluid communication with other shallow formations, and disturbance of the seafloor because of breakthrough of the shallow flow to the mudline (*RP 65* 2002). Besides water, other fluids can flow from shallow formations, such as mixtures of water, gas, and formation fines. Shallow gas causes fewer problems in deepwater wells compared to conventional water depths and compared to shallow water. The backpressure provided by the long seawater column reduces the gas expansion, so the wellbore erosion or failure caused by shallow gas is less likely (Rae and Di Lullo 2004). But the influx of gas into the cement while the cement is setting is obviously a problem.

SWFs can cause large economic losses because of well collapses and seabed subsidence, resulting in loss of wells and abandonment of the location (Rae and Di Lullo 2004).

Most frequently, the SWF is a result of the existence of an abnormally high pore pressure resulting from undercompacted and overpressured sands caused by rapid deposition (an underwater landslide known as a turbidite depositional event). Another reason can be a hydraulic communication with a deeper, higher-pressure formation, causing abnormal shallow pressures. Destabilization of gas hydrates or induced storage during drilling, casing, and cementing operations can also cause SWFs (*RP 65* 2002).

To avoid or to control SWF, it is recommended to follow some practices regarding site selection, drilling, and cementing. To prevent induced storage, it is advisable to minimize water loss, equivalent circulating density (ECD), mud weight, and surge pressures (Alberty et al. 1999).

It is advisable to use a computational simulator to design the cementing job better by adjusting the properties of the fluids, such as rheological parameters, densities, volumes, and flow rates, to get a better displacement efficiency and to ensure that the resulting pressures are within the pore- and fracture-pressure margins. The use of cementing best practices, including a good centralization, is fundamental to providing an effective primary-cementing job.

In addition to making certain that the rheological parameters are designed properly to cause an efficient displacement of the previous fluids pumped into the well, and using a proper cement-slurry density to avoid exceeding the formation fracturing pressure, the cement slurry should present certain characteristics to avoid SWF problems: fast liquid-to-solid transition, long-term sealing, and good control of fluid loss, free water, and sedimentation.

The rapid set of the cement slurry is a very important characteristic to avoid water flow during the gelation period, when the slurry begins to develop gel strength and loses the ability to fully transmit hydrostatic pressure, causing an underbalanced condition that can lead to fluid invasion.

To ensure that hydrostatic pressure is always transmitted to the formation, two slurries can be used with different thickening times, the lead slurry having a longer thickening time than that of the tail slurry.

SWF problems occur in several places, and there are many reported cases of SWF that occurred in Gulf of Mexico. In the Campos basin in Brazil, SWFs are not a problem, and therefore a conventional cementing job can be applied there. The conductor casing is in general driven (jetted) to approximately 50 m (164 ft), and no cement is used to support this casing. The subsequent casing (20-in. or 13⅜-in.), which is run into a competent formation, is usually cemented with a conventional extended cement slurry.

Hydrates. Gas hydrate is a solid with an ice-like structure and is formed from a mixture of water and natural gas. The water can come from two sources: (1) drilling fluid and (2) formation. The formation of gas hydrates depends on the gas composition, liquid-phase composition, temperature, and pressure. The high hydrostatic pressure and the low temperature found in deepwater wells increase the likelihood of hydrate formation in chokelines, risers, blowout preventers (BOPs), and subsea wellheads (Barker and Gomez 1989).

Hydrates belong to a group called clathrates, consisting of host molecules (water) that form a lattice structure to entrap the guest substances that are the gas molecules. Small molecules such as hydrocarbons C_1 to C_4, hydrogen sulfide (H_2S), and carbon dioxide (CO_2) can produce hydrates with water. Molecules that are too large to fit into the lattice cannot form hydrates. Because of the large volume of trapped gas, the decomposition of hydrates, by pressure decrease or by temperature increase, can generate a very large volume of gas.

Fig. 7.1.13 shows the combinations of pressure, temperature, and gas density (which is a function of gas composition) where the hydrates are stable. If gas hydrates are present before the cement job, it is necessary to reduce the risk of destabilization of gas hydrates by the use of cement slurries that generate low heat of hydration (Hampshire et al. 2004).

Cement Systems. Because of the conditions found in most deepwater wells, such as SWF, shallow gas flow, gas hydrates, and a narrow margin between pore and fracture pressures, the cement slurries to be used in these conditions have to exhibit properties that can ensure cement-sheath integrity. Some desirable properties include low density, short thickening time, fast liquid-to-solid transition, rapid development of mechanical properties, and low permeability.

In order to reduce the cement-slurry density, it is necessary to use extenders as additives. Besides the density reduction, the use of extenders allows one to reduce the amount of cement required to produce a given volume of slurry. The extenders can be classified as follows (Nelson 1990):

1. *Water extenders* are additives that allow the addition of water in excess without causing the settling of the cement slurry. Clays and water-viscosifying agents are water extenders.
2. *Low-density materials* are solids with density lower than that of the cement. The density of the cement slurry is lowered by the use of these materials in the slurry composition.
3. *Gaseous extenders* are nitrogen or air used to prepare foamed cements, reducing the density of the slurry.

Lightweight-cement systems include the following:

1. Foamed cement
2. Extended cement with clays
3. Extended cement with diatomaceous earth
4. Microsphere cement
5. Optimized-particle-size-distribution (OPSD) cement

Foamed Cement. (See Rae and Di Lullo 2004; Ravi et al. 2007; Reddy et al. 2002; White et al. 2000; Biezen and Ravi 1999; Davies and Cobbett 1981; Smith 2003.) Foamed cement consists of a mixture of

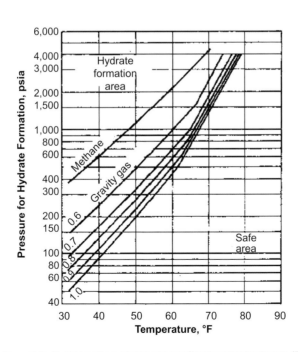

Fig. 7.1.13—Conditions favorable for formation of natural-gas/freshwater hydrates (Barker and Gomez 1989).

cement slurry and nitrogen that aims to obtain a lighter, stable cement system. This kind of system has been used to avoid SWF problems. The advantages of the foamed cement are very low densities [719 kg/m³ (6 lbm/gal) minimum] with relatively high strength, control of gas migration and SWF, and higher ductility compared to conventional cement systems. The disadvantages of this system are (1) that it is operationally more difficult to prepare the foamed slurry compared with conventional cement systems and (2) there is a requirement for more-precise control to ensure that the base cement slurry and nitrogen are mixed together in the correct proportions and homogeneously, which is becoming easier to achieve with automated mixing control.

Extended Cement With Clays. The most common clay used as a water extender is bentonite, mainly composed of sodium montmorillonite. Bentonite expands several times its original volume when placed in water, resulting in higher fluid viscosity, gel strength, and solids-suspending ability (Nelson 1990).

Extended Cement With Diatomaceous Earth. Diatomaceous earth is composed mostly of very small amorphous silica shells of organisms called diatoms. The density of this additive is much lower than that of the cement, and it requires a large quantity of water, resulting in the density reduction of the cement slurry.

Microsphere Cement. (See Rae and Di Lullo 2004; Smith 2003.) Hollow glass or ceramic microspheres can be added to the cement to obtain cement-slurry densities as low as 959 kg/m³ (8 lbm/gal). The crush strength of different microspheres varies in a broad range, and there are materials that can resist hydrostatic pressures greater than 414 MPa (60,000 psi). It is very important in selecting the material to consider its properties and the maximum pressure the cement slurry will be subjected to in the well.

The microspheres can be added to the cement forming a dry blend, and because of the large density difference between the materials, gravity segregation can occur, with the light microspheres separating to the top of the blend. This can cause problems of nonhomogeneity in the cement column.

OPSD Cement. (See Rae and Di Lullo 2004; Hampshire et al. 2004; Watson et al. 2003; Piot et al. 2001; Sørgård and Villar 2001.) Cement systems based on OPSD were used first by the construction industry. Part of the cement was replaced by aggregates (sand and gravels) and by other materials, such as fly ash, blast-furnace slag, or microsilica, increasing the compressive strength of the set concrete (Sørgård and Villar 2001). This technology was then brought to the oil industry to design cement slurries with optimized properties to be used in cement jobs.

This technology uses materials with different particle-size distributions in order to achieve a high packing density, reducing the voids among them. The water amount used in these systems is reduced compared to conventional slurries, and the higher solids content of OPSD cement is responsible for the low permeability and high compressive strength presented by this system. The OPSD cement can present a broad range of densities, depending on the density of the solids used in the design and on the amount used. Lightweight OPSD cement uses materials that present low density, such as microspheres.

Fig. 7.1.14 presents the relationship between compressive strength and density of three kinds of cement systems: water-extended cement, high-strength ceramic-bead cement, and foamed cement. The OPSD cement, which is not presented in Fig. 7.1.14, presents in general higher compressive strength than the ceramic-bead cement.

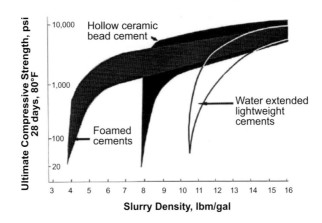

Fig. 7.1.14—Compressive strength vs. slurry density for different slurry compositions (Harness et al. 1992).

The cement slurries presenting very low densities have to be prepared with care because the density of the dry blend can be similar to the density of the water, making it impossible to control the ratio of dry blend to mix water by the mixture density. For small volumes, it is possible to prepare the slurry with known amounts of dry blend and water using a batching mixer, but for large volumes of slurry it is important to use appropriate equipment to ensure that the proper quantities of dry blend and water are mixed. This can be accomplished by measuring the flow of the mix water and of the slurry and calculating the solids content in the prepared mixture, enabling one to prepare an ultralight cement as designed.

Physical Properties of the Cement Systems. The cement slurry used in deepwater cementing jobs must exhibit certain properties, as already mentioned in the preceding sections. To avoid SWFs, the cement slurry must present

- Low density to avoid fracturing the formation that presents a low fracture gradient
- Fast liquid-to-solid transition and long-term sealing to avoid fluid or gas migration
- Good control of fluid loss so the slurry density will not increase to the point of causing lost circulation and alterations in the properties of the slurry, such as rheology and thickening time
- Zero free water and no sedimentation to avoid creating a path for gas or fluid and to avoid causing density differentials that may result in insufficient hydrostatic pressure to maintain well control (Kolstad et al. 2004)

When gas hydrates are present, the cement slurry must exhibit low heat of hydration to avoid destabilization of gas hydrates. To provide a long-term hydraulic seal and structural support, the cement slurries must exhibit flexibility and chemical resistance. These parameters must be evaluated considering the stress and strain the cement sheath will have to withstand and the aggressive fluids that will be in contact with the cement.

The heat generated during cement-slurry hydration (heat of hydration) should be considered to better estimate the temperature the cement will be submitted to. This increase in temperature because of hydration will speed the cement reaction, reducing the WOC time. A better understanding of this reaction rate can lead to savings in deepwater drilling operations by reducing idle rig time (Ravi et al. 1999; Romero and Loizzo 2000).

7.1.4 Deepwater Hydraulics—André Leibsohn Martins, Petrobras

Introduction. Hydrocarbon exploration in deepwater environments presents several particularities concerning hydraulics design. Because of the thin sediment coverage, rock formations frequently present low competency, and consequently, hydraulics should contemplate the narrow operational window between pore and fracture pressures. Additional difficulties may be faced in situations where the lower collapse pressure of the wellbore is higher than pore pressures and/or the upper collapse pressure of the wellbore is lower than fracture pressures.

In this scenario, a thorough understanding of the phenomena governing bottomhole pressure is necessary for high-cost ultradeepwater operations. Among other topics, the presence of solids in the annulus plays a major role in bottomhole-pressure prediction by two different mechanisms:

- Solids traveling in the annulus transmit hydrostatic pressure, which impacts bottomhole pressure directly. This effect increases with water depth as a result of the natural solids loading at the low-velocity annular flows through the riser. A common approach for predicting the impact of solids loading is to consider an average density of the fluid-cuttings mixture (ρ_m), as follows:

$$\rho_m = \rho_f \left(1 - C_s\right) + \rho_s C_s, \quad \dots\dots\dots\dots\dots\dots\dots\dots\dots\dots\dots\dots\dots\dots\dots\dots \quad (7.1.1)$$

where C_s is the solids concentration (vol%) and ρ_f and ρ_s are the fluid and cuttings density, respectively. **Fig. 7.1.15** illustrates the pressure increase caused by the presence of cuttings in a 10-lbm/gal mud for a complete range of water depths. **Fig. 7.1.16** shows the effect of cuttings loading on ECD in a typical deepwater well.

- Solids forming a cuttings bed in a highly inclined section may not transmit hydrostatic pressure but will restrict flow area, and also will accumulate near-annulus restrictions, resulting in pressure peaks.

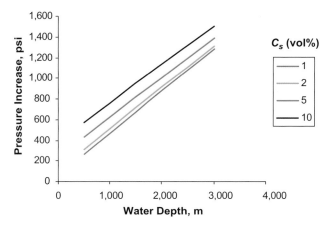

Fig. 7.1.15—Impact of cuttings concentration (C_s) at several water depths.

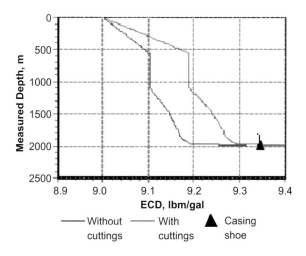

Fig. 7.1.16—Influence of the presence of solids on ECD in a typical deepwater well.

Several operational parameters affect solids concentration in deepwater wells. A discussion of their impact on bottomhole pressure follows.

- **Rate of Penetration (ROP).** ROP impacts solids concentration dramatically in different manners. In the high-angle sections, the tendency toward cuttings-bed formation increases, while at lower angles, cuttings loading will increase. In general, an increase in ROP will result at an increase of ECD in deep-water wells. Special care should be taken to keep solids concentration inside acceptable ranges and ECD inside the operational window. **Figs. 7.1.17 and 7.1.18** illustrate the impact of ROP on solids concentration in the inclined section and in the riser, respectively. Results are based on the solution of a mechanistic two-layer model described by Santana et al. (1998). Certainly, different models could capture such tendencies in different manners.
- **Well Depth.** An increase in measured depth directly affects the frictional terms and consequently increases ECD. This factor may be irrelevant for clean large-diameter holes where annular frictional pressure losses are negligible. For smaller-diameter holes (0.241-m, 0.216-m, and smaller), frictional pressure losses start to play an important role in total bottomhole pressures. **Fig. 7.1.19** (Martins et al. 2004) represents a discussion of the hydraulic limits for wells with a long horizontal section in deep-water environments. The figure represents a 0.216-m openhole horizontal section drilled with 5-in. drillpipe, considering the flow rates that result in two different hole-cleaning criteria: total cuttings-bed

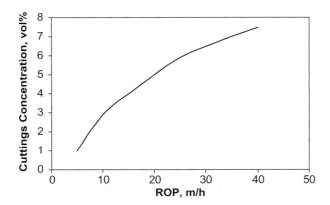

Fig. 7.1.17—Effect of ROP on cuttings concentration in a deviated well.

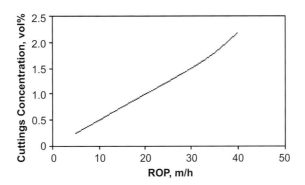

Fig. 7.1.18—Effect of ROP on cuttings concentration in the riser.

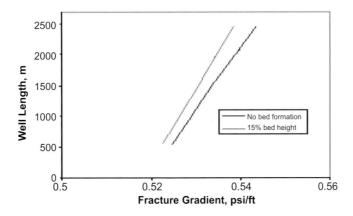

Fig. 7.1.19—Impact of the well length on fracture gradient.

removal and 15% of cuttings bed height in the annulus. The figure highlights that, in critical situations, pumping rates have to be reduced and the drilling activity should be conducted in the presence of a cuttings bed. In such situations, additional hole-cleaning procedures are needed, such as dedicated wiper trips or tandem pills. Wells with a long horizontal section may constitute economical drives for the exploitation of offshore heavy-oil fields (Vicente et al. 2003).

- **Pipe Rotation.** Low or zero pipe rotations stimulate bed formation in highly inclined wells, although a hole-cleaning problem may not be reflected by ECD measurements. On the other hand, high pipe rotations enhance solids resuspension, which immediately affects ECD. **Fig. 7.1.20** shows typical

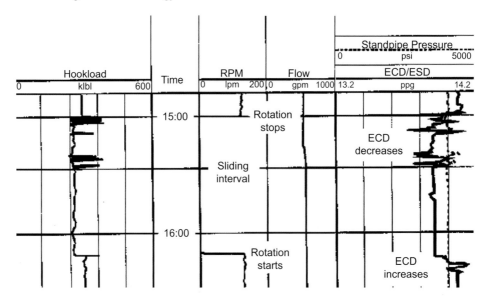

Fig. 7.1.20—Effect of pipe rotation on PMWD response.

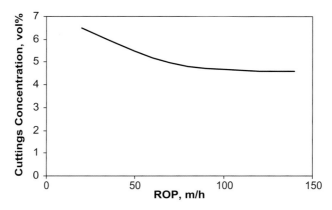

Fig. 7.1.21—Impact of pipe rotation on cuttings concentration.

pressure-measurement-while-drilling (PMWD) responses to pipe rotation. **Fig. 7.1.21** shows the impact of rotation on solids concentration based on the empirical model proposed by Sanchez et al. (1999).

- **Flow Rate.** Increasing flow rate will enhance hole cleaning immediately and, consequently, will reduce solids concentration. On the other hand, friction losses are directly proportional to flow rate. ECD may increase or decrease depending on the importance of both aspects. **Fig. 7.1.22** shows a minimum in the ECD curve. Whether this minimum will occur inside the operational range will depend on each specific well design.
- **Rheology.** The role of rheology in downhole pressure is complex and affects several events, including hole cleaning, frictional pressure losses, and pressure peaks after circulation stops. In dynamic conditions, highly pseudoplastic behavior is desired: High viscosity at low shear rates prevents cuttings sedimentation, while low viscosity at high shear rates enhances cuttings-bed resuspension and minimizes frictional pressure losses. **Fig. 7.1.23** illustrates the role of low- and high-shear viscosities in the drilling process.

Proper evaluation of rheological properties is fundamental to providing data for hydraulics simulators, which provide hydraulic design of deepwater operations. Evaluating rheological properties, which represent

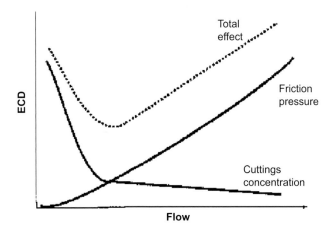

Fig. 7.1.22—Impact of the flow rate on ECD regarding hole cleaning and pressure drop effects.

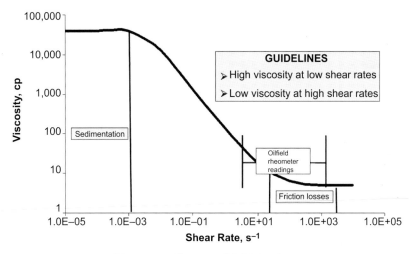

Fig. 7.1.23—The role of fluid rheology.

the fluid behavior in deepwater wells, is a critical topic. As highlighted in Fig. 7.1.23, the range of shear rates covered by a typical oilfield rheometer varies from 5 to 1,022 s^{-1}. It is desirable to obtain data in the widest possible range of shear rates in order to properly represent the fluid behavior in the several geometries that characterize the circulation system. **Table 7.1.4** and **Fig. 7.1.24** highlight the impact of the number of rheometer readings on the rheological parameters for a typical drilling fluid. Yield stresses can be obtained graphically by the intercept of the rheological curve with the shear-stress axis.

The choice of an accurate rheological model also affects hydraulics predictions. Dynamic viscosities obtained by Fann VG 35* readings represent properly the frictional-pressure-loss processes but fail to capture sedimentation phenomena. Low-shear-rate rheometers (*DV-III+ Rheometer* 1998) are available for laboratory use, but the reliability of their results is not achieved in floating vessels. **Table 7.1.5** highlights the difference in evaluating yield stresses using the conventional field device and laboratory low-shear rheometers for two typical fluids used in deepwater applications (Monteiro et al. 2005). The impact of the choice of a rheological model is also addressed in the table.

Figs. 7.1.25 and 7.1.26 illustrate the effects of fluid rheology on cuttings concentration and ECD, respectively, for a typical deepwater well. Oilfield rheometer readings decrease from Fluid A to Fluid C.

---

*Trademark of Fann Instrument Company, Houston.

TABLE 7.1.4—ESTIMATION OF RHEOLOGICAL PARAMETERS AS A FUNCTION OF THE NUMBER OF OILFIELD RHEOMETER READINGS

Rotation (rev/min)	Reading (°)	Shear Rate (s^{-1})	Shear Stress (lbf/ft^2)
3	7	5.109	7.46
6	8	10.218	8.53
100	32	170.3	34.11
200	48	340.6	51.17
300	65	510.9	69.29
600	90	1021.8	95.94

	μ_p [kg/(m·s)]	τ_0 (Pa per 1000 m^2)	Correlation Coefficient
Fit for two readings	2.503	64.16	1.00
Fit for six readings	4.217	19.17	0.95

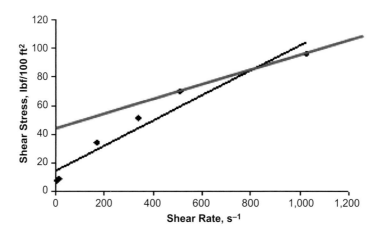

Fig. 7.1.24—Effect of the number of rheometer readings on rheological parameters.

TABLE 7.1.5—YIELD-STRESS VALUES EVALUATED BY DIFFERENT RHEOMETERS

Rheological Model	Fluids	Oilfield Rheometer (Pa)	Low-Shear-Rate Rheometer (Pa)
Bingham	Synthetic-based fluid	8.41	8.52
	Polymer-based fluid	9.40	4.30
Herschel-Bulkley	Synthetic-based fluid	5.99	7.67
	Polymer-based fluid	6.13	2.96

The hydraulics-optimization process is a compromise between solids concentration and downhole pressure. Unlike typical onshore and shallow-water applications, where fracture pressures are not limiting, optimum design for deepwater drilling prioritizes fluids with high viscosities at low shear rates and low viscosities at high shear rates which can guarantee proper cuttings transport with lower flow rates.

Fast and nonprogressive drilling-fluid gelation is desired to prevent drilled-solids sedimentation when the pumps are off, while avoiding excessive pressure peaks when circulation is resumed. Gelation tendencies are normally higher at low temperatures typical of deepwater risers. Whenever the gelled structure is formed, the energy required to break it will be higher, and consequently, a pressure peak is observed. This way, the gelled fluid induces pressure peaks when the pumps are turned on again after a static period. Pressure peaks can reach the fracture pressure, bringing risks to the operation.

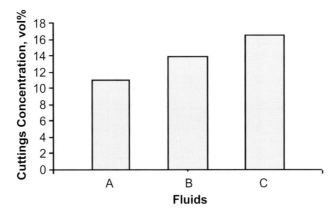

Fig. 7.1.25—Effect of fluid rheology on cuttings concentration.

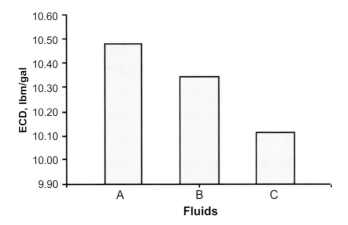

Fig. 7.1.26—Effect of fluid rheology on ECD.

Important parameters governing the startup circulation pressures are temperature, pressure, pumps-off time, and startup flow rate. **Fig. 7.1.27** shows, in a typical PMWD log, pressure increases when the circulation is resumed. Among other operational parameters, the log shows the fluid-flow rate (blue line in the central column) and the bottomhole pressure (green line in the right column). Such peaks can be minimized by simple startup operational procedures, such as rotating and reciprocating the drillstring before starting the pump and starting with low flow rates and gradually increasing to the desired flow rate. The idea is to break gelled structures before resuming total flow rates and, consequently, to minimize pressure peaks.

Although gelation tendencies can be estimated with the oilfield rheometer, important additional information can be gathered with different rheological experiments, including the small-amplitude oscillatory tests, creep-recovery experiments, or flow-initiation loops (Gandelmann et al. 2007). In all cases, evaluating properties at the low seafloor temperatures is essential.

Fluid Substitution. Several fluid-substitution operations take place while drilling and completing oil wells. Wells with a long horizontal section make such operations critical because of pressure limitations, long travel times, and huge volumes of fluid involved. Such facts contribute to fluid contamination, reducing the efficiency of several operations, and generating costs and disposal issues.

The quality of fluid displacement is controlled by the interface shape between the two fluids. Sharp profiles tend to result in fluid channeling, while flat profiles normally promote efficient displacement. The optimization of the fluid rheological properties is essential for the success of fluid-substitution operations. Several authors (Haut and Crook 1979, 1982; Sauer 1987; Lockyear and Hibbert 1989) state that the viscosity ratio between the displaced and displacing fluids, velocity profile in the eccentric annulus, wellbore angle, and fluid-density difference govern the process.

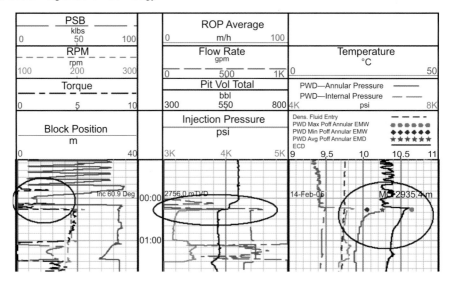

Fig. 7.1.27—Pressure peak when the circulation is resumed after a static period.

The dynamics of fluid substitution will be defined by the simulation of the two-phase flow where a Fluid A displaces a Fluid B of different physical properties. The solution of linear momentum equations for each phase plus the closure relations aims to capture the evolution of the shape of the interface between the fluids (Bittleston et al. 2002; Dutra et al. 2005).

Two fluid-substitution operations are critical in deepwater operations:

- Substitution of synthetic-based fluids by water-based fluids before the reservoir phase: Normally, this operation is critical when the interface reaches the riser annulus. Because of the low velocities in the riser, fluid contamination is a big issue. Since the displacing fluid is less viscous than the displaced fluid, efficiencies in the riser annulus are low, as highlighted by **Fig. 7.1.28** (Dutra et al. 2005). Maximizing flow rates (with additional flow through the booster line) and pumping a spacer pill are common alternatives to minimize synthetic-fluid contamination. **Fig. 7.1.29** illustrates the contamination resulting from the high viscosity ratio between displaced and displacing fluids. The yellow color represents the displaced fluid (synthetic-based), while the red color represents the displacing fluid (water-based).

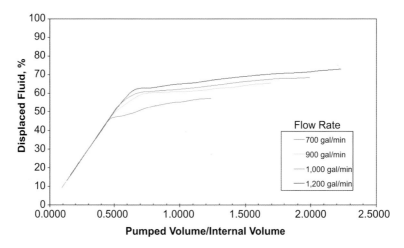

Fig. 7.1.28—Fluid displacement in the riser annulus.

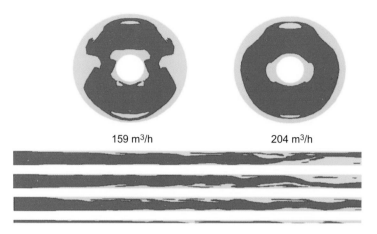

159 m³/h   204 m³/h

Fig. 7.1.29—Fluid contamination due to viscosity difference. The annulus is 0.508 m × 0.127 m. The displaced fluid is synthetic-based (yellow), and the displacing fluid is water-based (red).

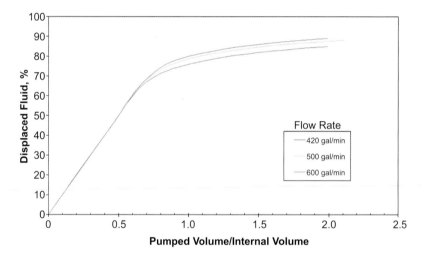

Fig. 7.1.30—Fluid displacement in a 0.216-m horizontal well.

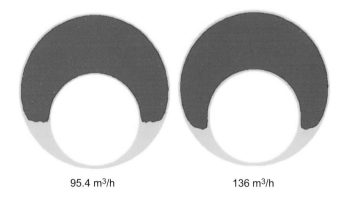

95.4 m³/h   136 m³/h

Fig. 7.1.31—Flow displacement pattern in eccentric horizontal section. The annulus is 0.216 m × 0.127 m. The displaced fluid is the drilling fluid (yellow), and the displacing fluid is the completion fluid (red).

- Substitution of drilling fluid by the completion fluid in the openhole long horizontal section: This is a critical situation in deepwater wells where flow rates are limited by the operational window and it is necessary to remove all the drilling fluid from the open hole to guarantee a solids-free environment to run the screen. **Fig. 7.1.30** (Dutra et al. 2005) shows the displacement percentage vs. volume required as a function of flow rate for an 8½-in. well. In this case, excess completion fluid should be pumped to guarantee total removal. Concentric strings and maximum flow rates allowed by the operational window enhance displacement efficiency. **Fig. 7.1.31** highlights the displacement pattern in the eccentric annulus. The yellow color represents the displaced fluid, while the red color represents the displacing fluid.

Nomenclature

C_s = solids concentration, vol%
ρ_f = fluid density, lbm/gal
ρ_s = cuttings density, lbm/gal

7.1.5 Deepwater Well Control—Antonio C.V.M. Lage, Petrobras

Introduction. The application of methods to repress fluid inflow and remove it from the well in a safe way is designated as well-control operations, as previously discussed in Chapter 4.2. It was properly stated that, regardless of the operational environment, well control is a key aspect and deserves special attention from all rig-crew members, supervision personnel, and engineers in charge of the designing activities. However, deepwater drilling has posed some additional technical challenges to this matter because of the relatively low wellbore fracture gradients. As a consequence of this particularity, well construction in deep water means managing the difficulties associated with narrow operational margins between the curves of pore pressure and formation fracture. In such conditions, conventional design criteria might be cost-prohibitive or might greatly reduce chances of reaching the target, mainly because of borehole size constraints. Besides that, controlling the well during kick events is difficult in deep water because of the risks of losing circulation.

The purpose of this section is to describe important concepts and calculations to design and deal with well-control situations in deep water.

Kick Tolerance. For the last 30 years, kick tolerance has been studied and improved by different authors (Pilkington 1975; Redmann 1991; Leach and Wand 1992; Lage et al. 1997). The main sources of motivation for those developments were not only the difficulties associated with the deepwater environment, but also some critical high-pressure/high-temperature (HP/HT) scenarios. In both cases, narrow operational margins between the curves of pore pressure and formation fracture are the technical challenges.

First, two fundamental definitions are presented within the scope of this subject: *tolerance*—a limit value, a maximum or a minimum, for a certain variable; and *margin*—the difference between the value of such a variable and its limit. Based on this, three parameters are defined:

1. Kick tolerance (ρ_{kt})—the maximum allowable pore pressure, expressed in equivalent mass density such that if a kick with a certain volume occurs at a particular depth with a specific drilling fluid, the well could be closed down and the kick circulated out safely—that is, not fracturing the weakest formation in the openhole section.
2. Kick safety margin ($\Delta\rho_{ksm}$)—the difference between the equivalent circulating density (ECD) estimated for the fracture or absorption pressure (ρ_f) at the weakest formation and the maximum ECD acting at this point ($\rho_{eq,\,cs}$) when the well is closed or while the kick is circulated out of the well (Eq. 7.1.7).
3. Pore pressure margin ($\Delta\rho_{kt}$)—the difference between the kick tolerance (ρ_{kt}) and the estimated pore pressure (ρ_p) of the producing formation, expressed in equivalent mass density (Eq. 7.1.6).

The simplest way to estimate kick tolerance is based on the hypothesis of a single-gas-bubble influx. However, even such a simplified model involves the definition of a well scenario and several assumptions related to the influx and control procedures. The most common method applied to evaluate kick tolerance assumes that gas occupies the annular space as a continuous block, and well closure is the most dangerous circumstance while controlling the kick (shut-in kick tolerance) (Redmann 1991).

Fig. 7.1.32 presents the main geometric variables of a directional well for deriving a kick-tolerance equation based on the single-gas-bubble influx. Because well closure is assumed to be the most-critical condition, the gas bubble is placed at the bottom of the well. In accordance with Fig. 7.1.32, L_{vk} is the vertical projection of the gas-bubble length, D_{vbh} is the true vertical depth of the bottom of the hole, and D_{vcs} is the true vertical depth of the casing shoe.

The kick-tolerance condition is obtained when the well-control incident leads to the fracture of the weakest formation of the open hole. Assuming that formation fracture occurs at the shoe, it is possible to state that the formation pressure minus the pressure corresponding to the hydrostatic column between the bottom of the well and the casing shoe is equal to fracture pressure, which is expressed as

$$\rho_{kt}D_{vbh}g - \rho_g L_{vk}g - \rho_{df}\left(D_{vbh} - D_{vcs} - L_{vk}\right)g = \rho_f D_{vcs}g \quad \dots\dots\dots\dots\dots\dots\dots\dots\dots\dots \quad (7.1.2)$$

On the basis of Eq. 7.1.2, it is possible to derive an expression for ρ_{kt}:

$$\rho_{kt} = \rho_{df} + \frac{D_{cs}}{D_{vbh}}\left(\rho_f - \rho_{df}\right) - \frac{L_{vk}}{D_{vcs}}\left(\rho_{df} - \rho_g\right). \quad \dots\dots\dots\dots\dots\dots\dots\dots \quad (7.1.3)$$

Considering an operational drilling scenario after performing the leakoff test at the casing shoe, all variables except L_{vk} on the right side of Eq. 7.1.3 are known. L_{vk} depends on well geometry and the volume of gas that entered the well before closure, which corresponds to the pit gain. However, if the well geometry is defined,

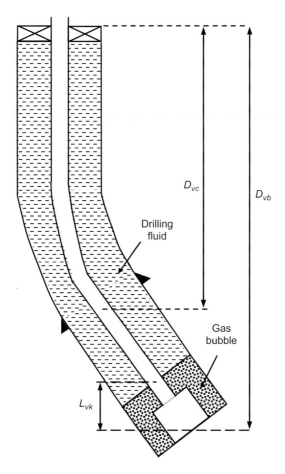

Fig. 7.1.32—Well geometry for deriving a kick tolerance equation based on a single gas bubble.

there is a straight-line relationship between the pit gain and L_{vk}. As a consequence, it is possible to plot ρ_{kt} as a function of the pit gain, which is presented in **Fig. 7.1.33**.

The slope of the straight line decreases when gas reaches the annular space between the open hole and the drillpipe. In contrast, after the top of the gas is at the casing shoe, ρ_{kt} remains constant.

Eq. 7.1.3 is very meaningful for the complete understanding of some important aspects related to kick tolerance. The analysis of partial derivatives leads to the comprehension of the tendencies while increasing or decreasing the variables, as shown in **Table 7.1.6**.

With a simple algorithm (Redmann 1991), it is possible to calculate the gas expansion during kick control. This improvement allows the calculation of ρ_{kt} when the top of the gas reaches the casing shoe, which is considered, in general, to be the weakest point in the open hole. Thus, the adopted value for ρ_{kt} would be the minimum between this one and the other obtained from Eq. 7.1.3. **Fig. 7.1.34** shows an example. It is noticeable that the initial shut-in condition dominates the minimum curve for smaller values of pit gain. On the other hand, the condition of gas at the casing shoe prevails for larger values of pit gain.

The initial shut-in condition usually defines the minimum value for ρ_{kt}. Expansion of the influx that travels from the bottom of the hole to the casing shoe is not, frequently, enough to compensate changes in annular

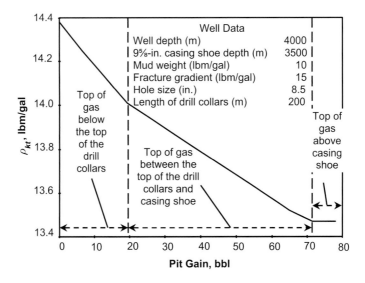

Fig. 7.1.33—Kick tolerance as a function of the pit gain (Lage et al. 1997).

TABLE 7.1.6—SENSITIVITY ANALYSIS OF ρ_{kt}

Variable	Partial Derivative	Signal	Behavior of ρ_{kt}
D_{vbh}	$\dfrac{L_{vk}(\rho_{df}-\rho_{g})-D_{vcs}(\rho_{g}-\rho_{f})}{D_{vbh}^{2}}$	<0	$\rho_{kt}>\rho_{df}\Rightarrow\rho_{kt}\downarrow$
		≥ 0	$\rho_{kt}\leq\rho_{df}\Rightarrow\rho_{kt}\uparrow$
ρ_{df}	$1-\dfrac{D_{vcs}+L_{vk}}{D_{vbh}}$	≥ 0	$\rho_{kt}\uparrow$
D_{vcs}	$\dfrac{\rho_{f}-\rho_{df}}{D_{vbh}}$	≥ 0	$\rho_{kt}\uparrow$
$==L_{vk}$	$\dfrac{\rho_{g}-\rho_{df}}{D_{vbh}}$	<0	$\rho_{kt}\downarrow$
ρ_{g}	$\dfrac{L_{vk}}{D_{vbh}}$	≥ 0	$\rho_{kt}\uparrow$

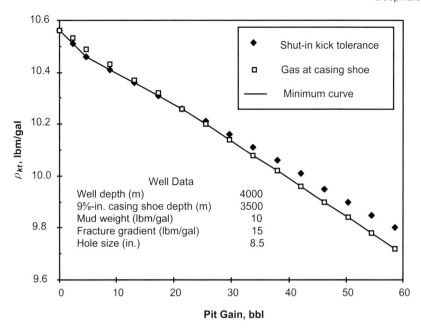

Fig. 7.1.34—Kick tolerance as a function of the pit gain for shut-in condition and gas reaching the casing shoe (Lage et al. 1997).

geometry. Part of the gas occupies the annulus between open hole and drill collars at the bottom, while at the casing shoe the influx is in the openhole/drillpipe annulus. **Table 7.1.7**, derived from the example presented in Fig. 7.1.33, clarifies this point. It is remarkable that the length of drill collars is shorter and the openhole section is longer than usual. Otherwise, the initial shut-in condition would dominate the minimum curve for the whole domain presented.

Previously, one of the most common ways of using kick tolerance (Pilkington 1975; Redmann 1991) was to link it to the weight of drilling fluid at a given well condition. Then, the so-called kick-tolerance margin, K, was defined as follows:

$$K = \rho_{kt} - \rho_{df} \qquad \qquad \qquad \qquad (7.1.4)$$

Notice that K decreases as the mud weight, ρ_{df}, approaches the kick tolerance, ρ_{kt}; thus, K is used as an indicator for the need of setting a new casing every time it reaches an established value, such as 59.91 kg/m³ (0.5 lbm/gal). However, this concept is valid only if the formation pore pressure is the unique basis for the definition of the mud weight. Still, other factors may require an increase in mud weight, such as borehole-stability problems, riser safety (RS) margin, or any other safety margin. In such circumstances, if K were the exclusive parameter for well design, a casing would be set at a depth less than necessary since K would fall prematurely below the limiting value.

TABLE 7.1.7—SHUT-IN CONDITION AND GAS AT CASING SHOE		
	Shut-In Condition	Gas at Casing Shoe
Influx volume, m³ (bbl)	3.40 (21.4)	14.3 (89.9)
L_{vk}, m	150	119
ρ_g, kg/m³ (lbm/gal)	233.7 (1.95)	79.56 (0.664)
ρ_{kt}, kg/m³ (lbm/gal)	1229 (10.26)	1229 (10.26)

An analysis of partial derivatives of Eqs. 7.1.3 and 7.1.4 make this point clear. First, Table 7.1.6 shows that ρ_{kt} may increase with ρ_{df}, differently from Eq. 7.1.4, where K decreases with ρ_{df}:

$$\frac{\partial K}{\partial \rho_{df}} = -\frac{D_{vcs} + L_{vk}}{D_{vbh}} < 0 \quad \dots\dots\dots\dots\dots\dots\dots\dots\dots\dots\dots\dots\dots\dots\dots \quad (7.1.5)$$

Therefore, an increment in the mud weight, because of any reason other than the pore pressure, could lead one to think that a casing should be set at that point, on the basis of the K value. This is not consistent, since ρ_{kt} would have increased at that same point. Consequently, the direct comparison between the kick tolerance and the estimated pore pressure is the best alternative to avoid this problem. Thus, the pore-pressure margin, $\Delta\rho_{kt}$, was defined as

$$\Delta\rho_{kt} = \rho_{kt} - \rho_p, \quad \dots\dots\dots\dots\dots\dots\dots\dots\dots\dots\dots\dots\dots\dots\dots\dots\dots\dots \quad (7.1.6)$$

where $\Delta\rho_{kt}$ expresses the safety margin between the maximum allowable reservoir pressure, on the basis of the loads imposed on the open hole and the expected or evaluated pore pressure ρ_p.

Fig. 7.1.35 illustrates a case history, Well ALS-47, that was drilled in 1993 at a water depth of 746 m. Fig. 7.1.35a presents pore pressure, two different mud-weight curves, and fracture gradient. Depending on the type of the drilling vessel used, moored or dynamically positioned, two approaches were possible. If a moored unit was assigned to drill the well, the pore-pressure estimation added to an overbalance safety margin (35.9 kg/m³ or 0.3 lbm/gal) would define the mud weight. Considering a minimum kick tolerance margin, K, of 59.9 kg/m³ (0.5 lbm/gal), Fig. 7.1.35b shows that the next casing would be set at 2450 m. Otherwise, if a dynamically positioned unit was adopted, the pore pressure evaluation added to the RS margin (see the RS margin in Fig. 7.1.35) would specify the mud weight. Under such circumstances, it would not be possible to drill ahead based on the K criterion, as shown in Fig. 7.1.35c.

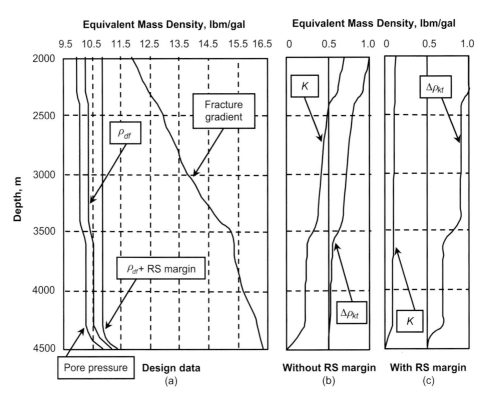

Fig. 7.1.35—Use of K and $\Delta\rho_{kt}$ in a deepwater drilling scenario (Lage et al. 1997).

As seen in the preceding example, K loses its meaning if ρ_{df} is not directly related to pore pressure. Thus, it is possible to have a small or even negative value for K without being in an unsafe well condition. This example characterizes the inconvenience of basing casing design exclusively on K analysis. If the weight of the drilling fluid is considerably higher than the pore pressure, this criterion will not provide proper results.

Figs. 7.1.35b and 7.1.35c also present $\Delta\rho_{kt}$ as a function of depth. The adoption of the RS margin in a dynamically positioned vessel (Fig. 7.1.35c) would lead to a heavier mud weight in relation to the one in a moored vessel (Fig. 7.1.35b) and would cause an increase in ρ_{kt} and $\Delta\rho_{kt}$. According to Fig. 7.1.35b, 4400 m is the proper depth for setting casing. With the application of heavier mud it is possible to set casing even deeper, at 4500 m (Fig. 7.1.35c). The $\Delta\rho_{kt}$ method is a better approach because it does not hide the relation between ρ_{kt} and pore pressure.

The pore-pressure margin is related to the formation pressure at the bottom of the well. Alternatively, a similar analysis can be performed on the basis of the stresses acting in front of the weakest formation, which is normally associated with the casing shoe. In this case, the kick safety margin ($\Delta\rho_{ksm}$) has to be used:

$$\Delta\rho_{ksm} = \rho_f - \rho_{eq,cs} . \quad\dots\dots\dots\dots\dots\dots\dots\dots\dots\dots\dots\dots\dots\dots\dots\dots\dots\dots \quad (7.1.7)$$

It can be shown from Eqs. 7.1.6 and 7.1.7 that both margins are related to each other by the following relation:

$$\Delta\rho_{kt} = \frac{D_{vcs}}{D_{vbh}}\left(\rho_f - \rho_{eq,cs}\right) = \frac{D_{vcs}}{D_{vbh}}\Delta\rho_{ksm} \quad\dots\dots\dots\dots\dots\dots\dots\dots\dots\dots\dots\dots\dots\dots\dots \quad (7.1.8)$$

From Eq. 7.1.8, it is clear that both margins always present the same sign; that is, they always give coherent results in terms of indicating the feasibility of closing the well and circulating the kick.

Finally, both margins can be calculated in terms of surface pressure. From Eq. 7.1.8, it can be shown that

$$D_{vcs}\left[(\rho_f - \rho_{df}) - (\rho_{eq,cs} - \rho_{df})\right] = D_{vbh}\left[(\rho_{kt} - \rho_{df}) - (\rho_f - \rho_{df})\right]. \quad\dots\dots\dots\dots\dots\dots\dots\dots \quad (7.1.9)$$

Therefore, considering the shut-in casing pressure (SICP) and the shut-in drillpipe pressure (SIDPP),

$$\text{SICP}_{max} - \text{SICP} = \text{SIDPP}_{max} - \text{SIDPP} \quad\dots\dots\dots\dots\dots\dots\dots\dots\dots\dots\dots\dots\dots\dots\dots\dots \quad (7.1.10)$$

Thus, it can be concluded that the surface-pressure margins in the annulus and inside the drillstring are exactly the same. Once more, the coherence of both the pore-pressure margin and the kick safety margin is demonstrated.

Two-Phase-Flow Modeling. Previous studies (Leach and Wand 1992; Nakagawa and Lage 1994; Rommetveit 1994) have shown that two-phase-flow modeling typically predicts less-conservative kick-tolerance values. The reason is that gas enters the well as bubbles, not as a continuous slug, and that it rises faster than mud (Lage et al. 1997). Depending on the properties of the mixture, gas and mud interact and establish different flow patterns. Bubble flow is the expected phase arrangement when the gas enters the well. Later, as the two-phase mixture flows up in the annulus while performing the well-control procedure, bubbles expand and coalesce. Sometimes they form Taylor bubbles (Nakagawa and Bourgoyne 1992), changing the phase arrangement to slug flow. However, in most of the situations, kick-tolerance calculation will deal only with bubble flow.

Fig. 7.1.36 physically compares the difference between both approaches, gas as a continuous slug and as a two-phase-flow mixture. Emphasis is given to the significant difference in kick-tolerance values and in the dispersion of the gas inside the well, with meaningful consequences on the gas arrival time and gas-flow rates at surface.

Kick-tolerance variations, affected by gas dispersion and slippage, are also shown in **Fig. 7.1.37.** The most conservative curve, Curve 1, is based on the assumption of gas as a continuous slug in the annulus, while the other curves (Curves 2, 3, and 4), are based on a two-phase-flow approach. The differences among the two-phase model curves are a result of different reservoir properties: the higher the reservoir productivity index (J) (Craft and Hawkins 1959), the lower the kick-tolerance values.

In addition, the two-phase-flow approach is feasible only through computer simulation. Because of the complex mathematical models, these simulators require the definition of more-detailed scenarios and

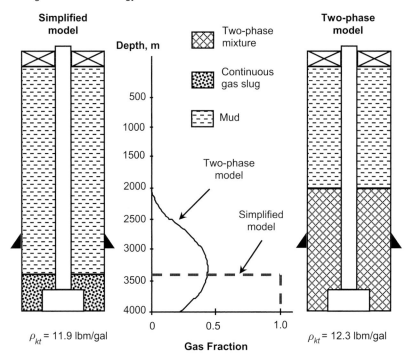

Fig. 7.1.36—Comparison of the gas distribution (Lage et al. 1997).

sophisticated numerical schemes, causing kick-tolerance calculations to be time-consuming. It is very important to gather precise information about well-control procedures, reservoir properties, and penetration rate. These data are important for predicting the reservoir production rate.

Finally, it is interesting to notice that, for smaller influx volumes, the kick-tolerance values calculated by the simplified model and two-phase simulators are basically the same.

Still in the same figure, there are three regions defined in relation to the position of the top of the two-phase mixture. In the first region, the gas is still below the top of the bottomhole assembly. In the second region, the gas has arrived at the annulus between the drillpipe and the open hole. Finally, the third region begins when the gas arrives at the casing shoe. It is clear that, as the reservoir productivity index decreases, the kick reaches the casing shoe with smaller volumes of influx. In other words, lower reservoir productivity index corresponds to higher dispersion of gas in the annulus. Consequently, with smaller volumes of influx, gas reaches the casing shoe causing a smaller hydrostatic loss and higher values of ρ_{kt}.

Some Case Histories. During the last 15 years, the oil industry has been concentrating efforts on deepwater exploration. Consequently, interesting case histories are available in the literature, such as that of Well PAS-25 (Lage et al. 1997) that was drilled close to the mouth of the Amazon River, in the north coast area of Brazil, in 1993. Because of logistic constraints, well design and all other planning activities demanded many hours from specialists, better methodologies, and sophisticated engineering tools.

Despite all the planning care, significant differences were noticed between presumed and measured data. As a consequence, execution of the well was not as expected. **Fig. 7.1.38a** shows the well plan for drilling the 0.31-m (12¼-in.) section. While drilling at 3200 m, the mud weight was increased from 1258 kg/m³ (10.5 lbm/gal) to 1306 kg/m³ (10.9 lbm/gal) to solve wellbore-stability problems. Under such conditions, several questions were raised related to the chances of safely reaching the final depth of this phase. Note that the kick tolerance calculated for a maximum acceptable pit gain of 7.95 m³ (50 bbl) using the simple model, Eq. 7.1.3, gives a small K of –0.02 lbm/gal. Using this value without questioning it, we would probably justify the premature setting of an extra casing string. However, the mud-logging unit available at the rig estimated a pore pressure value of 2.4 kg/m³ (10.2 lbm/gal), which meant that, at 3200 m, the available $\Delta\rho_{kt}$ was 81.5 kg/m³ (0.68 lbm/gal), a safe condition to continue drilling operations.

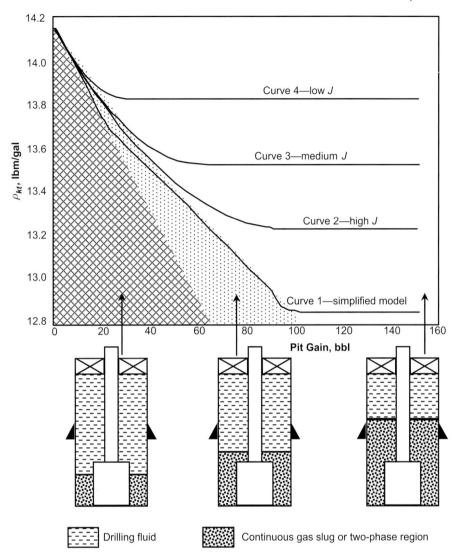

Fig. 7.1.37—Kick tolerance curves (Lage et al. 1997).

Fig. 7.1.39 shows both kick-tolerance curves for the well shown in Fig. 7.1.38a; the first one is based on the simplified model, and the other one is based on the two-phase model. The depth of the well, the mud weight, the pore pressure, and the acceptable $\Delta\rho_{kt}$ were 3514 m, 1306 kg/m^3 (10.9 lbm/gal), 1270 kg/m^3 (10.6 lbm/gal), and 35.9 kg/m^3 (0.3 lbm/gal), respectively. Note that for pit gains larger than 3.18 m^3 (20 bbl), the curves start to separate. Again for 7.95 m^3 (50 bbl) of pit gain, $\Delta\rho_{kt}$ is 33.6 kg/m^3 (0.28 lbm/gal) and 59.9 kg/m^3 (0.50 lbm/gal) for the simplified and two-phase models, respectively. Note that, while the simplified model indicates that drill operations should stop to set a casing string, the two-phase model suggests that drilling could go on. It is important to point out that in this case, the application of the described kick-tolerance method avoided stopping the well prematurely (3200 m), saving approximately USD 450,000 in casing string and the corresponding operational time.

In another case, this time in the Brazilian northeast coast area, well CES-111 had to be drilled in 1772 m of water. As shown in Fig. 7.1.38b, this well had a relatively shallow secondary target at 2300 m. Thus, it required a detailed analysis of the following points: (1) temperature influence on the friction pressure loss through the chokeline (FPLCL), (2) kick tolerance analysis with evaluation of the FPLCL, (3) simplified vs. two-phase kick tolerance models, and (4) the feasibility of drilling the 0.31-m (12-1/4-in.) section, from 2140 m to 2300 m, with the riser safety margin.

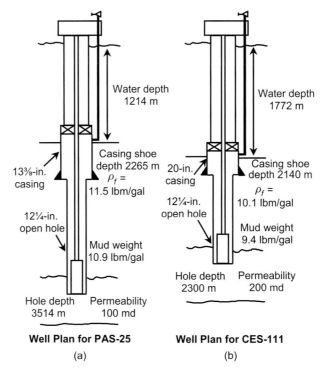

Fig. 7.1.38—Deepwater case histories (Lage et al. 1997).

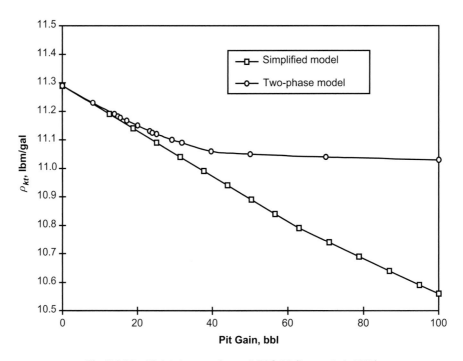

Fig. 7.1.39—Kick tolerance for well PAS-25 (Lage et al. 1997).

Fig. 7.1.40 shows kick-tolerance calculations determined under two temperature conditions, 60°C (140°F) and 4°C (40°F). The respective mud weights, viscosities, and yield points are (1) 1126 kg/m³ (9.4 lbm/gal), 15 cp, and 7.18 Pa (15 lbf per 100 ft² and (2) 1126 kg/m³ (9.4 lbm/gal), 30 cp, and 14.4 Pa (30 lbf per 100 ft²). The reason for considering both conditions is the significant effect of temperature variations on the mud

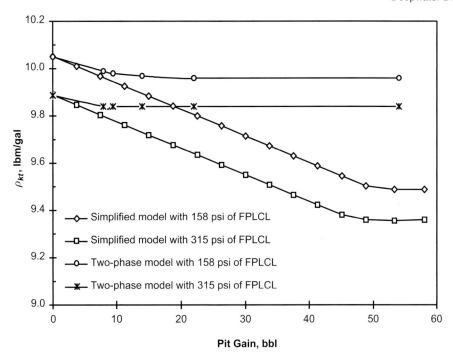

Fig. 7.1.40—Effect of temperature, mud properties, and friction pressure loss through the chokeline (FPLCL) on kick tolerance (Lage et al. 1997).

properties (Nakagawa and Lage 1994). For a typical fluid, when the temperature drops from 60°C (140°F) to 4°C (40°F), mud properties increase two-fold. Because temperature variations of this order are expected during the operations, their effect on the FPLCL and on kick tolerance had to be determined.

For 60°C (140°F), the FPLCL of 1089 kPa (158 psi) at the reduced flow rate of 6.31 L/s (100 gal/min) was not enough to affect kick-tolerance values because it was smaller than the calculated SICP of 1758 kPa (255 psi). However, for 4°C (40°F), the FPLCL of 2172 kPa (315 psi) induced a reduction in the kick tolerance as shown in Fig. 7.1.40. Anyway, considering an expected pore pressure equivalent to 1054 kg/m³ (8.8 lbm/gal) to 1078 kg/m³ (9.0 lbm/gal), kick tolerance is not of concern in that particular well.

The difference between using a simplified model and a two-phase model is also displayed in Fig. 7.1.40. The simplified model indicates that the gas would be at the casing shoe if the well were closed in with an influx of 8.27 m³ (52 bbl). On the other hand, more-complex simulators, using two-phase models, show that the influx reaches the casing shoe with shut-in volumes smaller than 2.39 m³ (15 bbl), and more important yet, this influx would be in the riser if the well were closed in with more than 8.59 m³ (54 bbl) of pit gain.

Finally, an aspect to be considered is the possibility of using the RS margin in this well. Taking into account the expected pore pressure, equivalent to 1054 kg/m³ (8.8 lbm/gal), the required mud weight would be 1172 kg/m³ (9.78 lbm/gal), including the RS margin. In this case, even with no gas influx inside the well, the circulation of the fluid through the chokeline would induce the casing-shoe fracture. The FPLCL of 1089 kPa (158 psi), caused by the reduced flow rate in the chokeline, is equivalent to 51.5 kg/m³ (0.43 lbm/gal) at the casing shoe. Therefore, the total equivalent pressure load at the casing shoe would be 1223 kg/m³ (10.21 lbm/gal), larger than the fracture gradient at that point, equivalent to 1210 kg/m³ (10.1 lbm/gal). Thus, the RS margin could not be adopted under those circumstances.

This planning evaluation illustrates how the precise quantification of the FPLCL is important for a drilling operation in deep waters. As a matter of fact, its relevance motivated a field test before drilling the 0.22-m (8½-in.) section of Well CES-111. The main targets of this initiative were (1) appraisal of the operational practices adopted to measure the FPLCL, (2) evaluation of the influence of the temperature, (3) improvement of the FPLCL computer predictions, and (4) verification of the advantages of flowing through kill and choke lines in parallel to decrease the FPLCL.

The 9⅝-in. casing was set at 3067 m, and the tests were performed before cutting cement. The FPLCL was measured by three different methods: (1) pumping mud through the drillstring and returning it through the chokeline, (2) pumping mud through the chokeline and returning it through the kill line, and (3) pumping mud through the drillstring and returning it through the choke and kill lines in parallel.

The measurements have shown that those three different approaches to evaluate the FPLCL have reached almost the same results. The differences among them were not significant. As a consequence, any of those procedures can be adopted to measure the FPLCL. In addition, the temperature of the drilling fluid in the return was approximately 11.6°C (53°F), regardless of the adopted measurement procedure.

The best fits between experimental data and theoretical predictions were obtained with the power-law model and considering the return temperature for measuring the rheological properties of the drilling fluid. **Fig. 7.1.41** shows experimental data and compares them to the computer predictions. In the laminar regime, less than 7.57 L/s (120 gal/min) in this particular application, the power-law model predicts an FPLCL closer to the experimental results. On the other hand, when the inertial forces dominate over the viscosity, in the turbulent regime, it does not make any difference to use the Bingham or the power-law model.

Furthermore, the application of both choke and kill lines in parallel for decreasing the FPLCL brings better results depending on the flow regime. If the use of only one of those lines to circulate leads to the turbulent regime, than having choke and kill lines in parallel will produce a significant advantage. On the other hand, if it leads to the laminar regime, than having the lines in parallel will not be advantageous in terms of decreasing the FPLCL. As an example, Fig. 7.1.41 shows that the flow-rate reduction from 12.6 L/s (200 gal/min) to 6.31 L/s (100 gal/min) produces a very significant decrease in the FPLCL. It drops from 430 to 140 psi, which represents one-third of the original value. On the other hand, if the flowing regime through a single line is laminar, this reduction will not be as significant as in the previous situation. Fig. 7.1.41 shows that the FPLCL drops from 965 kPa (140 psi) to 655 kPa (95 psi) when reducing the flowing rate from 6.31 L/s (100 gal/min) to 3.15 L/s (50 gal/min).

On the basis of the experience gained with Well BSS-70, drilled in 1993 in Santos basin **(Fig. 7.1.42)**, on the southwest coast of Brazil, an optimized project was developed for Well BSS-78. Some parts of this area are characterized by overpressurized sand/shale sequences surrounded by limestone intervals. Beginning near 3500 m, these sequences display an increasing pore-pressure profile that rapidly approaches the fracture-gradient curve, as shown in **Fig. 7.1.43** for Well BSS-78. Besides the high pore-pressure gradient, approximately 2157 kg/m³ (18 lbm/gal), the temperature at bottom reaches 171°C (340°F).

Fig. 7.1.43 also displays the casing-design program for BSS-78. The kick-tolerance criterion used to define casing-shoe depths was applied from bottom to top. The maximum allowable fracture gradient, derived from Eqs. 7.1.3 and 7.1.6 with a margin of 35.9 kg/m³ (0.3 lbm/gal) over pore pressure, was calculated along the well and compared to the expected fracture gradient. The points where both curves are equal determine the

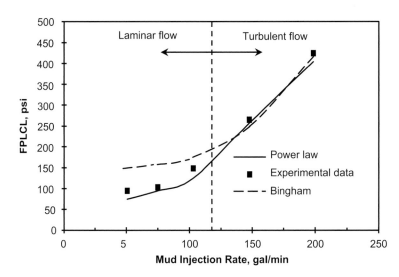

Fig. 7.1.41—Friction pressure loss through the chokeline (FPLCL) (Lage et al. 1997).

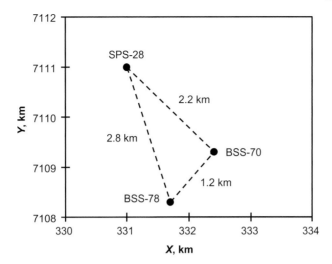

Fig. 7.1.42—BSS-78 location and reference wells (Lage et al. 1997).

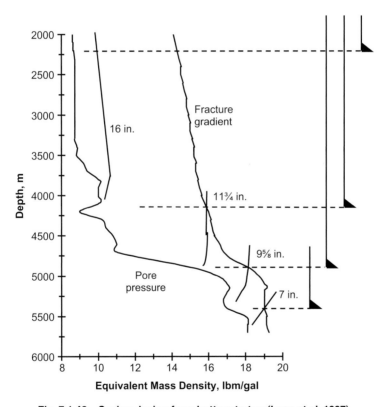

Fig. 7.1.43—Casing design from bottom to top (Lage et al. 1997).

casing-shoe depths. The 0.18-m (7-in.), 0.24-m (9⅝-in.), and 0.30-m (11¾-in.) casings were all set through this method. The other shallower casings were set to protect the long wellbore interval from excessive exposure to the drilling fluid, or to isolate shallower zones.

Besides the application of simplified kick-tolerance models for casing design, critical wells often require the use of complex kick simulators for designing one or more drilling sections. In general, it is not practical to use them to define the whole casing program, since the calculation of kick-tolerance curves is very time-consuming. Sometimes, however, they have to be used for checking alternative programs that could make the well feasible, or simply less expensive and safer. They can also be very helpful while drilling. For example, when the collected data for pore pressure and fracture gradient are different from those used for designing purposes, it might be necessary to re-evaluate the scenario and to decide about the possibility of drilling ahead in that phase.

It is important to point out the differences between two approaches to determine the casing depths: from bottom to top (Fig. 7.1.43) and from top to bottom **(Fig. 7.1.44)**. In addition, **Fig. 7.1.45** compares the results of casing design for those two different planning techniques. Depending on the scenario, one

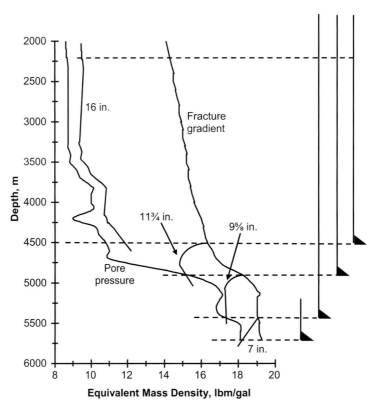

Fig. 7.1.44—Casing design from top to bottom (Lage et al. 1997).

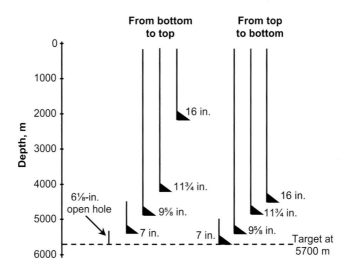

Fig. 7.1.45—Comparison of casing design approaches (Lage et al. 1997).

or the other could lead to safer and more economic design. With the first one, the casing lengths are minimized, since the procedure allows the calculation of the minimum depths for each casing, based on the limiting fracture gradient. In the second method, the casing lengths are maximized, having as constraint the maximum pore pressure allowed by the openhole section. In this particular example, the top-to-bottom approach leads to the elimination of a complete drilling section in comparison to the other planning approach. However, this economy would be achieved only by stretching other drilling sections, mainly the 0.41-m (16-in.) one, which must be drilled from 500 to 4500 m. As a matter of fact, this well-planning possibility brings about some operational risks related to the drilling problems associated with very long openhole sections.

Safe and economic casing design calls for the assumption of small pit gains in the kick-tolerance analysis. In the case of BSS-78, the narrow difference between pore pressure and fracture gradient required kick volumes smaller than 4.77 m³ (30 bbl) for making the well feasible. However, these predicted pit gains should also be realistic enough to allow the drilling team to perform shut-in operations. Therefore, a relation between the well-design premises and the operational procedures had to be established (Wilkie and Bernard 1981).

Shutting in the well quickly, after detecting a small pit gain, requires the arrangement of several preceding actions, such as specifying adequate pit-gain sensors, alerting the drilling operator, training the rig team, and defining specific procedures based on how critical a well is. In fact, the achievement of safer and more-economic casing programs requires the same arrangement.

Nomenclature

D = depth, L, m

D_{vbh} = true vertical depth of the bottom of the hole, m

D_{vcs} = true vertical depth of the casing shoe, m

J = productivity index, m⁻¹L⁴t, L/s/kPa (bbl/D/psi)

K = kick-tolerance margin, which is the difference between the kick tolerance (ρ_{kt}) and the mass density of the drilling fluid (ρ_{df}), expressed in equivalent mass density, mL⁻³, kg/m³ (lbm/gal)

L = length, L, m

L_{vk} = vertical projection of the gas-bubble length, m

ρ = mass density, mL⁻³, kg/m³ (lbm/gal)

ρ_{df} = mass density of the drilling fluid, mL⁻³, kg/m³ (lbm/gal)

ρ_{eq} = maximum equivalent mass density acting at a specific point in the well when it is closed or while the kick is being circulated out, mL⁻³, kg/m³ (lbm/gal)

$\rho_{eq,\,cs}$ = the maximum equivalent mass density acting at the casing shoe when the well is closed or while the kick is circulated out of the well, mL⁻³, kg/m³ (lbm/gal)

ρ_f = formation fracture or absorption pressure expressed in terms of the equivalent mass density, mL⁻³, kg/m³ (lbm/gal)

ρ_g = mass density of the influx, which is supposed to be gas, mL⁻³, kg/m³ (lbm/gal)

ρ_{kt} = kick tolerance, which expresses pore pressure in terms of the equivalent mass density, mL⁻³, kg/m³ (lbm/gal)

ρ_p = pore pressure of the formation expressed in terms of the equivalent mass density, mL⁻³, kg/m³ (lbm/gal)

$\Delta\rho_{ksm}$ = kick safety margin, which is the difference between the equivalent circulating density estimated for the fracture or absorption pressure (ρ_f) at the weakest formation and the maximum equivalent circulating density acting at this point ($\rho_{eq,\,cs}$) when the well is closed in or while the kick is circulated out of the well, expressed in equivalent mass density, mL⁻³, kg/m³ (lbm/gal)

$\Delta\rho_{kt}$ = pore-pressure margin, which is the difference between the kick tolerance (ρ_{kt}) and the estimated pore pressure (ρ_p) of the formation, expressed in equivalent mass density, mL⁻³, kg/m³ (lbm/gal)

Subscripts

df = drilling fluid

eq = equivalent

eq, cs = equivalent at casing shoe
   f = formation fracture
   g = gas
  kt = kick tolerance
ksm = kick safety margin
   p = pore
vbh = relative to the true vertical depth of the bottom of the hole
vcs = relative to the true vertical depth of the casing shoe
 vk = relative to the vertical projection of the gas-bubble length

7.1.6 Deepwater Fracture-Pressure Gradient—Clemente J.C. Gonçalves, José L. Falcão, and Luiz A.S. Rocha, Petrobras

Introduction. Fracture-pressure gradient is defined as the pressure gradient that will cause fracture of the formation. In other words, if the formation is exposed to a pressure higher than its fracture-pressure limit, the formation will break (fracture) and, possibly, lost circulation will occur.

Smaller tolerance between pore-pressure and fracture-pressure gradients, resulting in narrow pressure margins while drilling, is probably the most recognized deepwater challenge. The reduction of the fracture-pressure gradient observed in the deeper water is mainly because of the low stress regime as result of the reduction of the overburden-pressure gradient. In addition, the structurally weak, undercompacted, and unconsolidated sediments commonly found in the shallower portion of the underground can reduce the fracture gradient even further. Under these circumstances, the operational window formed by the pore-pressure and the fracture-pressure gradients will be reduced more and more as the water depth increases. Typical ways that a reduced operational window affects deepwater and ultradeepwater drilling are an excessive number of casing strings, small hole size at total depth (or inability to reach total depth), and fracturing the formation during kick-control operations.

Methods for estimating the fracture-pressure gradient can be classified as "direct" and "indirect." Direct methods rely on measuring the pressure required to fracture the rock and the pressure required to propagate the resulting fracture. They are generally based on a field procedure called a leakoff test (LOT) and use mud to pressurize the well until formation fracture is initiated. An LOT is a normal procedure in vertical wildcat wells where the formation-fracture gradient is not well established.

Indirect methods are based on analytical or numerical models and can be used to estimate fracture pressure along the entire well. Some of them are well known in the oil industry, others were built for specific areas, and all generally require data that are difficult to obtain.

Background. The fracture pressure is the upper limit below which the pressure in the well should be kept to avoid fracturing the formation and causing lost circulation. Although the most common approach when planning and drilling wells has been to assume leakoff pressure as the upper limit, other considerations sometimes are also taken into account and are discussed in the following, having **Fig. 7.1.46** as the basis.

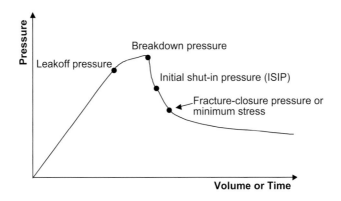

Fig. 7.1.46—Typical extended leakoff test.

The typical extended LOT displayed in Fig. 7.1.46 shows four pressure points:

- Leakoff pressure is the pressure at which plastic deformation starts occurring and is identified as the point where the curve is no longer a "straight line." Depending on the rock type, existing small fractures will open, and the formation will start taking fluid. Generally, typical LOTs are stopped after this pressure is reached, and this value is assumed as the fracture pressure.
- Breakdown pressure is the peak point of the graph, where rock tensile strength is overcome. Often, but not always, a peak pressure is observed before the formation fractures and the well starts taking fluid (Fjær et al. 1992).
- The initial shut-in pressure (generally referred to as ISIP) is the pressure recorded immediately after the pump is stopped. The pressure will fall off to a level at which it balances the formation stresses trying to close the fracture.
- Fracture-closure pressure is by definition equal to the minimum principal stress of the formation containing the fracture.

Although the actual upper limit is the breakdown pressure, it is common in the oil industry to assume leakoff pressure as the safe limit pressure for planning and execution purposes. To assume the fracture-closure pressure (minimum in-situ stress) as an upper limit is considered a conservative approach because results have shown that leakoff pressure is approximately 10% higher than the corresponding minimum in-situ stress (Fjær et al. 1992). The fact is that the term "fracture-pressure gradient" is not always clear. Sometimes it expresses either leakoff pressure or breakdown pressure. In some cases, it can also mean fracture-closure pressure. It is important to point out that LOTs in deviated wells may reflect the reduced breakdown pressure in such wells, but the minimum in-situ stress, which is independent of the orientation of the well, is still the same (Fjær et al. 1992). In this section, fracture-pressure gradient will mean leakoff pressure.

Methods Used To Estimate Fracture Gradients. The main idea of this section is to present methods, and a case history will be used as an example. A number of different fracture-gradient methods will be examined; they were classified according to their underlying assumptions and were divided into three groups.

Nine deepwater wells selected from the same basin were used in this study. Different methods, which will be explained subsequently, were calibrated with data obtained from six deepwater wells, referred to as "the six wells." The results were compared with the fracture-pressure gradients from the three remaining deepwater wells, referred to as "the three wells." Data that include overburden pressure, pore pressure, and LOTs of each well were used. It is important to say that all the LOTs used were reported as having been taken in shales. **Fig. 7.1.47** shows a map with the six wells (blue) used to calibrate the model and the three wells (red) used to verify the results. **Table 7.1.8** shows some well characteristics such as water depth, rotary-table elevation, final depth, LOT, and pore-pressure and overburden gradient.

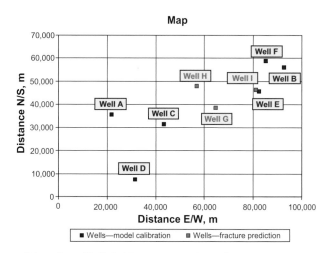

Fig. 7.1.47—Map showing well locations. Wells in blue were used to calibrate the model, and wells in red to verify it.

TABLE 7.1.8—WELL CHARACTERISTICS

	Well	Water Depth (m)	Final Depth (m)	Rotary-Table Elevation (m)	Test Depth (m)	LOT (lbm/gal) (Pa/m)	Pore Pressure (lbm/gal) (Pa/m)	Overburden (lbm/gal) (Pa/m)
Model calibration	A	1361	5020	14	1935	9.8 *11,516*	8.8 *10,341*	10.7 *12,574*
					2765	11.6 *13,631*	9.3 *10,928*	12.9 *15,159*
	B	1549	5528	24	2286	10.3 *12,103*	8.6 *10,106*	11.0 *12,926*
	C	1522	3157	18	2169	9.9 *11,633*	8.8 *10,341*	10.8 *12,691*
	D	1831	3898	18	2790	10.5 *12,339*	8.8 *10,341*	11.3 *13,279*
	E	1489	4529	24	2310	10.0 *11,751*	8.7 *10,223*	11.3 *13,279*
	F	1412	5347	14	1985	10.4 *12,221*	8.8 *10,341*	11.0 *12,926*
Fracture prediction	G	1490	4826	24	2395	10.3 *12,103*	8.5 *9988*	11.3 *13,279*
	H	1391	6024	24	2741	10.6 *12,456*	8.9 *10,458*	12.7 *14,924*
	I	1491	3675	14	2450	9.8 *11,516*	8.7 *10,223*	11.8 *13,866*

Basically, the procedures used to estimate the fracture-pressure gradient can be divided into three groups: methods based on the determination of the stresses around the wellbore, methods based on the minimum in-situ stress, and correlations developed for specific areas. Brief descriptions of each procedure and its applications are shown next.

Fracture Gradient Based on the Stresses Around the Wellbore.   The original compressive stress state acting on the rock before a well is drilled can be divided into three components. The first one is vertical and equal to the overburden. The other two are horizontal and generally unequal: σ_H (the maximum horizontal stress) and σ_h (the minimum horizontal stress). After the well is drilled, the mud replaces the drilled rock, resulting in a stress concentration around the wellbore. These redistributed stresses are known as the hoop stresses and are represented by

- σ_θ, which acts circumferentially around the wellbore wall
- σ_r, the radial stress
- σ_z, the axial stress, which acts parallel to the wellbore axis
- $\tau_{\theta z}$, an additional shear component, which is generated in deviated wells

Because most strength criteria are expressed in terms of the principal stresses, the hoop stresses must be converted into these stresses in order to be compared with the stress state that leads to rock failure. This is the underlying assumption used to estimate fracture gradient on the basis of the method described in this section.

Although easy to state, determining stresses around the wellbore is a quite difficult and complex problem. They are a function of several factors that include in-situ stresses, rock type, wellbore angle, mud properties, and pore pressure. In general, the solutions proposed by different authors (McLean and Addis 1990b) are based on assumptions that make the problem simple and feasible. Assuming rocks as homogeneous, isotropic, and perfectly linear-elastic materials is often the common departing point for most of models described in the literature. As known, rocks are neither homogeneous nor isotropic, so generally the results have a high uncertainty, although they help to explain certain phenomena. In addition, even though based on many simplifications, the models still require data that are difficult to obtain.

The determination of the stress state that causes rock to fail is also difficult to obtain because of a lack of data. Rock-failure criteria require a number of laboratory tests that generally are economically justified only for the reservoir rock. Despite all these restrictions, much work has been conducted to establish the stresses around the wellbore and to associate them with a given failure criterion. The relatively simple method described below (McLean and Addis 1990b; Fjær et al. 1992; Jaeger and Cook 1979) is based on a linear-elastic approach and can be used in vertical and directional wells. It has the following steps.

1. Determination of the in-situ stresses (original compressive stress state) σ_V, σ_H, and σ_h: although this is the most important aspect of this method, the lack of in-situ stress values usually leads the engineers to make assumptions such as

$$\sigma_V = \text{overburden pressure,} \qquad\qquad\qquad\qquad\qquad\qquad\qquad (7.1.11)$$

$$\sigma_H = \sigma_h, \qquad\qquad\qquad\qquad\qquad\qquad\qquad\qquad\qquad\qquad (7.1.12)$$

$$\text{and } K_{SAW} = \frac{\sigma_H - p_f}{\sigma_V - p_f}, \qquad\qquad\qquad\qquad\qquad\qquad\qquad (7.1.13)$$

where K_{SAW} is the stress ratio for the "stresses-around-the-wellbore" method.

2. Transposition of the in-situ stress tensor to a coordinate system with one of its axes parallel to the wellbore axis and another in a horizontal plane:

$$\sigma_x = \sigma_H \sin^2 \beta + \sigma_h \cos^2 \beta \qquad\qquad\qquad\qquad\qquad\qquad (7.1.14)$$

$$\sigma_y = \cos^2 \alpha (\sigma_H \cos^2 \beta + \sigma_h \sin^2 \beta) + \sigma_V \sin^2 \alpha \qquad\qquad (7.1.15)$$

$$\sigma_{zz} = \sin^2 \alpha (\sigma_H \cos^2 \beta + \sigma_h \sin^2 \beta) + \sigma_V \cos^2 \alpha \qquad\qquad (7.1.16)$$

$$\tau_{xy} = \cos \alpha \sin \beta \cos \beta (\sigma_H - \sigma_h) \qquad\qquad\qquad\qquad\qquad (7.1.17)$$

$$\tau_{yz} = \sin \alpha \cos \alpha (\sigma_V - \sigma_H \cos^2 \beta - \sigma_h \sin^2 \beta) \qquad\qquad\qquad (7.1.18)$$

$$\tau_{zx} = \sin \alpha \sin \beta \cos \beta (\sigma_h - \sigma_H) \qquad\qquad\qquad\qquad\qquad (7.1.19)$$

3. Determination of the hoop stresses at the wellbore wall because they will normally form the most critical stress state when using a linear-elastic approach:

$$\sigma_r = P_W \qquad\qquad\qquad\qquad\qquad\qquad\qquad\qquad\qquad\qquad (7.1.20)$$

$$\sigma_\theta = (\sigma_X + \sigma_Y - P_F) - 2(\sigma_X - \sigma_Y) \cos 2\theta$$
$$- 4\tau_{XY} \sin 2\theta - (P_W - P_F)\left[1 - \frac{h(1-2v)}{1-v}\right] \qquad (7.1.21)$$

$$\sigma_z = \sigma_{zz} - 2v\left[(\sigma_X - \sigma_Y)\cos 2\theta + 2\tau_{XY} \sin 2\theta\right]$$
$$+ \frac{h(1-2v)(P_W - P_F)}{1-v} \qquad\qquad\qquad\qquad\qquad (7.1.22)$$

$$\tau_{\theta z} = 2(\tau_{YZ} \cos \theta - \tau_{ZX} \sin \theta) \qquad\qquad\qquad\qquad\qquad\qquad (7.1.23)$$

$$p_f = P_F + h(P_W - P_F), \qquad\qquad\qquad\qquad\qquad\qquad\qquad (7.1.24)$$

where θ varies from 0 to 360°, representing points around the wellbore wall.

4. Conversion of the stresses at the wellbore wall to the three principal stresses, using the following equations:

$$\sigma_A = \frac{\sigma_\theta + \sigma_z}{2} + \sqrt{\left(\frac{\sigma_\theta - \sigma_z}{2}\right)^2 + \tau_{\theta z}^2} \quad \dots\dots\dots\dots\dots\dots\dots\dots\dots\dots\dots\dots\dots \quad (7.1.25)$$

$$\sigma_B = \frac{\sigma_\theta + \sigma_z}{2} - \sqrt{\left(\frac{\sigma_\theta - \sigma_z}{2}\right)^2 + \tau_{\theta z}^2} \quad \dots\dots\dots\dots\dots\dots\dots\dots\dots\dots\dots\dots\dots \quad (7.1.26)$$

$$\sigma_C = P_W \quad \dots\dots\dots\dots\dots\dots\dots\dots\dots\dots\dots\dots\dots\dots\dots\dots\dots\dots\dots \quad (7.1.27)$$

Further, let

$\sigma_1 = \max(\sigma_A, \sigma_B, \sigma_C)$
$\sigma_3 = \min(\sigma_A, \sigma_B, \sigma_C)$
$\sigma_2 = $ intermediate principal stress

These three expressions mean that σ_1 will be the maximum stress among σ_A, σ_B, and σ_C; σ_3 will be the minimum stress among σ_A, σ_B, and σ_C; and σ_2 will be the intermediate one.

5. Comparison between the computed minimum effective principal stress and the tensile strength: The failure is considered to have initiated when the minimum effective stress at the wall is less than the tensile strength of the formation:

$$\sigma_3 - p_f \leq -\left|\sigma_t\right|. \quad \dots\dots\dots\dots\dots\dots\dots\dots\dots\dots\dots\dots\dots\dots\dots\dots \quad (7.1.28)$$

The following procedure is to check if the fracture will propagate. The propagation criterion assumes that the minimum horizontal principal stress is less than the minimum effective stress, and it can be expressed as

$$\sigma_3 - p_f \geq \sigma_h. \quad \dots\dots\dots\dots\dots\dots\dots\dots\dots\dots\dots\dots\dots\dots\dots\dots\dots \quad (7.1.29)$$

Besides the assumptions shown in the preceding, for the application of this method it was necessary to assume zero for the tensile strength of the formation. The calibration of the model was achieved by varying the stress ratio until the calculated fracture gradient matched the LOT point for each of the six wells. The final result is shown in **Fig. 7.1.48**, where K_{SAW} is plotted against sediment depth. As can be seen, there are different values of stress ratio associated with different LOT sediment depths. At the end, to make future predictions, a curve fitting was used to represent the stress ratio for the entire basin.

Table 7.1.9 shows the results of the application of this method for the six wells used to calibrate the model. **Table 7.1.10** shows the predictions for the three wells.

Fracture-Gradient Methods Based on Minimum Stress. The "minimum-stress" methods are based on the statement that the vertical and the horizontal effective stresses are related by the following simple equation (Rocha and Bourgoyne 1996; Bowers 2001; Fjær et al. 1992):

$$K_{MS} = \frac{\sigma_h - p_f}{\sigma_V - p_f}, \quad \dots\dots\dots\dots\dots\dots\dots\dots\dots\dots\dots\dots\dots\dots\dots \quad (7.1.30)$$

where K_{MS} is the stress ratio for the minimum-stress method. The main assumption here is that fracture of the formation will take place when the tangential stress component equals the minimum in-situ stress:

$$\sigma_f = \sigma_h. \quad \dots\dots\dots\dots\dots\dots\dots\dots\dots\dots\dots\dots\dots\dots\dots\dots\dots\dots\dots \quad (7.1.31)$$

Substituting Eq. 7.1.31 into Eq. 7.1.30, the expression for the fracture gradient becomes

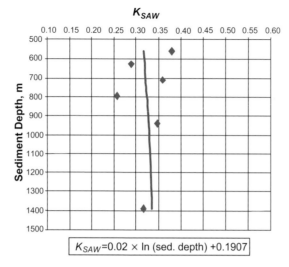

Fig. 7.1.48—Stress ratio K_{SAW} based on leakoff test for the "stresses around the wellbore" method.

TABLE 7.1.9—RESULTS OF THE CALIBRATION OF THE METHOD BASED ON THE SIX WELLS				
Well	Test Depth (m)	LOT (lbm/gal) (Pa/m)	Fracture (lbm/gal) (Pa/m)	Error (lbm/gal) (Pa/m)
A	1935	9.8 11,516	10.03 11,786	−0.23 −270
	2765	11.6 13,631	11.71 13,760	−0.11 −129
B	2286	10.3 12,103	10.12 11,892	0.18 212
C	2169	9.9 11,633	10.06 11,821	−0.16 −188
D	2790	10.5 12,339	10.41 12,233	0.09 106
E	2310	10.0 11,751	10.34 12,150	−0.34 −400
F	1985	10.4 12,221	10.11 11,880	0.29 341
Average error				−0.04 −47

TABLE 7.1.10—RESULTS OF THE PREDICTION OF FRACTURE GRADIENTS FOR THE THREE WELLS				
Well	Test Depth (m)	LOT (lbm/gal) (Pa/m)	Fracture (lbm/gal) (Pa/m)	Error (lbm/gal) (Pa/m)
G	2395	10.3 12,103	10.33 12,139	−0.03 −35
H	2741	10.6 12,456	11.38 13,373	−0.78 −917
I	2450	9.8 11,516	10.70 12,574	−0.90 −1,058
Average error				−0.57 −670

$$\sigma_f = K_{MS}\left(\sigma_V - p_f\right) + p_f. \quad \dots \quad (7.1.32)$$

As implied in the previous method, the direct use of Eq. 7.1.32 also implies that overburden- and pore-pressure gradients have already been estimated. The difference between the diverse methods based on minimum stress is the way K_{MS} has been calculated (Bowers 2001). Probably one of the simplest ways to obtain K_{MS} for the entire well is to correlate it with sediment depth. The procedure was as follows:

- K_{MS} was calculated for each of the six wells by use of Eq. 7.1.32 and Table 7.1.8. The results are shown in **Table 7.1.11**.
- K_{MS} was plotted against sediment depth and is shown in **Fig. 7.1.49**.
- A correlation for K_{MS} was obtained for the entire basin. Fig. 7.1.49 displays the correlation.

Table 7.1.11 shows the results of the application of this method for the six wells. Similarly, **Table 7.1.12** displays the predictions for the three wells.

Specific Correlations To Estimate Fracture Gradient. A simple approach to estimating the fracture-pressure gradient for a given area is through the use of specific correlations. The method used stems from the idea that a good correlation must take into account the effects of several factors such as well depth, water depth, existing stress state, and LOT measurements. As the overburden gradient is itself an in-situ stress, and as it is a function of well depth and water depth, it was chosen as the parameter to be correlated directly with available LOT data.

The idea behind the method was further applied in two scenarios: where overburden gradients and LOTs are available, and where only LOTs are on hand.

Scenarios Where Overburden Gradients and LOTs Are Available. **Fig. 7.1.50** shows the graphs of two direct correlations (linear and power) between LOT data and calculated overburden gradients shown in Table 7.1.8 for the six wells. These correlations, which can be used to predict the fracture gradient for this specific area, are as follows:

$$FG = 0.92 \times OBG \text{ (given in lbm/gal; } R^2 = 0.84)$$

$$FG = 1{,}633 \times (OBG/1175)^{0.83} \text{ (given in Pa/m; } R^2 = 0.86) \quad \dots \quad (7.1.33)$$

$$FG = 1.39 \times OBG^{0.83} \text{ (given in lbm/gal; } R^2 = 0.86) \quad \dots \quad (7.1.34)$$

TABLE 7.1.11—RESULTS OF THE CALIBRATION OF THE METHOD BASED ON THE SIX WELLS

Well	LOT (lbm/gal) (Pa/m)	Test Depth (m)	K_{MS}	Fracture (lbm/gal) (Pa/m)	Error (lbm/gal) (Pa/m)
A	9.8 11,516	1935	0.59	9.94 11,680	−0.14 −165
	11.6 13,631	2765	0.62	11.53 13,549	0.07 82
B	10.3 12,103	2286	0.6	10.01 11,763	0.29 341
C	9.9 11,633	2169	0.59	9.97 11,716	−0.07 −82
D	10.5 12,339	2790	0.61	10.31 12,115	0.19 223
E	10.0 11,751	2310	0.6	10.24 12,033	−0.24 −282
F	10.4 12,221	1985	0.59	10.08 11,845	0.32 376
Average error					0.06 71

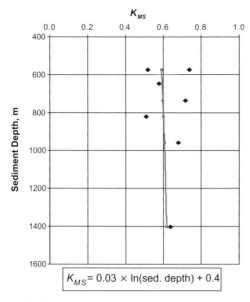

Fig. 7.1.49—Stress ratio K_{MS} based on leakoff test for the "minimum stress" method.

Well	LOT (lbm/gal) (Pa/m)	Test Depth (m)	K_{MS}	Fracture (lbm/gal) (Pa/m)	Error (lbm/gal) (Pa/m)
TABLE 7.1.12—RESULTS OF THE PREDICTION OF FRACTURE GRADIENTS FOR THE THREE WELLS					
G	10.3 12,103	2395	0.6	10.19 11,974	0.11 129
H	10.6 12,456	2741	0.62	11.20 13,161	−0.60 −705
I	9.8 11,516	2450	0.61	10.55 12,397	−0.75 −881
Average error					−0.41 −486

Table 7.1.13 shows the results of the application of the correlation from Eq. 7.1.33 for the six wells. Similarly, **Table 7.1.14** displays the predictions for the three wells.

Table 7.1.15 shows the results of the application of the correlation from Eq. 7.1.34 for the six wells. Similarly, **Table 7.1.16** displays the predictions for the three wells.

Scenarios Where Only LOTs Are Available. All the methods presented in the preceding were based on the assumption that information such as overburden gradient, pore-pressure gradient, LOT, and rock characteristics is readily accessible. However, because this is usually not the case, simple correlations based on available parameters can be quite useful for engineers.

The last method presented is described in the literature and is called pseudo-overburden (Rocha and Bourgoyne 1996). It uses LOT data only and is based on the following equation (Bourgoyne et al. 1986):

$$\sigma_{\text{pseudo}} = A\left[\rho_w z_w + \rho_g z_s - \frac{(\rho_g - \rho_{fl})\phi_0}{K_0}\left(1 - e^{-K_0 z_s}\right) \right], \quad \dots\dots\dots\dots\dots\dots\dots\dots\dots (7.1.35)$$

where

$A = 9.81$ (with σ_{pseudo} in Pa, ρ in kg/m^3, Z in m)

and $A = 0.1704$ (with σ_{pseudo} in psi, ρ in lbm/gal, Z in m).

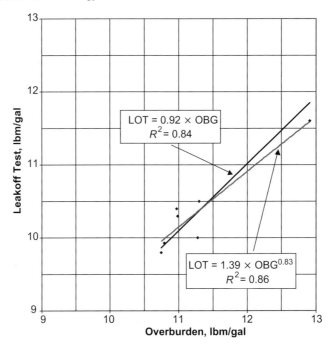

Fig. 7.1.50—Correlation between LOT and overburden gradient.

Well	Test Depth (m)	LOT (lbm/gal) (Pa/m)	Overburden (lbm/gal) (Pa/m)	Fracture (lbm/gal) (Pa/m)	Error (lbm/gal) (Pa/m)
		TABLE 7.1.13—FRACTURE-PRESSURE GRADIENTS BASED ON THE CORRELATION FG = 0.92 × OBG FOR THE SIX WELLS			
A	1935	9.8 *11,516*	10.7 *12,574*	9.84 *11,568*	−0.04 *−52*
	2765	11.6 *13,631*	12.9 *15,159*	11.87 *13,946*	−0.27 *−315*
B	2286	10.3 *12,103*	11.0 *12,926*	10.12 *11,892*	0.18 *212*
C	2169	9.9 *11,633*	10.8 *12,691*	9.94 *11,676*	−0.04 *−42*
D	2790	10.5 *12,339*	11.3 *13,279*	10.40 *12,216*	0.10 *122*
E	2310	10.0 *11,751*	11.3 *13,279*	10.40 *12,216*	−0.40 *−465*
F	1985	10.4 *12,221*	11.0 *12,926*	10.12 *11,892*	0.28 *329*
Average error					−0.03 *−30*

Eq. 7.1.35 can also be expressed in gradient form as follows:

$$G_{pseudo} = B \frac{\sigma_{pseudo}}{Z}, \quad\quad\quad\quad\quad\quad\quad\quad\quad\quad\quad\quad\quad\quad\quad\quad\quad (7.1.36)$$

where

$B = 1$ (with σ_{pseudo} in Pa, G_{pseudo} in Pa/m, Z in m)

and $B = 5.869$ (with σ_{pseudo} in psi, G_{pseudo} in lbm/gal, Z in m).

TABLE 7.1.14—FRACTURE-PRESSURE GRADIENTS BASED ON THE CORRELATION FG = 0.92 × OBG FOR THE THREE WELLS

Well	Test Depth (m)	LOT (lbm/gal) (Pa/m)	Overburden (lbm/gal) (Pa/m)	Fracture (lbm/gal) (Pa/m)	Error (lbm/gal) (Pa/m)
G	2395	10.3 *12,103*	11.3 *13,279*	10.40 *12,216*	−0.10 *−113*
H	2741	10.6 *12,456*	12.7 *14,924*	11.68 *13,730*	−1.08 *−1,274*
I	2450	9.8 *11,516*	11.8 *13,866*	10.86 *12,757*	−1.06 *−1,241*
Average error					−0.75 *−876*

TABLE 7.1.15—FRACTURE-PRESSURE GRADIENTS BASED ON THE CORRELATION FG = 1.39 × OBG$^{0.83}$ (lbm/gal) [FG = 1633 × (OBG/1175)$^{0.83}$ (Pa/m)] FOR THE SIX WELLS

Well	Test Depth (m)	LOT (lbm/gal) (Pa/m)	Overburden (lbm/gal) (Pa/m)	Fracture (lbm/gal) (Pa/m)	Error (lbm/gal) (Pa/m)
A	1935	9.8 *11,516*	10.7 *12,574*	9.97 *11,716*	−0.17 *−200*
	2765	11.6 *13,631*	12.9 *15,159*	11.61 *13,643*	−0.01 *−12*
B	2286	10.3 *12,103*	11.0 *12,926*	10.16 *11,939*	0.14 *165*
C	2169	9.9 *11,633*	10.8 *12,691*	10.01 *11,763*	−0.11 *−129*
D	2790	10.5 *12,339*	11.3 *13,279*	10.40 *12,221*	0.10 *118*
E	2310	10.0 *11,751*	11.3 *13,279*	10.38 *12,197*	−0.38 *−447*
F	1985	10.4 *12,221*	11.0 *12,926*	10.15 *11,927*	0.25 *294*
Average error					−0.03 *−30*

TABLE 7.1.16—FRACTURE-PRESSURE GRADIENTS BASED ON THE CORRELATION FG = 1.39 × OBG$^{0.83}$ (lbm/gal) [FG = 1633 × (OBG/1175)$^{0.83}$ (Pa/m)] FOR THE THREE WELLS

Well	Test Depth (m)	LOT (lbm/gal) (Pa/m)	Overburden (lbm/gal) (Pa/m)	Fracture (lbm/gal) (Pa/m)	Error (lbm/gal) (Pa/m)
G	2395	10.3 *12,103*	11.3 *13,279*	10.38 *12,197*	−0.08 *−94*
H	2741	10.6 *12,456*	12.7 *14,924*	11.42 *13,420*	−0.82 *−964*
I	2450	9.8 *11,516*	11.8 *13,866*	10.75 *12,632*	−0.95 *−1,116*
Average error					−0.62 *−725*

To apply the method we need to crossplot σ_{pseudo} vs. LOT and G_{pseudo} vs. LOT to find the parameters ϕ_0 and K_0. The simultaneous calibration of Eqs. 7.1.35 and 7.1.36 is accomplished in order to make the points of the pseudo-overburden stress and pseudo-overburden-stress gradient, calculated at each LOT location, to match the fracture data, and to make these points fall along a straight line through the origin with a slope of 1. After the match is obtained, the fracture-pressure gradient is assumed equal to the pseudo-overburden as follows:

$$FG = G_{pseudo} \quad \dots\dots\dots\dots\dots\dots\dots\dots\dots\dots\dots\dots\dots\dots\dots\dots\dots\dots\dots\dots\dots \quad (7.1.37)$$

Figs. 7.1.51 and 7.1.52 (in lbm/gal and psi, respectively) display pseudo-overburden curves vs. LOT. The average values used for ρ_w and ρ_g in the calculations, and the results obtained for ϕ_0 and K_0, are as follows:
$\rho_w = 1.03 \times 10^3 \text{ kg/m}^3$,
$\rho_g = 2.6 \times 10^3 \text{ kg/m}^3$,
$\phi_0 = 0.627$,
and $K_0 = 4.90 \times 10^{-4}$.

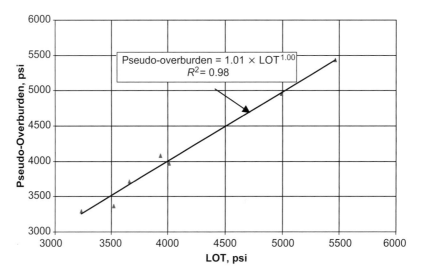

Fig. 7.1.51—Correlation between LOT and pseudo-overburden gradient.

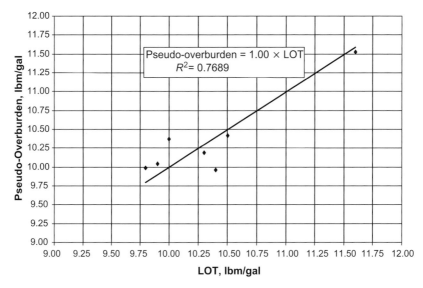

Fig. 7.1.52—Correlation between LOT and pseudo-overburden gradient.

Table 7.1.17 shows the results of the application of this method for the six wells. **Table 7.1.18** displays the predictions for the three wells.

Comparisons of Different Methods. **Table 7.1.19** summarizes the errors between fracture gradients estimated from different methods and actual LOT values for the six wells. **Table 7.1.20** displays fracture-gradient estimates for the same set of wells. Similarly, **Tables 7.1.21 and 7.1.22** are for the three wells.

Figs. 7.1.53, 7.1.54, and 7.1.55 refer to the three wells. They show pore pressure, overburden gradient, and fracture gradients estimated using the different methods described above.

An analysis of Tables 7.1.19 through 7.1.22, which refer to LOTs, indicates that

- All the methods presented absolute average errors smaller than 0.06 lbm/gal (70.5 Pa/m) for the six wells and smaller than 0.75 lbm/gal (876 Pa/m) for the three wells.
- The pseudo-overburden gradient method led to the smallest absolute average error for the six wells.
- The method based on the minimum stress gave the smallest average absolute error for the three wells.
- Well G presented the smallest absolute error among all the methods used.
- Wells H and I displayed the worst results, presenting error up to 1.08 lbm/gal (1274 Pa/m).
- An analysis of Figs. 7.1.53 through 7.1.55, which show plots of the different methods, indicates that

TABLE 7.1.17—RESULTS OF THE CALIBRATION OF THE METHOD BASED ON THE SIX WELLS

Well	LOT (lbm/gal) (Pa/m)	Test Depth (m)	Pseudo-OBG (lbm/gal) (Pa/m)	Error (lbm/gal) (Pa/m)
A	9.8 11,516	1935	9.99 11,739	−0.19 −223
	11.6 13,631	2765	11.52 13,537	0.08 94
B	10.3 12,103	2286	10.19 11,974	0.11 129
C	9.9 11,633	2169	10.04 11,798	−0.14 −165
D	10.5 12,339	2790	10.42 12,244	0.08 94
E	10.0 11,751	2310	10.37 12,186	−0.37 −435
F	10.4 12,221	1985	9.96 11,704	0.44 517
Average error				0.00 2

TABLE 7.1.18—RESULTS OF THE PREDICTION OF FRACTURE GRADIENTS FOR THE THREE WELLS

Well	LOT (lbm/gal) (Pa/m)	Test Depth (m)	Pseudo-OBG (lbm/gal) (Pa/m)	Error (lbm/gal) (Pa/m)
G	10.3 12,103	2395	10.53 12,374	−0.23 −270
H	10.6 12,456	2741	11.34 13,326	−0.74 −870
I	9.8 11,516	2450	10.62 12,480	−0.82 −964
Average error				−0.60 −701

TABLE 7.1.19—COMPARISONS OF THE ERRORS OBTAINED FROM DIFFERENT METHODS BASED ON THE SIX WELLS

Well Data		K_{SAW}	K_{MS}	Linear 0.92 × OBG	Power 1.39 × OBG$^{0.83}$	Pseudo-OBG
		Results for Different Procedures				
Well	LOT (lbm/gal) (Pa/m)	Error (lbm/gal) (Pa/m)	Error (lbm/gal) (Pa/m)	Error (lbm/gal) (Pa/m)	Error (lbm/gal) (Pa/m)	Error (lbm/gal) (Pa/m)
A	9.8 11,516	−0.23 −270	−0.14 −165	−0.04 −52	−0.17 −200	−0.19 −223
	11.6 13,631	−0.11 −129	0.07 82	−0.27 −315	−0.01 −12	0.08 94
B	10.3 12,103	0.18 212	0.29 341	0.18 212	0.14 165	0.11 129
C	9.9 11,633	−0.16 −188	−0.07 −82	−0.04 −42	−0.11 −129	−0.14 −165
D	10.5 12,339	0.09 106	0.19 223	0.10 122	0.10 118	0.08 94
E	10.0 11,751	−0.34 −400	−0.24 −282	−0.40 −465	−0.38 −447	−0.37 −435
F	10.4 12,221	0.29 341	0.32 376	0.28 329	0.25 294	0.44 517
Average error		−0.04 −47	0.06 71	−0.03 −30	−0.03 −30	0.00 2

TABLE 7.1.20—COMPARISONS OF LOTs AND FRACTURE GRADIENTS OBTAINED FROM DIFFERENT METHODS BASED ON THE SIX WELLS

Well Data		K_{SAW}	K_{MS}	Linear 0.92 × OBG	Power 1.39 × OBG$^{0.83}$	Pseudo-OBG
		Results for Different Procedures				
Well	LOT (lbm/gal) (Pa/m)	Error (lbm/gal) (Pa/m)	Error (lbm/gal) (Pa/m)	Error (lbm/gal) (Pa/m)	Error (lbm/gal) (Pa/m)	Error (lbm/gal) (Pa/m)
A	9.8 11,516	10.03 11,786	9.94 11,680	9.84 11,568	9.97 11,716	9.99 11,739
	11.6 13,631	11.71 13,760	11.53 13,549	11.87 13,946	11.61 13,643	11.52 13,537
B	10.3 12,103	10.12 11,892	10.01 11,763	10.12 11,892	10.16 11,939	10.19 11,974
C	9.9 11,633	10.06 11,821	9.97 11,716	9.94 11,676	10.01 11,763	10.04 11,798
D	10.5 12,339	10.41 12,233	10.31 12,115	10.40 12,216	10.40 12,221	10.42 12,244
E	10.0 11,751	10.34 12,150	10.24 12,033	10.40 12,216	10.38 12,197	10.37 12,186
F	10.4 12,221	10.11 11,880	10.08 11,845	10.12 11,892	10.15 11,927	9.96 11,704

- The largest differences among the methods occur either at small depths or at greater depths.
- All the methods show results very similar to each other near LOT locations.
- A worst and a best scenario for fracture-gradient estimates at each depth can be obtained by applying all the methods simultaneously.

TABLE 7.1.21—COMPARISONS OF THE ERRORS OBTAINED FROM DIFFERENT METHODS BASED ON THE THREE WELLS

Well Data		K_{SAW}	K_{MS}	Linear 0.92 × OBG	Power 1.39 × OBG$^{0.83}$	Pseudo-OBG
		Results for Different Procedures				
Well	LOT (lbm/gal) (Pa/m)	Error (lbm/gal) (Pa/m)	Error (lbm/gal) (Pa/m)	Error (lbm/gal) (Pa/m)	Error (lbm/gal) (Pa/m)	Error (lbm/gal) (Pa/m)
G	10.3 *12,103*	−0.03 *−36*	0.11 *129*	−0.10 *−113*	−0.08 *−94*	−0.23 *−270*
H	10.6 *12,456*	−0.78 *−917*	−0.60 *−705*	−1.08 *−1,274*	−0.82 *−964*	−0.74 *−870*
I	9.8 *11,516*	−0.90 *−1,058*	−0.75 *−881*	−1.06 *−1,241*	−0.95 *−1,116*	−0.82 *−964*
Average error		−0.57 *−670*	−0.41 *−486*	−0.75 *−876*	−0.62 *−725*	−0.60 *−701*

TABLE 7.1.22—COMPARISONS OF LOTs AND FRACTURE GRADIENTS OBTAINED FROM DIFFERENT METHODS BASED ON THE THREE WELLS

Well Data		K_{SAW}	K_{MS}	Linear 0.92 × OBG	Power 1.39 × OBG$^{0.83}$	Pseudo-OBG
		Results for Different Procedures				
Well	LOT (lbm/gal) (Pa/m)	Error (lbm/gal) (Pa/m)	Error (lbm/gal) (Pa/m)	Error (lbm/gal) (Pa/m)	Error (lbm/gal) (Pa/m)	Error (lbm/gal) (Pa/m)
G	10.3 *12,103*	10.33 *12,139*	10.19 *11,974*	10.40 *12,216*	10.38 *12,197*	10.53 *12,374*
H	10.6 *12,456*	11.38 *13,373*	11.2 *13,161*	11.68 *13,730*	11.42 *13,420*	11.34 *13,326*
I	9.8 *11,516*	10.7 *12,574*	10.55 *12,397*	10.86 *12,757*	10.75 *12,632*	10.62 *12,480*

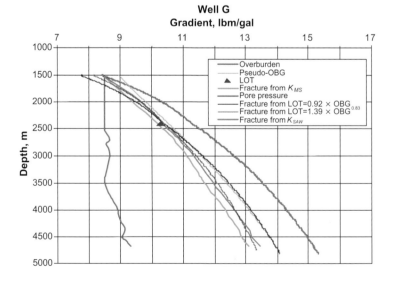

Fig. 7.1.53—Plot of gradients vs. depth shows fracture gradients from different procedures, pore pressure, and OBG of well G.

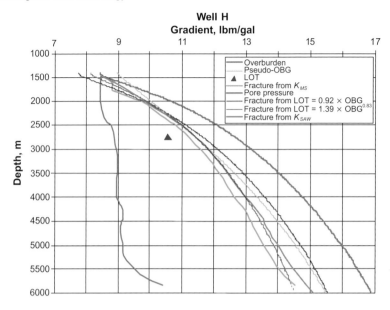

Fig. 7.1.54—Plot of gradients vs. depth shows fracture gradients from different procedures, pore pressure, and OBG of well H.

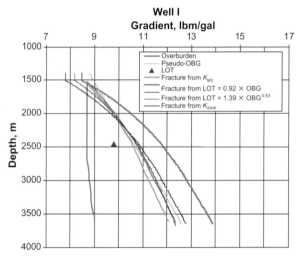

Fig. 7.1.55—Plot of gradients vs. depth shows fracture gradients from different procedures, pore pressure, and OBG of well I.

Nomenclature

G_{pseudo} = pseudo-overburden gradient
K = stress ratio
K_{MS} = stress ratio for minimum-stress method
K_{SAW} = stress ratio for stresses-around-the-wellbore method
p_f = pore-fluid pressure at the wellbore wall
K_0 = pseudodeclining porosity
P_F = far-field formation pressure
P_W = well pressure
z = depth
z_s = sediment depth

z_w = water depth
α = well inclination
β = well azimuth
ϕ_0 = pseudo surface porosity
ρ_{fl} = formation-fluid density
ρ_g = grain density
ρ_w = water density
σ_f = fracture stress
σ_h = minimum horizontal stress
σ_H = maximum horizontal stress
σ_{pseudo} = pseudo-overburden stress
σ_r = radial stress
σ_t = tensile strength
σ_V = vertical stress (overburden pressure)
σ_z = axial stress
σ_1 = max (σ_A, σ_B, σ_C)
σ_2 = intermediate principal stress
σ_3 = min (σ_A, σ_B, σ_C)
σ_θ = stress around wellbore wall
$\tau_{\theta z}$ = shear component
θ = angle around the wellbore

7.1.7 Special Topics in Deepwater Well Design—João Carlos R. Plácido and Emmanuel Franco Nogueira, Petrobras

Slender Well Technology. The design of a slender well (Saliés et al. 1998, 1999) [30-in. (0.762 m) × 13 ⅜-in. (0.340 m) × 9⅝-in. (0.244 m)] must combine information from drilling, completion, and workover. The goal of this technology is the reduction of the drilling-riser diameter, from 21 in. (0.533 m) to 15 in. (0.381 m), in order to use deepwater platforms of the second and third generation, with less deck capacity, but which are less expensive than the normally necessary fourth- and fifth-generation rigs.

Another advantage of slender technology is the reduction of drilling costs because of the elimination of one casing string, which is the primary objective of this technology. The idea is to change the conventional-well design [30-in. (0.762 m) × 20-in. (0.508 m) × 13⅜-in. (0.340 m) × 9⅝-in. (0.244 m) × optional 7-in. (0.178-m) liner] to a slender-well design [30-in. (0.762 m) × 13⅜-in. (0.340 m) × 9⅝-in. (0.244 m) × optional 7-in. (0.178-m) liner] by eliminating the 20-in. (0.508-m) casing. **Fig. 7.1.56** shows a schematic comparing a conventional-well and a slender-well design.

To eliminate the 20-in. (0.508-m) casing, the 17½-in. (0.445-m) phase has to go deeper in a riserless mode. Therefore, it is necessary to have a good knowledge of the area, which means that the pore-pressure profile and the fracture gradient must be well known. Because of this, slender technology is not recommended for exploratory wells.

One more advantage of slender technology is the reduction of the volume capacity of the drilling riser. In case of either a leak or an emergency disconnection, fewer environmental problems will result because a smaller amount of fluid will be discharged.

The slender wellhead can be 16¾ in. (0.425 m) or 18¾ in. (0.476 m), and it should withstand all forces exerted by the drilling riser. The wellhead is designed to receive two casing strings. The other strings must be liners. The high-pressure housing is located at the top of the 13⅜-in. (0.340-m) casing [instead of the 20-in. (0.508-m) casing, as used in conventional casing-string design].

In Brazil, slender wells have been drilled normally in three phases, as follows:

- Phase 1: Jetted or Drilled and Cemented. For Campos basin soil (low resistance), three joints of 30-in. (0.762-m) casing are normally necessary to support the weight of the next casing [13⅜ in. (0.340 m)]. In some areas, it is necessary to use four or even five joints. The phase can be either drilled or jetted. If drilled, a 26-in. (0.660-m) pilot bit and a 36-in. (0.914-m) hole opener are used to drill the phase; then

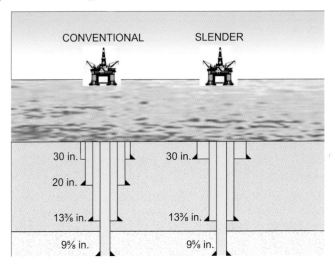

Fig. 7.1.56—Schematic of conventional and slender well. Courtesy of Petrobras.

the 30-in. (0.762-m) casing is set and cemented. The jetting operation is executed using a special tool called a Jet Cam (developed by Petrobras and Vetco), which allows a considerable rig-time savings because it is possible to drill ahead after the end of the jetting operation without having to trip out.

- Phase 2: Drilling Without Returns. In case of using jetting in the previous phase, Phase 2 starts just after landing the drilling guide base with the 30-in. (0.762-m) casing. The phase can be drilled using either a 16-in. (0.406-m) or 17 1/2-in. (0.445-m) bit. The length of this phase normally varies from 600 to 1200 m. Because of the smaller diameter of this phase, the rate of penetration is usually faster than that obtained with the 26-in. (0.660-m) bit used in the conventional-well design. Also, wellbore stability is normally better in smaller-diameter wells. To increase the resistance of the 13⅜-in. (0.340-m) casing, the pipe at the top is connected to a special stress joint at the bottom of the 16¾-in. (0.425 m) or 18¾-in. (0.476 m) high-pressure housing. This stress joint is welded to the bottom of the high-pressure housing, having a 13⅜-in. (0.340-m) buttress-double-seal (BDS) thread for the casing connection, allowing gradual changes in diameter for reducing the stress concentration. Because this stress joint must fit any kind of slender wellhead, it was necessary to adapt it to the standard and preloaded high-pressure housings. Next, the casing is cemented and the BOP is installed with the drilling riser.
- Phase 3: Drilling With Returns. A 12¼-in. (0.311-m) bit is used to drill this phase. The target zones should be reached in this phase, which will be drilled safely using conventional well-control equipment.

The structural strength of the wellhead system is a result of the low- and high-pressure housings and the structural casing, which, in this case, is the 36-in. (0.914-m) or 30-in. (0.762-m) casing. The loads generated by the riser system, BOP, and currents act on this wellhead system in the form of bending moments, compressive axial loads, and horizontal shear forces. On the other hand, the axial resistance is provided by the 36-in. (0.914-m) or 30-in. (0.762-m) casing together with the cement between the 30-in. (0.762-m) and 13⅜-in. (0.340-m) casing strings.

Slender technology was developed not only to reduce drilling cost, though that is a consequence. The main reason is the reduction of the riser diameter, allowing the use of the second- and third-generation offshore platforms with less deck capacity. As a premise, it was decided that the slender wellhead should permit drilling and workover operations using the 21-in. (0.533-m) or the 18⅝-in. (0.473-m) standard riser and also the 15-in. (0.381-m) slender riser. Consequently, a new slender wellhead of 16¾ in. (0.425 m) × 13 in. (0.330 m) × 10 ksi (68.9 MPa) was developed by Petrobras. The external shape of the high-pressure housing was kept the same as the standard 16¾-in. (0.425-m) × 10-ksi (68.9-MPa) wellhead. This new wellhead is designed to receive two casing hangers: the 9⅝ in. (0.244 m) and the 7⅝ in. (0.194 m), the last one being used only rarely. In a standard slender well, the 9⅝-in. (0.244-m) casing is used often for safety measures because it is difficult to evaluate the 13⅜-in. (0.340-m) casing wear. Otherwise, the 9⅝-in. (0.244-m) casing can be a liner.

The first slender well drilled in Brazil was in the Marlim field in March 1998, and it became the new Campos basin drilling record, taking 7.6 days from the jetting to the final depth. Such performance resulted not only from the slender well design—which was the main factor—but also from the excellent teamwork in the planning and execution of the well operations. Since then, more than 400 slender wells have been drilled in the Campos basin. A comparative analysis of the 65 best conventional wells in the Marlim field and the slender wells has shown that the latter technique resulted in drilling-time savings of approximately 17%.

Again, it is important to mention that the slender-well technology must be used only in known areas, since well safety is very important.

Also, following the same objective of reducing riser weight, some studies are being conducted with lighter alternative materials, such as titanium, aluminum, and composites. These new riser pipes must be submitted to several analyses before being approved. Stress, fatigue, vibration, and wear analyses should be performed. If these analyses are positive, the prototypes must be submitted to field tests.

Drilling With a Surface BOP (SBOP). The objective of using an SBOP is an efficient, safer, and faster ultradeepwater exploration program. The main idea is to avoid running the traditional and complex subsea BOP **(Figs. 7.1.57 and 7.1.58)**, which normally involves considerable downtime, with constant inherent electronic and hydraulics problems. Moreover, SBOP technology presents more advantages as the water depth increases.

The main advantage of using an SBOP is the cost savings. This technology significantly reduces the rig deck loading when compared to a conventional riser and allows the use of smaller rigs in deep water.

The SBOP concept (Azancot et al. 2002) **(Fig. 7.1.59)** uses at surface the same jackup or tension-leg-platform (TLP) BOP that allows easy maintenance, has no need for redundancy, and gives increased operational efficiency. In this case, it is not necessary to run the conventional subsea BOP with a riser; therefore, the downtime is reduced because the riser is a casing string rather than a special flanged joint.

The use of an SBOP in deep water means that lighter weights will be handled at the platform. Then, lower-cost third-generation rigs can replace the very expensive fourth- and fifth-generation rigs normally used for deep water.

Design guidelines were established for the following specific systems of an SBOP: mooring, riser, BOP, and tensioning.

The mooring system is a critical component of an SBOP on a floating rig and is the only aspect that differs from a TLP or a jackup drilling system. The integrity of this system has a very high impact on the riser integ-

Fig. 7.1.57—Conventional subsea BOP: two pipe rams, shear ram, and annular BOP. Courtesy of Petrobras.

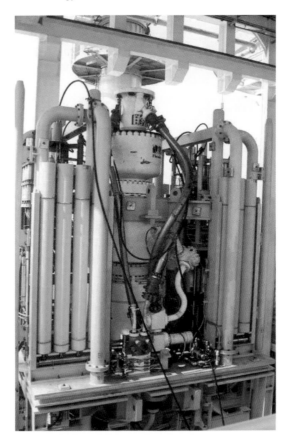

Fig. 7.1.58—Conventional subsea BOP: lower marine riser package (LMRP). Courtesy of Petrobras.

rity since bending stresses are related to the rig offset. The mooring is designed to maintain the vessel with minimal optimum-offset limits, reducing the bending stresses on the riser. The mooring system must meet the guidelines found in *RP 2SK* (2008).

The riser is the only barrier between the drilling fluid and the environment; therefore, its design and analysis are extremely important. The riser is subjected to specific loads from different sources, many of which combine to increase the overall load levels, and it must be designed to support extreme combinations of well pressure, current, waves, wind, and vessel offset. The riser must be submitted to static and dynamic analysis. The most critical points of the riser are the section near the seabed and the section just below the SBOP. At the seabed, the largest contributor to the load is the vessel offset, which creates large bending moments. The section just below the SBOP is subjected to large wave forces and also experiences loading because of the vessel displacements. All components below the SBOP must be checked against bending moments. At the surface, the tension is at the highest level. Therefore, the local state of stress must consider tension, bending moment, and internal pressure. Below the SBOP, the riser is subjected to cyclic forces generated by waves, which may cause fatigue damage mainly to the wellhead, pipe, and first connector below the SBOP. The riser must also be checked against vortex-induced vibrations (VIVs) because the alternating forces that result from vortex shedding can cause significant vibration if the frequency of vortex shedding gets close to one of the natural frequencies of the riser system. In order to determine the fatigue life of the riser components, the effects of the stress-amplification factor at the connector need to be evaluated. In summary, the riser must be designed to the following standards:

- API *RP 16Q* (2001), *Recommended Practice for Design, Selection, Operation, and Maintenance of Marine Drilling Riser Systems*

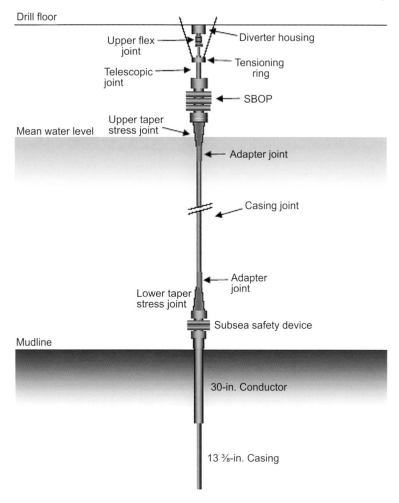

Fig. 7.1.59—Surface BOP system schematic. Courtesy of Petrobras.

- API *RP 2RD* (2006), *Recommended Practice for Design of Risers for Floating Production Systems and Tension Leg Platforms*
- Section VIII, Division 2 of the ASME *Boiler and Pressure Vessel Code* (2007)
- NACE *MR-01-75-98* (1998)

The SBOP system has several components: SBOP, ball joint, slip joint, tensioner ring, BOP frame, and wellhead. The elements are designed and manufactured in accordance with NACE Standard MR-01-75 sour services and API 6A specifications.

The SBOP is a conventional surface-type BOP stack, which is usually found on a jackup TLP, and it consists of two pipe rams, one shear ram, and one annular BOP. The arrangement of BOPs and the kill- and chokelines inlet and outlet allows maximum flexibility during well-control operations, which will be conducted in the same way as well-control operations on a jackup or TLP. The SBOP handles all well-control functions at the surface. The SBOP consists of the following components (bottom to top): 13⅝-in. (0.346-m) mechanical wellhead connector, 13⅝-in. (0.346-m) lower pipe rams, kill-line inlet port, 13⅝-in. (0.346-m) upper pipe rams, chokeline outlet port, 13⅝-in. (0.346-m) tandem boosted shear blind ram, and 13⅝-in. (0.346-m) annular BOP. Rams are usually rated to 10 ksi (68.9 MPa), and the annular BOP is usually rated to 5 ksi (34.5 MPa). The BOP frame must be placed in a position that minimizes the risk of clash with the moonpool and the wave impact on the BOP system.

The ball joint must accept an operational angular range compatible with the expected BOP displacements. The slip joint and stroke design must consider mainly the set-down operation, heave, and tide. The largest contribution comes from the set-down, which is based on the water depth and maximum offset.

The tensioning system is the most critical interface between the rig and the SBOP system. Most third-generation rigs that are not upgraded have a tensioning capacity on the order of 640 kip, which must be increased up to 1,280 kip. This system must be checked against the required tension for the designed riser system in different modes of operation, such as cementing and drilling. The rig structure must be checked with respect to the modification of the tensioning system.

A 13⅜-in. (0.340-m) conventional casing is used as a riser to connect the SBOP at the top to the seabed package. This riser is held at the surface by the tensioner ring using the tensioning system. The 13⅜-in. (0.340-m) casing presents much easier connection compared to the conventional riser flanges. Besides, the 13⅜-in. casing has higher pressure limits compared to the riser pipe. The 13⅜-in. (0.340-m) casing riser is rated to 7.5 ksi (51.7 MPa) API burst; however, it has been considered as a drilling riser to 5 ksi (34.5 MPa). Transition joints are located at the bottom of the riser, just above the SDS, and at the top of the riser, just below the SBOP. These joints avoid stress concentration between the riser and the SDS and SBOP, which are stiffer compared to the riser.

At the seabed, the conventional subsea package, which weighs approximately 650,000 lbm, is replaced by a subsea package weighing approximately 150,000 lbm. The seabed package consists of upper and lower connectors, and two shear rams (Fig. 7.1.59). This system, the SDS, is designed to shear the drillstring, seal off the well, and disconnect the riser in an emergency situation. The SDS is composed of the following main components (bottom to top): 18¾-in. wellhead connector, 13⅝-in. lower and upper shearing blind rams, and 13⅝-in. riser connector (facing upward). The shearing blind rams must be spaced out so that a drillpipe connector is able to fit between the two rams to ensure that at least one ram will cut the body of the pipe and not a connector. The SDS is equipped with a mechanical automatic closure system.

Regular well control is maintained by using the SBOP. It is not expected that any of the functions on the SDS will be operated regularly, although this capability is designed into the system. Because of this, there is no fixed hydraulic supply line from the surface to the SDS, which is controlled by a multiplex, electric-over-hydraulic, control system. There are two control panels for the SDS: the master control panel in the toolpusher's office and the driller panel in the driller's cabin. Signals from these panels are sent from the surface to the SDS in two ways: Acoustic signals are transmitted through the water from a surface-mounted transducer to two identical receivers and electronic control pods on the SDS, and electrical signals are transmitted through a wire attached to the riser. These signals actuate valves in the control package by sending hydraulic fluid to the appropriate function on the SDS.

For SBOP technology to be used safely and responsibly, a management system is necessary, which allows a consistent approach to the design and implementation of the SBOP system. It is a dynamic system built to learn from previous experience in order to extend this approach to more-severe scenarios, such as deep water. This approach reduces and controls risks by the quantitative-risk-analysis process. Identification and assessment of each critical system is achieved combining knowledge and experience by performing a hazard identification. A hazard operation (HAZOP) is also held before starting the operations. The purpose of the HAZOP is to identify the potential major accident hazards for a specific operation. For any new technology, a HAZOP is considered to be the main step of the process to manage the risk, gather experience and feedback, and train operational people. Operators that have already used an SBOP have considered this technology safer and more adequate than the traditional subsea system to drill in deep water for the following reasons: less risk of blowout, less downtime, larger offsets, lower-cost rig (third-generation), more-reliable well-control system, reduction of VIV and fatigue, better hydrate management, and less discharge to the environment during emergency disconnection.

There are already several documented case histories. SBOP technology was successfully proved in the Far East. Unocal has drilled approximately 140 wells in Kalimantan, the Andaman Sea, and the Sulu Sea; Chevron has drilled four wells in Bo Hai Bay, China; Shell Brunei has drilled three wells; and Shell Sarawak has drilled two wells. Nearly half of these wells were drilled in water deeper than 900 m.

In Brazil, this technology was applied by Shell and partners (Wintershall and Statoil) for the first time in a dynamically positioned drillship (Brander et al. 2004). The deepwater Well 1-SHEL-14RJS (Block BM-C-10 in the Campos basin) was drilled successfully by Shell in 2003. The well was drilled to beyond 5200 m total depth in 2887 m of water by the drillship *Stena Tay*. The well duration was 52.8 days from spud to release, with a total of 15 days of nonproductive time. Shell Brazil had already worked with *Stena Tay* for 2 years with a very good safety record.

References

10426-3, Petroleum and Natural Gas Industries—Cements and Materials for Well Cementing—Part 3: Testing of Deepwater Well Cement Formulations. 2003. Geneva, Switzerland: ISO.

Addis, M.A., Hanssen, T.H., Yassir, N., Willoughby, D.R., and Enever, J. 1998. A Comparison of Leak-Off Test and Extended Leak-Off Test Data for Stress Estimation. Paper SPE 47235 presented at SPE/ISRM Rock Mechanics in Petroleum Engineering, Trondheim, Norway, 8–10 July. DOI: 10.2118/47235-MS.

Alberty, M.W., Hafle, M.E., Mingle, J.C., and Byrd, T.M. 1999. Mechanisms of Shallow Waterflows and Drilling Practices for Intervention. *SPEDC* **14** (2): 123–129. SPE-56868-PA. DOI: 10.2118/56868-PA.

Altun, G., Langlinais, J., and Bourgoyne, A.T. 2001. Application of a New Model To Analyze Leak-Off Tests. *SPEDC* **16** (2): 108–116. SPE-72061-PA. DOI: 10.2118/72061-PA.

Azancot, P., Magne, E., and Zhang, J. 2002. Surface BOP—Management System and Design Guidelines. Paper SPE 74531 presented at the IADC/SPE Drilling Conference, Dallas, 26–28 February. DOI: 10.2118/74531-MS.

AzerMSA. 1999. The Development of the Oil and Gas Industry in Azerbaijan, http://members.tripod.com/azmsa/oil.html. Downloaded 13 October 2008.

Barker, J.W. and Gomez, R.K. 1989. Formation of Hydrates During Deepwater Drilling Operations. *JPT* **41** (3): 297–301; *Trans.*, AIME, **287**. SPE-16130-PA. DOI: 10.2118/16130-PA.

Berthezene, N., de Hemptinne, J.-C., Audibert, A., and Argillier, J.-F. 1999. Methane Solubility in Synthetic Oil-Based Drilling Muds. *J. Pet. Sci. Eng.* **23** (2): 71–81. DOI: 10.1016/S0920-4105(99)00008-X.

Biezen, E. and Ravi, K. 1999. Designing Effective Zonal Isolation for High-Pressure/High-Temperature and Low Temperature Wells. Paper SPE 57583 presented at the SPE/IADC Middle East Drilling Technology Conference, Abu Dhabi, UAE, 8–10 November. DOI: 10.2118/57583-MS.

Bittleston, S.H., Ferguson, J., and Frigaard, I.A. 2002. Mud Removal and Cement Placement During Primary Cementing of an Oil Well—Laminar Non-Newtonian Displacements in an Eccentric Annular Hele-Shaw Cell. *Journal of Engineering Mathematics* **43** (2–4): 229–253. DOI: 10.1023/A:1020370417367.

Boiler and Pressure Vessel Code, Section VIII, Division 2. 2007. New York City: ASME.

Bourgoyne, A.T., Chenevert, M.E., and Millhein, K.K. 1986. *Applied Drilling Engineering.* Textbook Series, SPE, Richardson, Texas.

Bowers, G. 2001. An Improved Methodology To Predict Predrill Pore Pressure in Deepwater Gulf of Mexico. Phase 1 Report, DEA 119 JIP, Knowledge Systems, Sugar Land, Texas, USA (June 2001).

Brander, G., Magne, E., Newman, T., Taklo, T., and Mitchell, C. 2004. Drilling in Brazil in 2887m Water Depth Using a Surface BOP System and a DP Vessel. Paper SPE 87113 presented at the IADC/SPE Drilling Conference, Dallas, 2–4 March. DOI: 10.2118/87113-MS.

Brantly, J.E. 1971. History of Oil Well Drilling, Chap. 31, 1358–1362. Houston: Gulf Publishing Co.

Bureau, N., de Hemptinne, J.-C., Audibert, A., and Herzhaft, B. 2002. Interactions Between Drilling Fluid and Reservoir Fluid. Paper SPE 77475 presented at the SPE Annual Technical Conference and Exhibition, San Antonio, Texas, USA, 29 September–2 October. DOI: 10.2118/77475-MS.

Calvert, D.G. and Griffin, T.J. 1998. Determination of Temperatures for Cementing in Wells Drilled in Deep Water. Paper SPE 39315 presented at the IADC/SPE Drilling Conference, Dallas, 3–6 March. DOI: 10.2118/39315-MS.

Chevron. 2007. About Chevron, http://www.chevron.com/about/.

Craft, B.C. and Hawkins, M.F. 1959. *Petroleum Reservoir Engineering*, 289. Upper Saddle River, New Jersey: Prentice Hall Press.

Davies, D.R. and Cobbett, J.S. 1981. Foamed Cement—A Cement With Many Applications. Paper SPE 9598 presented at the Middle East Technical Conference and Exhibition, Bahrain, 9–12 March. DOI: 10.2118/9598-MS.

Davison, J.M., Clary, S., Saasen, A., Allouche, M., Bodin, D., and Nguyen, V.-A. 1999. Rheology of Various Drilling Fluid Systems Under Deepwater Drilling Conditions and the Importance of Accurate Predictions of Downhole Fluid Hydraulics. Paper SPE 56632 presented at the SPE Annual Technical Conference and Exhibition, Houston, 3–6 October. DOI: 10.2118/56632-MS.

Dutra, E.S.S., Martins, A.L., Miranda, C.R. et al. 2005. Dynamics of Fluid Substitution While Drilling and Completing Long Horizontal-Section Wells. Paper SPE 94623 presented at the SPE Latin American and Caribbean Petroleum Engineering Conference, Rio de Janeiro, 20–23 June. DOI: 10.2118/94623-MS.

DV-III+ Rheometer Operator Manual. 1998. Middleboro, Massachusetts: Brookfield Engineering Laboratories.

Ebeltoft, H., Yousif, M., and Soergaard, E. 1997. Hydrate Control During Deep Water Drilling: Overview and New Drilling Fluids Formulations. Paper SPE 38567 presented at the SPE Annual Technical Conference and Exhibition, San Antonio, Texas, USA, 5–8 October. DOI: 10.2118/38567-MS.

Fjær, E., Holt, R.M., Horsrud, P., Raaen, A.M., and Risnes, R. 1992. *Petroleum Related Rock Mechanics.* Amsterdam: Elsevier B.V.

Gandelmann, R.A., Leal, R.A.F., Gonçalves, J.T., Aragão, A.F.L., Lomba, R.F., and Martins, A.L. 2007. Study on Gelation and Freezing Phenomena of Synthetic Drilling Fluid in Ultra Deep Water Environments. Paper SPE 105881 presented at the SPE/IADC Drilling Conference, Amsterdam, 20–22 February. DOI: 10.2118/105881-MS.

Hampshire, K., McFadyen, M., Ong, D., Mukoro, P., and Elmarsafawi, Y. 2004. Overcoming Deepwater Cementing Challenges in South China Sea, East Malaysia. Paper SPE 88012 presented at the IADC/SPE Asia Pacific Drilling Technology Conference and Exhibition, Kuala Lumpur, 13–15 September. DOI: 10.2118/88012-MS.

Harness, P.E., Sabins, F.L., and Griffith, J.E. 1992. New Technique Provides Better Low-Density-Cement Evaluation. Paper SPE 24050 presented at the SPE Western Regional Meeting, Bakersfield, California, 30 March–1 April. DOI: 10.2118/24050-MS.

Haut, R.C. and Crook, R.J. 1979. Primary Cementing: The Mud Displacement Process. Paper SPE 8253 presented at the SPE Annual Technical Conference and Exhibition, Las Vegas, Nevada, USA, 23–26 September. DOI: 10.2118/8253-MS.

Haut, R.C. and Crook, R.J. 1982. Laboratory Investigation of Lightweight, Low-Viscosity Cementing Spacer Fluids. *JPT* **34** (8): 1828–1834. SPE-10305-PA. DOI: 10.2118/10305-PA.

Jaeger, J.C. and Cook, N.G.W. 1979. *Fundamentals of Rock Mechanics,* third edition. London: Chapman and Hall.

Kolstad, E., Mozill, G., and Flores, J.C. 2004. Deepwater Isolation, Shallow-Water Flow Hazards Test Cement in Marco Polo—New Slurries Help Reduce the Known Risks. *Offshore* **64** (1): 76–80.

Kotkoskie, T.S., Al-Ubaidl, B., Wildeman, T.R., and Sloan, E.D. 1992. Inhibition of Gas Hydrates in Water-Based Drilling Muds. *SPEDE* **7** (2): 130–136. SPE-20437-PA. DOI: 10.2118/20437-PA.

Lage, A.C.V.M., Nakagawa, E.Y., and Rocha, L.A.S. 1997. Description and Application of New Criteria for Casing Setting Design. Paper OTC 8464 presented at the Offshore Technology Conference, Houston, 5–8 May.

Leach, C.P. and Wand, P.A. 1992. Use of a Kick Simulator as a Well Planning Tool. Paper SPE 24577 presented at the SPE Annual Technical Conference and Exhibition, Washington, DC, 4–7 October. DOI: 10.2118/24577-MS.

Lockyear, C.F. and Hibbert, A.P. 1989. Integrated Primary Cementing Study Defines Key Factors for Field Success. *JPT* **41** (12): 1320–1325; *Trans.,* AIME, **287**. SPE-18376-PA. DOI: 10.2118/18376-PA.

Marine Drilling Rigs 2003–2004. *World Oil* (December 2003): R3–R62.

Martins, A.L., Aragão, A.F.L., Calderon, A., Leal, R.A.F., Magalhães, J.V.M., and Silva, R.A. 2004. Hydraulic Limits for Drilling and Completing Long Horizontal Deepwater Wells. Paper SPE 86923 presented at the SPE International Thermal Operations and Heavy Oil Symposium and Western Regional Meeting, Bakersfield, California, USA, 16–18 March. DOI: 10.2118/86923-MS.

McLean, M.R. and Addis, M.A. 1990a. Wellbore Stability Analysis: A Review of Current Methods of Analysis and Their Field Application. Paper SPE 19941 presented at the SPE/IADC Drilling Conference, Houston, 27 February–2 March. DOI: 10.2118/19941-MS.

McLean, M.R. and Addis, M.A. 1990b. Wellbore Stability Analysis: The Effect of Strength Criteria on Mud Weight Recommendations. Paper SPE 20405 presented at the SPE Annual Technical Conference and Exhibition, New Orleans, 23–26 September. DOI: 10.2118/20405-MS.

Minerals Management Service (MMS). 2008. MMS Gulf of Mexico Region, http://www.gomr.mms.gov/index.html.

Monteiro, V.A.R., Brandão, E.M., Mello, E.C., and Martins, A.L. 2005. Drilling Fluid Rheology Optimization for Ultra Deep Water Wells in Campos Basin, Brazil. Paper No. 26 presented at the Offshore Mediterranean Conference, Ravenna, Italy, 16–18 March.

MR-01-75-98, Standard Material Requirements: Sulfide Stress Cracking Resistant Metallic Materials for Oilfield Equipment. 1998. Houston: NACE.

Nakagawa, E.Y. and Bourgoyne, A.T. Jr. 1992. Experimental Study of Gas Slip Velocity and Liquid Holdup in an Inclined Eccentric Annulus. In *Multiphase Flow in Wells and Pipelines: Presented at the Winter Annual Meeting of the American Society of Mechanical Engineers, Anaheim, California, November 8–13, 1992*, Vol. 144, ed. M.P. Sharma, 71–79. New York City: FED Series, ASME.

Nakagawa, E.Y. and Lage, A.C.V.M. 1994. Kick and Blowout Control Development for Deep Water Operations. Paper SPE 27497 presented at the SPE/IADC Drilling Conference, Dallas, 15–18 February. DOI: 10.2118/27497-MS.

Nelson, E.B. ed. 1990. *Well Cementing*. Oxford, UK: Developments in Petroleum Science Series, Elsevier.

O'Brien, T.B. 1981. Handling Gas in an Oil Mud Takes Special Precautions. *World Oil* (January): 22.

O'Bryan, P.L. 1983. The Experimental and Theoretical Study of Methane Solubility in an Oil-Base Drilling Fluid. MS thesis, Louisiana State University, Baton Rouge, Louisiana.

O'Bryan, P.L. and Bourgoyne, A.T. Jr. 1990. Swelling of Oil-Based Drilling Fluids Resulting From Dissolved Gas. *SPEDE* **5** (2): 149–155. SPE-16676-PA. DOI: 10.2118/16676-PA.

O'Bryan, P.L., Bourgoyne, A.T. Jr., Monger, T.G., and Kopcso, D.P. 1988. An Experimental Study of Gas Solubility in Oil-Based Drilling Fluids. *SPEDE* **3** (1): 33–42; *Trans.,* AIME, **285**. SPE-15414-PA. DOI: 10.2118/15414-PA.

Pilkington, P.E. 1975. Exploding the Myths About Kick Tolerance. *World Oil* (June): 59–62.

Piot, B., Ferri, A., Mananga, S.-P., Kalbare, C., and Viela, D. 2001. West Africa Deepwater Wells Benefit From Low-Temperature Cements. Paper SPE 67774 presented at the SPE/IADC Drilling Conference, Amsterdam, 27 February–1 March. DOI: 10.2118/67774-MS.

Rae, P. and Di Lullo, G. 2004. Lightweight Cement Formulations for Deep Water Cementing: Fact and Fiction. Paper SPE 91002 presented at the SPE Annual Technical Conference and Exhibition, Houston, 26–29 September. DOI: 10.2118/91002-MS.

Ravi, K., Biezen, E.N., Lightford, S.C., Hibbert, A., and Greaves, C. 1999. Deepwater Cementing Challenges. Paper SPE 56534 presented at the SPE Annual Technical Conference and Exhibition, Houston, 3–6 October. DOI: 10.2118/56534-MS.

Ravi, K., McMechan, D.E., Reddy, B.R., and Crook, R. 2007. A Comparative Study of Mechanical Properties of Density-Reduced Cement Compositions. *SPEDC* **22** (2): 119–126. SPE-90068-PA. DOI: 10.2118/90068-PA.

Reddy, B.R., Vargo, R., Sepulvado, B., and Weisinger, D. 2002. Value Created Through Versatile Additive Technology and Innovation for Zonal Isolation in Deepwater Environments. Paper SPE 77757 presented at the SPE Annual Technical Conference and Exhibition, San Antonio, Texas, USA, 29 September–2 October. DOI: 10.2118/77757-MS.

Redmann, K.P. Jr. 1991. Understanding Kick Tolerance and Its Significance in Drilling Planning and Execution. *SPEDE* **6** (4): 245–249. SPE-19991-PA. DOI: 10.2118/19991-PA.

Rocha, L.A. and Bourgoyne, A.T. 1996. A New Simple Method To Estimate Fracture Pressure Gradient. *SPEDC* **11** (3): 153–159. SPE-28710-PA. DOI: 10.2118/28710-PA.

Romero, J. and Loizzo, M. 2000. The Importance of Hydration Heat on Cement Strength Development for Deep Water Wells. Paper SPE 62894 presented at the SPE Annual Technical Conference and Exhibition, Dallas, 1–4 October. DOI: 10.2118/62894-MS.

Romero, J. and Touboul, E. 1998. Temperature Prediction for Deepwater Wells: A Field Validated Methodology. Paper SPE 49056 prepared for presentation at the SPE Annual Technical Conference and Exhibition, New Orleans, 27–30 September. DOI: 10.2118/49056-MS.

Rommetveit, R. 1994. Kick Simulator Improves Well Control Engineering and Planning. *Oil & Gas Journal* **92** (34): 64–71.

RP 10B, Recommended Practice for Testing Well Cements, 22nd edition. 1997. Washington, DC: API.

RP 16Q, Recommended Practice for Design, Selection, Operation and Maintenance of Marine Drilling Riser Systems. 2001. Washington, DC: API.

RP 2RD, Recommended Practice for Design of Risers for Floating Production Systems (FPSs) and Tension-Leg Platforms (TLPs), first edition. 2006. Washington, DC: API.

API RP 2SK, Recommended Practice for Design and Analysis of Stationkeeping Systems for Floating Structures, third edition. 2008. Washington, DC: API.

API RP 65, Cementing Shallow Water Flow Zones in Deep Water Wells. 2002. Washington, DC: API.

Saliés, J.B. et al. 1998. Slender Technology for Ultra-Deepwater in Campos Basin. Presented at the Deep Offshore Technology Conference (DOT) '98, New Orleans, 17–19 November.

Saliés, J.B., Nogueira, E.F., and Evandro, T.M.F. 1999. Evolution of Well Design in the Campos Basin Deep Water. Paper SPE 52785 presented at the SPE/IADC Drilling Conference, Amsterdam, 9–11 March. DOI: 10.2118/52785-MS.

Sanchez, R.A., Azar, J.J., Bassal, A.A., and Martins, A.L. 1999. Effect of Drillpipe Rotation on Hole Cleaning During Directional-Well Drilling. *SPEJ* **4** (2): 101–108. SPE-56406-PA. DOI: 10.2118/56406-PA.

Santana, M., Martins, A.L., and Sales, A. Jr. 1998. Advances in the Modeling of the Stratified Flow of Drilled Cuttings in High Horizontal Wells. Paper SPE 39890 presented at the International Petroleum Conference and Exhibition in Mexico, Villahermosa, Mexico, 3–5 March. DOI: 10.2118/39890-MS.

Sauer, C.W. 1987. Mud Displacement During Cementing: State of the Art. *JPT* **39** (9): 1091–1101. SPE-14197-PA. DOI: 10.2118/14197-PA.

Silva, C.T., Mariolani, J.R.L., Bonet, E.J., Lomba, R.F.T., Santos, O.L.A., and Ribeiro, P.R. 2004. Gas Solubility in Synthetic Fluids: A Well Control Issue. Paper SPE 91009 presented at the SPE Annual Technical Conference and Exhibition, Houston, 26–29 September. DOI: 10.2118/91009-MS.

Smith, D.K. 2003. *Cementing.* Monograph Series, SPE, Richardson, Texas **4**.

Sørgård, E. and Villar, J.P. 2001. Reducing the Environmental Impact by Replacing Chemistry With Physics. Paper SPE 66551 presented at the SPE/EPA/DOE Exploration and Production Environmental Conference, San Antonio, Texas, USA, 26–28 February. DOI: 10.2118/66551-MS.

Thomas, D.C., Lea, J.F. Jr., and Turek, E.A. 1984. Gas Solubility in Oil-Based Drilling Fluids: Effects on Kick Detection. *JPT* **36** (6): 959–968. SPE-11115-PA. DOI: 10.2118/11115-PA.

van Oort, E., Lee, J., Friedheim, J., and Toups, B. 2004. New Flat-Rheology Synthetic-Based Mud for Improved Deepwater Drilling. Paper SPE 90987 presented at the SPE Annual Technical Conference and Exhibition, Houston, 26–29 September. DOI: 10.2118/90987-MS.

Vicente, R., Sarica, C., and Ertekin, T. 2003. Horizontal Well Design Optimization: A Study of the Parameters Affecting Productivity and Flux Distribution of a Horizontal Well. Paper SPE 84194 presented at the SPE Annual Technical Conference and Exhibition, Denver, 5–8 October. DOI: 10.2118/84194-MS.

Ward, M., Granberry, V., Campos, M.R. et al. 2001. A Joint Industry Project To Assess Circulating Temperatures in Deepwater Wells. Paper SPE 71364 presented at the SPE Annual Technical Conference and Exhibition, New Orleans, 30 September–3 October. DOI: 10.2118/71364-MS.

Ward, M., Granberry, V., Campos, M.R. et al. 2003. A Joint Industry Project To Assess Circulating Temperatures in Deepwater Wells. *SPEDC* **18** (2): 133–137. SPE-83725-PA. DOI: 10.2118/83725-PA.

Watson, P., Kolstad, E., Borstmayer, R., Pope, T., and Reseigh, A. 2003. An Innovative Approach to Development Drilling in Deepwater Gulf of Mexico. Paper SPE 79809 presented at the SPE/IADC Drilling Conference, Amsterdam, 19–21 February. DOI: 10.2118/79809-MS.

White, J., Moore, S., and Miller, R.F. 2000. Foaming Cement as a Deterrent to Compaction Damage in Deepwater Production. Paper SPE 59136 presented at the IADC/SPE Drilling Conference, New Orleans, 23–25 February. DOI: 10.2118/59136-MS.

Wilkie, D.I. and Bernard, W.F. 1981. Detecting and Controlling Abnormal Pressure. *World Oil* **193** (1): 129–144.

Zhongming, C. and Novotny, R.J. 2003. Accurate Prediction of Wellbore Transient Temperature Profile Under Multiple Temperature Gradients: Finite Difference Approach and Case History. Paper SPE 84583 presented at the SPE Annual Technical Conference and Exhibition, Denver, 5–8 October. DOI: 10.2118/84583-MS.

SI Metric Conversion Factors

bbl	× 1.589 873	E – 01 = m^3
cp	× 1.0*	E – 03 = Pa·s
ft	× 3.048*	E – 01 = m

$$ft^2 \quad \times 9.290\ 304* \qquad E-02 = m^2$$
$$ft^3 \quad \times 2.831\ 685 \qquad E-02 = m^3$$
$$°F \qquad (°F-32)/1.8 \qquad\quad\ = °C$$
$$gal \quad \times 3.785\ 412 \qquad E-03 = m^3$$
$$in. \quad\ \times 2.54* \qquad\qquad E+00 = cm$$
$$ksi \quad \times 6.894\ 757 \qquad E+03 = kPa$$
$$lbf \quad \times 4.448\ 222 \qquad E+00 = N$$
$$lbm \ \times 4.535\ 924 \qquad E-01 = kg$$
$$mile \times 1.609\ 344* \qquad E+00 = km$$
$$psi \quad \times 6.894\ 757 \qquad E+00 = kPa$$

*Conversion factor is exact.

7.2 Deepwater Dual-Gradient Drilling—Brandee A. Elieff, Texas A&M University and Jerome J. Schubert, Texas A&M University

7.2.1 Dual-Gradient Drilling. To meet the world's increasing demand for energy, the search for oil and gas extends into increasingly hostile and challenging environments. Among these problematical environments are the deepwater regions of the world. As technology progresses, the definition of deep water encompasses greater and greater depths, and as the water depth increases, the associated technical, economic, and safety complexities increase proportionately. This has led to a high demand for new technologies throughout the oil field, but with a specific focus on improving drilling technologies. The industrywide goals are to increase accessibility to reserves, improve wellbore integrity, reduce overhead costs, and, most importantly, provide a safe working environment. Applying a dual-gradient technology to offshore drilling is not a new concept, but one that is being addressed with new fervor and can help meet all of these industry goals.

One of the many challenges faced when drilling deepwater offshore wells is the decreasing window between formation pore pressures and formation-fracture pressures. "In certain offshore areas with younger sedimentary deposits, the presence of a very narrow margin between formation pore pressure and fracture pressure creates tremendous drilling challenges with increasing water depths" (Rocha and Bourgoyne 1996). This occurrence is explained as being the result of the lower overburden pressures, due to the lower pressure gradient of seawater, than that which is exerted by typical sand/shale formations. The resulting situation is that the overburden and fracture pressures in an offshore well are significantly lower than those of an onshore well of a similar depth, and it is more difficult to maintain overpressured drilling techniques offshore without fracturing the formations (Johnson and Rowden 2001). Typically, the method for combating this problem has been to fortify the wellbore casing, by increasing the number of casing strings set in the well during drilling and completions operations. However, this can be extremely costly, both from a materials-cost perspective and a time-cost perspective. It has been proved that the number of casing strings set in a well can be reduced if the difference between the pore pressure and fracture pressure can be managed better. This has resulted in the development of new managed-pressure drilling (MPD) techniques. The International Association of Drilling Contractors (IADC) Underbalanced Operations Committee defines MPD as "An adaptive drilling process used to precisely control the annular pressure profile throughout the wellbore. The objectives are to ascertain the downhole pressure environment limits and to manage the annular hydraulic pressure profile accordingly. *The intention of MPD is to avoid continuous influx of formation fluids to the surface. Any influx incidental to the operation will be safely contained using an appropriate process*"(Drilling Contractor 2008). The objectives are to ascertain the downhole pressure environment limits and to manage the annular hydraulic pressure profile accordingly (MPD 2005; Grottheim 2005). One MPD technique that is being pursued for commercial use in deepwater environments is dual-gradient drilling.

7.2.2 Why Dual-Gradient Drilling? A dual-gradient system removes the mud-filled riser from the typical deepwater drilling system. In a conventional system, the annulus section of the riser is filled with mud, and below the seafloor the pressure within the annulus is so high that to avoid a pressure in the wellbore that exceeds the formation fracture pressure, it is necessary to set casing strings more frequently than is technically and economically desirable.

When using a dual-gradient drilling system, the riser is removed from the system (figuratively and/or literally depending upon the variation of the dual-gradient system). This allows the pressure at the seafloor to be lower (the seawater pressure gradient is lower than that of most drilling fluids) than in a conventional system, and this allows the driller to navigate more accurately in the pressure window between formation-fracture pressure and formation-pore pressure. As long as there is a safe margin (approximately 0.5 lbm/gal gradient) between the wellbore annular-pressure gradient and the fracture-pressure gradient, it is unnecessary to set casing strings as often as in the conventional system. An illustration of how the pressures are managed so that annular pressure remains above pore pressure at drilling depth but below fracture pressure at shallower depths in the well can be seen in **Fig. 7.2.1**.

Managing the pressure window between the formation-fracture and -pore pressures decreases the number of casing strings required to maintain wellbore integrity while drilling. A comparison between conventional deepwater-drilling casing requirements and dual-gradient deepwater-drilling casing requirements can be seen in **Figs. 7.2.2 and 7.2.3**.

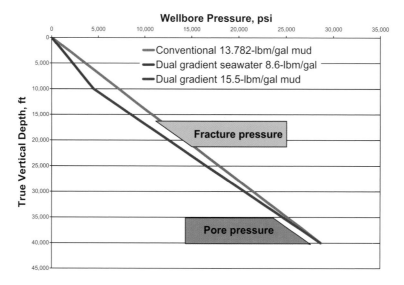

Fig. 7.2.1—Wellbore pressures in a dual-gradient system.

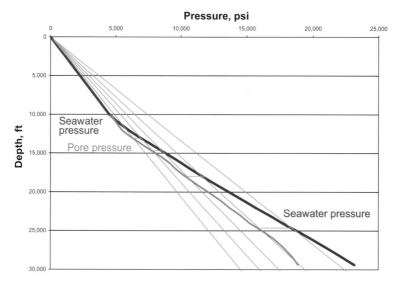

Fig. 7.2.2—Casing selection in a conventional system.

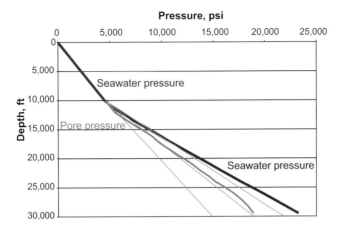

Fig. 7.2.3—Casing selection in a dual-gradient system.

When drilling conventionally in deepwater conditions, the riser is treated as part of the wellbore, and as the water depth increases the pressures within the wellbore change as though the depth of the well is increasing as well. However, when using the dual-gradient-drilling-system procedures, the depth of the water is no longer a factor affecting wellbore pressure. It is like "taking water out of the way."*

Many benefits are realized by employing dual-gradient drilling technology in a deepwater environment. A few of these benefits are

- Fewer required casing strings
- Larger production tubing (accommodates higher production rates)
- Improved well control and reduction of lost-circulation setbacks
- Lower costs, as the "water depth capabilities of smaller rigs may be extended" (Smith et al. 2001; Schumacher et al. 2001; Eggemeyer et al. 2001; Alford et al. 2005)

7.2.3 Dual-Gradient Drilling History and Evolution. The concept of dual-gradient drilling was first considered in the 1960s. At the time, the idea was to simply remove the riser, and therefore the technology was referred to as riserless drilling. The technology, however, was not pursued at the time because there was no driving economic or technical need for improving offshore drilling. As offshore drilling progressed into deeper water, the desire to improve project-development economics and technical characteristics resurrected the technology in the 1990s.

Beginning in 1996, four main projects began in an effort to improve deepwater drilling technology by implementing dual-gradient systems. The four projects were Shell Oil Company's Deep Vision project, Maurer Technology's Hollow Glass Spheres project, and the SubSea MudLift Drilling JIP (Schubert et al. 2003).

The most extensive study was the SubSea MudLift Drilling JIP that began in 1996 when a group of deepwater drilling contractors, operators, service companies, and a manufacturer gathered to discuss the merits of riserless or dual-gradient drilling. The result was an extensive system design, construction, and field test that would span 5 years. The main reason the group was interested in developing this technology was the promise it held to potentially reduce the necessary number of casing strings, specifically in the Gulf of Mexico, where high pore pressures and low formation strengths require operators to set casing strings often during drilling and completion operations (Smith et al. 2001; Schumacher et al. 2001; Eggemeyer et al. 2001).

The SubSea MudLift Drilling JIP was charged with the tasks of designing the hardware and the necessary procedures to effectively and safely operate the dual-gradient drilling system. Phase I of the project took place from September 1996 to April 1998 and cost approximately USD 1.05 million. Phase I was the Conceptual Engineering Phase, and the participants were to create a dual-gradient drilling design that was

---

*Personal communication of information from the SubSea MudLift Drilling Joint Industry Project (JIP) Phase III: Final Report, 2001, Houston.

feasible, considered well-control requirements, and was adaptable to a large rig fleet (not just a few specialized rigs). Phase I is considered to have been very successful and resulted in a design for drilling extended-reach, 12¼-in. holes at total depth (TD), in 10,000 ft of water. One of the most challenging design issues was how to lift the mud after it had been circulated through the wellbore.

Once circulated through the wellbore, the mud or drilling fluid is loaded with free gases, metal shavings, rock chips, and other drilling debris. What kind of pump is capable of pumping the mud from the seafloor back to the rig floor? The JIP answered this question in Phase I with the response of a positive-displacement diaphragm pump. However, no such pump existed that met the JIP's needs, so it was concluded that the JIP would have to design and build one. Other conclusions of Phase I were that this technology is more than feasible, but well-control procedures would need to be modified, and that a field test is necessary, specifically in the Gulf of Mexico where the driving need for this technology is based.

Phase II, or Component Design, Testing, Procedure, and Development, began in January 1998 and continued until April 2000 and cost approximately USD 12.65 million. The purpose of Phase II was to actually design, build, and test the subsea pumping system, create all the drilling operations and well-control procedures, and determine the best methods for incorporating the dual-gradient drilling technology onto existing drilling rigs. Phase II resulted in a proven, reliable seawater-driven diaphragm pumping system, drilling and well-control procedures capable of withstanding potential-equipment-failure cases, and an understanding that a system training program was necessary.

Phase III, or System Design, Fabrication, and Testing, began in January 2000 and was completed in November 2001 with a budget of USD 31.2 million. The purpose of Phase III was to validate the design of the technology through an actual field application. This goal was accomplished, and the first dual-gradient test well was spudded on 24 August 2001, and by 27 August 2001, the 20-in. casing had been run and cemented. On 29 August, the SubSea MudLift Drilling JIP system was finally put to test in the field. Although there were many problems initially (especially with the electrical system), "Once a problem was identified and repaired, it stayed repaired."**

Ultimately, 90% of the field-test objectives were met and considered successful. Although still requiring industry support, dual-gradient drilling was proved a viable and useful technology.

Another JIP began in 2000 and culminated with a successful test application in 2004. This was the development of AGR Ability Group's Riserless Mud Recovery (RMR) System. The system was designed and tested specifically for the application of drilling the tophole portion of a wellbore. The desired results were to increase control over shallow water and gas flows and to increase the depth of the surface-casing strings by reducing the number of typically selected casing seats. The RMR system was rated to a depth of 450 m of seawater, but was tested in only 330 m of seawater. The successful field test took place in December 2004 in the North Sea (Stave et al. 2005). The conclusions of this JIP were that using dual-gradient technology for tophole drilling results in

- Improved hole stability and reduced washouts
- Improved control over shallow gas and water flows
- Improved gas detection (because of accurate flow checks and improved mud-volume control)
- Preventing the accumulation of mud and cuttings on subsea templates and preventing the dispersion of drilling fluids into environmentally sensitive areas
- Reduced number of necessary surface-casing strings

The most current research being conducted in the dual-gradient drilling area is a project through the Offshore Technology Research Center (OTRC), a division of the National Science Foundation (NSF) that is a joint partnership between Texas A&M University and the University of Texas at Austin. The project the OTRC is pursuing, which is initially funded by the US Minerals Management Service (MMS), is called Application of Dual-Gradient Technology to Tophole Drilling. The purpose of the project is the design and testing of a dual-gradient drilling system geared specifically to drilling the tophole portion of the wellbore in a deepwater environment. Although this has already been achieved in shallow water,

---

** Personal communication of information from the SubSea MudLift Drilling JIP Phase III: Final Report, 2001, Houston.

this OTRC project is to focus on the application of a dual-gradient tophole drilling system (DGTHDS) in deep water. The driving factors for this project are the increasingly hazardous shallow hazards commonly found in deepwater environments, especially in the Gulf of Mexico. These shallow hazards are overpressured shallow gas zones, shallow water flows, and methane hydrates, which are jeopardizing drilling activities in deep water. It is hypothesized that a DGTHDS can control these shallow hazards while drilling in deep water. The project will explore increasing control over these hazards in two ways: One is in the increased well control available from a DGTHDS, and the second is to improve the wellbore integrity by setting surface casing deeper than in conventional drilling applications. Once the shallow hazards are controlled and the conductor and surface casing are set deeper, this will also allow for safer drilling of the intermediate-depth portions of the well and will ultimately reduce the number of casing strings used throughout the well.

7.2.4 Achieving the Dual-Gradient Condition. There are different methods used to achieve the dual-gradient condition when drilling offshore. Basically, a dual gradient is achieved when there are two different pressure gradients in the annulus, the volume between the wellbore inside diameter (ID) and the drillstring outside diameter (OD). The condition can be achieved by reducing the density of the drilling fluid in a portion of the wellbore or riser, removing the riser completely and allowing seawater to be the second gradient, or managing the level of the mud within the riser and allowing the second gradient within the riser to be that of another fluid (Herrmann and Shaughnessy 2001).

One method, nitrogen injection, is based on air-drilling procedures and underbalanced-drilling techniques. This technique uses nitrogen to reduce the weight of the mud in the riser (Schumacher et al. 2001). In an effort to reduce the amount of nitrogen required to lower the mud pressure gradient in the riser, a concentric riser system is considered the most economical. In this system, a casing string is placed inside the riser with a rotating blowout preventer (BOP) at the top of the riser (in the moonpool) to control the returning flow. The mud is held in the annulus between the casing string and the riser, and nitrogen is injected at the bottom of the riser into the annulus. Buoyancy causes the nitrogen to flow up the annulus, which reduces the density and pressure gradient of the drilling fluid as a result of nitrogen's liquid-holdup properties. The injection of nitrogen can reduce the weight of a 16.2-lbm/gal mud to 6.9 lbm/gal. This can be applied when the second gradient is desired to be even lower than that of seawater, which has a typical pressure gradient of 8.55 lbm/gal. The most noteworthy characteristic of this method of using nitrogen injection to create two gradients is that the formation is not underbalanced, as one might initially conclude. The cased hole is underbalanced to a depth, but below the casing, in the open hole, the wellbore is actually overbalanced, which prevents an influx of fluids from the formation into the wellbore. One serious concern with this method of creating a dual-density system is the uncertainty as to whether well control and kick recognition will be more difficult. In this case, the system is very dynamic, and well control and kick detection are definitely more complex, but not necessarily unsafe (Schubert 1999).

Another method of creating a dual-gradient system is to begin by drilling the upper portions of the well without a riser and simply returning the drilling mud to the seafloor. In this setup, the pressure inside the wellbore at the seafloor is the same as the pressure at the seafloor. In other words, the pressure gradient from the ocean surface to the seafloor is that of the seawater pressure gradient. Then, inside the wellbore a heavier-than-typical mud is used to maintain proper pressures while drilling. Once the initial spudding has taken place and the structural pipe has been set, the subsea BOP stack is installed with some variation on a typical system. The mud returns are moved from the wellhead by a rotating diverter to a subsea pump, which returns the mud to the rig floor through a 6-in.-ID return line. Drilling continues with this setup, and the remaining casing strings are set using this dual-gradient system where mud returns to the rig through a separate line (Schumacher et al. 2001). An illustration of this system can be seen in **Fig. 7.2.4**.

Initially, this method was regarded with skepticism because of the perceived difficulty of kick detection. However, with more-advanced technology, and the ability to monitor pressure in the subsea BOP accurately, kick detection and the detection of circulation loss is reliable and safe. In fact, it is possible for the riser to act as a trip tank in this system (Schubert 1999).

Another method of creating a dual-gradient system is similar to nitrogen injection. A US Department of Energy (DOE) project was conducted to test how the injection of hollow spheres into the mud returning through the riser can create a dual-gradient system. This system is similar to the nitrogen-injection method, but separating the gas from the mud at the rig floor is simplified because dissolved gas in the drilling fluid is

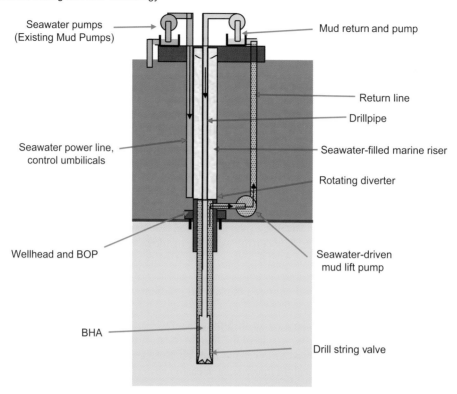

Fig. 7.2.4—Dual-gradient system.

not a concern. The glass spheres are separated from the mud and reinjected at the base of the riser. **Fig. 7.2.5** illustrates a typical hollow-glass-sphere-injection system.

7.2.5 Typical Dual-Gradient System and Components. The most commonly researched and pursued method of achieving a dual-gradient system is the riserless system, described in Section 7.2.4 and shown in Fig. 7.2.4. This system pumps the drilling mud through the drillstring, out the drill-bit nozzles, into the open hole, up the annulus, into the BOP stack, through the rotating head, into the subsea mud pump, and up the 6-in. return line to the rig floor. The mud is then cleaned at the rig floor and recycled back to the drillstring to be circulated again. The main components in this system that are unique to the dual-gradient system are the drillstring valve (DSV), the rotating head, the subsea mud pump, and the mud-return line.

Once the drilling mud flows up the annulus to the BOP it must be diverted so that it can be pumped up the return line. In the SubSea MudLift Drilling JIP this was accomplished through a rotating head referred to as the subsea rotating diverter (SRD). This SRD is capable of handling 6⅝-, 5½-, and 5-in. drillpipe and has a retrievable rotating seal rated to 500 psi, although, typically, the pressure difference across this seal is less than 50 psi. Once the mud is diverted to the subsea mud pump, the main concern is the handling of solids. This was addressed through the addition of a subsea rock-crusher assembly. Basically, as the returning mud passes through this assembly, any rock chips are crushed between two rotating cylinders with teeth. A photo of this rock-crusher assembly can be seen in **Fig. 7.2.6**.

Once the cuttings are crushed and processed through the unit, they have been reduced to small pieces. The crushed cuttings and mud are then passed through into the subsea mud pump. The requirements that the pump is subject to are very demanding. The pump must be able to pump up to 5 vol% mud cuttings, produce a flow rate between 10 and 1,800 gal/min, operate to a maximum pressure of 6,600 psi and within a temperature range between 28 and 180°F, and, finally, be able to pump 100% gas when the need arises to circulate a gas kick out of the well.

As mentioned in Section 7.2.3, the necessary result is a positive-displacement diaphragm pump that is hydraulically powered by seawater. The seawater providing hydraulic power is pumped from the rig floor,

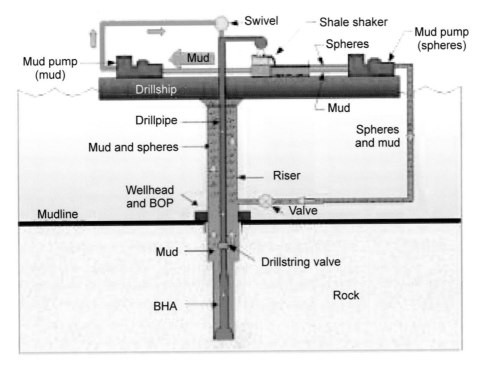

Fig. 7.2.5—Hollow-sphere-injection dual-gradient system (NETL 2003).

Fig. 7.2.6—Subsea rock-crusher assembly used in SubSea MudLift Drilling JIP. Courtesy of GE Oil & Gas Drilling and Production Systems. All rights reserved.

using conventional surface mud pumps, down an auxiliary line to the mud pump. **Fig. 7.2.7** is a cross section of the mechanisms at work within this diaphragm pump.

This pump also acts as a check valve by preventing the hydrostatic pressure of the drilling fluid within the return line from impacting on the pressure within the wellbore. This pump is normally run in an automatic mode, which means it is set to run at a constant inlet pressure, and the pump rate is automatically altered to maintain a constant inlet pump pressure. This allows the driller to change the surface mud-pumping rates as if the system were conventional (Kennedy 2001). The pump in normal drilling operations will operate at a setting of a constant inlet pressure. Should a well-control incident occur and a kick need to be circulated in order to maintain pressure on the wellbore, the pump will be switched to a setting of constant pump rate that equals the surface pump's pump rate.

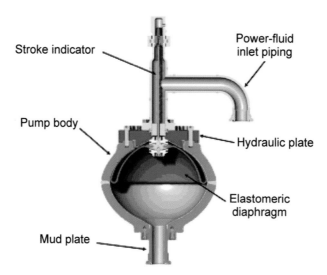

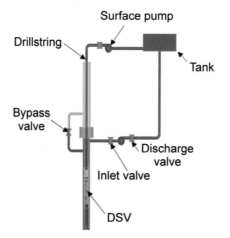

The last main component of the riserless dual-gradient drilling system is the DSV. The DSV was developed to control the U-tube effect, which is often encountered in drilling and completion operations. The U-tube effect is caused when the total hydrostatic pressure (HSP) of the fluid in the drillstring differs from the total HSP of the fluid in the annulus. In response, the fluid will flow through the drill-bit nozzles from the region (drillstring or annulus) with the higher HSP to the region with the lower HSP. In conventional operations the U-tube effect only occurs occasionally, and most commonly during cementing. However, in riserless dual-gradient drilling, the U-tube effect is always a factor because the HSP of the fluid in the drillstring is often greater than the sum of HSP of the fluid in the wellbore annulus and the HSP at the seafloor. The concern is that, when mud circulation is stopped to make or break a drillpipe connection, the mud within the drillstring will drain into the wellbore and up the annulus. The DSV assembly is placed in line with the drillstring, and when mud circulation is stopped, the DSV is closed to prevent the free fall of drilling fluid within the drillstring.* An illustration of the system with the DSV assembly in place can be seen in **Fig. 7.2.8**.

---

*Personal communication of information from the SubSea MudLift Drilling JIP Phase III: Final Report, 2001, Houston.

7.2.6 Dual-Gradient vs. Conventional Operations. There are several aspects of dual-gradient drilling that are different from those of conventional drilling operations. Regarding general drilling operations, a smaller rig may be used for applying dual-gradient technology than that which would be used conventionally. One reason is that in order to support a 21-in. riser (a common size used in conventional drilling), the rig must be large enough to support the weight of the riser. In a riserless dual-gradient drilling system, the weight hanging from the rig is reduced to that of the drillstring, the mud-return line, and the umbilical control lines. Also contributing to the large rig size necessary for conventional drilling are the deck-space limitations that are caused by the necessity of having large drilling-fluid volumes on hand. In a conventional drilling system, a large volume of mud is necessary in order to fill the riser. Another problem is that a high volume of mud is lost during the *pump-and-dump* method for drilling the tophole portion of the wellbore. In a DGTHDS, only the drillstring must be filled with mud, and the mud is returned to the rig floor where it is cleaned and recycled. This reduces the necessary deck space and the costs associated with supplying the necessary mud. Reducing the weight rating of the rig and the necessary deck space allows for the use of a smaller rig.

Another difference between a conventional drilling system and a dual-gradient drilling system is that removing the riser leaves only the drillstring to be affected by the forces exerted by the ocean currents. Since the diameter of the drillstring is considerably smaller than that of a 21-in. riser, the impact these forces have on drilling operations is reduced.

Perhaps the most time- and cost-saving benefit that results from the application of dual-gradient drilling over conventional drilling is how the necessary number of casing strings is reduced. This accomplishes two things: It allows for the final tubing size to be larger, which increases production flow rates, and it reduces the amount of time necessary to drill a deepwater well because less time is spent on completions.

From a safety perspective, the main differences between dual-gradient drilling and a conventional drilling system are the well-control procedures. Basically, a dual-gradient system as a managed-pressure drilling technique improves well control. A modified driller's method used in riserless dual-gradient drilling is described in Section 7.2.7.

The similarity between the two systems is that the drilling program is not altered significantly. Trips and connections are handled in the same manner, and the basic acts of drilling, such as bit selection and general rig procedures, are not altered (Schubert et al. 2003).

7.2.7 Dual-Gradient-Systems Well-Control Procedures. Well control is not simply something that must be implemented in the eventuality of a kick. Proper well control must be considered throughout all phases of drilling operations. This means from the initial planning, through the well completion, and into the abandonment stages. The basic purpose of proper well control is to prevent blowouts and create a quality wellbore. This is accomplished best through proper prediction of formation-pore and -fracture pressures, the design and use of the proper equipment (e.g., BOP, kick-detection devices, and casing), and proper kick-detection and kill procedures (Schubert et al. 2003; Hannegan and Wanzer 2003).

Taking a kick while drilling is common and must be prepared for. Quick kick detection and proper well-control response are imperative. Kicks may be detected through several different observations, and the driller must be aware of all inconsistencies experienced while drilling. The most common kick detection methods are: drilling breaks; a flow increase; mud-pit gains; a decrease in circulating pressure that is accompanied by an increase in pump speed within the surface pumps; if the well flows when the surface pumps are off, an increase in rotary torque and drag; and/or an increase in drillstring weight.

These kick-detection techniques are just as applicable, if not more so, in dual-gradient drilling as in conventional drilling. The major difference between dual-gradient drilling and conventional drilling is the U-tube effect. The U-tube effect occurs when drilling-mud circulation through the drillstring, up the annulus, and through the subsea mud pump is stopped. The U-tube effect causes the system to try to equalize the pressure difference between the HSP within the drillstring and the HSP in the annulus by draining the drilling fluid contained within the drillstring, through the drill-bit nozzles, into the annulus. Again, this occurs any time the HSP of the fluid in the drillstring differs from the HSP of the fluid in the annulus. The solution to the U-tube effect is simply a DSV, which is described in Section 7.2.5. There is, however, a benefit to the U-tube effect that occurs in dual-gradient drilling. This effect allows for lower circulating pressures by the rig pumps and makes small changes in pressures easier to detect. These pressure changes often serve as excellent kick detectors.

Another method of kick detection involves the inlet and outlet pressure of the subsea mud pump. When a kick enters the wellbore, the annular flow rate of the drilling fluid increases by an amount that is equal to that of the kick-influx rate. Generally, while drilling, the subsea mud pumps are set to operate in a constant-inlet-pressure mode. This means that if the rate of flow increases because of a kick influx, the pumping rate of the subsea mud pumps will increase automatically also, to maintain a constant subsea pump inlet pressure. This is an excellent indicator to the driller that a kick is occurring, and the driller can then take the measures necessary to stop the kick influx into the annulus.

Approximately half of all kicks occur while tripping the drillpipe into or out of the hole. The best method, which is also the earliest, of determining that a kick has taken place is to measure the volume of mud required to fill the hole after removing some of the pipe. This is usually done every five stands of drillpipe. If the mud required to fill the hole is less than the volume of the drillpipe removed, a kick has entered the wellbore. This is a kick detection employed by conventional drilling practices. In dual-gradient drilling, this kick-detection procedure must be considered for use both with a DSV and without a DSV. When operating without a DSV, an accurate determination of the amount of mud necessary to fill the wellbore is not possible until after the U-tube effect has ceased. When operating with a DSV, the volume of mud necessary to fill the hole is equal to the volume of a cylinder with a diameter equal to the OD of the pipe removed. The only major change from conventional operations is that more-frequent hole-fill intervals are necessary, and, if possible, continuous fill of the hole is even more desirable.

As soon as a kick is detected, one must take the necessary actions to stop the influx so that excessive casing pressures can be avoided. Excessive casing pressures can result in lost circulation, formation fracturing, and the worst-case scenario of a surface blowout. When a kick is initially detected, usually the response is to shut in the well by closing the BOP stack. When shutting in a dual-gradient drilling system, immediate shut-in should not be performed unless a DSV is in place. The DSV must be closed before shut-in to ensure that the HSP of the mud within the drillstring does not cause formation fracturing. If there is no DSV in place, it is necessary to allow the U-tube effect to take place and then to shut in the well by closing the BOP. When the U-tube effect is taking place it is difficult to prevent any additional influx from entering the wellbore. This is why it is recommended to deploy a DSV in all dual-gradient-drilling operations. A DSV allows immediate shut-in of the well, and killing procedures can then commence in a manner more similar to that of conventional drilling. However, the following procedures should be adhered to when the driller is not employing a complete shut-in scenario (i.e., no DSV) (Schubert et al. 2003; Schubert et al. 2002; Choe and Juvkam-Wold 1998; Forrest et al. 2001). This is known as a modified driller's method and is considered the most effective and most common in a dual-gradient system.

1. Slow the subsea pumps to the prekick rate (maintain the rig pumps at constant drilling rate).
2. Allow the drillpipe pressure to stabilize, and record this pressure and the circulating rate.
3. Continue circulating at the drillpipe pressure and rate recorded in Step 2 until kick fluids are circulated from the wellbore.
4. The constant drillpipe pressure is maintained by adjusting the subsea pump inlet pressure in a manner similar to adjusting the casing pressure with the adjustable choke on a conventional kill procedure.
5. After the kick fluids are circulated from the wellbore, a kill fluid of higher density is circulated in order toincrease the HSP imposed on the bottomhole.

Other methods such as the wait-and-weight method and the volumetric method are applicable to a riserless dual-gradient system. However, both of these methods require the use of a DSV. Although the DSV is applicable with the Modified Driller's Method, it is unnecessary, and it is always advisable to ensure that proper well control relies on as few pieces of equipment as possible.

7.2.8 Dual-Gradient Drilling Challenges. The main challenges that are associated with dual-gradient drilling are basically those that are associated with all new technologies. The technology has been designed, developed, and field tested successfully. The key now is to streamline the equipment and procedures to ensure that dual-gradient technology is seamlessly the next step forward in deepwater drilling.

In the field test of the SubSea MudLift Drilling JIP, the main delay while drilling the test hole was equipment-commissioning problems. The technology functioned successfully the way it was designed but had electrical and commissioning delays. Once these *kinks* were worked out of the system, the test hole was drilled with minimal delays.*

In order for the industry to embrace a new technology such as dual-gradient drilling, the kinks must be all worked out and the new technology must offer substantial benefits over conventional technologies.

An interesting point is that a dual-gradient system will need to be somewhat customized depending on water depth, temperatures above and below the mud line, formation pressures, ocean conditions, and a number of other conditions. However, even in conventional technology, no two wells are ever drilled with the exact same equipment or procedures. The difference is that personnel are familiar with how to alter conventional technology to fit with the current drilling environment. In order for personnel to become as familiar with dual-gradient technology as they are with conventional technology, training is a necessity.**

Eventually, dual-gradient technology will become a conventional technology and be one of the many tools in a driller's *toolbox*. The remaining obstacles are equipment commissioning, personnel training, and overcoming initial industry resistance.

7.2.9 Applying Dual Gradient to Tophole Drilling. Shallow hazards encountered while drilling the tophole portion of the wellbore are a problem, and controlling these shallow hazards has become a priority for exploration and production (E&P) companies operating in deepwater environments. The category of shallow hazards includes three main subcategories: methane hydrates, shallow gas zones, and shallow water flows. These hazards can be found in deepwater environments and generally between the mudline and approximately 5,000 ft below the mudline. Each of these hazards creates a different problem for E&P companies, which are pursuing oil and gas fields in deep water. Shallow hazards may appear to cause problems only during drilling and completion operations, but in reality they can have long-term ramifications that affect production long into the life of the field. Shallow hazards compromise the safety of operations, well control, wellbore integrity, and reservoir accessibility.

That is why it is surprising to find that the conventional method of drilling the tophole portion of the wellbore, *pump-and-dump,* is still used as the industry standard. Pump-and-dump is lacking in many ways, and dual-gradient technology can easily control shallow hazards with acceptable modifications to current drilling and completions equipment, drilling procedures, and well-control procedures.

Conventional Technology: Pump-and-Dump. The current pump-and-dump method used to drill the tophole portion of the wellbore in deep water is fairly basic. The mud is pumped down the drillstring, into the wellbore, up the annulus, and onto the seafloor. There is no BOP stack in place and there is no drilling-fluid return to the rig floor. The pump-and-dump method can cause several problems. These problems include, but are not limited to, limited well control, increased number of shallow casing strings, poor wellbore integrity, increased initial hole size (requiring larger rigs), loss of mud, and finally a negative environmental impact, which limits acceptable types of drilling fluids that meet regulations.

The pump-and-dump method offers few methods of kick detection and limited well-control methods when a kick does occur. Because the mud is not returned to the rig floor, there is limited downhole-pressure information available to the driller, and often the driller relies on visual kick-detection methods to determine when an influx has entered the wellbore. In an effort to avoid shallow hazards like hydrates and shallow gas zones, seismic data are carefully analyzed and the surface location of the rig may be moved to avoid these zones. This can result in the need for complicated directional wells that increase time, cost and risk required to drill. In the eventuality that these zones cannot be avoided, drillers have no proactive well-control methods in their toolbox. In the case of shallow water flows, these zones are generally allowed to produce until the formation pressure is reduced. Unfortunately, by the time this happens, erosion of the formation has often already occurred.

Dealing with these shallow hazards can increase the number of shallow casing strings, when compared to drilling in normally pressured zones. To ensure that the drilling fluid can be heavy enough to maintain overbalanced drilling, even when drilling through overpressured shallow gas zones, casing must be set often to prevent shallower parts of the wellbore from fracturing and causing lost circulation. Lost circulation can result

*Personal communication of information from the SubSea MudLift Drilling JIP Phase III: Final Report, 2001, Houston.
**Personal communication of information from the SubSea MudLift Drilling JIP Phase III: Final Report, 2001, Houston.

in stuck pipe or, worse, an underground blowout. Poor wellbore quality is also often the result of pump-and-dump. The pump-and-dump method limits the use of specialty drilling fluids that lift cuttings out of the hole at lower circulation rates. This means that in order to lift the cuttings with a less-specialized mud, the circulation rate is increased. This increased drilling-fluid-circulation rate can cause wellbore erosion, and the wellbore often becomes jaggedly shaped, which makes a high-quality cement job difficult to implement.

Aside from the technical, safety, and economical disadvantages to the pump-and-dump method, there is the obvious environmental impact, not to mention how the continuous loss of drilling fluid can become a high-cost constraint to the development of a field. The environmental restrictions placed on the types of acceptable drilling fluids can stop the driller from using the optimal fluid for the formation type and also prevent the addition of chemicals that minimize problems such as the formation of hydrates within equipment. The pump-and-dump method is not really a method at all. It is simply the standard rut that the industry has fallen into. It is obvious that a new method of tophole drilling is imperative. Applying dual-gradient-drilling technology to the tophole portion of the wellbore is likely to eliminate the majority, if not all, of these associated problems (Judge and Thethi 2003). Possibly the most important reasons that dual-gradient technology would be beneficial in tophole drilling are the control over shallow hazards, the improved well control, and the improved safety.

Comparing Dual-Gradient-Drilling Technology to Pump-and-Dump. The flow of the drilling fluid in a DGTHDS does not vary greatly from that in conventional riser drilling. It is, however, different from the pump-and-dump method. There are inherent benefits to this system over pump-and-dump, simply because the DGTHDS is a closed system. The amount of required mud is reduced because the drilling fluid is recycled and reused. Seafloor pollution is reduced, and because there is no environmental impact, the number of drilling-fluid types that meet regulations increases. It has been proved that selecting the proper drilling fluid can improve drilling operations significantly. Also important is how the closed system allows for the admission of backpressure to increase the wellbore annulus pressure. This allows the driller to maintain the proper wellbore annulus pressure with heavier mud at lower circulation rates. This prevents the wellbore erosion that is commonly associated with the pump-and-dump method. This additional pressure control also improves kick detection, offers proactive well-control methods, and ultimately reduces the number of required shallow casing strings.

Kick Detection. The DGTHDS offers more-accurate and faster kick-detection methods in addition to those that are already used during the pump-and-dump method. As discussed earlier in Section 7.2.5, in standard drilling mode the subsea pump is operated at a constant inlet pressure. When a kick enters the wellbore, the pump inlet pressure increases. In order to maintain a constant inlet pressure, the subsea pump responds by increasing its pumping rate to compensate for the additional inlet pressure created by the influx. This increase in pump rate is the first kick indicator. As the subsea pump increases its pumping rate, its outlet pressure increases and the levels in the mud pit increase. These are the second and third kick indicators, respectively. Finally, in response to the pressure changes within the wellbore, the surface pump pressure decreases—the fourth kick indicator. When a kick is detected, the system uses a modified driller's method to prevent further influx and circulate the kick safely out of hole.

Well Control Modified Driller's Method. As soon as the system detects a kick, the subsea pump is returned to the prekick rate and a constant-pumping-rate mode is maintained, which is equal to the surface pumping rate. This creates backpressure on the fluids within the wellbore annulus and increases bottomhole pressure until it is balanced with formation-pore pressure, and further influx is prevented. It is important to record the stabilized drillpipe pressure and the pumping rate. Circulation of the fluids is then continued, and the recorded drillpipe pressure is maintained at balance by changing the subsea pump rate. (This is similar to an adjustable choke in a conventional kill procedure.) Circulation is continued until kick fluids are removed from the wellbore. Once the kick fluids have been removed from the wellbore, a kill-weight mud is circulated to increase the HSP imposed on the bottomhole, and drilling can resume. The subsea pump rate increases, to maintain a constant inlet pressure, as the influx enters the wellbore. At the same time, the surface pump outlet pressure decreases. Once the kick is detected and well-control procedures commence, you can see the rate of the subsea pump return to the prekick rate, which is equal to that of the surface pump. It can also be seen how this causes the subsea pump inlet pressure and surface pump outlet pressure to increase.

DGTHDS Control of Methane Hydrates. As described earlier, methane hydrates affect drilling operations by forming within the equipment and by dissociating within the wellbore annulus. Dual-gradient technology applied to tophole drilling controls both of these problems caused by methane hydrates.

The introduction of a closed system allows for chemicals such as hydrate inhibitors to be added to the drilling fluid. These hydrate inhibitors have been proved to be very successful at preventing the formation of hydrates in drilling and production equipment.

In the case of drilling through dissociating hydrates, a significant well-control problem, dual-gradient technology offers the advantage of fast kick detection. When methane hydrates dissociate into the wellbore, the dual-gradient-drilling system reacts the same way as if a gas influx has entered the wellbore. The subsea pump inlet pressure will increase and the subsea pump rate will increase automatically to compensate. Then the pit-gain warning, the increased subsea pump outlet pressure, and the decreased surface pump outlet pressure will alert the driller to employ well-control methods. The subsea mud-return system supplies the driller with backpressure control over the formation that prevents the dissociating methane hydrates from causing other influxes. The dissociating methane hydrates can be circulated from the wellbore proactively and safely, and drilling can resume quickly.

DGTHDS Control of Shallow Gas Flows. A DGTHDS controls shallow gas flows the same way it controls dissociating methane hydrates: through effective kick detection and proactive well-control methods. Again, the gas influx into the wellbore is detected quickly, and the modified driller's method quickly circulates the kick from the wellbore and prevents further influx. The drilling-fluid weight is adjusted for the new formation-pore pressure, and drilling continues without the need to set dynamically selected casing seats.

DGTHDS Control of Shallow Water Flows. Shallow water flows are easier to control than methane-hydrate dissolution or gas kicks. Controlling these shallow water flows will allow the driller to prevent the erosion of the formation and ultimately will ensure that the operator will have a wellbore of high quality, because the casing seats are cemented securely to the formation (Roller 2003).

DGTHDS Control of Shallow Hazards. This is a new technology that is still in the research-and-development stage, but it has all the signs of benefiting the offshore drilling industry significantly and of being adopted as a conventional technology. The technical and safety benefits associated with this new technology far outweigh the inherent industry resistance to the implementation of a new technology. The benefits that the industry stands to gain from the implementation of dual-gradient technology range from financial to safety to environmental (Stave et al. 2005).

7.2.10 The Future of Dual-Gradient Drilling. Dual-gradient-drilling technology is not beyond our reach. This technology has been designed, engineered, and field tested for feasibility. This technology has been applied to the tophole portion (before surface casing is set) of a wellbore in a shallow-water environment and in a deepwater environment after surface casing have been set. In 2005, the OTRC launched a research project called Application of Dual-Gradient Technology to Tophole Drilling. The main objective of this project is to prove that dual-gradient technology will properly control shallow hazards (shallow gas and water flows and methane hydrates) that are encountered often when drilling in a deepwater environment. A dual-gradient system will aid this well control in two ways. First, it will allow for surface casing to be set deeper, which will improve wellbore integrity and intermediate-depth drilling. Second, as shallow hazards are encountered, the subsea mud-return system will enable the driller to have more-complete and safer well control.

Dual-gradient technology promises to improve safety and well control while drilling, decrease costs, improve wellbore quality, and reduce environmental impact. Even so, developing a new technology can be expensive and difficult to implement. The next step, which is paramount to implementing dual-gradient technology into commercial use, is to convince the industry end users (operators and service companies alike) that dual-gradient technology will significantly improve deepwater drilling operations (Elieff 2006).

References

Alford, S.E., Asko, A., Campbell, M., Aston, M.S., and Kvalvaag, E. 2005. Silicate-Based Fluid, Mud Recovery System Combine To Stabilize Surface Formations of Azeri Wells. Paper SPE 92769 presented at the SPE/IADC Drilling Conference, Amsterdam, 23–25 February. DOI: 10.2118/92769-MS.

Choe, J. and Juvkam-Wold, H.C. 1998. Well Control Aspects of Riserless Drilling. Paper SPE 49058 presented at the SPE Annual Technical Conference and Exhibition, New Orleans, 27–30 September. DOI: 10.2118/49058-MS.

Revised Defintions Clarify Distinction Between MPD, UB. Drilling Contractor. http://drillingcontractor.org/index/index.php?option=com_content&task=view&id=44&Itemid=1. Downloaded 28 January 2008.

Dual Gradient Drilling System Using Glass Hollow Spheres. 2003. NETL Project DE-AC26-02NT41641, US DOE/NETL, Tulsa (October 2002–February 2003).

Eggemeyer, J.C., Akins, M.E., Brainard, R.R. et al. 2001. SubSea MudLift Drilling: Design and Implementation of a Dual Gradient Drilling System. Paper SPE 71359 presented at the SPE Annual Technical Conference and Exhibition, New Orleans, 30 September–3 October. DOI: 10.2118/71359-MS.

Elieff, B.A. 2006. Top Hole Drilling With Dual Gradient Technology To Control Shallow Hazards. MS thesis, Texas A&M University, College Station, Texas.

Forrest, N., Bailey, T., and Hannegan, D. 2001. Subsea Equipment for Deep Water Drilling Using Dual Gradient Mud System. Paper SPE 67707 presented at the SPE/IADC Drilling Conference, Amsterdam, 27 February–1 March. DOI: 10.2118/67707-MS.

Grottheim, O.E. 2005. Development and Assessment of Electronic Manual for Well Control and Blowout Containment. MS thesis, Texas A&M University, College Station, Texas.

Hannegan, D.M. and Wanzer, G. 2003. Well Control Considerations—Offshore Applications of Underbalanced Drilling Technology. Paper SPE 79854 presented at the SPE/IADC Drilling Conference, Amsterdam, 19–21 February. DOI: 10.2118/79854-MS.

Herrmann, R.P. and Shaughnessy, J.M. 2001. Two Methods for Achieving a Dual Gradient in Deepwater. Paper SPE/IADC 67745 presented at the SPE/IADC Drilling Conference, Amsterdam, 27 February–1 March. DOI: 10.2118/67745-MS.

Johnson, M. and Rowden, M. 2001. Riserless Drilling Technique Saves Time and Money by Reducing Logistics and Maximizing Borehole Stability. Paper SPE 71752 presented at the SPE Annual Technical Conference and Exhibition, New Orleans, 30 September–3 October. DOI: 10.2118/71752-MS.

Judge, R.A. and Thethi, R. 2003. Deploying Dual Gradient Drilling Technology on a Purpose-Built Rig for Drilling Upper Hole Sections. Paper SPE 79808 presented at the SPE/IADC Drilling Conference, Amsterdam, 19–21 February. DOI: 10.2118/79808-MS.

Kennedy, J. ed. 2001. First Dual Gradient Drilling System Set for Field Test. *Drilling Contractor* (May/June): 20–23.

MPD Could Tap Huge Quantities of Methane Hydrate. 2005. *Drilling Contractor* (March/April): 32–33.

NETL Project DE-AC26-02NT41641.2003. Dual Gradient Drilling System Using Glass Hollow Spheres. US DOE/NETL, Tulsa, October 2002–February 2003.

Rocha, L.A. and Bourgoyne, A.T. 1996. A New Simple Method To Estimate Fracture Pressure Gradient. *SPEDC* **11** (3): 153–159. SPE-28710-PA. DOI: 10.2118/28710-PA.

Roller, P.R. 2003. Riserless Drilling Performance in a Shallow Hazard Environment. Paper SPE 79878 presented at the SPE/IADC Drilling Conference, Amsterdam, 19–21 February. DOI: 10.2118/79878-MS.

Schubert, J.J. 1999. Well Control Procedures for Riserless/Mudlift Drilling and Their Integration Into a Well Control Training Program. PhD dissertation, Texas A&M University, College Station, Texas.

Schubert, J.J., Juvkam-Wold, H.C., Weddle, C.E., and Alexander, C.H. 2002. HAZOP of Well Control Procedures Provides Assurance of the Safety of the SubSea MudLift Drilling System. Paper SPE 74482 presented at the IADC/SPE Drilling Conference, Dallas, 26–28 February. DOI: 10.2118/74482-MS.

Schubert, J.J., Juvkam-Wold, H.C., and Choe, J. 2003. Well Control Procedures for Dual Gradient Drilling as Compared to Conventional Riser Drilling. Paper SPE 79880 presented at the SPE/IADC Drilling Conference, Amsterdam, 19–21 February. DOI: 10.2118/79880-MS.

Schumacher, J.P., Dowell, J.D., Ribbeck, L.R., and Eggemeyer, J.C. 2001. Subsea Mudlift Drilling: Planning and Preparation for the First Subsea Field Test of a Full-Scale Dual Gradient Drilling System at Green Canyon 136, Gulf of Mexico. Paper SPE 71358 presented at the SPE Annual Technical Conference and Exhibition, New Orleans, 30 September–3 October. DOI: 10.2118/71358-MS.

Smith, K.L., Gault, A.D., Witt, D.E., and Weddle, C.E. 2001. SubSea MudLift Drilling Joint Industry Project: Delivering Dual Gradient Technology to Industry. Paper SPE 71357 presented at the SPE Annual Technical Conference and Exhibition, New Orleans, 30 September–3 October. DOI: 10.2118/71357-MS.

Stave, R., Farestveit, R., Høyland, S., Rochmann, P.O., and Rolland, N.L. 2005. Demonstration and Qualification of a Riserless Dual Gradient System. Paper OTC 17665 presented at the Offshore Technology Conference, Houston, 2–5 May.

SI Metric Conversion Factors

ft   × 3.048*   E – 01 = m
°F   (°F – 32)/1.8   = °C
°F   (°F + 459.67)/1.8   = K
gal × 3.785 412   E – 03 = m³
in.   × 2.54*   E + 00 = cm
lbm × 4.535 924   E – 01 = kg
psi × 6.894 757   E + 00 = kPa

*Conversion factor is exact.

7.3 Buoyant Wellhead—Terje Magnussen, Atlantis Deepwater Technology A/S

7.3.1 Deepwater Challenges. *Deepwater Exploration Drilling.* Many of the deepwater challenges are re-lated to the long distance between the floating drilling vessel and the top of the well [wellhead and blowout preventer (BOP)] and the working environment for vital well-control equipment at great water depths. The drilling riser, and the kill and choke lines, represent high loads on the drilling vessel, causing the vessel's capacity requirements to escalate. Consequently, the older drilling vessels cannot be used in deeper water. Riser buoyancy elements can be used to reduce the effective weight of the riser, but this is costly and may have some undesired effects on factors such as hydrodynamic behaviour and deck storage.

At water depths beyond 2000 m, very few units are able to operate. On the other hand, these units may not be able to operate in shallower water because of high financial and operational cost. In addition, several operational and safety-related challenges are related to having the BOP located a long distance below the drilling vessel:

- A gas kick may not be easy to detect because it will not expand much between the reservoir and the BOP. This may cause gas to be in the riser before it is detected and the BOP is closed.
- The very long choke and kill lines cause large pressure losses when kicks are circulated out, which complicates use of conventional kick-control methods.
- The large volume of mud in the riser will cause the total well mud volume to be large, which will affect
  - Bottom-up circulation time
  - Time to condition the mud system
  - The amount of cuttings in the riser, which may settle and block the BOP
  - The time needed to perform a controlled riser disconnect
- No riser margin exists, and the pressure underneath the BOP must contribute to maintaining pressure balance. If there is a possibility to have migrating, nonexpanding gas in the well below the BOP after shut-in, the well may collapse since the bottomhole pressure will increase. The result may be an under-ground blowout.

Deepwater-Field Development. When a deepwater discovery is ready for development, the operator is faced with several challenges, which are different from those in shallower water. The field-development and produc-tion technology for shallower water has been extended for use in deeper water. This development has included

- Alternative floating-production-platform designs
- Subsea production equipment for high seawater hydrostatic pressure and low temperature
- Control systems for deepwater production equipment
- Deepwater remotely operated vehicle (ROV) equipment
- Installation and suspension equipment for deepwater equipment and risers
- Risers for high axial and transverse loads
- Deepwater platform-mooring systems

However, the deepwater equipment is more complex and more expensive than similar shallow-water equip-ment, and the high loads, limited access, and lack of long-term experience make it difficult to maintain an acceptable reliability.

Well intervention is a major technical and economical challenge in deep water, and lack of well maintenance can easily jeopardize flow assurance. To maintain access to the wells, some deepwater-platform concepts such as tension leg platforms and spars use rigid risers with surface production trees. However, because of the vertical riser loads and hydrodynamic forces, such concepts can be applied only up to a maximum water depth. Further, all subsea-completed satellite wells connected to such platforms will suffer from lack of accessibility.

7.3.2 Principle of the Buoyant Wellhead. The aim of the buoyant wellhead is to eliminate the aforementioned challenges by making deepwater wells adopt the geometry of shallow-water wells such that the wellhead is located on the order of 200–500 m below surface. This can be achieved by using a buoy to suspend the wellhead and BOP at the desired depth so that the well may be treated as in shallow water and so that recognized shallow-water production concepts such as "floating production" may be applied. This principle is called "buoyant wellhead."

A floating rig drills the tophole riserless and installs the surface casing from the seabed. A similar-sized casing string is used to extend the well from the seabed to the buoy. The subseabed casing provides the reliable anchoring for the extension and buoy.

7.3.3 Anchoring, Buoyancy, and Forces. *Anchoring Principle.* **Fig. 7.3.1** shows the idealized relationship between the lateral force F acting on the buoy when offset at an angle α and a net buoyant force K_{net}, keeping it from further lateral movement caused by current. **Fig. 7.3.2** shows the main components when the extension casing is straight and its weight W is not taken into account.

$$F = K_{net} \times tg\ \alpha, \quad \dots\dots\dots\dots\dots\dots\dots\dots\dots\dots\dots\dots\dots\dots\dots\dots\dots\dots\dots (7.3.1)$$

where tg = tangens. The sea currents will push the buoy and extension casing aside until α has reached the value where F balances the total drag force from the currents. K_{net} is the net vertical tension at the seabed (K_{net} is the net buoyancy force from the buoy minus casing/fluid weight).

However, Eq. 7.3.1 is an idealized case, which is nearly correct at small offsets and high K_{net}. At larger angles, the weight of the extension casing and high-density internal fluids will cause the extension casing to take a bowed shape. This will affect α and complicate the calculation of $F \cdot K_{net}$ can be controlled from the surface vessel by injection or discharge of air to or from the buoy through an umbilical.

With the wellhead located at shallow depth, further drilling, completion, production, and well intervention can take place with equipment designed for such depth. The drilling, production, and intervention-operations procedures will also be affected positively because most of the well is below the mechanical well barrier elements.

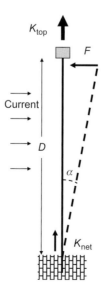

Fig. 7.3.1—Balance of forces.

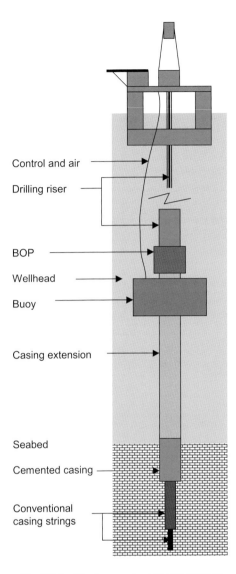

Fig. 7.3.2—The buoyant-wellhead system.

Net Buoyancy Force K_{net}. The net buoyancy force K_{net} is the vertical force in the casing at seabed. The pull in the casing at the wellhead, K_{top} is

$K_{top} = K_{net}$ plus the weight of casing plus the weight of fluid minus the buoyancy of casing/fluid in seawater.

Total Buoyancy of the Buoy. After the buoy has been installed, it carries the extension casing top tension K_{top}, its own weight, and the weight of all equipment on the well, including the BOP. K_{net} must be kept above a minimum value to prevent too much offset. Therefore, as new loads are added, the buoyancy of the buoy must be increased by injection of air. Such new loads will be casing strings and heavier drilling fluids and drillstring. Further, the buoyancy may also include a margin for emergency hang-off of the drillstring. When designing the buoy for dimensions, its volume should also include a margin for temporary loss of air in one or two compartments.

Extension-Casing Stress and Fatigue. The high vertical tension combined with offset caused by current and/or other external forces will cause bending stress in parts of the extension casing. Simulations show that the stress is highest at the seabed and at the bottom of the buoy. K_{net} must be kept high enough to prevent the bending at the seabed from reaching an excessive value.

Simulations may show that bending restrictors are needed at both top and bottom to prevent unacceptable stresses in the extension casing. Further, vortex-induced vibrations, generated when seawater flows past the extension casing, may cause vibrations in the extension casing. The hydrodynamic behavior of the buoy may also cause cyclic loads in the extension casing. All such high- and low-frequency vibrations and cyclic loads will expose the extension casing to fatigue effects, which must be controlled through design and operation.

7.3.4 Operational Effects on Drilling. *General.*

Typical well schematics for the long riser and the buoyant wellhead are shown in **Fig. 7.3.3.** Deepwater drilling and production with the wellhead at the seabed is still challenging, and more challenging the deeper the water.

When using the buoyant-wellhead principle, the wellhead remains at a shallow depth regardless of water depth, and there is little escalation in the complexity of the operation. However, the floating drilling or production unit will need a station-keeping system, which is designed for the full water depth (i.e., a fiber-rope taut-line mooring).

Well Control. With the buoyant seabed, the well control, which is a major concern in deepwater drilling, may be conducted as in shallow water. This is mainly because most of the well is underneath the BOP and because the well's geometry is not affected by the water depth. The possibility of detecting an influx before it reaches the BOP is very good, which is also the case in shallow water.

After an influx has been detected and the well has been shut in, the influx can be circulated out according to conventional kick-circulation practice, mainly because the choke and kill lines are short and represent a moderate circulation-pressure loss.

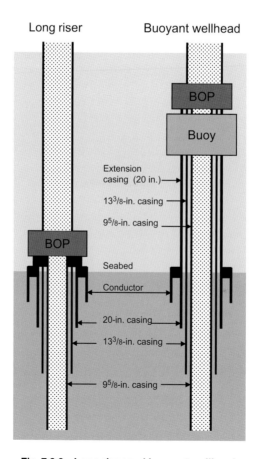

Fig. 7.3.3—Long riser and buoyant wellhead.

Riser Margin. The riser margin is the addition in mud density needed to make the downhole mud column and the seawater column together balance the formation pressure at any depth below the shoe of the surface casing in case of a riser leak or sudden disconnect. The riser margin cannot be maintained in deepwater drilling with the wellhead at the seabed. This means that the well will be underbalanced if the mud column in the riser is lost because of a severe leak deep in the riser or if the riser has to be pulled quickly in an emergency situation.

Only quick closing of the BOP may prevent the well from flowing. With the buoyant riser, the riser margin may be maintained as in shallow water because of the short drilling riser.

Riser Disconnects. With a long deepwater riser, a controlled disconnect of the riser takes a long time because of the time needed to displace the large riser mud volume with seawater and the time needed to hang off the drillstring in the wellhead/BOP. When resuming operation, reconnecting usually takes a long time because of the lack of lateral control of the lower marine riser package (LMRP). With the buoyant wellhead, the riser disconnect and reconnect will approximate those in shallow water.

Drilling-Mud and Hole Monitoring. With a long deepwater riser, a large portion of the drilling-mud volume will be in the riser, and the total mud volume will be large because of the large diameter of the riser compared to smaller casings. This large riser volume causes disadvantages compared to shallow-water drilling:

- The total consumption of mud and additives increases.
- The time needed for conditioning and circulating the mud increases.
- The time needed to circulate bottom-up for samples and well monitoring increases.
- The risk of large mud discharge to the environment increases.

With the buoyant wellhead, these disadvantages will not be present because the length of the riser is short.

7.3.5 Effect on Essential Well Equipment While Drilling. *Bending Stress on the Wellhead and BOP.*
When the wellhead and BOP are located at the seabed, the riser is exposed to significant drag forces. These forces are also transmitted to the BOP and into the wellhead and conductor where they may cause high bending forces. Therefore, with a long deepwater riser, the BOP, wellhead, and conductor need to be designed for higher stress than similar equipment used in shallow water. With the buoyant wellhead, the wellhead and BOP do not experience such stress.

Riser Loads. Because of high loads, the long deepwater riser needs to be designed for the purpose and cannot be an extension of a shallow-water riser. With the heavy deepwater riser, the riser-tensioning system must also be designed to suspend the high riser load. With the buoyant wellhead, the shallow-water riser can be used and the riser-tensioning system for shallow water can be used.

7.3.6 Economical Effects on Drilling.
In addition to the operational effects, there are also significant economical effects from the buoyant wellhead. The largest economical effect is obtained from savings on the drilling-vessel day rate. This is because of the significant difference in cost between shallow-water and deepwater drilling vessels. Further, in some cases, drilling time may be saved with a buoyant wellhead.

7.3.7 Effects on Field Development and Production.
The buoyant wellhead also can be used for field development and will have effects on design and drilling of production wells similar to those it has on exploration wells. Both single-well and multiple-well buoys can be used for production wells. The buoy may also accommodate subsea production equipment.

The buoyant wellhead will also have positive effects on production:

- Improved operational environment for vital subsea production equipment
- Easier well access for well intervention
- Easier access to vital subsea production equipment
- Less well and intervention cost
- Less risk of forming hydrates

Fig. 7.3.4—Launching the first ABS to sea.

7.3.8 The Atlantis Buoyant Wellhead. The patented Atlantis system is the only commercial buoyant-well-head system. The first unit for exploration drilling was constructed in 2002–2003 and went through sea trials. The manufacturer uses "Artificial Buoyant Seabed" (ABS) as the name of the buoy itself. (See **Fig. 7.3.4.**) Its first application was scheduled to occur in Chinese deep waters in 2009. Its main cylindrical steel body has a diameter of 16 m, and its height is approximately 7.50 m. In addition, buoyancy elements, which make the buoy float without any trapped air, are attached to the circumference of the main steel body.

Nomenclature

F = lateral force, kN
K_{net} = net buoyant force, kN
K_{top} = pull in the casing at the wellhead, kN
W = weight, kg
α = angle, degrees

General References

Magnussen, T. 1998. Subsea Installation. Norwegian Patent No. 303028.

SI Metric Conversion Factors

ft	× 3.048*	E – 01 = m
lbf	× 4.448 222	E + 00 = N
lbm	× 4.535 924	E – 01 = kg

*Conversion factor is exact.

Chapter 8

Design Considerations for High-Pressure/High-Temperature Wells

P.V. Suryanarayana, Blade Energy Partners, and **Knut Bjorkevoll,** Sintef

8.1 Introduction

High-pressure/high-temperature (HP/HT) wells are an unavoidable consequence of the ever-expanding search for new oil and gas resources around the world. The elevated pressures and temperatures in these wells make the design, drilling, and operation of such wells challenging. The basis of design itself has to be reconsidered when designing such wells—the design of each string in the well affects the design of other strings because loads like annular pressure buildup (APB) connect multiple strings in ways that are not commonly seen in conventional well design. HP/HT conditions create several nonstandard load scenarios, creating the need for nonstandard fluids and materials, advanced design approaches, and new procedures. Often, the design of these wells results in material or equipment requirements that are not easily met at the current state of the art.

In this chapter, we review the design considerations for HP/HT wells. We attempt to define the term *HP/HT* in this section, and note the distribution of HP/HT conditions around the world. An exhaustive literature survey of additional reading is provided for interested readers. The survey also serves to provide a theme for the material presented in this chapter. Methods and approaches to the estimation of temperature and pressure, which are obviously important in the design of HP/HT wells, are discussed next. The effects of HP/HT conditions on the properties and performance of fluids and materials are discussed, as are the nonstandard loads placed on wells. Operational implications of HP/HT conditions are discussed, particularly as relevant to well control. Other chapters in this text that provide additional detail on topics relevant to HP/HT well design are referenced where appropriate.

In summary, this chapter provides a broad coverage of the design considerations for HP/HT wells. This can necessarily result in limited treatment of some of the topics. The exhaustive reference list at the end of the chapter provides interested readers with a starting point for their more detailed excursions into this deeply challenging, and ultimately rewarding, aspect of oil and gas drilling.

8.1.1 Definition of HP/HT Conditions in Drilling. A universal definition of HP/HT wells has not been established; however, most operators consider a well with bottomhole temperature in excess of 300°F and surface shut-in pressure greater than 10,000 psi as an HP/HT application (Harrold et al. 2004). Wells with bottomhole temperatures exceeding 425°F and pressures above 15,000 psi are generally regarded as ultra-HP/HT wells (Baird et al. 1998) or extreme-HP/HT (X-HP/HT) wells.

Among regulatory bodies, the UK Health and Safety Executive has suggested that for an application to be classified as HP/HT, the undisturbed bottomhole temperature must exceed 300°F and have a pore pressure gradient in excess of 0.8 psi/ft or require the use of well-control equipment in excess of 10,000 psi working pressure (Seymour and MacAndrew 1993).

It is important to note that bottomhole temperature and pore (or surface) pressures are not the only indicators of HP/HT conditions. A third parameter that may be used in identifying situations in which thermal problems need special attention is the geothermal gradient. Beyond a geothermal gradient of 0.014°F/ft, thermal effects become serious enough to merit special attention during design.

8.1.2 Geographical Distribution of HP/HT Conditions. HP/HT and ultra-HP/HT fields are predominantly gas producers because of high temperatures coexisting with high pore pressures. Several of them are also offshore wells. These fields are concentrated in the Gulf of Mexico (GOM) and the North Sea. Indonesia, Offshore eastern India, West Africa, China, continental US, Yemen, and Kuwait are other key areas where high pressures and temperatures occur. **Fig. 8.1** illustrates the reservoir pressure and temperature conditions of several HP/HT and ultra-HP/HT fields around the world. The figure is based on an illustration presented by Baird et al. (1998) and has been updated with some of the more recent HP/HT fields.

MacAndrew et al. (1993) provide a good summary of the evolution and prevalence of HP/HT wells around the world. In the early 1980s, many gas wells were drilled in the Tuscaloosa trend in Louisiana, and in other southern US states where temperatures above 350°F and pressures above 16,000 psi were encountered. Most of the North Sea HP/HT wells are situated in the Central Graben—a series of upthrown and downthrown rocks. The Central Graben contains several Jurassic gas condensate prospects at 12,000 to 20,000 ft, with pressures of 18,000 psi and temperatures of 400°F. The Elgin and Franklin fields in the Central Graben area, considered to be the largest developed HP/HT fields in the world, contain huge quantities of rich natural gas condensates.

The Kristin field in Norway is the first HP/HT field to be developed subsea. The Thunder Horse project in the GOM, characterized by extremely high pressures, is one of the toughest HP/HT fields in the world to be drilled, at water depths greater than 6,000 ft. Despite these challenging conditions, the number of HP/HT wells is growing steadily. One major reason is that the high pressure in these reservoirs typically means

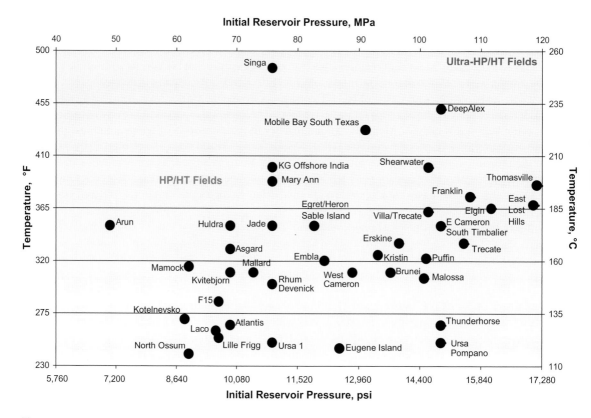

Fig. 8.1—Reservoir pressures and temperatures of HP/HT and ultra-HP/HT fields. After Baird et al. (1998). Image copyright Schlumberger. Used with permission.

production rates well beyond those possible from more normally pressured reservoirs (MacAndrew et al. 1993). Thus, for those who persevere, the prize is substantial.

8.1.3 Implications of HP/HT Conditions for Drilling and Well Construction. HP/HT conditions can significantly change the basis of design for wells. The higher pressures place significant demand on the well-construction materials used. The temperature of the wellbore is a function of its operating conditions (drilling, circulation, cementing, shut-in, production, injection, etc.). The temperature itself affects the properties of steel, cement, and fluids used. In particular, fluid rheology and pressure/volume/temperature (PVT) behavior affect both drilling operability and well control. The casing strings and the contents of the wellbore annuli respond to the temperature changes. In almost every instance, the parameter of interest is a temperature change, which is the difference between the temperatures during an initial or quiescent condition and some final operating condition. For example, a tubing string is installed during well completion. The packer is installed when the wellbore fluids are usually close to geothermal temperature. When the well starts producing, the tubing string (as well as the casing strings) are heated. The change in temperature causes thermal expansion of the tubing and casing. Depending on the packer (i.e., the amount of motion permitted by the packer at the tubing tail), its weight, and the internal and external fluids, the tubing string may buckle. Simultaneously, the fluid in the tubing annulus expands due to temperature increase and causes a pressure buildup (known as APB). An increase in temperature can also cause wellheads to move or *grow*, leading to surface facilities and wellbore concerns. Finally, post-drilling operations such as logging, testing, and completions are also affected. These phenomena, which are the result of temperature changes in the wellbore, have a serious impact on the loads imposed on the casing strings in particular and the integrity of the wellbore in general.

8.2 Literature Survey

Extensive literature is available on all aspects of HP/HT wells. Though these wells are the most challenging to drill, they are the most sought-after because of the prodigious amount of hydrocarbons that can be produced from these wells. Most papers on HP/HT applications regard bottomhole temperatures in excess of 300°F and surface shut-in pressures greater than 10,000 psi as HP/HT conditions (Harrold et al. 2004; Baird et al. 1998; Seymour and MacAndrew 1993). A map presented by Brownlee et al. (2005) indicates the concentration of HP/HT wells in the North Sea and the GOM. An updated version of this map is shown as Fig. 8.1. MacAndrew et al. (1993) provide valuable information regarding the development of HP/HT fields in these areas.

The traditional methods of design fail in the case of HP/HT wells, because high pressures and temperatures influence almost all aspects, including casing design, equipment, drilling fluids, cements, well control, connections, materials and completions. Shaughnessy et al. (2003) discuss the problems associated with ultradeep HP/HT wells and their possible solutions. The problems are mainly related to well control, drilling, and completions. Accurate prediction of both pressure and temperature is very critical in HP/HT wells. Swarbick (2002) presents the common methods of pore pressure prediction and the limitations of porosity-based prediction techniques. Esmersoy and Mallick (2004) discuss a new technique that can predict pressures ahead of the bit called the vertical seismic profile (VSP). Reliance on a single technique is not recommended for pore pressure prediction. Skomedal et al. (2002) present a study of the mechanical behavior of the reservoir rock when subjected to changes in pore pressure and stress. The depletion of the reservoir results in total stress reduction in the horizontal direction, leading to shear deformations. Maury and Idelovici (1995) present a case study that describes the loss in well stability when drilling in HP/HT conditions due to the transient thermal regime, which involves alternate cooling and heating of the wellbore. Cooling induces extra tensile, tangential, and axial stresses, which results in increase of mud weight, and warming has the opposite effect. All this influences the opening/shutting of fractures. Kelley et al. (2001), Sweatman et al. (2001), and Webb et al. (2001) discuss the formation pressure integrity of HP/HT wells and treatments to improve it. Causes of low borehole pressure integrity include natural in-situ stresses that cause weak points or flaws in rocks, drilling-induced stresses, and weakening of formation due to reactivity to certain drilling fluids.

Hasan and Kabir (2002) provide an excellent reference for the understanding of the wellbore thermal problem. Circulation of drilling fluids and production of formation fluids develop different thermal conditions in the well. Typical thermal properties of the formation and wellbore materials have been discussed by Hasan and Kabir (2002) and Marshal and Bentsen (1982). Prensky (1992) discusses methods for estimation of geothermal temperature, which is very important in casing design. The high geothermal gradients encountered in offshore fields for wells that generally are in close proximity to each other has been discussed by Sathuvalli et al. (2001). This can have a detrimental effect on mud properties, measurement-while-drilling (MWD) equipment, and kick tolerance. Geologic noise can cause errors in the estimation of the geothermal gradient (Beardsmore and Cull 2001).

Ramey's (1962) paper on wellbore heat transfer is one of the classic papers on wellbore thermal analysis, in which he developed a semi-analytic model for estimating the temperatures during injection and production. Most thermal simulators are based on Ramey's (1962) approach or a modified version of it. A similar problem has been addressed by Raymond (1969) and Willhite (1967). To calibrate the results obtained from thermal simulators, temperature can be measured from offset data, temperature probes, and thermometers. Vidick and Acock (1991) present the drawbacks of some of the flowing temperature measurement techniques.

High temperatures influence material properties. Berckenhoff and Wendt (2005) discuss the effect of high temperatures on elastomers. Elastomers tend to become soft at higher temperatures, and this can have a detrimental effect on the sealing integrity of the system. In addition to temperature, the presence of corrosive fluids affects material properties. Brownlee et al. (2005) have presented a good review of the selection procedures of materials for sour HP/HT wells used in the industry. Solid, high-strength corrosion-resistant alloys (CRAs) or clad CRAs are most suitable for this application. Nice et al. (2005) describe the development of a *mild sour service*, 125-ksi high-strength low-alloy steel grade for the Kristin field production casing.

PVT behavior of reservoir fluids is also affected by the HP/HT conditions. Volatility of heavy compounds increases with temperature. Gozalpour et al. (2005) present their work related to the increased volatility of water at higher temperatures for HP/HT fluids. Danesh (2002) conducted experiments to show that higher water content can result in increased viscosity of formation fluids. Fisk and Jamison (1989), Oakley et al. (2000), and Zamora et al. (2000) discuss the effect of high temperature on properties of mud. Mud density and viscosity typically reduce with increasing temperature. High-temperature gelation, fluid loss, and thermal degradation are some of the resulting effects. Assuming a constant density of the drilling fluid in the case of HP/HT wells is erroneous, as illustrated by Harris and Osisanya (2005). Rommetveit and Bjørkevoll (1997) present results from a simulator developed to predict pressure and temperature dependence of the mud density and its rheological properties. Wang and Su (2000) have proposed a pressure-temperature model for the estimation of equivalent static density (ESD) at HP/HT conditions. Results indicate that temperature gradient influences ESD to a great extent, and ESD variations with well pressures, equivalent circulating density (ECD), and surge and swab pressures have been reported. Saasen et al. (2002) discuss the advantages of cesium formate for the Huldra field in the North Sea, which experienced a kick due to barite sag in the oil-based mud (OBM) used. The advantages include no-sag potential, low ECD, quick thermal stabilization during flow checks, and fewer screen plugging risks. Romero and Loizzo (2000), MacAndrew et al. (1993), and Mansour et al. (1999) present their studies based on the deterioration of the mechanical properties of cements under the influence of temperature. High temperature affects cement's hydration, thickening time, stability, and compressive strength. Therefore, it is extremely important that cement is tested at the maximum expected temperatures before the cementing operation. Griffith et al. (2004) discuss the alternative of using foamed cement systems.

High temperatures affect connection performance by creating significant thermal loads, and this problem has been addressed by Bradley et al. (2005) and Carcagno (2005). Premium connections are necessary for HP/HT wells and must be designed using the stringent testing procedures in *ISO 13679* (2002).

High thermal loads are capable of causing buckling of unsupported sections. Paslay (1994), He and Kyllingstad (1995), Lea et al. (1995), Mitchell (1986), Sparks (1984), Hammerlindl (1980), Lubinski (1951), Lubinski et al. (1962), and Handelman (1946) have addressed the issue of buckling of constrained tubulars. The theory of buckling and experimental results have been discussed by Suryanarayana and McCann (1995). Handelman (1946), Lubinski (1951), Lubinski et al. (1962), and Sparks (1984) provide necessary equations and theory for calculation of effective forces for buckling and post-buckling analysis.

Wellhead movement (WHM) and APB are important considerations in casing design. Samuel and Gonzales (1999) present the optimization of multi-string casing design for the combination of annuli fluid expansion and wellhead growth. They introduce the concept of a dimensionless parameter, the wellhead growth index, which is defined as the ratio of the annulus fluid expansion of the casing to the actual volume of the exposed segment above the top of the cement. Adams (1991) presents the equations to calculate WHM and thermal stresses created due to thermal expansion of the strings using the elastic spring model. Halal and Mitchell (1993) lay emphasis on the multistring casing design that considers the elastic response of the whole casing system.

Adams and MacEachran (1994) discuss the impact of APB on casing design. The authors put forth a model that helps to relate the stresses developed in the composite system to heat up pressures. The real-life implications of APB problems, and the need for mitigation strategies other than an open shoe were first highlighted to the industry in a series of now-famous papers on the Marlin failures (Bradford et al. 2002; Ellis et al. 2002; Gosch et al. 2002). Loder et al. (2003) also present a case-study of mitigation of APB in a subsea-completed well. Payne et al. (2003), Sathuvalli et al. (2005), Ellis et al. (2004), and Belkin et al. (2005) discuss the mitigation strategies for APB and provide a PVT-based explanation for a selection of a strategy. The importance of transient analysis for APB mitigation strategy selection is highlighted by Oudeman and Kerem (2006). Mitigation strategies include nitrogen gas cushion, vacuum-insulated tubing (Azzola et al. 2004), syntactic foams, and rupture disks. Klever and Stewart (1998) and Adams et al. (2001) discuss the consideration of the tubular strength in addition to the assessment of loads for selection of a mitigation strategy.

Casing design for HP/HT wells often requires the use of limit state strength and probabilistic methods of load and strength estimation. Maes et al. (1995), Mason and Chandrashekhar (2005), and Payne et al. (2003) have considered the stochastic nature of loads and strengths in the tubular design. Limit states design and probabilistic use of limit state strength are also discussed in American Petroleum Institute (API) *TR 5C3/ISO-TR 10400* (2008).

Apart from the design of HP/HT wells, HP/HT conditions influence the operational aspects of HP/HT wells, including well control, drilling, cementing, surface facilities, well testing, and completions. Berckenhoff and Wendt (2005) and Young et al. (2005) discuss the implication of HP/HT conditions on material and equipment involved in well control. Rommetveit et al. (2003) emphasize the importance of advanced detection tools for kicks in HP/HT wells because mud volume can change as a result of increase in temperature. Mason and Chandrashekhar (2005) present a stochastic kick load model that accounts for the solubility of gas in OBM. Rudolf and Suryanarayana (1997) emphasize good tripping practice in HP/HT wells because tripping speed can cause swab pressures while tripping in, and this in combination with high temperature is capable of causing a kick. Berckenhoff and Wendt (2005) and Walton (2000) discuss the influence of HP/HT conditions on surface facilities, worsened by the presence of corrosive fluids in the system. Boscan et al. (2003) emphasize qualification testing of downhole equipment used for well testing and proper selection of equipment for surface and downhole conditions. Hahn et al. (2000) reemphasize the fact that rigorous design and testing procedures are required for completion equipment for HP/HT wells.

8.3 Estimation of Pressure

8.3.1 Pore and Fracture Pressure. Accurate pore pressure and fracture pressure estimation is extremely important in HP/HT wells because they are commonly drilled with a narrow margin between the mud and fracture gradients. In particular, estimation of the depth and magnitude of pressure transition zones is crucial. The accuracy of these estimates can impact casing setting depths, material and connections selection, mud design, and identification of potential zones where influx/losses can occur.

The most common basis for pore pressure estimation is seismic and geologic/basin modeling, and/or analysis of offset well data. Little may be known of the mechanisms that generated the overpressure conditions, and data quality is likely to be variable (Swarbick 2002). Particularly for HP/HT wells, conventional techniques prove to be unreliable. For instance, along the North Sea Central Graben axis, two different overburden pressure domains exist, separated by a low-porosity horizontal pressure seal (Ward et al. 1994).

Direct measurements of pore fluid pressures [e.g., wireline pressure tests such as repeated-formation testers (RFTs) and modular-formation dynamics testers (MDTs), or production tests such as drillstem tests (DSTs)] can be used to calibrate any prediction (Brownlee et al. 2005). Conventional real-time pore pressure prediction techniques typically relate measured porosity indicators (velocity from seismic, resistivity and density from logs) and real-time drilling parameters (drill rate, formation gas, etc.) for pore pressure prediction. However, these methods are highly constrained by lack of data ahead of the bit and do not provide the accuracy required to be fully effective tools. In addition, porosity-based pore prediction methods such as the Eaton ratio method and the equivalent depth method do not prove to be effective in the case of mixed-lithology reservoirs and for overpressure mechanisms other than disequilibrium compaction. For the prediction of pore pressure ahead of the bit, a technique called the look-ahead VSP, which measures the acoustic impedance, allows the estimation of lithological changes ahead of the bit. However, this approach alone cannot be used in cases in which pore pressure builds up gradually and is not associated with sharp impedance change (Esmersoy and Mallick 2004). For accurate pore pressure predictions, reliance on one particular method is not recommended, and a combination of techniques (MWD sonic, VSP, and seismic) will improve the confidence of the analysis. Accurate pore pressure prediction also allows for better control of pressure gradients (Kuyken and de Laange 1999).

Fracture pressure estimation usually is augmented by leakoff or formation integrity tests. Where offset wells are available, data from those wells can be used to obtain a reasonable basis for fracture pressure. Geomechanical modeling also can help in establishing the fracture closure stresses and overburden stresses in a given region, provided that adequate data is available to enable such modeling. Because some fluids develop significant gel strength after being exposed to high temperatures, one should be aware of gel effects when interpreting leakoff or formation integrity tests. When doing a test with a gelled-up fluid in the well, a significant fraction of the pressure that is imposed at surface may not reach the open hole. After the fluid has been circulated for some time, the gel will break down, and more of the surface pressure will add to the open-hole pressure (see Section 8.3.1).

Because considerable uncertainty remains in pressure estimation, it is important to consider pore and fracture pressure profiles from several data sources. The design then can be based on a range of pore and fracture pressure expectations. **Fig. 8.2** shows one example of pore and fracture pressure estimation and measurements based on different methods, including actual real-time data. The uncertainty and variability in pressure estimation using the different methods is apparent in Fig. 8.2.

8.3.2 Circulation Pressure. Steady-state and dynamic hydraulic calculations proceed in a manner similar to non-HP/HT wells, except that some effects that are normally ignored may become significant. This includes pressure and temperature dependence of fluid properties, which is discussed in some detail in Section 8.5.1, and gel effects, which are boosted for some fluids when exposed to high temperatures.

This introduces the concept of time-dependent rheology or thixotropy, which to the author's knowledge lacks an accurate mathematical description. However, investigations that have been done on drilling fluids (Bjørkevoll et al. 2003; Herzhaft et al. 2006) reveal some characteristic features that seem to be predictable on the basis of measurements with simple rheometers. The gel behavior seems to be well represented by a sum of two effects: a gel buildup that depends on time, and a gel decay that typically is represented by one or two exponential functions, with decay rate depending on shear rate. A critical shear rate exists, below which gel buildup dominates, and above which gel breakdown dominates. Near the critical shear rates, the gel may *hesitate* before the thins settle. It is recommended that the gel strength of actual drilling fluids be measured before doing operations in critical sections. The methods cited are used to determine whether gel effects can be significant and require extra care when starting pumps and when interpreting pressure tests. A high degree of gel also may reduce the reliability of downhole tools that are operated by pressure pulses.

8.4 Estimation of Temperature

Temperature estimation usually is based on measurements or estimates of geothermal data and on modeling of different flow conditions (production, drilling, cementing) to obtain temperature profiles. While temperature estimation is important in HP/HT wells, it is also the most disheartening aspect of HP/HT well design, because the high bottomhole temperatures and prodigious production rates usually mean that the well typically heats up significantly. In this section, we discuss the wellbore thermal problem and the different methods of temperature estimation.

8.4.1 The Thermal Problem. One of the most comprehensive books on fluid flow and heat transfer in wellbores is the text by Hasan and Kabir (2002). The basic heat transfer problem is depicted in **Fig. 8.3** for drilling (or circulation) and **Fig. 8.4** for production. During circulation, fluid at known temperature and rate enters the drillstring. Temperature continuity is maintained at the bit. Heat is transferred between the circulating fluid and the geothermal *sink*. Return rates are usually the same as injected rates. During production, fluid enters the production string at a known temperature and assumed rate, reaching the surface at some unknown temperature. (The situation is similar during injection, with the known temperature at surface.) Heat is lost from the wellbore to the geothermal sink. The return temperature and the temperature profile at any radius from the centerline are the results of interest.

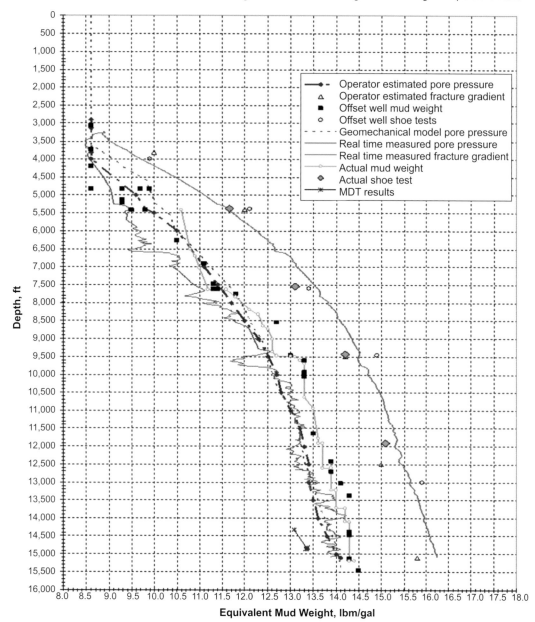

Fig. 8.2—Example of pore and fracture pressure estimation methods and uncertainty range.

Although wellbore temperatures are typically the highest during production in HP/HT wells, drilling condition results are also very important, especially for the outer tubulars. Drilling temperature profiles also impact the properties and performance of the drilling fluid itself, which is also a key design consideration.

Thermal Properties of Formation and Wellbore Materials. Thermal conductivity, specific heat, thermal diffusivity, and thermal anisotropy are typical thermal properties of interest. Typical values of these properties that are frequently used in thermal analysis are indicated in **Table 8.1.** For HP/HT wells in particular, it is important to use formation properties according to lithology, rather than to use generic constants. At a minimum, differentiation between sandstone and shale is important.

8.4.2 Geothermal Temperature Estimation. Geothermal temperature refers to the undisturbed temperature of the Earth that exists prior to operations such as drilling, production, or injection. The geothermal

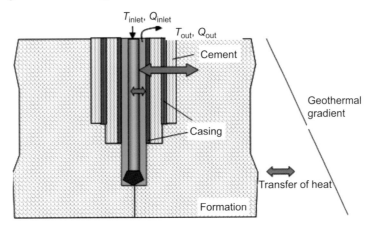

Fig. 8.3—The wellbore thermal problem while drilling.

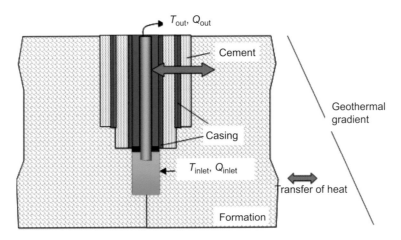

Fig. 8.4—The wellbore thermal problem while producing.

temperature is estimated from temperature logs and temperature data from offset wells or from the general region. Wireline-conveyed maximum-recording thermometers and continuous-reading thermistors are used to measure absolute temperatures, differential temperatures, and temperature gradients at depth (Prensky 1992). Geothermal temperature often is characterized by a geothermal temperature gradient, in °C/m or °F/ft, its magnitude varying widely across the world. Geothermal gradients typically lie between 0.022 and 0.033°C/m (0.012 and 0.018°F/ft). In regions where geothermal energy is harvested, the gradient can be as high as 0.05°C/m (0.027°F/ft).

Geothermal temperature estimation is critical in casing design and thermal modeling, especially for deepwater and HP/HT wells. Therefore, care must be exercised in obtaining accurate temperature data. If the gradients have to be estimated, it is recommended that sensitivities be conducted over a range of expected geothermal temperature gradients.

Particularly in offshore fields, wells are spaced sufficiently close for optimization of production facilities. Due to close proximity, diffusion of thermal energy between the wellbores has a pronounced effect on the undisturbed geothermal gradient of the formation (Sathuvalli et al. 2001). Higher geothermal gradients are observed at shallower depths for new wells when these wells are drilled in mature offshore fields in proximity to wells that have been on production for 6 months or longer. Drilling mud, cementing operations, and casing design of the new wells are strongly impacted by the higher geothermal gradients. The MWD equipment, drilling fluids, and motors that are used to drill the upper hole sections may not be rated for these higher

TABLE 8.1—TYPICAL VALUES OF THERMAL PROPERTIES OF FORMATION AND WELLBORE MATERIALS*

Formation/ Materials	Thermal Conductivity, k (Btu/hr-°F-ft)	Specific Heat, $C_{p,f}$ (Btu/lbm-°F)
Formation	1.3–3.33	0.2–0.625
Cement	0.38–0.5	0.4771
Formation oil	0.08–0.1	0.4–0.5
Formation gas	0.1–0.3	0.25
Water	0.36	0.997
Steel	30	0.09542
Packer fluid (clear brine)	0.35	0.9

*Hasan and Kabir (2002); Marshall and Bentsen (1982)

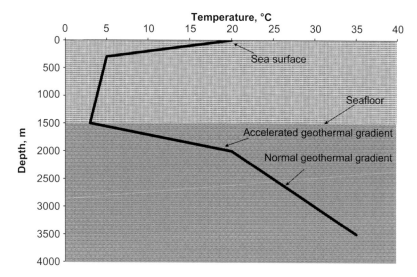

Fig. 8.5—Geothermal profile in offshore conditions.

temperatures. In addition to this, cooling (or heating) of the rock formation induces contraction (or expansion) and creates tangential compressive (or tensile) stresses around the wellbore; the altered temperatures will in turn affect mud density, kick tolerance, and ECD considerations.

Geothermal Temperatures in Offshore Conditions. Most of the sunlight is absorbed very near to the surface of the sea. The surface layer, frequently called the mixed layer, varies in depth according to location and season. Between approximately 300 m and 1 km, temperature decreases rapidly with depth. This region of steep temperature gradient is called the *thermocline*, beneath which there exists a deeper layer where temperature decreases very slowly with depth. The thermocline separates the warmer mixed layer above from the cooler deep layer below.

For subsea wells, geothermal temperature estimation must include this temperature reduction from ambient as a function of water depth, as well as the rapid increase in temperature to *normal* geothermal temperature below the mudline. A typical approach is given here and illustrated in **Fig. 8.5:**

- From surface to mudline, the temperature gradient characteristic of the waters in the region is used.
- The temperature at mudline is the temperature of water at the water depth to mudline. For most deepwater wells (>2,000 ft or 610 m water depth), the temperature at mudline is essentially constant at approximately 4°C (39°F), although not always.

- Below the mudline, a higher-than-normal (*accelerated*) temperature gradient is assumed, such that the temperature at 1,000 ft (305 m) below the mudline is the same as it would be if it were a land well.
- At a depth greater than 1,000 ft (305 m), the normal geothermal temperature gradient is used.

Errors in Geothermal Temperature Estimation. Some of the possible sources of error in geothermal temperature estimation are as follows:

1. Geologic noise associated with the data affects the signal-to-noise ratio. Regions of higher geothermal gradients generally yield noisier logs (Beardsmore and Cull 2001).
2. At higher temperatures, sensors are inefficient if not rated accordingly.
3. Error associated with transferring the data to the surface due to insulation failure or fluctuations in the cable resistance may affect the temperature readout.
4. Logs give erroneous readings if they are recorded when thermal equilibrium in the borehole is not achieved (nonequilibrated logs).

8.4.3 Estimation of Flowing Temperatures Using Thermal Models. Wellbore heat transfer is a mature discipline, and there exist standardized techniques to compute the temperatures in a wellbore. From the point of view of well design and analysis, it is necessary to determine the temperatures in the wellbore during production, injection, drilling, and shut-in. The temperatures during cementing usually can be obtained as a special case of circulation followed by shut-in.

The nature of heat transfer in a well during production or injection through tubing is discussed by Ramey (1962) in his classic paper. The heat transfer during circulation and drilling is addressed by Raymond (1969). Willhite (1967) provides a detailed summary of the key equations and correlations required to apply the approach described by Raymond. Despite the passage of time, and notwithstanding the development of sophisticated computer programs that determine wellbore temperatures, the papers by Ramey, Willhite, and Raymond are classic works, because they describe the salient physics underlying wellbore heat transfer. These papers must be read and understood by any engineer who wishes to understand the role of temperature changes and thermally induced loads in a wellbore.

There is comprehensive literature on heat transfer in pipes and annuli, most of which focuses on heat exchange with static or circulating Newtonian fluids such as air, water, or pure oils in regular geometries. A review of single-phase heat transfer in annuli, which is considerably more complicated than single-phase heat transfer in pipes, is given by Childs and Long (1996).

Wellbores are different from typical heat exchange experiments in several ways, especially when drilling: The length of the well is orders of magnitude larger than its diameter; fluids are frequently non-Newtonian with high viscosity; and the drillstring is irregular, eccentric, rotating, and vibrating. An important issue is to distinguish between different flow regimes. Purely laminar single-phase radial heat transfer is very poor and essentially conductive, and any regular *Taylor* vortices or more irregular turbulence will enhance radial heat transfer significantly. A mathematically correct approach to all flow regimes would be both beyond the current state of modeling and, to a large extent, irrelevant. A more pragmatic approach is sufficient for most practical cases, although an overview of the complexity of the problem is useful when evaluating the reliability of models used.

The mathematically simplest case is a static well with no forced convection and a given initial temperature distribution. If the fluid is sufficiently viscous or gelled up at a given temperature gradient, heat transfer will be purely conductive, and accurate calculations can be done if thermal properties of all materials are known. A lower viscosity or higher temperature gradient may induce natural convection, which is covered by published heat transfer correlations [e.g., see Appendix A in Hasan and Kabir (2002), which, however, unveils some uncertainty with respect to the quoted equations]. When using such expressions for gelled non-Newtonian fluids, one has to estimate the apparent viscosity that enters the calculations. This is best done at a typical shear rate for the natural convection in question.

Forced convection is mathematically very complex due to all the irregularities involved, but some degree of turbulence will normally occur and justify the use of the turbulent heat transfer correlations, several of which are well documented in the literature (Chapman 1974; Bejan 1984). This will probably be an overestimate of heat transfer when the pump rate is low and the drillstring is not rotating, and care must be taken to use appropriate heat transfer correlations.

A more detailed discussion of wellbore heat transfer is beyond the scope of this section. The reader is, however, urged to consult the work by Hasan and Kabir (2002) for an excellent discussion of a practical approach to this problem and for a bibliography of the seminal papers in the last few decades.

Semianalytic Models Using the Overall Heat Transfer Coefficient Concept. Ramey (1962) developed the now famous semianalytic model for the estimation of transient injection and production temperatures, which is the basis of most of the wellbore thermal models in use today. Several modifications to Ramey's approach have been proposed, and many thermal simulators in the production engineering and pipeline industry use some version of Ramey's approach.

In essence, Ramey's approach is very simple. He assumes azimuthal symmetry, 1D heat transfer (radial heat transfer dominates the problem), and poses the problem as a quasisteady 1D problem, with the transient behavior removed to the boundary between the wellbore and formation (last radial wellbore section at any given depth). The steady-state problem in 1D can be posed in terms of an overall heat transfer coefficient U_o [defined at a suitable radial location in the wellbore, typically the outside diameter (OD) of the tubing, U_{to}]. U_{to} can be found using the resistance analogy to the problem. Thus, in the steady-state region, over any axial elemental region ΔZ where the resistance does not change with depth (i.e., section over which geometry or fluids do not change with depth), the heat transfer Δq can be written as

$$\Delta q = U_o A_h \left(T_h - T_f \right) = U_{to} \, \pi d_{to} \Delta Z \left(T_h - T_f \right), \quad \dots\dots\dots\dots\dots\dots\dots\dots\dots\dots \quad (8.1)$$

where A_h is the heat transfer area at the location where U_o is defined (for instance, the outer surface area of tubing), T_h is the temperature at the boundary of the wellbore (at a radius $r = r_h$), and T_f is the temperature of the flowing fluid. The overall heat transfer coefficient can then be written in terms of the resistances of each radial component in the system over the selected axial region:

$$U_o = \frac{1}{\sum_{i=1}^{N} R'_i},$$

where R' represents the resistance of each component (e.g., steel, fluids, and cement).

The transient temperature distribution in the element ΔZ and region $r_h < r \leq \infty$ is governed by the 1D transient heat conduction equation,

$$\frac{\delta^2 T}{\delta r^2} + \frac{1}{r} \frac{\delta T}{\delta r} = \frac{1}{\alpha_e} \frac{\delta T}{\delta t}, \; r_h < r \leq \infty,$$

subject to the initial condition

$$T = aZ + b \; at \; t = 0 \; \text{(geothermal temperature, assumed linear)}$$

and boundary conditions

$$T = T_e = aZ + b, \, r \to \infty, t > 0 \; \text{(geothermal far from wellbore)}$$

and

$$-k_e \frac{\delta T}{\delta r} = \Delta q = U_o A_h \left(T_h - T_f \right) = U_{to} \, \pi d_{to} \Delta Z \left(T_h - T_f \right) \; \text{(continuity of heat flow)}.$$

This is a textbook problem in heat transfer, with the solution being given by

$$\Delta q = \frac{2 \pi k_e \left(T_h - T_e \right)}{f \left(\tau, \text{Bi} \right)} \Delta Z, \quad \dots\dots\dots\dots\dots\dots\dots\dots\dots\dots\dots\dots\dots\dots\dots\dots \quad (8.2)$$

where Bi is the Biot number, τ is the Fourier number, and the function f is obtained from the zeros of a Bessel function.

Because Δq is conserved, and, furthermore, because Δq is the heat lost by the flowing fluid as it flows up the tubing, we have

$$\Delta q = \frac{2\pi k_e \left(T_h - T_e\right)}{f\left(\tau, \mathrm{Bi}\right)} \Delta Z = U_{to} \pi d_{to} \Delta Z \left(T_h - T_f\right) = -Q_m C_{p,f} dT_f. \quad \dots \dots \dots \dots \dots \quad (8.3)$$

Eq. 8.3 results in a first-order ordinary differential equation (ODE) in T_f, whose analytical (nondimensional) solution is given by

$$T_f(Z,t) = aZ + b - aA(t) + \left(T_0 + aA(t) - b\right)e^{Z/A}, \quad \dots \dots \dots \dots \dots \dots \quad (8.4)$$

where

$$\frac{A}{L} = \mathrm{Pe}\left[\frac{1}{\mathrm{Bi}} - f\left(\tau, \mathrm{Bi}\right)\right]$$

$$\mathrm{Pe} = \frac{Q_m C_{p,f}}{2\pi k_e L}$$

$$\mathrm{Bi} = \frac{U_{to} r_{to}}{k_e}$$

$$\tau = \frac{\alpha_e t}{r_h^2}.$$

Pe is a Peclet number, representing the ratio of the heat carrying capacity of the flowing fluid to the heat removing capacity of the formation (high Pe means that fluid carries heat faster than the formation removes it); Bi is the Biot number, which is representative of the ratio of heat transfer within the wellbore to that in the formation; and, finally, τ, the Fourier number, is the nondimensional time.

Also, in the above equations, L is the length of the axial element over which Eq. 8.3 is applied and integrated, and T_0 is the boundary condition at a known Z.

Pe, Bi, and τ are the three nondimensional numbers that govern the problem. (Although Ramey does not use nondimensional numbers in the development presented in his 1962 work, it is convenient to do so, as shown here.) Their relative magnitude is indicative of the impact of temperature on the design of the wellbore. For instance, high Pe and Bi indicate that the outer strings and annuli of the wellbore can heat up significantly, while low Bi indicates that regardless of Pe, the outer strings and annuli do not heat up significantly.

In terms of these nondimensional numbers, Eq. 8.3 gives excellent results when $\tau > 0.01$ (in terms of time, approximately 1 hour or greater). Because times shorter than this are unlikely to be of interest in most HP/HT situations, this is more than adequate. Several improvements to this approach have been suggested, notably by Hasan and Kabir (2002), to improve its accuracy for shorter time.

Once T_f is found using Eq. 8.4, the temperature at any other radial location can be found easily by using the definition of Δq and its radial continuity.

It should be noted that because the solution to the natural convection problem requires knowledge of the temperature gradient across the enclosure (which is unknown), the problem therefore requires an iterative solution. Previous applications of the above approach show that convergence is very fast (3 to 5 iterations).

For the flowing fluid, the heat transfer coefficient can be obtained using a suitable correlation, which will require knowledge of the thermal properties of the fluid as a function of axial location. An accurate solution of this problem requires the use of a multiphase flow correlation (to obtain the fraction of liquid and gas at any location, and hence the density and viscosity at that location). A simple multiphase flow correlation [such as the one suggested by Hasan and Kabir (2002), for instance] can be used.

However, even a rough averaging of the properties may be reasonable for this problem, because the resistance of the flowing fluid is orders of magnitude smaller than that of the annular fluids and cement. Therefore, error in the flowing fluid heat transfer coefficient does not contribute greatly to the error in the temperature estimate.

Finite-Difference Methods: Use of Thermal Simulators. Most wellbore heat transfer analyses are conducted using wellbore thermal simulator programs. Given a well configuration, the simulator calculates the temperature distribution in the wellbore by using analytical solutions or numerical methods (or a combination of these methods) for different operational circumstances. The simulator treats the wellbore as an axisymmetric 2D finite-difference grid (Wooley 1980). Heat transfer from the flow stream to the wellbore is calculated by using the usual correlations for convective heat transfer. Natural convection in the annuli is modeled by using correlations such as those given by Dropkin and Somerscales (1965), which are used to determine a conductivity augmentation factor. Natural convection correlations used in modern thermal simulators are based on empirical work on short vertical annuli or between flat plates with aspect (length to gap) ratio on the order of 100. This can lead to errors in temperature estimation because natural convection in wellbores occurs in tall, vertical annuli. The problem of natural convection in tall vertical annuli, where multiple natural convection cells arise, is an unsolved problem to the best of the author's knowledge. Heat transfer in the tubing and cement sections is modeled by conduction. Heat transfer in the formation is calculated by solving the 2D heat conduction equations. Suitable boundary conditions (far away from the wellbore and at the wellbore-formation interface) are imposed. Generally, the distance at which the geothermal boundary condition is imposed is a user selection.

Because production flows are usually multiphase, solution of the simultaneous pressure/temperature problem requires the use of multiphase flow models and correlations. Most production engineering programs, which focus on pressure and rate calculations, incorporate a wide range of multiphase flow models, both mechanistic and empirical. In some of the wellbore thermal simulators used in drilling engineering, the focus is on temperature estimation, which results in a compromise where multiphase flow models are concerned. Some workers (Hasan and Kabir, 2002) have developed simple multiphase flow corrections to Ramey's approach and these models are usually applicable in systems with low to moderate gas/oil ratio. Needless to say, the selection of the appropriate multiphase flow model is critical in obtaining reasonable estimates of pressure and temperature. In particular, most correlations are suitable for vertical wells, and care should be exercised when dealing with horizontal and deviated wells.

The PVT behavior of production fluids (and as we shall see, drilling and completion fluids) is of importance when dealing with HP/HT conditions. The range of available PVT models depends upon the simulator under consideration, but most allow at least black-oil PVT models.

The output of a wellbore thermal simulator consists of the temperature in each string and annulus of the wellbore as a function of depth. In addition, the pressures in the flow stream are also calculated and provided as an output.

Errors in Flowing Temperature Estimation. Sometimes, flowing temperatures can be measured to calibrate models and assist in well design. The advantages and disadvantages of the techniques used to predict bottomhole circulating/static temperatures are discussed in **Table 8.2.**

8.5 Thermal Effects

8.5.1 Effect of Temperature on Material Properties. Several materials are used in a wellbore, and the elevated temperatures in HP/HT wells can affect the properties of these materials.

TABLE 8.2—TECHNIQUES USED TO MEASURE BOTTOMHOLE CIRCULATING/STATIC TEMPERATURES*	
Technique	Comment
API tables	Inaccurate; based on limited number of wells
Local knowledge	Inaccurate
Computer simulations	Need more validations over a large range of conditions
Temperature probe	Accurate; requires an assumption on the temperature; difficult to recover on surface among the cuttings
Downhole thermometers	Expensive; time-consuming
*Vidick and Acock (1991)	

Yield and Ultimate Strength. The material stress-strain curve is a function of temperature. Elevated-temperature stress-strain curves are used to obtain the yield strength of steels and alloys used in well construction. For casing material, the rate of decrease in yield strength with temperature is a function of the grade under consideration, and it must be obtained experimentally. For design purposes, linear deration of yield strength is commonly used, with the rate of deration usually approximately 0.02 to 0.05% per °F (0.04 to 0.09% per °C). A typical value used for yield strength deration is 0.03% per °F (0.054% per °C).

In design, the derated yield strength is used in place of minimum yield strength in the strength calculations to account for the temperature effect. The geothermal temperature at the depth of interest is used in the strength reduction calculation. It is common to derate yield strength in burst and von Mises equivalent stress (VME). In practice, collapse strength usually is not derated for temperature. However, it is appropriate to do so, because yield strength governs collapse in three of the four modes of collapse.

In limit state functions, ultimate strength often is used instead of yield strength. Ultimate strength of steels is not a strong function of temperature within the practical range of wellbore temperatures (<500°F or 260°C). *Therefore, it is not appropriate to derate ultimate strength for temperature.*

Thermal Properties. The thermal conductivity of materials typically decreases with temperature, and this should be considered when thermal elongation and thermal stresses are of interest.

Other Material Properties. The *fracture toughness*, an important property of materials governing brittle failure, is a function of material temperature. Because HP/HT wells often demand the use of non-steel alloys, fracture toughness should be determined for these materials under in-situ conditions of environment and temperature. Fracture toughness and environmentally assisted cracking are discussed in Section 8.5.3.

Sealing Elements and Elastomers. Elastomer properties are influenced by temperature. At low temperatures, the elastomers become hard and resistant to deformation, and at higher temperatures, they become soft and flow under pressure (Berckenhoff and Wendt 2005). The effect of temperature on sealing elements is one of the critical concerns in HP/HT operations. Although metal seals are an alternative to elastomers at elevated temperatures, they are much more expensive. Thermoplastics exhibit more stability at elevated temperatures and may be used as sealing elements.

8.5.2 Effect of Temperature on Fluid Properties. HP/HT conditions also affect the properties of the different fluids used in the construction of a well. Some of the fluids and the impact of temperature on the properties of the fluids are discussed in this section.

HP/HT Fluids. The composition of HP/HT fluids can be considerably different from that of conventional reservoir fluids. HP/HT conditions increase the volatility of heavy compounds, which results in gas condensates being richer in heavy compounds. Furthermore, volatility of water also increases with temperature; hence, water can become a major constituent of HP/HT fluids and therefore should be taken into account while studying the phase behavior of HP/HT fluids (Gozalpour et al. 2005). **Fig. 8.6** compares measured viscosity of a model volatile oil at a temperature of 392°F and at pressures above the bubblepoint, without water and with 5.4 mol% of water (Danesh 2002). A 20% increase in viscosity is observed in the presence of water. Therefore, higher water content in HP/HT fluids is an important consideration in their phase behavior, unlike conventional reservoir fluids. Many HP/HT fluids also have considerable solids content.

Drilling Fluids. Drilling fluids are complex fluids consisting of different chemical additives that must remain stable at high temperatures and pressures. Water-based drilling fluids often contain bentonite clay, and thermal instability is often associated with bentonite-based fluids. Oil-based fluids often contain organophilic clays that do not experience the large changes in physical properties with increasing temperature observed in bentonite-based fluids (Fisk and Jamison 1989). Many drilling fluid products are susceptible to thermal degradation at elevated temperatures. High-temperature gelation occurs in both OBM and water-based mud (WBM) in addition to high-temperature fluid loss (Oakley et al. 2000). In WBM, clay (bentonite) flocculation causes gelation, and the situation is further compounded by thermal degradation of thinners, a drop in pH, and an increase in filtrate loss. In OBM, the interaction of colloidal particles (clays and fluid-loss additives) and breakdown of emulsifiers may cause gelation. The rheology of the drilling mud needs to be controlled at HP/HT conditions because it may lead to poor hole cleaning, barite sag, and a nonuniform density profile in the annulus. Cesium formate is being recognized as a potential drilling fluid for HP/HT wells due to its no-sag potential and low solids content.

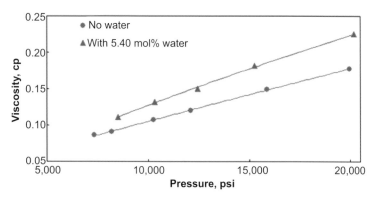

Fig. 8.6—Measured viscosities of volatile oil at different pressures with and without water at 392°F. This figure is a reproduction from Reservoir Fluid Studies. *Sharp IOR eNewsletter* **3 (September 2002): 4.2.2, Fig. 2 by Danesh, a 2002 copyright owned by Heriot-Watt University.**

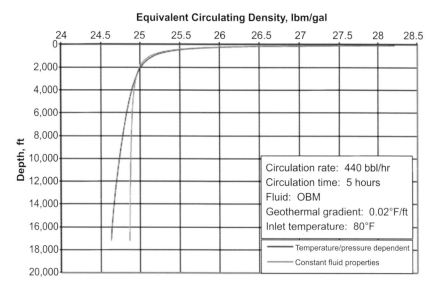

Fig. 8.7—Equivalent circulating density in a 17,200-ft well (Harris and Osisanya 2005).

In HP/HT wells, high-temperature conditions cause the fluid in the wellbore to expand, while high-pressure conditions cause fluid compression. These two conditions have opposing effects on the ECD of the drilling fluid and, therefore, on the estimation of the bottomhole pressure (Harris and Osisanya 2005). Hence, the temperature and pressure effects on ECD must be considered in HP/HT wells. **Fig. 8.7** illustrates the difference in ECD estimation with and without consideration of temperature and pressure. As depth increases, the ECD dependent on pressure and temperature continues to decrease relative to the ECD assuming constant fluid properties. This trend is a result of the more dominant effect of thermal expansion, as opposed to compression from increased pressure (Harris and Osisanya 2005).

Accurate prediction of ECD in HP/HT wells requires knowledge of the pressure and temperature dependence of the rheological properties of the drilling mud and the accurate temperature profile in the well (Maglione et al. 1996; Rommetveit and Bjørkevoll 1997). In the absence of laboratory data, models are available that predict pressure and temperature dependence of rheology. **Figs. 8.8 through 8.11** illustrate results from an HP/HT simulator for a 5000 m deep North Sea well and a water depth of approximately 100 m. Two different mud systems are used, WBM and OBM. The pressure dependence is shown to be more pronounced for OBM than WBM. The circulating temperature profiles for both muds are different, with the WBM being cooler than the OBM in the lower part of the well. This is due to the lower viscosity and higher specific heat of WBM. In Figs. 8.8 and 8.9, the density profiles for OBM and WBM vary with depth, with the density of

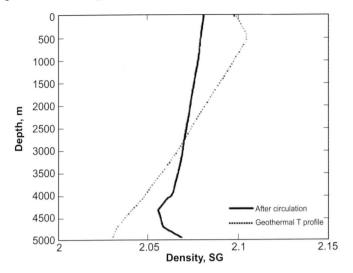

Fig. 8.8—Calculated mud density profile for WBM along the well with geothermal temperature profile and circulation temperature profile (Rommetveit and Bjørkevoll 1997).

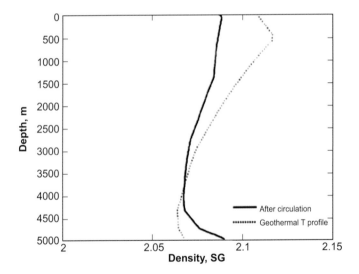

Fig. 8.9—Calculated mud density profile for OBM along the well with geothermal temperature profile and circulation temperature profile (Rommetveit and Bjørkevoll 1997).

WBM being more temperature-dependent compared to OBM. In Figs. 8.10 and 8.11, the equivalent viscosity profiles for OBM and WBM vary with depth. Mud rheology is not only temperature and pressure dependent, but also depends on shear history. In the upper riser section, the annulus is very wide and the equivalent viscosity is large due to low shear rates. Below the riser section and down to approximately 3000 m, shear rates are larger than in the riser, viscosity decreases with increasing temperature for both mud systems. In the lower part of the well, both the temperature and the shear rates are higher. The viscosity of the WBM is not largely impacted in the lower part of the annulus, while that of OBM decreases with increase in temperature. Clearly, calculation of density, viscosity, and pressure profiles is more challenging in HP/HT wells. (Note that similarly, the temperature will impact the thermal properties of packer fluids during production.)

PVT models for wellbore fluids have been developed, and, in particular, the Zamora et al. (2000) approach of characterizing the dependence of fluid density on temperature and pressure is commonly applied in thermal analyses of wellbores. They propose a quadratic, six-parameter curve fit for density in

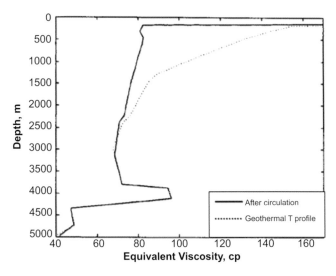

Fig. 8.10—Calculated equivalent viscosity profile for WBM along the well with geothermal temperature profile and circulation temperature profile (Rommetveit and Bjørkevoll 1997).

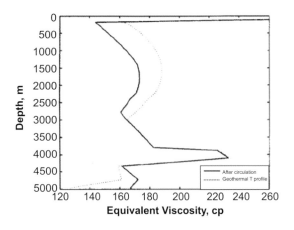

Fig. 8.11—Calculated equivalent viscosity profile for OBM along the well with geothermal temperature profile and circulation temperature profile (Rommetveit and Bjørkevoll 1997).

terms of temperature and pressure. The parameters of the equation are determined experimentally. Care should be exercised when extrapolating outside the experimental data; second-order terms may blow up to unrealistic values. It should be checked that the Zamora model fits the experimental data reasonably well. If not, a standard interpolation/extrapolation method can be considered, but this should be used with caution.

A number of correlation-based models have been published and can be used when good PVT data at actual pressures and temperatures are not available for the fluid(s) under consideration. Such models should be used with care, especially when pressure and/or temperatures outside the recommended range of the correlation are used. There are several PVT models for crudes (Glassø 1980; Standing 1977) and brines (Kemp and Thomas 1987), but for muds, it is general practice to use the Zamora et al. (2000) approach, or conduct tests to obtain the density and viscosity curves as a function of temperature and pressure. Zamora et al. (2000) propose a curve fit for density, ρ, in terms of temperature, T, and pressure, P:

$$\rho(P,T) = \rho_w \left[(a_o T + b_o) + (a_2 T + b_1)P + (a_2 T + b_2)P^2 \right]. \qquad (8.5)$$

The six coefficients of the equation, commonly known as the *Zamora* coefficients, are determined experimentally.

Cements. High-temperature cementing operations are generally performed using API Class G or Class H cements. High temperature escalates the hydration of cements and therefore decreases the thickening time. Cement hydration is highly dependent on temperature because the rate of the heat generation associated with hydration increases with temperature (Romero and Loizzo 2000). Because the bottomhole static temperatures are high in HP/HT wells, accurate prediction of bottomhole circulating temperatures is important to ensure that the cement sets at the right time. Factors such as rheology, fluid loss, stability, and compressive strength are influenced by temperature. The compressive strength of the set cement deteriorates with time, which is more likely once the expected bottomhole static temperature exceeds 225°F (MacAndrew et al. 1993). In addition to compressive strength of cement, the tensile and flexural strengths, elastic modulus, and Poisson's ratio are important to maximize the integrity of cement. The spacer for mud displacement, anti-gas-migration additive, and retarders must be selected based on the maximum expected downhole temperature (Mansour et al. 1999). Accurate temperature prediction therefore is extremely important to best simulate and test the cement slurries before the cementing operation is executed.

Foamed cement systems are emerging as alternatives to nonfoamed cement systems for HP/HT wells. Foamed systems can be optimized by varying the amount of nitrogen and base slurry composition to compensate for volume reduction and optimize mechanical properties (Griffith et al. 2004). The ability to control density also improves the possibility of a full cement column. In addition, foamed cement systems offer improved mud displacement, fluid loss, and rheology, as well as increased elasticity, better fracture toughness, and higher tensile strength.

8.5.3 Effect of Temperature on Connection Performance.

The thermal effects of high-temperature wells most likely subject the tubular connections to high compression loads during production and high external pressures due to annular fluid expansion (Bradley et al. 2005). These loads are applied in a 3D state, and connections design is based on triaxial performance criteria commonly referred to as VME stress. Finite-element analysis frequently is used as a design tool for API and proprietary casing and tubing threaded connection designs, along with the physical testing procedures documented in *ISO 13679* (2002). It is noted (Carcagno 2005) that the new *ISO 13679* (2002) exposes the connection to extreme loading conditions and thermal-pressure cycling to 180°C and is more demanding than previous editions of API *RP 5C5*, which has back-adopted *ISO 13679 (2002)*. Radial metal seals are the critical component for isolating or containing the high formation pressures encountered in HP/HT well drilling and completions and may be accompanied by resilient seals as a backup (Bradley et al. 2005). In essence, for HP/HT applications (which almost always imply gas is present), proprietary connections, tested to the CAL IV testing protocol of *ISO 13679* (2002), are usually required.

8.5.4 Effect of Temperature on Loads.

The standard load cases of interest in well design still apply, but the thermal effects lead to several special considerations when designing HP/HT wells. In addition, special load cases arise in HP/HT well design and need to be considered.

Thermal Elongation and Thermal Stresses. When tubulars are subjected to an increase in temperature, they elongate. In fully cemented sections, elongation is not possible, and a compressive axial thermal stress results. In unsupported sections of tubulars, depending upon the end conditions, some or all of the elongation is not permitted, resulting once again in a thermal stress. The thermal stress usually is ignored in conventional well design but can be critical in HP/HT well design.

The average temperature change across an uncemented string can be calculated from estimated initial and final temperature profiles as

$$\Delta T = \frac{1}{L_{uc}} \int_0^{L_{uc}} \left(T_{final} - T_{initial} \right) ds, \quad\dots\dots\dots\dots\dots\dots\dots\dots\dots\dots\dots\dots\dots\dots\dots \quad (8.6)$$

where L_{uc} denotes the length of the uncemented section of the casing.

The average thermal expansion of the casing string is then

$$\Delta L_T = \alpha_{steel} L_{uc} \Delta T, \quad\dots\dots\dots\dots\dots\dots\dots\dots\dots\dots\dots\dots\dots\dots\dots\dots\dots\dots\dots\dots\dots \quad (8.7)$$

where α_{steel} is the coefficient of thermal expansion of steel. This has a typical range of 6.5×10^{-6} to 6.9×10^{-6} per °F for steel. If this thermal expansion is suppressed, thermal stress is created in the string thusly:

$$\sigma_{\text{thermal}} = E\alpha_{\text{steel}}\Delta T, \quad \dots \dots \dots \dots \dots \dots \dots \dots \dots \dots \dots \dots \dots \dots \dots \dots \quad (8.8)$$

where E, the modulus of elasticity of steel, is usually taken to be 3×10^{7} psi. Both E and α_{steel} are functions of temperature, but in US units, their product is close to 200 psi/°F, leading to the well-known rule of thumb for thermal stress calculation, $\sigma_{\text{thermal}} = 200\,\Delta T$.

The corresponding thermal force in the string is given by

$$F_{\text{thermal}} = A_p E\alpha_{\text{steel}}\Delta T, \quad \dots \dots \dots \dots \dots \dots \dots \dots \dots \dots \dots \dots \dots \dots \dots \quad (8.9)$$

where A_p is the cross-sectional area of the casing.

Buckling of Unsupported Sections. The combination of hydrostatic forces and thermal forces can lead to an effective compressive load on unsupported casing or tubing sections, leading to buckling. Buckling of casing strings can also occur due to nonthermal loads such as formation subsidence and pressure changes in the annuli. Irrespective of the loads that cause buckling, it is necessary to understand the causes of buckling and determine if the buckled state of the tubular can compromise the integrity of the string or the tubular.

An unsupported casing string buckles when it experiences an *effective* compressive load greater than its critical buckling load. The critical buckling load is a function of the casing geometry, the nature of the hole (i.e., vertical, inclined, or curved), the clearance between the casing and the hole section, and the fluids inside and outside the casing.

Buckling of drillpipe and casing/tubing in constrained holes has been extensively studied by the drilling community. In fact, buckling of oil country tubular goods (OCTG) is a subset of the mechanics of tubulars subjected to hydrostatic, thermal, and mechanical loads. The readers are strongly urged to consult the works of Paslay (1994), He and Kyllingstad (1995), Lea et al. (1995), Mitchell (1986), Sparks (1984), Hammerlindl (1980), Lubinski (1951), Lubinski et al. (1962), Suryanarayana and McCann (1995), and Handelman (1946) for a discussion of issues related to the mechanics of constrained tubulars in general and issues of their elastic stability in particular. A detailed discussion of the theory of buckling and post-buckling behavior is beyond the scope of this chapter and is covered in Chapter 3.4.

The Effective Force. As described in the previous section, the tubular is said to be buckled if $|F_e| > |F_{cr}|$, where F_e is the effective compressive load and F_{cr} is the critical buckling load. This may happen over only a portion (usually the lower portion) of the unsupported pipe. A buckling neutral point is often defined, above which the pipe is in effective tension and therefore cannot buckle.

Note that in determination of buckling tendency, the effective force is used rather than the real force. In buckling and post-buckling analyses, therefore, a distinction between real and effective forces must be made. The physical basis for this distinction is described by Handelman (1946), Lubinski (1951), Lubinski et al. (1962), and Sparks (1984). In particular, the work by Sparks is an excellent description of the concept of effective force and its importance in all mechanical systems in which mechanical and fluid forces act. Pressure forces conveyed by fluids act on tubulars, but because fluids offer no flexural rigidity, bending and buckling behavior cannot be affected by these pressure forces. The effective force is essentially the force that exists when these forces are excluded,

$$F_e = F_{\text{real}} - p_i A_i + p_o A_o, \quad \dots \dots \dots \dots \dots \dots \dots \dots \dots \dots \dots \dots \dots \dots \quad (8.10)$$

where F_e is the effective compressive load, F_{real} is the real axial force, p_i and p_o are the internal and external pressures acting on the tubular respectively, A_i and A_o are the inside and outside surface areas of the tubular respectively. Buckling is influenced by forces caused by pressure as well as by real axial forces (e.g., weight, overpull or slackoff, thermal forces), and their combined effect is best described by using the effective force. Note, from Eq. 8.10, that the effective force can be negative (i.e., buckling is possible) even when the real force is positive (i.e., the stress state is tensile), depending upon the internal and external pressure and upon the inside and outside surface areas of the pipe. Conversely, even when the real force is compressive, buckling may not occur.

This is the reason why, although the compressive stress increases with depth in a purely hydrostatic environment (such as when a pipe is lowered into an infinitely deep sea), the pipe cannot buckle (the reader can verify that in this case, the effective compressive force at the free end of the pipe is zero regardless of the depth).

Post-Buckling Behavior. A post-buckling helical shape creates additional bending stress in the tubular. The post-buckling bending stress, $\sigma_{b,h}$, is given by (Lubinski et al. 1962)

$$\sigma_{b,h} = \pm \frac{d_o r_c}{4I} F_e, \dots\dots\dots\dots\dots\dots\dots\dots\dots\dots\dots\dots\dots\dots\dots\dots\dots \quad (8.11)$$

where F_e is the effective compressive force, d_o is the OD of the tubular, I is the moment of inertia of tubular cross-section with respect to its diameter and r_c is the radial clearance. Note that the bending stress is a function of the effective force acting on the tubular.

When a tubular is buckled helically, it is in contact with the constraining hole or, in the case of drilling through an unsupported section of casing, with the drillpipe. The compressive load on the string creates a normal contact force F_n between the buckled string and the constraining hole or the contacting drillpipe. This normal contact force is given by Mitchell (1982):

$$F_n = \frac{r_c F_e^2}{4EI}. \dots\dots\dots\dots\dots\dots\dots\dots\dots\dots\dots\dots\dots\dots\dots\dots\dots\dots\dots\dots\dots\dots\dots \quad (8.12)$$

Note that the normal contact force is the force per unit length of contact.

It is common to calculate an effective dogleg severity (DLS) for helically buckled pipe, as a function of the geometry, stiffness, and effective compressive force. The DLS is a useful characterization of buckling, especially in estimating wear when drilling through the buckled section. The DLS is given by

$$\text{DLS} = C_n \frac{r_c |F_e|}{2EI}, \dots\dots\dots\dots\dots\dots\dots\dots\dots\dots\dots\dots\dots\dots\dots\dots\dots\dots\dots\dots\dots \quad (8.13)$$

where C_n is a conversion constant, depending upon the system of units. For US customary units, $C_n = 68,755$, and the DLS is in degrees per 100 ft. The other variables are as defined earlier.

Implications for HP/HT Well Design. In HP/HT wells, due to the elevated geothermal temperature gradient, drilling fluid return temperatures can be very high, thus heating the upper, and usually unsupported, portions of the casing being drilled through. As a result of the heavier mud weight (and therefore higher internal pressure) and higher temperatures, the effective force on the unsupported casing section can become negative and can lead to buckling. Because the drillpipe contacts the buckled casing, the normal contact force that results can lead to wear of the casing and, hence, compromise of its integrity. This has to be checked in design. In practice, if the equivalent DLS of the buckled casing (Eq. 8.13) is lower than 2° per 100 ft, wear is unlikely to be a problem. If wear emerges as a concern, remedial measures such as wear protection or application of pretension may have to be considered.

WHM and Redistribution of Tubular Stresses. Temperature changes (and, to a certain extent, annular pressure changes) in a wellbore can cause length changes in the tubulars. Because at least some of these tubulars terminate at the wellhead, the tubulars are forced to move such that their axial displacements at the wellhead are the same. This phenomenon is known as WHM. Though WHM can occur during different stages of well construction, the term generally refers to the movement caused during production.

The movement of the wellhead is a response to the thermal forces exerted by the expansion of the strings (hung off the wellhead) and to the pressure-area forces in the annular spaces. This response is countered by the axial stiffness of the strings and the resistance of the structural member (usually the conductor or the surface casing on which the wellhead is installed).

The axial stiffness of the tubulars depends on the *point of fixity* (i.e., the lowest point of the tubular that experiences zero axial displacement). Typically, the point of fixity for the inner strings coincides with the cement top. Sometimes, depending on specific well geometry, assumptions regarding the point of fixity for the inner strings may have to be examined. However, choosing the point of fixity for the structural member (which usually is in contact with a cement sheath or with the formation), and therefore evaluating its resistance to the thermal forces exerted by the inner strings, poses some difficulty. WHM also is affected by the thermal expansion of the porous soil near the mudline (which is not as compacted as the deeper formations).

Though WHM models are available (Samuel and Gonzales 1999), the correlation between the observed growth in real wells and model predictions is less than satisfactory. The reasons for the poor correlation between the observed magnitudes of WHM and predicted values are likely as follows:

- Inadequate understanding/modeling of shallow soil resistance and interaction with the structural members, especially near the mudline
- Neglecting effects of thermal expansion of the soils and the pore fluids, given that the greatest temperature changes occur near the mudline during conditions of steady-state production
- Incomplete modeling (or neglect) of nonlinear soil-casing interaction that lead to ratcheting caused by cycles of prolonged periods of production and shutdown
- Effects of preloads on the strings
- Miscellaneous effects at the casing-cement interface and assumptions regarding the point of fixity

A complete model that accounts for these effects usually requires the use of finite-element methods. Therefore, if WHM is expected, a detailed model that considers the specific features of the wellbore, soil properties, and other attendant details must be considered.

APB. APB is a consequence of the difference between the volume change of the annular fluid and the volume change of the casing strings that form the annulus. Annulus fluid volume change may be caused by thermal expansion and/or addition/removal of fluid. The annulus changes its volume in response to fluid pressure and temperature changes while maintaining mechanical equilibrium at all times. Though the physics of APB is well understood and though there are well-documented methods of calculating APB, there is significant uncertainty in the assessment of the risk caused by APB. This is partly due to the fact that APB is a function of several variables, chief among them being fluid PVT response, uncertainties in inputs to the thermal models used to predict wellbore temperatures, second-order unknown variables such as wellbore elasticity, behavior of openhole sections, and long-term behavior of fluids sealed in inaccessible subsea annuli (Halal and Mitchell 1993; Payne et al. 2003). As a result, most APB assessments, while theoretically sound, must be regarded as a probable range of estimates. Therefore, casing strings and the wellbore must be designed so that the wellbore is robust enough to ride the uncertainties inherent in APB-induced loads.

Wellbore integrity issues in HP/HT and deepwater wells can seldom be addressed solely by increasing the tubular strengths (Payne et al. 2003). The differential collapse and burst loads on the various strings must be controlled by explicit APB management and active consideration of APB-induced loads on the basis of casing design. Wellbore integrity studies recommend at least two mitigation strategies (not counting an open casing shoe) to manage APB (Payne et al. 2003; Sathuvalli et al. 2005), and require that each annulus *stand on its own* (i.e., the strings forming any given annulus must survive APB in that annulus with the neighboring annuli vented). This recommendation is based on a consideration of loads likely to occur during the service life of the well, rather than those that occur immediately on well startup. Such considerations include sensitivity analyses of late-life wellbore thermal states, the possibility of converting producers to injectors, and accounting for the long-term effects of muds in annuli. In general, the overall APB management strategy is finalized by trial-and-error analyses of load scenarios in the presence of mitigation strategies, while simultaneously optimizing the overall risk-weighted cost. The current state of the art implies that rupture disks, nitrogen gas cushions, and solid syntactic foams are the methods of APB mitigation most often used (in increasing order of cost). Though vacuum-insulated tubing has been used [for example, in the BP Marlin wells in the GOM (Ellis et al. 2004)], it may not always present the most cost-effective solution. Other methods such as fluids with gas-laden but stable bubbles (and hence higher bulk compressibilities) have been suggested (Belkin et al. 2005). However, such methods are yet to be field-proven.

Design of a Mitigation Strategy. The design of an APB mitigation strategy is based on understanding the pressure-temperature response of the subject annulus and the mitigation device. The principles of how the more popular mitigation methods are used are detailed in Payne et al. (2003). Guidelines to design these mitigation methods are described in Sathuvalli et al. (2005). In general, irrespective of the mitigation strategy used, the design involves identifying an *operating point* at which the mitigation strategy is activated. This operating point (characterized by a pressure and temperature) is obtained by predicting the annular response to production heat-up. Usually, this involves routine (and sometimes tedious) but judiciously chosen thermal and APB sensitivity analyses.

Fig. 8.12 illustrates schematically the PVT response of the three commonly used mitigation strategies. The design of rupture disks is based on ensuring that disk placement in a string does not jeopardize its integrity to drilling loads. This usually implies verifying that the integrity of the subject string is not compromised until the differential pressures caused by APB-induced loads are exerted.

The crush pressure of syntactic foam is chosen by ensuring that it crushes at a predetermined pressure. This pressure must be clearly less than or equal to the allowable APB in the subject annulus. The allowable APB is a function of casing design. Because the crush pressure of syntactic foams usually decreases with increasing temperature, the design principle involves matching the crush pressure with the allowable APB. Once this operating point is determined, annular geometric and fluid expansion considerations are sufficient to determine size and volume requirements. It remains to ensure that the foam response outpaces the transient pressure increase in the annulus during startup. The rest of the design involves details of ensuring integrity under different load combinations.

The design of gas caps is straightforward and involves no new physical principles. The design is dictated more by practical and operational considerations. A general rule of thumb is that gas cap efficiency is maximal when its in-situ volume (during placement) is roughly 10% of the annular volume. However, the depth of the cement top and other considerations can influence the final parameters that define a gas cap.

A key aspect in the design of a mitigation strategy is the notion of allowable APB. The allowable APB is tied to the design basis of the well. For example, the most severe collapse load on the production casing occurs when the internal pressure in the tubing annulus (i.e., the annulus outside the production tubing) is the least. If the well is gas lifted late in its life, the possibility of near-vacuum conditions in the tubing annulus may have to be considered. Alternatively, if gas lift is not planned, the possibility of evacuation (of the packer fluid) to a depleted reservoir may have to be assessed. In short, a string-by-string analysis, to determine the worst load scenarios that the mitigation strategy must protect against, must be performed. This process, therefore, involves careful review of the design basis of the well (or field).

Simultaneously, the assessment of the loads is accompanied by an assessment of tubular strengths. Recent and ongoing studies have shown that the performance properties of the OCTG (if manufactured to well-toleranced

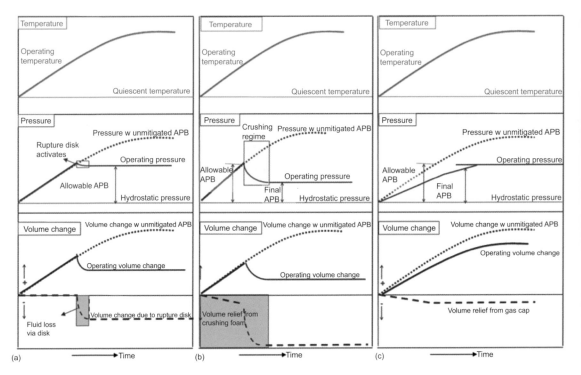

Fig. 8.12—Transient annular PVT responses with various APB mitigation strategies: (a) burst disks, (b) syntactic foam, and (c) nitrified spacers.

specifications) can exceed the specifications. In this light, ISO and API work groups are addressing the revision of OCTG performance properties (*ISO DIS 10400* 2004). Depending on the well parameters, a complete mitigation strategy may require an assessment of tubular strengths based on statistical analyses of mill data and advanced models to assess tubular performance properties (Klever and Stewart 1998; Adams et al. 2001). Such approaches are being used in several GOM deepwater developments on a case-by-case basis (Payne et al. 2003).

Thus, a complete design of a mitigation strategy will start with a review of the basis of the well design and with a string-by-string analysis to determine the allowable loads that the strings can sustain. This step also establishes the allowable APB magnitude in each annulus. The preceding results are used to design the mitigation strategy. As noted earlier, wellbore integrity studies suggest at least two mitigation strategies per annulus. This last step involves assessing the performance and interaction of the two mitigation strategies for every annulus.

Geomechanical Stress Redistribution and Resultant Loads. By definition the pore pressure is very high in HP/HT wells and, therefore, depletion during production can have a significant effect on the geomechanical stresses in the reservoir section. This is because the magnitude of depletion (in psi) is much greater in an HP/HT reservoir than in a conventional reservoir. The reduction in pore pressure is associated with a reduction in horizontal stress. Consequently, the fracture closure stress (lower bound of fracture pressure) reduces, sometimes significantly enough to affect future drilling and well design. This is a well-recognized phenomenon and is often discovered empirically during hydraulic fracturing of depleted reservoirs. It is common to define a linear stress-path coefficient, the ratio of the reduction in fracture pressure to that in pore pressure. Numerous published data (Salz 1977; Warpinski et al. 1991; Kristiansen 1998) indicate that reduction in fracture pressure can be 50% to 80% of the magnitude by which the pore pressure reduces. For example, Salz (1977) presents the following relationship between pore pressure and fracture gradient as a function of depletion, in the Vicksburg formation in south Texas. **Fig. 8.13** shows this relationship. The initial pore pressure gradient was in the range of 0.8 to 0.9 psi/ft, with the corresponding fracture pressure being approximately 1.0 to 1.05 psi/ft. With approximately 40% depletion, when the pore pressure drops to 0.5 psi/ft, the fracture gradient falls to 0.8 psi/ft, in this case.

Subsidence and compaction caused by substantial depletion are two other effects of interest in well design. These are discussed in Chapter 5.1, Geomechanics and Wellbore Stability. Suffice it to mention here that subsidence and compaction can lead to shear forces on wellbore tubulars, compromise the cement bond, and ultimately lead to failure of wells. Due to the larger degree of depletion typical in HP/HT wells, this is a particularly important concern.

Long-Term Impact of Temperature. *Corrosion and Environmentally Assisted Cracking.* CRAs are commonly used in HP/HT wells, especially for production tubing. Because many HP/HT reservoir fluids have H_2S and CO_2, and because fluids with high chloride content are used in such wells, corrosion and environmentally assisted cracking remain important concerns in the design of HP/HT wells. CO_2 corrosion generally worsens with temperature. Alloy selection should take the partial pressure of CO_2, chloride content, and temperature into account.

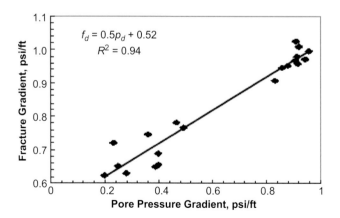

Fig. 8.13—Pore pressure and fracture gradient relationship in the Vicksburg formation, south Texas (Salz 1977).

The option of utilizing carbon steel tubing with inhibitors should be carefully considered as part of the overall tubular design. Defining the flow regime in the production tubing, along with characterization of the natural inhibition provided by the producing hydrocarbons, will define the regime of carbon steel applicability.

Because native pressure is high, the partial pressure of H_2S can be very high in HP/HT wells (Craig 1993). Therefore, although temperature does alleviate sulfide stress cracking (SSC), the higher partial pressure can lead to atomic hydrogen charging in susceptible alloy matrices, leading to acceleration of SSC. The threshold stress intensity factor (K_{ISSC}) is an important consideration in the design for SSC. *ISO DIS 10400* (2004) and *ISO 15156* (2003) [former *NACE Standard MR-0175* (2003)] describe methods to measure K_{ISSC}. The property, sometimes referred to as *fracture toughness*, is commonly measured using NACE method D, in fully saturated H_2S solution at atmospheric pressure (14.7 psia partial pressure) and temperature. Although this may be interpreted as *worst-case* value of K_{ISSC} for most applications, this may not be the case at elevated temperatures and partial pressures of H_2S. SSC is then correlated to the product of the mole fraction of H_2S and the system pressure (highest reservoir pressure) for design and material selection. This assumes ideal behavior. At higher pressures, the solubility of H_2S is less than indicated by ideal gas behavior. Because *dissolved* H_2S is implicated in SSC, it is important to determine the solubility of H_2S at elevated pressures. Nonideal gas equations (such as the ensemble Henry's law) can be used to determine the equivalent mole fraction of dissolved H_2S in such cases. It is recommended that tests be conducted at appropriate partial pressures of H_2S to determine the appropriate value of K_{ISSC} to be used in the design for SSC. *ISO DIS 10400* (2004) includes a brittle-burst equation based on fracture mechanics and failure assessment diagrams, which can be used with the measured value of K_{ISSC} to design for SSC.

Hysteresis in Stress-Strain Curves. The mechanical response of a metal depends not only on its current stress state but also on its deformation history (Xiang and Vlassak 2005). **Fig. 8.14** illustrates a typical stress-strain curve for metallic materials. The stress σ_f is the forward flow stress, σ_r at the start of reverse plastic flow is the reverse flow stress and σ_y is the yield stress. If $\sigma_r = \sigma_f$, the material hardens isotropically. For many metals, however, the reverse flow stress is found to be lower than the forward flow stress. This anisotropic flow behavior is known as the Bauschinger effect. The Bauschinger effect is normally associated with conditions in which the yield strength of a metal decreases when the direction of strain is changed (Han et al. 2005). The loss of strength due to this effect is of practical importance particularly if the working stress acts in a reverse direction compared to the manufacturing stress. It is also important when cyclic loading is imposed on the tubulars (such as in steam wells). In the more classical HP/HT applications, it is important to remember that design to limit states allows some wall plasticization. As a result, the Bauschinger effect is an important consideration in the extreme designs.

Thermal Creep. In the absence of a corrosive environment and at ordinary temperatures, say 32°F to 122°F, a properly designed member supports its static design load for an unlimited time (Boresi et al. 1993). But at elevated temperatures, a sustained load may produce inelastic strain in the material that increases with

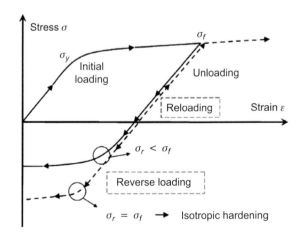

Fig. 8.14—Typical stress-strain curve of a metallic material that exhibits the Bauschinger effect (Xiang and Vlassak 2005). Reprinted with permission from Elsevier.

time. In such a situation, the material is said to creep, creep being defined as a time-dependent inelastic strain under sustained load and elevated temperature. If creep is sustained for a sufficiently long time, creep failure or fracture occurs. This phenomenon is dependent on the material and the environment. Therefore, as noted in *ASM Handbook* (1976), elevated-temperature behavior for various metals occurs over a wide range of temperature—for instance, at 700°F for low-alloy steels; 1000°F for austenitic iron-based high-temperature alloys; 400°F for aluminum alloys; and between 32 and 122°F for concrete and certain plastics (Boresi et al. 1993). In general, creep behavior is a function of the material, stress, temperature, time, stress history, and temperature history. Creep of materials at elevated temperatures is characterized by the fact that most of the deformation is irreversible and only a small part of the strain is recovered after the load is removed. Also, dependence of creep rate on stress is nonlinear. Therefore, the theory of creep of metals is patterned after the theory of plasticity in real situations (Boresi et al. 1993).

Thermal creep may become relevant in well design as the emerging development of X-HP/HT wells can expose low-alloy carbon steel casings to shut-in pressures over 18,000 psi, 50 to 450°F changing temperatures, and a sour environment containing H_2S, CO_2, and water. For such severe conditions, creep testing may be given consideration.

In general, however, thermal creep is not a major concern for steels at temperatures less than 700°F, but it could be an issue for CRAs.

8.6 Design of Wells for HP/HT Conditions

In general, as stated at the beginning of this chapter, HP/HT wells are unique in that

- The elevated temperatures and pressure place extraordinary demands on materials.
- Special load scenarios such as APB, WHM, subsidence and compaction, and buckling of unsupported strings become critical in well design. Moreover, the standard string-by-string design approach may no longer be adequate because design choices made for a certain string can impact the response of the other strings, and nonstandard loads lead to interaction between strings that is not common in conventional wells.
- Fluids and cements have to be designed with the HP/HT conditions in mind because their PVT and rheological behavior can be very different when compared to standard well designs.

Thus, the design of wells for HP/HT conditions requires special design approaches. This section covers some of the important implications of HP/HT conditions for well design.

8.6.1 Load Basis for Design. The standard load basis used in conventional well design still applies. However, deterministic load methodologies that work for conventional well design may be too conservative for HP/HT wells. The two most common loads where the deterministic load basis may be too conservative are the kick or well-control load case for drilling casings, and the tubing leak load case for production casing. Kick design usually considers a worst-case gas kick, with the annulus full of gas balancing the fracture pressure of the weakest exposed formation (usually the previous casing shoe). The tubing leak load case usually considers a leak in the tubing near the surface such that the highest shut-in pressure acts on the column of completion fluid. These load definitions usually lead to an estimation of the highest likely loads on a casing string. However, due to the variability and uncertainty in these loads, the probability of occurrence of the highest load magnitudes can be very low, especially given the standards of operational practices and quality assurance/quality control (QA/QC) typical in challenging wells. As a result, designers have begun questioning these load cases and turning to probabilistic methods for load estimation (Payne et al. 2003; Lewis 2004; Maes et al. 1995; Mason and Chandrashekhar 2005). While probabilistic consideration of loads involves the application of complex statistical methods, it is clear that there is a trend toward more-realistic estimation of loads when faced with challenging conditions such as high pressure and high temperature.

In HP/HT well design, new load cases such as APB will have to be considered. This often forces the designer to introduce mitigation measures in some of the strings or problem annuli. Often, the mitigation measures themselves change the design basis, and the standard loads will have to be rechecked to assure well integrity. In addition, trapped drilling fluids can degrade due to the settlement of solids, leading to reduced hydrostatic pressures in the annuli with time. This has to be accounted for in design.

8.6.2 Design Strength. Designing well casings requires the engineer to compare the estimated loads to a strength, which is based on tubular performance ratings and theories of strength. An important ramification of HP/HT conditions in well design is in the choice of design strength. In well design, the most common tubular performance ratings are the pipe body yield strength, the internal pressure rating (*burst* rating), and the external pressure rating (collapse rating). The ratings typically are based on minimum properties and theories of strength based on incipient yield criteria. These ratings historically have been governed by the procedures and equations described in the document API *TR 5C3/ISO 10400* (2008). Many of these equations date back to the late 1960s. For HP/HT wells, the use of these methods can imply impossibility of design.

With improvement in the manufacture of OCTG tubulars, development of more advanced theories of strength, and better control of the performance properties, the industry is reexamining the performance properties in API *TR 5C3/ISO 10400* (2008) and their applicability to OCTG tubulars, particularly when confronted with challenging well designs. *ISO DIS 10400* (2004) is one of the products of this trend, and it introduces limit states for burst, collapse, and brittle burst to the industry. An important consequence of using limit states is that more of the available pipe capacity is used by design. Equally important, designing to limit state strength implies allowing partial plasticization of the tubular. This can impact cyclic behavior of the material (i.e., when an HP/HT well is shut in and restarted). Hysteresis of the stress-strain curve can become important in designing to cyclic load conditions.

For the first time, *ISO DIS 10400* (2004) also allows probabilistic consideration of strength on the basis of measured material properties and dimensional parameters. Thus, if a designer takes the trouble to measure properties and dimensions and uses the resultant distributions in the limit state strength definitions, the result is a probability density function for strength, which can be used directly in design. This description of strength can be used regardless of whether the load is deterministic or probabilistic. In essence, it captures the enhancement in material performance properties at the design stage, thus leading to more rational design. In practice, the strength distribution can be compared to either a deterministic or probabilistic load, and the notional probability of failure (probability that the load exceeds strength) can be estimated and used as the basis of design. If the load is deterministic and representative of a *worst case*, then the probability of failure so estimated is an upper bound. **Fig. 8.15** illustrates the design process using both stochastic and deterministic loads.

Using probabilistic strength for design brings two additional burdens to the design process:

- Additional measurement and testing to define the actual strength distributions
- More rigorous QA/QC to ensure material conformance and enable adequate characterization of the strength distribution

Despite these burdens, as Payne et al. (2003) argue, "Advanced work in this area consistently identifies probabilistic engineering as pivotal to effective solutions of such design optimizations." Probabilistic design methods are here to stay and are probably necessary in the design of HP/HT wells.

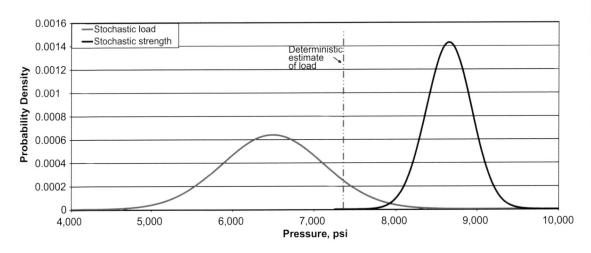

Fig. 8.15—Interference between load and strength, showing both deterministic and stochastic loads.

8.7 Operational Implications of HP/HT Wells

8.7.1 Well Control. HP/HT conditions offer an aggressive environment for the operation of well-control equipment. This is an environment in which metallic materials fall off in strength and elastomers lose their ability to maintain a long-term seal (Berckenhoff and Wendt 2005). Thermal cycling becomes an issue due to higher heat fluxes of HP/HT operation. Most material properties such as strength and elastic modulus decrease as temperature increases. The effect on materials due to the presence of H_2S and CO_2 needs to be considered. Elastomers are the most common sealing elements in internal well control equipment, and their properties vary widely with temperature. The higher-working-pressure environment demands that the well-control equipment grow in outside dimensions as well as total weight (Young et al. 2005).

Sections of equipment may be designed to operate at high temperatures while other sections may be designed to operate at low temperature, and this adds to the total height and weight of the equipment deployed (Berckenhoff and Wendt 2005). In addition, frequent elastomeric changes are expected as a part of preventive maintenance.

The frequency of well-control incidents is higher than one per HP/HT well, and an increasing number of these take place during completion (Rommetveit et al. 2003), the main reasons being

- Dynamically affected mud properties at high pressures and temperatures
- Narrow margin between pore and fracture pressure
- Hydrocarbon influx dissolution in OBM with infinite amounts of gas dissolving in the mud
- Barite sag, particularly in inclined and horizontal sections of the well

An advanced modeling tool that includes thermohydraulic modeling and kick modeling along with a transient well-control simulator is recommended (Rommetveit et al. 2003) to evaluate well control. In HP/HT wells, detection of kicks needs careful consideration because the change in mud volume may be due to the large variation in pressure and temperature conditions. Both absolute temperature and temperature variations increase with depth. Mud rheological properties along with density are dependent on temperature; therefore, temperature variations will effect a change in the mud volume, which may be detected as a kick. However, in a true sense, it is a *temperature kick*.

Another important phenomenon that is particularly evident in HP/HT wells is the so-called *wellbore breathing*. The term refers to the flow of fluid from the formation into the wellbore during noncirculating periods (such as during the making of a connection). During circulation in deep wells with reduced annular clearances, the friction loss (the ECD effect) can be significant, leading to considerable overbalance. During this period, some invasion of drilling fluid and near-wellbore pressure charging of the reservoir is likely. Because the bottomhole pressure drops during noncirculating periods (flowing friction being absent), there could be periods of underbalance with respect to near-wellbore formation conditions, leading to a flowback. The elastic contraction of the wellbore on reduction of bottomhole pressure also contributes to this effect. This flowback can be interpreted as a kick, although this is not a true kick. As soon as circulation resumes, or pressure balance is achieved, the effect ceases. This effect is more prevalent in tighter formations where near-wellbore charging and depletion are common. It also can lead to large-volume flowback during long-lasting noncirculation events such as trips. Typically, most of the flowback is of the invaded filtrate.

In the case of HP/HT wells, the established methods for calculating maximum anticipated surface pressure produce loads that cater to the worst-case scenario and may result in overdesign of wells. Established methods include the Fracture at Shoe with Gas to Surface method and the Minerals Management System (MMS) calculation method (Mason and Chandrashekhar 2005). The stochastic kick load model is emerging as a technique for HP/HT wells because it considers the entire physical system in which a kick is initiated and offers a higher level of confidence than any arbitrary method can. It calculates a range of possible design loads based on the spread of the stochastic variable's uncertainty. It accounts for the effect of gas solubility in oil-based drilling fluids and the effect it has on the physical environment.

8.7.2 Drilling and Cementing. Drilling-mud density and rheology are temperature-dependent and vary rapidly along the well, depending on the ongoing drilling operation. The mud expands or contracts as a result of these temperature variations, and this changes the effective volume of mud in the hole (Rommetveit et al. 2003). The hydrostatic pressure varies depending on the mud density. The frictional pressure changes as the rheology changes as a result of thermal effects. Hole cleaning is affected as a result of the changes in mud rheology. High temperatures in HP/HT wells cause thermal expansion of the fluid and a reduction in effective density (Rudolf and Suryanarayana 1997). Pressure transients occur during the tripping operation, with the swab pressures associated with tripping out and surge pressures with tripping in. However, it is possible to have significant swabs while tripping in (Rudolf and Suryanarayana 1997). This secondary swab effect when coupled with the elevated temperature and formation elasticity is capable of creating an under-balanced situation while tripping in, thereby inducing a kick. Therefore, it is extremely important to follow proper tripping practice. The mud weight should be adjusted to account for temperature and swab effects. The trip-in speed is an important operating parameter because this velocity can initiate the secondary swab effect that may result in a kick (Rudolf and Suryanarayana 1998). Therefore adequate mud weight and safe trip-in speeds must be considered as design parameters in the planning of HP/HT wells.

8.7.3 Surface Facilities. The temperature experienced by the surface equipment varies from seafloor temperature after initial tripping or long shut-ins to higher temperatures during circulation (Berckenhoff and Wendt 2005). The high fluid temperatures result in relatively high thermal expansion forces, which need to be accounted for in the design of the equipment. It also may be necessary to install coolers at surface to reduce the temperature of the circulation fluids. Thermal modeling and testing are extremely important before installation of the equipment. The design consequence of high-pressure conditions is the increased thickness and weight of the piping components, especially high-pressure valves and manifolds (Walton 2000). In addition, wellbore fluids contain compounds that may be corrosive, such as H_2S and CO_2, and their corrosion rates are a function of temperature. HP/HT wells demand high-volume surface gas-handling systems equipped with necessary sampling points, temperature sensors, and alarm systems. Mudline control demands sealing elements designed for aggressive conditions and equipment that can withstand a wide range of temperature.

8.7.4 Impact on Well Testing, Completions, and Formation Evaluation. In a well-testing operation, different tools and equipments are needed both downhole and at the surface. HP/HT test planning and equipment selection need careful consideration because of the demanding operating conditions. These conditions necessitate the use of permanent downhole packers instead of retrievable packers (Boscan et al. 2003). For HP/HT environments, the downhole valves and gauges must undergo extensive qualification testing at elevated-pressure and -temperature conditions. The rates and pressures experienced during well testing of HP/HT wells are extremely high. Surface testing equipment includes sampling points and a facility to separate and measure the fluids produced. It is recommended that critical parameters such as maximum expected temperature and pressure, expected flow rates, and possible production of corrosive fluids be predetermined from offset data before selection of the well-testing equipment.

Similar to the well-testing equipment, the completion equipment requires rigorous design and testing considerations. The production tubing selection in most cases is a CRA because HP/HT wells are most likely to contain H_2S (Hahn et al. 2000). After the selection of the CRA material, it needs to be tested for the various load conditions likely to be encountered. Care must be exercised when designing with CRAs because material properties used in design are different in CRAs. Premium connections are used for HP/HT wells to ensure continuous leak resistance. The need for a subsurface safety valve needs to be evaluated on the basis of the overall reliability of the valves in HP/HT applications (Hahn et al. 2001). Because HP/HT packers are expected to be a permanent installation, it is important that the packer material is corrosion-resistant. Also, the preferred completion fluid is clear brine because it has the advantage of being solids-free. Tubing-conveyed perforating systems are preferred because of the pressure control risks associated with electric wireline operations.

During the early 1960s and 1970s, hostile-environment logging (HEL) included HP/HT logging (Sarian and Gibson 2005). Downhole electronics limitations and signal degradation with temperature were the main concerns, compounded by reduced sensor efficiency with temperature. However, today, downhole sensor technology has improved significantly; for instance, standard wireline tools can operate continuously at 350°F and 20,000 psi.

Use of low-power components and efficient circuit design reduce the internal temperature increase due to the heat generated within the tool (Baird et al. 1998). Thermal flasks are used as HP/HT logging tool housings that protect the electronic circuits from high borehole temperatures. High-strength cable with electrically controlled weak points allows higher pull to release wireline jars and free logging tools. Planner software for tool sticking and holding time is used to predict a stuck-tool situation and prevent high-temperature failures (Sarian and Gibson 2005). Like all other HP/HT equipment, HP/HT logging tools are subject to far more rigorous maintenance programs than their standard counterparts.

Nomenclature

A_h = heat transfer area at the location where U_o is defined, ft²
A_p = cross-sectional area of the casing, in.²
A_i = inside surface area, in.²
A_o = outside surface area, in.²
Bi = Biot number
C_n = conversion constant
$C_{p,f}$ = specific heat, Btu/lbm-°F
d_{to} = outside diameter of tubular, ft
E = modulus of elasticity of steel, 3×10^7 psi
f = function
F_{cr} = critical force of buckling, lbf
F_e = effective force, lbf
F_{real} = real force, lbf
F_n = normal contact force, lbf
I = moment of inertia, in.⁴
k_e = thermal conductivity of earth (Btu/D-ft-°F)
K_{ISSC} = threshold stress intensity factor
L = length, ft
L_{uc} = length of uncemented section of the casing, ft
P = pressure, psi
p_i = internal pressure, psi
p_o = external pressure, psi
Pe = Peclet number
Q_{inlet} = heat flow in, Btu/lbm
Q_m = heat transferred from the surroundings, Btu/lbm
Q_{out} = heat flow out, Btu/lbm
r = radius, ft
r_h = radius at the boundary of the wellbore, ft
r_c = radial clearance, ft
R' = resistance
t = time, D
T = temperature, °F
$T_{initial}$ = initial temperature, °F
T_{final} = final temperature, °F
T_0 = reference temperature at a known Z, °F
T_e = temperature of the earth, °F
T_f = temperature of the flowing fluid, °F
T_h = temperature at the boundary of the wellbore (at a radius $r = r_h$), °F
τ = Fourier number
U_o = overall heat transfer coefficient, Btu/D-ft²-°F
U_{to} = overall heat transfer coefficient at the OD of the tubular, Btu/D-ft²-°F
Z = depth, ft
α_e = thermal diffusivity of earth, ft²/D
α_{steel} = coefficient of thermal expansion of steel, per °F
Δq = heat transfer rate, Btu/D

ΔZ = differential element of depth, ft

ρ = density, lbm/gal

σ_t = thermal stress, psi

σ_f = forward flow stress, psi

σ_r = reverse flow stress, psi

σ_y = yield stress, psi

$\sigma_{b,h}$ = post-buckling bending stress, psi

Acknowledgements

The authors gratefully acknowledge Anamika Gupta of Blade Energy Partners for the tremendous effort she has put into the writing of this chapter.

References

Adams, A. 1991. How To Design for Annulus Fluid Heat-Up. Paper SPE 22871 presented at the SPE Annual Technical Conference and Exhibition, Dallas, 6–9 October. DOI: 10.2118/22871-MS.

Adams, A.J. and MacEachran, A. 1994. Impact on Casing Design of Thermal Expansion of Fluids in Confined Annuli. *SPEDC* **9** (3): 210–216. SPE-21911-PA. DOI: 10.2118/21911-PA.

Adams, A.J., Moore, P.W., and Payne, M.L. 2001. On the Calibration of Design Collapse Strengths for Quenched and Tempered Pipe. Paper OTC 13048 presented at the Offshore Technology Conference, Houston, 30 April–3 May.

ASM Handbook. 1976. Russell Township, Ohio: ASM.

Azzola, J.H., Tselepidakis, D.P., Patillo, P.D. et al. 2004. Application of Vacuum Insulated Tubing To Mitigate Annular Pressure Buildup. Paper SPE 90232 presented at the SPE Annual Technical Conference and Exhibition, Houston, 26–29 September. DOI: 10.2118/90232-MS.

Baird, T., Drummond, R., Langseth, B., and Silipigno, L. 1998. High-Pressure, High-Temperature Well Logging, Perforating and Testing. *Oilfield Review* **10** (2): 51–67.

Beardsmore, G.R. and Cull, J.P. 2001. *Crustal Heat Flow: A Guide to Measurement and Modelling*. Cambridge, UK: Cambridge University Press.

Bejan, A. 1984. *Convection Heat Transfer*. New York: Wiley Publishing.

Belkin, A., Irving, M., O'Connor, R., Fosdick, M., Hoff, T., and Growcock, F.B. 2005. How Aphron Drilling Fluids Work. Paper SPE 96145 presented at the SPE Annual Technical Conference and Exhibition, Dallas, 9–12 October. DOI: 10.2118/96145-MS.

Berckenhoff, M. and Wendt, D. 2005. Design and Qualification Challenges for Mudline Well Control Equipment Intended for HP/HT Service. Paper SPE 97563 presented at the SPE High Pressure/High Temperature Sour Well Design Applied Technology Workshop, The Woodlands, Texas, 17–19 May. DOI: 10.2118/97563-MS.

Bjørkevoll, K.S., Rommetveit, R., Aas, B., Gjeralstveit, H., and Merlo, A. 2003. Transient Gel Breaking Model for Critical Wells Applications With Field Data Verification. Paper SPE 79843 presented at the SPE/IADC Drilling Conference, Amsterdam, 19–21 February. DOI: 10.2118/79843-MS.

Boresi, A.P., Schmidt, R.J., and Sidebottom, O.M. 1993. *Advanced Mechanics of Materials*, fifth edition. New York: Wiley and Sons.

Boscan, J., Almanza, E., and Wendler, C. 2003. Successful Well Testing Operations in High-Pressure/High Temperature Environment: Case Histories. Paper SPE 84096 presented at the SPE Annual Technical Conference and Exhibition, Denver, 5–8 October. DOI: 10.2118/84096-MS.

Bradley, A.B., Nagasaku, S., and Verger, E. 2005. Premium Connection Design, Testing, and Installation for HP/HT Sour Wells. Paper SPE 97585 presented at the SPE High Pressure/High Temperature Sour Well Design Applied Technology Workshop, The Woodlands, Texas, USA, 17–19 May. DOI: 10.2118/97585-MS.

Bradford, D.W., Fritchie, D.G., Gibson, D.H., Gosch, S.W., Pattillo, P.D., Sharp, J.W., and Taylor, C.E. 2002. Marlin Failure Analysis and Redesign: Part 1, Description of Failure. Paper SPE 74528 presented at the IADC/SPE Drilling Conference, Dallas, 26–28 February. DOI: 10.2118/74528-MS.

Brownlee, J.K., Flesner, K.O., Riggs, K.R., and Miglin, B.P. 2005. Selection and Qualification of Materials for HP/HT Wells. Paper SPE 97590 presented at the SPE High Pressure/High Temperature Sour Well Design Applied Technology Workshop, The Woodlands, Texas, 17–19 May. DOI: 10.2118/97590-MS.

Carcagno, G. 2005. The Design of Tubing and Casing Premium Connections for HT/HP Wells. Paper SPE 97584 presented at the SPE High Pressure/High Temperature Sour Well Design Applied Technology Workshop, The Woodlands, Texas, 17–19 May. DOI: 10.2118/97584-MS.

Chapman, A.J. 1974. *Heat Transfer*, third edition. New York: Macmillan Publishing.

Childs, P.R.N. and Long, C.A. 1996. A Review of Forced Convective Heat Transfer in Stationary and Rotating Annuli. *Proc. Instn. Mech. Eng. Part C: Journal of Mechanical Engineering Science* **210** (C2): 123–134. DOI: 10.1243/PIME_PROC_1996_210_179_02.

Craig, B.D. 1993. *Practical Oilfield Metallurgy and Corrosion*, second edition. Tulsa, Oklahoma: Pennwell Books.

Danesh, A. 2002. Reservoir Fluid Studies. *Sharp IOR eNewsletter* **3** (September 2002): 4.2.2.

Dropkin, D. and Somerscales, E. 1965. Heat Transfer by Natural Convection in Liquids Confined by Two Parallel Plates Which Are Inclined at Various Angles With Respect to the Horizontal. *Trans. ASME Journal of Heat Transfer* **87** (1): 77–84.

Ellis, R.C., Fritchie, D.G. Jr., Gibson, D.H., Gosch, S.W., and Pattillo, P.D. 2004. Marlin Failure Analysis and Redesign: Part 2—Redesign. *SPEDC* **19** (2): 112–119. SPE-88838-PA. DOI: 10.2118/88838-PA.

Esmersoy, C. and Mallick, S. 2004. Real-Time Pore-Pressure Prediction Ahead of the Bit. Final Report RPSEA-0016-04, Subcontract No. R-515, Sugar Land, Texas (January–September 2004) prepared for Research Partnership to Secure Energy for America (RPSEA). www.rpsea.org/attachments/wysiwyg/4/realtime_porepressure_sum.pdf

Fisk, J.V. and Jamison, D.E. 1989. Physical Properties of Drilling Fluids at High Temperatures and Pressures. *SPEDE* **4** (4): 341–346. SPE-17200-PA. DOI: 10.2118/17200-PA.

Glassø, Ø. 1980. Generalized Pressure-Volume-Temperature Correlations. *JPT* **32** (5): 785–795. SPE-8016-PA. DOI: 10.2118/8016-PA.

Gosch, S.W., Horne, D.J., Pattillo, P.D., Sharp, J.W., and Shah, P.C. 2002. Marlin Failure Analysis and Redesign: Part 3, VIT Completion with Real Time Monitoring. Paper SPE 74530 presented at the IADC/SPE Drilling Conference, Dallas, 26–28 February. DOI: 10.2118/74530-MS.

Gozalpour, F., Danesh, A., Fonseca, M., Todd, A.C., Tohidi, B., and Al-Syabi, Z. 2005. Physical and Rheological Behavior of High-Pressure/High-Temperature Fluids in Presence of Water. Paper SPE 94068 presented at the SPE Europec/EAGE Annual Conference, Madrid, Spain, 13–16 June. DOI: 10.2118/94068-MS.

Griffith, J.E., Lende, G., Ravi, K., Saasen, A., Nødland, N.E., and Jordal, O.H. 2004. Foam Cement Engineering and Implementation for Cement Sheath Integrity at High Temperature and High Pressure. Paper SPE 87194 presented at the IADC/SPE Drilling Conference, Dallas, 2–4 March. DOI: 10.2118/87194-MS.

Hahn, D.E., Pearson, R.M., and Hancock, S.H. 2000. Importance of Completion Design Considerations for Complex, Hostile, and HP/HT Wells in Frontier Areas. Paper SPE 59750 presented at the SPE/CERI Gas Technology Symposium, Calgary, 3–5 April. DOI: 10.2118/59750-MS.

Hahn, D.E., Burke, L.H., Mackenzie, S.F., and Archibald, K.H. 2001. Completion Design and Implementation in Challenging, Extreme HP/HT Wells in California. Paper SPE 71684 presented at the SPE Annual Technical Conference and Exhibition, New Orleans, 30 September–3 October. DOI: 10.2118/71684-MS.

Halal, A.S. and Mitchell, R.F. 1993. Casing Design for Trapped Annular Pressure Buildup. *SPEDE* **9** (2): 107–114. SPE-25694-PA. DOI: 10.2118/25694-PA.

Hammerlindl, D.J. 1980. Basic Fluid and Pressure Forces on Oilwell Tubulars. *JPT* **32** (1): 153–159. SPE-7594-PA. DOI: 10.2118/7594-PA.

Han, K., Van Tyne, C.J., and Levy, B.S. 2005. Effect of Strain and Strain Rate on the Bauschinger Effect Response of Three Different Steels. *Metallurgical and Materials Transactions* **36A** (9): 2379–2384. DOI: 10.1007/s11661-005-0110-7.

Handelman, G.H. 1946. Buckling Under Locally Hydrostatic Pressure. *Journal of Applied Mechanics* **13**: A198–A200.

Harris, O.O. and Osisanya, S.O. 2005. Evaluation of Equivalent Circulating Density of Drilling Fluids Under High-Pressure/High-Temperature Conditions. Paper SPE 97018 presented at the SPE Annual Technical Conference and Exhibition, Dallas, 9–12 October. DOI: 10.2118/97018-MS.

Harrold, D., Ringle, E., Taylor, M., Timte, S., and Wabnitz, F. 2004. Reliability by Design—HP/HT System Development. Paper OTC 16393 presented at the Offshore Technology Conference, Houston, 3–6 May.

Hasan, A.R. and Kabir, C.S. 2002. *Fluid Flow and Heat Transfer in Wellbores*. Richardson, Texas: SPE.

He, X. and Kyllingstad, A. 1995. Helical Buckling and Lock-Up Conditions for Coiled Tubing in Curved Wells. *SPEDC* **10** (1): 10–15. SPE-25370-PA. DOI: 10.2118/25370-PA.

Herzhaft, B., Ragouillaux, A., and Coussot, P. 2006. How To Unify Low-Shear-Rate Rheology and Gel Properties of Drilling Muds: A Transient Rheological and Structural Model for Complex Wells Applications. Paper SPE 99080 presented at the IADC/SPE Drilling Conference, Miami, Florida, 21–23 February. DOI: 10.2118/99080-MS.

ISO 13679, Petroleum and Natural Gas Industries—Procedures for Testing Casing and Tubing Connections. 2002. Geneva, Switzerland: ISO.

ISO 15156, Petroleum and Natural Gas Industries—Materials for Use in H_2S-Containing Environments in Oil and Gas Production, Part 1, Part 2, and Part 3. 2003. Geneva, Switzerland: ISO.

Kelley, S., Sweatman, R., and Heathman, J. 2001. Treatments Increase Formation Pressure Integrity in HTHP Wells. Paper AADE 01-NC-HO-42 prepared for presentation at the AADE National Drilling Conference, Houston, 27–29 March. http://www.aade.org/TechPapers/2001Papers/formation/AADE%2042.pdf. Downloaded 11 August 2008.

Kemp, N.P. and Thomas, D.C. 1987. Density Modeling for Pure and Mixed-Salt Brines as a Function of Composition, Temperature, and Pressure. Paper SPE 16079 presented at the SPE/IADC Drilling Conference, New Orleans, 15–18 March. DOI: 10.2118/16079-MS.

Klever, F.J. and Stewart, G. 1998. Analytical Burst Strength Prediction of OCTG With and Without Defects. Paper SPE 48329 presented at the SPE Applied Technology Workshop on Risk Based Design of Well Casing and Tubing, The Woodlands, Texas, 7–8 May. DOI: 10.2118/48329-MS.

Kristiansen,T. 1998. Geomechanical Characterization of the Overburden Above the Compacting Chalk Reservoir at Valhall. Paper SPE 47348 presented at the SPE/ISRM Rock Mechanics in Petroleum Engineering, Norway, 8-10 July. DOI: 10.2118/47348-MS.

Kuyken, C.W. and de Lange, F. 1999. Pore Pressure Prediction Allows for Tighter Pressure Gradient Control. *Offshore* **59** (12), 64–65.

Lea, J.F., Pattillo, P.D., and Studenmund, W.R. 1995. Interpretation of Calculated Forces on Sucker Rods. *SPEPF* **10** (1): 41–45; *Trans.*, AIME, **299**. SPE-25416-PA. DOI: 10.2118/25416-PA.

Lewis, D.B. 2004. Quantitative Risk Analysis Optimizes Well Design in Deep, HP/HT Projects. *The American Oil and Gas Reporter* (October 2004): 103–109.

Loder, T., Evans, J.H., and Griffith, J.E. 2003. Prediction and Effective Prevention Solution for Annular Pressure Buildup on Subsea Completed Wells—Case Study. Paper SPE 84270 presented at the SPE Annual Technical Conference and Exhibition, Denver, 5–8 October. DOI: 10.2118/84270-MS.

Lubinski, A. 1951. A Study of the Buckling of Rotary Drilling String. *API Drilling & Production Practice* (1951): 178–214.

Lubinski, A., Althouse, W.S., and Logan, J.L. 1962. Helical Buckling of Tubing Sealed in Packers. *JPT* **14** (6): 655–670; *Trans.*, AIME, **225**. SPE-178-PA. DOI: 10.2118/178-PA.

MacAndrew, R., Parry, N., Prieur, J., Wiggelman, J., Diggins, E., Guicheney, P., and Cameron, D. 1993. Drilling and Testing Hot, High Pressure Wells. *Oilfield Review* **5** (2): 15–32.

Maes, M.A., Gulati, K.C., McKenna, D.L., Brand, P.R., Lewis, D.B., and Johnson, R.C. 1995. Reliability-Based Casing Design. *Journal of Energy Resources Technology* **117** (2): 93–100. DOI: 10.1115/1.2835336.

Maglione, R., Gallino, G., Robotti, G., Romagnoli, R., di Torino, P., and Rommetveit, R. 1996. A Drilling Well as Viscometer: Studying the Effects of Well Pressure and Temperature on the Rheology of the Drilling Fluids. Paper SPE 36885 presented at the European Petroleum Conference, Milan, Italy, 22–24 October. DOI: 10.2118/36885-MS.

Mansour, S., Schulz, J., Haddad, G.S., and Helou, H. 1999. Cementing Under Extreme Conditions of High Pressure and High Temperature. Paper SPE 57582 presented at the SPE/IADC Middle East Drilling Technology Conference, Abu Dhabi, 8–10 November. DOI: 10.2118/57582-MS.

Marshall, D.W. and Bentsen, R.G. 1982. A Computer Model To Determine the Temperature Distribution in a Wellbore. *J. Cdn. Pet. Tech.* (January–February): 63–75.

Mason, S. and Chandrashekhar, S. 2005. Stochastic Kick Load Modeling. Paper SPE 97564 presented at the SPE High Pressure/High Temperature Sour Well Design Applied Technology Workshop, The Woodlands, Texas, 17–19 May. DOI: 10.2118/97564-MS.

Maury, V. and Idelovici, J.L. 1995. Safe Drilling of HP/HT Wells: The Role of the Thermal Regime in Loss and Gain Phenomena. Paper SPE 29428 presented at the SPE/IADC Drilling Conference, Amsterdam, 28 February–2 March. DOI: 10.2118/29428-MS.

Mitchell, R.F. 1982. Buckling Behavior of Well Tubing: The Packer Effect. *SPEJ* **22** (5): 616–624. SPE-9264-PA. DOI: 10.2118/9264-PA.

Mitchell, R.F. 1986. Simple Frictional Analysis of Helical Buckling of Tubing. *SPEDE* **1** (6): 457–465; *Trans.*, AIME, **281**. SPE-13064-PA. DOI: 10.2118/13064-PA.

NACE Standard MR-0175: Metals for Sulfide Stress Cracking and Stress Corrosion Cracking Resistance in Sour Oilfield Environments, 2003 revision. 2003. Houston: NACE.

Nice, P.I., Øksenvåg, S., Eiane, D.J., Ueda, M., and Loulergue, D. 2005. Development and Implementation of a High Strength "Mild Sour Service" Casing Grade Steel for the Kristin HP/HT Field. Paper SPE 97583 presented at the SPE High Pressure/High Temperature Sour Well Design Applied Technology Workshop, The Woodlands, Texas, 17–19 May. DOI: 10.2118/97583-MS.

Oakley, D.J., Morton, K., Eunson, A., Gilmour, A., Pritchard, D., and Valentine, A. 2000. Innovative Drilling Fluid Design and Rigorous Pre-Well Planning Enable Success in an Extreme HP/HT Well. Paper SPE 62729 presented at IADC/SPE Asia Pacific Drilling Technology, Kuala Lumpur, 11–13 September. DOI: 10.2118/62729-MS.

Oudeman, P. and Kerem, M. 2006. Transient Behavior of Annular Pressure Buildup in HP/HT Wells. *SPEDC* **21** (4): 234–241. SPE-88735-PA. DOI: 10.2118/88735-PA.

Paslay, P.R. 1994. Stress Analysis of Drillstrings. Paper SPE 27976 presented at the University of Tulsa Centennial Petroleum Engineering Symposium, Tulsa, 29–31 August. DOI: 10.2118/27976-MS.

Payne, M.L., Pattillo, P.D., Sathuvalli, U.B., and Miller, R.A. 2003. Advanced Topics for Critical Service Deepwater Well Design. Paper presented at the Deep Offshore Technology (DOT) Conference, Marseille, France, 19–21 November.

Prensky, S. 1992. Temperature Measurements in Boreholes: An Overview of Engineering and Scientific Applications. *The Log Analyst* **33** (3): 313–333.

Ramey, H.J. 1962. Wellbore Heat Transmission. *JPT* **14** (4): 427–435; *Trans.*, AIME, **225**. SPE-96-PA. DOI: 10.2118/96-PA.

Raymond, L.R. 1969. Temperature Distribution in a Circulating Drilling Fluid. *JPT* **21** (3): 333–341; *Trans.*, AIME, **246**. SPE-2320-PA. DOI: 10.2118/2320-PA.

Romero, J. and Loizzo, M. 2000. The Importance of Hydration Heat on Cement Strength Development for Deep Water Wells. Paper SPE 62894 presented at the SPE Annual Technical Conference and Exhibition, Dallas, 1–4 October. DOI: 10.2118/62894-MS.

Rommetveit, R. and Bjørkevoll, K.S. 1997. Temperature and Pressure Effects on Drilling Fluid Rheology and ECD in Very Deep Wells. Paper SPE/IADC 39282 presented at the SPE/IADC Middle East Drilling Technology Conference, Bahrain, 23–25 November. DOI: 10.2118/39282-MS.

Rommetveit, R., Fjelde, K.K., Aas, B., Day, N.F., Low, E., and Schwartz, D.H. 2003. HP/HT Well Control: An Integrated Approach. Paper OTC 15322 presented at the Offshore Technology Conference, Houston, 5–8 May.

Rudolf, R.L. and Suryanarayana, P.V. 1997. Kicks Caused by Tripping-In the Hole on Deep, High Temperature Wells. Paper SPE 38055 presented at the SPE Asia Pacific Oil and Gas Conference and Exhibition, Kuala Lumpur, 14–16 April. DOI: 10.2118/38055-MS.

Rudolf, R.L. and Suryanarayana, P.V. 1998. Field Validation of Swab Effects While Tripping-In the Hole on Deep, High Temperature Wells. Paper SPE 39395 presented at the IADC/SPE Drilling Conference, Dallas, 3–6 March. DOI: 10.2118/39395-MS.

Saasen, A., Jordal, O.H., Burkhead, D. et al. 2002. Drilling HT/HP Wells Using a Cesium-Formate-Based Drilling Fluid. Paper SPE 74541 presented at the IADC/SPE Drilling Conference, Dallas, 26–28 February. DOI: 10.2118/74541-MS.

Salz, L.B. 1977. Relationship Between Fracture Propagation Pressure and Pore Pressure. Paper SPE 6870 presented at the SPE Annual Technical Conference and Exhibition, Denver, 9–12 October. DOI: 10.2118/6870-MS.

Samuel, G.R. and Gonzales, A. 1999. Optimization of Multistring Casing Design With Wellhead Growth. Paper SPE 56762 presented at the SPE Annual Technical Conference and Exhibition, Houston, 3–6 October. DOI: 10.2118/56762-MS.

Sarian, S. and Gibson, A. 2005. Wireline Evaluation Technology in HP/HT Wells. Paper SPE 97571 presented at the SPE High Pressure/High Temperature Sour Well Design Applied Technology Workshop, The Woodlands, Texas, 17–19 May. DOI: 10.2118/97571-MS.

Sathuvalli, U.B., Suryanarayana, P.V., and Erpelding, P. 2001. Variations in Formation Temperature Due to Mature Field Production Impacts Drilling Performance. Paper SPE 67828 presented at the SPE/IADC Drilling Conference, Amsterdam, 27 February–1 March. DOI: 10.2118/67828-MS.

Sathuvalli, U.B., Payne, M.L., Patillo, P., Rahman, S., and Suryanarayana, P.V. 2005. Development of a Screening System To Identify Deepwater Wells at Risk for Annular Pressure Buildup. Paper SPE 92594 presented at the SPE/IADC Drilling Conference, Amsterdam, 23–25 February. DOI: 10.2118/92594-MS.

Seymour, K.P. and MacAndrew, R. 1993. The Design, Drilling, and Testing of a Deviated High-Temperature, High-Pressure Exploration Well in the North Sea. Paper OTC 7338 presented at the Offshore Technology Conference, Houston, 3–6 May.

Shaughnessy, J.M., Romo, L.A., and Soza, R.L. 2003. Problems of Ultradeep High-Temperature, High-Pressure Drilling. Paper SPE 84555 presented at the SPE Annual Technical Conference and Exhibition, Denver, 5–8 October. DOI: 10.2118/84555-MS.

Skomedal, E., Jostad, H.P., and Hettema, M.H. 2002. Effect of Pore Pressure and Stress Path on Rock Mechanical Properties for HP/HT Application. Paper SPE 78152 presented at the SPE/ISRM Rock Mechanics Conference, Irving, Texas, 20–23 October. DOI: 10.2118/78152-MS.

Sparks, C.P. 1984. The Influence of Tension, Pressure and Weight on Pipe and Riser Deformations and Stresses. *Journal of Energy Resources and Technology* **106** (1): 46–54.

Standing, M.B. 1977. *Volumetric and Phase Behavior of Oil Field Hydrocarbon Systems*. Richardson, Texas: SPE.

Suryanarayana, P.V.R. and McCann, R.C. 1995. An Experimental Study of Buckling and Post-Buckling of Laterally Constrained Rods. *ASME Journal of Energy Resources Technology* **117** (2): 115–124. DOI: 10.1115/1.2835327.

Swarbick, R.E. 2002. Challenges of Porosity-Based Pore Pressure Prediction. *CSEG Recorder* **27** (7): 75–77.

Sweatman, R., Kelley, S., and Heathman, J. 2001. Formation Pressure Integrity Treatments Optimize Drilling and Completion of HTHP Production Hole Sections. Paper SPE 68946 presented at the SPE European Formation Damage Conference, The Hague, 21–22 May. DOI: 10.2118/68946-MS.

TR 5C3/ISO 10400, Petroleum and Natural Gas Industries—Formulae and Calculation for Casing, Tubing, Drill Pipe and Line Properties, first edition. 2008. Washington, DC: API.

Vidick, B. and Acock, A. 1991. Minimizing Risks in High-Temperature/High-Pressure Cementing: The Quality Assurance/Quality Control Approach. Paper SPE 23074 presented at Offshore Europe, Aberdeen, 3–6 September. DOI: 10.2118/23074-MS.

Walton, D. 2000. Equipment and Material Selection To Cope With High-Pressure/High-Temperature Surface Conditions. Paper OTC 12122 presented at the Offshore Technology Conference, Houston, 1–4 May.

Wang, H. and Su, Y. 2000. High Temperature and High Pressure (HTHP) Mud P-T Behavior and Its Effect on Wellbore Pressure Calculations. Paper IADC/SPE 59266 presented at the IADC/SPE Drilling Conference, New Orleans, 23–25 February. DOI: 10.2118/59266-MS.

Ward, C.D., Coghill, K., and Broussard, M.D. 1994. The Application of Petrophysical Data To Improve Pore and Fracture Pressure Determination in North Sea Central Graben HPHT Wells. Paper SPE 28297 presented at the SPE Annual Technical Conference and Exhibition, New Orleans, 25–28 September. DOI: 10.2118/28297-MS.

Warpinski, N.R., Teufel, L.W., and Graf, D.C. 1991. Effect of Stress and Pressure on Gas Flow Through Natural Fractures. Paper SPE 22666 presented at the SPE Annual Technical Conference and Exhibition, Dallas, 6–9 October. DOI: 10.2118/22666-MS.

Webb, S., Anderson, T., Sweatman, R., and Vargo, R. 2001. New Treatments Substantially Increase LOT/FIT Pressures To Solve Deep HTHP Drilling Challenges. Paper SPE 71390 presented at the SPE Annual Technical Conference and Exhibition, New Orleans, 30 September–3 October. DOI: 10.2118/71390-MS.

Willhite, G.P. 1967. Over-All Heat Transfer Coefficients in Steam and Hot Water Injection Wells. *JPT* **19** (5): 607–615. SPE-1449-PA. DOI: 10.2118/1449-PA.

Wooley, G.R. 1980. Computing Downhole Temperatures in Circulation, Injection and Production Wells. *JPT* **32** (9): 1509–1522. SPE-8441-PA. DOI: 10.2118/8441-PA.

Xiang, Y. and Vlassak, J.J. 2005. Bauschinger Effect in Thin Metal Films. *Scripta Materialia* **53** (2): 177–182. DOI: 10.1016/j.scriptamat.2005.03.048.

Young, K., Alexander, C., Biel, R., and Shanks, E. 2005. Updated Design Methods for HP/HT Equipment. Paper SPE 97595 presented at SPE High Pressure/High Temperature Sour Well Design Applied Technology Workshop, The Woodlands, Texas, 17–19 May. DOI: 10.2118/97595-MS.

Zamora, M., Broussard, P.N., and Stephens, M.P. 2000. The Top 10 Mud-Related Concerns in Deepwater Drilling Operations. Paper SPE 59019 presented at the SPE International Petroleum Conference and Exhibition in Mexico, Villahermosa, Mexico, 1–3 February. DOI: 10.2118/59019-MS.

SI Metric Conversion Factors

bbl	× 1.589 873	E – 01	= m^3
Btu	× 1.055 056	E + 00	= kJ
Btu/hr	× 2.930 711	E – 04	= kW
Btu/(lbm-°F)	× 4.186 8*	E + 00	= kJ(kg·K)
cp	× 1.0*	E – 03	= Pa·s
ft	× 3.048*	E – 01	= m
°F	× (°F – 32)/1.8		= °C
gal	× 3.785 412	E – 03	= m^3
ksi	× 6.894 757	E + 03	= kPa
lbm	× 4.535 924	E – 01	= kg
psi	× 6.894 757	E + 00	= kPa

*Conversion factor is exact.

Chapter 9

Innovative Drilling Methods

9.1 Underbalanced Drilling Operations—Steve Nas and Deepak M. Gala, Weatherford

9.1.1 What Is Underbalanced Drilling? The official definition of underbalanced drilling (UBD) originates from the Alberta Energy Board and is also defined by the International Association of Drilling Contractors (IADC) Underbalanced Operations (UBO) committee* as:

> Drilling with the hydrostatic head of the drilling fluid intentionally designed to be lower than the pressure of the formations being drilled. The hydrostatic head of the fluid may naturally be less than the formation pressure, or it can be induced. The induced state may be created by adding natural gas, nitrogen, or air to the liquid phase of the drilling fluid. Whether the underbalanced status is induced or natural, the result may be an influx of formation fluids which must be circulated from the well and controlled at surface.

The words *intentionally designed* are important because this clarifies that the underbalanced process was designed as a part of the well design process.

In conventionally overbalanced, drilled wells a column of fluid of a certain density in the wellbore provides the primary well-control mechanism. The pressure at the bottom of the well is always designed to be higher than the pressure in the formation.

$$P_{bottomhole} = P_{hydrostatic} + P_{friction}$$

In underbalanced drilled wells, a less dense fluid has replaced this fluid column, and the pressure at the bottom of the well is now intentionally designed to be lower than the pressure in the formation at all times. This means that any reservoir formations containing fluids and gases that have sufficient porosity and permeability will start to produce when they are drilled underbalanced.

Overbalanced drilling:

$$P_{reservoir} < P_{bottomhole} = P_{hydrostatic} + P_{friction}$$

UBD:

$$P_{reservoir} > P_{bottomhole} = P_{hydrostatic} + P_{friction} + P_{choke}$$

In UBO, the fluid no longer acts as the primary well-control mechanism as it would do in conventional overbalanced drilling. Instead, the surface equipment used for UBD operations, such as the rotating control

---

* More information on the committee can be found at http://www.iadc.org/committees/ubo_mpd/index.html.

diverter (RCD) and choke manifold, has replaced the primary well control. The secondary well control in the form of the blowout-preventer (BOP) stack remains exactly the same as with conventional overbalanced operations. It is important that secondary well control equipment is not used for routine UBD operations; this equipment must remain a secondary barrier. As with overbalanced drilling, well control must be maintained at all times when drilling underbalanced. This means that both the well and the surface equipment must be designed in such a way that well control can always maintained. One of the main differences in UBD wells is that we plan for reservoir influx, whereas in overbalanced drilling, one intends to avoid influx.

When candidate wells for UBD are properly selected and the operation is properly planned and conducted, superior results can be expected in drilling performance, in production performance, and in safety records. Recent advances in reservoir characterization during UBO have resulted in discovery of previously unknown productive reservoir zones in existing mature reservoirs, leading to significant increases in reserves for the operators.

Managed-Pressure Drilling. Managed-pressure drilling is defined by the IADC UBO committee as follows:

> Managed Pressure Drilling (MPD) is an adaptive drilling process used to precisely control the annular pressure profile throughout the wellbore. The objectives are to ascertain the downhole pressure environment limits and to manage the annular hydraulic pressure profile accordingly. MPD is intended to avoid continuous influx of formation fluids to the surface. Any influx incidental to the operation will be safely contained using an appropriate process.

Notes added to this definition are:

1. The MPD process uses a collection of tools and techniques which may mitigate the risks and costs associated with drilling wells that have narrow downhole environmental limits, by proactively managing the annular hydraulic pressure profile.
2. MPD may include control of back pressure, fluid density, fluid rheology, annular fluid level, circulating friction, and hole geometry, or combinations thereof.
3. MPD may allow faster corrective action to deal with observed pressure variations. The ability to dynamically control annular pressures facilitates drilling of what might otherwise be economically unattainable prospects.

History of UBD. Initially all wells drilled with cable tool rigs were underbalanced. It's a known fact that as soon as the early pioneers struck oil the well blew out. In 1895, rotary drilling was introduced, and with that a fluid had to be circulated to carry cuttings out of the hole. In the 1920s, the first mud systems were introduced to provide viscosity to enhance cuttings transport. In 1928, the first BOPs were being used to control blowouts and to close wells in. Since then, wells have been drilled overbalanced. Air drilling was introduced in the 1950s for hard-rock applications, and foam drilling was introduced in the 1960s to allow loss zones to be drilled with returns.

In the 1980s, the first known underbalanced wells were drilled in the Austin chalk using low-pressure rotating seals and open flow lines into a waste pit, as in air drilling. These open flow lines are known as blooie lines. In an effort to improve the safety of flowing hydrocarbons into surface pits and to measure flow rates, the Canadians introduced the installation of separators in the return lines in the late 1980s and, with that, took the first steps into modern UBD technology.

The application and improvements in multiphase flow modeling and the use of more-advanced separation systems resulted in an increased uptake of UBD. The technology is still being improved as more wells are drilled, and the service providers are still advancing the UBD technology on a regular basis.

9.1.2 Why Drill Underbalanced? The reasons for performing an underbalanced operation can be broken down into three main categories:

- Minimizing pressure-related drilling problems
- Maximizing hydrocarbon recovery
- Characterizing the reservoir

Minimizing Pressure-Related Drilling Problems. Most drilling problems related to pressures can be minimized through the use of UBD. This makes the technology ideally suited for infill drilling in mature depleted reservoirs. This is also where the majority of UBD wells are currently being used.

Differential Sticking. The absence of an overbalanced pressure on the formation, combined with the lack of filter cake, prevents the drillstring from becoming differentially stuck.

Fluid Losses. In general, a reduction of the hydrostatic pressure in the annulus reduces the fluid losses into a reservoir formation. In UBD, the hydrostatic pressure is reduced to a level at which losses do not occur. This can be especially important in the protection of fractures in a reservoir.

Increased Penetration Rate. The lowering of the hydrostatic pressure and drilling with a solids-free fluid has a significant effect on penetration rate. This also has a positive impact on bit life. The rate-of-penetration (ROP) improvements are still a function of the formation type, porosity, and compressive strength and the relationship of weight on bit and RPM. It is difficult to state that the penetration rate will always increase significantly when drilling underbalanced. In some reservoirs, the increase in ROP is limited by other drilling factors.

Although drilling rates are faster when drilling underbalanced, due to the handling of hydrocarbons at surface, making connections and tripping can take significantly longer; this often eliminates any time gains made as a result of the increased penetration rate (Moore et al. 2004; Pinkstone et al. 2004; Tetley et al. 1999; Gedge 1999).

Maximizing Hydrocarbon Recovery. Although at first this was not a primary reason for the selection of UBD technology, results from early underbalanced drilled wells indicated that reservoir productivity improved significantly by drilling underbalanced and was an important driver in the application of UBD techniques. The reasons for the increased productivity can be traced to the fact that no invasion of solids or mud filtrates occurs into the reservoir formations. Because of the increased productivity of an underbalanced well combined with the ability to drill infill wells in depleted fields, the recovery of bypassed hydrocarbons is made possible. This can significantly extend the life of a field. The improved productivity of the wells also leads to a lower drawdown, which can, in turn, reduce water coning. Some long-term production results from underbalanced wells are now being proven. Although initial production from underbalanced wells may not indicate significantly increased production, the long-term production profiles show a much slower decline curve for underbalanced wells.

This phenomenon is attributed to production from poorer reservoir zones that have suffered less damage and can therefore contribute to overall well production for much longer (Bennion et al. 1998; Helio and Queiroz 2000; Luo et al. 2000a, 2000b; Hunt and Rester 2000; Labat et al. 2000; Stuczynski 2001; Pia et al. 2002a, 2000b; Culen et al. 2003; Devaul and Coy 2002; Sarssam et al. 2003; Kimery and McCaffrey 2004).

Reservoir Characterization. The ability to identify fractures and prolific reservoir zones, as well as productive zones initially believed to be nonproductive, while still drilling allows reservoir engineers to gain a better understanding of the reservoir. This ability in combination with the ability to steer wells real time into the more productive features in a reservoir and the ability to change completion designs based on the drilling results has led to a much better understanding of reservoirs, resulting in significantly better production from UBO. The industry has however not yet taken full advantage of this technology (Kneissl 2001; Biswas et al. 2003; Murphy, et al. 2005).

There are a number of specific advantages and disadvantages associated with UBD operations; these are summarized in **Table 9.1.1**.

9.1.3 IADC Classification System for UBO. A classification system has been developed by IADC to help in establishing the risks associated with underbalanced wells **(Table 9.1.2)**. The purpose of the IADC Well Classification System is to describe the overall risk, application category, and fluid system used in UBO and MPD. It also allows the industry to readily compare underbalanced wells and share the lessons learned. Wells are classified according to

- Risk level (0 to 5)
- Application category (A, B or C)
- Fluid system (1 to 5)

TABLE 9.1.1—ADVANTAGES AND DISADVANTAGES ASSOCIATED WITH UBD OPERATIONS

Advantages	Disadvantages
Increases ROP	Possible wellbore stability problems
Decreases formation damage	Increased drilling costs
Eliminates risk of differential sticking	Compatibility with conventional MWD systems
Reduces risk of loss circulation	Generally more complex drilling system
Improves bit life	Possible increased torque and drag
Allows drilling of depleted reservoir zones	Requires significantly more people on location
Allows reservoir characterization and well testing during drilling	

TABLE 9.1.2—IADC RISK CLASSIFICATION SYSTEM FOR UNDERBALANCED WELLS

Level 0	Performance enhancement only. No hydrocarbon-containing zones.
Level 1	Well incapable of natural flow to surface. Well is "inherently stable" and is a low-level risk from a well-control point of view.
Level 2	Well capable of natural flow to surface but enabling conventional well-kill methods and limited consequences in case of catastrophic equipment failure.
Level 3	Geothermal and non hydrocarbon production. Maximum shut-in pressures less than UBD equipment operating pressure rating. Catastrophic failure has immediate serious consequences.
Level 4	Hydrocarbon production. Maximum shut-in pressures less than UBD equipment operating pressure rating. Catastrophic failure has immediate serious consequences.
Level 5	Maximum projected surface pressures exceed UBD operating pressure rating but are below BOP stack rating. Catastrophic failure has immediate serious consequences.

This classification system provides a framework for defining minimum equipment requirements, specialized procedures, and safety management practices. For further information, refer to the IADC UBO HSE Planning Guidelines and other related documents.

Table 9.1.3 classifies the application category of the well. This indicates if wells are drilled using underbalanced, managed pressure, or mud cap drilling techniques. **Table 9.1.4** classifies the fluid system being used for the operation.

Example of Classification System Use. A well is being drilled from 10,000 ft to 12,000 ft using managed-pressure-drilling techniques. The pore pressure of the formation is 14.5 lbm/gal, and the fracture gradient is 16.5 lbm/gal. The design is predicated on using a 13.0-lbm/gal fluid and maintaining a balanced system with surface pressure. The rotating control device (RCD) and emergency shutdown (ESD) systems are rated at 5,000 psi.

From the above information:

TABLE 9.1.3—IADC CATEGORY CLASSIFICATION FOR UNDERBALANCED WELLS

Category A	Managed Pressure Drilling (MPD)—Drilling with returns to surface using an equivalent mud weight that is maintained at or above the openhole pore pressure.
Category B	Underbalanced Operations (UBO)—Performing operations with returns to surface using an equivalent mud weight that is maintained below the openhole pore pressure.
Category C	Mud-Cap Drilling—Drilling with a variable-length annular fluid column, which is maintained above a formation that is taking injected fluid and drilled cuttings without returns to surface.

TABLE 9.1.4—FLUID CLASSIFICATION FOR UNDERBALANCED WELLS

Gas drilling	1
Mist drilling	2
Foam drilling	3
Gasified liquid drilling	4
Liquid drilling	5

- Maximum allowable surface pressure (MASP) is the lesser of BHP minus gas to surface or frac at shoe minus gas to surface.
- MASPBHP = $12,000 \times 0.052 \times (14.5–2) = 7,800$ psi.
- MASPfrc = $10,000 \times 0.052 \times (16.5–2) = 7,540$ psi.

Because the maximum anticipated surface pressure exceeds the UBO/MPD equipment rating, the classification for the well would be Level 5, Category A, Fluid System 5 or 5A5.

All wells classified as Level 4 or Level 5 wells will require significant engineering and operational planning to ensure that they can be safely drilled. Well design should ensure that equipment provisions for the next risk level are available if possible.

9.1.4 Candidate Selection for UBD. A detailed assessment of the reservoir properties and the issues associated with the reservoir fluids and rock properties is recommended for all UBO. This is especially important for hole stability, well deliverability, and reservoir pressures.

Candidate selection for UBD has significantly improved in the past few years, and there are now some automated software systems available to assist with reservoir candidate selection. Some of these candidate-selection models also provide an estimate of the technical and economic risks for UBO. A manual selection chart is provided in **Fig. 9.1.1**.

It is important that the right candidate reservoir is selected for a UBD operation. A standardized approach to reservoir candidate selection should be deployed to ensure that all reservoirs are analyzed using the same standard parameters **(Fig 9.1.2)**. UBD avoids damage; it must be remembered that UBD does not improve a bad or a tight reservoir.

Table 9.1.5 shows reservoir types that will and will not benefit from UBD.

Well Types. The reservoir is the main objective for underbalanced technology, and once reservoir benefits have been determined, the well type can be reviewed for its suitability. UBD can be applied to new drills, existing wells that are being sidetracked. It can be used onshore and offshore and also can be applied to sour wells. Multilaterals can be drilled underbalanced, and the technology is already implementing UBO from floating rigs (Purvis and Smith 1998; Xiong and Shan 2003; Garrouch and Labbabidi 2003).

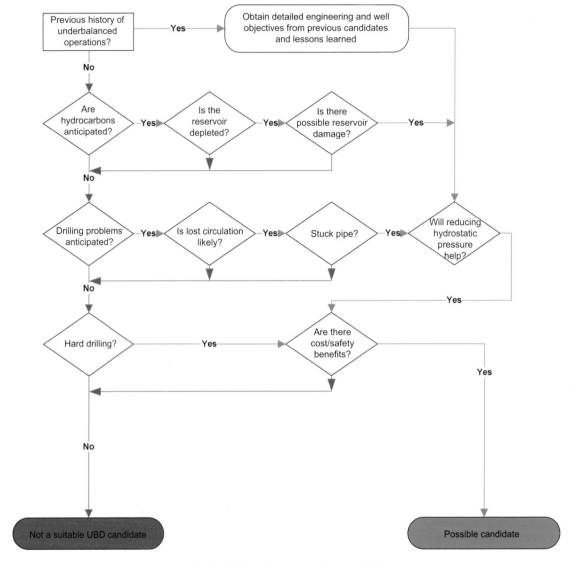

Fig. 9.1.1—Candidate selection for UBD.

Reservoir Considerations. UBD of a properly selected candidate reservoir can result in delivering some very productive wells because it eliminates the productivity-damaging effect of drilling mud (Xiong et al. 2003; Van der Werken 2005).

Formation Damage Mechanisms. Formation damage can occur as a result of many actions. However, the most common causes are drilling-, completion-, workover- and production-induced damage. Drilling, completion, and workover operations generally cause formation damage through the introduction of incompatible or dirty fluids to the formation. It has been stated that "the act of drilling is probably the most damaging process in the life of a reservoir."

Although less common than drilling- and completion-induced damage, well production also can cause formation damage. This type of damage usually occurs as a result of fluid incompatibilities or pressure-related alterations in the produced fluids. Regardless of cause, all formation damage is shown as follows:

- Change in absolute permeability of the reservoir
- Change in relative permeability of hydrocarbons
- Change in viscosity of formation fluid

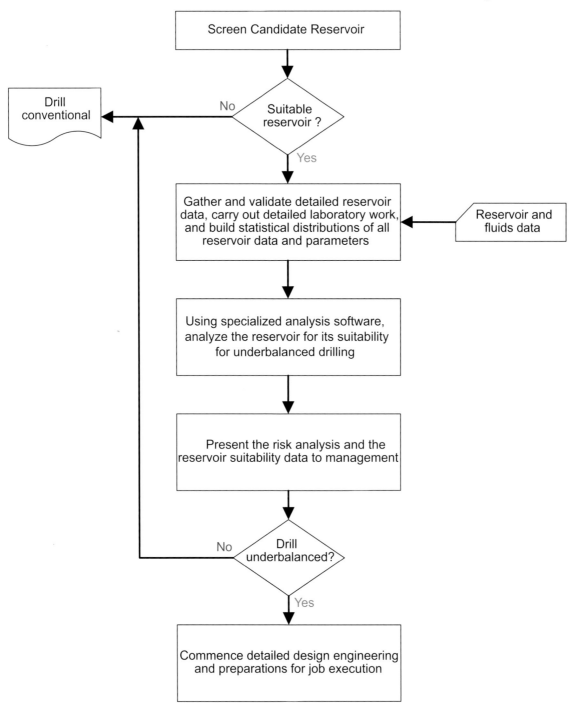

Fig. 9.1.2—Reservoir screening for underbalanced wells.

Formation damage can, sometimes, be cured by well-stimulation treatments, but it must be asked if prevention through the use of UBD is not better than the cure. Formation damage can be classified into four major groups:

- Mechanically induced formation damage
- Chemically induced formation damage
- Biologically induced formation damage
- Thermally induced formation damage

TABLE 9.1.5—RESERVOIR TYPES THAT WILL AND WILL NOT BENEFIT FROM UNDERBALANCED DRILLING	
Will Benefit	Will Not Benefit
Formations that usually suffer major formation damage during drilling or completion operations; wells with skin factors of 5 or higher	Wells in areas of very low conventional drilling cost
Formations that exhibit differential sticking tendencies	Wells drilled in areas of extremely high ROP (i.e., ROP > 1,000 ft/D)
Formations with zones with severe losses or fluid invasion from drilling or completion operations	Extremely-high-permeability wells
Wells with large macroscopic fractures	Ultralow-permeability wells
Low-permeability wells	Poorly consolidated formations
Wells with massive heterogeneous or highly laminated formations characterized by differing permeabilities, porosities, and pore throat throughput	Wells with low borehole stability
High-production reservoirs with low to medium permeabilities	Wells with loosely cemented laminar boundaries
Formation with rock fluid sensitivities	Wells that contain multiple zones with different pressure regimes
Formations that exhibit low ROP with overbalanced drilling	Reservoirs with interbedded shales, claystones or coal seams.

Mechanically induced formation damage can be subdivided further into three sections:

- Physical migration and entrainment of naturally occurring in-situ fines
- Entrainment and plugging due to the introduction of external solids into the formation
- Relative permeability effects such as phase trapping or hydrocarbon retention

Not all of these damage mechanisms are avoided through the use of UBD. Chemical causes of formation damage are normally the result of fluid incompatibilities, which may cause precipitation of solution salts or formation of emulsions with the reservoir fluids. The other main cause of chemically induced damage is the change in clay minerals lining the pore spaces. Biological formation damage is caused primarily by the introduction of bacteria into the reservoir as a result of certain chemical treatments. Thermal formation damage normally occurs only when gas or air is used to drill underbalanced in hard formations and glazing occurs due to the presence of fine cuttings combined with a lack of cooling. A thin but highly impermeable pottery-like glaze can be formed, which severely affects the performance of the well. Damage caused by drilling processes usually is collectively referred to as the skin factor of the well and is a term familiar from well-test analysis.

It must be highlighted that underbalanced wells still can be damaged. Depending on the process used, gas drilling can cause glazing of the formation, while water-based drilling fluids might cause significant imbibition effects, resulting in *filtrate* invasion. There is no protection from mudcake during overbalanced periods, resulting in extensive filtrate and solids invasion and a less efficient solids removal system—especially significant in horizontal wells where gravity and increased contact time of cuttings on the low side of the hole can increase damage (Brant et al. 1996; Helio and Queiroz 2000; Kimery and van der Werken 2004; Ding et al. 2006; Qutob 2004).

Hole Stability Considerations. Hole stability in UBD is an important issue. Shale and coal intervals commonly have borehole instability problems. They can be at a significantly higher pressure than the target formation to be drilled underbalanced. The underbalanced differential created in a shale formation can be significant enough to cause serious borehole sloughing. Shale studies of core samples can determine the potential for borehole instabilities and the size of the affected borehole area. Prediction models are available for borehole stability analysis (Falcao and Fonseca 2000; Parra et al. 2000; Hawkes et al. 2002; Wang and Lu 2002).

Economic Considerations. The economic basis for choosing UBD is more complex than merely suggesting higher well productivity. The high well productivity, combined with reservoir studies, can translate the improved well performance into a higher recovery factor and shorter time to recover the recoverable reserves. But the higher well construction cost due to the increased amount of equipment, personnel, and services required for an underbalanced well needs to be offset against potential production gains and long-term recovery rates from a reservoir. In some cases, UBD has allowed recovery of stranded reserves and allowed further infill drilling, thus postponing abandonment of a field by a number of years.

The economic considerations for UBD cannot be based solely on costs. The reservoir improvement and the reduction of pressure-related drilling problems all must be offset against the cost of implementing UBO. Costs will vary depending on the location, offshore or onshore, as well as with the complexity of the project. A standalone, single underbalanced well in a remote offshore location that requires generated nitrogen and gas recovery and a fully closed high-pressure separation system will significantly increase the well costs for UBD. Where as a multiwell project allows front-end engineering, mobilization, and equipment use to be optimized, enabling some of the costs to be offset against a higher number of wells.

9.1.5 Designing UBD Operations. The design of underbalanced wells follows a set pattern for most wells. The design process is similar to that of conventional wells, but there are some additional design steps required for UBD. Offset data collection and a good reservoir candidate-selection process are essential for good planning. A first UBD well in a field drilled with an operator that has never drilled underbalanced before, using a drilling contractor that has never drilled underbalanced with his rigs, can have a severe impact on the planning time required. Timings for planning of an underbalanced well are very much dependent on the well objectives and the complexity of the reservoir and the drilling operation. An offshore well from a production platform will require significantly longer planning time compared to a standalone land well (Giffin and Lyons 1999; Nas and Laird 2001; Rommetveit and Lage 2001; Lage et al. 2003). **Fig 9.1.3** shows the planning process for a typical UBD operation.

Fluid Selection. The fluid selection for UBO is a complex but very important step in the design of an underbalanced drilled well. The fluid types are classified into five major fluid types based primarily on the required equivalent circulating density (ECD): gas, mist, foam, gasified liquid, and liquid.

Fluid selection for UBD operations can be extremely complex. Fluid selection depends on reservoir characteristics, geophysical characteristics, well-fluid characteristics, well geometry, compatibility, hole cleaning, temperature stability, corrosion, drilling bottomhole assembly (BHA), data transmission, surface fluid handling and separation, formation lithology, health and safety, environmental impact, fluid source availability, and the primary objective for drilling underbalanced, all of which have to be taken into consideration before final fluid design.

One of the most important design criteria is the type of fluid, and the density that needs to be considered with regard to the anticipated pressures.

The last stage in fluid selection must be the optimization of the fluid with regard to hole cleaning, hole stability, and availability. Flow modeling using special multiphase flow models must be carried out to ensure that an underbalanced status can be achieved and maintained **(Table 9.1.6)**.

$$\text{Equivalent fluid density (lbm/gal)} = \frac{\text{Reservoir pressure (psi)}}{\text{Reservoir depth (ft TVD)} \times 0.052}$$

Gas. Gas drilling uses dry gas as the drilling medium. No liquids are added intentionally. The gas may be nitrogen, natural gas, or exhaust gas. Gas drilling is a common method for hard-rock drilling.

In reservoir applications, air cannot be used, and nitrogen or deoxygenated air systems would have to be used to avoid the formation of potentially explosive mixtures. Natural gas can be considered as a drilling

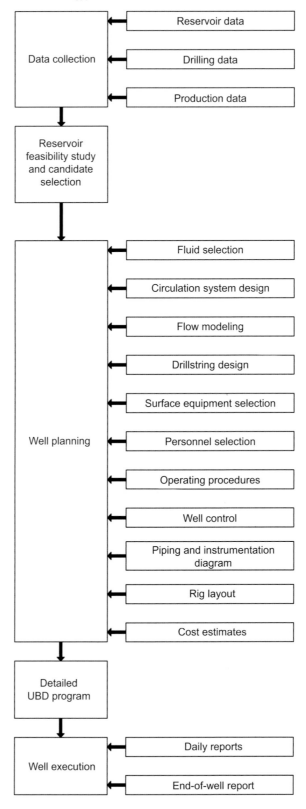

Fig. 9.1.3—Planning process for UBD.

TABLE 9.1.6—FLUID SELECTION FOR UNDERBALANCED DRILLING*

Equivalent Fluid Density	IADC Class	Fluid System
0 to 0.2 lbm/gal	1	Nitrogen/gas
0.2 to 0.6 lbm/gal	2	Mist
0.6 to 4 lbm/gal	3	Stable foam
4 to 7.5 lbm/gal	4	Foam or gasified fluid
7.5 to 8.5 lbm/gal	4	Native crude or diesel oil
8.5 to 10 lbm/gal	5	Single-phase fluids
10 to 12 lbm/gal	5	Brine
>12 lbm/gal	5	High-pressure UBD (this will require expert engineering)

* Divine 2003; Luo et al. 2000c

fluid, provided that a source of sufficient quantity and pressure of natural gas is present at or near the location. Stringent safety considerations must be adhered to when using hydrocarbon gas as a drilling medium. Gas drilling operations can be characterized as:

- Fast penetration rates
- Longer bit life
- Greater footage per bit
- Can handle only minimal liquid influx
- Possible slugging
- Possible mud rings in the presence of fluid ingress
- Reliance on high annular velocity to remove cuttings from the well

Mist. Mist drilling is gas drilling with injection of small quantities of fluid into the gas stream. A mist system will have less than 2.5 vol% of liquid content. In general, this technique needs to be used in areas where some formation water exists, which prevents the use of complete *dry gas* drilling.

Characteristics of mist drilling:

- Reliance on annular velocity to remove cuttings from the well
- Reduced formation of mud rings
- 30–40% more gas required than with dry air drilling
- Pressures generally higher than with dry gas drilling
- Slugging caused by incorrect air/liquid ratio (or gas/liquid ratio), with attendant pressure increase

Foam. Stable-foam drilling uses a homogeneous emulsion generated by mixing liquid, gas, and an emulsifying agent such as a surfactant. Stable foam contains a lot of gas, typically from 55 to 97 vol% gas. Several gases can be used to form foam. Nitrogen is the most common in most foam applications today because it is inert and environmentally friendly. Air can be used with foam if stability of the foam can be established. The foam structure encapsulates the oxygen and this eliminates many of the downhole fire risks.

Drilling with stable foam has some appeal due to the very low hydrostatic densities that can be generated with foam systems. Foam has good rheology and excellent cuttings transport properties, the fact that stable foam has some natural inherent viscosity as well as fluid loss control properties makes foam a very attractive drilling medium. During foam drilling, the volumes of liquid and gas injected into the well are carefully controlled. This ensures that foam forms when the liquid enters the gas stream, at the surface. The drilling fluid remains foam throughout its circulation path down the drillstring, up the annulus, and out of the well.

The more stable nature of foam also results in a much more continuous downhole pressure condition due to slower fluid and gas separation when the injection is stopped. In earlier foam systems, the amount of defoamer had to be tested carefully so that the foam was broken down before any fluid entered the separators. Recently developed stable-foam systems, including oil-based foam systems, are simpler to break, and the liquid can also be refoamed so that less foaming agent is required and a closed circulation system can be used. These systems, in general, rely on a chemical method to make and break the foam. The foam quality (the amount of gas) at surface used for drilling is normally between 80 and 95 vol%.

Characteristics of foam systems (Argillier et al. 1998):

- Extra fluid in the system reduces the influence of formation water.
- Stable foam has a very high carrying capacity.
- Improved cuttings transport reduces pump rates.
- Stable foam reduces slugging tendencies of the wellbore.
- Stable foam can withstand limited circulation stoppages without affecting the cuttings removal or ECD to any significant degree.
- Stable foam gives improved surface control and more stable downhole environment.
- The breaking of the foam at surface needs to be addressed at the design stage.
- If a recyclable foam system is used, more surface equipment is required.

Gasified Fluids. In fluid systems, the bottomhole pressure (BHP) in a well can be lowered through the introduction of gas into the fluid column. The fluid system can be water, crude oil, diesel, water-based mud, or oil-based mud. The gas is normally nitrogen or natural gas.

There are a number of methods that can be used to gasify a liquid system, and these methods are discussed in Section 9.1.8. The use of gas and liquid, as a circulation system in a well, complicates the flow behavior of a fluid, and the ratio of gas to liquid must be carefully calculated to ensure that a stable circulation system is used. If too much gas is used, slugging will occur. If not enough gas is used, the required BHP will be exceeded and the well may become overbalanced. Characteristics of gasified-fluid systems (Giffin and Lyons 2000):

- Extra fluid in the system will almost eliminate the influence of formation fluid unless incompatibilities occur.
- The fluid properties can be identified easily before starting the operation.
- Generally, less gas is required.
- Slugging of the gas and fluid must be managed correctly.
- Increased surface equipment will be required to store and clean the base fluid.
- Velocities, especially at surface, will be lower, reducing wear and erosion both downhole and to the surface equipment.

Single-Phase Fluids. Single-phase fluids are used if formation pressures are high enough to provide the desired underbalanced conditions. Many oil reservoirs are drilled with native crude that provides sufficient underbalanced conditions.

9.1.6 Gas Lift Systems. If gas injection is required to achieve the required underbalanced level, it offers not only a choice of what gas to use but also a choice of the way the gas is injected into the well.

Drillstring Injection. For drillstring injection systems, gas is injected at the standpipe manifold, where it mixes with the drilling fluid. One of the advantages of drillstring injection is that there is no special downhole equipment required in a well. Nonreturn valves have to be used in the drillstring to ensure that no flow returns up the drillstring. Using drillstring injection ensures that low BHPs can be achieved.

One of the disadvantages of drillstring injection is that the gas needs to be bled off prior to each connection. As a result of this, it can be difficult to obtain a stable circulation system and avoid pressure spikes at the reservoir if drilling is fast.

The other disadvantage is that the use of pulse-type measurement-while-drilling (MWD) tools is possible only to a maximum of 20 vol% gas. If higher gas volumes are required, electromagnetic MWD tools may have to be used.

Other issues that will need to be considered when using drillstring injection are the impregnation of the gas into any downhole rubber components. Positive-displacement motors (PDMs) need to be selected carefully when using drillstring gas injection to ensure that explosive decompression of the rubber is avoided.

In addition, downhole motor stalling with a compressible fluid system often results in overspeeding and subsequent failure of the motor once a stalled motor is picked up off bottom.

Annular Injection. Annular injection through a concentric string is used most commonly in underbalanced wells. Often a liner is run and cemented just above the target formation, this liner is tied back to surface, and the casing-liner annulus is used to inject gas into the well.

The main advantage of using an annulus to inject gas into the well is that gas injection can be continued during connections, thus creating a more stable BHP. Single-phase fluid is pumped down the drillstring, and this has the advantage that conventional MWD tools can be used (Mykytiw et al. 2003).

Parasite String Injection. A parasite string for gas injection is really used only in vertical wells. Often, two 1- or 2-in. coiled-tubing strings are connected and strapped to the casing string above the reservoir as the casing is run into the well. Gas is injected down the parasite string and into the drilling annulus (Westermark 1986).

The installation of a production casing string combined with the installation of two parasite strings can become a complicated operation. Some modifications to the wellhead system normally are required to provide surface connections to the parasite strings. Parasite strings are widely used in the USA, where they are used to avoid not only lost circulation but also differential sticking. Although losses can be cured, to avoid differential sticking of the drillstring, they run parasite strings.

Combined Annular and Drillstring Injection. Combining both the string injection and annular injection can also be done to provide the required BHPs. It makes surface operations a little more complex, but it has been successfully applied in a number of wells.

9.1.7 Hydraulics in UBO. The circulation system is the most critical element in UBD. It combines drilling and production operations, where well control becomes flow control. A properly designed circulation system manages pressures and does not allow extreme reservoir inflow rates or high annular flowing pressures at surface. The circulation system must effectively achieve an underbalanced state, provide adequate hole cleaning, power the BHA, and provide well control. At the preplanning phase, multiphase flow modeling is required to determine circulation system parameters. Multiphase flow calculations are among the most complex fluid-engineering calculations known in the industry. Compressible fluids change considerably with pressures and temperatures. Specialized software is required to ensure that the use of compressible fluids in drilling can be modeled. Both static and dynamic flow models are currently available, but the use of multiphase flow models requires significant training for engineers. Expert advice should be sought before undertaking any hydraulics calculations for an underbalanced well.

Depending on the fluids and gases used, different models may have to be used to verify the operating window for UBD. A multiphase model for UBD should include the following:

- Prediction of flow regime at any given point in the well
- Liquid holdup calculations at any given point in the well
- Frictional pressure loss calculations
- Thermal pressure/volume/temperature (PVT) calculations
- Hole cleaning and cuttings transport indications

Operating limits are determined before execution so that contingency plans for the circulating system can be put in place before starting the operations.

BHP Calculations. When designing underbalanced circulation systems, the BHP must be maintained below the reservoir pressure under static and dynamic conditions. Surface pressures and reservoir flow rates must be maintained as low as possible while, at the same time, maintaining control of the well. To ensure that proper hydraulics calculations are performed, a graph showing the UBD operating window is normally generated by the UBD engineer **(Fig 9.1.4)**. This graph shows liquid flow rate, gas-injection rate, and BHPs and provides wellsite personnel with a range of drilling parameters that can be used to maintain an underbalanced condition during drilling.

According to Saponja (1998), the pressure regime for UBD should be maintained in the friction-dominated part of the pressure window **(Fig 9.1.5)**. The hydrostatically dominated region depicts an unstable system because of a negatively sloped curve. In this region, small changes in gas-injection rate or inflow of formation

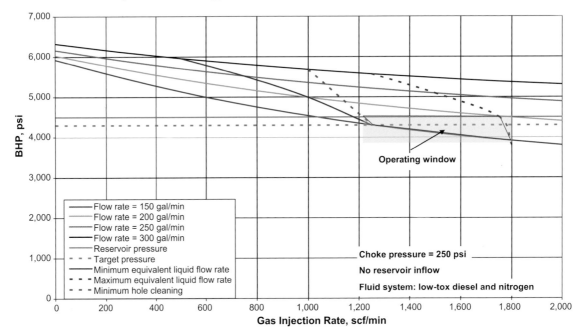

Fig. 9.1.4—Typical depiction of operating window for UBD.

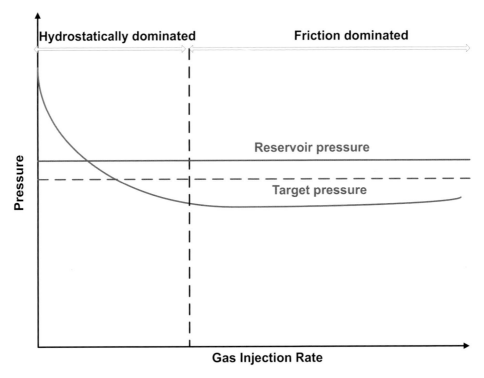

Fig. 9.1.5—Flow modeling for UBD.

gas dramatically change the BHP. Gas inflow not only reduces the BHP but also encourages more reservoir inflow, which can further reduce BHP. A circulation system operating in the friction-dominated region is more stable with changes in gas injection rates or gas inflow from the reservoir. Increasing the N_2 injection rate only results in small increases the BHP. Thus, a circulating system operating in the friction-dominated region can result beneficially in self-control of the inflow of formation gas (Saponja 1998; Bijleveld et al. 1998; Guo and Ghalambor 2002; Smith et al. 1998; Rommetveit et al. 2004).

Further flow modeling calculations that are normally performed are:

- Ensuring that sufficient flow rate is provided for a downhole motor; to achieve this, the equivalent liquid rate through the motor is calculated by combining liquid and gas rates.
- Hole cleaning while drilling underbalanced must be closely monitored. There is a reduced fluid rheology (a very thin, non-solids suspending fluid, often with turbulent two-phase flow) and, normally, an increased rate of penetration (ROP). Two-phase hole cleaning is largely dependent on the same criteria as for single phase. Hole cleaning efficiency and solids transport are primarily controlled by liquid phase velocities and solids concentration. Liquid velocity is the critical parameter controlling the system's ability to transport solids. From past experience, it has been concluded that a minimum liquid phase annular velocity of 180 to 250 ft/min is required in a wellbore with a deviation greater than 10° inclination.
- Annular friction pressure vs. gas injection rate. The annular friction pressure provides an indication about the pressure losses seen in the annulus. This is important in UBD operations as the annular friction is lost when making connections, resulting in pressure spikes at the reservoir. An indication of annular pressure losses will assist the engineer in designing and formulating connection procedures.
- Annular liquid holdup. This calculation is to understand what happens in the well once circulation is stopped for tripping or connections. Knowing the average percentage of gas and liquid in the annulus allows the engineers to calculate the liquid levels and the resulting BHPs when tripping.
- Injection pressures must be calculated to ensure that sufficient pump pressure capacity is provided to inject gas and liquids into the drillstring when circulating.

9.1.8 Downhole Equipment for UBD. No special drillstring tools are required for UBD. Dual downhole float valves or nonreturn valves are used in the bottom of the BHA to avoid backflow when making connections. If gas injection is used, a nonreturn valve at the top of the string helps in avoiding the need to bleed the gas from the entire drillstring when making connections.

Pressure measurement while drilling (PMWD) tools are often used to measure the BHP and these sensors have been invaluable in every UBD operation to date in which they have been included in the drillstring and operated without downtime.

In conventional drilling, data from the MWD and logging while drilling (LWD) tools are transmitted as pressure pulses through the fluid in the drillstring to a receiver at the top of the string. Several methods are used for mud-pulse data transmission. They all work well with single-phase fluids (incompressible fluids) in the string. When a compressible fluid is circulated through the string, the pressure pulses are dampened and information cannot be transmitted. Tools such as the electromagnetic tools that transmit low-frequency electromagnetic signals from the BHA through the Earth to surface receivers are often used in gasified or foam drilling when a compressible fluid is pumped down the drillstring.

Deployment Valves. A downhole deployment valve can be installed as an integral part of the casing or liner string above the openhole interval. The valve is designed to close the wellbore when the bit passes through it on tripping out of the hole, and it is opened when the bit is run back into the well. The valve isolates the wellbore from the formation pressure, thus allowing tripping without stripping or snubbing. This provides significant time savings during trips (Herbal et al. 2002; Muir 2004; Sutherland et al. 2005).

Drillstring Issues. No special issues are required for the drillstring when drilling underbalanced. Some operators insist on gas-tight connections or high-torque connections when specifying drillstrings for UBD. Any hardbanding on the pipe should be as smooth as possible to minimize wear on the rotating control head rubbers.

Elastomers. The use of elastomers in UBD must be carefully considered. In pressurized, gaseous environments, gases will dissolve into and permeate through elastomers. Especially at high BHPs, gas will absorb into elastomers.

Very large quantities of gas can be dissolved within an elastomer. This can be as high as 700 cm³ of gas per cubic centimeter of rubber at high pressures. This gas will try to escape out of the rubber once the pressure is reduced and often results in explosive decompression damage to the rubber components. Special elastomer compounds are now being developed for downhole tools and BOP components to ensure that gas impregnation is minimized.

Coiled Tubing or Jointed Pipe. Hole size and directional requirements and some economic considerations will determine if coiled tubing or jointed pipe is the optimal drillstring medium. For hole sizes larger than 6 in., jointed pipe will have to be used; for hole sizes smaller than 6 in., coiled tubing can be used, but the size of the coil required for a UBD operation is determined by many factors, and coiled tubing drilling experts should be consulted for advice on coil sizes. Generally, coiled tubing has several advantages over jointed pipe systems with regard to UBD. For jointed pipe, tripping under pressure and making connections are some of the issues that will need careful consideration (Nas 1999; Graham 1998; Luft and Wilde 1999; van Venrooy et al. 1999; Tinkham et al. 2000; Thatcher et al. 2000; Fraser and Ravensbergen 2002).

Gases for UBD. *Air.* Compressed air is the cheapest and, probably, the simplest to use because it requires compression only. It is used extensively by the mining industry and in water-well drilling. It has been widely used in the past in shallow-gas-well drilling. Its application in oil- and gas-well drilling is diminishing today due to the risk of bottomhole fires that may occur when the oxygen in the air is mixed with reservoir hydrocarbon gas.

Natural Gas. Natural gas has been used in UBO, but it normally is used in combination with annular gas injection. It solves the problems of corrosion and downhole fires, which are typical to air drilling. It has favorable properties when mixed with mud in gasified systems. Drillstring injection with natural gas is possible, but safety procedures will have to be implemented when making connections.

A source of natural gas at the right volumes and pressures is required if natural gas is to be used. In coiled-tubing operations, gas normally is not injected through the coil because of the potential issues associated with pinhole leaks in the reel of the coil at surface.

Nitrogen. Nitrogen is the preferred gas for UBD in the petroleum industry, particularly in offshore installations. Nitrogen is inert, nonflammable, nonexplosive, and noncorrosive. It has been used in underbalanced-well-cleaning operations for many years, and there is considerable experience in handling it in land and offshore operations.

Cryogenic or liquid nitrogen normally is supplied in 2,000-gal transport tanks. Liquid nitrogen is passed through the nitrogen converter, where the fluid is converted to gas. The gas is then injected into the string.

In 1995, a US patent was issued for a process to use membrane gas separation technology to drill oil and gas wells with nitrogen produced on site **(Fig. 9.1.6)** to replace higher-cost cryogenically produced nitrogen as an alternative gas source for UBD. A nitrogen generator filters the oxygen out of the air and produces nitrogen as a gas with a purity of approximately 97 vol%. The nitrogen is then compressed and injected into the well or drillstring (Chitty 1998).

Exhaust Gas. Some new-technology gas systems have been introduced in the form of exhaust gas systems. Propane gas is burned in an engine, and the exhaust is used as a lift gas in UBD. Exhaust gas from conventional diesel engines cannot be used in UBO because it still contains too much oxygen.

Well-Control Equipment. In UBO, the well is continuously pressurized, and the drillstring has to rotate and move axially through a seal at the top of the well. There are various designs of seal systems that can accommodate the rotation and the axial loads associated with UBD operations. These are basically an annular seal element in constant contact with the rotating drillstring, and it rotates together with the string. The seal element is mounted on bearings that allow rotation relative to the housing.

The conventional BOP comprising the annular BOP and various ram BOPs is still used as in conventional operations **(Fig. 9.1.7)**. The BOP and rig choke manifold normally are not used for routine UBO but are maintained as secondary well-control devices. There are two categories of rotating annulus seal elements, or RCDs, as they are now called:

- Passive seal or force-fit seal
- Active seal

The passive seal equipment uses the elasticity of the rubber element with added energy from the well pressure to maintain the seal around the drillstring. In the active seal equipment, the seal is energized by hydraulic pressure.

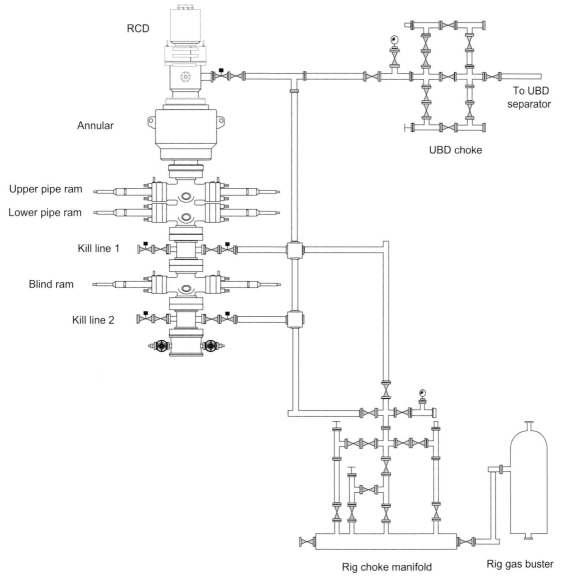

Oxygen and water vapor are fast gases. These quickly permeate through the membrane, allowing nitrogen to flow through the fiber bores as product.

Fig 9.1.6—Nitrogen membrane generation.

Fig. 9.1.7—Typical BOP stack for UBD.

The RCD is mounted on top of the conventional BOP and forms part of the principal barrier that contains well pressure during UBO. The conventional BOP stack, which is used in its open position just as in conventional drilling, remains the secondary well-control barrier that can be activated if required during well-control scenarios.

Passive RCDs. The passive RCDs use the elasticity of the rubber element with added energy from the well pressure to maintain the seal around the drillstring. The differential pressure enhances the force-fit contact between the pipe and the rubber element across the seal **(Fig. 9.1.8)**.

The seal elements are mounted on a bearing-supported assembly. The contact force between the seal and the rotating string creates a large enough friction force to rotate the bearing assembly. Because of the high friction between the drillstring and the rubber elements, large axial loads, downward or upward, are developed and transferred from the seal to the bearing. The highly loaded bearing generates heat, and the bearings are cooled and lubricated by circulating cooling oil around the system.

Active Rotating Control Devices. In active RCDs, the rubber seal element or the annular packer is inflated or energized by hydraulic pressure. As in passive RCDs, the bearing assembly is rotated by the grip force on the rotating pipe. The bearings are lubricated and cooled by circulating oil. In certain types of RCDs, the hydraulic module is equipped with an automatic control system that automatically regulates the energizing pressure. The hydraulic pressure and thus the sealing pressure increases automatically as the wellhead pressure increases **(Fig. 9.1.9)**. Most common RCDs used for UBD operations are rated to a static pressure of 5,000 psi and a rotating pressure of 2,500 psi.

Snubbing Systems. The hydrostatic pressure in the wellbore exerts axial force upward on the bottom of the drillstring—the higher the pressure, the higher the upward force. If the hydrostatic force is larger than the weight of the drillstring, the pipe will be forced out of the well (pipe-light). Therefore, a system is required to ensure that the drillstring is controlled when tripping out and pushed in when tripping into the well.

Fig. 9.1.8—Typical passive RCD configuration.

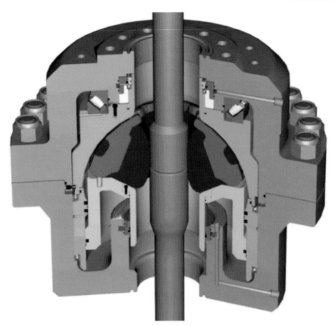

Fig. 9.1.9—Active rotating diverter.

If tripping is to be conducted underbalanced, a snubbing system will have to be installed on top of the rotating control head system. The most common systems used for UBD operations are rig-assist snubbing systems. A jack with a 10-ft stroke is used to push pipe into the hole or to trip pipe out of the hole. Once the weight of the string exceeds the upward force of the well, the snubbing system is switched to standby and the pipe is tripped in the hole using the drawworks. To facilitate snubbing, these so-called rig-assist snubbing units are installed on the rig floor when tripping pipe. On offshore jackup rigs where there is sufficient space under the rig floor, rig assist snubbing units can be installed below the rig floor, allowing the rig floor to be used in the conventional drilling way (Robichaux 1999; Aasen and Skaugen 2002; Schmigel and MacPherson 2003; Hannegan and Divine 2002; Hannegan and Wanzer 2003; Cantu et al. 2004).

Separation Systems. The separation system in UBD has to be tailored to the expected reservoir fluids. The separation system must be designed to handle the expected reservoir influx, and it must be able to handle any slugging efficiently **(Fig. 9.1.10)**.

The surface separation system in UBD can be compared with a process plant, and there are many similarities with the process industry. Fluid streams while drilling underbalanced are often described as four-phase flow, as the return flow comprises oil, water, gas, and solids. The challenge of any of the separation equipment for UBD is to separate the various phases of the return fluid stream into individual streams effectively and efficiently. Important factors in successful separation technology can be described as

- Having sufficient residence time (4–5 minutes) between entering and leaving the separator to allow phase separation
- Minimizing turbulence, particularly turbulence induced by a large volume of gas or too high flow velocities in the separator
- Avoiding agitation and remixing of the already separated phases
- Maintaining proper level control of the various phases to avoid carryover of liquid in the gas line and blowby of gas in the liquid lines
- Having enough capacity to accommodate irregular flow, pressure surges, and volume fluctuation in the inflow stream

Careful design of the surface separation system is required to ensure that a wide range of flow rates and reservoir fluids can be separated. Both horizontal and vertical separation systems are available for UBD.

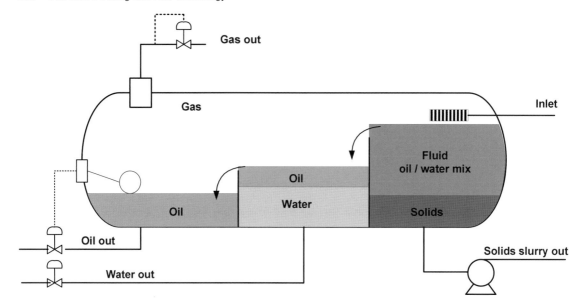

Fig. 9.1.10—Four-phase horizontal separator.

Vertical separators are more effective when the returns are predominantly liquid, while horizontal separators have higher and more efficient gas-handling capacities. In most cases, the location and reservoir characteristics combined with the drilling requirements will determine the optimum separation system to be used.

Solids Handling. Depending on the configuration of the separation system, solids are normally removed from the system at the first-stage separator and then transported and handled at surface. Solids can be pumped back to the shale shakers or they can be centrifuged as part of a separation system completely independent of the rig. In large horizontal separators, the solids sometimes can be left in the separator until the job is completed.

Data Acquisition. The data acquisition used on a UBD operation should provide as much information as possible about the reservoir and the drilling operation. Depending on the particular operation, monitored information may include the following parameters:

- The return stream flow rate, pressure, and temperature
- The injection stream flow rate, pressure, and temperature
- MWD downhole pressure and temperature
- Surface drilling data (e.g., hookloads, depth, ROP)
- Separator operating conditions (fluid levels, pressure, temperature)
- Choke position
- LWD information

Monitoring systems normally have audiovisual alarms that alert the operator in the data acquisition cabin when monitored parameters deviate from their normal operating range.

Reservoir characterization is a prime objective in UBD, and accurate reservoir flow data should be obtained.

Erosion. In some of the high-velocity, high-rate gas wells, the erosion of the pipework from solids can lead to washouts in the pipework. Inspection procedures for the pipework during and after UBO need to be implemented to ensure the safety of the installation.

Corrosion. The use of gases and production of reservoir fluids during the drilling process have led in some cases to corrosion of the drillstring and the installed well equipment. The introduction of oxygen as a result of impure nitrogen from nitrogen generation systems adds to the corrosion potential in underbalanced wells. It is difficult to accurately predict the corrosion potential of a UBD fluid system. Some type of corrosion monitoring and management program should be implemented for UBD operations.

Certain information can greatly enhance the effectiveness of a corrosion management program. Corrosion mechanism identification begins by analyzing the following items:

1. Reservoir fluid type and chemistry
2. Bottomhole temperature
3. BHP
4. Acid gas (H_2S or CO_2) concentrations
5. Electrical conductivity of fluids
6. Fluid velocity

Corrosion inhibitors always should be considered, even in oil-based fluid systems. A corrosion-monitoring program should be implemented as soon as operations commence. If membrane nitrogen is used, oxygen sensors should be used and monitored routinely by operations personnel.

9.1.9 Completing Underbalanced Wells. A lot of early underbalanced drilled wells were displaced to an overbalanced kill fluid prior to the installation of the completion. If the purpose of UBD is reservoir improvement, it is important that the reservoir is never exposed to overbalanced pressure with a nonreservoir fluid. Significant reductions in productivity have been observed in underbalanced wells that were killed for the installation of the completion. The reservoir should be isolated either through the installation of the permanent production packer and tailpipe with plugs or through the use of a downhole deployment valve or some kind of retrievable, inflatable plug system.

A number of technologies are now available for completion of underbalanced wells. A permanent packer and tailpipe with plugs can be installed in the well as a barrier, and this allows the completion to be run and installed in a conventional manner (Harting et al. 2003; Cavender and Restarick 2004; Cuthbertson et al. 2003).

Underbalanced Cementing. If cement is used in an underbalanced well, most of the gains achieved by the underbalanced process will be eliminated. Foam-based cements could be a solution, but there are very few case studies of underbalanced wells completed with lightweight cement slurries.

9.1.10 Health, Safety, and Environment (HSE) Issues in UBD. Some new hazards and new challenges related to UBD must be recognized when introducing UBD from an HSE perspective. Most of the hazards involved in UBO are generic to drilling, well operations, and production activities, and all can be handled by a variety of well-established measures.

The IADC UBO Committee has released a document to assist in managing HSE issues associated with UBD operations. It is recommended that this guidance document, *HSE Planning Guidelines*, be downloaded from the IADC website because it provides detailed guidance on the HSE issues associated with UBD operations.

Barriers and Well Control in UBD. It is widely believed that because underbalanced wells are flowing to surface, well control is not considered. Although underbalanced wells are indeed flowing to surface, the flow rate from the reservoir is carefully controlled, and well control is maintained by keeping surface pressures and reservoir in flow rates as low as possible. A matrix showing surface pressures and flow rates is normally compiled as part of the planning process to help in the well-control decision-making process **(Fig. 9.1.11)**.

If, while drilling an underbalanced well, a situation occurs that compromises rig or personnel safety, the well may have to be killed. This can be as a result of the following situations:

- Failure of pressure-control equipment
- Parted drillstring
- Unexpected H_2S
- General emergency

In some cases, killing an underbalanced well may be a complex process. Losses are almost always severe when killing a UBD well. A UBD well is normally killed using the driller's method, by circulating out the hydrocarbons in the well first. If an emergency kill is required, the fastest way to kill the well may have to be chosen. The emergency well-kill methods must be agreed and documented during the well-design stage of the project (Ramalho et al. 2006; Graham 2008).

		RCD Pressure Rating	
	Pressure 50% of dynamic rating	Pressure >75% of RCD dynamic rating	Pressure >90% of RCD dynamic rating
Low	Manageable	Increase BHCP	Shut in on rig BOPs
Medium	Increase BHCP	Increase BHCP	Shut in on rig BOPs
High	Shut in on rig BOPs	Shut in on rig BOPs	Shut in on rig BOPs

Reservoir inflow rate

Fig 9.1.11—UBD well-control matrix.

Emergency Systems for UBD. A full emergency shutdown system normally is installed on any UBD operation. Depending on the complexity of the project, this may include fire and gas detection on offshore platforms, or it may just be a simple warning siren on simple onshore low-pressure operations. The full requirements for safety systems must be reviewed during the planning phase (Knight et al. 2004; Arild et al. 2004).

References

Aasen, J. and Skaugen, E. 2002. Pipe Buckling at Surface in Underbalanced Drilling. Paper SPE 77241 presented at the IADC/SPE Asia Pacific Drilling Technology Conference, Jakarta, 8–11 September. DOI: 10.2118/77241-MS.

Argillier, J.-F., Saintpere, S., Hertzhaft, B., and Toure, A. 1998. Stability and Flowing Properties of Aqueous Foams for Underbalanced Drilling. Paper SPE 48982 presented at the SPE Annual Technical Conference and Exhibition, New Orleans, 27–30 September. DOI: 10.2118/48982-MS.

Arild, Ø., Nilsen, T., and Sandøy, M. 2004. Risk-Based Decision Support for Planning of an Underbalanced Drilling Operation. Paper SPE 91242 presented at the SPE/IADC Underbalanced Technology Conference and Exhibition, Houston, 11–12 October. DOI: 10.2118/91242-MS.

Bennion, D.B., Thomas, F.B., Bietz, R.F., and Bennion, D.W. 1998. Underbalanced Drilling: Praises and Perils. *SPEDC* **13** (4): 214–222. SPE-52889-PA. DOI: 10.2118/52889-PA.

Bijleveld, A.F., Koper, M., and Saponja, J. 1998. Development and Application of an Underbalanced Drilling Simulator. Paper SPE 39303 presented at the IADC/SPE Drilling Conference, Dallas, 3–6 March. DOI: 10.2118/39303-MS.

Biswas, D., Suryanarayana, P.V., Frink, P.J., and Rahman, S. 2003. An Improved Model to Predict Reservoir Characteristics During Underbalanced Drilling. Paper SPE 84176 presented at the SPE Annual Technical Conference and Exhibition, Denver, 5–8 October. DOI: 10.2118/84176-MS.

Brant, B.D., Brent, T.F., and Bietz, R.F. 1996. Formation Damage and Horizontal Wells—A Productivity Killer? Paper SPE 37138 presented at the International Conference on Horizontal Well Technology, Calgary, 18–20 November. DOI: 10.2118/37138-MS.

Cantu, J.A., May, J., and Shelton, J. 2004. Using Rotating Control Devices Safely in Today's Managed Pressure and Underbalanced Drilling Operations. Paper SPE 91583 presented at the SPE/IADC Underbalanced Technology Conference and Exhibition, Houston, 11–12 October. DOI: 10.2118/91583-MS.

Cavender, T.W. and Restarick, H.L. 2004. Well-Completion Techniques and Methodologies for Maintaining Underbalanced Conditions Throughout Initial and Subsequent Well Interventions. Paper SPE 90836 presented at the SPE Annual Technical Conference and Exhibition, Houston, 26–29 September. DOI: 10.2118/90836-MS.

Chitty, G.H. 1998. Corrosion Issues With Underbalanced Drilling in H_2S Reservoirs. Paper SPE 46039 presented at the SPE/ICoTA Coiled Tubing Roundtable, Houston, 15–16 March. DOI: 10.2118/46039-MS.

Culen, M.S., Harthi, S., and Hashimi, H. 2003. A Direct Comparison Between Conventional and Underbalanced Drilling Techniques in the Saih Rawl Field, Oman. Paper SPE 81629 presented at the IADC/SPE Underbalanced Technology Conference and Exhibition, Houston, 25–26 March. DOI: 10.2118/81629-MS.

Cuthbertson, R., Green, A., Dewar, J.A.G., and Truelove, B.D. 2003. Completion of an Underbalanced Well Using Expandable Sand Screen for Sand Control. Paper SPE 79792 presented at the SPE/IADC Drilling Conference, Amsterdam, 19–21 February. DOI: 10.2118/79792-MS.

Devaul, T. and Coy, A. 2002. Underbalanced Horizontal Drilling Yields Significant Productivity Gains in the Hugoton Field. Paper SPE 81632 presented at the IADC/SPE Underbalanced Technology Conference and Exhibition, Houston, 25–26 March. DOI: 10.2118/81632-MS.

Ding, Y., Herzhaft, B., and Renard, G. 2006. Near-Wellbore Formation Damage Effects on Well Performance: A Comparison Between Underbalanced and Overbalanced Drilling. *SPEPO* **21** (1): 51–57. SPE-86558. PA. DOI: 10.2118/86558-PA.

Divine, R. 2003. Planning Is Critical for Underbalanced Applications With Under-Experienced Operators. Paper SPE 81627 presented at the IADC/SPE Underbalanced Technology Conference and Exhibition, Houston, 25–26 March. DOI: 10.2118/81627-MS.

Falcao, J.L. and Fonseca, C.F. 2000. Underbalanced Horizontal Drilling: A Field Study of Wellbore Stability in Brazil. Paper SPE 64379 presented at the SPE Asia Pacific Oil and Gas Conference and Exhibition, Brisbane, Australia, 16–18 October. DOI: 10.2118/64379-MS.

Fraser, R.G. and Ravensbergen, J. 2002. Improving the Performance of Coiled Tubing Underbalanced Horizontal Drilling Operations. Paper SPE 74841 presented at the SPE/ICoTA Coiled Tubing Conference and Exhibition, Houston, 9–10 April. DOI: 10.2118/74841-MS.

Garrouch, A.A. and Labbabidi, H.M.S. 2003. Using Fuzzy Logic for UBD Candidate Selection. Paper SPE 81644 presented at the IADC/SPE Underbalanced Technology Conference and Exhibition, Houston, 25–26 March. DOI: 10.2118/81644-MS.

Gedge, B. 1999. Underbalanced Drilling Gains Acceptance in Europe and the International Arena. Paper SPE 52833 presented at the SPE/IADC Drilling Conference, Amsterdam, 9–11 March. DOI: 10.2118/52833-MS.

Giffin, D.R. and Lyons, W.C. 1999. Case Histories of Design and Implementation of Underbalanced Wells. Paper SPE 55606 presented at the SPE Rocky Mountain Regional Meeting, Gillette, Wyoming, USA, 15–18 May. DOI: 10.2118/55606-MS.

Giffin, D.R. and Lyons, W.C. 2000. Case Histories of Design and Implementation of Underbalanced Wells. Paper SPE 59166 presented at the IADC/SPE Drilling Conference, New Orleans, 23–25 February. DOI: 10.2118/59166-MS.

Gough, G. and Graham, R. 2008. Offshore Underbalanced Drilling—The Challenge at Surface. Paper SPE 112779 presented at the IADC/SPE Drilling Conference, Orlando, Florida, USA, 4–6 March. DOI: 10.2118/112779-MS.

Graham, R.A. 1998. Planning for Underbalanced Drilling Using Coiled Tubing. Paper SPE 46042 presented at the SPE/ICoTA Coiled Tubing Roundtable, Houston, 15–16 March. DOI: 10.2118/46042-MS.

Guo, B. and Ghalambor, A. 2002. An Innovation in Designing Underbalanced Drilling Flow Rates: A Gas-Liquid Rate Window (GLRW) Approach. Paper SPE 77237 presented at the IADC/SPE Asia Pacific Drilling Technology Conference, Jakarta, 8–11 September. DOI: 10.2118/77237-MS.

Hannegan, D. and Divine, R. 2002. Underbalanced Drilling—Perceptions and Realities of Today's Technology in Offshore Applications. Paper SPE 74448 presented at the IADC/SPE Drilling Conference, Dallas, 26–28 February. DOI: 10.2118/74448-MS.

Hannegan, D. and Wanzer, G. 2003. Well Control Considerations—Offshore Applications of Underbalanced Drilling Technology. Paper SPE 79854 presented at the SPE/IADC Drilling Conference, Amsterdam, 19–21 February. DOI: 10.2118/79854-MS.

Harting, T., Gent, J., and Anderson, T. 2003. Drilling Near Balance and Completing Open Hole To Minimize Formation Damage in a Sour Gas Reservoir. Paper SPE 81622 presented at the IADC/SPE Underbalanced Technology Conference and Exhibition, Houston, 25–26 March. DOI: 10.2118/81622-MS.

Hawkes, C.D., Smith, S.P., and McLellan, P.J. 2002. Coupled Modeling of Borehole Instability and Multiphase Flow for Underbalanced Drilling. Paper SPE 74447 presented at the IADC/SPE Drilling Conference, Dallas, 26–28 February. DOI: 10.2118/74447-MS.

Helio, S. and Queiroz, J. 2000. How Effective Is Underbalanced Drilling at Preventing Formation Damage? Paper SPE 58739 presented at the SPE International Symposium on Formation Damage Control, Lafayette, Louisiana, USA, 23–24 February. DOI: 10.2118/58739-MS.

Herbal, S., Grant, R., Grayson, B., Hosie, D., and Cuthbertson, B. 2002. Downhole Deployment Valve Addresses Problems Associated With Tripping Drill Pipe During Underbalanced Drilling Operations. Paper SPE 77240 presented at the IADC/SPE Asia Pacific Technology Conference, Jakarta, 8–11 September. DOI: 10.2118/77240-MS.

Hunt, J.L. and Rester, S. 2000. Reservoir Characterization During Underbalanced Drilling: A New Model. Paper SPE 59743 presented at the SPE/CERI Gas Technology Symposium, Calgary, 3–5 April. DOI: 10.2118/59743-MS.

Keshka, A., Al Rawahi, A.S., Murawwi, R.A., Al Yaaqeeb, K., Qutob, H., Boutalbi, S., and Villatoro, J. 2007. Reservoir Candidate Screening: The SURE Way to Successful Underbalanced Drilling Projects in ADCO. Paper SPE 108402 presented at the SPE Asia Pacific Oil and Gas Conference and Exhibition, Jakarta, 30 October–1 November. DOI: 10.2118/108402-MS.

Kimery, D. and McCaffrey, M. 2004. Underbalanced Drilling in Canada: Tracking the Long-Term Performance of Underbalanced Drilling Projects in Canada. Paper SPE 91593 presented at the SPE/IADC Underbalanced Technology Conference and Exhibition, Houston, 11–12 October. DOI: 10.2118/91593-MS.

Kimery, D. and van der Werken, T. 2004. Damage Interpretation of Properly and Improperly Drilled Underbalanced Horizontals in the Fractured Jean Marie Reservoir Using Novel Modeling and Methodology. Paper SPE 91607 presented at the SPE/IADC Underbalanced Technology Conference and Exhibition, Houston, 11–12 October. DOI: 10.2118/91607-MS.

Kneissl, W. 2001. Reservoir Characterization Whilst Underbalanced Drilling. Paper SPE 67690 presented at the SPE/IADC Drilling Conference, Amsterdam, 27 February–1 March. DOI: 10.2118/67690-MS.

Knight, J., Pickles, R., Smith, B., and Reynolds, M. 2004. HSE Training, Implementation, and Production Results for a Long-Term Underbalanced Coiled-Tubing Multilateral Drilling Project. Paper SPE 91581 presented at the SPE/IADC Underbalanced Technology Conference and Exhibition, Houston, 11–12 October. DOI: 10.2118/91581-MS.

Labat, C.P., Benoit, D.J., and Vining, P.R. 2000. Underbalanced Drilling at Its Limits Brings Life to Old Field. Paper SPE 62896 presented at the SPE Annual Technical Conference and Exhibition, Dallas, 1–4 October. DOI: 10.2118/62896-MS.

Lage, A.C.V.M., Sotomayor, G.P., Vargas, A.C. et al. 2003. Planning, Executing and Analyzing the Productive Life of the First Six Branches Multilateral Well Drilled Underbalanced in Brazil. Paper SPE 81620 presented at the IADC/SPE Underbalanced Technology Conference and Exhibition, Houston, 25–26 March. DOI: 10.2118/81620-MS.

Luft, H.B. and Wilde, G. 1999. Industry Guidelines for Underbalanced Coiled Tubing Drilling of Critical Sour Wells. Paper SPE 54483 presented at the SPE/ICoTA Coiled Tubing Roundtable, Houston, 25–26 May. DOI: 10.2118/54483-MS.

Luo, S., Hong, R., Meng, Y., Zhang, L., Li, Y., and Qin, C. 2000a. Underbalanced Drilling in High-Loss Formation Achieved Great Success—A Field Case Study. Paper SPE 59260 presented at the IADC/SPE Drilling Conference, New Orleans, 23–25 February. DOI: 10.2118/59260-MS.

Luo, S., Li, Y., Meng, Y., and Zhang, L. 2000b. A New Drilling Fluid for Formation Damage Control Used in Underbalanced Drilling. Paper SPE 59261 presented at the IADC/SPE Drilling Conference, New Orleans, 23–25 February. DOI: 10.2118/59261-MS.

Luo, S., Meng, Y., Tang, H., and Zhou, Y. 2000c. A New Drill-In Fluid Used for Successful Underbalanced Drilling. Paper SPE 58800 presented at the SPE International Symposium on Formation Damage Control, Lafayette, Louisiana, USA, 23–24 February. DOI: 10.2118/58800-MS.

Medley, G. and Stone, C.R. 2004. MudCap Drilling When? Techniques for Determining When To Switch From Conventional to Underbalanced Drilling. Paper SPE 91566 presented at the SPE/IADC Underbalanced Technology Conference and Exhibition, Houston, 11–12 October. DOI: 10.2118/91566-MS.

Moore, D.D., Bencheikh, A., and Chopty, J.R. 2004. Drilling Underbalanced in Hassi Messaoud. Paper SPE 91519 presented at the SPE/IADC Underbalanced Technology Conference and Exhibition, Houston, 11–12 October. DOI: 10.2118/91519-MS.

Murphy, D., Davidson, I., Kennedy, Busaidi, R., Wind, J., Mykytiw, C., and Arsenault, L. 2005. Applications of Underbalanced Drilling Reservoir Characterization for Water Shut Off in a Fractured Carbonate Reservoir—A Project Overview. Paper SPE 93695 presented at the SPE Middle East Oil and Gas Show and Conference, Bahrain, 12–15 March. DOI: 10.2118/93695-MS.

Mykytiw, C.G., Davidson, I.A., and Frink, P.J. 2003. Design and Operational Considerations To Maintain Underbalanced Conditions With Concentric Casing Injection. Paper SPE 81631 presented at the IADC/SPE Underbalanced Technology Conference and Exhibition, Houston, 25–26 March. DOI: 10.2118/81631-MS.

Nas, S. 1999. Underbalanced Drilling in a Depleted Gas Field Onshore UK With Coiled Tubing and Stable Foam. Paper SPE 52826 presented at the SPE/IADC Drilling Conference, Amsterdam, 9–11 March. DOI: 10.2118/52826-MS.

Nas, S. and Laird, A. 2001. Designing Underbalanced Thru Tubing Drilling Operations. Paper SPE 67829 presented at the SPE/IADC Drilling Conference, Amsterdam, 27 February–1 March. DOI: 10.2118/67829-MS.

Parra, J.G., Celis, E., and De Gennaro, S. 2000. Wellbore Stability Simulations for Underbalanced Drilling Operations in Highly Depleted Reservoirs. Paper SPE 65512 presented at the SPE/CIM International Conference on Horizontal Well Technology, Calgary, 6–8 November. DOI: 10.2118/65512-MS.

Pia, G., Fuller, T., Haselton, T., and Kirvelis, R. 2002a. Underbalanced—Undervalued? Direct Qualitative Comparison Proves the Technique! Paper SPE 74446 presented at the IADC/SPE Drilling Conference, Dallas, 26–28 February. DOI: 10.2118/74446-MS.

Pia, G., Fuller, T., Haselton, T., and Kirvelis, R. 2002b. Underbalanced Production Steering Delivers Record Productivity. Paper SPE 77529 presented at the SPE Annual Technical Conference and Exhibition, San Antonio, Texas, USA, 29 September–2 October. DOI: 10.2118/77529-MS.

Pinkstone, H., Timms, A., McMillan, S., Doll, R., and de Vries, H. 2004. Underbalanced Drilling of Fractured Carbonates in Northern Thailand Overcomes Conventional Drilling Problems Leading to a Major Gas Discovery. Paper SPE 90185 presented at the SPE/IADC Annual Technical Conference and Exhibition, Houston, 26–29 September. DOI: 10.2118/90185-MS.

Purvis, L. and Smith, D.D. 1998. Underbalanced Drilling in the Williston Basin. Paper SPE 39924 presented at the SPE Rocky Mountain Regional/Low-Permeability Reservoir Symposium, Denver, 5–8 April. DOI: 10.2118/39924-MS.

Qutob, H. 2004. Underbalanced Drilling; Remedy for Formation Damage, Lost Circulation, and Other Related Conventional Drilling Problems. Paper SPE 88698 presented at the Abu Dhabi International Conference and Exhibition, Abu Dhabi, 10–13 October. DOI: 10.2118/88698-MS.

Ramalho, J. and Davidson, I.A. 2006. Well-Control Aspects of Underbalanced Drilling Operations. Paper SPE 106367 presented at the IADC/SPE Asia Pacific Drilling Technology Conference and Exhibition, Bangkok, Thailand, 13–15 November. DOI: 10.2118/106367-MS.

Robichaux, D. 1999. Successful Use of the Hydraulic Workover Unit Method for Underbalanced Drilling. Paper SPE 52827 presented at the SPE/IADC Drilling Conference, Amsterdam, 9–11 March. DOI: 10.2118/52827-MS.

Rommetveit, R. and Lage, A.C.V.M. 2001. Designing Underbalanced and Lightweight Drilling Operations; Recent Technology Developments and Field Applications. Paper SPE 69449 presented at the SPE Latin American and Caribbean Petroleum Engineering Conference, Buenos Aires, 25–28 March. DOI: 10.2118/69449-MS.

Rommetveit, R., Fjelde, K.K., Frøyen, J., Bjørkevoll, K.S., Boyce, G., and Eck-Olsen, J. 2004. Use of Dynamic Modeling in Preparations for the Gullfaks C-5A Well. Paper SPE 91243 presented at the SPE/IADC Underbalanced Technology Conference and Exhibition, Houston, 11–12 October. DOI: 10.2118/91243-MS.

Saponja, J. 1998. Challenges With Jointed Pipe Underbalanced Operations. *SPEDC* **13** (2): 121–128. SPE-37066-PA. DOI: 10.2118/37066-PA.

Sarssam, M., Peterson, R., Ward, M., Elliott, D., and McMillan, S. 2003. Underbalanced Drilling for Production Enhancement in the Rasau Oil Field, Brunei. Paper SPE 85319 presented at the SPE/IADC Middle East Drilling Technology Conference and Exhibition, Abu Dhabi, 20–22 October. DOI: 10.2118/85319-MS.

Schmigel, K. and MacPherson, L. 2003. Snubbing Provides Options for Broader Application of Underbalanced Drilling Lessons. Paper SPE 81069 presented at the SPE Latin American and Caribbean Petroleum Engineering Conference, Port-of-Spain, Trinidad and Tobago, West Indies, 27–30 April. DOI: 10.2118/81069-MS.

Smith, S.P., Gregory, G.A., Munro, N., and Muqueem, M. 1998. Application of Multiphase Flow Methods to Underbalanced Horizontal Drilling. Paper SPE 51500 presented at the SPE International Conference on Horizontal Well Technology, Calgary, 1–4 November. DOI: 10.2118/51500-MS.

Stuczynski, M.C. 2001. Recovery of Lost Reserves Through Application of Underbalanced Drilling Techniques in the Safah Field. Paper SPE 72300 presented at the SPE/IADC Middle East Drilling Technology, Bahrain, 22–24 October. DOI: 10.2118/72300-MS.

Sutherland, I. and Grayson, B. 2005. DDV Reduces Time to Round-Trip Drillstring by Three Days, Saving £400,000. Paper SPE 92595 presented at the IADC/SPE Drilling Conference, Amsterdam, 23–25 February. DOI: 10.2118/92595-MS.

Tetley, N.P., Hazzard, V., and Neciri, T. 1999. Application of Diamond-Enhanced Insert Bits in Underbalanced Drilling. Paper SPE 56877 presented at the SPE Annual Technical Conference and Exhibition, Houston, 3–6 October. DOI: 10.2118/56877-MS.

Thatcher, D.A.A., Szutiak, G.A., and Lemay, M.M. 2000. Integration of Coiled Tubing Underbalanced Drilling Service To Improve Efficiency and Value. Paper SPE 60708 presented at the SPE/ ICoTA Coiled Tubing Roundtable, Houston, 5–6 April. DOI: 10.2118/60708-MS.

Timms, A., Muir, K., and Wuest, C. 2005. Downhole Deployment Valve—Case History. Paper SPE 93784 presented at the SPE Asia Pacific Oil and Gas Conference and Exhibition, Jakarta, 5–7 April. DOI: 10.2118/93784-MS.

Tinkham, S.K., Meek, D.E., and Staal, T.W. 2000. Wired BHA Applications in Underbalanced Coiled Tubing Drilling. Paper SPE 59161 presented at the IADC/SPE Drilling Conference, New Orleans, 23–25 February. DOI: 10.2118/59161-MS.

van der Werken, T., Boutalbi, S., and Kimery, D. 2005. Reservoir Screening Methodology for Horizontal Underbalanced Drilling Candidacy. Paper IPTC 10966 presented at the International Petroleum Technology Conference, Doha, Qatar, 21–23 November. DOI: 10.2523/10966-MS.

van Venrooy, J., van Beelen, N., Hoekstra, T., Fleck, A., Bell, G., and Weihe, A. 1999. Underbalanced Drilling With Coiled Tubing in Oman. Paper SPE 57571 presented at the SPE/IADC Middle East Drilling Technology Conference, Abu Dhabi, 8–10 November. DOI: 10.2118/57571-MS.

Vefring, E.H., Nygaard, G., Lorentzen, R.J., Nævdal, G., and Fjelde, K.K. 2003. Reservoir Characterization During UBD: Methodology and Active Tests. Paper SPE 81634 presented at the IADC/SPE Underbalanced Technology Conference and Exhibition, Houston, 25–26 March. DOI: 10.2118/81634-MS.

Wang, Y. and Lu, B. 2002. Fully Coupled Chemico-Geomechanics Model and Applications to Wellbore Stability in Shale Formation in an Underbalanced Field Conditions. Paper SPE 78978 presented at the SPE International Thermal Operations and Heavy Oil Symposium and International Horizontal Well Technology Conference, Calgary, 4–7 November. DOI: 10.2118/78978-MS.

Westermark, R.V. 1986. Drilling with a Parasite Aerating String in the Disturbed Belt, Gallatin County, Montana. Paper SPE 14734 presented at the IADC/SPE Drilling Conference, Dallas, 9–12 February. DOI: 10.2118/14734-MS.

Xiong, H. and Shan, D. 2003. Reservoir Criteria for Selecting Underbalanced Drilling Candidates. Paper SPE 81621 presented at the IADC/SPE Underbalanced Technology Conference and Exhibition, Houston, 25–26 March. DOI: 10.2118/81621-MS.

SI Metric Conversion Factors

bbl	× 1.589 873	E – 01 = m^3
ft	× 3.048*	E – 01 = m
ft^3	× 2.831 685	E – 02 = m^3
gal	× 3.785 412	E – 03 = m^3
in.	× 2.54*	E + 00 = cm
lbm	× 4.535 924	E – 01 = kg
psi	× 6.894 757	E + 00 = kPa

*Conversion factor is exact.

9.2 Casing While Drilling—Tommy M. Warren, Tescocorp

9.2.1 Introduction. Casing while drilling (CWD) is a process for using standard oilfield casing for the drill-string so that a well is simultaneously drilled and cased. Both surface and downhole tools and components are necessary to make this process possible. While many of the functions and activities are similar to the conventional drilling process in which drillpipe and drill collars are used, they are sufficiently different to warrant special engineering considerations.

When rotary drilling was first introduced in the early 1900s, casing was routinely used for the drillstring because *drillpipe* did not exist. The connections were not very robust, and over time, *drillpipe* evolved as stronger connections were developed and the resulting pipe was too expensive to purchase for each well.

The idea of drilling with casing re-emerged in the 1950s and 1960s. While there were many potential advantages of this technique, it was not commercially accepted because of limitations in materials and cutting tools that were available to use with it. This early work on drilling with casing was done in both the USA and Russia (Kammerer 1958; Brown 1971; Gelfgat and Alikin 1998). The development and common use of top-drives, polycrystalline diamond compact (PDC) bits, better pipe metallurgy, and stronger connection design, coupled with the need to drill through depleted zones, spawned another attempt to drill with casing in the late 1990s (Leturno 1993; Tessari and Madell 1999). These initiatives facilitated the process sufficiently so that it has become a successful commercial service.

The conventional drilling process for oil and gas uses a drillstring made up of drill collars and drillpipe to apply mechanical energy (rotary power and axial load) to the bit, as well as to provide a hydraulic conduit for the drilling fluid. Even when a downhole motor is used to supply the rotary energy, the drillstring remains essentially the same.

A conventional drillstring must be tripped out of the hole each time a component of the drilling assembly needs to be changed, the casing point is reached, or the borehole needs to be *conditioned*. Casing is then run into the well as a completely separate process to provide permanent access to the wellbore. CWD systems integrate the drilling and casing process to provide a more efficient well-construction system by eliminating these drillstring trips and allowing the well to be simultaneously drilled and cased.

Advantages of CWD. The fundamental reason to use a CWD system is that a significant fraction of the well cost for some wells can be eliminated (1) by using a drilling system where the casing is installed as the well is drilled, (2) by designing the surface and downhole equipment to effectively implement this process, and (3) by designing the well plan around this process.

Savings result from eliminating the costs of purchasing, handling, inspecting, transporting, and tripping the drillstring; reducing hole problems that are associated with tripping; reducing trouble time associated with lost circulation, wellbore instability, and well control; eliminating trouble time for running casing; and reducing rig equipment capital costs and operating costs.

The potential savings from reducing drillstring tripping and handling can be identified quite easily for any particular situation, but the savings from reducing hole problems are more difficult to quantify. There are many situations where problems such as lost circulation, well-control incidents, and borehole stability problems can be directly attributed to tripping the drillstring. Additionally, many routine practices such as making conditioning trips are performed solely for the purpose of preventing trouble time and thus are not captured as a potential savings that might result from using CWD.

In other cases, it is difficult to run the casing after the conventional drillstring is tripped out because of poor borehole quality. Some of these difficulties are related to borehole stability problems directly attributed to drillstring vibrations, while others are related more to the particular well geometry and formation conditions being drilled (Santos et al. 1999). The CWD system reduces these incidents by installing the casing immediately as the well is drilled.

CWD offers the opportunity to drive the casing setting depth deeper than may be obtained with the conventional drilling process. The need to drill with a sufficient mud weight to provide a trip margin before tripping out the drillstring to run casing is eliminated. The CWD process also mechanically enhances the wellbore wall filter cake to reduce lost circulation. The combination of better lost-circulation control and eliminating the need for a trip margin may allow the casing setting depth to be pushed deep enough to eliminate a string of casing.

In addition to reducing cost, the CWD process is safer than the conventional drilling process. Personnel exposure to pipe handling during tripping and casing running operations is reduced. These two operations have some of the highest accident rates of any operation conducted on a drilling rig. The CWD process also provides a circulation path to the bottom of the well at all times, which reduces risks associated with well-control operations. In order to trip out a conventional drillstring, the well must be static and remain static during the trip. Unfortunately, the tripping process tends to destabilize the pressure environment in the well due to surge and swab effects. If this disturbance causes an influx of fluid, there is no circulation path to restabilize the well when the drillstring is partially out of the well. Not only does CWD provide a circulation path to the bottom of the well at all times, it eliminates the surge and swab effects of tripping a conventional drillstring.

Managed-pressure drilling may be applied with the CWD system for drilling into an overpressured formation with a lower mud weight than would be required if drilled conventionally. With the appropriate well design, surface equipment, and planning, the dynamic friction of the flowing mud can be used for well control. At total depth (TD), the mud can be displaced with heavier-weight cement using a flow rate displacement schedule to cement the casing in place without breaking down shallow formations.

Downhole Equipment. The CWD process eliminates the conventional drillstring by using the casing itself as the hydraulic conduit and means of transmitting mechanical energy to the bit. The downhole drilling tools may consist of a nonretrievable bit attached directly to the casing or a retrievable and rerunnable bottomhole assembly (BHA).

The simplest nonretrievable system uses a crossover to attach a conventional bit to the bottom of the casing. An alternative way of using a conventional bit is to attach it with a *bit release tool* so that it can be jettisoned once the casing point is reached. This can allow logging tools to be run through the casing to log the bottom portion of the wellbore by pulling a few joints of casing after reaching TD, but it is practical for use on only the last string of casing run in the well.

Another type of nonretrievable system uses a *bit* that can be drilled out to allow the next section of casing to be run. **Fig. 9.2.1** shows two examples of drillout bits. The first consists of an aluminum core and thermally stable polycrystalline (TSP) diamond cutters that can be drilled up with a conventional PDC bit. The second is a steel shell with PDC cutters, but it requires special features on the bit used to drill through it.

A retrievable drilling assembly is required for cases where the bit may need to be changed before reaching casing point or a directional assembly is used. The retrievable BHA locks to the bottom of the casing and protrudes below the casing shoe to drill a hole of adequate size to allow the casing to pass freely. The top of the BHA fits into a landing sub (profile nipple) on the bottom of the casing in such a way that the BHA can be retrieved and replaced without needing to trip pipe out of the well **(Fig. 9.2.2)**.

The BHA components below the casing include a pilot bit and underreamer that are sized to pass through the drill casing and drill a hole that provides adequate clearance for the drill casing and subsequent cementing.

The BHA components also may include a downhole motor or other tools such as measurement-while-drilling (MWD) tools, coring equipment, or fishing tools needed to perform almost any operation that can be conducted with a conventional drillstring.

A drilling shoe may be used to stabilize the end of the casing and is sometimes dressed with either PDC cutters or tungsten carbide chips to ensure that a full-gauge hole is obtained ahead of the casing. The casing may include appropriate centralizers to prevent wear and to aid in cementing.

The retrievable BHA may be run on wireline or drillpipe, or it may be pumped into place. It is often pulled with a wireline, but for large-diameter tools it may be pulled with drillpipe. The tool-setting and -releasing

Defyer™   EZCase™

Fig. 9.2.1—Examples of drillout bits. Images courtesy of Weatherford (Defyer) and Baker Hughes (EZCase).

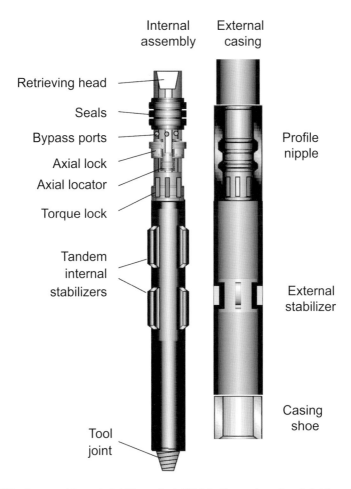

Fig. 9.2.2—Drill lock assembly and stabilizers lock BHA to the casing shoe joint (courtesy of Tesco).

function may be accomplished by wireline manipulations or with pump pressure assist, depending on the particular tool design.

Surface Equipment. In addition to the downhole equipment described above, the surface rig equipment must be selected to support the CWD process. In some cases, this may entail using a rig specifically designed for CWD, but in many cases, a conventional rig is used. In either case, the rig must be equipped with a top-drive to drill with casing effectively. The topdrive is often used to torque the casing to the coupling manufacturer's specifications, as well as to rotate the casing for drilling.

For systems that require the BHA to be run and retrieved with a wireline, the rig must be equipped with a wireline unit. In most cases, the wireline unit used for electric logging is not adequate because the wire will not handle the loads without risk of damaging the conductors. Wireline units with load-handling capacity up to 40,000 lbf are available for use with the CWD system.

A method of attaching the topdrive to the casing for drilling is also required. This can be as simple as using a crossover attached to the topdrive to screw into the box connection on the casing.

Alternatively, a casing drive assembly (CDA) provides a way of handling the casing that is faster and less likely to damage the connection threads. The CDA **(Fig. 9.2.3)** is a device that mounts below the topdrive and attaches to the casing without making a threaded connection. It has a spear-and-seal assembly that stabs into the casing inside diameter (ID) and a slip assembly that clamps to the casing. This allows the casing to be made up and drilled down without undergoing an additional make/break cycle on the casing threads.

The slip assembly generally is located on the casing ID for large-diameter casing and on the casing outside diameter (OD) for smaller casing. The CDA includes hydraulically extended single-joint elevators to facilitate picking casing up from the V-door.

9.2.2 CWD Engineering Considerations. In many ways, designing a CWD well is similar to designing a conventional well. Considerations such as borehole stability, well control, casing setting depths, directional planning, and bit selection are treated much as they are for conventional drilling. The well operator is normally responsible for this process, and the service provider designs the CWD process to operate within these constraints as much as possible.

One significant difference is that the casing may be subjected to different stresses in a CWD situation from those for conventional uses.

Engineering Process. The process of designing a CWD well begins much the same as a conventional well. The casing points and casing design are selected on the basis of wellbore stability, well control, and production requirements. The directional program for the well is designed to hit the selected targets, and the drilling fluid program is developed. Once a tentative conventional design process is completed, the design must be reviewed to assure that the CWD process can be used successfully to install the casing strings and maintain the required casing performance specifications.

Fig. 9.2.4 shows some of the interactions that affect the integrity of casing used for CWD. The three primary considerations for casing integrity (elastic loads, wear, and fatigue) are shown on the right side of

Fig. 9.2.3—Casing drive assembly (courtesy of Tesco).

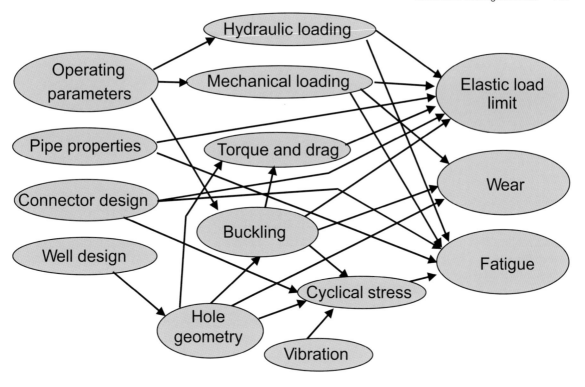

Fig. 9.2.4—Interactions affecting casing integrity for CWD applications.

Fig. 9.2.4, while the parameters that are under the operator's control (operating parameters, pipe properties, connector design, and well design) are shown on the left side. The overall casing integrity is controlled by a quite complex system of interactions that relate the parameters directly under operator control to the processes that degrade casing integrity.

The geometry of the fluid flow path provides another significant difference between CWD and conventional drilling. The flow path down the ID of the casing is large and unrestricted, so there is usually little pressure drop in the casing ID.

The CWD annulus is more restricted, so annular pressure losses may be higher than for drilling with a conventional drilling system. While the annular flow path is more restricted, it is also more uniform, so the annular velocities are nearly constant from the casing shoe to the surface. This provides the opportunity to clean the hole with relatively low flow rates, but the drilling fluid properties must be properly considered and adequate hydraulic energy must be provided to clean the bit and underreamer. These factors must also be considered in light of equivalent circulating density (ECD) limits imposed by the formation strength and pressure environment.

It is often advantageous to adjust portions of the conventional well plan to better implement the CWD process. For example, the hole size is normally dictated by the size of conventional bits. The optimal hole size for CWD applications may not be a standard bit size. When drilling with a retrievable CWD system, this is easily accommodated because the final hole size is dictated by the underreamer, which is relatively easy to customize.

Adjustments to the directional plan may be needed for optimal application of the CWD process. For example, the build rate while drilling with large-diameter casing may be less than is sometimes used for conventional drilling. Or it may be more efficient to perform the directional work higher in the hole to eliminate the need to use the smaller directional tools used with small-diameter casings.

Operational practices also may need to be adjusted to optimize the CWD process. Flow rates are generally less. The best response to lost circulation is normally to continue drilling at a slow rate while introducing loss control materials (LCM), rather than to stop drilling. Most hole-conditioning activities are eliminated with CWD, except when excessive *hydraulic lift* indicates that conditioning is needed.

It is the purpose of this chapter to provide insight and engineering tools to help plan and execute CWD wells. The chapter provides an explanation of the processes that affect CWD and is organized to complement and support the normal process one would follow in designing a CWD well.

9.2.3 Hydraulics for the Casing-Drilled Well. The hydraulic calculations for a well drilled with casing are much like the calculations for a conventionally drilled well. Most of these calculations can be made manually or with any of the available hydraulics software packages. However, a few of the calculations are specific to CWD and may not be available with general-purpose drilling hydraulics software. The discussion below concentrates on issues that are somewhat different for conventional drilling and CWD.

The relatively narrow annulus between the casing and borehole wall tends to increase the friction loss at conventional circulation rates, increases surge and swab effects, and in some cases may make the annular friction loss sensitive to drill-casing rotation speed. It also exerts a hydraulic lift force on the casing string.

9.2.4 Pressure Calculations. The pressure profile for the circulating system is typically *inverted* for a CWD well compared to a well drilled with drillpipe. Frictional pressure in the drillstring ID contributes a major portion of the total circulating pressure for conventional drilling systems. The friction pressure on the casing ID is much less significant because of the larger flow area of the casing ID compared to the drillpipe ID, while just the opposite is true for the annular friction pressure.

Frictional pressure in the annulus is more important than the other pressures in the CWD system because it directly affects the formation integrity and exerts a hydraulic lifting force on the casing. This pressure must be managed appropriately to produce a successful CWD service.

The frictional pressure in the annulus depends on the annular geometry, flow rates, fluid rheology, drill-string rotational speed, and eccentricity of the casing in the borehole. The rotational speed and eccentricity are generally not important for conventional drilling situations because of the large annulus, but they may be significant for the CWD annulus.

The casing tends to lie on the low side of the borehole, as shown in **Fig. 9.2.5** (assuming that it is not centralized). The eccentricity of the casing (*e*) affects both the frictional pressure in the annulus and the ability to remove cuttings. The fluid has a higher velocity in the thick part of the annulus, and the velocity decreases rapidly toward the areas where the annulus pinches out at the low side of the hole (Ooms et al. 1999). The overall result is that the annular friction pressure tends to be reduced approximately 30% when the pipe moves from being centered to fully eccentric.

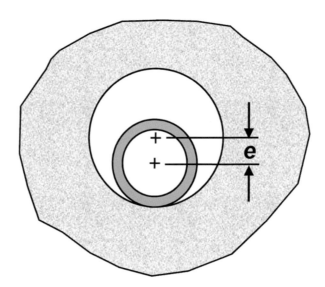

Fig. 9.2.5—Casing eccentricity in borehole.

The annular friction pressure increases with pipe rotation speed for cases where the annular gap is small compared to the pipe diameter. Some researchers (Ooms et al. 1999) claim that the pressure increases as rotational speed increases for *laminar flow*, but not for turbulent flow. Other work indicates that the effect continues into turbulent flow but is more significant for laminar flow.

It is not unusual for the annular flow regime for CWD situations to be in the transition region between laminar and turbulent. **Fig. 9.2.6** shows data measured with a 3.7-in. flush-joint pipe in a 4⅜-in. hole (ratio of casing diameter to hole diameter = 0.85) over the range of Reynolds numbers that are encountered in CWD (Bode et al. 1991).

The 100-rev/min data probably are more typical of CWD operating conditions. At this speed, the annular friction pressure increases 15 to 40% due to casing rotation. Other data and field observations have often shown a much smaller (often negligible) increase in friction pressure with increased rotary speed.

The friction pressure is also affected by the *roughness* of the conduit confining the flow if the flow regime is turbulent. The flow regime for most conventional drilling situations is either laminar or transitional; thus, the roughness of the borehole wall can be ignored.

However, the flow regime may be turbulent for CWD applications in which a low-viscosity and/or low-density fluid is used and is almost always turbulent when water is used as the drilling fluid. In these situations, the actual annular friction pressure may be higher than calculated because the roughness of the borehole wall is greater than the value used in the model to determine the friction factor for pipe flow.

When using CWD in situations where lost circulation is likely, the uncertainty in calculating annular friction pressure should be kept in mind. For critical situations, one may consider running an annular pressure-monitoring sub. Monitoring the hydraulic lift, as discussed later, is also an effective way to track the annular pressure in vertical wells, but is less effective in directional wells due to mechanical friction.

9.2.5 Casing Annulus. The frictional pressure in the annulus can be calculated and for typical CWD geometries is higher than for conventionally drilled wells. The sum of all pressure components acting in the annulus is typically reported as an equivalent circulating density (ECD). The ECD is simply the total pressure (at a point in the annulus near the bit) converted to a mud weight that would exert the same pressure at the same depth. For example, a 200-psi annular friction pressure with 9.0 lbm/gal mud at 8,000-ft true vertical depth (TVD) gives an ECD of 9.48 lbm/gal [9.0 + 200/(0.052 × 8,000)].

The ECD includes the effect of cuttings load in the annulus in addition to the pressure of the clean mud plus the friction. A decrease in the flow rate will decrease the friction but may increase the annular cuttings load at a fixed drilling rate. Thus, there may be an optimum flow rate that minimizes the ECD for any given well geometry, fluid properties, and drilling rate.

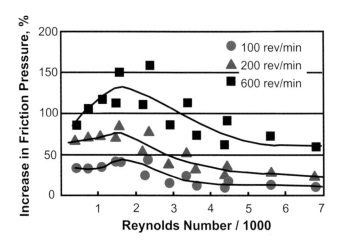

Fig. 9.2.6—Effect of rotary speed on annular pressure loss.

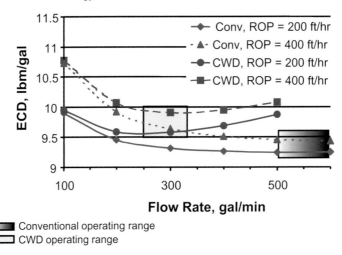

Fig. 9.2.7—Optimum flow rates for CWD and conventional application.

For example, **Fig. 9.2.7** shows a comparison of the ECD for a conventional well drilled with 6⅛-in. drill collars and 4-in. drillpipe and a CWD well drilled with 7-in. casing at a depth of 3,200 ft (mud density 8.9 lbm/gal). The ECD is minimized for CWD at a flow rate of approximately 300 gal/min compared to a conventional flow rate of 550 gal/min. Even though the CWD well might be drilled at 300 gal/min and the conventional well at 600 gal/min, the CWD well still has a 0.5-lbm/gal higher ECD.

Often, a conventionally drilled well will carry at least a 0.3-lbm/gal excess mud weight for trip margin to ensure that the well is static while the drillstring is tripped. Because no openhole tripping is required with the CWD system, the trip margin usually is eliminated for CWD wells. This often reduces the ECD to a level negligibly greater than that for the conventional drilling system.

Experience gained while drilling with casing has shown that less lost circulation occurs while drilling with casing than when drilling conventionally. This effect has not been fully explained, but it seems to be caused by the casing mechanically plastering drilled solids into the wall of the borehole. This plugs small fractures in the wall and reduces the effective permeability at the rock face. The plugging of microfractures with solid particles may also enhance the wellbore wall strength by the *stress cage* effect (Aston et al. 2004). Fluid flow into the wellbore is also inhibited, making well control easier for casing-drilled wells.

Even though it is difficult to quantify the effect, it is quite significant and is a major advantage of CWD. Some operators have documented a preference for using the CWD system over a conventional system for drilling lost-circulation-prone areas because of this benefit.

Reduction in Hookload. The annular friction exerts forces on the casing that reduce the hookload. The friction increases the pressure acting upward on the lower end of the casing and the annular flow exerts a drag force on the casing surface.

The total lift force can be calculated as explained below. The net effect, though, is that the annular friction makes the pipe behave very similar to the way it would if it were suspended in a more dense drilling fluid.

As shown in **Fig. 9.2.8**, the force acts as a concentrated force on the bottom of the casing and as a distributed drag force along its side. The pressure at the surface is zero (assuming the flowline is not plugged and the flow is not diverted through a choke), and the incremental pressure on bottom is the annular friction pressure.

The annular pressure acts on the end of the casing to generate a force given by

$$F_{end} = \Delta p_o \times \pi \times r_o^2, \dots\dots\dots\dots\dots\dots\dots\dots\dots\dots\dots\dots\dots\dots\dots\dots\dots\dots\dots \quad (9.2.1)$$

where F_{end} = the end force, Δp_o = the annular friction pressure, and r_o = the casing exterior radius.

The distributed drag force acting on the side of the casing is not as straightforward, but it can be derived as follows. **Fig. 9.2.9** shows a free-body diagram of the annular flow in a wellbore around the casing. The pressure drop times the annular cross section is equal to the sum of the drag forces on the borehole wall plus the drag force on the casing exterior wall, but only the drag on the casing wall affects the hookload.

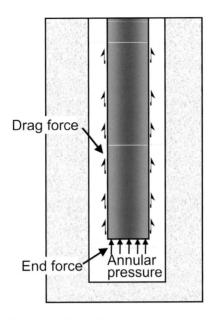

Fig. 9.2.8—Hydraulic lift force on casing.

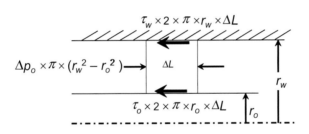

Fig. 9.2.9—Free body of fluid element in annular flow.

From this free body, the following equality can be written:

$$\Delta p_o \times \pi \times (r_w^2 - r_o^2) = \tau_w \times 2 \times \pi \times r_w \times \Delta L + \tau_o \times 2 \times \pi \times r_o \times \Delta L, \quad \dots\dots\dots\dots\dots\dots \quad (9.2.2)$$

where r_w = wellbore radius, r_o = casing exterior radius, τ_w = borehole wall shear stress, τ_o = casing wall shear stress, and ΔL = casing length.

The shear stress at the casing wall and wellbore wall may not be identical, but both are expressed by

$$\tau = \rho \times f_F \times v^2, \quad \dots\dots\dots\dots\dots\dots\dots\dots\dots\dots\dots\dots\dots\dots\dots\dots \quad (9.2.3)$$

where ρ = fluid density, f_F = friction factor, and v = average velocity. The average velocity and fluid density for both the wellbore wall and pipe wall are identical, but the friction factors may not be the same. This means that the shear stress at the wellbore wall is equal to the shear stress at the pipe wall times the ratio of the friction factors.

The drag force, F_d, on the borehole wall can then be written in terms of the casing wall parameters as

$$F_d = \tau_w f_o / f_w \, 2\pi \, (r_o / r_w) \, r_w \, \Delta L, \quad \dots\dots\dots\dots\dots\dots\dots\dots\dots\dots \quad (9.2.4)$$

where f_o = casing friction factor, and f_w = borehole wall friction factor.

The change in hookload that is caused by the flow rate increasing from zero to some other value is given by

$$\Delta HL = \tau_o \times 2 \times \pi \times r_o \times \Delta L, \quad \dots\dots\dots\dots\dots\dots\dots\dots\dots\dots\dots\dots \quad (9.2.5)$$

where ΔHL = change in hook load.

Eqs. 9.2.2, 9.2.4, and 9.2.5 can be combined to give the following expression for the annular friction pressure in terms of the change in hookload due to surface drag:

$$\Delta p_o = \Delta HL \times (1 + f_o/f_w \times r_w/r_o)/[\pi \times (r_w^2 - r_o^2)]. \quad \dots\dots\dots\dots\dots\dots\dots\dots\dots \quad (9.2.6)$$

Eqs. 9.2.1 and 9.2.6 then can be combined to give an expression for the annular pressure loss in terms of the total hookload change that includes both the end effect and the drag forces:

$$\Delta p_o = \Delta HL/[(\pi \times (r_w^2 - r_o^2)/(1 + f_o/f_w \times r_w/r_o) + \pi \times r_o^2] \quad \dots\dots\dots\dots\dots\dots\dots\dots \quad (9.2.7)$$

The friction factor for both the wellbore wall and casing wall depends on the Reynolds number (which is the same for both) and the relative roughness of each surface. As indicated in **Fig. 9.2.10**, the friction factor for laminar flow is independent of the roughness, but at a higher Reynolds number the friction factor can vary by a factor of three, depending on the surface roughness.

There is no significant roughness effect on friction factor for many CWD situations when the Reynolds number is less than 4,000. For cases in which water is used as the drilling fluid, the Reynolds number will be quite high and the drag effect on the casing will be a significantly smaller portion of the total hydraulic lift.

Fig. 9.2.11 shows the ratio of the friction for pipe compared to a rough wall ($\varepsilon/D = 1/30$), which might represent the roughness of the wellbore wall.

In cases where two different sizes of casing are used, the hydraulic lift acting on both must be calculated, and this will include not only the shear stresses and bottom end force but also a pressure-area force that acts on the lip created by the change in diameter.

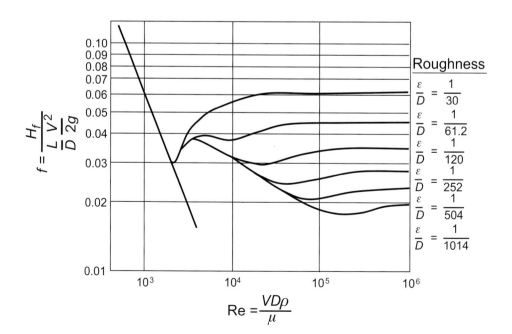

Fig. 9.2.10—Friction factors as a function of wall roughness.

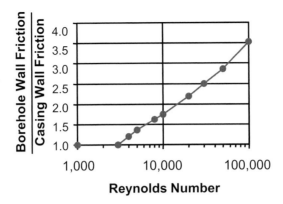

Fig. 9.2.11—Ratio of borehole wall friction to casing wall friction.

For example, consider a casing string made up of 1,000 ft of 5-in. casing and 6,500 ft of 4½-in. casing where the observed hydraulic lift is 8,000 lbf. The flow rate is 215 gal/min with mud properties that give a Reynolds number of 6,000 adjacent to the 5-in. casing and 5,000 adjacent to the 4½-in. casing. The corresponding velocities are 375 and 280 ft/min, respectively. Assume that the friction factors for the two sections are equal.

In this case, the annular friction pressure occurs over two sections (5-in. casing and 4½-in. casing). The annular friction pressure is proportional to the friction factor times the section length times the velocity squared. Thus, 78% of the friction loss occurs opposite the 4½-in. casing.

The upward end force must be adjusted for the downward force at the transition between the 5-in. and 4½-in. casing, and it is given by

$$F_{end} = \Delta p_o \times \pi \times [2.5^2 - 0.78 \times (2.5^2 - 2.25^2)] = \Delta p_o \times 16.72$$

The drag force is given by

$$F_d = \Delta p_o \times [0.22 \times \pi \times (3.125^2 - 2.5^2)/(1 + 3.125/2.5) + 0.78 \times \pi \times (3.125^2 - 2.25^2)$$
$$/ (1 + 3.125/2.25)]$$
$$F_d = \Delta p_o \times (1.08 + 4.82) = \Delta p_o \times 5.9$$

The annular pressure loss is then

$$\Delta p_o = 8,000/(16.72 + 5.9) = 354 \text{ psi}$$

The hydraulic lift force is easily observed at the rigsite and must be accounted for in order to apply the proper weight on bit (WOB) and to monitor hole cleaning and ECD. **Fig. 9.2.12** shows an example of the behavior of the pump pressure and hookload as the pumps are engaged and brought up to full speed. When the pumps are first brought on line, there is no change in hookload as the pipe is filled. Once the pipe is full, the pressure increases as the fluid begins to circulate down the casing and up the annulus. The hookload decreases as the flow rate is increased and pump pressure increases.

The hydraulic lift may be affected by the fluid properties as much as (or more than) it is by the flow rate. For example, comparing the hydraulic lift for Mud A (9.25 lbm/gal, 18 cp, and 12 lbf per 100 ft²) and Mud B (8.9 lbm/gal, 12 cp, and 5 lbf per 100 ft²) shows that the hydraulic lift for Mud A is much higher than for Mud B. The hydraulic lift is even much higher for Mud A at 250 gal/min than it is for Mud B at 300 gal/min (**Fig. 9.2.13**). Mud A was in fact used for a CWD operation through a depleted zone and was changed to Mud B to reduce the lost circulation and improve the hole cleaning and rate of penetration (ROP).

One of the most common mistakes made when planning a CWD well is to overdesign the mud. In general, *thin* muds perform much better for CWD operations than thicker muds. This is true even when lost-circulation

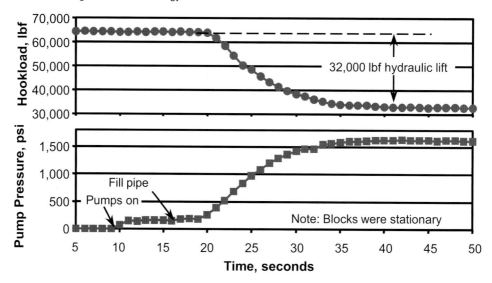

Fig. 9.2.12—Decrease in hookload as pumps are brought up to normal flow rate.

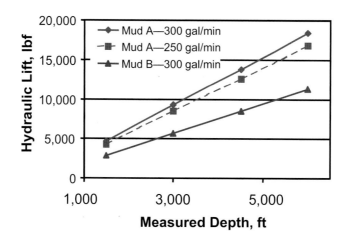

Fig. 9.2.13—Effect of mud properties on pumpoff.

zones are anticipated and in cases where a low-fluid-loss mud might be desired for conventional drilling. Not only does the thinner mud reduce the ECD because it reduces the shear stresses on the casing, but it allows a higher flow rate to clean the hole better.

Fig. 9.2.14 shows cakes of gumbo clay (note penny in center of photo) that were recovered over the shaker after working the pipe while drilling with Mud A discussed above. This shale apparently collected on the outside of the casing even though the annular velocity was quite high. Switching to Mud B not only reduced the hydraulic lift due to having thinner rheology, but also provided better hole cleaning and eliminated the production of the mudcakes.

The hydraulic lift should be monitored as a CWD well is drilled in order to monitor the hole cleaning. The hydraulic lift will naturally increase as the well is drilled deeper because the annular friction pressure increases. It also will increase as the mud weight increases and if the flow rate increases. Thus, a baseline should be developed for the theoretical hydraulic lift for the given operating and mud conditions so that poor cleaning conditions are easily identified.

A schedule of the anticipated hydraulic lift should be prepared from the planned flow rates and expected mud properties. It is helpful in detecting poor hole cleaning by comparing the expected hydraulic lift to the observed hydraulic lift as drilling progresses. This step generally is not needed for conventional well planning.

Fig. 9.2.14—Cakes of gumbo recovered while drilling with thick mud.

By plotting the observed hydraulic lift on the same graph as the expected hydraulic lift, evidence of poor hole cleaning shows up clearly. For example, **Fig. 9.2.15** shows the predicted hydraulic lift for a well drilled with 7-in. casing in south Texas compared to the observed hydraulic lift. The predictions match the observations reasonably well for most of the drilled interval. The hydraulic lift abruptly increased at approximately 3,000 ft. Working the pipe and backreaming at 3,500 ft brought the hydraulic lift back in line with expectations. Lost circulation can occur easily in cases in which the hydraulic lift increases and drilling continues with no remedial action.

One issue that field personnel often erroneously promote is the idea that hydraulic lift can be eliminated by doing something with the design of the casing shoe or centralizers. In most cases, the hydraulic lift is not

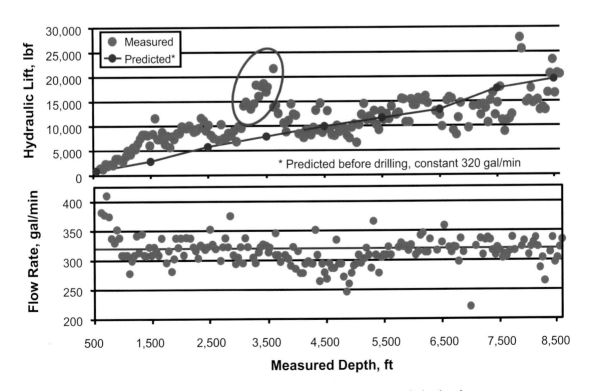

Fig. 9.2.15—Hydraulic lift can be monitored to indicate poor hole cleaning.

caused by a restriction of flow around the shoe or centralizers. It is simply due to the frictional pressure losses in the annulus due to the shear stresses acting on the hole wall and outside wall of the casing. But there may be instances in which the shoe or a stabilizer will ball up and increase the hydraulic lift.

The hydraulic lift is measured by observing the change in hookload as the pump is brought up to speed with the bit off bottom. It is possible for the hydraulic lift to increase once the bit is placed on bottom and starts to drill. If the ROP is high and the hole is circulated clean (usually at shallow depths) at a connection, the hydraulic lift will always increase as the annulus is loaded with cuttings. This will be accompanied by an increase in pump pressure. It is a good operating practice to pick up the drillstring and rezero the WOB any time the pump pressure increases.

Surge and Swab. Pipe movement affects the bottomhole pressure by (1) altering the effective annular flow rate due to fluid displacement, (2) fluid acceleration effects, and (3) fluid shear forces. All these effects may be more significant for a CWD well geometry than for a conventional well geometry.

At connections and while reaming, the casing will be lowered while circulating. In this case, the pipe displacement adds to the overall flow rate. For example, 7-in. casing displaces approximately 2 gal/ft in an 8½-in. hole (assuming a drilling float is in use); thus, the overall flow rate is increased by 2 gal/min times the pipe movement speed in ft/min. In addition to the increase in annular pressure due to this increased flow rate, the acceleration of the fluid caused by pipe movement also increases the pressure. For these reasons, it is a good practice to reduce the flow rate when the bit is off bottom and avoid sudden increases in pipe movement while drilling if there is any chance of inducing lost circulation.

Cuttings Removal. The annular velocities for CWD wells typically are much higher than for conventionally drilled wells. This higher velocity assists in transporting the cuttings but does not assure highly efficient cuttings removal. The combination of casing eccentricity and relatively low rotation speed of the casing for directional applications may let solids drop out on the low side of the casing. Thinner fluids may be more effective at cleaning solids from the low side of the hole than thicker muds. Periodically reciprocating the casing also helps to remove cuttings buildup.

The casing provides a large surface area and may be susceptible to balling when drilling gumbo-type formations. Stabilizers used on the casing may increase this tendency significantly. Again, thinner muds may be helpful in dealing with the gumbo issue.

9.2.6 Mechanical Loads.
The mechanical loading of casing from steady-state gravity and frictional forces, as well as dynamic loading caused by various modes of vibration, should be considered when designing a CWD application. The steady-state forces are relatively easy to predict with *torque-and-drag* calculations, while the dynamic forces are much less predictable.

Torsional forces usually are more critical than axial forces because of strength limitations of common casing connectors. In extreme cases with large-diameter casing, the torque rating of the topdrive also may need to be considered.

9.2.7 Drillstring Torque and Drag.
Most general-purpose drilling software packages include a module (such as the one described by Johancsik et al. 1984) that can be used to calculate torque and drag to provide the overall drillstring loading. These programs calculate the axial and normal forces, which along with Coulomb friction are integrated along the drillstring to provide the torque and drag forces. The friction factor used in the calculations is generally assumed to be independent of velocity; thus, the rotating torque should be independent of rotary speed as long as the drillstring is rotating at a constant rate.

For a given wellbore geometry, the difference in torque and drag for a CWD well compared to a conventionally drilled well is primarily related to the difference in average weight and diameter of the drillstring.

The average drillstring weight for a CWD well may be greater than for a conventional well, depending on the well depth, casing size, and conventional drillstring design. For example, **Fig. 9.2.16** shows that 23-lbm/ft 7-in. casing is very similar to a conventional drillstring composed of 4½-in. drillpipe and 20 6¼-in. drill collars, but 9⅝-in. casing is much heavier than this conventional drillstring. This example indicates that, in general, smaller casing is not much heavier than the drillstring that might be use to drill the same well, but larger casing may be considerably heavier. The effective diameter of the CWD drillstring (casing couplings) is generally slightly larger than that for a conventional drillstring (drillpipe tool joints and collars) for small

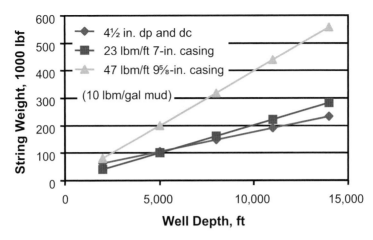

Fig. 9.2.16—String weight for casing and for drillpipe and drill collars (dp and dc).

casing and much larger for large casing. The combined effect of weight and diameter cause the drilling torque for a particular well drilled with large casing to often be greater than if it were drilled with a conventional drillstring.

The well path geometry is important in determining the torque because, first, the inclination is a key parameter in determining the normal force that causes friction. Second, the path determines the total curvature *and* location of the curvature that contributes an additional component to normal force due to the change in direction of tension as the drillstring is pulled around curved sections. In general, curved sections high in the well contribute more to torque and drag than the same curvature located deeper in the well.

Fig. 9.2.17 shows a comparison of the torque in an S-shaped well for three different casing sizes. The well path includes a build to 20° at 2,000 ft and a 1,000-ft hold section, and then drops to 10° at 1° per 100 ft and holds this inclination for the remainder of the well. Note that the casing weight from 4½ to 7 in. and from 7 to 9⅝ in. approximately doubles, but the torque more than doubles because of the increasing casing diameter.

It might be assumed that the torque would be very low because the inclination of the bottom 5,000 ft of the well is only 10°. It is true that in this portion of the well the normal force is low, but the axial force creating tension in the curved portion of the hole between 1,000 and 4,000 ft is high. The overall result is that the torque in the S-shaped, low-angle well can be quite high.

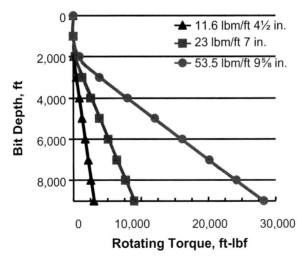

Fig. 9.2.17—Torque for an S-shaped well.

Effect of Tortuosity and BHA Design on Drillstring Torque. Accurately defining the well path for CWD applications is important because the well path plays such a significant role in determining torque and drag. The limited torque rating of casing couplings dictates that a reasonable estimate of torque be used when selecting casing connections. Furthermore, it is important to use practices that manage the torque while drilling a well to make the drilling operations as efficient as possible.

As discussed above, the well path geometry is a key parameter in determining the rotating torque for a drillstring. The well path is specified by directional surveys from either a real well or fictitious points used to define a well plan. In either case, the inclination and azimuth at discrete points along the path define the survey stations, and a smooth, circular arc delineates the well path between these survey stations.

Closely spaced survey stations are needed to adequately define the path for torque calculations, but it is common to find survey stations spaced 100 ft or more apart, particularly in *vertical* wells. For these situations, a quantity defined as *tortuosity* may be used to represent the amount of curvature between stations that is not measured.

For example, in **Fig. 9.2.18**, the path defined by the survey calculation model between station "a" and station "a + 1" is shown as the dashed line. If the stations are very far apart, the real path may be as shown by the solid line. In both cases the slopes (inclination and azimuth) at the survey stations are the same, but the two paths are much different.

Tortuosity is a parameter used to quantify this unmeasured curvature. It is a statistical parameter that specifies the average amount of unmeasured curvature between survey stations and has units of degrees per length, typically degrees per 100 ft.

The vertical CWD wells discussed by Shepard et al. (2002) provide a good example to demonstrate the effect of drillstring design on tortuosity and the effect of tortuosity on torque. These wells were drilled to approximately 9,500 ft with a full-hole bit and 11.6 lbm/ft 4½-in. casing. A total of 1,500 ft of slick 23.2-lbm/ft 5-in. casing was used between the 4½-in. casing and bit to provide additional weight.

Fig. 9.2.19 shows the inclination surveys for Wells 1 through 3 drilled with a slick assembly. Even though the maximum inclination observed in the wells was no more than 4°, the wells exhibited considerable variation in the inclination and much more torque when rotating off bottom than calculated (based on the surveyed well path).

The higher torque was more than could be expected from any reasonable variation in friction factor, and it was concluded that the higher torque was due to tortuosity between surveys. The drilling assembly was changed to a *packed* assembly **(Fig. 9.2.20)** in order to attempt to drill a *straighter* well path.

The lowermost stabilizer consisted of a conventional near-bit stabilizer that also served as a float sub and crossover from the rotary connection on the bit to the casing connection. The upper two spiral-blade stabilizers were rigid-body centralizer subs with full-gauge hardfaced blades.

This assembly was run in 11 additional wells in this same field, and the resulting inclinations are also shown in Fig. 9.2.19 for Wells 4 through 7. The variation in inclination within each well was significantly reduced.

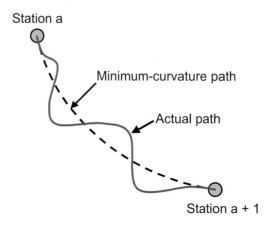

Fig. 9.2.18—Difference between actual well path and *surveyed* path.

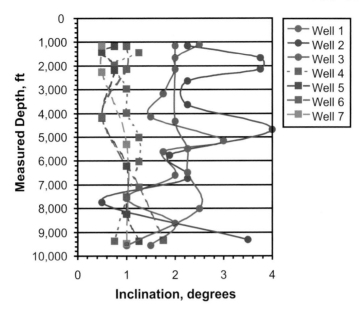

Fig. 9.2.19—Inclination surveys for Wamsutter wells drilled with 4½-in. and 5-in. casing and full-hole bits.

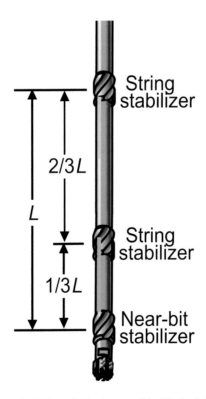

Fig. 9.2.20—Stabilized drillstring used for Wells 4 through 7.

The reduction in rotating torque was more significant than the reduction in inclination variability. **Fig. 9.2.21** shows a plot of the *off-bottom* torque for Well 1, drilled with the slick assembly. (Wells 2 and 3 were almost identical.) Using a typical openhole friction factor of 0.3 and cased-hole friction factor of 0.2, the torque was predicted to be only approximately 800 ft-lbf at TD (dashed curve). The measured torque was three to four times larger than the calculated torque when the borehole was assumed to be smooth. A tortuosity of 0.6° per 100 ft was required to match the actual measured torque.

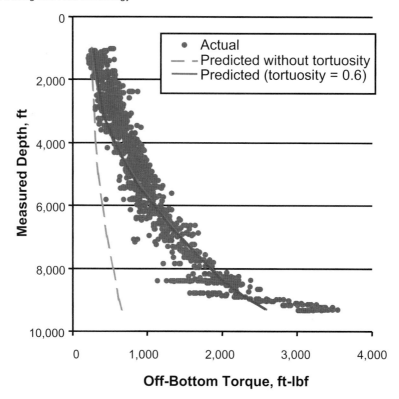

Fig. 9.2.21—Off-bottom torque without stabilized BHA.

The torque was reduced dramatically, as shown in **Fig. 9.2.22**, for the next four wells, drilled with the stabilized drilling assembly. For these wells, the torque is matched by using a tortuosity of 0.15° per 100 ft. The clear conclusion is that the stabilized assembly drilled a straighter and more vertical hole, and thus the torque was significantly less.

The observations from these wells can be applied to other CWD situations. First, the BHA can make quite a difference in the torque that is required to drill vertical wells. If the wells are shallow (less than 5,000 ft) the variation in torque probably doesn't matter, but for deeper wells, excessive off-bottom torque may adversely impact the maximum WOB that can be used by limiting the torque that is available for drilling.

Second, it is important to use an appropriate tortuosity in the torque and drag model when planning wells. For a properly stabilized rotary drilling assembly, it appears that a tortuosity of approximately 0.15° per 100 ft is reasonable, and for an unstabilized rotary assembly, a tortuosity of 0.6° per 100 ft is a reasonable choice. A review of the literature indicates that when using a steerable motor, reasonable values for the tortuosity are 0.8° per 100 ft for curved sections and 0.5° per 100 ft for tangent sections (Rezmer-Cooper et al. 1999; Weijermans et al. 2001).

Summary of Steady-State String Torque. For any particular situation, a torque-and-drag model must be run using the actual well geometry and drillstring definition to know what drillstring loading can be expected, but the main points about torque due to *steady-state rotating friction* can be summarized as follows:

- String torque is unaffected by rotary speed.
- String torque is zero for a truly vertical, straight well.
- For a nonvertical well, string torque increases with casing weight and diameter.
- String torque increases with higher inclinations.
- String torque increases with greater total hole curvature.
- A dogleg high in the hole increases torque more than the same dogleg at a lower point.
- Torque is affected by fluid type—lower with oil-based mud, higher with water-based mud, and still higher in air.

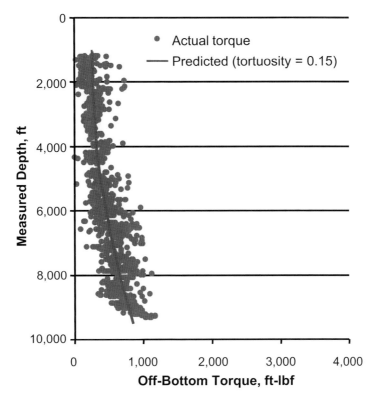

Fig. 9.2.22—Off-bottom torque with stabilized BHA.

- Torque is reduced as mud weight increases and would be maximum in an air-drilled hole.
- Realistic torque calculations can be made only when an accurate well survey is available.
- It is important that a reasonable estimate of tortuosity be used when planning a well.

Lateral Vibration. The most damaging vibration that has been observed while drilling with casing has been lateral vibration, typically referred to as whirl. This vibration can lead to fatigue failures of the casing couplings when it is allowed to persist. Thus, it is important to identify the onset of casing whirl and to know how to mitigate it.

As discussed above, the friction factor does not change appreciably as the sliding velocity changes, and the contact force does not change for constant rotation; thus, the torque is expected to be constant for any rotary speed. A small torque due to viscous force caused by the rotation of the casing in the drilling fluid is negligible.

But in some cases, the torque does increase substantially with increased rotary speed. *Any time this is observed, it should serve as a warning that destructive lateral vibration may be occurring at some point along the drillstring.* Lateral vibration is sometimes thought of as occurring in a plane aligned with the borehole axis. In fact, these vibrations most often display an interaction with the drillstring rotation as lateral displacements that orbit the borehole center.

Fig. 9.2.23 shows a cross section through the borehole with the drillstring in contact with the borehole wall. For CWD, this contact point may be at a centralizer, coupling, or point on the pipe body. The contact may be initiated by buckling. As the drillstring rotates, the initially small lateral force causes friction, which produces a traction force that tends to cause the contact point to walk around the hole in a counterclockwise direction. For common casing and hole dimensions, the casing will precess around the hole multiple times for each drillstring revolution.

This precession of the string around the hole (commonly referred to as whirl) generates a centrifugal force that increases the contact force. Whirl is less likely to be initiated at low rotation speeds, but once it is initiated, the increase in contact force causes the whirl to be self-sustaining and makes it continue, even if the rotary speed is reduced as shown in **Fig. 9.2.24**.

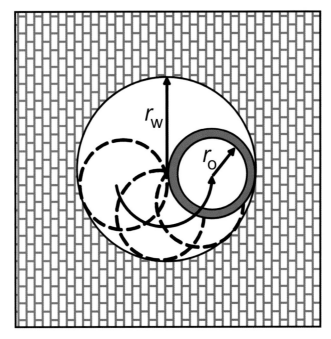

Fig. 9.2.23—Schematic of pipe contact with borehole wall.

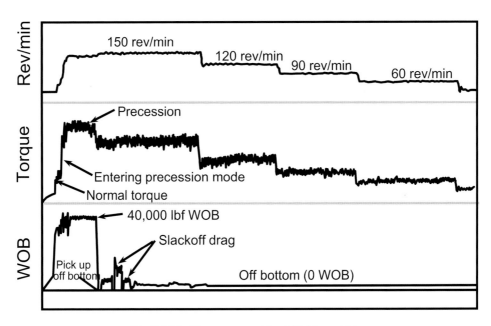

Fig. 9.2.24—Torque caused by drillstring whirl.

The data shown in Fig. 9.2.24 were collected while drilling with a conventional drillstring and an 8.5-in. hard-formation roller-cone bit. The WOB was increased to 40,000 lbf at the same time the rotary speed was increased to 150 rev/min. There was a short period during which the torque hesitated at approximately 2,500 ft-lbf, which is the normal drilling torque for these conditions, but then it abruptly jumped to 7,500 ft-lbf. After a short while, the bit was picked up off bottom and the torque decreased approximately 2,500 ft-lbf (bit torque), but it remained high. The rotary speed was then decreased in 30-rev/min increments, and the torque decreased as the square of the rotary speed. This type of torque-RPM interaction is almost always indicative of downhole lateral vibration of some portion of the drillstring.

Fig. 9.2.25 shows the off-bottom torque measured in two wells being drilled at similar depths with 4½-in. casing. The well showing the strong increase in torque as rotary speed increases experienced a fatigue failure of the casing after less than 1 million drillstring rotations. The casing in the well showing no increase in torque did not fail, even after more than 2 million rotations. Neither of the wells exhibited any discernible vibrations at the surface, but clearly the second well was experiencing lateral vibration.

Not only did the lateral vibration cause the connection to fatigue, but it significantly increased the wear of the casing couplings. The increased wear is due to both an increase in contact force and an increase in the number of effective rotations.

Whirl is most likely to occur at the bit end of the string, where the drillstring is in compression and may be buckled. Often whirl may be caused by drilling through an abrasive sand where the drillstring friction increases. It may be necessary to reduce the rotary speed while drilling several joints before it can be gradually increased without the whirl returning. Another technique that has been used to minimize this problem is to run a mud motor to power the bit so that the casing rotation can be kept low (less than 40 rev/min).

Lateral vibrations are more likely to occur when drilling with less-lubricating drilling fluids (e.g., water vs. a polymer mud). In some cases, pumping a viscous sweep has been effective at temporarily dissipating the tendency for the casing to whirl.

The equations shown below for calculating the whirl speed and the resulting torque demonstrate the effect of various parameters on the phenomenon. The important thing is to detect the detrimental whirling condition on the rig and respond to eliminate it. Without running some type of vibration-monitoring MWD, monitoring the surface torque is the only known method of detecting the onset of casing whirl.

Torque Caused by Whirl. If the casing moves around the hole with no slippage, it will orbit the hole at a speed given by

$$RPM_m = RPM_s \times d_o/(d_w - d_o), \quad \dots\dots\dots\dots\dots\dots\dots\dots\dots\dots\dots\dots (9.2.8)$$

where RPM_m = rate of precession of mass around hole (cycles per minute, cpm), RPM_s = string rotation speed (rev/min), d_w = diameter of hole (in.), and d_o = diameter of pipe at contact (in.).

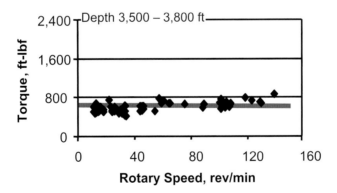

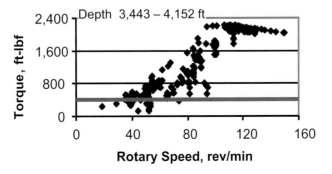

Fig. 9.2.25—Torque response for steady rotation and for whirl.

For example, if the hole diameter is 6.25 in., a 5.5-in. centralizer will precess around the hole at a speed of 880 cpm, even though the string rotary speed is only 120 rev/min. This increased speed will accumulate bending stress cycles at a rate of 7.3 times the drillstring rotary speed.

The increase in torque caused by whirl can be quite simply explained. Centrifugal force from the whirling drillstring provides the normal contact force needed to generate the friction for torque. The whirling mass includes both the steel of the drillstring and the internal volume of drilling fluid. The magnitude of the contact force due to whirl (written in drilling units) is given by

$$F_n = 0.000014 \times w \times (\text{RPM}_s \times d_o)^2/(d_w - d_o), \qquad \qquad \text{(9.2.9)}$$

where w = weight of precessing mass (lbm), RPM_s = string rotation speed (rev/min), d_w = diameter of hole (in.), and d_o = diameter of pipe at contact (in.). The resulting torque is

$$M = \mu_f \times F_n \times r_w/12, \qquad \qquad \text{(9.2.10)}$$

where M = torque due to whirl (ft-lbf), μ_f = coefficient of friction, F_n = contact force due to whirl (lbf), and r_w = hole radius (in.).

Fig. 9.2.26 shows a case with 4½-in. 11.6-lbm/ft casing in which the torque increased with rotary speed and indicated that a portion of the casing string was whirling. Also plotted along with the data is the calculated torque from Eq. 9.2.10, where the steady-state torque is taken to be 970 ft-lbf, the friction factor is 0.3, and it is assumed that 550 ft of casing is laterally vibrating. This particular casing length was used simply because it fit the data, but it also seemed reasonable given the particular drilling conditions. The vibration shown in this figure resulted in a fatigue failure of the casing at the last engaged thread in a connection box.

The increase in surface torque when the drillstring whirls is often the only indicator of this condition. Continuing to drill while the drillstring whirls leads to accelerated wear and to rapidly accumulating fatigue cycles. Both of these conditions are very detrimental to a successful CWD operation.

The best way to detect whirl is to closely monitor the torque and periodically plot the drilling torque trend and compare it to the torque-and-drag model predictions. The drilling torque normally will increase when weight is added to the bit and when soft formations are drilled that produce high drilling rates. These responses often mask the effect of whirl.

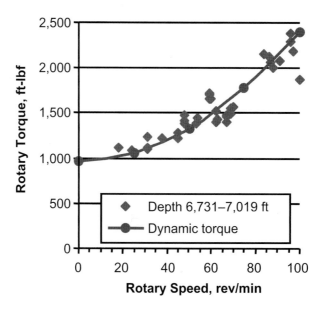

Fig. 9.2.26—Torque caused by lateral vibrations.

When the torque is high and drillstring whirl is suspected, the following procedure can be used to determine if the high torque is caused by drillstring whirl.

1. Note the on-bottom drilling torque and then pick the bit up off bottom without reducing the rotary speed.
2. Record the torque and then slowly reduce the rotary speed to zero.
3. Re-establish rotation at approximately 30 rev/min and note the torque.
4. If the torque remains high when the bit is picked up, gradually decreases as the rotational speed is reduced, and returns to a considerably lower value when rotation is reestablished, it is quite certain that the drillstring was whirling.

When whirl is detected, steps must be undertaken to eliminate it to prevent fatigue failures of the casing couplings. Once an incident of whirl is detected, the drillstring rotation should be reduced to zero and the drilling restarted at lower RPM. Simply reducing the RPM without fully stopping the rotation will often reduce the torque sufficiently so that the whirl is masked—but it may then continue undetected. The rotary speed may need to be reduced while drilling the next interval in order to keep whirl from restarting.

A whirling drillstring can give symptoms very similar to formation sloughing. **Fig. 9.2.27** shows an example of this. At a time of approximately 4 minutes, the torque almost doubled and had the appearance of being caused by whirl. Drilling continued for approximately 6 minutes, and the bit was picked up off bottom without decreasing the rotary speed. After pulling the entire joint of casing from below the rotary table with no decrease in torque, the string was then lowered. The bit appeared to take weight and the torque increased as if it were reaming through a bridge. After spending several minutes reaming back in, the topdrive rotation was stopped momentarily and then restarted at a low speed. The casing easily slid back to bottom without taking weight or requiring torque. The symptoms were similar to what would be expected from a sloughing formation, but the actual cause of the torque was whirl.

The main points relating to casing whirl are as follows:

- Torque that increases when the rotary speed increases is often indicative of whirl.
- Torque that remains high when the bit is picked up off bottom (at constant RPM) but returns to normal when the rotation speed is reduced to zero and restarted is proof of whirl.
- Often there is a threshold WOB and RPM to initiate drillstring whirl.
- Whirl often can be eliminated by reducing RPM and/or WOB.
- Whirl is more likely to occur when drilling hard, abrasive formations.

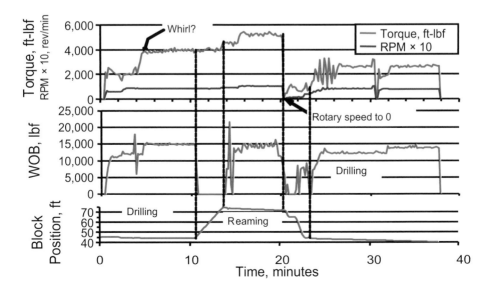

Fig. 9.2.27—Whirl while drilling with 4½-in. casing shows symptoms similar to formation sloughing.

- Whirl is sometimes associated with buckling.
- Whirl may be associated with the lateral resonant frequencies of the drillstring and in some situations may be avoided by quickly increasing the RPM to a level above the resonant frequency.
- Whirl is more likely in a drilling fluid with poor lubricating properties.
- Centralizers can increase the tendency to whirl, particularly if they are not smooth.
- Some bit designs are much more prone to initiating whirl than others.
- The tendency to whirl while drilling cement inside casing is much greater than when drilling formation and is very damaging to PDC cutters.
- An accumulation of cuttings in the annulus can initiate whirl, and reciprocating the casing to clean up the hole may reduce the initiation of whirl.
- The torque measured with a hydraulic topdrive should be corrected for fluid friction effects before using it to monitor for whirl.

Torsional Oscillation. A second type of vibration that is relatively common when drilling with casing in harder rock is torsional oscillation. While this type of vibration is not as destructive as casing whirl, it still may damage the connections by overtorquing them and may damage the bit by either allowing it to rotate backward or inducing a short-duration incident of bit whirl.

When drilling with conventional drillstrings, situations often are encountered in which the torque oscillates with a period (time for one complete cycle) of several seconds. This oscillation is caused by the drillstring behaving as a torsional pendulum, with the collars moving as a lumped mass and the drillpipe acting as a torsional spring. In severe cases, it may progress into *stick/slip*, where the bit comes to a full stop and the period exceeds that predicted for a torsional pendulum.

Many times when this phenomenon occurs, one will hear the driller say that the bit is *taking a bite* at the point of the oscillation when the torque is highest. In reality, the torque that is observed on the surface has more to do with the inertia of the drillstring than with the bit-rock interaction.

To understand this process, one must consider the drillstring as a rather long flexible spring. Ignoring the friction along the string, the drillstring will twist in proportion to the torque applied at its ends:

$$\alpha = ML/JG, \quad\dots\dots\dots\dots\dots\dots\dots\dots\dots\dots\dots\dots\dots\dots\dots\dots\dots\dots\dots\dots\dots\dots\dots \text{(9.2.11)}$$

where α = twist (radians), M = torque, L = drillstring length, J = polar moment of inertia, and G = shear modulus of material.

For a given applied torque, the twist increases linearly with depth but increases inversely with the pipe diameter raised *to the fourth power*. Thus, the twist magnitude is drastically less for large-diameter casing than for small-diameter casing or drillpipe.

One might naturally conclude that torsional vibration would be insignificant for CWD applications. Experience has indicated that this is not true, even though no substantial BHA mass is used. Torsional vibration has been experienced primarily in casing sizes of 7 in. or less.

The drillstring always exhibits some amount of twist while drilling. If the twist is constant, the bit will rotate at the same speed as the surface RPM. When the twist is increasing, the bit rotates slower than the surface speed, and when the twist is decreasing, the bit rotates faster than the surface speed.

The torque required to rotate a PDC bit at a constant WOB decreases as the rotary speed increases. This can (and often does) provide negative damping that will initiate and sustain torsional resonance of the entire casing string when drilling in relatively firm rock.

Not only does the bit speed oscillate, even when the surface rotational speed is constant, but the drillstring inertia amplifies the variations in torque at the bit. The surface torque oscillations are much higher than the bit torque oscillations.

Drillstring vibration software can determine the vibration frequency and the magnitude of force (torque), displacement (twist), and velocity (RPM) along the string. For example, consider a drillstring composed of 8,400 ft of 4½-in. 11.6-lbm/ft casing with 1,000 ft of 23.2-lbm/ft 5-in. casing on bottom. The fundamental natural frequency of the torsional vibration is 0.25 Hz, providing a period of 4 seconds between maximum peaks in torque.

Fig. 9.2.28 shows the mode shape of the torque along the drillstring at the fundamental frequency when a sinusoidal excitation of 1-ft-lbf amplitude is provided at the bit. The actual magnitude depends on the damping along the drillstring, and for this particular plot the damping was selected to provide a reasonable match for field observations. The torque magnitude is also influenced by the bit torque characteristics, which are not sinusoidal but are less symmetrical.

The mode shape simply shows the *magnitude* of the periodic vibration that occurs all along the casing string. For example, the torque vs. time is shown at three places along the drillstring for the torsional vibration.

At the fundamental frequency, which is the one often encountered, the torque amplitude gradually increases from a minimum at the bit to a maximum at the surface. A slight effect of the stiffer bottom 1,000 ft of casing can also be seen. As long as the topdrive is relatively stiff, no point in the drillstring will experience a torque higher than the surface torque. Unfortunately, the torque-measuring system commonly used on drilling rigs may not react fast enough to show the true torque peaks.

The actual torque vibration magnitude at any point along the drillstring is the amplification factor shown in Fig. 9.2.28 times the magnitude of the vibration at the bit. In other words, the simulations indicate that the surface-observed torque oscillations would be approximately 15 times higher than the bit torque oscillations.

Similarly, the change in bit velocity is shown in Fig. 9.2.28. Even though the torque on the bit is relatively constant, the bit velocity fluctuates by approximately 200 rev/min for a surface torque oscillation of 3,000 ft-lbf.

These simulations were conducted with the top of the casing *fixed*. While the topdrive is not totally stiff and will alter the torsional response somewhat, the simulations are good enough to show the principles of torsional vibration.

Torque oscillations may be less prevalent with the CWD system (particularly for larger sizes of casing) because the casing is stiffer than drillpipe and the mass of the collars is eliminated, but it is still possible to excite torsional vibrations in the casing. Small torsional vibrations at the bit may be *amplified* by the casing string if they occur at the right frequency and the nature of the system assures that the natural frequency is excited. The torsional forces result from the inertia of the speeding up and slowing down of the casing, rather than from sliding friction.

The period is shorter for torsional oscillation of casing than for a conventional drillstring. The period is generally less than 5 seconds and is independent of the casing size. The period depends only on casing length, which mode is excited, and the characteristics of the topdrive. The use of a heavier weight of casing on bottom will also alter the natural frequency.

There are three primary reasons to be concerned by torsional vibrations. First, for CWD applications, the casing connections may be overtorqued. Second, if the period is relatively long, the bit may be damaged by

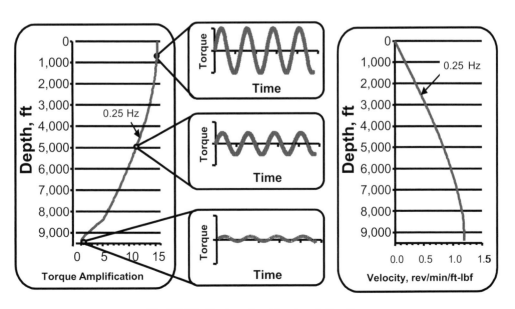

Fig. 9.2.28—First mode of torsional vibration.

intermittent whirl when the bit overspeeds. Third, there is a potential for the bit to travel backward due to torsional vibration. This can be very damaging for PDC bits because it places the diamond in tension and often causes spalling of the diamond cutter face.

Generally, torsional vibration is worse at low rotary speeds and high WOB. It is also more likely to occur when drilling hard rock with PDC bits and when drilling with a worn bit. In fact, the *high torque* that is most often reported when the bearings fail in a roller-cone bit is actually torsional oscillations excited by the friction of the dragging cone.

Speeding up the topdrive and reducing the WOB can often eliminate torsional vibration. In some cases, this may not be possible if the topdrive is not powered adequately. Any time torsional oscillations occur late in the life of a bit, consideration should be given to the possibility that they indicate it is time to pull the bit.

Fig. 9.2.29 shows the recorded torque while drilling with 11.6-lbm/ft 4½-in. casing with 1,000 ft of 23.2-lbm/ft 5-in. casing immediately above the bit. These data were collected while drilling at 8,200 ft with 11,500 lbf WOB and 50 rev/min and are typical of torsional oscillation seen while drilling with small casing in harder rock.

The main points relating to torsional oscillation of casing can be summarized as follows:

- Torsional oscillation is less likely to occur with casing than with a conventional drillstring, particularly for large-diameter casing.
- When it does occur, the period will be shorter than with a conventional drillstring.
- Torsional oscillations may indicate that the bit is worn.
- Reducing the WOB and increasing the RPM can often eliminate torsional oscillations.
- Torsional oscillations are damaging because they can overtorque the connections and may damage the bit.
- Torsional oscillations may be more likely if a section of heavier-weight casing is used on the bottom of the casing string.
- Torsional oscillations are more likely to occur as the well gets deeper.

Combined Vibration. The two types of vibration most often encountered while drilling with casing are lateral vibrations and torsional vibrations. The attributes of both are discussed above, and both should be avoided. Unfortunately, the actions taken to reduce one may increase the risk of the other.

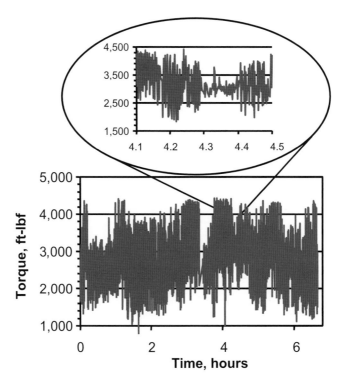

Fig. 9.2.29—Torsional oscillation recorder while drilling with 4½-in. casing.

Increasing the rotary speed reduces the risk of torsional vibrations, but it increases the risk of lateral vibrations. In some instances, it is not possible to pick a rotary speed that avoids both. Sometimes, when the rotary speed is increased to eliminate torsional vibrations, the torque smoothes out at a high value. One may think that the vibrations are eliminated, but in reality what has happened is that the torsional oscillations have been traded for drillstring whirl.

To determine if the vibration has really been eliminated, it is important to know the value of the normal frictional torque. This can be determined accurately only by picking the bit up off bottom, letting the drillstring come to rest, and then observing the torque when rotating at a low rotary speed (15 to 25 rev/min). If the drilling torque is higher than this value plus a reasonable bit torque, then there is a good chance that the lateral vibrations have been induced.

9.2.8 Buckling. One of the obvious differences between drilling with a conventional drillstring and CWD is that drill collars are not used to provide WOB. For years, drillers have been taught that they need to run drill collars to assure that the drillstring is not damaged from buckling. An obvious question, then, is, *how can the CWD process operate safely without using drill collars?* The following discussion explains some fundamental aspects of buckling and its impact on CWD.

Drillstring Buckling. The lower portion of a casing string that is used to apply weight to the bit is in compression. For straight boreholes, the critical buckling load (point at which the casing string may buckle) is determined by the stiffness of the pipe (*EI*), the lateral force of gravity (pipe weight and hole inclination), and the distance from the borehole wall (radial clearance).

As a casing string first begins to buckle in an inclined hole, it generally will deflect into a planar, sinusoidal shape (sinusoidal buckling). As the axial load increases, the string will transform into a helix, spiraling around the inside of the borehole. If the well is vertical, the buckling may immediately form a helix without passing through the sinusoidal buckling stage.

The critical buckling load can be calculated as follows for sinusoidal buckling (Dawson and Paslay 1984):

$$F_{crit} = \sqrt{4 \times w_1 \times \sin\theta \times E \times I / r}, \quad\dots\dots\dots\dots\dots\dots\dots\dots\dots\dots \quad (9.2.12)$$

where F_{crit} = critical buckling force, w_1 = buoyed weight per unit length, E = Young's modulus, I = moment of inertia, r = pipe/hole radial clearance, and θ = inclination.

Helical buckling generally occurs at a force given by $2.8 \times F_{crit}$ when the load is increasing. But after entering helical buckling, the pipe will not come out of helical buckling until the load is decreased to $1.4 \times F_{crit}$ (Mitchell 1996). Other references simply show a single value of $1.4 \times F_{crit}$ as the onset of helical buckling.

Fig. 9.2.30 shows the loads required to helically buckle 11.6-lbm/ft 4½-in. casing, 23-lbm/ft 7-in. casing, and 36-lbm/ft 9⅝-in. casing. These calculations were made using the more conservative $1.4 \times F_{crit}$ as the onset

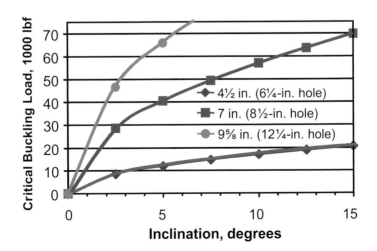

Fig. 9.2.30—Helical buckling force increases rapidly as casing size increases.

of helical buckling. The general conclusion from these calculations is that for most practical values of WOB used for CWD, buckling is insignificant for casing sizes larger than 7 in. if the inclination is at least 2°.

Buckling may influence two factors critical to drilling performance. First, the borehole wall contact force affects the torque required to rotate the drillstring and the wear that will be experienced by the casing. The location of the contact determines if the wear is localized to the casing couplings or also affects the casing body.

Second, buckling will cause a curvature in the casing that will affect the stress that the pipe experiences. If the stress level is high enough, the pipe will yield and fail. At a lower level, the stress may influence the rate at which the pipe fatigues. But in most CWD situations, the clearance between the wall of the borehole and the casing is small enough that the stress level from buckling is not excessive.

From a CWD engineering standpoint, it is important to manage the overall downhole drilling process to maintain casing integrity and drilling efficiency. One step in this process is to screen for whether buckling is significant for the particular well conditions. This can be done with hand calculations or with many of the generally available drilling engineering software packages.

Fig. 9.2.30 indicates that 4½-in. casing in a near-vertical hole would be buckled at low weights. The pipe deflection was calculated with a finite-element drillstring analysis program for WOB from 0 to 15,000 lbf for the 4½-in. casing with a full-gauge shoe and 30-ft shoe joint inside a 6¼-in. straight inclined hole. **Fig. 9.2.31** shows results of the finite-element model of the 4½-in. casing. The casing is not buckled at 4,000 lbf WOB, although the pipe body clearance is reduced to approximately 0.150 in. above the first collar. As the WOB increases, the deflection increases until the pipe begins to buckle at slightly less than 8,000 lbf.

The pitch length of the buckling (Fig. 9.2.31) decreases from approximately 130 ft at 8,000 lbf to approximately 90 ft near the shoe for 15,000 lbf. Even with 15,000 lbf applied WOB, the curvature in the pipe from buckling is only approximately 2° per 100 ft, which is very moderate for 4½-in. casing and results in a bending stress of only approximately 3,000 psi.

The stress due to buckling of the 4½-in. casing at 15,000 lbf WOB is much less than that required to yield the pipe and is also well below the endurance limit (minimum stress at which fatigue failures occur); thus, fatigue should not be a problem. The detrimental effect of the buckling, however, is the fact that the pipe body contacts the borehole wall and will experience wear if the pipe is run for very long in an abrasive formation, and that it may initiate whirl.

Fig. 9.2.32 shows the cumulative contact force for the bottom 300 ft of the casing. Not only does this figure show the contact force, but the discontinuity at 7,000 lbf clearly indicates the point at which buckling begins.

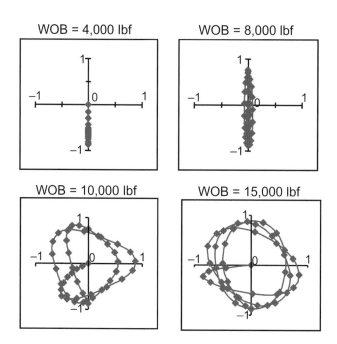

Fig. 9.2.31—End views of 4½-in. 9.5-lbm/ft casing at 0.5° inclination.

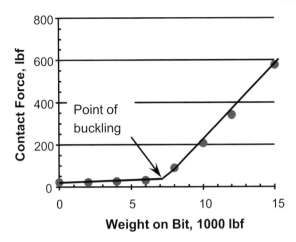

Fig. 9.2.32—Total side force for bottom 300 ft of 4½-in. casing at 0.5° inclination.

Weights up to approximately 6,000 lbf for these conditions can be run without significant concern for pipe body wear, as long as the hole is straight. If higher weights are needed for drilling, then the appropriate casing centralizers would be needed, or alternatively a motor could be used to minimize the wear. Even a slight increase in well inclination will significantly increase the permissible WOB that can be run; thus, a complete buckling analysis should be done using the actual well conditions. The prevention of buckling by using stabilizers on 7-in. and smaller casing is not very effective because more than one stabilizer may be required.

Effect of Torque. The critical buckling force is reduced by the application of torque to the casing—but not by much. Typically, the reduction in the critical buckling force will be only a few percent for torques approaching the yield limit of the pipe (He et al. 1995).

Once the pipe buckles, the application of torque will increase the contact force between the pipe and borehole wall. Again, this effect is relatively small, but in some cases it is not negligible. The contact force for helically buckled pipe is composed of an axial force component and a torque component. The combined contact force can be calculated with the following equation (He et al. 1995):

$$F_c = [r F_e^2 / (4EI)]\left(1 + \sqrt{M \times EI}\right), \quad \dots\dots\dots\dots\dots\dots\dots\dots\dots\dots\dots\dots\dots\dots \quad (9.2.13)$$

where F_c = lateral contact force and F_e = effective tension.

Curved Wellbores. It is normal practice to calculate a maximum load that can be imposed on a drillstring without buckling (critical buckling load) and operate at conditions that are below this load. For straight holes, the previously discussed method of calculating critical buckling load is well established, but there is less agreement on an accepted equation for buckling in curved holes.

The critical buckling load for casing in a curved hole when the inclination is increasing is higher than for a straight hole at the same average inclination. This is due to the axial compression forcing the pipe into the outside of the curve to assist gravity in holding the pipe firmly against the borehole wall. Conversely, the casing may become less stable when the inclination is decreasing at a low curvature.

If the inclination is decreasing at low curvatures, the critical buckling load decreases as the pipe become less stable near the point where it moves from the inside of the curve to the outside. At the point where the pipe is pushed to the outside of the curve, it again becomes stable and exceeds the critical load for a straight hole.

The model proposed by He and Kyllingstad (1995) is probably the best method of estimating the critical buckling loads in an inclined, curved borehole. It can be used to calculate both the sinusoidal and helical critical buckling loads for casing in curved holes:

$$F_{\text{crit}}^4 = \left(1 + a_{ni} \times F_{\text{crit}}\right)^2 \left(a_{n\theta} \times F_{\text{crit}}\right)^2, \quad \dots\dots\dots\dots\dots\dots\dots\dots\dots\dots\dots\dots\dots\dots\dots \quad (9.2.14)$$

where $a_{ni} = a_i \sqrt{\beta EI / r \times w \, \sin \theta}$, $a_{n\theta} = a_\theta \sqrt{\beta EI \sin \theta / r \times w_1}$, F_{crit} = critical buckling load, $\beta = 4$ for sinusoidal buckling, $\beta = 8$ for helical buckling, a_i = inclination build rate, a_θ = azimuth turn rate, a_{ni} = normalized inclination build rate, and $a_{n\theta}$ = normalized azimuth turn rate.

Fig. 9.2.33 shows sinusoidal critical buckling loads for both building (solid lines) and dropping (dashed lines) curvature for 4½-in. casing in a 6¼-in. hole. This small-diameter casing is probably the worst case for CWD. The casing will be buckled at low inclinations for both building and dropping curvatures. In a horizontal well, the critical buckling load is reduced for dropping curvature cases, but this is probably not very significant.

Hydraulic Lift Force. It was shown in the previous discussion of hydraulic effects that the annular friction creates a hydraulic lift on the casing that may be higher than the WOB. This force certainly causes axial compression in the lower end of the casing, but it may not contribute to buckling. This apparent contradiction is explained below.

The neutral point in a drillstring is the point that separates the portion of the string that may buckle from the portion that cannot buckle. This point often is believed to be the point where axial tension is zero, which is incorrect for a string suspended in a liquid. The neutral point is correctly defined as the point where the axial stress is equal to the average of the radial and tangential stress (Hammerlindl 1980).

This definition can best be explained by considering a string of casing suspended vertically in the ocean to a depth of 10,000 ft. We would not expect the casing to be buckled, and it is not. The hydrostatic pressure of the ocean clearly provides a compressive force acting on the end of the casing given by the cross-sectional area times 4,524 psi (10,000 ft × 0.052 × 8.7 lbm/gal). For hydrostatic conditions, the radial and tangential stresses are also 4,524 psi at the end of the casing. Thus, the neutral point is at the very end of the casing, where the axial stress is equal to the average of the radial and tangential stress.

We normally account for this effect by using the buoyed weight of the casing instead of the air weight to determine the neutral point when the only forces acting on the casing are a concentrated mechanical end force and hydrostatic pressure. The neutral point for this condition is simply the mechanical end force divided by the buoyed weight per foot of casing.

The radial and tangential stress must be included for conditions in which a concentrated end force is created by a nonhydrostatic pressure. The hydraulic lift force is caused by a pressure in the annulus that decreases from the bit to the surface. But this pressure also elevates the pressure on the inside of the casing and almost cancels out the effect of the annular pressure on the neutral point. Thus, even though an elevated annular pressure can exert a very significant compression force on the bottom of the casing, its contribution to buckling is negligible as long as the annular pressure is distributed along the string.

For example, consider a relatively tight clearance case with 7-in. casing being drilled below 9⅝-in. casing at 10,000 ft. For the particular drilling-fluid conditions being used on the well, a flow rate of 316 gal/min

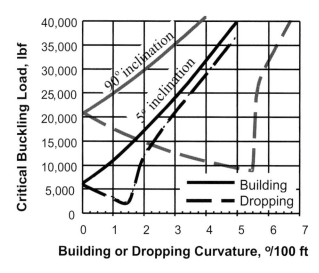

Fig. 9.2.33—Critical buckling loads for 4½-in. casing in 6¼-in. hole at 5° and 90° inclination.

provides a hydraulic lift force of 20,000 lbf, while 479 gal/min provides 40,000 lbf of hydraulic lift. With no flow, the neutral point is at 9,412 ft, with an applied WOB of 10,000 lbf. As shown in **Fig. 9.2.34**, increasing the flow rate sufficiently to cause 40,000 lbf of hydraulic lift only raised the neutral point to 9,217 ft. In other words, for this case, the neutral point moves approximately 58.8 ft per 1,000 lbf of mechanical force applied to the end of the casing, but only approximately 4.9 ft per 1,000 lbf of hydraulic lift.

However, if the annular pressure drop is caused by a localized restriction near the end of the casing (such as packing off around the shoe), the hydraulic lift will contribute much more significantly to buckling because the external pressure is no longer distributed along the length of the casing.

In practical situations such as that of the above example, where the pressure inside the pipe is different from outside the pipe, an *effective axial tension* is used to determine the neutral point and the degree of buckling if the pipe is buckled.

The effective tension for determining if the pipe is buckled and, if so, for determining the contact force is defined as

$$F_e = F - A_i \times P_i + A_o \times P_o, \dots\dots\dots\dots\dots\dots\dots\dots\dots\dots\dots\dots\dots \quad (9.2.15)$$

where F_e = effective tension, F = actual applied tension, A_i = internal pipe area, A_o = external pipe area, P_i = internal pipe pressure, and P_o = external pipe pressure.

The sign convention for the forces is for tension to be positive and compression negative. Thus, the effect of increasing internal pressure is to increase the tendency to buckle.

9.2.9 Fatigue in CWD Applications. The design of drillstring components has evolved over many years into the particular connection geometry, tube body dimensions, and material properties that currently are used for conventional drilling. A large body of experience is available to help select prudent and safe operating guidelines for drillstrings. Even so, fatigue failures account for 60 to 80% of the downhole tool failures experienced in drilling operations (Baryshnikov 1997).

Fatigue failures are caused by cyclical loading at stresses well below the elastic strength of the part. Under repeated loading, a crack develops at a point of localized high stress and propagates through the body until the remaining cross-sectional area is insufficient to support the static load.

Drillstring fatigue failures generally result from oscillating bending loads rather than axial or torsional loads. They are predominantly located in the lower portion of the drillstring rather than at the top where the static stress is highest. In many cases, a fatigue crack will result in a leak before the final rupture; thus, most of the *washouts* that are found in drillstrings are actually caused by fatigue cracks. These failures often are located either in the threaded portion of the connection or in the slip area of drillpipe and are most commonly referred to as *twistoffs* if allowed to continue until the pipe parts.

The *drillstring* design and operating practices for CWD are significantly different from those used with conventional drillpipe and collars. The prevalence of fatigue failures with conventional drillstrings makes it prudent to consider the implication of fatigue for CWD operations.

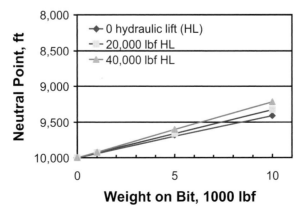

Fig. 9.2.34—Neutral point reduced by hydraulic lift.

Drillpipe Fatigue. **Fig. 9.2.35** shows the fatigue life for Grade D and E drillpipe. The fatigue data do not plot as a single line but rather as a band of failure. Most fatigue testing shows this type of scatter.

Fatigue life varies with many small imperfections in the material and surface finish and is not an intrinsic property of the material. The data shown in Fig. 9.2.35 indicate an endurance limit for the drillpipe of approximately 20,000 psi, which means that there is little chance of the pipe failing by fatigue if bending stresses are kept below 20,000 psi.

Casing Fatigue. The fatigue life expected from a steel part is directly related to the ultimate tensile strength of the material (Reed-Hill and Abbaschian 1991). **Table 9.2.1** shows a comparison of the strength specifications of casing and drillpipe. On the basis of this table, it might be expected that the fatigue life for K55 casing would look very similar to Fig. 9.2.35 for Grade D and E drillpipe.

Fig. 9.2.36 shows the results of fatigue tests of casing with a buttress thread-form connection compared to Grade D drillpipe. It is apparent that the performance of the casing fell slightly below the lower limit for the drillpipe. The points are plotted at a stress level determined by the pipe body bending stress and the connection tensile efficiency. These results are similar to test data available for 7⅝-in. N80 buttress connections. Based on these tests, an endurance limit for K55 buttress connections is estimated to be 12,000 psi.

It is likely that the actual local stress in the connection is higher than calculated by this method, but using the pipe body stress is an easy way to compare casing to drillpipe.

Stresses That Cause Fatigue. In order for a fatigue failure to occur, the part must be exposed to an alternating tensile stress. There are two common sources of cyclical tensile stresses in drillstrings. The first is bending stresses that result from rotating the pipe in a curved geometry, and the second is vibration.

When a tube is elastically deflected with a radius of curvature *R* (**Fig. 9.2.37**), an element of the tube on the inside of the curve is stressed with a compressive stress and an element of the tube on the outside of the curve

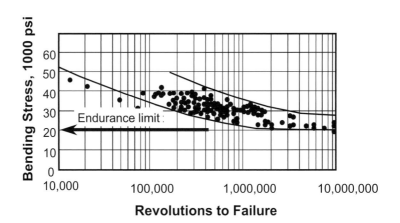

Fig. 9.2.35—Fatigue curve for Grade D and E drillpipe.

TABLE 9.2.1—COMPARISON OF CASING AND DRILLPIPE PROPERTIES			
	Yield Strength (psi)		Minimum Tensile Strength (psi)
	Min.	Max.	
Grade D drillpipe	55,000		95,000
Grade E drillpipe	75,000	105,000	100,000
Grade G drillpipe	105,000	135,000	115,000
Grade S drillpipe	135,000	165,000	145,000
Casing, J55	55,000	80,000	75,000
Casing, K55	55,000	80,000	95,000
Casing, N80	80,000	110,000	100,000
Casing, P110	110,000	140,000	125,000

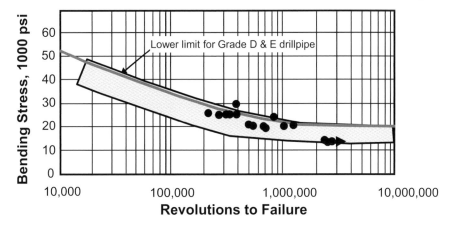

Fig. 9.2.36—Fatigue data from 4½-in. K55 casing.

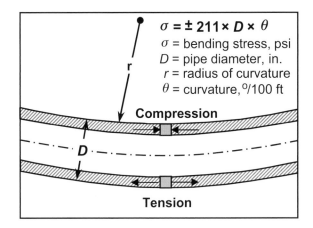

Fig. 9.2.37—Bending stress from pipe curvature.

is stressed with a tensile stress. As the pipe rotates, each point is alternately exposed to the tensile and compressive loads, which are added to any mean axial tension or compression in the tube.

Because the wall of the tube is relatively thin, the entire wall thickness is alternately exposed to tensile and compressive stresses. The outer fiber of the wall will be stressed slightly more than the inside of the wall, but not by much.

The magnitude of the alternating bending stress is proportional to the product of the pipe diameter and the dogleg severity. This dictates that larger casing must be limited to smaller curvatures in order to prevent fatigue failures.

Fig. 9.2.38 shows the alternating stresses for four sizes of casing that may be used in CWD for rotating through curvatures ranging from 0 to 15° per 100 ft. Using an arbitrary stress limit of 12,000 psi allows a maximum dogleg severity of 4, 6, 8, and 12° per 100 ft for 13⅜-in., 9⅝-in., 7-in., and 4½-in. casing, respectively.

In addition to the alternating bending stresses, the casing is also subjected to axial stresses from the hanging weight of the drillstring and from the internal pressure of the drilling fluid. These stresses increase the mean stress level in the pipe and reduce the alternating stress level that can be tolerated without fatigue (Reed-Hill and Abbaschian 1991). Adding the axial and pressure loads to the bending load can easily reduce the maximum dogleg severity that can be tolerated by 20 to 30% or more.

The maximum allowable operating stress for a CWD situation depends on the number of alternating stress cycles expected. If the conditions are such that the stress level is less than 12,000 psi, then the fatigue life will

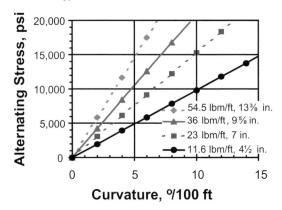

Fig. 9.2.38—Alternating stress from curved casing.

be almost infinite. There may be market-driven situations in which operating at stress levels above the endurance limit is acceptable, but it is much better to design operating parameters below the endurance limit.

Figs. 9.2.35 and 9.2.36 can be used to compare the fatigue life for casing and drillpipe at the same stress levels, but it should be kept in mind that for a given curvature, the casing is likely to be subjected to a higher alternating stress than the drillpipe that would be used to drill the same size hole. For example, the bending stress of 5½-in. casing is approximately 1.5 times that of 3½-in. drillpipe, and that for 7-in. casing is also approximately 1.5 times higher than for 4½-in. drillpipe.

Casing Connector Stresses. The above discussion of fatigue considers casing to have a uniform pipe body, but the casing connections provide discontinuities that may be the weakest points for fatigue. Reduced fatigue resistance from the casing connections can result from several factors, but the most significant is that the casing connector geometry may create stress risers where the local stresses are higher than expected from area considerations.

Any bending moment applied to the pipe body due to the pipe being rotated in a curvature is also applied to the casing connector, but may create an alternating stress on the casing coupling that is higher than on the pipe body.

When the connector is made up at the surface, the torque acting through the threads produce a jack-screw effect that places compression on the torque shoulder (if the connection includes a shoulder) but also creates a zone of relatively high tension on the coupling wall. The magnitude of this tensile stress depends on the makeup torque, the thread design, coupling wall thickness, joint compound friction coefficient, and amount of interference from the threads. This tensile load acts like a mean body stress that adds to the tensile stress for calculating the effective alternating stress for any degree of bending.

Connections may appear similar in terms of torque rating, but may be quite different in terms of actual well conditions that they can withstand. In selecting a connection for CWD applications, *the fatigue life and ultimate torque rating must be balanced.*

Fatigue testing is a critical part of qualifying various connectors for the CWD system. The small amount of testing that has been done to date shows the following:

1. It is possible to provide a casing connection that has adequate strength for CWD at a reasonable cost.
2. There is a wide range in fatigue resistance of connections that various vendors provide.
3. High torsional strength does not imply high fatigue resistance.

Adequate tests must be conducted to confidently provide a representative fatigue curve for each connector that is accepted for routine use in CWD. These curves allow an engineering analysis to provide a well design and casing design for any drilling situation.

Curvature for Stress Calculations. The number of bending cycles and the bending stress magnitude control fatigue. Bending stress, which is proportional to the pipe curvature, may result from curvature due to the pipe following the borehole curvature, the pipe bucking within the borehole, and/or the presence of lateral drillstring vibrations.

Borehole Curvature. The curvature caused by rotating the casing in a curved borehole is the simplest case to consider. The curvature of the casing is normally taken to be the same as the dogleg severity determined from the borehole surveys. The number of cycles experienced from rotating pipe in a curved borehole is equal to the number of revolutions of the pipe in the curve.

Buckling. The post-buckling curvature of the pipe is more important than whether the pipe is buckled. When the pipe buckles, its curvature becomes greater than the hole curvature. For a straight section of hole, the increased curvature can be calculated and is shown in **Fig. 9.2.39** for 4½-in., 7-in., and 9⅝-in. casing vs. axial loading, above the helical critical buckling load.

The increase in curvature is probably negligible for the two larger-sized casings, but it must be considered in fatigue calculations for the smaller casing run in near-vertical wells and in inclination-decreasing curvatures where helical buckling can occur at low weights. [There is no universally accepted way to calculate the additional curvature from buckling in a curved hole, but one method proposed by Schuh (1991) indicates increases in curvature quantitatively similar to those in Fig. 9.2.39.]

It is generally considered that the pipe rotates through the buckled curvature, but the buckled pipe may also rotate as a unit, similar to rotating a corkscrew. If this happens, the buckled curvature does not add to the fatigue loading, but will result in higher torque and a spiral-shaped wear on the outside of the casing.

The net effect of considering buckling is that it must be assumed that the casing may buckle in the lower portion of near-vertical wells and while drilling many zones with a negative curvature (decreasing inclination). Fortunately, the clearance between the casing and hole is small enough that the buckling will not contribute much to pipe fatigue for the larger sizes of casing.

Vibration. Cyclical tensile stresses also can result from either axial or lateral drillstring vibrations. Axial vibrations, such as bit bounce, impart an oscillating axial stress in the pipe but rarely cause casing fatigue.

Lateral vibration, such as drillstring whirl, can cause a significant bending stress near stabilizers and casing couplings. It is often a source of drillstring failure. The reversing stress frequency is generally several times the rotary speed, and the magnitude depends on both the whirl accelerations and the clearance between the pipe and the borehole wall. As discussed in Section 9.2.7, this type of vibration can be very damaging and procedures should be used to identify and prevent it.

Operational Practices Relating to Fatigue. Prudent operating practices for CWD are somewhat different from those for conventional drilling, and one of the most critical differences is taking proper care of the pipe to prevent fatigue failures.

The stress stored in the made-up connection will affect the fatigue life of the connection. Unfortunately, we do not have a way of evaluating how much a given variation will cause, so it is imperative that all the connections are made up to the proper specification.

Casing Handling. It is important that the casing be protected from circumferential slip and tong marks during handling and makeup. **Fig. 9.2.40** shows the stress concentration factor reported for die marks on

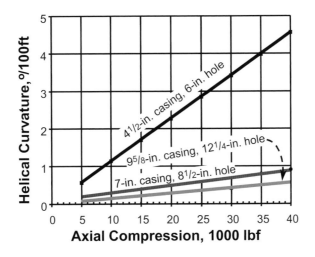

Fig. 9.2.39—Curvature due to helical buckling.

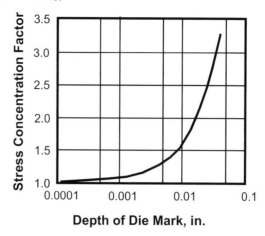

Fig. 9.2.40—Axial stress concentration factor for die marks on drillpipe.

drillpipe by Hossain et al. (1998). This stress concentration factor is multiplied by the bending stress to obtain the alternating stress for fatigue life calculations; thus, even a small score mark is quite significant. For example, a die mark 0.01 in. deep could decrease the safe stress from 12,000 psi to only 8,000 psi.

Monitoring the rotary speed may be more critical with the CWD system than with a conventional drilling operation. Fatigue is a direct accumulation of stress cycles, and rotating the pipe accounts for most of the fatigue-generating stress. While drilling, the rotary speed should be no higher than provides a linear increase in penetration rate. The rotating speed should be minimized when not drilling.

9.2.10 Directional Drilling With Casing. CWD technology can be applied to directional wells and provides benefits similar to those observed in vertical wells. Drilling directionally with casing generally requires a retrievable and rerunnable directional drilling assembly to allow the directional drilling tools to be recovered, to provide access below the casing, and to replace failed tools before reaching the casing point. The retrievable directional BHA provides all the functions of the directional tools used in a conventional directional BHA and in general is composed of conventional components. Directional trajectory control can be provided with a passive stabilized BHA, a steerable motor BHA, or a rotary steerable BHA.

Passive Stabilized Directional BHA. The appropriate placement of stabilizers on a CWD BHA can provide the capability to maintain, build, or drop inclination. The stabilization generally is applied to the drilling assembly in the pilot hole below the underreamer, where full-gauge stabilizers can be used that are also small enough to be tripped through the casing. **Fig. 9.2.41** shows a simple passive BHA that is used to drill a straight section of hole. Similarly, the BHA below the underreamer can consist of a pendulum assembly **(Fig. 9.2.42)** for reducing inclination or a building assembly for increasing the inclination. These assemblies are designed using the same principles that are used to design conventional passive directional assemblies (Bourgoyne et al. 1986). In some cases, the building assembly is constructed by placing the underreamer near the bit so that the bit axis will be more tilted in the hole and build inclination at a faster rate. In general, these assemblies are limited in versatility and are sensitive to operating and geological conditions.

Stabilizing the drilling assembly in the pilot hole between the pilot bit and underreamer provides the directional control and stabilizes the underreamer to protect it from lateral vibration. Stabilization in the pilot hole reduces the rotating torque by drilling a smoother hole than any configuration of casing stabilization. Undergauge stabilization may be used on the bottom of the casing to provide centralization for cementing and to control wear, but it has little effect on the directional trajectory.

Steerable Motor Directional BHA. Active trajectory control can be provided by a steerable motor assembly where both inclination and azimuth changes can be made in a controlled manner. The steerable motor CWD directional BHA **(Fig. 9.2.43)** normally consists of a pilot bit, underreamer, steerable mud motor, MWD tool, and nonmagnetic drill collar(s) extending below the casing (Warren et al. 2005). This is similar to the assembly that is commonly used for conventional directional drilling, except that an underreamer is

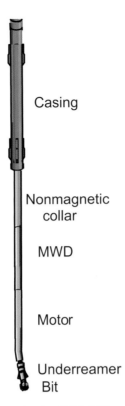

Casing

Nonmagnetic
collar

MWD

Motor

Underreamer
Bit

Fig. 9.2.41—Passive CWD drilling assembly.

Packed   Pendulum   Building

Fig. 9.2.42—Passive directional assemblies that may be used below the underreamer.

required and, for smaller casing sizes, the mud motor may be smaller than would be used for conventional directional work in the same size hole. A magnetic MWD tool is often used for steering and requires a section of nonmagnetic collars between it and the casing shoe. This extends the bit and underreamer to 80 to 120 ft below the casing shoe.

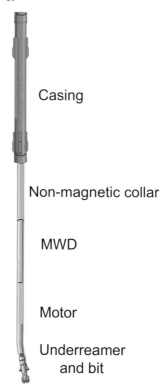

Casing

Non-magnetic collar

MWD

Motor

Underreamer
and bit

Fig. 9.2.43—Steerable motor directional CWD assembly.

Rotational power may be limited by the smaller motor required for some sizes of casing, but this is of little consequence for casing sizes larger than 7 in. The bend in the motor is also limited by the fact that the assembly must pass through the casing being used as the drillstring. In general, though, an adequate bend angle can be run to drill the maximum curvature that is safe to use when drilling with casing.

Orienting the motor is often very easy when drilling with casing because there is little twist between the surface and motor. The casing twist for a particular situation is much smaller than for drillpipe. This may allow the motor to be oriented without reciprocating the casing to work out stored twist as is normally required for orienting a motor run below drillpipe.

Stabilizing the bit and underreamer for directional wells when using a steerable motor (Fig. 9.2.43) is more difficult than when using a passive assembly. Because of the need to drill a portion of the hole without pipe rotation, the underreamer must be placed below the motor, which places it directly above the bit.

No full-gauge rigid stabilization can be placed on the BHA above the underreamer. For larger sizes of casing, this does not pose any problem; however, in smaller hole sizes, higher build rates are often desired, and the directional control is a little more difficult with an unstabilized steerable motor assembly. Running an expandable stabilizer above the top of the motor can eliminate this problem, but it introduces another component to the BHA.

Choosing the motor bend angle to run with the CWD system involves an additional consideration to that required for conventional drilling. The motor, bit, and underreamer must pass through the casing that may be only slightly larger than the motor housing. The bit to bend length is longer than normal because of the introduction of the underreamer to the assembly. This limits the bend angle to one that might be less than would be used to conventionally drill at the same desired curvature. In some cases, the smaller bend angle may increase the amount of sliding relative to rotating, but in most situations, the smaller bend will deliver an adequate build rate and is not as restrictive as might be expected.

In planning the build rate, the casing size and grade must be considered to minimize the potential for a fatigue failure as discussed above. **Table 9.2.2** shows suggested values for the maximum build rate that could be used for some typical casing strings.

TABLE 9.2.2—MAXIMUM ALLOWED CURVATURE FOR VARIOUS WEIGHTS AND GRADES OF CASING

Casing Size (in.)	Casing Weight (lbm/ft)	Casing Grade	Maximum Curvature (° / 100 ft)
5½	17	P110	13
7	23	L80	8
9⅝	36	J55	4.5
13⅜	54.5	J55	3

Torque considerations also affect the determination of whether a particular well is a good directional CWD candidate. The torque required while drilling with casing in directional wells has been similar to that observed in conventional wells, when adjusted for the casing weight and exterior diameter. For example, **Fig. 9.2.44** shows a comparison of the predicted and observed torque while drilling a directional well with 9⅝-in. (water-based mud) and 7-in. (oil-based mud) casing. The larger casing was used to drill to 2,400 ft while building the inclination to 15°, and then the 7-in. casing was used to drill a tangent section with a steerable motor. The torque for both strings was predicted adequately with a friction factor of 0.3 for water-based mud and 0.2 for oil-based mud. Both hole sections were drilled with stabilization above the motor.

In general, the off-bottom rotating torque is predicted quite well with friction factors of 0.25 to 0.35 for water-based mud, 0.20 to 0.25 for oil-based mud, and 0.15 to 0.20 for synthetic mud. It is important that an appropriate tortuosity be used in the calculations. The appropriate value of tortuosity will depend on the type of drilling assembly being used. Much higher tortuosity (0.8° per 100 ft) may be needed for unstabilized assemblies, while values as low as 0.15 may be adequate when drilling with a rotary steerable tool (RST).

Running a motor while drilling with casing introduces some challenges that are not apparent when running the motor below drillpipe and drill collars. There is an interaction between the motor pressure and casing elongation that affects motor performance when drilling with a positive-displacement mud motor. This presents a consideration that must be incorporated into motor selection and operating practices.

The internal casing pressure increases when the motor loading increases. This causes the casing to try to elongate and moves the neutral point higher in the drillstring, thus increasing the WOB. This in turn further increases the bit torque and motor loading. For example, **Fig. 9.2.45** shows that the increase in WOB for a given internal pressure change for 7-in. casing is approximately six times as much as it is for the 3½-in. drillpipe that would normally be used with the same size motor. This ignores the effect of the drill collars that would increase the difference.

This interaction of internal pressure and WOB affects the selection of the optimum motor for CWD and also affects the operation of any motor that is run with the casing. The importance of this interaction can be shown by comparing a 5½-in. 7/8-lobe motor to a 5-in. 6/7-lobe motor that is more commonly available for use below 7-in. casing. While both motors can deliver 2,500 ft-lbf of torque (**Fig. 9.2.46**), the 5-in. motor

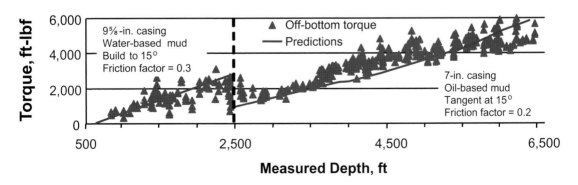

Fig. 9.2.44—Observed vs. predicted torque with 9⅝-in. casing and water-based mud, and with 7-in. casing and oil-based mud.

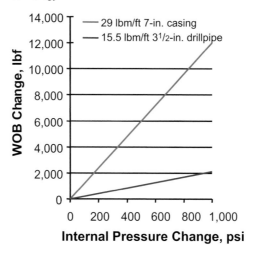

Fig. 9.2.45—Comparison of 7-in. casing and 3½-in. drillpipe.

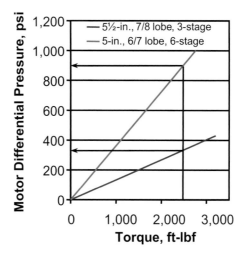

Fig. 9.2.46—A 5½-in. 7/8-lobe motor delivers adequate torque at a lower pressure than a 5-in. 6/7-lobe motor.

requires 900 psi compared to only 350 psi for the 5½-in. motor. This increased pressure adversely affects the drilling efficiency when drilling with the 7-in. casing.

Fig. 9.2.47 shows a comparison of the 5- and 5½-in. motor performance predicted with a mud motor drilling *simulator* that incorporates models for ROP, bit torque, motor characteristics, and pipe stretch (Warren et al. 2005). The smaller motor is approximately two times as sensitive to abrupt drillstring advancements as is the larger motor. Both motors are capable of developing the torque needed to drill with a pilot bit and underreamer with 7-in. casing, but the higher pressure of the 5-in. 6/7-lobe six-stage motor makes it more sensitive to any activity that can abruptly increase the motor loading. This abrupt motor loading may result from block movement at the surface, but it is more likely to result from stick/slip of the drillstring advancement in the borehole.

Fig. 9.2.48 shows field data that illustrate the motor-casing interaction while drilling with a 6¾-in. motor at 715 m in a 12¼-in. hole section. While sliding ahead at approximately 20 m/h, the motor stalled after slacking off only a little more than normal. The indicated drilling WOB was approximately 15,000 lbf, with a differential pressure of 150 psi. The pressure increased approximately 800 psi before the pumps were shut down, while the WOB increased approximately 21,000 lbf as the motor stalled.

Simulations of similar conditions **(Fig. 9.2.49)** show that the motor would continue to rotate with surface drillstring advances as large as 1.15 in. Observed advances as high as 0.9 in. were handled easily. An advance of 1.3 in. stalled the motor, created a total motor pressure drop of 950 psi, and increased the WOB by 22,000 lbf.

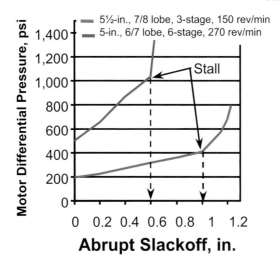

Fig. 9.2.47—Sensitivity of two similar motors to abrupt drillstring advancement.

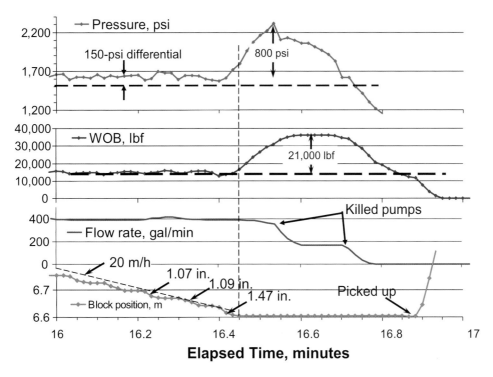

Fig. 9.2.48—Stall with 6¾-in. motor while sliding at 715 m.

The WOB increase when the motor stalls with 9⅝-in. casing is higher than when the motor stalls with the 7-in. casing. In fact, for the same depth and pressure increase, 40-lbm/ft 9⅝-in. casing will exhibit two times as much WOB increase as 29-lbm/ft 7-in. casing. Even so, operating a motor with 9⅝-in. casing is easier because the 6¾-in. motor has approximately 2.5 times as much torque capacity as the 5½-in. motor, while the 12¼-in. underreamer requires less than two times as much torque as the 8⅞-in. underreamer.

The casing characteristics combined with the bit characteristics define a torque demand (ft-lbf/psi) with respect to internal casing pressure changes. The motor performance curve defines a torque delivery (ft-lbf/psi) for the motor. As long as the torque delivery is substantially greater than the torque demand, the casing interaction is inconsequential. The ratio of torque delivery to motor differential pressure is more important in

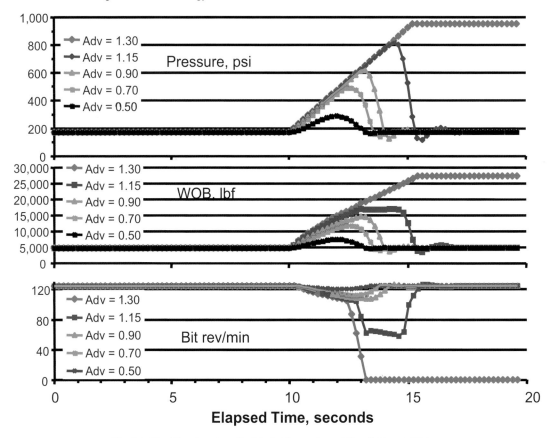

Fig. 9.2.49—Simulated stall with 6¾-in. motor while sliding at 715 m.

selecting a motor to use in CWD applications than is the absolute maximum torque that the motor will deliver. This almost always directs the user to a low-speed motor with a relatively high lobe count.

Both the simulator and drilling experience indicate that the casing elongation effect is of little consequence while drilling with a motor as long as the motor is selected properly, the WOB application is smooth, and the motor is operated at no more than half its rated differential pressure. As the loading increases and the WOB becomes erratic, problems with the motor stalling are encountered. These changes can occur from both surface control activities and inconsistent drag while sliding.

A second issue that occurs when drilling directionally with a steerable motor is that the casing may not slide as readily as drillpipe does. The casing may contact the wellbore wall with higher contact forces due to its greater weight, thus making sliding more difficult. One way of overcoming the frictional contact force is to rotate the casing, alternately clockwise and counterclockwise, to convert the axial frictional resistance to a circumferential force so that sliding is less inhibited. This process, known as *rocking*, is quite effective with casing because of the torsional stiffness of the casing, which allows a small degree of surface rotation to be effective.

Fig. 9.2.50 shows an example of sliding with 7-in. casing in a 15° tangent section of a directional well where rocking made a substantial difference in drilling performance. In this particular case, the motor available for the drilling was not optimized for the particular application and was thus very sensitive to stalling due to the pressure elongation effect. Without rocking, it was impossible to apply the WOB to the bit without stalling the motor. Rocking the casing approximately 90° to either side of the neutral position almost totally eliminated the stalling and significantly increased the ROP, as indicated by the slope of the block position data.

Rotary Steerable Directional BHA. Active trajectory control in directional wells is often much more efficient when using a RST for drilling conventional wells. The RST allows the trajectory to be continuously adjusted while rotating the drillstring and eliminates the need for sliding. A rotary steerable directional drilling assembly can be used below casing to provide the same benefits when drilling with casing as are achieved

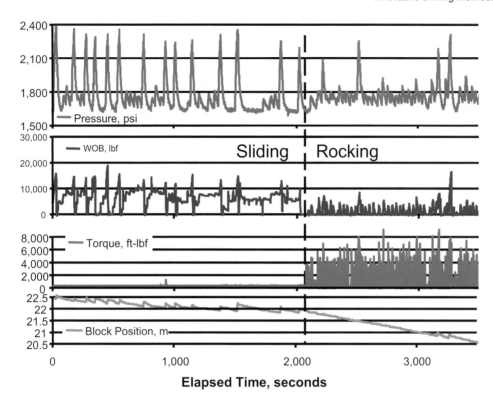

Fig. 9.2.50—Rocking 7-in. casing reduces motor stalling and increases the effective ROP.

when drilling conventionally. The RST technology also drills a smoother hole and reduces the possibility of accidentally introducing a localized dogleg that could cause a fatigue failure of the casing.

Fig. 9.2.51 shows a typical RST-retrievable BHA that is used when drilling with casing. The RST is placed immediately above the pilot bit to provide the trajectory control in the pilot hole. This allows conventional RSTs, MWD tools, and LWD tools to be operated in the hole size for which they were designed. This provides a complete complement of conventional tools to be selected for use with CWD without need for any special modifications.

The pilot hole is opened with an underreamer placed above the directional tools and immediately below the casing shoe. Often, stabilization (stabilizers or roller reamers) is positioned below the underreamer to stabilize it as it opens the hole to the final diameter.

Rotational power for the RST may be provided by a mud motor positioned in the bottom of the casing. Using the motor allows the casing to be rotated at low speed to minimize casing wear and fatigue of the casing connections. It is preferable that the motor be designed as a *straight hole motor* rather than a bent directional motor with the bend set to zero. This not only ensures that there is no residual bend in the motor, but it also allows a stronger drive shaft that is desirable when rotating the large mass of the rotary steerable BHA.

Selecting the motor is somewhat complicated by the fact that the rotational speed is generally limited to 150 rev/min to protect the underreamer. The flow rate may be limited by ECD considerations. Relatively high torque is needed to drive the BHA. The size of the motor is limited by the need for it to fit within the casing. Finally, the motor selection involves considerations of the casing elongation effect discussed above. When all these considerations are incorporated into the BHA design, there are relatively few motors available that can provide satisfactory performance.

When using a motor to drive the BHA, the MWD tool will be positioned below the motor. This means that the MWD tool turns any time the drilling fluid is circulated. Many MWD tools are designed to acquire survey data when the circulation of drilling fluid is resumed after making a connection. This type of MWD tool cannot be used in the rotary steerable CWD assembly that is motor-driven, but rather an MWD tool must be chosen that acquires the survey data in the *quiet* time when the pump is off and then transmits the data after

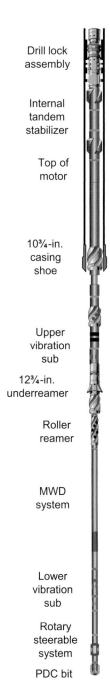

Fig. 9.2.51—Rotary steerable assembly for CWD.

pumping is resumed. The mud-pulse survey data are transmitted through the mud motor, but this has not been found to cause a significant degradation of the pressure signal. Electromagnetic MWD tools have been used in a few cases when the appropriate mud-pulse tools were not available.

Rotary steerable CWD tools have been used to drill both onshore and offshore wells that involve significant build and turn in the wellbore trajectory. **Fig. 9.2.52** shows one example of the application of RST technology to drill the 10¾- and 7¾-in. casing sections of a North Sea well. The well was successfully drilled and met all the objectives that originally drove the decision to drill with CWD. However, there were a number of directional tool failures in the well, and this is partially associated with the complexity of the tools that are used.

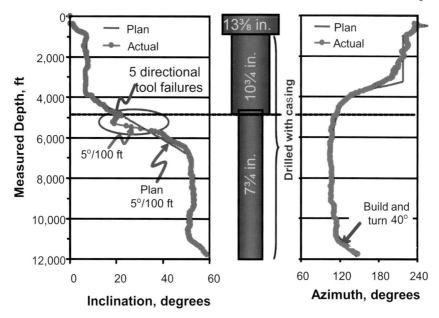

Fig. 9.2.52—North Sea well drilled with rotary steerable CWD tools.

The well shown in this example is one of the more complex wells drilled with CWD technology, but it would in no way be considered a very complex well in terms of conventional directional technology. The use of CWD technology in directional wells is just in its infancy, and its continued use will allow more complex wells to be drilled and more reliability to be gained with the drilling assemblies used in these wells. In fact, there is an effort under way in several of the major directional tool providers to develop directional tools specifically designed to be used with CWD. These tools may allow more of the BHA to be positioned inside the casing, where it is more protected and may allow the stick-out to be shorter.

Nomenclature

a_i = inclination build rate, °/100 ft
a_{ni} = normalized inclination build rate, dimensionless
$a_{n\theta}$ = normalized azimuth turn rate, dimensionless
a_θ = azimuth turn rate, °/100 ft
A_i = internal pipe area, in.^{2}
A_o = external pipe area, in.^{2}
d_o = external diameter of tubular at contact, in.
d_w = diameter of borehole, in.
D = conduit diameter, ft
E = Young's modulus, psi
f_F = friction factor, dimensionless
f_o = hydraulic friction factor, casing wall, dimensionless
f_w = hydraulic friction factor, borehole wall, dimensionless
F = applied tension, lbf
F_c = lateral contact force, lbf
F_{crit} = critical buckling force, lbf
F_d = drag force on borehole wall, lbf
F_e = effective tension, lbf
F_{end} = end force, lbf
F_n = contact force due to whirl, lbf
g = acceleration of gravity, 32.174 ft/sec^2

G = modulus of rigidity, lbf/in.^{2}
H_f = hydraulic head loss, ft
I = area moment of inertia, in.^{4}
J = polar moment of inertia, in.^{4}
L = drillstring length, ft
M = torque, ft-lbf
P_i = internal pipe pressure, psi
P_o = external pipe pressure, psi
r = pipe/hole radial clearance, in.
r_o = casing exterior radius, in.
r_w = wellbore radius, in.
Re = Reynolds number, dimensionless
RPM_m = rate of precession of mass around hole, cycles/min
RPM_s = string rotation speed, rev/min
v = average velocity, ft/sec
V = fluid velocity, ft/sec
w = weight of whirling mass, lbm
w_1 = buoyed weight per unit length, lbm/ft
α = twist, radians
β = 4 for sinusoidal buckling, β = 8 for helical buckling
ΔHL = change in hook load, lbf
ΔL = casing element length, in.
Δp_o = annular friction pressure, psi
ε = conduit wall roughness, in.
μ_f = coefficient of Coulomb friction, dimensionless
ρ = fluid density, lbm/gal
τ_o = casing wall shear stress
τ_w = borehole wall shear stress
θ = inclination

References

Aston, M.S., Alberty, M.W., McLean, M.R., de Jong, H.J., and Armagost, K. 2004. Drilling Fluids for Wellbore Strengthening. Paper SPE 87130 presented at the IADC/SPE Drilling Conference, Dallas, 2–4 March. DOI: 10.2118/87130-MS.

Baryshnikov, A. 1997. Downhole Tool Full-Scale Fatigue Test: Experience and Practice Recommendations. Paper presented at the ASME 8th Annual International Energy Week Conference and Exhibition, Houston, 28–30 January.

Bode, D.J., Noffke, R.B., and Nickens, H.V. 1991. Well Control Methods and Practices in Small-Diameter Wellbores. *JPT* **43** (11): 1380–1386; *Trans.*, AIME, **291**. SPE-19526-PA. DOI: 10.2118/19526-PA.

Bourgoyne, A.T., Chenevert, M.E., and Millheim, K.K. 1986. Principles of BHA. In *Applied Drilling Engineering*, Vol. 2, 426–443. Richardson, Texas: Textbook Series, SPE.

Brown, C.C. 1971. System for Rotary Drilling of Wells Using Casing as the Drillstring. US Patent No. 3,552,507.

Dawson, R. and Paslay, P.R. 1984. Drillpipe Buckling in Inclined Holes. *JPT* **36** (10): 1734–1738. SPE-11167-PA. DOI: 10.2118/11167-PA.

Gelfgat, M. and Alikin, R. 1998. Retractable Bits Development and Application. *Journal of Energy Resources Technology* **120** (2): 124–130. DOI: 10.1115/1.2795022.

Green, M.D., Thomesen, C.R., Wolfson, L., and Bern, P.A. 1999. An Integrated Solution of Extended-Reach Drilling Problems in the Niakuk Field, Alaska: Part II—Hydraulics, Cuttings Transport and PWD. Paper SPE 56564 presented at the SPE Annual Technical Conference and Exhibition, Houston, 3–6 October. DOI: 10.2118/56564-MS.

Hammerlindl, D.J. 1980. Basic Fluid and Pressure Forces on Oilwell Tubulars. *JPT* **32** (1): 153–159. SPE-7594-PA. DOI: 10.2118/7594-PA.

He, X. and Kyllingstad, A. 1995. Helical Buckling and Lock-Up Conditions for Coiled Tubing in Curved Wells. *SPEDC* **10** (1): 10–15. SPE-25370-PA. DOI: 10.2118/25370-PA.

He, X., Halsey, G.W., and Kyllingstad, A. 1995. Interactions Between Torque and Helical Buckling in Drilling. Paper SPE 30521 presented at the SPE Annual Technical Conference and Exhibition, Dallas, 22–25 October. DOI: 10.2118/30521-MS.

Hossain, M.M., Rahman, M.K., Rahman, S.S., Akgun, F., and Kinzel, H. 1998. Fatigue Life Evaluation: A Key To Avoid Drillpipe Failure Due to Die-Marks. Paper SPE 47789 presented at IADC/SPE Asia Pacific Drilling Technology, Jakarta, 7–9 September. DOI: 10.2118/47789-MS.

Johancsik, C.A., Friesen, D.B., and Dawson, R. 1984. Torque and Drag in Directional Wells—Prediction and Measurement. *JPT* **36** (6): 987–992. SPE-1380-PA. DOI: 10.2118/11380-PA.

Kammerer, A.W. Jr. 1958. Rotary Expansible Drill Bits. US Patent No. 2,822,149.

Leturno, R.E. 1993. Drilling With Casing and Retrievable Drill Bit. US Patent No. 5,271,472.

Mitchell, R.F. 1996. Buckling Analysis in Deviated Wells: A Practical Method. Paper SPE 36761 presented at the SPE Annual Technical Conference and Exhibition, Denver, 6–9 October. DOI: 10.2118/36761-MS.

Ooms, G., Burgerscentrum, J.M., and Kampman-Reinhartz, B.E. 1999. Influence of Drillpipe Rotation and Eccentricity on Pressure Drop Over Borehole During Drilling. Paper SPE 56638 presented at the SPE Annual Technical Conference and Exhibition, Houston, 3–6 October. DOI: 10.2118/56638-MS.

Reed-Hill, R.E. and Abbaschian, R. 1991. *Physical Metallurgy Principles*, third edition, Chap. 21. Belmont, California: Pws-Kent Series in Engineering, Cengage Learning.

Rezmer-Cooper, I., Chau, M., Hendricks, A. et al. 1999. Field Data Supports the Use of Stiffness and Tortuosity in Solving Complex Well Design Problems. Paper SPE 52819 presented at the SPE/IADC Drilling Conference, Amsterdam, 9–11 March. DOI: 10.2118/52819-MS.

Santos, H., Placido, J.C.R., and Wolter, C. 1999. Consequences and Relevance of Drillstring Vibration on Wellbore Stability. Paper SPE 52820 presented at the SPE/IADC Drilling Conference, Amsterdam, 9–11 March. DOI: 10.2118/52820-MS.

Schuh, F.J. 1991. The Critical Buckling Force and Stresses in Pipe in Inclined Curved Boreholes. Paper SPE 21942 presented at the SPE/IADC Drilling Conference, Amsterdam, 11–14 March. DOI: 10.2118/21942-MS.

Shepard, S.F., Reiley, R.H., and Warren, T.M. 2002. Casing Drilling Successfully Applied in Southern Wyoming. *World Oil* **223** (6): 33–41.

Tessari, R.M. and Madell, G. 1999. Casing Drilling—A Revolutionary Approach to Reducing Well Costs. Paper SPE 52789 presented at the SPE/IADC Drilling Conference, Amsterdam, 9–11 March. DOI: 10.2118/52789-MS.

Warren, T., Tessari, R., and Houtchens, B. 2004. Directional Casing While Drilling. Paper WOCWD-0430-01 presented at the World Oil Casing Drilling Technical Conference, Houston, 30–31 March.

Warren, T., Houtchens, B., and Madell, G. 2005. Directional Drilling With Casing. *SPEDC* **20** (1): 17–23. SPE-79914-PA. DOI: 10.2118/79914-PA.

Warren, T.M. 1984. Factors Affecting Torque for a Roller Cone Bit. *JPT* **36** (9): 1500–1508. SPE-11994-PA. DOI: 10.2118/11994-PA.

Warren, T.W., Oster, J.H., Sinor, L.A., and Chen, D.C.K. 1998. Shock Sub Performance Tests. Paper SPE 39323 presented at the IADC/SPE Drilling Conference, Dallas, 3–6 March. DOI: 10.2118/39323-MS.

Weijermans, P., Ruszka, J., Jamshidian, H., and Matheson, M. 2001. Drilling With Rotary Steerable System Reduces Wellbore Tortuosity. Paper SPE 67715 presented at the SPE/IADC Drilling Conference, Amsterdam, 27 February–1 March. DOI: 10.2118/67715-MS.

SI Metric Conversion Factors

cp	× 1.0*	E − 01 = Pa·s
cycles/sec	× 1.0*	E + 00 = Hz
ft	× 3.048*	E − 01 = m
ft/hr	× 8.466 667	E − 05 = m/s
ft/s	× 3.048*	E − 01 = m/s
ft-lbf	× 1.355 818	E + 00 = J

gal/min	× 6.309 020	E – 05 = m³/s
in	× 2.54*	E – 02 = m
in²	× 6.451 6*	E – 04 = m²
lbf	× 4.448 222	E – 00 = N
psi	× 6.894 757	E – 00 = kPa
lbm	× 4.535 924	E – 01 = kg
lbm/gal	× 9.977 633	E + 01 = kg/m³

*Conversion factor is exact.

9.3 Managed-Pressure Drilling—Don M. Hannegan, Weatherford International

9.3.1 Introduction. Managed-pressure drilling (MPD) is an advanced form of primary well control usually employing a closed and pressurizable circulating-drilling-fluids (mud) system that facilitates drilling with precise management of the wellbore-pressure profile. The primary objective of MPD is to optimize drilling processes by decreasing nonproductive time (NPT) and mitigating drilling hazards (Malloy et al. 2009).

The root concepts of MPD challenge conventional drilling wisdom, but because they *make sense*, the concepts have inspired exploration of ways in which they can be exploited to greater degrees when facing difficult wells. Therefore, the material presented here is not a recipe drilling program applicable to all prospects. It is, rather, an explanation of ingredients to consider applying when faced with drilling-related issues and hazards that act as barriers to conventional problem-solving methods.

MPD's specialized equipment and techniques to practice its variations safely and effectively have evolved since the mid-1960s on thousands of US land-drilling programs and are considered status quo by many who pioneered the root concepts. Compared to conventional drilling programs, MPD applications have established a commendable well-control incident track record.

Because MPD addresses NPT, the technology is of greatest potential benefit to offshore drilling programs in which cost of lost drilling time is much higher than onshore. Although MPD has been practiced safely and efficiently from all types of offshore rigs and producing the desired results in the process, it is still considered a relatively *new* technology in marine environments. Therefore, applications in marine environments will be the primary focus of this discussion.

It was not until 2003 that the enabling characteristics of the technology began to be more fully appreciated by offshore drilling decision makers. MPD is a technology that addresses a litany of drilling-related issues or barriers to conventional methods. The encounter of *drilling trouble zones* is undeniably on the increase. This is due in part to a requirement to drill in greater water depths and through depleted zones or reservoirs. And, as many would argue, most of the *easy* prospects in shallow and deep waters have already been drilled. Those remaining are more likely to be *hydraulically challenged*, requiring more precisely controlled management of the wellbore pressure profile to be drilled safely and efficiently.

A precursor to MPD technology is Chapter 9.1, Underbalanced Drilling Operations. Many of the tools required for the practice of MPD are illustrated in that section.

Drilling-related issues such as excessive mud cost, slow rate of penetration (ROP), wellbore ballooning/breathing, kick-detection limitations, difficulty in avoiding gross overbalance conditions, differentially stuck pipe, twistoffs, and resulting well-control issues contribute to defining the offshore industry's need for MPD technology. Kick-loss scenarios that frequently occur when drilling into narrow or relatively unknown downhole pressure environments also define a requirement to deviate from conventional methods. Excessive drilling flat time and health, safety, and environment (HSE) issues further indicate the necessity for a technology that addresses the root causes.

Most *drilling hazards* have one thing in common: they all may be addressed to some degree or other by drilling with more precise wellbore pressure management.

A key characteristic of MPD concepts is that they enable the drilling decision maker to view the circulating fluids system as one would a pressure vessel. With the mud in the hole at the time, a much wider range of adjustments to the equivalent mud weight (EMW) may be made with little or no interruption to drilling progress.

MPD technology is synergistic with and complements other drilling-hazard-mitigation technologies such as drilling with liner/casing and expandable tubulars (Galloway, 2003).

The predominance of the text speaks to drilling for conventional oil and gas. However, MPD technology has unique applications to coalbed methane, geothermal resources, and drilling for commercial quantities of methane hydrates.

9.3.2 Closed and Pressurizable Fluids Systems. A closed and pressurizable circulating mud system in its most basic configuration includes a rotating control device (RCD), dedicated drilling choke, and drillstring nonreturn valves [e.g., floats. (Bourgoyne 1997; Rehm 2003)]. The RCD is the key enabling tool for a closed-loop circulating fluids system, and the technologies based upon that concept have evolved in harmony with the evolution of its numerous onshore and offshore designs (Hannegan 2001).

Onshore and offshore applications of RCDs from fixed rigs such as jack-up and platform-mounted rigs often use *surface models* that usually are mounted atop or on the head of a typical blowout preventer (BOP) stack (Hannegan and Wanzer 2003). One RCD design allows its bearing and seal assembly to be remotely latched within a fixed rig's existing bell nipple. Another allows the RCD to be secured within the rig's existing marine diverter or within a dedicated annular or pipe ram BOP.

Floating rigs such as semisubmersibles and drillships use RCD designs that may be configured to be atop a typical marine riser in the moon pool area (Terwogt et al. 2005). A recently introduced design facilitates *docking* the RCD bearing and annular seal assembly in the upper marine riser system, typically under the upper telescoping slip joint. This design requires minimum modifications to the rig's conventional mud-return system and enables rapid transition from conventional returns to pressured returns, and vice versa. All RCD designs for floating rigs incorporate flexible flowlines to compensate for the relative movement between the rig and the riser.

Subsea RCD designs are applicable to riserless drilling, with or without riserless mud recovery, and to several variations of dual-gradient drilling with a marine riser system.

9.3.3 Background. In the 1960s, RCDs enabled the practice of drilling with compressible fluids (gas, air, mist, and foam) to flourish. Now referred to as performance drilling (PD) or simply *air drilling*, value is realized primarily in the form of improved penetration rates, longer life of drilling bits, and reduced overall costs of drilling the prospect.

In the 1990s, a relatively wide range of sizes, designs, and pressure containment capabilities of RCDs evolved to meet growing industry demand. Their usage became a key enabler of underbalanced drilling (UBD) with mud and nitrified fluids, and the benefit was improved productivity from easily damaged reservoirs.

Over time, other uses of the RCD evolved—uses other than *air drilling* and *underbalanced operations*. The industry learned to use the RCD to more precisely manipulate the annular hydraulic pressure profile when drilling with a conventional mud system. It also enabled one to drill safely with an EMW nearer the reservoir pore pressure. Although an influx of hydrocarbons during the drilling process is not invited, one is better prepared to safely and efficiently deal with any that may be incidental to the operation. In 2003, the assortment of techniques was recognized as a technology within itself and given the label *managed-pressure drilling* (Hannegan 2001).

Today, about 45% of all wells drilled onshore in the US use PD techniques. About 5% utilize UBD to improve drilling performance or well productivity, respectively. MPD techniques are used on another approximately 25% for the purpose of improving drillability, reducing NPT, and enhancing well control. The balance is drilled with a conventional open-to-atmosphere mud-return system. In other words, three of every four US land-drilling programs drill at least one section of the wellbore using an RCD.

9.3.4 UBD, PD, and MPD: Similarities and Differences. As mentioned, closed, pressurizable mud-return systems are required to practice UBD, PD, and MPD. This commonality has naturally produced the question, *specifically how do these three technologies differ?*

UBD is drilling with an EMW intentionally designed and maintained below adjacent wellbore reservoir pressures to invite fluid influx. The primary objective of UBD is to enhance well economics and asset exploitation strategies by reducing drilling-induced formation impairment, and by providing reservoir characterization. Typically a full surface complement of separation and flaring equipment is required to properly handle hydrocarbons produced during the drilling process. Nitrified fluids are required on some UBD prospects to achieve an EMW less than the pore pressure of the zone of interest (Bennion 1999).

PD is applying air, mist, or foam drilling-fluid systems to drill with subhydrostatic wellbore pressures to increase drilling ROP and extend bit life. The primary objective of PD is to optimize drilling economics. PD is mostly applicable to drilling in hard rock.

MPD does not invite influx, but one is equipped to contain any that may be incidental to the operation. The objective is to benefit from strengths of UBD technology and some tools developed to practice it safely, to more precisely manage the pressure profile throughout the wellbore. The primary focus is upon mitigating drilling hazards:

- Avoid exceeding the wellbore fracture gradient by drilling with a lighter-than-conventional-wisdom drilling fluid.
- Maintain the EMW above the pore pressure of the formation being drilled, even when the formation is perhaps hydrostatically underbalanced.
- Apply a method of backpressure when shut in to make jointed pipe connections [e.g., the shut-in bottomhole pressure (BHP) is very close if not identical to the BHP when circulating and drilling ahead].
- Contain and control any influx that may be incidental to the operation with a closed and pressurizable mud-return system.
- Apply conventional well-control principles.

In summary, a primary objective of UBD is to avoid damaging a pay zone's ability to produce. *UBD is generally reservoir-issue-related.* PD is for the primary purpose of improving penetration rate. *PD is generally drilling-performance-related.* A primary objective of MPD is to address a litany of drilling-related problems or barriers to economic drillability. *MPD is generally drilling-issue related.*

9.3.5 IADC Definition of MPD. Today, many thousands of onshore drilling programs and a rapidly growing number of offshore programs have proved that drilling with a closed and pressurizable mud-return system enables more precise wellbore pressure management. In recognition of the fact that interest in the various MPD techniques was growing quickly, a subcommittee of the International Association of Drilling Contractors (IADC) Underbalanced Operations (UBO) Committee was established in 2003 as a technology transference vehicle. In 2004, the name of the committee was changed to IADC UBO & MPD Committee. A deliverable was a definition that describes the essence of the technology. (IADC UBO & MPD Glossary of Terms 2008). Subsequent SPE Applied Technology Workshops, regulatory bodies, and the industry at large have further institutionalized the resulting definition below:

> MPD means an adaptive drilling process used to control precisely the annular pressure profile throughout the wellbore. The objectives are to ascertain the downhole pressure environment limits and to manage the annular hydraulic pressure profile accordingly. MPD is intended to avoid continuous influx of formation fluids to the surface. Any flow incidental to the operation will be safely contained using an appropriate process.

Technical Notes:

1. MPD process employs a collection of tools and techniques which may mitigate the risks and costs associated with drilling wells that have narrow downhole environmental limits, by proactively managing the annular hydraulic pressure profile.
2. MPD may include control of back pressure, fluid density, fluid rheology, annular fluid level, circulating friction, and hole geometry, or combinations thereof.
3. MPD may allow faster corrective action to deal with observed pressure variations. The ability to control annular pressures dynamically facilitates drilling of what might otherwise be economically unattainable prospects.

9.3.6 Methane Hydrates Drilling Applications of MPD. It is perhaps an oversight that temperature management was not mentioned in the above definition of MPD. For example, MPD technology is uniquely applicable to drilling for commercial quantities of methane hydrates (Hannegan et al. 2004). Methane hydrates are governed by Boyle's law, the pressure and temperature relationships of gases.

Methane hydrates exist in huge quantities on the slopes of continental shelves, where pressures and temperatures are suitable for methane and water to coexist. They are also found onshore in permafrost regions. They exist naturally as a frozen crystalline lattice structure consisting of molecules of water that have formed an open, cage-like lattice that encloses molecules of methane. Depending upon the purity of the methane hydrate, it can contain between 70 and 164 times its volume of free gas at standard temperature and pressure. Temperature management plus precise pressure management is required to drill for commercial quantities of methane hydrates without a well-control incident (Todd et al. 2006).

9.3.7 Conventional Hydraulics and Hydraulics Manipulation. To appreciate the potential for MPD, it is important for one to respect the limitations of conventional circulating-fluids systems when facing a difficult-to-drill prospect. A conventional mud-return system is typically open to the atmosphere under the rig floor. Mud and drilled cuttings gravity-flow away from beneath the rig floor to the shale shakers and mud pits.

BHP or EMW is determined by the sum of the mud weight hydrostatic head [$HH_{(MW)}$] and circulating annulus friction pressure (AFP):

$$BHP = HH_{(MW)} + AFP$$

AFP is present when drilling ahead and ceases to exist when the mud pumps are off, such as when making jointed pipe connections. Therefore, the dynamic BHP is significantly higher than the BHP when the fluids system is static. For drilling to progress, the EMW must remain within the formation pressure and fracture pressure, typically referred to as *the drilling window*. See **Fig. 9.3.1.**

Fig. 9.3.2 illustrates that when drilling within narrow or relatively unknown formation pore-pressure and fracture-gradient margins, kick-loss scenarios often occur. Shutting the rig's mud pumps off to make a jointed pipe connection may allow an influx of hydrocarbons into the wellbore, an event that interrupts drilling progress to be circulated out and produced to the surface. In this case, the HH alone results in a BHP that is less than the pore pressure of the formation at that depth. Alternatively, if the hydrostatic mud weight is near the

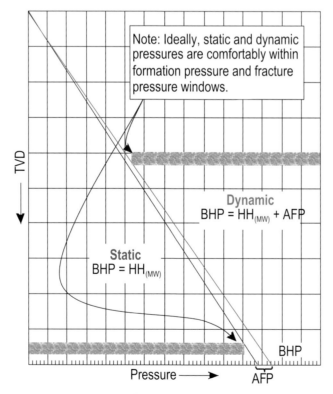

Fig. 9.3.1—Conventional circulation.

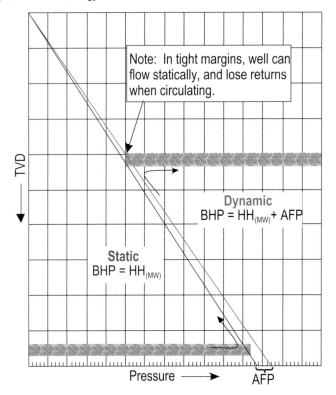

Fig. 9.3.2—Conventional circulation in tight margins.

fracture pressure, starting the rig's mud pumps to regain circulation may result in mud losses into the formation. In this event, the sum of HH and AFP results in an EMW that exceeds the fracture pressure. Excessive mud cost, differential sticking, risk of twistoff, resulting well-control issues, and/or loss of hole may occur.

A major limitation of conventional drilling is the fact that with the mud in the hole at the time, the only way to make easy adjustments to the effective wellbore pressure profile is by adjusting the speed of the rig's mud pumps.

9.3.8 Categories of MPD. There are two categories of MPD:

- *Reactive.* One is prepared to practice MPD as a contingency. A conventional-wisdom well construction and fluids program is planned, but the rig is equipped with at least an RCD, choke, and drillstring float(s) as a means to more safely and efficiently deal with, for example, unexpected downhole pressure environment limits (e.g., the mud in the hole at the time is not best suited for the drilling window encountered). For example, of the one-in-four US land-drilling programs practicing MPD, many are practicing the reactive-category MPD. As a means of preparing for unexpected developments, the drilling program is equipped or *tooled up* from the beginning to deal more efficiently and safely with downhole surprises. This, in part, explains why some underwriters require that wells they insure be drilled with a closed and pressurizable mud-return system.

    It should be noted that when an MPD contingency is incorporated into a conventional-wisdom drilling program; planning, training, HazId/HazOp processes, equipment pre-qualification and applicable regulatory permitting should be considered as if MPD were the primary program (US MMS 2008).

- *Proactive.* The drilling program is designed from the beginning with a casing, fluids, and openhole drilling plan and/or alternate plans that take full advantage of the ability to more precisely manage the wellbore pressure profile. This *walk the line* category of MPD technology offers the greatest benefit to both onshore and offshore drilling programs. Most offshore applications to date have been of this category. Of significance is the fact that a growing percentage of land MPD programs are transitioning from reactive to proactive MPD. This shift requires that the wells be pre-planned more thoroughly, but the benefits to the drilling program typically more than offset the cost of the additional MPD engineering and project management.

9.3.9 Variations of MPD. There are four key variations of MPD. Each is addressed in the context of the drilling hazards to which it has proved applicable. Occasionally, combinations of variations are practiced on the same challenging prospect. Combining several variations on the same prospect is expected to become more frequent as the technology becomes more status quo in the minds of drilling decision makers and as prospects become increasingly more difficult to drill.

1. ***Returns Flow Control.*** The objective is to drill with a closed annulus return system for HSE reasons only. For example, a conventional production platform drilling operation with an open-to-atmosphere system may allow explosive vapors to escape from drilled cuttings and trigger atmospheric monitors and/or automatically shut down production elsewhere on the platform. Other applications of this variation include toxicological ramifications of drilling with fluids emitting harmful vapors onto the rig floor, as a precaution wherever there is a risk of a shallow-gas hazards, and when drilling in populated areas. Typically only an RCD is added to the drilling operation to accomplish this variation.
2. ***Constant BHP (CBHP).*** This method is uniquely applicable to drilling in narrow or relatively unknown margins between the pore and fracture gradients. The objective is to maintain a constant EMW, whether the rig's mud pumps are on or off. Typically, a lighter-than-conventional-wisdom fluids program is implemented, nearer balanced, perhaps even hydrostatically underbalanced. When shut in to make jointed pipe connections, surface backpressure (BP) contributes to the HH pressure to maintain a desired degree of overbalance, preventing an influx of reservoir fluids.

   Fig. 9.3.3 illustrates a typical application of CBHP MPD; the choke is open when drilling ahead and closed when making jointed pipe connections. The amount of BP required when not circulating is roughly equivalent to the amount of AFP when circulating.

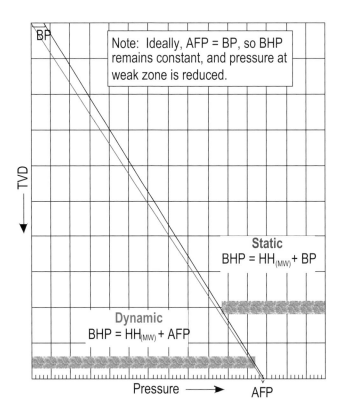

Fig. 9.3.3—CBHP MPD.

Fig. 9.3.4 illustrates a limitation to the application of BP on the uncased hole or casing shoe(s) above. In this case, the amount of BP applied would cause lost returns at a casing shoe.

3. ***Dual Gradient and Deepwater Dual Gradient.*** When a wellbore is exposed to two or more different depth-vs.-pressure gradients in its annulus returns path, it is referred to as *dual-gradient MPD*. See **Fig. 9.3.5**. This may be accomplished by injecting a less dense fluid from a parasite string attached to the casing or marine riser into the annulus at some predetermined depth. Introducing air, inert gas, light liquids, cuttings-free mud, or solids decreases the fluid density from that point to the surface. Offshore, this can be accomplished several ways. For example, a less dense medium can be injected into the marine riser as a means to adjust BHP without changing the base MW.

Another method of achieving a deepwater dual gradient is to artificially lift returns from the seabed back to the surface with subsea pumps and through dedicated return lines external to the drilling riser (Smith 2001; Eggemeyer 2001). In this case, the drilling riser may be filled with seawater to prevent collapse. The intent is not to reduce the EMW or effective BHP to a point less than formation pore pressure. Instead, the intent is most often to avoid gross overbalance and not exceed the fracture gradient. In both cases, the wellbore is exposed to two fluid densities: one below the point of injection or pumping, and one above that point. This form of MPD can be practiced with or without a subsea RCD, although there are advantages to having the subsea RCD (Forrest et al. 2001). In the case of gas injection into the riser, a surface RCD must be run.

4. ***Pressurized-Mudcap Drilling (PMCD).*** PMCD is a method of dealing with severe or near-total lost circulation. A predetermined column height of heavy mud, perhaps the rig's kill fluid, is pumped down the *backside* via the RCD's *fill-up line connection* (Terwogt 2005). This *mud cap* serves as an annulus barrier, preventing returns to surface. Varying amounts of surface BP is used to assure the ability of the mud cap to prevent returns to surface and to minimize changes to the mud cap column height as drilling progresses. See **Fig. 9.3.6**.

Typically a lighter and non-damaging fluid, perhaps seawater, is the sacrificial fluid used for PMCD. Mud and cuttings are *single passed* or forced into the otherwise troublesome zone above the loss circulation

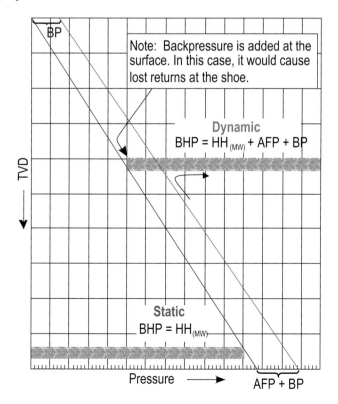

Fig. 9.3.4—Limitations of CBHP MPD.

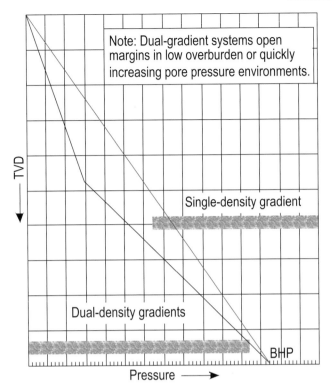

Note: Dual-gradient systems open margins in low overburden or quickly increasing pore pressure environments.

Single-density gradient

Dual-density gradients

BHP

TVD

Pressure ⟶

Fig. 9.3.5—Dual-gradient MPD.

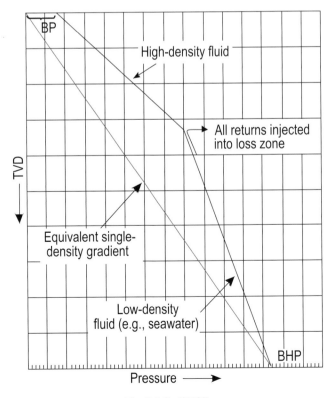

BP

High-density fluid

All returns injected into loss zone

Equivalent single-density gradient

Low-density fluid (e.g., seawater)

BHP

TVD

Pressure ⟶

Fig. 9.3.6—PMCD.

hazard. Lighter fluid increases the ROP. Mud sacrificed into the depleted zone is less expensive than conventional mud, and control of the well is enhanced. In the reservoir, there is less invasive damage to the well's ultimate productivity. PMCD may also be considered when the prospect is sour hydrocarbons and in deepwater applications requiring drilling through severely depleted formations to access deeper objectives. Once past the loss circulation zone and the section of interest drilled, some prospects may not successfully receive the kill fluid required to trip out the drillstring (Urselmann et al. 1999). This explains why many applications of PMCD incorporate the use of a casing isolation or downhole deployment valve (Sutherland et al. 2005).

9.3.10 MPD—Hydraulics and Hydraulics Manipulation. A closed-loop circulating fluids system is utilized in most applications of MPD. One may consider MPD to be a lesson learned from the process industries, which have been emphasizing for decades the many virtues of avoiding systems that are open to the atmosphere (Hannegan 2001).

MPD adds another variable element to the equation for determining effective BHP or EMW. One has the ability to apply desired amounts of hydraulic BP on the circulating-fluids system. When drilling with an essentially incompressible fluid, BP adjustments result in an almost instant change in effective BHP:

$$\text{MPD BHP} = \text{HH}_{(MW)} + \text{AFP} + \text{BP}$$

BP applications are of course critical to conventional well-control methods. This explains in part why MPD is considered an *advanced form of well control.*

In some adaptations of dual-gradient MPD, the BP is applied via speed of subsea annulus returns pumps, and not from the surface (Forrest et al. 2001).

9.3.11 The Offshore Environment. It is estimated that at least half of all offshore *prospects* are economically undrillable with conventional drilling equipment and methods. The percentage is destined to increase over time as reservoir pressures deplete and as requirements to drill in deeper water increase.

A major contributor to uneconomic drilling prospects is excessive costs stemming from drilling-related problems and barriers. Drilling-related situations such as lost circulation, occurrences of differentially stuck pipe, twistoffs, many mud density changes to total depth (TD), and kick-loss issues associated with narrow drilling windows contribute significantly to a growing number of prospects that are destined to exceed the drilling program's authorization for expenditure.

All of the aforementioned drilling-related challenges have one thing in common: They indicate a requirement for more precise wellbore pressure management, containment, and control, with fewer interruptions to drilling progress. Drilling in marine environments with more precise pressure management throughout the wellbore can address to varying degrees a significant number of these conventional challenges.

The Size of the Prize? A recent study has quantified the causes of NPT on drilling programs in the US Gulf of Mexico (GOM). Approximately 22% of total days from spud to TD objective are NPT. Lost circulation, resulting differentially stuck pipe and twistoffs, kick-loss scenarios associated with narrow downhole environmental limits, and flat time associated with interrupting the drilling program to increase or decrease the MW numerous times account for almost half of the total amount of NPT (Problem Incidents 2003). The balance is nondrilling-related, such as *wait on weather* or rig equipment failure. To varying degrees, the lost drilling days in the GOM and their causes are mirrored globally.

Perhaps as much as 10 to 15% of *borderline* drilling prospects could be drilled economically if half of the total NPT could be addressed with offshore applications of MPD technology. A number of otherwise undrillable prospects have been drilled successfully, increasing the operators' recoverable assets.

9.3.12 Opportunities With MPD. All variations of MPD involve the management of one or more of the following:

- The ability to quickly change the EMW of the mud in the hole at the time
- Use of otherwise negative characteristics of friction loss to an advantage
- Altering the fluid density
- Wellbore strengthening

9.3.13 Concepts Spawned by the Variations of MPD. Inspired by the ability to drill with a pressurized mud system, the industry has developed a number of specialty applications:

- **Riserless MPD.** This is *riserless pumping and dumping* with subsea well control. A subsea rotating device is used when establishing a subsea location via riserless drilling with seawater or other fluids compatible to be discharged onto the seafloor. The purpose of this technique typically is to establish deepwater locations by batch drilling. Because there is no marine riser and subsea BOP to buoy, smaller and less expensive rigs can be used to establish locations in water depths greater than those for which the rig was originally intended. A remotely operated vehicle (ROV) or subsea automatic choke adjusts BP at the flowline outlet of the subsea RCD. Closing the subsea choke increases BHP, virtually as if the subsea location were being drilled with a marine riser filled with mud and cuttings. Result: A degree of overbalance greater than the drilling fluid would impart and beneficial for subsea well control in the presence of shallow water flow or shallow gas hazards. See **Fig. 9.3.7.**
- **Dual-Gradient Riserless Drilling.** This is also known as *riserless mud recovery.* A subsea pump returns mud and cuttings to the rig for analysis and proper handling. Effective BHP may be adjusted via subsea annulus BP and speed of both the rig and subsea pump(s). See **Fig. 9.3.8.**
- **Compressible-Fluids MPD.** The concept of more precise wellbore pressure management has application to air, mist, foam, and gas drilling mediums. An example is a downhole air diverter subbed into the drillstring. The tool responds to a preset differential between drillstring and annulus pressure. An amount of cuttings-free compressed air is diverted into the annulus. Improved hole cleaning and a corresponding decrease in BHP increases the differential pressure across a percussion hammer, typically improving its performance. ROP increases and in some cases allows drilling to a greater depth in a *wet hole* than otherwise possible.
- **Deepwater Surface BOP Application of MPD.** The initial purpose of drilling from a moored semisubmersible or dynamically positioned drillship with a surface BOP was to enable wells to be drilled in water depths greater than the depth rating of the rig when using a subsea BOP stack (Gallager and Bond

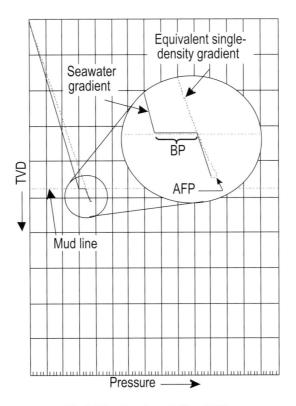

Fig. 9.3.7—Riserless drilling MPD.

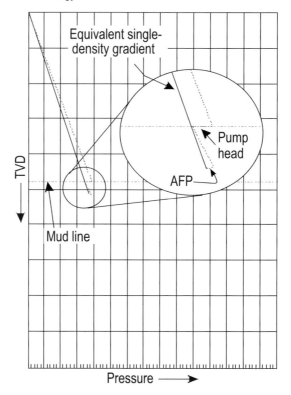

Fig. 9.3.8—Riserless dual-gradient MPD.

2001). However, drilling with a surface BOP enables many of the same MPD technologies otherwise available only to fixed rigs to be exploited in deep water. High pressure and usually smaller-diameter casing serves as the marine riser (Leach et al. 2002).

- *Continuous-Circulating Concentric Casing MPD.* This process involves more precise and almost instantaneous BHP management by using hydraulic friction control on the return annulus through continuous annular fluid circulation. By pumping additional volumes of drilling fluid through a concentric casing or drillstring, the bottomhole AFP is manipulated into seeing a more steady-state condition. By increasing the annular fluid rate down the concentric casing during connections by a volume equal to the normal standpipe rate, the downhole environment in the wellbore sees a more constant AFP. Pressure spikes typically associated with making jointed pipe connections may be eliminated or reduced significantly with this method.

- *Wellbore-Strengthening MPD.* In the early 1990s, work was done to investigate the impact of strengthening the wellbore by maintaining a sized solids content in the mud, which effectively plugged the microfractures that occurred in weaker formations as the mud density was increased. While this is not MPD in the sense of requiring a closed and pressurizable mud-return system, it achieves similar goals by widening the margin between pore pressure and fracture pressure in the wellbore. At the time of this writing, there is renewed interest in this technique.

- *Downhole Pumping MPD.* A newly emerging variation of MPD is through the use of a drilling-fluid-powered pump in the drillstring and within the casing that adds energy to the annulus fluid returns. Such an *ECD reduction tool* has the effect of creating a significant change in differential pressure at the point of the pump, reducing or eliminating the impact of the friction pressures on the BHP. See **Fig. 9.3.9.**

- *Applications of Hydraulic Flow Modeling and Process Control Computers.* Process industries (e.g., chemical, refineries, pulp and paper) have benefited from the use of closed and pressurizable systems for many decades. Today, few are found to be operating those systems without the aid of process-control computers. Greatly improved safety, more consistent product quality, reduction of waste, lower energy consumption, and a more positive environmental impact have resulted.

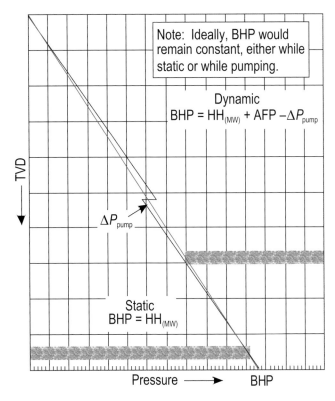

Fig. 9.3.9—Downhole pumping MPD.

The closed-loop characteristics of MPD have enabled the development of a technology that enables time- and temperature-corrected mass balance accuracy of measurement and analysis of flow and pressure profile data. The technology enables control of the circulating-fluids system with process-industries capability. Very small amounts of fluids influx and mud losses are detected, allowing the actual, not predicted, drilling window to be revealed and responded to safely, efficiently, and typically with less NPT. Such *micro-flux control* is particularly applicable to CBHP MPD and other drilling programs in which early kick detection is advantageous.

9.3.14 MPD Tool Chest. A number of technologies have evolved to enable the practice of MPD from all types of rigs, and others have been enabled or fostered by its variations and applications:

- Surface and subsea RCD designs for all types of rigs with pressure containment capabilities up to 5,000 psi and for all applications of MPD
- Wireline-retrievable drillstring floats
- Manual, semiautomatic, and process-controlled choke manifolds
- Casing isolation valves [downhole deployment valve (DDV)]
- Nitrogen production units
- Subsea mud-return pumps
- Surface mud logging
- Real-time pressure and flow-rate monitoring
- MPD hydraulics software modeling
- Continuous flow drillsstring subs (continuous flow valves)
- Continuous circulating systems
- ECD reduction tools
- MPD candidate selection

9.3.15 Status of Offshore Adoption of MPD. Of the three controlled pressure technologies (air drilling, UBD, and MPD), offshore drilling decision makers have embraced the last most robustly. Air drilling is not applicable in marine environments. UBD faces regulatory prohibition in some jurisdictions, lack of space on most offshore rigs for the required equipment, and limited ability to handle produced hydrocarbons. MPD, on the other hand, has caused many in the industry to rate it second only to horizontal and directional drilling as having promise to be one of the most influential technologies over the next 20 years.

With data from a significant number of case histories globally, a risk assessment study funded by a joint industry project (DEA 155 2008) concluded the following about applications of MPD from all types of off-shore rigs:

- "Properly applied, has a high probability of mitigating most, if not all, drilling-related risks."
- "Continues to demonstrate a bright future."
- "Is as safe as or safer than current conventional drilling techniques."
- "Is a sophisticated form of well control and deserves a balanced quality appraisal of risks–positive and negative."

9.3.16 Summary. Mother Nature has been generous in allowing the upstream industry to access copious quantities of hydrocarbon resources with the relatively simple and century-old hydraulics of a conventional rotary rig circulating-mud system the returns of which are open to the atmosphere under the rig floor. Conventional methods are rooted in the principle of drilling with weighted mud and its hydrostatic head pressure at that depth. The only way to change the effective wellbore pressure profile (i.e., EMW) with the mud in the hole at the time is to change the rate of circulation.

One of the biggest advantages of drilling with a pressurizable circulating-fluids system is its enabling of a step change in accuracy in flow rate and pressure monitoring. This allows real-time monitoring, as well as the ability to proactively control deviations from expected behavior of flow and pressure. This awareness fosters quick reaction that prevents normal drilling problems from escalating, therefore reducing NPT.

A tremendous number of improvements to drilling fluids, rigs, HSE issues, electronic sophistication, automation, materials, reservoir characterization, and analysis of prospects has evolved over the past 7 or 8 decades. However, most of the offshore industry today is still drilling with basic circulating-fluids system concepts that have been around since the first offshore well was drilled. More upstream technologists are arriving at the conclusion that this conventional wisdom has serious limitations when applied to the prospects that will have to be drilled in the future, those that are more *hydraulically challenged.*

Historically, many have associated drilling with closed and pressurizable circulating-fluids systems with UBD and air/mist/foam drilling and with a requirement for an RCD when drilling with jointed pipe is fairly obvious. In part, this explains why offshore-drilling decision makers have, until fairly recently, paid little attention to the RCD and consequently to the concept of drilling with pressurized mud systems in general.

MPD does not invite influx during the drilling process. Instead, the ability to apply varying degrees of BP at will is the key enabling technology to contain and safely control any influx that may be incidental to the MPD operation.

The use of an RCD, choke manifold, and drillstring floats has, more or less, become status quo in onshore operations, particular on US land programs. Downhole deployment valves are rapidly being accepted as an integral component of the basic *MPD kit* for their ability to assist in avoiding slow tripping times and swab/surge pressure issues. When practicing PMCD and once past the severe loss circulation hazard necessitating the method, a DDV is invaluable as a means of tripping out a drillstring within a cavity that would be very expensive or impossible to fill with kill fluid.

The practice of drilling sections of the wellbore with closed and pressurizable fluids systems has increased from 15% of US land-drilling programs in 1996 to 75%.

Compared to drilling with a conventional circulating-fluids system, a positive HSE and reportable well-control incident track record has been firmly established for MPD. A properly planned and executed MPD program actually has less potential for hydrocarbons to be produced to the surface during the drilling process than if the same prospect were to have been drilled conventionally.

A growing number of upstream technologists are of the opinion that MPD is the way that all wells should be drilled. The stage is set for the significant uptick in offshore MPD applications to continue its momentum in the minds of offshore-drilling decision makers.

9.3.17 Conclusion. Today, a growing percentage of prospects are economically undrillable with conventional, technology and the trend is unstoppable. MPD offers a means to increase economic drillability and do so short of producing formation fluids to the surface in the process. The practice of MPD has been documented to increase recoverable offshore assets. MPD has a commendable onshore track record that is being repeated in marine environments.

MPD technology is a more readily acceptable step change toward the inevitable requirement to drill underbalanced. The closed and pressurizable circulating-mud system characteristic of MPD is a necessary enabler of process-control operated pressure management.

References

API-16RCD, *Specification for Drill-Through Equipment-Rotating Control Devices*, first edition. 2005. Washington, DC: American Petroleum Institute.

Bennion, D.B. 1999. UBD Technology Offers Pathway to Solving Complex Reservoir Problem. *The American Oil & Gas Reporter* **42** (2): 49–54, 84.

Bourgoyne, A.T. Jr. 1997. Well Control Considerations for Underbalanced Drilling. SPE 38584 presented at the SPE Annual Technical Conference and Exhibition, San Antonio, Texas, 5–8 October. DOI: 10.2118/38584-MS.

DEA 155 Joint Industry Project: *A Probabilistic Approach to Risk Assessment of Managed Pressure Drilling in Offshore Applications.* 31 October 2008. Drilling Engineering Association: Final Report.

Eggemeyer, J.C., Akins, M.E., Brainard, R.R. et al. 2001. SubSea MudLift Drilling: Design and Implementation of a Dual Gradient Drilling System. Paper SPE 71359 presented at the SPE Annual Technical Conference and Exhibition, New Orleans, 30 September–3 October. DOI: 10.2118/71359-MS.

Forrest, N., Bailey, T., and Hannegan, D. 2001. Subsea Equipment for Deep Water Drilling Using Dual Gradient Mud System. Paper 67707 presented at the SPE/IADC Drilling Conference, Amsterdam, 27 February–1 March. DOI: 10.2118/67707-MS.

Gallagher, K.T. and Bond, D.F. 2001. Redefining the Environmental Envelope for Surface BOPs on a Semi-Submersible Drilling Unit. Paper 67709 presented at the SPE/IADC Drilling Conference, Amsterdam, 27 February–1 March. DOI: 10.2118/67709-MS.

Galloway, G. 2003. Rotary Drilling With Casing—A Field Proven Method of Reducing Wellbore Constructions Cost. Paper WOCD-0306-02 presented at the World Oil Casing Drilling Technical Conference, Houston, 6–7 March.

Hannegan, D. 2001. UBD Tools and Technology Solve Conventional Deepwater Drilling Problems. Paper AADE 01-NC-HO-01 presented at the AADE National Drilling Conference, Houston, 27–29 March.

Hannegan, D. and Wanzer, G. 2003. Well Control Considerations—Offshore Applications of Underbalanced Drilling Technology. Paper SPE 79854 presented at the SPE/IADC Drilling Conference, Amsterdam, 19–21 February. DOI: 10.2118/79854-MS.

IADC UBO & MPD Committee Meeting Minutes. 2004a. IADC Underbalanced Drilling Operations Committee Meeting–1st Quarter, Leiden, The Netherlands, 15–16 March.

IADC UBO & MPD Committee Meeting Minutes. 2004b. IADC Underbalanced Drilling Operations Committee Meeting, Houston, 13–14 December.

Leach, C., Dupal, K., Hakulin, C., Fossli, B., and Dech, J. 2002. Design of a Drilling Rig for 10,000 ft Water Depth Using a Pressured Riser. Paper SPE 74533 presented at the IADC/SPE Drilling Conference, Dallas, 26–28 February. DOI: 10.2118/74533-MS.

Malloy, Kenneth P., Stone, C. Rick, Medley, George H. Jr. 2009. Managed Pressure Drilling: What It Is and What It Is Not. IADC/SPE Paper 122281-PP presented at the IADC/SPE Managed Pressure Drilling and Underbalanced Operations Conference and Exhibition, San Antonio, Texas, 12–13 February.

MPD & UBO Glossary of Terms. International Association of Drilling Contractors, http://www.iadc.org/committees/ubo_mpd/Documents/UBO%20&%20MPD%20Glossary%20Jan08.pdf. Downloaded August 2008.

Problem Incidents—GOM Shelf Gas Wells: Wellbores Drilled 1993–2002; Water Depth <600 Feet. 2003. Houston: James K. Dodson Company.

Rehm, W. 2003. *Practical Underbalanced Drilling and Workover.* Austin, Texas: Petroleum Engineering Extension, University of Texas at Austin.

Reitsma, D. 2007. Development and Application of Combining a Real-Time Hydraulics Model and Automated Choke To Maintain a Relatively Constant Bottomhole Pressure While Drilling. Paper SPE 10708 presented at the International Petroleum Technology Conference, Doha, Qatar, 21–23 November.

Smith, K.L., Gault, A.D., Witt, D.E., and Weddle, C.E. 2001. SubSea MudLift Drilling Joint Industry Project: Delivering Dual Gradient Drilling Technology to Industry. Paper SPE 71357 presented at the SPE Annual Technical Conference and Exhibition, New Orleans, 30 September–3 October. DOI: 10.2118/71357-MS.

Sutherland, I. and Grayson, B. 2006. DDV Reduces Time To Round-Trip Drillstring by Three Days, Saving £400,000. Paper SPE 92595 presented at the SPE/IADC Drilling Conference, Amsterdam, 23–25 February. DOI: 10.2118/92595-MS.

Terwogt, J., Makiaho, L.B., van Beelen, N., Gedge, B.J., and Jenkins, J. 2005. Pressurized Mud Cap Drilling From a Semi-Submersible Drilling Rig. Paper SPE 92294 presented at the SPE/IADC Drilling Conference, Amsterdam, 23–25 February. DOI: 10.2118/92294-MS.

Urselmann, R., Cummins, J., Worral, R.N., and House, G. 1999. Pressurized Mud Cap Drilling: Efficient Drilling of High-Pressure Fractured Reservoirs. Paper SPE 52828 presented at the SPE/IADC Drilling Conference, Amsterdam, 9–11 March. DOI: 10.2118/52828-MS.

US Department of the Interior, Minerals Management Service, GoM Region. 15 May 2008. *Notice To Leaseholders No. 2008-G07.*

Todd, R., Hannegan, D., Harrall, S., 2006. New Technology Needs for Methane Hydrates Production. OTC Paper 18247-PP presented at the Offshore Technology Conference, Houston, 1–4 May.

Hannegan, D., Todd, R., Pritchard, D., Jonasson, B. 2004. MPD–eUniquely Applicable to Methane Hydrate Drilling. Paper SPE 91560-MS presented at SPE/IADC Underbalanced Technology Conference, Houston, 11-12 October. DOI: 10.2118/91560-MS.

9.4 Coiled-Tubing Drilling—Ian Retalic, Andy Laird, and Angus MacLeod, Senergy LEA

9.4.1 Background. Coiled tubing (CT) refers to any continuous string of tubing that can be deployed without making or breaking connections while tripping, and that is spooled on a drum, allowing continuous pumping into the string while deploying the pipe.

The development of modern CT has its roots in a secret Allied project during World War II. Project PLUTO, an acronym for *pipelines under the ocean,* deployed 23 *continuous tubing* pipelines across the English Channel to fuel the invasion forces. Of these, 17 were lead pipelines deployed to a total length of 30 miles. The remaining six pipelines were constructed by butt-welding 20-ft joints of 3-in. internal diameter (ID) steel pipe. These tubulars were spooled onto 40-ft-diameter reels and deployed by towing the reel behind a ship (**Fig. 9.4.1**).

It was not until 18 years later, however, that Bowen Tools (IRI International) and the California Oil Company developed a prototype CT workover unit with the injector head designed as a vertical, counterrotating chain drive system built onto a string of 1.315 in.-outside-diameter (OD) pipe (NOV CTES 2005).

Since its introduction in 1963, CT has been heralded as a technology that has the potential to revolutionize the oil field. Unfortunately, early mechanical failures, mainly due to quality issues with the manual butt-welding process and low-strength pipe, resulted in numerous fishing operations and promoted an industry-wide hesitancy to use the technology. The added high oil prices, and the oil industry's unwillingness to adopt change, have limited the growth of CT. Over the past 10–15 years, however, interest in CT has increased dramatically. The fall of oil prices in 1986 triggered the increased use of, and interest in, cost-saving technologies such as CT.

Worldwide experience of CT drilling (CTD), though rising, is still limited, and CTD is still considered by some as a *new technology.* CTD, therefore, has yet to mature. Consequently, CTD performance has been extremely varied throughout the world over the last 13 years or so. The projects that have the most continued success are those whereby a continuous learning curve can be established and strong management support exists.

Fig. 9.4.1—Project PLUTO coiled pipe (NOV CTES 2005).

The key areas where CTD has been used successfully are Canada (high CTD growth with more than 600 wells drilled or reentered each year), Alaska, Venezuela, Oman, and Sharja. All of these are land locations except Venezuela, where the operations occurred on Lake Maracaibo. The similarity between the operations was that there was an extended period of work, a large number of wells and little mobilization cost. This allowed for a considerable amount of time for learning to take effect.

9.4.2 Capabilities and Limitations. CT offers a number of advantages and unique capabilities over conventional jointed-pipe drilling methods, and it also has a number of disadvantages and limitations over jointed-pipe methods.

CT Capabilities.

- Drill and trip under pressure
- Faster trips (up to 4×)
- Continuous circulation while tripping pipe
- Continuous high-quality two-way telemetry between surface and downhole (with electrical line or fiber optics)
- Slimhole and through-tubing capability (4-in. tubing)
- Portable

CT Limitations.

- Cannot rotate
- Limited fishing capabilities (limited pull, no rotation, etc.)
- Limited reach in horizontal wells due to friction (no rotation)
- Possible (dependent on casing production string size) poor circulating hydraulics and limited hole-cleaning capability
- Limited tube life (due to fatigue through cold working)
- High maintenance
- Higher costs (additional equipment and crew)

Applications for CTD.

- Drill multilaterals from the motherbore
- Drill sidetracks without removing the completion

- Drill underbalanced wells and increase productivity by eliminating reservoir impairment
- Managed-pressure drilling
- Drill wells without a conventional rig
- Drill wells simultaneously with a conventional rig
- Low-cost, quick-move environments

Nonsuitable CTD Applications. There are number of applications for which CT is not best suited. Some of these applications are better drilled using other methods. These are

- Drilling extended-reach wells (in excess of 16,000 ft)
- Drilling 8½-in. or bigger hole sizes (due to hole cleaning and weight transfer issues)
- Drilling wells in unstable formations

9.4.3 Surface Equipment. There are a number of considerations to be accounted for when specifying the equipment requirements. The main requirements and the equipment affected are discussed in this section **(Fig. 9.4.2).**

CT Selection and Design. CT selection and design comes down to a few specific issues:

- Tubing forces
- Hydraulics
- CT fatigue
- Logistical constraints such as crane, roads, deck loading

Designing and selecting a string is an iterative process that generally results in a compromise between hydraulics, tubing forces, and fatigue, while complying with the maximum weight limitations of the location. A detailed sensitivity analysis needs to be performed using different CT diameters, grade, and wall thickness.

An analysis of the tubing forces will establish whether the CT can reach the total depth of the well without lockup. Lockup is the condition in which the coil has passed through the sinusoidal and helical buckling stages, and weight can no longer be transferred to the bottom of the string irrespective of the weight applied at surface.

For CTD, it is recommended to have a minimum of 2,000 lbf weight on bit (WOB) available and a minimum pickup weight of at least 15,000 lbf (80% of tensile yield).

The hydraulic calculations need to be carried out using a program that can account for the additional quantity of tubing at surface for determining surface and pump pressures. It is possible to some extent to be able to model this by increasing the length of the standpipe on a conventional mud hydraulics model; however, this cannot account for the tubing being spooled on a drum, and, therefore, the results will vary from a specific CT hydraulics model.

The optimum result for hydraulic calculations regarding CT size is to minimize the pump pressure and gooseneck pressure that might reduce the CT life cycles and surface equipment reliability. The number of cycles that can be imposed on a CT string depends on a number of factors. The primary parameters are the OD of the string, the wall thickness, the material grade, and the internal pressure while the tubing is traveling over the gooseneck. In general, the larger the CT OD, the smaller the number of cycles available. Similarly, the higher the pressure, the lower the number of cycles available.

All being equal, smaller-diameter CT will result in a higher pump pressure than larger tubing due to frictional loss inside the tubing. Though larger-diameter CT can provide better weight transfer properties and improved hydraulics, the life of larger CT can be substantially worse than smaller strings. This can be offset by the reduction in pump pressure through the coil, but in general, larger strings will always have a lower cycle threshold than a smaller-diameter string.

The most susceptible area for CT fatigue is over the gooseneck and at the reel. The higher the internal pressure while the tubing travels over the gooseneck or onto the reel, the greater the impact to the life of the tubing.

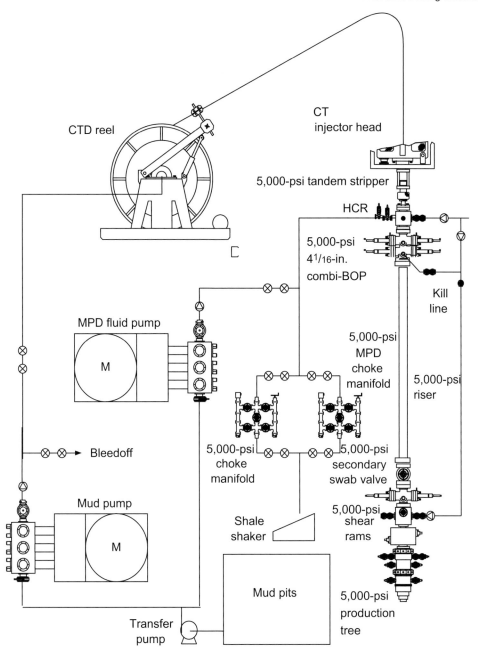

CTD reel

CT
injector head

5,000-psi tandem stripper

HCR

5,000-psi
4 1/16-in.
combi-BOP

Kill
line

MPD fluid pump

M

5,000-psi
MPD
choke
manifold

5,000-psi
riser

Bleedoff

5,000-psi
choke
manifold

5,000-psi
secondary
swab valve

Mud pump

Shale
shaker

5,000-psi
shear
rams

M

Mud pits

Transfer
pump

5,000-psi
production
tree

Fig. 9.4.2—Typical CT spread.

Ideally, when designing a CT string, thicker-walled tubing is placed at the top of the string to provide additional strength to increase the overpull available and to reduce the impact of coil fatigue. The CT contractor closely monitors the number of cycles during the operation, and a coil-cutting regime is used to minimize the life cycles at a particular point in the coil to avoid failure.

The weight of the tubing string and reel has to be taken into consideration for all offshore operations. The weight and overall dimensions need to be considered for land transportation. It may be that the tubing string needs to be tailored to suit the particular application if the logistical limits are approached. It is also important to model the tubing weight including any wireline cable and/or hydraulic umbilicals where installed within the coiled tubing (LEAding Edge Advantage International Ltd. 2002).

There is the potential to inject the cable on the rigsite if further weight reduction is necessary; however, the preference is to perform this offline prior to mobilizing. To assist with the weight reduction, a *drop-in drum* system can be utilized.

If the weight of the tubing cannot be reduced sufficiently, the coil can be transported to location in separate sections and either welded or connected, using a spoolable connector, on site. For offshore locations, the tubing can be spooled onto the platform from a supply boat.

Reel. The CT unit reel, in simplest terms, provides a means of storage for the tubing (**Fig. 9.4.3**). The capacities for a standard reel (width 87 in., core 112 in., and drum OD 180 in.) are listed in **Table 9.4.1.**

The injector head supplies all of the force required to run and pull the CT. The reel only supplies tension on the tubing between itself and the injector head. This provides the ability to smoothly feed the CT off from the reel while running in hole and properly spool the tubing onto the reel while pulling out of hole.

The reel drive system incorporates a motor and brake. This can provide the proper amount of backtension for a given operation or while the injector head control valve is in the neutral position.

During transportation or periods of inactivity, the reel generally is chained back to prevent movement of the tubing.

The reel incorporates a level-wind assembly, which is designed to automatically guide the CT on and off the reel. Within the control cabin, there is also a manual override facility to enable the operator to correct, or prevent, any improper spooling. The height of the level-wind assembly is fully adjustable in order to match the angle that the CT makes between the reel and the injector head. In most instances, there is also a mechanical depth counter positioned on the level-wind assembly, with a friction wheel that contacts the CT. The mechanical counter provides an odometer-type readout, which is used as a backup to the depth measurement system used on the injector head.

For a CT string with umbilicals or wireline installed, a rotating collector is installed to transfer either electrical data or hydraulic fluid from surface to the coil.

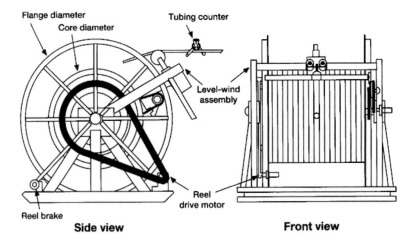

Fig. 9.4.3—Example CT reel (LEAding Edge Advantage Ltd. 2002).

TABLE 9.4.1—CAPACITIES FOR A STANDARD REEL*	
CT Size	Length
2 in.	25,937 ft
$2^3/_8$ in.	19,165 ft
$2^7/_8$ in.	12,408 ft
* Width 87 in., core 112 in., and drum OD 180 in.	

The final main component of the CT reel provides one of the most advantageous facets of the CTU. The reel swivel and manifold provides a pressure-tight rotating seal, which enables fluids to be pumped through the CT under pressure while running in hole or pulling out of hole.

The reel manifold design will vary, but as a minimum, it will include a valve within the reel core to provide the ability to isolate the CT at surface.

Gooseneck. Generally, a gooseneck with a minimum radius of 100 in. is required to reduce cycle fatigue and increase CT life. Smaller goosenecks can be used, but these can have a substantial effect on the life of the CT.

Injector Head. The main purpose of the CT injector head is to supply the effort and traction required to run and retrieve the CT in and out of the well. The injector head consists of two continuous opposing chains with gripper-type blocks matched to the CT size.

The chains normally are driven by two or four hydraulically powered sprocket drives. The CT operator is able to control the running speed of the CT in and out of the well by adjusting the hydraulic feed pressure.

Pressure Control Equipment. This refers to all equipment that will provide pressure containment in the event that the primary well control is insufficient and an influx of hydrocarbons is taken into the wellbore. Additionally, live well intervention is generally part of the operational process. As such, mitigation and control measures must be in place in the event of a leak at surface. The principal well-control equipment consists of

- CT stripper
- Blowout preventer (BOP) stack and annular BOP
- Choke manifold
- Hydraulically controlled regulator (HCR)

Additional well control equipment and pressure containment is provided by

- Tandem bottomhole assembly (BHA) check valves rated to 5,000 psi
- Surface check valves
- Lower shear/seal rams

The primary function of the CT BOP and stripper is to maintain control of the well at all times. BOPs and strippers must be properly maintained and kept ready to operate. Before the job, the BOPs and strippers must be inspected, function tested, and pressure tested.

All BOPs and strippers basically operate on the same design principle. Standard CT BOPs are available generally in quad-, combi-, or monoblocks **(Fig. 9.4.4).**

Each set of rams functions independently from the others; the operator manually selects the hydraulic controls at the operator's console. Many older BOPs are still rated for 5,000 psi, but more and more BOPs are rated for 10,000- or 15,000-psi working pressure, with some rated for 20,000 psi. The BOP working pressure is based on body design and the lower connection rating.

Where four rams are used, they should be configured in the following order:

- Ram 1: blind rams. The blind rams are designed to seal off the wellbore once the pipe is out of the well.
- Ram 2: shear rams. The shear or cutter rams are designed to cut the CT and/or wireline cables. (If hydraulic lines are installed in the coil, the shear rams may need to be upgraded or a booster unit installed on the end of the ram bonnet.)
- Ram 3: slip rams. The slip rams hold the pipe in either the pipe-light (snub) or pipe-heavy position.
- Ram 4: tubing rams. The tubing rams seal off the annular area around the CT, preventing wellbore fluids from reaching the surface or the slip inserts. Generally, however, the BOP rig-up incorporates combination rams. These allow a greater flexibility of well control without extending the height of the stack:
- Top rams: shear/seal rams. The ram insert is generally a dual-purpose shear/seal ram that can cut the CT and any wireline or umbilical and seal the wellbore simultaneously. Depending on the application, a booster unit may be fitted to the end of the ram bonnet, increasing the effective cutting force that the ram provides.

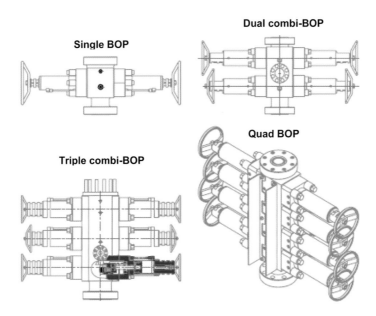

Fig. 9.4.4—BOP configurations (NOV CTES 2005).

- Middle rams: pipe/slip rams. The slip rams prevent the pipe from moving either up or down, depending on the forces involved. The insert includes a sealing mechanism that prevents wellbore fluids from bypassing the ram. The seal also prevents fluids from detrimentally interacting with the slip inserts.
- Bottom ram: shear/seal ram. This ram usually is identical to the top ram and usually is sited directly on top of the casing flange or wellhead. This provides a method of well control in the event of problems arising with the upper well control equipment.

An annular BOP is usually incorporated in the stack above the upper shear/seal ram. This allows sealing on any part of the BHA that has an OD different from that of the CT. Above the annular BOP usually just below the injector head, a stripper is located. This is a hydraulically energized, pressure-containing device. The stripper contains the well fluids and maintains wellbore integrity while tripping in or out with the CT. The stripper uses a large elastomer sealing element. The element is subject to wear and must be serviced periodically. The stripper may be rated for 5,000-, 10,000-, or 15,000-psi working pressures.

There may be instances, especially in high-pressure gas wells, in which strippers are run in tandem to provide adequate redundancy during the operation. This arrangement provides flexibility either to change out the worn stripper elements with the pipe in the hole or to trip back out of the hole. The upper stripper is the primary seal, allowing the lower stripper and annular BOP to provide a dual barrier to change out the elements where appropriate.

The kill line inlet enables fluids to be pumped into the annulus of the CT during normal operations or down the coil if in a situation where the tubing has been cut. This side outlet should not be used for taking returns.

Both the BOP and the stripper use elastomer sealing elements that are subject to mechanical wear and chemical degradation during normal operation. Some of the well fluids and gases may react with the sealing elements. Such a reaction reduces the effective life of the elastomers. The maintenance program must be arranged according to the wellbore fluids and activity. H_2S-certified equipment must be used when working on any well that contains H_2S. Verify that all valves and BOP parts are H_2S-certified when used in such applications.

In addition, positioning a gate valve as close to the tree as possible will provide the maximum length of riser to deploy tools into the well. (The lower shear/seal ram is the final well-control barrier and as such could not be used as a barrier to deploy the tool string.)

Power Pack. The power pack can be either a diesel or an electric power source. Zone II requirements must be met in all power packs, and all of the relevant emergency shutdowns will be fitted and tested prior to operations.

The power pack drives an array of hydraulic pumps, supplying each system or circuit with the required hydraulic pressure and flow rate. The power pack also provides the necessary components to safely and effectively provide a source of power and control to the related ancillary CT equipment.

Measuring Equipment. Dual weight sensors or diaphragms are required on the injector head to measure both pipe-light forces (snubbing forces) and pipe-heavy forces. Dual depth counters will be used, both at the injector head and at the reel to provide sufficient redundancy. The depth counters will be calibrated before use. Depth encoders with high pulses per revolution are available, which provide greater accuracy during milling and drilling. Where possible, these should be used. Often two depth encoders will be used, one from the CTD supplier and one from the BHA provider. The second depth encoder provides redundancy in case of failure of the primary encoder.

CTD Stack Support. A wellhead is not normally designed to support the full weight of the full BOP stack, the weight of the injector head, and the weight of CT at depth. A CTD jacking frame, mast, or crane should be used to support the weight of the CTD equipment through to the rig floor or ground level, eliminating issues of weight transfer to the tree.

Stack Configuration. The relevance for the surface equipment selection is the limited amount of weight that can be placed on top of the production tree. Because the tree generally remains in place during drilling operations, it is recommended that a method of support be provided for the BOP equipment, injector head, and weight of CT while in hole.

All of the BOP riser connections should be flanged up to the highest part of the secondary well-control equipment. In many cases, the complete riser is flanged, though quick unions can be used above the highest ram if required. If all the unions are flanged, a quick connect is required to make up the injector head to the top of the riser after BHA deployment. This connection will be pressure tested each time it is made up and prior to opening the riser to the well. To facilitate this with the least impact to the operation or the drilling fluid, a Cromar sub or Hydracon is utilized.

Both of these quick unions can be provided with a pressure test port sandwiched between two O-ring seals. Pressure testing takes place between these seals, minimizing the volume required and isolating the test fluid from the drilling fluid.

Mud Package Considerations. The mud system is one of the major components of the drilling package and requires careful selection. For offshore work, a detailed site survey is necessary to determine what is available and the suitability for slimhole drilling applications. The volumes required are significantly less than for conventional drilling operations.

Key equipment for the mud package includes

- Mud tanks
  - With surface lines to allow transfer of mud between the different tanks and provide flexibility while mixing
  - Trip tank (optional)
- Transfer pumps
- Fluid monitors
  - Gas monitors for H_2S and hydrocarbons
  - Pump stroke meters for flow
  - Flowmeters for flow in and out
  - Constant pit level monitoring using *radar* or sonic level sensors for improved accuracy
  - A pit volume totalizer with visual and audible alarms
- Mud-cleaning equipment
  - Preferably two variable-speed elliptical shakers
  - Centrifuge (when the mud system and formation dictates)
- Degasser
  - Poor-boy degasser
  - A vacuum degasser (optional)

- Mud pumps
  - Two mud pumps (minimum)
  - A standard CTD mud pump specification
- Pressure: 5,000 psi
- Pump rate: 3 bbl/min maximum
- Power: 370 hydraulic horsepower

The mud package controls should be in the CTD cabin. There should be a visual display unit on the mud package to allow the mud engineer to monitor the mud tanks and flow.

Data Acquisition. The data acquisition system will monitor and record all of the relevant CTD parameters for real-time monitoring, well reporting, and interpretation. It must be capable of setting alarms for different parameters and must provide both audible and visual warnings. The data that will be displayed both recorded and in real time will include

- CTD parameters: depth, weight, speed, CT life (real-time)
- Mud parameters: standpipe pressure, wellhead pressure, choke manifold pressure, flow in, flow out, pit volume totalizer, gain/loss
- Well control: BOP position indicator, accumulator pressure, stripper pressure, choke position
- BHA data: inclination, azimuth, gamma ray, bottomhole pressure (BHP), WOB, vibration

During operations, the operational data need to be routinely compared to previous trends and simulation results and updated as required.

Control Unit. The control unit of a modern CTD unit is highly sophisticated, compared with that of just a few years ago. In addition to having the capability to monitor downhole conditions and change tool settings, it often has a computerized capability to monitor CT conditions and update CT service-life predictions continually, to perform hydraulics calculations, and to predict borehole friction. The control cabin should contain all of the necessary controls and instruments to enable the complete CTD operation to be run from one central control station. As such, the control cab should have as a minimum the ability to perform the following functions:

- Control and monitor the operation of all CT operating functions.
- Control and monitor all downhole information and tools.
- Control the orientation of the drilling assembly.
- Control and monitor the operation of CT well-control equipment.
- Control and monitor pumps and surface volumes.
- Monitor and record the principal well and CT string parameters, including wellhead pressure, circulating pressure, tubing weight at the injector head, and tubing depth.

Additional attributes that are useful in a CTD cabin include

- Method for remotely controlling the backpressure of the well in an underbalanced or managed-pressure drilling scenario
- Method for controlling and monitoring any surface separation equipment

With some of the modern control cabins, a single operator can control the CT and pumping operation from the driller's chair in the control cabin. This type of *fly by wire* control eliminates the control panel and allows the operator to control both units from controls in the arms of the chair.

Platform and Utilities. If operating on a platform offshore, the interface with the utilities may be required, and general interface with the platform and production modules is extremely important. A detailed review to identify the interfaces required is essential at the beginning of the project planning. It may be necessary to implement a number of modifications to the infrastructure to accommodate the CTD spread (Fig. 9.4.2). If this is necessary, sufficient time and budget must be allocated early on to achieve this.

9.4.4 Downhole Equipment. The downhole equipment for CTD can be simple or quite complex. For a simple vertical-drilling project, it may consist of a bit, a downhole motor, and a few drill collars. It is much more elaborate for a directionally drilled well [**Fig. 9.4.5** (NOV CTES 2005)].

Tubing Connectors. Many CT operations require the use of a variety of tools on the bottom of the tubing, depending upon the particular application. To attach the tools to the end of the coil, a CT end connector is used. Several types are available:

- *Dimple/setscrew.* The operator uses a number of setscrews to attach the tubing connector to the end of the tubing. Usually the tubing is *dimpled* prior to this to allow small indentations, recesses for the setscrews to locate and bind into. A tool thread is used on the opposite end. Incorporated in the tool is either a single or double O-ring seal to maintain pressure integrity.
- *Swageloc.* The operator uses a Swageloc ferrule to connect the Swageloc tubing connector to the end of the tubing. A tool thread is used on the opposite end.
- *Roll-on.* The roll-on tubing connector has O-rings and recesses that allow the operator to roll-crimp the connector internally onto the tubing. The operator uses a pipe cutter with a roller instead of a blade to attach this connector to the tubing. A tool thread is used on the opposite end. The maximum connector OD is the same as the OD of the CT. These connectors generally are used where a slick assembly is required due to wellbore geometry restrictions or, in the case of a double roll-on, to attach two lengths of CT together temporarily. Roll-on tubing connectors should not be used for any operation that uses a downhole motor because the connector cannot withstand torque.
- *Slip/grapple.* This type of connector uses a grapple mechanism that is tightened onto the tubing during the makeup procedure. The connector is made in two parts, the upper containing a bowl that forces the grapple onto the outside of the CT. Later models incorporate small Allen screws to increase the torque capability of the connector. An O-ring is used on this connector to obtain a positive seal. A tool thread is used on the opposite end.
- *Modern dimple connector.* This connector is an internal connector whereby the CT is hydraulically dimpled into machined recesses in the connector. This type of connector provides excellent torque and yield characteristics and is the most common type used for CTD applications.
- *Backpressure Valves.* Generally a double check valve assembly is incorporated in the BHA just below the CT end connector, which prevents fluids from entering the coil from the wellbore. The main reason to incorporate check valves in the assembly is to prevent an inflow of formation fluids from reaching the surface if the CT develops a pinhole or there is a disconnect. There are two kinds of backpressure valves typically used:
- *Ball and seat.* The ball-and-seat valve is the basic type of check valve because of its simple construction, ease of maintenance, reliability, and cost. The main disadvantages are the restriction of flow area and restriction of mechanical access below the valve.
- *Flapper.* The flapper-type valve is used when applications may require that a ball or dart is pumped to actuate a tool function or when flow requirements do not allow a ball-and-seat check valve to be used. Typically a pair of these valves are included to provide a double barrier.

Disconnect. A disconnect sub is run on all CT runs, with very few exceptions. The disconnect is generally inserted in the tool string either directly below the double flapper check valve or just above the motor in some

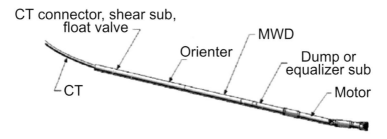

Fig. 9.4.5—Typical coiled-tubing drilling BHA.

drilling operations. Ideally, the disconnect should be run just below the check valves; however, due to the cost of many CTD assemblies, the disconnect can be run just below the MWD/orienter section. The disconnect will operate effectively only if the stuck point is below the disconnect. There are a number of methods for actuating the disconnect:

- **Hydraulic.** This type of disconnect can be used only when the CT has no electric cable, hydraulic lines, or other restrictions that would prevent a ball from being dropped and reaching the sub. Should the BHA become stuck, preventing the tubing from being retrieved, a ball can be dropped into the CT string at the reel and pumped down the coil to activate a piston in the disconnect. This then releases the bottom half of the disconnect sub. The bottom half of the hydraulic disconnect remains attached to the lower part of the drilling assembly and generally incorporates an external fishing neck that can be latched onto and retrieved with the correct fishing assembly. The proper ball/seat size in the disconnect should be chosen carefully to ensure that the ball will pass through any restrictions in the coil and items in the BHA above the disconnect sub.
- **Mechanical.** Mechanical disconnects are used either when a ball cannot be dropped through the coil or when the operation dictates that it is preferable not to run a ball-actuated sub. Mechanical disconnects can come in two distinct types: single-actuated disconnects and dual-actuated disconnects. The single-actuated tool relies purely on pull at the tool to activate. This predetermined force is set up at surface prior to tripping in and can be set either by using shear pins or shear rods or by incorporating a set of pretensioned springs. The dual-actuated tool requires a force at the tool to activate the primary release mechanism and then the addition of pressure or downward motion to initiate the second and final release from the tool string. The mechanical disconnect eliminates having to drop a ball through the tubing, but it is limited operationally because the release force is set up and fixed prior to tripping in, which effectively reduces the maximum force that can be applied downhole prior to the tool activating.
- **Electrical.** Essentially an electrical signal is sent from surface to the tool to activate it. The signal has to be shielded in such a way that accidental activation cannot occur. The tool also has to be in tension or compression prior to activation occurring.
- **Hydraulically surface-activated.** This type of tool is run with a pair of umbilicals inside the CT. The hydraulic umbilicals generally are used to orient the tool face of the drilling assembly in normal operations. To disconnect, pressure is applied down both umbilicals simultaneously, activating the disconnect sub.

Circulating Subs. Circulating subs are used to allow the well to be circulated without rotating the motor and bit. These allow high flow rates to be used after drilling a section and prior to tripping out or while tripping in to condition the well prior to drilling. These can be mechanical or electrical.

Jars. Jars are used routinely in rotary drilling operations to activate an upward or downward impact load, depending upon the design. The jars in rotary assemblies are used to free stuck pipe. For CTD operations, two types of jars are available: hydraulic and mechanical jars.

Hydraulic Jars. Hydraulic jars use a piston and oil to create a time delay. The time delay allows energy to be stored in stretched tubing or an accelerator. Once the energy is stored, the jar piston bypasses, creating a hammer-and-anvil effect that imparts an impact load to the end of the tool assembly. As the jar is pulled in tension, a piston moves through a restricted bore containing oil. This process enables the system to store energy. Continued upward pull moves the piston into a larger cylinder ID, which releases the stored energy in the accelerator or tubing and allows the mass to rapidly accelerate to the top of its stroke. This process creates an impact force that can be much greater than the tension that can be pulled on the tubing alone. The force supplied depends upon the force applied at the tool. These jars can be supplied either up-acting, down-acting, or bidirectional.

Mechanical Jars. Mechanical jars utilize the physical force from surface to activate either an upward or downward jarring motion. The difficulty with this is that the CT can act as a large spring, damping the force that can be applied downhole.

Generally, it is not very common for jars to be run in CTD operations because experience has shown that they are not very effective in freeing stuck pipe. In many CTD operations, it is the coil that is stuck either mechanically or differentially, and a jar in the BHA is ineffective.

Accelerators. Accelerators are run with jars, but similarly these normally are not run during CTD operations.

Logging Tools. The main function of the logging tools or MWD package is to measure the direction and gamma ray count while drilling. The package itself generally is contained inside a nonmagnetic collar (or monel) to prevent magnetic interference with the tool. These tools generally rely on one of four methods for the transmission of data between the tool string and surface: mud-pulse, electrical, electromagnetic, and fiber-optic.

Mud-Pulse Tools. These are based on conventional mud-pulse technology. Information is sent to and from the tool through the drilling fluid. This system usually is very reliable and eliminates the requirement for cable or umbilicals in the tubing, but it can suffer from some limitations. There is a limit on the amount of information that can be sent from the tool to surface. To communicate with the tool, the pump pressure has to be cycled, and depending on other components of the tool string, this may have alternative effects elsewhere. The tool also has a limitation in that it will operate only with a mud that does not include more than 20% gas in the base fluid, which therefore results in limited use in underbalanced environments. If the quantity of gas in the fluid exceeds this figure, the compressibility of the aerated mud dampens the signal to an extent that it cannot be read at surface. Pumping lost-circulation material through the tool sometimes can clog the pulsing mechanism, depending on the size of the material.

Electrical Tools. These require a continuous electrical connection to surface. This usually is provided by the inclusion of an electric cable inside the CT, using mono-, coaxial, or hepta- (seven conductors) cable. Due to the substantial increase in data transmission both to and from the tool, much greater control can be offered over the information requested from the tool, and more information can be sent to surface. This allows the tool string to include not just directional information and gamma ray but other downhole information. This additional information becomes especially important when drilling underbalanced.

Electromagnetic Tools. These tools provide an advantage in wells drilled underbalanced where mud-pulse telemetry is not effective (e.g., using air, foam, or mist). Because there are no moving parts, they have proved to be robust in demanding drilling conditions. The systems use bidirectional communications and offer the advantage of very high data rates, approximately four times faster than conventional MWD. However, there are depth (<12,000 ft) and overburden (salt) limitations, which need to be taken into consideration.

Fiber Optic Tools. Fiber optics are beginning to come into use. The data transmission rate exceeds that of electrical transmission, and the weight and dimensions of the cable are minimized, reducing the effect on shipping weights and pressure drop through the coil.

Currently, tools can host a number of data acquisition systems, such as

- Direction (azimuth)
- Angle (inclination)
- Gamma ray
- Internal pressure
- External (annulus) pressure
- Temperature
- WOB (positive and negative)
- Acceleration (vibration)
- Resistivity (limited availability and hole size)
- Torque

Downhole Motors. The choice of motor is key to the performance of the BHA. It must perform optimally at the lower flow rates and provide sufficient torque to turn the bit. The OD of the motor selected is a tradeoff between the horsepower requirement to drive the bit, the flow rate available, and the annular clearance required.

The following criteria constitute rules of thumb for CTD motor selection:

- Reliability: extended on-bottom and circulating hours to minimize any effect to the CT life and general success of the well
- Consistently achieves doglegs of 30–40° per 100 ft

- Optimum torque provided at the necessary flow rate
- Torque output between 300 and 500 ft-lbf
- Maximum differential across motor of less than 750 psi
- Minimum motor length to reduce the distance between bit and direction and inclination (D&I) package
- Rotational bit speed of 200–400 rev/min
- Resistant to anticipated bottomhole temperature
- Elastomers resistant to the drilling mud

There are essentially two main types of downhole motors available for slimhole drilling: a positive-displacement motor or a turbine motor.

Positive-Displacement Motor. The selection of this type of motor is dependent on the torque and speed output requirement, which is dependent on bit selection, ROP optimization, and available flow rates. To operate, the motor requires a rubber seal between the rotor and stator. The rubber components can be susceptible to damage from gaseous ingression and swelling when exposed to aromatics, especially at high temperatures. When selecting a motor, it is recommended to determine the effects of drilling fluid on motor reliability by performing swell tests on the elastomer components.

Turbine Motors. These run at very high speed, generally in excess of 900 rev/min. The motors do not have rubber components and as such are not detrimentally affected by either temperature or fluid type; however, the output speed severely limits bit choice to very high-speed bits, such as diamond-impregnated bits. There is the possibility of positioning a gearbox below the motor to reduce the speed output to that acceptable for polycrystalline diamond compact (PDC) bits, but this reduces the build rate ability of the motor. Due to the drive mechanism, the operating differential pressure across the motor usually exceeds 1,000 psi, which would result in excessive surface pressures and potentially would require flow rates to be reduced, affecting hole cleaning. Turbines are useful in very hard formations but are rarely used in conventional CTD applications. The main benefits of a positive-displacement motor over a turbine motor are

- Lower operating differential pressure
- Lower speed output: 200–500 rev/min, allowing the use of PDC bits
- Medium to high torque output in relation to motor dimensions
- Adjustable bent housing incorporated into the design of the motor

The lobe configuration of the motor will determine the torque and the rev/min. More lobes allow the rotor to turn at lower speeds but with higher torque. **Fig. 9.4.6** shows various positive-displacement motor (PDM) lobe configurations.

The maximum dimension of the motor usually is limited by the size of the open hole or by the minimum restriction in the existing wellbore. One item to be aware of during the selection process is that motors can have a larger maximum OD than the nominal stated size. This may be due to the size of the bearing assembly or wear pad.

To prevent stalling, the torque requirement is relative to the reactive torque provided by the bit. The bit should be designed to reduce excessive reactive torque. If the formation consists of various layers with varying geological characteristics, it is important to ensure that the bit can drill through a range of hard and soft rock types. It is important that the motor can provide sufficient torque to prevent excessive stalling.

Frequent motor stalls are very detrimental to the life of the coil, the motor, drilling efficiency, and potentially to the borehole stability because the BHP regime fluctuates with each cycle of the pumps.

The power rating used for the maximum output of the motor always should be the stall torque. If this is not provided, then twice the maximum operating torque should be used as a rule of thumb. The factors that determine the maximum torque output are as follows:

- Torsional rating of the CT
- Torque rating of the drilling assembly connections and CT connector
- Maximum pressure differential possible across the motor without exceeding surface limits

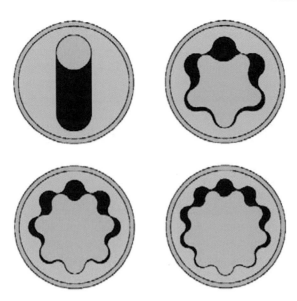

Fig. 9.4.6—PDM lobe configurations (NOV CTES 2005).

One of the primary functions of the motor is to provide the desired buildup rate from the kickoff point to the target. These doglegs can be fairly severe, with many wells requiring build rates in excess of 30° per 100 ft. There are inherent difficulties in achieving theoretical build rates due to variances in maintaining the required tool face and WOB, especially in formations that have not been drilled previously using CT. In many cases, if the build rate is not achieved, there may be no opportunity to correct the well path, necessitating a sidetrack of the open hole. This is a scenario associated with high risk and cost and therefore needs to be avoided.

To ensure that the required trajectory is drilled, the adjustable kick off (AKO) sub usually is set to provide a greater buildup rate than planned. To provide this flexibility, the motor selected should always be able to provide a theoretical dogleg severity (DLS) greater than that required by the well path. If the maximum setting of the motor AKO provides less than the required buildup rate, there are a couple of options that can be investigated.

An additional bent sub can be incorporated into the tool string just above the motor. The bent sub usually is preset with a 1.0–1.5° bend. This type of BHA can achieve dogleg severities greater than 45° per 100 ft without sacrificing the ability to pass through small restrictions, though a pass-through test is recommended. A wear pad or sleeve also can be used on the bearing assembly, increasing the build capability. Similarly, the restrictions within the wellbore need to be reviewed to ensure that there are no conflicts. It is essential to be able to alter the AKO setting on the rig because it provides the capability to react to the existing well profile as required.

The selected motor must be tolerant of the bottomhole environment and drilling fluid. As a result, the rubber components, especially the stator, must be compatible with the mud at the bottomhole temperature without excessive swelling or deterioration. The ideal motor would exhibit high potential maximum torque, low-pressure drop, short length, and high DLS capability. To achieve this requires a slight compromise. From the hydraulics modeling, the intention would be to determine which motor can be used whereby the maximum differential across the motor can be used while avoiding excessive surface pressures.

Orienting Tools. These are used to turn the bit in the desired direction because the tubing cannot be rotated. The orienting tool can be controlled from surface either hydraulically or electrically. The two main hydraulic methods for operating the orienter are drilling fluid and control lines.

Drilling Fluid. In this method drilling fluid is used to index the tool face. These tools require reducing or stopping the pump rate to reduce the differential across the indexing tool. This release of pressure allows a spring inside the tool to turn the tool face by 30–60°, depending on the type of tool used. The tool face can be indexed in only one direction, usually counterclockwise. The effect is that it is not possible to *wig-wag* either side of the required tool face; instead, the tool has to be cycled 300° or more to correct the trajectory. This can be quite time-consuming, especially because positive indication of the tool face is hard to ascertain at times until drilling commences and the effects of reactive torque are determined. The advantage of this type of orienter is that due to the simplicity of the internal mechanism, the tools are fairly reliable.

Control Lines. Control lines require one or more small umbilical cables to be inserted through the CT to allow hydraulic fluid to be pumped from a surface control unit to the orienter, which in turn regulates the positioning of the tool face. The advantage of this system is that the tool face can be adjusted infinitely while drilling without affecting any other drilling parameter. The orienter generally has a range of 400° movement from one stop to the other.

The electrical systems operate by two main methods, powering either a hydraulic pump or an electric motor downhole. In the first method, the electrical current powers a hydraulic pump in the orienting tool, which then forces a splined cylindrical mandrel to turn inside the orienter, altering the tool face. This tool has the advantage of direct control from surface; however, because the tool face ultimately is controlled by hydraulics, there is a maximum of around 400° between the stops, preventing continuous rotation. Alternatively, the current can be supplied directly to an electric motor, which then turns the tool face to the required angle.

This system allows the tool face to be turned continuously while drilling, which can result in a smoother borehole when tangential or horizontal sections are required.

Although any cables or umbilicals add weight to the CT string and reduce the effective ID, they provide the CT unit with a marked advantage over conventional MWD systems by way of providing two-way communication and high-quality real-time data from downhole instruments.

Pad-steering CT BHAs are being introduced to the market that allow continuous adjustment of the tool face and DLS, similar in capability to rotary steerable assemblies. These provide a significantly improved wellbore but have limitations on the DLS that can be drilled. Currently the DLS is limited to around 15° per 100 ft.

Bits. CT and slimhole drilling introduce certain limitations in the choice and design characteristics of the bits used. The size and specifications of the tubing restrict motor choice due to torsional limitations and low flow rates. As such, motors generally provide high speed and low torque. The small hole size affects technical specifications, annular clearances, and, therefore, bit type and design.

The primary characteristics that need to be reviewed during the bit design and selection include the reactive torque, steerability, durability, and ROP. For every application, these primary attributes need to be assessed to determine what the priorities are for the well or for particular sections of the wellbore (e.g., build and tangent sections).

Torque. Due to the small hole size and comparatively low flow through the CT, the bit must be designed to limit the amount of reactive torque while drilling. Aggressive bits provide greater instantaneous ROP but are more likely to stall the motor. This is harmful to the motor and results in additional time off bottom in reducing pump rates, picking off bottom, recommencing drilling, and adjusting the tool face. Aggressive bits also result in fluctuating reactive torques, which in turn are detrimental to maintaining a constant tool face. If high dogleg severities are required, fluctuations in tool face could result in not attaining the target horizon. As such, sacrificing ROP for greater tool face stability due to a constant reactive torque over a range of WOB values is more crucial.

Steerability. Good motor steerability, especially at the beginning of the openhole section, where the majority of any directional work is performed can be critical to meeting the directional requirements of the well. Conversely, this is also where the least information is available on the formation and drilling assembly compatibility. It is prudent to have a bit design available that provides excellent steerability, possibly at the expense of the other characteristics, to ensure that the desired well trajectory is attained. The profile of the bit is also important because this affects the stability of the bit, which can be important for tangent sections.

Durability. Any bit run must provide an effective bit life. The bit also must provide a reasonable penetration rate, and the hole must be in gauge. Due to the time required to perform a round trip to change out the drilling assembly, it is generally preferable to sacrifice some ROP for longer time on bottom drilling and an in-gauge hole.

ROP. In the majority of bit designs, ROP is important. Improved ROP reduces well cost and decreases the potential for additional trips by not achieving the required depth prior to a downhole tool failure or pulling the bit.

As in conventional operations, contingency bits always should be available. If there is no direct comparable offset information, a wide selection of bits is advisable, including possibly a selection of concentric and bicenter bits if there are concerns with the maximum extension possible and mechanical or geometrical concerns.

9.4.5 Fluids for CTD. There is much in the literature relating to CTD, but little that specifically addresses the issue of drilling fluids. It is evident, however, that increased interest is being shown in this subject, and some of the drilling-fluids service companies, particularly those with associated CT companies, are beginning to look closely at the fluid requirements for CTD.

The basic rheological requirements for a drilling fluid for CT operations are significantly different from those used in conventional rotary drilling, but they are akin to the requirements for slimhole applications. The essential difference from conventional drilling fluids is that in CT operations, small-diameter pipe and narrow annuli usually exist, which then require the use of a fluid specifically selected to minimize pressure losses and equivalent circulating densities (ECDs).

During the planning of a CT operation, the following factors must be considered:

- Tubing ID
- Length of CT
- Fluid type and rheology
- Average fluid temperature
- Fluid density
- Annular size
- Annular velocity
- Choke/wellhead pressure (if applicable)

Pump rate vs. tubing pressure drop can be calculated from this information by the use of hydraulics programs.

The following points must be considered when designing a fluid's rheology and hydraulics for CT operations:

- Flow rate must be high enough to transport cuttings.
- Pressure drops through the coil and up the annulus limit the maximum flow rate.
- Downhole motors and drilling assemblies have a maximum and minimum flow rate, often limiting the drilling-fluid flow rate.
- Some directional assemblies have limitations on solids that can be pumped through them (e.g., lost-circulation material such as calcium carbonate).

A big factor for cleaning deviated wells with CTD is that drillstring rotation is not possible. This limits mechanical agitation of the cuttings beds. Flow rates need to be increased to compensate for this. The ideal viscosity of the fluid can depend on the application. Thick fluids in laminar flow sometimes can improve transport, since a proportion of the solids will be carried in suspension; however, due to the nature of CTD, the tubing is always in slide mode. Because of this, once the solids have settled out of the mud, it can be very difficult to get the solids back into the fluid flow with a high-viscosity mud. It has been seen that low-viscosity mud can improve hole cleaning because the fluid remains in turbulent flow longer at reduced flow rates, thereby preventing the solids from settling around the tubing and resulting in a cleaner hole. Usually, high-viscosity sweeps are circulated occasionally to ensure efficient hole cleaning, but these must be used prudently to avoid detrimentally altering the rheology of the base drilling fluid. The selection of the appropriate rheology needs to take account of ECD restrictions to prevent formation breakdown.

Some ambiguity exists in the literature regarding the terminology relating to friction reduction in slimhole and CT applications. Because there is no pipe rotation used in CT operations, the requirement for a lubricant is related more to minimizing tubing lockup in larger casing diameters (i.e., larger than the hole being drilled), getting weight to bit, minimizing differential sticking, and facilitating trips in and out of the hole. Increased penetration, by CT, is essential for cost-effective drainage of horizontal wells. It is sometimes necessary to add lubricants to the drilling fluid to achieve the required performance. The additives used generally are either vegetable-oil-based or use small glass beads or the equivalent to reduce the friction coefficient between the casing or open hole and the CT. The use of such additives must be carefully considered in relation to formation compatibility because they may result in excessive impairment in the reservoir section.

Drag reduction is a term applied to the addition of a substance to a fluid to decrease the pressure losses that occur when that fluid flows in a pipe or annulus (for this reason, they are also known as *flow improvers*). This phenomenon also is referred to as friction reduction. This can lead to ambiguity because the phenomenon is unrelated to the lubricity functions of a drilling fluid—conventionally reduced by the addition of lubricants to reduce torque and drag of the drillstring, which, itself, is in intimate contact with the casing or formation.

Essentially, any fluid that is compatible with the elastomers used in downhole motors and elsewhere in the circulating system has the potential to be used as a drilling fluid for CT operations. The criteria that must be met are similar to those identified for slimhole drilling:

- Viscosity suitable for the application
- Density control
- Good shale inhibition (if shales exposed)
- Formation compatibility
- Good lubricity characteristics
- Acceptable cuttings-carrying capability
- Acceptable fluid-loss characteristics
- Acceptable temperature stability

The slimhole requirement of *no tendency for solids centrifugation* has little relevance in CTD because no pipe rotation occurs.

9.4.6 Slimhole Well Control. Well control in slimholes is achieved using the same principles as are used in any other size hole. However, because of the hole and drillstring geometry, the relative influence of the various factors may differ widely from what is observed in conventional-geometry wells.

Because of the inherent nature of CTD candidate wells, many opportunities have one or more complexities regarding selection of the mud weight. For example, many CTD candidates are depleted reservoirs that may contain normally pressurized zones or shale sections. To provide borehole stability, a higher mud weight than that required for well control may be necessary. CTD mud-weight selection is generally a compromise between

- Primary well control
- Differential sticking
- Hole stability
- Minimizing ECD effects
- Limiting formation damage

Conventional vs. Slimhole Well Control. Unlike with conventional wells, the driller's method always should be used in slimhole well-control situations because the wait-and-weight method gives no advantage due to the CT string volume being many times that of the openhole annular volume. For example, a 10,000-ft $2\frac{3}{8}$-in. string would have an internal volume of 41.5 bbl, while 500 ft of $3\frac{1}{4}$-in. open hole has a volume of 5.1 bbl. (NOV CTES 2005).

Because of the small annular clearances, sufficient cuttings-lift velocity and thus hole cleaning can be achieved with low mud-circulation rates. Even at low circulation rates, it is not uncommon to have turbulent flow conditions in the annulus. Turbulence assists in fragmentation and dispersion of gas bubbles and results in lower annulus kill pressures than would otherwise be expected.

Bottom-up and cuttings arrive relatively quickly after drilling a specific formation. Indications of approach to a high-pressure zone should be that much earlier, and there should be less risk of taking a kick.

System circulating friction pressure is relatively low, and unlike with standard wells, the annulus friction pressure may represent a larger proportion of the total. This means that small changes in annulus flow conditions may be reflected in a detectable change in total system pressure loss.

High annulus circulating friction pressure results in high bottomhole ECD. This can mask the fact that a high-pressure zone that cannot be overbalanced by the mud hydrostatic pressure may have been penetrated. The well may not kick until the circulation is stopped. The ECD must be known at all times for a range of circulation rates with the current mud in the hole. These data are needed for secondary well control.

Secondary well control in slimholes is recognized as being potentially hazardous. Therefore, it is absolutely imperative that primary control is maintained. Every possible indicator should be used to forewarn of possible entry to a high-pressure transition zone, and mud density should be adjusted accordingly. Additionally, only the very best practices should be used while tripping and in execution of all other operations.

Kick Detection for Slimholes. Since a small volume of influx fills a large height in the annulus, a small gas kick can have a very large impact on maximum well pressures. If a kick occurs, every effort must be made to limit the size of the influx by detecting it as early as possible. This can be achieved as follows:

- Train the crew (rapid response times are essential).
- Have sensitive flow-out measuring instrumentation.
- Perform accurate flow-in/flow-out measurement and comparison.
- Have a sensitive pit level indicator on a small-volume active pit.
- Have a sensitive pit volume totalizer.
- Perform continuous calculated/actual standpipe pressure comparison.
- Perform long and thorough flow checks after drilling a section and stopping circulation. The well may kick because the pore pressure gradient of the newly drilled formation could be higher than the hydrostatic mud gradient.
- Drilling breaks should be treated as potential kick situations. The well should be closed in immediately after circulation is stopped after each drilling break and the well observed for any pressure buildup.
- When circulating bottom-up, the well should be closely monitored for any kick indications (e.g., increased mud returns, pit level increase, change in pump pressure).
- Perform sensitive trip tank calibration and monitoring, with constant circulation across the wellhead while tripping.
- Pumping out of hole (inside the casing as well as in open hole) should be considered to avoid swabbing.

Early kick detection is critical to the success of any CTD program. Slimhole kick detection is an issue because openhole annular volume can be less than 0.006 bbl/ft. Typical openhole volumes are less than 15 bbl at total depth (TD). Kick detection with conventional systems typically should detect a <10-bbl influx with the right attention; however, 5 bbl in slimhole drilling means a large part of the open hole is already evacuated.

In conventional overbalanced drilling, it is normal practice to take slow circulation rates (SCR) after:

- Change of BHA
- Change of mud weight
- Deepening of the well over a predetermined depth (usually every 500 ft)
- Beginning of each tour by the driller
- Pump repairs

For CTD, this reasoning does not change, but the rates must be low enough to eliminate most, if not all, of any ECD effects. This may not be possible because many pumps, even those sized specially for CTD operations, may not be able to pump at less than 0.5 bbl/min. If rig pumps are used for CTD operations, these may not be able to pump at less than 1 bbl/min. The liners should be changed to provide a range in keeping with the operational requirements.

9.4.7 CTD Planning Considerations. The main consideration when planning any CTD project is to keep it simple and follow industry best practices for any drilling operation:

- Ensure that support and commitment are provided by all of the stakeholders.
- Ensure that a campaign of wells is performed to provide the best chance of implementing learning and seeing improvements, and to provide backup candidates in case one or more fall off the sequence.
- Start with the simplest candidate wells possible to *shake down* the crews and equipment.
- Ensure that the correct resources are available. The team needs to consist of both the right experience and sufficient personnel. Inadequate resources will lead to inadequate planning.
- Sufficient time must be allowed to correctly plan the operation. On the initial wells, this can add a considerable amount of time and effort to the project.
- Where possible, all complexities should be minimized until sufficient learning has been incorporated to allow greater technical challenges to be attempted.
- The operation should be set up within the drilling group rather than well services to ensure that the emphasis is on drilling rather than on being a CT operation.
- Agree on the project goals and objectives up front with the stakeholders, and ensure that they are clear, attainable, and measurable.
- Plan to perform as much of the work offline as possible to minimize cost and risk to the project.

It is important to be realistic in planning any project from scratch. A high-level time frame with milestones needs to be prepared early on and updated often. The plan should account for

- The internal approval process
- Peer reviews and risk assessments
- Contracts
- Procurement of long-lead items
- Platform or site modifications
- New-build equipment (specifically, the operation of the equipment)
- Design and engineering
- Testing, commissioning, and trials (especially if the design of the well is impacted by results of trial)
- Training
- Mobilization

As can be seen, there are a number of items that are on the critical path. The added complication is that there needs to be some level of commitment and exposure very early on in the project. Any commitment levels need to be identified and raised at the feasibility or concept review stage to ensure that suitable approval is obtained.

Subsurface Considerations. There are a number of issues that need to be reviewed to determine whether a well is a reasonable CTD candidate. A summary of the issues is listed here:

- Reservoir characteristics, fluids, and pressures
- Well type, age, and existing completion
- Well integrity
- Drilling and intervention history
- Targets: hole length, distance from motherbore, well path
- Production and completion requirements
- Number of candidate wells
- Formation

There are a number of considerations to take into account when planning any openhole section. The advantage of slimhole drilling, with the potential to be able to drill high doglegs, is the ability to enter directly into the reservoir section. One of the main disadvantages is that the slimhole capability precludes the ability to run any liners to isolate problem zones. Therefore, the well design needs to account for, and overcome, the following challenges:

- Shales, coals, and other formations of interest
- Pressures, stability, and reactivity

- Compartmentalization
- Depletion and high- or low-pressure zones
- Fractures and brecciated zones
- Abrupt changes in lithology
- Borehole stability
- Overburden

Because CTD is generally a mature-field development technique, it is likely that the reservoir will be under-pressured. The pressure difference between the overburden and the reservoir often can be very large prior to initial production, which is exacerbated by any pressure depletion. Therefore, it is advisable to plan the bore-hole such that the complete well path remains in the reservoir section.

The directional performance in the reservoir may be limited if the reservoir is significantly depressurized. As a general rule, if steering difficulties are experienced in the reservoir in larger hole sizes, they must be expected to be magnified in CTD wells.

Ideally, the trajectory should be lined up prior to entering the reservoir; however, if the proposed well is a reentry candidate, the casing exit is generally in the reservoir section, and, therefore, high doglegs are often unavoidable. It is only in new wells drilled specifically for CTD that the trajectory of the previous casing can be lined up for the final reservoir section.

The formation, lithology, structure, and dip all need to be accounted for in the well engineering phase, from the well path and steerability to differential sticking and well control.

Typically, the targets that are identified as CTD candidates are small pockets of hydrocarbons located close to an existing well. In general, a subsurface team would screen for targets of between 0.5 and 1.5 million BOE located within a 1,000- to 2,000-ft horizontal radius of an existing well, which can be reviewed by the drilling team for suitable candidates. Ideally, the candidate well will be close to or at the end of its productive life, to minimize any production regret. Issues to take into account include

- Requirement for zonal isolation
- Pressure prediction vs. shoe strength at window (no intermediate casing possible)
- Risk of economic failure in small targets

To ensure that the wellbore is placed in the optimum position, an agreed corridor to land the well must be developed. This corridor can be communicated to the operational staff by means of the controlled rig plots of the proposed trajectory. This *tolerance corridor* differs from the geological targets in that it remains a guide-line, and not a *contractual target*. This is especially important when designing wells close to the top of the reservoir, where there can be a degree of uncertainty regarding depths. The risks must be balanced between entering the top cap and not accessing stranded reserves. With improved resistivity tools that have a deeper reservoir penetration, proactive geosteering is becoming possible to ensure that the wellbore remains in the optimum location.

Good communication throughout the drilling process must be maintained between the drilling and subsur-face groups to ensure that the optimum well path is drilled. Altering the well path dramatically while drilling to remain within tight targets can result in a shorter well path or problems running the liner, and so the team must be able to react to variations in the plan.

As in standard well design, the geologist will define targets for the trajectory to intersect. In order to maximize drilling efficiency, the geologist should be challenged to provide targets as large as can be justified. Any reduction of steering will lead to a significant time savings. The majority of CTD assemblies cannot drill straight tangent sections. The well plan should account for this by maintaining a slight turn or build through-out.

A driller's target of 100 ft diameter would be considered a reasonable size at a measured depth of 12,000 to 14,000 ft. If the target is being intersected horizontally, a vertical target extent should be provided.

The reasons for the target size and shape should be clearly communicated so that the full team has an un-derstanding of the issues. This can be critical if the trajectory is approaching a target edge and decisions on how to proceed are required. The length of the section is likely to be limited by the CT string, hydraulics (ECD and surface pressures), and weight transfer considerations.

It is recommended, during the planning phase, to evaluate the well path to check accessibility and conventional CT reach and to check for wireline access for future activities.

Candidate Screening. The completions of the proposed donor wells should be screened to ensure that there are no restrictions that could prevent the BHA from passing through. The main considerations include the following:

- *Minimum drift through the completion.* For CTD applications, BHAs typically require a minimum drift size of around 3¾ in. through a minimum tubing size of 4½ in. Nipple profiles generally are not a problem because they can be milled out if required, but side pocket mandrels, sliding side doors, and the like cannot be milled out. To verify that there are no unforeseen restrictions in the well, it is recommended that a caliper or drift run be carried out. Another option is to carry out a pass-through test, which will provide an indication of whether the assembly will cause problems—this is especially useful where bicenter bits or high-angle AKO assemblies are proposed. Smaller 2⅜-in. drilling assemblies provide for access through 4-in. tubing and hole sizes below 3 in.; however, the tubing forces and hydraulics modeling become critical at these hole sizes and can limit the overall length that can be achieved (US Department of Energy 2003).
- *Age and condition of tubing.* The condition of the tubing can dictate whether a well can, or should, be drilled. The tubing must have pressure integrity and not have significant wear (e.g., from wireline runs). The condition of the tubing can be reviewed by looking at the history of the well and by carrying out a wireline caliper run or ultra-sonic imaging tool (USIT) log. It should be remembered that if the tubing is plastic-coated, then this will be removed while milling/drilling—which may affect the long-term life of the completion.
- *Producer or water injector.* Water injectors generally do not make good candidates because the tubing can be in a somewhat deteriorated condition. Any internal coating also can cause problems during drilling because it can strip off in chunks while drilling.
- *Scale/wax issues.* Scale and wax can cause restriction problems within the tubing. If this is known to occur in the candidate well, a wireline drift run, caliper, or gauge run should be considered in the planning stage. If necessary, a cleanout trip should be planned prior to running the whipstock.
- *Completion jewelry.* It is useful if the candidate wells have a wireline-retrievable safety valve in place. This allows the valve to be retrieved and a protector sleeve run. If a tubing-retrievable subsurface safety valve (TRSSSV) is in place, protection sleeves can be installed where deemed necessary, however there are advantages and disadvantages with respect to running a sleeve, which should be reviewed on a well-by-well basis.

Legislation and Health, Safety, and Environmental (HSE) Considerations. The introduction of CTD has an impact across all of the health, safety, and environment controls and provisions in place. It is essential that additional hazards introduced by the CTD equipment, operations, and personnel are identified and managed to reduce the residual risks to acceptable levels. Permits, consents, notifications, and consultations also must be achieved in a timely manner to facilitate the project schedule.

Key HSE focus areas should include

- Definition of the project HSE requirements, objectives, strategies, methods, responsibilities, and resources
- Hazard management (i.e., identification, analysis, assessment, definition of controls, and reassessment)
- Application of HSE principles in the technical design development of the drilling equipment
- Application of self-verification arrangements to the equipment and modifications
- Assessment of equipment and operational impact on the in-place HSE management system (e.g., emergency response) and physical arrangements (e.g., emergency shutdown systems), and the revisions required
- Review of asset Safety Case document and amendment where necessary
- Review, planning, and delivery of permits, consents, notifications, and consultation necessitated

- Interface with contractors to ensure capability and integration of their Health, Safety and Environmental Management System (HSEMS) and generation of a Safety Management System (SMS) interface document to enable project HSE objectives to be met
- Assessment and control of simultaneous operations
- Application of well-control policies to the selection of equipment, provision of procedures, and training
- Identification of training and awareness needs, followed by development and delivery of training material
- Recognition of project HSE auditing effort and integration with existing programs
- Application of HSE performance targets, monitoring, and reporting
- Identification of occupational health hazards and provision of controls to reduce residual risks to acceptable levels (LEAding Edge Advantage International Ltd. 2002).

References

An Introduction to Coiled Tubing—History, Applications, and Benefits. Intervention and Coiled Tubing Association (ICoTA), www.icota.com. Downloaded 27 August 2008.

Coiled Tubing Manual. 2008. Conroe: NOV CTES 18-20 and 551-556.

Introduction to Coiled Tubing Drilling. 2002. Aberdeen: Leading Edge Advantage International Ltd., 32.

US Department of Energy Microdrill Initiative: Initial Market Evaluation. 2003. Tulsa: Spears & Associates, 6.

SI Metric Conversion Factors

bbl	$\times$ 1.589 873	E $-$ 01 $=$ m^3
bbl/hr	$\times$ 4.416 314	E $-$ 05 $=$ m^3/s
ft	$\times$ 3.048*	E $-$ 01 $=$ m
ft-lbf	$\times$ 1.355 818	E $+$ 00 $=$ J
in.	$\times$ 2.54*	E $+$ 00 $=$ cm
lbf	$\times$ 4.448 222	E $+$ 00 $=$ N
psi	$\times$ 6.894 757	E $+$ 00 $=$ kPa

*Conversion factor is exact.

9.5 Novel Drilling Techniques—B.P. Jeffryes, Schlumberger Oilfield Research

9.5.1 Introduction. Discussion of novel or unusual drilling methods must first acknowledge the comprehensive reviews by Maurer (1968, 1980) and Eskin et al. (1995).

It is not easy to compare different drilling techniques because there are many different constraints or advantages that go beyond simple numerical metrics. However, one fundamental measure is the energy required to remove a given volume of rock. This is referred to as the drilling specific energy. For rocks, drilling experiments have shown little or no rate dependence in energy requirements; thus, to drill at twice the rate of penetration (ROP) requires approximately twice the power. Drilling specific energy is a also a convenient tool to go from small-scale laboratory rock-removal experiments to estimating power requirements for practical use in different hole sizes. A convenient unit for drilling specific energy is J/cm^3. Energy divided by volume is dimensionally the same as pressure, and thus 1 J/cm^3 = 1 MPa.

Any novel drilling method has many hurdles to overcome before it can compete economically and logistically with conventional rotary drilling. The energy necessary for rock destruction must of course be delivered to the cutting face, which will be at a considerable distance from the surface, but additionally the material removed must be brought to surface or otherwise removed, well control must be maintained, the rock-destruction means must be compatible with the subsurface temperature and pressure environment, and the resulting hole must be suitable for subsequent operations, such as logging, casing, and completing.

Realistic potential applications for novel methods tend to be in niches where conventional methods are highly constrained, such as drilling very hard rocks or small holes, or where adequate weight on bit cannot be provided.

The different methods described in this chapter by no means exhaust the methods that have been re-searched. Those chosen for inclusion have been selected on the basis of how close they are to practicality, or the level of recent research-and-development work in them.

Disk Bits. Of the methods to be considered here, disk bits are the closest to conventional oilfield drilling technology. Thin, hard, rolling discs are pressed forcefully in the rock face. The very high contact stress on the cutter edge fails the rock, producing large cuttings. Disk cutters are now standard technology for tun-neling in hard rock, leading to research sponsored by the Gas Research Institute (GRI) on applying disk technology in oilfield conditions (Friant 1997; Friant and Anderson 2000). An example disk bit is shown in **Fig. 9.5.1.** The five steel discs are each 3¼ in. in diameter.

Under atmospheric conditions, drilling hard rocks, disk bits outperform tricone bits significantly in drilling efficiency. Drilling welded tuff, Friant (1997) quotes 15–19 hp-hr/ton (100–130 J/cm^3) as the drilling specific energy for a disk bit, compared with 80–120 hp-hr/ton (550–800 J/cm^3) for a tricone bit. Just as for tricone bits, pressure and in-situ stress increase the drilling specific energy. Laboratory tests under pressure on welded tuff and on Indiana limestone showed reduced advantage for disk bits, especially in the limestone; however, disk bits still provided a higher ROP than tricone bits in comparable conditions.

Field trials using the bit shown in Fig. 9.5.1 were conducted at the GRI test well in Catoosa, Oklahoma (Plácido and Friant 2004). In the hard-rock intervals, ROPs were obtained similar to those that would be ex-pected for roller-cone bits, but with lower applied weights. For softer-rock intervals, weight had to be reduced to avoid bit plugging. During these tests, there were some mechanical problems with the disk cutter assem-blies. With high forces being applied to rolling bearings, such problems are to be expected; however, very low drilling vibrations were observed, potentially increasing the lifetimes and reliability of other components in the drillstring.

The best conditions for disk bits appear to be near-vertical wells (so that high weight can be applied easily), where the drilling time is dominated by the time taken to drill hard, low-permeability formations.

Laser Drilling. Cutting of materials using lasers is now established practice in industrial settings, so it is not surprising that its use for deep-hole drilling has also been explored. In the early 1970s, research was pur-sued on combining laser and conventional cutting means for hard-rock tunneling, and the first proposals for using downhole lasers for perforating were made in the late 1960s (Venghiattis 1969, Carstens and Brown 1971). Interest in lasers resurfaced in the 1990s, chiefly through the work of Ramona Graves and collaborators

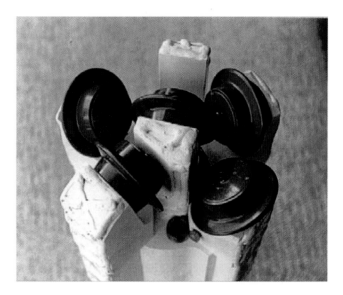

Fig. 9.5.1—An 8½-in. disk bit (Plácido and Friant 2004).

(Graves and O'Brian 1998; O'Brian et al. 1999; Graves et al. 2002). Experimental results that showed how laser drilling might be performed on an acceptable energy budget were published in Gahan et al. (2001). Normally, laser energy is associated with the very high energy densities required to melt or even vaporize materials; however, this will be very inefficient as a rock-removal process. Standard drilling methods reduce rock to macroscopic particles, and the minimum energy required for this process is only that necessary to create the free surface area of these particles—considerably less than the energy needed to physically transform the rock to a liquid. By tracking a pulsed laser over rock samples, while simultaneously decreasing the power density (through the use of a focusing lens and varying the distance from the laser to the sample), the authors showed that there was an optimal energy density for rock destruction that generated fractures and spalling through thermal processes. If the energy density was too high, the laser energy was used wastefully in vaporization and melting. If it was too low, the heat was conducted away without any destructive benefit. An example is shown in **Fig. 9.5.2. Fig. 9.5.3** shows the calculated specific energy for rock removal in shales.

While these results were obtained in laboratory conditions not representative of the downhole environment, more-recent experiments have been conducted under stress and pressure and with fluid-saturated rocks

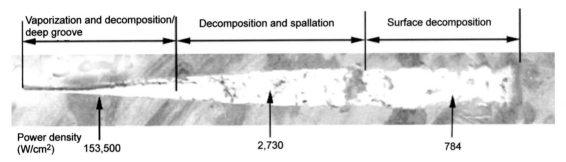

Fig. 9.5.2—Linear Nd:YAG laser track test samples with constant focal-position change rate for sandstone, identifying laser/rock reaction zones and calculated power densities (Gahan et al. 2001).

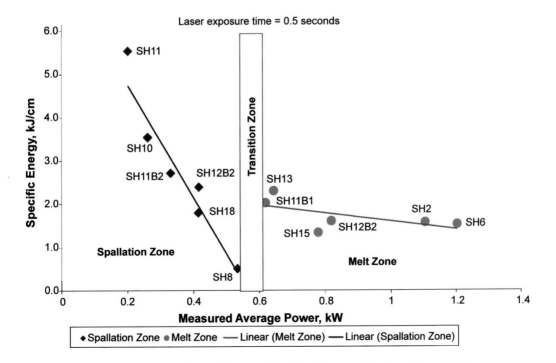

Fig. 9.5.3—Diagram showing change of material-removal method from spalling to melting of shale sample lased by Nd:YAG laser (Gahan et al. 2001).

(Gahan et al. 2005). The results do not show any fundamental barriers to laser rock destruction under these conditions, but the drilling specific energies measured were, in general, greater than 10 kJ/cm^3. To put this in perspective, to drill an 8½-in. hole at 120 ft/hr with a drilling specific energy of 10 kJ/cm^3 would require 1.5 MW of laser optical power.

Experiments on the effects of laser drilling on rock permeability have shown zero or negative skins (Gahan et al. 2004). When combined with the more achievable powers required for small-hole drilling (a 0.5-in. hole at 120 ft/hr and 10 kJ/cm^3 requires 5 kW of optical output), it is logical that recent work has centered on laser perforation. Fiber lasers, or fiber-coupled laser diodes, with outputs of this range could be engineered into a compact form that is geometrically suited to a borehole geometry, and the inevitable heat generated as a by-product could be dealt with through fluid cooling (an efficiency of converting electrical to optical energy of 25–30% is normal, though this reduces with temperature); however, even laser perforation is still a long way from being applied in the subsurface.

Electrohydraulic and Electropulse Drilling. Electrohydraulic drilling (also known as spark drilling) and electropulse drilling are both based on high-speed electrical discharge. A high voltage is created between two electrodes separated by an insulator, initiating dielectric breakdown and the creation of a plasma. Stored electric charge passes between the electrodes, discharging energy into the insulator. In electrohydraulic methods, the discharge is created in the fluid between the electrodes, and the resulting fluid pressure, impacting the rock face, creates damage. In electropulse drilling, the discharge takes place through the rock, directly destroying or damaging the material structure.

Since two electrodes, situated above a rock surface, will always have a shorter path connecting them through the liquid within which they are immersed rather than through the rock below them, one would expect liquid discharge always to take place unless the liquid was extremely resistant to dielectric breakdown. In fact, if the voltage rise time is sufficiently short, then even if the fluid is mildly ionic (such as normal tap water), the discharge will occur through the rock. To ensure this, the rise time for the voltage must be on the order of 100 ns. This direct application of energy to the rock interior gives a drilling specific energy that is much lower than that for other methods, with numbers as low as 180 J/cm^3 being reported for granite at atmospheric conditions (Inoue et al. 2000).

Fig. 9.5.4 shows clearly the geometry of an electropulse drilling head. A central conductor and electrode are joined to three circumferential electrodes, which are connected to the high-voltage-generating apparatus. The outer surface of the pipe is the ground, and three further electrodes are attached to this. When the central electrode is raised to a high voltage, the breakdown path through the rock will either be radial between the central electrode and a circumferential electrode, or be between two circumferential electrodes. In a vertical borehole, the drill bit will rest on the bottom, and the high-voltage generator repeatedly discharges, preferentially through any rock that is higher than the average position of the rock face, and thus the drill bit advances. The energy discharged per pulse depends on the electrical storage system behind the bit, but pulses in the region of 100–1000 J are normal (Martunovich and Fedorovich 2000).

The drilling head shown is not suited to drilling deep holes because there is no means for fluid circulation, but it is clear that the geometry allows for central fluid flow to take cuttings away from the electrodes.

Chapter 21 of Maurer (1980) provides extensive details on electrohydraulic-drilling research using very-high-energy pulses (typically discharging in excess of 1 kJ), finally concluding that the drilling rates achievable are not competitive with conventional rotary drilling. More recent work has concentrated on lower-energy pulses, near 100 J, and on optimizing the electrode arrangement to focus acoustic energy onto the rock face (Moeny and Small 1988; Moeny and Barrett 1999).

Neither electrohydraulic nor electropulse drilling require a weight on bit beyond that to take the cutting head close to the end of the borehole, and thus they are potentially suited to use on wireline or in geometries where weight is difficult to apply. They may also find a role as a method of enhancing conventional drilling using electrical energy generated downhole, a concept that goes back at least as far as 1968 (Smith 1970).

Thermal Drilling. There are (at least) two distinct methods of thermal drilling: melt drilling and spallation drilling. Melt drilling uses a very hot drilling head, manufactured out of a high-melting metal (such as molybdenum) or ceramic, to melt the rock surrounding the drilling head, with some weight on bit to move forward. In rock with adequate porosity (and silica content) a dense, glass-like layer is formed around the borehole, and no excess material need be transported uphole. For low-porosity rocks, *cuttings* consisting of filaments of glassy material must be taken away from the bit. The drilling-head temperature normally quoted

Fig. 9.5.4—A small electropulse drilling head (drill bit courtesy of the Research Institute of High Voltages, Tomsk Polytechnic University).

is approximately 1500°C. The biggest advantages of melt drilling are the absence (or reduction in volume) of cuttings and the presence of the vitrified layer around the borehole, acting as a kind of casing. While is it unlikely that this layer would ever be considered adequate as a long-term substitute for conventional casing materials, it could provide a short-term borehole lining for drilling unconsolidated or very fractured formations where borehole integrity is otherwise difficult to achieve. Drilling of carbonates is more problematic for melt drilling because of the generation of carbon dioxide and the high melting point of the resulting calcium oxide. Unsurprisingly, the drilling specific energy for melt drilling is high. Data quoted by Cort et al. (1994) for drilling a 3.5-in. hole converts to 13 kJ/cm^3, with little variation between rock types.

Melt-drilling research was carried out in both the USA and the former USSR at nuclear-weapons laboratories. The US research is summarized in Chapter 23 of Maurer (1980) and in Rowley and Neudecker (1986). A small-scale research effort was resumed at Los Alamos National Laboratories in the late 1990s as part of the microhole drilling project, aimed at mounting a melt-drilling head on coiled tubing (Bussod et al. 1998), and research was also undertaken at the St. Petersburg Gorny Institute (Soloviev et al. 1996).

Spallation drilling uses a downhole flame to heat the rock locally, fracturing the rock by the thermally created stress. In contrast to melt drilling, the normal volume of cuttings is produced. The method also has a higher energy efficiency, since the rock is merely broken, not melted. No drill bit in the conventional sense is employed, so issues such as bit wear do not arise; however, since there is no contact between the tool and the cutting face, the flame position must be controlled so that the correct standoff is obtained (Rauenzahn and Tester 1991). Hard, brittle rocks are actually those that break most readily under thermally induced strains, making spallation drilling an attractive option for these kinds of formations. Energy-efficiency calculations compared to other methods are not very informative, as the chemical energy is delivered directly through combustion, avoiding inevitable conversion losses present in other methods. Willams (1986) quotes a two-man test rig as achieving average penetration rates of 15.7 m/h through surface granite for a 20- to 25-cm-diameter hole, burning fuel oil at 5.7 L/min (thus approximately 5 L per foot of hole). Using literature values for the heat of combustion of liquid hydrocarbons, this is approximately 2–2.5 kJ/cm^3 as a drilling specific energy (ignoring the energy required to pump fuel and air to the burner).

There are of course complications. The fuel and air must be delivered to the burner, and if additional fluid is needed for cooling and cuttings lift, that also must be supplied. There is no control over hole shape or quality, and so a much more jagged and irregular hole is formed than with traditional rock-cutting methods. For particular situations, for instance surface-hole drilling through hard rocks, the advantages of spallation drilling could outweigh the drawbacks.

Projectile-Impact Drilling. Using high-velocity, dense particles to break rock goes back at least as far as the use of bullet perforators, and the concept of using steel shot in a drill bit goes back to at least 1952 (Deily 1955). During the 1960s and early 1970s, Gulf Oil conducted extensive testing and development of systems employing a variety of abrasive materials, including steel shot, with very-high-speed bit nozzles to maximize the destructive potential of the particles. Surface pressures were in the 8,000- to 14,500-psi range, and the drilling fluid was loaded with approximately 6 vol% of steel shot. Drag cutters were employed to remove the rock between the channels cut by the particles (Goodwin et al. 1968; Hasiba 1970; Juvkam-Wold 1975). Problems with erosion, and the combination of very high pump pressures and abrasive particles, led to the abandonment of the development program. The concept was revived by Curlett et al. (2002), who found that high penetration rates could be obtained at much lower particle velocities and concentrations than had been used earlier, and that required pump pressures were more in line with conventional oilfield equipment. Additionally, the use of polycrystalline-diamond-compact (PDC) cutters improved conventional rock-removal aspects of the design, and the lower velocities also increased the lifetime and survivability of the bit nozzles.

Fig. 9.5.5 shows the design of a projectile-impact drill bit, with two near-gauge nozzles drilling out a circumferential groove and a further nozzle close to the axis of the bit removing the central section. Laboratory tests at in-situ confining stresses and pressures showed that high rates of penetration in hard rock were potentially achievable (Rach 2007).

An advantage of projectile-impact drilling is the compatibility of the method with the normal mechanics of rotary drilling. So long as the particles can be added to and removed from the drilling fluid, and the formation-pressure window is wider than the difference between the bottomhole pressure with and without shot in the annular fluid, the method appears well suited to drilling vertical holes in hard rocks. However, while lowering the pressures required certainly increased the feasibility of operation, the difficulties of incorporating the steel shot into the fluid stream has so far proved to be the basic weakness of the technique.

Drilling Without a Borehole. For the drilling of production boreholes, it is self-evident that a channel to surface must be produced. For an exploration well, where the only product is data, so long as all the data required can be obtained while drilling, there is no need for a conduit to surface to be left behind. This is the idea behind the disposable drilling assembly shown in **Fig. 9.5.6** (taken from Stokka 2006), an idea currently being pursued commercially.

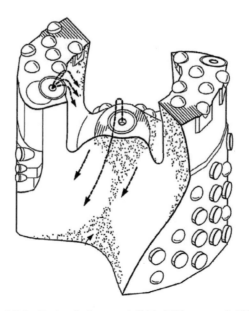

Fig. 9.5.5—Projectile-impact drill bit (Tibbetts et al. 2007).

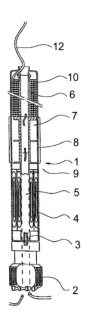

Fig. 9.5.6—Disposable drilling assembly (Stokka 2006).

The assembly unspools a power and communications cable as it proceeds. Cuttings removed at the bottom of the assembly are mechanically compacted either back into the borehole above the assembly, or by fracturing into the borehole walls; thus, instead of a hole to surface, a moving void is created that contains the assembly, connected to surface by the cable. This void will be at the same pressure as the surrounding rocks, and so it will be filled by connate fluid.

Measuring instruments contained in the assembly can thus measure not just the properties of the rock, but also those of the pore fluid. Since all the power must pass down the electrical cable, the power available at the bit will be low, but as the operational costs of the tool are low, this is not in itself a problem. It is envisaged that operators on a vessel (no drilling rig of course being required) could control a number of assemblies simultaneously.

There are of course considerable hurdles to making the concept a reality. All the components of the tool, including the drill bit, the umbilical, and other mechanical moving parts, must survive the entire distance drilled from surface to the target. All the measurements required must be decided before drilling starts, and there will be much higher positional uncertainty of those measurements once they have been taken than is the case in conventional drilling (though, potentially, the use of seismic receivers in the assembly allows them to be placed accurately with respect to seismic horizons).

Jet Drilling. The use of high-velocity fluid jets to cut materials is well established, either with clean fluids or incorporating small quantities of abrasives (Summers and Henry 1972). The technique is very suited to drilling small holes, since all the required fluid may be carried through a thin hose, which is required only to sustain internal pressure and not rotate or undergo torque or compression as a conventional drillstem would be. Additionally, the pressures involved (generally greater than 10 kpsi) make continuous tubing an attractive option. Conventional tool joints are not suited to these pressures, and neither are high-pressure connections appropriate to be made up and broken with the speed and frequency required during a drilling operation. Large-diameter coiled tubing is operationally inconvenient and has a lower pressure rating, which currently practically limits jet drilling to the small-hole environment. Thus, the oilfield application in which jet drilling has been brought to commercial practicality is in drilling short lateral holes out of a larger main borehole (usually vertical). The particular advantage of jet drilling is that the flexibility of the hose allows holes to be drilled perpendicularly out of the main borehole.

Jet-drilling heads normally comprise multiple nozzles. The orientation of these nozzles allows multiple objectives to be achieved. While forward-facing nozzles cut the rock, backward-facing nozzles help with washing away cuttings and can also provide forward thrust. Tangentially mounted nozzles generate torque, allowing the forward-facing-nozzle assembly to rotate, and side-mounted nozzles help in reaming a large hole and potentially can be used to control the cutting direction.

Commercial systems have been developed for drilling holes used to degas coal, either for gas production or as a prelude to mining (Trueman et al. 2005). The system requires an underreamed cavity (approximately 0.3 m in radius) at the level to be drilled, so that a hydraulic whipstock can be deployed to guide the drilling assembly into the formation. A flow of 234 L/min at 16 kpsi of pump pressure drilled a narrow hole for 192 m in 97 minutes. Instrumentation such as triaxial accelerometers and magnetometers may be incorporated into the drilling head, connected by wireline to the surface, allowing the trajectory position and direction to be monitored and the hole direction to be adjusted. This is accomplished by moving a rotating deflector into the path of one of the nozzles, which changes the drilling-head angle.

If the holes are to be drilled in an existing cased well, then two conditions must also be fulfilled. First, the system must be able to drill a hole through the existing casing; second, the full 90° of angle change must be achievable within the existing hole diameter. (The alternative is to mill out a window, with the consequent increase in time and cost.) Systems capable of achieving this have been built and used operationally (Landers 2000; Buset et al. 2001; Cirigliano and Blacutt 2007; Medvedev and Tencic 2007). In order to create the initial hole in the casing, a conventional mill bit on a downhole motor and flexible shaft is used, to be replaced by a jet bit on a flexible hose. Clearly, in order for the jet bit to enter the hole drilled previously, a guide is necessary, and so the drilling operation takes place through a fixed whipstock deployed on tubing, which ensures hole re-entry and induces both the flexible shaft and jet-bit hose to deflect through 90°. The whipstock may be detached and rotated or moved, allowing further holes to be drilled either at the same level or at different levels. The holes generated are up to 2 in. in diameter and 100 m in length, and 10-kpsi pumps are required. The whole operation of drilling four laterals at one depth takes approximately 24 hours (Cirigliano and Blacutt 2007). As well as the coiled-tubing unit and pumps, workover equipment is required in order to deploy the whipstock.

It has been suggested recently to use not water but supercritical carbon dioxide as the drilling fluid for jet drilling (Kolle 2000, 2002). As the fluid emerges through the drilling head, its volume increases, and this expansion process aids the penetration of the fluid into rock. Experimental tests show that drilling speeds comparable to those of water drilling may be achieved at lower pumping pressures.

Jet-Assisted Drilling. Operationally, having 10–15 kpsi of pressure present at surface is a considerable safety hazard during drilling operations. In order to generate high bit velocities without such high pressures at surface, an alternative is to use a downhole pressure intensifier. A pressure intensifier uses hydraulic energy from the majority of the flow to increase the pressure of the minority. For instance, approximately 7% of the flow has its pressure increased to 30 kpsi, at a cost to the surface system of an additional 1,500–2,000 psi of surface pressure (Veenhuizen et al. 1997; O'Hanlon et al. 1998). Since most of the flow is not at high pressure, such a system must be used to augment the cutting action of a drill bit breaking rock by another means.

In tests with jet-assisted tricone bits, an improvement in ROP of approximately 50% has been reported in one case (Veenhuizen et al. 1997), and an unquantified improvement in another (Santos et al. 2000). As shown in **Fig. 9.5.7**, the high-pressure nozzle is mounted with its aperture close to the rock face and is positioned so as to cut a slot circumferentially around the hole. This has the effect of relieving some of the stress on the rock inside the slot, aiding in its removal by the conventional cones. Of course, the reliability of the downhole pump is an issue, and good solids control must be maintained on the drilling fluid to avoid erosion.

Jet-assisted drilling has also been attempted using a PDC drag bit (Cohen et al. 2005), where in this case all the pressure is generated at surface and so all the drilling fluid emerges with a high velocity through the bit nozzles. The bit pressure drop is near 10 kpsi. The nozzles cut channels in the rock, and the cutters are positioned to remove the ridges between them. Laboratory testing for the bit showed ROP multiplied by a factor between two and four by the use of high-pressure fluid. Limited testing at depth showed ROP between two and six times faster than that of conventional drilling in offset wells, though because the drilling took place at a test center, it is not clear whether in drilling the offset wells any attempt had been made to optimize ROP. For the same reasons as were given with jet drilling, the high pressures at surface operationally limit the technique to small holes drilled with coiled tubing.

Acid Drilling. The idea of acid drilling was introduced in 2001 (Rae and Di Lullo 2001). The idea is very simple and consists of combining a fluid-jetting system and acid to create holes in an acid-soluble rock. While use of acid drilling is clearly restricted, in practice, to drilling carbonates, the simplicity of the technique is a

Fig. 9.5.7—A tricone bit, modified to take a high-pressure nozzle (Santos et al. 2000).

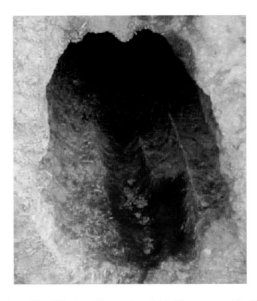

Fig. 9.5.8—Hole produced by 5⁵⁄₁₆-in., 10-nozzle acid jetting assembly (Portman et al. 2002).

great advantage—no returns to surface, no weight on bit, and no expensive downhole components. Safety and environmental issues make the use of jointless pipe (coiled tubing) essential.

Surface testing of simple jetting geometries showed that while a hole could be created using acid at much faster rates than just using water, to make a hole that the jetting assembly itself could enter required the correct nozzle geometry (Portman et al. 2002; Rae et al. 2004). Experiments showed that a number of small nozzles, evenly distributed in a circle, produced a clean, near-circular hole (Rae et al. 2004) **(Fig. 9.5.8)**.

Acid drilling has been deployed in the field on 1½-in. coiled tubing (Rae et al. 2007) and is used to create a number of short laterals off a main borehole. In Rae et al. (2007), a typical job is described. The objective is to stimulate production in low-producing wells. The tool is lowered to the required depth, the application of pump pressure activates a knuckle to aim the jet into the borehole wall, and the new hole is kicked off. Initially, approximately 15% hydrochloric acid is used, reducing to 10%. Sidetracks in excess of 100 ft in length have been drilled using this method. Acid requirements have averaged 1.25 bbl/ft of new borehole, and in terms of stimulation efficiency, it has required much less acid than conventional stimulation.

References

Buset, P., Riiber, M., and Eek, A. 2001. Jet Drilling Tool: Cost-Effective Lateral Drilling Technology for Enhanced Oil Recovery. Paper SPE 68504 presented at the SPE/ICoTA Coiled Tubing Roundtable, Houston, 7–8 March. DOI: 10.2118/68504-MS.

Bussod, G.Y., Dick, A.J., and Cort, G.E. 1998. Rock Melting Tool With Annealer Section. US Patent No. 5,735,355.

Carstens, J.P. and Brown, C.O. 1971. Rock Cutting by Laser. Paper SPE 3529 presented at the SPE Annual Meeting, New Orleans, 3–6 October. DOI: 10.2118/3529-MS.

Cirigliano, R.A. and Blacutt, J.F.T. 2007. First Experience in the Application of Radial Perforation Technology in Deep Wells. Paper SPE 107182 presented at the Latin American & Caribbean Petroleum Engineering Conference, Buenos Aires, 15–18 April. DOI: 10.2118/107182-MS.

Cohen, J.H., Deskins, G., and Rogers, J. 2005. High-Pressure Jet Kerf Drilling Shows Significant Potential To Increase ROP. Paper SPE 96557 presented at the SPE Annual Technical Conference and Exhibition, Dallas, 9–12 October. DOI: 10.2118/96557-MS.

Cort, G.E., Goff, S.J., Rowley, J.C., Neudecker, J.W., Dreesen, D.S., and Winchester, W. 1994. The Rock Melting Approach to Drilling. Technical report presented at the Drilling Technology Symposium, Energy-Sources Technology Conference and Exhibition, New Orleans, 23–26 January; Los Alamos National Laboratory, Report Number LA-UR--93-3191; CONF-940126—2.

Curlett, H.P., Sharp, D.P., and Gregory, M.A. 2002. Formation Cutting Method and System. US Patent No. 6,386,300.

Deily, F.H. 1955. Pellet Impact Core Drill. US Patent No. 2,724,575.

Eskin, M., Maurer, W.C., and Leviant, A. 1995. Former-USSR R&D on Novel Drilling Techniques. Houston: Maurer Engineering.

Friant, J.E. 1997. Disc Cutter Technology Applied to Drill Bits. Paper presented at the US DOE Natural Gas Conference, Houston, 24–27 March, Paper 2.3

Friant, J.E. and Anderson, M.A. 2000. Small Disc Cutter, and Drill Bits, Cutterheads, and Tunnel Boring Machines Employing Such Rolling Disc Cutters. US Patent No. 6,131,676.

Gahan, B.C., Parker, R.A., Batarseh, S., Figueroa, H., Reed, C.B., and Xu, Z. 2001. Laser Drilling: Determination of Energy Required To Remove Rock. Paper SPE 71466 presented at the SPE Annual Technical Conference and Exhibition, New Orleans, 30 September–3 October. DOI: 10.2118/71466-MS.

Gahan, B.C., Batarseh, S., Sharma, B., and Gowelly, S. 2004. Analysis of Efficient High-Power Fiber Lasers for Well Perforation. Paper SPE 90661 presented at the SPE Annual Technical Conference and Exhibition, Houston, 26–29 September. DOI: 10.2118/90661-MS.

Gahan, B.C., Batarseh, S., Watson, R., and Deeg, W. 2005. Effect of Downhole Pressure Conditions on High-Power Laser Perforation. Paper SPE 97093 presented at the SPE Annual Technical Conference and Exhibition, Dallas, 9–12 October. DOI: 10.2118/97093-MS.

Goodwin, R.J., Mori, E.A., Pekarek, J.L., and Schaub, P.W. 1968. Hydraulic Jet Drilling Method Using Ferrous Abrasives. US Patent No. 3,416,614.

Graves, R.M. and O'Brian, D.G. 1998. Star Wars Laser Technology Applied to Drilling and Completing Gas Wells. Paper SPE 49259 presented at the SPE Annual Technical Conference and Exhibition, New Orleans, 27–30 September. DOI: 10.2118/49259-MS.

Graves, R.M., Araya, A., Gahan, B.C., and Parker, R.A. 2002. Comparison of Specific Energy Between Drilling High Power Lasers and Other Methods. Paper SPE 77627 presented at the SPE Annual Technical Conference and Exhibition, San Antonio, Texas, USA, 29 September–2 October. DOI: 10.2118/77627-MS.

Hasiba, H.H. 1970. Relief Type Jet Bits. US Patent No. 3,548,959.

Inoue, H., Lisitsyn, I.V., Akiyama, H., and Nishizawa, I. 2000. Drilling of Hard Rocks by Pulsed Power. *IEEE Electrical Insulation Magazine* **16** (3): 19–25.

Juvkam-Wold, H.C. 1975. Drill Bit and Method of Drilling. US Patent No. 3,924,698.

Kolle, J.J. 2000. Coiled Tubing Drilling With Supercritical Carbon Dioxide. Paper SPE 65534 presented at the SPE/CIM International Conference on Horizontal Well Technology, Calgary, 6–8 November. DOI: 10.2118/65534-MS.

Kolle, J.J. 2002. Coiled Tubing Drilling With Supercritical Carbon Dioxide. US Patent No. 6,347,675.

Landers, C. 2000. Method and Apparatus for Horizontal Well Drilling. US Patent No. 6,125,949.

Martunovich, A.A. and Fedorovich, V.V. 2000. Electropulse Method of Holes Boring and Boring Machine. US Patent No. 6,164,388.

Maurer, W.C. 1968. *Novel Drilling Techniques*. Oxford, UK: Pergamon Press.

Maurer, W.C. 1980. *Advanced Drilling Techniques*. Tulsa: Petroleum Publishing Co.

Medvedev, P. and Tencic, M. 2007. Radial Formation Drilling: Economic Recovery of Remaining Reserves. *TNK-BP Innovator* **18** (October–November 2007): 15–17.

Moeny, W.M. and Barrett, D.M. 1999. Portable Electrohydraulic Mining Drill. US Patent No. 5,896,938.

Moeny, W.M. and Small, J.G. 1988. Focused Shock Spark Discharge Drill Using Multiple Electrodes. US Patent No. 4,741,405.

O'Brian, D.G., Graves, R.M., and O'Brian, E.A. 1999. StarWars Laser Technology for Gas Drilling and Completions in the 21st Century. Paper SPE 56625 presented at the SPE Annual Technical Conference and Exhibition, Houston, 3–6 October. DOI: 10.2118/56625-MS.

O'Hanlon, T.A., Kelley, D.P., and Veenhuizen, S.D. 1998. Downhole Pressure Intensifier and Drilling Assembly and Method. US Patent No. 5,787,998.

Plácido, J.C.R. and Friant, J.E. 2004. The Disc Bit—A Tool for Hard-Rock Drilling. *SPEDC* **19** (4): 205–211. SPE-79798-PA. DOI: 10.2118/79798-PA.

Portman, L., Rae, P., and Munir, A. 2002. Full-Scale Tests Prove It Practical To "Drill" Holes With Coiled Tubing Using Only Acid; No Motors, No Bits. Paper SPE 74824 presented at the SPE/ICoTA Coiled Tubing Conference and Exhibition, Houston, 9–10 April. DOI: 10.2118/74824-MS.

Rach, N.M. 2007. Particle Impact Drilling Blasts Away Hard Rock. *Oil & Gas Journal* **105** (6): 43–48.

Rae, P. and Di Lullo, G. 2001. Chemically Enhanced Drilling With Coiled Tubing in Carbonate Reservoirs. Paper SPE 68439 presented at the SPE/ICoTA Coiled Tubing Roundtable, Houston, 7–8 March. DOI: 10.2118/68439-MS.

Rae, P., Di Lullo, G., and Portman, L. 2004. Chemically Enhanced Drilling Methods. US Patent No. 6,772,847.

Rae, P., Di Lullo, G., Moss, P., and Portman, L. 2007. The Dendritic Well: A Simple Process Creates an Ideal Reservoir Drainage System. Paper SPE 108023 presented at the European Formation Damage Conference, Scheveningen, The Netherlands, 30 May–1 June. DOI: 10.2118/108023-MS.

Rauenzahn, R.M. and Tester, J.W. 1991. Numerical Simulation and Field Testing of Flame-Jet Thermal Spallation Drilling—Part I and II. *Intl. J. of Heat Mass Transfer* **34** (3): 795–818.

Rowley, J.C. and Neudecker, J.W. 1986. In Situ Rock Melting Applied to Lunar Base Construction and for Exploration Drilling and Coring on the Moon. In *Lunar Bases and Space Activities of the 21st Century*, ed. W.W. Mendall, 465–477. Houston: Lunar and Planetary Institute, NASA.

Santos, H., Placido, J.C.R., Oliviera, J.E., and Gamboa, L. 2000. Overcoming Hard Rock Drilling Challenges. Paper SPE 59182 presented at the IADC/SPE Drilling Conference, New Orleans, 23–25 February. DOI: 10.2118/59182-MS.

Smith, N.D. Jr. 1970. Shaped Spark Drill. US Patent No. 3,500,942.

Soloviev, G.N., Kudryashov, B.B., and Litvinenko, V.S. 1996. Method of Electrothermalmechanical Drilling and Device for Its Implementation. US Patent No. 5,479,994.

Stokka, S. 2006. Drilling Device. US Patent No. 7,093,673.

Summers, D.A. and Henry, R.L. 1972. Water Jet Cutting of Sedimentary Rock. *JPT* **24** (7): 797–802. SPE-3533-PA. DOI: 10.2118/3533-PA.

Tibbetts, G.A., Padgett, P.O., Curlett, H.B., Curlett, S.R., and Harder, N.J. 2007. Drill Bit. US Patent No. 7,258,176.

Trueman, R., Meyer, T.G.H., and Stockwell, M. 2005. Fluid Drilling System With Flexible Drill String and Retro Jets. US Patent No. 6,866,106.

Veenhuizen, S.D., Stang, D.L., Kelley, D.P., Duda, J.R., and Aslakson, J.K. 1997. Development and Testing of Downhole Pump for High-Pressure Jet-Assisted Drilling. Paper SPE 38581 presented at the SPE Annual Technical Conference and Exhibition, San Antonio, Texas, USA, 5–8 October. DOI: 10.2118/38581-MS.

Venghiattis, A.A. 1969. Well Perforating Apparatus and Method. US Patent No. 3,461,964.

Williams, R.E. 1986. The Thermal Spallation Drilling Process. *Geothermics* **15** (1): 17–22. DOI: 10.1016/0375-6505(86)90026-X.

SI Metric Conversion Factors

bbl	$\times$ 1.589 873	E $-$ 01 =	m^3
Btu	$\times$ 1.055 056	E + 00 =	kJ
ft	$\times$ 3.048*	E $-$ 01 =	m
°F	(°F $-$ 32)/1.8	=	°C
hp	$\times$ 7.460 43	E $-$ 01 =	kW
hp-hr	$\times$ 2.684 520	E + 00 =	MJ
in.	$\times$ 2.54*	E + 00 =	cm
kW-hr	$\times$ 3.6*	E + 00 =	J
psi	$\times$ 6.894 757	E + 00 =	kPa
ton	$\times$ 9.071 847	E $-$ 01 =	Mg

*Conversion factor is exact.

Chapter 10

Heat Transfer in Wells

10.1 Wellbore Thermal and Flow Simulation—Robert F. Mitchell, Halliburton, and Udaya B. Sathuvalli, Blade Energy Partners

10.1.1 Introduction. Casing loads result from running the casing, cementing the casing, subsequent drilling operations, production, and well-workover operations. The rationale for the design and analysis of casing and tubing (addressed in various degrees in Chapters 2, 3, and 8 of this book) show that casing loads can be classified broadly into pressure-induced and temperature-induced loads. Pressure-induced loads are produced by fluids within the casing, cement and fluids outside the casing, pressures imposed at the surface by drilling and workover operations, and pressures imposed by the formation during drilling and production. Temperature-induced loads result from thermal expansion of well tubulars, annular fluids, and the formation.

Though worst-case scenarios for many of these loads can be devised, a more accurate set of methods is necessary to determine the actual values of pressures and temperatures in the well. This is especially important in critical wells that are drilled in challenging environments—for example, in deepwater and high-pressure/high-temperature (HP/HT) fields. Greater water depths and completion depths in harsher environments impose significantly higher loads on tubulars. As a result, tubulars are required to operate very close to their structural operating limits.

The complex wells that are being drilled currently require the evaluation of a large number of loads to reflect the various scenarios that can occur during the life of the well. Casing design and analysis today are performed with computer programs, which generate the appropriate load sets (often custom-tailored for a particular operator), evaluate the results, and sometimes even determine a minimum-cost design automatically.

An accurate assessment of the thermal and mechanical loads is vital for an optimized well design. This chapter focuses on the thermal response of a wellbore and its components and describes methods to calculate this response. The determination of the thermal response requires the calculation of some or all of the following:

- Temperatures of the casing and tubing strings and risers
- Temperatures in the annuli between the strings
- Temperature of the formation (usually limited to a radial extent of 15 to 20 wellbore radii)
- Pressures in various flow streams
- Pressure changes in uncemented annuli (if they are present)

The above quantities are usually assessed for a given wellbore operation—for example, during drilling, reverse circulating, production, or injection. The results are used subsequently to determine the mechanical loads caused by the temperature changes.

Wellbore heat transfer is a mature discipline, and there exist standardized techniques to compute the temperatures in a wellbore. From the point of view of casing design and analysis, it is necessary to determine the temperatures in the wellbore during production, injection, drilling, and shut-in. The temperatures during cementing can usually be obtained as a special case of circulation followed by shut-in.

Among early papers on wellbore temperatures, heat transfer in a well during production or injection through tubing was discussed by Ramey (1962), and the heat transfer during circulation and drilling was investigated by Raymond (1969). The papers by Ramey (1962) and Raymond (1969) are valuable, since they describe the basic physics underlying wellbore heat transfer. The reader may consult Hasan and Kabir (2002) for a comprehensive discussion of all issues concerning wellbore heat transfer and for a bibliography of the seminal papers of the last few decades.

10.1.2 Effects of Temperature on Casing and Tubing Strings. The temperature of the wellbore is a function of its operating condition. The temperatures in the different sections of the wellbore will change, depending on the operation (e.g., drilling, circulation, cementing, shut-in, production, injection). The casing strings and the contents of the wellbore annuli respond to the temperature changes. In almost every instance, the parameter of interest is a temperature change, which is the difference between the temperatures during an initial or quiescent condition and some final operating condition. These temperature changes affect tubular design in the following ways.

Tubing Thermal Expansion. Changes in temperature will change the axial-force profile in the casing because of thermal contraction or expansion. For example, the reduction in length of the tubing (fixed at both ends) caused by pumping cool fluid into the wellbore during a stimulation job can be a critical axial design criterion. In contrast, reduction in tension during production because of thermal expansion can increase buckling and possibly result in compression at the wellhead.

Annular-Fluid-Expansion Pressure or Annular-Pressure Buildup (APB). Increases in temperature after the casing is landed will cause thermal expansion of fluids in sealed annuli, and this may result in significant pressure loads. This pressure change is known as APB. These pressure changes are a function of fluid pressure/volume/temperature (PVT) response, the mechanical stiffness of the annulus, and, of course, the temperature change. Methods to calculate APB are described by Halal and Mitchell (1994) and by Adams and MacEachran (1994). In land wells and in platform-completed wells, the loads caused by APB may not have to be considered explicitly if the pressures can be bled off. However, in subsea wells, where the outer annuli cannot be accessed after the hanger is landed, the pressure increase will influence the axial-load and differential-pressure profiles on the casing. In deepwater wells, especially those drilled in the Gulf of Mexico (GOM), APB-induced loads can pose a serious threat to well integrity and have led to several high-profile incidents such as the Marlin production-casing collapse (Bradford et al. 2004; Ellis et al. 2004; Gosch et al. 2004; Pattillo et al. 2006).

Temperature-Dependent Yield. Since changes in temperature affect the material yield strength, higher wellbore temperatures reduce the burst, collapse, axial, and triaxial ratings of the casing. A typical value used for derating the yield strength with temperature is 0.03% per °F (Steiner 1990).

Sour-Gas-Well Design. In sour environments, operating temperatures can determine what materials can be used at different depths in the wellbore.

Tubing Internal Pressures. Produced temperatures in gas wells will influence the gas gradient inside the tubing since gas density is a strong function of temperature and pressure.

10.1.3 Wellbore Temperatures and Flow Profiles. The first step in the analysis of thermal loads is the determination of the wellbore temperatures at a given operating condition. The temperatures and flow profiles in a wellbore are determined by using a computer program known as the wellbore thermal simulator.

The wellbore simulator combines techniques from a number of different engineering disciplines and predicts wellbore temperatures and flowing-fluid variables. The formulation of the computer model depends on the key features required for specific applications.

Given a well configuration, the simulator calculates the temperature distribution in the wellbore by using analytical solutions or numerical methods (or a combination of these methods) for different operational circumstances. Most simulators treat the wellbore as an axisymmetric finite-difference grid (Wooley 1980). Heat transfer from the flow stream to the wellbore is calculated by using the appropriate correlations for convective heat transfer. Heat transfer in the tubing and cement sections is modeled by conduction. Heat transfer in the formation is calculated by solving the 2D heat-conduction equations. Suitable boundary conditions (far away from the wellbore and at the wellbore/formation interface) are imposed.

The first need for the simulator is the best available properties and correlations to model the performance of the different fluids and wellbore materials. These correlations are coupled with the governing equations of mass, momentum, and energy in the simulator. The following model formulation discussion includes these governing equations. Specific references are cited to document the basis for the properties and correlations to be used in the simulator. The references should be reviewed for a more complete understanding of each correlation and its range of applicability.

10.1.4 Unique Aspects of Wellbore Heat Transfer. The next few subsections describe the theoretical basis for a wellbore simulator. These principles are based on adapting the principles of fluid flow and heat transfer to specific situations encountered in a typical wellbore. Before proceeding further, however, it is instructive to consider some aspects that require careful attention during wellbore temperature calculations. These include assumptions regarding the geothermal profile, the thermal interaction between the flowing multiphase fluid in the tubing and the wellbore, and the compositional model of the production fluid.

The geometry of the wellbore and the geothermal gradient have significant influence on the thermal response of the well. Since the wellbore geometry is characterized by a very large aspect ratio, the predominant direction of heat conduction is almost always radial. Heat transfer is directed radially outward during production, while it is directed radially inward during injection. Convected heat transfer, however, may be strongly vertical as a result of fluid flow. Operations such as circulation, drilling, and cementing tend to cool the lower section of the wellbore while heating the upper sections of the wellbore.

Geothermal Temperature. **Fig. 10.1.1** shows the geothermal profile based on temperature logs from exploration wells in a deepwater field in the GOM. The solid dark line appears to be a nearly linear fit to the data. However, a closer examination of the curves indicates that the temperature is not strictly linear with depth. The functional dependence of geothermal temperature with depth in this case is represented better by a bilinear curve, as shown in **Fig. 10.1.2**. In this illustration, the geothermal gradient is higher in the shallower section of the well, and it reduces below a certain distance from the mud line.

Fig. 10.1.2 illustrates the importance of using an accurate geothermal-temperature profile. The solid red and blue lines roughly represent the temperatures in the tubing and Annulus A during production and

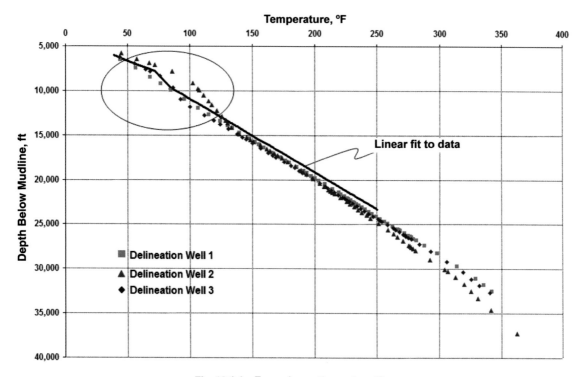

Fig. 10.1.1—Example geothermal profile.

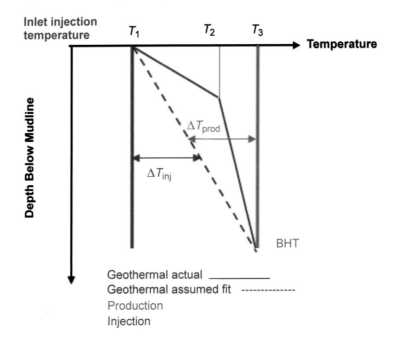

Fig. 10.1.2—The effect of the geothermal profile in assessing temperature changes.

injection, respectively. From the figure, the average change in temperature during production on the basis of an assumed linear geothermal profile (brown dashed line) is $(T_3 - T_1)/2$. This value reduces to $T_3 - (T_1 + T_2)/2$ when the bilinear geothermal profile (brown solid line) is assumed. Depending on the magnitude of T_2, the average temperature change is overestimated when a linear geothermal profile is assumed. A similar argument indicates that assuming a linear profile during injection underestimates the magnitude of temperature change at a given point in the wellbore. The thermal length changes of the tubulars and volume changes of the annular fluids are driven by the temperature changes, and errors in the geothermal-temperature gradient can have a significant effect on the final well design.

Fig. 10.1.3 illustrates this discussion for a real well. The solid lines are based on actual geothermal temperatures available (based on logs), while the dotted lines are based on an assumed linear fit to the available undisturbed temperature data. Most wellbore thermal simulators allow the user to enter the undisturbed temperature profile as a series of temperatures at different depths. When data are available (as is usually the case in development wells), it is preferable to use the actual geothermal temperatures rather than a linear profile that fits the data.

Multiphase Flow and Effects of Fluid Compositional Models. The analysis of multiphase flow during production is, perhaps, the most complicated aspect of wellbore heat transfer. The flow parameters and heat transfer in a flowing multiphase fluid differ from those in a single-phase fluid (Ramey 1962; Hasan et al. 1998). Since the temperatures are influenced by the relative magnitudes of gas, oil, and water that are produced, and by the nature of multiphase flow in the production tubing, special attention must be paid to this aspect of simulation. Most wellbore thermal simulators allow the user to specify a variety of multiphase correlations.

Fig. 10.1.4 shows the variation of steady-state wellhead temperature as a function of water content and flow rate. The relative magnitude of produced water increases toward the end of the life of a well, and significantly higher temperatures in the wellbore (and therefore the annuli) are expected. Finally, **Figs. 10.1.5 and 10.1.6** show the influence of the compositional model on the temperatures in various annuli of a typical subsea producer.

10.1.5 Key Wellbore-Simulation Capabilities. To address the wellbore operations of interest, a wellbore simulator must have a wide range of capabilities. These fall into the following four categories.

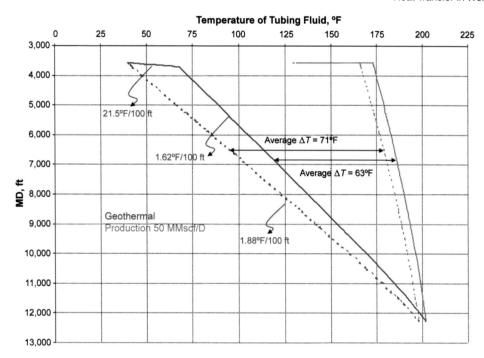

Fig. 10.1.3—Effect of geothermal gradient on tubing temperatures in a subsea well.

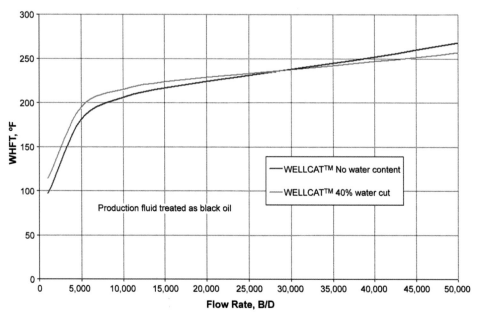

Fig. 10.1.4—Influence of flow rate and water content on wellhead temperature (WHT).

1. Transient effects
2. Flowing fluids
3. Wellbore geometry
4. Flow options

Many applications for operational design require an analysis of the transients. Drilling, cementing, fracturing, and production startup are all transient operations where fluid temperatures can change on the order of 100°F

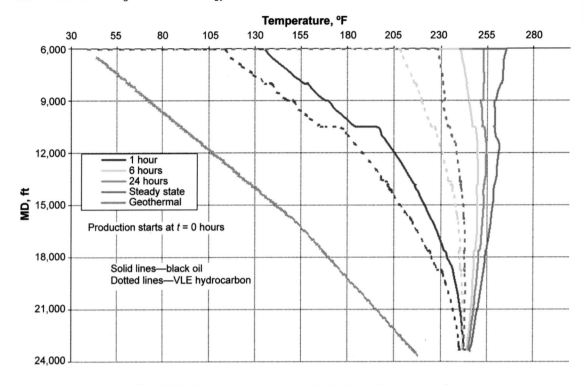

Fig. 10.1.5—Temperature in Annulus A of subsea deepwater well.

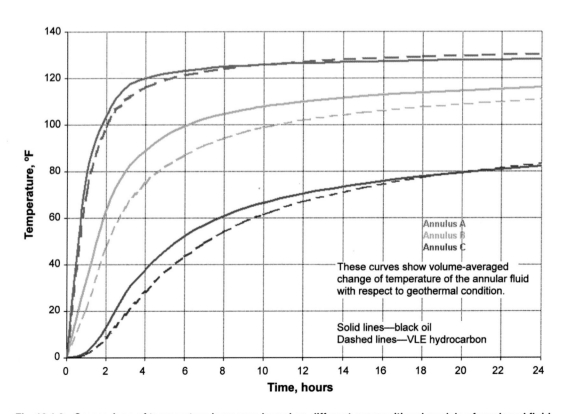

Fig. 10.1.6—Comparison of temperature increases based on different compositional models of produced fluid.

or more in a matter of minutes during flow in the well. While production and injection tend to heat or cool the wellbore, respectively, across its entire length, operations such as circulation and drilling cool the lower sections of the wellbore and heat the upper sections of the wellbore. These phenomena determine the severity of the loads on the casing strings and can influence well integrity significantly. For example, most deepwater and HP/HT wells experience large annular-pressure increases during production. The design of pressure-mitigation strategies such as burst disks and syntactic foams is linked intimately to the transient response of the wellbore annuli (Payne et al. 2003).

Fig. 10.1.7 illustrates the rate of change of the average annular temperature in a deepwater well. Note the rapid heat-up of the wellbore soon after production starts (temperature changes as high as 80°F/hr).

Table 10.1.1 shows the average temperature changes in the various strings and their respective annuli when drilling fluid is circulated at the bottom of the final hole section. The temperature changes are computed from the results of a wellbore thermal simulator, with the geothermal state as the reference condition. Negative numbers therefore indicate cooling, and positive numbers indicate heating.

Casing load analyses require the determination of data of the type shown in Fig. 10.1.7 and Table 10.1.1. The creation of such data is based on extensive sensitivity analyses that are best performed by a thermal simulator. Though the impact of the major variables can be studied qualitatively using analytical techniques such as those outlined by Ramey (1962), Raymond (1969), and Arnold (1990), a complete design requires a full numerical analysis.

Thus, a fully transient thermal response must be modeled in the flowing stream, the wellbore assembly, and the formation. The model must handle changing flow conditions, including changes in flow rate, inlet temperature and pressure, fluid type, and flow direction.

Oil-and gas-well operations involve fluids of many different types. The heat-transfer characteristics and temperature pressure coupling vary with fluid type. Oil- and water-based liquids and polymers behave differently from compressible and two-phase systems. Multiple fluids in the wellbore, including spacers and displacement fluids, are an important consideration. Temperature-dependent properties and wall coefficients

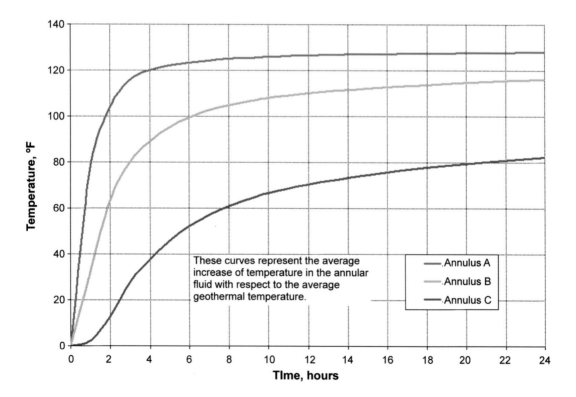

Fig. 10.1.7—Transient thermal response of wellbore annuli during production from a prolific deepwater well.

TABLE 10.1.1—THE EFFECT OF CIRCULATION AT THE TOTAL DEPTH (TD) OF THE $9^5/_8$-in. PRODUCTION CASING IN A TYPICAL DEEPWATER WELL*

	Circulation at TD				Drill From $9^5/_8$-in. Shoe to TD
	0.25 Days	1 Day	10 Days	40 Days	
Fluid	−12.95	−14.81	−20.48	−19.10	−19.10
Drillstring	−9.55	−11.72	−17.81	−16.36	−16.36
Drillstring annulus	−0.17	−2.83	−9.61	−8.02	−8.02
$9^5/_8$-in.	−8.61	−21.05	−38.88	−35.98	−35.98
$9^5/_8$-in. annulus	−6.01	−17.45	−34.62	−31.60	−31.60
$11^7/_8$-in.	−4.35	−8.80	−17.51	−15.56	−15.56
$11^7/_8$-in. annulus	−3.25	−7.57	−15.93	−13.97	−13.97
$13^5/_8$-in.	16.33	21.85	22.22	23.13	23.13
$13^5/_8$-in. annulus	10.23	18.16	20.15	21.08	21.08
16-in.	17.70	22.62	23.26	24.15	24.15
16-in. annulus	15.46	21.10	22.37	23.27	23.27
$17^7/_8$-in.	20.57	26.12	27.29	28.12	28.12
$17^7/_8$-in. annulus	18.46	24.65	26.39	27.24	27.24
22-in.	31.73	41.26	44.02	44.69	44.69
22-in. annulus	18.54	33.37	39.31	40.02	40.02
28-in.	3.74	18.24	30.90	31.58	31.58
28-in. annulus	1.73	12.48	25.87	26.56	26.56

* Entries in table indicate temperature changes in °F.

must be updated as temperatures and rheological properties change with time and depth. Even with drilling muds, the viscosity changes with temperature during the mud's circuit down the drillpipe and up the annulus, affecting the overall hydraulics of the system.

Flexibility in wellbore geometry is needed to accommodate different configurations such as deviated wells, liners, dual completions, and offshore risers. The geometry determines the cross-sectional flow area and the fluid velocity, which in turn govern the heat transfer. Temperatures during liner cementing are strongly influenced by the size of the liner and the annular clearance.

Flow options include production, injection, forward circulation, reverse circulation, drilling, and shut-in. Drilling is a special case of forward circulation, where the depth of circulation and the wellbore thermal resistance change as the well is drilled and casing is set. Flexibility in flow path and direction is needed to accommodate operations such as submersible pumps, hot-oil circulation, and multiple completions. The simulator must allow flow of any fluid in any annulus or pipe.

10.1.6 Model Formulation. The simulator should simultaneously solve wellbore heat-transfer and fluid-flow problems numerically and include the features specified in the preceding section. The importance of this linkage is obvious for compressible flow. However, this linkage is also necessary when temperature-dependent properties, such as viscosity, have a significant effect on the thermal and flow behavior.

10.1.7 Governing Equations. The wellbore simulator solves the energy equation for the wellbore and formation and solves the equations of mass, momentum, and energy conservation for each flow stream. The wellbore and formation are formulated as axisymmetric, transient-heat-transfer problems. Heat transfer in the wellbore consists of conduction, convection, and radiation. Heat transfer in the formation consists of conduction and convection. The flow streams are treated as 1D constant-area flow, though 3D-heat transfer effects are included along with the effects of discontinuous area changes. The mass- and momentum-balance equations are 1D, and pressure or density gradients in the direction normal to the flow are neglected.

The assumption of 1D flow means that the flow variables, such as density, velocity, and viscosity, are given as their average values over the cross-sectional area of the flow stream. The equations for frictional pressure drop and heat-transfer film coefficients used in the wellbore simulator are consistent with this convention. For instance, for flow in a tube, the frictional pressure drop is formulated in terms of the average velocity, density, and viscosity. The flow equations are given here in two sets: single-phase and two-phase flow. For further discussion of the basis of these equations and the 1D flow assumption, the texts by Zucro and Hoffman (1976) and Bird et al. (1960) are recommended.

The equations of mass and momentum conservation are solved subject to the assumption of quasisteady flow. This means that time variations of all variables except temperature are neglected in a time increment. In particular, this means that mass-accumulation effects are not considered in the mass-balance equation and that velocity is a function only of position in the momentum-balance equation. The steady-flow mass and momentum equations are solved for each time increment. These solutions are called quasisteady because they vary slightly between time increments because the temperature varies over the time increment.

The balance equations are written in a control volume form. The equations are written as integrals over a specified volume with specified surface areas rather than as partial-differential equations at a point. For the flow streams, the volume under consideration has cross-sectional area A and length Δz. The surface areas are the circular or annular cross-sectional areas A of the ends and the cylindrical lateral surface area. In the numerical solution of these equations, only the entrance and exit values of the flow variables are calculated. In order to evaluate the integrals, the variation of variables between the entrance and exit of each space increment may be needed. For example, the wellbore simulator may assume that density, velocity, viscosity, and thermal conductivity are constant and equal to the entrance conditions through the increment and that pressure and temperature vary linearly between entrance and exit. Experience has shown that these assumptions are reasonable except for compressible flow near critical flow conditions. When adequate computational performance is available, modern numerical techniques such as Romberg integration, adaptive Runge-Kutta, or Bulirsch-Stoer methods (Press et al. 1999) should be used to enhance accuracy and allow for coarser spatial increments for any type of flow.

Single-Phase Flow. The balance of mass for single-phase flow is given by

$$\dot{m} = \rho v A = \text{constant}, \dotfill (10.1.1)$$

where $\dot{m}$ is the mass-flow rate (kg/s), ρ is density (kg/m^3), v is the average velocity (m/s), A is the area (m^2), and quasisteady flow has been assumed (in this chapter, a dot above the symbol for a variable represents the derivative of the quantity with respect to time. For example, $\dot{m} = \dfrac{dm}{dt}$). In this context, this means that the partial derivatives of velocity and density with time have been neglected. In other words, the equation is solved with properties evaluated at the current fixed time. As temperatures change, the density will change, and hence velocity will vary with time. We are assuming this happens *slowly* compared to transient pressure conditions, so there is no mass accumulation or wave propagation. By Eq. 10.1.1, the mass-flow rate in any flow stream is constant. Note that where there is an interface between two different fluids, there may be a discontinuity in mass-flow rate, since we are assuming continuity of volume-flow rate at the interface.

The quasisteady momentum balance for single-phase flow has the form

$$\Delta P + \rho v \Delta v + \int_{\Delta z} \rho g \cos\phi \, dz \pm \int_{\Delta z} \frac{2 f \rho v^2}{D_h} \, dz = 0, \dotfill (10.1.2)$$

where P is the pressure (Pa), g is the acceleration due to gravity (m/s^2), ϕ is the angle of pipe (i.e., the wellbore) with vertical, f is the Fanning friction factor, D_h is the hydraulic diameter (m), v is the flow velocity (m/s), and Δz is the length of flow increment (m).

Eq. 10.1.2 balances the pressure-area forces acting on a control volume, bounded by two planes perpendicular to the flow direction and separated by a distance Δz along the flow direction, with the inertial, the gravitational, and the viscous forces (represented by the three terms from the right-hand side). The viscous force is calculated by use of correlations that relate the Fanning friction factor to the Reynolds number. The Fanning friction factor typically depends on the fluid density, velocity, viscosity, and type, and on the pipe roughness. The wellbore simulator should contain appropriate models for f—for instance, Newtonian fluids, power-law fluids, and polymer-based fluids.

The basic balance-of-energy equation for single-phase flow is given by

$$\int_{\Delta z} \rho \dot{\varepsilon}\, A\, dz = -\int_{\Delta z} p \frac{dv}{dz} A\, dz + \Phi + Q + R \quad \dots\dots\dots\dots\dots\dots\dots\dots\dots\dots\dots\dots\dots \quad (10.1.3)$$

where

ε = internal energy per unit mass, J/kg,
Φ = viscous dissipation, W,
Q = heat transferred into volume, W,
R = rate of volume energy added, W, and

$\dot{\varepsilon}$ = rate of change of internal energy = $\dfrac{\partial \varepsilon}{\partial t} + v \dfrac{\partial \varepsilon}{\partial z}$, W/kg.

Eq. 10.1.3 represents the first law of thermodynamics for a control volume (i.e., the increase in the internal energy of the control volume equals the sum of work performed and heat added). The right side of the equation represents the rate of change of internal energy of the control volume. The first and second terms on the right-hand side of Eq. 10.1.3 represent the pressure-volume work and viscous dissipation, respectively. The work performed by the pressure forces is usually significant when compressible flow is involved. The viscous-dissipation term Φ represents the work performed to overcome friction (i.e., the shear stresses at the fluid/pipe boundary). The last two terms on the right-hand side represent the heat loss and heat added to (generated in) the control volume, respectively. The term Q is usually written as the total heat flux into the control volume. The term R represents any sources of heat (e.g., cement heat of hydration) in the control volume. In most wellbore flow situations, R is set to zero. Eq. 10.1.3 can be rewritten in terms of enthalpy:

$$\int_{\Delta z} \rho \dot{h}\, A\, dz = \int_{\Delta z} v \frac{dp}{dz} A\, dz + \Phi + Q + R, \quad \dots\dots\dots\dots\dots\dots\dots\dots\dots\dots\dots\dots \quad (10.1.4)$$

where h = enthalpy = $\varepsilon + pAz$ (J/kg).

By choosing pressure and temperature as independent variables, and by using the following thermodynamic relationship for the dependence of enthalpy on pressure,

$$dh = C_p\, dT + (1 - \beta T) \frac{dp}{\rho},$$

Eq. 10.1.4 can be rewritten as

$$\int_{\Delta z} \rho C_p \frac{\partial T}{\partial t} A\, dz + \int_{\Delta z} \rho v C_p \frac{\partial T}{\partial z} A\, dz - \int_{\Delta z} v \beta T \frac{\partial p}{\partial z} A\, dz = \Phi + Q + R, \quad \dots\dots\dots\dots\dots \quad (10.1.5)$$

where

$C_p = \dfrac{\partial h}{\partial T}(P, T)$ = heat capacity at constant pressure, J/kg·K

$\beta = -\dfrac{1}{\rho} \dfrac{\partial \rho}{\partial T}(P, T)$ = isobaric coefficient of thermal expansion, 1/K

T = absolute temperature, K

Two-Phase Flow. The balance of mass for a two-component flow stream is given by

$$\dot{m}_1 + \dot{m}_2 = \dot{m} = \text{constant}, \quad \dots\dots\dots\dots\dots\dots\dots\dots\dots\dots\dots\dots\dots\dots\dots \quad (10.1.6)$$

where subscript 1 denotes component 1, subscript 2 denotes component 2, $\dot{m}$ is the total mass flux (kg/s), and quasisteady flow has been assumed. This equation means that the total mass flux of the system is constant and that the two phases can exchange mass as long as the mass-flux change of component 1 equals the negative of the mass-flux change of component 2. The two-phase-flow literature has many ways to define a density and velocity for each component of the flowing stream. The most common definitions are described here:

χ_a = volume fraction of component a, a = 1, 2 (liquid holdup is volume fraction of the liquid; 1 minus holdup is the volume fraction of the gas)

$\bar{\rho}_a$ = density of single-phase component a, a = 1, 2

$\rho_a = \chi_a \bar{\rho}_a$ = in-mixture density of component a, a = 1, 2

$v_a = \dot{m}_a/(\rho_a A)$ = local average velocity of component a, a = 1, 2

$v_{sa} = \dot{m}_a/(\bar{\rho}_a A)$ = superficial velocity of component a, a = 1, 2

$\dot{m}_a = \bar{\rho}_a v_{sa} A = \rho_a v_a A$, a = 1, 2

The balance of momentum for two-phase flow is given by the following equation:

$$\Delta P + \Delta P_{acc} + \int_{\Delta z}(\rho_1 + \rho_2)g\,\cos\varphi\,dz \pm \int_{\Delta z} F\,dz = 0, \quad\dots\dots\dots\dots\dots\dots\dots \text{(10.1.7)}$$

where

$$\Delta P_{acc} = \int_{\Delta z}\rho_1 v_1\frac{\partial v_1}{\partial z} + \rho_2 v_2\frac{\partial v_2}{\partial z}\,dz = \text{acceleration pressure drop, Pa, and}$$

F = two-phase frictional pressure drop per unit length, Pa/m.

The energy equation for two-phase flow has the following form:

$$\int_{\Delta z}\sum_{a=1}^{2}\left[\rho_a\frac{\partial \varepsilon_a}{\partial t} + \frac{\partial}{\partial z}(\rho_a v_a \varepsilon_a)\right]A\,dz = -\int_{\Delta z}\sum_{a=1}^{2}\left(P_{pa}\frac{\partial v_a}{\partial z}\right)A\,dz + \Phi + Q + R, \quad\dots\dots\dots\dots \text{(10.1.8)}$$

where P_{pa} is the partial pressure of a (Pa), ε_a is the internal energy of a (J/kg), and mechanical-energy terms that are second order or higher in the diffusion velocity have been neglected. Using the quasisteady assumptions, Eq. 10.1.8 can be rewritten in terms of enthalpy:

$$\int_{\Delta z}\sum_{a=1}^{2}\left[\rho_a\frac{\partial h_a}{\partial t} + \frac{\partial}{\partial z}(\rho_a v_a h_a)\right]A\,dz = \int_{\Delta z}\sum_{a=1}^{2}\left(v_a\frac{\partial P_{pa}}{\partial z}\right)A\,dz + \Phi + Q + R, \quad\dots\dots\dots\dots \text{(10.1.9)}$$

where h_a is the enthalpy of a (J/kg). The final form of the energy equation is written in terms of a mixture temperature T and partial pressure for each component:

$$\int_{\Delta z}\sum_{a=1}^{2}\left[\rho_a C_{pa}\frac{\partial T}{\partial t} + \rho_a v_a\left(C_{pa}\frac{\partial T}{\partial z} - T\frac{\beta_a}{\rho_a}\frac{\partial P_{pa}}{\partial z}\right) + h_a\frac{\partial}{\partial z}(\rho_a v_a)\right]A\,dz = \Phi + Q + R \quad\dots\dots \text{(10.1.10)}$$

Formation Heat Transfer. The energy equation for the formation is given by

$$\rho_f C\frac{\partial T}{\partial t} = \frac{1}{r}\frac{\partial}{\partial r}\left(rK_r\frac{\partial T}{\partial r}\right) + \frac{\partial}{\partial z}\left(K_z\frac{\partial T}{\partial z}\right) + q, \quad\dots\dots\dots\dots\dots\dots\dots\dots \text{(10.1.11)}$$

where ρ_f is formation density (kg/m³), C is formation heat capacity (J/kg·K), K_r is radial formation thermal conductivity (W/m·K), K_z is vertical formation thermal conductivity (W/m·K), q is volume heat added (W/m³), and heat transfer in the formation is assumed to be anisotropic Fourier heat conduction in cylindrical coordinates. Eq. 10.1.11 is modified by the inclusion of convected energy when flow in the formation is modeled:

$$\rho_f C\frac{\partial T}{\partial t} + \rho_{fl}v_{fl}C_{fl}\frac{\partial T}{\partial z} = \frac{1}{r}\frac{\partial}{\partial r}\left(rK_r\frac{\partial T}{\partial r}\right) + \frac{\partial}{\partial z}\left(K_z\frac{\partial T}{\partial z}\right) + q, \quad\dots\dots\dots\dots\dots\dots \text{(10.1.12)}$$

where ρ_{fl} is fluid density (kg/m³), v_{fl} is fluid velocity (m/s), and C_{fl} is fluid heat capacity (J/kg·K).

Boundary Conditions and Initial Conditions. The wellbore-simulator boundary conditions are fixed temperatures at the maximum radius of the model, at the maximum depth of the model, and in the environment above the surface. Heat transfer between the environment and the surface is given by a free-convection film coefficient (Chapman 1967).

Initial conditions for a given flow period are the temperature distribution corresponding to the results from the last flow period and the inlet pressure and temperature of the flowing streams. In the case of reservoir flow into the wellbore, the temperature at the maximum radius of the model is set equal to the initial reservoir temperature.

The initial temperature distribution for the first flow period is equal to the user-input undisturbed static gradient throughout the wellbore and formation. A restart option would save temperature conditions from one run to use as initial conditions for the next run.

Fluid Properties. The wellbore simulator uses a variety of fluid types. This section describes some fluid types that should be made available, and the sources to be used to determine fluid properties.

Drilling Fluids. Non-Newtonian Viscosities. While good viscosity correlations exist for pure water (Wagner and Kruse 1998; Hill et al. 1969) and a variety of hydrocarbons (Reid et al. 1977), there are very few viscosity correlations for water-based and oil-based drilling fluids. Older papers (Gray and Darley 1980; Annis 1967; Hiller 1963; Combs and Whitmire 1968; Houwen and Geehan 1986) give a good description of problems evaluating non-Newtonian viscosities. Alderman et al. (1988) and Sorelle et al. (1982) offer practical correlations for water- and oil-based fluids, but more research seems desirable.

Mud Density. Pressure and temperature dependence of mud density is calculated from the following relationship:

$$\rho(P,T) = \frac{\rho(P_a, T_r)}{1 - \dfrac{f_o \Delta \rho_o}{\rho_o(P,T)} - \dfrac{f_w \Delta \rho_w}{\rho_w(P,T)}}, \quad \dots\dots\dots\dots\dots\dots\dots\dots\dots \quad (10.1.13)$$

where

$\rho(P, T)$ = mud density at pressure P and temperature T, kg/m^3,
$\rho(P_a, T_r)$ = mud density at atmospheric pressure P_a and reference temperature T_r, kg/m^3,
f_o = volume fraction of oil at P_a and T_r,
ρ_o = density of oil phase, kg/m^3,
f_w = volume fraction of water at P_a and T_r,
ρ_w = density of water phase, kg/m^3,
$\Delta \rho_o = \rho_o(P, T) - \rho_o(P_a, T_r)$,
and $\Delta \rho_w = \rho_w(P, T) - \rho_w(P_a, T_r)$.

The formulas for calculating the water density should be based on the 1997 international standards (Wagner and Kruse 1998). Oil density may be based on the work of Sorelle et al. (1982), or on a model proposed for synthetic oils by Zamora et al. (2000), which has proved to correlate liquid density well and could be extended to other hydrocarbons. The Soave-Redlich-Kwong or Peng-Robinson equations of state (Reid et al. 1977) could be used for light oils.

Brine Density. The density of brines may be based on the thermodynamic electrolyte model of Kemp and Thomas (1987):

$$\rho(P,T) = \frac{1 + \sum_i m_i M_i}{(1/\rho_w) + \sum_i m_i \phi_i}, \quad \dots\dots\dots\dots\dots\dots\dots\dots\dots\dots\dots\dots \quad (10.1.14)$$

where

$\rho(P, T)$ = density of the brine, kg/m^3,
m_i = the concentration of the ith salt, mol/kg water,
M_i = molecular weight of the ith salt, kg/mol,
ρ_w = density of pure water, kg/m^3,
and ϕ_i = apparent molal volume of the ith salt, m^3/mol.

Because of the importance of NaCl brine, the correlation of Rodgers and Pitzer (1982) is recommended in place of Kemp and Thomas (1987). The reader is urged to consult recent literature for brine properties to supplement these references.

Steam and Water Properties. Water-based fluids are perhaps the most common fluids found in wellbores. For this reason, we recommend the formulation developed for the Industrial Standard IAPWS-IF97 (Wagner and Kruse 1998). Older correlations by Hill et al. (1969) are still valid. Viscosity, thermal conductivity, and surface tension for steam and water may be calculated from correlations given in appendices of Wagner and Kruse (1998).

Production Fluids. The properties of hydrocarbon mixtures have been researched extensively for both the upstream and downstream parts of the industry. An in-depth discussion of this subject is far beyond the scope of this chapter. The reader is referred to Reid et al. (1977), which will lead the reader to some of the (massive) literature on the subject. For a more hydrocarbon-production point of view, there is the monograph by Whitson and Brule (2000).

There are two common approaches to characterizing production hydrocarbons: (1) compositional thermodynamic models and (2) *black-oil* models. For gases and light hydrocarbon fluids, called condensates, the properties are calculated by means of a thermodynamic-based compositional model, as discussed by Reid et al. (1977) and Whitson and Brule (2000). For modeling the PVT and thermodynamic behavior of real gases, two representative PVT models are the Soave-Redlich-Kwong and Peng-Robinson equations of state (Reid et al. 1977). These models are particularly convenient because the pressure dependence is cubic in specific volume, which makes calculation of volume from pressure simple. Full-thermodynamic vapor/liquid equilibrium (VLE) calculations are then used to model hydrocarbon behavior for condensate wells. In these models, thermodynamic equilibrium is used to determine the gas composition, liquid composition, and liquid fraction of the mixture. Once these compositions are known, transport properties, such as viscosity and thermal conductivity, are determined from correlations.

Because these calculations are so complex and difficult, a simplified model has been developed for heavier oils. In this model, a residual liquid phase is always present, and the gas composition is fixed. VLE is reached by dissolving the gas in the liquid phase, as defined by correlations for oil-dissolved-gas/oil ratio, bubble-point, and formation volume factor (liquid density). Typical black-oil models are Beggs and Vasquez (1980), Standing (1981), and Brill (1999). These black-oil models are usually not thermodynamically complete, so other sources may be needed. For example, the oil heat capacity was found in Katz (1959). Viscosity correlations may be found in other sources, such as Beggs and Robinson (1975).

An intermediate-weight hydrocarbon mixture, called a volatile oil, is usually modeled with a compositional model, though these calculations can be very difficult.

Because of the difficulty of modeling condensates and heavy oils, an alternative approach for wellbore thermal modeling is to use a dedicated hydrocarbon model to generate tables of properties, which can then be interpolated by the thermal code.

10.1.8 Flow Correlations. The simulator needs to calculate pressure drop from both single- and two-phase-flow correlations.

Single-Phase-Flow Correlations. Single-phase-flow frictional pressure drop is calculated with the Fanning friction factor, as shown in Eq. 10.1.2. The friction factor can be derived for laminar flow in tubing and annuli for Newtonian fluids in basic fluid-mechanics textbooks (Bird et al. 1960). The friction factor can be derived for laminar flow in tubing and annuli for non-Newtonian-fluid models with the same equation form, but with the Reynolds number redefined (Gray and Darley 1980; Savins 1958; Govier 1977; Economides et al. 1998). Turbulent-flow friction factors for Newtonian fluids were correlated experimentally by Colebrook, as summarized by Govier (1977). Developed for power-law fluids, the Dodge and Metzner (1959) correlation for turbulent flow is commonly used for all non-Newtonian fluids. As an alternative to Dodge and Metzner, there is the correlation of Randall and Anderson (1982) for pressure drops in tubing and annuli.

Two-Phase-Flow Correlations. Two-phase-flow pressure-drop correlations predict the effects of hydrostatic pressure change and frictional pressure drop. Hydrostatic pressure change is given by

$$\Delta P_h = \left(\overline{\rho}_L H_L + \overline{\rho}_v H_v \right) g \cos\phi, \quad \text{...} \quad (10.1.15)$$

where

$\bar{\rho}_L$ = the liquid phase density, kg/m³,

H_L = the liquid holdup,

$\bar{\rho}_v$ = the vapor phase density, kg/m³,

$H_v = 1 - H_L$,

and ϕ = pipe-inclination angle from the vertical.

The liquid holdup corresponds to χ_L as defined in Section 10.1.7 and represents the volume fraction occupied by the liquid. Since the remaining fraction is occupied by the vapor, $H_v = 1 - H_L$. The liquid holdup is usually determined from a comprehensive correlation that depends on the properties of the liquid and vapor phases, the flow rate of each phase, the pipe diameter, and the pipe inclination. The frictional-pressure-drop correlation has the form

$$\Delta P_f = 2f_{tp}\, \rho_m v_m^{\,2}/D_h, \quad \dots\dots\dots\dots\dots\dots\dots\dots\dots\dots\dots\dots\dots\dots\dots\dots\dots\dots\dots\dots\dots\dots\dots\dots\dots\dots \quad (10.1.16)$$

where f_{tp} is two-phase friction factor, ρ_m is some density definition for mixture (kg/m³), and v_m is some mixture-velocity definition (m/s). The definitions of density and velocity of the mixture used in Eq. 10.1.16 vary from correlation to correlation, and sometimes vary within a single correlation.

One commonly used set of two-phase-flow correlations was developed by Brill (1999). Of the two-phase-flow correlations currently available, this set is most general regarding pipe inclination and flow direction, but its accuracy of 17.8% is not as good as more specialized correlations. However, this set of correlations was found to be accurate for the vertical downward flow of steam. A set of two-phase-flow correlations developed by Orkiszewski (1967) provides an alternative to the Brill (1999) correlations for vertical oil/gas/water production. The Orkiszewski correlation is accurate to approximately 8.6% but is applicable only to vertical upward flow. Other commonly used two-phase-flow correlations are Gray (1974), Hagedorn and Brown (1964), and Duns and Ros (1963). A new method, called *mechanistic modeling*, is currently being developed (Shoham 2006).

10.1.9 Heat-Transfer Correlations. For fluids, both flowing and shut-in, correlations are needed to handle six conditions: (1) conduction, (2) forced convection, (3) free convection, (4) radiation, (5) combined free and forced convection, and (6) turbulent forced convection.

The main source for the conduction and forced-convection correlations is the standard textbook material summarized in Chapman (1967, 333–336, 339). The best free-convection correlation for the annuli was developed by Dropkin and Somerscales (1965). Radiation correlations may be taken from Willhite (1967). The textbook by White (1984, 353–355, 288) presents correlations for situations where combined free and forced convection and turbulent forced convection apply.

10.1.10 Solution Method. The numerical solution process consists of two principal steps.

1. The reduction of the differential equations of energy and momentum conservation to algebraic equations.
2. The solution of these algebraic equations subject to the appropriate boundary conditions.

Development of Algebraic Equations. Several techniques are used to develop the algebraic equations. In all cases, the wellbore and formation are subdivided into small volumes. Associated with these volumes are nodal points. The resulting mesh of nodal points takes the following form.

1. The mesh should be rectangular with variable vertical and radial nodal-point spacing.
2. The radial mesh spacing within the wellbore is governed by the well completion. At each depth, radial nodal points are arranged so that there is a node in each annulus and in each casing.
3. Radial points outside the wellbore are distributed logarithmically with the refined mesh near the wellbore and the coarser mesh near the boundaries.

In the wellbore, the heat-balance technique is used to derive the coefficients. This method starts with the integral energy equation (Eq. 10.1.5) for single-phase flow or Eq. 10.1.10 for two-phase flow. Appropriate choices for the heat fluxes are made (either film coefficients for flow or conduction for solids), and the integration is performed. The alternative method would be to start with the differential energy equation and apply finite-difference techniques. The heat-balance method was chosen because of the variety of heat-transfer models necessary to simulate wellbore heat transfer (Lunardini 1981). Alternatively, finite-difference techniques are used to generate coefficients in the formation (Peaceman 1977). Note that for solid materials, these two formulations are equivalent. Both formulations allow arbitrary variation in radial and vertical grid spacing. A definite advantage to these techniques is the narrow bandwidth of the resulting algebraic equations that allows a very efficient solution algorithm.

Fig. 10.1.8 illustrates the heat balance for cell j, k, where j is the radial index associated with r_j, and k is the vertical index associated with z_k. Heat flux into cell j, k is given by

$$Q^{j,k}_{j+1,k} = R^{j,k}_{j+1,k} \left[T_{j+1,k} - T_{j,k} \right], \qquad \dotfill \qquad (10.1.17)$$

where $\Re$ is the heat transfer coefficient in the appropriate direction, as given by the indices. The total heat flux into cell j, k is then

$$Q^{j,k} = Q^{j,k}_{j-1,k} + Q^{j,k}_{j+1,k} + Q^{j,k}_{j,k-1} + Q^{j,k}_{j,k+1} + Q_{j,k}, \qquad \dotfill \qquad (10.1.18)$$

where $Q_{j,k}$ is the heat generated within cell j, k. The transient heat balance is therefore

$$\bar{C}_{j,k} \frac{dT_{j,k}}{dt} = Q^{j,k}, \qquad \dotfill \qquad (10.1.19)$$

where $\bar{C}_{j,k} = 2\pi \iint \rho c_p r \, dr \, dz$ is the total heat capacity of cell j, k.

The following heat transfer effects are used in the above formulation to evaluate $\Re$.

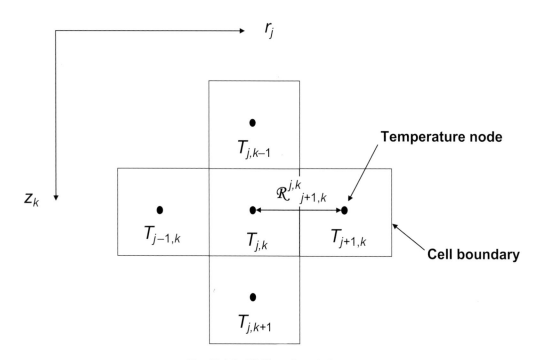

Fig. 10.1.8—Wellbore heat balance.

1. Flowing fluids.
   a. Vertical free and forced convection.
   b. Vertical and radial heat conduction.
   c. Change of phase.
2. Wellbore.
   a. Vertical and radial heat conduction.
   b. Natural convection in annuli.
   c. Radiation in annuli.
3. Formation.
   a. Vertical and radial heat conduction.
   b. Change of phase.
      • Melting of frozen soils.
      • Desiccation of wet soils.

Using finite differences for the time derivative,

$$\frac{dT}{dt} \cong \frac{T^n - T^{n-1}}{\Delta t}, \quad \dots\dots\dots\dots\dots\dots\dots\dots\dots\dots\dots\dots\dots\dots\dots\dots\dots \quad (10.1.20)$$

the following governing equation results:

$$T_{j,k}^n = A_{j,k} T_{j-1,k}^m + B_{j,k} T_{j+1,k}^m + C_{j,k} T_{j,k-1}^m + D_{j,k} T_{j,k+1}^m \\ + E_{j,k} T_{j,k}^{n-1} + F_{j,k}, \quad \dots\dots\dots\dots\dots\dots\dots\dots\dots\dots\dots\dots \quad (10.1.21)$$

where the superscript n indicates the nth timestep, superscript m may be either n or $n-1$, and the coefficients $A_{j,k}$ through $F_{j,k}$ are determined by rearranging Eqs. 10.1.17 through 10.1.19. For a more complete description of the development of these equations, see Wooley (1980), Chapman (1967), and Willhite (1967).

Solution of Algebraic Equations. The solution of the algebraic equations (Eq. 10.1.21) uses conventional reservoir-engineering techniques. As previously stated, the index m may be either n or $n-1$. If we choose $m = n-1$, the formulation is called explicit. This formulation is quite easy to solve, since everything on the right side of Eq. 10.1.21 is known. However, a major consideration for solving transient problems is stability (Peaceman 1977). It is important to recognize that explicit formulations are stable up to a maximum timestep size. Because the maximum timestep size is usually governed by convected heat, typically the time for a flowing fluid to traverse the smallest cell, explicit calculations may take a long time to execute.

If the index m equals n, the formulation is called implicit. Implicit solutions provide stability; however, the penalty for an implicit formulation is increased computational difficulty and some loss of accuracy.

For most applications, we recommend that a fully implicit calculation be made, using the alternating-direction implicit (ADI) algorithm (Peaceman 1977). Because of the narrow bandwidth of the equations, a tridiagonal solution method is used to solve the implicit equations. This solution technique has been used extensively for many applications, including reservoir engineering, because of its high efficiency. An alternative solution method is Gauss-Seidel iteration (Wooley 1980); however, while it works well for liquid systems and is easy to formulate, it becomes less effective than the ADI algorithm for more-complex systems.

Nomenclature

A = area, m^2
C = formation heat capacity, J/kg·K
C_{fl} = heat capacity of fluid, J/kg·K
C_p = heat capacity at constant pressure, J/kg·K
D_h = hydraulic diameter, m
f = Fanning friction coefficient

f_o = volume fraction of oil

f_{tp} = two-phase fanning friction coefficient

f_w = volume fraction of water

F = two-phase frictional pressure drop per unit length, Pa/m

g = acceleration of gravity, m/s^2

h = enthalpy per unit mass, J/kg

h_a = enthalpy per unit mass for component a, J/kg

H_L = liquid holdup

H_v = vapor volume fraction

K_r = formation radial thermal conductivity, W/m·K

K_z = formation vertical thermal conductivity, W/m·K

$\dot{m}$ = mass-flow rate, kg/s

$\dot{m}_a$ = mass-flow rate of component a, kg/s

m_i = concentration of ith salt, mol/kg water

M_i = molecular weight of ith salt, kg/mol

P = pressure, Pa

P_a = atmospheric pressure, Pa

P_{pa} = partial pressure of component a, Pa

Q = heat transferred into volume, W

q = volume heat added, W/m^3

R = rate of energy added, W

$\Re$ = heat transfer coefficient

r = radius, m

T = absolute temperature, K

T_r = reference temperature, K

v = flow velocity, m/s

v_a = local average velocity of component a, m/s

v_{fl} = fluid velocity, m/s

v_m = flow velocity of mixture, m/s

v_{sa} = superficial velocity of component a, m/s

z_k = length of increment, m

β = isobaric coefficient of thermal expansion, 1/K

ΔP_{acc} = acceleration pressure drop, Pa

ΔP_f = mixture frictional pressure drop, Pa

Δz = length of flow increment, m

ε = internal energy per unit mass, J/kg

ε_a = internal energy per unit mass for component a, J/kg

ρ = density, kg/m^3

$\bar{\rho}_a$ = density of single-phase component a, kg/m^3

$\bar{\rho}_L$ = the liquid-phase density, kg/m^3

ρ_f = formation density, kg/m^3

ρ_{fl} = fluid density, kg/m^3

ρ_m = mixture fluid density, kg/m^3

ρ_o = density of oil phase, kg/m^3

$\bar{\rho}_v$ = the vapor phase density, kg/m^3

ρ_w = density of water phase, kg/m^3

ϕ = inclination of flow, radians

ϕ_i = apparent molal volume of ith salt, m^3/mol

Φ = viscous dissipation, W

χ_a = volume fraction of component a

χ_L = volume fraction of liquid phase

References

Adams, A.J. and MacEachran, A. 1994. Impact on Casing Design of Thermal Expansion of Fluids in Confined Annuli. *SPEDC* **9** (3): 210–216. SPE-21911-PA. DOI: 10.2118/21911-PA.

Alderman, N.J., Gavignet, A., Guillot, D., and Maitland, G.C. 1988. High-Temperature, High-Pressure Rheology of Water-Based Muds. Paper SPE 18035 presented at the SPE Annual Technical Conference and Exhibition, Houston, 2–5 October. DOI: 10.2118/18035-MS.

Annis, M.R. 1967. High-Temperature Flow Properties of Water-Base Drilling Fluids. *JPT* **19** (8): 1074–1080; *Trans.*, AIME, **240**. SPE-1698-PA. DOI: 10.2118/1698-PA.

Arnold, F.C. 1990. Temperature Variation in a Circulating Wellbore Fluid. *Journal of Energy Resources Technology* **112** (2): 79–83. DOI: 10.1115/1.2905726.

Beggs, H.D. and Robinson, J.R. 1975. Estimating the Viscosity of Crude Oil Systems. *JPT* **27** (9): 1140–1141. SPE-5434-PA. DOI: 10.2118/5434-PA.

Beggs, H.D. and Vasquez, M. 1980. Correlations for Fluid Physical Property Prediction. *JPT* **32** (6): 968–970. SPE-6719-PA. DOI: 10.2118/6719-PA.

Bird, R.B., Steward, W.E., and Lightfoot, E.N. 1960. *Transport Phenomena*. New York City: Wiley and Sons Publishing.

Bradford, D.W., Fritchie, D.G. Jr., Gibson, D.H. et al. 2004. Marlin Failure Analysis and Redesign: Part 1—Description of Failure. *SPEDC* **19** (2): 104–111. SPE-88814-PA. DOI: 10.2118/88814-PA.

Brill, J.P. 1999. *Multiphase Flow in Wells*. Monograph Series, SPE, Richardson, Texas, **17**.

Chapman, A.J. 1967. *Heat Transfer*, second edition. New York City: Macmillan Press.

Combs, G.D. and Whitmire, L.D. 1968. Capillary Viscometer Simulates Bottom Hole Conditions. *Oil and Gas Journal* (30 September): 108–113.

Dodge, D.W. and Metzner, A.B. 1959. Turbulent Flow of Non-Newtonian Systems. *AIChE Journal* **5** (2): 189–204. DOI: 10.1002/aic.690050214.

Dropkin, E. and Somerscales, E. 1965. Heat Transfer by Natural Convection by Fluids Confined by Two Parallel Plates Which Are Inclined at Various Angles With Respect to the Horizon. *ASME Journal of Heat Transfer* **87**: 77.

Duns, H. and Ros., N.C.J. 1963. Vertical flow of gas and liquid mixtures in wells. *Proc.*, 6th World Petroleum Congress, Frankfurt am Main, Germany, Section II, 451–465.

Economides, M.J., Watters, L.T., and Dunn-Norman, S. 1998. *Petroleum Well Construction*. New York City: John Wiley and Sons.

Ellis, R.C., Fritchie, D.G. Jr., Gibson, D.H., Gosch, S.W., and Pattillo, P.D. 2004. Marlin Failure Analysis and Redesign: Part 2—Redesign. *SPEDC* **19** (2): 112–119. SPE-88838-PA. DOI: 10.2118/88838-PA.

Gosch, S.W., Horne, D.J., Pattillo, P.D., Sharp, J.W., and Shah, P.C. 2004. Marlin Failure Analysis and Redesign: Part 3—VIT Completion With Real-Time Monitoring. *SPEDC* **19** (2): 120–128. SPE-88839-PA. DOI: 10.2118/88839-PA.

Govier, G.W. 1977. *The Flow of Complex Mixtures in Pipes*. Melbourne, Florida, USA: Krieger Publishing Co.

Gray, G.R. and Darley, H.C.H. 1980. *Composition and Properties of Oil Well Drilling Fluids*, fourth edition. Houston: Gulf Publishing Co.

Gray, H.E. 1974. Vertical Flow Correlation in Gas Wells. In *User Manual for API 14B Subsurface Controlled Safety Valve Sizing Computer Program*, Appendix B. Washington, DC: API.

Hagedorn, A.R. and Brown, K.E. 1964. The Effect of Liquid Viscosity in Two-Phase Vertical Flow. *JPT* **16** (2): 203–210; *Trans.*, AIME, **231**. SPE-733-PA. DOI: 10.2118/733-PA.

Halal, A.S. and Mitchell, R.F. 1994. Casing Design for Trapped Annulus Pressure Buildup. *SPEDC* **9** (2): 107–114. SPE-25694-PA. DOI: 10.2118/25694-PA.

Hasan, A.R. and Kabir, C.S. 2002. *Fluid Flow and Heat Transfer in Wellbores*. Richardson, Texas, USA: SPE.

Hasan, A.R., Kabir, C.S., and Wang, X. 1998. Wellbore Two-Phase Flow and Heat Transfer During Transient Testing. *SPEJ* **3** (2): 174–180. SPE-38946-PA. DOI: 10.2118/38946-PA.

Hill, P.G., Keenan, J.H., Moore, J.G., and Keyes, F.C. 1969. *Steam Tables: Thermodynamic Properties of Water Including Vapor, Liquid and Solid*. New York: John Wiley and Sons.

Hiller, H.K. 1963. Rheological Measurements on Clay Suspensions and Drilling Fluids at High Temperatures and Pressures. *JPT* **15** (7): 779–788. SPE-489-PA. DOI: 10.2118/489-PA.

Houwen, O.H. and Geehan, T. 1986. Rheology of Oil-Base Muds. Paper SPE 15416 presented at the SPE Annual Technical Conference and Exhibition, New Orleans, 5–8 October. DOI: 10.2118/15416-MS.

Katz, D.L. 1959. *Handbook of Natural Gas Engineering.* New York City: McGraw-Hill Higher Education.

Kemp, N.P. and Thomas, D.C. 1987. Density Modeling for Pure and Mixed-Salt Brines as a Function of Composition, Temperature, and Pressure. Paper SPE 16079 presented at the SPE/IADC Drilling Conference, New Orleans, 15–18 March. DOI: 10.2118/16079-MS.

Lunardini, V.J. 1981. *Heat Transfer in Cold Climates.* New York City: Van Nostrand Reinhold Co.

Orkiszewski, J. 1967. Predicting Two-Phase Pressure Drops in Vertical Pipe. *JPT* **19** (6): 829–838. SPE-1546-PA. DOI: 10.2118/1546-PA.

Pattillo, P.D., Cocales, B.W., and Morey, S.C. 2006. Analysis of an Annular Pressure Buildup Failure During Drill Ahead. *SPEDC* **21** (4): 242–247. SPE-89775-PA. DOI: 10.2118/89775-PA.

Payne, M., Pattillo, P.D., Sathuvalli, U.B., and Miller, R.A. 2003. Advanced Topics for Critical Service Deepwater Well Design. Paper presented at the Deep Offshore Technology Conference, Marseilles, France, 19–21 November.

Peaceman, D. 1977. *Fundamentals of Numerical Reservoir Simulation*, first edition. Oxford, UK: Elsevier Publishing.

Press, W.H., Flannery, B.P., Teukolsky, S.A., and Vetterling, W.T. 1999. *Numerical Recipes in FORTRAN 77: The Art of Scientific Computing*, Vol. 1. New York City: Cambridge University Press.

Ramey, H.J. Jr. 1962. Wellbore Heat Transmission. *JPT* **14** (4): 427–435; *Trans.*, AIME, **225**. SPE-96-PA. DOI: 10.2118/96-PA.

Randall, B.V. and Anderson, D.B. 1982. Flow of Mud During Drilling Operations. *JPT* **34** (7): 1414–1420. SPE-9444-PA. DOI: 10.2118/9444-PA.

Raymond, L.R. 1969. Temperature Distribution in a Circulating Drilling Fluid. *JPT* **21** (3): 333–341; *Trans.*, AIME, **246**. SPE-2320-PA. DOI: 10.2118/2320-PA.

Reid, R.C., Prausnitz, J.M., and Sherwood, T.K. 1977. *The Properties of Gases and Liquids*, third edition. New York City: McGraw-Hill.

Rogers, P.S.Z. and Pitzer, K.S. 1982. Volumetric Properties of Aqueous Sodium Chloride Solutions. *J. Phys. Chem. Ref. Data* **11** (1): 15–81.

Savins, J.G. 1958. Generalized Newtonian (Pseudoplastic) Flow in Stationary Pipes and Annuli. *Trans.*, AIME **213**: 325–332. SPE-1151-G.

Shoham, O. 2006. *Mechanistic Modeling of Gas-Liquid Two-Phase Flow in Pipes.* Richardson, Texas, USA: SPE.

Sorelle, R.R., Jardiolin, R.A., Buckley, P., and Barrios, J.R. 1982. Mathematical Field Model Predicts Downhole Density Changes in Static Drilling Fluids. Paper SPE Paper 11118 presented at the SPE Annual Technical Conference and Exhibition, New Orleans, 26–29 September. DOI: 10.2118/11118-MS.

Standing, M.B. 1981. *Volumetric and Phase Behavior of Oil Field Hydrocarbon Systems.* Richardson, Texas: Society of Petroleum Engineers of AIME.

Steiner, R. 1990. *ASM Handbook Volume 1: Properties and Selection: Irons, Steels, and High-Performance Alloys*, tenth edition. Materials Park, Ohio, USA: ASM International.

Wagner, W. and Kruse, A. 1998. *Properties of Water and Steam: The Industrial Standard IAPWS-IF97 for the Thermodynamic Properties and Supplementary Equations for Other Properties.* Heidelberg, Germany: Springer-Verlag.

White, F. 1984. *Heat Transfer.* White Plains, New York: Addison Wesley Educational Publishers.

Whitson, C. and Brulé, M. 2000. *Phase Behavior.* Monograph Series, SPE, Richardson, Texas, **20**.

Willhite, G.P. 1967. Overall Heat Transfer Coefficients in Steam and Hot Water Injection Wells. *JPT* **19** (5): 607–615. SPE-1449-PA. DOI: 10.2118/1449-PA.

Wooley, G.R. 1980. Computing Downhole Temperatures in Circulation, Injection and Production Wells. *JPT* **32** (9): 1509–1522. SPE-8441-PA. DOI: 10.2118/8441-PA.

Zamora, M., Broussard, P.N., and Stephens, M.P. 2000. The Top 10 Mud-Related Concerns in Deepwater Drilling Operations. Paper SPE 59019 presented at the SPE International Petroleum Conference and Exhibition in Mexico, Villahermosa, Mexico, 1–3 February. DOI: 10.2118/59019-MS.

Zucro, M.J. and Hoffman, J.D. 1976. *Gas Dynamics.* New York City: Wiley and Sons Publishing.

SI Metric Conversion Factors
°F × (°F − 32)/1.8   = °C
°F × (°F + 459.67)/1.8 = K

10.2 Analytical Approach to Temperature-Induced Well Problems—Eirik Karstad, University of Stavanger

10.2.1 Introduction. During various well operations, heat is transported to or from the rock. During water-injection operations, considerable cooling takes place, with known effects of aiding fracturing processes. During flow testing of wells, on the other hand, a considerable amount of heat is transported up to the well-head. Also, during ordinary drilling, heat is transported up to the surface.

It has become clear that this heat transport is significant in many respects. Mud ballooning was long believed to be volumetric variations due to borehole flexing, but now it is evident that the drilling fluid expands and contracts due to variations in pressure and temperature. The physical parameters are measured at surface conditions, and it is clear that significant errors arise by using simple static models. In critical high-pressure/high-temperature (HP/HT) wells, the difference in physical parameters between surface conditions and downhole conditions may be critical and may confuse the interpretation of well kicks.

In this work, time-dependent temperature behavior in the borehole is modeled (Karstad 1999). This is actually an exact analytic model of a countercurrent heat exchanger applied on the physical conditions that exist in HP/HT wells. Also, we present a new analytical model for the density/pressure/temperature dependence of drilling fluids. These models are combined to study the variation in effective mud density and variation in mud volume during drilling operations.

Uses of Temperature Data in the Petroleum Industry. Temperatures in a well are important for many aspects of drilling, completion, production, and injection. Some well-known applications that require knowledge of accurate downhole temperature are (Kabir et al. 1996; Hermanrud 1988)

1. Cement composition, placement, and setting time
2. Drilling mud and annulus fluid composition (fluid temperature, as a function of both depth and elapsed time, dictates the intrinsic fluid properties such as viscosity and density)
3. Determining equivalent circulating density (ECD)
4. Determining equivalent static density (ESD)
5. Packer design and selection
6. Logging tool design and log interpretation
7. Wax deposition in production tubing
8. Thermal stresses in casings and tubing
9. Permafrost thaw and refreezing
10. Wellhead and production equipment design
11. Drill bit design
12. Elastomer and seal selection
13. Maximum allowable pumping rate
14. Pressure/volume/temperature (PVT) modeling of hydrocarbons
15. Reservoir modeling and volume estimates
16. Detection of zones with poor casing cement
17. Correlation between wells
18. Understanding temperature-dependent geological processes such as cementation/dissolution of minerals, altering of maturation indicators, and generation of hydrocarbons

Applications concerning reservoir and geological aspects require knowledge of the undisturbed formation temperature (also referred to as *virgin* or *true* temperature). Others, such as cement composition or thermal stresses in casings, mainly require the knowledge of the well fluid temperature as a function of depth and circulation time. However, since the temperature actually encountered in the well is directly related to the virgin formation temperature (VFT), it is obvious that the VFT is important in all aspects of temperature modeling of downhole applications.

This study focuses mainly on the investigation of the temperature in the borehole. The main objective has been to estimate the time-dependent temperature behavior of the well fluid as a function of depth and circulation time. This is used later for determining parameters and investigating how the temperature affects the well fluid density during drilling operations.

Literature Review. Many models for analyzing temperature behavior and temperature effects in rock and borehole are found in the literature.

VFT. Accurate determination of VFT is important in hydrocarbon exploration because these temperatures are vital to the calculation of source rock temperatures in a sedimentary basin, and therefore to the calculation of the depth to the oil window.

VFTs can be calculated with high precision from drillstem tests (Hermanrud et al. 1991) when some crucial requirements are fulfilled. However, these procedures are very time-consuming compared to the more widely used log-derived temperatures (LDTs). If three or more temperature measurements are recorded at the same depth, estimates of the formation temperature can presumably be made.

Numerous methods for VFT determination from LDT methods have been presented through the last decades. We present a summary of the different LDT methods that have emerged:

Line Source Model. The simplest and most frequently used procedure to calculate the VFT is the Horner plot method. The method of Horner originally was devised for the study of pressure buildup in drillstem tests, and not for the determination of VFTs. The Horner plot method and other line source models have been widely discussed (Bullard 1947; Edwardson et al. 1962; Luheshi 1983). This model generally will estimate VFTs that are too low for data taken short times after circulation stops.

Exponential Decay Method. The exponential decay method assumes that the formation temperature recovery is exponential with time. The method is discussed by many authors (Oxburgh et al. 1972), none of whom present a sound physical argument for this behavior.

One-Medium, Zero-Circulation-Time Models. The one-medium, zero-circulation-time models are based on the following initial conditions in the borehole:

$$
\begin{aligned}
T &= T_{\text{mud}} \quad t = 0 \\
T &= \text{VFT} \quad t < 0
\end{aligned}
\qquad\qquad\qquad (10.2.1)
$$

Models have been suggested by Middleton (1982).

Two-Medium, Zero-Circulation-Time Models. In this method, thermal variation following an instantaneous temperature drop caused by mud circulation is evaluated for different thermal properties of rock and drilling mud. This problem has been treated by Oxburgh et al. (1972) and Middleton (1982).

Two-Medium, Nonzero-Circulation-Time Models. Two-medium models with nonzero circulating time have been applied numerically by Lee (1982) and Luheshi (1983), and analytically by Cao et al. (1988). With nonzero circulation, the mud circulation is considered to have the effect of either maintaining the borehole mud at constant temperature or supplying a constant amount of heat per unit length per unit time to the borehole.

The development of LDT methods generally has moved from models based on fairly simple physical descriptions to more sophisticated models that give a more accurate description of the prevailing conditions in a borehole. However, it is not obvious that models that give a good physical description of the borehole are best for estimating the VRT. The general trend seems to be that the sophisticated models are more accurate but have large standard deviations, while the simpler models based on the line source concept give temperatures consistently too low but with smaller standard deviations.

Circulating Fluid Temperature. As pointed out, there are many important aspects of well operations that are strongly affected by the temperature. This appears particularly when drilling deeper wells and drilling in hostile environments. Therefore, to evaluate effects associated with higher temperatures, knowledge of the temperature of the flowing fluids and formation face is needed.

Reliable predictive temperature tools are limited, and it is shown that the most common method, American Petroleum Institute (API) thickening schedules, often over- or underestimate the true temperature. Although measurement-while-drilling (MWD) systems are more frequently used to measure downhole pressures and temperatures, we cannot overemphasize the need for computing the entire wellbore by simulating the transient process.

Much of the classical work in this area was developed by Ramey (1962). He presented approximate methods for predicting the temperature of either a single-phase incompressible liquid or a single-phase perfect gas flowing in injection and production wells.

The circulating mud temperature is a function of many variables such as well depth, circulation rate, formation properties, inlet mud temperature, and wellbore and drillpipe size. Numerous investigations have been made to determine how these variables affect the flowing temperature distribution. Two approaches have emerged: numerical (Raymond 1969; Corre et al. 1984) and analytical (Kabir et al. 1996; Holmes and Swift 1970). The numerical models often require extensive data input and also may be time-consuming.

Problems involving dynamic flow conditions generally exhibit numerical instabilities at early times unless very small timestep sizes are used. Still, accurate estimates of circulating fluid temperatures can be obtained by assuming steady heat flow in the conduit and annulus through which fluid is circulated and transient heat conduction within the formation.

All the works cited above focus on onshore wells. However, at substantial water depths, it is obvious that the cooling of the riser affects the fluid temperature. Water currents can cause significant cooling of the riser and may mask any geothermal gradient effects of the mud within.

Another variable, which has been neglected, is the supply of energy to the well from rotation of the drillstring and from the hydraulic pumping system. It is known that friction between the circulating fluid and the wellbore/casing generates heat, but it is not included in any earlier analytical models.

Effects of Pressure and Temperature on the Density of Drilling Fluids. One of the most important factors fundamental to well control and cost control in drilling is the maintenance of the correct hydrostatic head of drilling fluid in the well. As a result, more-precise methods of evaluating and measuring downhole pressures exerted by the drilling fluid column have become increasingly important in today's drilling industry.

Liquids expand when heat is applied and are compressed by pressure. Therefore, the density of the fluid decreases with increasing temperature, but increases with increasing pressure. As a drilling fluid is pumped downhole, the temperature and pressure effects change its density.

Several models for calculating downhole pressure and density are found in the literature. Compositional material-balance models (Peters et al. 1990) are most widely used, and with great success. These models predict the downhole density from knowledge of the mud composition, the density of the mud constituents at ambient temperature and pressure, and the density of the liquid constituents at elevated pressure and temperature. The actual density is determined through numerical integration.

However, analysis of data from laboratory studies conducted for HP/HT wells demonstrates discrepancies between measured densities and those calculated through use of a compositional model. It is known that if a significant amount of chemicals is present in the mud, some chemical interaction can cause changes in the solid-fluid system. In these cases, the compositional models cannot be used. Thus, for many applications, an accurate analytical model would be advantageous.

On the basis of empirical correlations, there have been presented various models to describe downhole mud density with analytic equations (Sorelle et al. 1982; Kutasov 1988). Studies by Babu (1996) show that the empirical model suggested by Kutasov (1988) represents measured data more accurately than other models for many mud compositions.

One important advantage of the empirical equations is that a minimum of input data is required. However, reliable laboratory density measurements at elevated pressures and temperatures are needed for accurate downhole density predictions. Also, none of the empirical models give an accurate physical description of the effects of pressure and temperature on the density of drilling fluids.

Mathematical Approach. Approaches to temperature modeling can be divided into two main methods: *numerical* and *analytical*. The focus in this book is in essence on the analytical approach (Karstad 1999). The two main reasons are as follows:

First, numerical methods may have limited value for parametric studies and studying the overall system behavior. In contrast, the analytic methods can overcome these problems, but they may be restricted to simpler geometries and contain simplifying assumptions.

Second, as stated earlier, numerical models often require extensive data input and may also be time-consuming. Though the sophisticated numerical models give a very accurate description of the prevailing conditions in a borehole, it does not follow that the more complex models are best suited for estimating, for example, the circulating fluid temperature or downhole fluid density. Introducing a new parameter makes a model more flexible, but the accuracy may not increase. One might suppose that there exists some optimum between the number of parameters and the resulting accuracy. This is shown in **Fig. 10.2.1**. By introducing a new parameter, we also introduce more uncertainty, depending on the accuracy of the new parameter.

We give a short introduction to the method of computing the accuracy of results calculated from data of known accuracy. Suppose we want the accuracy of quantity N.

$$N = f(\mu_1, \mu_2, \ldots, \mu_n)$$

μ_i = measured data

$\pm\Delta\mu_i$ = error of measurement μ_i

$\qquad\qquad\qquad\qquad\qquad\qquad\qquad\qquad\qquad$ (10.2.2)

The error in the measured data will cause an error, ΔN, in the computed quantity N. By performing a Taylor series expansion, and assuming that $\Delta\mu_i$ is small such that $\Delta\mu_i^2$ is negligible, the absolute error, Err_{abs}, may then be written as follows:

$$Err_{abs} = (N + \Delta N) - N$$
$$= \Delta N \qquad\qquad\qquad\qquad\qquad\qquad\qquad (10.2.3)$$
$$= \left|\Delta\mu_1 \frac{\partial f}{\partial \mu_1}\right| + \left|\Delta\mu_2 \frac{\partial f}{\partial \mu_2}\right| + \cdots + \left|\Delta\mu_n \frac{\partial f}{\partial \mu_n}\right|$$

We observe from Eq. 10.2.3 that the error in N is dependent not only on the error in the measured data, but also on the sensitivity of N to that particular measurement. A small error in μ_i may be significant if N is very sensitive to μ_i, or a large error in μ_k may be negligible if N is very insensitive to μ_k.

The disadvantage of introducing many parameters, as often is the case in numerical models, is that we introduce many errors. In addition, we may not even know the magnitude of the error or how sensitive our model is to that particular parameter. It becomes difficult to keep first-order effects distinct from second-order effects. However, it must be stated that there are many good numerical simulators available. Analytical and numerical methods do not exclude each other.

10.2.2 Temperature Model. The physical problem is to determine the temperature profile of the circulating fluid under various operational modes. For this reason, both forward circulation and reverse circulation will be analyzed.

Numerous contributions on the subject have been presented. But common for most of them is that they focus on onshore wells, and they do not study the effect of changes in well depth and mud-in temperature, yet there is no doubt that these effects have an impact on the fluid temperature. Also, the supply of energy to the well from rotation of the drillstring and from the hydraulic pumping system have been largely neglected.

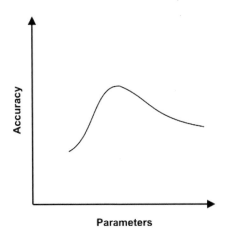

Fig. 10.2.1—Schematic view of possible correlation between number of parameters and accuracy in modeling.

In this section, we present models for calculating circulating fluid temperature. The models are used to study the temperature behavior during drilling operations. We also develop explicit solutions (with respect to circulating time) of the temperature behavior at the bottom of the hole and at the return mud flowline. Finally, we present new models for calculating the circulating fluid temperature in offshore applications, where we take into account the cooling effect of having a riser and the heating effect from energy input due to drillstring rotation and mud pumping.

Representation of the Problem. Because heat conduction in the formation is a rather slow process, the fluid temperature is always in a transient state and is strongly sensitive to the flow rate.

When circulating drilling fluids, the bottom part of the well is usually being cooled. The drilling fluid at bottom is heated in the process, resulting in a heat transfer to the upper part of the well and also heat lost from the well. In fact, the hydraulic system can be described as a countercurrent heat exchanger with a moving outer boundary. This is represented schematically in **Fig. 10.2.2**.

This leads to a nonlinear heating of the mud system. A typical picture of temperature profiles downhole is shown in **Fig. 10.2.3**. The virgin temperature profile of the formation, T_f, represents the temperature that exists in the formation before the well is drilled. This is not measured directly, but is important to be determined both as a reference temperature and for analysis work related to hydrocarbon PVT behavior (see Introduction).

The cold mud pumped down inside the drillstring is being heated by contact with the pipe wall, while the return mud on the outside of the pipe is exposed both to the outside of the pipe and the annulus. For this reason, the inside of the pipe (T_d) and the outside of the pipe (T_a) have different temperatures.

The temperature profile for the formation at the borehole wall (T_{wb}) is different from the virgin temperature (T_f). This difference depends on the heat conduction properties of the rock formation and the openhole time exposure.

At the borehole wall, there is a heat exchange between the formation and the drilling mud, resulting in a temperature profile in the mud in the annulus defined by T_a. Because the drillstring is a countercurrent heat exchanger, the mud temperature inside the drillstring (T_d) is usually lower than T_a. We observe that the total heat exchange process results in four different temperature profiles.

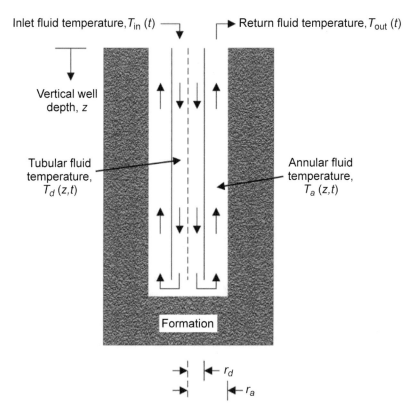

Fig. 10.2.2—Schematic view of circulating fluid system.

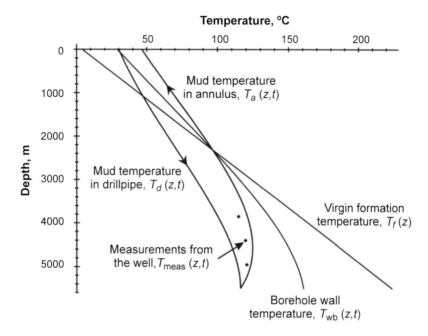

Fig. 10.2.3—Downhole temperatures.

Recently, the industry started to measure temperatures during drilling. A temperature measurement probe is placed somewhere in the bottomhole assembly (BHA). Since this probe is located between the inside and the outside of the string, the measured temperature probably reflects some average between T_a and T_d. By thermal insulation on the inside, the probe may approximately read T_a. It is important to observe that a temperature measurement in the well may not directly apply to one of the curves shown in Fig. 10.2.3.

Formation/Casing Heat Conduction. Calculating the circulating fluid temperature involves the heat transfer from the wellbore/casing and the formation surrounding the well. From geometrical consideration, it is plausible to assume radial heat transfer in the formation surrounding the borehole. Heat conduction from the wellbore/casing to the surrounding formation may then be written as (Ramey 1962)

$$q_f = \frac{2\pi k_f}{f(t_D)}(T_f - T_{wb})dz, \quad \dots\dots\dots\dots\dots\dots\dots\dots\dots\dots\dots\dots\dots \quad (10.2.4)$$

where $f(t_D)$ is a dimensionless time function that depends on the formation-borehole heat transfer boundary conditions and describes how the transient heat flow from the formation to the wellbore changes with time. The solution was first presented by Carslaw and Jaeger (1959).

To determine sufficiently accurate values of the dimensionless time function, a rigorous treatment of the transient heat flow from the formation to the wellbore is required. Many different models exist that can be used to estimate $f(t_D)$ (Ramey 1962; Hasan and Kabir 1994a, 1994b).

In our study, we have focused on the circulating fluid temperature and therefore have chosen to use the existing models. In field applications, we have used the Hasan and Kabir model (1994a, 1994b), which assumes a constant heat flux at the wellbore-formation interface.

Dimensionless Time Function. For constant heat flux in a cylindrical-source well, the Hasan and Kabir formulation yields

$$f(t_D) = \left(1.1281\sqrt{t_D}\right)\left(1 - 0.3\sqrt{t_D}\right) \quad \text{if } 10^{-10} \le t_D \le 1.5, \quad \dots\dots\dots\dots\dots\dots\dots \quad (10.2.5)$$

$$f(t_D) = \left(0.4063 + 0.5\ln t_D\right)\left(1 + \frac{0.6}{t_D}\right) \quad \text{if } t_D \ge 1.5, \quad \dots\dots\dots\dots\dots\dots\dots\dots\dots \quad (10.2.6)$$

where the dimensionless time, t_D, is given by

$$t_D = \frac{\alpha_h t}{r_w^2} \times 3600 \quad \dots\dots\dots\dots\dots\dots\dots\dots\dots\dots\dots\dots\dots\dots\dots\dots\dots \quad (10.2.7)$$

(note that the number 3,600 is only to make the units consistent when the circulation time is measured in hours) and the thermal diffusivity is defined as

$$\alpha_h = \frac{k_f}{\rho_f c_f}. \quad \dots\dots\dots\dots\dots\dots\dots\dots\dots\dots\dots\dots\dots\dots\dots\dots\dots \quad (10.2.8)$$

Circulating Fluid Temperature. On the basis of the latest contributions to the subject of calculating the circulating fluid temperature, we here present analytical solutions for calculating the flowing fluid temperature and analyze temperature measurements during the drilling operation.

The usual procedure for flow problems of this type is to solve the total-energy and mechanical-energy equations simultaneously to yield both temperature and pressure distributions. However, the solution can be approximated by some simplifying assumptions:

- The circulating fluid is incompressible and is circulated at a constant rate.
- Axial conduction of heat in the circulating fluid is negligible compared with axial convection.
- Radial temperature gradients in the drillpipe, casing, or annulus are negligible (i.e., during circulation, the borehole mud is considered to be well mixed and can be treated as a perfect conductor).
- The geothermal gradient is constant.
- The physical properties of the circulating fluid and the formation (density, heat transfer coefficients, specific heat, and thermal conductivity) do not vary with temperature.
- The formation is radially symmetric and can be assumed to be infinite with respect to heat flow.
- Viscous flow energy, rotational energy, and drill bit energy are all neglected.
- The heat transfer in the wellbore is steady-state; the flows of heat in the tubing and annulus are rapid compared to the flow of heat in the formation.
- Transient heat transfer takes place in the formation surrounding the borehole.

Considering the steady flow of a single-phase fluid, we now set up an energy balance over a differential element, dz, of the well as shown in **Fig. 10.2.4**. Here we have chosen z to be positive in the downward direction, and the fluid is flowing down the drillstring and up through the annulus.

Forward Circulation—Fluid Flow Down the Drillstring. Heat enters the element by convection in the drillstring, q_d at z, and in the annulus, q_a at $z + dz$. Next, there will be conductive heat transfer through the drillstring, q_{ad}, and between the annulus and the formation, q_f. The energy balance gives us

$$q_d(z) - q_d(z+dz) = -q_{ad} \quad , \quad \dots\dots\dots\dots\dots\dots\dots\dots\dots\dots\dots\dots\dots\dots \quad (10.2.9)$$
$$q_a(z+dz) - q_a(z) = q_{ad} - q_f$$

where

$$q_d(z) = w C_{fl} T_d(z)$$
$$q_a(z) = w C_{fl} T_a(z)$$
$$q_{ad}(z) = 2\pi r_d U_d (T_a - T_d) dz \quad \dots\dots\dots\dots\dots\dots\dots\dots\dots\dots\dots\dots \quad (10.2.10)$$
$$q_f(z) = 2\pi r_w U_a (T_{wb} - T_a) dz$$

In Eq. 10.2.10, the wellbore interface temperature, T_{wb}, is related to the annular fluid temperature by the overall heat transfer coefficient, U_a. Thus, by combining Eqs. 10.2.10 and 10.2.4, we can eliminate the unknown T_{wb}, and q_f may be written as a function of the dimensionless time function. The function $f(t_D)$ describes how the transient heat flow from the formation to the wellbore changes with time.

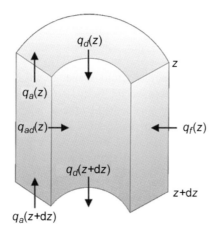

Fig. 10.2.4—Schematic view of heat balance (forward circulation).

The formation temperature, T_f, usually is approximated to vary linearly with depth:

$$T_f(z) = T_{sf} + g_G z. \quad \dots\dots\dots\dots\dots\dots\dots\dots\dots\dots\dots\dots\dots\dots\dots\dots \quad (10.2.11)$$

Now, by combining Eqs. 10.2.9 through 10.2.11, we can express T_d in terms of an inhomogeneous second-order linear differential equation (Kabir et al. 1996; Karstad and Aadnoy 1997). The boundary conditions of the forward-circulation case are that the tubing fluid temperature equals the inlet fluid temperature at the wellhead (i.e., $T_d = T_{in}$ at $z = 0$), and the heat exchange between the tubing and annular fluid is zero at the bottom of the hole (i.e., $T_d = T_a$ or $dT_d/dz = 0$ at $z = D$). Solving the differential equation gives the following expressions for the annular and tubing fluid temperature profiles:

$$\begin{aligned} T_d(z,t) &= \alpha e^{\lambda_1 z} + \beta e^{\lambda_2 z} + g_G z - B g_G + T_{sf} \\ T_a(z,t) &= (1 + \lambda_1 B)\alpha e^{\lambda_1 z} + (1 + \lambda_2 B)\beta e^{\lambda_2 z} + g_G z + T_{sf} \end{aligned}, \quad \dots\dots\dots\dots\dots\dots\dots\dots \quad (10.2.12)$$

where

$$A = \frac{wC_{fl}}{2\pi r_w U_a}\left(1 + \frac{r_w U_a f(t_D)}{k_f}\right), \quad \dots\dots\dots\dots\dots\dots\dots\dots\dots\dots\dots\dots\dots \quad (10.2.13)$$

$$B = \frac{wC_{fl}}{2\pi r_d U_d}$$

$$\begin{aligned} \lambda_1 &= \frac{1}{2A}\left(1 - \sqrt{1 + \frac{4A}{B}}\right) \\ \lambda_2 &= \frac{1}{2A}\left(1 + \sqrt{1 + \frac{4A}{B}}\right) \end{aligned}, \quad \dots\dots\dots\dots\dots\dots\dots\dots\dots\dots\dots \quad (10.2.14)$$

$$\alpha = -\frac{(T_{in} + B g_G - T_{sf})\lambda_2 e^{\lambda_2 D} + g_G}{\lambda_1 e^{\lambda_1 D} - \lambda_2 e^{\lambda_2 D}} \quad \dots\dots\dots\dots\dots\dots\dots\dots\dots\dots\dots \quad (10.2.15)$$

$$\beta = \frac{(T_{in} + B g_G - T_{sf})\lambda_1 e^{\lambda_1 D} + g_G}{\lambda_1 e^{\lambda_1 D} - \lambda_2 e^{\lambda_2 D}}$$

This is the general solution of the countercurrent heat exchanger consideration of a circulating well for forward circulation. Further mathematical details may be found in Kabir et al. (1996) and Karstad (1999).

The temperature profiles in Fig. 10.2.3 are calculated using data from **Table 10.2.1** applied to Eq. 10.2.12.

As we can observe in Fig. 10.2.3, the maximum fluid temperature is not at the bottom of the hole. This has been discussed in many previous works (Holmes and Swift 1970; Kabir et al. 1996).

When the fluid flows down the drillstring, it is heated by the annular fluid and reaches its maximum at the bottom of the hole. As the fluid starts to flow up the annulus, it receives heat from the surrounding formation. This gain in heat is larger than the heat it loses to the fluid in the drillstring. Thus, the annular fluid temperature continues to increase. The rate of heat gain decreases with decreasing depth, and at some depth away from the bottom of the hole, the net heat gained by the fluid equals the net heat loss. The maximum fluid temperature is reached.

This maximum value is found by differentiating Eq. 10.2.12 with respect to the well depth, z, and setting equal to zero:

$$\frac{dT_a}{dz}\left(z\left(T_{max}\right)\right) = \left(1+\lambda_1 B\right)\alpha\lambda_1 e^{\lambda_1 z\left(T_{max}\right)}$$

$$+ \left(1+\lambda_2 B\right)\beta\lambda_2 e^{\lambda_2 z\left(T_{max}\right)} + g_G \quad \dots\dots\dots\dots\dots\dots\dots\dots\dots\dots\dots\dots\dots \quad (10.2.16)$$

$$= 0$$

Because $z(T_{max})$ is implicit in Eq. 10.2.16, a numerical solution of $z(T_{max})$ has to be sought (e.g., by using the Newton-Raphson method).

Reverse Circulation—Fluid Flow Down the Annulus. In many drilling and well workover situations, flow down the annulus and up the tubing may be the case. Therefore, analytical solutions for the reverse-circulation case are desirable. These are easily obtained by reversing the fluid flow on the differential element and following the procedure for forward circulation. Noting that the flow direction is reversed, an energy balance yields

$$q_d\left(z+dz\right)-q_d\left(z\right)=-q_{ad}$$
$$q_a\left(z\right)-q_a\left(z+dz\right)=q_{ad}-q_f \quad , \dots\dots\dots\dots\dots\dots\dots\dots\dots\dots\dots\dots\dots \quad (10.2.17)$$

where the boundary conditions for the reverse circulation require that $T_a^{rev} = T_{in}$ at the wellhead ($z = 0$) and $dT_d^{rev}/dz = 0$ at the bottom of the hole ($z = D$). Now, using the steps from Eqs. 10.2.10 through 10.2.12, a general solution for the temperature distributions may be determined:

TABLE 10.2.1—WELL AND MUD DATA*	
Drillstem outside diameter (in.)	$6^5/_8$
Drill bit size (in.)	$8^3/_8$
Well depth (m)	4572
Circulation rate (m³/s)	0.013249
Inlet mud temperature (°C)	23.889
Mud specific heat (J/kg·°C)	1676
Mud density (kg/m³)	1198.264
Mud thermal conductivity (W/m·°C)	1.730
Formation thermal conductivity (W/m·°C)	2.250
Formation specific heat (J/kg·°C)	838
Surface Earth temperature (°C)	15.278
Geothermal gradient (°C/m)	0.02315
*Karstad and Aadnoy (1998); Holmes and Swift (1970); Kabir et al. (1996)	

$$T_d^{rev}(z,t) = \eta e^{-\lambda_1 z} + \xi e^{-\lambda_2 z} + g_G z + B g_G + T_{sf}$$
$$T_a^{rev}(z,t) = (1 + \lambda_1 B) \eta e^{-\lambda_1 z} + (1 + \lambda_2 B) \xi e^{-\lambda_2 z} + g_G z + T_{sf}$$

$$\text{..........................} \quad (10.2.18)$$

where

$$\eta = \frac{(T_{sf} - T_{in}) \lambda_2 e^{-\lambda_2 D} + g_G (1 + \lambda_2 B)}{\lambda_1 (1 + \lambda_2 B) e^{-\lambda_1 D} - \lambda_2 (1 + \lambda_1 B) e^{-\lambda_2 D}}$$

$$\text{...............................} \quad (10.2.19)$$

$$\xi = -\frac{(T_{sf} - T_{in}) \lambda_1 e^{-\lambda_1 D} + g_G (1 + \lambda_1 B)}{\lambda_1 (1 + \lambda_2 B) e^{-\lambda_1 D} - \lambda_2 (1 + \lambda_1 B) e^{-\lambda_2 D}}$$

Using data from Table 10.2.1, we demonstrate the temperature profiles during reverse circulation. This is shown in **Fig. 10.2.5**.

In the upper part of the well, the fluid looses heat to the colder surrounding formation, while it gains heat from the heated fluid in the drillstring. Then, at some distance down the wellbore, the formation temperature exceeds the fluid temperature in the annulus. From there, the fluid temperature rises as it flows downward. At the bottom of the hole, the fluid flow starts to flow up the drillstring and thereby reverses its direction. The fluid in the drillstring is now at a higher temperature than the countercurrent fluid in the annulus. Heat exchange is, therefore, from the drillstring to the annulus as it flows upward. Consequently, the bottomhole temperature is the maximum temperature in the well, and in general we should observe monotonic fluid temperature change in both conduits (there can sometimes be a local minimum in the annular fluid temperature near the surface).

Compared to the forward-circulation case, we have a greater heat loss from the well. This is also what we would expect intuitively. During reverse circulation there is a large contact area between the formation and the fluid that is being pumped into the well. Hence, heat transfer to the circulating fluid is more effective than during forward circulation, and we can expect a higher bottomhole temperature. Also, near the surface, the heat loss from the upward flowing fluid is less during reverse circulation.

Temperature Behavior During Drilling. Published models (Kabir et al. 1996; Holmes and Swift 1970; Corre et al. 1984) are all based on the assumption of a constant well depth and a constant mud-in temperature. For many applications within production technology, this is usually sufficient. The Circulating Fluid Temperature section defines such a model. However, due to an increasing mud-pit temperature during drilling,

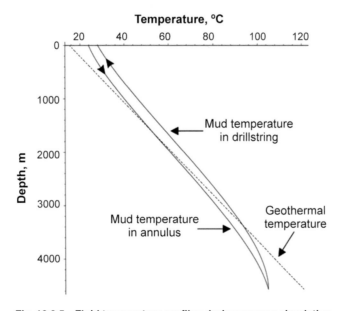

Fig. 10.2.5—Fluid temperature profiles during reverse circulation.

one often is faced with an increasing mud-in temperature. In this section, we will expand the model to be able to handle both an increasing well depth and a variable mud-in temperature (Karstad and Aadnoy 1997).

The model developed gives the fluid temperature profile in the drillstring and in the annulus for a circulating well at constant depth. To model an increasing well depth, the model must be modified. The true vertical well depth during drilling is given by

$$D(t) = D_0 + \text{ROP} \cdot t, \quad\dotfill\quad (10.2.20)$$

where ROP is the rate of penetration, and t is the time drilled. Also, by assuming the drilling process to be rapid compared to the heat flow in the formation, we assume that the instantaneous bottomhole temperature beneath the drill bit equals the VFT, $T_{wb} = T_f$ (note that this is equivalent to neglecting the time function, $f(t_D) = 0$, in Eq. 10.2.13). Thus, in the following calculations, we use

$$A = \frac{w C_{fl}}{2\pi r_w U_a}. \quad\dotfill\quad (10.2.21)$$

Similar to the increasing well depth, the variable mud-in temperature is given by

$$T_{in} = T_{in}^0 + g_T t, \quad\dotfill\quad (10.2.22)$$

where g_T is the rate of change of the mud-in temperature.

Since the temperatures at the drill bit and the mud return flowline are monitored during drilling, these are the most interesting depths to model. Using data from Table 10.2.1, a number of simulations were used to derive explicit temperature models for the bottomhole temperature and mud-return flowline temperature during drilling.

Mud-Return Temperature. First, we will consider the return flowline temperature while drilling. Letting the well depth z be zero in Eq. 10.2.12 reduces to the following:

$$T_{out}(0,t) = T_{in} + B g_G + B(\alpha \lambda_1 + \beta \lambda_2). \quad\dotfill\quad (10.2.23)$$

Assuming an ROP of 5 to 25 m/h, and varying the mud flow rate, the mud-return temperature was observed to obey the following equation:

$$T_{out}(0,t) = T_{out}(0,0) - B\lambda_1 g_T t. \quad\dotfill\quad (10.2.24)$$

Since the mud-out temperature usually is higher than the mud-in temperature, the active mud pit gradually will heat up. This will again result in an increase in the mud-in temperature, and from Eq. 10.2.23, the mud-out temperature will further increase.

For offshore HP/HT applications, the active mud-pit temperature may increase from approximately 10 to 60°C. This temperature increase depends not only on the return flow temperature, but also on the total specific heat of, and the heat loss from, the active mud pit. In other words, the mud pit itself may be modeled as a thermodynamic system. Hasan et al. (1996) established an exponential relationship between mud pit temperature and time that has been verified by actual data. Further development of heat transfer models for the mud pit is not included in this study.

Assuming a constant mud-in temperature, it furthermore was found that the mud-out temperature was nearly constant during drilling, or

$$T_{out}(0,t) = T_{out}(0,0). \quad\dotfill\quad (10.2.25)$$

Bottomhole Temperature. Similarly, we observe the bottomhole temperature during drilling:

$$T_{bh}(D,t) = \alpha e^{\lambda_1 D} + \beta e^{\lambda_2 D} + g_G D - B g_G + T_{sf}. \quad\dotfill\quad (10.2.26)$$

Again, we use the data in Table 10.2.1 to show the influence of drilling on the bottomhole temperature of the mud. Performing a number of simulations, it was discovered that the modeled bottomhole temperature showed a distinct linear trend with increasing depth. The following equation resulted from this analysis:

$$T_{bh}(D,t) = T_{bh}(D_0,0) + g_G \mathrm{ROP} \cdot t + g_T \left(1 - \frac{\lambda_1}{\lambda_2} \right) e^{\lambda_1 D_0} t. \dots \dots \dots \quad (10.2.27)$$

Again, studying the case of maintaining the mud-in temperature constant, for this condition, the following relationship was found:

$$T_{bh}(D,t) = T_{bh}(D_0,0) + g_G \mathrm{ROP} \cdot t. \dots \dots \dots \quad (10.2.28)$$

During drilling, viscous flow energy, rotational energy, and drill bit energy are believed to have a strong effect on the bottomhole temperature. However, this most likely will result in a perturbed offset in Eq. 10.2.27, and only minor changes in the geothermal temperature gradient, g_G. Thus, drilling with a constant ROP and constant mud-in temperature gives us direct information on the geothermal temperature gradient, g_G.

Effect of Changes in Intrinsic Fluid Properties on Circulating-Fluid Temperature Profiles. Physical parameters (mud density, mud specific heat, heat transfer coefficients, mud thermal conductivity, formation specific heat, formation thermal conductivity, formation density) that are included in the temperature model usually can be considered as constants. However, for the severe changes in temperature during HP/HT operations, this may not be a satisfactory assumption. Therefore, we conducted a study to analyze what effect a temperature dependence of these variables would have on the circulating fluid temperature.

The study showed that disturbances in the physical parameters caused by temperature variations have little impact on the fluid temperature profile. Possible changes are noticeable only near the bottom of the hole. A 5% change in model constants (which are compositions of physical parameters) results in less than 1% change in bottomhole temperature.

We therefore conclude that the effect of changes in intrinsic fluid properties on circulating fluid temperature profiles are minimal and can be neglected.

Effects of Riser, Hydraulics, and Rotation. In this model, we have expanded the temperature model presented earlier to include the effect of having a riser in offshore applications as well as input of energy due to drillstring rotation and the hydraulic pumping system. Since most of the mathematical details are described earlier in this section, we will here confine ourselves to the bare necessities.

First, we introduce some additional assumptions regarding the well and mud system:

- Input of energy due to drillstring rotation and the hydraulic pumping system is evenly distributed throughout the well. We do not assume any heating effect in the riser due to the cooling from the seawater.
- Seawater temperature remains time-invariant.

When performing the energy balance on the differential element in Fig. 10.2.4, we now will have to include input of energy, q_e, to the system:

$$\begin{aligned} q_d(z) - q_d(z+\mathrm{d}z) &= -q_{ad} \\ q_a(z+\mathrm{d}z) - q_a(z) &= q_{ad} - q_f - q_e \end{aligned}, \dots \dots \dots \quad (10.2.29)$$

where

$$q_e = \frac{\dot{E}_{rot} + \dot{E}_{pump}}{D} \, \mathrm{d}z. \dots \dots \dots \quad (10.2.30)$$

The temperature profiles now can be solved by following the same steps as performed in Eqs. 10.2.9 through 10.2.15. The initial and boundary conditions are now as follows:

- The tubing fluid temperature equals the inlet fluid temperature at the wellhead (i.e., $T_d = T_{in}$ at $z = 0$).

- Heat exchange between the fluid in the drillstring and annulus is zero at the bottom of the hole (i.e., $dT_d/dz = 0$ at $z = D$).
- The fluid temperature function is continuous at seabed level [i.e., $T_d(D_{sb}) = T_{dr}(D_{sb})$ and $T_{as}(D_{sb}) = T_{ar}(D_{sb})$].

Solving the requirement of continuity analytically will give very complex expressions, and it is therefore usually solved by numerical methods.

Because both the water and rock sections are parts of the same countercurrent heat exchanger consideration of a circulating well, the temperature expressions will be similar to Eq. 10.2.12. For detailed results and expressions, see Karstad (1999).

To demonstrate the effects of including a riser and input of energy from drillstring rotation and the hydraulics system, we simulate an offshore well using data from **Table 10.2.2**.

In **Fig. 10.2.6**, we have plotted the downhole temperature profiles for an offshore well. The effect of having a riser included is clearly demonstrated by the curvature of the annulus temperature in the water section.

We cannot observe any significant cooling effect on the mud flowing down the drillstring. This is due to the moderate heat loss in the annulus for this particular case. At very large water depths (e.g., 1000 m), we would notice that the temperature profile of the mud in the drillstring would have a more vertical trend in the riser section.

In **Fig. 10.2.7**, we compare temperature profiles obtained with and without input of energy to the flowing fluid system. Clearly, input of energy due to pumping and rotation leads to increased temperature. We can see that if we neglect input of energy to the well, we would underestimate bottomhole temperature by as much as 7°C. Since the energy that is put into the system is accumulated over time, this effect would increase as circulation continues.

TABLE 10.2.2—WELL AND MUD DATA FOR SYNTHETIC OFFSHORE WELL (Karstad and Aadnoy 1999)	
Water Data	
Water depth (m)	500
Water temperature at surface (°C)	10
Water temperature at seabed (°C)	4
Mud Data	
Inlet mud temperature (°C)	20
Mud thermal conductivity (W/m·°C)	1.730
Mud specific heat (J/kg·°C)	1676
Mud density (SG)	2.04
Formation Data	
Surface Earth temperature (°C)	4
Geothermal gradient (°C/m)	0.04
Formation thermal conductivity (W/m·°C)	2.250
Formation specific heat (J/kg·°C)	838
Formation density (kg/m^3)	2643
Well Data	
Drillpipe outside diameter (in.)	5
Drillpipe inside diameter (in.)	4.408
Drill bit size (in.)	8^3/$_8$
Well depth (m)	4000
Circulation rate (L/min)	2000
Rate of energy input from rotation and pumping (kW)	1500

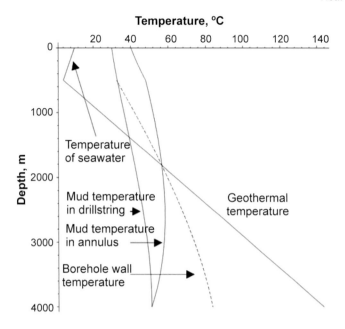

Fig. 10.2.6—Influence of riser upon circulating fluid temperature (Karstad 1999b).

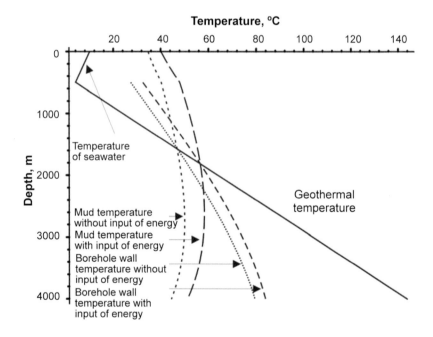

Fig. 10.2.7—Temperature profiles with and without input of energy.

In **Fig. 10.2.8**, we have a different scenario. Here we show the temperature profiles including an input of 1500 kW from pumping and rotation at two different times during circulation. Even though we have continuous input of energy from rotation and pumping, the fluid temperature is decreasing as we circulate.

One practical consequence of this result is seen during drilling of HP/HT wells. During temporary shut-in of the well, one often observes that the pressure increases under the wellhead (Aadnoy 1996). The reason is that during shut-in, the mud gradually is heated up toward geothermal temperature, expands, and reduces in density (Babu 1996). In other words, the static mud density decreases. The density/pressure/temperature dependence of drilling fluid is further discussed in the next section.

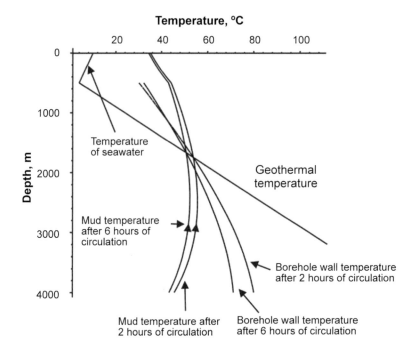

Fig. 10.2.8—Time-dependent behavior of temperature profiles.

However, the behavior of the circulating mud temperature is dependent on many factors, especially the mud flow rate. For a different well with different properties, we may observe the temperature to increase with circulation time (Gao et al. 1998).

Summary. In this section, the equations that govern the transport of heat from the formation to the drilling rigs are defined. A countercurrent heat exchange model gives the heat transfer in the borehole.

To improve the application of the total thermodynamic model, parametric studies are performed. These results are fitted to simple explicit equations, which describe the total thermodynamic model with reasonable accuracy.

The temperature model for calculating circulating fluid temperature also was improved to include the effect of having a riser in offshore applications, and input of energy due to drillstring rotation and the hydraulic pumping system.

Through this research, the following understanding emerged:

- During rapid drilling, the VFT gradient is directly determined from bottomhole temperature measurements. The condition is that the drilling rate exceeds the temperature conduction rate in the rock.
- If the mud-in temperature is constant during rapid drilling, the measured temperature gradient is equal to the VFT gradient. If the mud-in temperature changes, a correction must be applied.
- Any time the flow rate or rotary speed changes, the bottomhole temperature deviates from its previous trend. For interpretation, ensure that the depth interval under consideration was drilled under constant conditions.
- Circulating wellbore fluid when the bit is not drilling leads to cooling of the bottom of the well and increasing mud-out temperature. The cooling effect is strongly dependent on the circulating time and on the flow rate.
- Circulation at a constant well depth leads to a higher mud-out temperature at surface.
- Using the explicit equations derived in this section and temperature measurements at the top and bottom of the well, the model parameters can be directly determined.
- The riser has a definite cooling effect on the mud.
- Neglecting input of energy to the well would result in an underestimation of the bottomhole temperature. The effect will increase as circulation continues.

10.2.3 Density Model. During drilling, often it is observed that the return mud volume varies to some extent, giving either a too low or a too high return rate. Also, when the pumps are turned off, a mud backflow results. This phenomenon is often called borehole ballooning or breathing. Different interpretations of this phenomenon include ballooning shales, fractures being opened and closed, and compressibility of the drilling fluid (Gill 1989; Aadnoy 1996; Ward and Clark 1999). Unfortunately, these effects often complicate interpretation of the pore pressure, as seen in several critical HP/HT wells in the North Sea.

Under critical circumstances, it may be difficult to separate the ballooning effect from responses from taking a kick, or having a mud loss. And since the operational margins in HP/HT wells are very small, a misdiagnosis may lead to unwarranted well-control procedures that can be very costly. Therefore, borehole stability, well control, and interpretation of well signals are key issues related to this phenomenon. Since the change in temperature is nearly always transient or pseudo steady-state, the temperature effect is critical to the analysis of mud volume variations.

This section shows the results of a larger analytical study of the ballooning phenomenon. We present an analytical model for the density-pressure-temperature (DPT) dependence for drilling fluids (Karstad and Aadnoy 1998). By coupling this to the temperature model developed, we can study the temperature effects on mud-out volume variations.

Mathematical Development. Suppose that the density is a function of pressure and temperature. The change in density with depth then can be expressed as follows:

$$\frac{d\rho}{dz}(p,T) = \frac{\partial \rho}{\partial p}\frac{dp}{dz} + \frac{\partial \rho}{\partial T}\frac{dT}{dz}. \qquad (10.2.31)$$

In this equation, we make two important assumptions. We assume that both pressure and temperature are functions of the depth, z, only.

Coefficient of Isothermal Compressibility. The effective compressibility, c_{eff}, of drilling fluid composed of water, oil, and solids having compressibilities c_w, c_o, and c_{sd}, respectively, is given by

$$c_{eff} = c_w f_w + c_o f_o + c_{sd} f_{sd}, \qquad (10.2.32)$$

where f_w, f_o, and f_{sd} denote the volume fractions of water, oil, and solids. Also, by definition, the coefficient of isothermal compressibility is given by

$$c_{eff} = \frac{1}{\rho}\frac{\partial \rho}{\partial p}. \qquad (10.2.33)$$

However, when the mud is subjected to a higher temperature and pressure, the volume changes. Since water, oil, and solids have different compressibilities, the respective volume fractions will change also. Therefore, the effective compressibility of the drilling fluid will be a function of both temperature and pressure. Studies by Peters et al. (1990) show that for the pressures and temperatures of interest, c_{eff} is a weak function of both pressure and temperature. For first-order effects, we can write

$$c_{eff} = c_0\left[1 + c_1(p - p_0) + c_2(T - T_0)\right]. \qquad (10.2.34)$$

Then, by combining Eqs. 10.2.33 and 10.2.34, we obtain

$$\frac{\partial \rho}{\partial p} = c_0 \rho\left[1 + c_1(p - p_0) + c_2(T - T_0)\right]. \qquad (10.2.35)$$

Coefficient of Thermal Expansion. The arguments that we used for the coefficient of isothermal compressibility also may be used for the coefficient of thermal expansion. The isobaric coefficient of thermal expansion is defined as

$$\alpha_{eff} = \frac{1}{\rho}\frac{\partial \rho}{\partial T} \qquad (10.2.36)$$

and again using the first-order relation similar to Eq. 10.2.34,

$$\alpha_{eff} = \alpha_0 \left[1 + \alpha_1 (p - p_0) + \alpha_2 (T - T_0) \right]. \quad\ldots\ldots\ldots\ldots\ldots\ldots\ldots\ldots\ldots\ldots \quad (10.2.37)$$

This is in agreement with studies by Kutasov (1988), which show that the coefficient of thermal expansion of water- and oil-based mud and brine can be expressed as a linear function of temperature. By combining Eqs. 10.2.36 and 10.2.37, we get

$$\frac{\partial \rho}{\partial T} = \alpha_0 \rho \left[1 + \alpha_1 (p - p_0) + \alpha_2 (T - T_0) \right]. \quad\ldots\ldots\ldots\ldots\ldots\ldots\ldots\ldots \quad (10.2.38)$$

Density as a Function of Pressure and Temperature. Eqs. 10.2.35 and 10.2.38 now can be substituted into Eq. 10.2.31. Inspection of the properties of Eqs. 10.2.35 and 10.2.38 show that the second order effects $c_0 c_2$ and $a_0 a_2$ are in the same order of magnitude and can, with minor loss of accuracy, be replaced by an average gradient. The following simplification is therefore introduced:

$$c_0 c_2 \approx \alpha_0 \alpha_2 \approx \frac{c_0 c_2 + \alpha_0 \alpha_2}{2}. \quad\ldots\ldots\ldots\ldots\ldots\ldots\ldots\ldots\ldots\ldots \quad (10.2.39)$$

The mathematical details of the procedure of evaluating the density as a function of pressure and temperature are given in Karstad (1999). The resulting equation of state is given by

$$\rho = \rho_0 e^{\Gamma(p,T)}, \quad\ldots\ldots\ldots\ldots\ldots\ldots\ldots\ldots\ldots\ldots\ldots\ldots\ldots\ldots \quad (10.2.40)$$

where

$$\begin{aligned}
\Gamma(p,T) = \ &\gamma_p \left(\frac{p - p_0}{p_{max} - p_0} \right) + \gamma_{pp} \left(\frac{p - p_0}{p_{max} - p_0} \right)^2 \\
&+ \gamma_T \left(\frac{T - T_0}{T_{max} - T_0} \right) + \gamma_{TT} \left(\frac{T - T_0}{T_{max} - T_0} \right)^2. \quad\ldots\ldots\ldots\ldots\ldots\ldots\ldots \quad (10.2.41) \\
&+ \gamma_{pT} \left(\frac{T - T_0}{T_{max} - T_0} \right) \left(\frac{p - p_0}{p_{max} - p_0} \right)
\end{aligned}$$

The dimensionless constants γ_p, γ_{pp}, γ_T, γ_{TT}, and γ_{pT} are essentially unknown and must be determined for different muds from density measurements at elevated pressures and temperatures. We have normalized the model constants with bottomhole pressure, p_{max}, and temperature, T_{max}. However, we do not know the exact pressure and temperature, so we choose the following values:

$$p_{max} = p_0 + \rho_0 g D \quad\ldots\ldots\ldots\ldots\ldots\ldots\ldots\ldots\ldots\ldots\ldots\ldots \quad (10.2.42)$$

and

$$T_{max} = T_{sf} + g_s D_{sb} + g_G (D - D_{sb}). \quad\ldots\ldots\ldots\ldots\ldots\ldots\ldots\ldots \quad (10.2.43)$$

Thus, the constants γ_i are dimensionless and are scaled to avoid very small numerical values. Note that the choice of p_m and T_m is not rigorous; the important message is to choose a suitable value with regard to the pressures and temperatures encountered in the well.

If we know the pressure and temperature profiles of the well, we may calculate the mud density profile. Note that we have assumed surface conditions (p_{sf}, T_{sf}) to equal initial conditions (p_0, T_0). This may not be a satisfactory approximation if the surface temperature is very high. This can be accounted for by using the equation of state to calculate the density at surface conditions.

Comparison to Earlier Models. *Resemblance to Empirical Models.* In earlier models, which present an equation of state for the downhole density of static drilling fluids, the equations are obtained from curve-fitting methods. Our model is derived from pure analytical considerations. Therefore, it is interesting to observe the similarities to earlier models.

We observe that if we let the parameters $\gamma_{pp} = 0$ and $\gamma_{pT} = 0$, the general density equation is reduced and becomes identical to the empirical relation suggested by Kutasov (1988). Further, if we expand our equation of state into a Taylor series and neglect all second-order terms and higher, we get

$$\rho = \rho_0 \left[1 + \tilde{\gamma}_p \left(p - p_0 \right) + \tilde{\gamma}_T \left(T - T_0 \right) \right]. \qquad \text{(10.2.44)}$$

This equals the linear model suggested by Sorelle et al. (1982). This shows that models that earlier have been considered as pure empirical correlations can be derived easily from our analytical solution of the density/pressure/temperature dependence. Or, from another point of view, this verifies our model selection. This is further verified when we test our model against measured data from different muds.

Model Applied to Measured Data. We have tested the model against measured densities at elevated pressures and temperatures, and we have compared the results to the predictions made by other models.

The constants γ_i are evaluated for the measured data of 16 different muds. Muds 1 through 6 from Mc-Mordie et al. (1982), Muds 7 through 12 from Peters et al. (1990), Mud 13 from Sorelle et al. (1982), and Muds 14 through 16 from Isambourg et al. (1996). The constants yielded from linear regression analysis performed on the measured data are given in **Table 10.2.3**, where we have calculated the density ρ_0 such that p_0 and T_0 equal standard conditions ($p_0^{\text{gauge}} = 0$ Pa, $T_0 = 15°C$).

If we compare the average errors with those of other models, we observe a significant improvement. A comparison of different average errors is given in **Table 10.2.4**. In **Fig. 10.2.9**, we show some of the results from Table 10.2.4 in a bar chart for better visualization.

However, for the water-based muds, we observe the difference in average errors obtained by our model and Kutasov's (1988) to be very small. As we can observe in Table 10.2.3, this is expected. The constants γ_{pp} and γ_{pT} are generally lower for the water-based muds. This means that there is less coupling between the pressure and temperature effects and that in this range of pressures the effective coefficient of compressibility may be assumed constant (i.e., a very weak function of pressure). The result is that the Karstad and Aadnoy (1998) model is reduced to equal the empirical model suggested by Kutasov (1988).

TABLE 10.2.3—EMPIRICAL CONSTANTS FOR DIFFERENT MUDS

Mud	Base of Mud	ρ_0 (kg/m³)	$\gamma_p \times 10^2$	$\gamma_{pp} \times 10^3$	$\gamma_T \times 10^2$	$\gamma_{TT} \times 10^3$	$\gamma_{pT} \times 10^3$	Average Error
1	Water	1296	2.945	−4.854	−6.624	−29.245	15.347	0.16
2	Water	1643	2.945	−7.219	−4.993	−34.911	9.559	0.11
3	Water	2171	3.356	−6.686	−3.490	−27.259	8.428	0.11
4	Diesel Oil 2	1326	7.082	−34.951	−11.676	0.531	44.181	0.14
5	Diesel Oil 2	1721	6.560	−26.573	−9.042	−4.708	33.801	0.08
6	Diesel Oil 2	2171	5.575	−24.492	−8.091	−0.860	32.052	0.08
7	Diesel Oil 2	1328	4.292	−10.070	−10.637	9.183	20.255	0.04
8	Diesel Oil 2	2047	3.490	−8.537	−7.468	3.587	16.851	0.04
9	Mineral Oil A	1327	5.049	−11.611	−10.361	4.095	22.495	0.06
10	Mineral Oil A	2047	3.654	−8.970	−7.383	−1.906	17.676	0.04
11	Mineral Oil B	1328	5.216	−12.485	−11.145	7.770	24.293	0.06
12	Mineral Oil B	2048	3.827	−9.811	−7.956	2.440	19.368	0.05
13	Diesel Oil 2	849	5.896	−19.997	−11.271	−3.531	14.833	0.05
14	Low-toxicity oil	825	6.203	−32.776	−6.633	−28.898	42.746	0.17
15	Low-toxicity oil	2256	3.802	−11.541	−8.126	−6.372	16.815	0.04
16	Water	2212	2.383	−1.469	−2.818	−34.916	0.790	0.01

TABLE 10.2.4—COMPARISON OF AVERAGE ERRORS FOR DIFFERENT MODELS					
Mud No.	Base of Mud	ρ_0 (kg/m³)	Sorelle et al. (1982)	Compositional	Babu (1996)	Analytical
1	Water	1296	1.31	0.38	0.19	0.16
2	Water	1643	1.15	0.41	0.12	0.11
3	Water	2171	1.43	0.80	0.11	0.11
4	Diesel Oil 2	1326	1.35	NA	0.25	0.14
5	Diesel Oil 2	1721	1.29	NA	0.18	0.08
6	Diesel Oil 2	2171	0.49	NA	0.17	0.08
7	Diesel Oil 2	1328	0.50	0.27	0.27	0.04
8	Diesel Oil 2	2047	NA	0.23	0.23	0.04
9	Mineral Oil A	1327	NA	0.28	0.31	0.06
10	Mineral Oil A	2047	NA	0.16	0.24	0.04
11	Mineral Oil B	1328	NA	0.17	0.33	0.06
12	Mineral Oil B	2048	NA	0.10	0.27	0.05
13	Diesel Oil 2	849	0.40	NA	0.25	0.05
14	Low-toxicity oil	825	NA	NA	0.25	0.17
15	Low-toxicity oil	2256	NA	NA	0.07	0.04
16	Water	2212	NA	NA	0.01	0.01

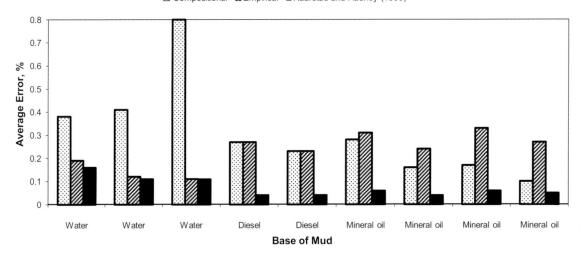

Fig. 10.2.9—Comparing accuracy of different density models.

Model Sensitivity to Input Parameters. As described in the Mathematical Approach section, the error in an expression is dependent on the error in each input parameter. In this section, we calculate the error obtained in predicted density and compare the contribution from each of the input parameters. The downhole fluid density can be written as follows:

$$\rho = f\left(\rho_0, p, T, \gamma_p, \gamma_{pp}, \gamma_T, \gamma_{TT}, \gamma_{pT}\right). \qquad (10.2.45)$$

Recalling Eq. 10.2.3, the absolute error in ρ is now represented by the partial derivatives. For this particular example, we assume that the temperature is 200°C, the pressure is 1.05×10^8 Pa, and there is an error of 5% in each of the input parameters, $\rho_0, p, T, \gamma_p, \gamma_{pp}, \gamma_T, \gamma_{TT}$, and γ_{pT}. Next, we compute the error contribution from each term using Mud 8 in Table 10.2.3. The absolute error from each of the parameters becomes

$$\left| \Delta\rho_0 \frac{\partial\rho}{\partial\rho_0} \right| = 0.05\rho \qquad \left| \Delta p \frac{\partial\rho}{\partial p} \right| = 0.002\rho$$

$$\left| \Delta T \frac{\partial\rho}{\partial T} \right| = 0.002\rho \qquad \left| \Delta\gamma_p \frac{\partial\rho}{\partial\gamma_p} \right| = 0.002\rho$$

$$\left| \Delta\gamma_{pp} \frac{\partial\rho}{\partial\gamma_{pp}} \right| = 0.0007\rho \qquad \left| \Delta\gamma_T \frac{\partial\rho}{\partial\gamma_T} \right| = 0.004\rho \qquad \text{......................................} \quad (10.2.46)$$

$$\left| \Delta\gamma_{TT} \frac{\partial\rho}{\partial\gamma_{TT}} \right| = 0.0002\rho \qquad \left| \Delta\gamma_{pT} \frac{\partial\rho}{\partial\gamma_{pT}} \right| = 0.001\rho$$

From these equations, it is obvious that an error in the measurement of the surface density is the most critical with regard to predicting the downhole density. A 5% error in surface density results in a 5% relative error in predicted density and is responsible for more than 80% of the total error. We consider this a first-order effect.

A 5% error in either pressure or temperature measurements results in approximately 0.2% relative error in predicted density. This is approximately 3% of the total error and may be considered a second-order effect.

However, one of the most interesting results we find is when we look at the sensitivity to the constants γ_i. The first-order terms, γ_p and γ_T, contribute 0.2 and 0.4% relative error in predicted density, respectively. These are of the same order as for pressure and temperature. The second-order terms γ_{pp} and γ_{TT} are less significant and contribute only 0.07 and 0.02% relative error in predicted density, or alternatively 0.1 and 0.03% of the total error, respectively. This may be regarded as a third-order effect. In contrast to this, the second-order term γ_{pT} represents 0.1% relative error in predicted density, which is significantly more than for the other second-order terms. The fact that the relative error obtained from γ_{pT} is an order of magnitude larger than for γ_{pp} and γ_{TT} indicates that the coupling effect between pressure and temperature is important.

Thus, this sensitivity analysis justifies the introduction of crosscoupling between pressure and temperature in the coefficient of isothermal compressibility and the coefficient of thermal expansion.

Conditions for Invariable Density. As the temperature and pressure of the mud column increases with depth, the mud experiences two opposing effects. Increase in temperature tends to decrease the mud density due to thermal expansion, while the increase in pressure tends to increase the mud density due to compressibility. Therefore, for a particular temperature profile, we expect these two opposing effects to cancel out. We define this as the constant-density temperature profile, $T(\rho_0)$. The result would be a constant density, ρ_0, throughout the interval. If the real mud temperature lies above $T(\rho_0)$, the mud density is lower than the surface density, and the thermal expansion effect is predominant. If the temperature is lower, the mud density is higher than that at surface, and the compressibility effect is predominant. Hence, if we can determine $T(\rho_0)$, we can compare it with the real temperature profile in the well and immediately determine which effect is predominant, thermal expansion or compressibility. A constant density implies that

$$\Gamma(p,T) = 0 \text{ and } (p - p_0) = g\rho_0 z. \quad \text{...} \quad (10.2.47)$$

$T(\rho_0)$ now may be determined by solving Eq. 10.2.47 for the temperature, T. $T(\rho_0)$ is the condition for constant density throughout the interval. **Fig. 10.2.10** shows this temperature profile for Muds 1, 7, and 10 in Table 10.2.3.

Prediction of Downhole Density Behavior. In **Fig. 10.2.11**, we can see this concept applied to a well. For a hypothetical 4600-m well, using Mud 8 in Table 10.2.3, Peters et al. (1990) concluded that no a priori prediction can be made about how the downhole mud densities will vary with depth. By plotting different static temperature profiles together with $T(\rho_0)$, we can predict the density variation immediately. The middle curve shows the temperature profile that gives constant mud density throughout the well. The bottomhole temperature is now 99°C.

With a bottomhole temperature of 177°C, the mud density will be lower than the surface density, and with a bottomhole temperature of 57°C, the mud density will be higher than the surface density. The measurements by Peters et al. (1990) show exactly the same results.

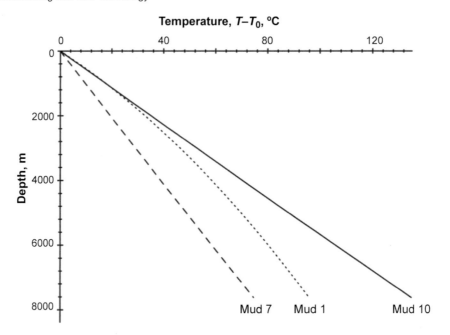

Fig. 10.2.10—Curves show temperature profiles that result in a constant density at any depth of the well for different fluids.

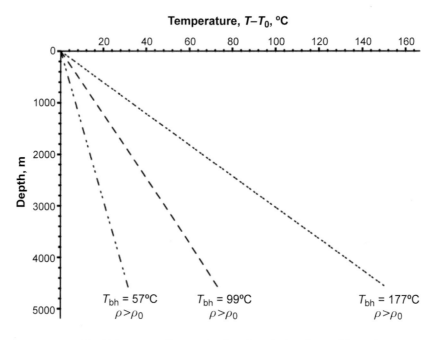

Fig. 10.2.11—Predicted behavior of static density profile.

For other mud temperature profiles, it may not be that easy to predict the density behavior. In **Fig. 10.2.12,** we can see the mud density profiles of Muds 7 and 10 compared to the surface density when using well data from Holmes and Swift (1970) and the resulting temperature profile after a short period of circulation. In the upper part, the density will be lower than the surface density. The pressure at some depth, z, in the circulated mud is therefore lower than the pressure in the constant-density mud. As a result, the circulated mud will have

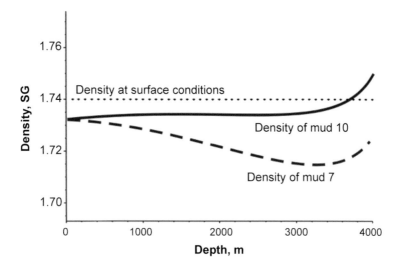

Fig. 10.2.12—Density profiles of different muds shortly after initiation of circulation.

a lower density than the surface density, even below the point of intersection between the two curves. Thus, it is not straightforward to determine at which depth the compressibility effect cancels out the effect of thermal expansion, or whether the average density is higher or lower than the surface density. However, we may say that the effect of thermal expansion (compressibility) is predominant if the area between $T(\rho_0)$ and actual mud temperature profile is visibly larger on the upper (lower) side of the point of intersection.

Static Pressure Profile. By using our equation of state, we now can calculate the pressure profile for a specific mud and a given temperature profile. The change in pressure with depth, z, is given by

$$\frac{dp}{dz} = \rho g = \rho_0 g e^{\Gamma(p,T)}, \qquad\qquad\qquad\qquad\qquad (10.2.48)$$

where the surface condition is given by

$$p\left(z=0\right)=p_{sf} \text{ and } T\left(z=0\right)=T_{sf}. \qquad\qquad\qquad (10.2.49)$$

The temperature profile, $T(z)$, is considered to be known. An analytical solution for this nonlinear initial value problem is not easily obtained. Therefore, we solve it numerically by using a classical Runge-Kutta method.

However, under certain conditions, we can approximate the equation so that an analytical solution may be obtained. As noted, the equation of state for water-based muds is well approximated by the simpler model of Kutasov (1988). If the well has been left static for some time, it is reasonable to assume that the temperature increases linearly with depth. Then

$$T = g_G z + T_{sf}, \qquad\qquad\qquad\qquad\qquad\qquad (10.2.50)$$

and Eq. 10.2.48 may be solved analytically.

Water-Based Muds and Linear Temperature Variation. Assuming Eq. 10.2.50 and surface conditions such that $p_{sf}=p_0$ and $T_{sf}=T_0$, the equation of state is reduced and Eq. 10.2.48 can be written as a separable differential equation. Solving this with appropriate manipulations yields (see Karstad, 1999, for mathematical details)

$$p - p_0 = \frac{1}{\hat{\lambda}_p} \ln \frac{1}{1-F\left(z\right)}, \qquad\qquad\qquad\qquad (10.2.51)$$

where $F(z)$ is given by

$$F(z) = \frac{g\hat{\gamma}_p \rho_0}{g_G \sqrt{-\gamma_{TT}}} \frac{\sqrt{\pi}}{2} e^{\eta^2} \left[\operatorname{erf}(z_D) - \operatorname{erf}(-\eta) \right]. \dots\dots\dots\dots\dots\dots \quad (10.2.52)$$

η is a dimensionless constant,

$$\eta = \frac{\hat{\gamma}_T}{2\sqrt{-\hat{\gamma}_{TT}}}, \dots\dots\dots\dots\dots\dots\dots\dots\dots\dots\dots \quad (10.2.53)$$

and z_D is a dimensionless depth variable,

$$z_D = g_G \sqrt{-\hat{\gamma}_{TT}} \, z - \eta. \dots\dots\dots\dots\dots\dots\dots\dots\dots \quad (10.2.54)$$

Equivalent Density Concept. For field applications, it is helpful to compare a complex well fluid column to an equivalent single-fluid column. Therefore, we calculate an equivalent density, ρ_{eq}, which is defined by

$$\rho_{eq} = \frac{p - p_0}{gD}. \dots\dots\dots\dots\dots\dots\dots\dots\dots\dots\dots \quad (10.2.55)$$

Hence, the pressure is represented as a pressure gradient expressed in density units. ρ_{eq} should always be referenced at a specified depth.

By studying Eq. 10.2.48, we observe that the equivalent density can be calculated in an alternative way. Integration yields

$$p - p_0 = g \int_0^D \rho \, dz \dots\dots\dots\dots\dots\dots\dots\dots\dots \quad (10.2.56)$$

Substituting this into Eq. 10.2.55 gives us

$$\rho_{eq} = \frac{1}{D} \int_0^D \rho \, dz. \dots\dots\dots\dots\dots\dots\dots\dots\dots \quad (10.2.57)$$

This shows that the equivalent density equals the average mud density of the well.

Also, we distinguish between equivalent densities for a circulating and a noncirculating well. Due to fluid circulation, the bottomhole pressure is greater than the hydrostatic pressure of the mud. The major contribution is from the frictional pressure loss in the annulus, but compressibility of the mud also may be a dominating parameter, especially in deep wells. For this scenario, the equivalent density is represented by the ECD.

When the well is not circulating, we have no frictional pressure loss. Any variations in density now are due to thermal expansion and compression of the mud. To incorporate this density/pressure/temperature behavior of the mud, we define an ESD. For a shallow well where pressure and temperature effects are not substantial, the ESD is very close to the surface density. Also, the ESD always has a lower value than the ECD.

Effect of Density Variations on Mud Gain/Loss. The return mud volume is often observed to vary to some extent, giving either a too-low or a too-high return rate. This phenomenon is referred to as ballooning. A thorough discussion on the evaluation of ballooning is given by Aadnoy (1996).

Because the mud density is strongly temperature-dependent, volume gain or losses due to thermal effects may be substantial. However, in contrast to the short-lived pressure effects, thermal effects are much slower. It takes many hours before the temperature has stabilized.

Since the volume of the well is constant at a given time, the process is defined as isochoric—that is, a constant-volume process. Thus, any pit gain or loss is the result of a gain or loss of mass to the mud in the well. The volume is given by

$$V = \pi D \left(r_{c,i}^2 - r_{d,o}^2 + r_{d,i}^2 \right), \dots\dots\dots\dots\dots\dots\dots \quad (10.2.58)$$

where $r_{c,i}$, $r_{d,o}$, and $r_{d,i}$ are the inner radius of the hole or casing, outer radius of the drillstring, and inner radius of drillstring, respectively. At time t, the mass of the mud system is

$$m_t = \rho_{eq,t} V, \quad \dots\dots\dots\dots\dots\dots\dots\dots\dots\dots\dots\dots\dots \quad (10.2.59)$$

where $\rho_{eq,t}$ is the ESD (or average density) at time t. At a later time, t', the mass has changed to $m_{t'}$. We can write the change in mass as

$$\Delta m = \Delta \rho_{eq} V. \quad \dots\dots\dots\dots\dots\dots\dots\dots\dots\dots\dots \quad (10.2.60)$$

This change of mass is recognized as the pit gain or loss, and we can represent it as surface volume units,

$$\Delta V = \frac{V}{\rho_0 D} \int_0^D (\rho_{t'} - \rho_t)\, dz, \quad \dots\dots\dots\dots\dots\dots\dots\dots \quad (10.2.61)$$

or, alternatively,

$$\frac{\Delta V}{V} = \frac{\Delta \rho_{eq}}{\rho_0}, \quad \dots\dots\dots\dots\dots\dots\dots\dots\dots\dots \quad (10.2.62)$$

where ΔV is the pit gain or loss, and ρ_0 is the surface density. Eq. 10.2.62 is valid for both circulating and noncirculating conditions by exchanging ρ_{eq} with ECD or ESD, respectively. This simple model agrees with the method proposed by Aadnoy (1996).

Coupling Between Pit Gain/Loss and Mud Density. In the planning of an HP/HT well, Aadnoy (1996) performed a larger study about ballooning. The following elements were included in the study:

* Pressure drop vs. flow-rate modeling
* Temperature-dependent mud behavior
* Pressure-related volume expansion of casing and open hole

Each of these elements was built into a complex numerical simulator. Using data given in **Table 10.2.5**, the ECD was predicted.

In the following, we will apply the same data in our equation describing the coupling between pit gain/loss and equivalent mud density (Eq 10.2.62).

Case 1. For the 12¼-in. (31.12-cm) hole evaluated in Case 1, we obtain an equivalent density of 1.727 SG. This is in reasonable agreement with the simulator-predicted equivalent mud density of 1.72 SG.

Case 2. For the 8½-in. (21.59-cm) hole evaluated in Case 2, we obtain an equivalent density of 2.095 SG. Again, this is close to the simulator-predicted equivalent mud density of 2.08 SG.

These two examples show that the variation in mud density is by far the most important factor in mud ballooning. Also, we have shown that the variation in equivalent mud density can be calculated by using the simple model derived in this section (Eq. 10.2.62).

Transient Density Behavior During Circulation. In this particular example, we are considering only short-time behavior that is relevant for an ordinary drilling operation. By using the temperature profiles, we have analyzed the time-dependent density behavior. Throughout the simulations, we use Mud 8 in Table 10.2.3.

TABLE 10.2.5—DATA FROM COMPUTER SIMULATIONS		
	Case 1	Case 2
Hole diameter (in.)	12¼	8½
Flow rate (L/min)	4000	2500
Depth of well (m)	4700	5000
Capacity of well (m^3/m)	1/15.78	1/41.77
Maximum pit gain (L)	12,000	1400
Mud density at surface (SG)	1.8	2.12
Predicted effective density (SG)	1.72	2.08

Well 1. First we consider an onshore well, and we do not include any external supply of energy to the well. Input data for the simulations are given in **Table 10.2.6**.

We have calculated what the downhole mud density profile would look like after some period of circulation. In **Fig. 10.2.13**, we have plotted the surface density (2.04 SG) together with

- Density profile at geothermal conditions (ESD = 1.99 SG)
- Density profile shortly after circulation begins (ESD = 1.99 SG)
- Density profile after 12 hours of circulation (ESD = 2.07 SG)

From the figure, we can see that the mud density increases as circulation continues. This is explained by the continuous cooling of the well during circulation. In this case, the ESD has changed by 0.08 SG during the 12 hours of circulation. The resulting volume variation would be an apparent mud loss of 5.06 m^3.

TABLE 10.2.6—WELL AND MUD DATA FOR SYNTHETIC ONSHORE HP/HT WELL (Karstad and Aadnoy 1998)	
Drillstem outside diameter (in.)	5
Drill bit size (in.)	8^3/$_8$
Well depth (m)	5500
Circulation rate (L/min)	1500
Inlet mud temperature (°C)	30
Mud thermal conductivity (W/m·°C)	1.730
Mud specific heat (J/kg·°C)	1676
Mud density (SG)	2.04
Formation thermal conductivity (W/m·°C)	2.250
Formation specific heat (J/kg·°C)	838
Formation density (SG)	2.643
Surface Earth temperature (°C)	4
Geothermal gradient (°C/m)	0.040

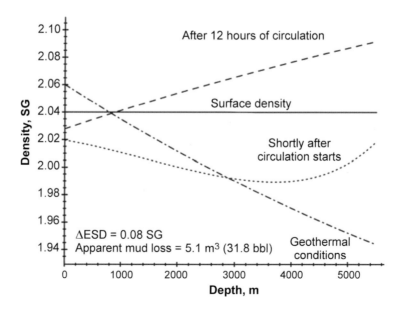

Fig. 10.2.13—Examples of different density profiles in a well.

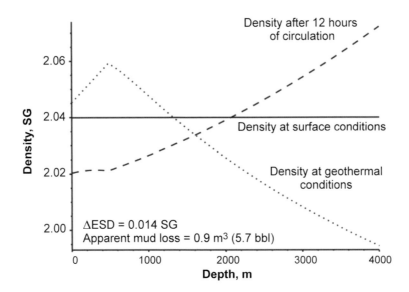

Fig. 10.2.14—Examples of density profiles at different circulation times in offshore well.

Well 2. Now we consider an offshore well. Also, in addition to including the effect of having a riser, we include input of energy due to drillstring rotation and the hydraulic pumping system. For simulations, we use the well data given in Table 10.2.6.

The resulting density profiles are shown in **Fig. 10.2.14.**

- Surface density (2040 kg/m^3)
- Density profile at geothermal conditions (ESD = 2027 kg/m^3)
- Density profile after 12 hours of circulation (ESD = 2041 kg/m^3)

For this well, the ESD has changed by only 0.014 SG during the 12 hours of circulation, with an apparent volumetric mud loss of 0.9 m^3. The much smaller change in ESD for this example is in line with what could be expected. Since the temperature of the seawater surrounding the riser does not change, it will have a dampening effect on temperature variations in the mud. This again will result in damped variations of mud density. The input of energy from pumping and rotation give a similar dampening effect. A continuous supply of energy will result in less cooling of the mud during circulation, and the change in mud density will be less. Hence, both effects counteract thermal contraction during circulation, and less change in ESD is expected.

However, it is important to note that the variation in ESD depends on many factors. In some wells where pressure and temperature effects tend to cancel out, the changes in ESD are minor, while in other wells the conditions result in significant changes in ESD.

By studying the equation of state, it is found that the flow rate has a strong effect on how the density changes. A higher circulation rate cools the well more, and therefore the equivalent density will stabilize more quickly and at a higher value. For most practical purposes, the equivalent density has stabilized within 12 hours of circulation.

Effect of Density Variations During Shut-In. When the pumps are stopped, the frictional losses inside the hydraulic system cease and the mud pressure decreases. This allows for volume changes of the mud, the casing, and the rock of the open hole. However, since both the casing and the rock are relatively stiff bodies, any volume changes due to expansion of casing and open hole amounts to approximately 10% of the total volume variation (Aadnoy 1996; Karstad 1999). Mud volume variation due to the compressibility of the mud is much more significant. The pressure effect is almost instantaneous, while the temperature effect is much slower.

The mud will expand due to heating of the well, and the apparent loss now will start to flow back out of the hole. This effect has been observed in HP/HT wells that are shut in after a well-control incident. Changes in shut-in pressures have been interpreted as a direct measurement of the pore pressure in the reservoir.

In reality, density variations due to temperature changes may significantly change the shut-in pressure, and should always be taken into account.

Summary. In this section we have presented a new analytical model for the density/pressure/temperature dependence of drilling fluids. The new equation of state is coupled to a transient temperature model. The result is the effective fluid density and wellbore pressure, taking effects of compression and thermal expansion into account.

A field example demonstrates the error caused by compressibility and thermal effects. Assuming a surface density of 2.04 SG, the effective density at geothermal conditions will be 1.99 SG. During 12 hours of circulation, it increases to 2.07 SG. This would lead to an apparent mud loss of 5 m^3. This demonstrates potential errors in, for example, pore pressure prediction.

Through this research, the following understanding emerged:

- The fluid density model derived takes into account coupling between temperature and pressure, and it is shown that it is more accurate than the models in use today.
- Measuring volume variations in the return mud can provide important information for calibration of the models and also can be used to give an estimate of the effective bottomhole pressure.
- The equivalent density changes more rapidly for high circulation rates, and it tends to stabilize at a higher value. For most practical purposes, the equivalent density has stabilized within 12 hours of circulation.
- Circulating wellbore fluid when the bit is not drilling leads to cooling of the bottom of the well and increasing mud return temperature. The cooling effect is strongly dependent on the circulating time and on the flow rate.

10.2.4 Energy Balance Approach. Due to frictional forces, the rotational energy of the drillstring and the hydraulic pump energy are transformed into heat downhole. This supply of energy is counteracting the otherwise cooling effect the circulating fluid has on the formation. The knowledge of accurate temperature with circulating time is important for determination of thermal stresses and equivalent mud density, design of cementing programs and logging tools, and log interpretation. In light of this, it is obvious that including the supply of energy is essential for temperature modeling.

This effect has not been included in earlier analytical models. Some numerical models include energy source terms as a function of location and time (Corre et al. 1984). However, as pointed out earlier, the complexity of numerical models makes them less suited for parametric studies or analyzing the system behavior.

In addition to including the supply of energy in the temperature model, we also have performed an energy balance of the circulating well. This was done to model the rate of heat that is supplied to or transported from the well. Knowing the direction and magnitude of the heat transport to/from the well provides important information on the downhole temperature and density behavior.

As shown in Fig. 10.2.3, the formation is being cooled in the lower section and heated in the upper section. When the net heat transport is from the well to the surface, the cooling effect in the lower section is the dominating effect. The result is that the temperature profile will drift away from the geothermal temperature, and the average fluid temperature will decrease. As the temperature decreases, the ECD will increase. Alternatively, the ESD or the static pressure tends to decrease when changed from circulating to static condition.

In case of a net heat supply to the well, the opposite occurs. The average fluid temperature increases, and the ECD decreases. Depending on the history, we even can get a situation in which the ESD increases when changed from circulating to static condition.

Conservation of Energy: Control Volume Approach. The expression for the energy balance applied to a control volume is developed from the first law of thermodynamics. First, we define a control volume as shown in **Fig. 10.2.15**.

It is difficult to model the heat loss in the riser, so the control volume is chosen not to include the water section of the well. Also, because the accumulation of energy in the formation happens within a few radii away from the wellbore, we choose the surface of the control volume to be sufficiently far away from the wellbore that geothermal conditions may be assumed for all times. Hence, there will be no heat exchange between the control volume and the surrounding formations.

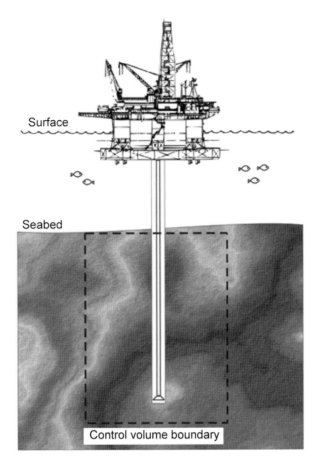

Fig. 10.2.15—Conservation of energy for a control volume, applied to a circulating well.

The first law of thermodynamics may now be written as follows (Welty et al. 1984):

$$
\begin{Bmatrix} \text{Rate of addition} \\ \text{of heat to control} \\ \text{volume from its} \\ \text{surroundings} \end{Bmatrix} - \begin{Bmatrix} \text{Rate of work} \\ \text{done by control} \\ \text{volume on its} \\ \text{surroundings} \end{Bmatrix} =
$$

$$
\begin{Bmatrix} \text{Rate of energy} \\ \text{out of control} \\ \text{volume due to} \\ \text{fluid flow} \end{Bmatrix} - \begin{Bmatrix} \text{Rate of energy} \\ \text{into control} \\ \text{volume due to} \\ \text{fluid flow} \end{Bmatrix} + \begin{Bmatrix} \text{Rate of accumu-} \\ \text{lation of energy} \\ \text{within control} \\ \text{volume} \end{Bmatrix} \quad \text{.........................} \quad (10.2.63)
$$

Eq. 10.2.63 is in general an integral expression. Fortunately, the geometry of the control surface enables us, in a straightforward manner, to reduce the integral expression to a simple scalar expression.

Rate of Accumulation of Energy Within Control Volume. Energy storage is a volumetric phenomenon and can be expressed by the following equation:

$$
\begin{Bmatrix} \text{Rate of accumulation} \\ \text{of energy within} \\ \text{control volume} \end{Bmatrix} = \frac{\partial}{\partial t} \iiint_{c.v.} e\rho dV, \quad \text{.................................} \quad (10.2.64)
$$

where e is the specific energy, or energy per unit mass. In general terms, the specific energy includes the potential energy, gz, due to the position of the fluid continuum in the gravitational field; the kinetic energy of the fluid, $v^2/2$, due to its velocity; and the internal energy, Υ, of the fluid due to its thermal state:

$$e = gz + \frac{v^2}{2} + \Upsilon. \quad\dots\dots\dots\dots\dots\dots\dots\dots\dots\dots\dots\dots\dots\dots\dots\dots\dots \quad (10.2.65)$$

The change in internal energy, $\Delta\Upsilon$, generally can be divided into thermal energy, intermolecular forces, chemical energy, and nuclear energy. In applications of interest for our study, chemical and nuclear effects are neglected. Also, we assume that there is no phase change in the circulating fluid system, which implies that latent energy effects need not be considered. Hence, if kinetic and potential energy effects may be neglected, as is almost always the case in heat transfer analysis, changes in energy storage are due only to changes in the internal thermal energy.

Since the rate of accumulation of energy is generally unknown, we represent it by $Q_{R,\mathrm{acc}}$. The sign of this term defines whether energy is removed from or supplied to the control volume. If it is positive, the net rate of energy is into the control volume, and vice versa.

Rate of Energy Into and Out of Control Volume Due to Fluid Flow. The net efflux of energy from the control surface may be defined by the following integral (Welty et al. 1984):

$$\iint_{c.s.} e\rho(\mathbf{v}\cdot\mathbf{n})dA = e_{a,sb}\rho_{a,sb}v_{a,sb}A_a - e_{d,sb}\rho_{d,sb}v_{d,sb}A_d, \quad\dots\dots\dots\dots\dots\dots\dots \quad (10.2.66)$$

where $\mathbf{v}$ is the fluid velocity vector and $\mathbf{n}$ is the outwardly directed normal vector.

We now assume that we have no loss or gain of fluids in the control volume during circulation. Thus, we have conservation of mass within the fluid system. During drilling, this is not quite true because of the addition of drill cuttings to the fluid in the annulus. However, the amount of drill cuttings added may be considered insignificant compared to the mass flow rate of the circulating mud. Hence, the mass flow rate in the annulus equals the mass flow rate in the drillpipe. Since the fluid enters and leaves the control volume at the same depth, the potential energy contribution cancels out. The net rate of energy out of the control volume is then given by

$$\begin{Bmatrix} \text{Rate of energy out} \\ \text{of control volume} \\ \text{due to fluid flow} \end{Bmatrix} - \begin{Bmatrix} \text{Rate of energy into} \\ \text{control volume due} \\ \text{to fluid flow} \end{Bmatrix}, \quad\dots\dots\dots\dots\dots\dots\dots\dots\dots\dots\dots \quad (10.2.67)$$

$$= \frac{w}{2}\left(v_{a,sb}^2 - v_{d,sb}^2\right) + w\left(\Upsilon_{a,sb} - \Upsilon_{d,sb}\right)$$

where w is the mass flow rate.

Rate of Addition of Heat to Control Volume From Surroundings. Accumulation of energy in the formation takes place within a few radii away from the wellbore. Therefore, we choose the surface of the control volume to be sufficiently far away from the wellbore so that geothermal conditions may be assumed for all times. Hence, there will be no heat exchange between the control volume and the surrounding formations. Also, we neglect any axial conduction of heat in the circulating fluid. The result of this is that there will be no addition of heat to the control volume:

$$\begin{Bmatrix} \text{Rate of addition of} \\ \text{heat to control volume} \\ \text{from its surroundings} \end{Bmatrix} = \frac{dQ}{dt} = 0. \quad\dots\dots\dots\dots\dots\dots\dots\dots\dots\dots\dots \quad (10.2.68)$$

Rate of Work Done by Control Volume on Surroundings. The last term of the first-law expression to be considered is the work-rate or power term, dW/dt. This term includes three types of work:

1. *Shaft work*, W_s, is work done by the control volume on its surroundings that could cause a shaft to rotate. Alternatively, it can accomplish the raising of a weight through a distance.
2. *Flow work*, $W\sigma$, is that done on the surroundings to overcome normal stresses on the control surface where there is fluid flow.
3. *Shear work*, $W\tau$, is performed on the surroundings to overcome shear stresses at the control surface.

The stress vector has components σ_{ii} and τ_{ij} in the directions normal and tangential to the surface, respectively. The work done by the control volume on the surroundings in the presence of the stress vector is given by

$$-\iint_{c.s.} (\mathbf{v} \cdot \mathbf{S}) dA, \quad \dots\dots\dots\dots\dots\dots\dots\dots\dots\dots\dots\dots\dots \quad (10.2.69)$$

where $\mathbf{S}$ is the stress vector. The normal stress components of $\mathbf{S}$ may be written as $\sigma_{ii}\mathbf{n}$. Further, we may say simply that the normal stress term is the sum of pressure effects and viscous effects. The shear work and the work done to overcome the viscous portion of the normal stress are both unavailable to do mechanical work. If we assume axial flow, the fluid flow will be normal to the control surface. It is then reasonable to neglect the work rate accomplished in overcoming viscous effects at the control surface. Integrated over the control surface, the sum of flow work and shear work may then be written as follows (Karstad 1999):

$$\frac{dW_\sigma}{dt} + \frac{dW_\tau}{dt} = -\iint_{c.s.} (\mathbf{v} \cdot \mathbf{S}) dA$$

$$= w\left(\frac{P_{a,sb}}{\rho_{a,sb}} - \frac{P_{d,sb}}{\rho_{d,sb}}\right). \quad \dots\dots\dots\dots\dots\dots\dots\dots\dots\dots\dots \quad (10.2.70)$$

During circulation, the rig delivers power to the fluid through the mud pump and the rotation of the drillstring. There is friction between the fluid and the rotating drillstring and between the formation and the drill bit. Thus, due to frictional forces, the rotational energy of the drillstring, $\dot{E}_{rot}$, and the hydraulic pump energy, $\dot{E}_{pump}$, are transformed into heat in the

- Mud
- Casing and borehole wall
- Riser
- Drillstring, BHA, and drill bit
- Formation beneath drill bit

and also into the kinetic energy of the mud. The mud will not flow purely in the axial direction, but will flow as a spiral up through the annulus. We assume that both the increase in temperature and the kinetic energy are entirely due to mechanical work done on the control volume by the surroundings. The total work rate then becomes

$$\begin{Bmatrix} \text{Rate of work done} \\ \text{by control volume} \\ \text{on its surroundings} \end{Bmatrix} = \frac{dW_s}{dt} + \frac{dW_\sigma}{dt} + \frac{dW_\tau}{dt}$$

$$= -\dot{E}_{rot} - \dot{E}_{pump} + w\left(\frac{P_{a,sb}}{\rho_{a,sb}} - \frac{P_{d,sb}}{\rho_{d,sb}}\right). \quad \dots\dots\dots\dots\dots\dots \quad (10.2.71)$$

Total Energy Balance. Eq. 10.2.63 can now be written as follows:

$$Q_{R,acc} = -\dot{E}_{rot} - \dot{E}_{pump}$$

$$-w\left[\left(h_{a,sb} - h_{d,sb}\right) + \frac{w^2}{2}\left(\frac{1}{\rho_{a,sb}^2 A_a^2} - \frac{1}{\rho_{d,sb}^2 A_d^2}\right)\right], \quad \dots\dots\dots\dots\dots \quad (10.2.72)$$

where $Q_{R,\text{acc}}$ is the heat flow to the control volume, and h is the specific enthalpy of the drilling fluid (by definition, $h = U + P/\rho$). If $Q_{R,\text{acc}} > 0$, we have a net input of energy into the control volume, and consequently the fluid is heating the formation. Conversely, if $Q_{R,\text{acc}} < 0$, we have a net loss of energy from the control volume, and the fluid is cooling the formation.

Effect of Riser. The control volume is chosen to be valid for both onshore and offshore wells. From Fig. 10.2.15, it is obvious that for an onshore well the heat flow to the control volume will be the same as the heat flow to the well. For the offshore scenario, we have to take into account the heat transfer between the riser and the surrounding water. We define $Q_{R,\text{acc}} = \Phi + \Phi_r$, where Φ can be interpreted as the net input of energy to the well, and Φ_r is the rate of heat transfer between the riser and the surrounding water. Φ_r is by definition negative when heat is lost from the riser to the water.

It is reasonable to assume that the riser is an energy sink. For almost all cases, heat is expected to be transferred from the mud system to the seawater. Thus, accumulated heat in our control volume in an offshore well will always be less than accumulated heat in an equivalent control volume for an onshore well.

The rate of net input of energy to the well then is found by replacing all values in Eq. 10.2.72 with surface values. As shown in Section 10.2.3, the drilling fluids cannot be considered incompressible. Downhole mud density is strongly dependent on depth (i.e., pressure) and temperature (e.g., circulation time). However, the heat transfer properties of the drilling fluid can largely be assumed constant.

$$\Phi = -\dot{E}_{\text{rot}} - \dot{E}_{\text{pump}}$$
$$-w\left[(P_{\text{out}} - P_{\text{in}}) + \rho_0 C_{fl} (T_{\text{out}} - T_{\text{in}}) + \frac{q^2 \rho}{2}\left(\frac{1}{A_a^2} - \frac{1}{A_d^2} \right) \right]. \quad\quad\quad (10.2.73)$$

Using Eq. 10.2.73, we now can determine whether the heat flow, Φ, is out of or into the well. Depending on the heat loss in the riser, we have three distinct scenarios:

1. Heat efflux from the well and heat efflux from the formation
2. Heat influx to the well and heat efflux from the formation
3. Heat influx to the well and heat influx to the formation

Heat Efflux From Well and Heat Efflux From Formation. This is the most common scenario. There is a net flow of energy from the formation to the surface.

$$\Phi < 0 \Rightarrow Q_{R,\text{acc}} < 0. \quad\quad\quad\quad\quad\quad (10.2.74)$$

The flowing mud has a cooling effect on the formation. The return mud is hotter than the mud in, and the average mud temperature is lower than the average formation temperature.

This course of events can be demonstrated by the well described in Table 10.2.2. Inserting the data into Eq. 10.2.73 shows that there is a net heat flow out of the well. Accordingly, we expect the fluid temperature on average to be lower than the formation temperature. Inspecting Figs. 10.2.6 and 10.2.8 confirms this. The fluid temperature is significantly lower than the virgin formation temperature and is decreasing as circulation carries on.

An important consequence of this is that the net heat flow rate from the well, Φ, is directly related to the change in average fluid temperature. With good estimates of the heat loss in the riser, Φ_r, it is possible to determine the change in average fluid temperature from surface measurements. This is useful for estimates of the ECD.

Heat Influx to Well and Heat Efflux From Formation. When $\Phi > 0$, there is a net flow of energy into the well.

$$\Phi > 0 \text{ and } |\Phi| < |\Phi_r| \Rightarrow Q_{R,\text{acc}} < 0. \quad\quad\quad\quad\quad (10.2.75)$$

However, since the heat loss from the riser exceeds the energy input at surface, we still have a net flow of energy out of the formation. This is most likely in very deep water.

Heat Influx to Well and Heat Influx to Formation. When the flow of energy into the well exceeds heat loss in the riser, there is a net flow of energy from the surface to the formation.

$$\Phi > 0 \text{ and } |\Phi| > |\Phi_r| \Rightarrow Q_{R,acc} > 0. \qquad\qquad (10.2.76)$$

This results in an increasing average mud temperature, and on average the mud is heating the formation. This does not happen very often. Usually, the convection of heat to surface from the circulating fluid by far exceeds the input of energy from pumping and rotation. But at low flow rates, a net input of energy to the formation may occur.

Summary. An offshore well is modeled in terms of the first law of thermodynamics. It is shown that the riser gives a heat loss that significantly cools the drilling mud. Also, it is shown that including energy input due to drillstring rotation and the mud pumping system may lead to an increase in the temperature of the well.

The control volume approach is a new approach to modeling the effect of the circulating-fluid system. The method can be a useful tool for calibration of circulating-mud temperature models. Further development of the method also may be used for analyzing hole cleaning.

Nomenclature

A = area, m^3
c = coefficient of isothermal compressibility, Pa^{-1}
C = specific heat, kJ/kg·°C
D = true vertical depth of well, m
$f(t_D)$ = dimensionless time function
e = specific energy, J/kg
Err_{abs} = absolute error
$\dot{E}_{pump}$ = input of energy from mud pump, W
$\dot{E}_{rot}$ = input of energy from drillstring rotation, W
g = acceleration due to gravity, m/s^2
g_G = geothermal gradient, °C/m
g_s = temperature gradient in seawater, °C/m
g_T = rate of change of mud-in temperature, °C/hr
h = specific enthalpy, J/kg
k = conductivity, W/m·°C
m = mass, kg
$\mathbf{n}$ = normal vector
N = arbitrary quantity
p, P = pressure, Pa
q = heat flow, kJ/s (when subscript is used)
q = flow rate, m^3/s (without subscript)
Q = heat, J
Q_R, Φ = heat flow rate, J/s (W)
ρ = fluid density, kg/m^3
r_d = drillpipe radius, m
r_w = borehole radius, m
R = radius, m
$\mathbf{S}$ = stress vector, kg/m·s^2
t = circulation time, hours
t_D = dimensionless time
T = temperature, °C
U = overall heat transfer coefficient, W/m^2·°C

v = fluid velocity, m/s

$\mathbf{v}$ = fluid velocity vector, m/s

V = volume of the well, m^3

w = mass flow rate of fluid, m^3/h

W = work, J

Y = specific internal energy, J/kg

z = vertical depth, m

α_h = thermal diffusivity, m^2/s

α = coefficient of thermal expansion, °C^{-1}

Subscripts

0 = initial conditions

a = annulus

abs = absolute

acc = accumulated

ad = from annulus to drillstring

ar = annulus, in rock section

as = annulus, in water section

bh = bottomhole

$c.s.$ = surface of control volume

$c.v.$ = control volume

d = drillstring

D = implies that quantity is dimensionless

dr = drillstring, in rock section

ds = drillstring, in water section

e = external input

eff = effective

eq = equivalent

f = formation

fl = fluid

i,j,k,n = arbitrary number

in = mud inlet

max = maximum

$meas$ = measurement

o = oil

out = mud outlet

pump = due to pump

r = riser

rel = relative

rev = reverse

rot = due to rotation

s = water section

sb = seabed

sd = solids

sf = surface

σ = pertains to normal stresses

τ = pertains to shear stresses

μ = pertains to viscous stresses

t = circulation time

w = water

wb = wellbore interface

References

Aadnoy, B.S. 1996. Evaluation of Ballooning in Deep Wells. In *Modern Well Design*, Appendix B. Rotterdam, Netherlands: Balkema.

Babu, D.R. 1996. Effects of P-p-T Behavior of Muds on Static Pressures During Deep Well Drilling—Part 2: Static Pressures. *SPEDC* **11** (2): 91–97. SPE-27419-PA. DOI: 10.2118/27419-PA.

Bullard, E.C. 1947. The Time Necessary for a Bore Hole To Attain Temperature Equilibrium. *Monthly Notices Roy. Astron. Soc., Geophys. Suppl.* **5**: 127–130.

Cao, S., Hermanrud, C., and Lerche, I. 1988. Formation Temperature Estimation by Inversion of Borehole Measurements. *Geophysics* **53** (7): 979–988. DOI: 10.1190/1.1442534.

Carslaw, H.S. and Jaeger, J.C. 1959. *Conduction of Heat in Solids*, second edition. London: Oxford Science Publications.

Corre, B., Eymard, R., and Guenot, A. 1984. Numerical Computation of Temperature Distribution in a Wellbore While Drilling. Paper SPE 13208 presented at the SPE Annual Conference and Exhibition, Houston, 16–19 September. DOI: 10.2118/13208-MS.

Edwardson, M.J., Girner, H.M., Parkinson, H.R., Williams, C.D., and Matthews, C.S. 1962. Calculation of Formation Temperature Disturbances Caused by Mud Circulation. *JPT* **14** (4): 416–426. SPE-124-PA. DOI: 10.2118/124-PA.

Gao, E., Estensen, O., MacDonald, C., and Castle, S. 1998. Critical Requirements for Successful Fluid Engineering in HP/HT Wells: Modeling Tools, Design Procedures and Bottomhole Pressure Management in the Field. Paper SPE 50581 presented at the European Petroleum Conference, The Hague, 20–22 October. DOI: 10.2118/50581-MS.

Gill, J.A. 1989. How Borehole Ballooning Alters Drilling Responses. *Oil and Gas Journal* **87**: 43–52.

Hasan, A.R. and Kabir, C.S. 1994a. Aspects of Wellbore Heat Transfer During Two-Phase Flow. *SPEPF* **9** (3): 211–216. SPE-22948-PA. DOI: 10.2118/22948-PA.

Hasan, A.R. and Kabir, C.S. 1994b. Static Reservoir Temperature Determination From Transient Data After Mud Circulation. *SPEDC* **9** (1): 17–23. SPE-24085-PA. DOI: 10.2118/24085-PA.

Hasan, A.R., Kabir, C.S., and Ameen, M.M. 1996. A Fluid Circulating Temperature Model for Workover Operations. *SPEJ* **1** (2): 133–144. SPE-27848-PA. DOI: 10.2118/27848-PA.

Hermanrud, C. 1988. Determination of Formation Temperature From Downhole Measurements. PhD dissertation, University of South Carolina, Columbia, South Carolina.

Hermanrud, C., Lerche, I., and Meisingset, K.K. 1991. Determination of Virgin Rock Temperature From Drillstem Tests. *JPT* **43** (9): 1126–1131. SPE-19464-PA. DOI: 10.2118/19464-PA.

Holmes, C.S. and Swift, S.C. 1970. Calculation of Circulating Mud Temperatures. *JPT* **22** (6): 670–674. SPE-2318-PA. DOI: 10.2118/2318-PA.

Isambourg, P., Anfinsen, B.T., and Marken, C. 1996. Volumetric Behavior of Drilling Muds at High Pressure and High Temperature. Paper SPE 36830 presented at the European Petroleum Conference, Milan, Italy, 22–24 October. DOI: 10.2118/36830-MS.

Kabir, C.S., Hasan, A.R., Kouba, G.E., and Ameen, M.M. 1996. Determining Circulating Fluid Temperature in Drilling, Workover, and Well-Control Operations. *SPEDC* **11** (2): 74–79. SPE-24581-PA. DOI: 10.2118/24581-PA.

Karstad, E. 1999. Time-Dependent Temperature Behavior in Rock and Borehole. PhD dissertation, Stavanger University College, Stavanger, Norway.

Karstad, E. and Aadnoy, B.S. 1997. Analysis of Temperature Measurements During Drilling. Paper SPE 38603 presented at the SPE Annual Technical Conference and Exhibition, San Antonio, Texas, 5–8 October. DOI: 10.2118/38603-MS.

Karstad, E. and Aadnoy, B.S. 1998. Density Behavior of Drilling Fluids During High Pressure High Temperature Drilling Operations. Paper SPE 47806 presented at IADC/SPE Asia Pacific Drilling Technology, Jakarta, 7–9 September. DOI: 10.2118/47806-MS.

Karstad, E. and Aadnoy, B.S. 1999. Optimization of Mud Temperature and Fluid Models in Offshore Applications. Paper SPE 56939 presented at the Offshore Europe Oil Exhibition and Conference, Aberdeen, 7–10 September. DOI: 10.2118/56939-MS.

Kutasov, I.M. 1988. Empirical Correlation Determines Downhole Mud Density. *Oil and Gas Journal* **86**: 61–63.

Lee, T.-C. 1982. Estimation of Formation Temperature and Thermal Property From Dissipation of Heat Generated by Drilling. *Geophysics* **47** (11): 1577–1584. DOI: 10.1190/1.1441308.

Luheshi, M.N. 1983. Estimation of Formation Temperature From Borehole Measurements. *Geophysical Journal (Royal Astronomical Society)* **74**: 747–776.

McMordie, W.C. Jr., Bland, R.G., and Hauser, J.M.1982. Effect of Temperature and Pressure on the Density of Drilling Fluids. Paper SPE 11114 presented at the SPE Annual Technical Conference and Exhibition, New Orleans, 26–29 September. DOI: 10.2118/11114-MS.

Middleton, M.F. 1982. Bottom-Hole Temperature Stabilization With Continued Circulation of Drilling Mud. *Geophysics* **47** (12): 1716–1723. DOI: 10.1190/1.1441321.

Oxburgh, E.R., Richardson, S.W., Turcotte, D.L., and Hsui, A. 1972. Equilibrium Bore Hole Temperatures From Observation of Thermal Transients During Drilling. *Earth and Planetary Science Letters* **14** (1): 47–49. DOI: 10.1016/0012-821X(72)90077-5.

Peters, E.J., Chenevert, M.E., and Zhang, C. 1990. A Model for Predicting the Density of Oil-Base Muds at High Pressures and Temperatures. *SPEDE* **5** (2): 141–148; *Trans.*, AIME, **289**. SPE-18036-PA. DOI: 10.2118/18036-PA.

Ramey, H.J. Jr. 1962. Wellbore Heat Transmission. *JPT* **14** (4): 427–435; *Trans.*, AIME, **225**. SPE-96-PA. DOI: 10.2118/96-PA.

Raymond, L.R. 1969. Temperature Distribution in a Circulating Drilling Fluid. *JPT* **21** (3): 333–341; *Trans.*, AIME, **246**. SPE-2320-PA. DOI: 10.2118/2320-PA.

Sorelle, R.R., Jardiolin, R.A., Buckley, P., and Barrios, J.R. 1982. Mathematical Field Model Predicts Downhole Density Changes in Static Drilling Fluids. Paper SPE 11118 presented at the SPE Annual Technical Conference and Exhibition, New Orleans, 26–29 September. DOI: 10.2118/11118-MS.

Ward, C. and Clark, R. 1999. Anatomy of a Ballooning Borehole Using PWD. In *Overpressures in Petroleum Exploration*, Proc. from the Pau, France workshop held on 7–8 April 1998, ed. A. Mitchell and D. Grauls. Paris: Editions Technip.

Welty, J.R., Wicks, C.E., and Wilson, R.E. 1984. *Fundamentals of Momentum, Heat, and Mass Transfer*, third edition. New York: John Wiley and Sons.

AUTHOR INDEX

SUBJECT INDEX